FUNDAMENTALS OF APPLIED ELECTROMAGNETICS

Seventh Edition

Global Edition

Fawwaz T. Ulaby
University of Michigan, Ann Arbor

Umberto Ravaioli
University of Illinois, Urbana–Champaign

Pearson

Boston · Columbus · Indianapolis · New York · San Francisco · Hoboken · Amsterdam
Cape Town · Dubai · London · Madrid · Milan · Munich · Paris · Montreal · Toronto
Delhi · Mexico City · Sau Paula · Sydney · Hong Kong · Seoul · Singapore · Taipei · Tokyo

Library of Congress Cataloging-in-Publication Data on File

Vice President and Editorial Director, ECS: *Marcia J. Horton*
Head of Learning Asset Acquisition, Global Edition: *Laura Dent*
Acquisitions Editor: *Julie Bai*
Editorial Assistant: *Sandra Rodriguez*
Acquisitions Editor, Global Edition: *Murchana Borthakur*
Associate Project Editor, Golbal Edition: *Binita Roy*
Managing Editor: *Scott Disanno*
Production Editor: *Rose Kernan*
Art Director: *Marta Samsel*
Art Editor: *Gregory Dulles*
Manufacturing Manager: *Mary Fischer*
Manufacturing Buyer: *Maura Zaldivar-Garcia*
Senior Manufacturing Controller, Production, Global Edition: *Trudy Kimber*
Product Marketing Manager: *Bram Van Kempen*
Field Marketing Manager: *Demetrius Hall*
Marketing Assistant: *Jon Bryant*
Cover Designer: *Lumina Datamatics*

Pearson Education Limited
Edinburgh Gate
Harlow
Essex CM20 2JE
England

and Associated Companies throughout the world

Visit us on the World Wide Web at:
www.pearsonglobaleditions.com

Authorized adaptation from the United States edition, entitled Fundamentals of Applied Electromagnetics, 7th edition, ISBN 978-0-13-335681-6, by Fawwaz T. Ulaby and Umberto Ravaioli, published by Pearson Education ©2015.

ISBN 10: 1-292-08244-5
ISBN 13: 978-1-292-08244-8

British Library Cataloguing-in-Publication Data
A catalogue record for this book is available from the British Library

10 9 8 7 6 5 4 3 2 1

Typeset in Times by Paul Mailhot

Printed and bound by Courier Kendallville in The United States of America

We dedicate this book to
Jean and Ann Lucia.

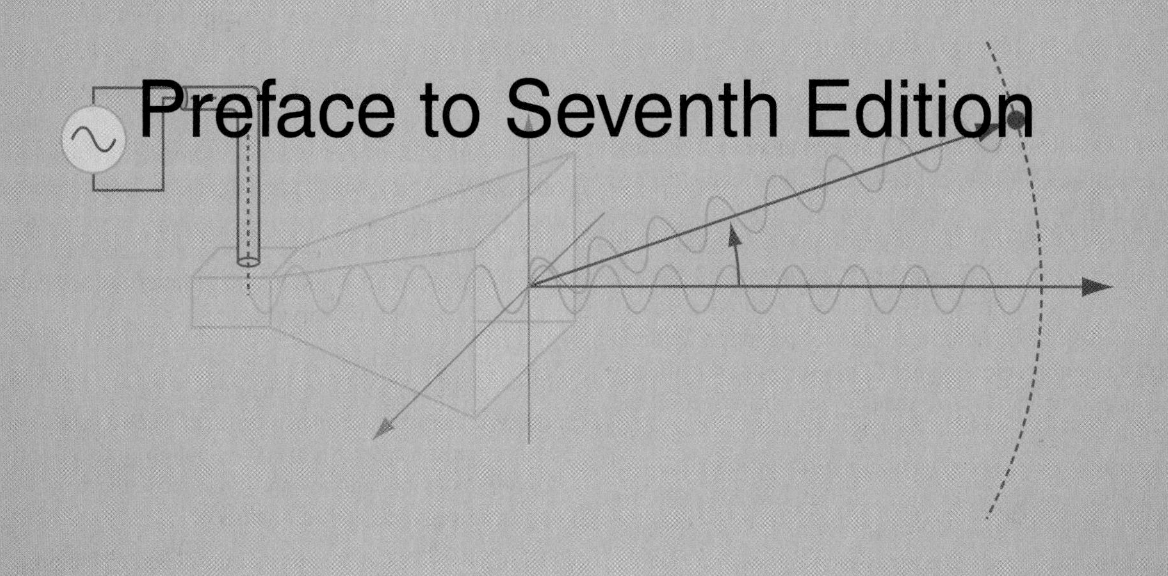

Preface to Seventh Edition

Building on the core content and style of its predecessor, this seventh edition (7/e) of *Applied Electromagnetics* introduces new features designed to help students develop a deeper understanding of electromagnetic concepts and applications. Prominent among them is a set of 52 web-based simulation modules that allow the user to interactively analyze and design transmission line circuits; generate spatial patterns of the electric and magnetic fields induced by charges and currents; visualize in 2-D and 3-D space how the gradient, divergence, and curl operate on spatial functions; observe the temporal and spatial waveforms of plane waves propagating in lossless and lossy media; calculate and display field distributions inside a rectangular waveguide; and generate radiation patterns for linear antennas and parabolic dishes. These are valuable learning tools; we encourage students to use them and urge instructors to incorporate them into their lecture materials and homework assignments.

Additionally, by enhancing the book's graphs and illustrations, and by expanding the scope of topics of the Technology Briefs, additional bridges between electromagnetic fundamentals and their countless engineering and scientific applications are established. In summary:

NEW TO THIS EDITION

- A set of 10 additional interactive simulation modules, bringing the total to 52
- Updated Technology Briefs
- Enhanced figures and images
- New/updated end-of-chapter problems
- The interactive modules and Technology Briefs can be found at the Student Website on http://www.pearsonglobaleditions.com/Ulaby.

ACKNOWLEDGMENTS

As authors, we were blessed to have worked on this book with the best team of professionals: Richard Carnes, Leland Pierce, Janice Richards, Rose Kernan, and Paul Mailhot. We are exceedingly grateful for their superb support and unwavering dedication to the project.

We enjoyed working on this book. We hope you enjoy learning from it.

FAWWAZ T. ULABY
UMBERTO RAVAIOLI

CONTENT

The book begins by building a bridge between what should be familiar to a third-year electrical engineering student and the electromagnetics (EM) material covered in the book. Prior to enrolling in an EM course, a typical student will have taken one or more courses in circuits. He or she should be familiar with circuit analysis, Ohm's law, Kirchhoff's current and voltage laws, and related topics. Transmission lines constitute a *natural* bridge between electric circuits and electromagnetics. Without having to deal with vectors or fields, the student uses already familiar concepts to learn about wave motion, the reflection and transmission of power, phasors, impedance matching, and many of the properties of wave propagation in a guided structure. All of these newly learned concepts will prove invaluable later (in Chapters 7 through 9) and will facilitate the learning of how plane waves propagate in free space and in material media. Transmission lines are covered in Chapter 2, which is preceded in Chapter 1 with reviews of complex numbers and phasor analysis.

The next part of the book, contained in Chapters 3 through 5, covers vector analysis, electrostatics, and magnetostatics. The electrostatics chapter begins with Maxwell's equations for the time-varying case, which are then specialized to electrostatics and magnetostatics, thereby providing the student with an overall framework for what is to come and showing him or her why electrostatics and magnetostatics are special cases of the more general time-varying case.

Chapter 6 deals with time-varying fields and sets the stage for the material in Chapters 7 through 9. Chapter 7 covers plane-wave propagation in dielectric and conducting media, and Chapter 8 covers reflection and transmission at discontinuous boundaries and introduces the student to fiber optics, waveguides and resonators.

In Chapter 9, the student is introduced to the principles of radiation by currents flowing in wires, such as dipoles, as well as

Suggested Syllabi

	Chapter	Two-semester Syllabus 6 credits (42 contact hours per semester) Sections	Hours	One-semester Syllabus 4 credits (56 contact hours) Sections	Hours
1	Introduction: Waves and Phasors	All	4	All	4
2	Transmission Lines	All	12	2-1 to 2-8 and 2-11	8
3	Vector Analysis	All	8	All	8
4	Electrostatics	All	8	4-1 to 4-10	6
5	Magnetostatics	All	7	5-1 to 5-5 and 5-7 to 5-8	5
	Exams		3		2
		Total for first semester	42		
6	Maxwell's Equations for Time-Varying Fields	All	6	6-1 to 6-3, and 6-6	3
7	Plane-wave Propagation	All	7	7-1 to 7-4, and 7-6	6
8	Wave Reflection and Transmission	All	9	8-1 to 8-3, and 8-6	7
9	Radiation and Antennas	All	10	9-1 to 9-6	6
10	Satellite Communication Systems and Radar Sensors	All	5	None	—
	Exams		3		1
		Total for second semester	40	Total	56
	Extra Hours		2		0

to radiation by apertures, such as a horn antenna or an opening in an opaque screen illuminated by a light source.

To give the student a taste of the wide-ranging applications of electromagnetics in today's technological society, Chapter 10 concludes the book with overview presentations of two system examples: satellite communication systems and radar sensors.

The material in this book was written for a two-semester sequence of six credits, but it is possible to trim it down to generate a syllabus for a one-semester four-credit course. The accompanying table provides syllabi for each of these two options.

MESSAGE TO THE STUDENT

The web-based interactive modules of this book were developed with you, the student, in mind. Take the time to use them in conjunction with the material in the textbook. Video animations can show you how fields and waves propagate in time and space, how the beam of an antenna array can be made to scan electronically, and examples of how current is induced in a circuit under the influence of a changing magnetic field. The modules are a useful resource for self-study. You can find them at the Student Website link on http://www.pearsonglobaleditions.com/Ulaby. Use them!

ACKNOWLEDGMENTS

Special thanks are due to reviewers for their valuable comments and suggestions. They include Constantine Balanis of Arizona State University, Harold Mott of the University of Alabama, David Pozar of the University of Massachusetts, S. N. Prasad of Bradley University, Robert Bond of New Mexico Institute of Technology, Mark Robinson of the University of Colorado at Colorado Springs, and Raj Mittra of the University of Illinois. I appreciate the dedicated efforts of the staff at Prentice Hall and I am grateful for their help in shepherding this project through the publication process in a very timely manner.

FAWWAZ T. ULABY

List of Technology Briefs

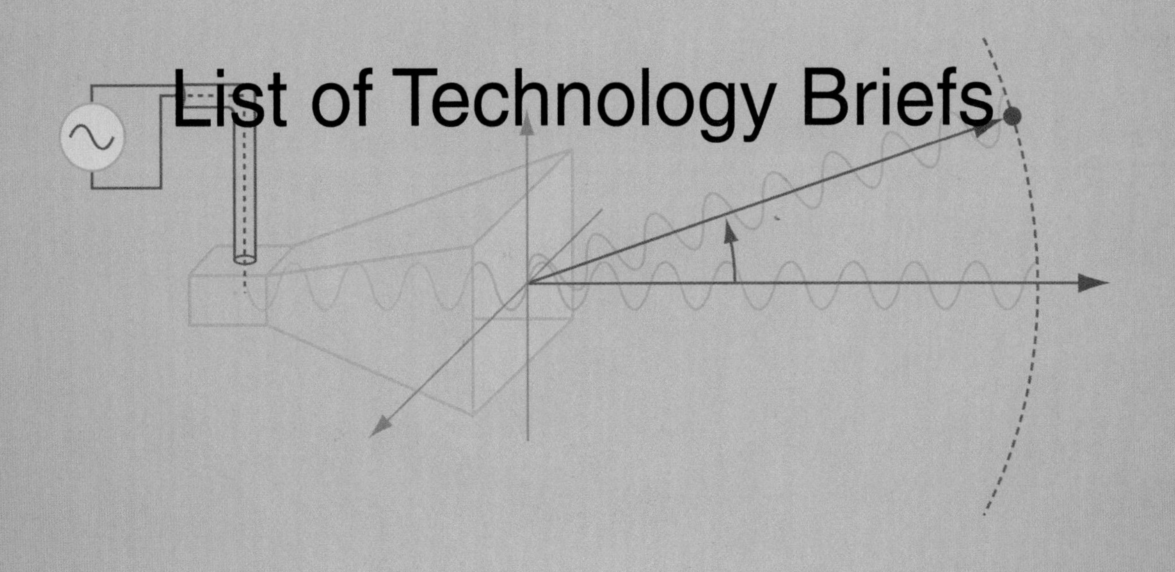

TB1	LED Lighting	42
TB2	Solar Cells	60
TB3	Microwave Ovens	104
TB4	EM Cancer Zappers	134
TB5	Global Positioning System	172
TB6	X-Ray Computed Tomography	186
TB7	Resistive Sensors	218
TB8	Supercapacitors as Batteries	236
TB9	Capacitive Sensors	240
TB10	Electromagnets	278
TB11	Inductive Sensors	290
TB12	EMF Sensors	314
TB13	RFID Systems	344
TB14	Liquid Crystal Display (LCD)	358
TB15	Lasers	390
TB16	Bar-Code Readers	404
TB17	Health Risks of EM Fields	446

Contents

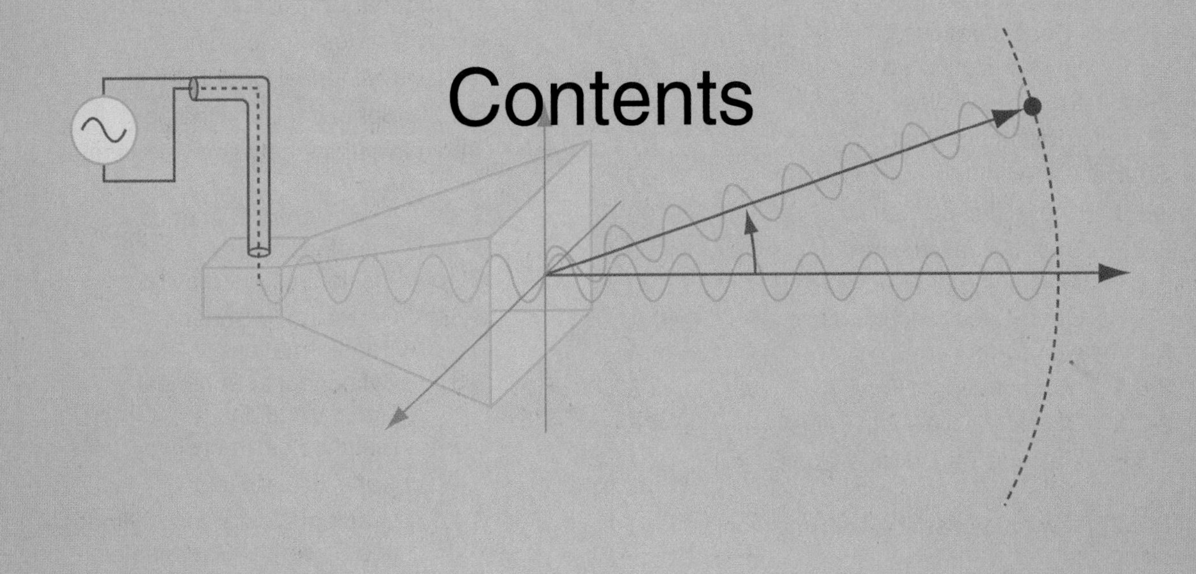

Preface 5

List of Technology Briefs 9

List of Modules 17

Photo Credits 19

Chapter 1 Introduction: Waves and Phasors 23

1-1 Historical Timeline 25

1-1.1 EM in the Classical Era 25

1-1.2 EM in the Modern Era 25

1-2 Dimensions, Units, and Notation 33

1-3 The Nature of Electromagnetism 34

1-3.1 The Gravitational Force: A Useful Analogue 34

1-3.2 Electric Fields 35

1-3.3 Magnetic Fields 37

1-3.4 Static and Dynamic Fields 38

1-4 Traveling Waves 40

1-4.1 Sinusoidal Waves in a Lossless Medium 41

TB1 LED Lighting 42

1-4.2 Sinusoidal Waves in a Lossy Medium 50

1-5 The Electromagnetic Spectrum 52

1-6 Review of Complex Numbers 54

1-7 Review of Phasors 58

1-7.1 Solution Procedure 59

TB2 Solar Cells 60

1-7.2 Traveling Waves in the Phasor Domain 65

Chapter 1 Summary 65

Problems 66

Chapter 2 Transmission Lines 70

2-1 General Considerations 71

2-1.1 The Role of Wavelength 71

2-1.2 Propagation Modes 73

2-2 Lumped-Element Model 74

2-3 Transmission-Line Equations 78

2-4 Wave Propagation on a Transmission Line 79

2-5 The Lossless Microstrip Line 82

2-6 The Lossless Transmission Line: General Considerations 87
2-6.1 Voltage Reflection Coefficient 88
2-6.2 Standing Waves 92
2-7 Wave Impedance of the Lossless Line 97
2-8 Special Cases of the Lossless Line 100
2-8.1 Short-Circuited Line 100
2-8.2 Open-Circuited Line 103
2-8.3 Application of Short-Circuit/ Open-Circuit Technique 103
TB3 Microwave Ovens 104
2-8.4 Lines of Length $l = n\lambda/2$ 106
2-8.5 Quarter-Wavelength Transformer 106
2-8.6 Matched Transmission Line: $Z_L = Z_0$ 107
2-9 Power Flow on a Lossless Transmission Line 108
2-9.1 Instantaneous Power 108
2-9.2 Time-Average Power 109
2-10 The Smith Chart 110
2-10.1 Parametric Equations 111
2-10.2 Wave Impedance 114
2-10.3 SWR, Voltage Maxima and Minima 115
2-10.4 Impedance to Admittance Transformations 118
2-11 Impedance Matching 123
2-11.1 Lumped-Element Matching 124
2-11.2 Single-Stub Matching 130
2-12 Transients on Transmission Lines 133
TB4 EM Cancer Zappers 134
2-12.1 Transient Response 137
2-12.2 Bounce Diagrams 140
Chapter 2 Summary 144
Problems 146

Chapter 3 Vector Analysis 155

3-1 Basic Laws of Vector Algebra 156
3-1.1 Equality of Two Vectors 157
3-1.2 Vector Addition and Subtraction 157
3-1.3 Position and Distance Vectors 158
3-1.4 Vector Multiplication 158
3-1.5 Scalar and Vector Triple Products 161
3-2 Orthogonal Coordinate Systems 162
3-2.1 Cartesian Coordinates 163
3-2.2 Cylindrical Coordinates 164
3-2.3 Spherical Coordinates 167
3-3 Transformations between Coordinate Systems 169
3-3.1 Cartesian to Cylindrical Transformations 169
TB5 Global Positioning System 172
3-3.2 Cartesian to Spherical Transformations 174
3-3.3 Cylindrical to Spherical Transformations 175
3-3.4 Distance between Two Points 175
3-4 Gradient of a Scalar Field 176
3-4.1 Gradient Operator in Cylindrical and Spherical Coordinates 177
3-4.2 Properties of the Gradient Operator 178
3-5 Divergence of a Vector Field 180
3-6 Curl of a Vector Field 184
TB6 X-Ray Computed Tomography 186
3-6.1 Vector Identities Involving the Curl 188
3-6.2 Stokes's Theorem 188
3-7 Laplacian Operator 189
Chapter 3 Summary 191
Problems 193

Chapter 4 Electrostatics 200

4-1 Maxwell's Equations 201
4-2 Charge and Current Distributions 202
4-2.1 Charge Densities 202
4-2.2 Current Density 203
4-3 Coulomb's Law 204
4-3.1 Electric Field due to Multiple Point Charges 205
4-3.2 Electric Field due to a Charge Distribution 206
4-4 Gauss's Law 209
4-5 Electric Scalar Potential 211
4-5.1 Electric Potential as a Function of Electric Field 211
4-5.2 Electric Potential Due to Point Charges 213

4-5.3 Electric Potential Due to Continuous Distributions 213
4-5.4 Electric Field as a Function of Electric Potential 214
4-5.5 Poisson's Equation 215
4-6 Conductors 217
TB7 Resistive Sensors 218
4-6.1 Drift Velocity 220
4-6.2 Resistance 221
4-6.3 Joule's Law 222
4-7 Dielectrics 223
4-7.1 Polarization Field 224
4-7.2 Dielectric Breakdown 225
4-8 Electric Boundary Conditions 225
4-8.1 Dielectric-Conductor Boundary 229
4-8.2 Conductor-Conductor Boundary 230
4-9 Capacitance 232
4-10 Electrostatic Potential Energy 235
TB8 Supercapacitors as Batteries 236
TB9 Capacitive Sensors 240
4-11 Image Method 245
Chapter 4 Summary 247
Problems 248

Chapter 5 Magnetostatics 257

5-1 Magnetic Forces and Torques 259
5-1.1 Magnetic Force on a Current-Carrying Conductor 250
5-1.2 Magnetic Torque on a Current-Carrying Loop 263
5-2 The Biot–Savart Law 266
5-2.1 Magnetic Field due to Surface and Volume Current Distributions 266
5-2.2 Magnetic Field of a Magnetic Dipole 270
5-2.3 Magnetic Force Between Two Parallel Conductors 272
5-3 Maxwell's Magnetostatic Equations 273
5-3.1 Gauss's Law for Magnetism 273
5-3.2 Ampère's Law 274
TB10 Electromagnets 278
5-4 Vector Magnetic Potential 281
5-5 Magnetic Properties of Materials 282
5-5.1 Electron Orbital and Spin Magnetic Moments 283
5-5.2 Magnetic Permeability 283
5-5.3 Magnetic Hysteresis of Ferromagnetic Materials 284
5-6 Magnetic Boundary Conditions 286
5-7 Inductance 287
5-7.1 Magnetic Field in a Solenoid 287
5-7.2 Self-Inductance 289
TB11 Inductive Sensors 290
5-7.3 Mutual Inductance 292
5-8 Magnetic Energy 293
Chapter 5 Summary 294
Problems 296

Chapter 6 Maxwell's Equations for Time-Varying Fields 303

6-1 Faraday's Law 304
6-2 Stationary Loop in a Time-Varying Magnetic Field 306
6-3 The Ideal Transformer 310
6-4 Moving Conductor in a Static Magnetic Field 311
TB12 EMF Sensors 314
6-5 The Electromagnetic Generator 316
6-6 Moving Conductor in a Time-Varying Magnetic Field 318
6-7 Displacement Current 319
6-8 Boundary Conditions for Electromagnetics 321
6-9 Charge-Current Continuity Relation 321
6-10 Free-Charge Dissipation in a Conductor 324
6-11 Electromagnetic Potentials 324
6-11.1 Retarded Potentials 325
6-11.2 Time-Harmonic Potentials 326
Chapter 6 Summary 329
Problems 330

Chapter 7 Plane-Wave Propagation 335

7-1 Time-Harmonic Fields 337
- 7-1.1 Complex Permittivity 337
- 7-1.2 Wave Equations 338

7-2 Plane-Wave Propagation in Lossless Media 338
- 7-2.1 Uniform Plane Waves 339
- 7-2.2 General Relation Between **E** and **H** 341

TB13 RFID Systems 344

7-3 Wave Polarization 346
- 7-3.1 Linear Polarization 347
- 7-3.2 Circular Polarization 348
- 7-3.3 Elliptical Polarization 350

7-4 Plane-Wave Propagation in Lossy Media 353
- 7-4.1 Low-Loss Dielectric 355
- 7-4.2 Good Conductor 356

TB14 Liquid Crystal Display (LCD) 358

7-5 Current Flow in a Good Conductor 361

7-6 Electromagnetic Power Density 365
- 7-6.1 Plane Wave in a Lossless Medium 365
- 7-6.2 Plane Wave in a Lossy Medium 366
- 7-6.3 Decibel Scale for Power Ratios 367

Chapter 7 Summary 368

Problems 370

Chapter 8 Wave Reflection and Transmission 374

8-1 Wave Reflection and Transmission at Normal Incidence 375
- 8-1.1 Boundary between Lossless Media 376
- 8-1.2 Transmission-Line Analogue 378
- 8-1.3 Power Flow in Lossless Media 379
- 8-1.4 Boundary between Lossy Media 381

8-2 Snell's Laws 384

8-3 Fiber Optics 387

8-4 Wave Reflection and Transmission at Oblique Incidence 389

TB15 Lasers 390
- 8-4.1 Perpendicular Polarization 392
- 8-4.2 Parallel Polarization 396
- 8-4.3 Brewster Angle 397

8-5 Reflectivity and Transmissivity 398

8-6 Waveguides 402

TB16 Bar-Code Readers 404

8-7 General Relations for **E** and **H** 405

8-8 TM Modes in Rectangular Waveguide 406

8-9 TE Modes in Rectangular Waveguide 410

8-10 Propagation Velocities 411

8-11 Cavity Resonators 414
- 8-11.1 Resonant Frequency 415
- 8-11.2 Quality Factor 415

Chapter 8 Summary 417

Problems 419

Chapter 9 Radiation and Antennas 425

9-1 The Hertzian Dipole 428
- 9-1.1 Far-Field Approximation 430
- 9-1.2 Power Density 431

9-2 Antenna Radiation Characteristics 432
- 9-2.1 Antenna Pattern 433
- 9-2.2 Beam Dimensions 434
- 9-2.3 Antenna Directivity 436
- 9-2.4 Antenna Gain 438
- 9-2.5 Radiation Resistance 438

9-3 Half-Wave Dipole Antenna 439
- 9-3.1 Directivity of $\lambda/2$ Dipole 441
- 9-3.2 Radiation Resistance of $\lambda/2$ Dipole 441
- 9-3.3 Quarter-Wave Monopole Antenna 442

9-4 Dipole of Arbitrary Length 442

9-5 Effective Area of a Receiving Antenna 444

TB17 Health Risks of EM Fields 446

9-6 Friis Transmission Formula 449

9-7 Radiation by Large-Aperture Antennas 451

9-8 Rectangular Aperture with Uniform Aperture Distribution 454
- 9-8.1 Beamwidth 455
- 9-8.2 Directivity and Effective Area 456

9-9 Antenna Arrays 456

9-10 N-Element Array with Uniform Phase Distribution 464

9-11 Electronic Scanning of Arrays 466
9-11.1 Uniform-Amplitude Excitation 467
9-11.2 Array Feeding 467
Chapter 9 Summary 472
Problems 474

Chapter 10 Satellite Communication Systems and Radar Sensors 479

10-1 Satellite Communication Systems 480
10-2 Satellite Transponders 482
10-3 Communication-Link Power Budget 484
10-4 Antenna Beams 485
10-5 Radar Sensors 486
10-5.1 Basic Operation of a Radar System 486
10-5.2 Unambiguous Range 487
10-5.3 Range and Angular Resolutions 488
10-6 Target Detection 489
10-7 Doppler Radar 491
10-8 Monopulse Radar 492
Chapter 10 Summary 495
Problems 496

Appendix A Symbols, Quantities, Units, and Abbreviations 497

Appendix B Material Constants of Some Common Materials 501

Appendix C Mathematical Formulas 505

Appendix D Answers to Selected Problems 507

Bibliography 513

Index 515

List of Modules

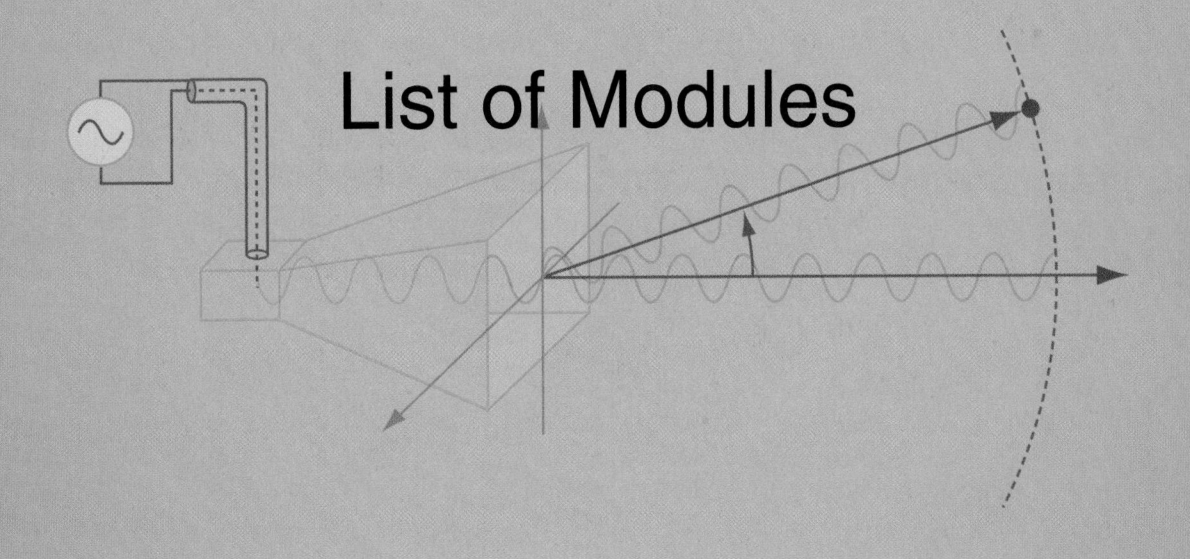

1.1 Sinusoidal Waveforms 49
1.2 Traveling Waves 51
1.3 Phase Lead/Lag 53
2.1 Two-Wire Line 82
2.2 Coaxial Cable 83
2.3 Lossless Microstrip Line 86
2.4 Transmission-Line Simulator 95
2.5 Wave and Input Impedance 100
2.6 Interactive Smith Chart 123
2.7 Quarter-Wavelength Transformer 131
2.8 Discrete Element Matching 132
2.9 Single-Stub Tuning 133
2.10 Transient Response 143
3.1 Vector Addition and Subtraction 167
3.2 Gradient 180
3.3 Divergence 184
3.4 Curl 190
4.1 Fields due to Charges 216
4.2 Charges in Adjacent Dielectrics 239
4.3 Charges above Conducting Plane 231
4.4 Charges near Conducting Sphere 232
5.1 Electron Motion in Static Fields 260
5.2 Magnetic Fields due to Line Sources 268
5.3 Magnetic Field of a Current Loop 271
5.4 Magnetic Force Between Two Parallel Conductors 273
6.1 Circular Loop in Time-varying Magnetic Field 309
6.2 Rotating Wire Loop in Constant Magnetic Field 318
6.3 Displacement Current 322
7.1 Linking E to H 343
7.2 Plane Wave 346
7.3 Polarization I 353
7.4 Polarization II 354
7.5 Wave Attenuation 361
7.6 Current in a Conductor 364
8.1 Normal Incidence on Perfect Conductor 384
8.2 Multimode Step-Index Optical Fiber 389
8.3 Oblique Incidence 401
8.4 Oblique Incidence in Lossy Medium 402
8.5 Rectangular Waveguide 415
9.1 Hertzian Dipole ($l \ll \lambda$) 432
9.2 Linear Dipole Antenna 444
9.3 Detailed Analysis of Linear Antenna 445
9.4 Large Parabolic Reflector 457
9.5 Two-dipole Array 462
9.6 Detailed Analysis of Two-Dipole Array 463
9.7 N-Element Array 469
9.8 Uniform Dipole Array 471

Photo Credits

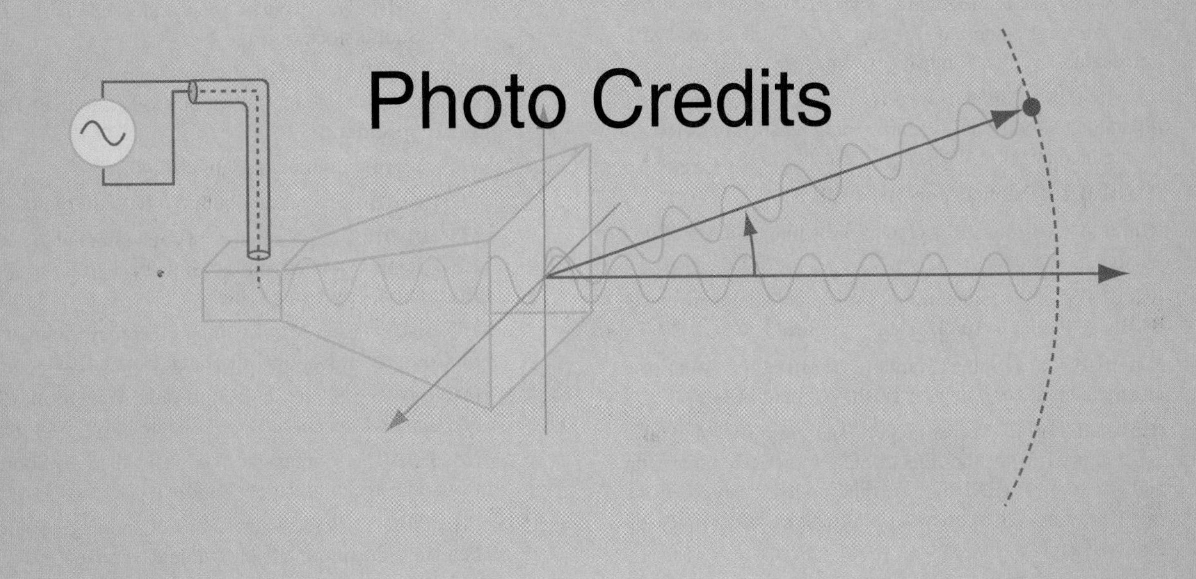

Page 24 (Fig 01-01): Line Art: 2-D LCD array, Source: Fawwaz Ulaby

Page 26 (Ch 01-01A): Thales of Miletus (624–546 BC), Photo Researchers, Inc./Science Source

Page 26 (Ch 01-01B): Isaac Newton, Mary Evans/Science Source

Page 26 (Ch 01-01C): Benjamin West, Benjamin Franklin Drawing Electricity from the Sky, Painting/Alamy

Page 26 (Ch 01-01D): Replica of the Voltaic pile invented by Alessandro Volta 1800, Clive Streeter/DK Images

Page 26 (Ch 01-01E): Hans Christian Ørsted, Danish Physicist, Science Source

Page 26 (Ch 01-01F): Andre-Marie Ampere, Nickolae/Fotolia

Page 27 (Ch 01-01G): Michael Faraday, Nicku/Shutterstock

Page 27 (Ch 01-01H): James Clerk Maxwell (1831–1879), SPL/Science Source

Page 27 (Ch 01-01I): Heinrich Rudolf Hertz, Science Source

Page 27 (Ch 01-01J): Nicola Tesla, Bain News Service/NASA

Page 27 (Ch 01-01K): Early X-Ray of Hand, Bettmann/Corbis

Page 27 (Ch 01-01M): Albert Einstein, Science Source

Page 28 (Ch 01-02A): Telegraph, Morse apparatus, vintage engraved illustration, Morphart Creation/Shutterstock

Page 28 (Ch 01-02B): Thomas Alva Edison With His 'Edison Effect' Lamps, Education Images/Getty Images, Inc.

Page 28 (Ch 01-02C): Replica of an early type of telephone made by Scottish-born telephony pioneer Alexander Graham Bell (1847–1922), Science & Society Picture Library/Getty Images

Page 28 (Ch 01-02D): Guglielmo Marconi, Pach Brothers/Library of Congress Prints and Photographs Division [LC-USZ62-39702]

Page 28 (Ch 01-02E): De Forest seated at his invention, the radio-telephone, called the Audion, Jessica Wilson/Science Source

Page 28 (Ch 01-02F): The staff of KDKA broadcast reports of the 1920 presidential election, Bettmann/Corbis

Page 29 (Ch 01-02G): This bottle-like object is a Cathode Ray tube which forms the receiver of the new style television invented by Dr. Vladimir Zworykin, Westinghouse research engineer, who is holding it, Bettmann/Corbis

Page 29 (Ch 01-02H): Radar in operation in the Second World War, Library of Congress Department of Prints and Photographs [LC-USZ62-101012]

Page 29 (Ch 01-02I): Shockly, Brattain, and Bardeen with an apparatus used in the early investigations which led to the invention of the transistor, Photo Researchers, Inc./Science Source

Page 29 (Ch 01-02J): A Photograph of Jack Kilby's Model of the First Working Integrated Circuit Ever Built circa 1958, Fotosearch/Archive Photos/Getty Images

Page 29 (Ch 01-02K): Shown here is the 135-foot rigidized inflatable balloon satellite undergoing tensile stress test in a dirigible hanger at Weekesville, North Carolina, NASA

Page 29 (Ch 01-02L): Pathfinder on Mars, JPL/NASA

Page 30 (Ch 01-03A): Abacus isolated on white, Sikarin Supphatada/Shutterstock

Page 30 (Ch 01-03B): Pascaline; a mechanical calculator invented by Blaise Pascal in 1642, Science Source

Page 30 (Ch 01-03C): Original Caption: Portrait of American electrical engineer Vannevar Bush, Bettmann/Corbis

Page 30 (Ch 01-03D): J. Presper Eckert and John W. Mauchly, are pictured with the Electronic Numerical Integrator and Computer (ENIAC) in this undated photo from the University of Pennsylvania Archives, University of Pennsylvania/AP images

Page 30 (Ch 01-03E): Description: DEC PDP-1 computer, on display at the Computer History Museum, USA, Volker Steger/Science Source

Page 31 (Ch 01-03F): Classic Antique Red LED Diode Calculator, James Brey/E+/Getty Images

Page 31 (Ch 01-03G): Apple I computer. This was released in April 1976 at the Homebrew Computer Club, USA, Volker Steger/Science

Page 31 (Ch 01-03H): UNITED STATES—DECEMBER 07: The IBM Personal Computer System was introduced to the market in early 1981, SSPL/Getty Images, Inc.

Page 31 (Ch 01-03I): NEW YORK, UNITED STATES: Chess enthusiasts watch World Chess champion Garry Kasparov on a television monitor as he holds his head in his hands, Stan Honda/Getty Images, Inc.

Page 32 (Fig 01-02A): The Very Large Array of Radio Telescopes, VLA, NRAO/NASA

Page 32 (Fig 01-02B): SCaN's Benefits to Society—Global Positioning System, Jet Propulsion Laboratory/NASA

Page 32 (Fig 01-02C): Motor, ABB

Page 32 (Fig 01-02D and Page 338 (Fig TF14-04)): TV on white background, Fad82/Fotolia

Page 32 (Fig 01-02E): Nuclear Propulsion Through Direct Conversion of Fusion Energy, John Slough/NASA

Page 32 (Fig 01-02F): Tracking station has bird's eye view on VAFB, Ashley Tyler/US Air Force

Page 32 (Fig 01-02G): Glass Fiber Cables, Kulka/Zefa/Corbis

Page 32 (Fig 01-02H): Electromagnetic sensors, HW Group

Page 32 (Fig 01-02I): Touchscreen smartphone, Oleksiy Mark/Shutterstock

Page 32 (Fig 01-02J): Line Art: Electromagnetics is at the heart of numerous systems and applications:, Source: Based on IEEE Spectrum

Page 42 (TF 01-01a): Lightbulb, Chones/Fotolia

Page 42 (TF 01-01b): Fluorescent bulb, Wolf1984/Fotolia

Page 42 (TF 01-01c): 3d render of an unbranded screw-in LED lamp, isolated on a white background, Marcello Bortolino/Getty Images, Inc.

Page 43 (TF 01-03): Line Art: Lighting efficiency, Source: Based on Courtesy of National Research Council, 2009

Page 49 (Mod 01-01): Screenshot: Sinusoidal Waveforms, Source: © Pearson Education, Upper Saddle River, New Jersey

Page 51 (Mod 01-02): Screenshot: TravelingWaves, Source: © Pearson Education, Upper Saddle River, New Jersey

Page 53 (Mod02-04): Screenshot: Phase Lead/Lag, Source: © Pearson Education, Upper Saddle River, New Jersey

Page 55 (Fig 01-17): Line Art: Individual bands of the radio spectrum and their primary allocations in the US. Student Website, Source: U.S. Department of Commerce

Page 82 (Mod 02-01): Screenshot: Two-Wire Line, Source: © Pearson Education, Upper Saddle River, New Jersey

Page 83 (Mod 02-02): Screenshot: Coaxial Cable, Source: © Pearson Education, Upper Saddle River, New Jersey

Page 84 (Fig 02-10a): Line Art: Microstrip line: longitudinal view, Source: Prof. Gabriel Rebeiz, U. California at San Diego

Page 84 (Fig 02-10b): Line Art: Microstrip line: Cross-sectional view, Source: Prof. Gabriel Rebeiz, U. California at San Diego

Page 84 (Fig 02-10c): Circuit board, Gabriel Reibeiz

Page 88 (Mod02-03): Screenshot: Lossless Microstrip Line, Source: © Pearson Education, Upper Saddle River, New Jersey

Page 95 (Mod02-04): Screenshot: Transmission-Line Simulator, Source: © Pearson Education, Upper Saddle River, New Jersey

Page 100 (Mod 02-05): Screenshot: Wave and Input Impedance, Source: © Pearson Education, Upper Saddle River, New Jersey

Page 105 (TF 03-02): Microwave oven cavity, Pearson Education, Inc.

Page 123 (Mod 02-06): Screenshot: Interactive Smith Chart, Source: © Pearson Education, Upper Saddle River, New Jersey

Page 131 (Mod 02-07): Screenshot: Quarter-Wavelength Transformer, Source: © Pearson Education, Upper Saddle River, New Jersey

Page 132 (Mod 02-08): Screenshot: Discrete Element Matching, Source: © Pearson Education, Upper Saddle River, New Jersey

Page 133 (Mod 02-09): Screenshot: Single-Stub Tuning, Source: © Pearson Education, Upper Saddle River, New Jersey

Page 134 (TF 04-01): Microwave ablation for cancer liver treatment, Radiological Society of North America (RSNA)

Page 135 (TF 04-02): Setup for a percutaneous microwave ablation procedure shows three single microwave applicators connected to three microwave generators, Radiological Society of North America (RSNA)

Page 134 (TF 04-03): Line Art: Bryan Christie Design LLC

Page 143 (Mod 02-10): Screenshot: Transient Response, Source: © Pearson Education, Upper Saddle River, New Jersey

Page 165 (Mod 03-01): Screenshot: Vector Addition and Subtraction, Source: © Pearson Education, Upper Saddle River, New Jersey

Page 172 (TF 05-01): Touchscreen smartphone with GPS navigation isolated on white reflective background, Oleksiy Mark/Shutterstock

Page 172 (TF 05-02): SCaN's Benefits to Society—Global Positioning System, Jet Propulsion Laboratory/NASA

Page 173 (TF 05-03): SUV, Konstantin/Fotolia

Page 180 (Mod 03-02): Screenshot: Gradient, Source: Graphics created with Wolfram Matematica®

Page 184 (Mod 03-03): Screenshot: Divergence, Source: Graphics created with Wolfram Matematica®

Page 186 (TF 06-01): X-ray of pelvis and spinal column, Cozyta/Getty Images, Inc.

Page 186 (TF 06-02): CT scan advance technology for medical diagnosis, Tawesit/Fotolia

Page 187 (TF 06-03c): Digitally enhanced CT scan of a normal brain in transaxial (horizontal) section, Scott Camazine/Science Source

Page 190 (Mod 03-04): Screenshot: Curl, Source: Graphics created with Wolfram Matematica

Page 216 (Mod 04-01): Screenshot: Fields due to Charges, Source: © Pearson Education, Upper Saddle River, New Jersey

Page 229 (Mod 04-02): Screenshot: Charges in Adjacent Dielectrics, Source: © Pearson Education, Upper Saddle River, New Jersey

Page 231 (Mod 04-03): Screenshot: Charges above Conducting Plane, Source: © Pearson Education, Upper Saddle River, New Jersey

Page 232 (Mod 04-04): Screenshot: Charges near Conducting Sphere, Source: © Pearson Education, Upper Saddle River, New Jersey

Page 236 (TF 08-01): Various electrolytic capacitors, David J. Green/Alamy

Page 236 (TF08-02A): High-speed train in motion, Metlion/Fotolia

Page 236 (TF08-02B): Cordless Drill, Derek Hatfield/Shutterstock

Page 236 (TF08-02C): The 2006 BMW X3 Concept Gasoline Electric Hybrid uses high-performance capacitors (or "Super Caps") to store and supply electric energy to the vehicle's Active Transmission, Passage/Car Culture/Corbis

Page 236 (TF 08-02D): LED Electric torch—laser Pointer isolated on white background, Artur Synenko/Shutterstock

Page 244 (TF 09-06): Line Art: Bryan Christie Design, LLC

Page 244 (TF 09-07): Line Art: Fingerprint representation, Source: Courtesy of Dr. M. Tartagni, University of Bologna, Italy

Page 260 (Mod 05-01): Screenshot: Electron Motion in Static Fields, Source: © Pearson Education, Upper Saddle River, New Jersey

Page 268 (Mod 05-02): Screenshot: Magnetic Fields due to Line Sources, Source: © Pearson Education, Upper Saddle River, New Jersey

Page 271 (Mod 05-03): Screenshot: Magnetic Field of a Current Loop, Source: © Pearson Education, Upper Saddle River, New Jersey

Page 273 (Mod 05-04): Screenshot: Magnetic Force Between Two Parallel Plates, Source: © Pearson Education, Upper Saddle River, New Jersey

Page 280 (TF 10-05A): CHINA—JUNE 20: A maglev train awaits departure in Shanghai, China, on Saturday, June 20, 2009, Qilai Shen/Bloomberg/Getty Images

Page 280 (TF 10-5b and c): Line Art: Magnetic trains—(b) internal workings of the Maglev train, Source: Amy Mast, Maglev trains are making history right now. Flux, volume 3 issue 1, National High Magnetic Field Laboratory

Page 309 (Mod 06-01): Screenshot: Circular Loop in Time-varying Magnetic Field, Source: Copyright © by Pearson Education, Upper Saddle River, New Jersey

Page 318 (Mod 06-02): Screenshot: Rotating Wire Loop in Constant Magnetic Field, Source: Copyright © by Pearson Education, Upper Saddle River, New Jersey

Page 322 (Mod 06-02): Screenshot: Displacement Current, Source: Copyright © by Pearson Education, Upper Saddle River, New Jersey

Page 343 (Mod 07-01): Screenshot: Linking E to H, Source: © Pearson Education, Upper Saddle River, New Jersey

Page 344 (TF 13-01): Jersey cow on pasture, Lakeview Images/Shutterstock

Page 345 (TF 13-2): Line Art: How an RFID system works is illustrated through this EZ-Pass example: Tag, Source: Prof. C. F. Huang

Page 346 (Mod 07-02): Screenshot: Plane Wave, Source: © Pearson Education, Upper Saddle River, New Jersey

Page 353 (Mod 07-03): Screenshot: Polarization I, Source: © Pearson Education, Upper Saddle River, New Jersey

Page 354 (Mod 07-04): Screenshot: Polarization II, Source: © Pearson Education, Upper Saddle River, New Jersey

Page 361 (Mod 07-05): Screenshot: Wave Attenuation, Source: © Pearson Education, Upper Saddle River, New Jersey

Page 364 (Mod 07-06): Screenshot: Current in Conductor, Source: © Pearson Education, Upper Saddle River, New Jersey

Page 384 (Mod 08-01): Screenshot: Normal Incidence on Perfect Conductor, Source: © Pearson Education, Upper Saddle River, New Jersey

Page 389 (Mod 08-02): Screenshot: Multimode Step-Index Optical Fiber, Source: © Pearson Education, Upper Saddle River, New Jersey

Page 390 (TF 15-01A): Optical Computer Mouse, William Whitehurst/Cusp/Corbis

Page 390 (TF 15-01B): Laser eye surgery, Will & Deni McIntyre/Science Source

Page 390 (TF 15-01C): Laser Star Guide, NASA

Page 390 (TF 15-01D): Laser: TRUMPF GmbH + Co. KG

Page 401 (Mod 08-03): Screenshot: Oblique Incidence, Source: © Pearson Education, Upper Saddle River, New Jersey

Page 402 (Mod 08-04): Screenshot: Oblique Incidence in Lossy Medium, Source: © Pearson Education, Upper Saddle River, New Jersey

Page 415 (Mod 08-05): Screenshot: Rectangular Waveguide, Source: © Pearson Education, Upper Saddle River, New Jersey

Page 432 (Mod 09-01): Screenshot: Hertzian Dipole ($l \ll \lambda$), Source: © Pearson Education, Upper Saddle River, New Jersey

Page 445 (Mod 09-03): Screenshot: Detailed Analysis of Linear Antenna, Source: © Pearson Education, Upper Saddle River, New Jersey

Page 446 (TF 17-01A): Smiling woman using computer, Edbockstock/Fotolia

Page 446 (TF 17-01B): Vector silhouette of Power lines and electric pylons, Ints Vikmanis/Alamy

Page 446 (TF 17-01C): Telecommunications tower, Poliki/Fotolia

Page 457 (Mod 09-04): Screenshot: Large Parabolic Reflector, Source: © Pearson Education, Upper Saddle River, New Jersey

Page 458 (Fig 09-25): The AN/FPS-85 Phased Array Radar Facility in the Florida panhandle, near the city of Freeport, NASA

Page 462 (Mod 09-05): Screenshot: Two-dipole Array, Source: © Pearson Education, Upper Saddle River, New Jersey

Page 469 (Mod 09-07): Screenshot: N-Element Array, Source: © Pearson Education, Upper Saddle River, New Jersey

Page 471 (Mod 09-08): Screenshot: Uniform Dipole Array, Source: © Pearson Education, Upper Saddle River, New Jersey

Page 486 (Text 10-01): 1. Dipoles and helices at VHF...steering and scanning. (79 words/212 pages), Source: R. G. Meadows and A. J. Parsons, Satellite Communications, Hutchinson Publishers, London, 1989

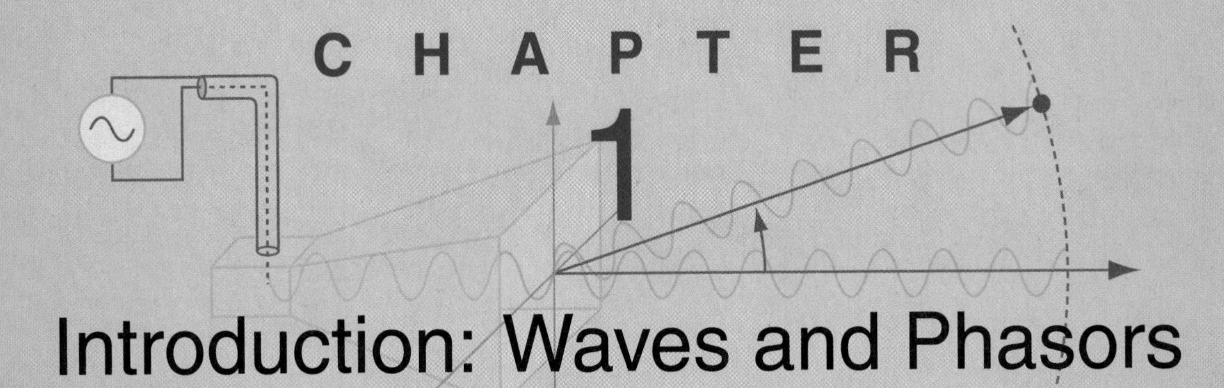

CHAPTER 1

Introduction: Waves and Phasors

Chapter Contents

Overview, 24
1-1 Historical Timeline, 25
1-2 Dimensions, Units, and Notation, 33
1-3 The Nature of Electromagnetism, 34
1-4 Traveling Waves, 40
TB1 LED Lighting, 42
1-5 The Electromagnetic Spectrum, 52
1-6 Review of Complex Numbers, 54
1-7 Review of Phasors, 58
TB2 Solar Cells, 60
Chapter 1 Summary, 65
Problems, 66

Objectives

Upon learning the material presented in this chapter, you should be able to:

1. Describe the basic properties of electric and magnetic forces.
2. Ascribe mathematical formulations to sinusoidal waves traveling in both lossless and lossy media.
3. Apply complex algebra in rectangular and polar forms.
4. Apply the phasor-domain technique to analyze circuits driven by sinusoidal sources.

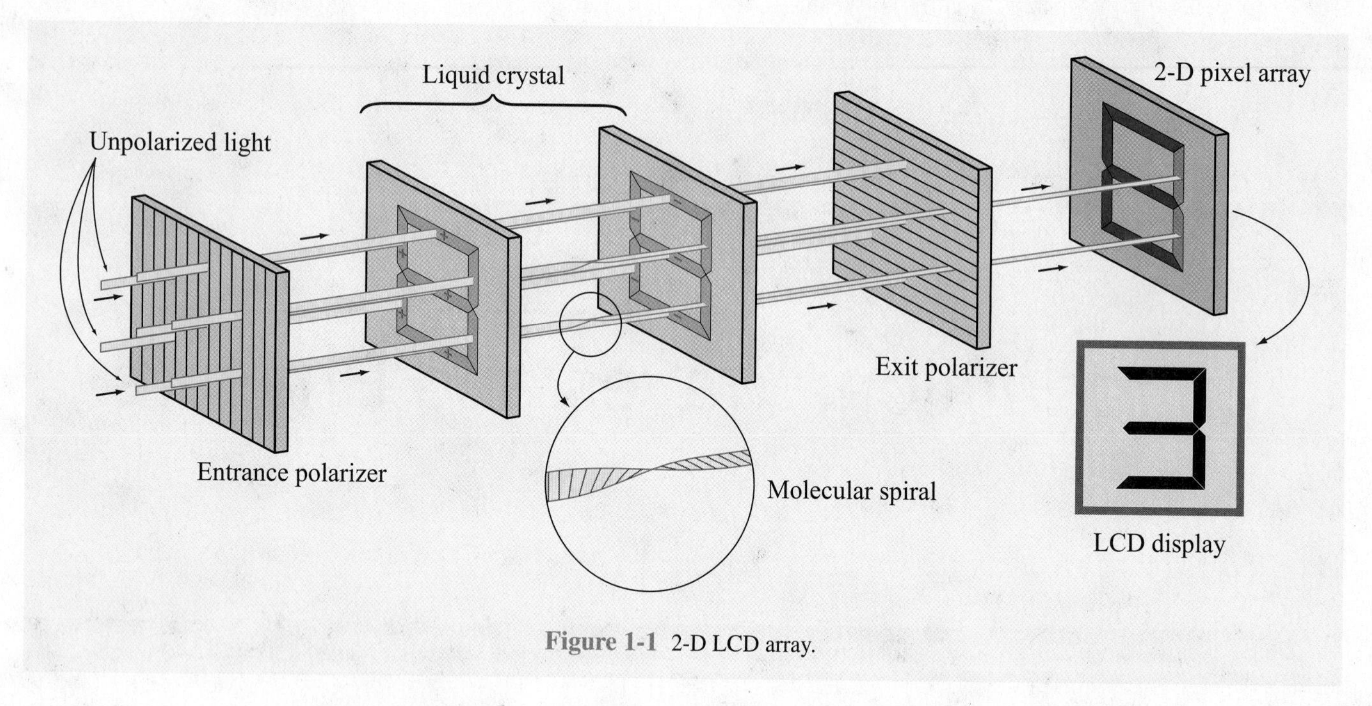

Figure 1-1 2-D LCD array.

Overview

Liquid crystal displays have become integral parts of many electronic consumer products, ranging from alarm clocks and cell phones to laptop computers and television systems. LCD technology relies on special electrical and optical properties of a class of materials known as ***liquid crystals***, which are neither pure solids nor pure liquids but rather a hybrid of both. The molecular structure of these materials is such that when light travels through them, the polarization of the emerging light depends on whether or not a voltage exists across the material. Consequently, when no voltage is applied, the exit surface appears bright, and conversely, when a voltage of a certain level is applied across the LCD material, no light passes through it, resulting in a dark pixel. In-between voltages translate into a range of grey levels. By controlling the voltages across individual pixels in a two-dimensional array, a complete image can be displayed (**Fig. 1-1**). Color displays are composed of three subpixels with red, green, and blue filters.

► The polarization behavior of light in an LCD is a prime example of how electromagnetics is at the heart of electrical and computer engineering. ◄

The subject of this book is applied electromagnetics (EM), which encompasses the study of both static and dynamic electric and magnetic phenomena and their engineering applications. Primary emphasis is placed on the fundamental properties of dynamic (time-varying) electromagnetic fields because of their greater relevance to practical problems in many applications, including wireless and optical communications, radar, bioelectromagnetics, and high-speed microelectronics. We study wave propagation in guided media, such as coaxial transmission lines, optical fibers and waveguides; wave reflection and transmission at interfaces between dissimilar media; radiation by antennas; and several other related topics. The concluding chapter is intended to illustrate a few aspects of applied EM through an examination of design considerations associated with the use and operation of radar sensors and satellite communication systems.

We begin this chapter with a chronology of the history of electricity and magnetism. Next, we introduce the fundamental electric and magnetic field quantities of electromagnetics, as well as their relationships to each other and to the electric charges and currents that generate them. These relationships constitute the underpinnings of the study of electromagnetic phenomena. Then, in preparation for the material presented in Chapter 2, we provide short reviews of three topics: traveling waves, complex numbers, and phasors, all useful in solving time-harmonic problems.

1-1 Historical Timeline

The history of EM may be divided into two overlapping eras. In the *classical era*, the fundamental laws of electricity and magnetism were discovered and formulated. Building on these formulations, the *modern era* of the past 100 years ushered in the birth of the field of applied EM, the topic of this book.

1-1.1 EM in the Classical Era

Chronology 1-1 provides a timeline for the development of electromagnetic theory in the classical era. It highlights those discoveries and inventions that have impacted the historical development of EM in a very significant way, even though the selected discoveries represent only a small fraction of those responsible for our current understanding of electromagnetics. As we proceed through the book, some of the names highlighted in Chronology 1-1, such as those of Coulomb and Faraday, will appear again later as we discuss the laws and formulations named after them.

The attractive force of magnetite was reported by the Greeks some 2800 years ago. It was also a Greek, *Thales of Miletus*, who first wrote about what we now call static electricity: he described how rubbing amber caused it to develop a force that could pick up light objects such as feathers. The term "*electric*" first appeared in print around 1600 in a treatise on the (electric) force generated by friction, authored by the physician to Queen Elizabeth I, *William Gilbert*.

About a century later, in 1733, *Charles-François du Fay* introduced the notion that electricity involves two types of "fluids," one "positive" and the other "negative," and that like-fluids repel and opposite-fluids attract. His notion of a fluid is what we today call electric charge. The invention of the capacitor in 1745, originally called the *Leyden jar*, made it possible to store significant amounts of electric charge in a single device. A few years later, in 1752, *Benjamin Franklin* demonstrated that lightning is a form of electricity. He transferred electric charge from a cloud to a Leyden jar via a silk kite flown in a thunderstorm. The collective eighteenth-century knowledge about electricity was integrated in 1785 by *Charles-Augustin de Coulomb*, in the form of a mathematical formulation characterizing the electrical force between two charges in terms of their strengths and polarities and the distance between them.

The year 1800 is noted for the development of the first electric battery by *Alessandro Volta*, and 1820 was a banner year for discoveries about how electric currents induce magnetism. This knowledge was put to good use by *Joseph Henry*, who developed one of the earliest electromagnets and dc (direct current) electric motors. Shortly thereafter, *Michael Faraday* built the first electric generator (the converse of the electric motor). Faraday, in essence, demonstrated that a changing magnetic field induces an electric field (and hence a voltage). The converse relation, namely that a changing electric field induces a magnetic field, was first proposed by *James Clerk Maxwell* in 1864 and then incorporated into his four (now) famous equations in 1873.

> ► Maxwell's equations represent the foundation of classical electromagnetic theory. ◄

Maxwell's theory, which predicted the existence of electromagnetic waves, was not fully accepted by the scientific community at that time, not until verified experimentally by means of radio waves by *Heinrich Hertz* in the 1880s. X-rays, another member of the EM family, were discovered in 1895 by *Wilhelm Röntgen*. In the same decade, *Nikola Tesla* was the first to develop the ac (alternating current) motor, considered a major advance over its predecessor, the dc motor.

Despite the advances made in the 19th century in our understanding of electricity and magnetism and how to put them to practical use, it was not until 1897 that the fundamental carrier of electric charge, the electron, was identified and its properties quantified by *Joseph Thomson*. The ability to eject electrons from a material by shining electromagnetic energy, such as light, on it is known as the *photoelectric effect*.

> ► To explain the photoelectric effect, *Albert Einstein* adopted the quantum concept of energy that had been advanced a few years earlier (1900) by *Max Planck*. Symbolically, this step represents the bridge between the classical and modern eras of electromagnetics. ◄

1-1.2 EM in the Modern Era

Electromagnetics plays a role in the design and operation of every conceivable electronic device, including the diode, transistor, integrated circuit, laser, display screen, bar-code reader, cell phone, and microwave oven, to name but a few. Given the breadth and diversity of these applications (Fig. 1-2), it is far more difficult to construct a meaningful timeline for the modern era than for the classical era. That said, one can develop timelines for specific technologies and link their milestone innovations to EM. Chronologies 1-2 and 1-3 present timelines for the development of telecommunications and computers,

Chronology 1-1: TIMELINE FOR ELECTROMAGNETICS IN THE CLASSICAL ERA

Electromagnetics in the Classical Era

ca. 900 BC — Legend has it that while walking across a field in northern Greece, a shepherd named Magnus experiences a pull on the iron nails in his sandals by the black rock he is standing on. The region was later named Magnesia and the rock became known as magnetite [a form of iron with permanent magnetism].

ca. 600 BC — Greek philosopher **Thales** describes how amber, after being rubbed with cat fur, can pick up feathers [static electricity].

ca. 1000 — Magnetic compass used as a navigational device.

1600 — **William Gilbert** (English) coins the term electric after the Greek word for amber (*elektron*), and observes that a compass needle points north-south because the Earth acts as a bar magnet.

1671 — **Isaac Newton** (English) demonstrates that white light is a mixture of all the colors.

1733 — **Charles-François du Fay** (French) discovers that electric charges are of two forms, and that like charges repel and unlike charges attract.

1745 — **Pieter van Musschenbroek** (Dutch) invents the Leyden jar, the first electrical capacitor.

1752 — **Benjamin Franklin** (American) invents the lightning rod and demonstrates that lightning is electricity.

1785 — **Charles-Augustin de Coulomb** (French) demonstrates that the electrical force between charges is proportional to the inverse of the square of the distance between them.

1800 — **Alessandro Volta** (Italian) develops the first electric battery.

1820 — **Hans Christian Oersted** (Danish) demonstrates the interconnection between electricity and magnetism through his discovery that an electric current in a wire causes a compass needle to orient itself perpendicular to the wire.

1820 — **Andre-Marie Ampère** (French) notes that parallel currents in wires attract each other and opposite currents repel.

1820 — **Jean-Baptiste Biot** (French) and **Félix Savart** (French) develop the Biot-Savart law relating the magnetic field induced by a wire segment to the current flowing through it.

Chronology 1-1: TIMELINE FOR ELECTROMAGNETICS IN THE CLASSICAL ERA (continued)

Electromagnetics in the Classical Era

1827 **Georg Simon Ohm** (German) formulates Ohm's law relating electric potential to current and resistance.

1827 **Joseph Henry** (American) introduces the concept of inductance, and builds one of the earliest electric motors. He also assisted Samual Morse in the development of the telegraph.

1831 **Michael Faraday** (English) discovers that a changing magnetic flux can induce an electromotive force.

1835 **Carl Friedrich Gauss** (German) formulates Gauss's law relating the electric flux flowing through an enclosed surface to the enclosed electric charge.

Gauss' Law for Electricity

$$\Phi_E = \oint \vec{E} \cdot d\vec{A} = \frac{q_{inside}}{\varepsilon_0}$$

1873 **James Clerk Maxwell** (Scottish) publishes his Treatise on Electricity and Magnetism in which he unites the discoveries of Coulomb, Oersted, Ampère, Faraday, and others into four elegantly constructed mathematical equations, now known as Maxwell's Equations.

1887 **Heinrich Hertz** (German) builds a system that can generate electromagnetic waves (at radio frequencies) and detect them.

1888 **Nikola Tesla** (Croatian-American) invents the ac (alternating current) electric motor.

1895 **Wilhelm Röntgen** (German) discovers X-rays. One of his first X-ray images was of the bones in his wife's hands. [1901 Nobel prize in physics.]

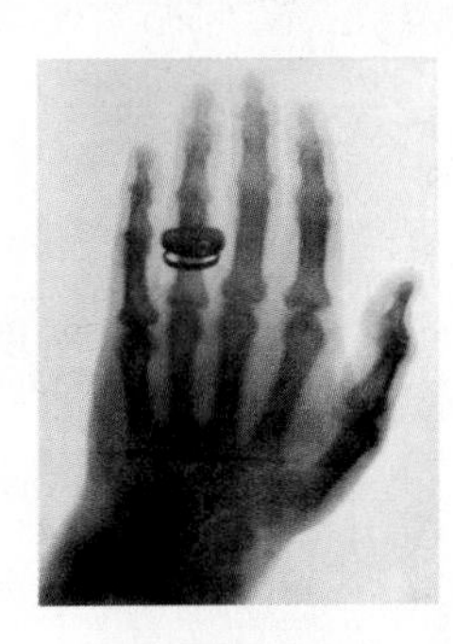

1897 **Joseph John Thomson** (English) discovers the electron and measures its charge-to-mass ratio. [1906 Nobel prize in physics.]

1905 **Albert Einstein** (German-American) explains the photoelectric effect discovered earlier by Hertz in 1887. [1921 Nobel prize in physics.]

Chronology 1-2: TIMELINE FOR TELECOMMUNICATIONS

Telecommunications

1825 **William Sturgeon** (English) develops the multiturn electromagnet.

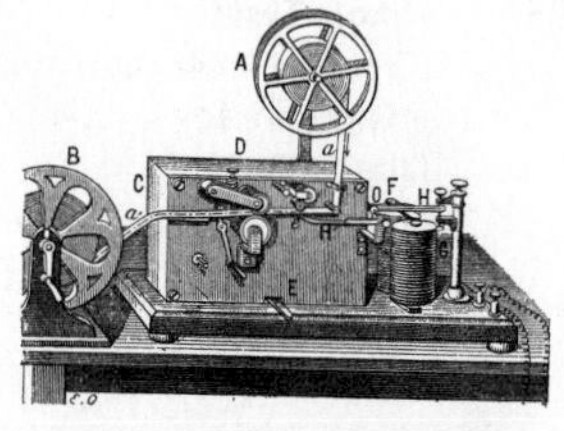

1837 **Samuel Morse** (American) patents the electromagnetic telegraph, using a code of dots and dashes to represent letters and numbers.

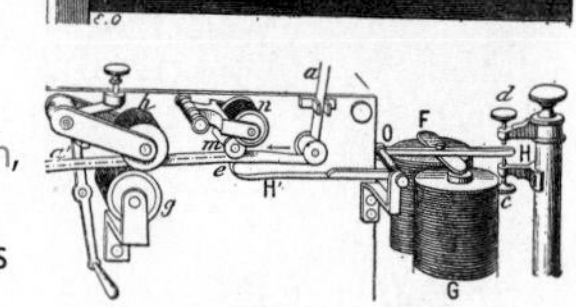

1872 **Thomas Edison** (American) patents the electric typewriter.

1876 **Alexander Graham Bell** (Scottish-American) invents the telephone, the rotary dial becomes available in 1890, and by 1900, telephone systems are installed in many communities.

1887 **Heinrich Hertz** (German) generates radio waves and demonstrates that they share the same properties as light.

1887 **Emil Berliner** (American) invents the flat gramophone disc, or record.

1896 **Guglielmo Marconi** (Italian) files his first of many patents on wireless transmission by radio. In 1901, he demonstrates radio telegraphy across the Atlantic Ocean. [1909 Nobel prize in physics, shared with Karl Braun (German).]

1897 **Karl Braun** (German) invents the cathode ray tube (CRT). [1909 Nobel prize with Marconi.]

1902 **Reginald Fessenden** (American) invents amplitude modulation for telephone transmission. In 1906, he introduces AM radio broadcasting of speech and music on Christmas Eve.

1912 **Lee De Forest** (American) develops the triode tube amplifier for wireless telegraphy. Also in 1912, the wireless distress call issued by the *Titanic* was heard 58 miles away by the ocean liner *Carpathia*, which managed to rescue 705 *Titanic* passengers 3.5 hours later.

1919 **Edwin Armstong** (American) invents the superheterodyne radio receiver.

1920 Birth of commercial radio broadcasting; Westinghouse Corporation establishes radio station KDKA in Pittsburgh, Pennsylvania.

Chronology 1-2: TIMELINE FOR TELECOMMUNICATIONS (continued)

Telecommunications

1923 

Vladimir Zworykin (Russian-American) invents television. In 1926, John Baird (Scottish) transmits TV images over telephone wires from London to Glasgow. Regular TV broadcasting began in Germany (1935), England (1936), and the United States (1939).

1926 Transatlantic telephone service between London and New York.

1932 First microwave telephone link, installed (by Marconi) between Vatican City and the Pope's summer residence.

1933 **Edwin Armstrong** (American) invents frequency modulation (FM) for radio transmission.

1935 **Robert Watson-Watt** (Scottish) invents radar.

1938 **H. A. Reeves** (American) invents pulse code modulation (PCM).

1947 **William Shockley**, **Walter Brattain**, and **John Bardeen** (all Americans) invent the junction transistor at Bell Labs. [1956 Nobel prize in physics.]

1955 Pager is introduced as a radio communication product in hospitals and factories.

1955 **Narinder Kapany** (Indian-American) demonstrates the optical fiber as a low-loss, light-transmission medium.

1958 **Jack Kilby** (American) builds first integrated circuit (IC) on germanium and, independently, **Robert Noyce** (American) builds first IC on silicon.

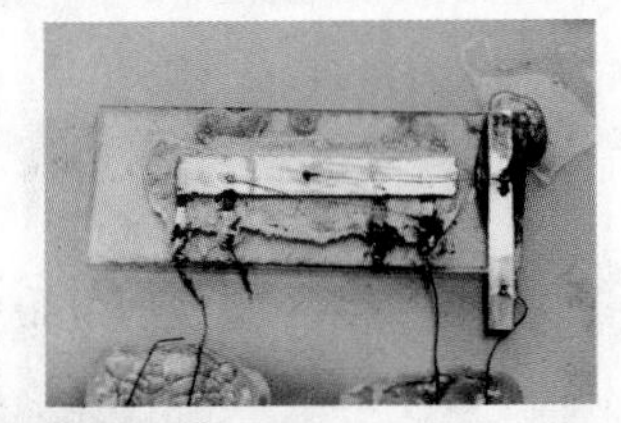

1960 Echo, the first passive communication satellite is launched, and successfully reflects radio signals back to Earth. In 1963, the first communication satellite is placed in geosynchronous orbit.

1969 ARPANET is established by the U.S. Department of Defense, to evolve later into the Internet.

1979 Japan builds the first cellular telephone network:

- 1983 cellular phone networks start in the United States.
- 1990 electronic beepers become common.
- 1995 cell phones become widely available.
- 2002 cell phone supports video and Internet.

1984 Worldwide Internet becomes operational.

1988 First transatlantic optical fiber cable between the U.S. and Europe.

1997 Mars Pathfinder sends images to Earth.

2004 Wireless communication supported by many airports, university campuses, and other facilities.

2012 Smartphones worldwide exceed 1 billion.

Chronology 1-3: TIMELINE FOR COMPUTER TECHNOLOGY

Computer Technology

ca 1100 BC **Abacus** is the earliest known calculating device.

1614 **John Napier** (Scottish) develops the logarithm system.

1642 **Blaise Pascal** (French) builds the first adding machine using multiple dials.

1671 **Gottfried von Leibniz** (German) builds calculator that can do both addition and multiplication.

1820 **Charles Xavier Thomas de Colmar** (French) builds the Arithmometer, the first mass-produced calculator.

1885 **Dorr Felt** (American) invents and markets a key-operated adding machine (and adds a printer in 1889).

1930 **Vannevar Bush** (American) develops the differential analyzer, an analog computer for solving differential equations.

1941 **Konrad Zuze** (German) develops the first programmable digital computer, using binary arithmetic and electric relays.

1945 **John Mauchly** and **J. Presper Eckert** develop the ENIAC, the first all-electronic computer.

1950 **Yoshiro Nakama** (Japanese) patents the floppy disk as a magnetic medium for storing data.

1956 **John Backus** (American) develops FORTRAN, the first major programming language.

```
C     FORTRAN PROGRAM FOR
PRINTING A TABLE OF CUBES
      DO 5 I = 1, 64
      ICUBE = I * I * I
      PRINT 2, I, ICUBE
    2 FORMAT (1H , I3,I7)
    5 CONTINUE
      STOP
```

1958 Bell Labs develops the modem.

1960 Digital Equipment Corporation introduces the first minicomputer, the PDP-1, to be followed with the PDP-8 in 1965.

1964 IBM's 360 mainframe becomes the standard computer for major businesses.

1965 **John Kemeny** and **Thomas Kurtz** (both American) develop the BASIC computer language.

```
PRINT
FOR Counter = 1 TO Items
  PRINT  USING "##."; Counter;
  LOCA TE , ItemColumn
  PRINT  Item$(Counter);
  LOCA TE , PriceColumn
 PRINT  Price$(Counter)
NEXT  Counter
```

Chronology 1-3: TIMELINE FOR COMPUTER TECHNOLOGY (continued)

Computer Technology

1968 **Douglas Engelbart** (American) demonstrates a word-processor system, the mouse pointing device and the use of "windows."

1971 Texas Instruments introduces the pocket calculator.

1971 **Ted Hoff** (American) invents the Intel 4004, the first computer microprocessor.

1976 IBM introduces the laser printer.

1976 Apple Computer sells Apple I in kit form, followed by the fully assembled Apple II in 1977 and the Macintosh in 1984.

1980 Microsoft introduces the MS-DOS computer disk operating system. Microsoft Windows is marketed in 1985.

1981 IBM introduces the PC.

1989 **Tim Berners-Lee** (British) invents the World Wide Web by introducing a networked hypertext system.

1991 Internet connects to 600,000 hosts in more than 100 countries.

1995 Sun Microsystems introduces the Java programming language.

1996 **Sabeer Bhatia** (Indian-American) and **Jack Smith** (American) launch Hotmail, the first webmail service.

1997 IBM's Deep Blue computer defeats World Chess Champion Garry Kasparov.

2002 The billionth personal computer was sold, second billion reached in 2007.

2010 iPad introduced in 2010.

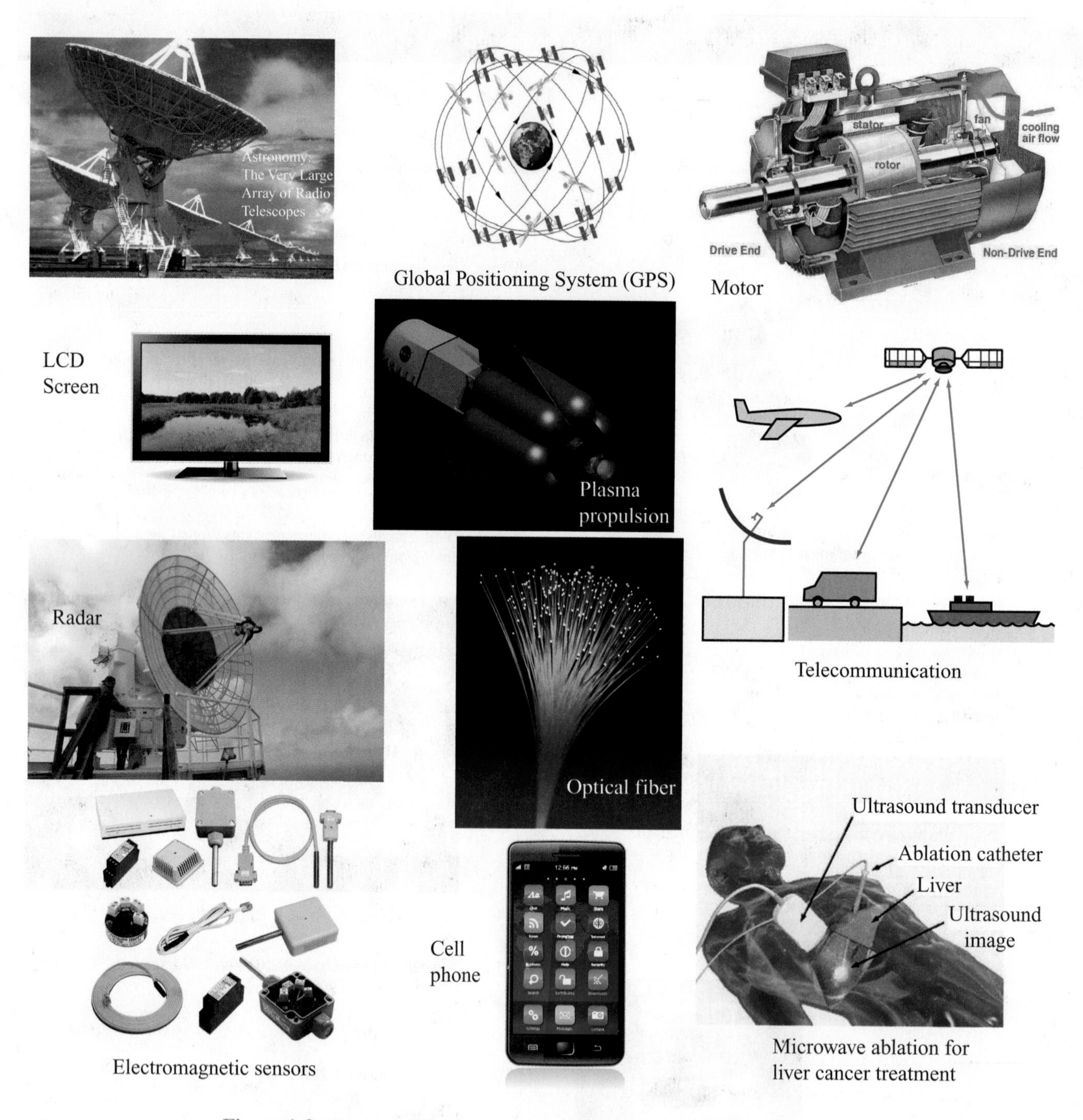

Figure 1-2 Electromagnetics is at the heart of numerous systems and applications.

Table 1-1 **Fundamental SI units.**

Dimension	Unit	Symbol
Length	meter	m
Mass	kilogram	kg
Time	second	s
Electric charge	coulomb	C
Temperature	kelvin	K
Amount of substance	mole	mol
Luminous intensity	candela	cd

technologies that have become integral parts of today's societal infrastructure. Some of the entries in these chronologies refer to specific inventions, such as the telegraph, the transistor, and the laser. The operational principles and capabilities of some of these technologies are highlighted in special sections called *Technology Briefs*, scattered throughout the book.

1-2 Dimensions, Units, and Notation

The ***International System of Units***, abbreviated ***SI*** after its French name *Système Internationale*, is the standard system used in today's scientific literature for expressing the units of physical quantities. Length is a ***dimension*** and meter is the ***unit*** by which it is expressed relative to a reference standard. The SI system is based on the units for the seven ***fundamental dimensions*** listed in **Table 1-1**. The units for all other dimensions are regarded as ***secondary*** because they are based on, and can be expressed in terms of, the seven fundamental units. Appendix A contains a list of quantities used in this book, together with their symbols and units.

For quantities ranging in value between 10^{-18} and 10^{18}, a set of prefixes, arranged in steps of 10^3, are commonly used to denote multiples and submultiples of units. These prefixes, all of which were derived from Greek, Latin, Spanish, and Danish terms, are listed in **Table 1-2**. A length of 5×10^{-9} m, for example, may be written as 5 nm.

In EM we work with scalar and vector quantities. In this book we use a medium-weight italic font for symbols denoting scalar quantities, such as R for resistance, and a boldface roman font for symbols denoting vectors, such as $\mathbf{E}$ for the electric field vector. A vector consists of a magnitude (scalar) and a direction, with the direction usually denoted by a unit vector. For example,

$$\mathbf{E} = \hat{\mathbf{x}} E, \tag{1.1}$$

Table 1-2 **Multiple and submultiple prefixes.**

Prefix	Symbol	Magnitude
exa	E	10^{18}
peta	P	10^{15}
tera	T	10^{12}
giga	G	10^{9}
mega	M	10^{6}
kilo	k	10^{3}
milli	m	10^{-3}
micro	μ	10^{-6}
nano	n	10^{-9}
pico	p	10^{-12}
femto	f	10^{-15}
atto	a	10^{-18}

where E is the magnitude of $\mathbf{E}$ and $\hat{\mathbf{x}}$ is its direction. A symbol denoting a unit vector is printed in boldface with a circumflex (ˆ) above it.

Throughout this book, we make extensive use of ***phasor representation*** in solving problems involving electromagnetic quantities that vary sinusoidally in time. Letters denoting phasor quantities are printed with a tilde (∼) over the letter. Thus, $\widetilde{\mathbf{E}}$ is the phasor electric field vector corresponding to the instantaneous electric field vector $\mathbf{E}(t)$. This notation is discussed in more detail in Section 1-7.

Notation Summary

- **Scalar quantity:** medium-weight italic, such as C for capacitance.
- **Units:** medium-weight roman, as in V/m for volts per meter.
- **Vector quantities:** boldface roman, such as $\mathbf{E}$ for electric field vector
- **Unit vectors:** boldface roman with circumflex (ˆ) over the letter, as in $\hat{\mathbf{x}}$.
- **Phasors:** a tilde (∼) over the letter; $\widetilde{E}$ is the phasor counterpart of the sinusoidally time-varying scalar field $E(t)$, and $\widetilde{\mathbf{E}}$ is the phasor counterpart of the sinusoidally time-varying vector field $\mathbf{E}(t)$.

1-3 The Nature of Electromagnetism

Our physical universe is governed by four fundamental forces of nature:

- The ***nuclear force***, which is the strongest of the four, but its range is limited to ***subatomic scales***, such as nuclei.
- The ***electromagnetic force*** exists between all charged particles. It is the dominant force in ***microscopic*** systems, such as atoms and molecules, and its strength is on the order of 10^{-2} that of the nuclear force.
- The ***weak-interaction force***, whose strength is only 10^{-14} that of the nuclear force. Its primary role is in interactions involving certain radioactive elementary particles.
- The ***gravitational force*** is the weakest of all four forces, having a strength on the order of 10^{-41} that of the nuclear force. However, it often is the dominant force in ***macroscopic*** systems, such as the solar system.

This book focuses on the electromagnetic force and its consequences. Even though the electromagnetic force operates at the atomic scale, its effects can be transmitted in the form of electromagnetic waves that can propagate through both free space and material media. The purpose of this section is to provide an overview of the basic ***framework of electromagnetism***, which consists of certain fundamental laws governing the electric and magnetic fields induced by static and moving electric charges, the relations between the electric and magnetic fields, and how these fields interact with matter. As a precursor, however, we will take advantage of our familiarity with the gravitational force by describing some of its properties because they provide a useful analogue to those of the electromagnetic force.

1-3.1 The Gravitational Force: A Useful Analogue

According to Newton's law of gravity, the gravitational force $\mathbf{F}_{g_{21}}$ acting on mass m_2 due to a mass m_1 at a distance R_{12} from m_2 (**Fig. 1-3**) is given by

$$\mathbf{F}_{g_{21}} = -\hat{\mathbf{R}}_{12} \frac{Gm_1m_2}{R_{12}^2} \quad \text{(N)}, \tag{1.2}$$

where G is the universal gravitational constant, $\hat{\mathbf{R}}_{12}$ is a unit vector that points from m_1 to m_2, and the unit for force is newton (N). The negative sign in Eq. (1.2) accounts for the fact that the gravitational force is attractive. Conversely, $\mathbf{F}_{g_{12}} = -\mathbf{F}_{g_{21}}$, where $\mathbf{F}_{g_{12}}$ is the force acting on mass m_1 due to the gravitational pull of mass m_2. Note that the first subscript of $\mathbf{F}_g$ denotes the mass experiencing the force and the second subscript denotes the source of the force.

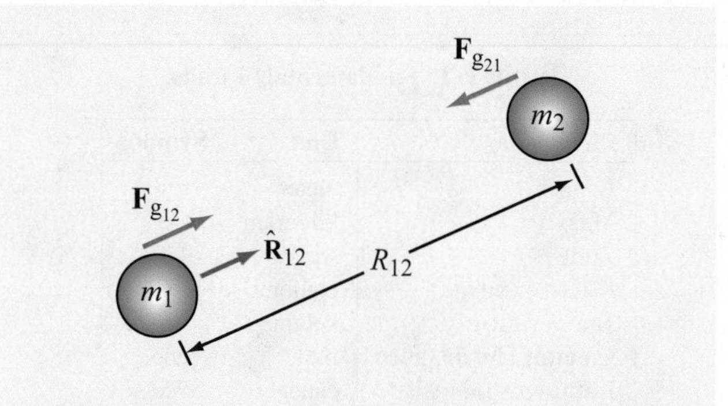

Figure 1-3 Gravitational forces between two masses.

> ▶ The force of gravitation acts at a distance. ◀

The two objects do not have to be in direct contact for each to experience the pull by the other. This phenomenon of action at a distance has led to the concept of ***fields***. An object of mass m_1 induces a ***gravitational field*** $\boldsymbol{\psi}_1$ (**Fig. 1-4**) that does not physically emanate from the object, yet its influence exists at every point in space such that if another object of mass m_2 were to exist at a distance R_{12} from the object of mass m_1, then

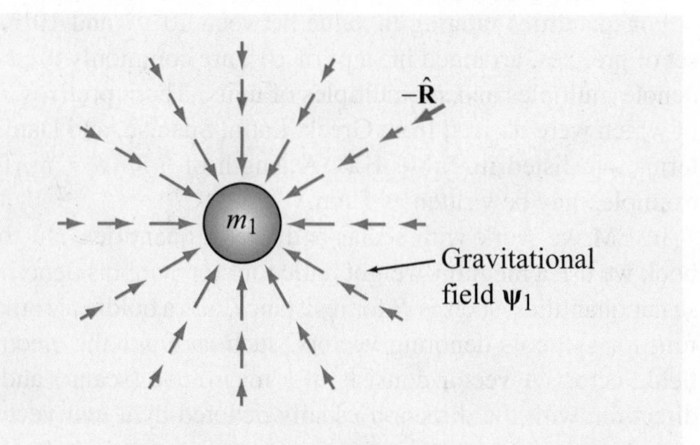

Figure 1-4 Gravitational field $\boldsymbol{\psi}_1$ induced by a mass m_1.

the object of mass m_2 would experience a force acting on it equal to

$$\mathbf{F}_{g_{21}} = \boldsymbol{\psi}_1 m_2, \qquad (1.3)$$

where

$$\boldsymbol{\psi}_1 = -\hat{\mathbf{R}}\,\frac{Gm_1}{R^2} \quad \text{(N/kg)}. \qquad (1.4)$$

In Eq. (1.4) $\hat{\mathbf{R}}$ is a unit vector that points in the radial direction away from object m_1, and therefore $-\hat{\mathbf{R}}$ points toward m_1. The force due to $\boldsymbol{\psi}_1$ acting on a mass m_2, for example, is obtained from the combination of Eqs. (1.3) and (1.4) with $R = R_{12}$ and $\hat{\mathbf{R}} = \hat{\mathbf{R}}_{12}$. The field concept may be generalized by defining the gravitational field $\boldsymbol{\psi}$ at any point in space such that when a test mass m is placed at that point, the force $\mathbf{F}_g$ acting on it is related to $\boldsymbol{\psi}$ by

$$\boldsymbol{\psi} = \frac{\mathbf{F}_g}{m}. \qquad (1.5)$$

The force $\mathbf{F}_g$ may be due to a single mass or a collection of many masses.

1-3.2 Electric Fields

The electromagnetic force consists of an electrical component $\mathbf{F}_e$ and a magnetic component $\mathbf{F}_m$.

► The electrical force $\mathbf{F}_e$ is similar to the gravitational force, but with two major differences:

(1) the source of the electrical field is electric charge, not mass, and

(2) even though both types of fields vary inversely as the square of the distance from their respective sources, electric charges may have positive or negative polarity, resulting in a force that may be attractive or repulsive. ◄

We know from atomic physics that all matter contains a mixture of neutrons, positively charged protons, and negatively charged electrons, with the fundamental quantity of charge being that of a single electron, usually denoted by the letter e. The unit by which electric charge is measured is the coulomb (C), named in honor of the eighteenth-century French scientist Charles Augustin de Coulomb (1736–1806). The magnitude of e is

$$e = 1.6 \times 10^{-19} \quad \text{(C)}. \qquad (1.6)$$

The charge of a single electron is $q_e = -e$, and that of a proton is equal in magnitude but opposite in polarity: $q_p = e$.

► Coulomb's experiments demonstrated that:

(1) two like charges repel one another, whereas two charges of opposite polarity attract,

(2) the force acts along the line joining the charges, and

(3) its strength is proportional to the product of the magnitudes of the two charges and inversely proportional to the square of the distance between them. ◄

These properties constitute what today is called ***Coulomb's law***, which can be expressed mathematically as

$$\mathbf{F}_{e_{21}} = \hat{\mathbf{R}}_{12}\,\frac{q_1 q_2}{4\pi\epsilon_0 R_{12}^2} \quad \text{(N)} \quad \textbf{(in free space)}, \qquad (1.7)$$

where $\mathbf{F}_{e_{21}}$ is the ***electrical force*** acting on charge q_2 due to charge q_1 when both are in ***free space*** (vacuum), R_{12} is the distance between the two charges, $\hat{\mathbf{R}}_{12}$ is a unit vector pointing from charge q_1 to charge q_2 (**Fig. 1-5**), and ϵ_0 is a universal constant called the ***electrical permittivity of free space*** [$\epsilon_0 = 8.854 \times 10^{-12}$ farad per meter (F/m)]. The two charges are assumed to be isolated from all other charges. The force $\mathbf{F}_{e_{12}}$ acting on charge q_1 due to charge q_2 is equal to force $\mathbf{F}_{e_{21}}$ in magnitude, but opposite in direction: $\mathbf{F}_{e_{12}} = -\mathbf{F}_{e_{21}}$.

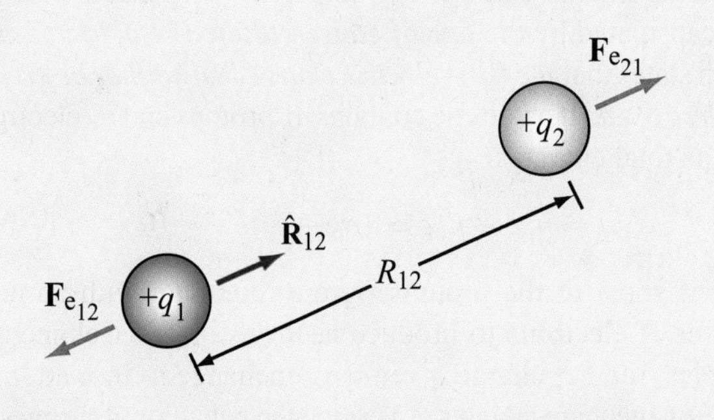

Figure 1-5 Electric forces on two positive point charges in free space.

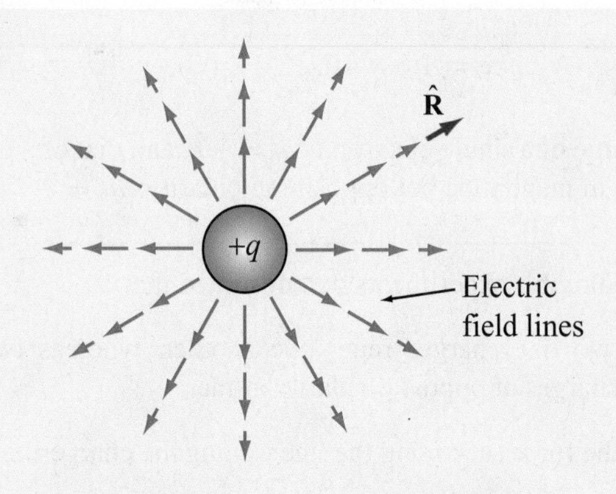

Figure 1-6 Electric field **E** due to charge q.

The expression given by Eq. (1.7) for the electrical force is analogous to that given by Eq. (1.2) for the gravitational force, and we can extend the analogy further by defining the existence of an ***electric field intensity*** **E** due to any charge q as

$$\mathbf{E} = \hat{\mathbf{R}} \frac{q}{4\pi\epsilon_0 R^2} \quad \text{(V/m)} \quad \textbf{(in free space)}, \tag{1.8}$$

where R is the distance between the charge and the observation point, and $\hat{\mathbf{R}}$ is the radial unit vector pointing away from the charge. **Figure 1-6** depicts the electric-field lines due to a positive charge. For reasons that will become apparent in later chapters, the unit for **E** is volt per meter (V/m).

▶ If any point charge q' is present in an electric field **E** (due to other charges), the point charge will experience a force acting on it equal to $\mathbf{F}_e = q'\mathbf{E}$. ◀

Electric charge exhibits two important properties. The first is encapsulated by the ***law of conservation of electric charge***, which states that *the (net) electric charge can neither be created nor destroyed*. If a volume contains n_p protons and n_e electrons, then its total charge is

$$q = n_p e - n_e e = (n_p - n_e)e \quad \text{(C)}. \tag{1.9}$$

Even if some of the protons were to combine with an equal number of electrons to produce neutrons or other elementary particles, the net charge q remains unchanged. In matter, the quantum mechanical laws governing the behavior of the protons inside the atom's nucleus and the electrons outside it do not allow them to combine.

▶ The second important property of electric charge is embodied by the ***principle of linear superposition***, which states that *the total vector electric field at a point in space due to a system of point charges is equal to the vector sum of the electric fields at that point due to the individual charges*. ◀

This seemingly simple concept allows us in future chapters to compute the electric field due to complex distributions of charge without having to be concerned with the forces acting on each individual charge due to the fields by all of the other charges.

The expression given by Eq. (1.8) describes the field induced by an electric charge residing in free space. Let us now consider what happens when we place a positive point charge in a material composed of atoms. In the absence of the point charge, the material is electrically neutral, with each atom having a positively charged nucleus surrounded by a cloud of electrons of equal but opposite polarity. Hence, at any point in the material not occupied by an atom the electric field **E** is zero. Upon placing a point charge in the material, as shown in **Fig. 1-7**, the atoms experience forces that cause them to become distorted. The center of symmetry of the electron cloud is altered with respect to the nucleus, with one pole of the atom becoming positively charged relative to the other pole. Such a polarized atom is called an ***electric dipole***, and the distortion process is called ***polarization***. The degree of polarization depends on the distance between the atom and the isolated point charge, and the orientation of the dipole is such that the axis connecting

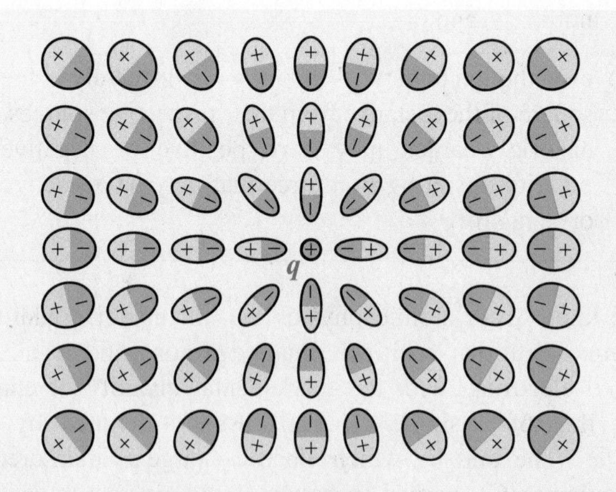

Figure 1-7 Polarization of the atoms of a dielectric material by a positive charge q.

its two poles is directed toward the point charge, as illustrated schematically in Fig. 1-7. The net result of this polarization process is that the electric fields of the dipoles of the atoms (or molecules) tend to counteract the field due to the point charge. Consequently, the electric field at any point in the material is different from the field that would have been induced by the point charge in the absence of the material. To extend Eq. (1.8) from the free-space case to any medium, we replace the permittivity of free space ϵ_0 with ϵ, where ϵ is the permittivity of the material in which the electric field is measured and is therefore characteristic of that particular material. Thus,

$$\mathbf{E} = \hat{\mathbf{R}} \frac{q}{4\pi\epsilon R^2} \quad \text{(V/m)}. \tag{1.10}$$

(material with permittivity ϵ)

Often, ϵ is expressed in the form

$$\epsilon = \epsilon_r \epsilon_0 \quad \text{(F/m)}, \tag{1.11}$$

where ϵ_r is a dimensionless quantity called the ***relative permittivity*** or ***dielectric constant*** of the material. For vacuum, $\epsilon_r = 1$; for air near Earth's surface, $\epsilon_r = 1.0006$; and the values of ϵ_r for materials that we have occasion to use in this book are tabulated in Appendix B.

In addition to the electric field intensity **E**, we often find it convenient to also use a related quantity called the ***electric flux density*** **D**, given by

$$\mathbf{D} = \epsilon \mathbf{E} \quad \text{(C/m}^2\text{)}, \tag{1.12}$$

with unit of coulomb per square meter (C/m^2).

▶ These two electric quantities, **E** and **D**, constitute one of two fundamental pairs of electromagnetic fields. The second pair consists of the magnetic fields discussed next. ◀

1-3.3 Magnetic Fields

As early as 800 B.C., the Greeks discovered that certain kinds of stones exhibit a force that attracts pieces of iron. These stones are now called ***magnetite*** (Fe_3O_4) and the phenomenon

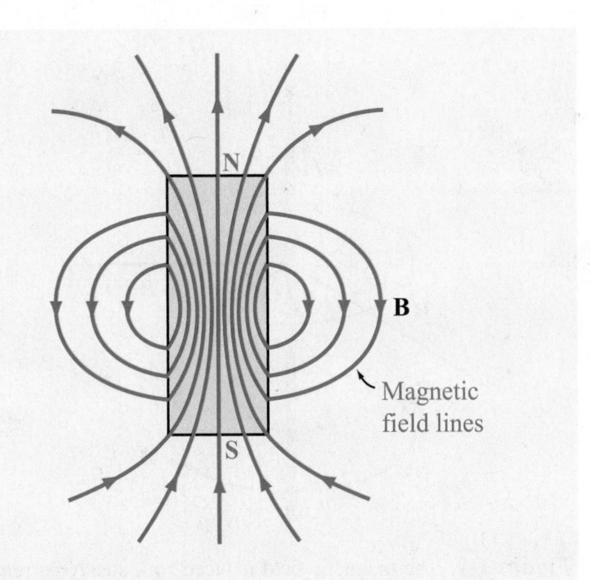

Figure 1-8 Pattern of magnetic field lines around a bar magnet.

they exhibit is known as ***magnetism***. In the thirteenth century, French scientists discovered that when a needle was placed on the surface of a spherical natural magnet, the needle oriented itself along different directions for different locations on the magnet. By mapping the directions indicated by the needle, it was determined that the magnetic force formed magnetic-field lines that encircled the sphere and appeared to pass through two points diametrically opposite to each other. These points, called the ***north and south poles*** of the magnet, were found to exist for every magnet, regardless of its shape. The magnetic-field pattern of a bar magnet is displayed in Fig. 1-8. It was also observed that like poles of different magnets repel each other and unlike poles attract each other.

▶ The attraction-repulsion property for magnets is similar to the electric force between electric charges, except for one important difference: ***electric charges can be isolated, but magnetic poles always exist in pairs.*** ◀

If a permanent magnet is cut into small pieces, no matter how small each piece is, it will always have a north and a south pole.

The magnetic lines surrounding a magnet represent the ***magnetic flux density*** **B**. A magnetic field not only exists around permanent magnets but can also be created by electric current. This connection between electricity and magnetism was discovered in 1819 by the Danish scientist Hans Oersted

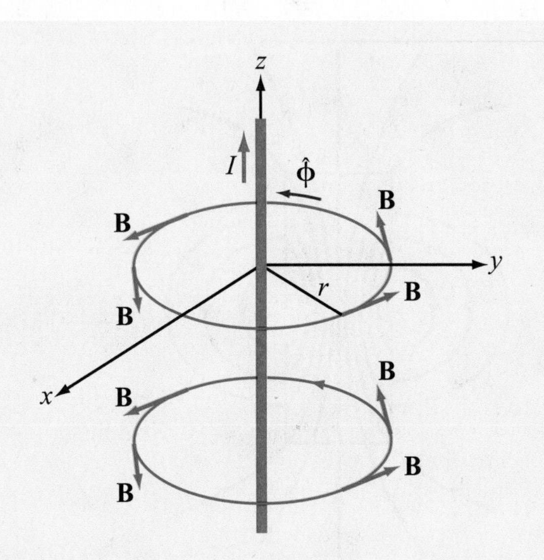

Figure 1-9 The magnetic field induced by a steady current flowing in the z direction.

(1777–1851), who observed that an electric current in a wire caused a compass needle placed in its vicinity to deflect and that the needle turned so that its direction was always perpendicular to the wire and to the radial line connecting the wire to the needle. From these observations, he deduced that the current-carrying wire induced a magnetic field that formed closed circular loops around the wire (Fig. 1-9). Shortly after Oersted's discovery, French scientists Jean Baptiste Biot and Felix Savart developed an expression that relates the magnetic flux density **B** at a point in space to the current I in the conductor. Application of their formulation, known today as the ***Biot–Savart law***, to the situation depicted in Fig. 1-9 for a very long wire residing in free space leads to the result that the ***magnetic flux density*** **B** induced by a constant current I flowing in the z direction is given by

$$\mathbf{B} = \hat{\boldsymbol{\phi}} \frac{\mu_0 I}{2\pi r} \qquad \text{(T)}, \tag{1.13}$$

where r is the radial distance from the current and $\hat{\boldsymbol{\phi}}$ is an azimuthal unit vector expressing the fact that the magnetic field direction is tangential to the circle surrounding the current (Fig. 1-9). The magnetic field is measured in tesla (T), named in honor of Nikola Tesla (1856–1943), a Croatian-American electrical engineer whose work on transformers made it possible to transport electricity over long wires without too much loss. The quantity μ_0 is called the ***magnetic permeability of free space*** [$\mu_0 = 4\pi \times 10^{-7}$ henry per meter (H/m)], and it is analogous to the electric permittivity ϵ_0. In fact, as we will see in Chapter 2, the product of ϵ_0 and μ_0 specifies c, the ***velocity of light in free space***:

$$c = \frac{1}{\sqrt{\mu_0 \epsilon_0}} = 3 \times 10^8 \qquad \text{(m/s)}. \tag{1.14}$$

We noted in Section 1-3.2 that when an electric charge q' is subjected to an electric field **E**, it experiences an electric force $\mathbf{F}_\mathrm{e} = q'\mathbf{E}$. Similarly, if a charge q' resides in the presence of a magnetic flux density **B**, it experiences a ***magnetic force*** $\mathbf{F}_\mathrm{m}$, but only if the charge is in motion and its velocity **u** is in a direction not parallel (or anti-parallel) to **B**. In fact, as we learn in more detail in Chapter 5, $\mathbf{F}_\mathrm{m}$ points in a direction perpendicular to both **B** and **u**.

To extend Eq. (1.13) to a medium other than free space, μ_0 should be replaced with μ, the ***magnetic permeability*** of the material in which **B** is being observed. The majority of natural materials are ***nonmagnetic***, meaning that they exhibit a magnetic permeability $\mu = \mu_0$. For ferromagnetic materials, such as iron and nickel, μ can be much larger than μ_0. The magnetic permeability μ accounts for ***magnetization*** properties of a material. In analogy with Eq. (1.11), μ of a particular material can be defined as

$$\mu = \mu_\mathrm{r} \mu_0 \qquad \text{(H/m)}, \tag{1.15}$$

where μ_r is a dimensionless quantity called the ***relative magnetic permeability*** of the material. The values of μ_r for commonly used ferromagnetic materials are given in Appendix B.

▶ We stated earlier that **E** and **D** constitute one of two pairs of electromagnetic field quantities. The second pair is **B** and the ***magnetic field intensity*** **H**, which are related to each other through μ:

$$\mathbf{B} = \mu \mathbf{H}. \tag{1.16}$$

1-3.4 Static and Dynamic Fields

In EM, the time variable t, or more precisely if and how electric and magnetic quantities vary with time, is of crucial importance. Before we elaborate further on the significance

Table 1-3 The three branches of electromagnetics.

Branch	Condition	Field Quantities (Units)
Electrostatics	Stationary charges ($\partial q/\partial t = 0$)	Electric field intensity **E** (V/m) Electric flux density **D** (C/m^2) $\mathbf{D} = \epsilon\mathbf{E}$
Magnetostatics	Steady currents ($\partial I/\partial t = 0$)	Magnetic flux density **B** (T) Magnetic field intensity **H** (A/m) $\mathbf{B} = \mu\mathbf{H}$
Dynamics (time-varying fields)	Time-varying currents ($\partial I/\partial t \neq 0$)	**E**, **D**, **B**, and **H** (**E**, **D**) coupled to (**B**, **H**)

of this statement, it will prove useful to define the following time-related adjectives unambiguously:

- *static*—describes a quantity that does not change with time. The term *dc* (i.e., direct current) is often used as a synonym for static to describe not only currents but other electromagnetic quantities as well.
- *dynamic*—refers to a quantity that does vary with time, but conveys no specific information about the character of the variation.
- *waveform*—refers to a plot of the magnitude profile of a quantity as a function of time.
- *periodic*—a quantity is periodic if its waveform repeats itself at a regular interval, namely its period T. Examples include the sinusoid and the square wave. By application of the Fourier series analysis technique, any periodic waveform can be expressed as the sum of an infinite series of sinusoids.
- *sinusoidal*—also called *ac* (i.e., alternating current), describes a quantity that varies sinusoidally (or cosinusoidally) with time.

In view of these terms, let us now examine the relationship between the electric field **E** and the magnetic flux density **B**. Because **E** is governed by the charge q and **B** is governed by $I = dq/dt$, one might expect that **E** and **B** must be somehow related to each other. They may or may not be interrelated, depending on whether I is static or dynamic.

Let us start by examining the dc case in which I remains constant with time. Consider a small section of a beam of charged particles, all moving at a constant velocity. The moving charges constitute a dc current. The electric field due to that section of the beam is determined by the total charge q contained in it. The magnetic field does not depend on q, but rather on the rate of charge (current) flowing through that section. Few charges moving very fast can constitute the same current as many charges moving slowly. In these two cases the induced magnetic field is the same because the current I is the same, but the induced electric field is quite different because the numbers of charges are not the same.

Electrostatics and ***magnetostatics*** refer to the study of EM under the specific, respective conditions of stationary charges and dc currents. They represent two *independent* branches, so characterized because the induced electric and magnetic fields do not couple to each other. ***Dynamics***, the third and more general branch of electromagnetics, involves ***time-varying fields*** induced by time-varying sources, that is, currents and associated charge densities. If the current associated with the beam of moving charged particles varies with time, then the amount of charge present in a given section of the beam also varies with time, and vice versa. As we see in Chapter 6, the electric and magnetic fields become coupled to each other in that case.

► A time-varying electric field generates a time-varying magnetic field, and vice versa. ◄

Table 1-3 provides a summary of the three branches of electromagnetics.

The electric and magnetic properties of materials are characterized by the parameters ϵ and μ, respectively. A third

Table 1-4 **Constitutive parameters of materials.**

Parameter	Units	Free-Space Value
Electrical permittivity ϵ	F/m	$\epsilon_0 = 8.854 \times 10^{-12}$ $\approx \frac{1}{36\pi} \times 10^{-9}$
Magnetic permeability μ	H/m	$\mu_0 = 4\pi \times 10^{-7}$
Conductivity σ	S/m	0

fundamental parameter is also needed, the ***conductivity*** of a material σ, which is measured in siemens per meter (S/m). The conductivity characterizes the ease with which charges (electrons) can move freely in a material. If $\sigma = 0$, the charges do not move more than atomic distances and the material is said to be a ***perfect dielectric***. Conversely, if $\sigma = \infty$, the charges can move very freely throughout the material, which is then called a ***perfect conductor***.

▶ The parameters ϵ, μ, and σ are often referred to as the ***constitutive parameters*** of a material (**Table 1-4**). A medium is said to be ***homogeneous*** if its constitutive parameters are constant throughout the medium. ◀

Concept Question 1-1: What are the four fundamental forces of nature and what are their relative strengths?

Concept Question 1-2: What is Coulomb's law? State its properties.

Concept Question 1-3: What are the two important properties of electric charge?

Concept Question 1-4: What do the electrical permittivity and magnetic permeability of a material account for?

Concept Question 1-5: What are the three branches and associated conditions of electromagnetics?

1-4 Traveling Waves

Waves are a natural consequence of many physical processes: waves manifest themselves as ripples on the surfaces of oceans and lakes; sound waves constitute pressure disturbances that travel through air; mechanical waves modulate stretched strings; and electromagnetic waves carry electric and magnetic fields through free space and material media as microwaves, light, and X-rays. All these various types of waves exhibit a number of common properties, including:

- ***Moving waves carry energy.***
- ***Waves have velocity;*** it takes time for a wave to travel from one point to another. Electromagnetic waves in vacuum travel at a speed of 3×10^8 m/s, and sound waves in air travel at a speed approximately a million times slower, specifically 330 m/s. Sound waves cannot travel in vacuum.
- ***Many waves exhibit a property called linearity.*** Waves that do not affect the passage of other waves are called ***linear*** because they can pass right through each other. The total of two linear waves is simply the sum of the two waves as they would exist separately. Electromagnetic waves are linear, as are sound waves. When two people speak to one another, the sound waves they generate do not interact with one another, but simply pass through each other. Water waves are approximately linear; the expanding circles of ripples caused by two pebbles thrown into two locations on a lake surface do not affect each other. Although the interaction of the two circles may exhibit a complicated pattern, it is simply the linear superposition of two independent expanding circles.

Waves are of two types: ***transient waves*** caused by sudden disturbances and ***continuous periodic waves*** generated by a repetitive source. We encounter both types of waves in this book, but most of our discussion deals with the propagation of continuous waves that vary sinusoidally with time.

An essential feature of a propagating wave is that it is a self-sustaining disturbance of the medium through which it travels. If this disturbance varies as a function of one space variable, such as the vertical displacement of the string shown in **Fig. 1-10**, we call the wave ***one-dimensional***. The vertical displacement varies with time and with the location along the length of the string. Even though the string rises up into a second dimension, the wave is only one-dimensional because the disturbance varies with only one space variable.

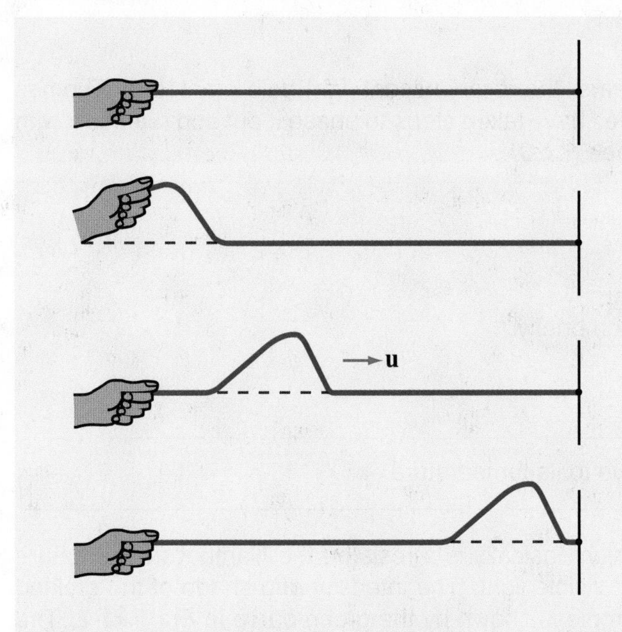

Figure 1-10 A one-dimensional wave traveling on a string.

A ***two-dimensional wave*** propagates out across a surface, like the ripples on a pond [**Fig. 1-11(a)**], and its disturbance can be described by two space variables. And by extension, a ***three-dimensional wave*** propagates through a volume and its disturbance may be a function of all three space variables. Three-dimensional waves may take on many different shapes; they include ***plane waves***, ***cylindrical waves***, and ***spherical waves***. A plane wave is characterized by a disturbance that at a given point in time has uniform properties across an infinite plane perpendicular to its direction of propagation [**Fig. 1-11(b)**]. Similarly, for cylindrical and spherical waves, the disturbances are uniform across cylindrical and spherical surfaces [**Figs. 1-11(b)** and **(c)**].

In the material that follows, we examine some of the basic properties of waves by developing mathematical formulations that describe their functional dependence on time and space variables. To keep the presentation simple, we limit our discussion to sinusoidally varying waves whose disturbances are functions of only one space variable, and we defer the discussion of more complicated waves to later chapters.

1-4.1 Sinusoidal Waves in a Lossless Medium

Regardless of the mechanism responsible for generating them, all linear waves can be described mathematically in common terms.

> ▶ A medium is said to be ***lossless*** if it does not attenuate the amplitude of the wave traveling within it or on its surface. ◀

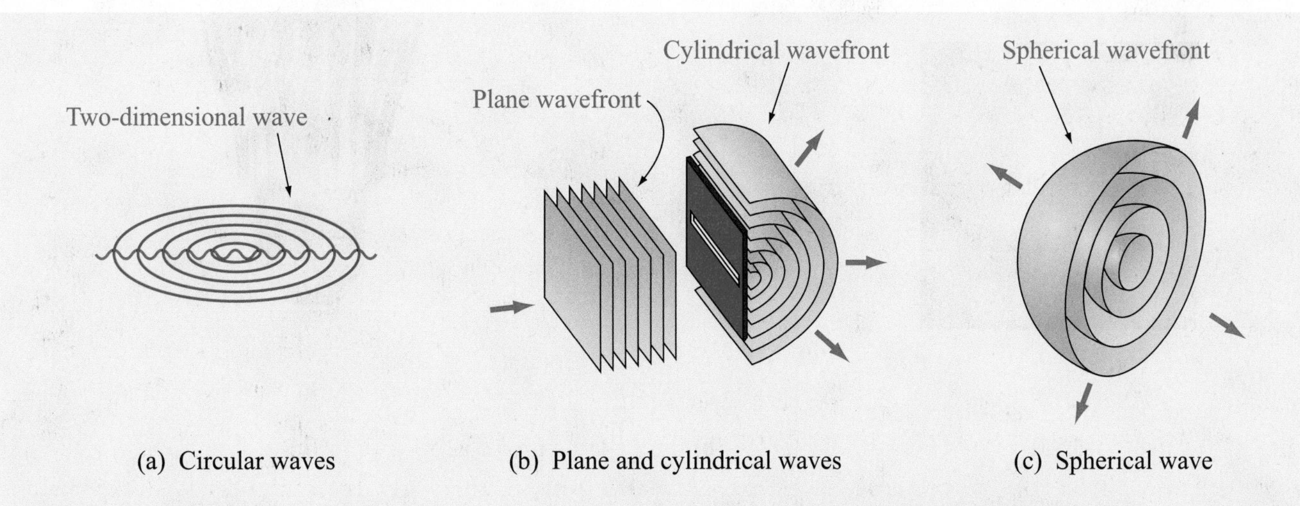

Figure 1-11 Examples of two-dimensional and three-dimensional waves: (a) circular waves on a pond, (b) a plane light wave exciting a cylindrical light wave through the use of a long narrow slit in an opaque screen, and (c) a sliced section of a spherical wave.

Technology Brief 1: LED Lighting

After lighting our homes, buildings, and streets for over 100 years, the incandescent light bulb created by Thomas Edison (1879) will soon become a relic of the past. Many countries have taken steps to phase it out and replace it with a much more energy-efficient alternative: the ***light-emitting diode (LED)***.

Light Sources

The three dominant sources of electric light are the incandescent, fluorescent, and LED light bulbs (**Fig. TF1-1**). We examine each briefly.

Incandescent Light Bulb

▶ ***Incandescence*** is the emission of light from a hot object due to its temperature. ◀

By passing electric current through a thin tungsten filament, which basically is a resistor, the filament's temperature rises to a very high level, causing the filament to glow and emit visible light. The intensity and shape of the emitted spectrum depends on the filament's temperature. A typical example is shown by the green curve in **Fig. TF1-2**. The tungsten spectrum is similar in shape to that of sunlight (yellow curve in **Fig. TF1-2**), particularly in the blue and green parts of the spectrum (400–550 nm). Despite the relatively strong (compared with sunlight) yellow light emitted by incandescent sources, the quasi-white light they produce has a quality that the human eye finds rather comfortable.

(a) (b) (c)

Figure TF1-1 (a) Incandescent light bulb; (b) fluorescent mercury vapor lamp; (c) white LED.

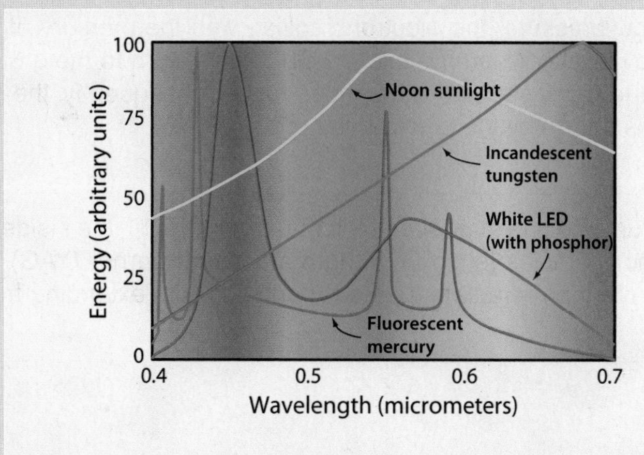

Figure TF1-2 Spectra of common sources of visible light.

▶ The incandescent light bulb is significantly less expensive to manufacture than the fluorescent and LED light bulbs, but it is far inferior with regard to energy efficacy and operational lifetime (**Fig. TF1-7**). ◀

Of the energy supplied to an incandescent light bulb, only about 2% is converted into light, with the remainder wasted as heat! In fact, the incandescent light bulb is the weakest link in the overall conversion sequence from coal to light (**Fig. TF1-3**).

Fluorescent Light Bulb

To ***fluoresce*** means to emit radiation in consequence to incident radiation of a shorter wavelength. By passing a stream of electrons between two electrodes at the ends of a tube [**Fig. TF1-1(b)**] containing mercury gas (or the noble gases

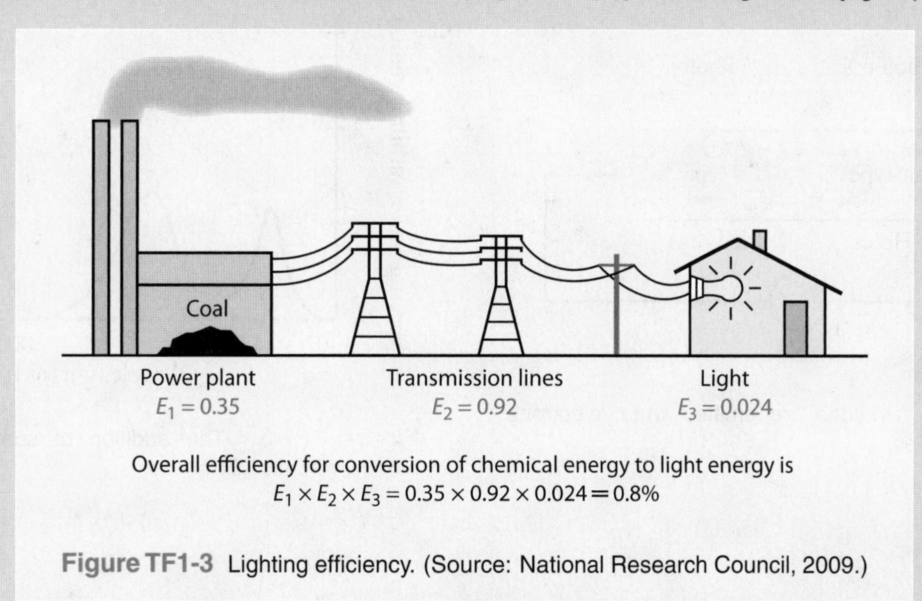

Figure TF1-3 Lighting efficiency. (Source: National Research Council, 2009.)

neon, argon, and xenon) at very low pressure, the electrons collide with the mercury atoms, causing them to excite their own electrons to higher energy levels. When the excited electrons return to the ground state, they emit photons at specific wavelengths, mostly in the ultraviolet part of the spectrum. Consequently, the spectrum of a mercury lamp is concentrated into narrow lines, as shown by the blue curve in **Fig. TF1-2**.

▶ To broaden the mercury spectrum into one that resembles that of white light, the inside surface of the fluorescent light tube is coated with phosphor particles [such as yttrium aluminum garnet (YAG) doped with cerium]. The particles absorb the UV energy and then reradiate it as a broad spectrum extending from blue to red; hence the name ***fluorescent***. ◀

Light-Emitting Diode

The LED contained inside the polymer jacket in **Fig. TF1-1(c)** is a p-n junction diode fabricated on a semiconductor chip. When a voltage is applied in a forward-biased direction across the diode (**Fig. TF1-4**), current flows through the junction and some of the streaming electrons are captured by positive charges (holes). Associated with each electron-hole recombining act is the release of energy in the form of a photon.

▶ The wavelength of the emitted photon depends on the diode's semiconductor material. The materials most commonly used are aluminum gallium arsenide (AlGaAs) to generate red light, indium gallium nitride (InGaN) to generate blue light, and aluminum gallium phosphide (AlGaP) to generate green light. In each case, the emitted energy is confined to a narrow spectral band. ◀

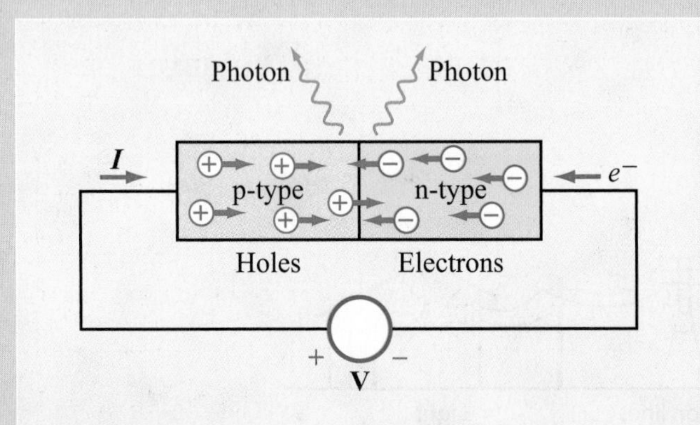

Figure TF1-4 Photons are emitted when electrons combine with holes.

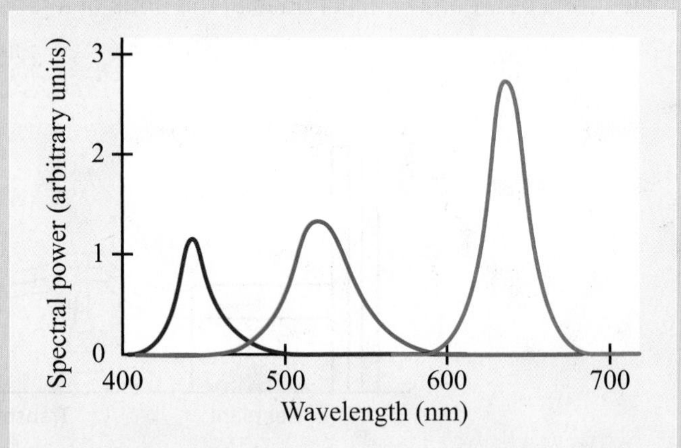

Figure TF1-5 The addition of spectra from three monochromatic LEDs.

Two basic techniques are available for generating white light with LEDs: (a) RGB and (b) blue/conversion. The RGB approach involves the use of three monochromatic LEDs whose primary colors (red, green, and blue) are mixed to generate an approximation of a white-light spectrum. An example is shown in Fig. TF1-5. The advantage of this approach is that the relative intensities of the three LEDs, can be controlled independently, thereby making it possible to "tune" the shape of the overall spectrum so as to generate an esthetically pleasing color of "white." The major shortcoming of the RGB technique is cost; manufacturing three LEDs instead of just one.

With the blue LED/phosphor conversion technique, a blue LED is used with phosphor powder particles suspended in the epoxy resin that encapsulates it. The blue light emitted by the LED is absorbed by the phosphor particles and then reemitted as a broad spectrum (Fig. TF1-6). To generate high-intensity light, several LEDs are clustered into a single enclosure.

Comparison

► ***Luminous efficacy*** (LE) is a measure of how much light in lumens is produced by a light source for each watt of electricity consumed by it. ◄

Of the three types of light bulbs we discussed, the incandescent light bulb is by far the most inefficient and its useful lifespan is the shortest (Fig. TF1-7). For a typical household scenario, the 10-year cost—including electricity and replacement cost—is several times smaller for the LED than for the alternatives.

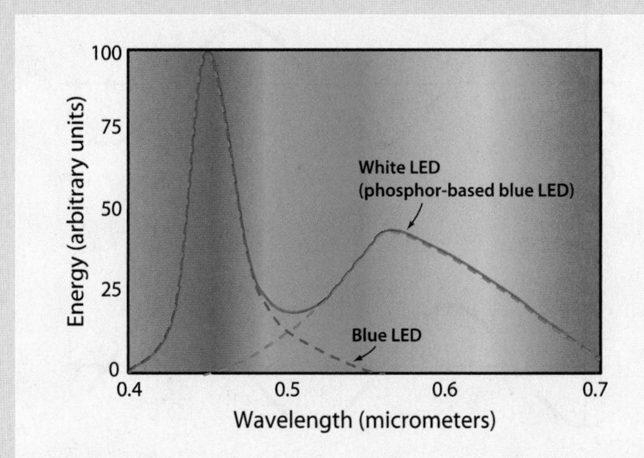

Figure TF1-6 Phosphor-based white LED emission spectrum.

Parameter	Type of Light Bulb			
	Incandescent	Fluorescent	White LED	
			Circa 2010	Circa 2025
Luminous Efficacy (lumens/W)	~12	~40	~70	~150
Useful Lifetime (hours)	~1000	~20,000	~60,000	~100,000
Purchase Price	~$1.50	~$5	~$10	~$5
Estimated Cost over 10 Years	~$410	~$110	~$100	~$40

Figure TF1-7 Even though the initial purchase price of a white LED is several times greater than that of the incandescent light bulb, the total 10-year cost of using the LED is only one-fourth of the incandescent's (in 2010) and is expected to decrease to one-tenth by 2025.

By way of an example, let us consider a wave traveling on a lake surface, and let us assume for the time being that frictional forces can be ignored, thereby allowing a wave generated on the water surface to travel indefinitely with no loss in energy. If y denotes the height of the water surface relative to the mean height (undisturbed condition) and x denotes the distance of wave travel, the functional dependence of y on time t and the spatial coordinate x has the general form

$$y(x,t) = A\cos\left(\frac{2\pi t}{T} - \frac{2\pi x}{\lambda} + \phi_0\right) \quad \text{(m)}, \tag{1.17}$$

where A is the ***amplitude*** of the wave, T is its ***time period***, λ is its ***spatial wavelength***, and ϕ_0 is a ***reference phase***. The quantity $y(x,t)$ can also be expressed in the form

$$y(x,t) = A\cos\phi(x,t) \quad \text{(m)}, \tag{1.18}$$

where

$$\phi(x,t) = \left(\frac{2\pi t}{T} - \frac{2\pi x}{\lambda} + \phi_0\right) \quad \text{(rad)}. \tag{1.19}$$

The angle $\phi(x,t)$ is called the ***phase*** of the wave, and it should not be confused with the reference phase ϕ_0, which is constant with respect to both time and space. Phase is measured by the same units as angles, that is, radians (rad) or degrees, with 2π radians = 360°.

Let us first analyze the simple case when $\phi_0 = 0$:

$$y(x,t) = A\cos\left(\frac{2\pi t}{T} - \frac{2\pi x}{\lambda}\right) \quad \text{(m)}. \tag{1.20}$$

The plots in **Fig. 1-12** show the variation of $y(x,t)$ with x at $t = 0$ and with t at $x = 0$. The wave pattern repeats itself at a spatial period λ along x and at a temporal period T along t.

If we take time snapshots of the water surface, the height profile $y(x,t)$ would exhibit the sinusoidal patterns shown in **Fig. 1-13**. In all three profiles, which correspond to three

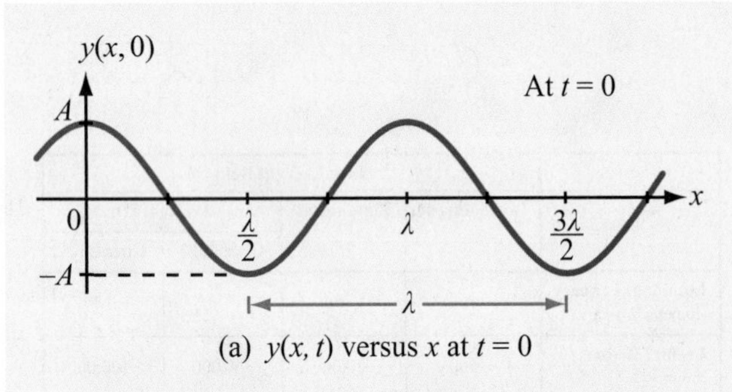

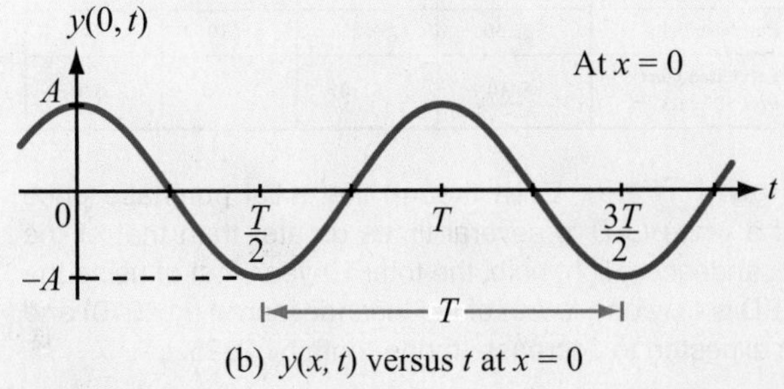

Figure 1-12 Plots of $y(x,t) = A\cos\left(\frac{2\pi t}{T} - \frac{2\pi x}{\lambda}\right)$ as a function of (a) x at $t = 0$ and (b) t at $x = 0$.

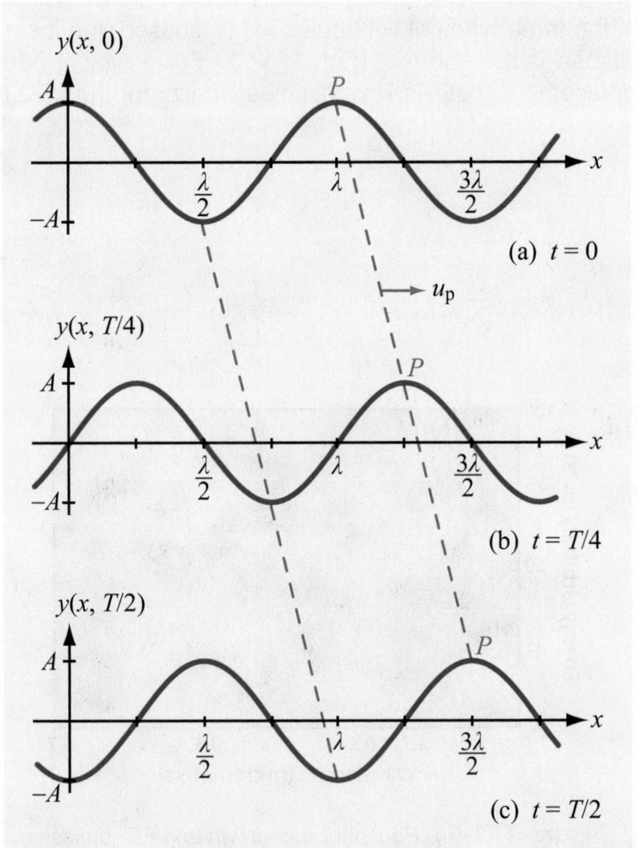

Figure 1-13 Plots of $y(x,t) = A\cos\left(\frac{2\pi t}{T} - \frac{2\pi x}{\lambda}\right)$ as a function of x at (a) $t = 0$, (b) $t = T/4$, and (c) $t = T/2$. Note that the wave moves in the $+x$ direction with a velocity $u_p = \lambda/T$.

different values of t, the spacing between peaks is equal to the wavelength λ, even though the patterns are shifted relative to one another because they correspond to different observation times. Because the pattern advances along the $+x$ direction at progressively increasing values of t, $y(x, t)$ is called a wave traveling in the $+x$ direction. If we track a given point on the wave, such as the peak P, and follow it in time, we can measure the ***phase velocity*** of the wave. At the peaks of the wave pattern, the phase $\phi(x, t)$ is equal to zero or multiples of 2π radians. Thus,

$$\phi(x, t) = \frac{2\pi t}{T} - \frac{2\pi x}{\lambda} = 2n\pi, \quad n = 0, 1, 2, \ldots \quad (1.21)$$

Had we chosen any other fixed height of the wave, say y_0, and monitored its movement as a function of t and x, this again would have been equivalent to setting the phase $\phi(x, t)$ constant such that

$$y(x, t) = y_0 = A \cos\left(\frac{2\pi t}{T} - \frac{2\pi x}{\lambda}\right), \quad (1.22)$$

or

$$\frac{2\pi t}{T} - \frac{2\pi x}{\lambda} = \cos^{-1}\left(\frac{y_0}{A}\right) = \text{constant}. \quad (1.23)$$

The apparent velocity of that fixed height is obtained by taking the time derivative of Eq. (1.23),

$$\frac{2\pi}{T} - \frac{2\pi}{\lambda}\frac{dx}{dt} = 0, \quad (1.24)$$

which gives the ***phase velocity*** u_p as

$$u_p = \frac{dx}{dt} = \frac{\lambda}{T} \quad \text{(m/s)}. \quad (1.25)$$

► The phase velocity, also called the ***propagation velocity***, is ***the velocity of the wave pattern*** as it moves across the water surface. ◄

The water itself mostly moves up and down; when the wave moves from one point to another, the water does not move physically along with it.

The ***frequency*** of a sinusoidal wave, f, is the reciprocal of its time period T:

$$f = \frac{1}{T} \quad \text{(Hz)}. \quad (1.26)$$

Combining the preceding two equations yields

$$u_p = f\lambda \quad \text{(m/s)}. \quad (1.27)$$

The wave frequency f, which is measured in cycles per second, has been assigned the unit (Hz), named in honor of the German physicist Heinrich Hertz (1857–1894), who pioneered the development of radio waves.

Using Eq. (1.26), Eq. (1.20) can be rewritten in a more compact form as

$$\begin{aligned} y(x, t) &= A \cos\left(2\pi f t - \frac{2\pi}{\lambda} x\right) \\ &= A \cos(\omega t - \beta x), \end{aligned} \quad (1.28)$$

(wave moving along $+x$ direction)

where ω is the ***angular velocity*** of the wave and β is its ***phase constant*** (or ***wavenumber***), defined as

$$\omega = 2\pi f \quad \text{(rad/s)}, \quad (1.29a)$$

$$\beta = \frac{2\pi}{\lambda} \quad \text{(rad/m)}. \quad (1.29b)$$

In terms of these two quantities,

$$u_p = f\lambda = \frac{\omega}{\beta}. \quad (1.30)$$

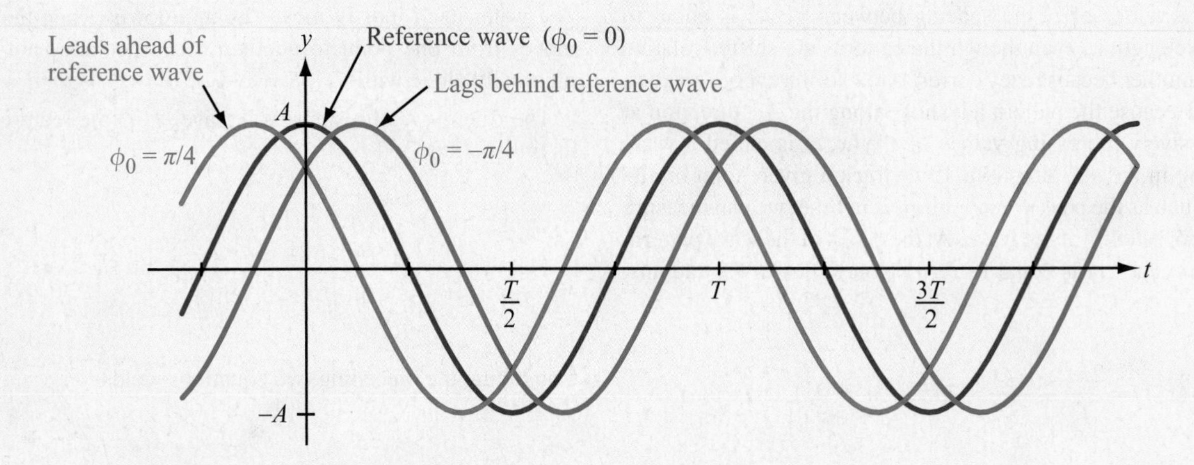

Figure 1-14 Plots of $y(0, t) = A\cos[(2\pi t/T) + \phi_0]$ for three different values of the reference phase ϕ_0.

So far, we have examined the behavior of a wave traveling in the $+x$ direction. To describe a wave traveling in the $-x$ direction, we reverse the sign of x in Eq. (1.28):

$$y(x, t) = A\cos(\omega t + \beta x). \tag{1.31}$$

(wave moving along $-x$ direction)

▶ The direction of wave propagation is easily determined by inspecting the signs of the t and x terms in the expression for the phase $\phi(x, t)$ given by Eq. (1.19): if one of the signs is positive and the other is negative, then the wave is traveling in the positive x direction, and if both signs are positive or both are negative, then the wave is traveling in the negative x direction. The constant phase reference ϕ_0 has no influence on either the speed or the direction of wave propagation. ◀

We now examine the role of the phase reference ϕ_0 given previously in Eq. (1.17). If ϕ_0 is not zero, then Eq. (1.28) should be written as

$$y(x, t) = A\cos(\omega t - \beta x + \phi_0). \tag{1.32}$$

A plot of $y(x, t)$ as a function of x at a specified t or as a function of t at a specified x is shifted in space or time, respectively, relative to a plot with $\phi_0 = 0$ by an amount proportional to ϕ_0. This is illustrated by the plots shown in **Fig. 1-14**. We observe that when ϕ_0 is positive, $y(t)$ reaches its peak value, or any other specified value, sooner than when $\phi_0 = 0$. Thus, the wave with $\phi_0 = \pi/4$ is said to ***lead*** the wave with $\phi_0 = 0$ by a ***phase lead*** of $\pi/4$; and similarly, the wave with $\phi_0 = -\pi/4$ is said to ***lag*** the wave with $\phi_0 = 0$ by a ***phase lag*** of $\pi/4$. A wave function with a negative ϕ_0 takes longer to reach a given value of $y(t)$, such as its peak, than the zero-phase reference function.

▶ When its value is positive, ϕ_0 signifies a phase lead in time, and when it is negative, it signifies a phase lag. ◀

Exercise 1-1: Consider the red wave shown in **Fig. E1.1**. What is the wave's (a) amplitude, (b) wavelength, and (c) frequency, given that its phase velocity is 6 m/s?

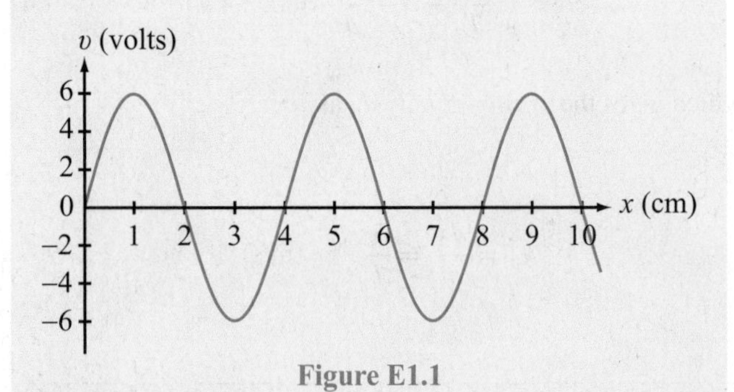

Figure E1.1

Answer: (a) $A = 6$ V, (b) $\lambda = 4$ cm, (c) $f = 150$ Hz.

Module 1.1 Sinusoidal Waveforms Learn how the shape of the waveform is related to the amplitude, frequency, and reference phase angle of a sinusoidal wave.

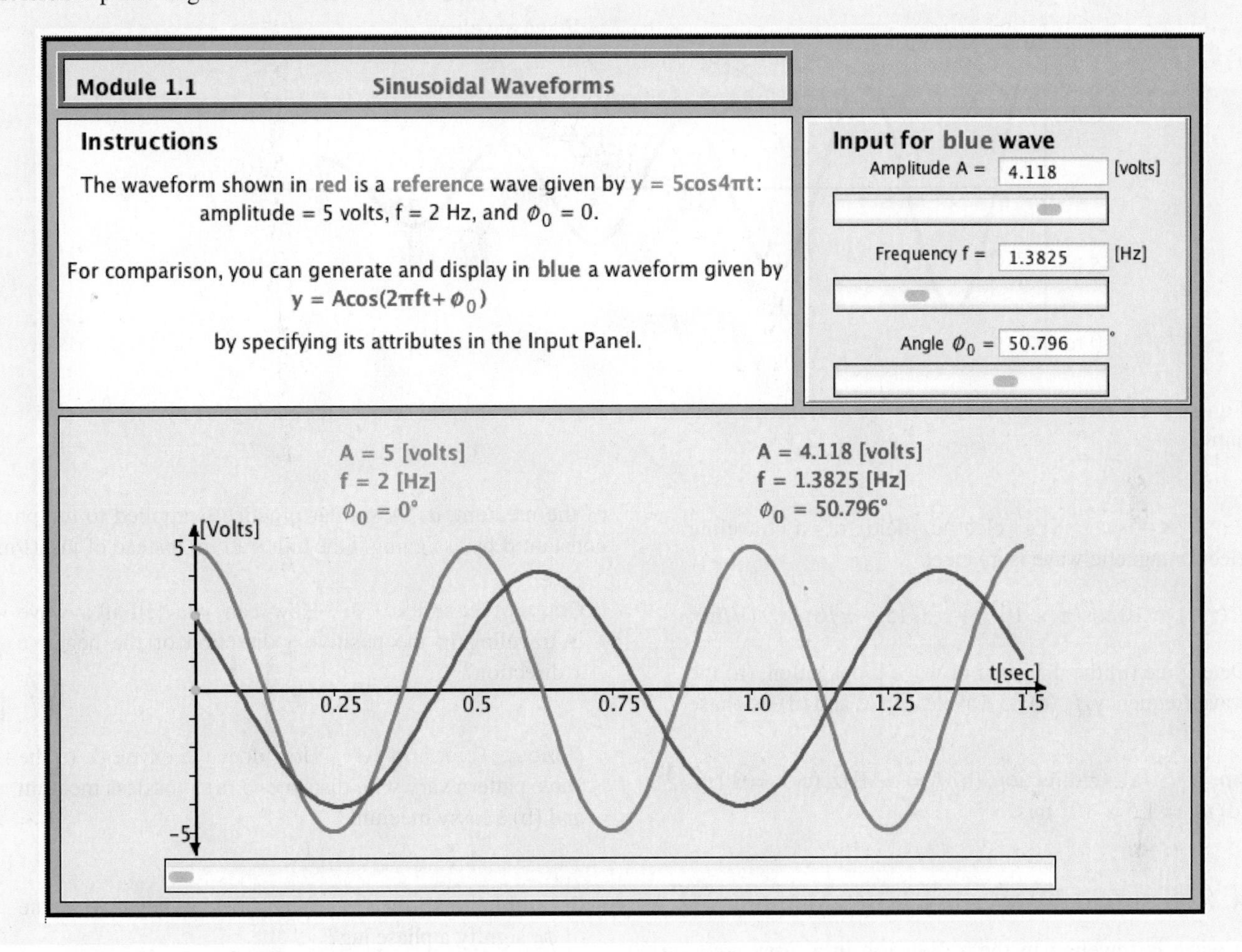

Exercise 1-2: The wave shown in red in Fig. E1.2 is given by $\upsilon = 5\cos 2\pi t/8$. Of the following four equations:

(1) $\upsilon = 5\cos(2\pi t/8 - \pi/4)$,

(2) $\upsilon = 5\cos(2\pi t/8 + \pi/4)$,

(3) $\upsilon = -5\cos(2\pi t/8 - \pi/4)$,

(4) $\upsilon = 5\sin 2\pi t/8$,

(a) which equation applies to the green wave? (b) which equation applies to the blue wave?

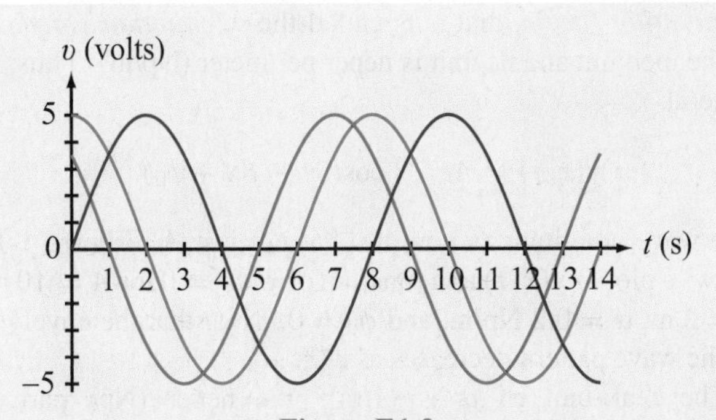

Figure E1.2

Answer: (a) #2, (b) #4.

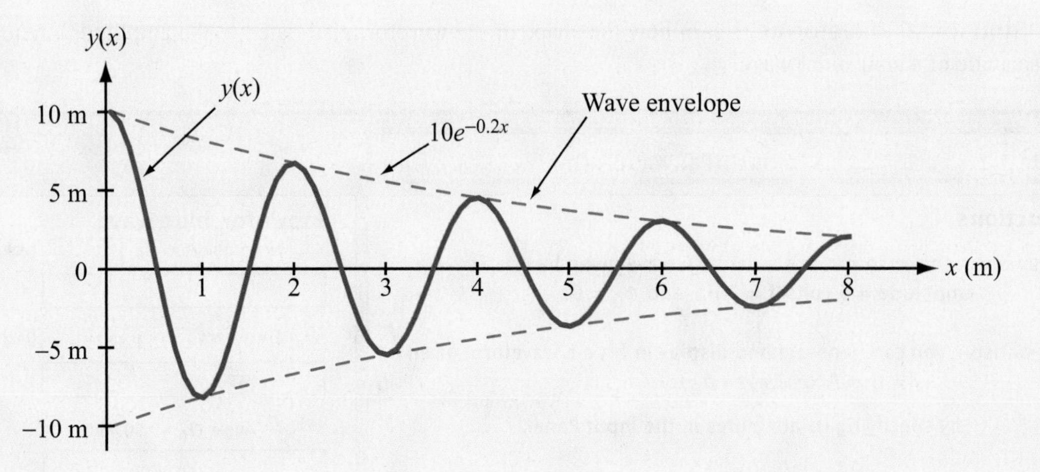

Figure 1-15 Plot of $y(x) = (10e^{-0.2x} \cos \pi x)$ meters. Note that the envelope is bounded between the curve given by $10e^{-0.2x}$ and its mirror image.

Exercise 1-3: The electric field of a traveling electromagnetic wave is given by

$$E(z, t) = 10 \cos(\pi \times 10^7 t + \pi z/15 + \pi/6) \quad \text{(V/m)}.$$

Determine (a) the direction of wave propagation, (b) the wave frequency f, (c) its wavelength λ, and (d) its phase velocity u_p.

Answer: (a) $-z$ direction, (b) $f = 5$ MHz, (c) $\lambda = 30$ m, (d) $u_p = 1.5 \times 10^8$ m/s.

1-4.2 Sinusoidal Waves in a Lossy Medium

If a wave is traveling in the x direction in a ***lossy medium***, its amplitude decreases as $e^{-\alpha x}$. This factor is called the ***attenuation factor***, and α is called the ***attenuation constant*** of the medium and its unit is neper per meter (Np/m). Thus, in general,

$$y(x, t) = Ae^{-\alpha x} \cos(\omega t - \beta x + \phi_0). \tag{1.33}$$

The wave amplitude is now $Ae^{-\alpha x}$, not just A. **Figure 1-15** shows a plot of $y(x, t)$ as a function of x at $t = 0$ for $A = 10$ m, $\lambda = 2$ m, $\alpha = 0.2$ Np/m, and $\phi_0 = 0$. Note that the envelope of the wave pattern decreases as $e^{-\alpha x}$.

The real unit of α is (1/m); the neper (Np) part is a dimensionless, artificial adjective traditionally used as a reminder that the unit (Np/m) refers to the attenuation constant of the medium, α. A similar practice is applied to the phase constant β by assigning it the unit (rad/m) instead of just (l/m).

Concept Question 1-6: How can you tell if a wave is traveling in the positive x direction or the negative x direction?

Concept Question 1-7: How does the envelope of the wave pattern vary with distance in (a) a lossless medium and (b) a lossy medium?

Concept Question 1-8: Why does a negative value of ϕ_0 signify a phase lag?

Example 1-1: Sound Wave in Water

An acoustic wave traveling in the x direction in a fluid (liquid or gas) is characterized by a differential pressure $p(x, t)$. The unit for pressure is newton per square meter (N/m^2). Find an expression for $p(x, t)$ for a sinusoidal sound wave traveling in the positive x direction in water, given that the wave frequency is 1 kHz, the velocity of sound in water is 1.5 km/s, the wave amplitude is 10 N/m^2, and $p(x, t)$ was observed to be at its maximum value at $t = 0$ and $x = 0.25$ m. Treat water as a lossless medium.

Module 1.2 Traveling Waves Learn how the shape of a traveling wave is related to its frequency and wavelength, and to the attenuation constant of the medium.

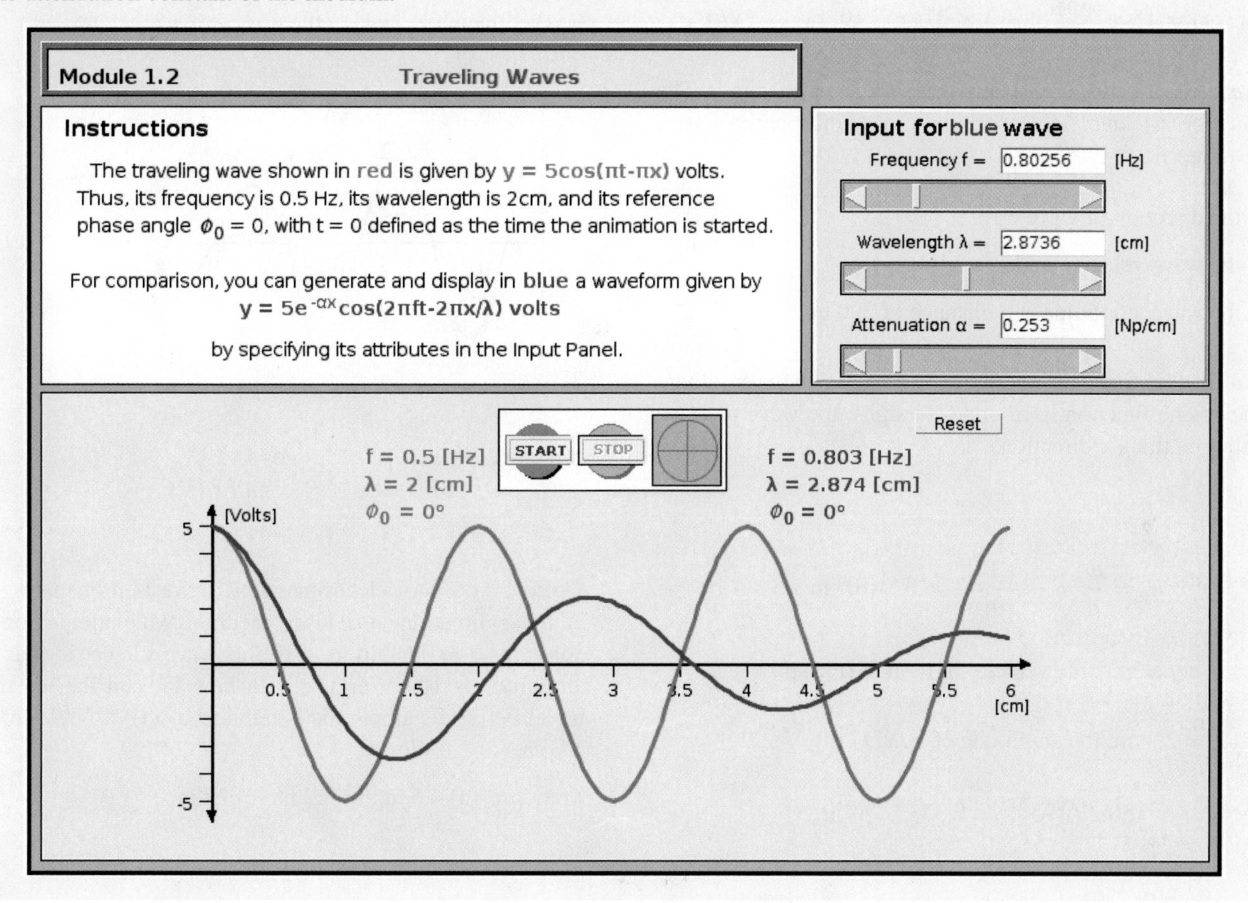

Solution: According to the general form given by Eq. (1.17) for a wave traveling in the positive x direction,

$$p(x,t) = A \cos\left(\frac{2\pi}{T}t - \frac{2\pi}{\lambda}x + \phi_0\right) \quad (\text{N/m}^2).$$

The amplitude $A = 10$ N/m^2, $T = 1/f = 10^{-3}$ s, and from $u_\text{p} = f\lambda$,

$$\lambda = \frac{u_\text{p}}{f} = \frac{1.5 \times 10^3}{10^3} = 1.5 \text{ m}.$$

Hence,

$$p(x,t) = 10\cos\left(2\pi \times 10^3 t - \frac{4\pi}{3}x + \phi_0\right) \quad (\text{N/m}^2).$$

Since at $t = 0$ and $x = 0.25$ m, $p(0.25, 0) = 10$ N/m^2, we have

$$10 = 10\cos\left(\frac{-4\pi}{3}\,0.25 + \phi_0\right) = 10\cos\left(\frac{-\pi}{3} + \phi_0\right),$$

which yields the result $(\phi_0 - \pi/3) = \cos^{-1}(1)$, or $\phi_0 = \pi/3$. Hence,

$$p(x,t) = 10\cos\left(2\pi \times 10^3 t - \frac{4\pi}{3}x + \frac{\pi}{3}\right) \quad (\text{N/m}^2).$$

Example 1-2: Power Loss

A laser beam of light propagating through the atmosphere is

characterized by an electric field given by

$$E(x, t) = 150e^{-0.03x} \cos(3 \times 10^{15}t - 10^7 x) \qquad \text{(V/m)},$$

where x is the distance from the source in meters. The attenuation is due to absorption by atmospheric gases. Determine

(a) the direction of wave travel,

(b) the wave velocity, and

(c) the wave amplitude at a distance of 200 m.

Solution: (a) Since the coefficients of t and x in the argument of the cosine function have opposite signs, the wave must be traveling in the $+x$ direction.

(b)

$$u_p = \frac{\omega}{\beta} = \frac{3 \times 10^{15}}{10^7} = 3 \times 10^8 \text{ m/s},$$

which is equal to c, the velocity of light in free space.

(c) At $x = 200$ m, the amplitude of $E(x, t)$ is

$$150e^{-0.03 \times 200} = 0.37 \qquad \text{(V/m)}.$$

Exercise 1-4: Consider the red wave shown in Fig. E1.4. What is the wave's (a) amplitude (at $x = 0$), (b) wavelength, and (c) attenuation constant?

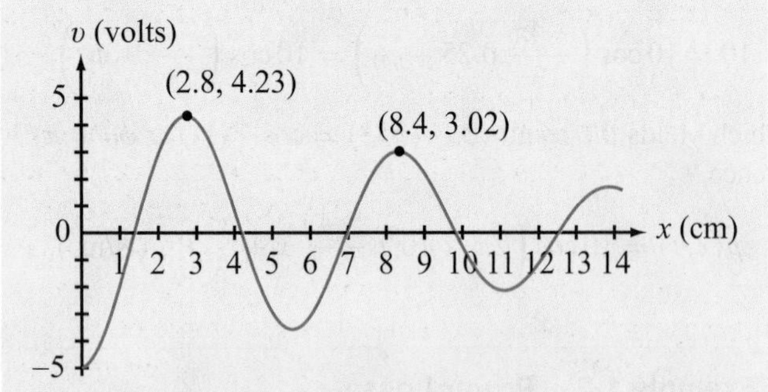

Figure E1.4

Answer: (a) 5 V, (b) 5.6 cm, (c) $\alpha = 0.06$ Np/cm.

Exercise 1-5: The red wave shown in Fig. E1.5 is given by $\upsilon = 5\cos 4\pi x$ (V). What expression is applicable to (a) the blue wave and (b) the green wave?

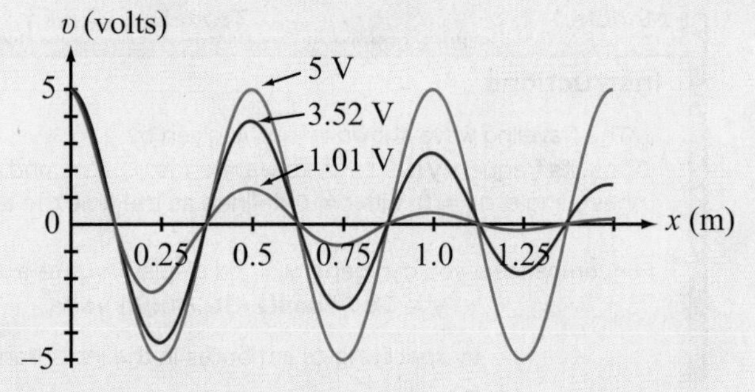

Figure E1.5

Answer: (a) $\upsilon = 5e^{-0.7x} \cos 4\pi x$ (V), (b) $\upsilon = 5e^{-3.2x} \cos 4\pi x$ (V).

Exercise 1-6: An electromagnetic wave is propagating in the z direction in a lossy medium with attenuation constant $\alpha = 0.5$ Np/m. If the wave's electric-field amplitude is 100 V/m at $z = 0$, how far can the wave travel before its amplitude is reduced to (a) 10 V/m, (b) 1 V/m, (c) 1 μV/m?

Answer: (a) 4.6 m, (b) 9.2 m, (c) 37 m.

1-5 The Electromagnetic Spectrum

Visible light belongs to a family of waves arranged according to frequency and wavelength along a continuum called the ***electromagnetic spectrum*** (Fig. 1-16). Other members of this family include gamma rays, X rays, infrared waves, and radio waves. Generically, they all are called EM waves because they share the following fundamental properties:

- A ***monochromatic*** (single frequency) EM wave consists of electric and magnetic fields that oscillate at the same frequency f.

- The phase velocity of an EM wave propagating in vacuum is a universal constant given by the velocity of light c, defined earlier by Eq. (1.14).

Module 1.3 Phase Lead/Lag Examine sinusoidal waveforms with different values of the reference phase constant ϕ_0.

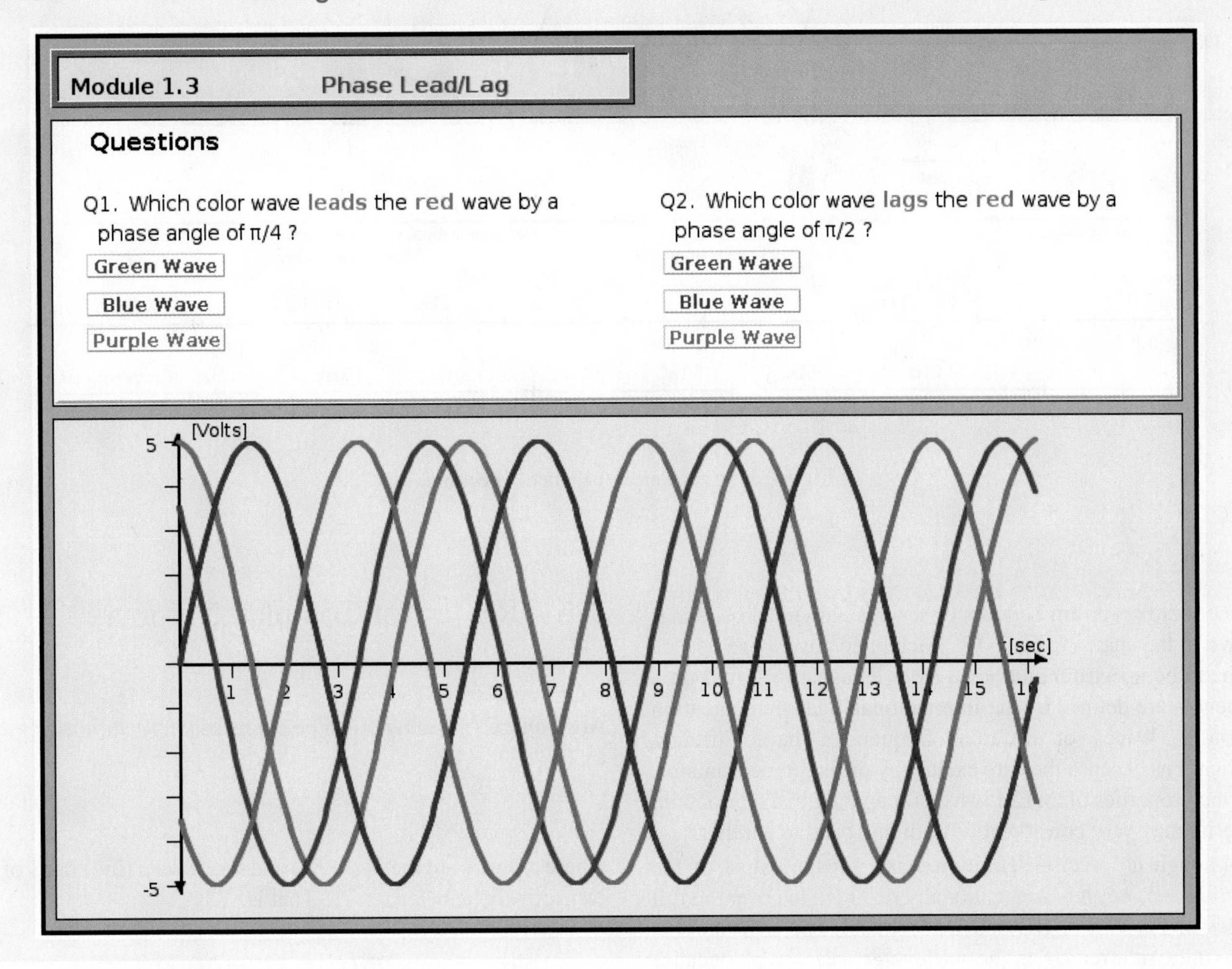

- In vacuum, the wavelength λ of an EM wave is related to its oscillation frequency f by

$$\lambda = \frac{c}{f} . \quad (1.34)$$

Whereas all monochromatic EM waves share these properties, each is distinguished by its own wavelength λ, or equivalently by its own oscillation frequency f.

The visible part of the EM spectrum shown in **Fig. 1-16** covers a very narrow wavelength range extending between $\lambda = 0.4$ μm (violet) and $\lambda = 0.7$ μm (red). As we move progressively toward shorter wavelengths, we encounter the ultraviolet, X-ray, and gamma-ray bands, each so named because of historical reasons associated with the discovery of waves with those wavelengths. On the other side of the visible spectrum lie the infrared band and then the microwave part of the radio region. Because of the link between λ and f given by Eq. (1.34), each of these spectral ranges may be specified in terms of its wavelength range or its frequency range. In practice, however, a wave is specified in terms of its wavelength λ if $\lambda < 1$ mm, which encompasses all parts of the EM spectrum except for the radio region, and the wave is specified in terms of its frequency f if $\lambda > 1$ mm (i.e., in the radio region). A wavelength of 1 mm corresponds to a frequency of 3×10^{11} Hz = 300 GHz in free space.

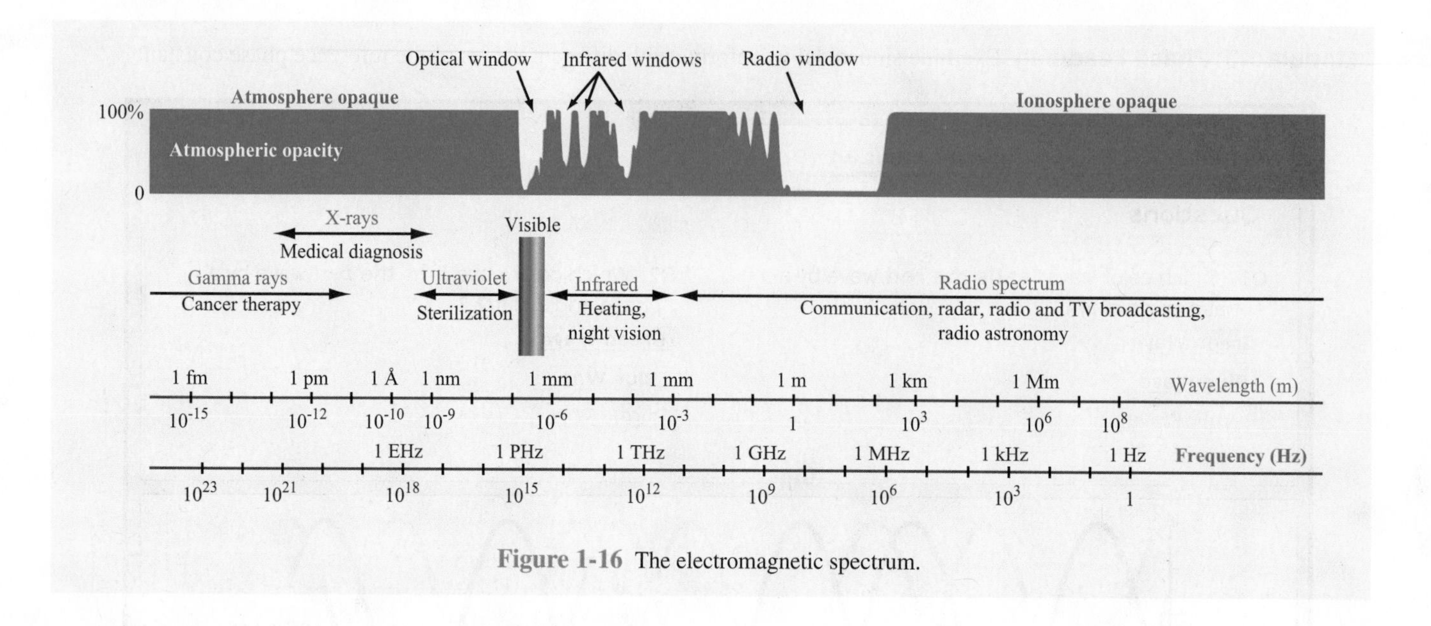

Figure 1-16 The electromagnetic spectrum.

The radio spectrum consists of several individual bands, as shown in the chart of **Fig. 1-17**. Each band covers one decade of the radio spectrum and has a letter designation based on a nomenclature defined by the International Telecommunication Union. Waves of different frequencies have different applications because they are excited by different mechanisms, and the properties of an EM wave propagating in a nonvacuum material may vary considerably from one band to another.

Although no precise definition exists for the extent of the ***microwave band***, it is conventionally regarded to cover the full ranges of the UHF, SHF, and EHF bands. The EHF band is sometimes referred to as the ***millimeter-wave band*** because the wavelength range covered by this band extends from 1 mm (300 GHz) to 1 cm (30 GHz).

Concept Question 1-9: What are the three fundamental properties of EM waves?

Concept Question 1-10: What is the range of frequencies covered by the microwave band?

Concept Question 1-11: What is the wavelength range of the visible spectrum? What are some of the applications of the infrared band?

1-6 Review of Complex Numbers

Any ***complex number*** z can be expressed in ***rectangular form*** as

$$z = x + jy, \tag{1.35}$$

where x and y are the ***real*** ($\mathfrak{Re}$) and ***imaginary*** ($\mathfrak{Im}$) parts of z, respectively, and $j = \sqrt{-1}$. That is,

$$x = \mathfrak{Re}(z), \qquad y = \mathfrak{Im}(z). \tag{1.36}$$

Alternatively, z may be cast in ***polar form*** as

$$z = |z|e^{j\theta} = |z|\angle\theta \tag{1.37}$$

where $|z|$ is the magnitude of z, θ is its phase angle, and $\angle\theta$ is a useful shorthand representation for $e^{j\theta}$. Applying ***Euler's identity***,

$$e^{j\theta} = \cos\theta + j\sin\theta, \tag{1.38}$$

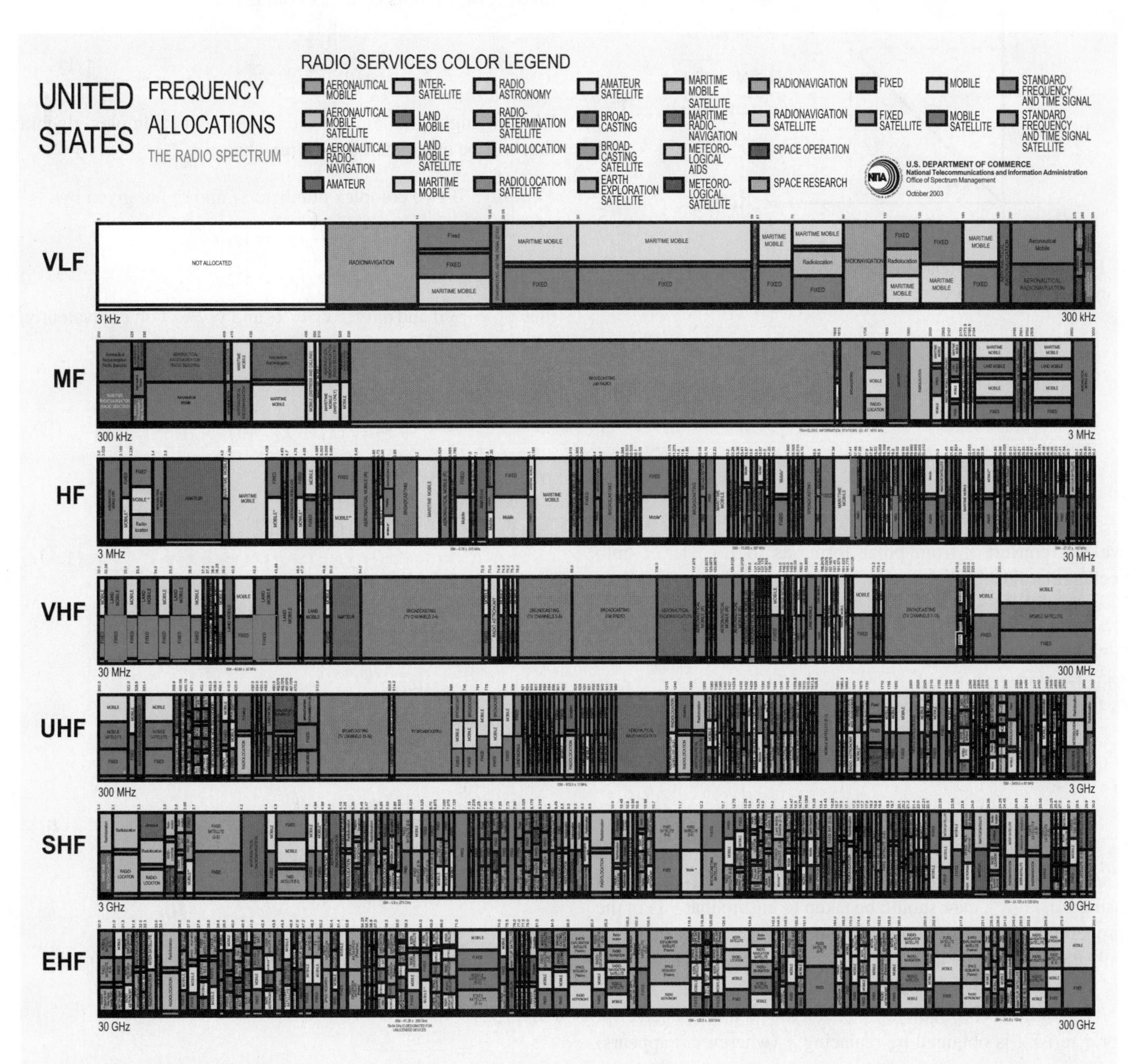

Figure 1-17 Individual bands of the radio spectrum and their primary allocations in the US. [See expandable version on book website: em.eecs.umich.edu.]

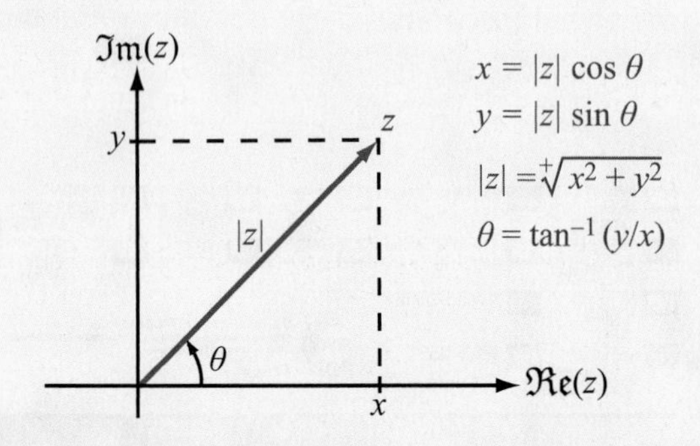

Figure 1-18 Relation between rectangular and polar representations of a complex number $z = x + jy = |z|e^{j\theta}$.

we can convert z from polar form, as in Eq. (1.37), into rectangular form,

$$z = |z|e^{j\theta} = |z| \cos\theta + j|z| \sin\theta. \tag{1.39}$$

This leads to the relations

$$x = |z| \cos\theta, \qquad y = |z| \sin\theta, \tag{1.40}$$

$$|z| = \sqrt[+]{x^2 + y^2}, \qquad \theta = \tan^{-1}(y/x). \tag{1.41}$$

The two forms are illustrated graphically in **Fig. 1-18**. When using Eq. (1.41), care should be taken to ensure that θ is in the proper quadrant. Also note that, since $|z|$ is a positive quantity, only the positive root in Eq. (1.41) is applicable. This is denoted by the + sign above the square-root sign.

The ***complex conjugate*** of z, denoted with a star superscript (or asterisk), is obtained by replacing j (wherever it appears) with $-j$, so that

$$z^* = (x + jy)^* = x - jy = |z|e^{-j\theta} = |z|\angle{-\theta}. \tag{1.42}$$

The magnitude $|z|$ is equal to the positive square root of the product of z and its complex conjugate:

$$|z| = \sqrt[+]{z\,z^*}. \tag{1.43}$$

We now highlight some of the properties of complex algebra that will be encountered in future chapters.

Equality: If two complex numbers z_1 and z_2 are given by

$$z_1 = x_1 + jy_1 = |z_1|e^{j\theta_1}, \tag{1.44}$$

$$z_2 = x_2 + jy_2 = |z_2|e^{j\theta_2}, \tag{1.45}$$

then $z_1 = z_2$ if and only if $x_1 = x_2$ and $y_1 = y_2$ or, equivalently, $|z_1| = |z_2|$ and $\theta_1 = \theta_2$.

Addition:

$$z_1 + z_2 = (x_1 + x_2) + j(y_1 + y_2). \tag{1.46}$$

Multiplication:

$$\begin{aligned} z_1 z_2 &= (x_1 + jy_1)(x_2 + jy_2) \\ &= (x_1x_2 - y_1y_2) + j(x_1y_2 + x_2y_1), \end{aligned} \tag{1.47a}$$

or

$$\begin{aligned} z_1 z_2 &= |z_1|e^{j\theta_1} \cdot |z_2|e^{j\theta_2} \\ &= |z_1||z_2|e^{j(\theta_1+\theta_2)} \\ &= |z_1||z_2|[\cos(\theta_1 + \theta_2) + j \sin(\theta_1 + \theta_2)]. \end{aligned} \tag{1.47b}$$

Division: For $z_2 \neq 0$,

$$\begin{aligned} \frac{z_1}{z_2} &= \frac{x_1 + jy_1}{x_2 + jy_2} \\ &= \frac{(x_1 + jy_1)}{(x_2 + jy_2)} \cdot \frac{(x_2 - jy_2)}{(x_2 - jy_2)} \\ &= \frac{(x_1x_2 + y_1y_2) + j(x_2y_1 - x_1y_2)}{x_2^2 + y_2^2}, \end{aligned} \tag{1.48a}$$

or

$$\begin{aligned} \frac{z_1}{z_2} &= \frac{|z_1|e^{j\theta_1}}{|z_2|e^{j\theta_2}} \\ &= \frac{|z_1|}{|z_2|} e^{j(\theta_1-\theta_2)} \\ &= \frac{|z_1|}{|z_2|}[\cos(\theta_1 - \theta_2) + j \sin(\theta_1 - \theta_2)]. \end{aligned} \tag{1.48b}$$

Powers: For any positive integer n,

$$z^n = (|z|e^{j\theta})^n$$
$$= |z|^n e^{jn\theta} = |z|^n(\cos n\theta + j \sin n\theta), \quad (1.49)$$

$$z^{1/2} = \pm|z|^{1/2}e^{j\theta/2}$$
$$= \pm|z|^{1/2}[\cos(\theta/2) + j \sin(\theta/2)]. \quad (1.50)$$

Useful Relations:

$$-1 = e^{j\pi} = e^{-j\pi} = 1\angle 180^\circ,$$
$$j = e^{j\pi/2} = 1\angle 90^\circ, \quad (1.51)$$
$$-j = -e^{j\pi/2} = e^{-j\pi/2} = 1\angle -90^\circ, \quad (1.52)$$
$$\sqrt{j} = (e^{j\pi/2})^{1/2} = \pm e^{j\pi/4} = \frac{\pm(1+j)}{\sqrt{2}}, \quad (1.53)$$
$$\sqrt{-j} = \pm e^{-j\pi/4} = \frac{\pm(1-j)}{\sqrt{2}}. \quad (1.54)$$

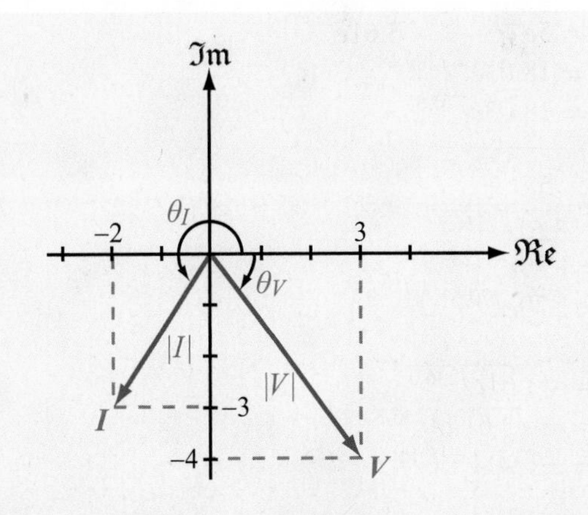

Figure 1-19 Complex numbers V and I in the complex plane (Example 1-3).

Example 1-3: Working with Complex Numbers

Given two complex numbers

$$V = 3 - j4,$$
$$I = -(2 + j3),$$

(a) express V and I in polar form, and find (b) VI, (c) VI^*, (d) V/I, and (e) $\sqrt{I}$.

Solution:

(a) $|V| = \sqrt[+]{VV^*}$
$= \sqrt[+]{(3-j4)(3+j4)} = \sqrt[+]{9+16} = 5,$
$\theta_V = \tan^{-1}(-4/3) = -53.1^\circ,$
$V = |V|e^{j\theta_V} = 5e^{-j53.1^\circ} = 5\angle -53.1^\circ,$

$|I| = \sqrt[+]{2^2 + 3^2} = \sqrt[+]{13} = 3.61.$

Since $I = (-2 - j3)$ is in the third quadrant in the complex plane [Fig. 1-19],

$$\theta_I = 180^\circ + \tan^{-1}\left(\tfrac{3}{2}\right) = 236.3^\circ,$$
$$I = 3.61\angle 236.3^\circ.$$

(b) $VI = 5e^{-j53.1^\circ} \times 3.61e^{j236.3^\circ}$
$= 18.03e^{j(236.3^\circ - 53.1^\circ)}$
$= 18.03e^{j183.2^\circ}.$

(c) $VI^* = 5e^{-j53.1^\circ} \times 3.61e^{-j236.3^\circ}$
$= 18.03e^{-j289.4^\circ}$
$= 18.03e^{j70.6^\circ}$.

(d) $\dfrac{V}{I} = \dfrac{5e^{-j53.1^\circ}}{3.61e^{j236.3^\circ}}$
$= 1.39e^{-j289.4^\circ}$
$= 1.39e^{j70.6^\circ}$.

(e) $\sqrt{I} = \sqrt{3.61e^{j236.3^\circ}}$
$= \pm\sqrt{3.61}\, e^{j236.3^\circ/2}$
$= \pm 1.90e^{j118.15^\circ}$.

Exercise 1-7: Express the following complex functions in polar form:

$$z_1 = (4 - j3)^2,$$
$$z_2 = (4 - j3)^{1/2}.$$

Answer: $z_1 = 25\angle{-73.7^\circ}$, $z_2 = \pm\sqrt{5}\angle{-18.4^\circ}$. [See (EM) (the "(EM)" symbol refers to the book website: em.eecs.umich.edu).]

Exercise 1-8: Show that $\sqrt{2j} = \pm(1 + j)$. (See (EM).)

1-7 Review of Phasors

Phasor analysis is a useful mathematical tool for solving problems involving linear systems in which the excitation is a periodic time function. Many engineering problems are cast in the form of linear integro-differential equations. If the excitation, more commonly known as the ***forcing function***, varies sinusoidally with time, the use of phasor notation to represent time-dependent variables allows us to convert a linear integro-differential equation into a linear equation with no sinusoidal functions, thereby simplifying the method of solution. After solving for the desired variable, such as the voltage or current in a circuit, conversion from the phasor domain back to the time domain provides the desired result.

The phasor technique can also be used to analyze linear systems when the forcing function is a (nonsinusoidal) periodic time function, such as a square wave or a sequence of pulses.

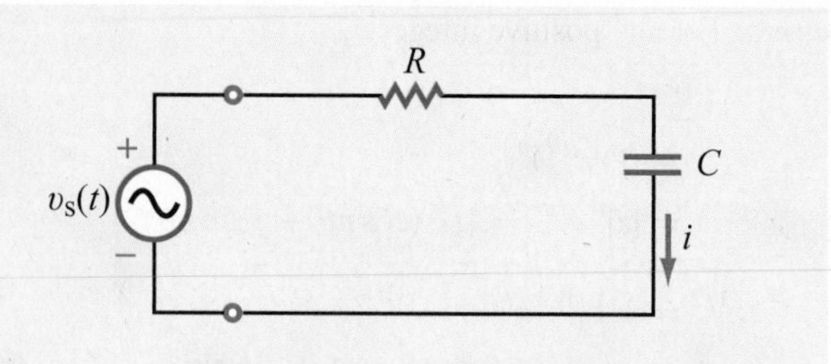

Figure 1-20 RC circuit connected to a voltage source $\upsilon_s(t)$.

By expanding the forcing function into a Fourier series of sinusoidal components, we can solve for the desired variable using phasor analysis for each Fourier component of the forcing function separately. According to the principle of superposition, the sum of the solutions due to all of the Fourier components gives the same result as one would obtain had the problem been solved entirely in the time domain without the aid of the Fourier representation. The obvious advantage of the phasor–Fourier approach is simplicity. Moreover, in the case of nonperiodic source functions, such as a single pulse, the functions can be expressed as Fourier integrals, and a similar application of the principle of superposition can be used as well.

The simple RC circuit shown in **Fig. 1-20** contains a sinusoidally time-varying voltage source given by

$$\upsilon_s(t) = V_0 \sin(\omega t + \phi_0), \tag{1.55}$$

where V_0 is the amplitude, ω is the angular frequency, and ϕ_0 is a reference phase. Application of Kirchhoff's voltage law gives the following loop equation:

$$R\,i(t) + \frac{1}{C}\int i(t)\,dt = \upsilon_s(t). \tag{1.56}$$

(time domain)

Our objective is to obtain an expression for the current $i(t)$. We can do this by solving Eq. (1.56) in the time domain, which is somewhat cumbersome because the forcing function $\upsilon_s(t)$ is a sinusoid. Alternatively, we can take advantage of the phasor-domain solution technique as follows.

1-7.1 Solution Procedure

Step 1: Adopt a cosine reference

To establish a phase reference for all time-varying currents and voltages in the circuit, the forcing function is expressed as a cosine (if not already in that form). In the present example,

$$\begin{aligned} \upsilon_s(t) &= V_0 \sin(\omega t + \phi_0) \\ &= V_0 \cos\left(\frac{\pi}{2} - \omega t - \phi_0\right) \\ &= V_0 \cos\left(\omega t + \phi_0 - \frac{\pi}{2}\right), \end{aligned} \tag{1.57}$$

where we used the properties $\sin x = \cos(\pi/2 - x)$ and $\cos(-x) = \cos x$.

Step 2: Express time-dependent variables as phasors

Any cosinusoidally time-varying function $z(t)$ can be expressed as

$$z(t) = \mathfrak{Re}\left[\widetilde{Z}\, e^{j\omega t}\right], \tag{1.58}$$

where $\widetilde{Z}$ is a time-independent function called the ***phasor*** of the ***instantaneous*** function $z(t)$. To distinguish instantaneous quantities from their phasor counterparts, a tilde ($\sim$) is added over the letter denoting a phasor. The voltage $\upsilon_s(t)$ given by Eq. (1.57) can be cast in the form

$$\begin{aligned} \upsilon_s(t) &= \mathfrak{Re}\left[V_0 e^{j(\omega t + \phi_0 - \pi/2)}\right] \\ &= \mathfrak{Re}\left[V_0 e^{j(\phi_0 - \pi/2)} e^{j\omega t}\right] = \mathfrak{Re}\left[\widetilde{V}_s e^{j\omega t}\right], \end{aligned} \tag{1.59}$$

where $\widetilde{V}_s$ consists of the expression inside the square bracket that multiplies $e^{j\omega t}$,

$$\widetilde{V}_s = V_0 e^{j(\phi_0 - \pi/2)}. \tag{1.60}$$

The phasor $\widetilde{V}_s$, corresponding to the time function $\upsilon_s(t)$, contains amplitude and phase information but is independent of the time variable t. Next we define the unknown variable $i(t)$ in terms of a phasor $\tilde{I}$,

$$i(t) = \mathfrak{Re}(\tilde{I} e^{j\omega t}), \tag{1.61}$$

and if the equation we are trying to solve contains derivatives or integrals, we use the following two properties:

$$\begin{aligned} \frac{di}{dt} &= \frac{d}{dt}\left[\mathfrak{Re}(\tilde{I} e^{j\omega t})\right] \\ &= \mathfrak{Re}\left[\frac{d}{dt}(\tilde{I} e^{j\omega t})\right] = \mathfrak{Re}[j\omega \tilde{I} e^{j\omega t}], \end{aligned} \tag{1.62}$$

and

$$\begin{aligned} \int i\; dt &= \int \mathfrak{Re}(\tilde{I} e^{j\omega t})\; dt \\ &= \mathfrak{Re}\left(\int \tilde{I} e^{j\omega t}\; dt\right) = \mathfrak{Re}\left(\frac{\tilde{I}}{j\omega}\, e^{j\omega t}\right). \end{aligned} \tag{1.63}$$

Thus, differentiation of the time function $i(t)$ is equivalent to multiplication of its phasor $\tilde{I}$ by $j\omega$, and integration is equivalent to division by $j\omega$.

Step 3: Recast the differential / integral equation in phasor form

Upon using Eqs. (1.59), (1.61), and (1.63) in Eq. (1.56), we have

$$R\, \mathfrak{Re}(\tilde{I} e^{j\omega t}) + \frac{1}{C}\, \mathfrak{Re}\left(\frac{\tilde{I}}{j\omega}\, e^{j\omega t}\right) = \mathfrak{Re}(\widetilde{V}_s e^{j\omega t}). \tag{1.64}$$

Combining all three terms under the same real-part ($\mathfrak{Re}$) operator leads to

$$\mathfrak{Re}\left\{\left[\left(R + \frac{1}{j\omega C}\right)\tilde{I} - \widetilde{V}_s\right] e^{j\omega t}\right\} = 0. \tag{1.65a}$$

Technology Brief 2: Solar Cells

A ***solar cell*** is a photovoltaic device that converts solar energy into electricity. The conversion process relies on the ***photovoltaic effect***, which was first reported by 19-year-old Edmund Bequerel in 1839 when he observed that a platinum electrode produced a small current if exposed to light. The photovoltaic effect is often confused with the ***photoelectric effect***; they are interrelated, but not identical (**Fig. TF2-1**).

The photoelectric effect explains the mechanism responsible for why an electron is ejected by a material in consequence to a photon incident upon its surface [**Fig. TF2-1(a)**]. For this to happen, the photon energy E (which is governed by its wavelength through $E = hc/\lambda$, with h being Planck's constant and c the velocity of light) has to exceed the binding energy with which the electron is held by the material. For his 1905 quantum-mechanical model of the photoelectric effect, Albert Einstein was awarded the 1921 Nobel Prize in physics.

Whereas a single material is sufficient for the photoelectric effect to occur, at least two adjoining materials with different electronic properties (to form a junction that can support a voltage across it) are needed to establish a ***photovoltaic current*** through an external load [**Fig. TF2-1(b)**]. Thus, the two effects are governed by the same quantum-mechanical rules associated with how photon energy can be used to liberate electrons away from their hosts, but the followup step of what happens to the liberated electrons is different in the two cases.

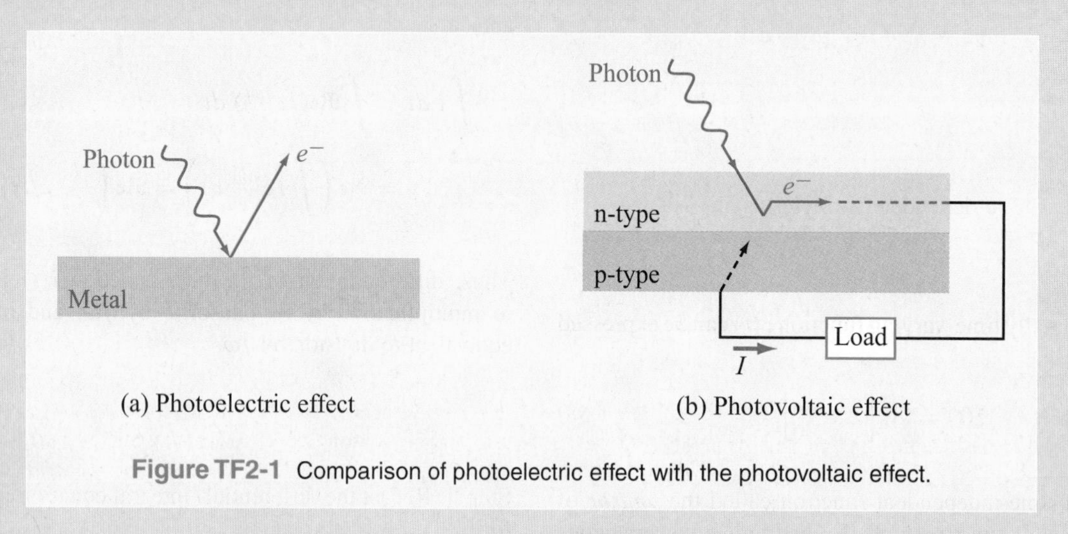

Figure TF2-1 Comparison of photoelectric effect with the photovoltaic effect.

The PV Cell

Today's photovoltaic (PV) cells are made of semiconductor materials. The basic structure of a PV cell consists of a ***p-n junction*** connected to a load (**Fig. TF2-2**).

Typically, the n-type layer is made of silicon doped with a material that creates an abundance of negatively charged atoms, and the p-type layer also is made of silicon but doped with a different material that creates an abundance of holes (atoms with missing electrons). The combination of the two layers induces an electric field across the junction, so when an incident photon liberates an electron, the electron is swept under the influence of the electric field through the n-layer and out to the external circuit connected to the load.

The ***conversion efficiency*** of a PV cell depends on several factors, including the fraction of the incident light that gets absorbed by the semiconductor material, as opposed to getting reflected by the n-type front surface or transmitted through to the back conducting electrode. To minimize the reflected component, an antireflective coating usually is inserted between the upper glass cover and the n-type layer (**Fig. TF2-2**).

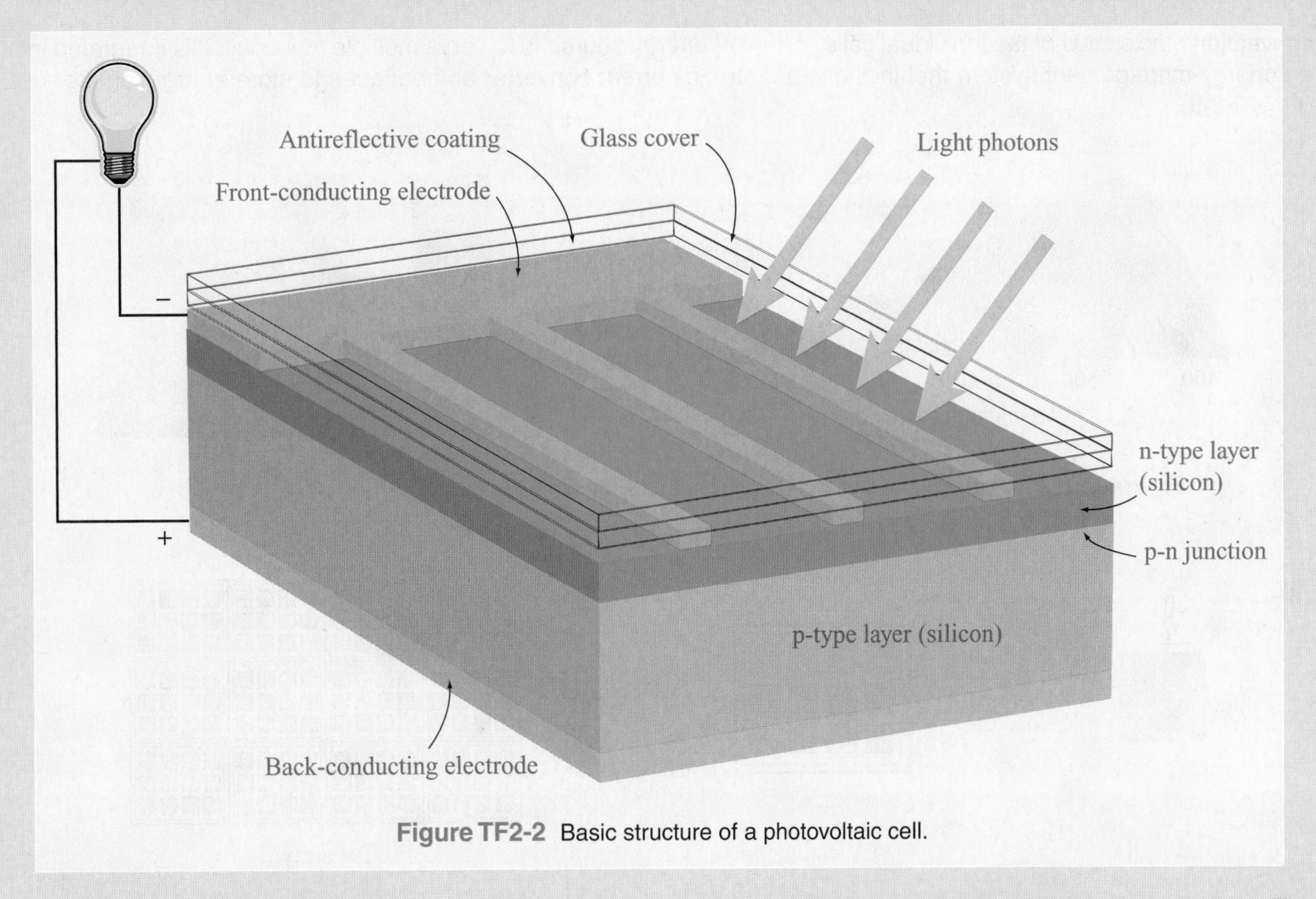

Figure TF2-2 Basic structure of a photovoltaic cell.

The PV cell shown in **Fig. TF2-2** is called a ***single-junction cell*** because it contains only one p-n junction. The semiconductor material is characterized by a quantity called its ***band gap energy***, which is the amount of energy needed to free an electron away from its host atom. Hence, for that to occur, the wavelength of the incident photon (which, in turn, defines its energy) has to be such that the photon's energy exceeds the band gap of the material. Solar energy extends over a broad spectrum, so only a fraction of the solar spectrum (photons with energies greater than the band gap) is absorbed by a single-junction material. To overcome this limitation, multiple p-n layers can be cascaded together to form a ***multijunction PV device*** (**Fig. TF2-3**). The cells usually are arranged such that the top cell has the highest band gap energy, thereby capturing the high-energy (short-wavelength) photons, followed by the cell with the next lower band gap, and so on.

► The multijunction technique offers an improvement in conversion efficiency of 2–4 times over that of the single-junction cell. However, the fabrication cost is significantly greater as well. ◄

Modules, Arrays, and Systems

A ***photovoltaic module*** consists of multiple PV cells connected together so as to supply electrical power at a specified voltage level, such as 12 or 24 V. The combination of multiple modules generates a ***PV array*** (**Fig. TF2-4**). The amount of generated power depends on the intensity of the intercepted sunlight, the total area of the module or array, and the conversion efficiencies of the individual cells. If the PV energy source is to serve multiple functions, it is integrated into an energy-management system that includes a dc-to-ac current converter and batteries to store energy for later use (**Fig. TF2-5**).

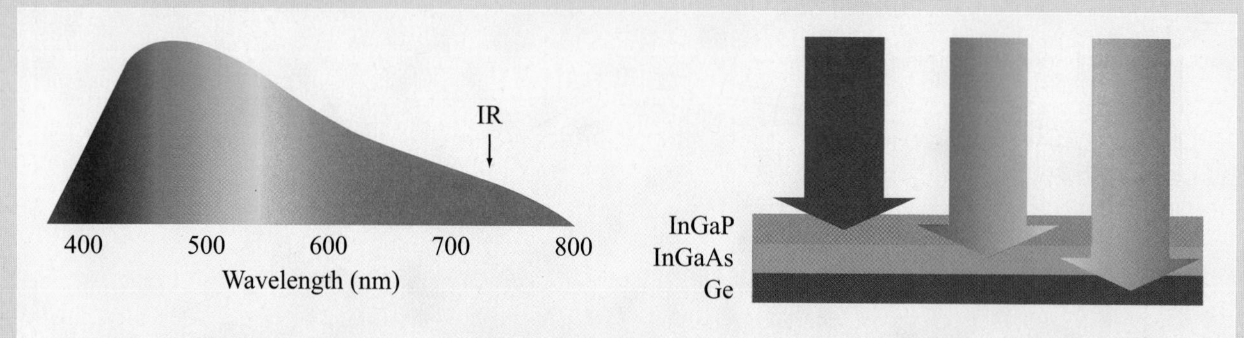

Figure TF2-3 In a multijunction PV device, different layers absorb different parts of the light spectrum.

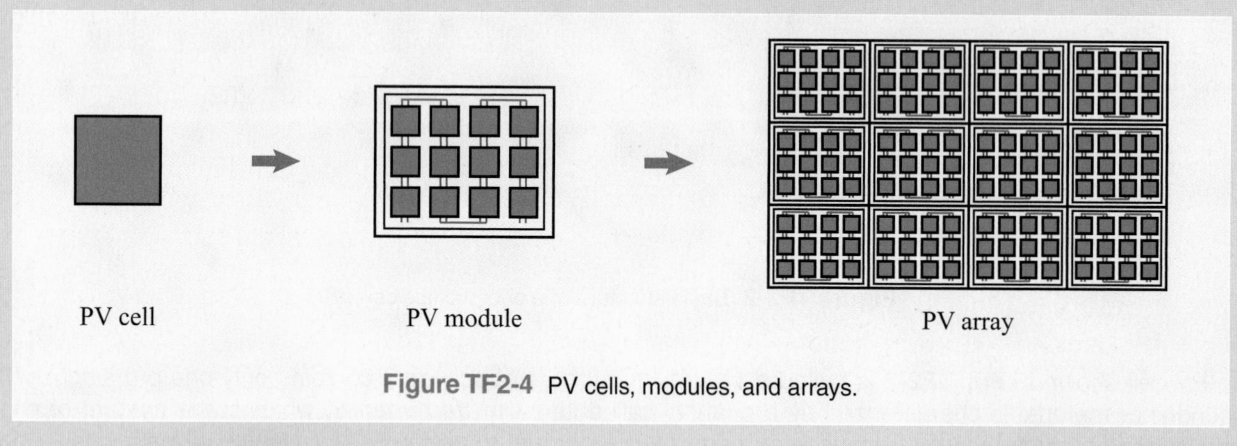

Figure TF2-4 PV cells, modules, and arrays.

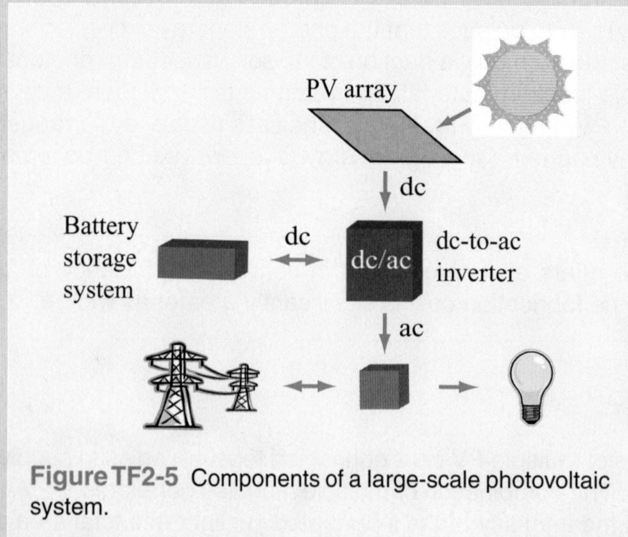

Figure TF2-5 Components of a large-scale photovoltaic system.

Had we adopted a sine reference—instead of a cosine reference—to define sinusoidal functions, the preceding treatment would have led to the result

$$\mathfrak{Im}\left\{\left[\left(R + \frac{1}{j\omega C}\right)\tilde{I} - \widetilde{V}_s\right]e^{j\omega t}\right\} = 0. \qquad (1.65b)$$

Since both the real and imaginary parts of the expression inside the curly brackets are zero, the expression itself must be zero. Moreover, since $e^{j\omega t} \neq 0$, it follows that

$$\tilde{I}\left(R + \frac{1}{j\omega C}\right) = \widetilde{V}_s \quad \textbf{(phasor domain)}. \qquad (1.66)$$

The time factor $e^{j\omega t}$ has disappeared because it was contained in all three terms. Equation (1.66) is the phasor-domain equivalent of Eq. (1.56).

Step 4: Solve the phasor-domain equation

From Eq. (1.66) the phasor current $\tilde{I}$ is given by

$$\tilde{I} = \frac{\widetilde{V}_s}{R + 1/(j\omega C)}. \qquad (1.67)$$

Before we apply the next step, we need to convert the right-hand side of Eq. (1.67) into the form $I_0 e^{j\theta}$ with I_0 being a real quantity. Thus,

$$\begin{aligned}\tilde{I} &= V_0 e^{j(\phi_0 - \pi/2)}\left[\frac{j\omega C}{1 + j\omega RC}\right] \\ &= V_0 e^{j(\phi_0 - \pi/2)}\left[\frac{\omega C e^{j\pi/2}}{\sqrt{1 + \omega^2 R^2 C^2}\, e^{j\phi_1}}\right] \\ &= \frac{V_0 \omega C}{\sqrt{1 + \omega^2 R^2 C^2}}\, e^{j(\phi_0 - \phi_1)}, \end{aligned} \qquad (1.68)$$

where we have used the identity $j = e^{j\pi/2}$. The phase angle associated with $(1 + j\omega RC)$ is $\phi_1 = \tan^{-1}(\omega RC)$ and lies in the first quadrant of the complex plane.

Table 1-5 **Time-domain sinusoidal functions $z(t)$ and their cosine-reference phasor-domain counterparts $\widetilde{Z}$, where $z(t) = \mathfrak{Re}\,[\widetilde{Z}e^{j\omega t}]$.**

$z(t)$		$\widetilde{Z}$
$A\cos\omega t$	$\leftrightarrow$	A
$A\cos(\omega t + \phi_0)$	$\leftrightarrow$	$Ae^{j\phi_0}$
$A\cos(\omega t + \beta x + \phi_0)$	$\leftrightarrow$	$Ae^{j(\beta x + \phi_0)}$
$Ae^{-\alpha x}\cos(\omega t + \beta x + \phi_0)$	$\leftrightarrow$	$Ae^{-\alpha x}e^{j(\beta x + \phi_0)}$
$A\sin\omega t$	$\leftrightarrow$	$Ae^{-j\pi/2}$
$A\sin(\omega t + \phi_0)$	$\leftrightarrow$	$Ae^{j(\phi_0 - \pi/2)}$
$\frac{d}{dt}(z(t))$	$\leftrightarrow$	$j\omega\widetilde{Z}$
$\frac{d}{dt}[A\cos(\omega t + \phi_0)]$	$\leftrightarrow$	$j\omega Ae^{j\phi_0}$
$\int z(t)\,dt$	$\leftrightarrow$	$\frac{1}{j\omega}\widetilde{Z}$
$\int A\sin(\omega t + \phi_0)\,dt$	$\leftrightarrow$	$\frac{1}{j\omega}Ae^{j(\phi_0 - \pi/2)}$

Step 5: Find the instantaneous value

To find $i(t)$, we simply apply Eq. (1.61). That is, we multiply the phasor $\tilde{I}$ given by Eq. (1.68) by $e^{j\omega t}$ and then take the real part:

$$\begin{aligned} i(t) &= \mathfrak{Re}\left[\tilde{I}e^{j\omega t}\right] \\ &= \mathfrak{Re}\left[\frac{V_0 \omega C}{\sqrt{1 + \omega^2 R^2 C^2}}\, e^{j(\phi_0 - \phi_1)} e^{j\omega t}\right] \\ &= \frac{V_0 \omega C}{\sqrt{1 + \omega^2 R^2 C^2}} \cos(\omega t + \phi_0 - \phi_1). \end{aligned} \qquad (1.69)$$

In summary, we converted all time-varying quantities into the phasor domain, solved for the phasor $\tilde{I}$ of the desired instantaneous current $i(t)$, and then converted back to the time domain to obtain an expression for $i(t)$. **Table 1-5** provides a summary of some time-domain functions and their phasor-domain equivalents.

Example 1-4: RL Circuit

The voltage source of the circuit shown in **Fig. 1-21** is given by

$$v_s(t) = 5\sin(4 \times 10^4 t - 30°) \quad \text{(V)}. \tag{1.70}$$

Obtain an expression for the voltage across the inductor.

Solution: The voltage loop equation of the RL circuit is

$$Ri + L\frac{di}{dt} = v_s(t). \tag{1.71}$$

Before converting Eq. (1.71) into the phasor domain, we express Eq. (1.70) in terms of a cosine reference:

$$\begin{aligned} v_s(t) &= 5\sin(4 \times 10^4 t - 30°) \\ &= 5\cos(4 \times 10^4 t - 120°) \quad \text{(V)}. \end{aligned} \tag{1.72}$$

The coefficient of t specifies the angular frequency as $\omega = 4 \times 10^4$ (rad/s). Per the second entry in **Table 1-5**, The voltage phasor corresponding to $v_s(t)$ is

$$\widetilde{V}_s = 5e^{-j120°} \quad \text{(V)},$$

and the phasor equation corresponding to Eq. (1.71) is

$$R\tilde{I} + j\omega L\tilde{I} = \widetilde{V}_s. \tag{1.73}$$

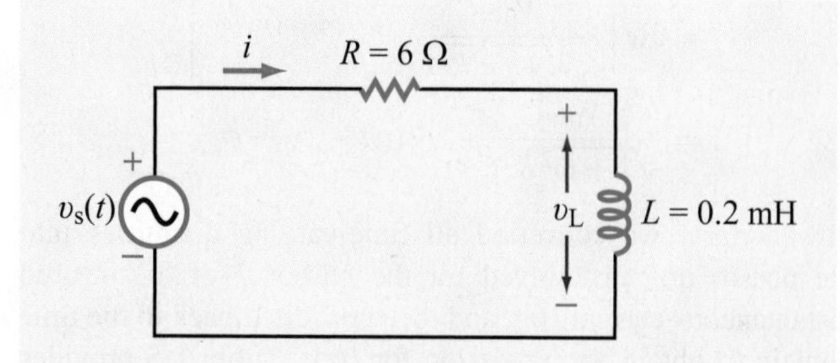

Figure 1-21 RL circuit (Example 1-4).

Solving for the current phasor $\tilde{I}$, we have

$$\begin{aligned} \tilde{I} &= \frac{\widetilde{V}_s}{R + j\omega L} \\ &= \frac{5e^{-j120°}}{6 + j4 \times 10^4 \times 2 \times 10^{-4}} \\ &= \frac{5e^{-j120°}}{6 + j8} = \frac{5e^{-j120°}}{10e^{j53.1°}} = 0.5e^{-j173.1°} \quad \text{(A)}. \end{aligned}$$

The voltage phasor across the inductor is related to $\tilde{I}$ by

$$\begin{aligned} \widetilde{V}_L &= j\omega L\tilde{I} \\ &= j4 \times 10^4 \times 2 \times 10^{-4} \times 0.5e^{-j173.1°} \\ &= 4e^{j(90° - 173.1°)} = 4e^{-j83.1°} \quad \text{(V)}, \end{aligned}$$

and the corresponding instantaneous voltage $v_L(t)$ is therefore

$$\begin{aligned} v_L(t) &= \mathfrak{Re}\left[\widetilde{V}_L e^{j\omega t}\right] \\ &= \mathfrak{Re}\left[4e^{-j83.1°} e^{j4\times 10^4 t}\right] \\ &= 4\cos(4 \times 10^4 t - 83.1°) \quad \text{(V)}. \end{aligned}$$

Concept Question 1-12: Why is the phasor technique useful? When is it used? Describe the process.

Concept Question 1-13: How is the phasor technique used when the forcing function is a nonsinusoidal periodic waveform, such as a train of pulses?

Exercise 1-9: A series RL circuit is connected to a voltage source given by $v_s(t) = 150\cos\omega t$ (V). Find (a) the phasor current $\tilde{I}$ and (b) the instantaneous current $i(t)$ for $R = 400\ \Omega$, $L = 3$ mH, and $\omega = 10^5$ rad/s.

Answer: (a) $\tilde{I} = 150/(R + j\omega L) = 0.3\angle{-36.9°}$ (A), (b) $i(t) = 0.3\cos(\omega t - 36.9°)$ (A). (See EM.)

Exercise 1-10: A phasor voltage is given by $\widetilde{V} = j5$ V. Find $v(t)$.

Answer: $v(t) = 5\cos(\omega t + \pi/2) = -5\sin\omega t$ (V). (See EM.)

1-7.2 Traveling Waves in the Phasor Domain

According to Table 1-5, if we set $\phi_0 = 0$, its third entry becomes

$$A\cos(\omega t + \beta x) \quad \leftrightarrow \quad Ae^{j\beta x}. \qquad (1.74)$$

From the discussion associated with Eq. (1.31), we concluded that $A\cos(\omega t + \beta x)$ describes a wave traveling in the negative x direction.

► In the phasor domain, a wave of amplitude A traveling in the positive x direction in a lossless medium with phase constant β is given by the negative exponential $Ae^{-j\beta x}$, and conversely, a wave traveling in the negative x direction is given by $Ae^{j\beta x}$. Thus, the sign of x in the exponential is opposite to the direction of travel. ◄

Chapter 1 Summary

Concepts

- Electromagnetics is the study of electric and magnetic phenomena and their engineering applications.
- The International System of Units consists of the six fundamental dimensions listed in Table 1-1. The units of all other physical quantities can be expressed in terms of the six fundamental units.
- The four fundamental forces of nature are the nuclear, weak-interaction, electromagnetic, and gravitational forces.
- The source of the electric field quantities **E** and **D** is the electric charge q. In a material, **E** and **D** are related by $\mathbf{D} = \epsilon\mathbf{E}$, where ϵ is the electrical permittivity of the material. In free space, $\epsilon = \epsilon_0 \approx (1/36\pi) \times 10^{-9}$ (F/m).
- The source of the magnetic field quantities **B** and **H** is the electric current I. In a material, **B** and **H** are related by $\mathbf{B} = \mu\mathbf{H}$, where μ is the magnetic permeability of the medium. In free space, $\mu = \mu_0 = 4\pi \times 10^{-7}$ (H/m).
- Electromagnetics consists of three branches: (1) electrostatics, which pertains to stationary charges, (2) magnetostatics, which pertains to dc currents, and (3) electrodynamics, which pertains to time-varying currents.
- A traveling wave is characterized by a spatial wavelength λ, a time period T, and a phase velocity $u_\text{p} = \lambda/T$.
- An electromagnetic (EM) wave consists of oscillating electric and magnetic field intensities and travels in free space at the velocity of light $c = 1/\sqrt{\epsilon_0\mu_0}$. The EM spectrum encompasses gamma rays, X-rays, visible light, infrared waves, and radio waves.
- Phasor analysis is a useful mathematical tool for solving problems involving time-periodic sources.

Mathematical and Physical Models

Electric field due to charge q in free space

$$\mathbf{E} = \hat{\mathbf{R}}\,\frac{q}{4\pi\epsilon_0 R^2}$$

Magnetic field due to current I in free space

$$\mathbf{B} = \hat{\boldsymbol{\phi}}\,\frac{\mu_0 I}{2\pi r}$$

Plane wave $\quad y(x,t) = Ae^{-\alpha x}\cos(\omega t - \beta x + \phi_0)$

- $\alpha = 0$ in lossless medium
- phase velocity $u_\text{p} = f\lambda = \frac{\omega}{\beta}$
- $\omega = 2\pi f;\ \beta = 2\pi/\lambda$
- $\phi_0 =$ phase reference

Complex numbers

- Euler's identity

$$e^{j\theta} = \cos\theta + j\sin\theta$$

- Rectangular-polar relations

$$x = |z|\cos\theta, \qquad y = |z|\sin\theta,$$

$$|z| = \sqrt[+]{x^2 + y^2}\,, \qquad \theta = \tan^{-1}(y/x)$$

Phasor-domain equivalents

Table 1-5

Important Terms

Provide definitions or explain the meaning of the following terms:

angular velocity ω
attenuation constant α
attenuation factor
Biot–Savart law
complex conjugate
complex number
conductivity σ
constitutive parameters
continuous periodic wave
Coulomb's law
dielectric constant
dynamic
electric dipole
electric field intensity **E**
electric flux density **D**
electric polarization
electrical force
electrical permittivity ϵ
electrodynamics
electrostatics
EM spectrum
Euler's identity
forcing function
fundamental dimensions
instantaneous function
law of conservation of electric charge
LCD
liquid crystal
lossless or lossy medium
magnetic field intensity **H**
magnetic flux density **B**
magnetic force
magnetic permeability μ
magnetostatics
microwave band
monochromatic
nonmagnetic materials
perfect conductor
perfect dielectric
periodic
phase
phase constant (wave number) β
phase lag and lead
phase velocity (propagation velocity) u_p
phasor
plane wave
principle of linear superposition
reference phase ϕ_0
relative permittivity or dielectric constant ϵ_r
SI system of units
static
transient wave
velocity of light c
wave amplitude
wave frequency f
wave period T
waveform
wavelength λ

PROBLEMS

Section 1-4: Traveling Waves

*1.1 A harmonic wave traveling along a string is generated by an oscillator that completes 360 vibrations per minute. If it is observed that a given crest, or maximum, travels 300 cm in 10 s, what is the wavelength?

1.2 For the pressure wave described in Example 1-1, plot the following:

(a) $p(x, t)$ versus x at $t = 0$

(b) $p(x, t)$ versus t at $x = 0$

Be sure to use appropriate scales for x and t so that each of your plots covers at least two cycles.

*1.3 A 2 kHz sound wave traveling in the x direction in air was observed to have a differential pressure $p(x, t) = 30$ N/m^2 at $x = 0$ and $t = 25$ μs. If the reference phase of $p(x, t)$ is 36°, find a complete expression for $p(x, t)$. The velocity of sound in air is 330 m/s.

1.4 A wave traveling along a string is given by

$$y(x, t) = 2 \sin(4\pi t + 10\pi x) \quad \text{(cm)},$$

where x is the distance along the string in meters and y is the vertical displacement. Determine: (a) the direction of wave travel, (b) the reference phase ϕ_0, (c) the frequency, (d) the wavelength, and (e) the phase velocity.

1.5 Two waves, $y_1(t)$ and $y_2(t)$, have identical amplitudes and oscillate at the same frequency, but $y_2(t)$ leads $y_1(t)$ by a phase angle of 60°. If

$$y_1(t) = 4 \cos(2\pi \times 10^3 t),$$

write the expression appropriate for $y_2(t)$ and plot both functions over the time span from 0 to 2 ms.

*1.6 The height of an ocean wave is described by the function

$$y(x, t) = 1.5 \sin(0.5t - 0.6x) \quad \text{(m)}.$$

Determine the phase velocity and wavelength, and then sketch $y(x, t)$ at $t = 2s$ over the range from $x = 0$ to $x = 2\lambda$.

*Answer(s) available in Appendix D.

1.7 A wave traveling along a string in the $+x$ direction is given by

$$y_1(x,t) = A\cos(\omega t - \beta x),$$

where $x = 0$ is the end of the string, which is tied rigidly to a wall, as shown in **Fig. P1.7**.

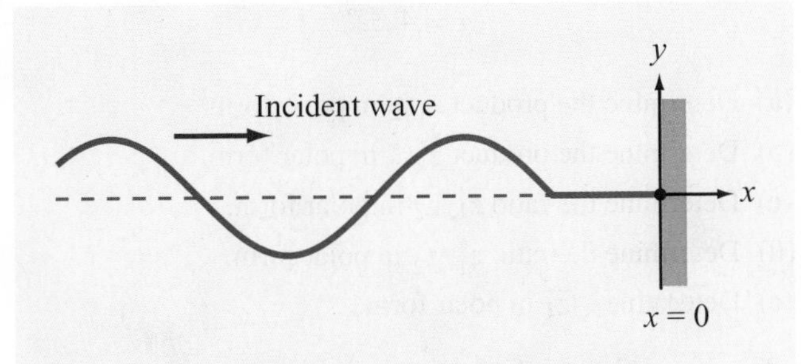

Figure P1.7 Wave on a string tied to a wall at $x = 0$ (Problem 1.7).

When wave $y_1(x,t)$ arrives at the wall, a reflected wave $y_2(x,t)$ is generated. Hence, at any location on the string, the vertical displacement y_s is the sum of the incident and reflected waves:

$$y_s(x,t) = y_1(x,t) + y_2(x,t).$$

(a) Write an expression for $y_2(x,t)$, keeping in mind its direction of travel and the fact that the end of the string cannot move.

(b) Generate plots of $y_1(x,t)$, $y_2(x,t)$ and $y_s(x,t)$ versus x over the range $-2\lambda \le x \le 0$ at $\omega t = \pi/4$ and at $\omega t = \pi/2$.

1.8 Two waves on a string are given by the following functions:

$$y_1(x,t) = 4\cos(20t - 30x) \qquad \text{(cm)}$$

$$y_2(x,t) = -4\cos(20t + 30x) \qquad \text{(cm)}$$

where x is in centimeters. The waves are said to interfere constructively when their superposition $|y_s| = |y_1 + y_2|$ is a maximum, and they interfere destructively when $|y_s|$ is a minimum.

*(a) What are the directions of propagation of waves $y_1(x,t)$ and $y_2(x,t)$?

(b) At $t = (\pi/50)$ s, at what location x do the two waves interfere constructively, and what is the corresponding value of $|y_s|$?

(c) At $t = (\pi/50)$ s, at what location x do the two waves interfere destructively, and what is the corresponding value of $|y_s|$?

1.9 Give expressions for $y(x,t)$ for a sinusoidal wave traveling along a string in the negative x direction, given that $y_{max} = 40$ cm, $\lambda = 30$ cm, $f = 10$ Hz, and

(a) $y(x,0) = 0$ at $x = 0$

(b) $y(x,0) = 0$ at $x = 3.75$ cm

***1.10** Given two waves characterized by

$$y_1(t) = 3\cos\omega t$$

$$y_2(t) = 3\sin(\omega t + 60°)$$

does $y_2(t)$ lead or lag $y_1(t)$ and by what phase angle?

1.11 The vertical displacement of a string is given by the harmonic function:

$$y(x,t) = 4\cos(16\pi t - 20\pi x) \qquad \text{(m)},$$

where x is the horizontal distance along the string in meters. Suppose a tiny particle were attached to the string at $x = 5$ cm. Obtain an expression for the vertical velocity of the particle as a function of time.

***1.12** An oscillator that generates a sinusoidal wave on a string completes 40 vibrations in 50 s. The wave peak is observed to travel a distance of 1.4 m along the string in 5 s. What is the wavelength?

1.13 The voltage of an electromagnetic wave traveling on a transmission line is given by

$$\upsilon(z,t) = 5e^{-\alpha z}\sin(4\pi \times 10^9 t - 20\pi z) \qquad \text{(V)},$$

where z is the distance in meters from the generator.

(a) Find the frequency, wavelength, and phase velocity of the wave.

(b) At $z = 4$ m, the amplitude of the wave was measured to be 2 V. Find α.

***1.14** A certain electromagnetic wave traveling in seawater was observed to have an amplitude of 98.02 (V/m) at a depth of 10 m, and an amplitude of 81.87 (V/m) at a depth of 100 m. What is the attenuation constant of seawater?

1.15 A laser beam traveling through fog was observed to have an intensity of 1 (μW/m^2) at a distance of 2 m from the laser gun and an intensity of 0.2 (μW/m^2) at a distance of 3 m. Given that the intensity of an electromagnetic wave is proportional to the square of its electric-field amplitude, find the attenuation constant α of fog.

Section 1-5: Complex Numbers

1.16 Complex numbers z_1 and z_2 are given

$$z_1 = 3 - j2$$
$$z_2 = -4 + j3$$

(a) Express z_1 and z_2 in polar form.

(b) Find $|z_1|$ by first applying Eq. (1.41) and then by applying Eq. (1.43).

*__(c)__ Determine the product $z_1 z_2$ in polar form.

(d) Determine the ratio z_1/z_2 in polar form.

(e) Determine z_1^3 in polar form.

1.17 Evaluate each of the following complex numbers and express the result in rectangular form:

(a) $z_1 = 8e^{j\pi/3}$

*__(b)__ $z_2 = \sqrt{3}\, e^{j3\pi/4}$

(c) $z_3 = 2e^{-j\pi/2}$

(d) $z_4 = j^3$

(e) $z_5 = j^{-4}$

(f) $z_6 = (1 - j)^3$

(g) $z_7 = (1 - j)^{1/2}$

1.18 Complex numbers z_1 and z_2 are given by

$$z_1 = -3 + j2$$
$$z_2 = 1 - j2$$

Determine (a) $z_1 z_2$, (b) z_1/z_2^*, (c) z_1^2, and (d) $z_1 z_1^*$, all in polar form.

1.19 If $z = -2 + j4$, determine the following quantities in polar form:

(a) $1/z$

(b) z^3

*__(c)__ $|z|^2$

(d) $\mathfrak{Im}\{z\}$

(e) $\mathfrak{Im}\{z^*\}$

1.20 Find complex numbers $t = z_1 + z_2$ and $s = z_1 - z_2$, both in polar form, for each of the following pairs:

(a) $z_1 = 2 + j3$ and $z_2 = 1 - j2$

(b) $z_1 = 3$ and $z_2 = j3$

(c) $z_1 = 3\angle 30^\circ$ and $z_2 = 3\angle -30^\circ$

*__(d)__ $z_1 = 3\angle 30^\circ$ and $z_2 = 3\angle -150^\circ$

1.21 Complex numbers z_1 and z_2 are given by

$$z_1 = 5\angle -60^\circ$$
$$z_2 = 4\angle 45^\circ.$$

(a) Determine the product $z_1 z_2$ in polar form.

(b) Determine the product $z_1 z_2^*$ in polar form.

(c) Determine the ratio z_1/z_2 in polar form.

(d) Determine the ratio z_1^*/z_2^* in polar form.

(e) Determine $\sqrt{z_1}$ in polar form.

*__1.22__ If $z = 3 + j5$, find the value of $\ln(z)$.

1.23 If $z = 3e^{j\pi/6}$, find the value of e^z.

1.24 If $z = 3 - j4$, find the value of e^z.

Section 1-6: Phasors

*__1.25__ A voltage source given by

$$v_s(t) = 25\cos(2\pi \times 10^3 t - 30^\circ) \quad \text{(V)}$$

is connected to a series RC load as shown in **Fig. 1-20**. If $R = 1$ MΩ and $C = 200$ pF, obtain an expression for $v_c(t)$, the voltage across the capacitor.

1.26 Find the instantaneous time sinusoidal functions corresponding to the following phasors:

(a) $\widetilde{V} = -5e^{j\pi/3}$ (V)

(b) $\widetilde{V} = j6e^{-j\pi/4}$ (V)

(c) $\widetilde{I} = (6 + j8)$ (A)

*__(d)__ $\widetilde{I} = -3 + j2$ (A)

(e) $\widetilde{I} = j$ (A)

(f) $\widetilde{I} = 2e^{j\pi/6}$ (A)

1.27 Find the phasors of the following time functions:

(a) $v(t) = 6\cos(\omega t - \pi/6)$ (V)

(b) $v(t) = 12\sin(\omega t - \pi/4)$ (V)

(c) $i(x, t) = 5e^{-2x}\sin(\omega t + \pi/6)$ (A)

*__(d)__ $i(t) = -2\cos(\omega t - 3\pi/4)$ (A)

(e) $i(t) = 4\sin(\omega t + \pi/3) + 7\cos(\omega t - \pi/6)$ (A)

1.28 A series RLC circuit is connected to a generator with a voltage $v_s(t) = V_0 \cos(\omega t + \pi/3)$ (V).

(a) Write the voltage loop equation in terms of the current $i(t)$, R, L, C, and $v_s(t)$.

(b) Obtain the corresponding phasor-domain equation.

(c) Solve the equation to obtain an expression for the phasor current $\tilde{I}$.

1.29 The voltage source of the circuit shown in **Fig. P1.29** is given by

$$v_s(t) = 25 \cos(4 \times 10^4 t - 45°) \quad \text{(V)}.$$

Obtain an expression for $i_L(t)$, the current flowing through the inductor.

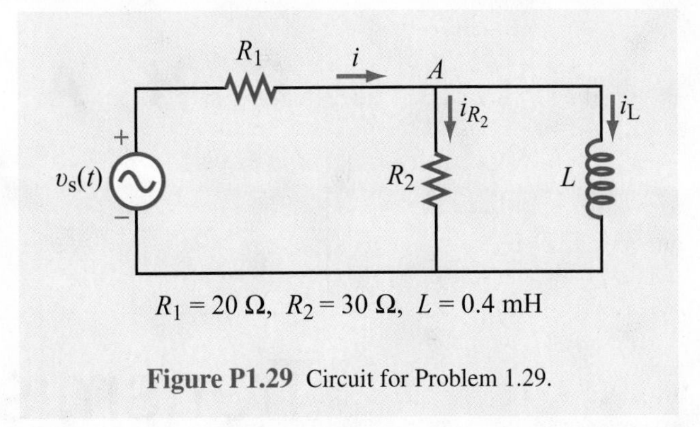

Figure P1.29 Circuit for Problem 1.29.

C H A P T E R

2

Transmission Lines

Chapter Contents

2-1 General Considerations, 71
2-2 Lumped-Element Model, 74
2-3 Transmission-Line Equations, 78
2-4 Wave Propagation on a Transmission Line, 79
2-5 The Lossless Microstrip Line, 82
2-6 The Lossless Transmission Line: General Considerations, 87
2-7 Wave Impedance of the Lossless Line, 97
2-8 Special Cases of the Lossless Line, 100
TB3 Microwave Ovens, 104
2-9 Power Flow on a Lossless Transmission Line, 108
2-10 The Smith Chart, 110
2-11 Impedance Matching, 123
2-12 Transients on Transmission Lines, 133
TB4 EM Cancer Zappers, 134
Chapter 2 Summary, 144
Problems, 146

Objectives

Upon learning the material presented in this chapter, you should be able to:

1. Calculate the line parameters, characteristic impedance, and propagation constant of coaxial, two-wire, parallel-plate, and microstrip transmission lines.
2. Determine the reflection coefficient at the load-end of the transmission line, the standing-wave pattern, and the locations of voltage and current maxima and minima.
3. Calculate the amount of power transferred from the generator to the load through the transmission line.
4. Use the Smith chart to perform transmission-line calculations.
5. Analyze the response of a transmission line to a voltage pulse.

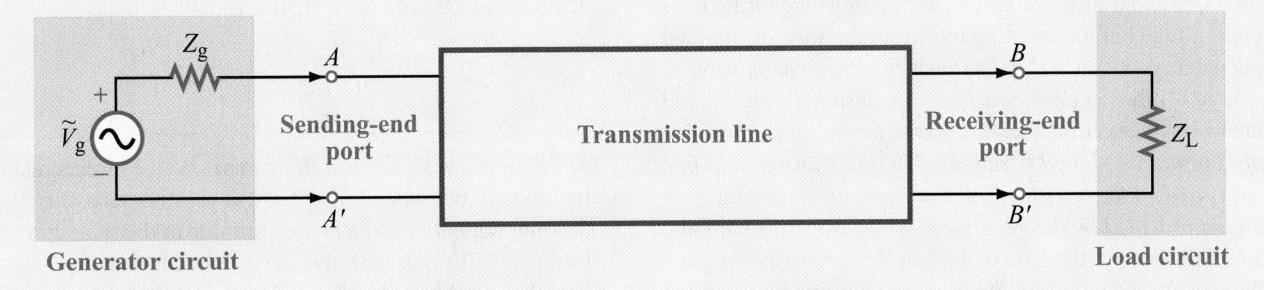

Figure 2-1 A transmission line is a two-port network connecting a generator circuit at the sending end to a load at the receiving end.

2-1 General Considerations

In most electrical engineering curricula, the study of electromagnetics is preceded by one or more courses on electrical circuits. In this book, we use this background to build a bridge between circuit theory and electromagnetic theory. The bridge is provided by transmission lines, the topic of this chapter. By modeling transmission lines in the form of equivalent circuits, we can use Kirchhoff's voltage and current laws to develop wave equations whose solutions provide an understanding of wave propagation, standing waves, and power transfer. Familiarity with these concepts facilitates the presentation of material in later chapters.

Although the notion of ***transmission lines*** may encompass all structures and media that serve to transfer energy or information between two points, including nerve fibers in the human body and fluids and solids that support the propagation of mechanical pressure waves, this chapter focuses on transmission lines that guide electromagnetic signals. Such transmission lines include telephone wires, coaxial cables carrying audio and video information to TV sets or digital data to computer monitors, microstrips printed on microwave circuit boards, and optical fibers carrying light waves for the transmission of data at very high rates.

Fundamentally, a transmission line is a two-port network, with each port consisting of two terminals, as illustrated in Fig. 2-1. One of the ports, the line's sending end, is connected to a source (also called the generator). The other port, the line's receiving end, is connected to a load. The source connected to the transmission line's sending end may be any circuit generating an output voltage, such as a radar transmitter, an amplifier, or a computer terminal operating in transmission mode. From circuit theory, a dc source can be represented by a Thévenin-equivalent ***generator circuit*** consisting of a generator voltage V_g in series with a generator resistance R_g, as shown in Fig. 2-1. In the case of alternating-current (ac) signals, the generator circuit is represented by a voltage phasor $\tilde{V}_g$ and an impedance Z_g.

The load circuit, or simply the ***load***, may be an antenna in the case of radar, a computer terminal operating in the receiving mode, the input terminals of an amplifier, or any output circuit whose input terminals can be represented by an equivalent load impedance Z_L.

2-1.1 The Role of Wavelength

In low-frequency circuits, circuit elements usually are interconnected using simple wires. In the circuit shown in Fig. 2-2, for example, the generator is connected to a simple RC load via a pair of wires. In view of our definition in the preceding paragraphs of what constitutes a transmission line, we pose the following question: Is the pair of wires between terminals AA' and terminals BB' a transmission line? If so, under what set of circumstances should we explicitly treat the

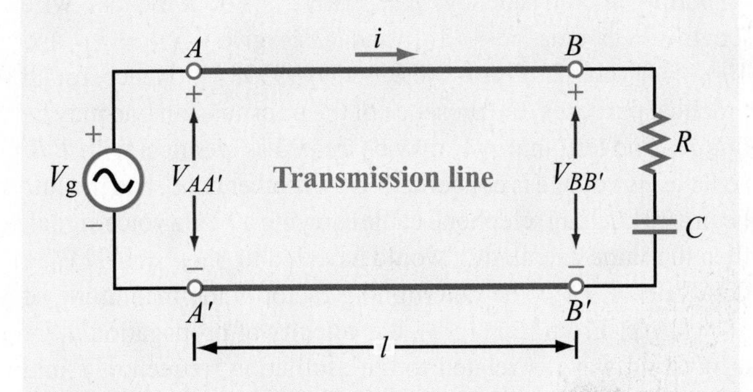

Figure 2-2 Generator connected to an RC circuit through a transmission line of length l.

pair of wires as a transmission line, as opposed to ignoring their presence altogether and treating the circuit as only an RC-load connected to a generator $\widetilde{V}_g$? The answer to the first question is: *yes, the pair of wires does constitute a transmission line.* And the answer to the second question is: *the factors that determine whether or not we should treat the wires as a transmission line are governed by the length of the line l and the frequency f of the signal provided by the generator.* (As we will see later, the determining factor is the ratio of the length l to the wavelength λ of the wave propagating on the transmission line between the source and load terminals AA' and BB', respectively.) If the generator voltage is cosinusoidal in time, then the voltage across the input terminals AA' is

$$V_{AA'} = V_g(t) = V_0 \cos \omega t \qquad \text{(V)}, \tag{2.1}$$

where $\omega = 2\pi f$ is the angular frequency, and if we assume that the current flowing through the wires travels at the speed of light, $c = 3 \times 10^8$ m/s, then the voltage across the output terminals BB' will have to be delayed in time relative to that across AA' by the travel delay-time l/c. Thus, assuming no ohmic losses in the transmission line and ignoring other transmission line effects discussed later in this chapter,

$$\begin{aligned} V_{BB'}(t) = V_{AA'}(t - l/c) &= V_0 \cos[\omega(t - l/c)] \\ &= V_0 \cos(\omega t - \phi_0), \end{aligned} \tag{2.2}$$

with

$$\phi_0 = \frac{\omega l}{c} \qquad \text{(rad)}. \tag{2.3}$$

Thus, the time delay associated with the length of the line l manifests itself as a constant phase shift ϕ_0 in the argument of the cosine. Let us compare $V_{BB'}$ to $V_{AA'}$ at $t = 0$ for an ultralow-frequency electronic circuit operating at a frequency $f = 1$ kHz. For a typical wire length $l = 5$ cm, Eqs. (2.1) and (2.2) give $V_{AA'} = V_0$ and $V_{BB'} = V_0 \cos(2\pi f l/c) = 0.999999999998\ V_0$. Hence, for all practical purposes, the presence of the transmission line may be ignored and terminal AA' may be treated as identical with BB' so far as its voltage is concerned. On the other hand, had the line been a 20 km long telephone cable carrying a 1 kHz voice signal, then the same calculation would have led to $V_{BB'} = 0.91 V_0$, a deviation of 9%. The determining factor is the magnitude of $\phi_0 = \omega l/c$. From Eq. (1.27), the velocity of propagation u_p of a traveling wave is related to the oscillation frequency f and the wavelength λ by

$$u_p = f\lambda \qquad \text{(m/s)}.$$

In the present case, $u_p = c$. Hence, the phase delay

$$\phi_0 = \frac{\omega l}{c} = \frac{2\pi f l}{c} = 2\pi \frac{l}{\lambda} \quad \text{radians}. \tag{2.4}$$

► When l/λ is very small, transmission-line effects may be ignored, but when $l/\lambda \gtrsim 0.01$, it may be necessary to account not only for the phase shift due to the time delay, but also for the presence of ***reflected*** signals that may have been bounced back by the load toward the generator. ◄

Power loss on the line and ***dispersive*** effects may need to be considered as well.

► A dispersive transmission line is one on which the wave velocity is not constant as a function of the frequency f. ◄

This means that the shape of a rectangular pulse, which through Fourier analysis can be decomposed into many sinusoidal waves of different frequencies, gets distorted as it travels down the line because its different frequency components do not propagate at the same velocity (Fig. 2-3). Preservation of pulse shape is very important in high-speed data transmission, not only between

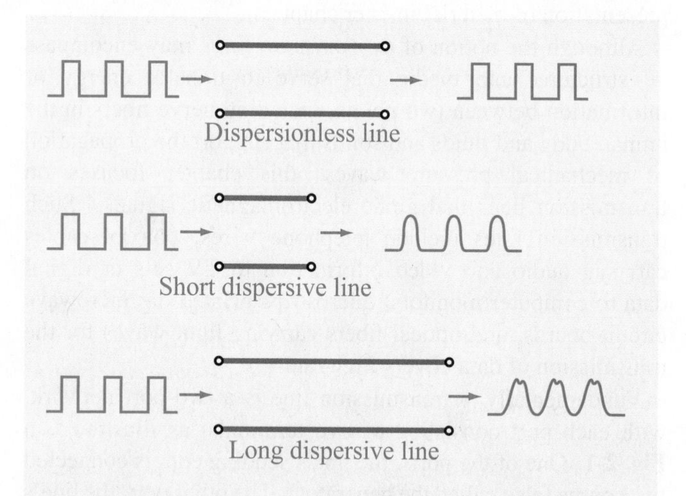

Figure 2-3 A dispersionless line does not distort signals passing through it regardless of its length, whereas a dispersive line distorts the shape of the input pulses because the different frequency components propagate at different velocities. The degree of distortion is proportional to the length of the dispersive line.

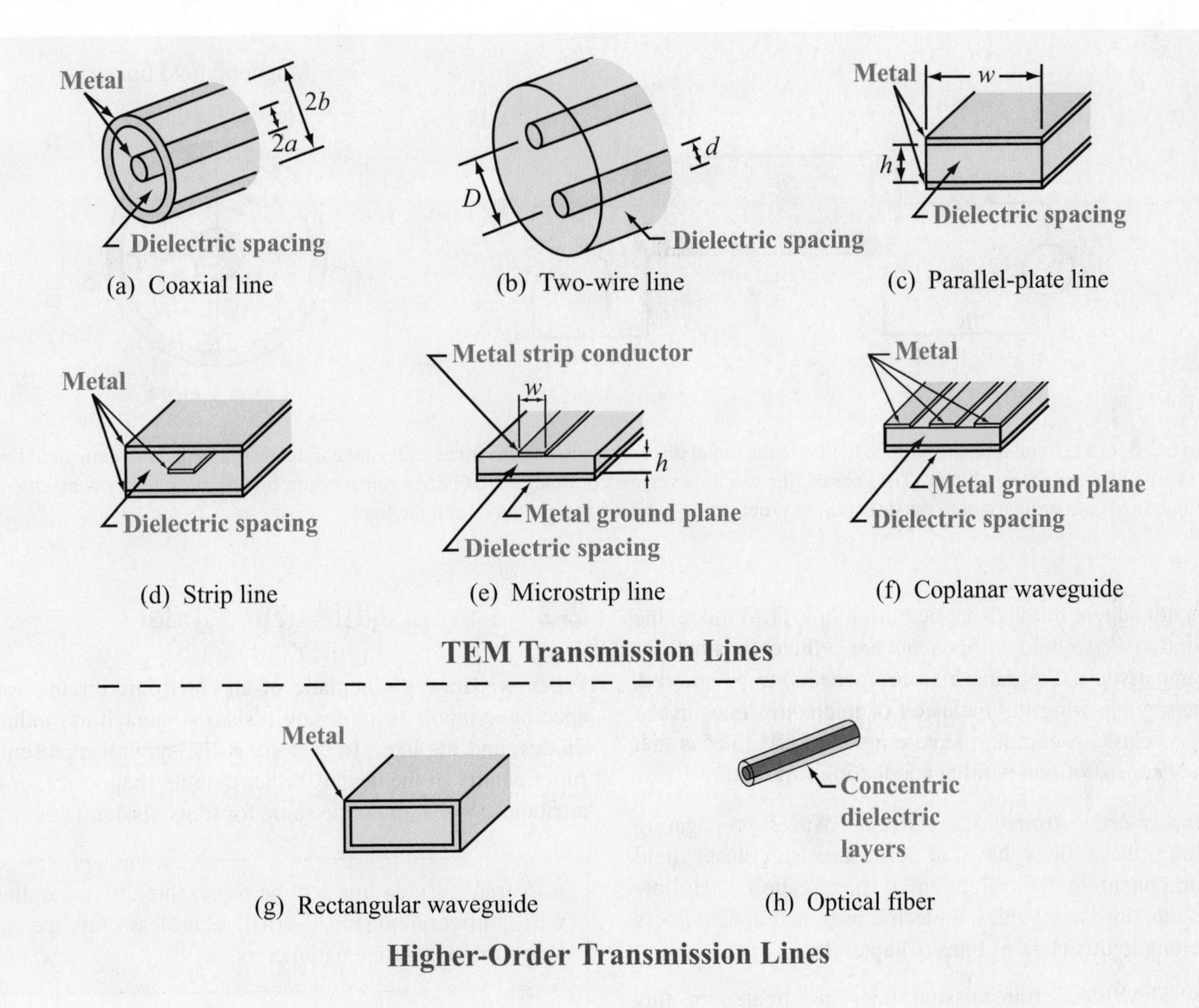

Figure 2-4 A few examples of transverse electromagnetic (TEM) and higher-order transmission lines.

terminals, but also across transmission line segments fabricated within high-speed integrated circuits. At 10 GHz, for example, the wavelength is $\lambda = 3$ cm in air but only on the order of 1 cm in a semiconductor material. Hence, even lengths between devices on the order of millimeters become significant, and their presence has to be accounted for in the design of the circuit.

2-1.2 Propagation Modes

A few examples of common types of transmission lines are shown in **Fig. 2-4**. Transmission lines may be classified into two basic types:

- ***Transverse electromagnetic (TEM) transmission lines:*** Waves propagating along these lines are characterized by electric and magnetic fields that are entirely ***transverse*** to the direction of propagation. Such an orthogonal configuration is called a ***TEM mode***. A good example is the coaxial line shown in **Fig. 2-5**: the electric field is in the radial direction between the inner and outer conductors, while the magnetic field circles the inner conductor, and neither has a component along the line axis (the direction of wave propagation). Other TEM transmission lines include the two-wire line and the parallel-plate line, both shown in **Fig. 2-4**. Although the fields present on a microstrip line

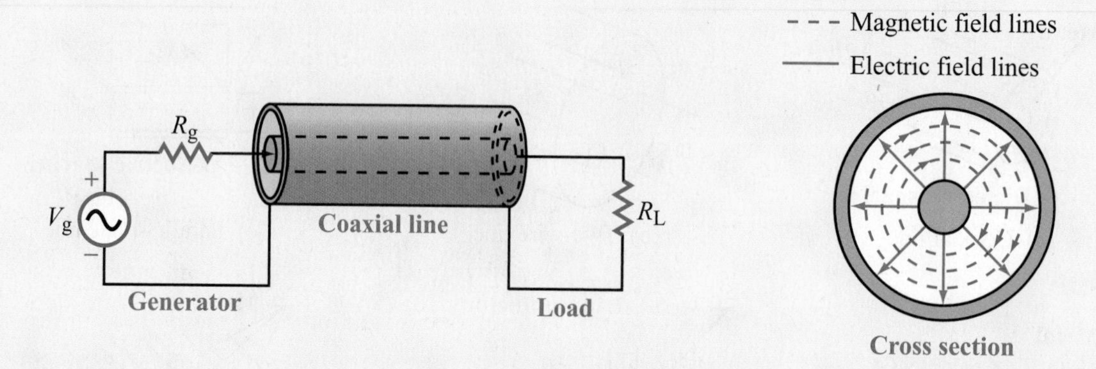

Figure 2-5 In a coaxial line, the electric field is in the radial direction between the inner and outer conductors, and the magnetic field forms circles around the inner conductor. The coaxial line is a transverse electromagnetic (TEM) transmission line because both the electric and magnetic fields are orthogonal to the direction of propagation between the generator and the load.

do not adhere to the exact definition of a TEM mode, the nontransverse field components are sufficiently small (in comparison to the transverse components) to be ignored, thereby allowing the inclusion of microstrip lines in the TEM class. A common feature among TEM lines is that they consist of two parallel conducting surfaces.

- ***Higher-order transmission lines:*** Waves propagating along these lines have at least one significant field component in the direction of propagation. Hollow conducting waveguides, dielectric rods, and optical fibers belong to this class of lines (Chapter 8).

Only TEM-mode transmission lines are treated in this chapter. This is because they are more commonly used in practice and, fortunately, less mathematical rigor is required for treating them than is required for lines that support higher-order modes. We start our treatment by representing the transmission line in terms of a lumped-element circuit model, and then we apply Kirchhoff's voltage and current laws to derive a pair of equations governing their behavior, known as the ***telegrapher's equations***. By combining these equations, we obtain wave equations for the voltage and current at any location along the line. Solution of the wave equations for the sinusoidal steady-state case leads to a set of formulas that can be used for solving a wide range of practical problems. In the latter part of this chapter we introduce a graphical tool known as the ***Smith chart***, which facilitates the solution of transmission-line problems without having to perform laborious calculations involving complex numbers.

2-2 Lumped-Element Model

When we draw a schematic of an electronic circuit, we use specific symbols to represent resistors, capacitors, inductors, diodes, and the like. In each case, the symbol represents the functionality of the device, rather than its shape, size, or other attributes. We shall do the same for transmission lines.

▶ A transmission line will be represented by a parallel-wire configuration [Fig. 2-6(a)], regardless of its specific shape or constitutive parameters. ◀

Thus, Fig. 2-6(a) may represent a coaxial line, a two-wire line, or any other TEM line.

Drawing again on our familiarity with electronic circuits, when we analyze a circuit containing a transistor, we mimic the functionality of the transistor by an equivalent circuit composed of sources, resistors, and capacitors. We apply the same approach to the transmission line by orienting the line along the z direction, subdividing it into differential sections each of length Δz [Fig. 2-6(b)] and then representing each section by an equivalent circuit, as illustrated in Fig. 2-6(c). This representation, often called the ***lumped-element circuit*** model, consists of four basic elements, with values that henceforth will be called the ***transmission line parameters***. These are:

- R': The combined ***resistance*** of both conductors per unit length, in Ω/m,

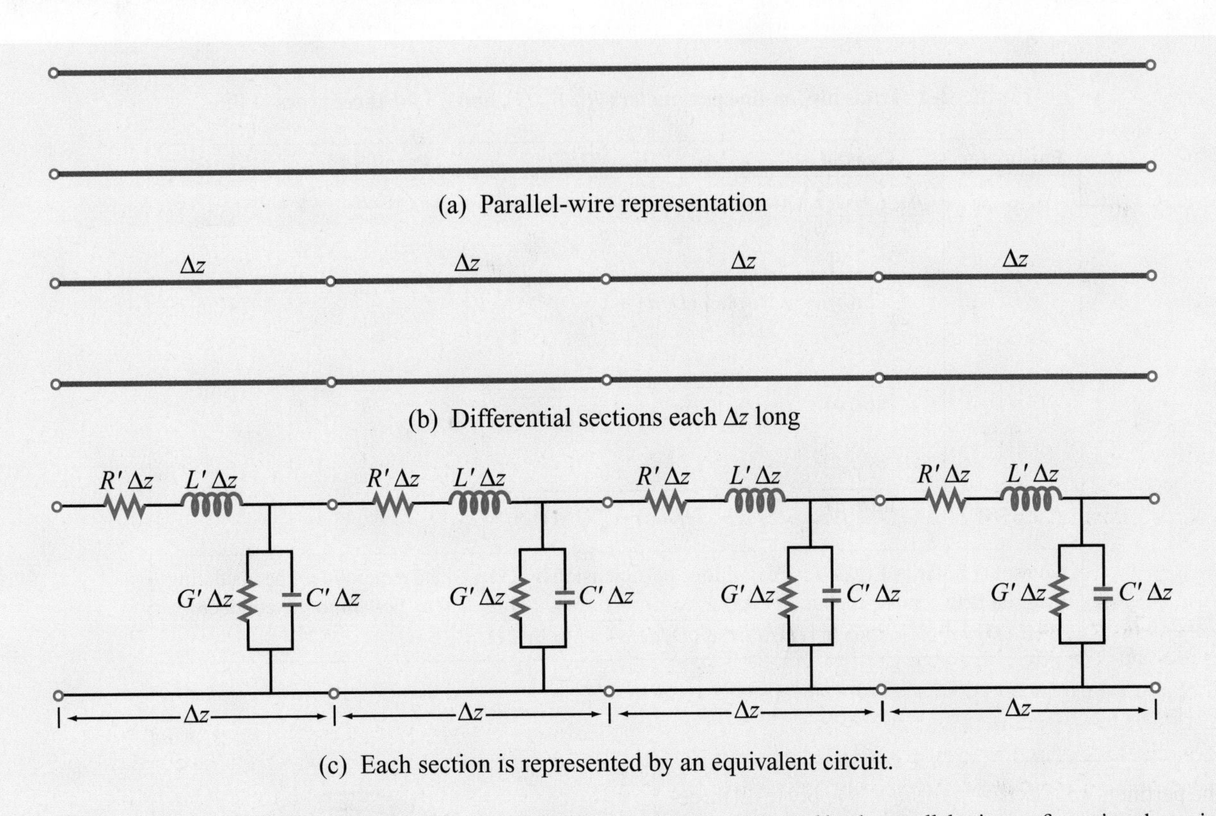

Figure 2-6 Regardless of its cross-sectional shape, a TEM transmission line is represented by the parallel-wire configuration shown in (a). To obtain equations relating voltages and currents, the line is subdivided into small differential sections (b), each of which is then represented by an equivalent circuit (c).

- L': The combined ***inductance*** of both conductors per unit length, in H/m,
- G': The ***conductance*** of the insulation medium between the two conductors per unit length, in S/m, and
- C': The ***capacitance*** of the two conductors per unit length, in F/m.

Whereas the four line parameters are characterized by different formulas for different types of transmission lines, the equivalent model represented by **Fig. 2-6(c)** is equally applicable to all TEM transmission lines. *The prime superscript is used as a reminder that the line parameters are differential quantities whose units are per unit length.*

Expressions for the line parameters R', L', G', and C' are given in Table 2-1 for the three types of TEM transmission lines diagrammed in parts (a) through (c) of **Fig. 2-4**. For each of these lines, the expressions are functions of two sets of parameters: (1) geometric parameters defining the cross-sectional dimensions of the given line and (2) the electromagnetic constitutive parameters of the conducting and insulating materials. The pertinent geometric parameters are:

- *Coaxial line [**Fig. 2-4(a)**]:*

  a = outer radius of inner conductor, m
  b = inner radius of outer conductor, m

- *Two-wire line [**Fig. 2-4(b)**]:*

  d = diameter of each wire, m
  D = spacing between wires' centers, m

- *Parallel-plate line [**Fig. 2-4(c)**]:*

  w = width of each plate, m
  h = thickness of insulation between plates, m

Table 2-1 **Transmission-line parameters R', L', G', and C' for three types of lines.**

Parameter	Coaxial	Two-Wire	Parallel-Plate	Unit
R'	$\frac{R_s}{2\pi}\left(\frac{1}{a}+\frac{1}{b}\right)$	$\frac{2R_s}{\pi d}$	$\frac{2R_s}{w}$	Ω/m
L'	$\frac{\mu}{2\pi}\ln(b/a)$	$\frac{\mu}{\pi}\ln\left[(D/d)+\sqrt{(D/d)^2-1}\right]$	$\frac{\mu h}{w}$	H/m
G'	$\frac{2\pi\sigma}{\ln(b/a)}$	$\frac{\pi\sigma}{\ln\left[(D/d)+\sqrt{(D/d)^2-1}\right]}$	$\frac{\sigma w}{h}$	S/m
C'	$\frac{2\pi\epsilon}{\ln(b/a)}$	$\frac{\pi\epsilon}{\ln\left[(D/d)+\sqrt{(D/d)^2-1}\right]}$	$\frac{\epsilon w}{h}$	F/m

Notes: (1) Refer to Fig. 2-4 for definitions of dimensions. (2) μ, ϵ, and σ pertain to the insulating material between the conductors. (3) $R_s = \sqrt{\pi f \mu_c/\sigma_c}$. (4) μ_c and σ_c pertain to the conductors. (5) If $(D/d)^2 \gg 1$, then $\ln\left[(D/d)+\sqrt{(D/d)^2-1}\right] \approx \ln(2D/d)$.

► The pertinent ***constitutive parameters*** apply to all three lines and consist of two groups:

(1) μ_c and σ_c are the magnetic permeability and electrical conductivity of the conductors, and

(2) ϵ, μ, and σ are the electrical permittivity, magnetic permeability, and electrical conductivity of the insulation material separating them. ◄

Appendix B contains tabulated values for these constitutive parameters for various materials. For the purposes of the present chapter, we need not concern ourselves with the derivations leading to the expressions in Table 2-1. The techniques necessary for computing R', L', G', and C' for the general case of an arbitrary two-conductor configuration are presented in later chapters.

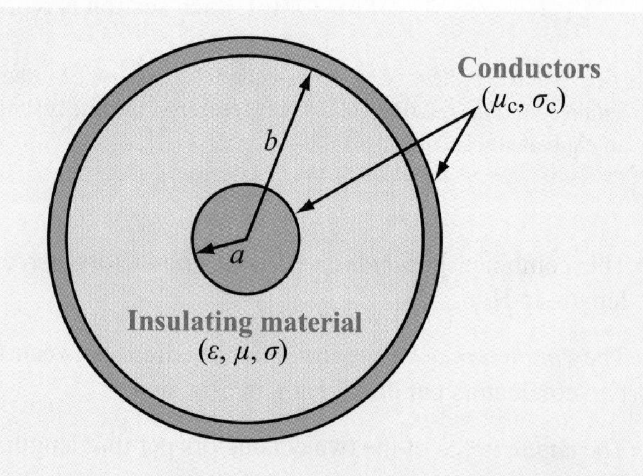

Figure 2-7 Cross section of a coaxial line with inner conductor of radius a and outer conductor of radius b. The conductors have magnetic permeability μ_c and conductivity σ_c, and the spacing material between the conductors has permittivity ϵ, permeability μ, and conductivity σ.

The lumped-element model shown in Fig. 2-6(c) reflects the physical phenomena associated with the currents and voltages on any TEM transmission line. It consists of two in-series elements, R' and L', and two shunt elements, G' and C'. To explain the lumped-element model, consider a small section of a coaxial line, as shown in Fig. 2-7. The line consists of inner and outer conductors of radii a and b separated by a material with permittivity ϵ, permeability μ, and conductivity σ. The two metal conductors are made of a material with conductivity σ_c and permeability μ_c. When a voltage source is connected across the terminals connected to the two conductors at the sending end of the line, currents flow through the conductors, primarily along the outer surface of the inner conductor and the inner surface of the outer conductor. The line resistance R' accounts for the combined resistance per unit length of the inner and outer conductors. The expression for R' is derived in Chapter 7 and is given by Eq. (7.96) as

$$R' = \frac{R_s}{2\pi}\left(\frac{1}{a} + \frac{1}{b}\right) \quad \textbf{(coax line)} \quad (\Omega/\text{m}), \tag{2.5}$$

where R_s, which represents the ***surface resistance*** of the conductors, is given by Eq. (7.92a) as

$$R_s = \sqrt{\frac{\pi f \mu_c}{\sigma_c}} \quad (\Omega). \tag{2.6}$$

The surface resistance depends not only on the material properties of the conductors (σ_c and μ_c), but also on the frequency f of the wave traveling on the line.

► For a ***perfect conductor*** with $\sigma_c = \infty$ or a high-conductivity material such that $(f\mu_c/\sigma_c) \ll 1$, R_s approaches zero, and so does R'. ◄

Next, let us examine the line inductance L', which accounts for the joint inductance of both conductors. Application of Ampère's law in Chapter 5 to the definition of inductance leads to the following expression [Eq. (5.99)] for the inductance per unit length of a coaxial line:

$$L' = \frac{\mu}{2\pi} \ln\left(\frac{b}{a}\right) \quad \textbf{(coax line)} \quad (\text{H/m}). \tag{2.7}$$

The line conductance G' accounts for current flow between the outer and inner conductors, made possible by the conductivity σ of the insulator. It is precisely because the current flow is from one conductor to the other that G' appears as a *shunt* element in the lumped-element model. For the coaxial line, the conductance per unit length is given by Eq. (4.76) as

$$G' = \frac{2\pi\sigma}{\ln(b/a)} \quad \textbf{(coax line)} \quad (\text{S/m}). \tag{2.8}$$

► If the material separating the inner and outer conductors is a ***perfect dielectric*** with $\sigma = 0$, then $G' = 0$. ◄

The last line parameter on our list is the line capacitance C'. When equal and opposite charges are placed on any two noncontacting conductors, a voltage difference develops between them. Capacitance is defined as the ratio of the charge to the voltage difference. For the coaxial line, the capacitance per unit length is given by Eq. (4.117) as

$$C' = \frac{2\pi\epsilon}{\ln(b/a)} \quad \textbf{(coax line)} \quad (\text{F/m}). \tag{2.9}$$

All TEM transmission lines share the following useful relations:

$$L'C' = \mu\epsilon \quad \textbf{(all TEM lines)}, \tag{2.10}$$

and

$$\frac{G'}{C'} = \frac{\sigma}{\epsilon} \quad \textbf{(all TEM lines)}. \tag{2.11}$$

If the insulating medium between the conductors is air, the transmission line is called an ***air line*** (e.g., coaxial air line or two-wire air line). For an air line, $\epsilon = \epsilon_0 = 8.854 \times 10^{-12}$ F/m, $\mu = \mu_0 = 4\pi \times 10^{-7}$ H/m, $\sigma = 0$, and $G' = 0$.

Concept Question 2-1: What is a transmission line? When should transmission-line effects be considered, and when may they be ignored?

Concept Question 2-2: What is the difference between dispersive and nondispersive transmission lines? What is the practical significance of dispersion?

Concept Question 2-3: What constitutes a TEM transmission line?

Concept Question 2-4: What purpose does the lumped-element circuit model serve? How are the line parameters R', L', G', and C' related to the physical and electromagnetic constitutive properties of the transmission line?

Exercise 2-1: Use **Table 2-1** to evaluate the line parameters of a two-wire air line with wires of radius 1 mm, separated by a distance of 2 cm. The wires may be treated as perfect conductors with $\sigma_c = \infty$.

Answer: $R' = 0$, $L' = 1.20$ (μH/m), $G' = 0$, $C' = 9.29$ (pF/m). (See EM.)

Exercise 2-2: Calculate the transmission line parameters at 1 MHz for a coaxial air line with inner and outer conductor diameters of 0.6 cm and 1.2 cm, respectively. The conductors are made of copper (see Appendix B for μ_c and σ_c of copper).

Answer: $R' = 2.07 \times 10^{-2}$ (Ω/m), $L' = 0.14$ (μH/m), $G' = 0$, $C' = 80.3$ (pF/m). (See EM.)

2-3 Transmission-Line Equations

A transmission line usually connects a source on one end to a load on the other. Before considering the complete circuit, however, we will develop general equations that describe the voltage across and current carried by the transmission line as a function of time t and spatial position z. Using the lumped-element model of **Fig. 2-6(c)**, we begin by considering a differential length Δz as shown in **Fig. 2-8**. The quantities $\upsilon(z,t)$ and $i(z,t)$ denote the instantaneous voltage and current at the left end of the differential section (node N), and similarly $\upsilon(z+\Delta z,t)$ and $i(z+\Delta z,t)$ denote the same quantities at node $(N+1)$, located at the right end of the section. Application of Kirchhoff's voltage law accounts for the voltage drop across the series resistance $R'\Delta z$ and inductance $L'\Delta z$:

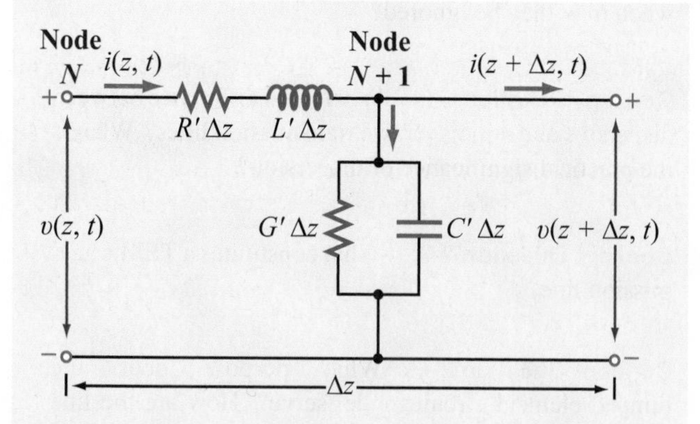

Figure 2-8 Equivalent circuit of a two-conductor transmission line of differential length Δz.

$$\upsilon(z,t) - R'\Delta z\, i(z,t) - L'\,\Delta z\,\frac{\partial\, i(z,t)}{\partial t} - \upsilon(z+\Delta z,t) = 0. \tag{2.12}$$

Upon dividing all terms by Δz and rearranging them, we obtain

$$-\left[\frac{\upsilon(z+\Delta z,\,t) - \upsilon(z,t)}{\Delta z}\right] = R'\,i(z,t) + L'\,\frac{\partial i(z,t)}{\partial t}. \tag{2.13}$$

In the limit as $\Delta z \to 0$, Eq. (2.13) becomes a differential equation:

$$-\frac{\partial \upsilon(z,t)}{\partial z} = R'\,i(z,t) + L'\,\frac{\partial\, i(z,t)}{\partial t}. \tag{2.14}$$

Similarly, Kirchhoff's current law accounts for current drawn from the upper line at node $(N+1)$ by the parallel conductance $G'\,\Delta z$ and capacitance $C'\,\Delta z$:

$$i(z,t) - G'\,\Delta z\,\upsilon(z+\Delta z,t) - C'\,\Delta z\,\frac{\partial \upsilon(z+\Delta z,t)}{\partial t} - i(z+\Delta z,t) = 0. \tag{2.15}$$

Upon dividing all terms by Δz and taking the limit $\Delta z \to 0$, Eq. (2.15) becomes a second-order differential equation:

$$-\frac{\partial i(z,t)}{\partial z} = G'\,\upsilon(z,t) + C'\,\frac{\partial \upsilon(z,t)}{\partial t}. \tag{2.16}$$

The first-order differential equations (2.14) and (2.16) are the time-domain forms of the ***transmission-line equations***, known as the ***telegrapher's equations***.

Except for the last section of this chapter, our primary interest is in sinusoidal steady-state conditions. To that end, we make use of the phasor representation with a cosine reference, as outlined in Section 1-7. Thus, we define

$$\upsilon(z,t) = \mathfrak{Re}[\widetilde{V}(z)\,e^{j\omega t}], \tag{2.17a}$$

$$i(z,t) = \mathfrak{Re}[\widetilde{I}(z)\,e^{j\omega t}], \tag{2.17b}$$

where $\widetilde{V}(z)$ and $\widetilde{I}(z)$ are the phasor counterparts of $\upsilon(z,t)$ and $i(z,t)$, respectively, each of which may be real or complex. Upon substituting Eqs. (2.17a) and (2.17b) into Eqs. (2.14) and

(2.16), and utilizing the property given by Eq. (1.62) that $\partial/\partial t$ in the time domain is equivalent to multiplication by $j\omega$ in the phasor domain, we obtain the following pair of equations:

$$-\frac{d\widetilde{V}(z)}{dz} = (R' + j\omega L')\,\widetilde{I}(z), \tag{2.18a}$$

$$-\frac{d\widetilde{I}(z)}{dz} = (G' + j\omega C')\,\widetilde{V}(z). \tag{2.18b}$$

(telegrapher's equations in phasor form)

2-4 Wave Propagation on a Transmission Line

The two first-order coupled equations (2.18a) and (2.18b) can be combined to give two second-order uncoupled wave equations, one for $\widetilde{V}(z)$ and another for $\widetilde{I}(z)$. The wave equation for $\widetilde{V}(z)$ is derived by first differentiating both sides of Eq. (2.18a) with respect to z, resulting in

$$-\frac{d^2\widetilde{V}(z)}{dz^2} = (R' + j\omega L')\frac{d\widetilde{I}(z)}{dz}. \tag{2.19}$$

Then, upon substituting Eq. (2.18b) for $d\widetilde{I}(z)/dz$, Eq. (2.19) becomes

$$\frac{d^2\widetilde{V}(z)}{dz^2} - (R' + j\omega L')(G' + j\omega C')\,\widetilde{V}(z) = 0, \tag{2.20}$$

or

$$\frac{d^2\widetilde{V}(z)}{dz^2} - \gamma^2\,\widetilde{V}(z) = 0, \tag{2.21}$$

(wave equation for $\widetilde{V}(z)$)

where

$$\gamma = \sqrt{(R' + j\omega L')(G' + j\omega C')}. \tag{2.22}$$

(propagation constant)

Application of the same steps to Eqs. (2.18a) and (2.18b) in reverse order leads to

$$\frac{d^2\widetilde{I}(z)}{dz^2} - \gamma^2\,\widetilde{I}(z) = 0. \tag{2.23}$$

(wave equation for $\widetilde{I}(z)$)

The second-order differential equations (2.21) and (2.23) are called ***wave equations*** for $\widetilde{V}(z)$ and $\widetilde{I}(z)$, respectively, and γ is called the ***complex propagation constant*** of the transmission line. As such, γ consists of a real part α, called the ***attenuation constant*** of the line with units of Np/m, and an imaginary part β, called the ***phase constant*** of the line with units of rad/m. Thus,

$$\gamma = \alpha + j\beta \tag{2.24}$$

with

$$\alpha = \mathfrak{Re}(\gamma) = \mathfrak{Re}\left(\sqrt{(R' + j\omega L')(G' + j\omega C')}\right) \quad \text{(Np/m)}, \tag{2.25a}$$

(attenuation constant)

$$\beta = \mathfrak{Im}(\gamma) = \mathfrak{Im}\left(\sqrt{(R' + j\omega L')(G' + j\omega C')}\right) \quad \text{(rad/m)}. \tag{2.25b}$$

(phase constant)

In Eqs. (2.25a) and (2.25b), we choose the square-root solutions that give positive values for α and β. For passive transmission lines, α is either zero or positive. Most transmission lines, and all those considered in this chapter, are of the passive type. The gain region of a laser is an example of an active transmission line with a negative α.

The wave equations (2.21) and (2.23) have traveling wave solutions of the following form:

$$\widetilde{V}(z) = V_0^+ e^{-\gamma z} + V_0^- e^{\gamma z} \quad \text{(V)}, \tag{2.26a}$$

$$\widetilde{I}(z) = I_0^+ e^{-\gamma z} + I_0^- e^{\gamma z} \quad \text{(A)}. \tag{2.26b}$$

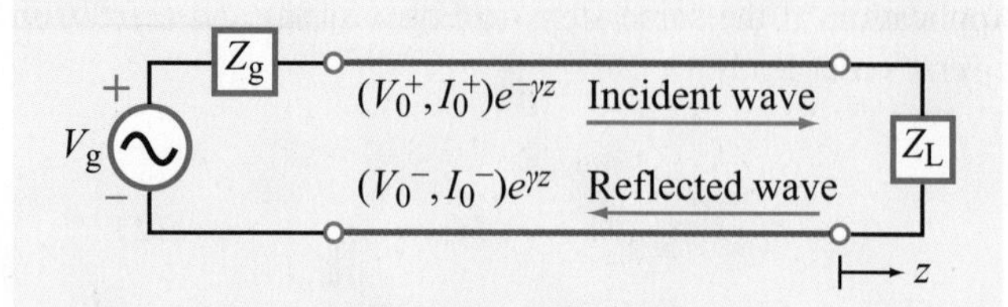

Figure 2-9 In general, a transmission line can support two traveling waves, an incident wave (with voltage and current amplitudes (V_0^+, I_0^+)) traveling along the $+z$ direction (towards the load) and a reflected wave (with (V_0^-, I_0^-)) traveling along the $-z$ direction (towards the source).

As shown later, the $e^{-\gamma z}$ term represents a wave propagating in the $+z$ direction while the $e^{\gamma z}$ term represents a wave propagating in the $-z$ direction (Fig. 2-9). Verification that these are indeed valid solutions is easily accomplished by substituting the expressions given by Eqs. (2.26a) and (2.26b), as well as their second derivatives, into Eqs. (2.21) and (2.23).

In their present form, the solutions given by Eqs. (2.26a) and (2.26b) contain four unknowns, the wave amplitudes (V_0^+, I_0^+) of the $+z$ propagating wave and (V_0^-, I_0^-) of the $-z$ propagating wave. We can easily relate the current wave amplitudes, I_0^+ and I_0^-, to the voltage wave amplitudes, V_0^+ and V_0^-, by using Eq. (2.26a) in Eq. (2.18a) and then solving for the current $\tilde{I}(z)$. The process leads to

$$\tilde{I}(z) = \frac{\gamma}{R' + j\omega L'}[V_0^+ e^{-\gamma z} - V_0^- e^{\gamma z}]. \qquad (2.27)$$

Comparison of each term with the corresponding term in Eq. (2.26b) leads us to conclude that

$$\frac{V_0^+}{I_0^+} = Z_0 = \frac{-V_0^-}{I_0^-}, \qquad (2.28)$$

where

$$Z_0 = \frac{R' + j\omega L'}{\gamma} = \sqrt{\frac{R' + j\omega L'}{G' + j\omega C'}} \quad (\Omega), \qquad (2.29)$$

is called the ***characteristic impedance*** of the line.

► It should be noted that Z_0 is equal to the ratio of the voltage amplitude to the current amplitude for each of the traveling waves individually (with an additional minus sign in the case of the $-z$ propagating wave), but it is not equal to the ratio of the total voltage $\tilde{V}(z)$ to the total current $\tilde{I}(z)$, unless one of the two waves is absent. ◄

It seems reasonable that the voltage-to-current ratios of the two waves V_0^+/I_0^+ and V_0^-/I_0^-, are both related to the same quantity, namely Z_0, but it is not immediately obvious as to why one of the ratios is the negative of the other. The explanation, which is available in more detail in Chapter 7, is based on a directional rule that specifies the relationships between the directions of the electric and magnetic fields of a TEM wave and its direction of propagation. On a transmission line, the voltage is related to the electric field **E** and the current is related to the magnetic field **H**. To satisfy the directional rule, reversing the direction of propagation requires reversal of the direction (or polarity) of I relative to V. Hence, $V_0^-/I_0^- = -V_0^+/I_0^+$.

In terms of Z_0, Eq. (2.27) can be cast in the form

$$\tilde{I}(z) = \frac{V_0^+}{Z_0} e^{-\gamma z} - \frac{V_0^-}{Z_0} e^{\gamma z}. \qquad (2.30)$$

According to Eq. (2.29), the characteristic impedance Z_0 is determined by the angular frequency ω of the wave traveling along the line and the four line parameters (R', L', G', and C'). These, in turn, are determined by the line geometry and its constitutive parameters. Consequently, the combination of Eqs. (2.26a) and (2.30) now contains only two unknowns, namely V_0^+ and V_0^-, as opposed to four.

In later sections, we apply boundary conditions at the source and load ends of the transmission line to obtain expressions for the remaining wave amplitudes V_0^+ and V_0^-. In general, each is a complex quantity characterized by a magnitude and a phase angle:

$$V_0^+ = |V_0^+|e^{j\phi^+}, \qquad (2.31a)$$

$$V_0^- = |V_0^-|e^{j\phi^-}. \qquad (2.31b)$$

After substituting these definitions in Eq. (2.26a) and using Eq. (2.24) to decompose γ into its real and imaginary parts, we can convert back to the time domain to obtain an expression for $\upsilon(z,t)$, the instantaneous voltage on the line:

$$\begin{aligned}\upsilon(z,t) &= \mathfrak{Re}(\widetilde{V}(z)e^{j\omega t}) \\ &= \mathfrak{Re}\left[\left(V_0^+ e^{-\gamma z} + V_0^- e^{\gamma z}\right) e^{j\omega t}\right] \\ &= \mathfrak{Re}[|V_0^+|e^{j\phi^+} e^{j\omega t} e^{-(\alpha+j\beta)z} \\ &\quad + |V_0^-|e^{j\phi^-} e^{j\omega t} e^{(\alpha+j\beta)z}] \\ &= |V_0^+|e^{-\alpha z}\cos(\omega t - \beta z + \phi^+) \\ &\quad + |V_0^-|e^{\alpha z}\cos(\omega t + \beta z + \phi^-). \end{aligned} \tag{2.32}$$

From our review of waves in Section 1-4, we recognize the first term in Eq. (2.32) as a wave traveling in the $+z$ direction (the coefficients of t and z have opposite signs) and the second term as a wave traveling in the $-z$ direction (the coefficients of t and z are both positive). Both waves propagate with a phase velocity u_p given by Eq. (1.30):

$$u_\text{p} = f\lambda = \frac{\omega}{\beta}. \tag{2.33}$$

Because the wave is *guided* by the transmission line, λ often is called the ***guide wavelength***. The factor $e^{-\alpha z}$ accounts for the attenuation of the $+z$ propagating wave, and the factor $e^{\alpha z}$ accounts for the attenuation of the $-z$ propagating wave.

▶ The presence of two waves on the line propagating in opposite directions produces a *standing wave*. ◀

To gain a physical understanding of what that means, we shall first examine the relatively simple but important case of a ***lossless line*** ($\alpha = 0$) and then extend the results to the more general case of a ***lossy transmission line*** ($\alpha \neq 0$). In fact, we shall devote the next several sections to the study of lossless transmission lines because in practice many lines can be designed to exhibit very low-loss characteristics.

Example 2-1: Air Line

An ***air line*** is a transmission line in which air separates the two conductors, which renders $G' = 0$ because $\sigma = 0$. In addition, assume that the conductors are made of a material with high conductivity so that $R' \approx 0$. For an air line with a characteristic impedance of 50 Ω and a phase constant of 20 rad/m at 700 MHz, find the line inductance L' and the line capacitance C'.

Solution: The following quantities are given:

$$Z_0 = 50\ \Omega,$$
$$\beta = 20 \text{ rad/m},$$
$$f = 700 \text{ MHz} = 7 \times 10^8 \text{ Hz}.$$

With $R' = G' = 0$, Eqs. (2.25b) and (2.29) reduce to

$$\beta = \mathfrak{Im}\left[\sqrt{(j\omega L')(j\omega C')}\right] = \mathfrak{Im}\left(j\omega\sqrt{L'C'}\right) = \omega\sqrt{L'C'},$$

$$Z_0 = \sqrt{\frac{j\omega L'}{j\omega C'}} = \sqrt{\frac{L'}{C'}}.$$

The ratio of β to Z_0 is

$$\frac{\beta}{Z_0} = \omega C',$$

or

$$C' = \frac{\beta}{\omega Z_0} = \frac{20}{2\pi \times 7 \times 10^8 \times 50} = 9.09 \times 10^{-11} \text{ (F/m)}$$
$$= 90.9 \text{ (pF/m)}.$$

From $Z_0 = \sqrt{L'/C'}$, it follows that

$$L' = Z_0^2 C' = (50)^2 \times 90.9 \times 10^{-12} = 2.27 \times 10^{-7} \text{ (H/m)}$$
$$= 227 \text{ (nH/m)}.$$

Exercise 2-3: Verify that Eq. (2.26a) indeed provides a solution to the wave equation (2.21). (See EM.)

Exercise 2-4: A two-wire air line has the following line parameters: $R' = 0.404$ (mΩ/m), $L' = 2.0$ (μH/m), $G' = 0$, and $C' = 5.56$ (pF/m). For operation at 5 kHz, determine (a) the attenuation constant α, (b) the phase constant β, (c) the phase velocity u_p, and (d) the characteristic impedance Z_0. (See EM.)

Answer: (a) $\alpha = 3.37 \times 10^{-7}$ (Np/m), (b) $\beta = 1.05 \times 10^{-4}$ (rad/m), (c) $u_\text{p} = 3.0 \times 10^8$ (m/s), (d) $Z_0 = (600 - j1.9)\ \Omega = 600\angle{-0.18^\circ}\ \Omega$.

Module 2.1 Two-Wire Line The input data specifies the geometric and electric parameters of a two-wire transmission line. The output includes the calculated values for the line parameters, characteristic impedance Z_0, and attenuation and phase constants, as well as plots of Z_0 as a function of d and D.

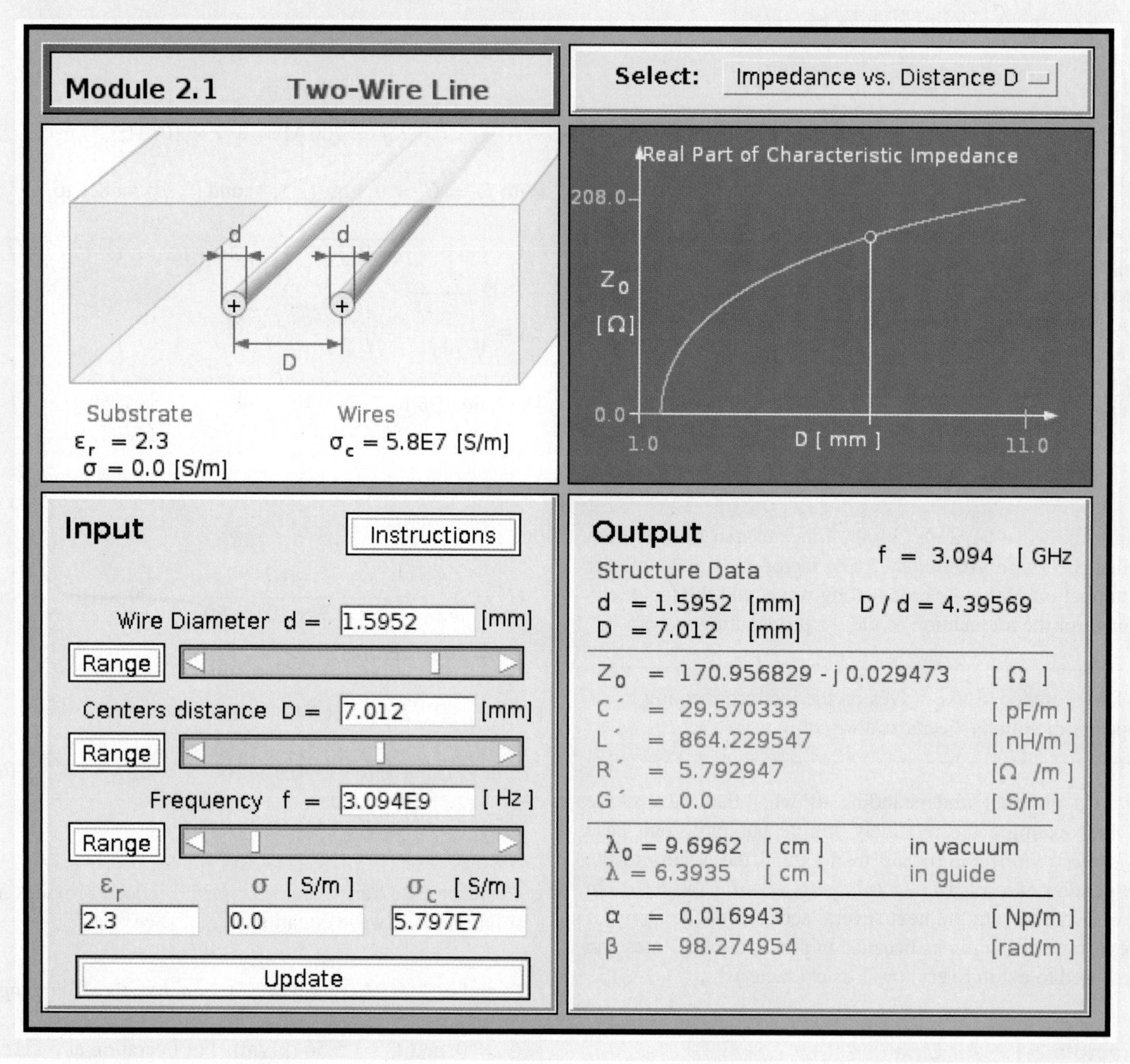

2-5 The Lossless Microstrip Line

Because its geometry is well suited for fabrication on printed circuit boards, the microstrip line is the most common interconnect configuration used in RF and microwave circuits. It consists of a narrow, very thin strip of copper (or another good conductor) printed on a dielectric substrate overlaying a ground plane (**Fig. 2-10(a)**). The presence of charges of opposite polarity on its two conducting surfaces gives rise to electric field lines between them (**Fig. 2-10(b)**). Also, the flow of current

Module 2.2 Coaxial Cable Except for changing the geometric parameters to those of a coaxial transmission line, this module offers the same output information as Module 2.1.

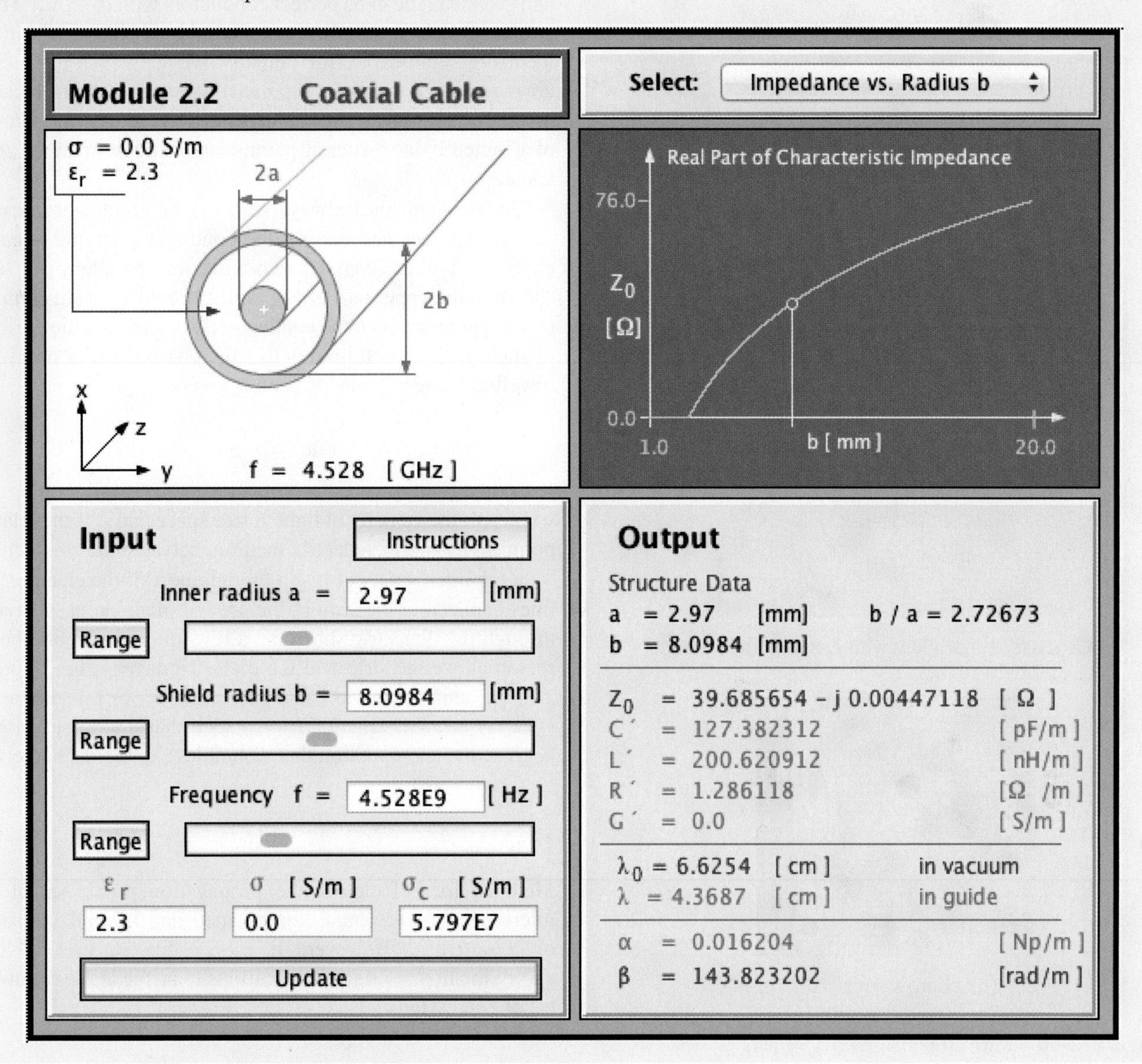

through the conductors (when part of a closed circuit) generates magnetic field loops around them, as illustrated in Fig. 2-10(b) for the narrow strip. Even though the patterns of **E** and **B** are not everywhere perfectly orthogonal, they are approximately so in the region between the conductors, which is where the **E** and **B** fields are concentrated the most. Accordingly, the microstrip line is considered a ***quasi-TEM*** transmission line, which allows us to describe its voltages and currents in terms of the one-dimensional TEM model of Section 2-4, namely Eqs. (2.26) through (2.33).

The microstrip line has two geometric parameters: the width of the *elevated* strip, w, and the thickness (height) of

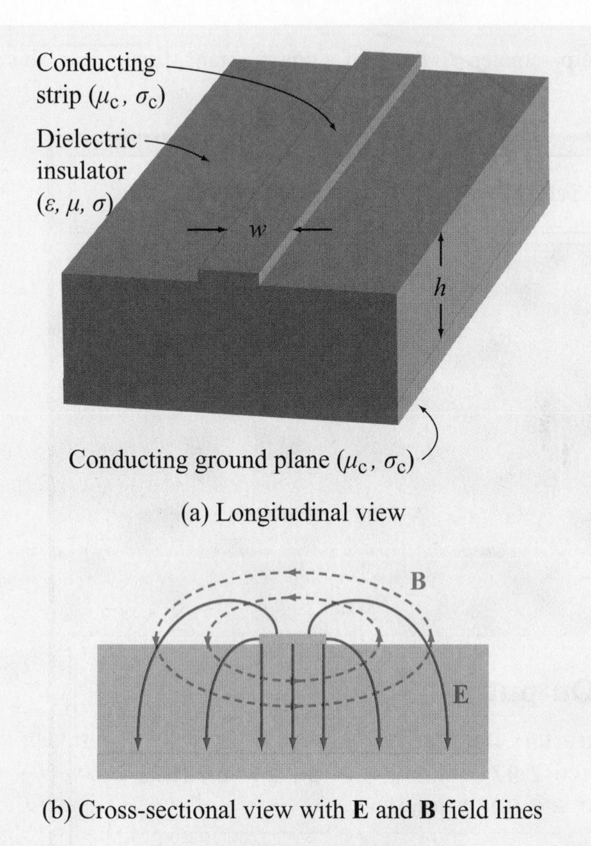

(a) Longitudinal view

(b) Cross-sectional view with **E** and **B** field lines

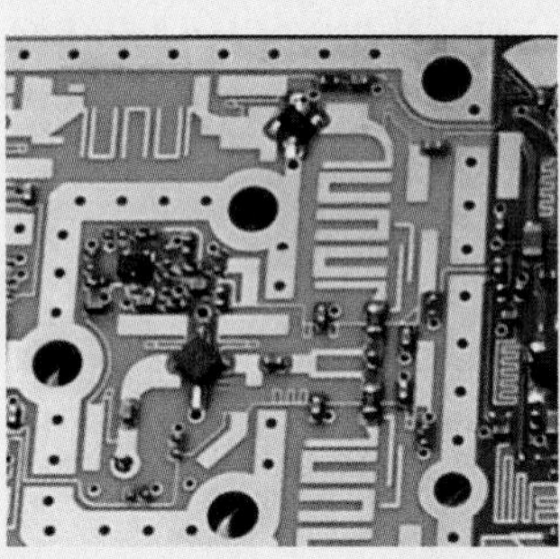

(c) Microwave circuit

Figure 2-10 Microstrip line: (a) longitudinal view, (b) cross-sectional view, and (c) circuit example. (Courtesy of Prof. Gabriel Rebeiz, U. California at San Diego.)

the dielectric layer, h. We will ignore the thickness of the conducting strip because it has a negligible influence on the propagation properties of the microstrip line, so long as the strip thickness is much smaller than the width w, which is almost always the case in practice. Also, we assume the substrate material to be a perfect dielectric with $\sigma = 0$ and the metal strip and ground plane to be perfect conductors with $\sigma_c \approx \infty$. These two assumptions simplify the analysis considerably without incurring significant error. Finally, we set $\mu = \mu_0$, which is always true for the dielectric materials used in the fabrication of microstrip lines. These simplifications reduce the number of geometric and material parameters to three, namely w, h, and ϵ.

Electric field lines always start on the conductor carrying positive charges and end on the conductor carrying negative charges. For the coaxial, two-wire, and parallel-plate lines shown in the upper part of **Fig. 2-4**, the field lines are confined to the region between the conductors. A characteristic attribute of such transmission lines is that the phase velocity of a wave traveling along any one of them is given by

$$u_p = \frac{c}{\sqrt{\epsilon_r}}, \tag{2.34}$$

where c is the velocity of light in free space and ϵ_r is the relative permittivity of the dielectric medium between the conductors.

In the microstrip line, even though most of the electric field lines connecting the strip to the ground plane do pass directly through the dielectric substrate, a few go through both the air region above the strip and the dielectric layer [**Fig. 2-10(b)**]. This nonuniform mixture can be accounted for by defining an ***effective relative permittivity*** ϵ_{eff} such that the phase velocity is given by an expression that resembles Eq. (2.34), namely

$$u_p = \frac{c}{\sqrt{\epsilon_{\text{eff}}}}. \tag{2.35}$$

Methods for calculating the propagation properties of the microstrip line are quite complicated and beyond the scope of this text. However, it is possible to use curve-fit approximations to rigorous solutions to arrive at the following set of expressions:†

$$\epsilon_{\text{eff}} = \frac{\epsilon_r + 1}{2} + \left(\frac{\epsilon_r - 1}{2}\right)\left(1 + \frac{10}{s}\right)^{-xy}, \tag{2.36}$$

where s is the ***width-to-thickness ratio***,

$$s = \frac{w}{h}, \tag{2.37}$$

†D. H. Schrader, *Microstrip Circuit Analysis*, Prentice Hall, 1995, pp. 31–32.

and x and y are intermediate variables given by

$$x = 0.56\left[\frac{\epsilon_r - 0.9}{\epsilon_r + 3}\right]^{0.05}, \tag{2.38a}$$

$$y = 1 + 0.02\ln\left(\frac{s^4 + 3.7\times10^{-4}s^2}{s^4 + 0.43}\right) + 0.05\ln(1 + 1.7\times10^{-4}s^3). \tag{2.38b}$$

The characteristic impedance of the microstrip line is given by

$$Z_0 = \frac{60}{\sqrt{\epsilon_{eff}}}\ln\left\{\frac{6 + (2\pi - 6)e^{-t}}{s} + \sqrt{1 + \frac{4}{s^2}}\right\}, \tag{2.39}$$

with

$$t = \left(\frac{30.67}{s}\right)^{0.75}. \tag{2.40}$$

Figure 2-11 displays plots of Z_0 as a function of s for various types of dielectric materials.

The corresponding line and propagation parameters are given by

$$R' = 0 \quad \text{(because } \sigma_c = \infty\text{)}, \tag{2.41a}$$

$$G' = 0 \quad \text{(because } \sigma = 0\text{)}, \tag{2.41b}$$

$$C' = \frac{\sqrt{\epsilon_{eff}}}{Z_0 c}, \tag{2.41c}$$

$$L' = Z_0^2 C', \tag{2.41d}$$

$$\alpha = 0 \quad \text{(because } R' = G' = 0\text{)}, \tag{2.41e}$$

$$\beta = \frac{\omega}{c}\sqrt{\epsilon_{eff}}. \tag{2.41f}$$

The preceding expressions allow us to compute the values of Z_0 and the other propagation parameters when given values for ϵ_r, h, and w. This is exactly what is needed in order to analyze a circuit containing a microstrip transmission line. To perform the reverse process, namely to design a microstrip line by selecting values for its w and h such that their ratio yields the required value of Z_0 (to satisfy design specifications), we need to express s in terms of Z_0. The expression for Z_0 given by Eq. (2.39) is rather complicated, so inverting it to obtain an expression for s in terms of Z_0 is rather difficult. An alternative option is to generate a family of curves similar to those displayed in **Fig. 2-11** and to use them to estimate s for a specified value of Z_0. A logical extension of the graphical approach is to generate curve-fit expressions that provide high-accuracy estimates of s. The error associated with the following formulas is less than 2%:

(a) For $Z_0 \le (44 - 2\epsilon_r)\ \Omega$,

$$s = \frac{w}{h} = \frac{2}{\pi}\left\{(q-1) - \ln(2q-1) + \frac{\epsilon_r - 1}{2\epsilon_r}\left[\ln(q-1) + 0.29 - \frac{0.52}{\epsilon_r}\right]\right\} \tag{2.42}$$

with

$$q = \frac{60\pi^2}{Z_0\sqrt{\epsilon_r}},$$

and

(b) for $Z_0 \ge (44 - 2\epsilon_r)\ \Omega$,

$$s = \frac{w}{h} = \frac{8e^p}{e^{2p} - 2}, \tag{2.43a}$$

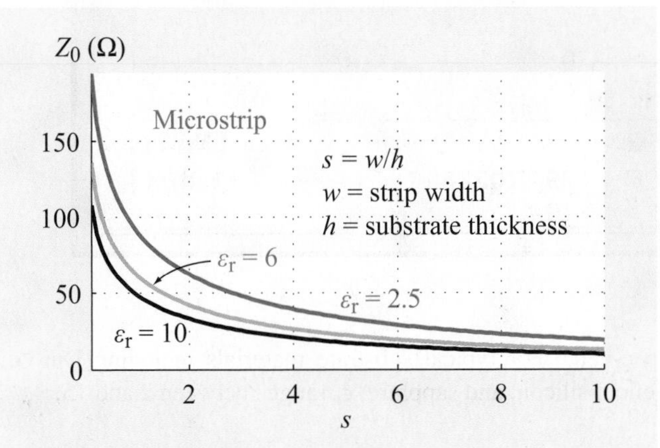

Figure 2-11 Plots of Z_0 as a function of s for various types of dielectric materials.

Module 2.3 Lossless Microstrip Line The output panel lists the values of the transmission-line parameters and displays the variation of Z_0 and ϵ_{eff} with h and w.

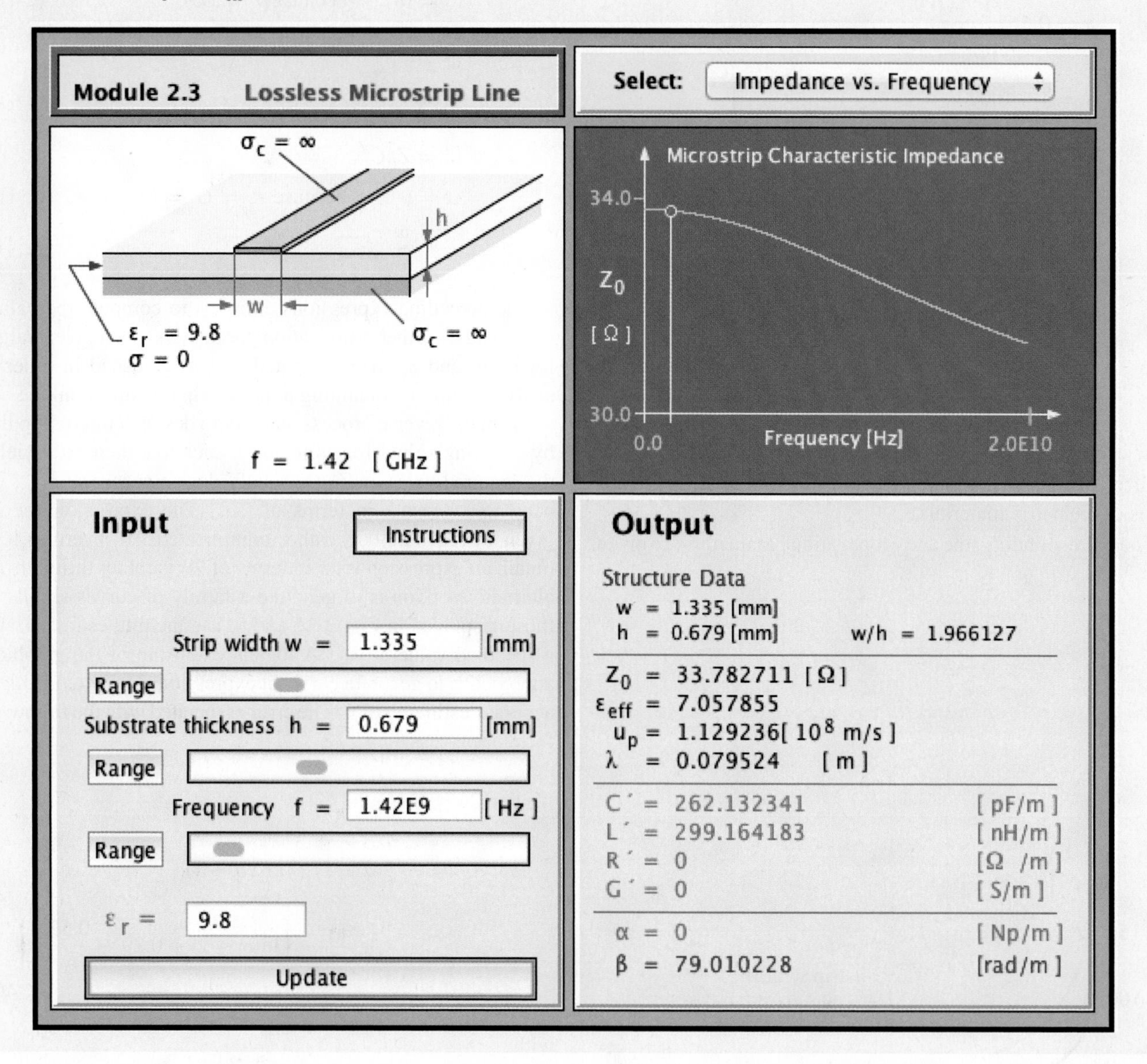

with

$$p = \sqrt{\frac{\epsilon_r + 1}{2} \frac{Z_0}{60}} + \left(\frac{\epsilon_r - 1}{\epsilon_r + 1}\right)\left(0.23 + \frac{0.12}{\epsilon_r}\right). \quad (2.43b)$$

The foregoing expressions presume that ϵ_r, the relative permittivity of the dielectric substrate, has already been specified. For typical substrate materials including Duroid, Teflon, silicon, and sapphire, ϵ_r ranges between 2 and 15.

Example 2-2: Microstrip Line

A 50 Ω microstrip line uses a 0.5 mm thick sapphire substrate with $\epsilon_r = 9$. What is the width of its copper strip?

Solution: Since $Z_0 = 50 > 44 - 18 = 32$, we should use Eq. (2.43):

$$p = \sqrt{\frac{\epsilon_r + 1}{2}} \times \frac{Z_0}{60} + \left(\frac{\epsilon_r - 1}{\epsilon_r + 1}\right)\left(0.23 + \frac{0.12}{\epsilon_r}\right)$$

$$= \sqrt{\frac{9+1}{2}} \times \frac{50}{60} + \left(\frac{9-1}{9+1}\right)\left(0.23 + \frac{0.12}{9}\right) = 2.06,$$

$$s = \frac{w}{h} = \frac{8e^p}{e^{2p} - 2} = \frac{8e^{2.06}}{e^{4.12} - 2} = 1.056.$$

Hence,

$$w = sh = 1.056 \times 0.5 \text{ mm} = 0.53 \text{ mm}.$$

To check our calculations, we use $s = 1.056$ to calculate Z_0 to verify that the value we obtained is indeed equal or close to 50 Ω. With $\epsilon_r = 9$, Eqs. (2.36) to (2.40) yield

$$x = 0.55, \qquad y = 0.99,$$

$$t = 12.51, \qquad \epsilon_{\text{eff}} = 6.11,$$

$$Z_0 = 49.93\ \Omega.$$

The calculated value of Z_0 is, for all practical purposes, equal to the value specified in the problem statement.

2-6 The Lossless Transmission Line: General Considerations

According to the preceding section, a transmission line is fully characterized by two fundamental parameters, its propagation constant γ and its characteristic impedance Z_0, both of which are specified by the angular frequency ω and the line parameters R', L', G', and C'. In many practical situations, the transmission line can be designed to exhibit low ohmic losses by selecting conductors with very high conductivities and dielectric materials (separating the conductors) with negligible conductivities. As a result, R' and G' assume very small values such that $R' \ll \omega L'$ and $G' \ll \omega C'$. These conditions allow us to set $R' = G' \approx 0$ in Eq. (2.22), which yields

$$\gamma = \alpha + j\beta = j\omega\sqrt{L'C'}, \qquad (2.44)$$

which in turn implies that

$$\alpha = 0 \qquad \textbf{(lossless line)},$$

$$\beta = \omega\sqrt{L'C'} \qquad \textbf{(lossless line)}. \qquad (2.45)$$

For the characteristic impedance, application of the lossless line conditions to Eq. (2.29) leads to

$$Z_0 = \sqrt{\frac{L'}{C'}} \qquad \textbf{(lossless line)}, \qquad (2.46)$$

which now is a real number. Using the lossless line expression for β [Eq. (2.45)], we obtain the following expressions for the guide wavelength λ and the phase velocity u_p:

$$\lambda = \frac{2\pi}{\beta} = \frac{2\pi}{\omega\sqrt{L'C'}}, \qquad (2.47)$$

$$u_p = \frac{\omega}{\beta} = \frac{1}{\sqrt{L'C'}}. \qquad (2.48)$$

Upon using Eq. (2.10), Eqs. (2.45) and (2.48) may be rewritten as

$$\beta = \omega\sqrt{\mu\epsilon} \qquad \text{(rad/m)}, \qquad (2.49)$$

$$u_p = \frac{1}{\sqrt{\mu\epsilon}} \qquad \text{(m/s)}, \qquad (2.50)$$

where μ and ϵ are, respectively, the magnetic permeability and electrical permittivity of the insulating material separating the conductors. Materials used for this purpose are usually characterized by a permeability $\mu_0 = 4\pi \times 10^{-7}$ H/m (the permeability of free space). Also, the permittivity ϵ is often specified in terms of the relative permittivity ϵ_r defined as

$$\epsilon_r = \epsilon/\epsilon_0, \qquad (2.51)$$

where $\epsilon_0 = 8.854 \times 10^{-12}$ F/m $\approx (1/36\pi) \times 10^{-9}$ F/m is the permittivity of free space (vacuum). Hence, Eq. (2.50) becomes

$$u_p = \frac{1}{\sqrt{\mu_0\epsilon_r\epsilon_0}} = \frac{1}{\sqrt{\mu_0\epsilon_0}} \cdot \frac{1}{\sqrt{\epsilon_r}} = \frac{c}{\sqrt{\epsilon_r}}, \qquad (2.52)$$

where $c = 1/\sqrt{\mu_0\epsilon_0} = 3 \times 10^8$ m/s is the velocity of light in free space. If the insulating material between the conductors is air, then $\epsilon_r = 1$ and $u_p = c$. In view of Eq. (2.51) and the relationship between λ and u_p given by Eq. (2.33), the wavelength is given by

$$\lambda = \frac{u_p}{f} = \frac{c}{f}\frac{1}{\sqrt{\epsilon_r}} = \frac{\lambda_0}{\sqrt{\epsilon_r}}, \tag{2.53}$$

where $\lambda_0 = c/f$ is the wavelength in air corresponding to a frequency f. Note that, because both u_p and λ depend on ϵ_r, the choice of the type of insulating material used in a transmission line is dictated not only by its mechanical properties, but by its electrical properties as well.

According to Eq. (2.52), if ϵ_r of the insulating material is independent of f (which usually is the case for commonly used TEM lines), the same independence applies to u_p.

► If sinusoidal waves of different frequencies travel on a transmission line with the same phase velocity, the line is called ***nondispersive***. ◄

This is an important feature to consider when digital data are transmitted in the form of pulses. A rectangular pulse or a series of pulses is composed of many Fourier components with different frequencies. If the phase velocity is the same for all frequency components (or at least for the dominant ones), then the pulse's shape does not change as it travels down the line. In contrast, the shape of a pulse propagating in a dispersive medium becomes progressively distorted, and the pulse length increases (stretches out) as a function of the distance traveled in the medium (**Fig. 2-3**), thereby imposing a limitation on the maximum data rate (which is related to the length of the individual pulses and the spacing between adjacent pulses) that can be transmitted through the medium without loss of information.

Table 2-2 provides a list of the expressions for γ, Z_0, and u_p for the general case of a lossy line and for several types of lossless lines. The expressions for the lossless lines are based on the equations for L' and C' given in **Table 2-1**.

Exercise 2-5: For a lossless transmission line, $\lambda = 20.7$ cm at 1 GHz. Find ϵ_r of the insulating material.

Answer: $\epsilon_r = 2.1$. (See EM.)

Exercise 2-6: A lossless transmission line uses a dielectric insulating material with $\epsilon_r = 4$. If its line capacitance is $C' = 10$ (pF/m), find (a) the phase velocity u_p, (b) the line inductance L', and (c) the characteristic impedance Z_0.

Answer: (a) $u_p = 1.5\times 10^8$ (m/s), (b) $L' = 4.45$ (μH/m), (c) $Z_0 = 667.1\ \Omega$. (See EM.)

2-6.1 Voltage Reflection Coefficient

With $\gamma = j\beta$ for the lossless line, Eqs. (2.26a) and (2.30) for the total voltage and current become

$$\widetilde{V}(z) = V_0^+ e^{-j\beta z} + V_0^- e^{j\beta z}, \tag{2.54a}$$

$$\widetilde{I}(z) = \frac{V_0^+}{Z_0} e^{-j\beta z} - \frac{V_0^-}{Z_0} e^{j\beta z}. \tag{2.54b}$$

These expressions contain two unknowns, V_0^+ and V_0^-. According to Section 1-7.2, an exponential factor of the form $e^{-j\beta z}$ is associated with a wave traveling in the positive z direction, from the source (sending end) to the load (receiving end). Accordingly, we refer to it as the ***incident wave***, with V_0^+ as its voltage amplitude. Similarly, the term containing $V_0^- e^{j\beta z}$ represents a ***reflected wave*** with voltage amplitude V_0^-, traveling along the negative z direction, from the load to the source.

To determine V_0^+ and V_0^-, we need to consider the lossless transmission line in the context of the complete circuit, including a generator circuit at its input terminals and a load at its output terminals, as shown in **Fig. 2-12**. The line, of length l, is terminated in an arbitrary ***load impedance*** Z_L.

► For convenience, the reference of the spatial coordinate z is chosen such that $z = 0$ corresponds to the location of the load. ◄

At the sending end, at $z = -l$, the line is connected to a sinusoidal voltage source with phasor voltage $\widetilde{V}_g$ and internal impedance Z_g. Since z points from the generator to the load, positive values of z correspond to locations beyond the load, and therefore are irrelevant to our circuit. In future sections, we will find it more convenient to work with a spatial dimension that also starts at the load, but whose direction is opposite of z. We shall call it the ***distance from the load*** d and define it as $d = -z$, as shown in **Fig. 2-12**.

Table 2-2 Characteristic parameters of transmission lines.

	Propagation Constant $\gamma = \alpha + j\beta$	Phase Velocity u_p	Characteristic Impedance Z_0
General case	$\gamma = \sqrt{(R' + j\omega L')(G' + j\omega C')}$	$u_p = \omega/\beta$	$Z_0 = \sqrt{\dfrac{(R' + j\omega L')}{(G' + j\omega C')}}$
Lossless $(R' = G' = 0)$	$\alpha = 0,\ \beta = \omega\sqrt{\epsilon_r}/c$	$u_p = c/\sqrt{\epsilon_r}$	$Z_0 = \sqrt{L'/C'}$
Lossless coaxial	$\alpha = 0,\ \beta = \omega\sqrt{\epsilon_r}/c$	$u_p = c/\sqrt{\epsilon_r}$	$Z_0 = (60/\sqrt{\epsilon_r})\ln(b/a)$
Lossless two-wire	$\alpha = 0,\ \beta = \omega\sqrt{\epsilon_r}/c$	$u_p = c/\sqrt{\epsilon_r}$	$Z_0 = (120/\sqrt{\epsilon_r}) \cdot \ln[(D/d) + \sqrt{(D/d)^2 - 1}]$ $Z_0 \approx (120/\sqrt{\epsilon_r})\ln(2D/d)$, if $D \gg d$
Lossless parallel-plate	$\alpha = 0,\ \beta = \omega\sqrt{\epsilon_r}/c$	$u_p = c/\sqrt{\epsilon_r}$	$Z_0 = (120\pi/\sqrt{\epsilon_r})\,(h/w)$

Notes: (1) $\mu = \mu_0$, $\epsilon = \epsilon_r\epsilon_0$, $c = 1/\sqrt{\mu_0\epsilon_0}$, and $\sqrt{\mu_0/\epsilon_0} \approx (120\pi)\ \Omega$, where ϵ_r is the relative permittivity of insulating material. (2) For coaxial line, a and b are radii of inner and outer conductors. (3) For two-wire line, d = wire diameter and D = separation between wire centers. (4) For parallel-plate line, w = width of plate and h = separation between the plates.

The phasor voltage across the load, $\widetilde{V}_L$, and the phasor current through it, $\widetilde{I}_L$, are related by the load impedance Z_L as

$$Z_L = \frac{\widetilde{V}_L}{\widetilde{I}_L}. \tag{2.55}$$

The voltage $\widetilde{V}_L$ is the total voltage on the line $\widetilde{V}(z)$ given by Eq. (2.54a), and $\widetilde{I}_L$ is the total current $\widetilde{I}(z)$ given by Eq. (2.54b), both evaluated at $z = 0$:

$$\widetilde{V}_L = \widetilde{V}(z{=}0) = V_0^+ + V_0^-, \tag{2.56a}$$

$$\widetilde{I}_L = \widetilde{I}(z{=}0) = \frac{V_0^+}{Z_0} - \frac{V_0^-}{Z_0}. \tag{2.56b}$$

Using these expressions in Eq. (2.55), we obtain

$$Z_L = \left(\frac{V_0^+ + V_0^-}{V_0^+ - V_0^-}\right) Z_0. \tag{2.57}$$

Solving for V_0^- gives

$$V_0^- = \left(\frac{Z_L - Z_0}{Z_L + Z_0}\right) V_0^+. \tag{2.58}$$

▶ The ratio of the amplitudes of the reflected and incident voltage waves at the load is known as the ***voltage reflection coefficient*** Γ. ◀

From Eq. (2.58), it follows that

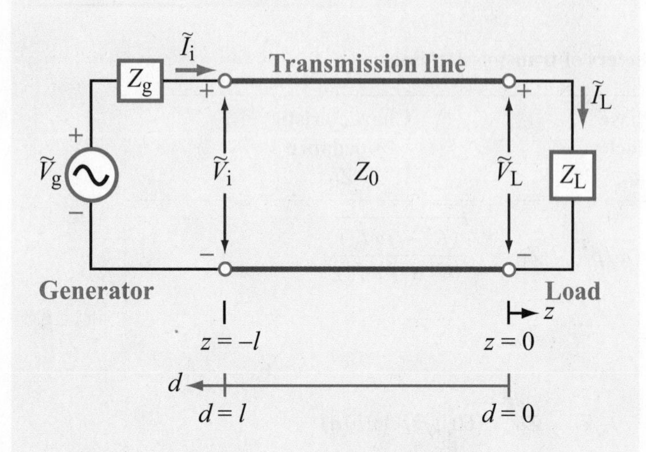

Figure 2-12 Transmission line of length l connected on one end to a generator circuit and on the other end to a load Z_L. The load is located at $z = 0$ and the generator terminals are at $z = -l$. Coordinate d is defined as $d = -z$.

$$\Gamma = \frac{V_0^-}{V_0^+} = \frac{Z_L - Z_0}{Z_L + Z_0} = \frac{Z_L/Z_0 - 1}{Z_L/Z_0 + 1} = \frac{z_L - 1}{z_L + 1} \quad \textbf{(dimensionless)}, \tag{2.59}$$

where

$$z_L = \frac{Z_L}{Z_0} \tag{2.60}$$

is the ***normalized load impedance***. In many transmission-line problems, we can streamline the necessary computation by normalizing all impedances in the circuit to the characteristic impedance Z_0. *Normalized impedances are denoted by lowercase letters.*

In view of Eq. (2.28), the ratio of the current amplitudes is

$$\frac{I_0^-}{I_0^+} = -\frac{V_0^-}{V_0^+} = -\Gamma. \tag{2.61}$$

▶ We note that whereas the ratio of the voltage amplitudes is equal to Γ, the ratio of the current amplitudes is equal to $-\Gamma$. ◀

The reflection coefficient Γ is governed by a single parameter, the normalized load impedance z_L. As indicated by Eq. (2.46), Z_0 of a lossless line is a real number. However, Z_L is in general a complex quantity, as in the case of a series RL circuit, for example, for which $Z_L = R + j\omega L$. Hence, in general Γ also is complex and given by

$$\Gamma = |\Gamma| e^{j\theta_r}, \tag{2.62}$$

where $|\Gamma|$ is the magnitude of Γ and θ_r is its phase angle. Note that $|\Gamma| \leq 1$.

▶ A load is said to be ***matched*** to a transmission line if $Z_L = Z_0$ because then there will be no reflection by the load ($\Gamma = 0$ and $V_0^- = 0$). ◀

On the other hand, when the load is an open circuit ($Z_L = \infty$), $\Gamma = 1$ and $V_0^- = V_0^+$, and when it is a short circuit ($Z_L = 0$), $\Gamma = -1$ and $V_0^- = -V_0^+$ (**Table 2-3**).

Example 2-3: Reflection Coefficient of a Series RC Load

A 100 Ω transmission line is connected to a load consisting of a 50 Ω resistor in series with a 10 pF capacitor. Find the reflection coefficient at the load for a 100 MHz signal.

Solution: The following quantities are given (**Fig. 2-13**):

$$R_L = 50\ \Omega, \qquad C_L = 10\ \text{pF} = 10^{-11}\ \text{F},$$

$$Z_0 = 100\ \Omega, \qquad f = 100\ \text{MHz} = 10^8\ \text{Hz}.$$

Table 2-3 **Magnitude and phase of the reflection coefficient for various types of load. The normalized load impedance $z_L = Z_L/Z_0 = (R + jX)/Z_0 = r + jx$, where $r = R/Z_0$ and $x = X/Z_0$ are the real and imaginary parts of z_L, respectively.**

Reflection Coefficient $\Gamma = |\Gamma|e^{j\theta_r}$

Load	$\lvert\Gamma\rvert$	θ_r
Z_0 $Z_L = (r + jx)Z_0$	$\left[\dfrac{(r-1)^2 + x^2}{(r+1)^2 + x^2}\right]^{1/2}$	$\tan^{-1}\left(\dfrac{x}{r-1}\right) - \tan^{-1}\left(\dfrac{x}{r+1}\right)$
Z_0 Z_0	0 (no reflection)	irrelevant
Z_0 (short)	1	$\pm 180°$ (phase opposition)
Z_0 (open)	1	0 (in-phase)
Z_0 $jX = j\omega L$	1	$\pm 180° - 2\tan^{-1} x$
Z_0 $jX = \dfrac{-j}{\omega C}$	1	$\pm 180° + 2\tan^{-1} x$

The normalized load impedance is

$$z_L = \frac{Z_L}{Z_0} = \frac{R_L - j/(\omega C_L)}{Z_0}$$

$$= \frac{1}{100}\left(50 - j\frac{1}{2\pi \times 10^8 \times 10^{-11}}\right)$$

$$= (0.5 - j1.59)\ \Omega.$$

From Eq. (2.59), the voltage reflection coefficient is

$$\Gamma = \frac{z_L - 1}{z_L + 1} = \frac{0.5 - j1.59 - 1}{0.5 - j1.59 + 1}$$

$$= \frac{-0.5 - j1.59}{1.5 - j1.59} = \frac{-1.67e^{j72.6°}}{2.19e^{-j46.7°}} = -0.76e^{j119.3°}.$$

This result may be converted into the form of Eq. (2.62) by replacing the minus sign with $e^{-j180°}$. Thus,

$$\Gamma = 0.76e^{j119.3°}e^{-j180°} = 0.76e^{-j60.7°} = 0.76\angle{-60.7°},$$

or

$$|\Gamma| = 0.76, \qquad \theta_r = -60.7°.$$

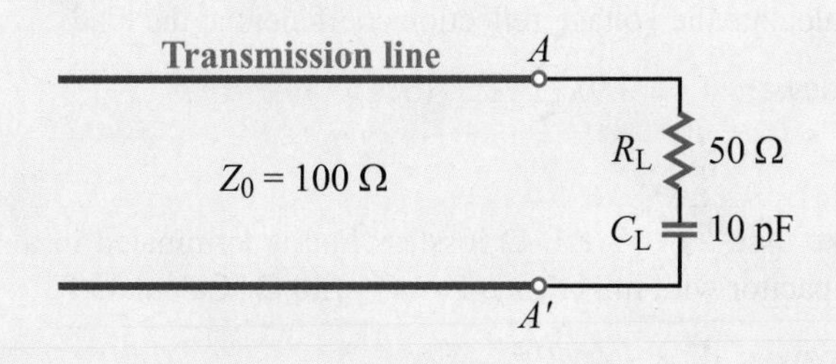

Figure 2-13 RC load (Example 2-3).

Example 2-4: $|\Gamma|$ for Purely Reactive Load

Show that $|\Gamma| = 1$ for a lossless line connected to a purely reactive load.

Solution: The load impedance of a purely reactive load is

$$Z_L = jX_L.$$

From Eq. (2.59), the reflection coefficient is

$$\begin{aligned}\Gamma &= \frac{Z_L - Z_0}{Z_L + Z_0} \\ &= \frac{jX_L - Z_0}{jX_L + Z_0} \\ &= \frac{-(Z_0 - jX_L)}{(Z_0 + jX_L)} \\ &= \frac{-\sqrt{Z_0^2 + X_L^2}\, e^{-j\theta}}{\sqrt{Z_0^2 + X_L^2}\, e^{j\theta}} \\ &= -e^{-j2\theta},\end{aligned}$$

where $\theta = \tan^{-1} X_L/Z_0$. Hence

$$\begin{aligned}|\Gamma| &= |-e^{-j2\theta}| \\ &= [(e^{-j2\theta})(e^{-j2\theta})^*]^{1/2} \\ &= 1.\end{aligned}$$

Exercise 2-7: A 50 Ω lossless transmission line is terminated in a load with impedance $Z_L = (30 - j200)$ Ω. Calculate the voltage reflection coefficient at the load.

Answer: $\Gamma = 0.93\angle{-27.5^\circ}$. (See EM.)

Exercise 2-8: A 150 Ω lossless line is terminated in a capacitor with impedance $Z_L = -j30$ Ω. Calculate Γ.

Answer: $\Gamma = 1\angle{-157.4^\circ}$. (See EM.)

2-6.2 Standing Waves

Using the relation $V_0^- = \Gamma V_0^+$ in Eqs. (2.54a) and (2.54b) yields

$$\widetilde{V}(z) = V_0^+(e^{-j\beta z} + \Gamma e^{j\beta z}), \tag{2.63a}$$

$$\widetilde{I}(z) = \frac{V_0^+}{Z_0}(e^{-j\beta z} - \Gamma e^{j\beta z}). \tag{2.63b}$$

These expressions now contain only one, yet to be determined, unknown, V_0^+. Before we proceed to solve for V_0^+, however, let us examine the physical meaning underlying these expressions. We begin by deriving an expression for $|\widetilde{V}(z)|$, the magnitude of $\widetilde{V}(z)$. Upon using Eq. (2.62) in Eq. (2.63a) and applying the relation $|\widetilde{V}(z)| = [\widetilde{V}(z)\,\widetilde{V}^*(z)]^{1/2}$, where $\widetilde{V}^*(z)$ is the complex conjugate of $\widetilde{V}(z)$, we have

$$\begin{aligned}|\widetilde{V}(z)| &= \Big\{\Big[V_0^+(e^{-j\beta z} + |\Gamma|e^{j\theta_r}e^{j\beta z})\Big] \\ &\quad\cdot \Big[(V_0^+)^*(e^{j\beta z} + |\Gamma|e^{-j\theta_r}e^{-j\beta z})\Big]\Big\}^{1/2} \\ &= |V_0^+|\Big[1 + |\Gamma|^2 + |\Gamma|(e^{j(2\beta z+\theta_r)} + e^{-j(2\beta z+\theta_r)})\Big]^{1/2} \\ &= |V_0^+|\Big[1 + |\Gamma|^2 + 2|\Gamma|\cos(2\beta z + \theta_r)\Big]^{1/2},\end{aligned} \tag{2.64}$$

where we have used the identity

$$e^{jx} + e^{-jx} = 2\cos x \tag{2.65}$$

for any real quantity x. To express the magnitude of $\widetilde{V}$ as a function of d instead of z, we replace z with $-d$ on the right-hand side of Eq. (2.64):

$$|\widetilde{V}(d)| = |V_0^+|\Big[1 + |\Gamma|^2 + 2|\Gamma|\cos(2\beta d - \theta_r)\Big]^{1/2}. \tag{2.66}$$

By applying the same steps to Eq. (2.63b), a similar expression can be derived for $|\widetilde{I}(d)|$, the magnitude of the current $\widetilde{I}(d)$:

$$|\widetilde{I}(d)| = \frac{|V_0^+|}{Z_0}\,[1 + |\Gamma|^2 - 2|\Gamma|\cos(2\beta d - \theta_r)]^{1/2}. \tag{2.67}$$

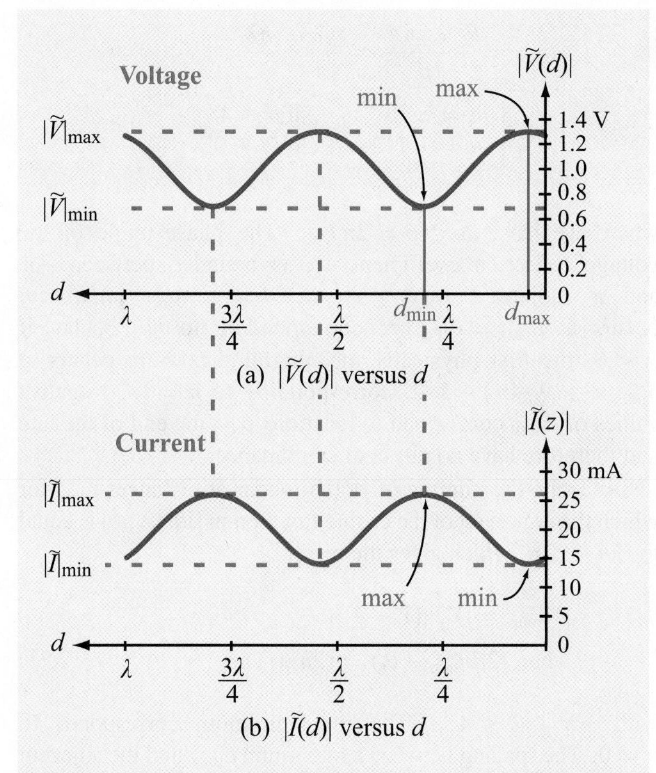

Figure 2-14 Standing-wave pattern for (a) $|\widetilde{V}(d)|$ and (b) $|\widetilde{I}(d)|$ for a lossless transmission line of characteristic impedance $Z_0 = 50\ \Omega$, terminated in a load with a reflection coefficient $\Gamma = 0.3e^{j30°}$. The magnitude of the incident wave $|V_0^+| = 1$ V. The standing-wave ratio is $S = |\widetilde{V}|_{max}/|\widetilde{V}|_{min} = 1.3/0.7 = 1.86$.

The variations of $|\widetilde{V}(d)|$ and $|\widetilde{I}(d)|$ as a function of d and the position on the line relative to the load (at $d = 0$), are illustrated in **Fig. 2-14** for a line with $|V_0^+| = 1$ V, $|\Gamma| = 0.3$, $\theta_r = 30°$, and $Z_0 = 50\ \Omega$. The sinusoidal patterns are called ***standing waves*** and are caused by the *interference* of the two traveling waves. The *maximum value* of the ***standing-wave pattern*** of $|\widetilde{V}(d)|$ corresponds to the position on the line at which the incident and reflected waves are ***in-phase*** [$2\beta d - \theta_r = 2n\pi$ in Eq. (2.66)] and therefore add constructively to give a value equal to $(1 + |\Gamma|)|V_0^+| = 1.3$ V. The *minimum value* of $|\widetilde{V}(d)|$ occurs when the two waves interfere destructively, which occurs when the incident and reflected waves are in ***phase-opposition*** [$2\beta d - \theta_r = (2n+1)\pi$]. In this case, $|\widetilde{V}(d)| = (1-|\Gamma|)|V_0^+| = 0.7$ V.

▶ Whereas the repetition period is λ for the incident and reflected waves considered individually, the repetition period of the standing-wave pattern is $\lambda/2$. ◀

The standing-wave pattern describes the spatial variation of the magnitude of $\widetilde{V}(d)$ as a function of d. If one were to observe the variation of the instantaneous voltage as a function of time at location $d = d_{max}$ in **Fig. 2-14**, that variation would be as $\cos \omega t$ and would have an amplitude equal to 1.3 V [i.e., $\upsilon(t)$ would oscillate between -1.3 V and $+1.3$ V]. Similarly, the instantaneous voltage $\upsilon(d, t)$ at any location d will be sinusoidal with amplitude equal to $|\widetilde{V}(d)|$ at that d. Interactive Module 2.4† provides a highly recommended simulation tool for gaining better understanding of the standing-wave patterns for $\widetilde{V}(d)$ and $\widetilde{I}(d)$ and the dynamic behavior of $\upsilon(d, t)$ and $i(d, t)$.

Close inspection of the voltage and current standing-wave patterns shown in **Fig. 2-14** reveals that the two patterns are in phase opposition (when one is at a maximum, the other is at a minimum, and vice versa). This is a consequence of the fact that the third term in Eq. (2.66) is preceded by a plus sign, whereas the third term in Eq. (2.67) is preceded by a minus sign.

The standing-wave patterns shown in **Fig. 2-14** are for $\Gamma = 0.3\,e^{j30°}$. The peak-to-peak variation of the pattern ($|\widetilde{V}|_{min}$ to $|\widetilde{V}|_{max}$) depends on $|\Gamma|$, which in general can vary between 0 and 1. For the special case of a matched line with $Z_L = Z_0$, we have $|\Gamma| = 0$ and $|\widetilde{V}(d)| = |V_0^+|$ for all values of d, as shown in **Fig. 2-15(a)**.

▶ With no reflected wave present, there are no interference and no standing waves. ◀

The other end of the $|\Gamma|$ scale, at $|\Gamma| = 1$, corresponds to when the load is a short circuit ($\Gamma = -1$) or an open circuit ($\Gamma = 1$). The standing-wave patterns for those two cases are shown in **Figs. 2-15(b)** and **(c)**; both exhibit maxima of $2|V_0^+|$ and minima equal to zero, but the two patterns are spatially shifted relative to each other by a distance of $\lambda/4$. A purely reactive load (capacitor or inductor) also satisfies the condition $|\Gamma| = 1$, but θ_r is generally neither zero nor 180° (**Table 2-3**). Exercise 2.9 examines the standing-wave pattern for a lossless line terminated in an inductor.

†At em.eecs.umich.edu

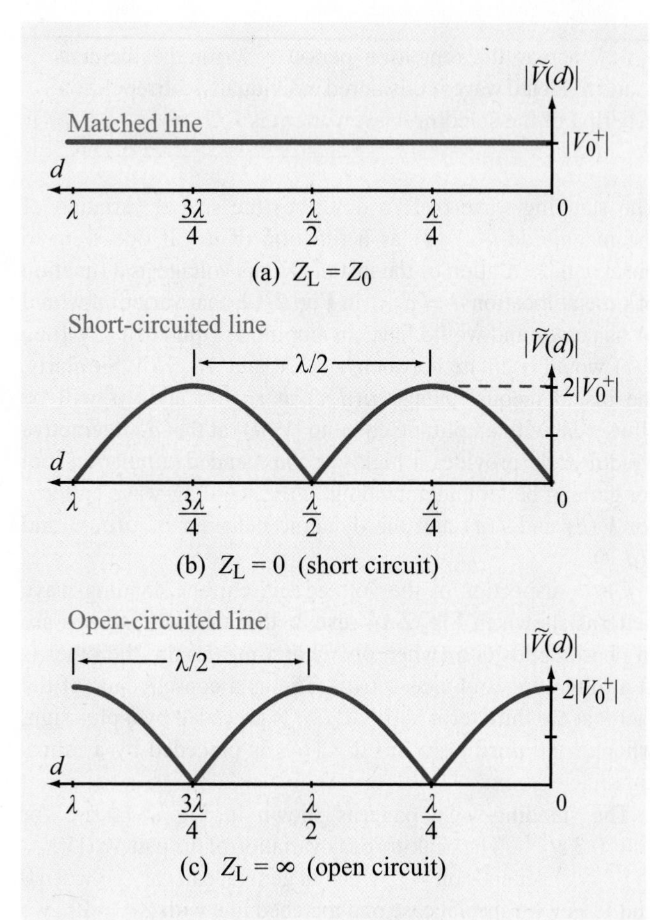

Figure 2-15 Voltage standing-wave patterns for (a) a matched load, (b) a short-circuited line, and (c) an open-circuited line.

Now let us examine the maximum and minimum values of the voltage magnitude. From Eq. (2.66), $|\widetilde{V}(d)|$ is a maximum when the argument of the cosine function is equal to zero or a multiple of 2π. Let us denote d_{max} as the distance from the load at which $|\widetilde{V}(d)|$ is a maximum. It then follows that

$$|\widetilde{V}(d)| = |\widetilde{V}|_{\text{max}} = |V_0^+|[1 + |\Gamma|], \tag{2.68}$$

when

$$2\beta d_{\text{max}} - \theta_r = 2n\pi, \tag{2.69}$$

with $n = 0$ or a positive integer. Solving Eq. (2.69) for d_{max}, we have

$$d_{\text{max}} = \frac{\theta_r + 2n\pi}{2\beta} = \frac{\theta_r \lambda}{4\pi} + \frac{n\lambda}{2}, \quad \begin{cases} n = 1, 2, \ldots & \text{if } \theta_r < 0, \\ n = 0, 1, 2, \ldots & \text{if } \theta_r \geq 0, \end{cases} \tag{2.70}$$

where we have used $\beta = 2\pi/\lambda$. The phase angle of the voltage reflection coefficient, θ_r, is bounded between $-\pi$ and π radians. If $\theta_r \geq 0$, the ***first voltage maximum*** occurs at $d_{\text{max}} = \theta_r\lambda/4\pi$, corresponding to $n = 0$, but if $\theta_r < 0$, the first physically meaningful maximum occurs at $d_{\text{max}} = (\theta_r\lambda/4\pi) + \lambda/2$, corresponding to $n = 1$. Negative values of d_{max} correspond to locations past the end of the line and therefore have no physical significance.

Similarly, the minima of $|\widetilde{V}(d)|$ occur at distances d_{min} for which the argument of the cosine function in Eq. (2.66) is equal to $(2n+1)\pi$, which gives the result

$$|\widetilde{V}|_{\text{min}} = |V_0^+|[1 - |\Gamma|], \quad \text{when } (2\beta d_{\text{min}} - \theta_r) = (2n+1)\pi, \tag{2.71}$$

with $-\pi \leq \theta_r \leq \pi$. The first minimum corresponds to $n = 0$. The spacing between a maximum d_{max} and the adjacent minimum d_{min} is $\lambda/4$. Hence, the ***first minimum*** occurs at

$$d_{\text{min}} = \begin{cases} d_{\text{max}} + \lambda/4, & \text{if } d_{\text{max}} < \lambda/4, \\ d_{\text{max}} - \lambda/4, & \text{if } d_{\text{max}} \geq \lambda/4. \end{cases} \tag{2.72}$$

► The locations on the line corresponding to voltage maxima correspond to current minima, and vice versa. ◄

The ratio of $|\widetilde{V}|_{\text{max}}$ to $|\widetilde{V}|_{\text{min}}$ is called the ***voltage standing-wave ratio*** S, which from Eqs. (2.68) and (2.71) is given by

$$S = \frac{|\widetilde{V}|_{\text{max}}}{|\widetilde{V}|_{\text{min}}} = \frac{1 + |\Gamma|}{1 - |\Gamma|} \quad \textbf{(dimensionless)}. \tag{2.73}$$

This quantity, which often is referred to by its acronym, VSWR, or the shorter acronym ***SWR***, provides a measure of the mismatch between the load and the transmission line; for a matched load with $\Gamma = 0$, we get $S = 1$, and for a line with $|\Gamma| = 1$, $S = \infty$.

Module 2.4 Transmission-Line Simulator Upon specifying the requisite input data—including the load impedance at $d = 0$ and the generator voltage and impedance at $d = l$—this module provides a wealth of output information about the voltage and current waveforms along the trasmission line. You can view plots of the standing wave patterns for voltage and current, the time and spatial variations of the instantaneous voltage $\upsilon(d,t)$ and current $i(d,t)$, and other related quantities.

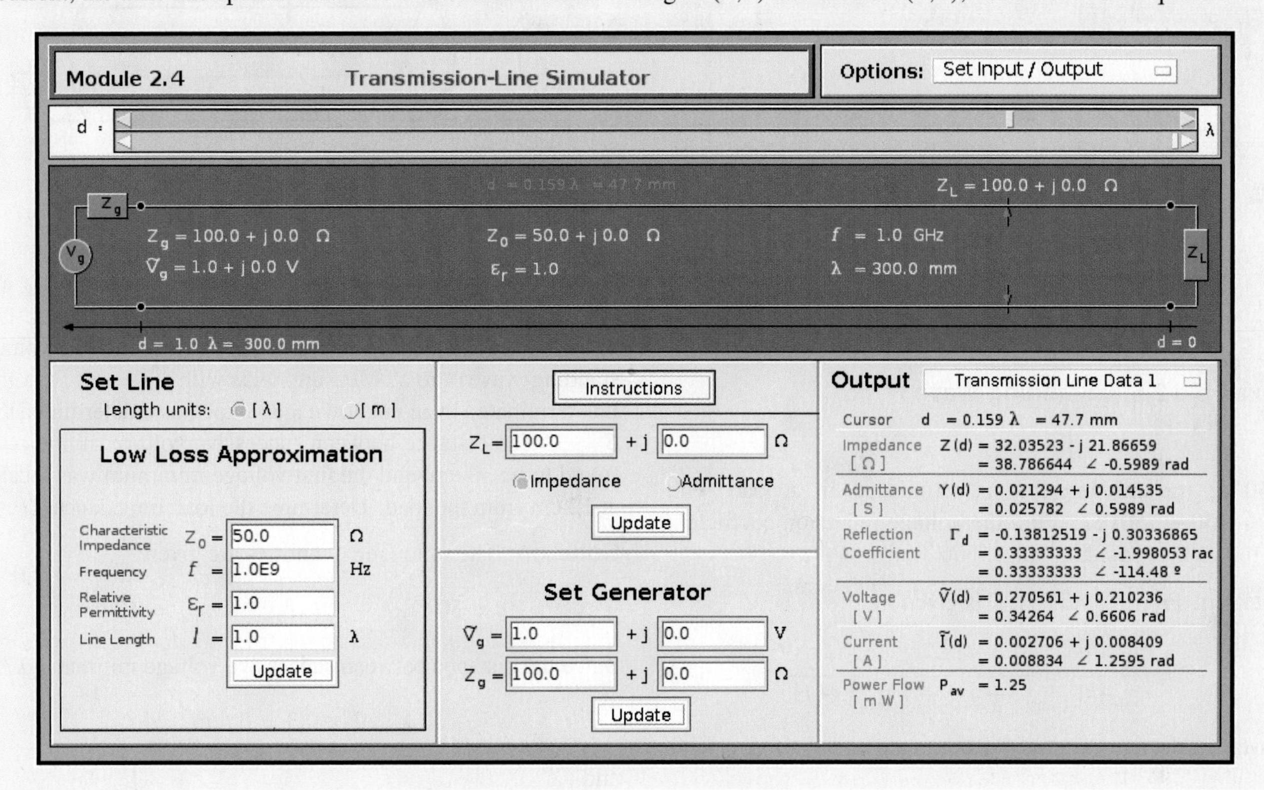

Concept Question 2-5: The attenuation constant α represents ohmic losses. In view of the model given in Fig. 2-6(c), what should R' and G' be in order to have no losses? Verify your expectation through the expression for α given by Eq. (2.25a).

Concept Question 2-6: How is the wavelength λ of the wave traveling on the transmission line related to the free-space wavelength λ_0?

Concept Question 2-7: When is a load matched to a transmission line? Why is it important?

Concept Question 2-8: What is a standing-wave pattern? Why is its period $\lambda/2$ and not λ?

Concept Question 2-9: What is the separation between the location of a voltage maximum and the adjacent current maximum on the line?

Exercise 2-9: Use Module 2.4 to generate the voltage and current standing-wave patterns for a 50 Ω line of length 1.5λ, terminated in an inductance with $Z_L = j140\ \Omega$.

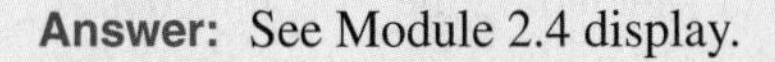

Answer: See Module 2.4 display.

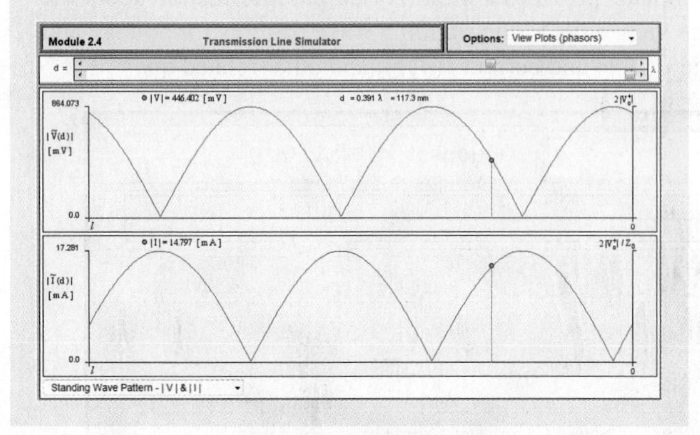

Example 2-5: Standing-Wave Ratio

A 50 Ω transmission line is terminated in a load with $Z_L = (100 + j50)$ Ω. Find the voltage reflection coefficient and the voltage standing-wave ratio.

Solution: From Eq. (2.59), Γ is given by

$$\Gamma = \frac{z_L - 1}{z_L + 1} = \frac{(2 + j1) - 1}{(2 + j1) + 1} = \frac{1 + j1}{3 + j1}.$$

Converting the numerator and denominator to polar form yields

$$\Gamma = \frac{1.414 e^{j45°}}{3.162 e^{j18.4°}} = 0.45 e^{j26.6°}.$$

Using the definition for S given by Eq. (2.73), we have

$$S = \frac{1 + |\Gamma|}{1 - |\Gamma|} = \frac{1 + 0.45}{1 - 0.45} = 2.6.$$

Example 2-6: Measuring Z_L

A ***slotted-line*** probe is an instrument used to measure the unknown impedance of a load, Z_L. A coaxial slotted line contains a narrow longitudinal slit in the outer conductor. A small probe inserted in the slit can be used to sample the magnitude of the electric field and, hence, the magnitude $|\widetilde{V}(d)|$ of the voltage on the line (Fig. 2-16). By moving the probe along the length of the slotted line, it is possible to measure $|\widetilde{V}|_{max}$ and $|\widetilde{V}|_{min}$ and the distances from the load at which they occur. Use of Eq. (2.73), namely $S = |\widetilde{V}|_{max}/|\widetilde{V}|_{min}$, provides the voltage standing-wave ratio S. Measurements with a $Z = 50$ Ω slotted line terminated in an unknown load impedance determined that $S = 3$. The distance between successive voltage minima was found to be 30 cm, and the first voltage minimum was located at 12 cm from the load. Determine the load impedance Z_L.

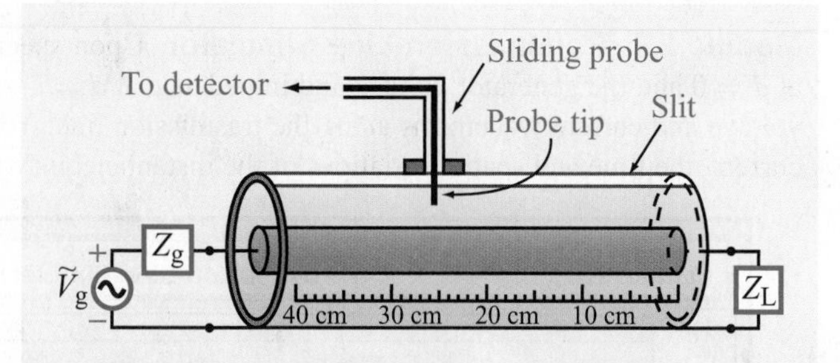

Figure 2-16 Slotted coaxial line (Example 2-6).

Solution: The following quantities are given:

$$Z_0 = 50\ \Omega, \qquad S = 3, \qquad d_{min} = 12\ \text{cm}.$$

Since the distance between successive voltage minima is $\lambda/2$,

$$\lambda = 2 \times 0.3 = 0.6\ \text{m},$$

and

$$\beta = \frac{2\pi}{\lambda} = \frac{2\pi}{0.6} = \frac{10\pi}{3} \qquad \text{(rad/m)}.$$

From Eq. (2.73), solving for $|\Gamma|$ in terms of S gives

$$|\Gamma| = \frac{S - 1}{S + 1} = \frac{3 - 1}{3 + 1} = 0.5.$$

Next, we use the condition given by Eq. (2.71) to find θ_r:

$$2\beta d_{min} - \theta_r = \pi, \qquad \text{for } n = 0 \text{ (first minimum)},$$

which gives

$$\theta_r = 2\beta d_{min} - \pi$$
$$= 2 \times \frac{10\pi}{3} \times 0.12 - \pi = -0.2\pi\ \text{(rad)} = -36°.$$

Hence,

$$\Gamma = |\Gamma| e^{j\theta_r} = 0.5 e^{-j36°} = 0.405 - j0.294.$$

Solving Eq. (2.59) for Z_L, we have

$$Z_L = Z_0 \left[\frac{1+\Gamma}{1-\Gamma}\right]$$

$$= 50 \left[\frac{1+0.405-j0.294}{1-0.405+j0.294}\right] = (85-j67)\ \Omega.$$

Exercise 2-10: If $\Gamma = 0.5\angle{-60^\circ}$ and $\lambda = 24$ cm, find the locations of the voltage maximum and minimum nearest to the load.

Answer: $d_{max} = 10$ cm, $d_{min} = 4$ cm. (See EM.)

Exercise 2-11: A 140 Ω lossless line is terminated in a load impedance $Z_L = (280+j182)$ Ω. If $\lambda = 72$ cm, find (a) the reflection coefficient Γ, (b) the voltage standing-wave ratio S, (c) the locations of voltage maxima, and (d) the locations of voltage minima.

Answer: (a) $\Gamma = 0.5\angle{29^\circ}$, (b) $S = 3.0$, (c) $d_{max} = 2.9\text{ cm} + n\lambda/2$, (d) $d_{min} = 20.9\text{ cm} + n\lambda/2$, where $n = 0, 1, 2, \ldots$. (See EM.)

2-7 Wave Impedance of the Lossless Line

The standing-wave patterns indicate that on a mismatched line the voltage and current magnitudes are oscillatory with position along the line and in phase opposition with each other. Hence, the voltage to current ratio, called the ***wave impedance*** $Z(d)$, must vary with position also. Using Eqs. (2.63a) and (2.63b) with $z = -d$,

$$Z(d) = \frac{\tilde{V}(d)}{\tilde{I}(d)}$$

$$= \frac{V_0^+[e^{j\beta d} + \Gamma e^{-j\beta d}]}{V_0^+[e^{j\beta d} - \Gamma e^{-j\beta d}]} Z_0$$

$$= Z_0 \left[\frac{1+\Gamma e^{-j2\beta d}}{1-\Gamma e^{-j2\beta d}}\right]$$

$$= Z_0 \left[\frac{1+\Gamma_d}{1-\Gamma_d}\right] \quad (\Omega), \tag{2.74}$$

where we define

$$\Gamma_d = \Gamma e^{-j2\beta d} = |\Gamma| e^{j\theta_r} e^{-j2\beta d} = |\Gamma| e^{j(\theta_r - 2\beta d)} \tag{2.75}$$

as the ***phase-shifted voltage reflection coefficient***, meaning that Γ_d has the same magnitude as Γ, but its phase is shifted by $2\beta d$ relative to that of Γ.

► $Z(d)$ is the ratio of the total voltage (incident- and reflected-wave voltages) to the total current at any point d on the line, in contrast with the characteristic impedance of the line Z_0, which relates the voltage and current of each of the two waves individually ($Z_0 = V_0^+/I_0^+ = -V_0^-/I_0^-$). ◄

In the circuit of Fig. 2-17(a), at terminals BB' at an arbitrary location d on the line, $Z(d)$ is the wave impedance of the line when "looking" to the right (i.e., towards the load). Application of the equivalence principle allows us to replace the segment to the right of terminals BB' with a lumped impedance of value $Z(d)$, as depicted in Fig. 2-17(b). From the standpoint of the input circuit to the left of terminals BB', the two circuit configurations are electrically identical.

Of particular interest in many transmission-line problems is the ***input impedance*** at the source end of the line, at $d = l$,

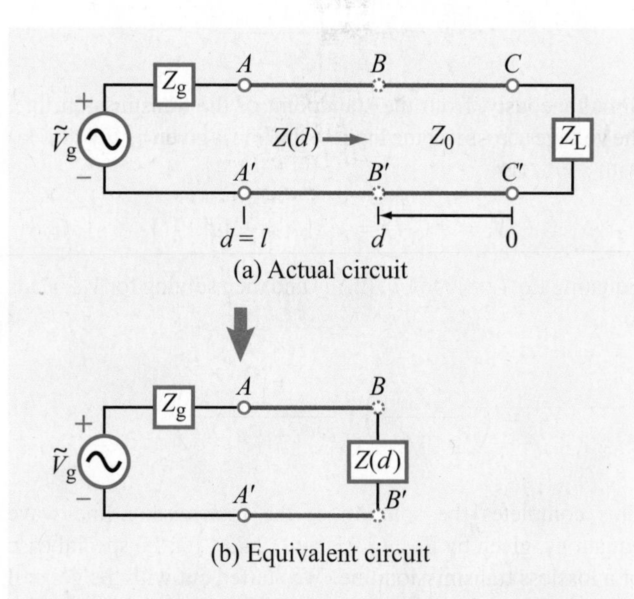

Figure 2-17 The segment to the right of terminals BB' can be replaced with a discrete impedance equal to the wave impedance $Z(d)$.

which is given by

$$Z_{\text{in}} = Z(l) = Z_0 \left[\frac{1+\Gamma_l}{1-\Gamma_l}\right]. \tag{2.76}$$

with

$$\Gamma_l = \Gamma e^{-j2\beta l} = |\Gamma| e^{j(\theta_r - 2\beta l)}. \tag{2.77}$$

By replacing Γ with Eq. (2.59) and using the relations

$$e^{j\beta l} = \cos\beta l + j\sin\beta l, \tag{2.78a}$$

$$e^{-j\beta l} = \cos\beta l - j\sin\beta l, \tag{2.78b}$$

Eq. (2.76) can be written in terms of z_L as

$$\begin{aligned} Z_{\text{in}} &= Z_0 \left(\frac{z_L \cos\beta l + j\sin\beta l}{\cos\beta l + j z_L \sin\beta l}\right) \\ &= Z_0 \left(\frac{z_L + j\tan\beta l}{1 + j z_L \tan\beta l}\right). \end{aligned} \tag{2.79}$$

From the standpoint of the generator circuit, the transmission line can be replaced with an impedance Z_{in}, as shown in Fig. 2-18. The phasor voltage across Z_{in} is given by

$$\widetilde{V}_i = \widetilde{I}_i Z_{\text{in}} = \frac{\widetilde{V}_g Z_{\text{in}}}{Z_g + Z_{\text{in}}}, \tag{2.80}$$

Simultaneously, from the standpoint of the transmission line, the voltage across it at the input of the line is given by Eq. (2.63a) with $z = -l$:

$$\widetilde{V}_i = \widetilde{V}(-l) = V_0^+[e^{j\beta l} + \Gamma e^{-j\beta l}]. \tag{2.81}$$

Equating Eq. (2.80) to Eq. (2.81) and then solving for V_0^+ leads to

$$V_0^+ = \left(\frac{\widetilde{V}_g Z_{\text{in}}}{Z_g + Z_{\text{in}}}\right)\left(\frac{1}{e^{j\beta l} + \Gamma e^{-j\beta l}}\right). \tag{2.82}$$

This completes the solution of the transmission-line wave equations, given by Eqs. (2.21) and (2.23), for the special case of a lossless transmission line. We started out with the general solutions given by Eq. (2.26), which included four unknown amplitudes, V_0^+, V_0^-, I_0^+, and I_0^-. We then determined that $Z_0 = V_0^+/I_0^+ = -V_0^-/I_0^-$, thereby reducing the unknowns to the two voltage amplitudes only. Upon applying the boundary condition at the load, we were able to relate V_0^- to V_0^+ through Γ, and, finally, by applying the boundary condition at the source, we obtained an expression for V_0^+.

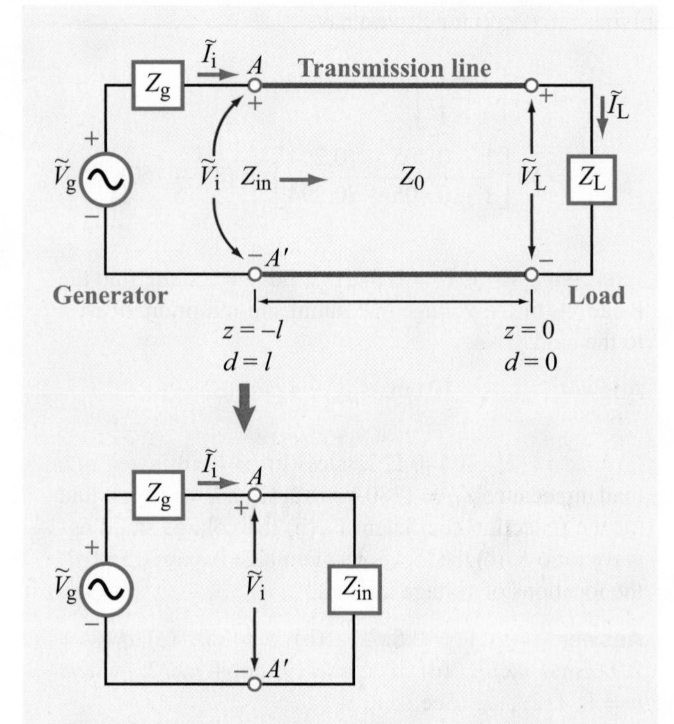

Figure 2-18 At the generator end, the terminated transmission line can be replaced with the input impedance of the line Z_{in}.

Example 2-7: Complete Solution for $v(z,t)$ and $i(z,t)$

A 1.05 GHz generator circuit with series impedance $Z_g = 10\ \Omega$ and voltage source given by

$$v_g(t) = 10\sin(\omega t + 30^\circ) \quad \text{(V)}$$

is connected to a load $Z_L = (100 + j50)\ \Omega$ through a 50 Ω, 67 cm long lossless transmission line. The phase velocity of the line is $0.7c$, where c is the velocity of light in a vacuum. Find $v(z,t)$ and $i(z,t)$ on the line.

Solution: From the relationship $u_p = \lambda f$, we find the wavelength

$$\lambda = \frac{u_p}{f} = \frac{0.7 \times 3 \times 10^8}{1.05 \times 10^9} = 0.2 \text{ m},$$

and

$$\beta l = \frac{2\pi}{\lambda} l = \frac{2\pi}{0.2} \times 0.67 = 6.7\pi = 0.7\pi = 126°,$$

where we have subtracted multiples of 2π. The voltage reflection coefficient at the load is

$$\Gamma = \frac{Z_L - Z_0}{Z_L + Z_0} = \frac{(100 + j50) - 50}{(100 + j50) + 50} = 0.45e^{j26.6°}.$$

With reference to Fig. 2-18, the input impedance of the line, given by Eq. (2.76), is

$$\begin{aligned} Z_{in} &= Z_0 \left(\frac{1 + \Gamma_l}{1 - \Gamma_l} \right) \\ &= Z_0 \left(\frac{1 + \Gamma e^{-j2\beta l}}{1 - \Gamma e^{-j2\beta l}} \right) \\ &= 50 \left(\frac{1 + 0.45e^{j26.6°} e^{-j252°}}{1 - 0.45e^{j26.6°} e^{-j252°}} \right) = (21.9 + j17.4)\ \Omega. \end{aligned}$$

Rewriting the expression for the generator voltage with the cosine reference, we have

$$\begin{aligned} \upsilon_g(t) &= 10 \sin(\omega t + 30°) \\ &= 10 \cos(90° - \omega t - 30°) \\ &= 10 \cos(\omega t - 60°) \\ &= \mathfrak{Re}[10e^{-j60°} e^{j\omega t}] = \mathfrak{Re}[\widetilde{V}_g e^{j\omega t}] \qquad (\text{V}). \end{aligned}$$

Hence, the phasor voltage $\widetilde{V}_g$ is given by

$$\widetilde{V}_g = 10\, e^{-j60°} = 10\angle{-60°} \qquad (\text{V}).$$

Application of Eq. (2.82) gives

$$\begin{aligned} V_0^+ &= \left(\frac{\widetilde{V}_g Z_{in}}{Z_g + Z_{in}} \right) \left(\frac{1}{e^{j\beta l} + \Gamma e^{-j\beta l}} \right) \\ &= \left[\frac{10e^{-j60°}(21.9 + j17.4)}{10 + 21.9 + j17.4} \right] \\ &\quad \cdot (e^{j126°} + 0.45e^{j26.6°} e^{-j126°})^{-1} \\ &= 10.2e^{j159°} \qquad (\text{V}). \end{aligned}$$

Using Eq. (2.63a) with $z = -d$, the phasor voltage on the line is

$$\begin{aligned} \widetilde{V}(d) &= V_0^+(e^{j\beta d} + \Gamma e^{-j\beta d}) \\ &= 10.2e^{j159°}(e^{j\beta d} + 0.45e^{j26.6°} e^{-j\beta d}), \end{aligned}$$

and the corresponding instantaneous voltage $\upsilon(d, t)$ is

$$\begin{aligned} \upsilon(d, t) &= \mathfrak{Re}[\widetilde{V}(d)\, e^{j\omega t}] \\ &= 10.2 \cos(\omega t + \beta d + 159°) \\ &\quad + 4.55 \cos(\omega t - \beta d + 185.6°) \qquad (\text{V}). \end{aligned}$$

Similarly, Eq. (2.63b) leads to

$$\begin{aligned} \widetilde{I}(d) &= 0.20e^{j159°}(e^{j\beta d} - 0.45e^{j26.6°} e^{-j\beta d}), \\ i(d, t) &= 0.20 \cos(\omega t + \beta d + 159°) \\ &\quad + 0.091 \cos(\omega t - \beta d + 185.6°) \qquad (\text{A}). \end{aligned}$$

Module 2.5 Wave and Input Impedance The wave impedance, $Z(d) = \widetilde{V}(d)/\widetilde{I}(d)$, exhibits a cyclical pattern as a function of position along the line. This module displays plots of the real and imaginary parts of $Z(d)$, specifies the locations of the voltage maximum and minimum nearest to the load, and provides other related information.

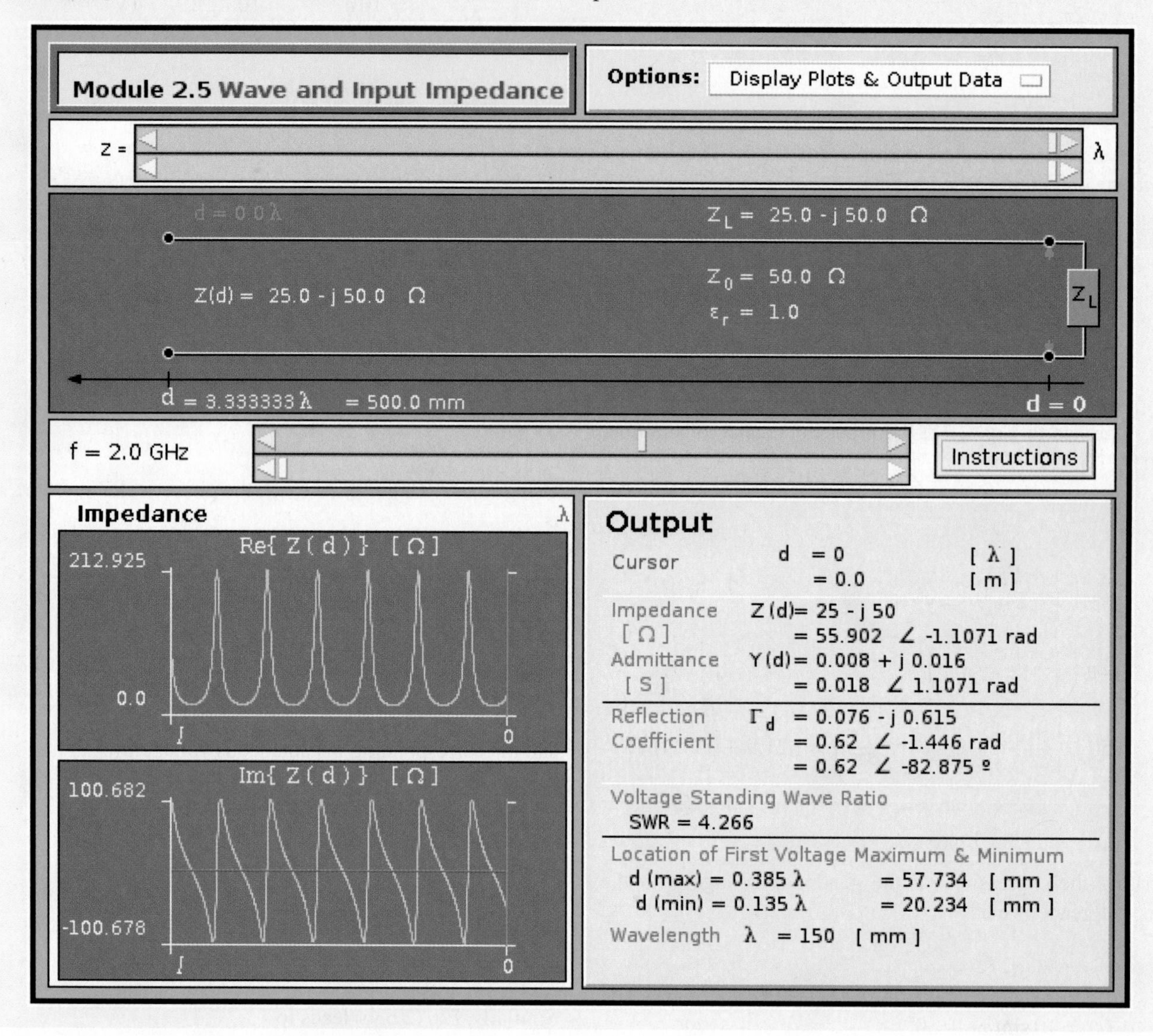

2-8 Special Cases of the Lossless Line

We often encounter situations involving lossless transmission lines with particular terminations or lines whose lengths lead to particularly useful line properties. We now consider some of these special cases.

2-8.1 Short-Circuited Line

The transmission line shown in **Fig. 2-19(a)** is terminated in a short circuit, $Z_L = 0$. Consequently, the voltage reflection coefficient defined by Eq. (2.59) is $\Gamma = -1$, and the voltage standing-wave ratio given by Eq. (2.73) is $S = \infty$. With

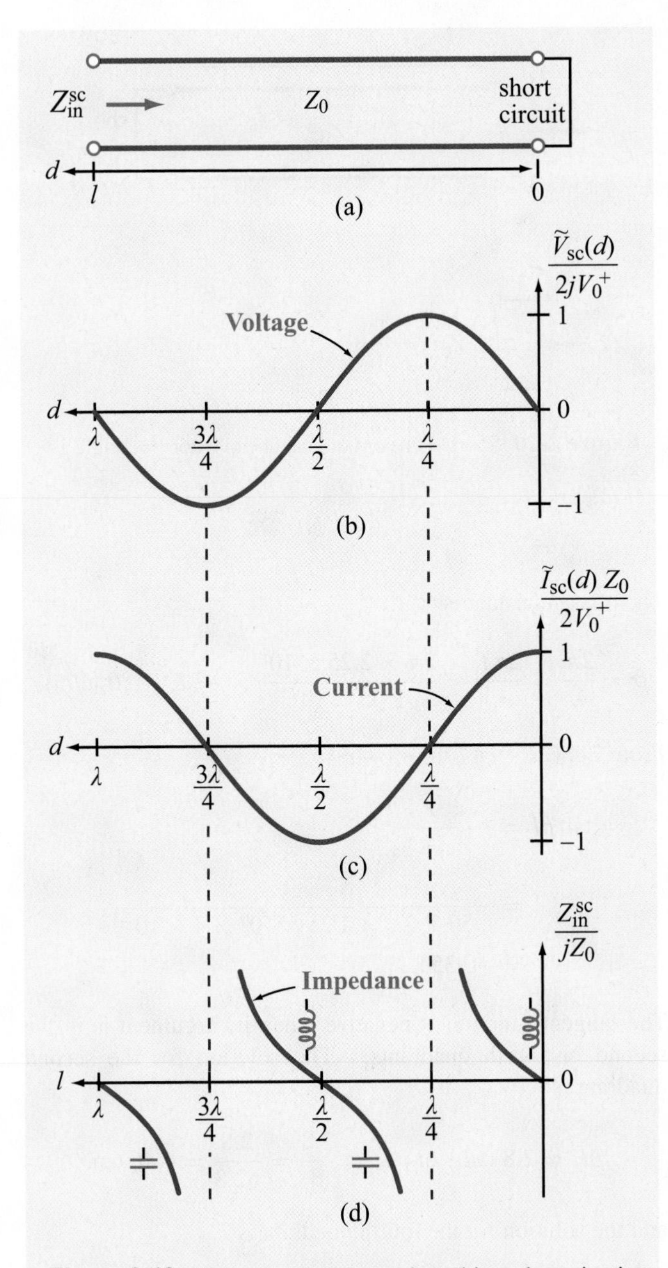

Figure 2-19 Transmission line terminated in a short circuit: (a) schematic representation, (b) normalized voltage on the line, (c) normalized current, and (d) normalized input impedance.

$z = -d$ and $\Gamma = -1$ in Eqs. (2.63a) and (2.63b), and $\Gamma = -1$ in Eq. (2.74), the voltage, current, and wave impedance on a short-circuited lossless transmission line are given by

$$\widetilde{V}_{sc}(d) = V_0^+[e^{j\beta d} - e^{-j\beta d}] = 2jV_0^+ \sin\beta d, \qquad (2.83a)$$

$$\widetilde{I}_{sc}(d) = \frac{V_0^+}{Z_0}[e^{j\beta d} + e^{-j\beta d}] = \frac{2V_0^+}{Z_0}\cos\beta d, \qquad (2.83b)$$

$$Z_{sc}(d) = \frac{\widetilde{V}_{sc}(d)}{\widetilde{I}_{sc}(d)} = jZ_0 \tan\beta d. \qquad (2.83c)$$

The voltage $\widetilde{V}_{sc}(d)$ is zero at the load ($d = 0$), as it should be for a short circuit, and its amplitude varies as $\sin\beta d$. In contrast, the current $\widetilde{I}_{sc}(d)$ is a maximum at the load and it varies as $\cos\beta d$. Both quantities are displayed in **Fig. 2-19** as a function of d.

Denoting Z_{in}^{sc} as the ***input impedance of a short-circuited line*** of length l,

$$Z_{in}^{sc} = \frac{\widetilde{V}_{sc}(l)}{\widetilde{I}_{sc}(l)} = jZ_0 \tan\beta l. \qquad (2.84)$$

A plot of Z_{in}^{sc}/jZ_0 versus l is shown in **Fig. 2-19(d)**. For the short-circuited line, if its length is less than $\lambda/4$, its impedance is equivalent to that of an inductor, and if it is between $\lambda/4$ and $\lambda/2$, it is equivalent to that of a capacitor.

In general, the input impedance Z_{in} of a line terminated in an arbitrary load has a real part, called the ***input resistance*** R_{in}, and an imaginary part, called the ***input reactance*** X_{in}:

$$Z_{in} = R_{in} + jX_{in}. \qquad (2.85)$$

In the case of the short-circuited lossless line, the input impedance is purely reactive ($R_{in} = 0$). If $\tan\beta l \geq 0$, the line appears inductive to the source, acting like an equivalent inductor L_{eq} whose impedance equals Z_{in}^{sc}. Thus,

$$j\omega L_{eq} = jZ_0 \tan\beta l, \quad \text{if } \tan\beta l \geq 0, \qquad (2.86)$$

or

$$L_{eq} = \frac{Z_0 \tan\beta l}{\omega} \quad \text{(H)}. \qquad (2.87)$$

The minimum line length l that would result in an input impedance Z_{in}^{sc} equivalent to that of an inductor with inductance L_{eq} is

$$l = \frac{1}{\beta}\tan^{-1}\left(\frac{\omega L_{eq}}{Z_0}\right) \quad \text{(m)}. \qquad (2.88)$$

Similarly, if $\tan\beta l \leq 0$, the input impedance is capacitive, in which case the line acts like an equivalent capacitor with capacitance C_{eq} such that

$$\frac{1}{j\omega C_{eq}} = jZ_0 \tan\beta l, \qquad \text{if } \tan\beta l \leq 0, \tag{2.89}$$

or

$$C_{eq} = -\frac{1}{Z_0\omega \tan\beta l} \qquad \text{(F)}. \tag{2.90}$$

Since l is a positive number, the shortest length l for which $\tan\beta l \leq 0$ corresponds to the range $\pi/2 \leq \beta l \leq \pi$. Hence, the minimum line length l that would result in an input impedance Z_{in}^{sc} equivalent to that of a capacitor of capacitance C_{eq} is

$$l = \frac{1}{\beta}\left[\pi - \tan^{-1}\left(\frac{1}{\omega C_{eq} Z_0}\right)\right] \qquad \text{(m)}. \tag{2.91}$$

► These results imply that, through proper choice of the length of a short-circuited line, we can make them into equivalent capacitors and inductors of any desired reactance. ◄

Such a practice is indeed common in the design of microwave circuits and high-speed integrated circuits, because making an actual capacitor or inductor often is much more difficult than fabricating a shorted microstrip transmission line on a circuit board.

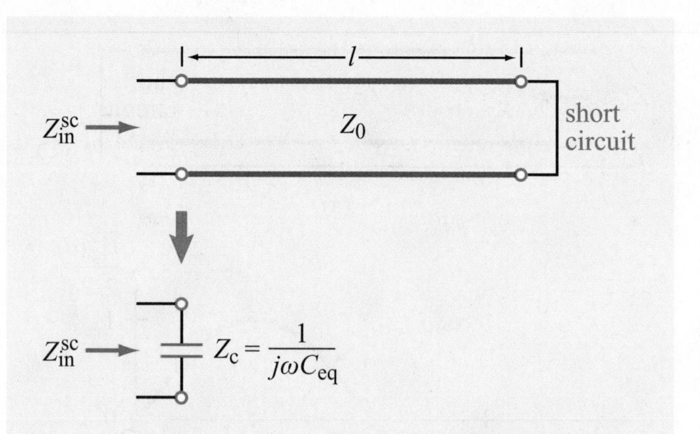

Figure 2-20 Shorted line as equivalent capacitor (Example 2-8).

Example 2-8: Equivalent Reactive Elements

Choose the length of a shorted 50 Ω lossless transmission line (**Fig. 2-20**) such that its input impedance at 2.25 GHz is identical to that of a capacitor with capacitance $C_{eq} = 4$ pF. The wave velocity on the line is $0.75c$.

Solution: We are given

$$u_p = 0.75c = 0.75 \times 3 \times 10^8 = 2.25 \times 10^8 \text{ m/s},$$

$$Z_0 = 50\ \Omega,$$

$$f = 2.25 \text{ GHz} = 2.25 \times 10^9 \text{ Hz},$$

$$C_{eq} = 4 \text{ pF} = 4 \times 10^{-12} \text{ F}.$$

The phase constant is

$$\beta = \frac{2\pi}{\lambda} = \frac{2\pi f}{u_p} = \frac{2\pi \times 2.25 \times 10^9}{2.25 \times 10^8} = 62.8 \qquad \text{(rad/m)}.$$

From Eq. (2.89), it follows that

$$\tan\beta l = -\frac{1}{Z_0\omega C_{eq}}$$
$$= -\frac{1}{50 \times 2\pi \times 2.25 \times 10^9 \times 4 \times 10^{-12}}$$
$$= -0.354.$$

The tangent function is negative when its argument is in the second or fourth quadrants. The solution for the second quadrant is

$$\beta l_1 = 2.8 \text{ rad} \quad \text{or} \quad l_1 = \frac{2.8}{\beta} = \frac{2.8}{62.8} = 4.46 \text{ cm},$$

and the solution for the fourth quadrant is

$$\beta l_2 = 5.94 \text{ rad} \quad \text{or} \quad l_2 = \frac{5.94}{62.8} = 9.46 \text{ cm}.$$

We also could have obtained the value of l_1 by applying Eq. (2.91). The length l_2 is greater than l_1 by exactly $\lambda/2$. In fact, any length $l = 4.46 \text{ cm} + n\lambda/2$, where n is a positive integer, also is a solution.

2-8.2 Open-Circuited Line

With $Z_L = \infty$, as illustrated in **Fig. 2-21(a)**, we have $\Gamma = 1$, $S = \infty$, and the voltage, current, and input impedance are given by

$$\widetilde{V}_{oc}(d) = V_0^+[e^{j\beta d} + e^{-j\beta d}] = 2V_0^+ \cos \beta d, \tag{2.92a}$$

$$\widetilde{I}_{oc}(d) = \frac{V_0^+}{Z_0}[e^{j\beta d} - e^{-j\beta d}] = \frac{2jV_0^+}{Z_0} \sin \beta d, \tag{2.92b}$$

$$Z_{in}^{oc} = \frac{\widetilde{V}_{oc}(l)}{\widetilde{I}_{oc}(l)} = -jZ_0 \cot \beta l. \tag{2.93}$$

Plots of these quantities are displayed in **Fig. 2-21** as a function of d.

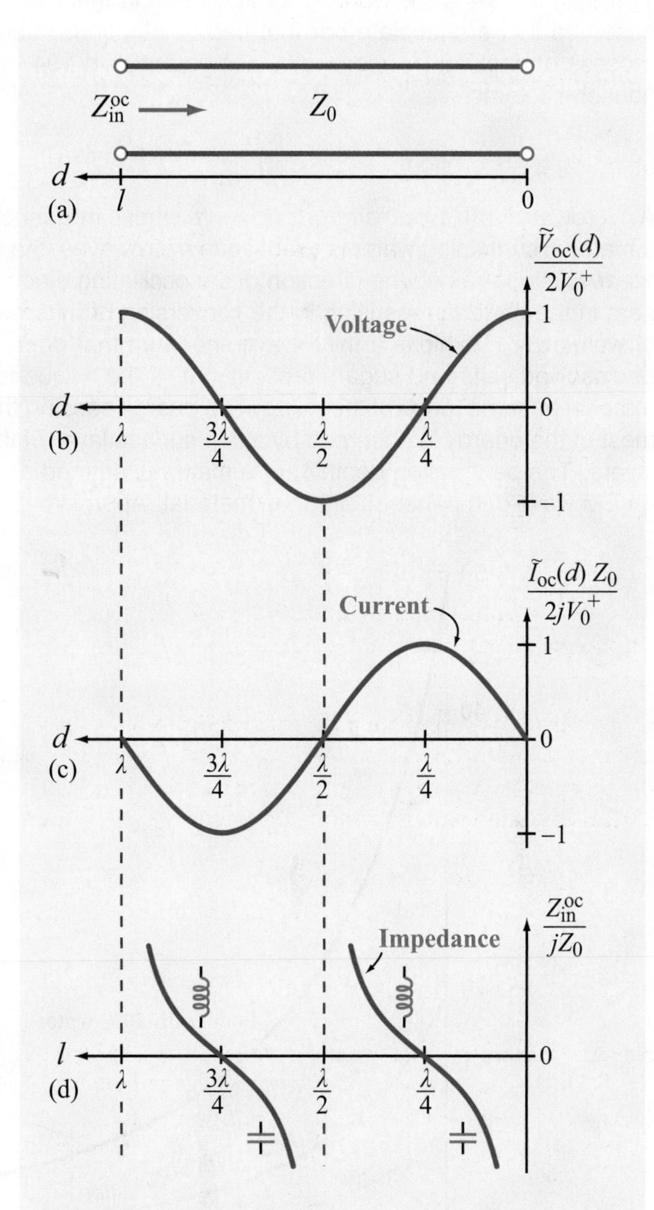

Figure 2-21 Transmission line terminated in an open circuit: (a) schematic representation, (b) normalized voltage on the line, (c) normalized current, and (d) normalized input impedance.

2-8.3 Application of Short-Circuit/ Open-Circuit Technique

A network analyzer is a radio-frequency (RF) instrument capable of measuring the impedance of any load connected to its input terminal. When used to measure (1) Z_{in}^{sc}, the input impedance of a lossless line when terminated in a short circuit, and (2) Z_{in}^{oc}, the input impedance of the line when terminated in an open circuit, the combination of the two measurements can be used to determine the characteristic impedance of the line Z_0 and its phase constant β. Indeed, the product of the expressions given by Eqs. (2.84) and (2.93) gives

$$Z_0 = \sqrt[+]{Z_{in}^{sc}\, Z_{in}^{oc}}\,, \tag{2.94}$$

and the ratio of the same expressions leads to

$$\tan \beta l = \sqrt{\frac{-Z_{in}^{sc}}{Z_{in}^{oc}}}\,. \tag{2.95}$$

Because of the π phase ambiguity associated with the tangent function, the length l should be less than or equal to $\lambda/2$ to provide an unambiguous result.

Technology Brief 3: Microwave Ovens

Percy Spencer, while working for Raytheon in the 1940s on the design and construction of ***magnetrons*** for radar, observed that a chocolate bar that had unintentionally been exposed to microwaves had melted in his pocket. The process of cooking by microwave was patented in 1946 and by the 1970s, microwave ovens had become standard household items.

Microwave Absorption

A microwave is an ***electromagnetic wave*** whose frequency lies in the 300 MHz–300 GHz range (see **Fig. 1-16**.) When a material containing water is exposed to microwaves, the water molecule reacts by rotating itself so as to align its own ***electric dipole*** along the direction of the oscillating electric field of the microwave. The rapid vibration motion creates heat in the material, resulting in the conversion of microwave energy into thermal energy. The absorption coefficient of water, $\alpha(f)$, exhibits a microwave spectrum that depends on the temperature of the water and the concentration of dissolved salts and sugars present in it. If the frequency f is chosen such that $\alpha(f)$ is high, the water-containing material absorbs much of the microwave energy passing through it and converts it to heat. However, it also means that most of the energy is absorbed by a thin surface layer of the material, with not much energy remaining to heat deeper layers. The penetration depth δ_p of a material, defined as $\delta_p = 1/2\alpha$, is a measure of how deep the power carried by an EM wave can penetrate into the material. Approximately 95% of the microwave energy incident upon a material is

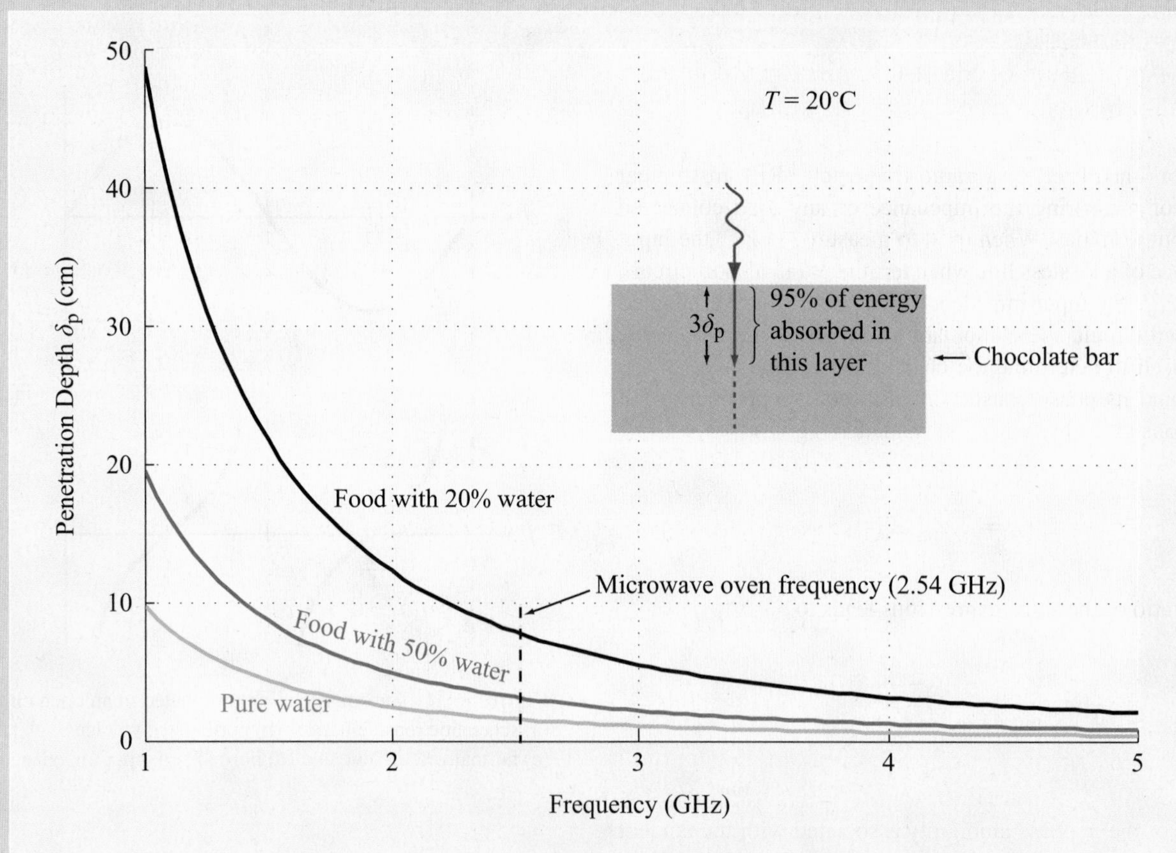

Figure TF3-1 Penetration depth as a function of frequency (1–5 GHz) for pure water and two foods with different water contents.

absorbed by the surface layer of thickness $3\delta_p$. **Figure TF3-1** displays calculated spectra of δ_p for pure water and two materials with different water contents.

> ▶ The frequency most commonly used in microwave ovens is 2.54 GHz. The magnitude of δ_s at 2.54 GHz varies beween $\sim$2 cm for pure water and 8 cm for a material with a water content of only 20%. ◀

This is a practical range for cooking food in a microwave oven; at much lower frequencies, the food is not a good absorber of energy (in addition to the fact that the design of the magnetron and the oven cavity become problematic), and at much higher frequencies, the microwave energy cooks the food very unevenly (mostly the surface layer). Whereas microwaves are readily absorbed by water, fats, and sugars, they can penetrate through most ceramics, glass, or plastics without loss of energy, thereby imparting little or no heat to those materials.

Oven Operation

To generate high-power microwaves ($\sim$700 watts) the microwave oven uses a ***magnetron tube*** (**Fig. TF3-2**), which requires the application of a voltage on the order of 4000 volts. The typical household voltage of 115 volts is increased to the required voltage level through a ***high-voltage transformer***. The microwave energy generated by the magnetron is transferred into a cooking chamber designed to contain the microwaves within it through the use of metal surfaces and safety Interlock switches.

> ▶ Microwaves are reflected by metal surfaces, so they can bounce around the interior of the chamber or be absorbed by the food, but not escape to the outside. ◀

If the oven door is made of a glass panel, a ***metal screen*** or a layer of conductive mesh is attached to it to ensure the necessary shielding; microwaves cannot pass through the metal screen if the mesh width is much smaller than the wavelength of the microwave ($\lambda \approx 12$ cm at 2.5 GHz). In the chamber, the microwave energy establishes a ***standing-wave pattern***, which leads to an uneven distribution. This is mitigated by using a rotating ***metal stirrer*** that disperses the microwave energy to different parts of the chamber.

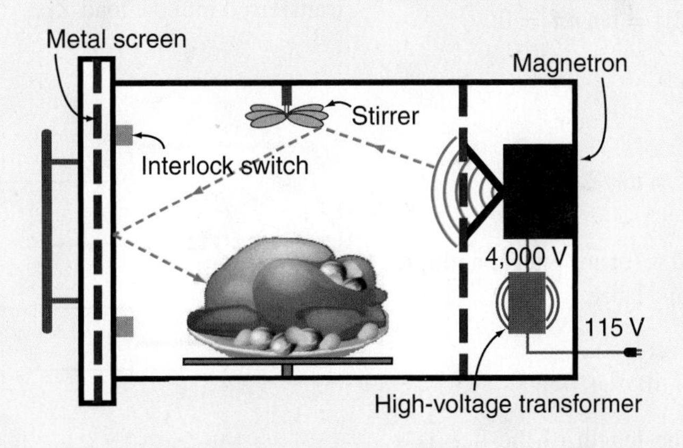

Figure TF3-2 Microwave oven cavity.

Example 2-9: Measuring Z_0 and β

Find Z_0 and β of a 57 cm long lossless transmission line whose input impedance was measured as $Z_{\rm in}^{\rm sc} = j40.42\ \Omega$ when terminated in a short circuit and as $Z_{\rm in}^{\rm oc} = -j121.24\ \Omega$ when terminated in an open circuit. From other measurements, we know that the line is between 3 and 3.25 wavelengths long.

Solution: From Eqs. (2.94) and (2.95),

$$Z_0 = \sqrt[+]{Z_{\rm in}^{\rm sc}\, Z_{\rm in}^{\rm oc}} = \sqrt{(j40.42)(-j121.24)} = 70\ \Omega,$$

$$\tan\beta l = \sqrt{\frac{-Z_{\rm in}^{\rm sc}}{Z_{\rm in}^{\rm oc}}} = \sqrt{\frac{1}{3}}\,.$$

Since l is between 3λ and 3.25λ, $\beta l = (2\pi l/\lambda)$ is between 6π radians and $(13\pi/2)$ radians. This places βl in the first quadrant (0 to $\pi/2$) radians. Hence, the only acceptable solution for $\tan\beta\ell = \sqrt{1/3}$ is $\beta l = \pi/6$ radians. This value, however, does not include the 2π multiples associated with the integer λ multiples of l. Hence, the true value of βl is

$$\beta l = 6\pi + \frac{\pi}{6} = 19.4 \qquad \text{(rad)},$$

in which case

$$\beta = \frac{19.4}{0.57} = 34 \qquad \text{(rad/m)}.$$

2-8.4 Lines of Length $l = n\lambda/2$

If $l = n\lambda/2$, where n is an integer,

$$\tan\beta l = \tan\left[(2\pi/\lambda)\,(n\lambda/2)\right] = \tan n\pi = 0.$$

Consequently, Eq. (2.79) reduces to

$$Z_{\rm in} = Z_{\rm L}, \qquad \text{for } l = n\lambda/2, \qquad (2.96)$$

which means that a half-wavelength line (or any integer multiple of $\lambda/2$) does not modify the load impedance.

2-8.5 Quarter-Wavelength Transformer

Another case of interest is when the length of the line is a quarter-wavelength (or $\lambda/4 + n\lambda/2$, where $n = 0$ or a positive integer), corresponding to $\beta l = (2\pi/\lambda)(\lambda/4) = \pi/2$. From Eq. (2.79), the input impedance becomes

$$Z_{\rm in} = \frac{Z_0^2}{Z_{\rm L}}, \qquad \text{for } l = \lambda/4 + n\lambda/2. \qquad (2.97)$$

The utility of such a quarter-wave transformer is illustrated by Example 2-10.

Example 2-10: $\lambda/4$ Transformer

A 50 Ω lossless transmission line is to be matched to a resistive load impedance with $Z_{\rm L} = 100\ \Omega$ via a quarter-wave section as shown in Fig. 2-22, thereby eliminating reflections along the feedline. Find the required characteristic impedance of the quarter-wave transformer.

Solution: To eliminate reflections at terminal AA', the input impedance $Z_{\rm in}$ looking into the quarter-wave line should be equal to Z_{01}, the characteristic impedance of the feedline. Thus, $Z_{\rm in} = 50\ \Omega$. From Eq. (2.97),

$$Z_{\rm in} = \frac{Z_{02}^2}{Z_{\rm L}}\,,$$

or

$$Z_{02} = \sqrt{Z_{\rm in}\, Z_{\rm L}} = \sqrt{50 \times 100} = 70.7\ \Omega.$$

Whereas this eliminates reflections on the feedline, it does not eliminate them on the $\lambda/4$ line. However, since the lines are lossless, all the power incident on AA' will end up getting transferred into the load $Z_{\rm L}$.

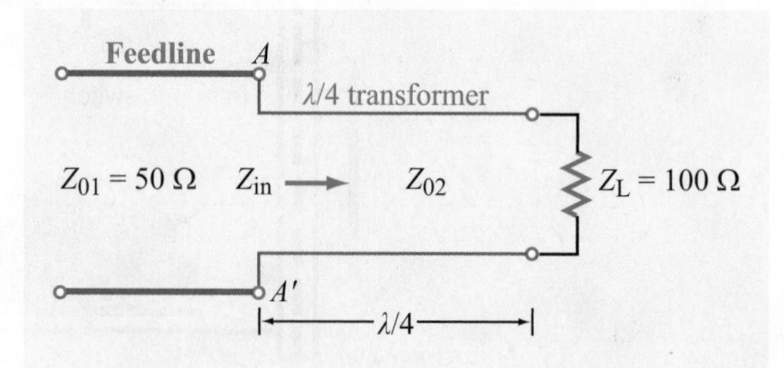

Figure 2-22 Configuration for Example 2-10.

In this example, Z_L is purely resistive. To apply the $\lambda/4$ transformer technique to match a transmission line to a load with a complex impedance, a slightly more elaborate procedure is required (Section 2-11).

2-8.6 Matched Transmission Line: $Z_L = Z_0$

For a matched lossless transmission line with $Z_L = Z_0$, (1) the input impedance $Z_{in} = Z_0$ for all locations d on the line, (2) $\Gamma = 0$, and (3) all the incident power is delivered to the load, regardless of the line length l. A summary of the properties of standing waves is given in **Table 2-4**.

Table 2-4 Properties of standing waves on a lossless transmission line.

Voltage maximum	$\lvert\widetilde{V}\rvert_{max} = \lvert V_0^+\rvert[1 + \lvert\Gamma\rvert]$
Voltage minimum	$\lvert\widetilde{V}\rvert_{min} = \lvert V_0^+\rvert[1 - \lvert\Gamma\rvert]$
Positions of voltage maxima (also positions of current minima)	$d_{max} = \frac{\theta_r\lambda}{4\pi} + \frac{n\lambda}{2}, \quad n = 0, 1, 2, \ldots$
Position of first maximum (also position of first current minimum)	$d_{max} = \begin{cases} \frac{\theta_r\lambda}{4\pi}, & \text{if } 0 \leq \theta_r \leq \pi \\ \frac{\theta_r\lambda}{4\pi} + \frac{\lambda}{2}, & \text{if } -\pi \leq \theta_r \leq 0 \end{cases}$
Positions of voltage minima (also positions of current maxima)	$d_{min} = \frac{\theta_r\lambda}{4\pi} + \frac{(2n+1)\lambda}{4}, \quad n = 0, 1, 2, \ldots$
Position of first minimum (also position of first current maximum)	$d_{min} = \frac{\lambda}{4}\left(1 + \frac{\theta_r}{\pi}\right)$
Input impedance	$Z_{in} = Z_0\left(\frac{z_L + j\tan\beta l}{1 + jz_L\tan\beta l}\right) = Z_0\left(\frac{1+\Gamma_l}{1-\Gamma_l}\right)$
Positions at which Z_{in} is real	at voltage maxima and minima
Z_{in} at voltage maxima	$Z_{in} = Z_0\left(\frac{1+\lvert\Gamma\rvert}{1-\lvert\Gamma\rvert}\right)$
Z_{in} at voltage minima	$Z_{in} = Z_0\left(\frac{1-\lvert\Gamma\rvert}{1+\lvert\Gamma\rvert}\right)$
Z_{in} of short-circuited line	$Z_{in}^{sc} = jZ_0\tan\beta l$
Z_{in} of open-circuited line	$Z_{in}^{oc} = -jZ_0\cot\beta l$
Z_{in} of line of length $l = n\lambda/2$	$Z_{in} = Z_L, \quad n = 0, 1, 2, \ldots$
Z_{in} of line of length $l = \lambda/4 + n\lambda/2$	$Z_{in} = Z_0^2/Z_L, \quad n = 0, 1, 2, \ldots$
Z_{in} of matched line	$Z_{in} = Z_0$
$\lvert V_0^+\rvert$ = amplitude of incident wave; $\Gamma = \lvert\Gamma\rvert e^{j\theta_r}$ with $-\pi < \theta_r < \pi$; θ_r in radians; $\Gamma_l = \Gamma e^{-j2\beta l}$.	

Concept Question 2-10: What is the difference between the characteristic impedance Z_0 and the input impedance Z_{in}? When are they the same?

Concept Question 2-11: What is a quarter-wave transformer? How can it be used?

Concept Question 2-12: A lossless transmission line of length l is terminated in a short circuit. If $l < \lambda/4$, is the input impedance inductive or capacitive?

Concept Question 2-13: What is the input impedance of an infinitely long line?

Concept Question 2-14: If the input impedance of a lossless line is inductive when terminated in a short circuit, will it be inductive or capacitive when the line is terminated in an open circuit?

Exercise 2-12: A 50 Ω lossless transmission line uses an insulating material with $\epsilon_r = 2.25$. When terminated in an open circuit, how long should the line be for its input impedance to be equivalent to a 10-pF capacitor at 50 MHz?

Answer: $l = 9.92$ cm. (See EM.)

Exercise 2-13: A 300 Ω feedline is to be connected to a 3 m long, 150 Ω line terminated in a 150 Ω resistor. Both lines are lossless and use air as the insulating material, and the operating frequency is 50 MHz. Determine (a) the input impedance of the 3 m long line, (b) the voltage standing-wave ratio on the feedline, and (c) the characteristic impedance of a quarter-wave transformer were it to be used between the two lines in order to achieve $S = 1$ on the feedline. (See EM.)

Answer: (a) $Z_{in} = 150$ Ω, (b) $S = 2$, (c) $Z_0 = 212.1$ Ω.

2-9 Power Flow on a Lossless Transmission Line

Our discussion thus far has focused on the voltage and current attributes of waves propagating on a transmission line. Now we examine the flow of power carried by the incident and reflected waves. We begin by reintroducing Eqs. (2.63a) and (2.63b) with $z = -d$:

$$\widetilde{V}(d) = V_0^+(e^{j\beta d} + \Gamma e^{-j\beta d}), \tag{2.98a}$$

$$\widetilde{I}(d) = \frac{V_0^+}{Z_0}(e^{j\beta d} - \Gamma e^{-j\beta d}). \tag{2.98b}$$

In these expressions, the first terms represent the incident-wave voltage and current, and the terms involving Γ represent the reflected-wave voltage and current. The time-domain expressions for the voltage and current at location d from the load are obtained by transforming Eq. (2.98) to the time domain:

$$\begin{aligned} \upsilon(d,t) &= \mathfrak{Re}[\widetilde{V}e^{j\omega t}] \\ &= \mathfrak{Re}[|V_0^+|e^{j\phi^+}(e^{j\beta d} + |\Gamma|e^{j\theta_r}e^{-j\beta d})e^{j\omega t}] \\ &= |V_0^+|[\cos(\omega t + \beta d + \phi^+) \\ &\qquad + |\Gamma|\cos(\omega t - \beta d + \phi^+ + \theta_r)], \end{aligned} \tag{2.99a}$$

$$\begin{aligned} i(d,t) &= \frac{|V_0^+|}{Z_0}[\cos(\omega t + \beta d + \phi^+) \\ &\qquad - |\Gamma|\cos(\omega t - \beta d + \phi^+ + \theta_r)], \end{aligned} \tag{2.99b}$$

where we used the relations $V_0^+ = |V_0^+|e^{j\phi^+}$ and $\Gamma = |\Gamma|e^{j\theta_r}$, both introduced earlier as Eqs. (2.31a) and (2.62), respectively.

2-9.1 Instantaneous Power

The instantaneous power carried by the transmission line is equal to the product of $\upsilon(d,t)$ and $i(d,t)$:

$$\begin{aligned} P(d,t) &= \upsilon(d,t)\, i(d,t) \\ &= |V_0^+|[\cos(\omega t + \beta d + \phi^+) \\ &\qquad + |\Gamma|\cos(\omega t - \beta d + \phi^+ + \theta_r)] \\ &\quad \times \frac{|V_0^+|}{Z_0}[\cos(\omega t + \beta d + \phi^+) \\ &\qquad - |\Gamma|\cos(\omega t - \beta d + \phi^+ + \theta_r)] \\ &= \frac{|V_0^+|^2}{Z_0}[\cos^2(\omega t + \beta d + \phi^+) \\ &\qquad - |\Gamma|^2\cos^2(\omega t - \beta d + \phi^+ + \theta_r)] \quad \text{(W)}. \end{aligned} \tag{2.100}$$

Per our earlier discussion in connection with Eq. (1.31), if the signs preceding ωt and βd in the argument of the cosine term are both positive or both negative, then the cosine term represents a wave traveling in the negative d direction. Since d points from the load to the generator, the first term in Eq. (2.100) represents the ***instantaneous incident power*** traveling towards the load. This is the power that would be delivered to the load in the absence of wave reflection (when $\Gamma = 0$). Because βd is preceded by a minus sign in the argument of the cosine of the second term in Eq. (2.100), that term represents the ***instantaneous reflected power*** traveling in the $+d$ direction, away from the load. Accordingly, we label these two power components

$$P^{\mathrm{i}}(d,t) = \frac{|V_0^+|^2}{Z_0}\cos^2(\omega t + \beta d + \phi^+) \quad \text{(W)}, \qquad (2.101a)$$

$$P^{\mathrm{r}}(d,t) = -|\Gamma|^2\frac{|V_0^+|^2}{Z_0}\cos^2(\omega t - \beta d + \phi^+ + \theta_{\mathrm{r}}) \quad \text{(W)}. \qquad (2.101b)$$

Using the trigonometric identity

$$\cos^2 x = \tfrac{1}{2}(1 + \cos 2x),$$

the expressions in Eq. (2.101) can be rewritten as

$$P^{\mathrm{i}}(d,t) = \frac{|V_0^+|^2}{2Z_0}[1 + \cos(2\omega t + 2\beta d + 2\phi^+)], \qquad (2.102a)$$

$$P^{\mathrm{r}}(d,t) = -|\Gamma|^2\frac{|V_0^+|^2}{2Z_0}[1 + \cos(2\omega t - 2\beta d + 2\phi^+ + 2\theta_{\mathrm{r}})]. \qquad (2.102b)$$

We note that in each case, the instantaneous power consists of a dc (non–time-varying) term and an ac term that oscillates at an angular frequency of 2ω.

▶ The power oscillates at twice the rate of the voltage or current. ◀

2-9.2 Time-Average Power

From a practical standpoint, we usually are more interested in the time-average power flowing along the transmission line, $P_{\mathrm{av}}(d)$, than in the instantaneous power $P(d,t)$. To compute $P_{\mathrm{av}}(d)$, we can use a time-domain approach or a computationally simpler phasor-domain approach. For completeness, we consider both.

Time-domain approach

The time-average power is equal to the instantaneous power averaged over one time period $T = 1/f = 2\pi/\omega$. For the incident wave, its time-average power is

$$P^{\mathrm{i}}_{\mathrm{av}}(d) = \frac{1}{T}\int_0^T P^{\mathrm{i}}(d,t)\,dt = \frac{\omega}{2\pi}\int_0^{2\pi/\omega} P^{\mathrm{i}}(d,t)\,dt. \qquad (2.103)$$

Upon inserting Eq. (2.102a) into Eq. (2.103) and performing the integration, we obtain

$$P^{\mathrm{i}}_{\mathrm{av}} = \frac{|V_0^+|^2}{2Z_0} \quad \text{(W)}, \qquad (2.104)$$

which is identical with the dc term of $P^{\mathrm{i}}(d,t)$ given by Eq. (2.102a). A similar treatment for the reflected wave gives

$$P^{\mathrm{r}}_{\mathrm{av}} = -|\Gamma|^2\frac{|V_0^+|^2}{2Z_0} = -|\Gamma|^2 P^{\mathrm{i}}_{\mathrm{av}}. \qquad (2.105)$$

▶ The average reflected power is equal to the average incident power, diminished by a multiplicative factor of $|\Gamma|^2$. ◀

Note that the expressions for $P^{\mathrm{i}}_{\mathrm{av}}$ and $P^{\mathrm{r}}_{\mathrm{av}}$ are independent of d, which means that the time-average powers carried by the incident and reflected waves do not change as they travel along the transmission line. This is as expected, because the transmission line is lossless.

The ***net average power flowing towards (and then absorbed by) the load*** shown in **Fig. 2-23** is

$$P_{\mathrm{av}} = P^{\mathrm{i}}_{\mathrm{av}} + P^{\mathrm{r}}_{\mathrm{av}} = \frac{|V_0^+|^2}{2Z_0}[1 - |\Gamma|^2] \quad \text{(W)}. \qquad (2.106)$$

Phasor-domain approach

For any propagating wave with voltage and current phasors $\widetilde{V}$ and $\widetilde{I}$, a useful formula for computing the time-average power is

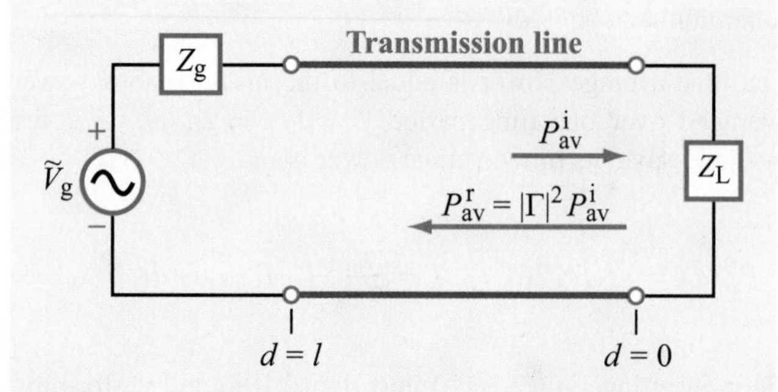

Figure 2-23 The time-average power reflected by a load connected to a lossless transmission line is equal to the incident power multiplied by $|\Gamma|^2$.

$$P_{av} = \tfrac{1}{2}\mathfrak{Re}\left[\tilde{V} \cdot \tilde{I}^*\right], \tag{2.107}$$

where $\tilde{I}^*$ is the complex conjugate of $\tilde{I}$. Application of this formula to Eqs. (2.98a) and (2.98b) gives

$$\begin{aligned}
P_{av} &= \frac{1}{2}\,\mathfrak{Re}\Bigg[V_0^+(e^{j\beta d} + \Gamma e^{-j\beta d}) \\
&\qquad \cdot \frac{V_0^{+*}}{Z_0}(e^{-j\beta d} - \Gamma^* e^{j\beta d})\Bigg] \\
&= \frac{1}{2}\,\mathfrak{Re}\left[\frac{|V_0^+|^2}{Z_0}(1 - |\Gamma|^2 + \Gamma e^{-j2\beta d} - \Gamma^* e^{j2\beta d})\right] \\
&= \frac{|V_0^+|^2}{2Z_0}\{[1 - |\Gamma|^2] \\
&\qquad + \mathfrak{Re}\,[|\Gamma| e^{-j(2\beta d - \theta_r)} - |\Gamma| e^{j(2\beta d - \theta_r)}]\} \\
&= \frac{|V_0^+|^2}{2Z_0}\{[1 - |\Gamma|^2] \\
&\qquad + |\Gamma|[\cos(2\beta d - \theta_r) - \cos(2\beta d - \theta_r)]\} \\
&= \frac{|V_0^+|^2}{2Z_0}[1 - |\Gamma|^2],
\end{aligned} \tag{2.108}$$

which is identical to Eq. (2.106).

Exercise 2-14: For a 50 Ω lossless transmission line terminated in a load impedance $Z_L = (100 + j50)$ Ω, determine the fraction of the average incident power reflected by the load.

Answer: 20%. (See EM.)

Exercise 2-15: For the line of Exercise 2-14, what is the magnitude of the average reflected power if $|V_0^+| = 1$ V?

Answer: $P^r_{av} = 2$ (mW). (See EM.)

Concept Question 2-15: According to Eq. (2.102b), the instantaneous value of the reflected power depends on the phase of the reflection coefficient θ_r, but the average reflected power given by Eq. (2.105) does not. Explain.

Concept Question 2-16: What is the average power delivered by a lossless transmission line to a reactive load?

Concept Question 2-17: What fraction of the incident power is delivered to a matched load?

Concept Question 2-18: Verify that

$$\frac{1}{T}\int_0^T \cos^2\left(\frac{2\pi t}{T} + \beta d + \phi\right) dt = \frac{1}{2},$$

regardless of the values of d and ϕ, so long as neither is a function of t.

2-10 The Smith Chart

The ***Smith chart***, developed by P. H. Smith in 1939, is a widely used graphical tool for analyzing and designing transmission-line circuits. Even though it was originally intended to facilitate calculations involving complex impedances, the Smith chart has become an important avenue for comparing and characterizing the performance of microwave circuits. As the material in this and the next section demonstrates, use of the Smith chart not only avoids tedious manipulations of complex numbers, but it also allows an engineer to design impedance-matching circuits with relative ease.

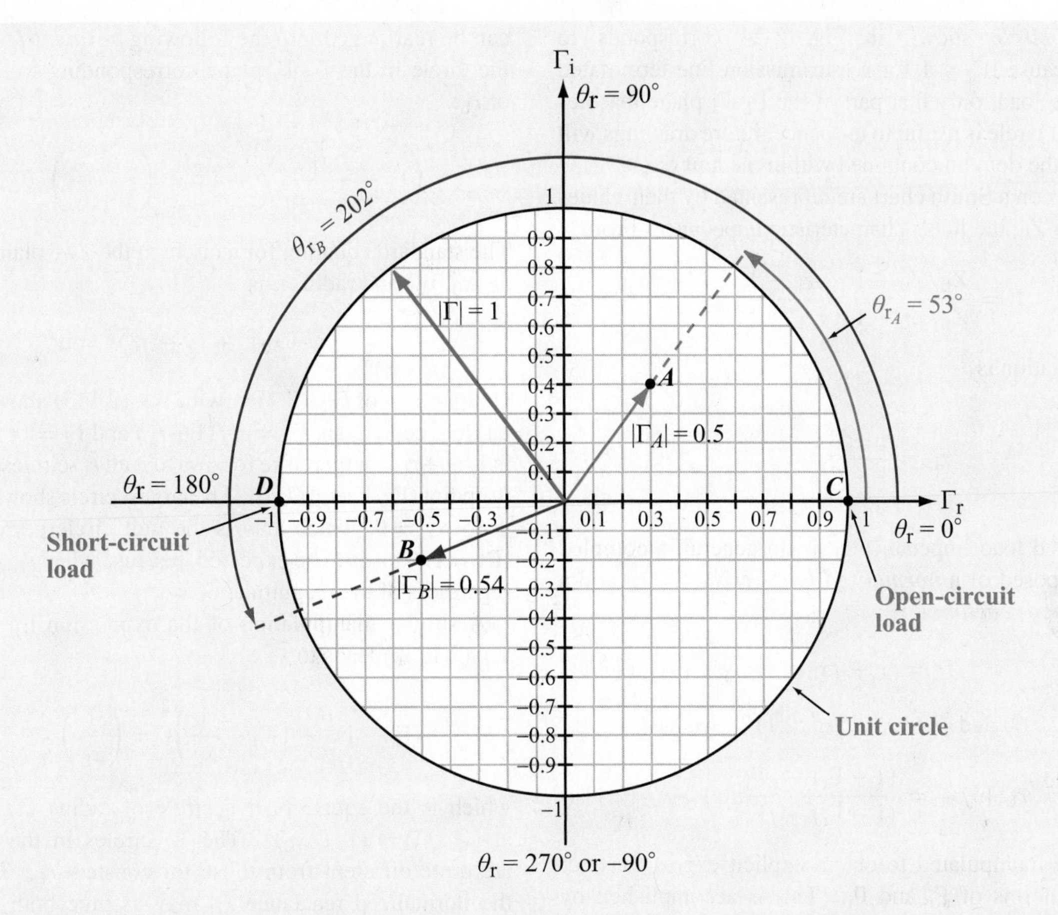

Figure 2-24 The complex Γ plane. Point A is at $\Gamma_A = 0.3 + j0.4 = 0.5e^{j53°}$, and point B is at $\Gamma_B = -0.5 - j0.2 = 0.54e^{j202°}$. The unit circle corresponds to $|\Gamma| = 1$. At point C, $\Gamma = 1$, corresponding to an open-circuit load, and at point D, $\Gamma = -1$, corresponding to a short circuit.

2-10.1 Parametric Equations

The reflection coefficient Γ is, in general, a complex quantity composed of a magnitude $|\Gamma|$ and a phase angle θ_r or, equivalently, a real part Γ_r and an imaginary part Γ_i,

$$\Gamma = |\Gamma|e^{j\theta_r} = \Gamma_r + j\Gamma_i, \quad (2.109)$$

where

$$\Gamma_r = |\Gamma|\cos\theta_r, \quad (2.110a)$$

$$\Gamma_i = |\Gamma|\sin\theta_r. \quad (2.110b)$$

The Smith chart lies in the complex Γ plane. In **Fig. 2-24**, point A represents a reflection coefficient $\Gamma_A = 0.3 + j0.4$ or, equivalently,

$$|\Gamma_A| = [(0.3)^2 + (0.4)^2]^{1/2} = 0.5$$

and

$$\theta_{r_A} = \tan^{-1}(0.4/0.3) = 53°.$$

Similarly, point B represents $\Gamma_B = -0.5 - j0.2$, or $|\Gamma_B| = 0.54$ and $\theta_{r_B} = 202°$ [or, equivalently, $\theta_{r_B} = (360° - 202°) = -158°$].

▶ When both Γ_r and Γ_i are negative, θ_r is in the third quadrant in the Γ_r–Γ_i plane. Thus, when using $\theta_r = \tan^{-1}(\Gamma_i/\Gamma_r)$ to compute θ_r, it may be necessary to add or subtract 180° to obtain the correct value of θ_r. ◀

The ***unit circle*** shown in **Fig. 2-24** corresponds to $|\Gamma| = 1$. Because $|\Gamma| \leq 1$ for a transmission line terminated with a passive load, only that part of the Γ_r–Γ_i plane that lies within the unit circle is useful to us; hence, future drawings will be limited to the domain contained within the unit circle.

Impedances on a Smith chart are represented by their values normalized to Z_0, the line's characteristic impedance. From

$$\Gamma = \frac{Z_L/Z_0 - 1}{Z_L/Z_0 + 1} = \frac{z_L - 1}{z_L + 1}, \tag{2.111}$$

the inverse relation is

$$z_L = \frac{1 + \Gamma}{1 - \Gamma}. \tag{2.112}$$

The normalized load impedance z_L is, in general, a complex quantity composed of a ***normalized load resistance*** r_L and a ***normalized load reactance*** x_L:

$$z_L = r_L + jx_L. \tag{2.113}$$

Using Eqs. (2.109) and (2.113) in Eq. (2.112), we have

$$r_L + jx_L = \frac{(1 + \Gamma_r) + j\Gamma_i}{(1 - \Gamma_r) - j\Gamma_i}, \tag{2.114}$$

which can be manipulated to obtain explicit expressions for r_L and x_L in terms of Γ_r and Γ_i. This is accomplished by multiplying the numerator and denominator of the right-hand side of Eq. (2.114) by the complex conjugate of the denominator and then separating the result into real and imaginary parts. These steps lead to

$$r_L = \frac{1 - \Gamma_r^2 - \Gamma_i^2}{(1 - \Gamma_r)^2 + \Gamma_i^2}, \tag{2.115a}$$

$$x_L = \frac{2\Gamma_i}{(1 - \Gamma_r)^2 + \Gamma_i^2}. \tag{2.115b}$$

Equation (2.115a) implies that there exist many combinations of values for Γ_r and Γ_i that yield the same value for the normalized load resistance r_L. For example, $(\Gamma_r, \Gamma_i) = (0.33, 0)$ gives $r_L = 2$, as does $(\Gamma_r, \Gamma_i) = (0.5, 0.29)$, as well as an infinite number of other combinations. In fact, if we were to plot in the Γ_r–Γ_i plane all possible combinations of Γ_r and Γ_i corresponding to $r_L = 2$, we would obtain the circle labeled $r_L = 2$ in **Fig. 2-25**. Similar circles can be obtained for other values of r_L. After some algebraic manipulations, Eq. (2.115a) can be rearranged into the following ***parametric equation*** for the circle in the Γ_r–Γ_i plane corresponding to a given value of r_L:

$$\left(\Gamma_r - \frac{r_L}{1 + r_L}\right)^2 + \Gamma_i^2 = \left(\frac{1}{1 + r_L}\right)^2. \tag{2.116}$$

The standard equation for a circle in the x–y plane with center at (x_0, y_0) and radius a is

$$(x - x_0)^2 + (y - y_0)^2 = a^2. \tag{2.117}$$

Comparison of Eq. (2.116) with Eq. (2.117) shows that the r_L circle is centered at $\Gamma_r = r_L/(1+r_L)$ and $\Gamma_i = 0$, and its radius is $1/(1+r_L)$. It therefore follows that all r_L-circles pass through the point $(\Gamma_r, \Gamma_i) = (1, 0)$. The largest circle shown in **Fig. 2-25** is for $r_L = 0$, which also is the unit circle corresponding to $|\Gamma| = 1$. This is to be expected, because when $r_L = 0$, $|\Gamma| = 1$ regardless of the magnitude of x_L.

A similar manipulation of the expression for x_L given by Eq. (2.115b) leads to

$$(\Gamma_r - 1)^2 + \left(\Gamma_i - \frac{1}{x_L}\right)^2 = \left(\frac{1}{x_L}\right)^2, \tag{2.118}$$

which is the equation of a circle of radius $(1/x_L)$ centered at $(\Gamma_r, \Gamma_i) = (1, 1/x_L)$. The x_L circles in the Γ_r–Γ_i plane are quite different from those for constant r_L. To start with, the normalized reactance x_L may assume both positive and negative values, whereas the normalized resistance cannot be negative (negative resistances cannot be realized in passive circuits). Hence, Eq. (2.118) represents two families of circles, one for positive values of x_L and another for negative ones. Furthermore, as shown in **Fig. 2-25**, only part of a given circle falls within the bounds of the $|\Gamma| = 1$ unit circle.

The families of circles of the two parametric equations given by Eqs. (2.116) and (2.118) plotted for selected values of r_L and x_L constitute the Smith chart shown in **Fig. 2-26**. The Smith chart provides a graphical evaluation of Eqs. (2.115a and b) and their inverses. For example, point P in **Fig. 2-26** represents a normalized load impedance $z_L = 2 - j1$, which corresponds to a voltage reflection coefficient $\Gamma = 0.45 \exp(-j26.6^\circ)$. The magnitude $|\Gamma| = 0.45$ is obtained by dividing the length of the line between the center of the Smith chart and the point P by the length of the line between the center of the Smith chart and the edge of the unit circle (the radius of the unit circle corresponds to $|\Gamma| = 1$). The perimeter of the Smith chart contains three concentric scales. The innermost scale is labeled ***angle of***

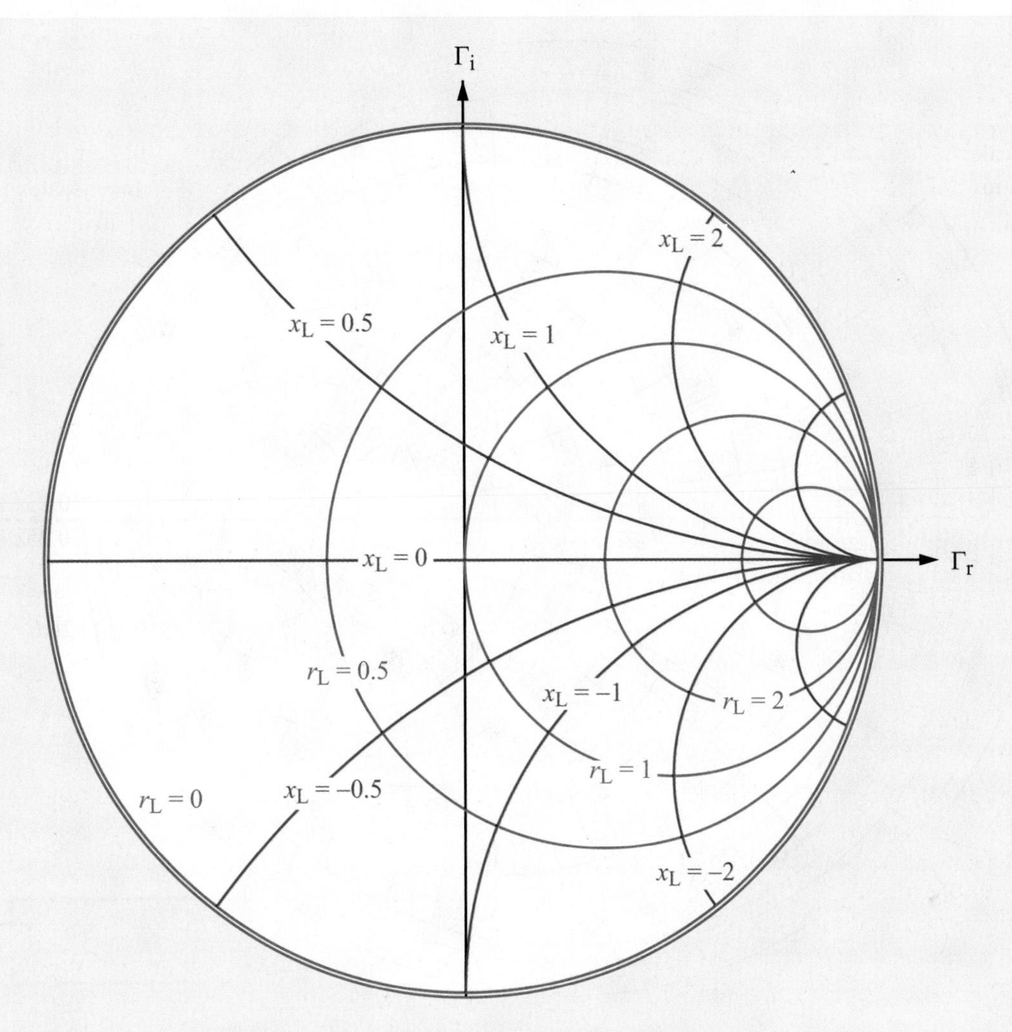

Figure 2-25 Families of r_L and x_L circles within the domain $|\Gamma| \leq 1$.

reflection coefficient in degrees. This is the scale for θ_r. As indicated in **Fig. 2-26**, $\theta_r = -26.6°$ (-0.46 rad) for point P. The meanings and uses of the other two scales are discussed next.

Exercise 2-16: Use the Smith chart to find the values of Γ corresponding to the following normalized load impedances: (a) $z_L = 2 + j0$, (b) $z_L = 1 - j1$, (c) $z_L = 0.5 - j2$, (d) $z_L = -j3$, (e) $z_L = 0$ (short circuit), (f) $z_L = \infty$ (open circuit), (g) $z_L = 1$ (matched load).

Answer: (a) $\Gamma = 0.33$, (b) $\Gamma = 0.45\angle{-63.4°}$, (c) $\Gamma = 0.83\angle{-50.9°}$, (d) $\Gamma = 1\angle{-36.9°}$, (e) $\Gamma = -1$, (f) $\Gamma = 1$, (g) $\Gamma = 0$. (See EM.)

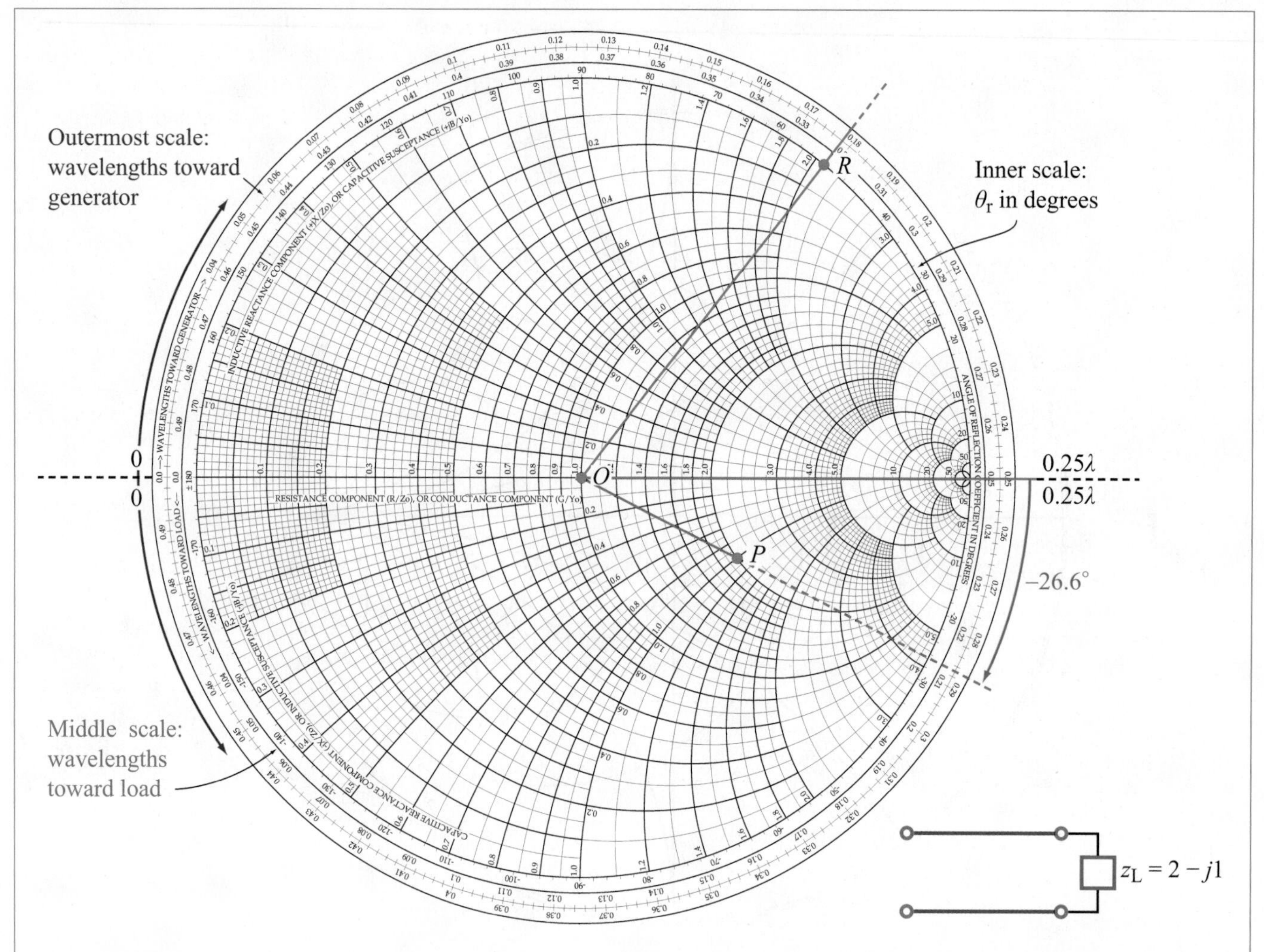

Figure 2-26 Point P represents a normalized load impedance $z_L = 2 - j1$. The reflection coefficient has a magnitude $|\Gamma| = \overline{OP}/\overline{OR} = 0.45$ and an angle $\theta_r = -26.6°$. Point R is an arbitrary point on the $r_L = 0$ circle (which also is the $|\Gamma| = 1$ circle).

2-10.2 Wave Impedance

From Eq. (2.74), the ***normalized wave impedance*** looking toward the load at a distance d from the load is

$$z(d) = \frac{Z(d)}{Z_0} = \frac{1 + \Gamma_d}{1 - \Gamma_d}, \tag{2.119}$$

where

$$\Gamma_d = \Gamma e^{-j2\beta d} = |\Gamma| e^{j(\theta_r - 2\beta d)} \tag{2.120}$$

is the *phase-shifted* voltage reflection coefficient. The form of Eq. (2.119) is identical with that for z_L given by Eq. (2.112):

$$z_L = \frac{1 + \Gamma}{1 - \Gamma}. \tag{2.121}$$

This similarity in form suggests that if Γ is transformed into Γ_d, z_L gets transformed into $z(d)$. On the Smith chart, the transformation from Γ to Γ_d is achieved by maintaining $|\Gamma|$ constant and *decreasing* its phase θ_r by $2\beta d$, which corresponds to a *clockwise* rotation (on the Smith chart) over an angle of $2\beta d$ radians. A complete rotation around the Smith chart corresponds to a phase change of 2π in Γ. The length d corresponding to this phase change satisfies

$$2\beta d = 2\,\frac{2\pi}{\lambda}\,d = 2\pi, \tag{2.122}$$

from which it follows that $d = \lambda/2$.

▶ The outermost scale around the perimeter of the Smith chart (Fig. 2-26), called the ***wavelengths toward generator*** (WTG) scale, has been constructed to denote movement on the transmission line toward the generator, in units of the wavelength λ. That is, d is measured in wavelengths, and one complete rotation corresponds to $d = \lambda/2$. ◀

In some transmission-line problems, it may be necessary to move from some point on the transmission line toward a point closer to the load, in which case the phase of Γ must be increased, which corresponds to rotation in the counterclockwise direction. For convenience, the Smith chart contains a third scale around its perimeter (in between the θ_r scale and the WTG scale) for accommodating such an operation. It is called the ***wavelengths toward load*** (WTL) scale.

To illustrate how the Smith chart is used to find $Z(d)$, consider a 50 Ω lossless transmission line terminated in a load impedance $Z_L = (100 - j50)$ Ω. Our objective is to find $Z(d)$ at a distance $d = 0.1\lambda$ from the load. The normalized load impedance is $z_L = Z_L/Z_0 = 2 - j1$, and is marked by point A on the Smith chart in Fig. 2-27. On the WTG scale, point A is located at 0.287λ. Next, we construct a circle centered at $(\Gamma_r, \Gamma_i) = (0, 0)$ and passing through point A. Since the center of the Smith chart is the intersection point of the Γ_r and Γ_i axes, all points on this circle have the same value of $|\Gamma|$. This ***constant-$|\Gamma|$ circle*** is also a ***constant-SWR circle***. This follows from the relation between the voltage standing-wave ratio (SWR) and $|\Gamma|$, namely

$$S = \frac{1 + |\Gamma|}{1 - |\Gamma|}. \tag{2.123}$$

▶ A constant value of $|\Gamma|$ corresponds to a constant value of S, and vice versa. ◀

As was stated earlier, to transform z_L to $z(d)$, we need to maintain $|\Gamma|$ constant, which means staying on the SWR circle, while decreasing the phase of Γ by $2\beta d$ radians. This is equivalent to moving a distance $d = 0.1\lambda$ toward the generator on the WTG scale. Since point A is located at 0.287λ on the WTG scale, $z(d)$ is found by moving to location $0.287\lambda + 0.1\lambda = 0.387\lambda$ on the WTG scale. A radial line through this new position on the WTG scale intersects the SWR circle at point B. This point represents $z(d)$, and its value is $z(d) = 0.6 - j0.66$. Finally, we unnormalize $z(d)$ by multiplying it by $Z_0 = 50$ Ω to get $Z(d) = (30 - j33)$ Ω. This result can be verified analytically using Eq. (2.119). The points between points A and B on the SWR circle represent different locations along the transmission line.

If a line is of length l, its ***input impedance*** is $Z_{in} = Z_0\, z(l)$, with $z(l)$ determined by rotating a distance l from the load along the WTG scale.

Exercise 2-17: Use the Smith chart to find the normalized input impedance of a lossless line of length l terminated in a normalized load impedance z_L for each of the following combinations: (a) $l = 0.25\lambda$, $z_L = 1 + j0$, (b) $l = 0.5\lambda$, $z_L = 1 + j1$, (c) $l = 0.3\lambda$, $z_L = 1 - j1$, (d) $l = 1.2\lambda$, $z_L = 0.5 - j0.5$, (e) $l = 0.1\lambda$, $z_L = 0$ (short circuit), (f) $l = 0.4\lambda$, $z_L = j3$, (g) $l = 0.2\lambda$, $z_L = \infty$ (open circuit).

Answer: (a) $z_{in} = 1 + j0$, (b) $z_{in} = 1 + j1$, (c) $z_{in} = 0.76 + j0.84$, (d) $z_{in} = 0.59 + j0.66$, (e) $z_{in} = 0 + j0.73$, (f) $z_{in} = 0 + j0.72$, (g) $z_{in} = 0 - j0.32$. (See EM.)

2-10.3 SWR, Voltage Maxima and Minima

Consider a load with $z_L = 2 + j1$. **Figure 2-28** shows a Smith chart with an SWR circle drawn through point A, representing z_L. The SWR circle intersects the real (Γ_r) axis at two points, labeled P_{max} and P_{min}. At both points $\Gamma_i = 0$ and $\Gamma = \Gamma_r$. Also, on the real axis, the imaginary part of the load impedance $x_L = 0$. From the definition of Γ,

$$\Gamma = \frac{z_L - 1}{z_L + 1}, \tag{2.124}$$

it follows that points P_{max} and P_{min} correspond to

$$\Gamma = \Gamma_r = \frac{r_0 - 1}{r_0 + 1} \qquad (\text{for } \Gamma_i = 0), \tag{2.125}$$

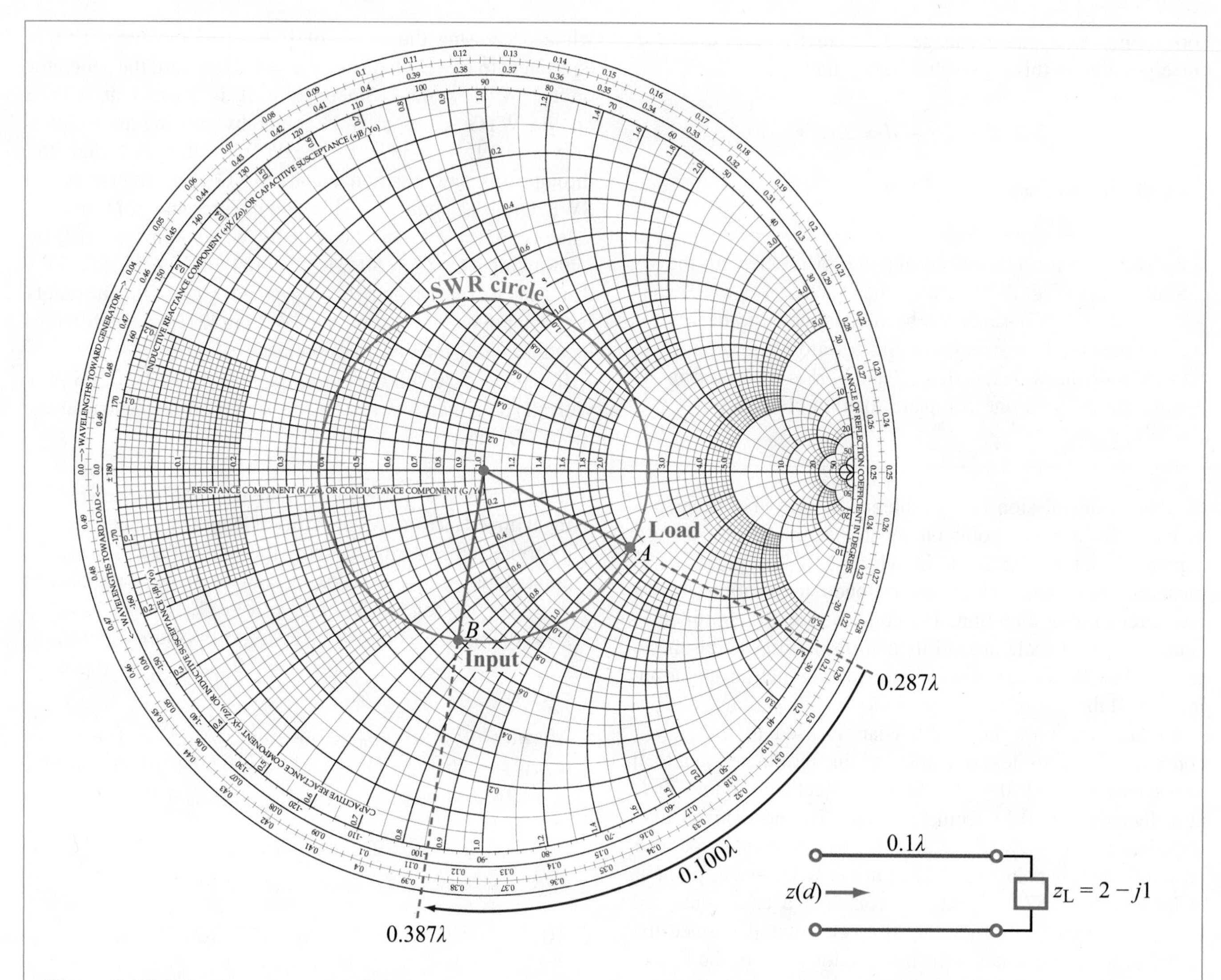

Figure 2-27 Point A represents a normalized load $z_L = 2 - j1$ at 0.287λ on the WTG scale. Point B represents the line input at $d = 0.1\lambda$ from the load. At B, $z(d) = 0.6 - j0.66$.

where r_0 is the value of r_L where the SWR circle intersects the Γ_r axis. Point P_{min} corresponds to $r_0 < 1$ and P_{max} corresponds to $r_0 > 1$. Rewriting Eq. (2.123) for $|\Gamma|$ in terms of S, we have

$$|\Gamma| = \frac{S-1}{S+1}\,. \tag{2.126}$$

For point P_{max}, $|\Gamma| = \Gamma_r$; hence

$$\Gamma_r = \frac{S-1}{S+1}\,. \tag{2.127}$$

The similarity in form of Eqs. (2.125) and (2.127) suggests that S equals the value of the normalized resistance r_0. By definition $S \geq 1$, and at point P_{max}, $r_0 > 1$, which further satisfies the similarity condition. In Fig. 2-28, $r_0 = 2.6$ at P_{max}; hence $S = 2.6$.

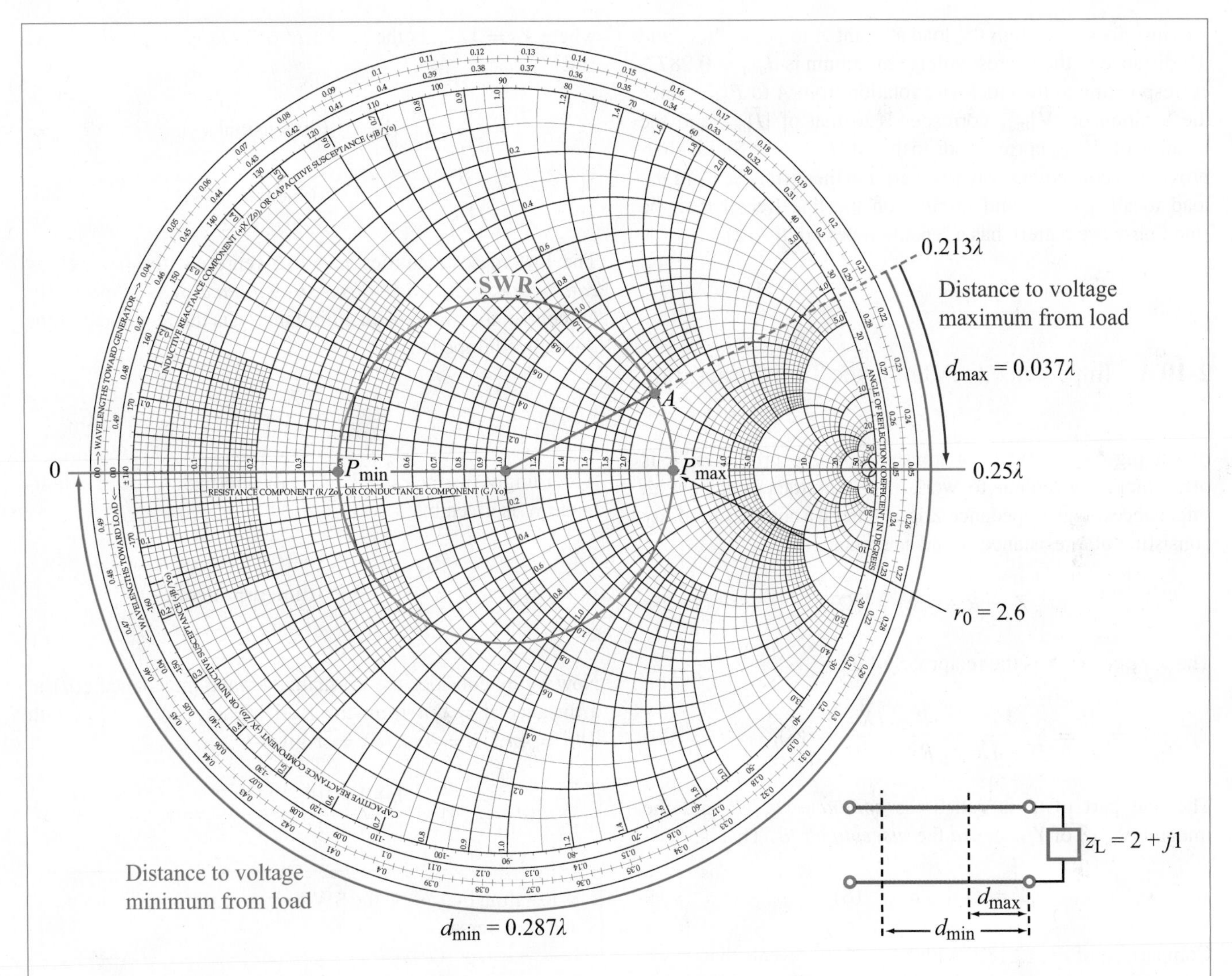

Figure 2-28 Point A represents a normalized load with $z_L = 2 + j1$. The standing wave ratio is $S = 2.6$ (at P_{max}), the distance between the load and the first voltage maximum is $d_{max} = (0.25 - 0.213)\lambda = 0.037\lambda$, and the distance between the load and the first voltage minimum is $d_{min} = (0.037 + 0.25)\lambda = 0.287\lambda$.

► S is numerically equal to the value of r_0 at P_{max}, the point at which the SWR circle intersects the real Γ axis to the right of the chart's center. ◄

Points P_{min} and P_{max} also represent locations on the line at which the magnitude of the voltage $|\widetilde{V}|$ is a minimum and a maximum, respectively. This is easily demonstrated by considering Eq. (2.120) for Γ_d. At point P_{max}, the total phase of Γ_d, that is, $(\theta_r - 2\beta d)$, equals zero or $-2n\pi$ (with n being a positive integer), which is the condition corresponding to $|\widetilde{V}|_{max}$, as indicated by Eq. (2.69). Similarly, at P_{min} the total phase of Γ_d equals $-(2n+1)\pi$, which is the condition for $|\widetilde{V}|_{min}$. Thus, for the transmission line represented by the SWR circle shown in Fig. 2-28, the distance between the load and the nearest voltage maximum is $d_{max} = 0.037\lambda$, obtained by moving clockwise from the load at point A to point $P_{\max}$, and the distance to the nearest voltage minimum is $d_{\min} = 0.287\lambda$, corresponding to the clockwise rotation from A to $P_{\min}$. Since the location of $|\widetilde{V}|_{\max}$ corresponds to that of $|\widetilde{I}|_{\min}$ and the location of $|\widetilde{V}|_{\min}$ corresponds to that of $|\widetilde{I}|_{\max}$, the Smith chart provides a convenient way to determine the distances from the load to all maxima and minima on the line (recall that the standing-wave pattern has a repetition period of $\lambda/2$).

2-10.4 Impedance to Admittance Transformations

In solving certain types of transmission-line problems, it is often more convenient to work with admittances than with impedances. Any impedance Z is in general a complex quantity consisting of a resistance R and a reactance X:

$$Z = R + jX \qquad (\Omega). \tag{2.128}$$

The ***admittance*** Y is the reciprocal of Z:

$$Y = \frac{1}{Z} = \frac{1}{R + jX} = \frac{R - jX}{R^2 + X^2} \qquad (\text{S}). \tag{2.129}$$

The real part of Y is called the ***conductance*** G, and the imaginary part of Y is called the ***susceptance*** B. That is,

$$Y = G + jB \qquad (\text{S}). \tag{2.130}$$

Comparison of Eq. (2.130) with Eq. (2.129) reveals that

$$G = \frac{R}{R^2 + X^2} \qquad (\text{S}), \tag{2.131a}$$

$$B = \frac{-X}{R^2 + X^2} \qquad (\text{S}). \tag{2.131b}$$

A normalized impedance z is defined as the ratio of Z to Z_0, the characteristic impedance of the line. The same concept applies to the definition of the ***normalized admittance*** y; that is,

$$y = \frac{Y}{Y_0} = \frac{G}{Y_0} + j\frac{B}{Y_0} = g + jb \qquad (\text{dimensionless}), \tag{2.132}$$

where $Y_0 = 1/Z_0$ is the ***characteristic admittance of the line*** and

$$g = \frac{G}{Y_0} = GZ_0 \qquad (\text{dimensionless}), \tag{2.133a}$$

$$b = \frac{B}{Y_0} = BZ_0 \qquad (\text{dimensionless}). \tag{2.133b}$$

The lowercase quantities g and b represent the ***normalized conductance*** and ***normalized susceptance*** of y, respectively. Of course, the normalized admittance y is the reciprocal of the normalized impedance z,

$$y = \frac{Y}{Y_0} = \frac{Z_0}{Z} = \frac{1}{z}\,. \tag{2.134}$$

Accordingly, using Eq. (2.121), the normalized load admittance y_{L} is given by

$$y_{\text{L}} = \frac{1}{z_{\text{L}}} = \frac{1 - \Gamma}{1 + \Gamma} \qquad \textbf{(dimensionless)}. \tag{2.135}$$

Now let us consider the normalized wave impedance $z(d)$ at a distance $d = \lambda/4$ from the load. Using Eq. (2.119) with $2\beta d = 4\pi d/\lambda = 4\pi\lambda/4\lambda = \pi$ gives

$$z(d = \lambda/4) = \frac{1 + \Gamma e^{-j\pi}}{1 - \Gamma e^{-j\pi}} = \frac{1 - \Gamma}{1 + \Gamma} = y_{\text{L}}. \tag{2.136}$$

► Rotation by $\lambda/4$ on the SWR circle transforms z into y, and vice versa. ◄

In **Fig. 2-29**, the points representing z_{L} and y_{L} are diametrically opposite to each other on the SWR circle. In fact, such a transformation on the Smith chart can be used to determine any normalized admittance from its corresponding normalized impedance, and vice versa.

The Smith chart can be used with normalized impedances or with normalized admittances. As an impedance chart, the Smith chart consists of r_{L} and x_{L} circles, the resistance and reactance of a normalized load impedance z_{L}, respectively. When used as an admittance chart, the r_{L} circles become g_{L} circles and the x_{L} circles become b_{L} circles, where g_{L} and b_{L} are the conductance and susceptance of the normalized load admittance y_{L}, respectively.

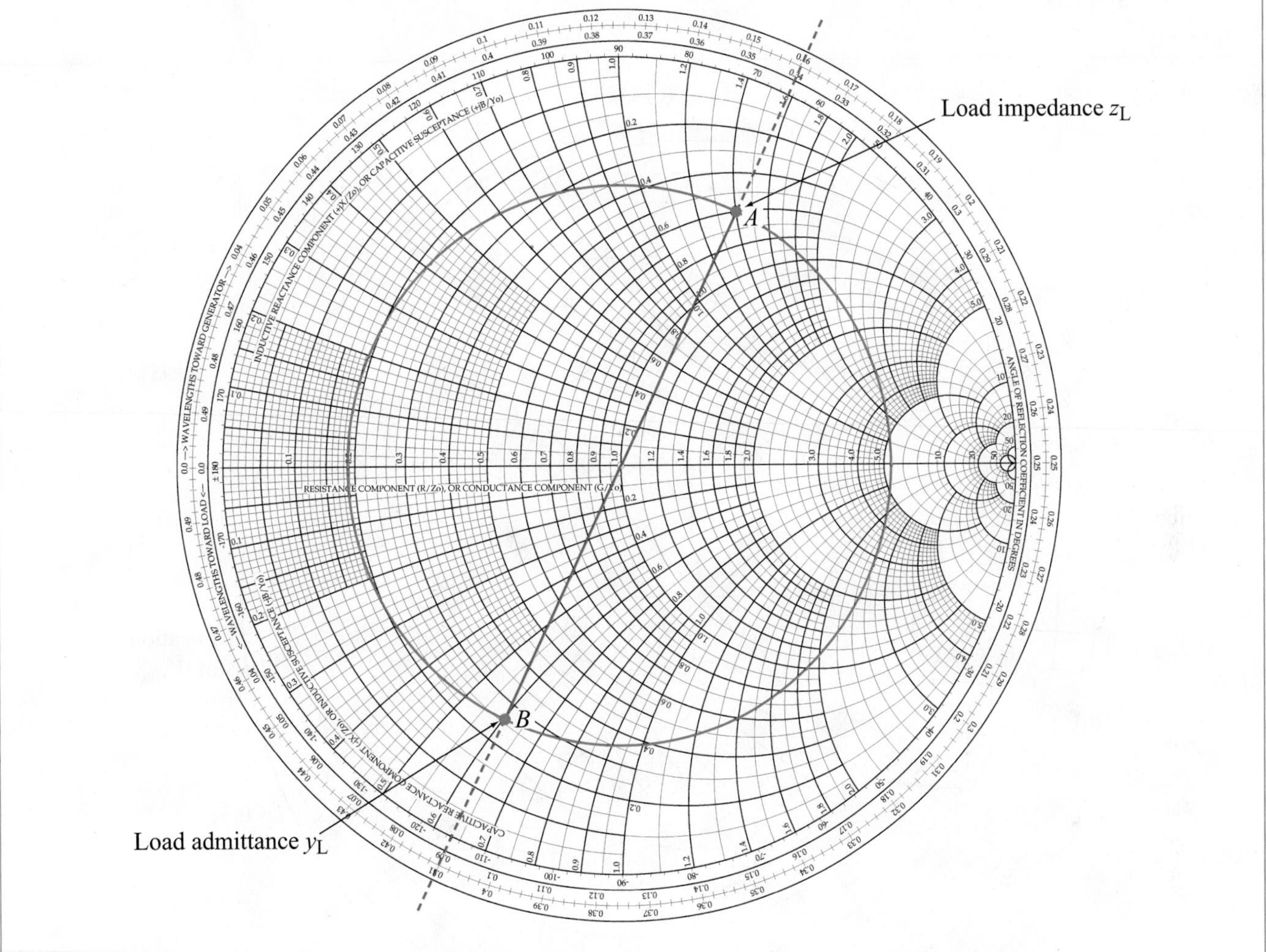

Figure 2-29 Point A represents a normalized load $z_L = 0.6 + j1.4$. Its corresponding normalized admittance is $y_L = 0.25 - j0.6$, and it is at point B.

Example 2-11: Smith-Chart Calculations

A 50 Ω lossless transmission line of length 3.3λ is terminated by a load impedance $Z_L = (25 + j50)$ Ω. Use the Smith chart to find (a) the voltage reflection coefficient, (b) the voltage standing-wave ratio, (c) the distances of the first voltage maximum and first voltage minimum from the load, (d) the input impedance of the line, and (e) the input admittance of the line.

Solution: (a) The normalized load impedance is

$$z_L = \frac{Z_L}{Z_0} = \frac{25 + j50}{50} = 0.5 + j1,$$

which is marked as point A on the Smith chart in Fig. 2-30. A radial line is drawn from the center of the chart at point O through point A to the outer perimeter of the chart. The line crosses the scale labeled "angle of reflection coefficient in degrees" at $\theta_r = 83°$. Next, measurements are made to determine lengths $\overline{OA}$ and $\overline{OO'}$, of the lines between O and A and between points O and O', respectively, where O' is an

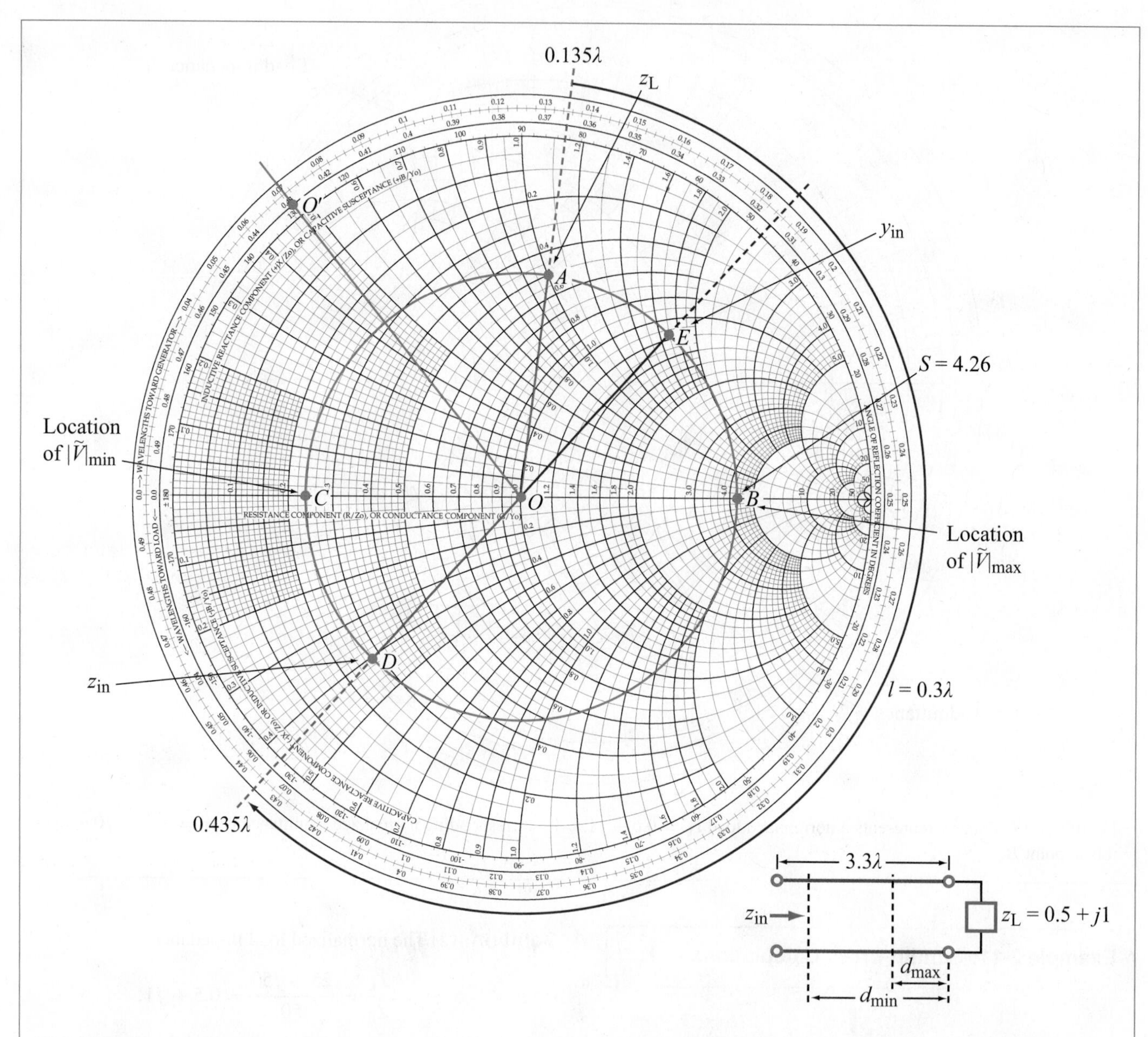

Figure 2-30 Solution for Example 2-11. Point A represents a normalized load $z_L = 0.5 + j1$ at 0.135λ on the WTG scale. At A, $\theta_r = 83°$ and $|\Gamma| = \overline{OA}/\overline{OO'} = 0.62$. At B, the standing-wave ratio is $S = 4.26$. The distance from A to B gives $d_{max} = 0.115\lambda$ and from A to C gives $d_{min} = 0.365\lambda$. Point D represents the normalized input impedance $z_{in} = 0.28 - j0.40$, and point E represents the normalized input admittance $y_{in} = 1.15 + j1.7$.

arbitrary point on the $r_L = 0$ circle. The length $\overline{OO'}$ is equal to the radius of the $|\Gamma| = 1$ circle. The magnitude of Γ is then obtained from $|\Gamma| = \overline{OA}/\overline{OO'} = 0.62$. Hence,

$$\Gamma = 0.62\angle 83^\circ. \quad (2.137)$$

(b) The SWR circle passing through point A crosses the Γ_r axis at points B and C. The value of r_L at point B is 4.26, from which it follows that

$$S = 4.26.$$

(c) The first voltage maximum is at point B on the SWR circle, which is at location 0.25λ on the WTG scale. The load, represented by point A, is at 0.135λ on the WTG scale. Hence, the distance between the load and the first voltage maximum is

$$d_{max} = (0.25 - 0.135)\lambda = 0.115\lambda.$$

The first voltage minimum is at point C. Moving on the WTG scale between points A and C gives

$$d_{min} = (0.5 - 0.135)\lambda = 0.365\lambda,$$

which is 0.25λ past d_{max}.

(d) The line is 3.3λ long; subtracting multiples of 0.5λ leaves 0.3λ. From the load at 0.135λ on the WTG scale, the input of the line is at $(0.135 + 0.3)\lambda = 0.435\lambda$. This is labeled as point D on the SWR circle, and the normalized impedance is

$$z_{in} = 0.28 - j0.40,$$

which yields

$$Z_{in} = z_{in} Z_0 = (0.28 - j0.40)50 = (14 - j20)\ \Omega.$$

(e) The normalized input admittance y_{in} is found by moving 0.25λ on the Smith chart to the image point of z_{in} across the circle, labeled point E on the SWR circle. The coordinates of point E give

$$y_{in} = 1.15 + j1.7,$$

and the corresponding input admittance is

$$Y_{in} = y_{in} Y_0 = \frac{y_{in}}{Z_0} = \frac{1.15 + j1.7}{50} = (0.023 + j0.034)\ \text{S}.$$

Example 2-12: Determining Z_L Using the Smith Chart

This problem is similar to Example 2-6, except that now we demonstrate its solution using the Smith chart.

Given that the voltage standing-wave ratio $S = 3$ on a 50 Ω line, that the first voltage minimum occurs at 5 cm from the load, and that the distance between successive minima is 20 cm, find the load impedance.

Solution: The distance between successive minima equals $\lambda/2$. Hence, $\lambda = 40$ cm. In wavelength units, the first voltage minimum is at

$$d_{min} = \frac{5}{40} = 0.125\lambda.$$

Point A on the Smith chart in Fig. 2-31 corresponds to $S = 3$. Using a compass, the constant S circle is drawn through point A. Point B corresponds to locations of voltage minima. Upon moving 0.125λ from point B toward the load on the WTL scale (counterclockwise), we arrive at point C, which represents the location of the load. The normalized load impedance at point C is

$$z_L = 0.6 - j0.8.$$

Multiplying by $Z_0 = 50\ \Omega$, we obtain

$$Z_L = 50(0.6 - j0.8) = (30 - j40)\ \Omega.$$

Concept Question 2-19: The outer perimeter of the Smith chart represents what value of $|\Gamma|$? Which point on the Smith chart represents a matched load?

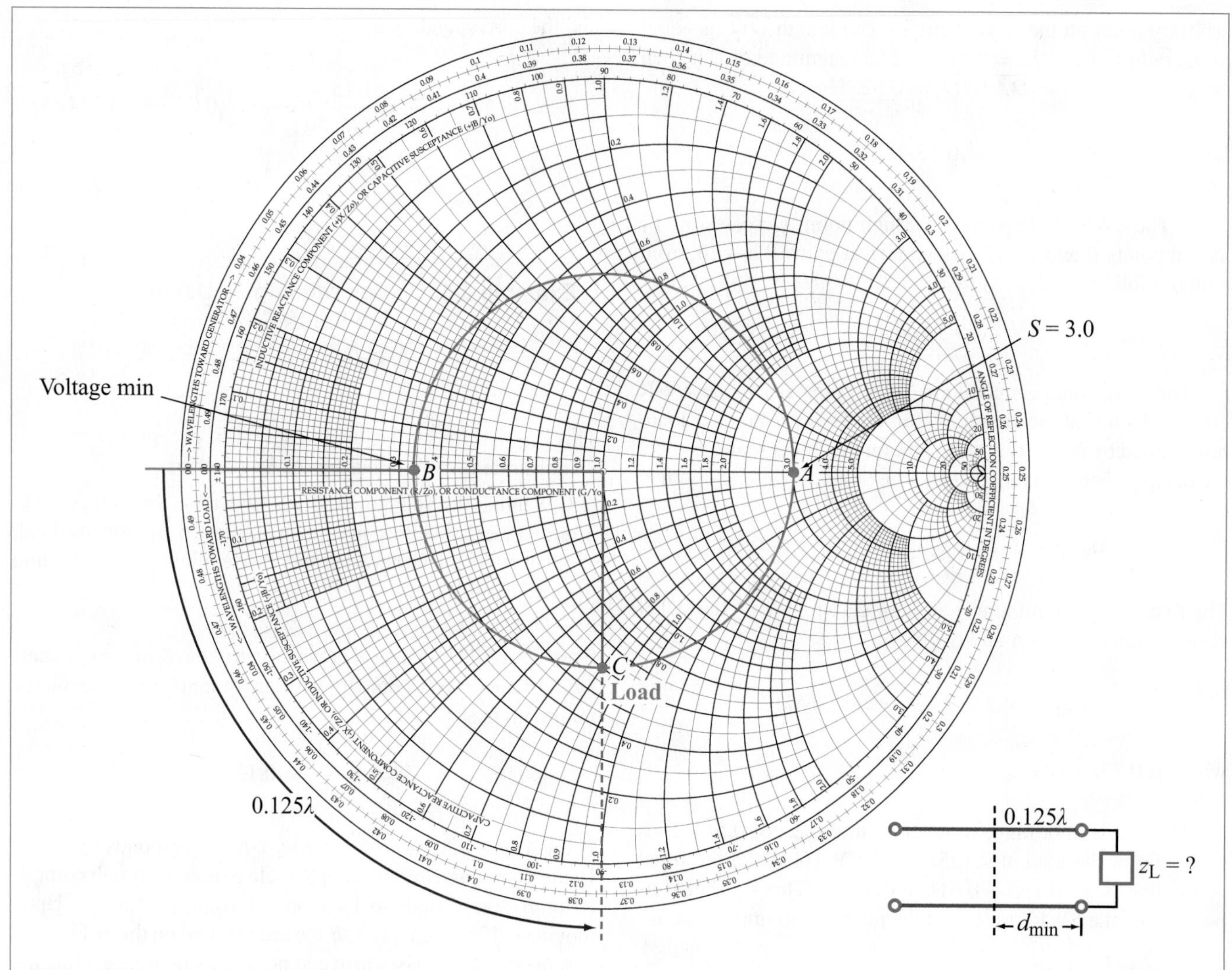

Figure 2-31 Solution for Example 2-12. Point A denotes that $S = 3$, point B represents the location of the voltage minimum, and point C represents the load at 0.125λ on the WTL scale from point B. At C, $z_L = 0.6 - j0.8$.

Concept Question 2-20: What is an SWR circle? What quantities are constant for all points on an SWR circle?

Concept Question 2-21: What line length corresponds to one complete rotation around the Smith chart? Why?

Concept Question 2-22: Which points on the SWR circle correspond to locations of voltage maxima and minima on the line and why?

Concept Question 2-23: Given a normalized impedance z_L, how do you use the Smith chart to find the corresponding normalized admittance $y_L = 1/z_L$?

Module 2.6 Interactive Smith Chart Locate the load on the Smith chart; display the corresponding reflection coefficient and SWR circle; "move" to a new location at a distance d from the load, and read the wave impedance $Z(d)$ and phase-shifted reflection coefficient Γ_d; perform impedance to admittance transformations and vice versa; and use all of these tools to solve transmission-line problems via the Smith chart.

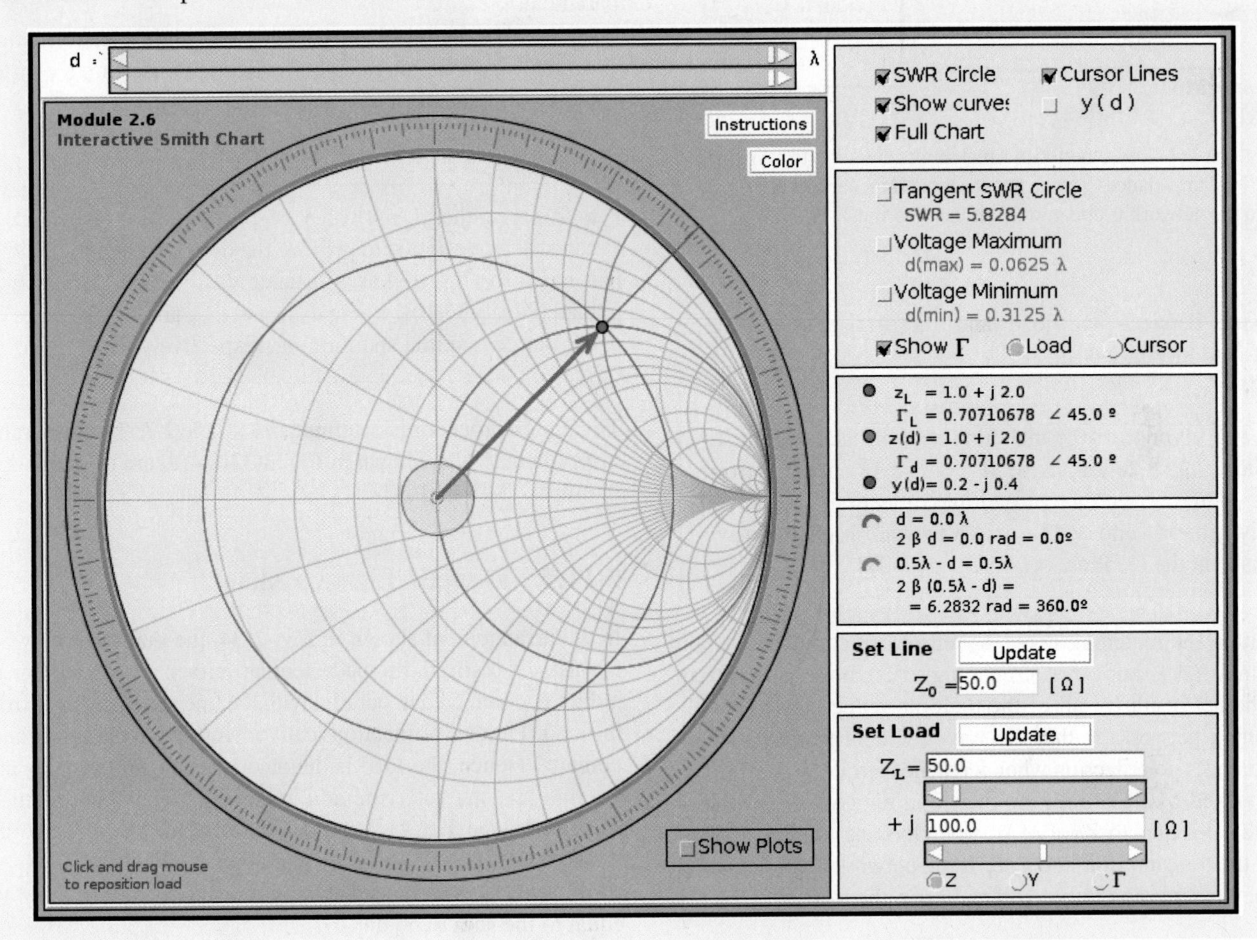

2-11 Impedance Matching

A transmission line usually connects a generator circuit at one end to a load at the other. The load may be an antenna, a computer terminal, or any circuit with an equivalent input impedance Z_L.

▶ The transmission line is said to be matched to the load when its characteristic impedance $Z_0 = Z_L$, in which case waves traveling on the line towards the load are not reflected back to the source. ◀

Since the primary use of a transmission line is to transfer power or transmit coded signals (such as digital data), a matched load ensures that all of the power delivered to the transmission line by the source is transferred to the load (and no *echoes* are relayed back to the source).

The simplest solution to matching a load to a transmission line is to design the load circuit such that its impedance $Z_L = Z_0$. Unfortunately, this may not be possible in practice because the load circuit may have to satisfy other requirements. An alternative solution is to place an ***impedance-matching***

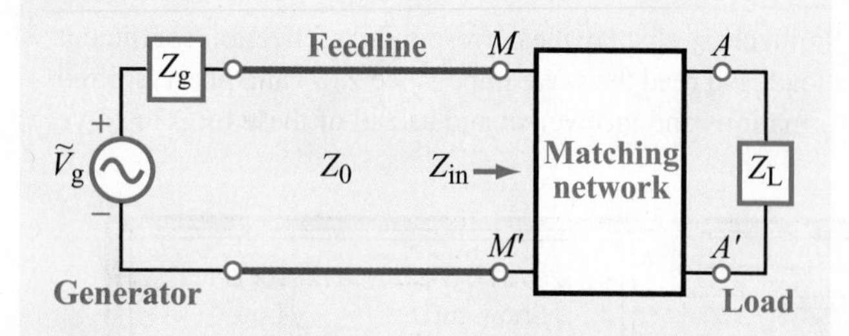

Figure 2-32 The function of a matching network is to transform the load impedance Z_L such that the input impedance Z_{in} looking into the network is equal to Z_0 of the feedline.

network between the load and the transmission line as illustrated in Fig. 2-32.

► The purpose of the matching network is to eliminate reflections at terminals MM' for waves incident from the source. Even though multiple reflections may occur between AA' and MM', only a forward traveling wave exists on the feedline. ◄

Within the matching network, reflections can occur at both terminals (AA' and MM'), creating a standing-wave pattern, but the net result (of all of the multiple reflections within the matching network) is that the wave incident from the source experiences no reflection when it reaches terminals MM'. This is achieved by designing the matching network to exhibit an impedance equal to Z_0 at MM' when looking into the network from the transmission line side. If the network is lossless, then all the power going into it will end up in the load.

► Matching networks may consist of lumped elements, such as capacitors and inductors (but not resistors, because resistors incur ohmic losses), or of sections of transmission lines with appropriate lengths and terminations. ◄

The matching network, which is intended to match a load impedance $Z_L = R_L + jX_L$ to a lossless transmission line with characteristic impedance Z_0, may be inserted either in series (between the load and the feedline) as in Fig. 2-33(a) and (b) or in parallel [Fig. 2-33(c) to (e)]. In either case, the network has to transform the real part of the load impedance from R_L (at the load) to Z_0 at MM' in Fig. 2-32 and transform the reactive part from X_L (at the load) to zero at MM'. To achieve these two transformations, the matching network must have at least two degrees of freedom (that is, two adjustable parameters).

If $X_L = 0$, the problem reduces to a single transformation, in which case matching can be realized by inserting a quarter-wavelength transformer (Section 2-8.5) next to the load [Fig. 2-33(a)].

► For the general case where $X_L \neq 0$, a $\lambda/4$ transformer can still be designed to provide the desired matching, but it has to be inserted at a distance d_{max} or d_{min} from the load [Fig. 2-33(b)], where d_{max} and d_{min} are the distances to voltage maxima and minima, respectively. ◄

The design procedure is outlined in Module 2.7. The in-parallel insertion networks shown in Fig. 2-33(c)–(e) are the subject of Examples 2-13 and 2-14.

2-11.1 Lumped-Element Matching

In the arrangement shown in Fig. 2-34, the matching network consists of a single lumped element, either a capacitor or an inductor, connected in parallel with the line at a distance d from the load. Parallel connections call for working in the admittance domain. Hence, the load is denoted by an admittance Y_L and the line has characteristic admittance Y_0. The shunt element has admittance Y_s. At MM', Y_d is the admittance due to the transmission-line segment to the right of MM'. The input admittance Y_{in} (referenced at a point just to the left of MM') is equal to the sum of Y_d and Y_s:

$$Y_{in} = Y_d + Y_s. \tag{2.138}$$

In general Y_d is complex and Y_s is purely imaginary because it represents a reactive element (capacitor or inductor). Hence, Eq. (2.138) can be written as

$$Y_{in} = (G_d + jB_d) + jB_s = G_d + j(B_d + B_s). \tag{2.139}$$

When all quantities are normalized to Y_0, Eq. (2.139) becomes

$$y_{in} = g_d + j(b_d + b_s). \tag{2.140}$$

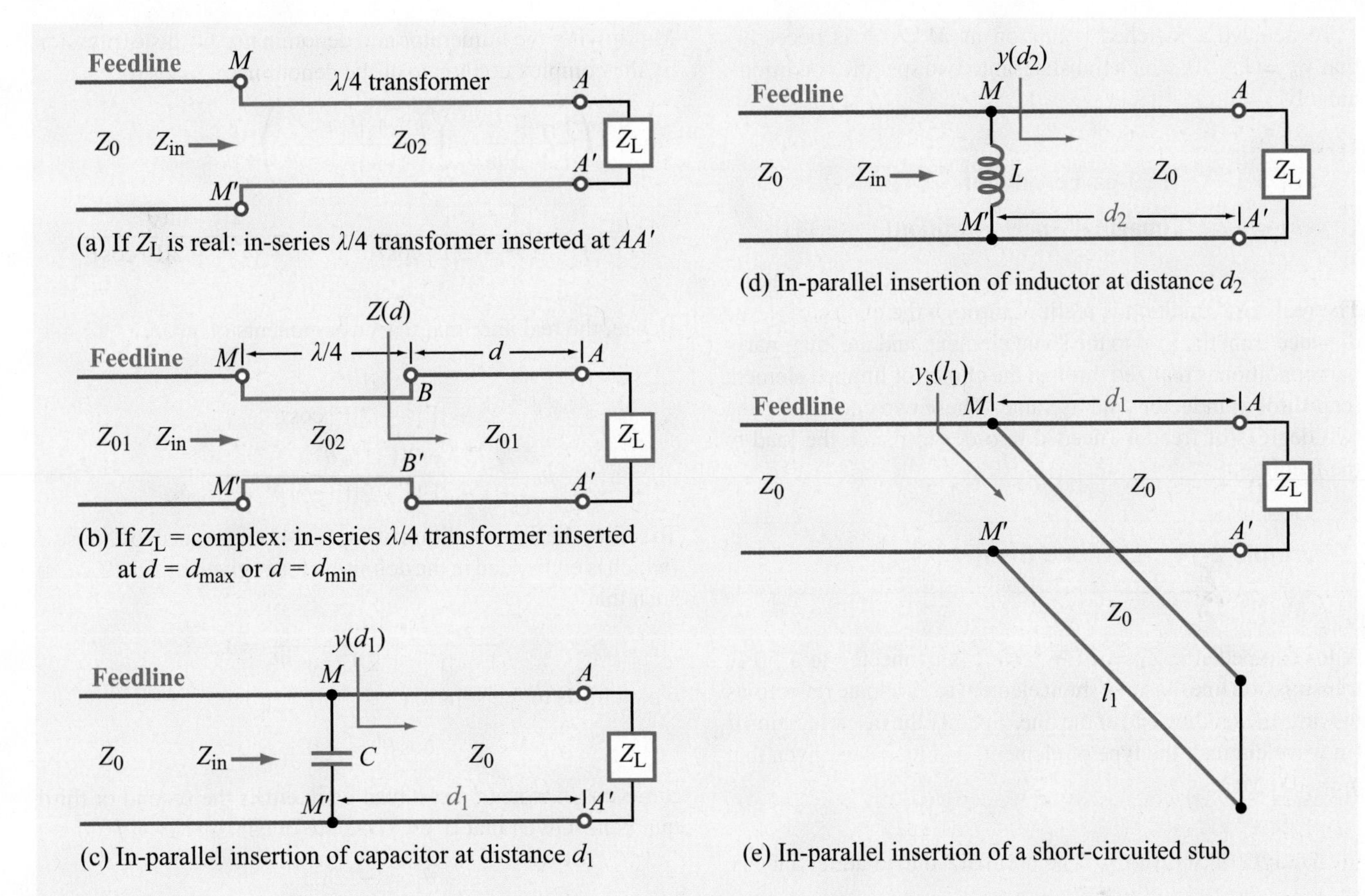

Figure 2-33 Five examples of in-series and in-parallel matching networks.

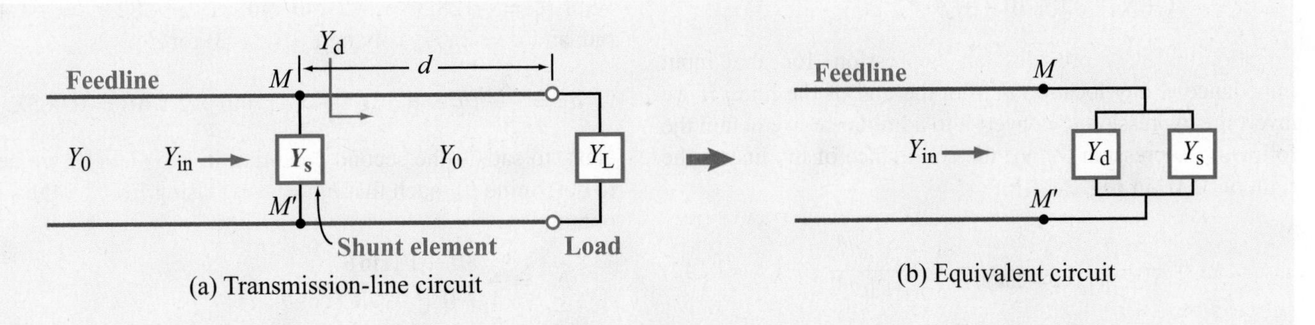

Figure 2-34 Inserting a reactive element with admittance Y_s at MM' modifies Y_d to Y_{in}.

To achieve a matched condition at MM', it is necessary that $y_{\text{in}} = 1 + j0$, which translates into two specific conditions, namely

$$g_{\text{d}} = 1 \quad \textbf{(real-part condition)}, \tag{2.141a}$$

$$b_{\text{s}} = -b_{\text{d}} \quad \textbf{(imaginary-part condition)}. \tag{2.141b}$$

The real-part condition is realized through the choice of d, the distance from the load to the shunt element, and the imaginary-part condition is realized through the choice of lumped element (capacitor or inductor) and its value. These two choices are the two degrees of freedom needed in order to match the load to the feedline.

Example 2-13: Lumped Element

A load impedance $Z_{\text{L}} = 25 - j50\ \Omega$ is connected to a 50 Ω transmission line. Insert a shunt element to eliminate reflections towards the sending end of the line. Specify the insert location d (in wavelengths), the type of element, and its value, given that $f = 100$ MHz.

(a) Analytical Solution: The normalized load admittance is

$$y_{\text{L}} = \frac{Z_0}{Z_{\text{L}}} = 50\left(\frac{1}{25 - j50}\right) = 0.4 + j0.8.$$

Upon replacing z_{L} with $1/y_{\text{L}}$ in Eq. (2.124), the reflection coefficient at the load becomes

$$\Gamma = \frac{1 - y_{\text{L}}}{1 + y_{\text{L}}} = \frac{1 - (0.4 + j0.8)}{1 + (0.4 + j0.8)} = 0.62e^{-j82.9^\circ}.$$

Equation (2.119) provides an expression for the input impedance at any location d from the end of the line. If we invert the expression to convert it to admittance, we obtain the following expression for y_{d}, the admittance of the line to the right of MM' in Fig. 2-34(a):

$$y_{\text{d}} = \frac{1 - |\Gamma|e^{j(\theta_{\text{r}} - 2\beta d)}}{1 + |\Gamma|e^{j(\theta_{\text{r}} - 2\beta d)}} = \frac{1 - |\Gamma|e^{j\theta'}}{1 + |\Gamma|e^{j\theta'}}, \tag{2.142}$$

where

$$\theta' = \theta_{\text{r}} - 2\beta d. \tag{2.143}$$

Multiplying the numerator and denominator of this expression by the complex conjugate of the denominator leads to

$$\begin{aligned} y_{\text{d}} &= \left(\frac{1 - |\Gamma|e^{j\theta'}}{1 + |\Gamma|e^{j\theta'}}\right)\left(\frac{1 + |\Gamma|e^{-j\theta'}}{1 + |\Gamma|e^{-j\theta'}}\right) \\ &= \frac{1 - |\Gamma|^2}{1 + |\Gamma|^2 + 2|\Gamma|\cos\theta'} - j\,\frac{2|\Gamma|\sin\theta'}{1 + |\Gamma|^2 + 2|\Gamma|\cos\theta'}. \end{aligned} \tag{2.144}$$

Hence, the real and imaginary components of y_{d} are

$$g_{\text{d}} = \frac{1 - |\Gamma|^2}{1 + |\Gamma|^2 + 2|\Gamma|\cos\theta'}, \tag{2.145a}$$

$$b_{\text{d}} = \frac{-2|\Gamma|\sin\theta'}{1 + |\Gamma|^2 + 2|\Gamma|\cos\theta'}. \tag{2.145b}$$

To satisfy the first condition of Eq. (2.141a), we need to choose d (which is embedded in the definition for θ' given by Eq. (2.143)) such that

$$\frac{1 - |\Gamma|^2}{1 + |\Gamma|^2 + 2|\Gamma|\cos\theta'} = 1,$$

which leads to the solution

$$\cos\theta' = -|\Gamma|. \tag{2.146}$$

Since $\cos\theta'$ is negative, θ' can be in either the second or third quadrant. Given that $|\Gamma| = 0.62$, we obtain

$$\theta'_1 = -128.3^\circ,$$

or

$$\theta'_2 = +128.3^\circ.$$

Each value of θ' offers a possible solution for d. We shall label them d_1 and d_2.

Solution for $\boldsymbol{d_1}$ **[Fig. 2-35(a)]**

With $\theta'_1 = -128.3^\circ = -2.240$ rad, $\theta_{\text{r}} = -82.9^\circ = -1.446$ rad, and $\beta = 2\pi/\lambda$, solving Eq. (2.143) for d gives

$$d_1 = \frac{\lambda}{4\pi}(\theta_{\text{r}} - \theta'_1) = \frac{\lambda}{4\pi}(-1.446 + 2.240) = 0.063\lambda.$$

Next, to satisfy the second condition in Eq. (2.141), we need to determine b_{s_1} such that $b_{s_1} = -b_{\text{d}}$. Using Eq. (2.145b), we obtain

$$\begin{aligned} b_{s_1} &= \frac{2|\Gamma|\sin\theta'}{1 + |\Gamma|^2 + 2|\Gamma|\cos\theta'} \\ &= \frac{2 \times 0.62\sin(-128.3^\circ)}{1 + 0.62^2 + 2 \times 0.62\cos(-128.3^\circ)} = -1.58. \end{aligned}$$

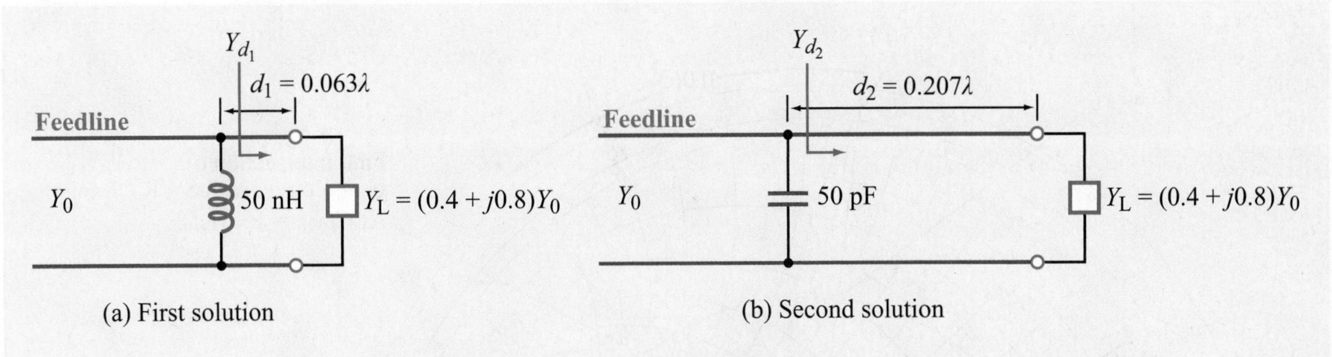

Figure 2-35 Solutions for Example 2-13.

The corresponding impedance of the lumped element is

$$Z_{s_1} = \frac{1}{Y_{s_1}} = \frac{1}{y_{s_1} Y_0} = \frac{Z_0}{jb_{s_1}} = \frac{Z_0}{-j1.58} = \frac{jZ_0}{1.58} = j31.62\ \Omega.$$

Since the value of Z_{s_1} is positive, the element to be inserted should be an inductor and its value should be

$$L = \frac{31.62}{\omega} = \frac{31.62}{2\pi \times 10^8} = 50\ \text{nH}.$$

The results of this solution have been incorporated into the circuit of **Fig. 2-35(a)**.

Solution for d_2 **[Fig. 2-35(b)]**

Repeating the procedure for $\theta_2' = 128.3°$ leads to

$$d_2 = \frac{\lambda}{4\pi}(-1.447 - 2.239) = -0.293\lambda.$$

A negative value for d_2 is physically meaningless because that would place it to the right of the load, but since we know that impedances repeat themselves every $\lambda/2$, we simply need to add $\lambda/2$ to the solution:

$$d_2 \text{ (physically realizable)} = -0.293\lambda + 0.5\lambda = 0.207\lambda.$$

The associated value for b_s is $+1.58$. Hence

$$Z_{s_2} = -j31.62\ \Omega,$$

which is the impedance of a capacitor with

$$C = \frac{1}{31.62\omega} = 50\ \text{pF}.$$

Figure 2-35(b) displays the circuit solution for d_2 and C.

(b) Smith-chart solution:

The normalized load impedance is

$$z_L = \frac{Z_L}{Z_0} = \frac{25 - j50}{50} = 0.5 - j1,$$

which is represented by point A on the Smith chart of **Fig. 2-36**. Next, we draw the constant S circle through point A. As alluded to earlier, to perform the matching task, it is easier to work with admittances than with impedances. The normalized load admittance y_L is represented by point B, obtained by rotating point A over 0.25λ, or equivalently by drawing a line from point A through the chart center to the image of point A on the S circle. The value of y_L at B is

$$y_L = 0.4 + j0.8,$$

and it is located at position 0.115λ on the WTG scale. In the admittance domain, the r_L circles become g_L circles, and the x_L circles become b_L circles. To achieve matching, we need to move from the load toward the generator a distance d such that the normalized input admittance y_d of the line terminated in the load (**Fig. 2-34**) has a real part of 1. This condition is satisfied by either of the two ***matching points*** C and D on the Smith charts of **Figs. 2-36** and **2-37**, respectively, corresponding to intersections of the S circle with the $g_L = 1$ circle. Points C and D represent two possible solutions for the distance d in **Fig. 2-34(a)**.

Solution for Point C **(Fig. 2-36):** At C,

$$y_d = 1 + j1.58,$$

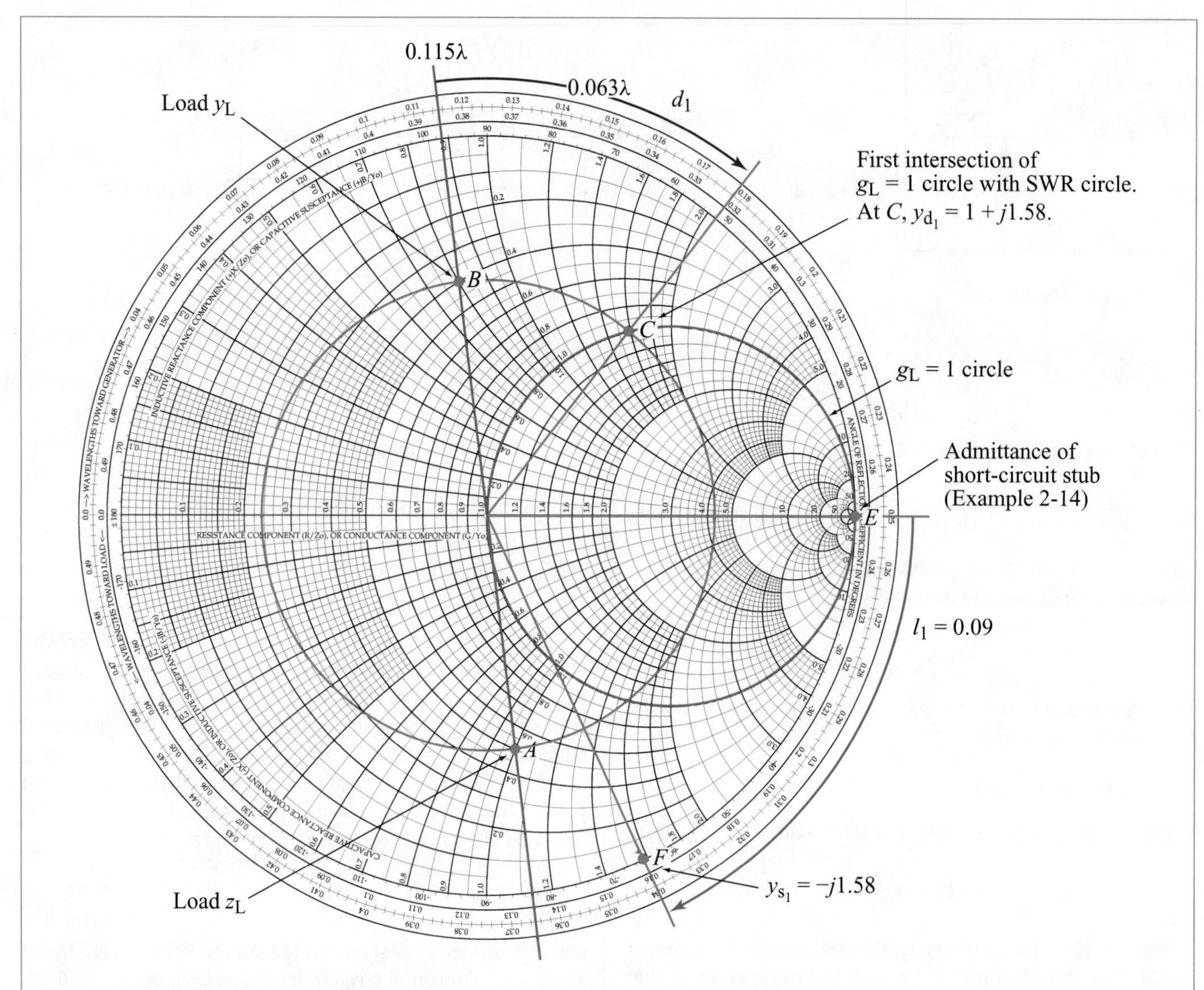

Figure 2-36 Solution for point C of Examples 2-13 and 2-14. Point A is the normalized load with $z_L = 0.5 - j1$; point B is $y_L = 0.4 + j0.8$. Point C is the intersection of the SWR circle with the $g_L = 1$ circle. The distance from B to C is $d_1 = 0.063\lambda$. The length of the shorted stub (E to F) is $l_1 = 0.09\lambda$ (Example 2-14).

which is located at 0.178λ on the WTG scale. The distance between points B and C is

$$d_1 = (0.178 - 0.115)\lambda = 0.063\lambda.$$

Looking from the generator toward the parallel combination of the line connected to the load and the shunt element, the normalized input admittance at terminals MM' is

$$y_{in} = y_s + y_d,$$

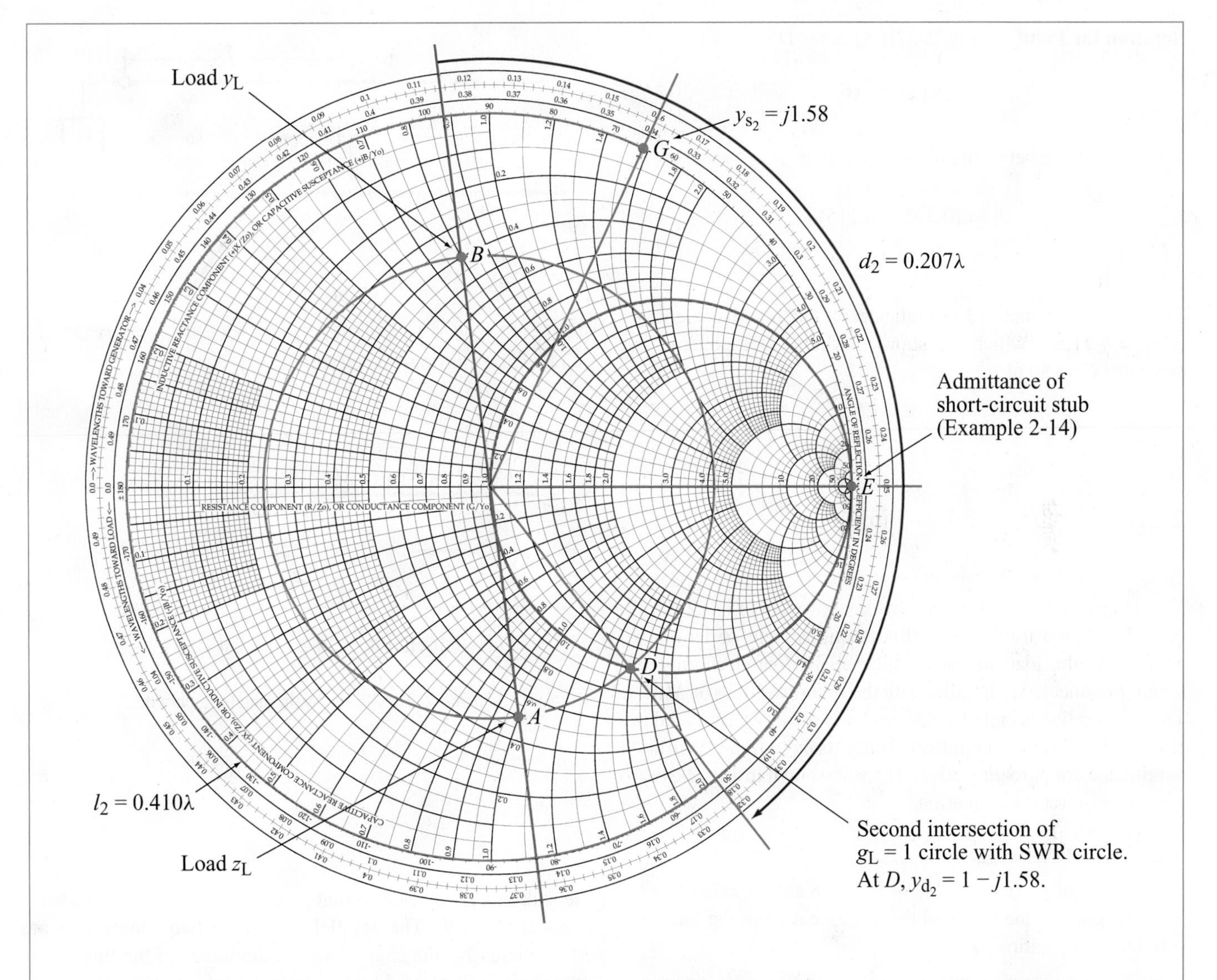

Figure 2-37 Solution for point D of Examples 2-13 and 2-14. Point D is the second point of intersection of the SWR circle and the $g_L = 1$ circle. The distance B to D gives $d_2 = 0.207\lambda$, and the distance E to G gives $l_2 = 0.410\lambda$ (Example 2-14).

where y_s is the normalized input admittance of the shunt element. To match the feed line to the parallel combination, we need $y_{in} = 1 + j0$. Thus,

$$1 + j0 = y_s + 1 + j1.58,$$

or

$$y_s = -j1.58.$$

This is the same result obtained earlier in the analytical solution, which led to choosing an inductor $L = 50$ nH.

Solution for Point D **(Fig. 2-37):** At point D,

$$y_d = 1 - j1.58,$$

and the distance between points B and D is

$$d_2 = (0.322 - 0.115)\lambda$$
$$= 0.207\lambda.$$

The needed normalized admittance of the reactive element is $y_s = +j1.58$, which, as shown earlier, corresponds to a capacitor $C = 50$ pF.

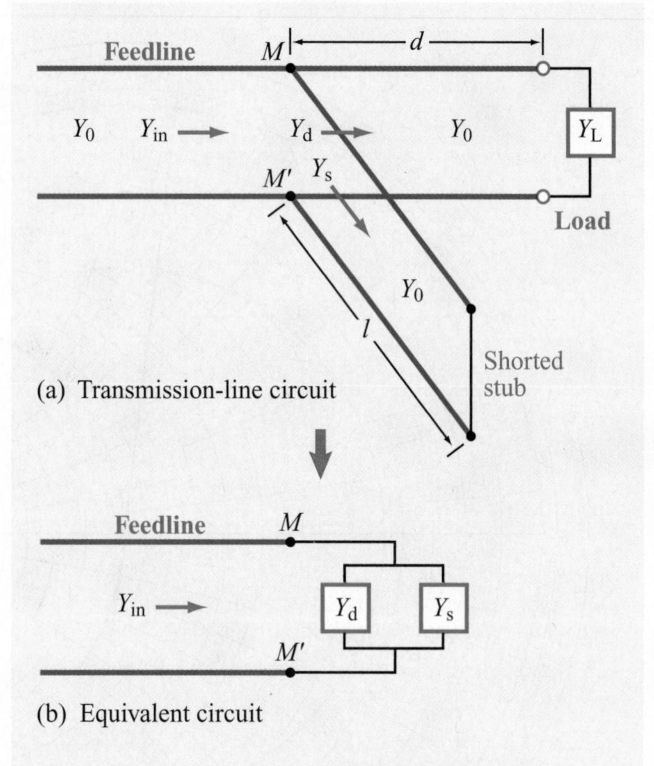

Figure 2-38 Shorted-stub matching network.

2-11.2 Single-Stub Matching

The single-stub matching network shown in Fig. 2-38(a) consists of two transmission line sections, one of length d connecting the load to the feedline at MM' and another of length l connected in parallel with the other two lines at MM'. This second line is called a ***stub***, and it is usually terminated in either a short or open circuit, and hence its input impedance and admittance are purely reactive. The stub shown in Fig. 2-38(a) has a short-circuit termination.

▶ The required two degrees of freedom are provided by the length l of the stub and the distance d from the load to the stub position. ◀

Because at MM' the stub is added in parallel to the line (which is why it is called a ***shunt*** stub), it is easier to work with admittances than with impedances. The matching procedure consists of two steps. In the first step, the distance d is selected so as to transform the load admittance $Y_L = 1/Z_L$ into an admittance of the form $Y_d = Y_0 + jB$, when looking toward the load at MM'. Then, in the second step, the length l of the stub line is selected so that its input admittance Y_s at MM' is equal to $-jB$. The parallel sum of the two admittances at MM' yields Y_0, the characteristic admittance of the line. The procedure is illustrated by Example 2-14.

Example 2-14: Single-Stub Matching

Repeat Example 2-13, but use a shorted stub (instead of a lumped element) to match the load impedance $Z_L = (25 - j50)\ \Omega$ to the $50\ \Omega$ transmission line.

Module 2.7 Quarter-Wavelength Transformer This module allows you to go through a multi-step procedure to design a quarter-wavelength transmission line that, when inserted at the appropriate location on the original line, presents a matched load to the feedline.

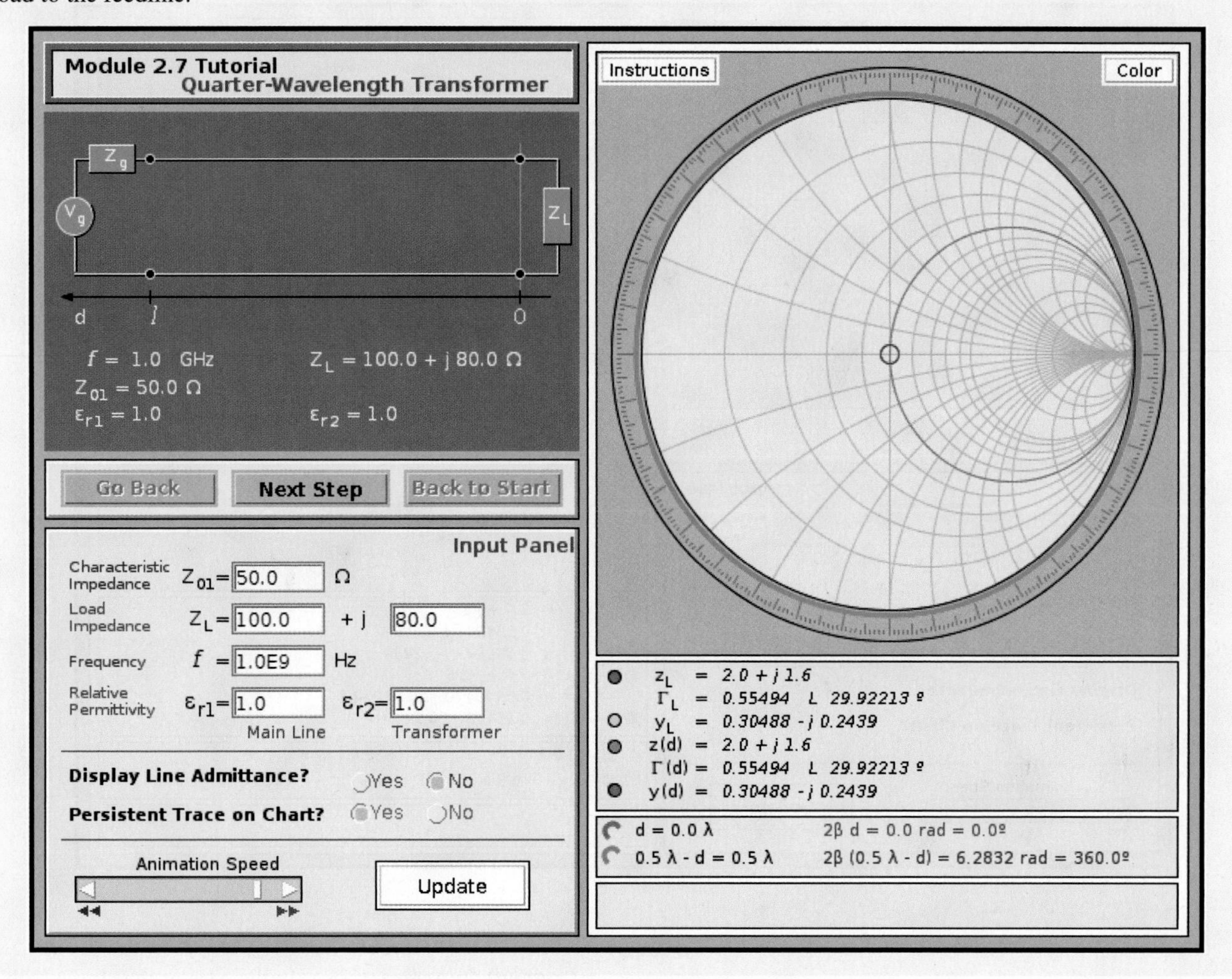

Solution: In Example 2-13, we demonstrated that the load can be matched to the line via either of two solutions:

(1) $d_1 = 0.063\lambda$, and $y_{s_1} = jb_{s_1} = -j1.58$,

(2) $d_2 = 0.207\lambda$, and $y_{s_2} = jb_{s_2} = j1.58$.

The locations of the insertion points, at distances d_1 and d_2, remain the same, but now our task is to select corresponding lengths l_1 and l_2 of shorted stubs that present the required admittances at their inputs.

To determine l_1, we use the Smith chart in Fig. 2-36. The normalized admittance of a short circuit is $-j\infty$, which is represented by point E on the Smith chart, with position 0.25λ on the WTG scale. A normalized input admittance of $-j1.58$ is located at point F, with position 0.34λ on the WTG scale.

Module 2.8 Discrete Element Matching For each of two possible solutions, the module guides the user through a procedure to match the feedline to the load by inserting a capacitor or an inductor at an appropriate location along the line.

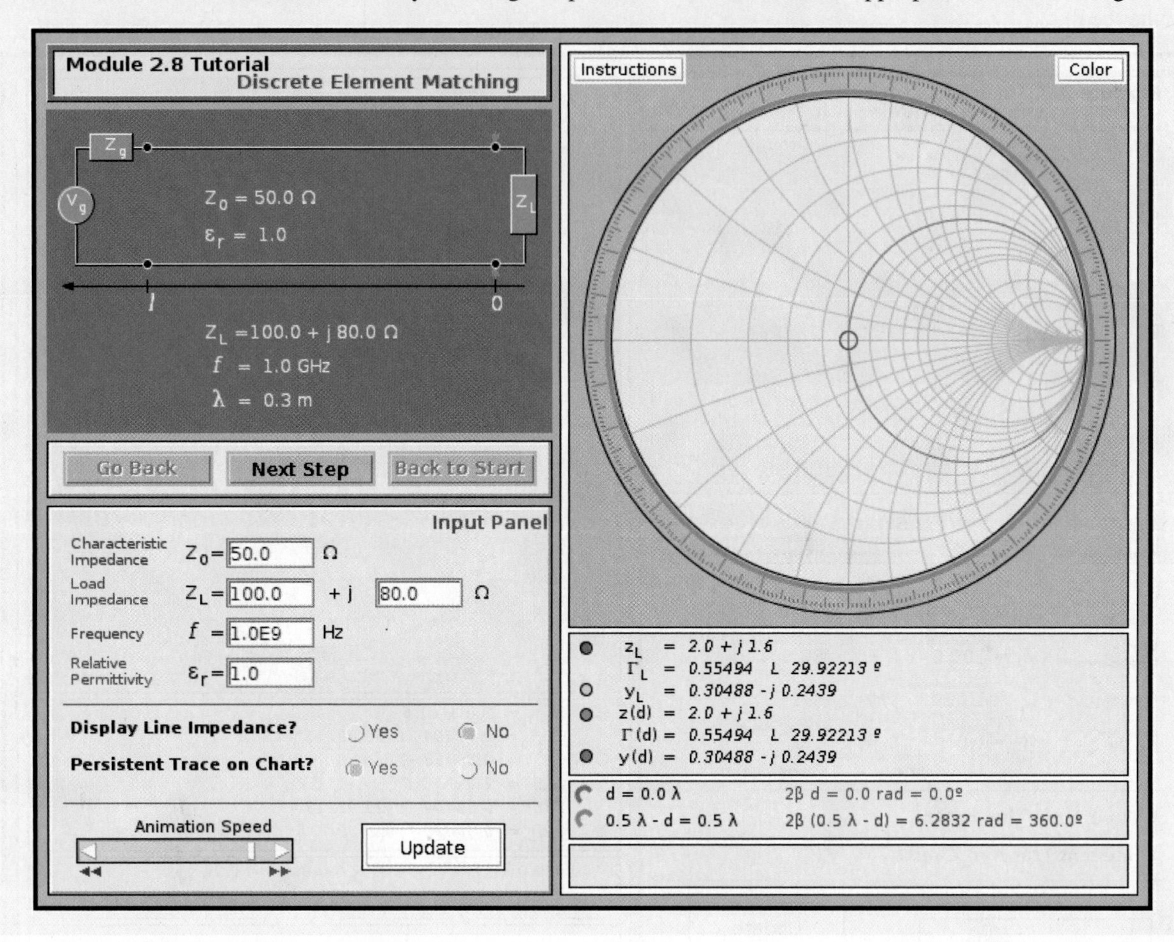

Hence,

$$l_1 = (0.34 - 0.25)\lambda = 0.09\lambda.$$

Similarly, $y_{s_2} = j1.58$ is represented by point G with position 0.16λ on the WTG scale of the Smith chart in Fig. 2-37. Rotating from point E to point G involves a rotation of 0.25λ plus an additional rotation of 0.16λ or

$$l_2 = (0.25 + 0.16)\lambda = 0.41\lambda.$$

Concept Question 2-24: To match an arbitrary load impedance to a lossless transmission line through a matching network, what is the required minimum number of degrees of freedom that the network should provide?

Concept Question 2-25: In the case of the single-stub matching network, what are the two degrees of freedom?

Concept Question 2-26: When a transmission line is matched to a load through a single-stub matching network, no waves are reflected toward the generator. What happens to the waves reflected by the load and by the shorted stub when they arrive at terminals MM' in Fig. 2-38?

Module 2.9 Single-Stub Tuning Instead of inserting a lumped element to match the feedline to the load, this module determines the length of a shorted stub that can accomplish the same goal.

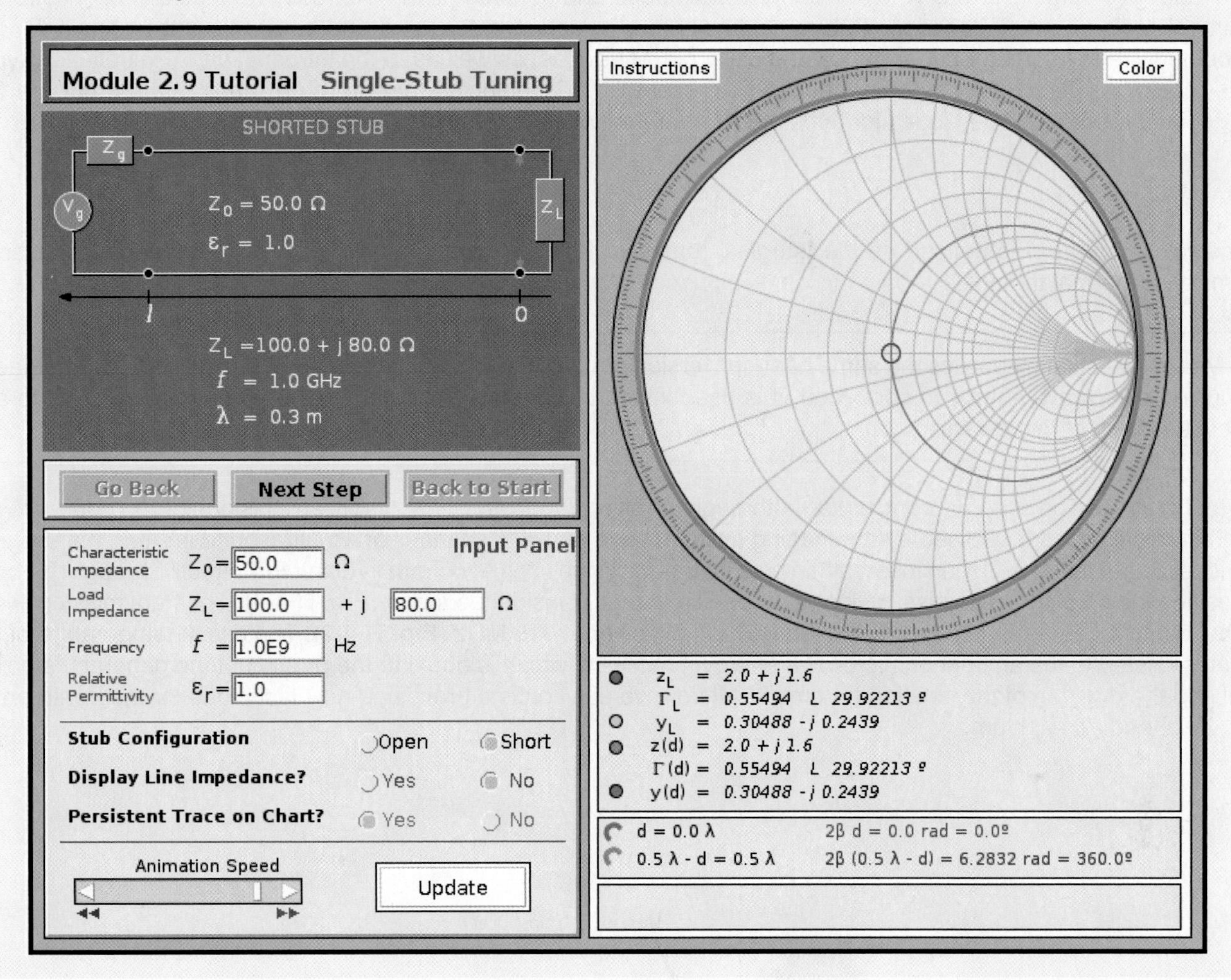

2-12 Transients on Transmission Lines

Thus far, our treatment of wave propagation on transmission lines has focused on the analysis of single-frequency, time-harmonic signals under steady-state conditions. The impedance-matching and Smith chart techniques we developed, while useful for a wide range of applications, are inappropriate for dealing with digital or wideband signals that exist in digital chips, circuits, and computer networks. For such signals, we need to examine the transient transmission-line response instead.

► The ***transient response*** of a voltage pulse on a transmission line is a time record of its back and forth travel between the sending and receiving ends of the line, taking into account all the multiple reflections (echoes) at both ends. ◄

Let us start by considering the case of a single rectangular pulse of amplitude V_0 and duration τ, as shown in **Fig. 2-39(a)** (page 137). The amplitude of the pulse is zero prior to $t = 0$, V_0 over the interval $0 \leq t \leq \tau$, and zero afterwards. The pulse

Technology Brief 4: EM Cancer Zappers

From laser eye surgery to 3-D X-ray imaging, EM sources and sensors have been used as medical diagnostic and treatment tools for many decades. Future advances in information processing and other relevant technologies will undoubtedly lead to greater performance and utility of EM devices, as well as to the introduction of entirely new types of devices. This Technology Brief introduces two recent EM technologies that are still in their infancy, but are fast developing into serious techniques for the surgical treatment of cancer tumors.

Microwave Ablation

In medicine, ***ablation*** is defined as the "surgical removal of body tissue," usually through the direct application of chemical or thermal therapies.

> ► Microwave ablation applies the same heat-conversion process used in a microwave oven (see TB3), but instead of using microwave energy to cook food, it is used instead to destroy cancerous tumors by exposing them to a focused beam of microwaves. ◄

The technique can be used ***percutaneously*** (through the skin), ***laparoscopically*** (via an incision), or ***intraoperatively*** (open surgical access). Guided by an imaging system, such as a CT scanner or an ultrasound imager, the surgeon can localize the tumor and then insert a thin coaxial transmission line (~ 1.5 mm in diameter) directly through the body to position the tip of the transmission line (a probe-like antenna) inside the tumor (**Fig. TF4-1**). The transmission line is connected to a generator capable of delivering 60 W of power at 915 MHz (**Fig. TF4-2**). The rise in temperature of the tumor is related to the amount of microwave energy it receives, which is equal to the product of the generator's power level and the duration of the ablation treatment. Microwave ablation is a promising new technique for the treatment of liver, lung, and adrenal tumors.

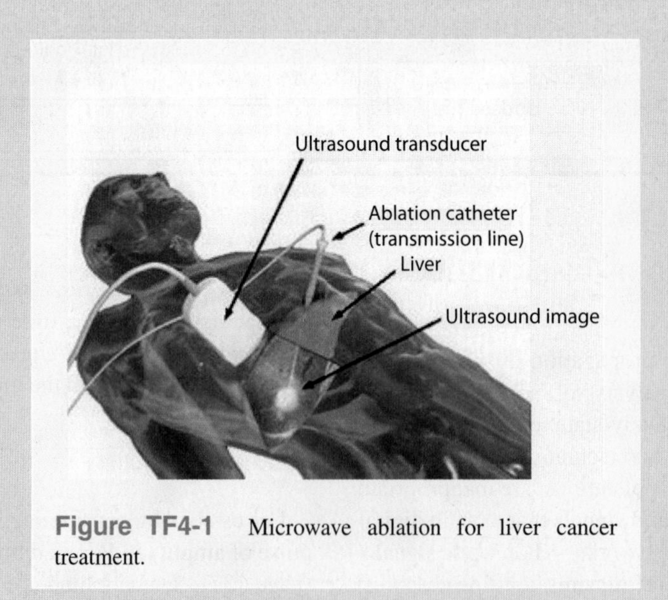

Figure TF4-1 Microwave ablation for liver cancer treatment.

Figure TF4-2 Photograph of the setup for a percutaneous microwave ablation procedure in which three single microwave applicators are connected to three microwave generators.

High-Power Nanosecond Pulses

Bioelectrics is an emerging field focused on the study of how electric fields behave in biological systems. Of particular recent interest is the desire to understand how living cells might respond to the application of extremely short pulses (on the order of nanoseconds (10^{-9} s), and even as short as picoseconds (10^{-12} s)) with exceptionally high voltage and current amplitudes.

> ▶ The motivation is to treat cancerous cells by ***zapping*** them with high-power pulses. The pulse power is delivered to the cell via a transmission line, as illustrated by the example in **Fig. TF4-3**. ◀

Note that the pulse is about 200 ns long, and its voltage and current amplitudes are approximately 3,000 V and 60 A, respectively. Thus, the peak power level is about 180,000 W! However, the total energy carried by the pulse is only $(1.8 \times 10^5) \times (2 \times 10^{-7}) = 0.0036$ Joules. Despite the low energy content, the very high voltage appears to be very effective in destroying malignant tumors (in mice, so far), with no regrowth.

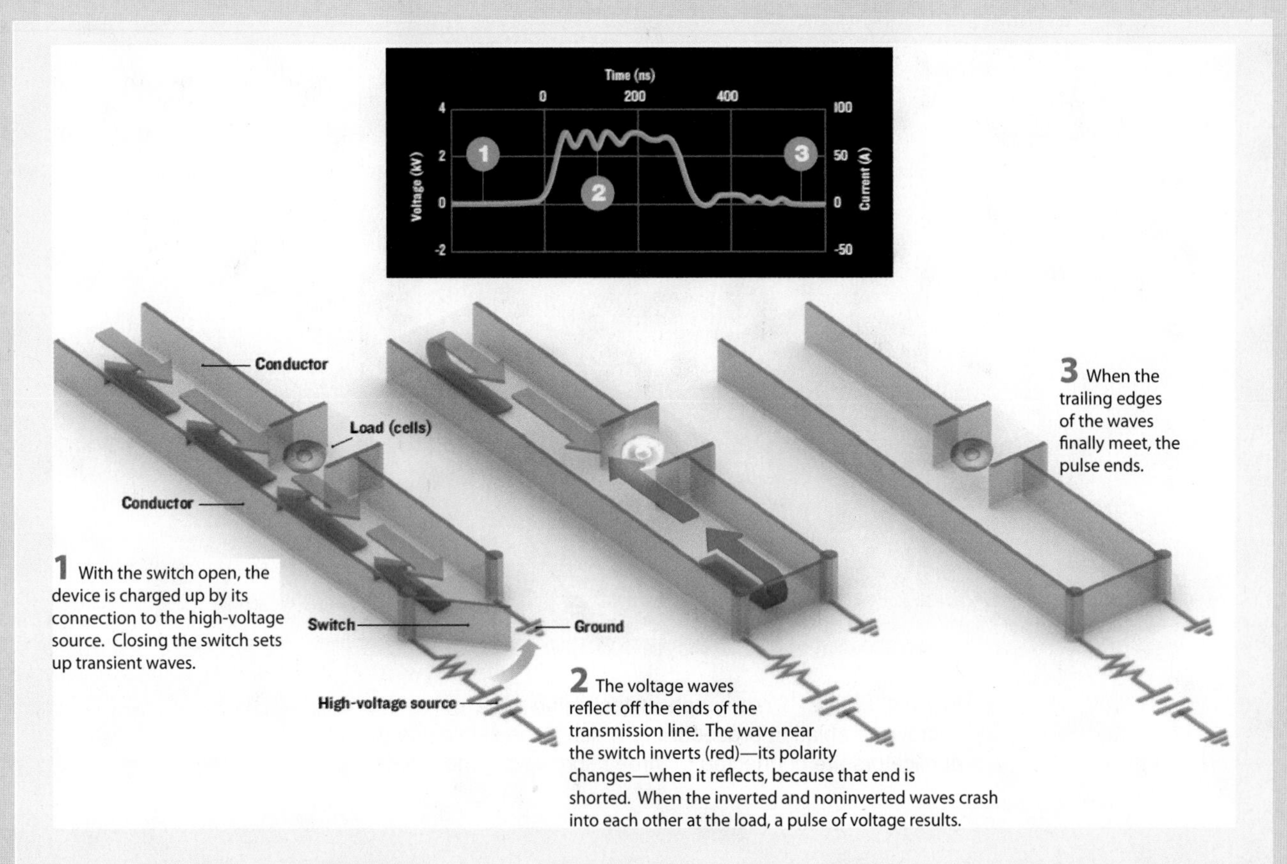

Figure TF4-3 High-voltage nanosecond pulse delivered to tumor cells via a transmission line. The cells to be shocked by the pulse sit in a break in one of the transmission-line conductors.

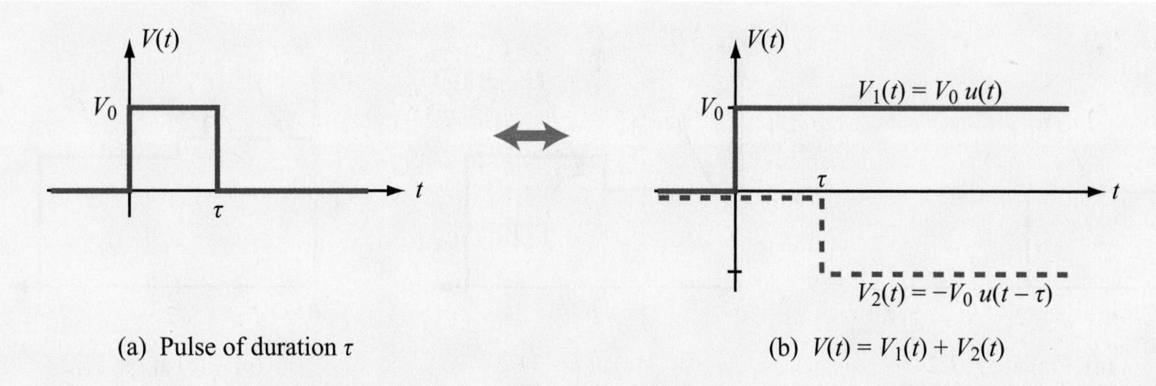

Figure 2-39 A rectangular pulse $V(t)$ of duration τ can be represented as the sum of two step functions of opposite polarities displaced by τ relative to each other.

can be described mathematically as the sum of two unit step functions:

$$V(t) = V_1(t) + V_2(t) = V_0\,u(t) - V_0\,u(t-\tau), \qquad (2.147)$$

where the unit step function $u(x)$ is

$$u(x) = \begin{cases} 1 & \text{for } x > 0, \\ 0 & \text{for } x < 0. \end{cases} \qquad (2.148)$$

The first component, $V_1(t) = V_0\,u(t)$, represents a dc voltage of amplitude V_0 that is switched on at $t = 0$ and retains that value indefinitely, and the second component, $V_2(t) = -V_0\,u(t-\tau)$, represents a dc voltage of amplitude $-V_0$ that is switched on at $t = \tau$ and remains that way indefinitely. As can be seen from **Fig. 2-39(b)**, the sum $V_1(t) + V_2(t)$ is equal to V_0 for $0 < t < \tau$ and equal to zero for $t < 0$ and $t > \tau$. This representation of a pulse in terms of two step functions allows us to analyze the transient behavior of the pulse on a transmission line as the superposition of two dc signals. Hence, if we can develop basic tools for describing the transient behavior of a single step function, we can apply the same tools for each of the two components of the pulse and then add the results to obtain the response to $V(t)$.

2-12.1 Transient Response to a Step Function

The circuit shown in **Fig. 2-40(a)** (page 137) consists of a generator, composed of a dc voltage source V_g and a series resistance R_g, connected to a lossless transmission line of length l and characteristic impedance Z_0. The line is terminated in a purely resistive load R_L at $z = l$.

▶ Note that whereas in previous sections, $z = 0$ was defined as the location of the load, now it is more convenient to define it as the location of the source. ◀

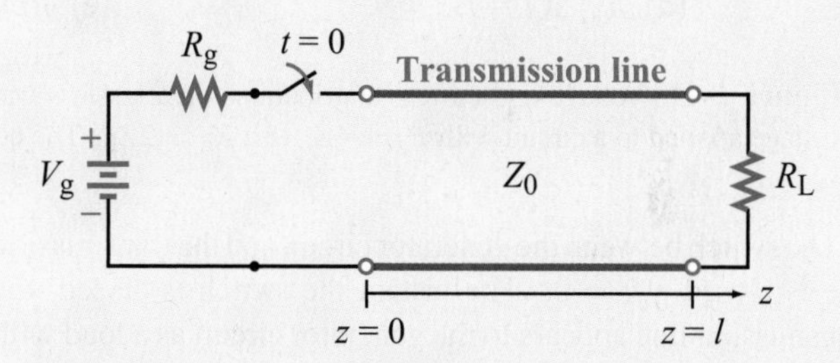

(a) Transmission-line circuit

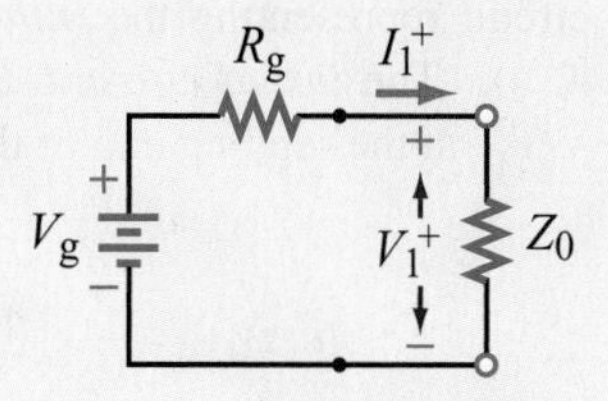

(b) Equivalent circuit at $t = 0^+$

Figure 2-40 At $t = 0^+$, immediately after closing the switch in the circuit in (a), the circuit can be represented by the equivalent circuit in (b).

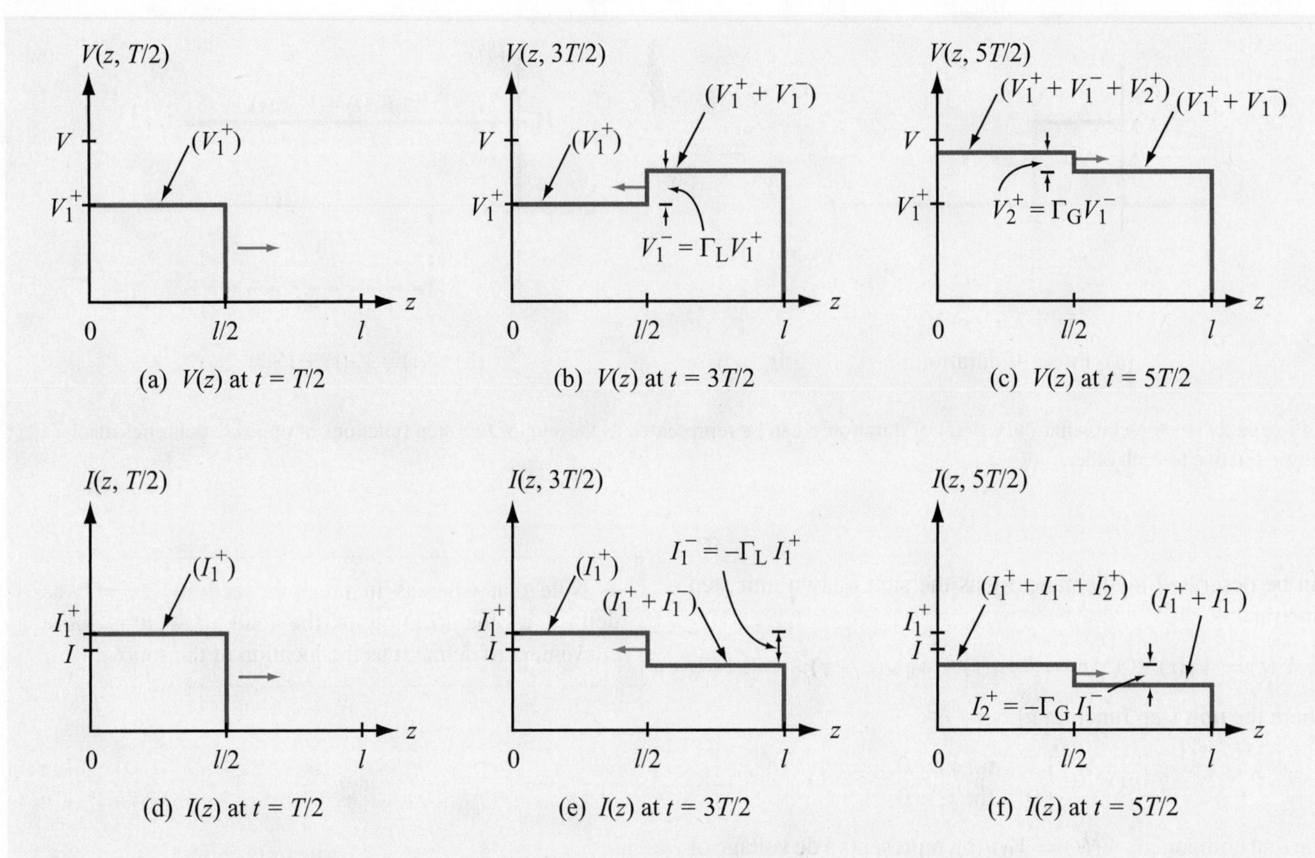

Figure 2-41 Voltage and current distributions on a lossless transmission line at $t = T/2$, $t = 3T/2$, and $t = 5T/2$, due to a unit step voltage applied to a circuit with $R_g = 4Z_0$ and $R_L = 2Z_0$. The corresponding reflection coefficients are $\Gamma_L = 1/3$ and $\Gamma_g = 3/5$.

The switch between the generator circuit and the transmission line is closed at $t = 0$. The instant the switch is closed, the transmission line appears to the generator circuit as a load with impedance Z_0. This is because, in the absence of a signal on the line, the input impedance of the line is unaffected by the load impedance R_L. The circuit representing the ***initial condition*** is shown in **Fig. 2-40(b)**. The ***initial current*** I_1^+ and corresponding ***initial voltage*** V_1^+ at the sending end of the transmission line are given by

$$I_1^+ = \frac{V_g}{R_g + Z_0}, \tag{2.149a}$$

$$V_1^+ = I_1^+ Z_0 = \frac{V_g Z_0}{R_g + Z_0}. \tag{2.149b}$$

The combination of V_1^+ and I_1^+ constitutes a wave that travels along the line with velocity $u_p = 1/\sqrt{\mu\epsilon}$, immediately after the switch is closed. The plus-sign superscript denotes the fact that the wave is traveling in the $+z$ direction. The transient response of the wave is shown in **Fig. 2-41** at each of three instances in time for a circuit with $R_g = 4Z_0$ and $R_L = 2Z_0$. The first response is at time $t_1 = T/2$, where $T = l/u_p$ is the time it takes the wave to travel the full length of the line. By time t_1, the wave has traveled halfway down the line; consequently, the voltage on the first half of the line is equal to V_1^+, while the voltage on the second half is still zero [**Fig. 2-41(a)**]. At $t = T$, the wave reaches the load at $z = l$, and because $R_L \neq Z_0$, the mismatch generates a reflected wave with amplitude

$$V_1^- = \Gamma_L V_1^+, \tag{2.150}$$

where

$$\Gamma_L = \frac{R_L - Z_0}{R_L + Z_0} \tag{2.151}$$

is the reflection coefficient of the load. For the specific case illustrated in Fig. 2-41, $R_L = 2Z_0$, which leads to $\Gamma_L = 1/3$. After this first reflection, the voltage on the line consists of the sum of two waves: the initial wave V_1^+ and the reflected wave V_1^-. The voltage on the transmission line at $t_2 = 3T/2$ is shown in Fig. 2-41(b); $V(z, 3T/2)$ equals V_1^+ on the first half of the line $(0 \le z < l/2)$, and $(V_1^+ + V_1^-)$ on the second half $(l/2 \le z \le l)$.

At $t = 2T$, the reflected wave V_1^- arrives at the sending end of the line. If $R_g \neq Z_0$, the mismatch at the sending end generates a reflection at $z = 0$ in the form of a wave with voltage amplitude V_2^+ given by

$$V_2^+ = \Gamma_g V_1^- = \Gamma_g \Gamma_L V_1^+, \tag{2.152}$$

where

$$\Gamma_g = \frac{R_g - Z_0}{R_g + Z_0} \tag{2.153}$$

is the reflection coefficient of the generator resistance R_g. For $R_g = 4Z_0$, we have $\Gamma_g = 0.6$. As time progresses after $t = 2T$, the wave V_2^+ travels down the line toward the load and adds to the previously established voltage on the line. Hence, at $t = 5T/2$, the total voltage on the first half of the line is

$$V(z, 5T/2) = V_1^+ + V_1^- + V_2^+ = (1 + \Gamma_L + \Gamma_L \Gamma_g) V_1^+ \quad (0 \le z < l/2), \tag{2.154}$$

while on the second half of the line the voltage is only

$$V(z, 5T/2) = V_1^+ + V_1^- = (1 + \Gamma_L) V_1^+ \quad (l/2 \le z \le l). \tag{2.155}$$

The voltage distribution is shown in Fig. 2-41(c).

So far, we have examined the transient response of the voltage wave $V(z, t)$. The associated transient response of the current $I(z, t)$ is shown in Figs. 2-41(d)–(f). The current behaves similarly to the voltage $V(z, t)$, except for one important difference. Whereas at either end of the line the reflected voltage is related to the incident voltage by the reflection coefficient at that end, the reflected current is related to the incident current by the negative of the reflection coefficient. This property of wave reflection is expressed by Eq. (2.61). Accordingly,

$$I_1^- = -\Gamma_L I_1^+, \tag{2.156a}$$

$$I_2^+ = -\Gamma_g I_1^- = \Gamma_g \Gamma_L I_1^+, \tag{2.156b}$$

and so on.

▶ The multiple-reflection process continues indefinitely, and the ultimate value that $V(z, t)$ reaches as t approaches $+\infty$ is the same at all locations on the transmission line. ◀

It is given by

$$\begin{aligned} V_\infty &= V_1^+ + V_1^- + V_2^+ + V_2^- + V_3^+ + V_3^- + \cdots \\ &= V_1^+[1 + \Gamma_L + \Gamma_L\Gamma_g + \Gamma_L^2\Gamma_g + \Gamma_L^2\Gamma_g^2 + \Gamma_L^3\Gamma_g^2 + \cdots] \\ &= V_1^+[(1 + \Gamma_L)(1 + \Gamma_L\Gamma_g + \Gamma_L^2\Gamma_g^2 + \cdots)] \\ &= V_1^+(1 + \Gamma_L)[1 + x + x^2 + \cdots], \end{aligned} \tag{2.157}$$

where $x = \Gamma_L \Gamma_g$. The series inside the square bracket is the geometric series of the function

$$\frac{1}{1 - x} = 1 + x + x^2 + \cdots \quad \text{for } |x| < 1. \tag{2.158}$$

Hence, Eq. (2.157) can be rewritten in the compact form

$$V_\infty = V_1^+ \frac{1 + \Gamma_L}{1 - \Gamma_L \Gamma_g}. \tag{2.159}$$

Upon replacing V_1^+, Γ_L, and Γ_g with Eqs. (2.149b), (2.151), and (2.153), and simplifying the resulting expression, we obtain

$$V_\infty = \frac{V_g R_L}{R_g + R_L}. \tag{2.160}$$

The voltage V_∞ is called the ***steady-state voltage*** on the line, and its expression is exactly what we should expect on the basis of dc analysis of the circuit in Fig. 2-40(a), wherein we treat the transmission line as simply a connecting wire between the generator circuit and the load. The corresponding ***steady-state current*** is

$$I_\infty = \frac{V_\infty}{R_L} = \frac{V_g}{R_g + R_L}. \tag{2.161}$$

2-12.2 Bounce Diagrams

Keeping track of the voltage and current waves as they bounce back and forth on the line is a rather tedious process. The ***bounce diagram*** is a graphical presentation that allows us

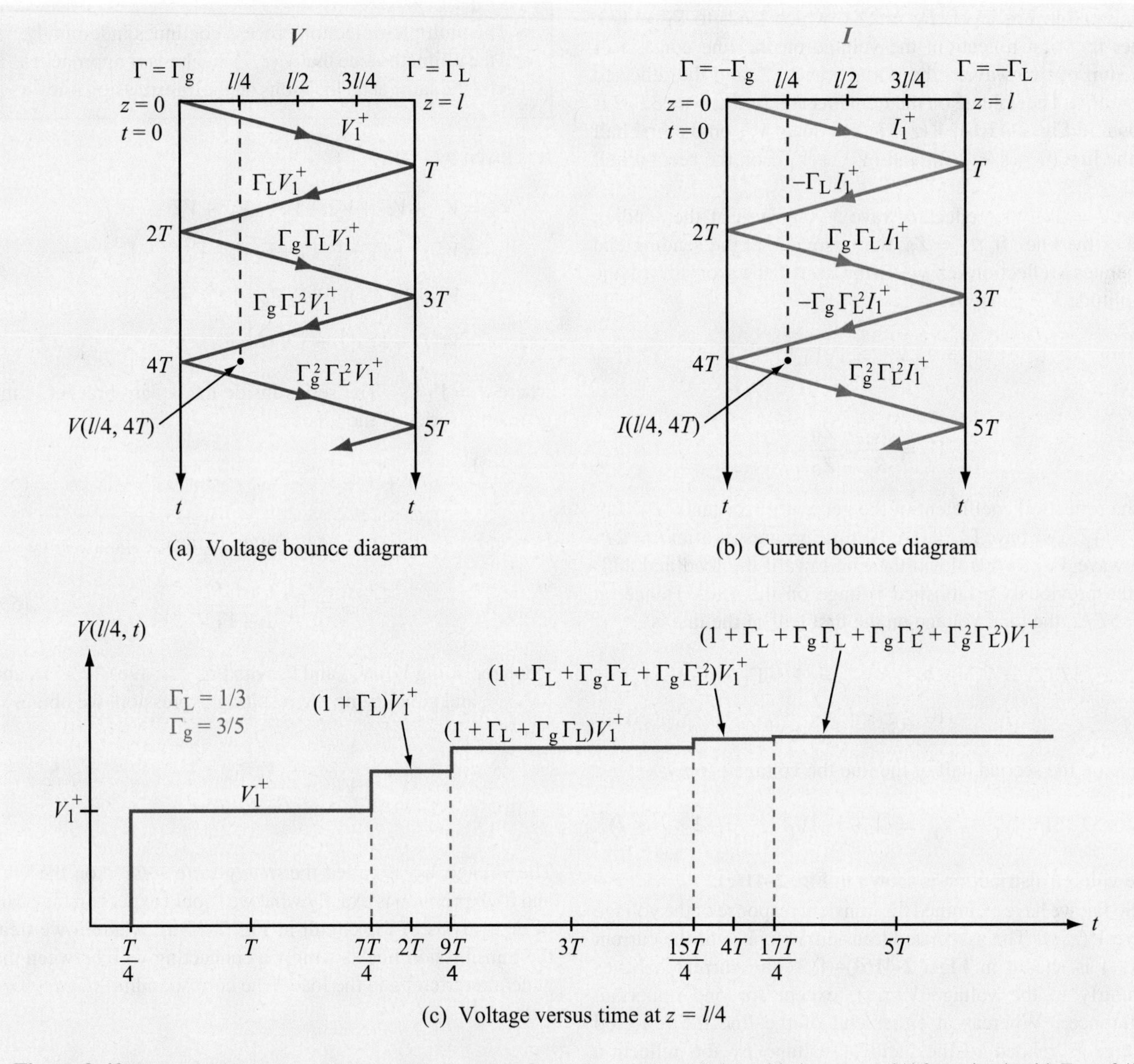

Figure 2-42 Bounce diagrams for (a) voltage and (b) current. In (c), the voltage variation with time at $z = l/4$ for a circuit with $\Gamma_g = 3/5$ and $\Gamma_L = 1/3$ is deduced from the vertical dashed line at $l/4$ in (a).

to accomplish the same goal, but with relative ease. The horizontal axes in **Figs. 2-42(a)** and **(b)** represent position along the transmission line, while the vertical axes denote time. **Figures 2-42(a)** and **(b)** pertain to $V(z, t)$ and $I(z, t)$, respectively. The bounce diagram in **Fig. 2-42(a)** consists of a zigzag line indicating the progress of the voltage wave on the line. The incident wave V_1^+ starts at $z = t = 0$ and travels in the $+z$ direction until it reaches the load at $z = l$ at time $t = T$. At the very top of the bounce diagram, the reflection coefficients are indicated by $\Gamma = \Gamma_g$ at the generator end and by $\Gamma = \Gamma_L$ at the load end. At the end of the first straight-line segment of the zigzag line, a second line is drawn to represent the reflected voltage wave $V_1^- = \Gamma_L V_1^+$. The amplitude of each new straight-line segment equals the product of the amplitude of the preceding straight-line segment and the reflection coefficient at that end of the line. The bounce diagram for the current $I(z,t)$ in **Fig. 2-42(b)** adheres to the same principle except for the reversal of the signs of Γ_L and Γ_g at the top of the bounce diagram.

Using the bounce diagram, the total voltage (or current) at any point z_1 and time t_1 can be determined by drawing a vertical line through point z_1, then adding the voltages (or currents) of all the zigzag segments intersected by that line between $t = 0$ and $t = t_1$. To find the voltage at $z = l/4$ and $T = 4T$, for example, we draw a dashed vertical line in **Fig. 2-42(a)** through $z = l/4$ and we extend it from $t = 0$ to $t = 4T$. The dashed line intersects four line segments. The total voltage at $z = l/4$ and $t = 4T$ therefore is

$$
\begin{aligned}
V(l/4, 4T) &= V_1^+ + \Gamma_L V_1^+ + \Gamma_g \Gamma_L V_1^+ + \Gamma_g \Gamma_L^2 V_1^+ \\
&= V_1^+ (1 + \Gamma_L + \Gamma_g \Gamma_L + \Gamma_g \Gamma_L^2).
\end{aligned}
$$

The time variation of $V(z,t)$ at a specific location z can be obtained by plotting the values of $V(z,t)$ along the (dashed) vertical line passing through z. **Figure 2-42(c)** shows the variation of V as a function of time at $z = l/4$ for a circuit with $\Gamma_g = 3/5$ and $\Gamma_L = 1/3$.

Example 2-15: Pulse Propagation

The transmission-line circuit of **Fig. 2-43(a)** is excited by a rectangular pulse of duration $\tau = 1$ ns that starts at $t = 0$. Establish the waveform of the voltage response at the load, given that the pulse amplitude is 5 V, the phase velocity is c, and the length of the line is 0.6 m.

Solution: The one-way propagation time is

$$
T = \frac{l}{c} = \frac{0.6}{3 \times 10^8} = 2 \text{ ns}.
$$

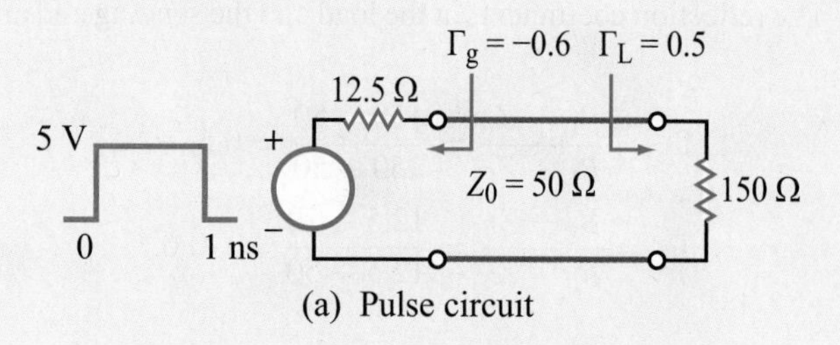

(a) Pulse circuit

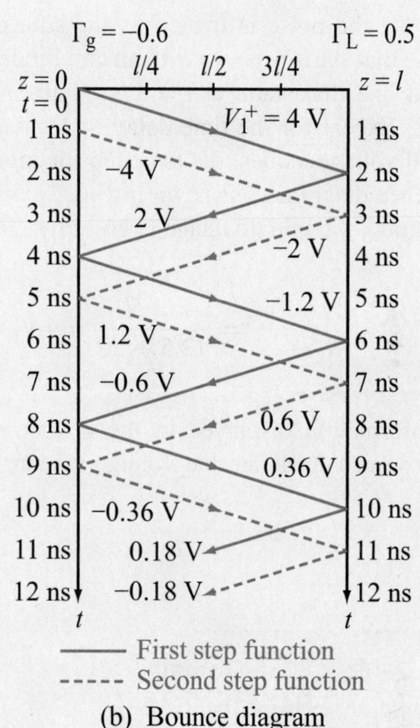

(b) Bounce diagram

V_L (V): 6 V, 4 V, 2 V, −2 V, −4 V; t (ns): 1 2 3 4 5 6 7 8 9 10 11 12; 0.54 V; −1.8 V

(c) Voltage waveform at the load

Figure 2-43 Example 2-15.

The reflection coefficients at the load and the sending end are

$$\Gamma_{\rm L} = \frac{R_{\rm L} - Z_0}{R_{\rm L} + Z_0} = \frac{150 - 50}{150 + 50} = 0.5,$$

$$\Gamma_{\rm g} = \frac{R_{\rm g} - Z_0}{R_{\rm g} + Z_0} = \frac{12.5 - 50}{12.5 + 50} = -0.6.$$

By Eq. (2.147), the pulse is treated as the sum of two step functions, one that starts at $t = 0$ with an amplitude $V_{10} = 5$ V and a second one that starts at $t = 1$ ns with an amplitude $V_{20} = -5$ V. Except for the time delay of 1 ns and the sign reversal of all voltage values, the two step functions generate identical bounce diagrams, as shown in **Fig. 2-43(b)**. For the first step function, the initial voltage is given by

$$V_1^+ = \frac{V_{01} Z_0}{R_{\rm g} + Z_0} = \frac{5 \times 50}{12.5 + 50} = 4 \text{ V}.$$

Using the information displayed in the bounce diagram, it is straightforward to generate the voltage response shown in **Fig. 2-43(c)**.

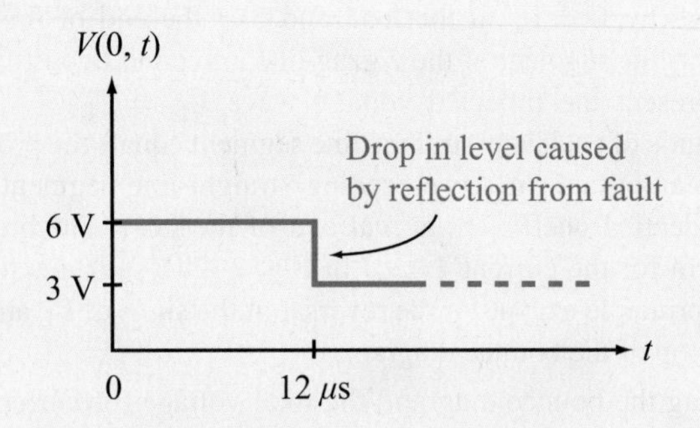

(a) Observed voltage at the sending end

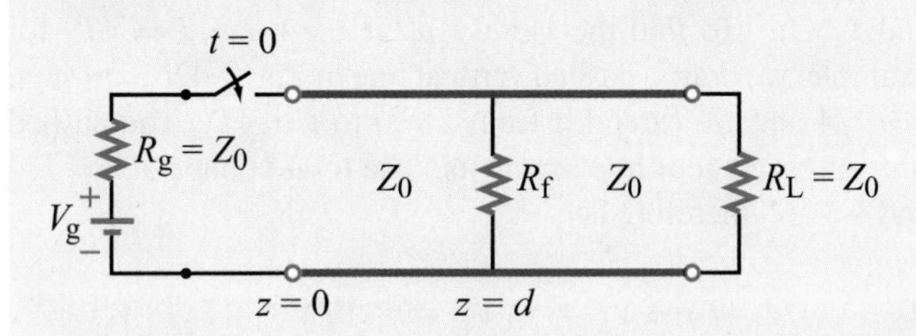

(b) The fault at $z = d$ is represented by a fault resistance $R_{\rm f}$

Figure 2-44 Time-domain reflectometer of Example 2-16.

Example 2-16: Time-Domain Reflectometer

A time-domain reflectometer (TDR) is an instrument used to locate faults on a transmission line. Consider, for example, a long underground or undersea cable that gets damaged at some distance d from the sending end of the line. The damage may alter the electrical properties or the shape of the cable, causing it to exhibit at the fault location an impedance $R_{\rm Lf}$. A TDR sends a step voltage down the line, and by observing the voltage at the sending end as a function of time, it is possible to determine the location of the fault and its severity.

If the voltage waveform shown in **Fig. 2-44(a)** is seen on an oscilloscope connected to the input of a 75 Ω matched transmission line, determine (a) the generator voltage, (b) the location of the fault, and (c) the fault shunt resistance. The line's insulating material is Teflon with $\epsilon_{\rm r} = 2.1$.

Solution: (a) Since the line is properly matched, $R_{\rm g} = R_{\rm L} = Z_0$. In **Fig. 2-44(b)**, the fault located a distance d from the sending end is represented by a shunt resistance $R_{\rm f}$. For a matched line, Eq. (2.149b) gives

$$V_1^+ = \frac{V_{\rm g} Z_0}{R_{\rm g} + Z_0} = \frac{V_{\rm g} Z_0}{2Z_0} = \frac{V_{\rm g}}{2}.$$

According to **Fig. 2-44(a)**, $V_1^+ = 6$ V. Hence,

$$V_{\rm g} = 2V_1^+ = 12 \text{ V}.$$

(b) The propagation velocity on the line is

$$u_{\rm p} = \frac{c}{\sqrt{\epsilon_{\rm r}}} = \frac{3 \times 10^8}{\sqrt{2.1}} = 2.07 \times 10^8 \text{ m/s}.$$

For a fault at a distance d, the round-trip time delay of the echo is

$$\Delta t = \frac{2d}{u_{\rm p}}.$$

Module 2.10 Transient Response For a lossless line terminated in a resistive load, the module simulates the dynamic response, at any location on the line, to either a step or pulse waveform sent by the generator.

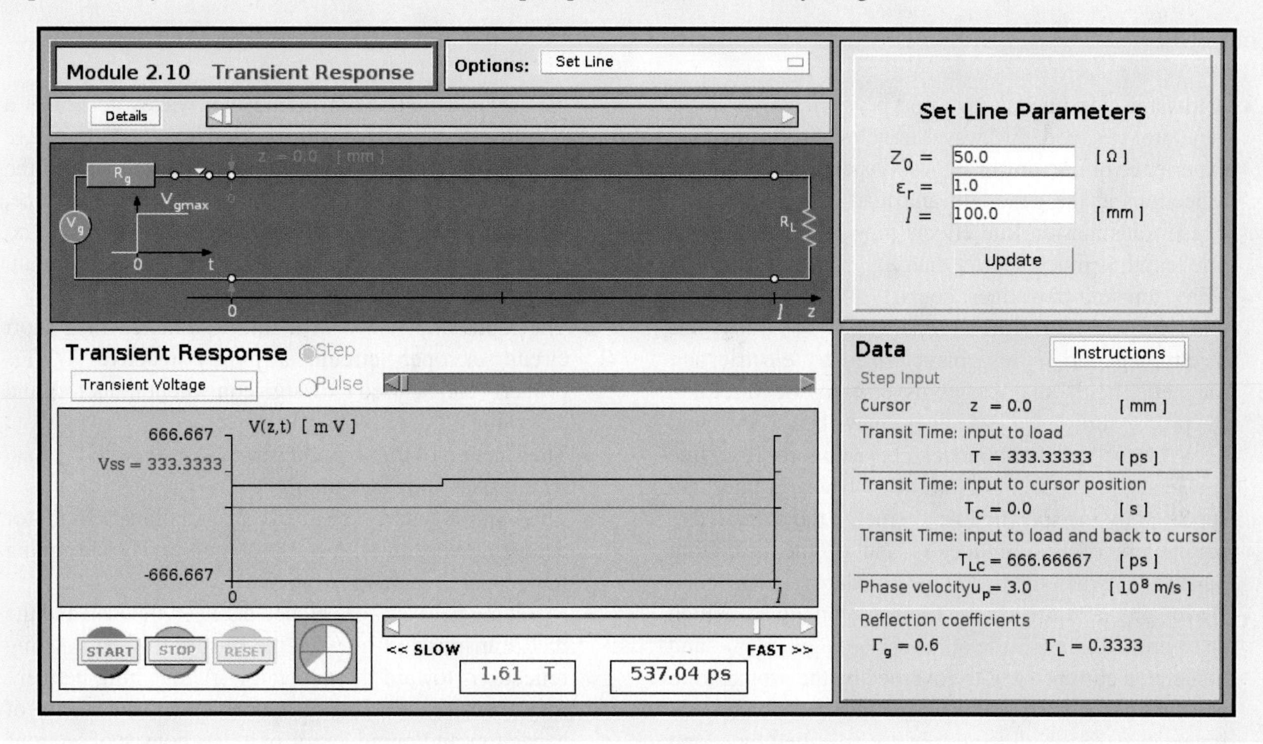

From Fig. 2-44(a), $\Delta t = 12\ \mu$s. Hence,

$$d = \frac{\Delta t}{2} u_{\mathrm{p}} = \frac{12 \times 10^{-6}}{2} \times 2.07 \times 10^{8} = 1,242 \text{ m}.$$

(c) The change in level of $V(0, t)$ shown in Fig. 2-44(a) represents V_1^-. Thus,

$$V_1^- = \Gamma_{\mathrm{f}} V_1^+ = -3 \text{ V},$$

or

$$\Gamma_{\mathrm{f}} = \frac{-3}{6} = -0.5,$$

where Γ_{f} is the reflection coefficient due to the fault load R_{Lf} that appears at $z = d$.

From Eq. (2.59),

$$\Gamma_{\mathrm{f}} = \frac{R_{\mathrm{Lf}} - Z_0}{R_{\mathrm{Lf}} + Z_0},$$

which leads to $R_{\mathrm{Lf}} = 25\ \Omega$. This fault load is composed of the fault shunt resistance R_{f} and the characteristic impedance Z_0 of the line to the right of the fault:

$$\frac{1}{R_{\mathrm{Lf}}} = \frac{1}{R_{\mathrm{f}}} + \frac{1}{Z_0},$$

so the shunt resistance must be 37.5 Ω.

Concept Question 2-27: What is transient analysis used for?

Concept Question 2-28: The transient analysis presented in this section was for a step voltage. How does one use it for analyzing the response to a pulse?

Concept Question 2-29: What is the difference between the bounce diagram for voltage and the bounce diagram for current?

Chapter 2 Summary

Concepts

- A transmission line is a two-port network connecting a generator to a load. EM waves traveling on the line may experience ohmic power losses, dispersive effects, and reflections at the generator and load ends of the line. These transmission-line effects may be ignored if the line length is much shorter than λ.
- TEM transmission lines consist of two conductors that can support the propagation of transverse electromagnetic waves characterized by electric and magnetic fields that are transverse to the direction of propagation. TEM lines may be represented by a lumped-element model consisting of four line parameters (R', L', G', and C') whose values are specified by the specific line geometry, the constitutive parameters of the conductors and of the insulating material between them, and the angular frequency ω.
- Wave propagation on a transmission line, which is represented by the phasor voltage $\widetilde{V}(z)$ and associated current $\widetilde{I}(z)$, is governed by the propagation constant of the line, $\gamma = \alpha + j\beta$, and its characteristic impedance Z_0. Both γ and Z_0 are specified by ω and the four line parameters.
- If $R' = G' = 0$, the line becomes lossless ($\alpha = 0$). A lossless line is generally nondispersive, meaning that the phase velocity of a wave is independent of the frequency.
- In general, a line supports two waves, an incident wave supplied by the generator and another wave reflected by the load. The sum of the two waves generates a standing-wave pattern with a period of $\lambda/2$. The voltage standing-wave ratio S, which is equal to the ratio of the maximum to minimum voltage magnitude on the line, varies between 1 for a matched load ($Z_L = Z_0$) to ∞ for a line terminated in an open circuit, a short circuit, or a purely reactive load.
- The input impedance of a line terminated in a short circuit or open circuit is purely reactive. This property can be used to design equivalent inductors and capacitors.
- The fraction of the incident power delivered to the load by a lossless line is equal to $(1 - |\Gamma|^2)$.
- The Smith chart is a useful graphical tool for analyzing transmission-line problems and for designing impedance-matching networks.
- Matching networks are placed between the load and the feed transmission line for the purpose of eliminating reflections toward the generator. A matching network may consist of lumped elements in the form of capacitors and/or inductors, or it may consist of sections of transmission lines with appropriate lengths and terminations.
- Transient analysis of pulses on transmission lines can be performed using a bounce-diagram graphical technique that tracks reflections at both the load and generator ends of the transmission line.

Mathematical and Physical Models

TEM Transmission Lines

$$L'C' = \mu\epsilon$$

$$\frac{G'}{C'} = \frac{\sigma}{\epsilon}$$

$$\alpha = \mathfrak{Re}(\gamma) = \mathfrak{Re}\left(\sqrt{(R' + j\omega L')(G' + j\omega C')}\right) \quad \text{(Np/m)}$$

$$\beta = \mathfrak{Im}(\gamma) = \mathfrak{Im}\left(\sqrt{(R' + j\omega L')(G' + j\omega C')}\right) \quad \text{(rad/m)}$$

$$Z_0 = \frac{R' + j\omega L'}{\gamma} = \sqrt{\frac{R' + j\omega L'}{G' + j\omega C'}} \quad (\Omega)$$

$$\Gamma = \frac{z_L - 1}{z_L + 1}$$

Mathematical and Physical Models (continued)

Step Function Transient Response

$$V_1^+ = \frac{V_g Z_0}{R_g + Z_0}$$

$$V_\infty = \frac{V_g R_L}{R_g + R_L}$$

$$\Gamma_g = \frac{R_g - Z_0}{R_g + Z_0}$$

$$\Gamma_L = \frac{R_L - Z_0}{R_L + Z_0}$$

Lossless Line

$$\alpha = 0$$

$$\beta = \omega\sqrt{L'C'}$$

$$Z_0 = \sqrt{\frac{L'}{C'}}$$

$$u_p = \frac{1}{\sqrt{\mu\epsilon}} \quad \text{(m/s)}$$

$$\lambda = \frac{u_p}{f} = \frac{c}{f}\,\frac{1}{\sqrt{\epsilon_r}} = \frac{\lambda_0}{\sqrt{\epsilon_r}}$$

$$d_{\max} = \frac{\theta_r\lambda}{4\pi} + \frac{n\lambda}{2}$$

$$d_{\min} = \frac{\theta_r\lambda}{4\pi} + \frac{(2n+1)\lambda}{4}$$

$$S = \frac{1+|\Gamma|}{1-|\Gamma|}$$

$$P_{av} = \frac{|V_0^+|^2}{2Z_0}\,[1 - |\Gamma|^2]$$

Important Terms

Provide definitions or explain the meaning of the following terms:

admittance Y
air line
attenuation constant α
bounce diagram
characteristic impedance Z_0
coaxial line
complex propagation constant γ
conductance G
current maxima and minima
dispersive transmission line
distortionless line
effective relative permittivity ϵ_{eff}
guide wavelength λ
higher-order transmission lines
impedance matching
in-phase
input impedance Z_{in}
load impedance Z_L
lossless line
lumped-element model
matched transmission line
matching network
microstrip line
normalized impedance
normalized load reactance x_L
normalized load resistance r_L
open-circuited line
optical fiber
parallel-plate line
perfect conductor
perfect dielectric
phase constant β
phase opposition
phase-shifted reflection coefficient Γ_d
quarter-wave transformer
short-circuited line
single-stub matching
slotted line
Smith chart
standing wave
standing-wave pattern
surface resistance R_s
susceptance B
SWR circle
telegrapher's equations
TEM transmission lines
time-average power P_{av}
transient response
transmission-line parameters
two-wire line
unit circle
voltage maxima and minima
voltage reflection coefficient Γ
voltage standing-wave ratio (VSWR or SWR) S
wave equations
wave impedance $Z(d)$
waveguide
WTG and WTL

PROBLEMS

Sections 2-1 to 2-4: Transmission-Line Model

2.1 A two-wire copper transmission line is embedded in a dielectric material with $\epsilon_r = 2.6$ and $\sigma = 2 \times 10^{-6}$ S/m. Its wires are separated by 3 cm, and their radii are 1 mm each.

(a) Calculate the line parameters R', L', G', and C' at 2 GHz.

(b) Compare your results with those based on Module 2.1. Include a printout of the screen display.

2.2 A transmission line of length l connects a load to a sinusoidal voltage source with an oscillation frequency f. Assuming that the velocity of wave propagation on the line is c, for which of the following situations is it reasonable to ignore the presence of the transmission line in the solution of the circuit:

*(a) $l = 30$ cm, $f = 20$ kHz

(b) $l = 50$ km, $f = 60$ Hz

*(c) $l = 30$ cm, $f = 600$ MHz

(d) $l = 2$ mm, $f = 100$ GHz

2.3 Show that the transmission-line model shown in **Fig. P2.3** yields the same telegrapher's equations given by Eqs. (2.14) and (2.16).

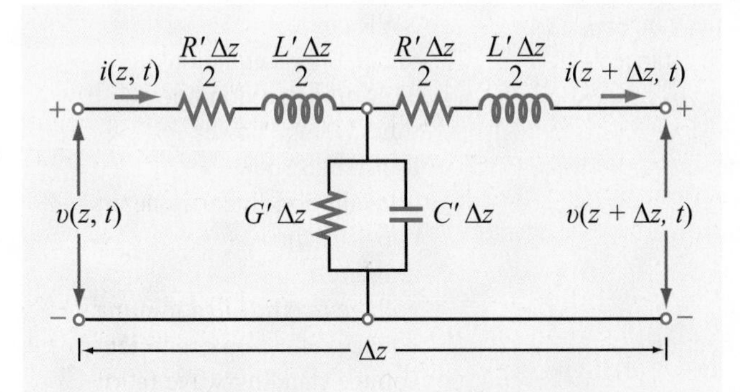

Figure P2.3 Transmission-line model for Problem 2.3.

*2.4 A 1 GHz parallel-plate transmission line consists of 2.4 cm wide copper strips separated by a 0.3 cm thick layer of polystyrene. Appendix B gives $\mu_c = \mu_0 = 4\pi \times 10^{-7}$ (H/m) and $\sigma_c = 5.8 \times 10^7$ (S/m) for copper, and $\epsilon_r = 2.6$ for polystyrene. Use **Table 2-1** to determine the line parameters of the transmission line. Assume that $\mu = \mu_0$ and $\sigma \approx 0$ for polystyrene.

2.5 For the parallel-plate transmission line of Problem 2.4, the line parameters are given by $R' = 2$ Ω/m, $L' = 335$ nH/m, $G' = 0$, and $C' = 344$ pF/m. Find α, β, u_p, and Z_0 at 1 GHz.

2.6 A coaxial line with inner and outer conductor diameters of 0.5 cm and 1 cm, respectively, is filled with an insulating material with $\epsilon_r = 4.5$ and $\sigma = 10^{-3}$ S/m. The conductors are made of copper.

(a) Calculate the line parameters at 1 GHz.

(b) Compare your results with those based on Module 2.2. Include a printout of the screen display.

*2.7 Find α, β, u_p, and Z_0 for the coaxial line of Problem 2.6. Verify your results by applying Module 2.2. Include a printout of the screen display.

2.8 Find α, β, u_p, and Z_0 for the two-wire line of Problem 2.1. Compare results with those based on Module 2.1. Include a printout of the screen display.

Section 2-5: The Lossless Microstrip Line

2.9 A lossless microstrip line uses a 1 mm wide conducting strip over a 1 cm thick substrate with $\epsilon_r = 2.5$. Determine the line parameters ϵ_{eff}, Z_0, and β at 10 GHz. Compare your results with those obtained by using Module 2.3. Include a printout of the screen display.

*2.10 Use Module 2.3 to design a 100 Ω microstrip transmission line. The substrate thickness is 1.8 mm and its $\epsilon_r = 2.3$. Select the strip width w, and determine the guide wavelength λ at $f = 5$ GHz. Include a printout of the screen display.

2.11 A 50 Ω microstrip line uses a 0.6 mm alumina substrate with $\epsilon_r = 9$. Use Module 2.3 to determine the required strip width w. Include a printout of the screen display.

2.12 Generate a plot of Z_0 as a function of strip width w, over the range from 0.05 mm to 5 mm, for a microstrip line fabricated on a 0.7 mm thick substrate with $\epsilon_r = 9.8$.

*Answer(s) available in Appendix D.

Section 2-6: The Lossless Transmission Line: General Considerations

2.13 In addition to not dissipating power, a lossless line has two important features: (1) it is dispersionless (u_p is independent of frequency); and (2) its characteristic impedance Z_0 is purely real. Sometimes, it is not possible to design a transmission line such that $R' \ll \omega L'$ and $G' \ll \omega C'$, but it is possible to choose the dimensions of the line and its material properties so as to satisfy the condition

$$R'C' = L'G' \qquad \text{(distortionless line)}$$

Such a line is called a ***distortionless*** line, because despite the fact that it is not lossless, it nonetheless possesses the previously mentioned features of the lossless line. Show that for a distortionless line,

$$\alpha = R'\sqrt{\frac{C'}{L'}} = \sqrt{R'G'}\,,$$

$$\beta = \omega\sqrt{L'C'}\,,$$

$$Z_0 = \sqrt{\frac{L'}{C'}}\,.$$

***2.14** For a distortionless line [see Problem 2.13] with $Z_0 = 50\ \Omega$, $\alpha = 10$ (mNp/m), and $u_p = 2.5 \times 10^8$ (m/s), find the line parameters and λ at 100 MHz.

2.15 Find α and Z_0 of a distortionless line whose $R' = 8\ \Omega$/m and $G' = 2 \times 10^{-4}$ S/m.

***2.16** A transmission line operating at 125 MHz has $Z_0 = 40\ \Omega$, $\alpha = 0.01$ (Np/m), and $\beta = 0.75$ rad/m. Find the line parameters R', L', G', and C'.

***2.17** Polyethylene with $\epsilon_r = 2.25$ is used as the insulating material in a lossless coaxial line with a characteristic impedance of 50 Ω. The radius of the inner conductor is 1.2 mm.

(a) What is the radius of the outer conductor?

(b) What is the phase velocity of the line?

2.18 Using a slotted line, the voltage on a lossless transmission line was found to have a maximum magnitude of 1.5 V and a minimum magnitude of 0.5 V. Find the magnitude of the load's reflection coefficient.

2.19 A 50 Ω lossless transmission line is terminated in a load with impedance $Z_L = (30 - j50)\ \Omega$. The wavelength is 8 cm. Determine:

(a) The reflection coefficient at the load.

(b) The standing-wave ratio on the line.

(c) The position of the voltage maximum nearest the load.

(d) The position of the current maximum nearest the load.

(e) Verify quantities in parts (a)–(d) using Module 2.4. Include a printout of the screen display.

2.20 A 300 Ω lossless air transmission line is connected to a complex load composed of a resistor in series with an inductor, as shown in Fig. P2.20. At 5 MHz, determine: (a) Γ, (b) S, (c) location of voltage maximum nearest to the load, and (d) location of current maximum nearest to the load.

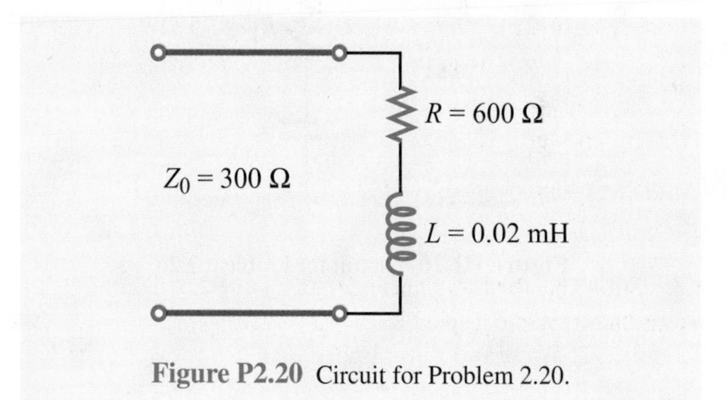

Figure P2.20 Circuit for Problem 2.20.

2.21 Using a slotted line, the following results were obtained: distance of first minimum from the load = 4 cm; distance of second minimum from the load = 14 cm; voltage standing-wave ratio = 1.5. If the line is lossless and $Z_0 = 50\ \Omega$, find the load impedance.

***2.22** On a 150 Ω lossless transmission line, the following observations were noted: distance of first voltage minimum from the load = 3 cm; distance of first voltage maximum from the load = 9 cm; $S = 2$. Find Z_L.

***2.23** A load with impedance $Z_L = (50 - j50)\ \Omega$ is to be connected to a lossless transmission line with characteristic impedance Z_0, with Z_0 chosen such that the standing-wave ratio is the smallest possible. What should Z_0 be?

2.24 A 50 Ω lossless line terminated in a purely resistive load has a voltage standing-wave ratio of 2. Find all possible values of Z_L.

2.25 Apply Module 2.4 to generate plots of the voltage standing-wave pattern for a 50 Ω line terminated in a load impedance $Z_L = (100 - j50)$ Ω. Set $V_g = 1$ V, $Z_g = 50$ Ω, $\epsilon_r = 2.25$, $l = 40$ cm, and $f = 1$ GHz. Also determine S, d_{max}, and d_{min}.

2.26 A 50 Ω lossless transmission line is connected to a load composed of a 75 Ω resistor in series with a capacitor of unknown capacitance (**Fig. P2.26**). If at 10 MHz the voltage standing-wave ratio on the line was measured to be 3, determine the capacitance C.

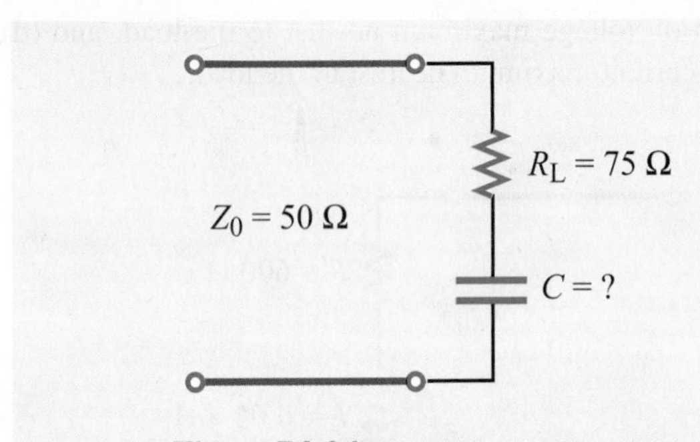

Figure P2.26 Circuit for Problem 2.26.

Section 2-7: Wave and Input Impedance

2.27 Show that the input impedance of a quarter-wavelength–long lossless line terminated in a short circuit appears as an open circuit.

2.28 A lossless transmission line of electrical length $l = 0.35\lambda$ is terminated in a load impedance as shown in **Fig. P2.28**. Find Γ, S, and Z_{in}. Verify your results using Modules 2.4 or 2.5. Include a printout of the screen's output display.

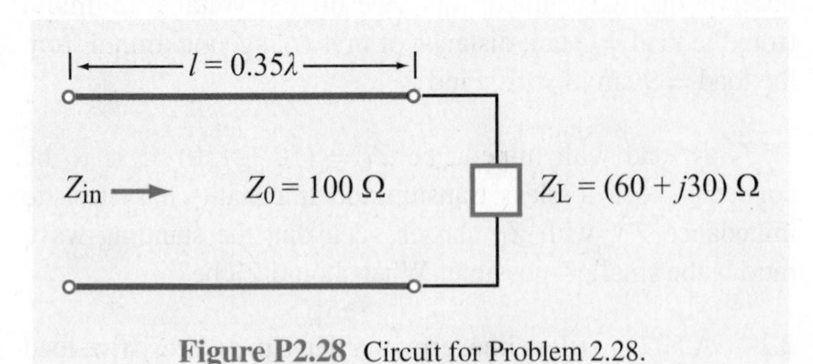

Figure P2.28 Circuit for Problem 2.28.

*__2.29__ At an operating frequency of 300 MHz, a lossless 50 Ω air-spaced transmission line 2.5 m in length is terminated with an impedance $Z_L = (40 + j20)$ Ω. Find the input impedance.

2.30 Show that at the position where the magnitude of the voltage on the line is a maximum, the input impedance is purely real.

2.31 A 6 m section of 150 Ω lossless line is driven by a source with

$$v_g(t) = 5\cos(8\pi \times 10^7 t - 30^\circ) \quad \text{(V)}$$

and $Z_g = 150$ Ω. If the line, which has a relative permittivity $\epsilon_r = 2.25$, is terminated in a load $Z_L = (150 - j50)$ Ω, determine:

(a) λ on the line.

*__(b)__ The reflection coefficient at the load.

(c) The input impedance.

(d) The input voltage $\widetilde{V}_i$.

(e) The time-domain input voltage $v_i(t)$.

(f) Quantities in (a) to (d) using Modules 2.4 or 2.5.

2.32 A voltage generator with

$$v_g(t) = 5\cos(2\pi \times 10^9 t) \text{ V}$$

and internal impedance $Z_g = 50$ Ω is connected to a 50 Ω lossless air-spaced transmission line. The line length is 5 cm, and the line is terminated in a load with impedance $Z_L = (100 - j100)$ Ω. Determine:

*__(a)__ Γ at the load.

(b) Z_{in} at the input to the transmission line.

(c) The input voltage $\widetilde{V}_i$ and input current $\widetilde{I}_i$.

(d) The quantities in (a)–(c) using Modules 2.4 or 2.5.

2.33 Two half-wave dipole antennas, each with an impedance of 75 Ω, are connected in parallel through a pair of transmission lines, and the combination is connected to a feed transmission line, as shown in **Fig. P2.33**. All lines are 50 Ω and lossless.

*__(a)__ Calculate Z_{in_1}, the input impedance of the antenna-terminated line, at the parallel juncture.

(b) Combine Z_{in_1} and Z_{in_2} in parallel to obtain Z'_L, the effective load impedance of the feedline.

(c) Calculate Z_{in} of the feedline.

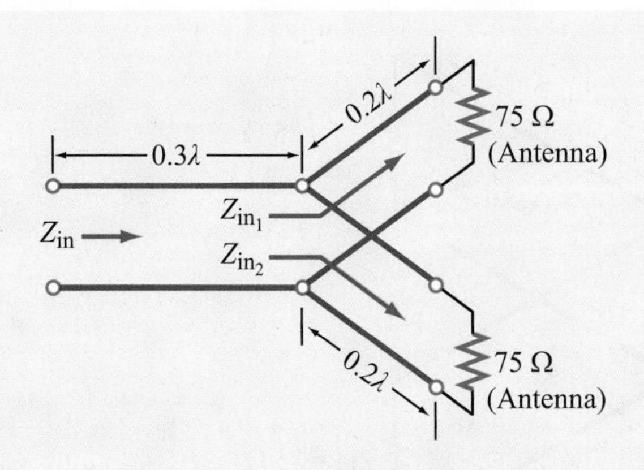

Figure P2.33 Circuit for Problem 2.33.

2.34 A 50 Ω lossless line is terminated in a load impedance $Z_L = (30 - j20)$ Ω.

(a) Calculate Γ and S.

(b) It has been proposed that by placing an appropriately selected resistor across the line at a distance d_{max} from the load (as shown in **Fig. P2.34(b)**), where d_{max} is the distance from the load of a voltage maximum, then it is possible to render $Z_i = Z_0$, thereby eliminating reflection back to the end. Show that the proposed approach is valid and find the value of the shunt resistance.

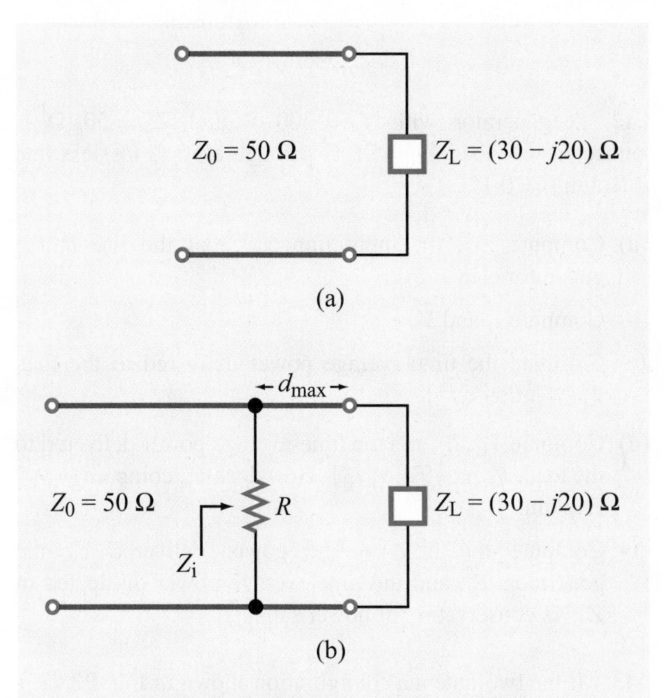

Figure P2.34 Circuit for Problem 2.34.

*__2.35__ For the lossless transmission line circuit shown in **Fig. P2.35**, determine the equivalent series lumped-element circuit at 400 MHz at the input to the line. The line has a characteristic impedance of 50 Ω, and the insulating layer has $\epsilon_r = 2.25$.

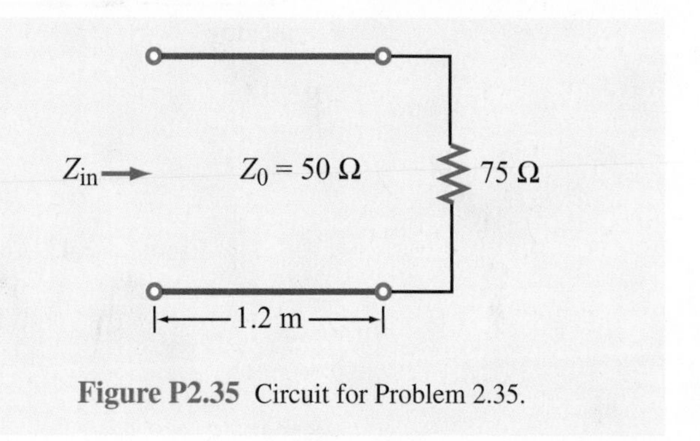

Figure P2.35 Circuit for Problem 2.35.

Section 2-8: Special Cases of the Lossless Line

*__2.36__ A lossless transmission line is terminated in a short circuit. How long (in wavelengths) should the line be for it to appear as an open circuit at its input terminals?

2.37 At an operating frequency of 300 MHz, it is desired to use a section of a lossless 50 Ω transmission line terminated in a short circuit to construct an equivalent load with reactance $X = 40$ Ω. If the phase velocity of the line is $0.75c$, what is the shortest possible line length that would exhibit the desired reactance at its input? Verify your result using Module 2.5.

2.38 The input impedance of a 36 cm long lossless transmission line of unknown characteristic impedance was measured at 1 MHz. With the line terminated in a short circuit, the measurement yielded an input impedance equivalent to an inductor with inductance of 0.064 μH, and when the line was open-circuited, the measurement yielded an input impedance equivalent to a capacitor with capacitance of 40 pF. Find Z_0 of the line, the phase velocity, and the relative permittivity of the insulating material.

*__2.39__ A 75 Ω resistive load is preceded by a $\lambda/4$ section of a 50 Ω lossless line, which itself is preceded by another $\lambda/4$ section of a 100 Ω line. What is the input impedance?

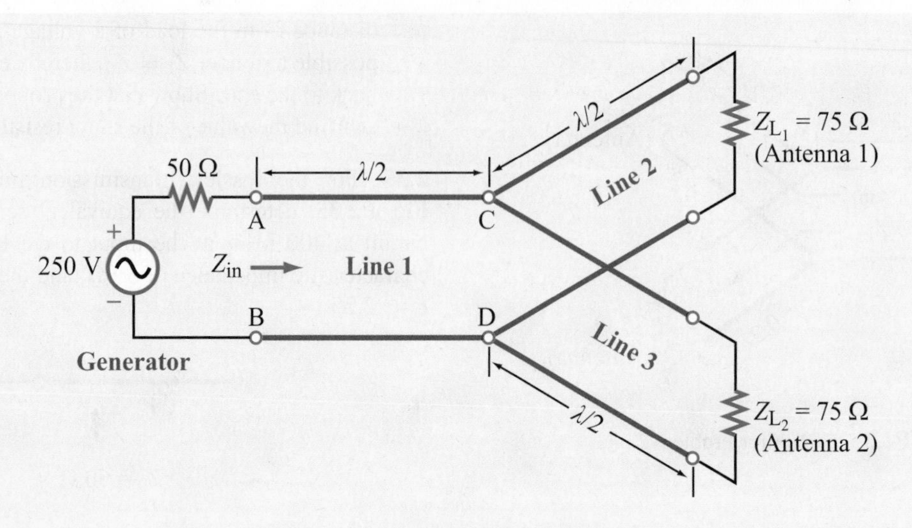

Figure P2.43 Antenna configuration for Problem 2.43.

Compare your result with that obtained through two successive applications of Module 2.5.

2.40 A 100 MHz FM broadcast station uses a 300 Ω transmission line between the transmitter and a tower-mounted half-wave dipole antenna. The antenna impedance is 73 Ω. You are asked to design a quarter-wave transformer to match the antenna to the line.

(a) Determine the electrical length and characteristic impedance of the quarter-wave section.

(b) If the quarter-wave section is a two-wire line with $D = 2.5$ cm, and the wires are embedded in polystyrene with $\epsilon_r = 2.6$, determine the physical length of the quarter-wave section and the radius of the two wire conductors.

2.41 A 50 Ω lossless line of length $l = 0.375\lambda$ connects a 300 MHz generator with $\widetilde{V}_g = 300$ V and $Z_g = 50$ Ω to a load Z_L. Determine the time-domain current through the load for:

(a) $Z_L = (50 - j50)$ Ω

*__(b)__ $Z_L = 50$ Ω

(c) $Z_L = 0$ (short circuit)

For (a), verify your results by deducing the information you need from the output products generated by Module 2.4.

Section 2-9: Power Flow on a Lossless Transmission Line

2.42 A generator with $\widetilde{V}_g = 300$ V and $Z_g = 50$ Ω is connected to a load $Z_L = 75$ Ω through a 50 Ω lossless line of length $l = 0.15\lambda$.

*__(a)__ Compute Z_{in}, the input impedance of the line at the generator end.

(b) Compute $\widetilde{I}_i$ and $\widetilde{V}_i$.

(c) Compute the time-average power delivered to the line, $P_{in} = \frac{1}{2}\mathfrak{Re}[\widetilde{V}_i \widetilde{I}_i^*]$.

(d) Compute $\widetilde{V}_L$, $\widetilde{I}_L$, and the time-average power delivered to the load, $P_L = \frac{1}{2}\mathfrak{Re}[\widetilde{V}_L \widetilde{I}_L^*]$. How does P_{in} compare to P_L? Explain.

(e) Compute the time-average power delivered by the generator, P_g, and the time-average power dissipated in Z_g. Is conservation of power satisfied?

2.43 If the two-antenna configuration shown in **Fig. P2.43** is connected to a generator with $\widetilde{V}_g = 250$ V and $Z_g = 50$ Ω, how much average power is delivered to each antenna?

*2.44 For the circuit shown in **Fig. P2.44**, calculate the average incident power, the average reflected power, and the average power transmitted into the infinite 100 Ω line. The $\lambda/2$ line is lossless and the infinitely long line is slightly lossy. (Hint: The input impedance of an infinitely long line is equal to its characteristic impedance so long as $\alpha \neq 0$.)

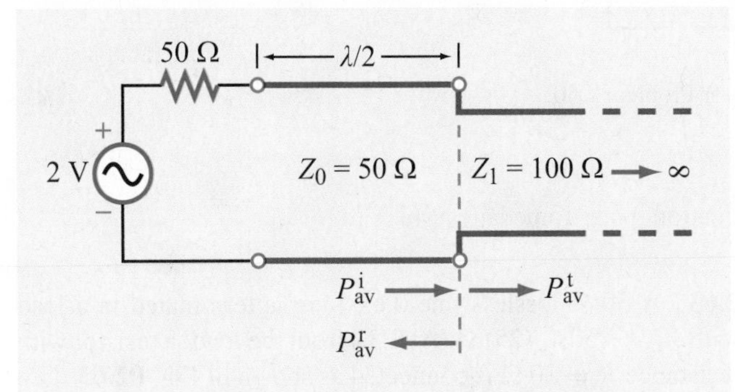

Figure P2.44 Circuit for Problem 2.44.

2.45 The circuit shown in **Fig. P2.45** consists of a 100 Ω lossless transmission line terminated in a load with $Z_L = (50 + j100)$ Ω. If the peak value of the load voltage was measured to be $|\widetilde{V}_L| = 12$ V, determine:

*(a) the time-average power dissipated in the load,

(b) the time-average power incident on the line,

(c) the time-average power reflected by the load.

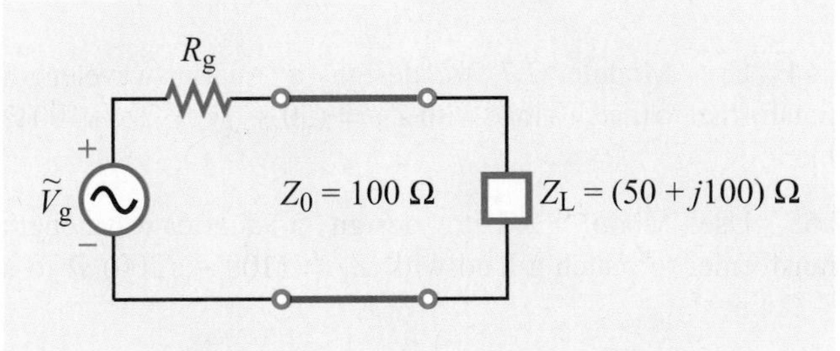

Figure P2.45 Circuit for Problem 2.45.

2.46 An antenna with a load impedance

$$Z_L = (75 + j25)\ \Omega$$

is connected to a transmitter through a 50 Ω lossless transmission line. If under matched conditions (50 Ω load) the transmitter can deliver 20 W to the load, how much power can it deliver to the antenna? Assume that $Z_g = Z_0$.

Section 2-10: The Smith Chart

2.47 Use the Smith chart to find the normalized load impedance corresponding to a reflection coefficient of

(a) $\Gamma = 0.5$

(b) $\Gamma = 0.5\angle 60^\circ$

(c) $\Gamma = -1$

(d) $\Gamma = 0.3\angle -30^\circ$

(e) $\Gamma = 0$

(f) $\Gamma = j$

2.48 Use the Smith chart to find the reflection coefficient corresponding to a load impedance of

(a) $Z_L = 3Z_0$

*(b) $Z_L = (2 - j2)Z_0$

(c) $Z_L = -j2Z_0$

(d) $Z_L = 0$ (short circuit)

2.49 Repeat Problem 2.48 using Module 2.6.

*2.50 Use the Smith chart to determine the input impedance Z_{in} of the two-line configuration shown in **Fig. P2.50**.

2.51 Repeat Problem 2.50 using Module 2.6.

2.52 A lossless 50 Ω transmission line is terminated in a load with $Z_L = (50 + j25)$ Ω. Use the Smith chart to find the following:

(a) The reflection coefficient Γ.

*(b) The standing-wave ratio.

(c) The input impedance at 0.35λ from the load.

(d) The input admittance at 0.35λ from the load.

(e) The shortest line length for which the input impedance is purely resistive.

(f) The position of the first voltage maximum from the load.

*2.53 On a lossless transmission line terminated in a load $Z_L = 100$ Ω, the standing-wave ratio was measured to be 2.5. Use the Smith chart to find the two possible values of Z_0.

2.54 Repeat Problem 2.53 using Module 2.6.

*2.55 A lossless 50 Ω transmission line is terminated in a short circuit. Use the Smith chart to determine:

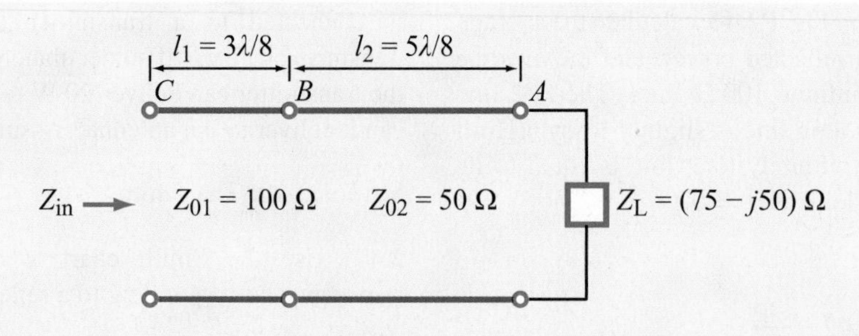

Figure P2.50 Circuit for Problem 2.50.

(a) The input impedance at a distance 2.3λ from the load.

(b) The distance from the load at which the input admittance is $Y_{in} = -j0.04$ S.

2.56 Repeat Problem 2.55 using Module 2.6.

***2.57** Use the Smith chart to find y_L if $z_L = 1.5 - j0.7$.

2.58 A lossless 100 Ω transmission line $3\lambda/8$ in length is terminated in an unknown impedance. If the input impedance is $Z_{in} = -j2.5\ \Omega$,

(a) Use the Smith chart to find Z_L.

(b) Verify your results using Module 2.6.

2.59 A 75 Ω lossless line is 0.6λ long. If $S = 1.8$ and $\theta_r = -60°$, use the Smith chart to find $|\Gamma|$, Z_L, and Z_{in}.

2.60 Repeat Problem 2.59 using Module 2.6.

***2.61** Using a slotted line on a 50 Ω air-spaced lossless line, the following measurements were obtained: $S = 1.6$ and $|\widetilde{V}|_{max}$ occurred only at 10 cm and 24 cm from the load. Use the Smith chart to find Z_L.

2.62 At an operating frequency of 5 GHz, a 50 Ω lossless coaxial line with insulating material having a relative permittivity $\epsilon_r = 2.25$ is terminated in an antenna with an impedance $Z_L = 150\ \Omega$. Use the Smith chart to find Z_{in}. The line length is 30 cm.

Section 2-11: Impedance Matching

***2.63** A 50 Ω lossless line 0.6λ long is terminated in a load with $Z_L = (50 + j25)\ \Omega$. At 0.3λ from the load, a resistor with resistance $R = 30\ \Omega$ is connected as shown in **Fig. P2.63**. Use the Smith chart to find Z_{in}.

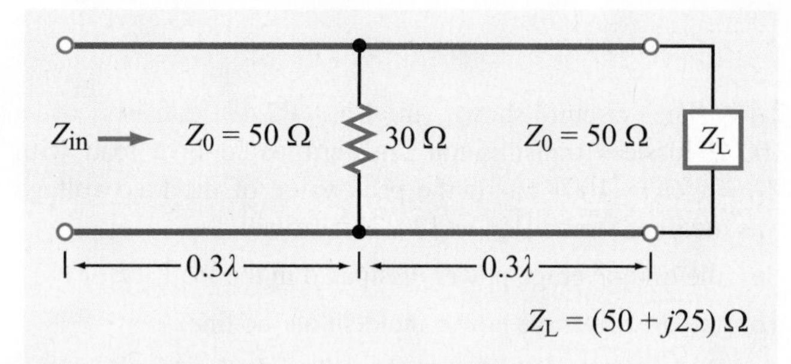

Figure P2.63 Circuit for Problem 2.63.

2.64 Use Module 2.7 to design a quarter-wavelength transformer to match a load with $Z_L = (50 + j10)\ \Omega$ to a 100 Ω line.

2.65 Use Module 2.7 to design a quarter-wavelength transformer to match a load with $Z_L = (100 - j200)\ \Omega$ to a 50 Ω line.

2.66 A 200 Ω transmission line is to be matched to a computer terminal with $Z_L = (50 - j25)\ \Omega$ by inserting an appropriate reactance in parallel with the line. If $f = 800$ MHz and $\epsilon_r = 4$, determine the location nearest to the load at which inserting:

(a) A capacitor can achieve the required matching, and the value of the capacitor.

(b) An inductor can achieve the required matching, and the value of the inductor.

2.67 Repeat Problem 2.66 using Module 2.8.

2.68 A 50 Ω lossless line is to be matched to an antenna with $Z_L = (75 - j20)$ Ω using a shorted stub. Use the Smith chart to determine the stub length and distance between the antenna and stub.

***2.69** Repeat Problem 2.68 for a load with

$$Z_L = (100 + j50)\ \Omega.$$

2.70 Repeat Problem 2.68 using Module 2.9.

2.71 Repeat Problem 2.69 using Module 2.9.

2.72 Determine Z_{in} of the feed line shown in Fig. P2.72. All lines are lossless with $Z_0 = 50$ Ω.

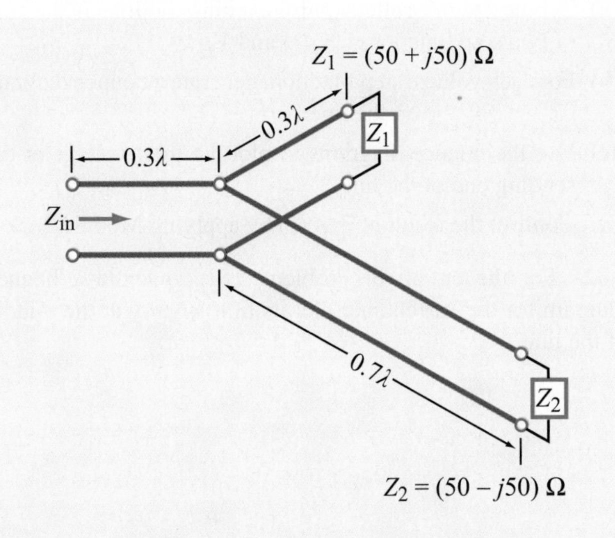

Figure P2.72 Network for Problem 2.72.

***2.73** Repeat Problem 2.72 for the case where all three transmission lines are $3\lambda/4$ in length.

2.74 A 25 Ω antenna is connected to a 75 Ω lossless transmission line. Reflections back toward the generator can be eliminated by placing a shunt impedance Z at a distance l from the load (Fig. P2.74). Determine the values of Z and l.

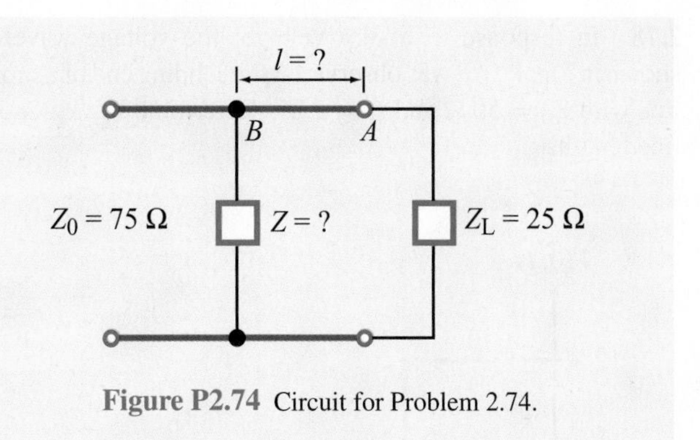

Figure P2.74 Circuit for Problem 2.74.

Section 2-12: Transients on Transmission Lines

2.75 In response to a step voltage, the voltage waveform shown in Fig. P2.75 was observed at the sending end of a lossless transmission line with $R_g = 50$ Ω, $Z_0 = 50$ Ω, and $\epsilon_r = 4$. Determine the following:

(a) The generator voltage.

(b) The length of the line.

(c) The load impedance.

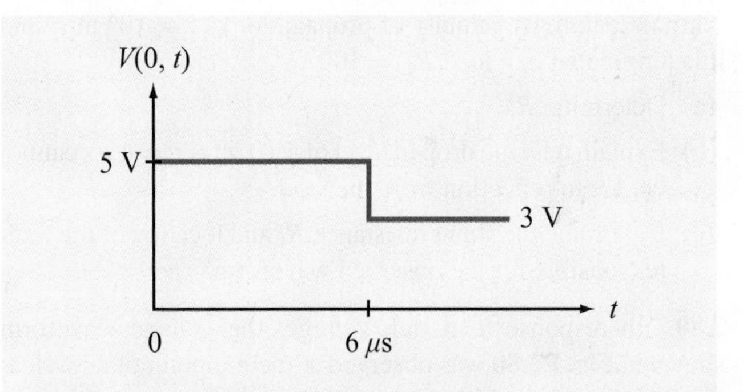

Figure P2.75 Voltage waveform for Problems 2.75 and 2.79.

2.76 Generate a bounce diagram for the voltage $V(z,t)$ for a 1 m long lossless line characterized by $Z_0 = 50$ Ω and $u_p = 2c/3$ (where c is the velocity of light) if the line is fed by a step voltage applied at $t = 0$ by a generator circuit with $V_g = 60$ V and $R_g = 100$ Ω. The line is terminated in a load $R_L = 25$ Ω. Use the bounce diagram to plot $V(t)$ at a point midway along the length of the line from $t = 0$ to $t = 25$ ns.

2.77 Repeat Problem 2.76 for the current $I(z,t)$ on the line.

*2.78 In response to a step voltage, the voltage waveform shown in **Fig. P2.78** was observed at the sending end of a shorted line with $Z_0 = 50\ \Omega$ and $\epsilon_r = 2.25$. Determine V_g, R_g, and the line length.

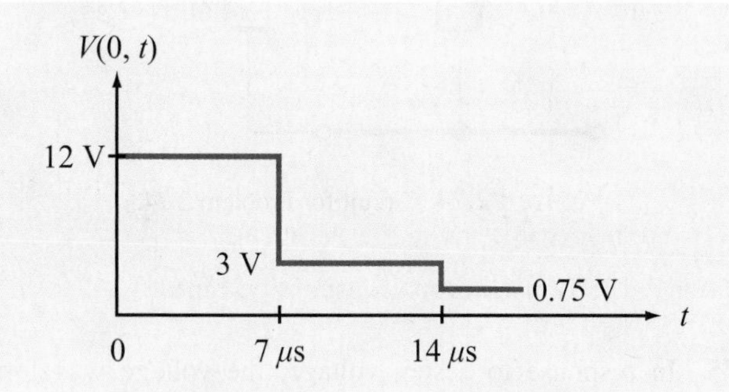

Figure P2.78 Voltage waveform of Problem 2.78.

2.79 Suppose the voltage waveform shown in **Fig. P2.75** was observed at the sending end of a 50 Ω transmission line in response to a step voltage introduced by a generator with $V_g = 15$ V and an unknown series resistance R_g. The line is 1 km in length, its velocity of propagation is 2×10^8 m/s, and it is terminated in a load $R_L = 100\ \Omega$.

(a) Determine R_g.

(b) Explain why the drop in level of $V(0, t)$ at $t = 6\ \mu$s cannot be due to reflection from the load.

(c) Determine the shunt resistance R_f and location of the fault responsible for the observed waveform.

*2.80 In response to a step voltage, the voltage waveform shown in **Fig. P2.80** was observed at the midpoint of a lossless transmission line with $Z_0 = 50\ \Omega$ and $u_p = 1 \times 10^8$ m/s. Determine: (a) the length of the line, (b) Z_L, (c) R_g, and (d) V_g.

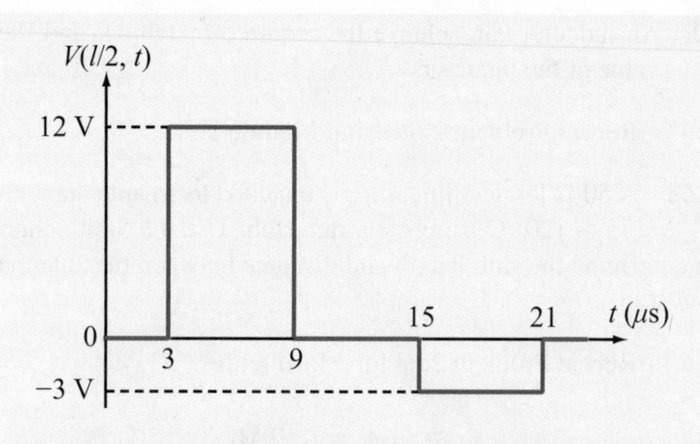

Figure P2.80 Circuit for Problem 2.80.

2.81 A generator circuit with $V_g = 200$ V and $R_g = 25\ \Omega$ was used to excite a 75 Ω lossless line with a rectangular pulse of duration $\tau = 0.4\ \mu$s. The line is 200 m long, its $u_p = 2 \times 10^8$ m/s, and it is terminated in a load $R_L = 125\ \Omega$.

(a) Synthesize the voltage pulse exciting the line as the sum of two step functions, $V_{g_1}(t)$ and $V_{g_2}(t)$.

(b) For each voltage step function, generate a bounce diagram for the voltage on the line.

(c) Use the bounce diagrams to plot the total voltage at the sending end of the line.

(d) Confirm the result of part (c) by applying Module 2.10.

2.82 For the circuit of Problem 2.81, generate a bounce diagram for the current and plot its time history at the middle of the line.

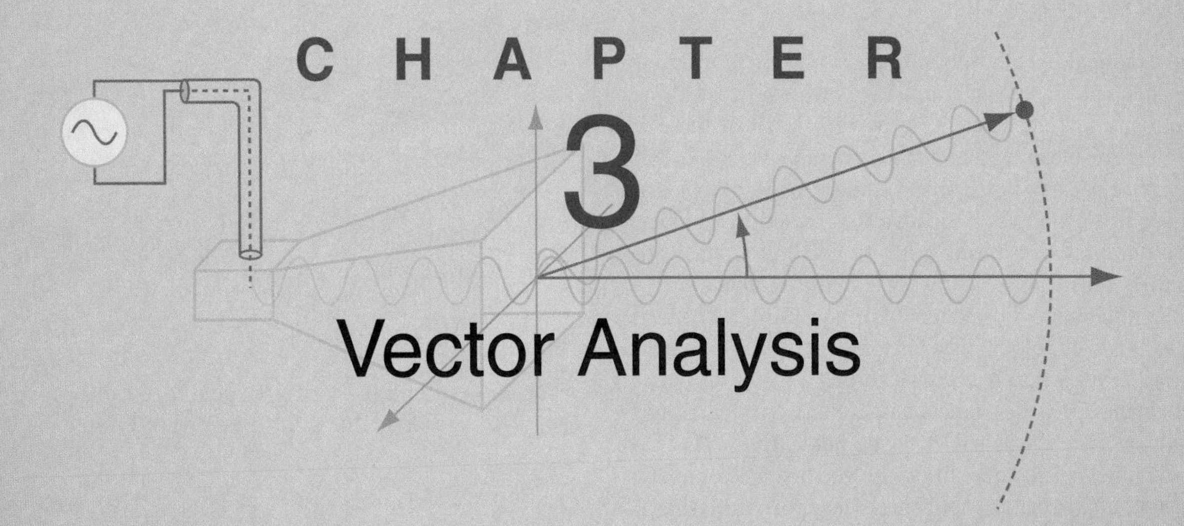

Vector Analysis

Chapter Contents

	Overview, 156
3-1	Basic Laws of Vector Algebra, 156
3-2	Orthogonal Coordinate Systems, 162
3-3	Transformations between Coordinated Systems, 169
TB5	Global Positioning System, 172
3-4	Gradient of a Scalar Field, 176
3-5	Divergence of a Vector Field, 180
3-6	Curl of a Vector Field, 184
TB6	X-Ray Computed Tomography, 186
3-7	Laplacian Operator, 189
	Chapter 3 Summary, 191
	Problems, 193

Objectives

Upon learning the material presented in this chapter, you should be able to:

1. Use vector algebra in Cartesian, cylindrical, and spherical coordinate systems.
2. Transform vectors between the three primary coordinate systems.
3. Calculate the gradient of a scalar function and the divergence and curl of a vector function in any of the three primary coordinate systems.
4. Apply the divergence theorem and Stokes's theorem.

Overview

In our examination of wave propagation on a transmission line in Chapter 2, the primary quantities we worked with were voltage, current, impedance, and power. Each of these is a ***scalar*** quantity, meaning that it can be completely specified by its magnitude, if it is a positive real number, or by its magnitude and phase angle if it is a negative or a complex number (a negative number has a positive magnitude and a phase angle of π (rad)). This chapter is concerned with vectors. A ***vector*** has a magnitude and a direction. The speed of an object is a scalar, whereas its velocity is a vector.

Starting in the next chapter and throughout the succeeding chapters in the book, the primary electromagnetic quantities we deal with are the electric and magnetic fields, **E** and **H**. These, and many other related quantities, are vectors. Vector analysis provides the mathematical tools necessary for expressing and manipulating vector quantities in an efficient and convenient manner. To specify a vector in three-dimensional space, it is necessary to specify its components along each of the three directions.

► Several types of coordinate systems are used in the study of vector quantities, the most common being the Cartesian (or rectangular), cylindrical, and spherical systems. A particular coordinate system is usually chosen to best suit the geometry of the problem under consideration. ◄

Vector algebra governs the laws of addition, subtraction, and "multiplication" of vectors. The rules of vector algebra and vector representation in each of the aforementioned orthogonal coordinate systems (including vector transformation between them) are two of the three major topics treated in this chapter. The third topic is ***vector calculus***, which encompasses the laws of differentiation and integration of vectors, the use of special vector operators (gradient, divergence, and curl), and the application of certain theorems that are particularly useful in the study of electromagnetics, most notably the divergence and Stokes's theorems.

3-1 Basic Laws of Vector Algebra

A vector is a mathematical object that resembles an arrow. Vector **A** in **Fig. 3-1** has ***magnitude*** (or length) $A = |\mathbf{A}|$ and ***unit vector*** $\hat{\mathbf{a}}$:

$$\mathbf{A} = \hat{\mathbf{a}}|\mathbf{A}| = \hat{\mathbf{a}}A. \tag{3.1}$$

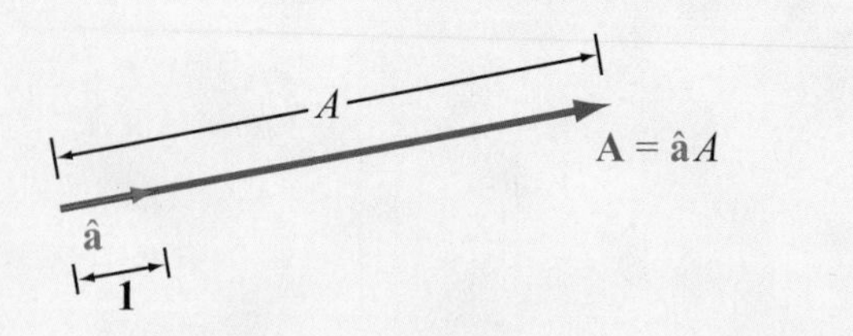

Figure 3-1 Vector $\mathbf{A} = \hat{\mathbf{a}}A$ has magnitude $A = |\mathbf{A}|$ and points in the direction of unit vector $\hat{\mathbf{a}} = \mathbf{A}/A$.

The unit vector $\hat{\mathbf{a}}$ has a magnitude of one ($|\hat{\mathbf{a}}| = 1$), and points from **A**'s tail or anchor to its head or tip. From Eq. (3.1),

$$\hat{\mathbf{a}} = \frac{\mathbf{A}}{|\mathbf{A}|} = \frac{\mathbf{A}}{A}. \tag{3.2}$$

In the Cartesian (or rectangular) coordinate system shown in **Fig. 3-2(a)**, the x, y, and z coordinate axes extend along directions of the three mutually perpendicular unit vectors $\hat{\mathbf{x}}$, $\hat{\mathbf{y}}$, and $\hat{\mathbf{z}}$, also called ***base vectors***. The vector **A** in **Fig. 3-2(b)** may be decomposed as

$$\mathbf{A} = \hat{\mathbf{x}}A_x + \hat{\mathbf{y}}A_y + \hat{\mathbf{z}}A_z, \tag{3.3}$$

where A_x, A_y, and A_z are **A**'s scalar components along the x-, y-, and z axes, respectively. The component A_z is equal to the perpendicular projection of **A** onto the z axis, and similar definitions apply to A_x and A_y. Application of the Pythagorean theorem, first to the right triangle in the x–y plane to express the hypotenuse A_r in terms of A_x and A_y, and then again to the vertical right triangle with sides A_r and A_z and hypotenuse A, yields the following expression for the magnitude of **A**:

$$A = |\mathbf{A}| = \sqrt[+]{A_x^2 + A_y^2 + A_z^2}. \tag{3.4}$$

Since A is a nonnegative scalar, only the positive root applies. From Eq. (3.2), the unit vector $\hat{\mathbf{a}}$ is

$$\hat{\mathbf{a}} = \frac{\mathbf{A}}{A} = \frac{\hat{\mathbf{x}}A_x + \hat{\mathbf{y}}A_y + \hat{\mathbf{z}}A_z}{\sqrt[+]{A_x^2 + A_y^2 + A_z^2}}. \tag{3.5}$$

Occasionally, we use the shorthand notation $\mathbf{A} = (A_x, A_y, A_z)$ to denote a vector with components A_x, A_y, and A_z in a Cartesian coordinate system.

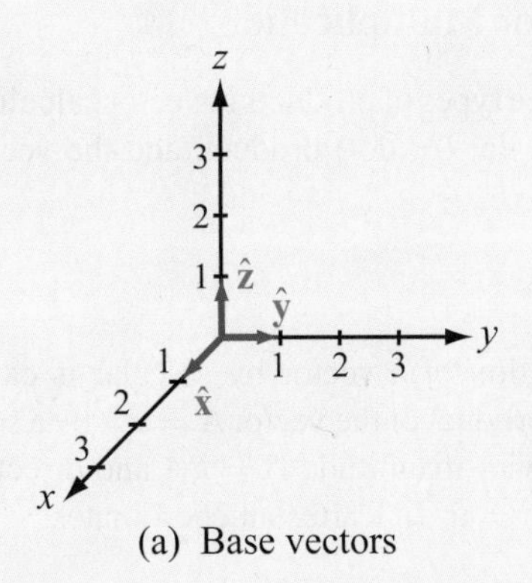

(a) Base vectors

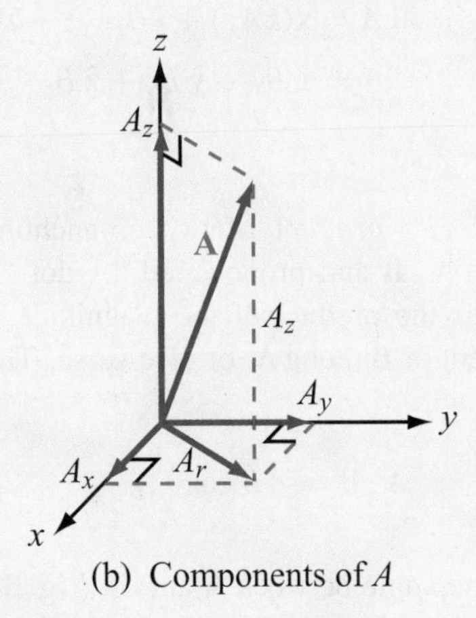

(b) Components of A

Figure 3-2 Cartesian coordinate system: (a) base vectors $\hat{\mathbf{x}}$, $\hat{\mathbf{y}}$, and $\hat{\mathbf{z}}$, and (b) components of vector **A**.

3-1.1 Equality of Two Vectors

Two vectors **A** and **B** are equal if they have equal magnitudes and identical unit vectors. Thus, if

$$\mathbf{A} = \hat{\mathbf{a}}A = \hat{\mathbf{x}}A_x + \hat{\mathbf{y}}A_y + \hat{\mathbf{z}}A_z, \tag{3.6a}$$

$$\mathbf{B} = \hat{\mathbf{b}}B = \hat{\mathbf{x}}B_x + \hat{\mathbf{y}}B_y + \hat{\mathbf{z}}B_z, \tag{3.6b}$$

then $\mathbf{A} = \mathbf{B}$ if and only if $A = B$ and $\hat{\mathbf{a}} = \hat{\mathbf{b}}$, which requires that $A_x = B_x$, $A_y = B_y$, and $A_z = B_z$.

► Equality of two vectors does not necessarily imply that they are identical; in Cartesian coordinates, two displaced parallel vectors of equal magnitude and pointing in the same direction are equal, but they are identical only if they lie on top of one another. ◄

3-1.2 Vector Addition and Subtraction

The sum of two vectors **A** and **B** is a vector $\mathbf{C} = \hat{\mathbf{x}}C_x + \hat{\mathbf{y}}C_y + \hat{\mathbf{z}}C_z$, given by

$$\begin{aligned}\mathbf{C} &= \mathbf{A} + \mathbf{B} \\ &= (\hat{\mathbf{x}}A_x + \hat{\mathbf{y}}A_y + \hat{\mathbf{z}}A_z) + (\hat{\mathbf{x}}B_x + \hat{\mathbf{y}}B_y + \hat{\mathbf{z}}B_z) \\ &= \hat{\mathbf{x}}(A_x + B_x) + \hat{\mathbf{y}}(A_y + B_y) + \hat{\mathbf{z}}(A_z + B_z) \\ &= \hat{\mathbf{x}}C_x + \hat{\mathbf{y}}C_y + \hat{\mathbf{z}}C_z.\end{aligned} \tag{3.7}$$

► Hence, vector addition is commutative:

$$\mathbf{C} = \mathbf{A} + \mathbf{B} = \mathbf{B} + \mathbf{A}. \tag{3.8}$$

Graphically, vector addition can be accomplished by either the parallelogram or the head-to-tail rule (**Fig. 3-3**). Vector **C** is the diagonal of the parallelogram with sides **A** and **B**. With the head-to-tail rule, we may either add **A** to **B** or **B** to **A**. When **A** is added to **B**, it is repositioned so that its tail starts at the tip of **B**, while keeping its length and direction unchanged. The sum vector **C** starts at the tail of **B** and ends at the tip of **A**.

Subtraction of vector **B** from vector **A** is equivalent to the addition of **A** to negative **B**. Thus,

$$\begin{aligned}\mathbf{D} &= \mathbf{A} - \mathbf{B} \\ &= \mathbf{A} + (-\mathbf{B}) \\ &= \hat{\mathbf{x}}(A_x - B_x) + \hat{\mathbf{y}}(A_y - B_y) + \hat{\mathbf{z}}(A_z - B_z).\end{aligned} \tag{3.9}$$

Graphically, the same rules used for vector addition are also applicable to vector subtraction; the only difference is that the arrowhead of $(-\mathbf{B})$ is drawn on the opposite end of the line segment representing the vector **B** (i.e., the tail and head are interchanged).

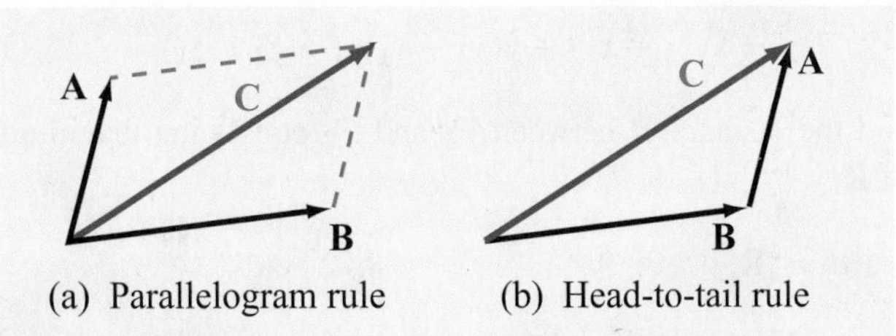

Figure 3-3 Vector addition by (a) the parallelogram rule and (b) the head-to-tail rule.

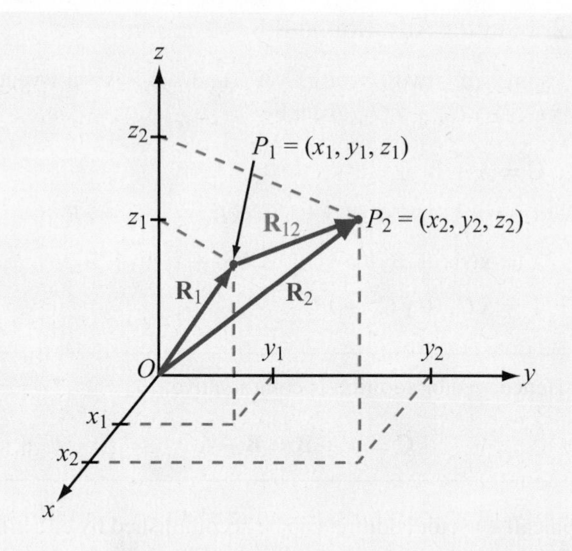

Figure 3-4 Distance vector $\mathbf{R}_{12} = \overrightarrow{P_1P_2} = \mathbf{R}_2 - \mathbf{R}_1$, where $\mathbf{R}_1$ and $\mathbf{R}_2$ are the position vectors of points P_1 and P_2, respectively.

3-1.3 Position and Distance Vectors

The ***position vector*** of a point P in space is the vector from the origin to P. Assuming points P_1 and P_2 are at (x_1, y_1, z_1) and (x_2, y_2, z_2) in **Fig. 3-4**, their position vectors are

$$\mathbf{R}_1 = \overrightarrow{OP_1} = \hat{\mathbf{x}}x_1 + \hat{\mathbf{y}}y_1 + \hat{\mathbf{z}}z_1, \tag{3.10a}$$

$$\mathbf{R}_2 = \overrightarrow{OP_2} = \hat{\mathbf{x}}x_2 + \hat{\mathbf{y}}y_2 + \hat{\mathbf{z}}z_2, \tag{3.10b}$$

where point O is the origin. The ***distance vector*** from P_1 to P_2 is defined as

$$\begin{aligned}\mathbf{R}_{12} &= \overrightarrow{P_1P_2} \\ &= \mathbf{R}_2 - \mathbf{R}_1 \\ &= \hat{\mathbf{x}}(x_2 - x_1) + \hat{\mathbf{y}}(y_2 - y_1) + \hat{\mathbf{z}}(z_2 - z_1),\end{aligned} \tag{3.11}$$

and the distance d between P_1 and P_2 equals the magnitude of $\mathbf{R}_{12}$:

$$\begin{aligned}d &= |\mathbf{R}_{12}| \\ &= [(x_2 - x_1)^2 + (y_2 - y_1)^2 + (z_2 - z_1)^2]^{1/2}.\end{aligned} \tag{3.12}$$

Note that the first and second subscripts of $\mathbf{R}_{12}$ denote the locations of its tail and head, respectively (**Fig. 3-4**).

3-1.4 Vector Multiplication

There exist three types of products in vector calculus: the simple product, the scalar (or dot) product, and the vector (or cross) product.

Simple product

The multiplication of a vector by a scalar is called a ***simple product***. The product of the vector $\mathbf{A} = \hat{\mathbf{a}}A$ by a scalar k results in a vector $\mathbf{B}$ with magnitude $B = kA$ and direction the same as $\mathbf{A}$. That is, $\hat{\mathbf{b}} = \hat{\mathbf{a}}$. In Cartesian coordinates,

$$\begin{aligned}\mathbf{B} = k\mathbf{A} = \hat{\mathbf{a}}kA &= \hat{\mathbf{x}}(kA_x) + \hat{\mathbf{y}}(kA_y) + \hat{\mathbf{z}}(kA_z) \\ &= \hat{\mathbf{x}}\,B_x + \hat{\mathbf{y}}\,B_y + \hat{\mathbf{z}}\,B_z.\end{aligned} \tag{3.13}$$

Scalar or dot product

The ***scalar*** (or ***dot***) ***product*** of two co-anchored vectors $\mathbf{A}$ and $\mathbf{B}$, denoted $\mathbf{A} \cdot \mathbf{B}$ and pronounced "A dot B," is defined geometrically as the product of the magnitude of $\mathbf{A}$ and the scalar component of $\mathbf{B}$ along $\mathbf{A}$, or vice versa. Thus,

$$\mathbf{A} \cdot \mathbf{B} = AB\cos\theta_{AB}, \tag{3.14}$$

where θ_{AB} is the angle between $\mathbf{A}$ and $\mathbf{B}$ (**Fig. 3-5**) measured *from* the tail of $\mathbf{A}$ *to* the tail of $\mathbf{B}$. Angle θ_{AB} is assumed to be in the range $0 \leq \theta_{AB} \leq 180°$. The scalar product of $\mathbf{A}$ and $\mathbf{B}$ yields a scalar whose magnitude is less than or equal to the products of their magnitudes (equality holds when $\theta_{AB} = 0$) and whose sign is positive if $0 < \theta_{AB} < 90°$ and negative if $90° < \theta_{AB} < 180°$. When $\theta_{AB} = 90°$, $\mathbf{A}$ and $\mathbf{B}$ are orthogonal, and their dot product is zero. The quantity $A\cos\theta_{AB}$ is the

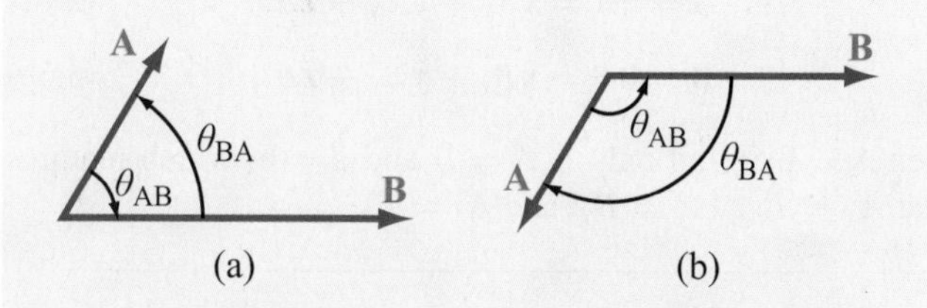

Figure 3-5 The angle θ_{AB} is the angle between $\mathbf{A}$ and $\mathbf{B}$, measured from $\mathbf{A}$ to $\mathbf{B}$ between vector tails. The dot product is positive if $0 \leq \theta_{AB} < 90°$, as in (a), and it is negative if $90° < \theta_{AB} \leq 180°$, as in (b).

scalar component of **A** along **B**. Similarly $B\cos\theta_{BA}$ is the scalar component of **B** along **A**.

The dot product obeys both the commutative and distributive properties of multiplication:

$$\mathbf{A}\cdot\mathbf{B} = \mathbf{B}\cdot\mathbf{A}, \tag{3.15a}$$

(commutative property)

$$\mathbf{A}\cdot(\mathbf{B}+\mathbf{C}) = \mathbf{A}\cdot\mathbf{B}+\mathbf{A}\cdot\mathbf{C}, \tag{3.15b}$$

(distributive property)

The commutative property follows from Eq. (3.14) and the fact that $\theta_{AB} = \theta_{BA}$. The distributive property expresses the fact that the scalar component of the sum of two vectors along a third one equals the sum of their respective scalar components.

The dot product of a vector with itself gives

$$\mathbf{A}\cdot\mathbf{A} = |\mathbf{A}|^2 = A^2, \tag{3.16}$$

which implies that

$$A = |\mathbf{A}| = \sqrt[+]{\mathbf{A}\cdot\mathbf{A}}\,. \tag{3.17}$$

Also, θ_{AB} can be determined from

$$\theta_{AB} = \cos^{-1}\left[\frac{\mathbf{A}\cdot\mathbf{B}}{\sqrt[+]{\mathbf{A}\cdot\mathbf{A}}\;\sqrt[+]{\mathbf{B}\cdot\mathbf{B}}}\right]. \tag{3.18}$$

Since the base vectors $\hat{\mathbf{x}}$, $\hat{\mathbf{y}}$, and $\hat{\mathbf{z}}$ are each orthogonal to the other two, it follows that

$$\hat{\mathbf{x}}\cdot\hat{\mathbf{x}} = \hat{\mathbf{y}}\cdot\hat{\mathbf{y}} = \hat{\mathbf{z}}\cdot\hat{\mathbf{z}} = 1, \tag{3.19a}$$

$$\hat{\mathbf{x}}\cdot\hat{\mathbf{y}} = \hat{\mathbf{y}}\cdot\hat{\mathbf{z}} = \hat{\mathbf{z}}\cdot\hat{\mathbf{x}} = 0. \tag{3.19b}$$

If $\mathbf{A} = (A_x, A_y, A_z)$ and $\mathbf{B} = (B_x, B_y, B_z)$, then

$$\mathbf{A}\cdot\mathbf{B} = (\hat{\mathbf{x}}A_x + \hat{\mathbf{y}}A_y + \hat{\mathbf{z}}A_z)\cdot(\hat{\mathbf{x}}B_x + \hat{\mathbf{y}}B_y + \hat{\mathbf{z}}B_z). \tag{3.20}$$

Use of Eqs. (3.19a) and (3.19b) in Eq. (3.20) leads to

$$\mathbf{A}\cdot\mathbf{B} = A_xB_x + A_yB_y + A_zB_z. \tag{3.21}$$

Vector or cross product

The ***vector*** (or ***cross***) ***product*** of two vectors **A** and **B**, denoted $\mathbf{A}\times\mathbf{B}$ and pronounced "*A* cross *B*," yields a vector defined as

$$\mathbf{A}\times\mathbf{B} = \hat{\mathbf{n}}\,AB\sin\theta_{AB}, \tag{3.22}$$

where $\hat{\mathbf{n}}$ is a *unit vector normal to the plane containing* **A** *and* **B** [Fig. 3-6(a)]. The magnitude of the cross product, $AB|\sin\theta_{AB}|$, equals the area of the parallelogram defined by the two vectors. The direction of $\hat{\mathbf{n}}$ is governed by the following ***right-hand rule*** [Fig. 3-6(b)]: $\hat{\mathbf{n}}$ points in the direction of the right thumb when the fingers rotate from **A** to **B** through the angle θ_{AB}. Note that, since $\hat{\mathbf{n}}$ is perpendicular to the plane containing **A** and **B**, $\mathbf{A}\times\mathbf{B}$ is perpendicular to both vectors **A** and **B**.

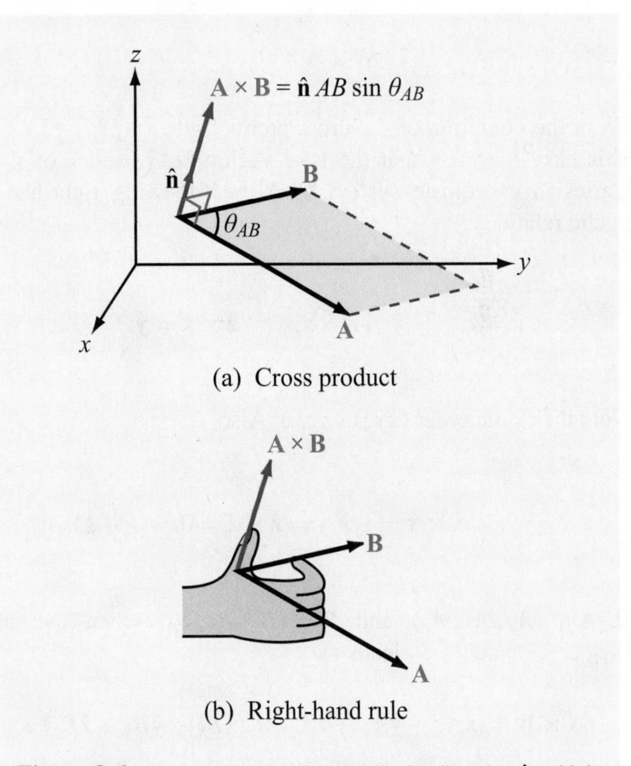

Figure 3-6 Cross product $\mathbf{A}\times\mathbf{B}$ points in the direction $\hat{\mathbf{n}}$, which is perpendicular to the plane containing **A** and **B** and defined by the right-hand rule.

The cross product is anticommutative and distributive:

$$\mathbf{A} \times \mathbf{B} = -\mathbf{B} \times \mathbf{A} \quad \text{(anticommutative)}. \tag{3.23a}$$

The anticommutative property follows from the application of the right-hand rule to determine $\hat{\mathbf{n}}$. The distributive property follows from the fact that the area of the parallelogram formed by **A** and (**B** + **C**) equals the sum of those formed by (**A** and **B**) and (**A** and **C**):

$$\mathbf{A} \times (\mathbf{B} + \mathbf{C}) = \mathbf{A} \times \mathbf{B} + \mathbf{A} \times \mathbf{C}, \quad \text{(distributive)} \tag{3.23b}$$

The cross product of a vector with itself vanishes. That is,

$$\mathbf{A} \times \mathbf{A} = 0. \tag{3.24}$$

From the definition of the cross product given by Eq. (3.22), it is easy to verify that the base vectors $\hat{\mathbf{x}}$, $\hat{\mathbf{y}}$, and $\hat{\mathbf{z}}$ of the Cartesian coordinate system obey the following right-hand cyclic relations:

$$\hat{\mathbf{x}} \times \hat{\mathbf{y}} = \hat{\mathbf{z}}, \qquad \hat{\mathbf{y}} \times \hat{\mathbf{z}} = \hat{\mathbf{x}}, \qquad \hat{\mathbf{z}} \times \hat{\mathbf{x}} = \hat{\mathbf{y}}. \tag{3.25}$$

Note the cyclic order $(xyzxyz\ldots)$. Also,

$$\hat{\mathbf{x}} \times \hat{\mathbf{x}} = \hat{\mathbf{y}} \times \hat{\mathbf{y}} = \hat{\mathbf{z}} \times \hat{\mathbf{z}} = 0. \tag{3.26}$$

If $\mathbf{A} = (A_x, A_y, A_z)$ and $\mathbf{B} = (B_x, B_y, B_z)$, then use of Eqs. (3.25) and (3.26) leads to

$$\begin{aligned} \mathbf{A} \times \mathbf{B} &= (\hat{\mathbf{x}}A_x + \hat{\mathbf{y}}A_y + \hat{\mathbf{z}}A_z) \times (\hat{\mathbf{x}}B_x + \hat{\mathbf{y}}B_y + \hat{\mathbf{z}}B_z) \\ &= \hat{\mathbf{x}}(A_yB_z - A_zB_y) + \hat{\mathbf{y}}(A_zB_x - A_xB_z) \\ &\quad + \hat{\mathbf{z}}(A_xB_y - A_yB_x). \end{aligned} \tag{3.27}$$

The cyclical form of the result given by Eq. (3.27) allows us to express the cross product in the form of a determinant:

$$\mathbf{A} \times \mathbf{B} = \begin{vmatrix} \hat{\mathbf{x}} & \hat{\mathbf{y}} & \hat{\mathbf{z}} \\ A_x & A_y & A_z \\ B_x & B_y & B_z \end{vmatrix}. \tag{3.28}$$

Example 3-1: Vectors and Angles

In Cartesian coordinates, vector **A** points from the origin to point $P_1 = (2, 3, 3)$, and vector **B** is directed from P_1 to point $P_2 = (1, -2, 2)$. Find:

(a) vector **A**, its magnitude A, and unit vector $\hat{\mathbf{a}}$,
(b) the angle between **A** and the y axis,
(c) vector **B**,
(d) the angle θ_{AB} between **A** and **B**, and
(e) perpendicular distance from the origin to vector **B**.

Solution: (a) Vector **A** is given by the position vector of $P_1 = (2, 3, 3)$ (Fig. 3-7). Thus,

$$\mathbf{A} = \hat{\mathbf{x}}2 + \hat{\mathbf{y}}3 + \hat{\mathbf{z}}3,$$

$$A = |\mathbf{A}| = \sqrt{2^2 + 3^2 + 3^2} = \sqrt{22}\,,$$

$$\hat{\mathbf{a}} = \frac{\mathbf{A}}{A} = (\hat{\mathbf{x}}2 + \hat{\mathbf{y}}3 + \hat{\mathbf{z}}3)/\sqrt{22}\,.$$

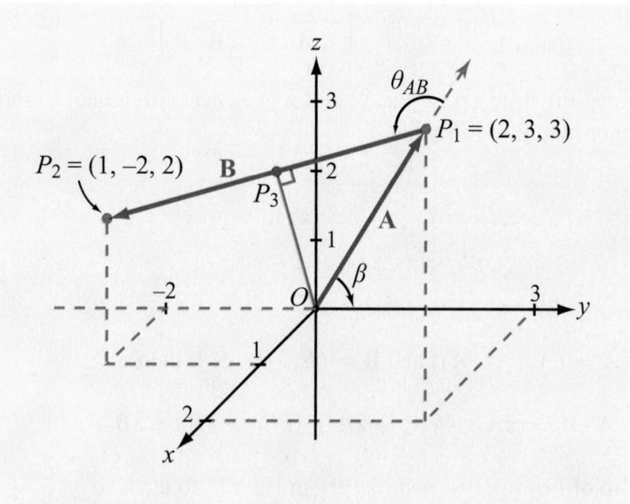

Figure 3-7 Geometry of Example 3-1.

(b) The angle β between **A** and the y axis is obtained from

$$\mathbf{A} \cdot \hat{\mathbf{y}} = |\mathbf{A}||\hat{\mathbf{y}}| \cos \beta = A \cos \beta,$$

or

$$\beta = \cos^{-1}\left(\frac{\mathbf{A} \cdot \hat{\mathbf{y}}}{A}\right) = \cos^{-1}\left(\frac{3}{\sqrt{22}}\right) = 50.2°.$$

(c)

$$\mathbf{B} = \hat{\mathbf{x}}(1-2) + \hat{\mathbf{y}}(-2-3) + \hat{\mathbf{z}}(2-3) = -\hat{\mathbf{x}} - \hat{\mathbf{y}}5 - \hat{\mathbf{z}}.$$

(d)

$$\theta_{AB} = \cos^{-1}\left[\frac{\mathbf{A} \cdot \mathbf{B}}{|\mathbf{A}||\mathbf{B}|}\right] = \cos^{-1}\left[\frac{(-2-15-3)}{\sqrt{22}\,\sqrt{27}}\right] = 145.1°.$$

(e) The perpendicular distance between the origin and vector **B** is the distance $|\overrightarrow{OP_3}|$ shown in **Fig. 3-7**. From right triangle OP_1P_3,

$$|\overrightarrow{OP_3}| = |\mathbf{A}| \sin(180° - \theta_{AB}) = \sqrt{22}\,\sin(180° - 145.1°) = 2.68.$$

Exercise 3-1: Find the distance vector between $P_1 = (1, 2, 3)$ and $P_2 = (-1, -2, 3)$ in Cartesian coordinates.

Answer: $\overrightarrow{P_1P_2} = -\hat{\mathbf{x}}2 - \hat{\mathbf{y}}4$. (See EM.)

Exercise 3-2: Find the angle θ_{AB} between vectors **A** and **B** of Example 3-1 from the cross product between them.

Answer: $\theta_{AB} = 145.1°$. (See EM.)

Exercise 3-3: Find the angle between vector **B** of Example 3-1 and the z axis.

Answer: 101.1°. (See EM.)

Exercise 3-4: Vectors **A** and **B** lie in the y-z plane and both have the same magnitude of 2 (**Fig. E3.4**). Determine (a) $\mathbf{A} \cdot \mathbf{B}$ and (b) $\mathbf{A} \times \mathbf{B}$.

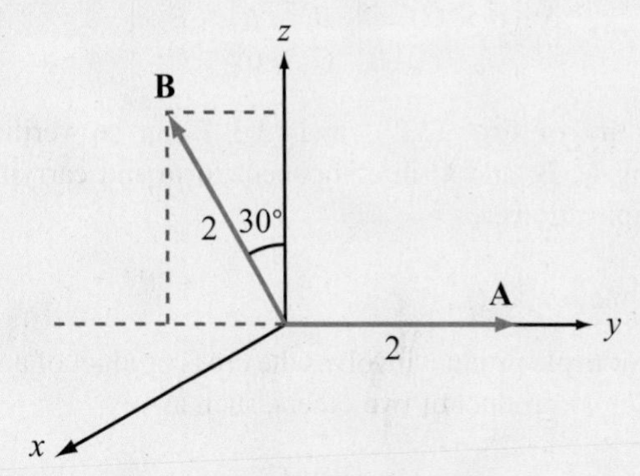

Figure E3.4

Answer: (a) $\mathbf{A} \cdot \mathbf{B} = -2$; (b) $\mathbf{A} \times \mathbf{B} = \hat{\mathbf{x}}\,3.46$. (See EM.)

Exercise 3-5: If $\mathbf{A} \cdot \mathbf{B} = \mathbf{A} \cdot \mathbf{C}$, does it follow that $\mathbf{B} = \mathbf{C}$?

Answer: No. (See EM.)

3-1.5 Scalar and Vector Triple Products

When three vectors are multiplied, not all combinations of dot and cross products are meaningful. For example, the product

$$\mathbf{A} \times (\mathbf{B} \cdot \mathbf{C})$$

does not make sense because $\mathbf{B} \cdot \mathbf{C}$ is a scalar, and the cross product of the vector **A** with a scalar is not defined under the rules of vector algebra. Other than the product of the form $\mathbf{A}(\mathbf{B} \cdot \mathbf{C})$, the only two meaningful products of three vectors are the scalar triple product and the vector triple product.

Scalar triple product

The dot product of a vector with the cross product of two other vectors is called a scalar triple product, so named because the result is a scalar. A scalar triple product obeys the following cyclic order:

$$\mathbf{A} \cdot (\mathbf{B} \times \mathbf{C}) = \mathbf{B} \cdot (\mathbf{C} \times \mathbf{A}) = \mathbf{C} \cdot (\mathbf{A} \times \mathbf{B}). \qquad (3.29)$$

The equalities hold as long as the cyclic order $(ABCABC\ldots)$ is preserved. The scalar triple product of vectors

$\mathbf{A} = (A_x, A_y, A_z)$, $\mathbf{B} = (B_x, B_y, B_z)$, and $\mathbf{C} = (C_x, C_y, C_z)$ can be expressed in the form of a 3×3 determinant:

$$\mathbf{A} \cdot (\mathbf{B} \times \mathbf{C}) = \begin{vmatrix} A_x & A_y & A_z \\ B_x & B_y & B_z \\ C_x & C_y & C_z \end{vmatrix}. \quad (3.30)$$

The validity of Eqs. (3.29) and (3.30) can be verified by expanding **A**, **B**, and **C** in component form and carrying out the multiplications.

Vector triple product

The vector triple product involves the cross product of a vector with the cross product of two others, such as

$$\mathbf{A} \times (\mathbf{B} \times \mathbf{C}). \quad (3.31)$$

Since each cross product yields a vector, the result of a vector triple product is also a vector. The vector triple product does not obey the associative law. That is,

$$\mathbf{A} \times (\mathbf{B} \times \mathbf{C}) \neq (\mathbf{A} \times \mathbf{B}) \times \mathbf{C}, \quad (3.32)$$

which means that it is important to specify which cross multiplication is to be performed first. By expanding the vectors **A**, **B**, and **C** in component form, it can be shown that

$$\mathbf{A} \times (\mathbf{B} \times \mathbf{C}) = \mathbf{B}(\mathbf{A} \cdot \mathbf{C}) - \mathbf{C}(\mathbf{A} \cdot \mathbf{B}), \quad (3.33)$$

which is known as the "bac-cab" rule.

Example 3-2: Vector Triple Product

Given $\mathbf{A} = \hat{\mathbf{x}} - \hat{\mathbf{y}} + \hat{\mathbf{z}}2$, $\mathbf{B} = \hat{\mathbf{y}} + \hat{\mathbf{z}}$, and $\mathbf{C} = -\hat{\mathbf{x}}2 + \hat{\mathbf{z}}3$, find $(\mathbf{A} \times \mathbf{B}) \times \mathbf{C}$ and compare it with $\mathbf{A} \times (\mathbf{B} \times \mathbf{C})$.

Solution:

$$\mathbf{A} \times \mathbf{B} = \begin{vmatrix} \hat{\mathbf{x}} & \hat{\mathbf{y}} & \hat{\mathbf{z}} \\ 1 & -1 & 2 \\ 0 & 1 & 1 \end{vmatrix} = -\hat{\mathbf{x}}3 - \hat{\mathbf{y}} + \hat{\mathbf{z}}$$

and

$$(\mathbf{A} \times \mathbf{B}) \times \mathbf{C} = \begin{vmatrix} \hat{\mathbf{x}} & \hat{\mathbf{y}} & \hat{\mathbf{z}} \\ -3 & -1 & 1 \\ -2 & 0 & 3 \end{vmatrix} = -\hat{\mathbf{x}}3 + \hat{\mathbf{y}}7 - \hat{\mathbf{z}}2.$$

A similar procedure gives $\mathbf{A} \times (\mathbf{B} \times \mathbf{C}) = \hat{\mathbf{x}}2 + \hat{\mathbf{y}}4 + \hat{\mathbf{z}}$. The fact that the results of two vector triple products are different demonstrates the inequality stated in Eq. (3.32).

Concept Question 3-1: When are two vectors equal and when are they identical?

Concept Question 3-2: When is the position vector of a point identical to the distance vector between two points?

Concept Question 3-3: If $\mathbf{A} \cdot \mathbf{B} = 0$, what is θ_{AB}?

Concept Question 3-4: If $\mathbf{A} \times \mathbf{B} = 0$, what is θ_{AB}?

Concept Question 3-5: Is $\mathbf{A}(\mathbf{B} \cdot \mathbf{C})$ a vector triple product?

Concept Question 3-6: If $\mathbf{A} \cdot \mathbf{B} = \mathbf{A} \cdot \mathbf{C}$, does it follow that $\mathbf{B} = \mathbf{C}$?

3-2 Orthogonal Coordinate Systems

A three-dimensional coordinate system allows us to uniquely specify locations of points in space and the magnitudes and directions of vectors. Coordinate systems may be orthogonal or nonorthogonal.

► An ***orthogonal coordinate system*** is one in which coordinates are measured along locally mutually perpendicular axes. ◄

Nonorthogonal systems are very specialized and seldom used in solving practical problems. Many orthogonal coordinate systems have been devised, but the most commonly used are

- the Cartesian (also called rectangular),
- the cylindrical, and
- the spherical coordinate system.

Why do we need more than one coordinate system? Whereas a point in space has the same location and an object has the same shape regardless of which coordinate system is used to describe them, the solution of a practical problem can be greatly facilitated by the choice of a coordinate system that best fits the geometry under consideration. The following subsections examine the properties of each of the aforementioned orthogonal systems, and Section 3-3 describes how a point or vector may be transformed from one system to another.

3-2.1 Cartesian Coordinates

The Cartesian coordinate system was introduced in Section 3-1 to illustrate the laws of vector algebra. Instead of repeating these laws for the Cartesian system, we summarize them in **Table 3-1**. Differential calculus involves the use of differential lengths, areas, and volumes. In Cartesian coordinates a ***differential length vector*** (**Fig. 3-8**) is expressed as

$$d\mathbf{l} = \hat{\mathbf{x}}\, dl_x + \hat{\mathbf{y}}\, dl_y + \hat{\mathbf{z}}\, dl_z = \hat{\mathbf{x}}\, dx + \hat{\mathbf{y}}\, dy + \hat{\mathbf{z}}\, dz, \quad (3.34)$$

where $dl_x = dx$ is a differential length along $\hat{\mathbf{x}}$, and similar interpretations apply to $dl_y = dy$ and $dl_z = dz$.

A ***differential area vector*** $d\mathbf{s}$ is a vector with magnitude ds equal to the product of two differential lengths (such as dl_y and dl_z), and direction specified by a unit vector along the third direction (such as $\hat{\mathbf{x}}$). Thus, for a differential area vector in the y–z plane,

$$d\mathbf{s}_x = \hat{\mathbf{x}}\, dl_y\, dl_z = \hat{\mathbf{x}}\, dy\, dz \qquad (y\text{–}z \text{ plane}), \quad (3.35a)$$

Table 3-1 **Summary of vector relations.**

	Cartesian Coordinates	Cylindrical Coordinates	Spherical Coordinates
Coordinate variables	x, y, z	r, ϕ, z	R, θ, ϕ
Vector representation $\mathbf{A} =$	$\hat{\mathbf{x}}A_x + \hat{\mathbf{y}}A_y + \hat{\mathbf{z}}A_z$	$\hat{\mathbf{r}}A_r + \hat{\boldsymbol{\phi}}A_\phi + \hat{\mathbf{z}}A_z$	$\hat{\mathbf{R}}A_R + \hat{\boldsymbol{\theta}}A_\theta + \hat{\boldsymbol{\phi}}A_\phi$
Magnitude of A $\lvert\mathbf{A}\rvert =$	$\sqrt[+]{A_x^2 + A_y^2 + A_z^2}$	$\sqrt[+]{A_r^2 + A_\phi^2 + A_z^2}$	$\sqrt[+]{A_R^2 + A_\theta^2 + A_\phi^2}$
Position vector $\overrightarrow{OP_1} =$	$\hat{\mathbf{x}}x_1 + \hat{\mathbf{y}}y_1 + \hat{\mathbf{z}}z_1$, for $P(x_1, y_1, z_1)$	$\hat{\mathbf{r}}r_1 + \hat{\mathbf{z}}z_1$, for $P(r_1, \phi_1, z_1)$	$\hat{\mathbf{R}}R_1$, for $P(R_1, \theta_1, \phi_1)$
Base vectors properties	$\hat{\mathbf{x}}\cdot\hat{\mathbf{x}} = \hat{\mathbf{y}}\cdot\hat{\mathbf{y}} = \hat{\mathbf{z}}\cdot\hat{\mathbf{z}} = 1$ $\hat{\mathbf{x}}\cdot\hat{\mathbf{y}} = \hat{\mathbf{y}}\cdot\hat{\mathbf{z}} = \hat{\mathbf{z}}\cdot\hat{\mathbf{x}} = 0$ $\hat{\mathbf{x}}\times\hat{\mathbf{y}} = \hat{\mathbf{z}}$ $\hat{\mathbf{y}}\times\hat{\mathbf{z}} = \hat{\mathbf{x}}$ $\hat{\mathbf{z}}\times\hat{\mathbf{x}} = \hat{\mathbf{y}}$	$\hat{\mathbf{r}}\cdot\hat{\mathbf{r}} = \hat{\boldsymbol{\phi}}\cdot\hat{\boldsymbol{\phi}} = \hat{\mathbf{z}}\cdot\hat{\mathbf{z}} = 1$ $\hat{\mathbf{r}}\cdot\hat{\boldsymbol{\phi}} = \hat{\boldsymbol{\phi}}\cdot\hat{\mathbf{z}} = \hat{\mathbf{z}}\cdot\hat{\mathbf{r}} = 0$ $\hat{\mathbf{r}}\times\hat{\boldsymbol{\phi}} = \hat{\mathbf{z}}$ $\hat{\boldsymbol{\phi}}\times\hat{\mathbf{z}} = \hat{\mathbf{r}}$ $\hat{\mathbf{z}}\times\hat{\mathbf{r}} = \hat{\boldsymbol{\phi}}$	$\hat{\mathbf{R}}\cdot\hat{\mathbf{R}} = \hat{\boldsymbol{\theta}}\cdot\hat{\boldsymbol{\theta}} = \hat{\boldsymbol{\phi}}\cdot\hat{\boldsymbol{\phi}} = 1$ $\hat{\mathbf{R}}\cdot\hat{\boldsymbol{\theta}} = \hat{\boldsymbol{\theta}}\cdot\hat{\boldsymbol{\phi}} = \hat{\boldsymbol{\phi}}\cdot\hat{\mathbf{R}} = 0$ $\hat{\mathbf{R}}\times\hat{\boldsymbol{\theta}} = \hat{\boldsymbol{\phi}}$ $\hat{\boldsymbol{\theta}}\times\hat{\boldsymbol{\phi}} = \hat{\mathbf{R}}$ $\hat{\boldsymbol{\phi}}\times\hat{\mathbf{R}} = \hat{\boldsymbol{\theta}}$
Dot product $\mathbf{A}\cdot\mathbf{B} =$	$A_xB_x + A_yB_y + A_zB_z$	$A_rB_r + A_\phi B_\phi + A_zB_z$	$A_RB_R + A_\theta B_\theta + A_\phi B_\phi$
Cross product $\mathbf{A}\times\mathbf{B} =$	$\begin{vmatrix} \hat{\mathbf{x}} & \hat{\mathbf{y}} & \hat{\mathbf{z}} \\ A_x & A_y & A_z \\ B_x & B_y & B_z \end{vmatrix}$	$\begin{vmatrix} \hat{\mathbf{r}} & \hat{\boldsymbol{\phi}} & \hat{\mathbf{z}} \\ A_r & A_\phi & A_z \\ B_r & B_\phi & B_z \end{vmatrix}$	$\begin{vmatrix} \hat{\mathbf{R}} & \hat{\boldsymbol{\theta}} & \hat{\boldsymbol{\phi}} \\ A_R & A_\theta & A_\phi \\ B_R & B_\theta & B_\phi \end{vmatrix}$
Differential length $d\mathbf{l} =$	$\hat{\mathbf{x}}\,dx + \hat{\mathbf{y}}\,dy + \hat{\mathbf{z}}\,dz$	$\hat{\mathbf{r}}\,dr + \hat{\boldsymbol{\phi}}r\,d\phi + \hat{\mathbf{z}}\,dz$	$\hat{\mathbf{R}}\,dR + \hat{\boldsymbol{\theta}}R\,d\theta + \hat{\boldsymbol{\phi}}R\sin\theta\,d\phi$
Differential surface areas	$d\mathbf{s}_x = \hat{\mathbf{x}}\,dy\,dz$ $d\mathbf{s}_y = \hat{\mathbf{y}}\,dx\,dz$ $d\mathbf{s}_z = \hat{\mathbf{z}}\,dx\,dy$	$d\mathbf{s}_r = \hat{\mathbf{r}}r\,d\phi\,dz$ $d\mathbf{s}_\phi = \hat{\boldsymbol{\phi}}\,dr\,dz$ $d\mathbf{s}_z = \hat{\mathbf{z}}r\,dr\,d\phi$	$d\mathbf{s}_R = \hat{\mathbf{R}}R^2\sin\theta\,d\theta\,d\phi$ $d\mathbf{s}_\theta = \hat{\boldsymbol{\theta}}R\sin\theta\,dR\,d\phi$ $d\mathbf{s}_\phi = \hat{\boldsymbol{\phi}}R\,dR\,d\theta$
Differential volume $d\mathcal{V} =$	$dx\,dy\,dz$	$r\,dr\,d\phi\,dz$	$R^2\sin\theta\,dR\,d\theta\,d\phi$

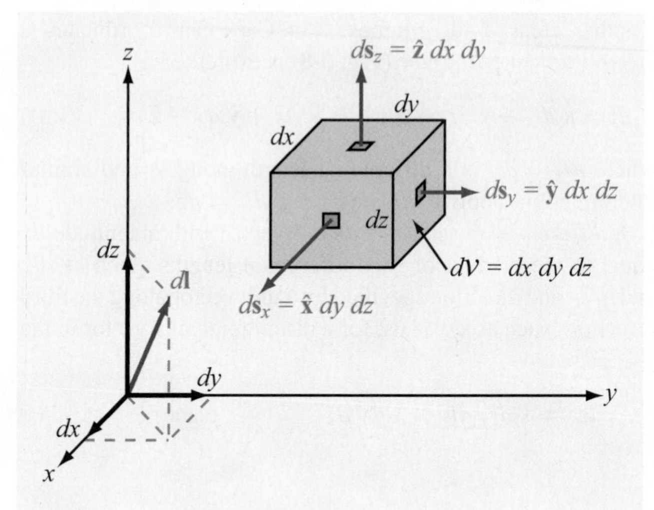

Figure 3-8 Differential length, area, and volume in Cartesian coordinates.

with the subscript on $d\mathbf{s}$ denoting its direction. Similarly,

$$d\mathbf{s}_y = \hat{\mathbf{y}}\, dx\, dz \qquad (x\text{–}z \text{ plane}), \tag{3.35b}$$

$$d\mathbf{s}_z = \hat{\mathbf{z}}\, dx\, dy \qquad (x\text{–}y \text{ plane}). \tag{3.35c}$$

A ***differential volume*** equals the product of all three differential lengths:

$$d\mathcal{V} = dx\, dy\, dz. \tag{3.36}$$

3-2.2 Cylindrical Coordinates

The cylindrical coordinate system is useful for solving problems involving structures with cylindrical symmetry, such as calculating the capacitance per unit length of a coaxial transmission line. In the cylindrical coordinate system, the location of a point in space is defined by three variables, r, ϕ, and z (**Fig. 3-9**). The coordinate r is the ***radial distance*** in the x–y plane, ϕ is the ***azimuth angle*** measured from the positive x axis, and z is as previously defined in the Cartesian coordinate system. Their ranges are $0 \le r < \infty$, $0 \le \phi < 2\pi$, and $-\infty < z < \infty$. Point $P(r_1, \phi_1, z_1)$ in **Fig. 3-9** is located

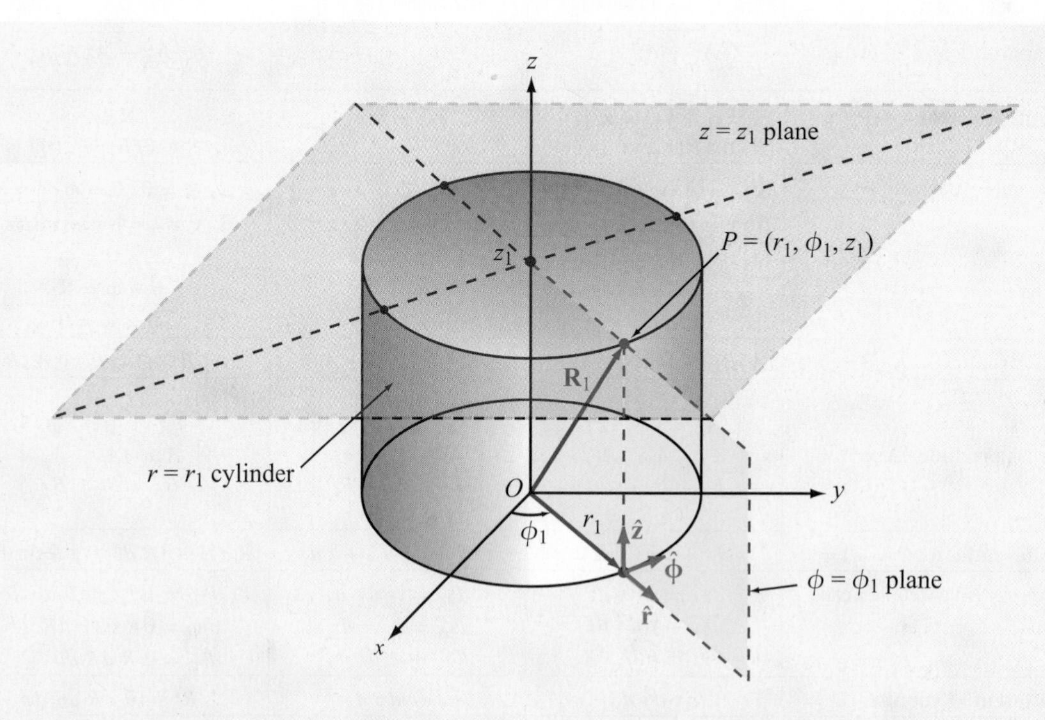

Figure 3-9 Point $P(r_1, \phi_1, z_1)$ in cylindrical coordinates; r_1 is the radial distance from the origin in the x–y plane, ϕ_1 is the angle in the x–y plane measured from the x axis toward the y axis, and z_1 is the vertical distance from the x–y plane.

at the intersection of three surfaces. These are the cylindrical surface defined by $r = r_1$, the vertical half-plane defined by $\phi = \phi_1$ (which extends outwardly from the z axis), and the horizontal plane defined by $z = z_1$.

▶ The mutually perpendicular base vectors are $\hat{\mathbf{r}}$, $\hat{\boldsymbol{\phi}}$, and $\hat{\mathbf{z}}$, with $\hat{\mathbf{r}}$ pointing away from the origin along r, $\hat{\boldsymbol{\phi}}$ pointing in a direction tangential to the cylindrical surface, and $\hat{\mathbf{z}}$ pointing along the vertical. Unlike the Cartesian system, in which the base vectors $\hat{\mathbf{x}}$, $\hat{\mathbf{y}}$, and $\hat{\mathbf{z}}$ are independent of the location of P, in the cylindrical system both $\hat{\mathbf{r}}$ and $\hat{\boldsymbol{\phi}}$ are functions of ϕ. ◀

The base unit vectors obey the following right-hand cyclic relations:

$$\hat{\mathbf{r}} \times \hat{\boldsymbol{\phi}} = \hat{\mathbf{z}}, \qquad \hat{\boldsymbol{\phi}} \times \hat{\mathbf{z}} = \hat{\mathbf{r}}, \qquad \hat{\mathbf{z}} \times \hat{\mathbf{r}} = \hat{\boldsymbol{\phi}}, \tag{3.37}$$

and like all unit vectors, $\hat{\mathbf{r}} \cdot \hat{\mathbf{r}} = \hat{\boldsymbol{\phi}} \cdot \hat{\boldsymbol{\phi}} = \hat{\mathbf{z}} \cdot \hat{\mathbf{z}} = 1$, and $\hat{\mathbf{r}} \times \hat{\mathbf{r}} = \hat{\boldsymbol{\phi}} \times \hat{\boldsymbol{\phi}} = \hat{\mathbf{z}} \times \hat{\mathbf{z}} = 0$.

In cylindrical coordinates, a vector is expressed as

$$\mathbf{A} = \hat{\mathbf{a}}|\mathbf{A}| = \hat{\mathbf{r}}A_r + \hat{\boldsymbol{\phi}}A_\phi + \hat{\mathbf{z}}A_z, \tag{3.38}$$

where A_r, A_ϕ, and A_z are the components of $\mathbf{A}$ along the $\hat{\mathbf{r}}$, $\hat{\boldsymbol{\phi}}$, and $\hat{\mathbf{z}}$ directions. The magnitude of $\mathbf{A}$ is obtained by applying Eq. (3.17), which gives

$$|\mathbf{A}| = \sqrt[+]{\mathbf{A} \cdot \mathbf{A}} = \sqrt[+]{A_r^2 + A_\phi^2 + A_z^2}\,. \tag{3.39}$$

The position vector $\overrightarrow{OP}$ shown in **Fig. 3-9** has components along r and z only. Thus,

$$\mathbf{R}_1 = \overrightarrow{OP} = \hat{\mathbf{r}}r_1 + \hat{\mathbf{z}}z_1. \tag{3.40}$$

The dependence of $\mathbf{R}_1$ on ϕ_1 is implicit through the dependence of $\hat{\mathbf{r}}$ on ϕ_1. Hence, when using Eq. (3.40) to denote the position vector of point $P(r_1, \phi_1, z_1)$, it is necessary to specify that $\hat{\mathbf{r}}$ is at ϕ_1.

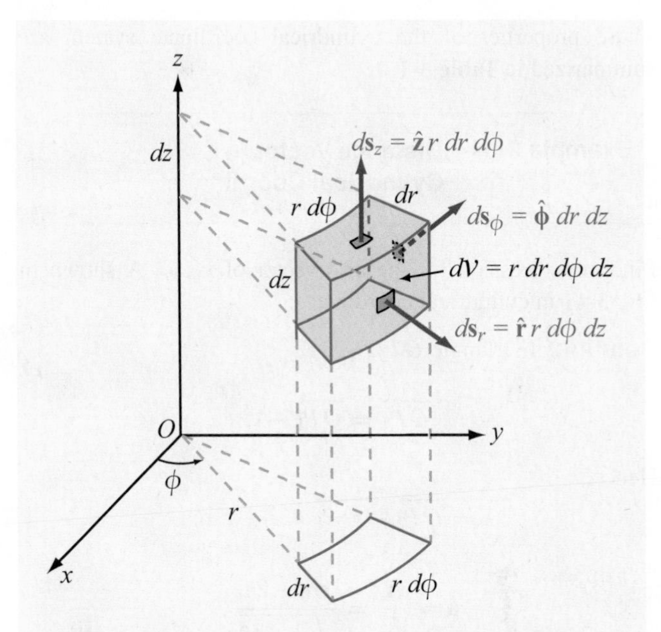

Figure 3-10 Differential areas and volume in cylindrical coordinates.

Figure 3-10 shows a differential volume element in cylindrical coordinates. The differential lengths along $\hat{\mathbf{r}}$, $\hat{\boldsymbol{\phi}}$, and $\hat{\mathbf{z}}$ are

$$dl_r = dr, \qquad dl_\phi = r\,d\phi, \qquad dl_z = dz. \tag{3.41}$$

Note that the differential length along $\hat{\boldsymbol{\phi}}$ is $r\,d\phi$, not just $d\phi$. The differential length $d\mathbf{l}$ in cylindrical coordinates is given by

$$d\mathbf{l} = \hat{\mathbf{r}}\,dl_r + \hat{\boldsymbol{\phi}}\,dl_\phi + \hat{\mathbf{z}}\,dl_z = \hat{\mathbf{r}}\,dr + \hat{\boldsymbol{\phi}}r\,d\phi + \hat{\mathbf{z}}\,dz. \tag{3.42}$$

As was stated previously for the Cartesian coordinate system, the product of any pair of differential lengths is equal to the magnitude of a vector differential surface area with a surface normal pointing along the direction of the third coordinate. Thus,

$$d\mathbf{s}_r = \hat{\mathbf{r}}\,dl_\phi\,dl_z = \hat{\mathbf{r}}r\,d\phi\,dz \quad (\phi\text{–}z \textbf{ cylindrical surface}), \tag{3.43a}$$

$$d\mathbf{s}_\phi = \hat{\boldsymbol{\phi}}\,dl_r\,dl_z = \hat{\boldsymbol{\phi}}\,dr\,dz \quad (r\text{–}z \textbf{ plane}), \tag{3.43b}$$

$$d\mathbf{s}_z = \hat{\mathbf{z}}\,dl_r\,dl_\phi = \hat{\mathbf{z}}r\,dr\,d\phi \quad (r\text{–}\phi \textbf{ plane}). \tag{3.43c}$$

The differential volume is the product of the three differential lengths,

$$d\mathcal{V} = dl_r\,dl_\phi\,dl_z = r\,dr\,d\phi\,dz. \tag{3.44}$$

These properties of the cylindrical coordinate system are summarized in **Table 3-1**.

Example 3-3: Distance Vector in Cylindrical Coordinates

Find an expression for the unit vector of vector **A** shown in **Fig. 3-11** in cylindrical coordinates.

Solution: In triangle OP_1P_2,

$$\overrightarrow{OP_2} = \overrightarrow{OP_1} + \mathbf{A}.$$

Hence,

$$\mathbf{A} = \overrightarrow{OP_2} - \overrightarrow{OP_1} = \hat{\mathbf{r}} r_0 - \hat{\mathbf{z}} h,$$

and

$$\hat{\mathbf{a}} = \frac{\mathbf{A}}{|\mathbf{A}|} = \frac{\hat{\mathbf{r}} r_0 - \hat{\mathbf{z}} h}{\sqrt{r_0^2 + h^2}} .$$

We note that the expression for **A** is independent of ϕ_0. This implies that all vectors from point P_1 to any point on the circle defined by $r = r_0$ in the x–y plane are equal in the cylindrical coordinate system, which is not true. The ambiguity can be resolved by specifying that **A** passes through a point whose $\phi = \phi_0$.

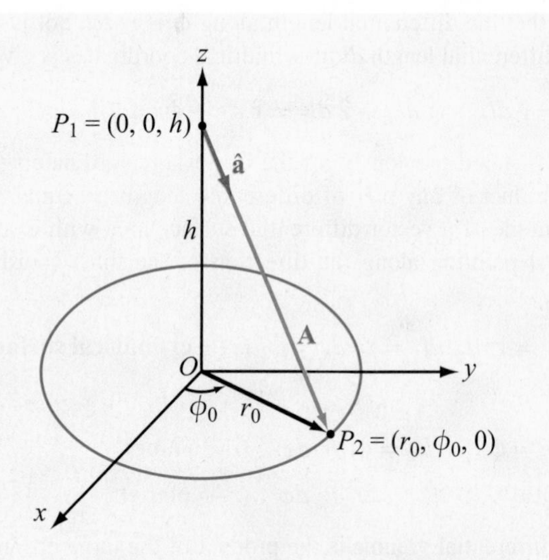

Figure 3-11 Geometry of Example 3-3.

Example 3-4: Cylindrical Area

Find the area of a cylindrical surface described by $r = 5$, $30° \leq \phi \leq 60°$, and $0 \leq z \leq 3$ (**Fig. 3-12**).

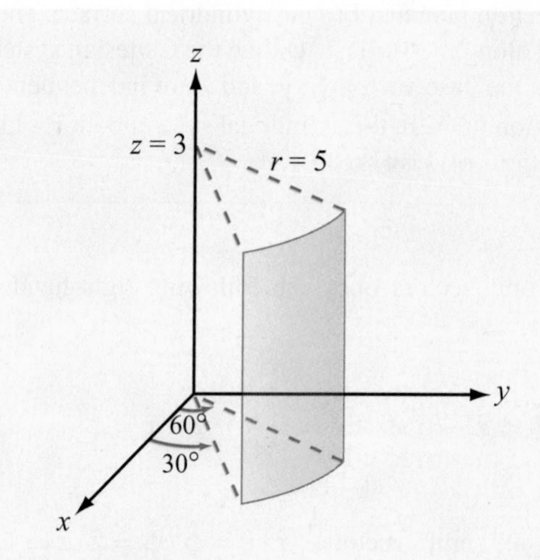

Figure 3-12 Cylindrical surface of Example 3-4.

Solution: The prescribed surface is shown in **Fig. 3-12**. Use of Eq. (3.43a) for a surface element with constant r gives

$$S = r \int_{\phi=30°}^{60°} d\phi \int_{z=0}^{3} dz = 5\phi \Big|_{\pi/6}^{\pi/3} z \Big|_0^3 = \frac{5\pi}{2} .$$

Note that ϕ had to be converted to radians before evaluating the integration limits.

Exercise 3-6: A circular cylinder of radius $r = 5$ cm is concentric with the z axis and extends between $z = -3$ cm and $z = 3$ cm. Use Eq. (3.44) to find the cylinder's volume.

Answer: 471.2 cm^3. (See EM.)

Module 3.1 Vector Addition and Subtraction Display two vectors in rectangular or cylindrical coordinates, and compute their sum and difference.

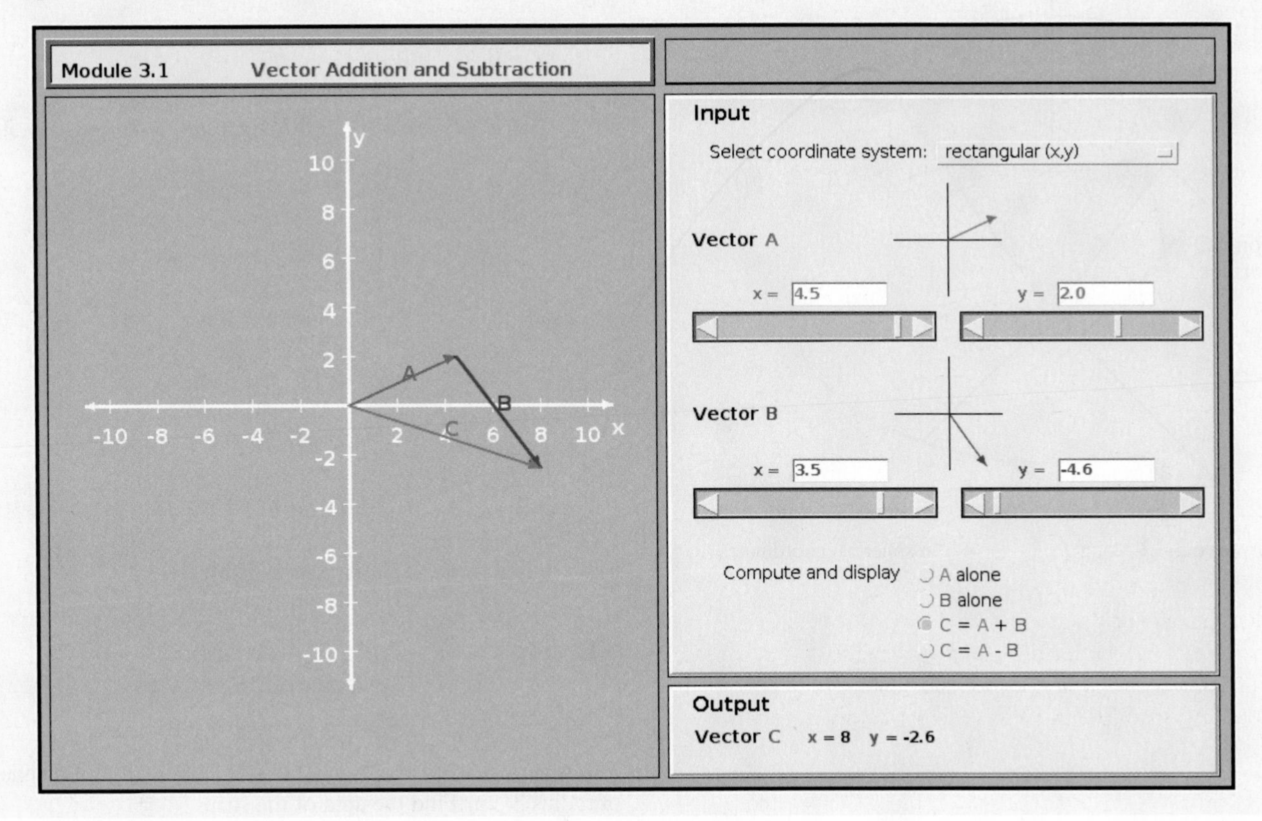

3-2.3 Spherical Coordinates

In the spherical coordinate system, the location of a point in space is uniquely specified by the variables R, θ, and ϕ (Fig. 3-13). The range coordinate R, which measures the distance from the origin to the point, describes a sphere of radius R centered at the origin. The *zenith angle* θ is measured from the positive z axis and it describes a conical surface with its apex at the origin, and the azimuth angle ϕ is the same as in cylindrical coordinates. The ranges of R, θ, and ϕ are $0 \le R < \infty$, $0 \le \theta \le \pi$, and $0 \le \phi < 2\pi$. The base vectors $\hat{\mathbf{R}}$, $\hat{\boldsymbol{\theta}}$, and $\hat{\boldsymbol{\phi}}$ obey the following right-hand cyclic relations:

$$\hat{\mathbf{R}} \times \hat{\boldsymbol{\theta}} = \hat{\boldsymbol{\phi}}, \quad \hat{\boldsymbol{\theta}} \times \hat{\boldsymbol{\phi}} = \hat{\mathbf{R}}, \quad \hat{\boldsymbol{\phi}} \times \hat{\mathbf{R}} = \hat{\boldsymbol{\theta}}. \tag{3.45}$$

A vector with components A_R, A_θ, and A_ϕ is written as

$$\mathbf{A} = \hat{\mathbf{a}}|\mathbf{A}| = \hat{\mathbf{R}}A_R + \hat{\boldsymbol{\theta}}A_\theta + \hat{\boldsymbol{\phi}}A_\phi, \tag{3.46}$$

and its magnitude is

$$|\mathbf{A}| = \sqrt[+]{\mathbf{A}\cdot\mathbf{A}} = \sqrt[+]{A_R^2 + A_\theta^2 + A_\phi^2}\,. \tag{3.47}$$

The position vector of point $P(R_1, \theta_1, \phi_1)$ is simply

$$\mathbf{R}_1 = \overrightarrow{OP} = \hat{\mathbf{R}}R_1, \tag{3.48}$$

while keeping in mind that $\hat{\mathbf{R}}$ is implicitly dependent on θ_1 and ϕ_1.

As shown in Fig. 3-14, the differential lengths along $\hat{\mathbf{R}}$, $\hat{\boldsymbol{\theta}}$, and $\hat{\boldsymbol{\phi}}$ are

$$dl_R = dR, \quad dl_\theta = R\,d\theta, \quad dl_\phi = R\sin\theta\,d\phi. \tag{3.49}$$

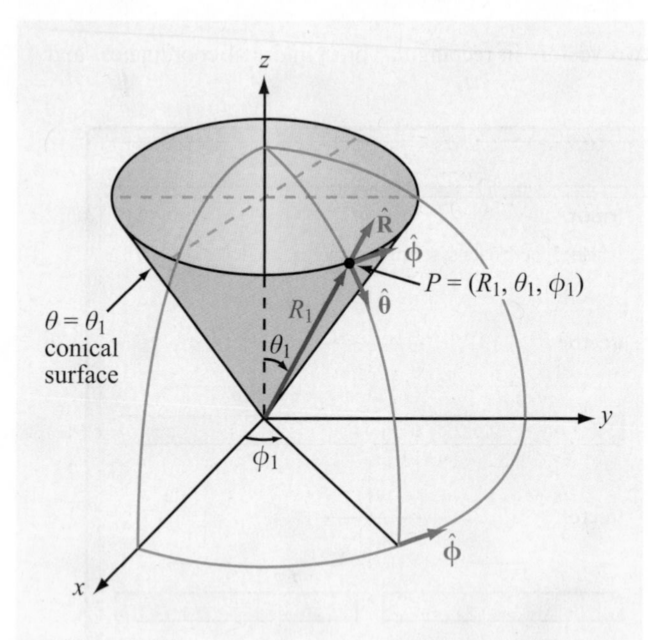

Figure 3-13 Point $P(R_1, \theta_1, \phi_1)$ in spherical coordinates.

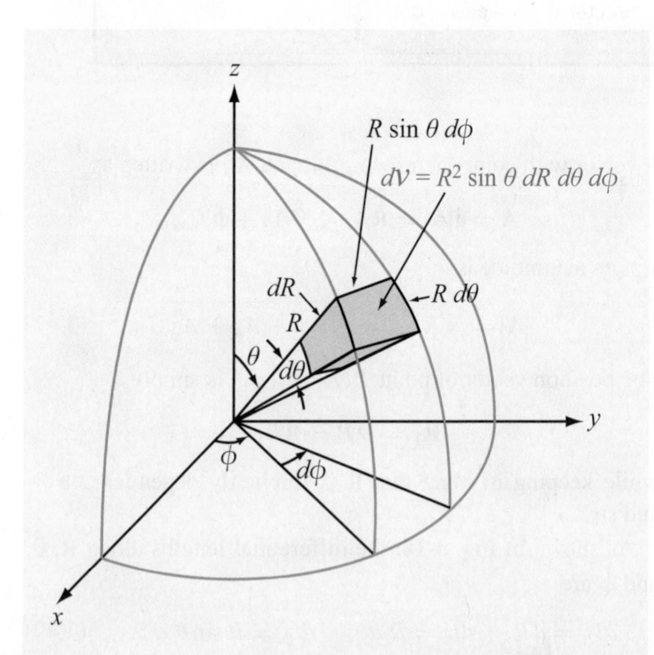

Figure 3-14 Differential volume in spherical coordinates.

Hence, the expressions for the vector differential length $d\mathbf{l}$, the vector differential surface $d\mathbf{s}$, and the differential volume $d\mathcal{V}$ are

$$d\mathbf{l} = \hat{\mathbf{R}}\, dl_R + \hat{\boldsymbol{\theta}}\, dl_\theta + \hat{\boldsymbol{\phi}}\, dl_\phi$$

$$= \hat{\mathbf{R}}\, dR + \hat{\boldsymbol{\theta}} R\, d\theta + \hat{\boldsymbol{\phi}} R \sin\theta\, d\phi, \qquad (3.50a)$$

$$d\mathbf{s}_R = \hat{\mathbf{R}}\, dl_\theta\, dl_\phi = \hat{\mathbf{R}} R^2 \sin\theta\, d\theta\, d\phi \qquad (3.50b)$$

(θ–ϕ spherical surface),

$$d\mathbf{s}_\theta = \hat{\boldsymbol{\theta}}\, dl_R\, dl_\phi = \hat{\boldsymbol{\theta}} R \sin\theta\, dR\, d\phi \qquad (3.50c)$$

(R–ϕ conical surface),

$$d\mathbf{s}_\phi = \hat{\boldsymbol{\phi}}\, dl_R\, dl_\theta = \hat{\boldsymbol{\phi}} R\, dR\, d\theta \quad (R\text{–}\theta \text{ plane}), \qquad (3.50d)$$

$$d\mathcal{V} = dl_R\, dl_\theta\, dl_\phi = R^2 \sin\theta\, dR\, d\theta\, d\phi. \qquad (3.50e)$$

These relations are summarized in **Table 3-1**.

Example 3-5: Surface Area in Spherical Coordinates

The spherical strip shown in **Fig. 3-15** is a section of a sphere of radius 3 cm. Find the area of the strip.

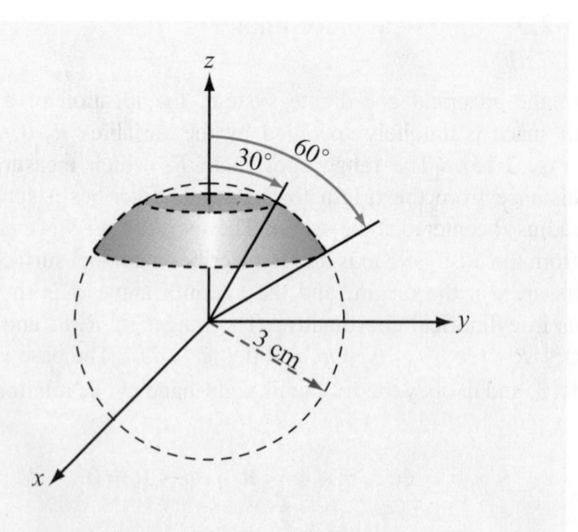

Figure 3-15 Spherical strip of Example 3-5.

Solution: Use of Eq. (3.50b) for the area of an elemental spherical area with constant radius R gives

$$
\begin{aligned}
S &= R^2 \int_{\theta=30^\circ}^{60^\circ} \sin\theta \, d\theta \int_{\phi=0}^{2\pi} d\phi \\
&= 9(-\cos\theta)\Big|_{30^\circ}^{60^\circ} \; \phi\Big|_0^{2\pi} \qquad (\text{cm}^2) \\
&= 18\pi(\cos 30^\circ - \cos 60^\circ) = 20.7 \text{ cm}^2.
\end{aligned}
$$

Example 3-6: Charge in a Sphere

A sphere of radius 2 cm contains a volume charge density $\rho_{\rm v}$ given by

$$\rho_{\rm v} = 4\cos^2\theta \qquad (\text{C/m}^3).$$

Find the total charge Q contained in the sphere.

Solution:

$$
\begin{aligned}
Q &= \int_{\mathcal{V}} \rho_{\rm v} \, d\mathcal{V} \\
&= \int_{\phi=0}^{2\pi} \int_{\theta=0}^{\pi} \int_{R=0}^{2\times10^{-2}} (4\cos^2\theta) R^2 \sin\theta \, dR \, d\theta \, d\phi \\
&= 4 \int_0^{2\pi} \int_0^{\pi} \left(\frac{R^3}{3}\right)\Big|_0^{2\times10^{-2}} \sin\theta\cos^2\theta \, d\theta \, d\phi \\
&= \frac{32}{3} \times 10^{-6} \int_0^{2\pi} \left(-\frac{\cos^3\theta}{3}\right)\Big|_0^{\pi} d\phi \\
&= \frac{64}{9} \times 10^{-6} \int_0^{2\pi} d\phi \\
&= \frac{128\pi}{9} \times 10^{-6} = 44.68 \qquad (\mu\text{C}).
\end{aligned}
$$

Note that the limits on R were converted to meters prior to evaluating the integral on R.

3-3 Transformations between Coordinate Systems

The position of a given point in space of course does not depend on the choice of coordinate system. That is, its location is the same irrespective of which specific coordinate system is used to represent it. The same is true for vectors. Nevertheless, certain coordinate systems may be more useful than others in solving a given problem, so it is essential that we have the tools to "translate" the problem from one system to another. In this section, we shall establish the relations between the variables (x, y, z) of the Cartesian system, (r, ϕ, z) of the cylindrical system, and (R, θ, ϕ) of the spherical system. These relations will then be used to transform expressions for vectors expressed in any one of the three systems into expressions applicable in the other two.

3-3.1 Cartesian to Cylindrical Transformations

Point P in **Fig. 3-16** has Cartesian coordinates (x, y, z) and cylindrical coordinates (r, ϕ, z). Both systems share the coordinate z, and the relations between the other two pairs of coordinates can be obtained from the geometry in **Fig. 3-16**. They are

$$r = \sqrt[+]{x^2 + y^2}, \qquad \phi = \tan^{-1}\left(\frac{y}{x}\right), \tag{3.51}$$

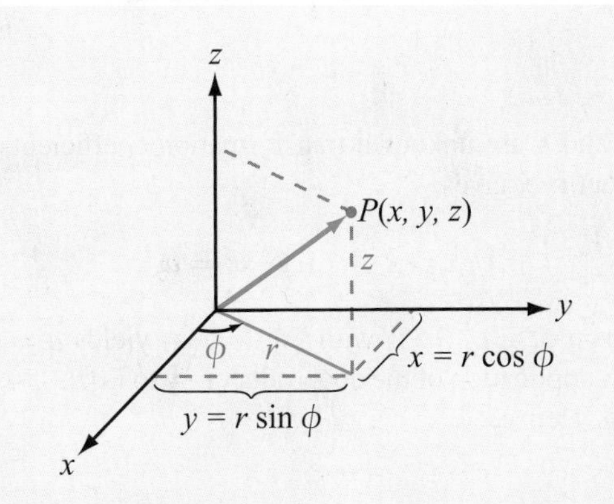

Figure 3-16 Interrelationships between Cartesian coordinates (x, y, z) and cylindrical coordinates (r, ϕ, z).

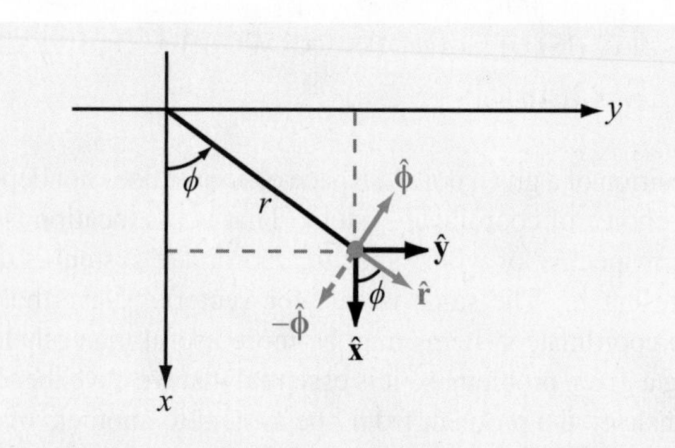

Figure 3-17 Interrelationships between base vectors ($\hat{\mathbf{x}}$, $\hat{\mathbf{y}}$) and ($\hat{\mathbf{r}}$, $\hat{\boldsymbol{\phi}}$).

and the inverse relations are

$$x = r\cos\phi, \qquad y = r\sin\phi. \tag{3.52}$$

Next, with the help of Fig. 3-17, which shows the directions of the unit vectors $\hat{\mathbf{x}}$, $\hat{\mathbf{y}}$, $\hat{\mathbf{r}}$, and $\hat{\boldsymbol{\phi}}$ in the x–y plane, we obtain the following relations:

$$\hat{\mathbf{r}}\cdot\hat{\mathbf{x}} = \cos\phi, \qquad \hat{\mathbf{r}}\cdot\hat{\mathbf{y}} = \sin\phi, \tag{3.53a}$$

$$\hat{\boldsymbol{\phi}}\cdot\hat{\mathbf{x}} = -\sin\phi, \qquad \hat{\boldsymbol{\phi}}\cdot\hat{\mathbf{y}} = \cos\phi. \tag{3.53b}$$

To express $\hat{\mathbf{r}}$ in terms of $\hat{\mathbf{x}}$ and $\hat{\mathbf{y}}$, we write $\hat{\mathbf{r}}$ as

$$\hat{\mathbf{r}} = \hat{\mathbf{x}}a + \hat{\mathbf{y}}b, \tag{3.54}$$

where a and b are unknown transformation coefficients. The dot product $\hat{\mathbf{r}}\cdot\hat{\mathbf{x}}$ gives

$$\hat{\mathbf{r}}\cdot\hat{\mathbf{x}} = \hat{\mathbf{x}}\cdot\hat{\mathbf{x}}a + \hat{\mathbf{y}}\cdot\hat{\mathbf{x}}b = a. \tag{3.55}$$

Comparison of Eq. (3.55) with Eq. (3.53a) yields $a = \cos\phi$. Similarly, application of the dot product $\hat{\mathbf{r}}\cdot\hat{\mathbf{y}}$ to Eq. (3.54) gives $b = \sin\phi$. Hence,

$$\hat{\mathbf{r}} = \hat{\mathbf{x}}\cos\phi + \hat{\mathbf{y}}\sin\phi. \tag{3.56a}$$

Repetition of the procedure for $\hat{\boldsymbol{\phi}}$ leads to

$$\hat{\boldsymbol{\phi}} = -\hat{\mathbf{x}}\sin\phi + \hat{\mathbf{y}}\cos\phi. \tag{3.56b}$$

The third base vector $\hat{\mathbf{z}}$ is the same in both coordinate systems. By solving Eqs. (3.56a) and (3.56b) simultaneously for $\hat{\mathbf{x}}$ and $\hat{\mathbf{y}}$, we obtain the following inverse relations:

$$\hat{\mathbf{x}} = \hat{\mathbf{r}}\cos\phi - \hat{\boldsymbol{\phi}}\sin\phi, \tag{3.57a}$$

$$\hat{\mathbf{y}} = \hat{\mathbf{r}}\sin\phi + \hat{\boldsymbol{\phi}}\cos\phi. \tag{3.57b}$$

The relations given by Eqs. (3.56a) to (3.57b) are not only useful for transforming the base vectors ($\hat{\mathbf{x}}$, $\hat{\mathbf{y}}$) into ($\hat{\mathbf{r}}$, $\hat{\boldsymbol{\phi}}$), and vice versa, they also can be used to transform the components of a vector expressed in either coördinate system into its corresponding components expressed in the other system. For example, a vector $\mathbf{A} = \hat{\mathbf{x}}A_x + \hat{\mathbf{y}}A_y + \hat{\mathbf{z}}A_z$ in Cartesian coordinates can be described by $\mathbf{A} = \hat{\mathbf{r}}A_r + \hat{\boldsymbol{\phi}}A_\phi + \hat{\mathbf{z}}A_z$ in cylindrical coordinates by applying Eqs. (3.56a) and (3.56b). That is,

$$A_r = A_x\cos\phi + A_y\sin\phi, \tag{3.58a}$$

$$A_\phi = -A_x\sin\phi + A_y\cos\phi, \tag{3.58b}$$

and, conversely,

$$A_x = A_r\cos\phi - A_\phi\sin\phi, \tag{3.59a}$$

$$A_y = A_r\sin\phi + A_\phi\cos\phi. \tag{3.59b}$$

The transformation relations given in this and the following two subsections are summarized in Table 3-2.

Table 3-2 Coordinate transformation relations.

Transformation	Coordinate Variables	Unit Vectors	Vector Components
Cartesian to cylindrical	$r = \sqrt[+]{x^2 + y^2}$ $\phi = \tan^{-1}(y/x)$ $z = z$	$\hat{\mathbf{r}} = \hat{\mathbf{x}} \cos\phi + \hat{\mathbf{y}} \sin\phi$ $\hat{\boldsymbol{\phi}} = -\hat{\mathbf{x}} \sin\phi + \hat{\mathbf{y}} \cos\phi$ $\hat{\mathbf{z}} = \hat{\mathbf{z}}$	$A_r = A_x \cos\phi + A_y \sin\phi$ $A_\phi = -A_x \sin\phi + A_y \cos\phi$ $A_z = A_z$
Cylindrical to Cartesian	$x = r\cos\phi$ $y = r\sin\phi$ $z = z$	$\hat{\mathbf{x}} = \hat{\mathbf{r}} \cos\phi - \hat{\boldsymbol{\phi}} \sin\phi$ $\hat{\mathbf{y}} = \hat{\mathbf{r}} \sin\phi + \hat{\boldsymbol{\phi}} \cos\phi$ $\hat{\mathbf{z}} = \hat{\mathbf{z}}$	$A_x = A_r \cos\phi - A_\phi \sin\phi$ $A_y = A_r \sin\phi + A_\phi \cos\phi$ $A_z = A_z$
Cartesian to spherical	$R = \sqrt[+]{x^2 + y^2 + z^2}$ $\theta = \tan^{-1}[\sqrt[+]{x^2 + y^2}/z]$ $\phi = \tan^{-1}(y/x)$	$\hat{\mathbf{R}} = \hat{\mathbf{x}} \sin\theta\cos\phi + \hat{\mathbf{y}} \sin\theta\sin\phi + \hat{\mathbf{z}} \cos\theta$ $\hat{\boldsymbol{\theta}} = \hat{\mathbf{x}} \cos\theta\cos\phi + \hat{\mathbf{y}} \cos\theta\sin\phi - \hat{\mathbf{z}} \sin\theta$ $\hat{\boldsymbol{\phi}} = -\hat{\mathbf{x}} \sin\phi + \hat{\mathbf{y}} \cos\phi$	$A_R = A_x \sin\theta\cos\phi + A_y \sin\theta\sin\phi + A_z \cos\theta$ $A_\theta = A_x \cos\theta\cos\phi + A_y \cos\theta\sin\phi - A_z \sin\theta$ $A_\phi = -A_x \sin\phi + A_y \cos\phi$
Spherical to Cartesian	$x = R\sin\theta\cos\phi$ $y = R\sin\theta\sin\phi$ $z = R\cos\theta$	$\hat{\mathbf{x}} = \hat{\mathbf{R}} \sin\theta\cos\phi + \hat{\boldsymbol{\theta}} \cos\theta\cos\phi - \hat{\boldsymbol{\phi}} \sin\phi$ $\hat{\mathbf{y}} = \hat{\mathbf{R}} \sin\theta\sin\phi + \hat{\boldsymbol{\theta}} \cos\theta\sin\phi + \hat{\boldsymbol{\phi}} \cos\phi$ $\hat{\mathbf{z}} = \hat{\mathbf{R}} \cos\theta - \hat{\boldsymbol{\theta}} \sin\theta$	$A_x = A_R \sin\theta\cos\phi + A_\theta \cos\theta\cos\phi - A_\phi \sin\phi$ $A_y = A_R \sin\theta\sin\phi + A_\theta \cos\theta\sin\phi + A_\phi \cos\phi$ $A_z = A_R \cos\theta - A_\theta \sin\theta$
Cylindrical to spherical	$R = \sqrt[+]{r^2 + z^2}$ $\theta = \tan^{-1}(r/z)$ $\phi = \phi$	$\hat{\mathbf{R}} = \hat{\mathbf{r}} \sin\theta + \hat{\mathbf{z}} \cos\theta$ $\hat{\boldsymbol{\theta}} = \hat{\mathbf{r}} \cos\theta - \hat{\mathbf{z}} \sin\theta$ $\hat{\boldsymbol{\phi}} = \hat{\boldsymbol{\phi}}$	$A_R = A_r \sin\theta + A_z \cos\theta$ $A_\theta = A_r \cos\theta - A_z \sin\theta$ $A_\phi = A_\phi$
Spherical to cylindrical	$r = R\sin\theta$ $\phi = \phi$ $z = R\cos\theta$	$\hat{\mathbf{r}} = \hat{\mathbf{R}} \sin\theta + \hat{\boldsymbol{\theta}} \cos\theta$ $\hat{\boldsymbol{\phi}} = \hat{\boldsymbol{\phi}}$ $\hat{\mathbf{z}} = \hat{\mathbf{R}} \cos\theta - \hat{\boldsymbol{\theta}} \sin\theta$	$A_r = A_R \sin\theta + A_\theta \cos\theta$ $A_\phi = A_\phi$ $A_z = A_R \cos\theta - A_\theta \sin\theta$

Example 3-7: Cartesian to Cylindrical Transformations

Given point $P_1 = (3, -4, 3)$ and vector $\mathbf{A} = \hat{\mathbf{x}}2 - \hat{\mathbf{y}}3 + \hat{\mathbf{z}}4$, defined in Cartesian coordinates, express P_1 and $\mathbf{A}$ in cylindrical coordinates and evaluate $\mathbf{A}$ at P_1.

Solution: For point P_1, $x = 3$, $y = -4$, and $z = 3$. Using Eq. (3.51), we have

$$r = \sqrt[+]{x^2 + y^2} = 5,$$

$$\phi = \tan^{-1}\frac{y}{x} = -53.1° = 306.9°,$$

and z remains unchanged. Hence, $P_1 = P_1(5, 306.9°, 3)$ in cylindrical coordinates.

The cylindrical components of vector $\mathbf{A} = \hat{\mathbf{r}}A_r + \hat{\boldsymbol{\phi}}A_\phi + \hat{\mathbf{z}}A_z$ can be determined by applying Eqs. (3.58a) and (3.58b):

$$A_r = A_x \cos\phi + A_y \sin\phi = 2\cos\phi - 3\sin\phi,$$

$$A_\phi = -A_x \sin\phi + A_y \cos\phi = -2\sin\phi - 3\cos\phi,$$

$$A_z = 4.$$

Hence,

$$\mathbf{A} = \hat{\mathbf{r}}(2\cos\phi - 3\sin\phi) - \hat{\boldsymbol{\phi}}(2\sin\phi + 3\cos\phi) + \hat{\mathbf{z}}4.$$

At point P, $\phi = 306.9°$, which gives

$$\mathbf{A} = \hat{\mathbf{r}}3.60 - \hat{\boldsymbol{\phi}}0.20 + \hat{\mathbf{z}}4.$$

Technology Brief 5: Global Positioning System

The Global Positioning System (GPS), initially developed in the 1980s by the U.S. Department of Defense as a navigation tool for military use, has evolved into a system with numerous civilian applications, including vehicle tracking, aircraft navigation, map displays in automobiles and hand-held cell phones (**Fig. TF5-1**), and topographic mapping. The overall GPS comprises three segments. The ***space segment*** consists of 24 satellites (**Fig. TF5-2**), each circling Earth every 12 hours at an orbital altitude of about 12,000 miles and transmitting continuous coded time signals. All satellite transmitters broadcast coded messages at two specific frequencies: 1.57542 GHz and 1.22760 GHz. The ***user segment*** consists of hand-held or vehicle-mounted receivers that determine their own locations by receiving and processing multiple satellite signals. The third segment is a network of five ***ground stations***, distributed around the world, that monitor the satellites and provide them with updates on their precise orbital information.

> ► GPS provides a location inaccuracy of about 30 m, both horizontally and vertically, but it can be improved to within 1 m by ***differential GPS***. (See final section.) ◄

Principle of Operation

The ***triangulation technique*** allows the determination of the location (x_0, y_0, z_0) of any object in 3-D space from knowledge of the distances d_1, d_2, and d_3 between that object and three other independent points in space of known locations (x_1, y_1, z_1) to (x_3, y_3, z_3). In GPS, the distances are established by measuring the times it takes the signals to travel from the satellites to the GPS receivers, and then multiplying them by the speed of light $c = 3 \times 10^8$ m/s. Time synchronization is achieved by using atomic clocks. The satellites use very precise clocks, accurate to 3 nanoseconds (3×10^{-9} s), but receivers use less accurate, inexpensive, ordinary quartz clocks. Consequently, the receiver clock may have an unknown ***time offset error*** t_0 relative to the satellite clocks. *To correct for the time error of a GPS receiver, a signal from a fourth satellite is needed.*

The GPS receiver of the automobile in **Fig. TF5-3** is at distances d_1 to d_4 from the GPS satellites. Each satellite sends a message identifying its orbital coordinates (x_1, y_1, z_1) for satellite 1, and so on for the other satellites, together with a binary-coded sequence common to all satellites. The GPS receiver generates the same binary sequence (**Fig. TF5-3**),

Figure TF5-1 iPhone map feature.

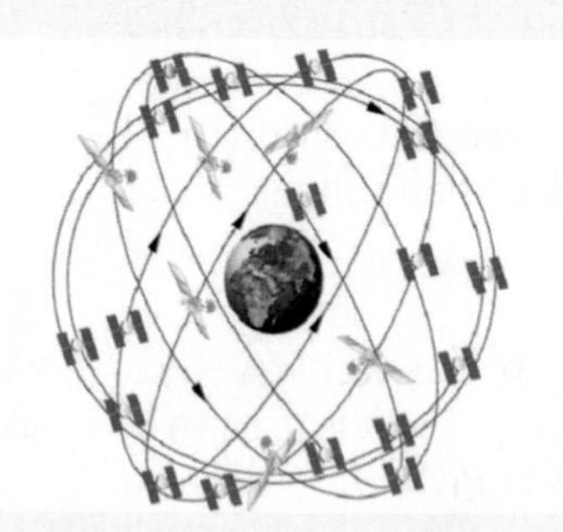

Figure TF5-2 GPS nominal satellite constellation. Four satellites in each plane, 20,200 km altitudes, 55° inclination.

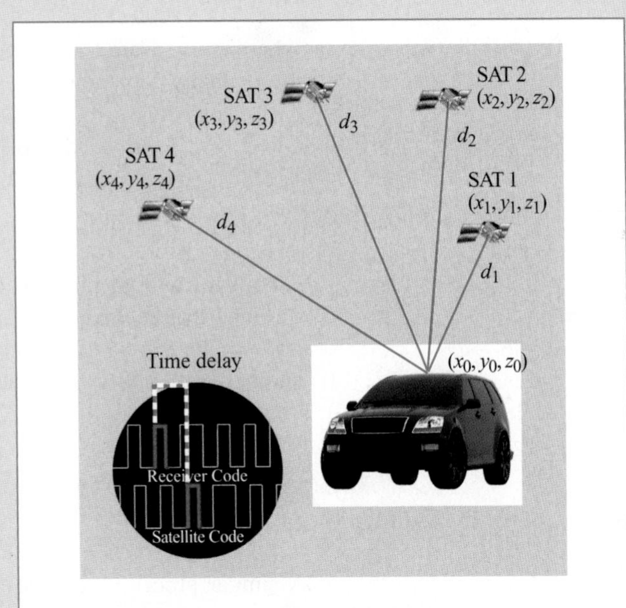

Figure TF5-3 Automobile GPS receiver at location (x_0, y_0, z_0).

and by comparing its code with the one received from satellite 1, it determines the time t_1 corresponding to travel time over the distance d_1. A similar process applies to satellites 2 to 4, leading to four equations:

$$
\begin{aligned}
d_1^2 &= (x_1 - x_0)^2 + (y_1 - y_0)^2 + (z_1 - z_0)^2 = c\,[(t_1 + t_0)]^2 \\
d_2^2 &= (x_2 - x_0)^2 + (y_2 - y_0)^2 + (z_2 - z_0)^2 = c\,[(t_2 + t_0)]^2 \\
d_3^2 &= (x_3 - x_0)^2 + (y_3 - y_0)^2 + (z_3 - z_0)^2 = c\,[(t_3 + t_0)]^2 \\
d_4^2 &= (x_4 - x_0)^2 + (y_4 - y_0)^2 + (z_4 - z_0)^2 = c\,[(t_4 + t_0)]^2 .
\end{aligned}
$$

The four satellites report their coordinates (x_1, y_1, z_1) to (x_4, y_4, z_4) to the GPS receiver, and the time delays t_1 to t_4 are measured directly by the receiver. The unknowns are (x_0, y_0, z_0), the coordinates of the GPS receiver, and the time offset of its clock t_0. Simultaneous solution of the four equations provides the desired location information.

Differential GPS

The 30 m GPS position inaccuracy is attributed to several factors, including ***time-delay errors*** (due to the difference between the speed of light and the actual signal speed in the troposphere) that depend on the receiver's location on Earth, delays due to signal reflections by tall buildings, and satellites' locations misreporting errors.

► ***Differential GPS***, or ***DGPS***, uses a stationary reference receiver at a location with known coordinates. ◄

By calculating the difference between its location on the basis of the GPS estimate and its true location, the ***reference receiver*** establishes coordinate correction factors and transmits them to all DGPS receivers in the area. Application of the correction information usually reduces the location inaccuracy down to about 1 m.

3-3.2 Cartesian to Spherical Transformations

From Fig. 3-18, we obtain the following relations between the Cartesian coordinates (x, y, z) and the spherical coordinates (R, θ, ϕ):

$$R = \sqrt[+]{x^2 + y^2 + z^2}, \quad (3.60a)$$

$$\theta = \tan^{-1}\left[\frac{\sqrt[+]{x^2 + y^2}}{z}\right], \quad (3.60b)$$

$$\phi = \tan^{-1}\left(\frac{y}{x}\right). \quad (3.60c)$$

The converse relations are

$$x = R \sin\theta \cos\phi, \quad (3.61a)$$

$$y = R \sin\theta \sin\phi, \quad (3.61b)$$

$$z = R \cos\theta. \quad (3.61c)$$

The unit vector $\hat{\mathbf{R}}$ lies in the $\hat{\mathbf{r}}$–$\hat{\mathbf{z}}$ plane. Hence, it can be expressed as a linear combination of $\hat{\mathbf{r}}$ and $\hat{\mathbf{z}}$ as follows:

$$\hat{\mathbf{R}} = \hat{\mathbf{r}}a + \hat{\mathbf{z}}b, \quad (3.62)$$

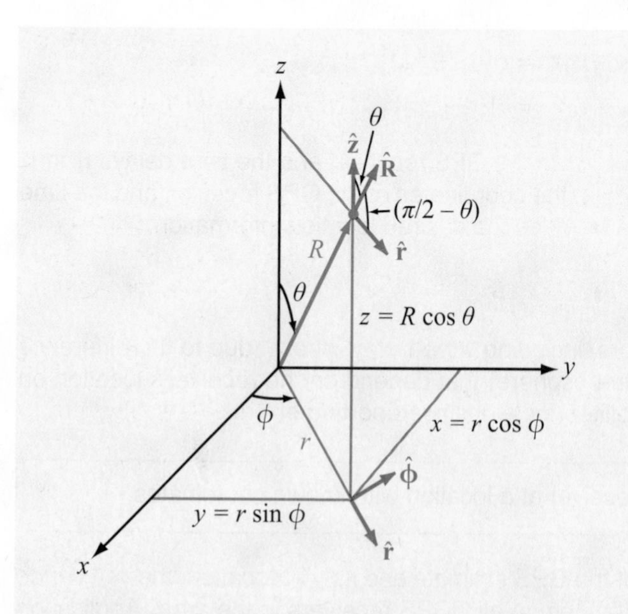

Figure 3-18 Interrelationships between (x, y, z) and (R, θ, ϕ).

where a and b are transformation coefficients. Since $\hat{\mathbf{r}}$ and $\hat{\mathbf{z}}$ are mutually perpendicular,

$$\hat{\mathbf{R}} \cdot \hat{\mathbf{r}} = a, \quad (3.63a)$$

$$\hat{\mathbf{R}} \cdot \hat{\mathbf{z}} = b. \quad (3.63b)$$

From Fig. 3-18, the angle between $\hat{\mathbf{R}}$ and $\hat{\mathbf{r}}$ is the complement of θ and that between $\hat{\mathbf{R}}$ and $\hat{\mathbf{z}}$ is θ. Hence, $a = \hat{\mathbf{R}} \cdot \hat{\mathbf{r}} = \sin\theta$ and $b = \hat{\mathbf{R}} \cdot \hat{\mathbf{z}} = \cos\theta$. Upon inserting these expressions for a and b in Eq. (3.62) and replacing $\hat{\mathbf{r}}$ with Eq. (3.56a), we have

$$\hat{\mathbf{R}} = \hat{\mathbf{x}} \sin\theta \cos\phi + \hat{\mathbf{y}} \sin\theta \sin\phi + \hat{\mathbf{z}} \cos\theta. \quad (3.64a)$$

A similar procedure can be followed to obtain the following expression for $\hat{\boldsymbol{\theta}}$:

$$\hat{\boldsymbol{\theta}} = \hat{\mathbf{x}} \cos\theta \cos\phi + \hat{\mathbf{y}} \cos\theta \sin\phi - \hat{\mathbf{z}} \sin\theta. \quad (3.64b)$$

Finally $\hat{\boldsymbol{\phi}}$ is given by Eq. (3.56b) as

$$\hat{\boldsymbol{\phi}} = -\hat{\mathbf{x}} \sin\phi + \hat{\mathbf{y}} \cos\phi. \quad (3.64c)$$

Equations (3.64a) through (3.64c) can be solved simultaneously to give the following expressions for $(\hat{\mathbf{x}}, \hat{\mathbf{y}}, \hat{\mathbf{z}})$ in terms of $(\hat{\mathbf{R}}, \hat{\boldsymbol{\theta}}, \hat{\boldsymbol{\phi}})$:

$$\hat{\mathbf{x}} = \hat{\mathbf{R}} \sin\theta \cos\phi + \hat{\boldsymbol{\theta}} \cos\theta \cos\phi - \hat{\boldsymbol{\phi}} \sin\phi, \quad (3.65a)$$

$$\hat{\mathbf{y}} = \hat{\mathbf{R}} \sin\theta \sin\phi + \hat{\boldsymbol{\theta}} \cos\theta \sin\phi + \hat{\boldsymbol{\phi}} \cos\phi, \quad (3.65b)$$

$$\hat{\mathbf{z}} = \hat{\mathbf{R}} \cos\theta - \hat{\boldsymbol{\theta}} \sin\theta. \quad (3.65c)$$

Equations (3.64a) to (3.65c) can also be used to transform Cartesian components (A_x, A_y, A_z) of vector $\mathbf{A}$ into their spherical counterparts (A_R, A_θ, A_ϕ), and vice versa, by replacing $(\hat{\mathbf{x}}, \hat{\mathbf{y}}, \hat{\mathbf{z}}, \hat{\mathbf{R}}, \hat{\boldsymbol{\theta}}, \hat{\boldsymbol{\phi}})$ with $(A_x, A_y, A_z, A_R, A_\theta, A_\phi)$.

Example 3-8: Cartesian to Spherical Transformation

Express vector $\mathbf{A} = \hat{\mathbf{x}}(x+y) + \hat{\mathbf{y}}(y-x) + \hat{\mathbf{z}}z$ in spherical coordinates.

Solution: Using the transformation relation for A_R given in **Table 3-2**, we have

$$\begin{aligned} A_R &= A_x \sin\theta\cos\phi + A_y \sin\theta\sin\phi + A_z\cos\theta \\ &= (x+y)\sin\theta\cos\phi + (y-x)\sin\theta\sin\phi + z\cos\theta. \end{aligned}$$

Using the expressions for x, y, and z given by Eq. (3.61c), we have

$$\begin{aligned} A_R &= (R\sin\theta\cos\phi + R\sin\theta\sin\phi)\sin\theta\cos\phi \\ &\quad + (R\sin\theta\sin\phi - R\sin\theta\cos\phi)\sin\theta\sin\phi + R\cos^2\theta \\ &= R\sin^2\theta\,(\cos^2\phi + \sin^2\phi) + R\cos^2\theta \\ &= R\sin^2\theta + R\cos^2\theta = R. \end{aligned}$$

Similarly,

$$\begin{aligned} A_\theta &= (x+y)\cos\theta\cos\phi + (y-x)\cos\theta\sin\phi - z\sin\theta, \\ A_\phi &= -(x+y)\sin\phi + (y-x)\cos\phi, \end{aligned}$$

and following the procedure used with A_R, we obtain

$$A_\theta = 0, \qquad A_\phi = -R\sin\theta.$$

Hence,

$$\mathbf{A} = \hat{\mathbf{R}}A_R + \hat{\boldsymbol{\theta}}A_\theta + \hat{\boldsymbol{\phi}}A_\phi = \hat{\mathbf{R}}R - \hat{\boldsymbol{\phi}}R\sin\theta.$$

3-3.3 Cylindrical to Spherical Transformations

Transformations between cylindrical and spherical coordinates can be realized by combining the transformation relations of the preceding two subsections. The results are given in **Table 3-2**.

3-3.4 Distance between Two Points

In Cartesian coordinates, the distance d between two points $P_1 = (x_1, y_1, z_1)$ and $P_2 = (x_2, y_2, z_2)$ is given by Eq. (3.12) as

$$d = |\mathbf{R}_{12}| = [(x_2-x_1)^2 + (y_2-y_1)^2 + (z_2-z_1)^2]^{1/2}. \tag{3.66}$$

Upon using Eq. (3.52) to convert the Cartesian coordinates of P_1 and P_2 into their cylindrical equivalents, we have

$$\begin{aligned} d &= \left[(r_2\cos\phi_2 - r_1\cos\phi_1)^2 \right. \\ &\quad \left. + (r_2\sin\phi_2 - r_1\sin\phi_1)^2 + (z_2-z_1)^2\right]^{1/2} \\ &= \left[r_2^2 + r_1^2 - 2r_1r_2\cos(\phi_2-\phi_1) + (z_2-z_1)^2\right]^{1/2}. \end{aligned}$$

(cylindrical) (3.67)

A similar transformation using Eqs. (3.61a-c) leads to an expression for d in terms of the spherical coordinates of P_1 and P_2:

$$\begin{aligned} d = \{&R_2^2 + R_1^2 - 2R_1R_2[\cos\theta_2\cos\theta_1 \\ &+ \sin\theta_1\sin\theta_2\cos(\phi_2-\phi_1)]\}^{1/2}. \end{aligned} \tag{3.68}$$

(spherical)

Concept Question 3-7: Why do we use more than one coordinate system?

Concept Question 3-8: Why is it that the base vectors $(\hat{\mathbf{x}}, \hat{\mathbf{y}}, \hat{\mathbf{z}})$ are independent of the location of a point, but $\hat{\mathbf{r}}$ and $\hat{\boldsymbol{\phi}}$ are not?

Concept Question 3-9: What are the cyclic relations for the base vectors in (a) Cartesian coordinates, (b) cylindrical coordinates, and (c) spherical coordinates?

Concept Question 3-10: How is the position vector of a point in cylindrical coordinates related to its position vector in spherical coordinates?

Exercise 3-7: Point $P = (2\sqrt{3}, \pi/3, -2)$ is given in cylindrical coordinates. Express P in spherical coordinates.

Answer: $P = (4, 2\pi/3, \pi/3)$. (See EM.)

Exercise 3-8: Transform vector

$$\mathbf{A} = \hat{\mathbf{x}}(x + y) + \hat{\mathbf{y}}(y - x) + \hat{\mathbf{z}}z$$

from Cartesian to cylindrical coordinates.

Answer: $\mathbf{A} = \hat{\mathbf{r}}r - \hat{\boldsymbol{\phi}}r + \hat{\mathbf{z}}z$. (See EM.)

3-4 Gradient of a Scalar Field

When dealing with a scalar physical quantity whose magnitude depends on a single variable, such as the temperature T as a function of height z, the rate of change of T with height can be described by the derivative dT/dz. However, if T is also a function of x and y, its spatial rate of change becomes more difficult to describe because we now have to deal with three separate variables. The differential change in T along x, y, and z can be described in terms of the partial derivatives of T with respect to the three coordinate variables, but it is not immediately obvious as to how we should combine the three partial derivatives so as to describe the spatial rate of change of T along a specified direction. Furthermore, many of the quantities we deal with in electromagnetics are vectors, and therefore both their magnitudes and directions may vary with spatial position. To this end, we introduce three fundamental operators to describe the differential spatial variations of scalars and vectors; these are the ***gradient***, ***divergence***, and ***curl*** operators. The gradient operator applies to scalar fields and is the subject of the present section. The other two operators, which apply to vector fields, are discussed in succeeding sections.

Suppose that $T_1 = T(x, y, z)$ is the temperature at point $P_1 = (x, y, z)$ in some region of space, and $T_2 = T(x + dx,\ y + dy,\ z + dz)$ is the temperature at a nearby point $P_2 = (x + dx,\ y + dy,\ z + dz)$ (Fig. 3-19). The differential distances dx, dy, and dz are the components of the differential distance vector $d\mathbf{l}$. That is,

$$d\mathbf{l} = \hat{\mathbf{x}}\, dx + \hat{\mathbf{y}}\, dy + \hat{\mathbf{z}}\, dz. \tag{3.69}$$

From differential calculus, the temperature difference between points P_1 and P_2, $dT = T_2 - T_1$, is

$$dT = \frac{\partial T}{\partial x}\, dx + \frac{\partial T}{\partial y}\, dy + \frac{\partial T}{\partial z}\, dz. \tag{3.70}$$

Because $dx = \hat{\mathbf{x}} \cdot d\mathbf{l}$, $dy = \hat{\mathbf{y}} \cdot d\mathbf{l}$, and $dz = \hat{\mathbf{z}} \cdot d\mathbf{l}$, Eq. (3.70) can be rewritten as

$$\begin{aligned} dT &= \hat{\mathbf{x}}\frac{\partial T}{\partial x} \cdot d\mathbf{l} + \hat{\mathbf{y}}\frac{\partial T}{\partial y} \cdot d\mathbf{l} + \hat{\mathbf{z}}\frac{\partial T}{\partial z} \cdot d\mathbf{l} \\ &= \left[\hat{\mathbf{x}}\frac{\partial T}{\partial x} + \hat{\mathbf{y}}\frac{\partial T}{\partial y} + \hat{\mathbf{z}}\frac{\partial T}{\partial z}\right] \cdot d\mathbf{l}. \end{aligned} \tag{3.71}$$

The vector inside the square brackets in Eq. (3.71) relates the change in temperature dT to a vector change in direction $d\mathbf{l}$. This vector is called the ***gradient*** of T, or ***grad*** T for short, and denoted ∇T:

$$\nabla T = \text{grad } T = \hat{\mathbf{x}}\frac{\partial T}{\partial x} + \hat{\mathbf{y}}\frac{\partial T}{\partial y} + \hat{\mathbf{z}}\frac{\partial T}{\partial z}. \tag{3.72}$$

Equation (3.71) can then be expressed as

$$dT = \nabla T \cdot d\mathbf{l}. \tag{3.73}$$

The symbol ∇ is called the ***del*** or ***gradient operator*** and is defined as

$$\nabla = \hat{\mathbf{x}}\frac{\partial}{\partial x} + \hat{\mathbf{y}}\frac{\partial}{\partial y} + \hat{\mathbf{z}}\frac{\partial}{\partial z} \qquad \textbf{(Cartesian)}. \tag{3.74}$$

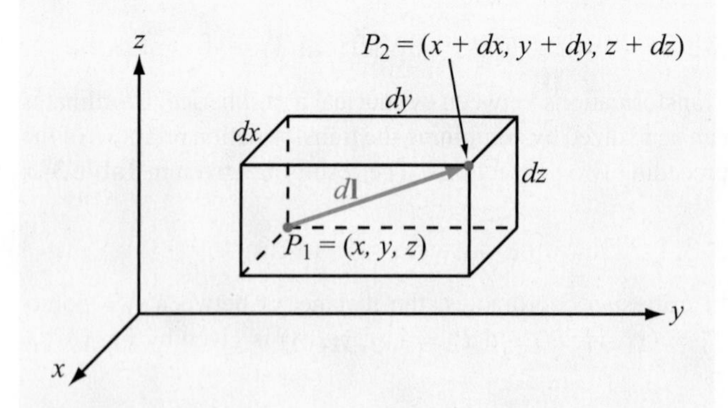

Figure 3-19 Differential distance vector $d\mathbf{l}$ between points P_1 and P_2.

▶ Whereas the gradient operator itself has no physical meaning, it attains a physical meaning once it operates on a scalar quantity, and the result of the operation is a vector with magnitude equal to the maximum rate of change of the physical quantity per unit distance and pointing in the direction of maximum increase. ◀

With $d\mathbf{l} = \hat{\mathbf{a}}_l \, dl$, where $\hat{\mathbf{a}}_l$ is the unit vector of $d\mathbf{l}$, the ***directional derivative*** of T along $\hat{\mathbf{a}}_l$ is

$$\frac{dT}{dl} = \nabla T \cdot \hat{\mathbf{a}}_l. \tag{3.75}$$

We can find the difference $(T_2 - T_1)$, where $T_1 = T(x_1, y_1, z_1)$ and $T_2 = T(x_2, y_2, z_2)$ are the values of T at points $P_1 = (x_1, y_1, z_1)$ and $P_2 = (x_2, y_2, z_2)$, not necessarily infinitesimally close to one another, by integrating both sides of Eq. (3.73). Thus,

$$T_2 - T_1 = \int_{P_1}^{P_2} \nabla T \cdot d\mathbf{l}. \tag{3.76}$$

Example 3-9: Directional Derivative

Find the directional derivative of $T = x^2 + y^2 z$ along direction $\hat{\mathbf{x}}2 + \hat{\mathbf{y}}3 - \hat{\mathbf{z}}2$ and evaluate it at $(1, -1, 2)$.

Solution: First, we find the gradient of T:

$$\begin{aligned}\nabla T &= \left(\hat{\mathbf{x}}\frac{\partial}{\partial x} + \hat{\mathbf{y}}\frac{\partial}{\partial y} + \hat{\mathbf{z}}\frac{\partial}{\partial z}\right)(x^2 + y^2 z) \\ &= \hat{\mathbf{x}}2x + \hat{\mathbf{y}}2yz + \hat{\mathbf{z}}y^2.\end{aligned}$$

We denote $\mathbf{l}$ as the given direction,

$$\mathbf{l} = \hat{\mathbf{x}}2 + \hat{\mathbf{y}}3 - \hat{\mathbf{z}}2.$$

Its unit vector is

$$\hat{\mathbf{a}}_l = \frac{\mathbf{l}}{|\mathbf{l}|} = \frac{\hat{\mathbf{x}}2 + \hat{\mathbf{y}}3 - \hat{\mathbf{z}}2}{\sqrt{2^2 + 3^2 + 2^2}} = \frac{\hat{\mathbf{x}}2 + \hat{\mathbf{y}}3 - \hat{\mathbf{z}}2}{\sqrt{17}}.$$

Application of Eq. (3.75) gives

$$\begin{aligned}\frac{dT}{dl} &= \nabla T \cdot \hat{\mathbf{a}}_l = (\hat{\mathbf{x}}2x + \hat{\mathbf{y}}2yz + \hat{\mathbf{z}}y^2) \cdot \left(\frac{\hat{\mathbf{x}}2 + \hat{\mathbf{y}}3 - \hat{\mathbf{z}}2}{\sqrt{17}}\right) \\ &= \frac{4x + 6yz - 2y^2}{\sqrt{17}}.\end{aligned}$$

At $(1, -1, 2)$,

$$\left.\frac{dT}{dl}\right|_{(1,-1,2)} = \frac{4 - 12 - 2}{\sqrt{17}} = \frac{-10}{\sqrt{17}}.$$

3-4.1 Gradient Operator in Cylindrical and Spherical Coordinates

Even though Eq. (3.73) was derived using Cartesian coordinates, it should have counterparts in other coordinate systems. To convert Eq. (3.72) into cylindrical coordinates (r, ϕ, z), we start by restating the coordinate relations

$$r = \sqrt{x^2 + y^2}, \qquad \tan\phi = \frac{y}{x}. \tag{3.77}$$

From differential calculus,

$$\frac{\partial T}{\partial x} = \frac{\partial T}{\partial r}\frac{\partial r}{\partial x} + \frac{\partial T}{\partial \phi}\frac{\partial \phi}{\partial x} + \frac{\partial T}{\partial z}\frac{\partial z}{\partial x}. \tag{3.78}$$

Since z is orthogonal to x and $\partial z/\partial x = 0$, the last term in Eq. (3.78) vanishes. From the coordinate relations given by Eq. (3.77), it follows that

$$\frac{\partial r}{\partial x} = \frac{x}{\sqrt{x^2 + y^2}} = \cos\phi, \tag{3.79a}$$

$$\frac{\partial \phi}{\partial x} = -\frac{1}{r}\sin\phi. \tag{3.79b}$$

Hence,

$$\frac{\partial T}{\partial x} = \cos\phi\frac{\partial T}{\partial r} - \frac{\sin\phi}{r}\frac{\partial T}{\partial \phi}. \tag{3.80}$$

This expression can be used to replace the coefficient of $\hat{\mathbf{x}}$ in Eq. (3.72), and a similar procedure can be followed to obtain an expression for $\partial T/\partial y$ in terms of r and ϕ. If, in addition, we use the relations $\hat{\mathbf{x}} = \hat{\mathbf{r}}\cos\phi - \hat{\boldsymbol{\phi}}\sin\phi$ and $\hat{\mathbf{y}} = \hat{\mathbf{r}}\sin\phi + \hat{\boldsymbol{\phi}}\cos\phi$ [from Eqs. (3.57a) and (3.57b)], then Eq. (3.72) becomes

$$\nabla T = \hat{\mathbf{r}}\frac{\partial T}{\partial r} + \hat{\boldsymbol{\phi}}\frac{1}{r}\frac{\partial T}{\partial \phi} + \hat{\mathbf{z}}\frac{\partial T}{\partial z}. \tag{3.81}$$

Hence, the gradient operator in cylindrical coordinates can be expressed as

$$\hat{\mathbf{r}}\frac{\partial}{\partial r} + \hat{\boldsymbol{\phi}}\frac{1}{r}\frac{\partial}{\partial \phi} + \hat{\mathbf{z}}\frac{\partial}{\partial z} \quad \textbf{(cylindrical)}. \tag{3.82}$$

A similar procedure leads to the following expression for the gradient in spherical coordinates:

$$\nabla = \hat{\mathbf{R}}\frac{\partial}{\partial R} + \hat{\boldsymbol{\theta}}\frac{1}{R}\frac{\partial}{\partial \theta} + \hat{\boldsymbol{\phi}}\frac{1}{R\sin\theta}\frac{\partial}{\partial \phi}. \tag{3.83}$$

(spherical)

3-4.2 Properties of the Gradient Operator

For any two scalar functions U and V, the following relations apply:

$$(1) \quad \nabla(U+V) = \nabla U + \nabla V, \tag{3.84a}$$

$$(2) \quad \nabla(UV) = U\,\nabla V + V\,\nabla U, \tag{3.84b}$$

$$(3) \quad \nabla V^n = nV^{n-1}\,\nabla V, \quad \text{for any } n. \tag{3.84c}$$

Example 3-10: Calculating the Gradient

Find the gradient of each of the following scalar functions and then evaluate it at the given point.

(a) $V_1 = 24V_0\cos(\pi y/3)\sin(2\pi z/3)$ at $(3, 2, 1)$ in Cartesian coordinates,

(b) $V_2 = V_0 e^{-2r}\sin 3\phi$ at $(1, \pi/2, 3)$ in cylindrical coordinates,

(c) $V_3 = V_0(a/R)\cos 2\theta$ at $(2a, 0, \pi)$ in spherical coordinates.

Solution: (a) Using Eq. (3.72) for ∇,

$$\begin{aligned}\nabla V_1 &= \hat{\mathbf{x}}\frac{\partial V_1}{\partial x} + \hat{\mathbf{y}}\frac{\partial V_1}{\partial y} + \hat{\mathbf{z}}\frac{\partial V_1}{\partial z} \\ &= -\hat{\mathbf{y}}8\pi V_0\sin\frac{\pi y}{3}\sin\frac{2\pi z}{3} + \hat{\mathbf{z}}16\pi V_0\cos\frac{\pi y}{3}\cos\frac{2\pi z}{3} \\ &= 8\pi V_0\left[-\hat{\mathbf{y}}\sin\frac{\pi y}{3}\sin\frac{2\pi z}{3} + \hat{\mathbf{z}}2\cos\frac{\pi y}{3}\cos\frac{2\pi z}{3}\right].\end{aligned}$$

At (3, 2, 1),

$$\begin{aligned}\nabla V_1 &= 8\pi V_0\left[-\hat{\mathbf{y}}\sin^2\frac{2\pi}{3} + \hat{\mathbf{z}}2\cos^2\frac{2\pi}{3}\right] \\ &= \pi V_0\left[-\hat{\mathbf{y}}6 + \hat{\mathbf{z}}4\right].\end{aligned}$$

(b) The function V_2 is expressed in terms of cylindrical variables. Hence, we need to use Eq. (3.82) for ∇:

$$\begin{aligned}\nabla V_2 &= \left(\hat{\mathbf{r}}\frac{\partial}{\partial r} + \hat{\boldsymbol{\phi}}\frac{1}{r}\frac{\partial}{\partial \phi} + \hat{\mathbf{z}}\frac{\partial}{\partial z}\right) V_0 e^{-2r}\sin 3\phi \\ &= -\hat{\mathbf{r}}2V_0 e^{-2r}\sin 3\phi + \hat{\boldsymbol{\phi}}(3V_0 e^{-2r}\cos 3\phi)/r \\ &= \left[-\hat{\mathbf{r}}2\sin 3\phi + \hat{\boldsymbol{\phi}}\frac{3\cos 3\phi}{r}\right] V_0 e^{-2r}.\end{aligned}$$

At $(1, \pi/2, 3)$, $r = 1$ and $\phi = \pi/2$. Hence,

$$\begin{aligned}\nabla V_2 &= \left[-\hat{\mathbf{r}}2\sin\frac{3\pi}{2} + \hat{\boldsymbol{\phi}}3\cos\frac{3\pi}{2}\right] V_0 e^{-2} \\ &= \hat{\mathbf{r}}2V_0 e^{-2} = \hat{\mathbf{r}}0.27V_0.\end{aligned}$$

(c) As V_3 is expressed in spherical coordinates, we apply Eq. (3.83) to V_3:

$$\begin{aligned}\nabla V_3 &= \left(\hat{\mathbf{R}}\frac{\partial}{\partial R} + \hat{\boldsymbol{\theta}}\frac{1}{R}\frac{\partial}{\partial \theta} + \hat{\boldsymbol{\phi}}\frac{1}{R\sin\theta}\frac{\partial}{\partial \phi}\right) V_0\left(\frac{a}{R}\right)\cos 2\theta \\ &= -\hat{\mathbf{R}}\frac{V_0 a}{R^2}\cos 2\theta - \hat{\boldsymbol{\theta}}\frac{2V_0 a}{R^2}\sin 2\theta \\ &= -[\hat{\mathbf{R}}\cos 2\theta + \hat{\boldsymbol{\theta}}2\sin 2\theta]\frac{V_0 a}{R^2}.\end{aligned}$$

At $(2a, 0, \pi)$, $R = 2a$ and $\theta = 0$, which yields

$$\nabla V_3 = -\hat{\mathbf{R}}\frac{V_0}{4a}.$$

Exercise 3-9: Given $V = x^2y + xy^2 + xz^2$, (a) find the gradient of V, and (b) evaluate it at $(1, -1, 2)$.

Answer: (a) $\nabla V = \hat{\mathbf{x}}(2xy + y^2 + z^2) + \hat{\mathbf{y}}(x^2 + 2xy) + \hat{\mathbf{z}}2xz$, (b) $\nabla V\big|_{(1,-1,2)} = \hat{\mathbf{x}}3 - \hat{\mathbf{y}} + \hat{\mathbf{z}}4$. (See EM.)

Exercise 3-10: Find the directional derivative of $V = rz^2 \cos 2\phi$ along the direction of $\mathbf{A} = \hat{\mathbf{r}}2 - \hat{\mathbf{z}}$ and evaluate it at $(1, \pi/2, 2)$.

Answer: $(dV/dl)\big|_{(1,\pi/2,2)} = -4/\sqrt{5}$. (See EM.)

Exercise 3-11: The power density radiated by a star [Fig. E3.11(a)] decreases radially as $S(R) = S_0/R^2$, where R is the radial distance from the star and S_0 is a constant. Recalling that the gradient of a scalar function denotes the maximum rate of change of that function per unit distance and the direction of the gradient is along the direction of maximum increase, generate an arrow representation of ∇S.

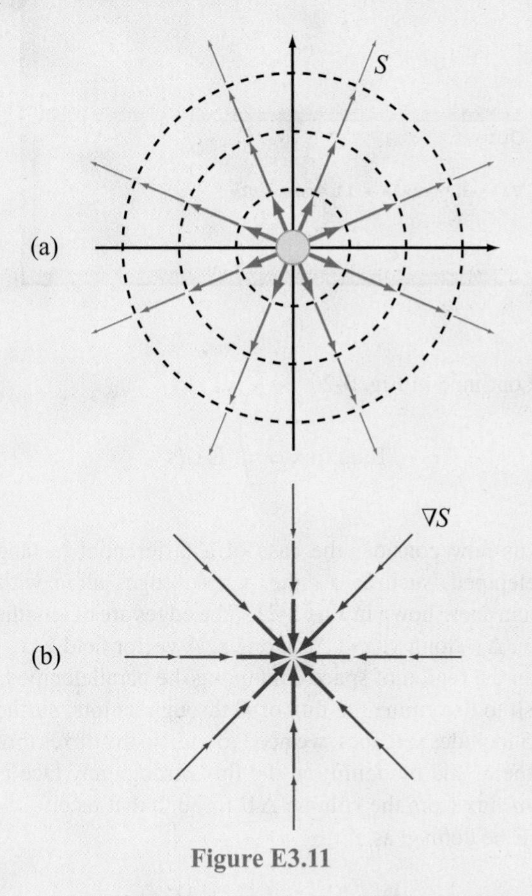

Figure E3.11

Answer: $\nabla S = -\hat{\mathbf{R}}\, 2S_0/R^3$ (Fig. 3.11(b)). (See EM.)

Exercise 3-12: The graph in Fig. E3.12(a) depicts a gentle change in atmospheric temperature from T_1 over the sea to T_2 over land. The temperature profile is described by the function

$$T(x) = T_1 + (T_2 - T_1)/(e^{-x} + 1),$$

where x is measured in kilometers and $x = 0$ is the sea–land boundary. (a) In which direction does ∇T point and (b) at what value of x is it a maximum?

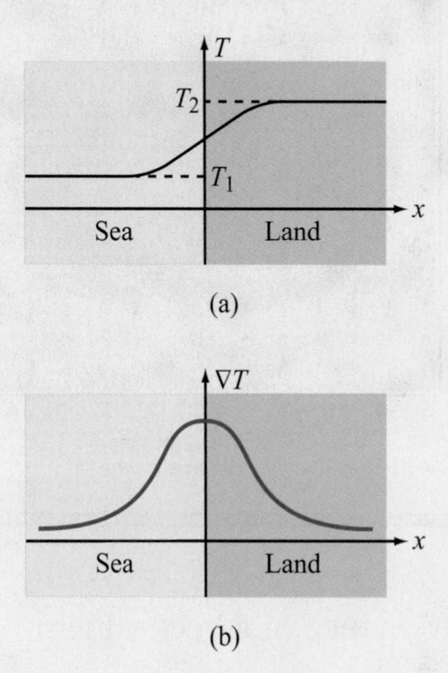

Figure E3.12

Answer: (a) $+\hat{\mathbf{x}}$; (b) at $x = 0$.

$$T(x) = T_1 + \frac{T_2 - T_1}{e^{-x} + 1},$$

$$\nabla T = \hat{\mathbf{x}} \frac{\partial T}{\partial x}$$

$$= \hat{\mathbf{x}} \frac{e^{-x}(T_2 - T_1)}{(e^{-x} + 1)^2}.$$

(See EM.)

Module 3.2 Gradient Select a scalar function $f(x, y, z)$, evaluate its gradient, and display both in an appropriate 2-D plane.

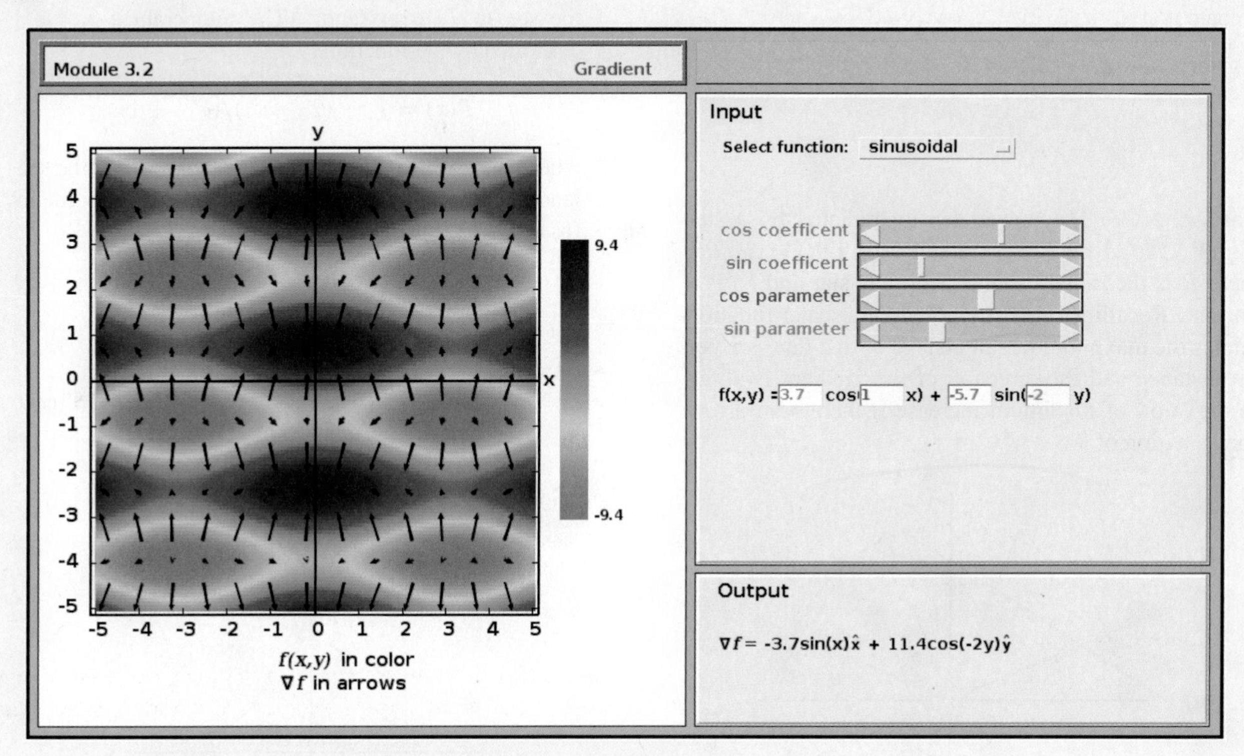

3-5 Divergence of a Vector Field

From our brief introduction of Coulomb's law in Chapter 1, we know that an isolated, positive point charge q induces an electric field $\mathbf{E}$ in the space around it, with the direction of $\mathbf{E}$ being outward away from the charge. Also, the strength (magnitude) of $\mathbf{E}$ is proportional to q and decreases with distance R from the charge as $1/R^2$. In a graphical presentation, a vector field is usually represented by ***field lines***, as shown in **Fig. 3-20**. The arrowhead denotes the direction of the field at the point where the field line is drawn, and the length of the line provides a qualitative depiction of the field's magnitude.

At a surface boundary, ***flux density*** is defined as the amount of outward flux crossing a unit surface ds:

$$\text{Flux density of } \mathbf{E} = \frac{\mathbf{E} \cdot d\mathbf{s}}{|d\mathbf{s}|} = \frac{\mathbf{E} \cdot \hat{\mathbf{n}}\, ds}{ds} = \mathbf{E} \cdot \hat{\mathbf{n}}, \tag{3.85}$$

where $\hat{\mathbf{n}}$ is the normal to $d\mathbf{s}$. The ***total flux*** outwardly crossing a closed surface S, such as the enclosed surface of the imaginary sphere outlined in **Fig. 3-20**, is

$$\text{Total flux} = \oint_S \mathbf{E} \cdot d\mathbf{s}. \tag{3.86}$$

Let us now consider the case of a differential rectangular parallelepiped, such as a cube, whose edges align with the Cartesian axes shown in **Fig. 3-21**. The edges are of lengths Δx along x, Δy along y, and Δz along z. A vector field $\mathbf{E}(x, y, z)$ exists in the region of space containing the parallelepiped, and we wish to determine the flux of $\mathbf{E}$ through its total surface S. Since S includes six faces, we need to sum up the fluxes through all of them, and by definition the flux through any face is the *outward* flux from the volume $\Delta \mathcal{V}$ through that face.

Let $\mathbf{E}$ be defined as

$$\mathbf{E} = \hat{\mathbf{x}} E_x + \hat{\mathbf{y}} E_y + \hat{\mathbf{z}} E_z. \tag{3.87}$$

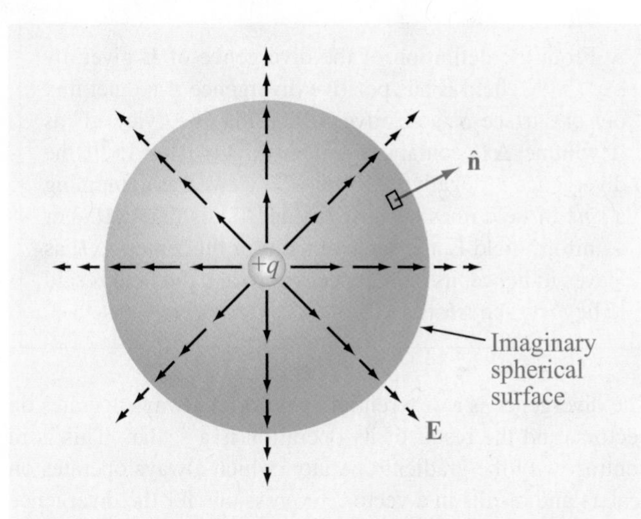

Figure 3-20 Flux lines of the electric field **E** due to a positive charge q.

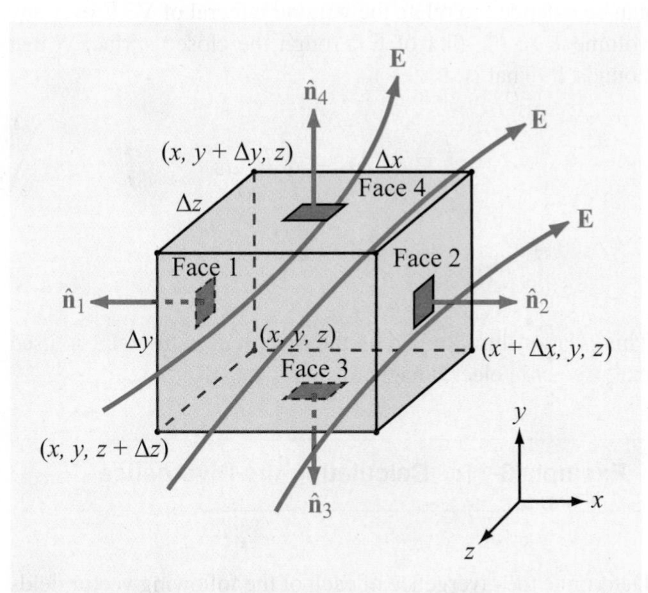

Figure 3-21 Flux lines of a vector field **E** passing through a differential rectangular parallelepiped of volume $\Delta v = \Delta x \, \Delta y \, \Delta z$.

The area of the face marked 1 in **Fig. 3-21** is $\Delta y \, \Delta z$, and its unit vector $\hat{\mathbf{n}}_1 = -\hat{\mathbf{x}}$. Hence, the outward flux F_1 through face 1 is

$$\begin{aligned} F_1 &= \int_{\text{Face 1}} \mathbf{E} \cdot \hat{\mathbf{n}}_1 \, ds \\ &= \int_{\text{Face 1}} (\hat{\mathbf{x}} E_x + \hat{\mathbf{y}} E_y + \hat{\mathbf{z}} E_z) \cdot (-\hat{\mathbf{x}}) \, dy \, dz \\ &\approx -E_x(1) \, \Delta y \, \Delta z, \end{aligned} \tag{3.88}$$

where $E_x(1)$ is the value of E_x at the center of face 1. Approximating E_x over face 1 by its value at the center is justified by the assumption that the differential volume under consideration is very small.

Similarly, the flux out of face 2 (with $\hat{\mathbf{n}}_2 = \hat{\mathbf{x}}$) is

$$F_2 = E_x(2) \, \Delta y \, \Delta z, \tag{3.89}$$

where $E_x(2)$ is the value of E_x at the center of face 2. Over a differential separation Δx between the centers of faces 1 and 2, $E_x(2)$ is related to $E_x(1)$ by

$$E_x(2) = E_x(1) + \frac{\partial E_x}{\partial x} \, \Delta x, \tag{3.90}$$

where we have ignored higher-order terms involving $(\Delta x)^2$ and higher powers because their contributions are negligibly small when Δx is very small. Substituting Eq. (3.90) into Eq. (3.89) gives

$$F_2 = \left[E_x(1) + \frac{\partial E_x}{\partial x} \, \Delta x \right] \Delta y \, \Delta z. \tag{3.91}$$

The sum of the fluxes out of faces 1 and 2 is obtained by adding Eqs. (3.88) and (3.91),

$$F_1 + F_2 = \frac{\partial E_x}{\partial x} \, \Delta x \, \Delta y \, \Delta z. \tag{3.92a}$$

Repeating the same procedure to each of the other face pairs leads to

$$F_3 + F_4 = \frac{\partial E_y}{\partial y}\,\Delta x\,\Delta y\,\Delta z, \tag{3.92b}$$

$$F_5 + F_6 = \frac{\partial E_z}{\partial z}\,\Delta x\,\Delta y\,\Delta z. \tag{3.92c}$$

The sum of fluxes F_1 through F_6 gives the total flux through surface S of the parallelepiped:

$$\oint_S \mathbf{E}\cdot d\mathbf{s} = \left(\frac{\partial E_x}{\partial x} + \frac{\partial E_y}{\partial y} + \frac{\partial E_z}{\partial z}\right)\Delta x\,\Delta y\,\Delta z$$

$$= (\text{div}\,\mathbf{E})\,\Delta \mathcal{V}, \tag{3.93}$$

where $\Delta\mathcal{V} = \Delta x\,\Delta y\,\Delta z$ and div **E** is a scalar function called the ***divergence*** of **E**, specified in Cartesian coordinates as

$$\text{div}\,\mathbf{E} = \frac{\partial E_x}{\partial x} + \frac{\partial E_y}{\partial y} + \frac{\partial E_z}{\partial z}. \tag{3.94}$$

► By shrinking the volume $\Delta\mathcal{V}$ to zero, we define the ***divergence*** of **E** at a point as the net outward flux per unit volume over a closed incremental surface. ◄

Thus, from Eq. (3.93), we have

$$\text{div}\,\mathbf{E} \triangleq \lim_{\Delta\mathcal{V}\to 0} \frac{\oint_S \mathbf{E}\cdot d\mathbf{s}}{\Delta\mathcal{V}}, \tag{3.95}$$

where S encloses the elemental volume $\Delta\mathcal{V}$. Instead of denoting the divergence of **E** by div **E**, it is common practice to denote it as $\nabla\cdot\mathbf{E}$. That is,

$$\nabla\cdot\mathbf{E} = \text{div}\,\mathbf{E} = \frac{\partial E_x}{\partial x} + \frac{\partial E_y}{\partial y} + \frac{\partial E_z}{\partial z} \tag{3.96}$$

for a vector **E** in Cartesian coordinates.

► From the definition of the divergence of **E** given by Eq. (3.95), field **E** has positive divergence if the net flux out of surface S is positive, which may be "viewed" as if volume $\Delta\mathcal{V}$ contains a *source* of field lines. If the divergence is negative, $\Delta\mathcal{V}$ may be viewed as containing a *sink* of field lines because the net flux is into $\Delta\mathcal{V}$. For a uniform field **E**, the same amount of flux enters $\Delta\mathcal{V}$ as leaves it; hence, its divergence is zero and the field is said to be ***divergenceless***. ◄

The divergence is a differential operator; it always operates on vectors, and the result of its operation is a scalar. This is in contrast with the gradient operator, which always operates on scalars and results in a vector. Expressions for the divergence of a vector in cylindrical and spherical coordinates are provided on the inside back cover of this book.

The divergence operator is distributive. That is, for any pair of vectors $\mathbf{E}_1$ and $\mathbf{E}_2$,

$$\nabla\cdot(\mathbf{E}_1 + \mathbf{E}_2) = \nabla\cdot\mathbf{E}_1 + \nabla\cdot\mathbf{E}_2. \tag{3.97}$$

If $\nabla\cdot\mathbf{E} = 0$, the vector field **E** is called ***divergenceless***.

The result given by Eq. (3.93) for a differential volume $\Delta\mathcal{V}$ can be extended to relate the volume integral of $\nabla\cdot\mathbf{E}$ over any volume $\mathcal{V}$ to the flux of **E** through the closed surface S that bounds $\mathcal{V}$. That is,

$$\int_{\mathcal{V}} \nabla\cdot\mathbf{E}\,d\mathcal{V} = \oint_S \mathbf{E}\cdot d\mathbf{s}. \tag{3.98}$$

(divergence theorem)

This relationship, known as the ***divergence theorem***, is used extensively in electromagnetics.

Example 3-11: Calculating the Divergence

Determine the divergence of each of the following vector fields and then evaluate them at the indicated points:

(a) $\mathbf{E} = \hat{\mathbf{x}}3x^2 + \hat{\mathbf{y}}2z + \hat{\mathbf{z}}x^2z$ at $(2, -2, 0)$;
(b) $\mathbf{E} = \hat{\mathbf{R}}(a^3\cos\theta/R^2) - \hat{\boldsymbol{\theta}}(a^3\sin\theta/R^2)$ at $(a/2, 0, \pi)$.

Solution: (a)

$$\begin{aligned}\nabla \cdot \mathbf{E} &= \frac{\partial E_x}{\partial x} + \frac{\partial E_y}{\partial y} + \frac{\partial E_z}{\partial z} \\ &= \frac{\partial}{\partial x}(3x^2) + \frac{\partial}{\partial y}(2z) + \frac{\partial}{\partial z}(x^2 z) \\ &= 6x + 0 + x^2 \\ &= x^2 + 6x.\end{aligned}$$

At (2, −2, 0), $\nabla \cdot \mathbf{E}\Big|_{(2,-2,0)} = 16.$

(b) From the expression given on the inside of the back cover of the book for the divergence of a vector in spherical coordinates, it follows that

$$\begin{aligned}\nabla \cdot \mathbf{E} &= \frac{1}{R^2}\frac{\partial}{\partial R}(R^2 E_R) + \frac{1}{R\sin\theta}\frac{\partial}{\partial \theta}(E_\theta \sin\theta) \\ &\quad + \frac{1}{R\sin\theta}\frac{\partial E_\phi}{\partial \phi} \\ &= \frac{1}{R^2}\frac{\partial}{\partial R}(a^3\cos\theta) + \frac{1}{R\sin\theta}\frac{\partial}{\partial \theta}\left(-\frac{a^3\sin^2\theta}{R^2}\right) \\ &= 0 - \frac{2a^3\cos\theta}{R^3} \\ &= -\frac{2a^3\cos\theta}{R^3}.\end{aligned}$$

At $R = a/2$ and $\theta = 0$, $\nabla \cdot \mathbf{E}\Big|_{(a/2,0,\pi)} = -16.$

Exercise 3-13: Given $\mathbf{A} = e^{-2y}(\hat{\mathbf{x}}\sin 2x + \hat{\mathbf{y}}\cos 2x)$, find $\nabla \cdot \mathbf{A}$.

Answer: $\nabla \cdot \mathbf{A} = 0$. (See EM.)

Exercise 3-14: Given $\mathbf{A} = \hat{\mathbf{r}} r\cos\phi + \hat{\boldsymbol{\phi}} r\sin\phi + \hat{\mathbf{z}}3z$, find $\nabla \cdot \mathbf{A}$ at (2, 0, 3).

Answer: $\nabla \cdot \mathbf{A} = 6$. (See EM.)

Exercise 3-15: If $\mathbf{E} = \hat{\mathbf{R}}AR$ in spherical coordinates, calculate the flux of $\mathbf{E}$ through a spherical surface of radius a, centered at the origin.

Answer: $\oint_S \mathbf{E} \cdot d\mathbf{s} = 4\pi A a^3$. (See EM.)

Exercise 3-16: Verify the divergence theorem by calculating the volume integral of the divergence of the field $\mathbf{E}$ of Exercise 3.15 over the volume bounded by the surface of radius a.

Exercise 3-17: The arrow representation in Fig. E3.17 represents the vector field $\mathbf{A} = \hat{\mathbf{x}}\,x - \hat{\mathbf{y}}\,y$. At a given point in space, $\mathbf{A}$ has a positive divergence $\nabla \cdot \mathbf{A}$ if the net flux flowing outward through the surface of an imaginary infinitesimal volume centered at that point is positive, $\nabla \cdot \mathbf{A}$ is negative if the net flux is into the volume, and $\nabla \cdot \mathbf{A} = 0$ if the same amount of flux enters into the volume as leaves it. Determine $\nabla \cdot \mathbf{A}$ everywhere in the x–y plane.

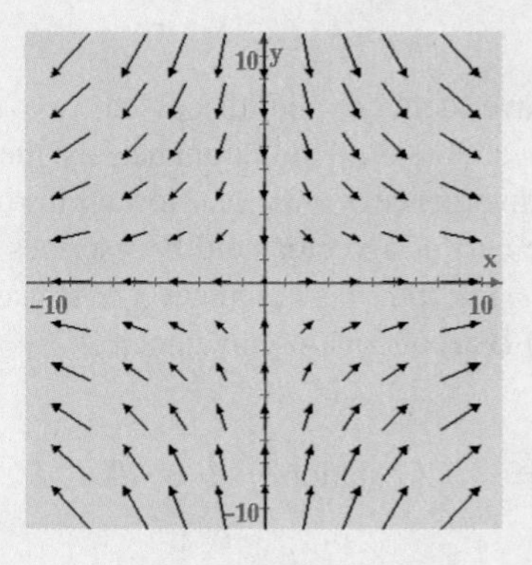

Figure E3.17

Answer: $\nabla \cdot \mathbf{A} = 0$ everywhere. (See EM.)

Module 3.3 Divergence Select a vector function $\mathbf{f}(x, y, z)$, evaluate its divergence, and display both in an appropriate 2-D plane.

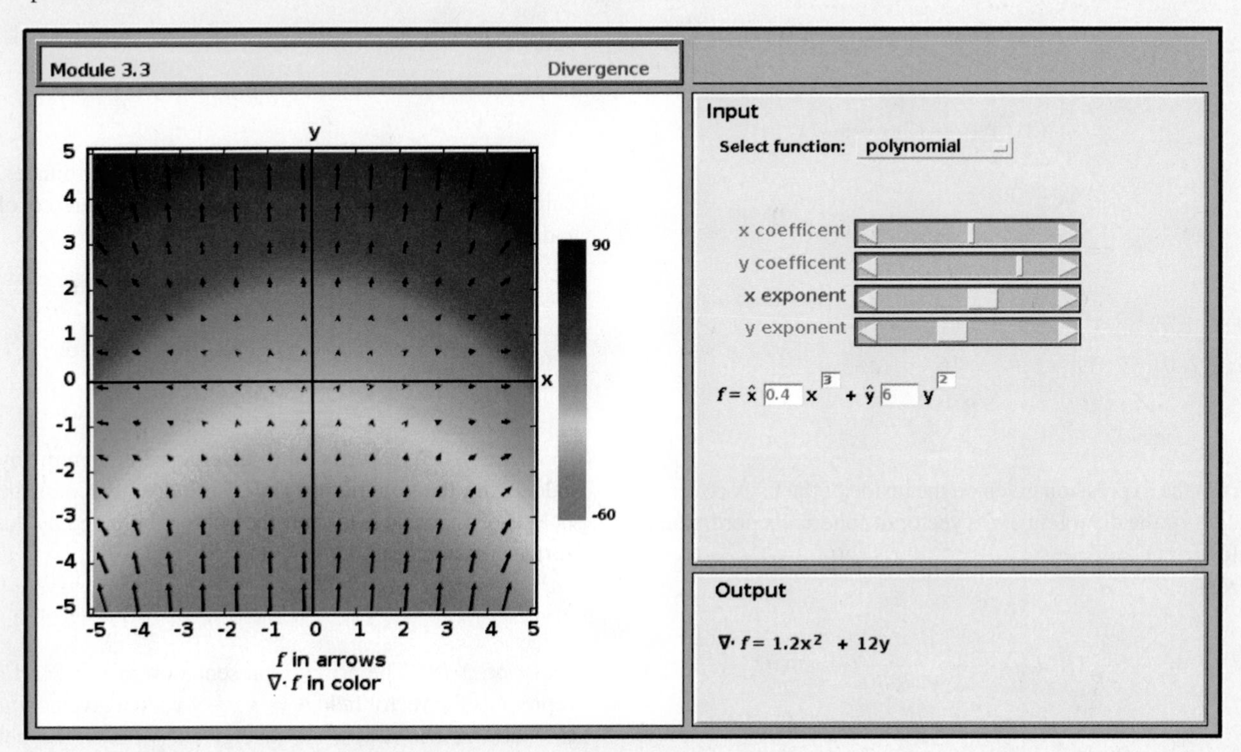

3-6 Curl of a Vector Field

So far we have defined and discussed two of the three fundamental operators used in vector analysis, the gradient of a scalar and the divergence of a vector. Now we introduce the ***curl operator***. The curl of a vector field $\mathbf{B}$ describes its rotational property, or ***circulation***. The circulation of $\mathbf{B}$ is defined as the line integral of $\mathbf{B}$ around a closed contour C;

$$\text{Circulation} = \oint_C \mathbf{B} \cdot d\mathbf{l}. \tag{3.99}$$

To gain a physical understanding of this definition, we consider two examples. The first is for a uniform field $\mathbf{B} = \hat{\mathbf{x}}B_0$, whose field lines are as depicted in Fig. 3-22(a). For the rectangular contour $abcd$ shown in the figure, we have

$$\begin{aligned}\text{Circulation} &= \int_a^b \hat{\mathbf{x}}B_0 \cdot \hat{\mathbf{x}}\,dx + \int_b^c \hat{\mathbf{x}}B_0 \cdot \hat{\mathbf{y}}\,dy \\ &\quad + \int_c^d \hat{\mathbf{x}}B_0 \cdot \hat{\mathbf{x}}\,dx + \int_d^a \hat{\mathbf{x}}B_0 \cdot \hat{\mathbf{y}}\,dy \\ &= B_0\,\Delta x - B_0\,\Delta x = 0, \end{aligned} \tag{3.100}$$

where $\Delta x = b - a = c - d$ and, because $\hat{\mathbf{x}} \cdot \hat{\mathbf{y}} = 0$, the second and fourth integrals are zero. According to Eq. (3.100), *the circulation of a uniform field is zero.*

Next, we consider the magnetic flux density $\mathbf{B}$ induced by an infinite wire carrying a dc current I. If the current is in free space and it is oriented along the z direction, then, from Eq. (1.13),

$$\mathbf{B} = \hat{\boldsymbol{\phi}}\frac{\mu_0 I}{2\pi r}, \tag{3.101}$$

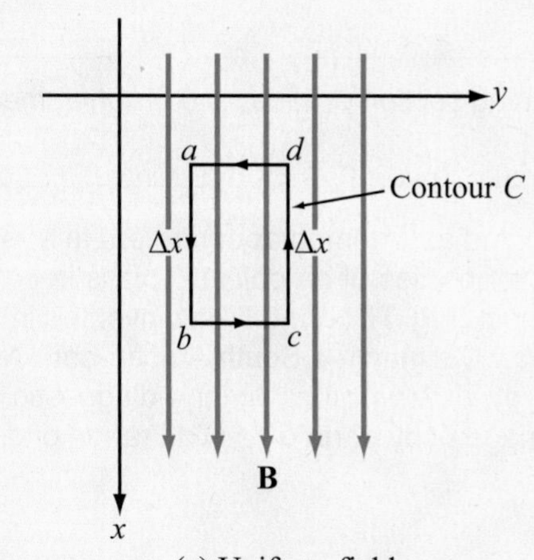

(a) Uniform field

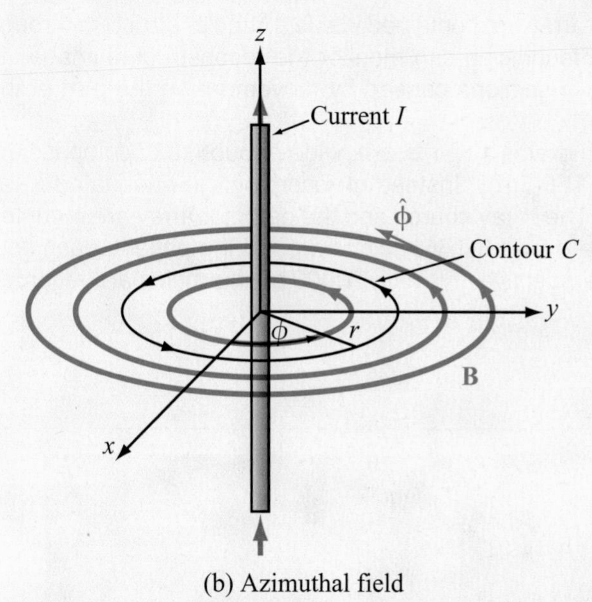

(b) Azimuthal field

Figure 3-22 Circulation is zero for the uniform field in (a), but it is not zero for the azimuthal field in (b).

where μ_0 is the permeability of free space and r is the radial distance from the current in the x–y plane. The direction of **B** is along the azimuth unit vector $\hat{\boldsymbol{\phi}}$. The field lines of **B** are concentric circles around the current, as shown in Fig. 3-22(b). For a circular contour C of radius r centered at the origin in the x–y plane, the differential length vector $d\mathbf{l} = \hat{\boldsymbol{\phi}} r\, d\phi$, and the circulation of **B** is

$$\text{Circulation} = \oint_C \mathbf{B} \cdot d\mathbf{l}$$

$$= \int_0^{2\pi} \hat{\boldsymbol{\phi}} \frac{\mu_0 I}{2\pi r} \cdot \hat{\boldsymbol{\phi}} r\, d\phi = \mu_0 I. \tag{3.102}$$

In this case, the circulation is not zero. However, had the contour C been in the x–z or y–z planes, $d\mathbf{l}$ would not have had a $\hat{\boldsymbol{\phi}}$ component, and the integral would have yielded a zero circulation. Clearly, the circulation of **B** depends on the choice of contour and the direction in which it is traversed. To describe the circulation of a tornado, for example, we would like to choose our contour such that the circulation of the wind field is maximum, and we would like the circulation to have both a magnitude and a direction, with the direction being toward the tornado's vortex. The ***curl operator*** embodies these properties. The curl of a vector field **B**, denoted curl **B** or $\nabla \times \mathbf{B}$, is defined as

$$\nabla \times \mathbf{B} = \text{curl}\, \mathbf{B}$$

$$= \lim_{\Delta s \to 0} \frac{1}{\Delta s} \left[\hat{\mathbf{n}} \oint_C \mathbf{B} \cdot d\mathbf{l} \right]_{\max}. \tag{3.103}$$

► Curl **B** is the circulation of **B** per unit area, with the area Δs of the contour C being oriented such that the circulation is maximum. ◄

The direction of curl **B** is $\hat{\mathbf{n}}$, the unit normal of Δs, defined according to the right-hand rule: with the four fingers of the right hand following the contour direction $d\mathbf{l}$, the thumb points along $\hat{\mathbf{n}}$ (Fig. 3-23). When we use the notation $\nabla \times \mathbf{B}$ to denote curl **B**, it should *not* be interpreted as the cross product of ∇ and **B**.

Technology Brief 6: X-Ray Computed Tomography

► The word ***tomography*** is derived from the Greek words *tome*, meaning section or slice, and *graphia*, meaning writing. ◄

Computed tomography, also known as ***CT scan*** or ***CAT scan*** (for computed axial tomography), refers to a technique capable of generating 3-D images of the X-ray attenuation (absorption) properties of an object. This is in contrast to the traditional, X-ray technique that produces only a 2-D profile of the object (**Fig. TF6-1**). CT was invented in 1972 by British electrical engineer ***Godfrey Hounsfeld*** and independently by ***Allan Cormack***, a South African-born American physicist. The two inventors shared the ***1979 Nobel Prize in Physiology or Medicine***. Among diagnostic imaging techniques, CT has the decided advantage in having the sensitivity to image body parts on a wide range of densities, from soft tissue to blood vessels and bones.

Principle of Operation

In the system shown in **Fig. TF6-2**, the X-ray source and detector array are contained inside a circular structure through which the patient is moved along a conveyor belt. A CAT scan technician can monitor the reconstructed images to insure that they do not contain artifacts such as streaks or blurry sections caused by movement on the part of the patient during the measurement process.

A CT scanner uses an ***X-ray source*** with a narrow slit that generates a ***fan-beam***, wide enough to encompass the extent of the body, but only a few millimeters in thickness [**Fig. TF6-3(a)**]. Instead of recording the attenuated X-ray beam on film, it is captured by an array of some 700 ***detectors***. The X-ray source and the detector array are mounted on a circular frame that rotates in steps of a fraction of a degree over a full 360° circle around the patient, each time recording an X-ray attenuation profile from a different angular perspective. Typically, 1,000 such profiles are recorded

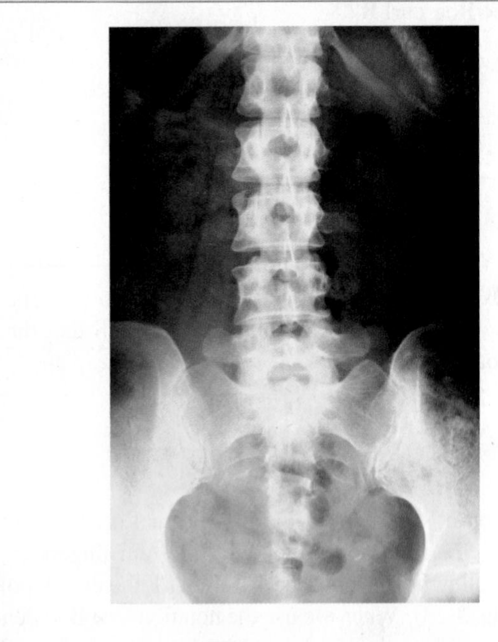

Figure TF6-1 2-D X-ray image.

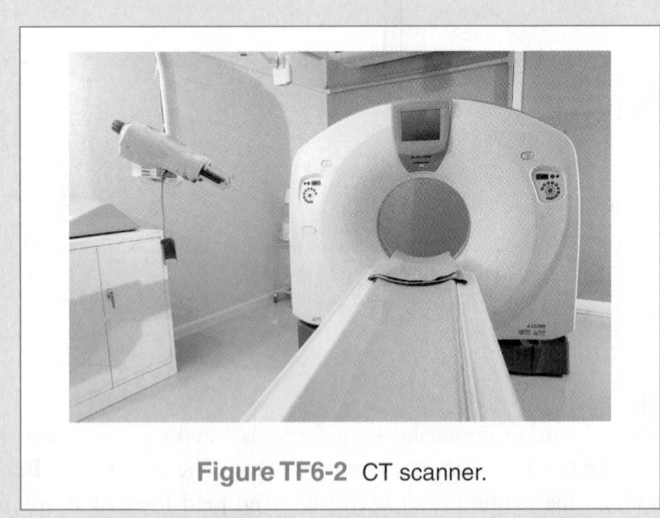

Figure TF6-2 CT scanner.

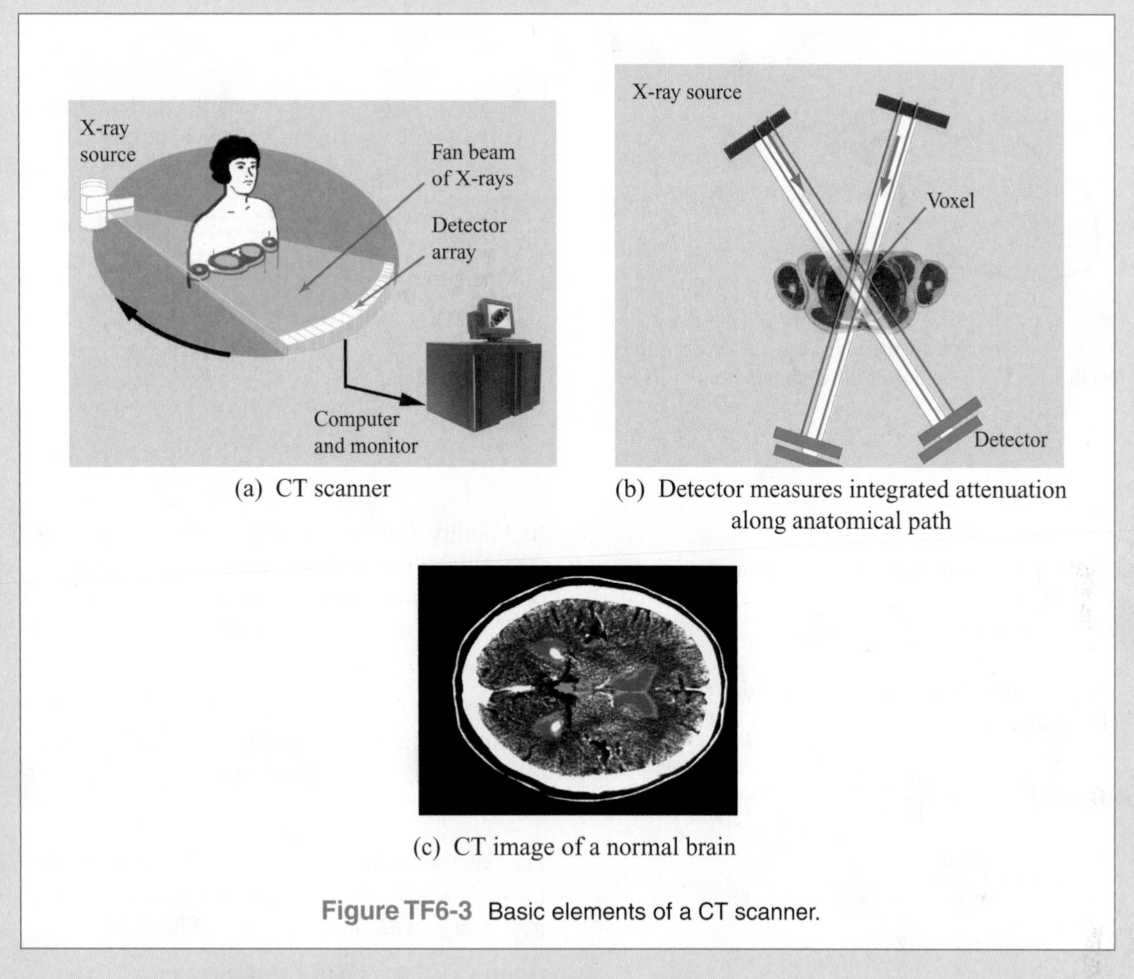

(a) CT scanner

(b) Detector measures integrated attenuation along anatomical path

(c) CT image of a normal brain

Figure TF6-3 Basic elements of a CT scanner.

per each thin traverse slice of anatomy. In today's technology, this process is completed in less than 1 second. To image an entire part of the body, such as the chest or head, the process is repeated over multiple slices (layers), which typically takes about 10 seconds to complete.

Image Reconstruction

For each anatomical slice, the CT scanner generates on the order of 7×10^5 measurements (1,000 angular orientations $\times$ 700 detector channels). Each measurement represents the integrated path attenuation for the narrow beam between the X-ray source and the detector [**Fig. TF6-3(b)**], and each volume element ***(voxel)*** contributes to 1,000 such measurement beams.

▶ Commercial CT machines use a technique called ***filtered back-projection*** to "reconstruct" an image of the attenuation rate of each voxel in the anatomical slice and, by extension, for each voxel in the entire body organ. This is accomplished through the application of a sophisticated matrix inversion process. ◀

A sample CT image of the brain is shown in **Fig. TF6-3(c)**.

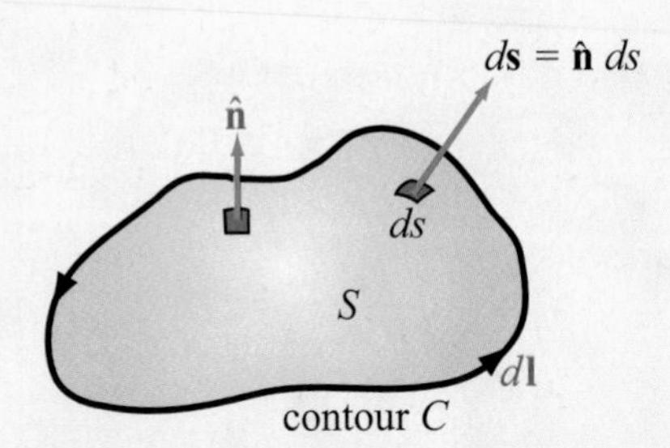

Figure 3-23 The direction of the unit vector $\hat{\mathbf{n}}$ is along the thumb when the other four fingers of the right hand follow $d\mathbf{l}$.

For a vector $\mathbf{B}$ specified in Cartesian coordinates as

$$\mathbf{B} = \hat{\mathbf{x}}B_x + \hat{\mathbf{y}}B_y + \hat{\mathbf{z}}B_z, \tag{3.104}$$

it can be shown, through a rather lengthy derivation, that Eq. (3.103) leads to

$$\begin{aligned}\nabla \times \mathbf{B} &= \hat{\mathbf{x}}\left(\frac{\partial B_z}{\partial y} - \frac{\partial B_y}{\partial z}\right) + \hat{\mathbf{y}}\left(\frac{\partial B_x}{\partial z} - \frac{\partial B_z}{\partial x}\right) \\ &\quad + \hat{\mathbf{z}}\left(\frac{\partial B_y}{\partial x} - \frac{\partial B_x}{\partial y}\right) \\ &= \begin{vmatrix} \hat{\mathbf{x}} & \hat{\mathbf{y}} & \hat{\mathbf{z}} \\ \dfrac{\partial}{\partial x} & \dfrac{\partial}{\partial y} & \dfrac{\partial}{\partial z} \\ B_x & B_y & B_z \end{vmatrix}. \end{aligned} \tag{3.105}$$

Expressions for $\nabla \times \mathbf{B}$ are given on the inside back cover of the book for the three orthogonal coordinate systems considered in this chapter.

3-6.1 Vector Identities Involving the Curl

For any two vectors $\mathbf{A}$ and $\mathbf{B}$ and scalar V,

$$(1) \quad \nabla \times (\mathbf{A} + \mathbf{B}) = \nabla \times \mathbf{A} + \nabla \times \mathbf{B}, \tag{3.106a}$$

$$(2) \quad \nabla \cdot (\nabla \times \mathbf{A}) = 0, \tag{3.106b}$$

$$(3) \quad \nabla \times (\nabla V) = 0. \tag{3.106c}$$

3-6.2 Stokes's Theorem

► ***Stokes's theorem*** converts the surface integral of the curl of a vector over an open surface S into a line integral of the vector along the contour C bounding the surface S. ◄

For the geometry shown in **Fig. 3-23**, ***Stokes's theorem*** states

$$\int_S (\nabla \times \mathbf{B}) \cdot d\mathbf{s} = \oint_C \mathbf{B} \cdot d\mathbf{l}. \tag{3.107}$$

(Stokes's theorem)

Its validity follows from the definition of $\nabla \times \mathbf{B}$ given by Eq. (3.103). If $\nabla \times \mathbf{B} = 0$, the field $\mathbf{B}$ is said to be ***conservative*** or ***irrotational*** because its circulation, represented by the right-hand side of Eq. (3.107), is zero, irrespective of the contour chosen.

Example 3-12: Verification of Stokes's Theorem

For vector field $\mathbf{B} = \hat{\mathbf{z}}\cos\phi/r$, verify Stokes's theorem for a segment of a cylindrical surface defined by $r = 2$, $\pi/3 \le \phi \le \pi/2$, and $0 \le z \le 3$ (**Fig. 3-24**).

Solution: Stokes's theorem states that

$$\int_S (\nabla \times \mathbf{B}) \cdot d\mathbf{s} = \oint_C \mathbf{B} \cdot d\mathbf{l}.$$

Left-hand side: With $\mathbf{B}$ having only a component $B_z = \cos\phi/r$, use of the expression for $\nabla \times \mathbf{B}$ in cylindrical coordinates from the inside back cover of the book gives

$$\begin{aligned}\nabla \times \mathbf{B} &= \hat{\mathbf{r}}\left(\frac{1}{r}\frac{\partial B_z}{\partial \phi} - \frac{\partial B_\phi}{\partial z}\right) + \hat{\boldsymbol{\phi}}\left(\frac{\partial B_r}{\partial z} - \frac{\partial B_z}{\partial r}\right) \\ &\quad + \hat{\mathbf{z}}\frac{1}{r}\left(\frac{\partial}{\partial r}(rB_\phi) - \frac{\partial B_r}{\partial \phi}\right) \\ &= \hat{\mathbf{r}}\frac{1}{r}\frac{\partial}{\partial \phi}\left(\frac{\cos\phi}{r}\right) - \hat{\boldsymbol{\phi}}\frac{\partial}{\partial r}\left(\frac{\cos\phi}{r}\right) \\ &= -\hat{\mathbf{r}}\frac{\sin\phi}{r^2} + \hat{\boldsymbol{\phi}}\frac{\cos\phi}{r^2}.\end{aligned}$$

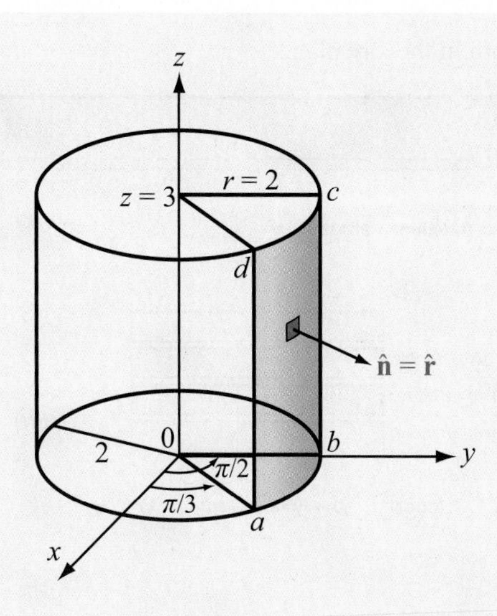

Figure 3-24 Geometry of Example 3-12.

The integral of $\nabla \times \mathbf{B}$ over the specified surface S is

$$\int_S (\nabla \times \mathbf{B}) \cdot d\mathbf{s}$$

$$= \int_{z=0}^{3} \int_{\phi=\pi/3}^{\pi/2} \left(-\hat{\mathbf{r}} \frac{\sin\phi}{r^2} + \hat{\boldsymbol{\phi}} \frac{\cos\phi}{r^2} \right) \cdot \hat{\mathbf{r}} r \, d\phi \, dz$$

$$= \int_{0}^{3} \int_{\pi/3}^{\pi/2} -\frac{\sin\phi}{r} \, d\phi \, dz = -\frac{3}{2r} = -\frac{3}{4}.$$

Right-hand side: The surface S is bounded by contour $C = abcd$ shown in **Fig. 3-24**. The direction of C is chosen so that it is compatible with the surface normal $\hat{\mathbf{r}}$ by the right-hand rule. Hence,

$$\oint_C \mathbf{B} \cdot d\mathbf{l} = \int_a^b \mathbf{B}_{ab} \cdot d\mathbf{l} + \int_b^c \mathbf{B}_{bc} \cdot d\mathbf{l} + \int_c^d \mathbf{B}_{cd} \cdot d\mathbf{l} + \int_d^a \mathbf{B}_{da} \cdot d\mathbf{l},$$

where $\mathbf{B}_{ab}$, $\mathbf{B}_{bc}$, $\mathbf{B}_{cd}$, and $\mathbf{B}_{da}$ are the field $\mathbf{B}$ along segments ab, bc, cd, and da, respectively. Over segment ab, the dot product of $\mathbf{B}_{ab} = \hat{\mathbf{z}}\,(\cos\phi)/2$ and $d\mathbf{l} = \hat{\boldsymbol{\phi}} r \, d\phi$ is zero, and the same is true for segment cd. Over segment bc, $\phi = \pi/2$; hence, $\mathbf{B}_{bc} = \hat{\mathbf{z}}(\cos\pi/2)/2 = 0$. For the last segment, $\mathbf{B}_{da} = \hat{\mathbf{z}}(\cos\pi/3)/2 = \hat{\mathbf{z}}/4$ and $d\mathbf{l} = \hat{\mathbf{z}}\, dz$. Hence,

$$\oint_C \mathbf{B} \cdot d\mathbf{l} = \int_d^a \left(\hat{\mathbf{z}} \frac{1}{4} \right) \cdot \hat{\mathbf{z}} \, dz = \int_3^0 \frac{1}{4} \, dz = -\frac{3}{4},$$

which is the same as the result obtained by evaluating the left-hand side of Stokes's equation.

Exercise 3-18: Find $\nabla \times \mathbf{A}$ at $(2, 0, 3)$ in cylindrical coordinates for the vector field

$$\mathbf{A} = \hat{\mathbf{r}} 10 e^{-2r} \cos\phi + \hat{\mathbf{z}} 10 \sin\phi.$$

Answer: (See EM.)

$$\nabla \times \mathbf{A} = \left(\hat{\mathbf{r}} \frac{10\cos\phi}{r} + \frac{\hat{\mathbf{z}}\, 10 e^{-2r}}{r} \sin\phi \right) \bigg|_{(2,0,3)} = \hat{\mathbf{r}} 5.$$

Exercise 3-19: Find $\nabla \times \mathbf{A}$ at $(3, \pi/6, 0)$ in spherical coordinates for the vector field $\mathbf{A} = \hat{\boldsymbol{\theta}} 12 \sin\theta$.

Answer: (See EM.)

$$\nabla \times \mathbf{A} = \hat{\boldsymbol{\phi}} \frac{12\sin\theta}{R} \bigg|_{(3,\pi/6,0)} = \hat{\boldsymbol{\phi}} 2.$$

3-7 Laplacian Operator

In later chapters, we sometimes deal with problems involving multiple combinations of operations on scalars and vectors. A

Module 3.4 Curl Select a vector $\mathbf{f}(x, y)$, evaluate its curl, and display both in the x-y plane.

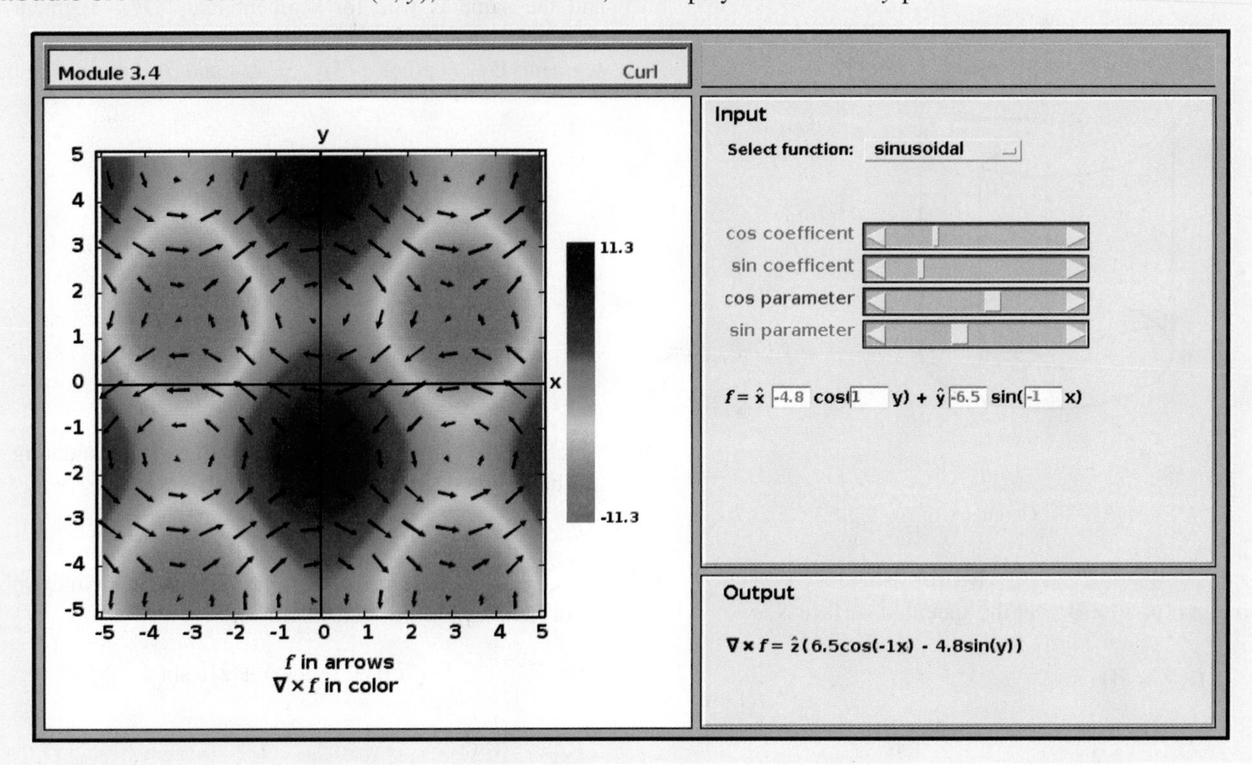

frequently encountered combination is the divergence of the gradient of a scalar. For a scalar function V defined in Cartesian coordinates, its gradient is

$$\begin{aligned}\nabla V &= \hat{\mathbf{x}}\frac{\partial V}{\partial x} + \hat{\mathbf{y}}\frac{\partial V}{\partial y} + \hat{\mathbf{z}}\frac{\partial V}{\partial z} \\ &= \hat{\mathbf{x}}A_x + \hat{\mathbf{y}}A_y + \hat{\mathbf{z}}A_z = \mathbf{A},\end{aligned} \tag{3.108}$$

where we defined a vector $\mathbf{A}$ with components $A_x = \partial V/\partial x$, $A_y = \partial V/\partial y$, and $A_z = \partial V/\partial z$. The divergence of ∇V is

$$\begin{aligned}\nabla \cdot (\nabla V) = \nabla \cdot \mathbf{A} &= \frac{\partial A_x}{\partial x} + \frac{\partial A_y}{\partial y} + \frac{\partial A_z}{\partial z} \\ &= \frac{\partial^2 V}{\partial x^2} + \frac{\partial^2 V}{\partial y^2} + \frac{\partial^2 V}{\partial z^2}.\end{aligned} \tag{3.109}$$

For convenience, $\nabla \cdot (\nabla V)$ is called the ***Laplacian*** of V and is denoted by $\nabla^2 V$ (the symbol ∇^2 is pronounced "del square"). That is,

$$\nabla^2 V = \nabla \cdot (\nabla V) = \frac{\partial^2 V}{\partial x^2} + \frac{\partial^2 V}{\partial y^2} + \frac{\partial^2 V}{\partial z^2}. \tag{3.110}$$

As we can see from Eq. (3.110), the Laplacian of a scalar function is a scalar. Expressions for $\nabla^2 V$ in cylindrical and spherical coordinates are given on the inside back cover of the book.

The Laplacian of a scalar can be used to define the Laplacian of a vector. For a vector $\mathbf{E}$ specified in Cartesian coordinates as

$$\mathbf{E} = \hat{\mathbf{x}}E_x + \hat{\mathbf{y}}E_y + \hat{\mathbf{z}}E_z, \tag{3.111}$$

the Laplacian of **E** is

$$\nabla^2 \mathbf{E} = \left(\frac{\partial^2}{\partial x^2} + \frac{\partial^2}{\partial y^2} + \frac{\partial^2}{\partial z^2}\right)\mathbf{E}$$
$$= \hat{\mathbf{x}}\,\nabla^2 E_x + \hat{\mathbf{y}}\,\nabla^2 E_y + \hat{\mathbf{z}}\,\nabla^2 E_z. \quad (3.112)$$

Thus, in Cartesian coordinates the Laplacian of a vector is a vector whose components are equal to the Laplacians of the vector components. Through direct substitution, it can be shown that

$$\nabla^2 \mathbf{E} = \nabla(\nabla \cdot \mathbf{E}) - \nabla \times (\nabla \times \mathbf{E}). \quad (3.113)$$

Concept Question 3-11: What do the magnitude and direction of the gradient of a scalar quantity represent?

Concept Question 3-12: Prove the validity of Eq. (3.84c) in Cartesian coordinates.

Concept Question 3-13: What is the physical meaning of the divergence of a vector field?

Concept Question 3-14: If a vector field is solenoidal at a given point in space, does it necessarily follow that the vector field is zero at that point? Explain.

Concept Question 3-15: What is the meaning of the transformation provided by the divergence theorem?

Concept Question 3-16: How is the curl of a vector field at a point related to the circulation of the vector field?

Concept Question 3-17: What is the meaning of the transformation provided by Stokes's theorem?

Concept Question 3-18: When is a vector field "conservative"?

Chapter 3 Summary

Concepts

- Vector algebra governs the laws of addition, subtraction, and multiplication of vectors, and vector calculus encompasses the laws of differentiation and integration of vectors.
- In a right-handed orthogonal coordinate system, the three base vectors are mutually perpendicular to each other at any point in space, and the cyclic relations governing the cross products of the base vectors obey the right-hand rule.
- The dot product of two vectors produces a scalar, whereas the cross product of two vectors produces another vector.
- A vector expressed in a given coordinate system can be expressed in another coordinate system through the use of transformation relations linking the two coordinate systems.
- The fundamental differential functions in vector calculus are the gradient, the divergence, and the curl.
- The gradient of a scalar function is a vector whose magnitude is equal to the maximum rate of increasing change of the scalar function per unit distance, and its direction is along the direction of maximum increase.
- The divergence of a vector field is a measure of the net outward flux per unit volume through a closed surface surrounding the unit volume.
- The divergence theorem transforms the volume integral of the divergence of a vector field into a surface integral of the field's flux through a closed surface surrounding the volume.
- The curl of a vector field is a measure of the circulation of the vector field per unit area Δs, with the orientation of Δs chosen such that the circulation is maximum.
- Stokes's theorem transforms the surface integral of the curl of a vector field into a line integral of the field over a contour that bounds the surface.
- The Laplacian of a scalar function is defined as the divergence of the gradient of that function.

Mathematical and Physical Models

Distance between Two Points

$$d = [(x_2 - x_1)^2 + (y_2 - y_1)^2 + (z_2 - z_1)^2]^{1/2}$$

$$d = \left[r_2^2 + r_1^2 - 2r_1r_2\cos(\phi_2 - \phi_1) + (z_2 - z_1)^2\right]^{1/2}$$

$$d = \left\{R_2^2 + R_1^2 - 2R_1R_2[\cos\theta_2\cos\theta_1 + \sin\theta_1\sin\theta_2\cos(\phi_2 - \phi_1)]\right\}^{1/2}$$

Coordinate Systems Table 3-1

Coordinate Transformations Table 3-2

Vector Products

$$\mathbf{A}\cdot\mathbf{B} = AB\cos\theta_{AB}$$

$$\mathbf{A}\times\mathbf{B} = \hat{\mathbf{n}}\,AB\sin\theta_{AB}$$

$$\mathbf{A}\cdot(\mathbf{B}\times\mathbf{C}) = \mathbf{B}\cdot(\mathbf{C}\times\mathbf{A}) = \mathbf{C}\cdot(\mathbf{A}\times\mathbf{B})$$

$$\mathbf{A}\times(\mathbf{B}\times\mathbf{C}) = \mathbf{B}(\mathbf{A}\cdot\mathbf{C}) - \mathbf{C}(\mathbf{A}\cdot\mathbf{B})$$

Divergence Theorem

$$\int_{\mathcal{V}} \nabla\cdot\mathbf{E}\,d\mathcal{V} = \oint_S \mathbf{E}\cdot d\mathbf{s}$$

Vector Operators

$$\nabla T = \hat{\mathbf{x}}\frac{\partial T}{\partial x} + \hat{\mathbf{y}}\frac{\partial T}{\partial y} + \hat{\mathbf{z}}\frac{\partial T}{\partial z}$$

$$\nabla\cdot\mathbf{E} = \frac{\partial E_x}{\partial x} + \frac{\partial E_y}{\partial y} + \frac{\partial E_z}{\partial z}$$

$$\nabla\times\mathbf{B} = \hat{\mathbf{x}}\left(\frac{\partial B_z}{\partial y} - \frac{\partial B_y}{\partial z}\right) + \hat{\mathbf{y}}\left(\frac{\partial B_x}{\partial z} - \frac{\partial B_z}{\partial x}\right) + \hat{\mathbf{z}}\left(\frac{\partial B_y}{\partial x} - \frac{\partial B_x}{\partial y}\right)$$

$$\nabla^2 V = \frac{\partial^2 V}{\partial x^2} + \frac{\partial^2 V}{\partial y^2} + \frac{\partial^2 V}{\partial z^2}$$

(see back cover for cylindrical and spherical coordinates)

Stokes's Theorem

$$\int_S (\nabla\times\mathbf{B})\cdot d\mathbf{s} = \oint_C \mathbf{B}\cdot d\mathbf{l}$$

Important Terms

Provide definitions or explain the meaning of the following terms:

azimuth angle
base vectors
Cartesian coordinate system
circulation of a vector
conservative field
cross product
curl operator
cylindrical coordinate system
differential area vector
differential length vector
differential volume
directional derivative
distance vector
divergenceless
divergence operator
divergence theorem
dot product
field lines
flux density
flux lines
gradient operator
irrotational field
Laplacian operator
magnitude
orthogonal coordinate system
position vector
radial distance r
range R
right-hand rule
scalar product
scalar quantity
simple product
solenoidal field
spherical coordinate system
Stokes's theorem
vector product
vector quantity
unit vector
zenith angle

PROBLEMS

Section 3-1: Basic Laws of Vector Algebra

3.1 Given vectors $\mathbf{A} = \hat{\mathbf{x}}2 - \hat{\mathbf{y}}3 + \hat{\mathbf{z}}$, $\mathbf{B} = \hat{\mathbf{x}}2 - \hat{\mathbf{y}} + \hat{\mathbf{z}}3$, and $\mathbf{C} = \hat{\mathbf{x}}2 + \hat{\mathbf{y}}1 - \hat{\mathbf{z}}1$, show that $\mathbf{C}$ is perpendicular to both $\mathbf{A}$ and $\mathbf{B}$.

*__3.2__ Vector $\mathbf{A}$ starts at point $(1, -1, 3)$ and ends at point $(2, -1, 0)$. Find a unit vector in the direction of $\mathbf{A}$.

*__3.3__ In Cartesian coordinates, the three corners of a triangle are $P_1 = (0, 4, 4)$, $P_2 = (4, -4, 4)$, and $P_3 = (2, 2, -4)$. Find the area of the triangle.

3.4 Given $\mathbf{A} = \hat{\mathbf{x}}2 - \hat{\mathbf{y}}3 + \hat{\mathbf{z}}1$ and $\mathbf{B} = \hat{\mathbf{x}}B_x + \hat{\mathbf{y}}2 + \hat{\mathbf{z}}B_z$:

(a) Find B_x and B_z if $\mathbf{A}$ is parallel to $\mathbf{B}$.

(b) Find a relation between B_x and B_z if $\mathbf{A}$ is perpendicular to $\mathbf{B}$.

3.5 Given vectors $\mathbf{A} = \hat{\mathbf{x}} + \hat{\mathbf{y}}2 - \hat{\mathbf{z}}3$, $\mathbf{B} = \hat{\mathbf{x}}2 - \hat{\mathbf{y}}4$, and $\mathbf{C} = \hat{\mathbf{y}}2 - \hat{\mathbf{z}}4$, find the following:

*__(a)__ A and $\hat{\mathbf{a}}$

(b) The component of $\mathbf{B}$ along $\mathbf{C}$

(c) θ_{AC}

(d) $\mathbf{A} \times \mathbf{C}$

*__(e)__ $\mathbf{A} \cdot (\mathbf{B} \times \mathbf{C})$

(f) $\mathbf{A} \times (\mathbf{B} \times \mathbf{C})$

(g) $\hat{\mathbf{x}} \times \mathbf{B}$

*__(h)__ $(\mathbf{A} \times \hat{\mathbf{y}}) \cdot \hat{\mathbf{z}}$

3.6 Given $\mathbf{A} = \hat{\mathbf{x}}(x + 2y) - \hat{\mathbf{y}}(y + 3z) + \hat{\mathbf{z}}(3x - y)$, determine a unit vector parallel to $\mathbf{A}$ at point $P = (1, -1, 2)$.

3.7 Given vectors $\mathbf{A} = \hat{\mathbf{x}}2 - \hat{\mathbf{y}} + \hat{\mathbf{z}}3$ and $\mathbf{B} = \hat{\mathbf{x}}3 - \hat{\mathbf{z}}2$, find a vector $\mathbf{C}$ whose magnitude is 9 and whose direction is perpendicular to both $\mathbf{A}$ and $\mathbf{B}$.

3.8 By expansion in Cartesian coordinates, prove:

(a) The relation for the scalar triple product given by Eq. (3.29).

(b) The relation for the vector triple product given by Eq. (3.33).

*__3.9__ Find an expression for the unit vector directed toward the origin from an arbitrary point on the line described by $x = 1$ and $z = -2$.

*__3.10__ Find a unit vector parallel to either direction of the line described by

$$2x + z = 6.$$

3.11 Find an expression for the unit vector directed toward the point P located on the z axis at a height h above the x–y plane from an arbitrary point $Q = (x, y, -5)$ in the plane $z = -5$.

*__3.12__ A given line is described by

$$x + 2y = 4.$$

Vector $\mathbf{A}$ starts at the origin and ends at point P on the line such that $\mathbf{A}$ is orthogonal to the line. Find an expression for $\mathbf{A}$.

3.13 Two lines in the x–y plane are described by the following expressions:

Line 1	$x + 2y = -6.$
Line 2	$3x + 4y = 8.$

Use vector algebra to find the smaller angle between the lines at their intersection point.

3.14 Show that, given two vectors $\mathbf{A}$ and $\mathbf{B}$,

(a) The vector $\mathbf{C}$ defined as the vector component of $\mathbf{B}$ in the direction of $\mathbf{A}$ is given by

$$\mathbf{C} = \hat{\mathbf{a}}(\mathbf{B} \cdot \hat{\mathbf{a}}) = \frac{\mathbf{A}(\mathbf{B} \cdot \mathbf{A})}{|\mathbf{A}|^2}$$

where $\hat{\mathbf{a}}$ is the unit vector of $\mathbf{A}$.

(b) The vector $\mathbf{D}$ defined as the vector component of $\mathbf{B}$ perpendicular to $\mathbf{A}$ is given by

$$\mathbf{D} = \mathbf{B} - \frac{\mathbf{A}(\mathbf{B} \cdot \mathbf{A})}{|\mathbf{A}|^2}.$$

*__3.15__ A certain plane is described by

$$2x + 3y + 4z = 16.$$

Find the unit vector normal to the surface in the direction away from the origin.

3.16 Given $\mathbf{B} = \hat{\mathbf{x}}(z - 3y) + \hat{\mathbf{y}}(2x - 3z) - \hat{\mathbf{z}}(x + y)$, find a unit vector parallel to $\mathbf{B}$ at point $P = (1, 0, -1)$.

*__3.17__ Find a vector $\mathbf{G}$ whose magnitude is 4 and whose direction is perpendicular to both vectors $\mathbf{E}$ and $\mathbf{F}$, where $\mathbf{E} = \hat{\mathbf{x}} + \hat{\mathbf{y}}\,2 - \hat{\mathbf{z}}\,2$ and $\mathbf{F} = \hat{\mathbf{y}}\,3 - \hat{\mathbf{z}}\,6$.

* Answer(s) available in Appendix D.

3.18 A given line is described by the equation:

$$y = x - 1.$$

Vector **A** starts at point $P_1 = (0, 2)$ and ends at point P_2 on the line, at which **A** is orthogonal to the line. Find an expression for **A**.

3.19 Vector field **E** is given by

$$\mathbf{E} = \hat{\mathbf{R}}\, 5R\cos\theta - \hat{\boldsymbol{\theta}}\frac{12}{R}\sin\theta\cos\phi + \hat{\boldsymbol{\phi}}3\sin\phi.$$

Determine the component of **E** tangential to the spherical surface $R = 3$ at point $P = (3, 30°, 60°)$.

3.20 When sketching or demonstrating the spatial variation of a vector field, we often use arrows, as in **Fig. P3.20**, wherein the length of the arrow is made to be proportional to the strength of the field and the direction of the arrow is the same as that of the field's. The sketch shown in **Fig. P3.20**, which represents the vector field $\mathbf{E} = \hat{\mathbf{r}}r$, consists of arrows pointing radially away from the origin and their lengths increasing linearly in proportion to their distance away from the origin. Using this arrow representation, sketch each of the following vector fields:

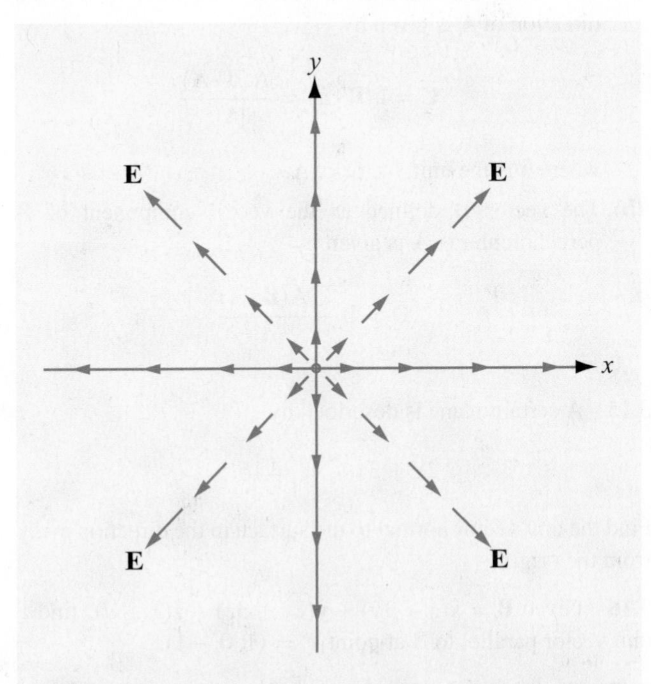

Figure P3.20 Arrow representation for vector field $\mathbf{E} = \hat{\mathbf{r}}\, r$ (Problem 3.20).

(a) $\mathbf{E}_1 = -\hat{\mathbf{x}}y$

(b) $\mathbf{E}_2 = \hat{\mathbf{y}}x$

(c) $\mathbf{E}_3 = \hat{\mathbf{x}}x + \hat{\mathbf{y}}y$

(d) $\mathbf{E}_4 = \hat{\mathbf{x}}x + \hat{\mathbf{y}}2y$

(e) $\mathbf{E}_5 = \hat{\boldsymbol{\phi}}r$

(f) $\mathbf{E}_6 = \hat{\mathbf{r}}\sin\phi$

3.21 Use arrows to sketch each of the following vector fields:

(a) $\mathbf{E}_1 = \hat{\mathbf{x}}x - \hat{\mathbf{y}}y$

(b) $\mathbf{E}_2 = -\hat{\boldsymbol{\phi}}$

(c) $\mathbf{E}_3 = \hat{\mathbf{y}}\,(1/x)$

(d) $\mathbf{E}_4 = \hat{\mathbf{r}}\cos\phi$

Sections 3-2 and 3-3: Coordinate Systems

3.22 Convert the coordinates of the following points from cylindrical to Cartesian coordinates:

(a) $P_1 = (2, \pi/4, -4)$

(b) $P_2 = (3, 0, -4)$

(c) $P_3 = (4, \pi, 4)$

3.23 Convert the coordinates of the following points from Cartesian to cylindrical and spherical coordinates:

***(a)** $P_1 = (1, 2, 0)$

(b) $P_2 = (0, 0, 2)$

(c) $P_3 = (1, 1, 3)$

***(d)** $P_4 = (-2, 2, -2)$

3.24 Convert the coordinates of the following points from spherical to cylindrical coordinates:

***(a)** $P_1 = (5, 0, 0)$

(b) $P_2 = (5, 0, \pi)$

(c) $P_3 = (3, \pi/2, 0)$

3.25 Use the appropriate expression for the differential surface area ds to determine the area of each of the following surfaces:

(a) $r = 3;\ 0 \le \phi \le \pi/3;\ -2 \le z \le 2$

(b) $2 \le r \le 5;\ \pi/2 \le \phi \le \pi;\ z = 0$

***(c)** $2 \le r \le 5;\ \phi = \pi/4;\ -2 \le z \le 2$

(d) $R = 2;\ 0 \le \theta \le \pi/3;\ 0 \le \phi \le \pi$

(e) $0 \le R \le 5;\ \theta = \pi/3;\ 0 \le \phi \le 2\pi$

Also sketch the outline of each surface.

3.26 A section of a sphere is described by $0 \le R \le 2$, $0 \le \theta \le 90°$, and $30° \le \phi \le 90°$. Find the following:

(a) The surface area of the spherical section.

(b) The enclosed volume.

Also sketch the outline of the section.

3.27 Find the volumes described by the following:

*(a) $2 \le r \le 5$; $\pi/2 \le \phi \le \pi$; $0 \le z \le 2$

(b) $0 \le R \le 5$; $0 \le \theta \le \pi/3$; $0 \le \phi \le 2\pi$

Also sketch the outline of each volume.

3.28 A vector field is given in cylindrical coordinates by

$$\mathbf{E} = \hat{\mathbf{r}}r\cos\phi + \hat{\boldsymbol{\phi}}r\sin\phi + \hat{\mathbf{z}}z^3.$$

Point $P = (2, \pi, 3)$ is located on the surface of the cylinder described by $r = 2$. At point P, find:

(a) The vector component of **E** perpendicular to the cylinder.

(b) The vector component of **E** tangential to the cylinder.

3.29 At a given point in space, vectors **A** and **B** are given in spherical coordinates by

$$\mathbf{A} = \hat{\mathbf{R}}4 + \hat{\boldsymbol{\theta}}2 - \hat{\boldsymbol{\phi}},$$
$$\mathbf{B} = -\hat{\mathbf{R}}2 + \hat{\boldsymbol{\phi}}3.$$

Find:

(a) The scalar component, or projection, of **B** in the direction of **A**.

(b) The vector component of **B** in the direction of **A**.

(c) The vector component of **B** perpendicular to **A**.

***3.30** Given vectors

$$\mathbf{A} = \hat{\mathbf{r}}(\cos\phi + 3z) - \hat{\boldsymbol{\phi}}(2r + 4\sin\phi) + \hat{\mathbf{z}}(r - 2z)$$
$$\mathbf{B} = -\hat{\mathbf{r}}\sin\phi + \hat{\mathbf{z}}\cos\phi$$

find

(a) θ_{AB} at $(2, \pi/2, 0)$

(b) A unit vector perpendicular to both **A** and **B** at $(2, \pi/3, 1)$

3.31 Determine the distance between the following pairs of points:

*(a) $P_1 = (1, 1, 2)$ and $P_2 = (0, 2, 3)$

(b) $P_3 = (2, \pi/3, 1)$ and $P_4 = (4, \pi/2, 3)$

(c) $P_5 = (3, \pi, \pi/2)$ and $P_6 = (4, \pi/2, \pi)$

3.32 Find the distance between the following pairs of points:

(a) $P_1 = (1, 2, 3)$ and $P_2 = (-2, -3, -2)$ in Cartesian coordinates.

(b) $P_3 = (1, \pi/4, 3)$ and $P_4 = (3, \pi/4, 4)$ in cylindrical coordinates.

(c) $P_5 = (4, \pi/2, 0)$ and $P_6 = (3, \pi, 0)$ in spherical coordinates.

3.33 Transform the vector

$$\mathbf{A} = \hat{\mathbf{R}}\sin^2\theta\cos\phi + \hat{\boldsymbol{\theta}}\cos^2\phi - 3\hat{\boldsymbol{\phi}}\sin\phi$$

into cylindrical coordinates and then evaluate it at $P = (2, \pi/2, \pi/2)$.

3.34 Transform the following vectors into cylindrical coordinates and then evaluate them at the indicated points:

(a) $\mathbf{A} = \hat{\mathbf{x}}(x + y)$ at $P_1 = (1, 2, 3)$

(b) $\mathbf{B} = \hat{\mathbf{x}}(y - x) + \hat{\mathbf{y}}(x - y)$ at $P_2 = (1, 0, 2)$

*(c) $\mathbf{C} = \hat{\mathbf{x}}y^2/(x^2 + y^2) - \hat{\mathbf{y}}x^2/(x^2 + y^2) + \hat{\mathbf{z}}4$ at $P_3 = (1, -1, 2)$

(d) $\mathbf{D} = \hat{\mathbf{R}}\sin\theta + \hat{\boldsymbol{\theta}}\cos\theta + \hat{\boldsymbol{\phi}}\cos^2\phi$ at $P_4 = (2, \pi/2, \pi/4)$

*(e) $\mathbf{E} = \hat{\mathbf{R}}\cos\phi + \hat{\boldsymbol{\theta}}\sin\phi + \hat{\boldsymbol{\phi}}\sin^2\theta$ at $P_5 = (3, \pi/2, \pi)$

3.35 Transform the following vectors into spherical coordinates and then evaluate them at the indicated points:

(a) $\mathbf{A} = \hat{\mathbf{x}}y^2 + \hat{\mathbf{y}}xz + \hat{\mathbf{z}}4$ at $P_1 = (1, -1, 2)$

(b) $\mathbf{B} = \hat{\mathbf{y}}(x^2 + y^2 + z^2) - \hat{\mathbf{z}}(x^2 + y^2)$ at $P_2 = (-1, 0, 2)$

*(c) $\mathbf{C} = \hat{\mathbf{r}}\cos\phi - \hat{\boldsymbol{\phi}}\sin\phi + \hat{\mathbf{z}}\cos\phi\sin\phi$ at $P_3 = (2, \pi/4, 2)$

(d) $\mathbf{D} = \hat{\mathbf{x}}y^2/(x^2 + y^2) - \hat{\mathbf{y}}x^2/(x^2 + y^2) + \hat{\mathbf{z}}4$ at $P_4 = (1, -1, 2)$

Sections 3-4 to 3-7: Gradient, Divergence, and Curl Operators

3.36 Find the gradient of the following scalar functions:

(a) $T = 3/(x^2 + z^2)$

(b) $V = xy^2z^4$

(c) $U = z\cos\phi/(1 + r^2)$

(d) $W = e^{-R}\sin\theta$

*(e) $S = 4x^2e^{-z} + y^3$

(f) $N = r^2 \cos^2 \phi$

(g) $M = R \cos\theta \sin\phi$

3.37 For each of the following scalar fields, obtain an analytical solution for ∇T and generate a corresponding arrow representation.

(a) $T = 10 + x$, for $-10 \le x \le 10$

*(b) $T = x^2$, for $-10 \le x \le 10$

(c) $T = 100 + xy$, for $-10 \le x \le 10$

(d) $T = x^2y^2$, for $-10 \le x, y \le 10$

(e) $T = 20 + x + y$, for $-10 \le x, y \le 10$

(f) $T = 1 + \sin(\pi x/3)$, for $-10 \le x \le 10$

*(g) $T = 1 + \cos(\pi x/3)$, for $-10 \le x \le 10$

(h) $T = 15 + r\cos\phi$, for $\begin{cases} 0 \le r \le 10 \\ 0 \le \phi \le 2\pi. \end{cases}$

(i) $T = 15 + r\cos^2\phi$, for $\begin{cases} 0 \le r \le 10 \\ 0 \le \phi \le 2\pi. \end{cases}$

***3.38** The gradient of a scalar function T is given by

$$\nabla T = \hat{\mathbf{z}} e^{-4z}.$$

If $T = 10$ at $z = 0$, find $T(z)$.

***3.39** For the scalar function $V = xy^2 - z^2$, determine its directional derivative along the direction of vector $\mathbf{A} = (\hat{\mathbf{x}} - \hat{\mathbf{y}}z)$ and then evaluate it at $P = (1, -1, 2)$.

3.40 Follow a procedure similar to that leading to Eq. (3.82) to derive the expression given by Eq. (3.83) for ∇ in spherical coordinates.

3.41 Evaluate the line integral of $\mathbf{E} = \hat{\mathbf{x}}\,x - \hat{\mathbf{y}}\,y$ along the segment P_1 to P_2 of the circular path shown in Fig. P3.41.

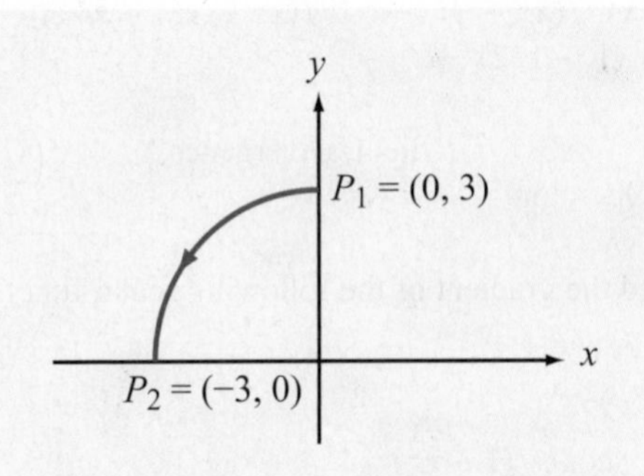

Figure P3.41 Problem 3.41.

***3.42** For the scalar function $U = \frac{1}{R}\sin^2\theta$, determine its directional derivative along the range direction $\hat{\mathbf{R}}$ and then evaluate it at $P = (5, \pi/4, \pi/2)$.

3.43 For the scalar function $T = \frac{1}{2}e^{-r/5}\cos\phi$, determine its directional derivative along the radial direction $\hat{\mathbf{r}}$ and then evaluate it at $P = (2, \pi/4, 4)$.

3.44 Each of the following vector fields is displayed in Fig. P3.44 in the form of a vector representation. Determine $\nabla \cdot \mathbf{A}$ analytically and then compare the result with your expectations on the basis of the displayed pattern.

(a) $\mathbf{A} = -\hat{\mathbf{x}}\cos x \sin y + \hat{\mathbf{y}}\sin x \cos y$, for $-\pi \le x, y \le \pi$

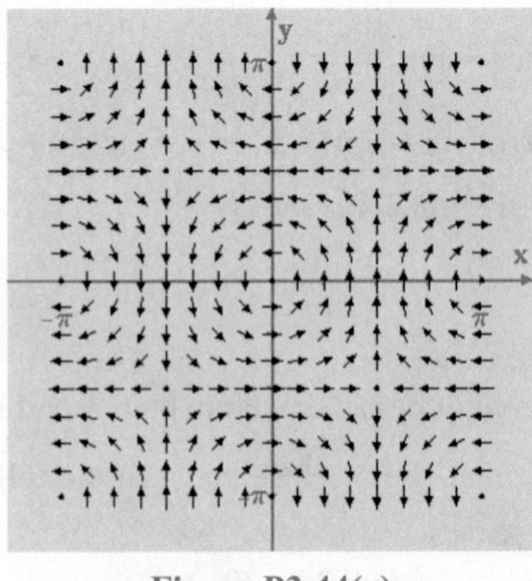

Figure P3.44(a)

(b) $\mathbf{A} = -\hat{\mathbf{x}} \sin 2y + \hat{\mathbf{y}} \cos 2x$, for $-\pi \le x, y \le \pi$

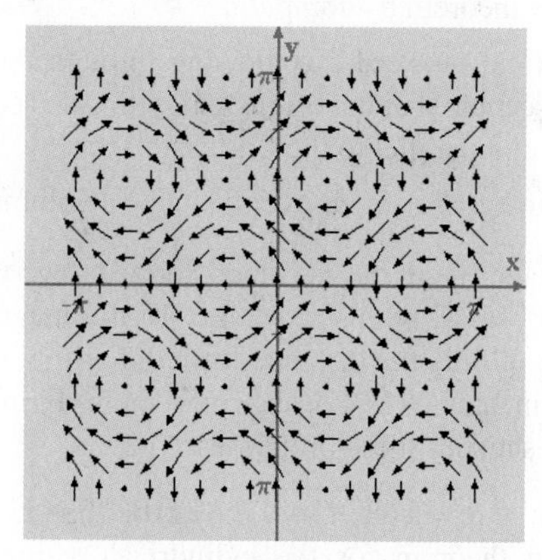

Figure P3.44(b)

(c) $\mathbf{A} = -\hat{\mathbf{x}} xy + \hat{\mathbf{y}} y^2$, for $-10 \le x, y \le 10$

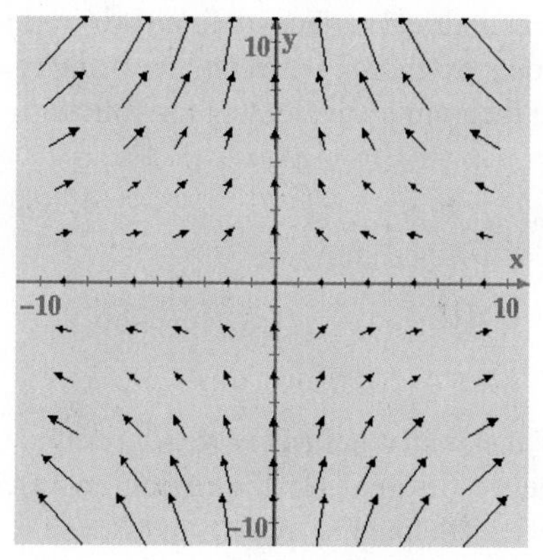

Figure P3.44(c)

(d) $\mathbf{A} = -\hat{\mathbf{x}} \cos x + \hat{\mathbf{y}} \sin y$, for $-\pi \le x, y \le \pi$

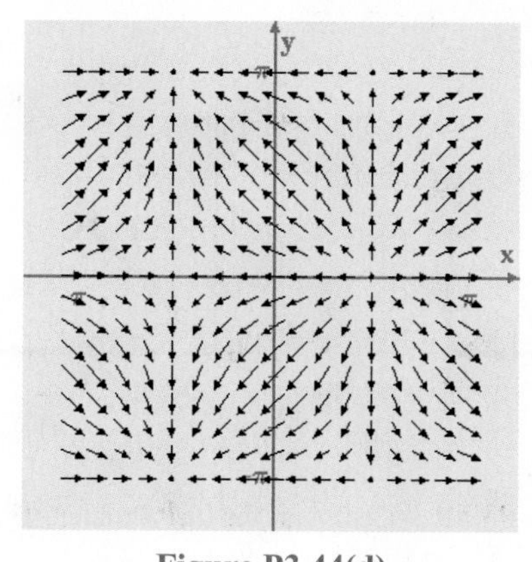

Figure P3.44(d)

(e) $\mathbf{A} = \hat{\mathbf{x}} x$, for $-10 \le x \le 10$

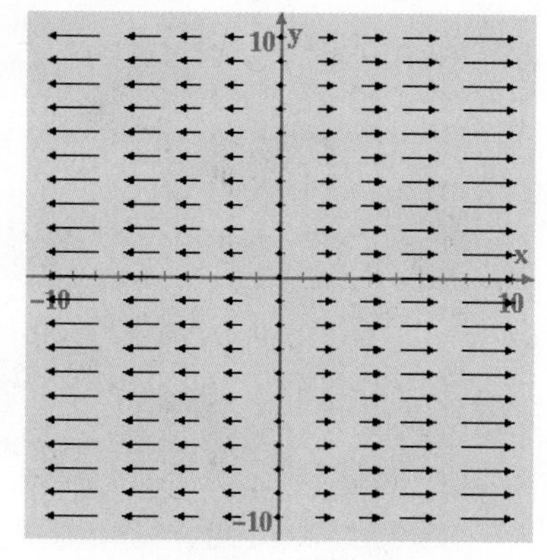

Figure P3.44(e)

(f) $\mathbf{A} = \hat{\mathbf{x}} xy^2$, for $-10 \le x, y \le 10$

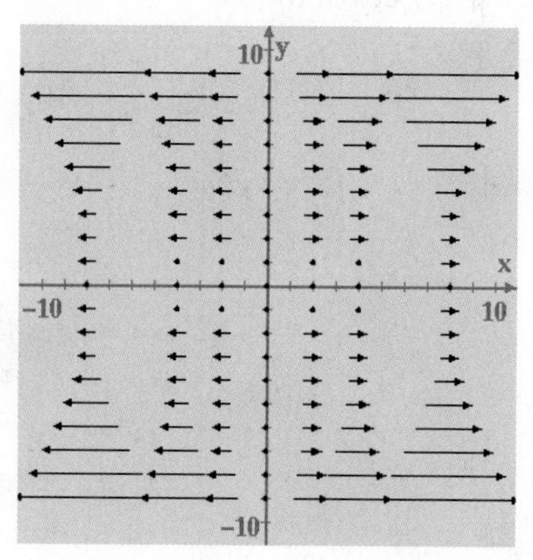

Figure P3.44(f)

(g) $\mathbf{A} = \hat{\mathbf{x}} xy^2 + \hat{\mathbf{y}} x^2 y$, for $-10 \le x, y \le 10$

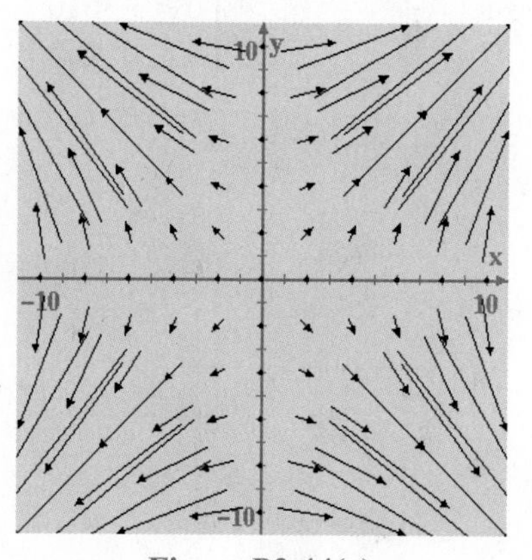

Figure P3.44(g)

(h) $\mathbf{A} = \hat{\mathbf{x}} \sin\left(\frac{\pi x}{10}\right) + \hat{\mathbf{y}} \sin\left(\frac{\pi y}{10}\right)$, for $-10 \le x, y \le 10$

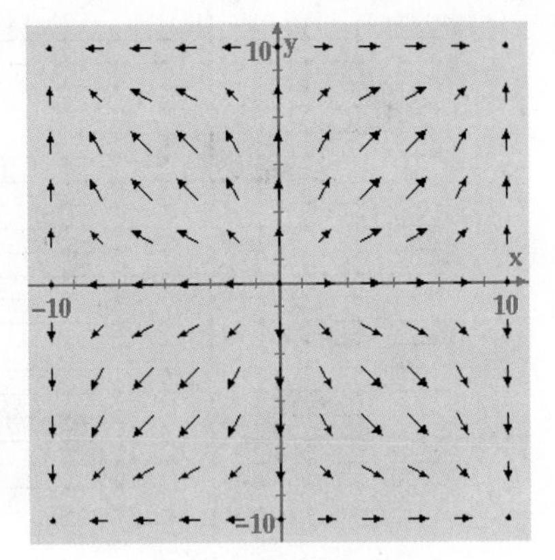

Figure P3.44(h)

(i) $\mathbf{A} = \hat{\mathbf{r}}\, r + \hat{\boldsymbol{\phi}}\, r \cos\phi$, for $\begin{cases} 0 \le r \le 10 \\ 0 \le \phi \le 2\pi. \end{cases}$

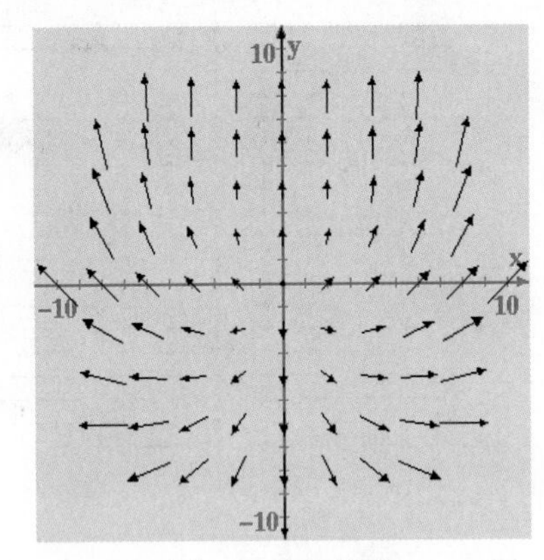

Figure P3.44(i)

(j) $\mathbf{A} = \hat{\mathbf{r}}\, r^2 + \hat{\boldsymbol{\phi}}\, r^2 \sin\phi$, for $\begin{cases} 0 \le r \le 10 \\ 0 \le \phi \le 2\pi. \end{cases}$

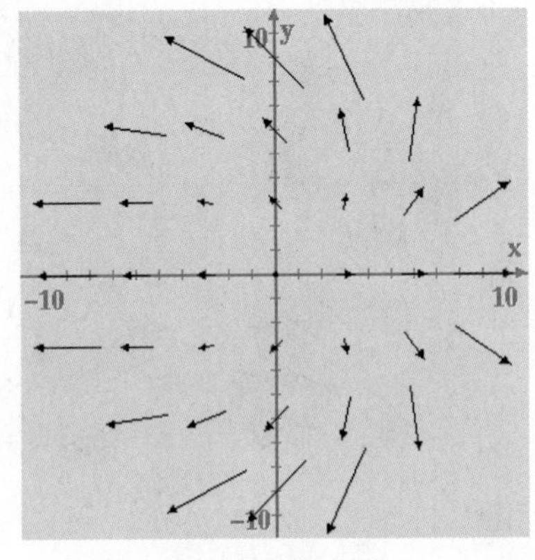

Figure P3.44(j)

3.45 For the vector field $\mathbf{E} = \hat{\mathbf{x}}xz - \hat{\mathbf{y}}yz^2 - \hat{\mathbf{z}}xy$, verify the divergence theorem by computing

(a) The total outward flux flowing through the surface of a cube centered at the origin and with sides equal to 2 units each and parallel to the Cartesian axes.

(b) The integral of $\nabla \cdot \mathbf{E}$ over the cube's volume.

*__3.46__ Vector field $\mathbf{E}$ is characterized by the following properties: (a) $\mathbf{E}$ points along $\hat{\mathbf{R}}$; (b) the magnitude of $\mathbf{E}$ is a function of only the distance from the origin; (c) $\mathbf{E}$ vanishes at the origin; and (d) $\nabla \cdot \mathbf{E} = 6$, everywhere. Find an expression for $\mathbf{E}$ that satisfies these properties.

3.47 For the vector field $\mathbf{E} = \hat{\mathbf{r}}10e^{-r} - \hat{\mathbf{z}}3z$, verify the divergence theorem for the cylindrical region enclosed by $r = 2$, $z = 0$, and $z = 4$.

*__3.48__ A vector field $\mathbf{D} = \hat{\mathbf{r}}r^3$ exists in the region between two concentric cylindrical surfaces defined by $r = 1$ and $r = 2$, with both cylinders extending between $z = 0$ and $z = 5$. Verify the divergence theorem by evaluating the following:

(a) $\oint_S \mathbf{D} \cdot d\mathbf{s}$

(b) $\int_{\mathcal{V}} \nabla \cdot \mathbf{D}\, d\mathcal{V}$

3.49 For the vector field $\mathbf{D} = \hat{\mathbf{R}}3R^2$, evaluate both sides of the divergence theorem for the region enclosed between the spherical shells defined by $R = 1$ and $R = 2$.

3.50 For the vector field $\mathbf{E} = \hat{\mathbf{x}}xy - \hat{\mathbf{y}}(x^2 + 2y^2)$, calculate

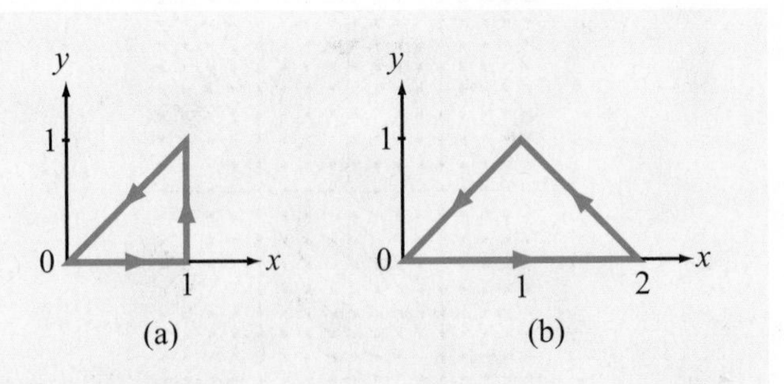

Figure P3.50 Contours for (a) Problem 3.50 and (b) Problem 3.51.

(a) $\oint_C \mathbf{E} \cdot d\mathbf{l}$ around the triangular contour shown in **Fig. P3.50(a)**.

(b) $\int_S (\nabla \times \mathbf{E}) \cdot d\mathbf{s}$ over the area of the triangle.

3.51 Repeat Problem 3.50 for the contour shown in **Fig. P3.50(b)**.

3.52 Verify Stokes's theorem for the vector field

$$\mathbf{B} = (\hat{\mathbf{r}} r \cos\phi + \hat{\boldsymbol{\phi}} \sin\phi)$$

by evaluating the following:

(a) $\oint_C \mathbf{B} \cdot d\mathbf{l}$ over the semicircular contour shown in **Fig. P3.52(a)**.

(b) $\int_S (\nabla \times \mathbf{B}) \cdot d\mathbf{s}$ over the surface of the semicircle.

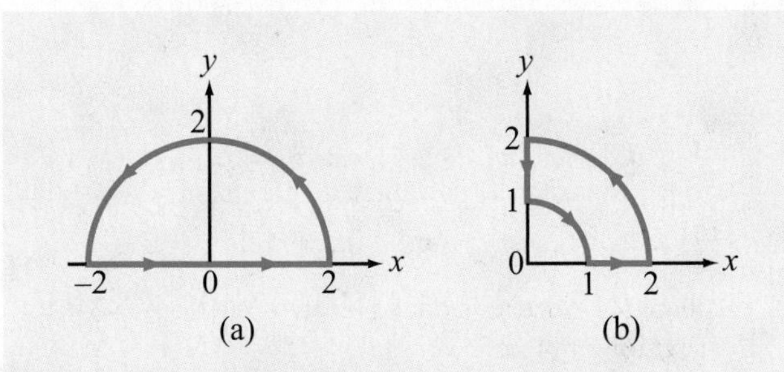

Figure P3.52 Contour paths for (a) Problem 3.52 and (b) Problem 3.53.

3.53 Repeat Problem 3.52 for the contour shown in **Fig. P3.52(b)**.

3.54 Verify Stokes's theorem for the vector field $\mathbf{A} = \hat{\mathbf{R}} \cos\theta + \hat{\boldsymbol{\phi}} \sin\theta$ by evaluating it on the hemisphere of unit radius.

3.55 Verify Stokes's theorem for the vector field $\mathbf{B} = (\hat{\mathbf{r}} \cos\phi + \hat{\boldsymbol{\phi}} \sin\phi)$ by evaluating:

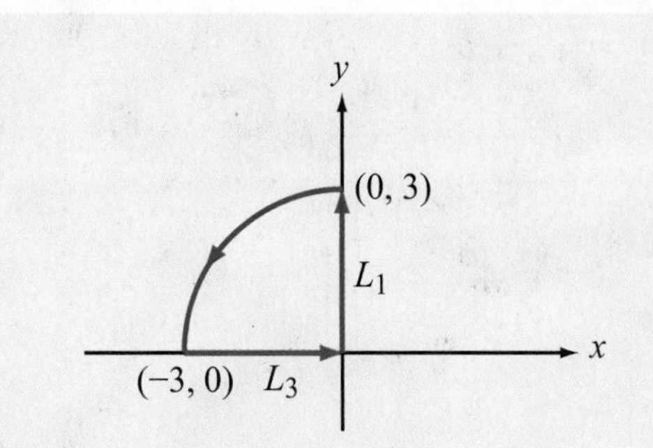

Figure P3.55 Problem 3.55.

(a) $\oint_C \mathbf{B} \cdot d\boldsymbol{\ell}$ over the path comprising a quarter section of a circle, as shown in **Fig. P3.55**, and

(b) $\int_S (\nabla \times \mathbf{B}) \cdot d\mathbf{s}$ over the surface of the quarter section.

3.56 Determine if each of the following vector fields is solenoidal, conservative, or both:

*(a) $\mathbf{A} = \hat{\mathbf{x}} x^2 - \hat{\mathbf{y}} 2xy$

(b) $\mathbf{B} = \hat{\mathbf{x}} x^2 - \hat{\mathbf{y}} y^2 + \hat{\mathbf{z}} 2z$

(c) $\mathbf{C} = \hat{\mathbf{r}}(\sin\phi)/r^2 + \hat{\boldsymbol{\phi}}(\cos\phi)/r^2$

*(d) $\mathbf{D} = \hat{\mathbf{R}}/R$

(e) $\mathbf{E} = \hat{\mathbf{r}}\left(3 - \frac{r}{1+r}\right) + \hat{\mathbf{z}} z$

(f) $\mathbf{F} = (\hat{\mathbf{x}} y + \hat{\mathbf{y}} x)/(x^2 + y^2)$

(g) $\mathbf{G} = \hat{\mathbf{x}}(x^2 + z^2) - \hat{\mathbf{y}}(y^2 + x^2) - \hat{\mathbf{z}}(y^2 + z^2)$

*(h) $\mathbf{H} = \hat{\mathbf{R}}(Re^{-R})$

3.57 Find the Laplacian of the following scalar functions:

(a) $V_1 = 10r^3 \sin 2\phi$

(b) $V_2 = (2/R^2) \cos\theta \sin\phi$

3.58 Find the Laplacian of the following scalar functions:

(a) $V = 4xy^2z^3$

(b) $V = xy + yz + zx$

*(c) $V = 3/(x^2 + y^2)$

(d) $V = 5e^{-r} \cos\phi$

(e) $V = 10e^{-R} \sin\theta$

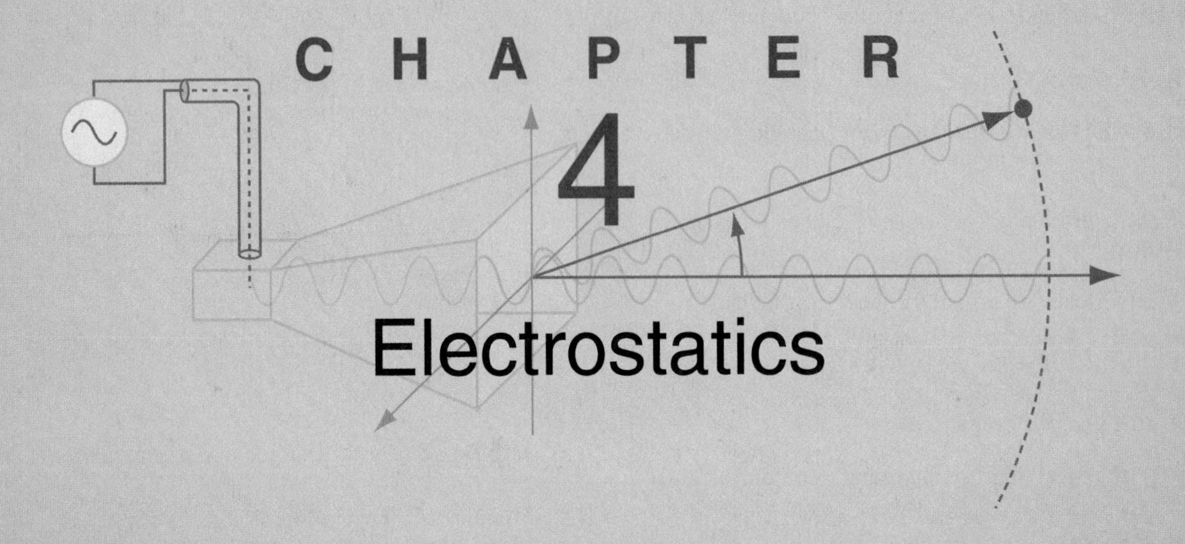

Electrostatics

Chapter Contents

4-1 Maxwell's Equations, 201
4-2 Charge and Current Distributions, 202
4-3 Coulomb's Law, 204
4-4 Gauss's Law, 209
4-5 Electric Scalar Potential, 211
4-6 Conductors, 217
TB7 Resistive Sensors, 218
4-7 Dielectrics, 223
4-8 Electric Boundary Conditions, 225
4-9 Capacitance, 232
4-10 Electrostatic Potential Energy, 235
TB8 Supercapacitors as Batteries, 236
TB9 Capacitive Sensors, 240
4-11 Image Method, 245
Chapter 4 Summary, 247
Problems, 248

Objectives

Upon learning the material presented in this chapter, you should be able to:

1. Evaluate the electric field and electric potential due to any distribution of electric charges.
2. Apply Gauss's law.
3. Calculate the resistance R of any shaped object, given the electric field at every point in its volume.
4. Describe the operational principles of resistive and capacitive sensors.
5. Calculate the capacitance of two-conductor configurations.

4-1 Maxwell's Equations

The modern theory of electromagnetism is based on a set of four fundamental relations known as ***Maxwell's equations***:

$$\nabla \cdot \mathbf{D} = \rho_{\mathrm{v}}, \tag{4.1a}$$

$$\nabla \times \mathbf{E} = -\frac{\partial \mathbf{B}}{\partial t}, \tag{4.1b}$$

$$\nabla \cdot \mathbf{B} = 0, \tag{4.1c}$$

$$\nabla \times \mathbf{H} = \mathbf{J} + \frac{\partial \mathbf{D}}{\partial t}. \tag{4.1d}$$

Here $\mathbf{E}$ and $\mathbf{D}$ are the ***electric field intensity*** and ***flux density***, interrelated by $\mathbf{D} = \epsilon \mathbf{E}$ where ϵ is the electrical permittivity; $\mathbf{H}$ and $\mathbf{B}$ are ***magnetic field intensity*** and ***flux density***, interrelated by $\mathbf{B} = \mu \mathbf{H}$ where μ is the magnetic permeability; ρ_{v} is the electric charge density per unit volume; and $\mathbf{J}$ is the current density per unit area. The fields and fluxes $\mathbf{E}$, $\mathbf{D}$, $\mathbf{B}$, $\mathbf{H}$ were introduced in Section 1-3, and ρ_{v} and $\mathbf{J}$ will be discussed in Section 4-2. Maxwell's equations hold in any material, including free space (vacuum). In general, all of the above quantities may depend on spatial location and time t. In the interest of readability, we will not, however, explicitly reference these dependencies [as in $\mathbf{E}(x, y, z, t)$] except when the context calls for it. By formulating these equations, published in a classic treatise in 1873, James Clerk Maxwell established the first unified theory of electricity and magnetism. His equations, deduced from experimental observations reported by Coulomb, Gauss, Ampère, Faraday, and others, not only encapsulate the connection between the electric field and electric charge and between the magnetic field and electric current, but also capture the bilateral coupling between electric and magnetic fields and fluxes. Together with some auxiliary relations, Maxwell's equations comprise the fundamental tenets of electromagnetic theory.

Under ***static*** conditions, none of the quantities appearing in Maxwell's equations are functions of time (i.e., $\partial/\partial t = 0$). *This happens when all charges are permanently fixed in space, or, if they move, they do so at a steady rate so that* ρ_{v} *and* $\mathbf{J}$ *are constant in time.* Under these circumstances, the time derivatives of $\mathbf{B}$ and $\mathbf{D}$ in Eqs. (4.1b) and (4.1d) vanish, and Maxwell's equations reduce to

Electrostatics

$$\nabla \cdot \mathbf{D} = \rho_{\mathrm{v}}, \tag{4.2a}$$

$$\nabla \times \mathbf{E} = 0. \tag{4.2b}$$

Magnetostatics

$$\nabla \cdot \mathbf{B} = 0, \tag{4.3a}$$

$$\nabla \times \mathbf{H} = \mathbf{J}. \tag{4.3b}$$

Maxwell's four equations separate into two uncoupled pairs, with the first pair involving only the electric field and flux $\mathbf{E}$ and $\mathbf{D}$ and the second pair involving only the magnetic field and flux $\mathbf{H}$ and $\mathbf{B}$.

▶ Electric and magnetic fields become ***decoupled*** in the static case. ◀

This allows us to study electricity and magnetism as two distinct and separate phenomena, as long as the spatial distributions of charge and current flow remain constant in time. We refer to the study of electric and magnetic phenomena under static conditions as ***electrostatics*** and ***magnetostatics***, respectively. Electrostatics is the subject of the present chapter, and in Chapter 5 we learn about magnetostatics. The experience gained through studying electrostatic and magnetostatic phenomena will prove invaluable in tackling the more involved material in subsequent chapters, which deal with time-varying fields, charge densities, and currents.

We study electrostatics not only as a prelude to the study of time-varying fields, but also because it is an important field in its own right. Many electronic devices and systems are based on the principles of electrostatics. They include x-ray machines, oscilloscopes, ink-jet electrostatic printers, liquid crystal displays, copy machines, micro-electromechanical switches and accelerometers, and many solid-state–based control devices. Electrostatic principles also guide the design of medical diagnostic sensors, such as the electrocardiogram, which records the heart's pumping pattern, and the electroencephalogram, which records brain activity, as well as the development of numerous industrial applications.

4-2 Charge and Current Distributions

In electromagnetics, we encounter various forms of electric charge distributions. When put in motion, these charge distributions constitute current distributions. Charges and currents may be distributed over a volume of space, across a surface, or along a line.

4-2.1 Charge Densities

At the atomic scale, the charge distribution in a material is discrete, meaning that charge exists only where electrons and nuclei are and nowhere else. In electromagnetics, we usually are interested in studying phenomena at a much larger scale, typically three or more orders of magnitude greater than the spacing between adjacent atoms. At such a macroscopic scale, we can disregard the discontinuous nature of the charge distribution and treat the net charge contained in an elemental volume $\Delta \mathcal{V}$ as if it were uniformly distributed within. Accordingly, we define the ***volume charge density*** ρ_v as

$$\rho_\mathrm{v} = \lim_{\Delta \mathcal{V} \to 0} \frac{\Delta q}{\Delta \mathcal{V}} = \frac{dq}{d\mathcal{V}} \qquad (\mathrm{C/m^3}), \tag{4.4}$$

where Δq is the charge contained in $\Delta \mathcal{V}$. In general, ρ_v depends on spatial location (x, y, z) and t; thus, $\rho_\mathrm{v} = \rho_\mathrm{v}(x, y, z, t)$. Physically, ρ_v represents the average charge per unit volume for a volume $\Delta \mathcal{V}$ centered at (x, y, z), with $\Delta \mathcal{V}$ being large enough to contain a large number of atoms, yet small enough to be regarded as a point at the macroscopic scale under consideration. The variation of ρ_v with spatial location is called its ***spatial distribution***, or simply its ***distribution***. The total charge contained in volume $\mathcal{V}$ is

$$Q = \int_{\mathcal{V}} \rho_\mathrm{v} \, d\mathcal{V} \qquad (\mathrm{C}). \tag{4.5}$$

In some cases, particularly when dealing with conductors, electric charge may be distributed across the surface of a material, in which case the quantity of interest is the ***surface charge density*** ρ_s, defined as

$$\rho_\mathrm{s} = \lim_{\Delta s \to 0} \frac{\Delta q}{\Delta s} = \frac{dq}{ds} \qquad (\mathrm{C/m^2}), \tag{4.6}$$

where Δq is the charge present across an elemental surface area Δs. Similarly, if the charge is, for all practical purposes, confined to a line, which need not be straight, we characterize its distribution in terms of the ***line charge density*** ρ_ℓ, defined as

$$\rho_\ell = \lim_{\Delta l \to 0} \frac{\Delta q}{\Delta l} = \frac{dq}{dl} \qquad (\mathrm{C/m}). \tag{4.7}$$

Example 4-1: Line Charge Distribution

Calculate the total charge Q contained in a cylindrical tube oriented along the z axis as shown in **Fig. 4-1(a)**. The line charge density is $\rho_\ell = 2z$, where z is the distance in meters from the bottom end of the tube. The tube length is 10 cm.

Solution: The total charge Q is

$$Q = \int_0^{0.1} \rho_\ell \, dz = \int_0^{0.1} 2z \, dz = z^2 \Big|_0^{0.1} = 10^{-2} \text{ C}.$$

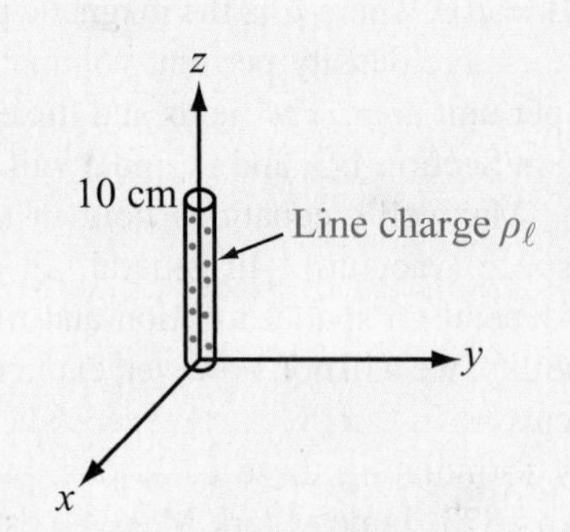

(a) Line charge distribution

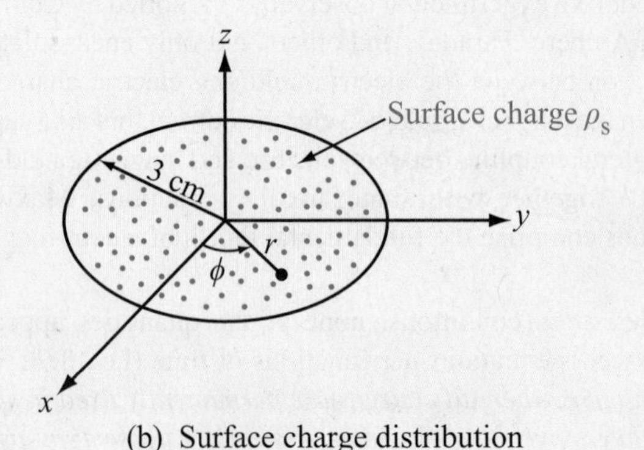

(b) Surface charge distribution

Figure 4-1 Charge distributions for Examples 4-1 and 4-2.

Example 4-2: Surface Charge Distribution

The circular disk of electric charge shown in **Fig. 4-1(b)** is characterized by an azimuthally symmetric surface charge density that increases linearly with r from zero at the center to 6 C/m^2 at $r = 3$ cm. Find the total charge present on the disk surface.

Solution: Since ρ_s is symmetrical with respect to the azimuth angle ϕ, it depends only on r and is given by

$$\rho_s = \frac{6r}{3 \times 10^{-2}} = 2 \times 10^2 r \qquad (\text{C/m}^2),$$

where r is in meters. In polar coordinates, an elemental area is $ds = r\, dr\, d\phi$, and for the disk shown in **Fig. 4-1(b)**, the limits of integration are from 0 to 2π (rad) for ϕ and from 0 to 3×10^{-2} m for r. Hence,

$$\begin{aligned} Q &= \int_S \rho_s\, ds \\ &= \int_{\phi=0}^{2\pi} \int_{r=0}^{3\times 10^{-2}} (2 \times 10^2 r) r\, dr\, d\phi \\ &= 2\pi \times 2 \times 10^2 \left. \frac{r^3}{3} \right|_0^{3\times 10^{-2}} = 11.31 \quad (\text{mC}). \end{aligned}$$

Exercise 4-1: A square plate residing in the x–y plane is situated in the space defined by -3 m $\leq x \leq 3$ m and -3 m $\leq y \leq 3$ m. Find the total charge on the plate if the surface charge density is $\rho_s = 4y^2$ (μC/m^2).

Answer: $Q = 0.432$ (mC). (See EM.)

Exercise 4-2: A thick spherical shell centered at the origin extends between $R = 2$ cm and $R = 3$ cm. If the volume charge density is $\rho_v = 3R \times 10^{-4}$ (C/m^3), find the total charge contained in the shell.

Answer: $Q = 0.61$ (nC). (See EM.)

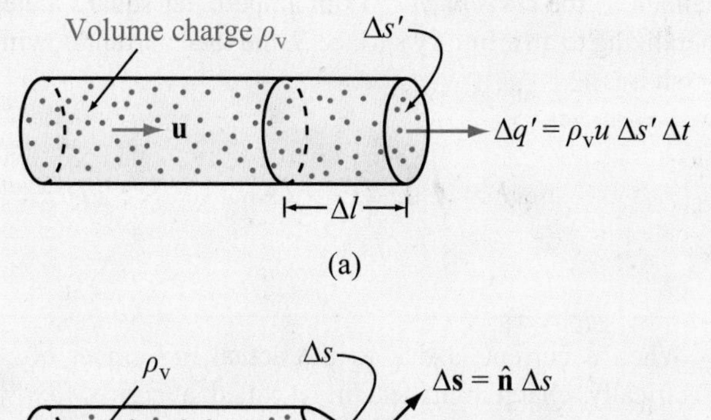

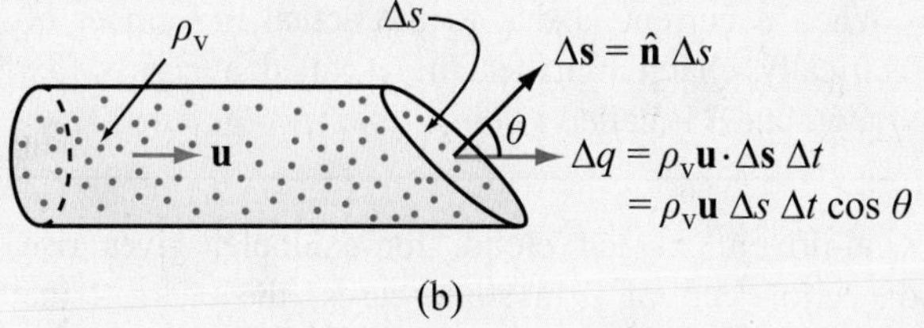

Figure 4-2 Charges with velocity **u** moving through a cross section $\Delta s'$ in (a) and Δs in (b).

4-2.2 Current Density

Consider a tube with volume charge density ρ_v [**Fig. 4-2(a)**]. The charges in the tube move with velocity **u** along the tube axis. Over a period Δt, the charges move a distance $\Delta l = u\, \Delta t$. The amount of charge that crosses the tube's cross-sectional surface $\Delta s'$ in time Δt is therefore

$$\Delta q' = \rho_v\, \Delta \mathcal{V} = \rho_v\, \Delta l\, \Delta s' = \rho_v u\, \Delta s'\, \Delta t. \qquad (4.8)$$

Now consider the more general case where the charges are flowing through a surface Δs with normal $\hat{\mathbf{n}}$ not necessarily parallel to **u** [**Fig. 4-2(b)**]. In this case, the amount of charge Δq flowing through Δs is

$$\Delta q = \rho_v \mathbf{u} \cdot \Delta \mathbf{s}\, \Delta t, \qquad (4.9)$$

where $\Delta \mathbf{s} = \hat{\mathbf{n}}\, \Delta s$ and the corresponding total current flowing in the tube is

$$\Delta I = \frac{\Delta q}{\Delta t} = \rho_v \mathbf{u} \cdot \Delta \mathbf{s} = \mathbf{J} \cdot \Delta \mathbf{s}, \qquad (4.10)$$

where

$$\mathbf{J} = \rho_v \mathbf{u} \qquad (\text{A/m}^2) \qquad (4.11)$$

is defined as the *current density* in ampere per square meter. Generalizing to an arbitrary surface S, the total current flowing through it is

$$I = \int_S \mathbf{J} \cdot d\mathbf{s} \qquad \text{(A).} \tag{4.12}$$

▶ When a current is due to the actual movement of electrically charged matter, it is called a ***convection current***, and **J** is called a ***convection current density***. ◀

A wind-driven charged cloud, for example, gives rise to a convection current. In some cases, the charged matter constituting the convection current consists solely of charged particles, such as the electron beam of a scanning electron microscope or the ion beam of a plasma propulsion system.

When a current is due to the movement of charged particles relative to their host material, **J** is called a ***conduction current density***. In a metal wire, for example, there are equal amounts of positive charges (in atomic nuclei) and negative charges (in the electron shells of the atoms). None of the positive charges and few of the negative charges can move; only those electrons in the outermost electron shells of the atoms can be pushed from one atom to the next if a voltage is applied across the ends of the wire.

▶ This movement of electrons from atom to atom constitutes a ***conduction current***. The electrons that emerge from the wire are not necessarily the same electrons that entered the wire at the other end. ◀

Conduction current, which is discussed in more detail in Section 4-6, obeys Ohm's law, whereas convection current does not.

Concept Question 4-1: What happens to Maxwell's equations under static conditions?

Concept Question 4-2: How is the current density **J** related to the volume charge density ρ_v?

Concept Question 4-3: What is the difference between convection and conduction currents?

4-3 Coulomb's Law

One of the primary goals of this chapter is to develop dexterity in applying the expressions for the ***electric field intensity* E** and associated ***electric flux density* D** induced by a specified distribution of charge. Our discussion will be limited to electrostatic fields induced by stationary charge densities.

We begin by reviewing the expression for the electric field introduced in Section 1-3.2 on the basis of the results of Coulomb's experiments on the electrical force between charged bodies. ***Coulomb's law***, which was first introduced for electrical charges in air and later generalized to material media, implies that:

(1) An isolated charge q induces an electric field **E** at every point in space, and at any specific point P, **E** is given by

$$\mathbf{E} = \hat{\mathbf{R}} \frac{q}{4\pi\epsilon R^2} \qquad \text{(V/m),} \tag{4.13}$$

where $\hat{\mathbf{R}}$ is a unit vector pointing from q to P (**Fig. 4-3**), R is the distance between them, and ϵ is the electrical permittivity of the medium containing the observation point P.

(2) In the presence of an electric field **E** at a given point in space, which may be due to a single charge or a distribution of charges, the force acting on a test charge q' when placed at P, is

$$\mathbf{F} = q'\mathbf{E} \qquad \text{(N).} \tag{4.14}$$

With **F** measured in newtons (N) and q' in coulombs (C), the unit of **E** is (N/C), which will be shown later in Section 4-5 to be the same as volt per meter (V/m).

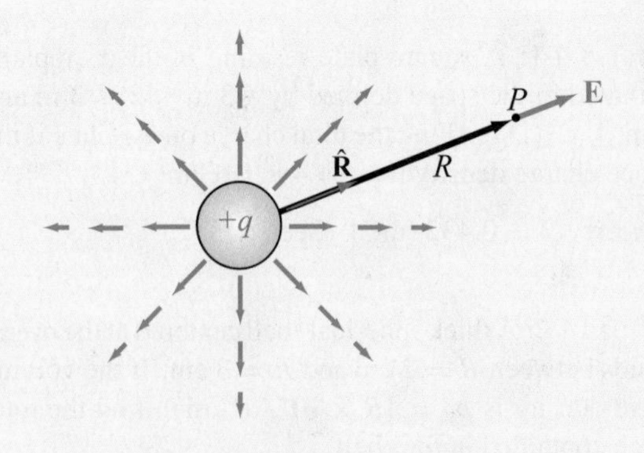

Figure 4-3 Electric-field lines due to a charge q.

For a material with electrical permittivity ϵ, the electric field quantities **D** and **E** are related by

$$\mathbf{D} = \epsilon \mathbf{E} \qquad (4.15)$$

with

$$\epsilon = \epsilon_r \epsilon_0, \qquad (4.16)$$

where

$$\epsilon_0 = 8.85 \times 10^{-12} \approx (1/36\pi) \times 10^{-9} \qquad \text{(F/m)}$$

is the electrical permittivity of free space, and $\epsilon_r = \epsilon/\epsilon_0$ is called the ***relative permittivity*** (or ***dielectric constant***) of the material. For most materials and under a wide range of conditions, ϵ is independent of both the magnitude and direction of **E** [as implied by Eq. (4.15)].

▶ If ϵ is independent of the magnitude of **E**, then the material is said to be ***linear*** because **D** and **E** are related linearly, and if it is independent of the direction of **E**, the material is said to be ***isotropic***. ◀

Materials usually do not exhibit nonlinear permittivity behavior except when the amplitude of **E** is very high (at levels approaching dielectric breakdown conditions discussed later in Section 4-7), and anisotropy is present only in certain materials with peculiar crystalline structures. Hence, except for unique materials under very special circumstances, the quantities **D** and **E** are effectively redundant; for a material with known ϵ, knowledge of either **D** or **E** is sufficient to specify the other in that material.

4-3.1 Electric Field Due to Multiple Point Charges

The expression given by Eq. (4.13) for the field **E** due to a single point charge can be extended to multiple charges. We begin by considering two point charges, q_1 and q_2 with position vectors $\mathbf{R}_1$ and $\mathbf{R}_2$ (measured from the origin in Fig. 4-4). The electric field **E** is to be evaluated at a point P with position vector **R**. At P, the electric field $\mathbf{E}_1$ due to q_1 alone is given by Eq. (4.13) with R, the distance between q_1 and P, replaced with $|\mathbf{R} - \mathbf{R}_1|$ and the unit vector $\hat{\mathbf{R}}$ replaced with $(\mathbf{R} - \mathbf{R}_1)/|\mathbf{R} - \mathbf{R}_1|$. Thus,

$$\mathbf{E}_1 = \frac{q_1(\mathbf{R} - \mathbf{R}_1)}{4\pi\epsilon|\mathbf{R} - \mathbf{R}_1|^3} \qquad \text{(V/m)}. \qquad (4.17a)$$

Similarly, the electric field at P due to q_2 alone is

$$\mathbf{E}_2 = \frac{q_2(\mathbf{R} - \mathbf{R}_2)}{4\pi\epsilon|\mathbf{R} - \mathbf{R}_2|^3} \qquad \text{(V/m)}. \qquad (4.17b)$$

▶ The electric field obeys the principle of linear superposition. ◀

Hence, the total electric field **E** at P due to q_1 and q_2 together is

$$\mathbf{E} = \mathbf{E}_1 + \mathbf{E}_2 = \frac{1}{4\pi\epsilon}\left[\frac{q_1(\mathbf{R} - \mathbf{R}_1)}{|\mathbf{R} - \mathbf{R}_1|^3} + \frac{q_2(\mathbf{R} - \mathbf{R}_2)}{|\mathbf{R} - \mathbf{R}_2|^3}\right]. \qquad (4.18)$$

Generalizing the preceding result to the case of N point charges, the electric field **E** at point P with position vector **R** due to charges $q_1, q_2, \ldots, q_N$ located at points with position vectors $\mathbf{R}_1, \mathbf{R}_2, \ldots, \mathbf{R}_N$, equals the vector sum of the electric fields induced by all the individual charges, or

$$\mathbf{E} = \frac{1}{4\pi\epsilon}\sum_{i=1}^{N}\frac{q_i(\mathbf{R} - \mathbf{R}_i)}{|\mathbf{R} - \mathbf{R}_i|^3} \qquad \text{(V/m)}. \qquad (4.19)$$

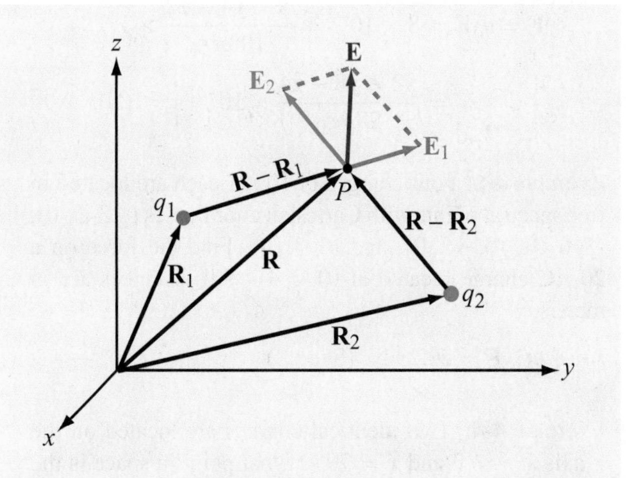

Figure 4-4 The electric field **E** at P due to two charges is equal to the vector sum of $\mathbf{E}_1$ and $\mathbf{E}_2$.

Example 4-3: Electric Field Due to Two Point Charges

Two point charges with $q_1 = 2 \times 10^{-5}$ C and $q_2 = -4 \times 10^{-5}$ C are located in free space at points with Cartesian coordinates $(1, 3, -1)$ and $(-3, 1, -2)$, respectively. Find (a) the electric field $\mathbf{E}$ at $(3, 1, -2)$ and (b) the force on a 8×10^{-5} C charge located at that point. All distances are in meters.

Solution: (a) From Eq. (4.18), the electric field $\mathbf{E}$ with $\epsilon = \epsilon_0$ (free space) is

$$\mathbf{E} = \frac{1}{4\pi\epsilon_0}\left[q_1 \frac{(\mathbf{R} - \mathbf{R}_1)}{|\mathbf{R} - \mathbf{R}_1|^3} + q_2 \frac{(\mathbf{R} - \mathbf{R}_2)}{|\mathbf{R} - \mathbf{R}_2|^3}\right] \quad \text{(V/m)}.$$

The vectors $\mathbf{R}_1$, $\mathbf{R}_2$, and $\mathbf{R}$ are

$$\mathbf{R}_1 = \hat{\mathbf{x}} + \hat{\mathbf{y}}3 - \hat{\mathbf{z}},$$

$$\mathbf{R}_2 = -\hat{\mathbf{x}}3 + \hat{\mathbf{y}} - \hat{\mathbf{z}}2,$$

$$\mathbf{R} = \hat{\mathbf{x}}3 + \hat{\mathbf{y}} - \hat{\mathbf{z}}2.$$

Hence,

$$\mathbf{E} = \frac{1}{4\pi\epsilon_0}\left[\frac{2(\hat{\mathbf{x}}2 - \hat{\mathbf{y}}2 - \hat{\mathbf{z}})}{27} - \frac{4(\hat{\mathbf{x}}6)}{216}\right] \times 10^{-5}$$

$$= \frac{\hat{\mathbf{x}} - \hat{\mathbf{y}}4 - \hat{\mathbf{z}}2}{108\pi\epsilon_0} \times 10^{-5} \quad \text{(V/m)}.$$

(b) The force on q_3 is

$$\mathbf{F} = q_3\mathbf{E} = 8 \times 10^{-5} \times \frac{\hat{\mathbf{x}} - \hat{\mathbf{y}}4 - \hat{\mathbf{z}}2}{108\pi\epsilon_0} \times 10^{-5}$$

$$= \frac{\hat{\mathbf{x}}2 - \hat{\mathbf{y}}8 - \hat{\mathbf{z}}4}{27\pi\epsilon_0} \times 10^{-10} \quad \text{(N)}.$$

Exercise 4-3: Four charges of 10 μC each are located in free space at points with Cartesian coordinates $(-3, 0, 0)$, $(3, 0, 0)$, $(0, -3, 0)$, and $(0, 3, 0)$. Find the force on a 20-μC charge located at $(0, 0, 4)$. All distances are in meters.

Answer: $\mathbf{F} = \hat{\mathbf{z}}0.23$ N. (See EM.)

Exercise 4-4: Two identical charges are located on the x axis at $x = 3$ and $x = 7$. At what point in space is the net electric field zero?

Answer: At point $(5, 0, 0)$. (See EM.)

Exercise 4-5: In a hydrogen atom the electron and proton are separated by an average distance of 5.3×10^{-11} m. Find the magnitude of the electrical force F_e between the two particles, and compare it with the gravitational force F_g between them.

Answer: $F_e = 8.2 \times 10^{-8}$ N, and $F_g = 3.6 \times 10^{-47}$ N. (See EM.)

4-3.2 Electric Field Due to a Charge Distribution

We now extend the results obtained for the field due to discrete point charges to continuous charge distributions. Consider a volume $\mathcal{V}'$ that contains a distribution of electric charge with volume charge density ρ_v, which may vary spatially within $\mathcal{V}'$ (Fig. 4-5). The differential electric field at a point P due to a differential amount of charge $dq = \rho_v \, d\mathcal{V}'$ contained in a differential volume $d\mathcal{V}'$ is

$$d\mathbf{E} = \hat{\mathbf{R}}' \frac{dq}{4\pi\epsilon R'^2} = \hat{\mathbf{R}}' \frac{\rho_v \, d\mathcal{V}'}{4\pi\epsilon R'^2}, \tag{4.20}$$

where $\mathbf{R}'$ is the vector from the differential volume $d\mathcal{V}'$ to point P. Applying the principle of linear superposition, the

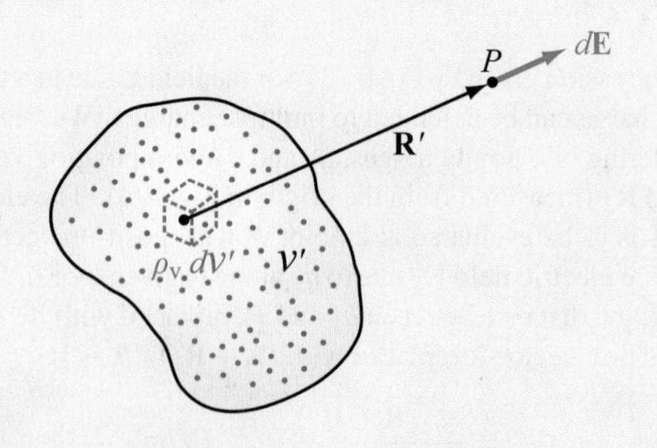

Figure 4-5 Electric field due to a volume charge distribution.

total electric field **E** is obtained by integrating the fields due to all differential charges in $\mathcal{V}'$. Thus,

$$\mathbf{E} = \int_{\mathcal{V}'} d\mathbf{E} = \frac{1}{4\pi\epsilon} \int_{\mathcal{V}'} \hat{\mathbf{R}}' \, \frac{\rho_\text{v} \, d\mathcal{V}'}{R'^2} \, . \tag{4.21a}$$

(volume distribution)

It is important to note that, in general, both R' and $\hat{\mathbf{R}}'$ vary as a function of position over the integration volume $\mathcal{V}'$.

If the charge is distributed across a surface S' with surface charge density ρ_s, then $dq = \rho_\text{s} \, ds'$, and if it is distributed along a line l' with a line charge density ρ_ℓ, then $dq = \rho_\ell \, dl'$. Accordingly, the electric fields due to surface and line charge distributions are

$$\mathbf{E} = \frac{1}{4\pi\epsilon} \int_{S'} \hat{\mathbf{R}}' \, \frac{\rho_\text{s} \, ds'}{R'^2} \, , \tag{4.21b}$$

(surface distribution)

$$\mathbf{E} = \frac{1}{4\pi\epsilon} \int_{l'} \hat{\mathbf{R}}' \, \frac{\rho_\ell \, dl'}{R'^2} \, . \tag{4.21c}$$

(line distribution)

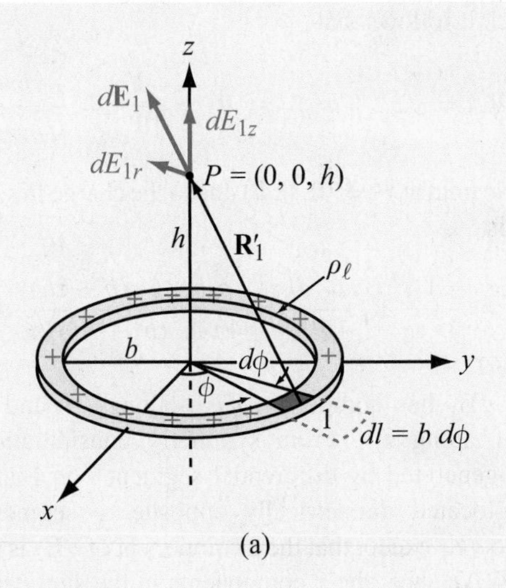

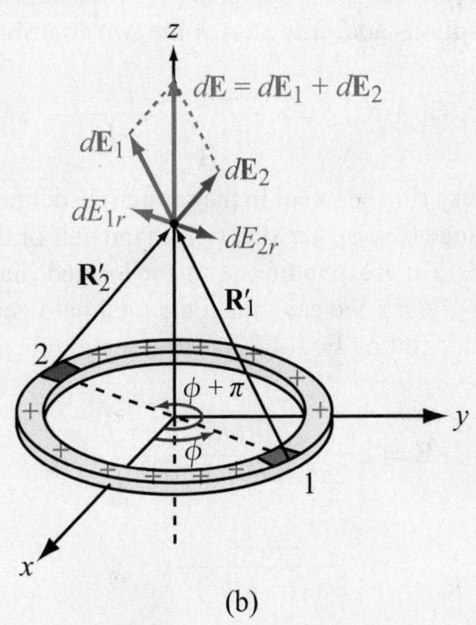

Figure 4-6 Ring of charge with line density ρ_ℓ. (a) The field $d\mathbf{E}_1$ due to infinitesimal segment 1 and (b) the fields $d\mathbf{E}_1$ and $d\mathbf{E}_2$ due to segments at diametrically opposite locations (Example 4-4).

Example 4-4: Electric Field of a Ring of Charge

A ring of charge of radius b is characterized by a uniform line charge density of positive polarity ρ_ℓ. The ring resides in free space and is positioned in the x–y plane as shown in Fig. 4-6. Determine the electric field intensity **E** at a point $P = (0, 0, h)$ along the axis of the ring at a distance h from its center.

Solution: We start by considering the electric field generated by a differential ring segment with cylindrical coordinates $(b, \phi, 0)$ in Fig. 4-6(a). The segment has length $dl = b \, d\phi$ and contains charge $dq = \rho_\ell \, dl = \rho_\ell b \, d\phi$. The distance vector $\mathbf{R}'_1$ from segment 1 to point $P = (0, 0, h)$ is

$$\mathbf{R}'_1 = -\hat{\mathbf{r}} b + \hat{\mathbf{z}} h,$$

from which it follows that

$$R_1' = |\mathbf{R}_1'| = \sqrt{b^2 + h^2}\,, \qquad \hat{\mathbf{R}}_1' = \frac{\mathbf{R}_1'}{|\mathbf{R}_1'|} = \frac{-\hat{\mathbf{r}}b + \hat{\mathbf{z}}h}{\sqrt{b^2 + h^2}}\,.$$

The electric field at $P = (0, 0, h)$ due to the charge in segment 1 therefore is

$$d\mathbf{E}_1 = \frac{1}{4\pi\epsilon_0}\,\hat{\mathbf{R}}_1'\,\frac{\rho_\ell\,dl}{R_1'^2} = \frac{\rho_\ell b}{4\pi\epsilon_0}\,\frac{(-\hat{\mathbf{r}}b + \hat{\mathbf{z}}h)}{(b^2 + h^2)^{3/2}}\,d\phi.$$

The field $d\mathbf{E}_1$ has component dE_{1r} along $-\hat{\mathbf{r}}$ and component dE_{1z} along $\hat{\mathbf{z}}$. From symmetry considerations, the field $d\mathbf{E}_2$ generated by differential segment 2 in Fig. 4-6(b), which is located diametrically opposite to segment 1, is identical to $d\mathbf{E}_1$ except that the $\hat{\mathbf{r}}$ component of $d\mathbf{E}_2$ is opposite that of $d\mathbf{E}_1$. Hence, the $\hat{\mathbf{r}}$ components in the sum cancel and the $\hat{\mathbf{z}}$ contributions add. The sum of the two contributions is

$$d\mathbf{E} = d\mathbf{E}_1 + d\mathbf{E}_2 = \hat{\mathbf{z}}\,\frac{\rho_\ell bh}{2\pi\epsilon_0}\,\frac{d\phi}{(b^2 + h^2)^{3/2}}\,. \tag{4.22}$$

Since for every ring segment in the semicircle defined over the azimuthal range $0 \le \phi \le \pi$ (the right-hand half of the circular ring) there is a corresponding segment located diametrically opposite at $(\phi + \pi)$, we can obtain the total field generated by the ring by integrating Eq. (4.22) over a semicircle as

$$\begin{aligned}
\mathbf{E} &= \hat{\mathbf{z}}\,\frac{\rho_\ell bh}{2\pi\epsilon_0(b^2 + h^2)^{3/2}}\int_0^\pi d\phi \\
&= \hat{\mathbf{z}}\,\frac{\rho_\ell bh}{2\epsilon_0(b^2 + h^2)^{3/2}} \\
&= \hat{\mathbf{z}}\,\frac{h}{4\pi\epsilon_0(b^2 + h^2)^{3/2}}\,Q,
\end{aligned} \tag{4.23}$$

where $Q = 2\pi b\rho_\ell$ is the total charge on the ring.

Example 4-5: Electric Field of a Circular Disk of Charge

Find the electric field at point P with Cartesian coordinates $(0, 0, h)$ due to a circular disk of radius a and uniform charge density ρ_s residing in the x–y plane (Fig. 4-7). Also, evaluate $\mathbf{E}$ due to an infinite sheet of charge density ρ_s by letting $a \to \infty$.

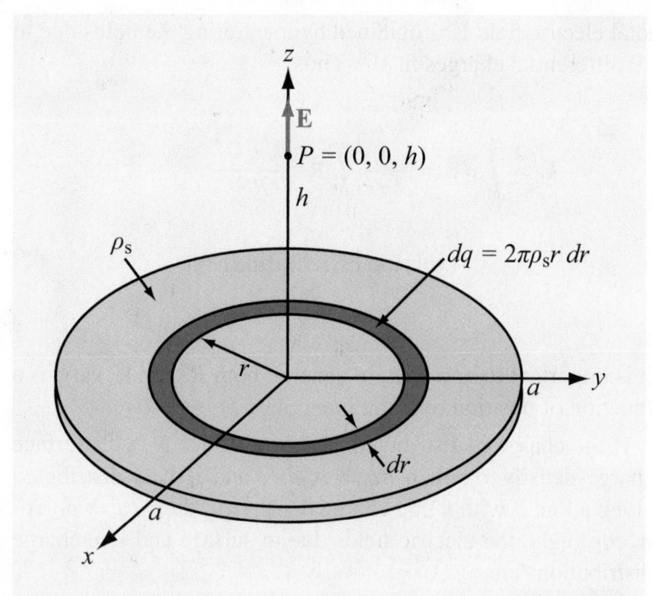

Figure 4-7 Circular disk of charge with surface charge density ρ_s. The electric field at $P = (0, 0, h)$ points along the z direction (Example 4-5).

Solution: Building on the expression obtained in Example 4-4 for the on-axis electric field due to a circular ring of charge, we can determine the field due to the circular disk by treating the disk as a set of concentric rings. A ring of radius r and width dr has an area $ds = 2\pi r\,dr$ and contains charge $dq = \rho_s\,ds = 2\pi\rho_s r\,dr$. Upon using this expression in Eq. (4.23) and also replacing b with r, we obtain the following expression for the field due to the ring:

$$d\mathbf{E} = \hat{\mathbf{z}}\,\frac{h}{4\pi\epsilon_0(r^2 + h^2)^{3/2}}\,(2\pi\rho_s r\,dr).$$

The total field at P is obtained by integrating the expression over the limits $r = 0$ to $r = a$:

$$\begin{aligned}
\mathbf{E} &= \hat{\mathbf{z}}\,\frac{\rho_s h}{2\epsilon_0}\int_0^a \frac{r\,dr}{(r^2 + h^2)^{3/2}} \\
&= \pm\hat{\mathbf{z}}\,\frac{\rho_s}{2\epsilon_0}\left[1 - \frac{|h|}{\sqrt{a^2 + h^2}}\right],
\end{aligned} \tag{4.24}$$

with the plus sign for $h > 0$ (P above the disk) and the minus sign when $h < 0$ (P below the disk).

For an infinite sheet of charge with $a = \infty$,

$$\mathbf{E} = \pm\hat{\mathbf{z}}\,\frac{\rho_s}{2\epsilon_0}. \tag{4.25}$$

(infinite sheet of charge)

We note that for an infinite sheet of charge $\mathbf{E}$ is the same at all points above the x–y plane, and a similar statement applies for points below the x–y plane.

Concept Question 4-4: When characterizing the electrical permittivity of a material, what do the terms *linear* and *isotropic* mean?

Concept Question 4-5: If the electric field is zero at a given point in space, does this imply the absence of electric charges?

Concept Question 4-6: State the principle of linear superposition as it applies to the electric field due to a distribution of electric charge.

Exercise 4-6: An infinite sheet with uniform surface charge density ρ_s is located at $z = 0$ (x–y plane), and another infinite sheet with density $-\rho_s$ is located at $z = 2$ m, both in free space. Determine $\mathbf{E}$ everywhere.

Answer: $\mathbf{E} = 0$ for $z < 0$; $\mathbf{E} = \hat{\mathbf{z}}\rho_s/\epsilon_0$ for $0 < z < 2$ m; and $\mathbf{E} = 0$ for $z > 2$ m. (See EM.)

4-4 Gauss's Law

In this section, we use Maxwell's equations to confirm the expressions for the electric field implied by Coulomb's law, and propose alternative techniques for evaluating electric fields induced by electric charge. To that end, we restate Eq. (4.1a):

$$\nabla \cdot \mathbf{D} = \rho_v, \tag{4.26}$$

(differential form of Gauss's law)

which is referred to as the differential form of ***Gauss's law***. The adjective "differential" refers to the fact that the divergence operation involves spatial derivatives. As we see shortly, Eq. (4.26) can be converted to an integral form. When solving electromagnetic problems, we often go back and forth between equations in differential and integral form, depending on which of the two happens to be the more applicable or convenient to use. To convert Eq. (4.26) into integral form, we multiply both sides by $d\mathcal{V}$ and evaluate their integrals over an arbitrary volume $\mathcal{V}$:

$$\int_{\mathcal{V}} \nabla \cdot \mathbf{D}\, d\mathcal{V} = \int_{\mathcal{V}} \rho_v\, d\mathcal{V} = Q. \tag{4.27}$$

Here, Q is the total charge enclosed in $\mathcal{V}$. The divergence theorem, given by Eq. (3.98), states that the volume integral of the divergence of any vector over a volume $\mathcal{V}$ equals the total outward flux of that vector through the surface S enclosing $\mathcal{V}$. Thus, for the vector $\mathbf{D}$,

$$\int_{\mathcal{V}} \nabla \cdot \mathbf{D}\, d\mathcal{V} = \oint_{S} \mathbf{D} \cdot d\mathbf{s}. \tag{4.28}$$

Comparison of Eq. (4.27) with Eq. (4.28) leads to

$$\oint_{S} \mathbf{D} \cdot d\mathbf{s} = Q. \tag{4.29}$$

(integral form of Gauss's law)

▶ The integral form of Gauss's law is illustrated diagrammatically in Fig. 4-8; for each differential surface element $d\mathbf{s}$, $\mathbf{D} \cdot d\mathbf{s}$ is the electric field flux flowing outward of $\mathcal{V}$ through $d\mathbf{s}$, and the total flux through surface S equals the enclosed charge Q. The surface S is called a ***Gaussian surface***. ◀

The integral form of Gauss's law can be applied to determine $\mathbf{D}$ due to a single isolated point charge q by enclosing the latter with a closed, spherical, Gaussian surface S of arbitrary radius R centered at q (Fig. 4-9). From symmetry considerations and assuming that q is positive, the direction of $\mathbf{D}$ must be radially outward along the unit vector $\hat{\mathbf{R}}$, and D_R, the magnitude of $\mathbf{D}$, must be the same at all points on S. Thus, at any point on S,

$$\mathbf{D} = \hat{\mathbf{R}} D_R, \tag{4.30}$$

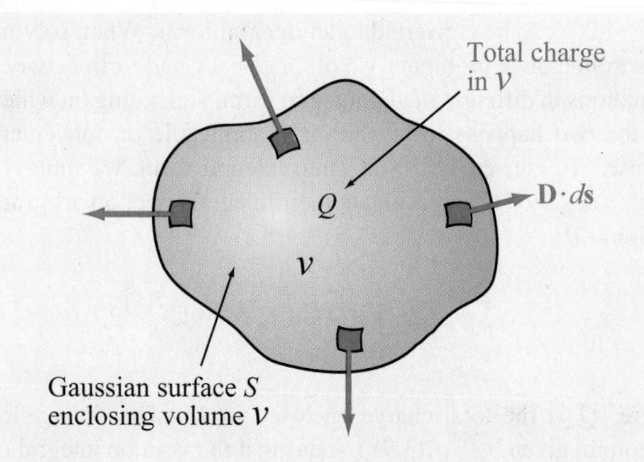

Figure 4-8 The integral form of Gauss's law states that the outward flux of **D** through a surface is proportional to the enclosed charge Q.

and $d\mathbf{s} = \hat{\mathbf{R}}\, ds$. Applying Gauss's law gives

$$\oint_S \mathbf{D} \cdot d\mathbf{s} = \oint_S \hat{\mathbf{R}} D_R \cdot \hat{\mathbf{R}}\, ds = \oint_S D_R\, ds = D_R(4\pi R^2) = q. \tag{4.31}$$

Solving for D_R and then inserting the result in Eq. (4.30) gives the following expression for the electric field **E** induced by an isolated point charge in a medium with permittivity ϵ:

$$\mathbf{E} = \frac{\mathbf{D}}{\epsilon} = \hat{\mathbf{R}}\, \frac{q}{4\pi\epsilon R^2} \qquad \text{(V/m)}. \tag{4.32}$$

This is identical with Eq. (4.13) obtained from Coulomb's law; after all, Maxwell's equations incorporate Coulomb's law. For this simple case of an isolated point charge, it does not matter whether Coulomb's law or Gauss's law is used to obtain the expression for **E**. However, it does matter as to which approach we follow when we deal with multiple point charges or continuous charge distributions. Even though Coulomb's law can be used to find **E** for any specified distribution of charge, Gauss's law is easier to apply than Coulomb's law, but its utility is limited to symmetrical charge distributions.

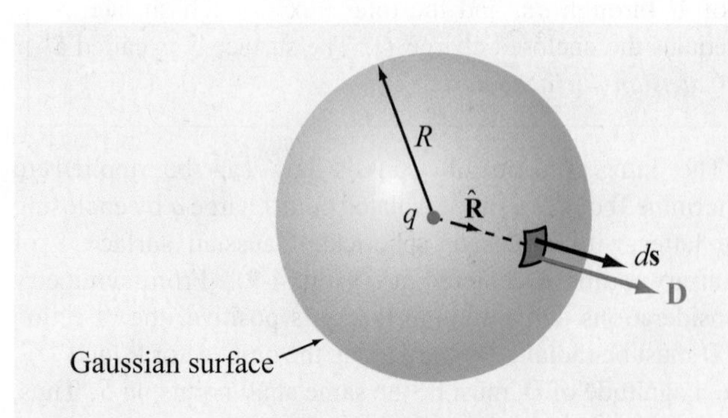

Figure 4-9 Electric field **D** due to point charge q.

► Gauss's law, as given by Eq. (4.29), provides a convenient method for determining the flux density **D** when the charge distribution possesses symmetry properties that allow us to infer the variations of the magnitude and direction of **D** as a function of spatial location, thereby facilitating the integration of **D** over a cleverly chosen Gaussian surface. ◄

Because at every point on the surface the direction of $d\mathbf{s}$ is along its outward normal, only the normal component of **D** at the surface contributes to the integral in Eq. (4.29). To successfully apply Gauss's law, the surface S should be chosen such that, from symmetry considerations, across each subsurface of S, **D** is constant in magnitude and its direction is either normal or purely tangential to the subsurface. These aspects are illustrated in Example 4-6.

Example 4-6: Electric Field of an Infinite Line Charge

Use Gauss's law to obtain an expression for **E** due to an infinitely long line with uniform charge density ρ_ℓ that resides along the z axis in free space.

Solution: Since the charge density along the line is uniform, infinite in extent and residing along the z axis, symmetry considerations dictate that **D** is in the radial $\hat{\mathbf{r}}$ direction and cannot depend on ϕ or z. Thus, $\mathbf{D} = \hat{\mathbf{r}}\, D_r$. Therefore, we construct a finite cylindrical Gaussian surface of radius r and height h, concentric around the line of charge (**Fig. 4-10**). The total charge contained within the cylinder is $Q = \rho_\ell h$. Since **D** is along $\hat{\mathbf{r}}$, the top and bottom surfaces of the cylinder do not contribute to the surface integral on the left-hand side of Eq. (4.29); that is, only the curved surface contributes to the

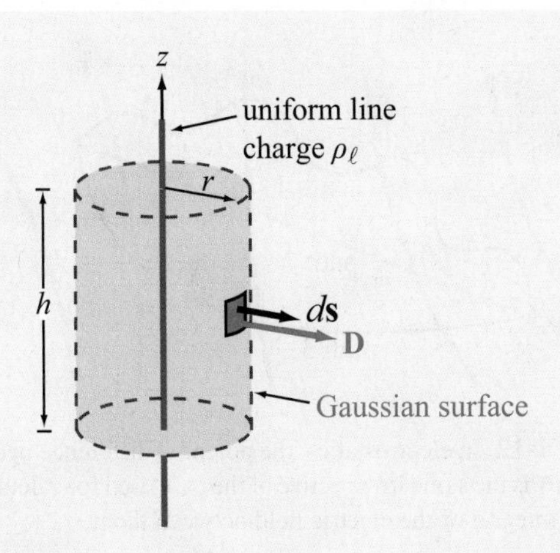

Figure 4-10 Gaussian surface around an infinitely long line of charge (Example 4-6).

integral. Hence,

$$\int_{z=0}^{h}\int_{\phi=0}^{2\pi} \hat{\mathbf{r}}D_r \cdot \hat{\mathbf{r}}r\,d\phi\,dz = \rho_\ell h$$

or

$$2\pi h D_r r = \rho_\ell h,$$

which yields

$$\mathbf{E} = \frac{\mathbf{D}}{\epsilon_0} = \hat{\mathbf{r}}\,\frac{D_r}{\epsilon_0} = \hat{\mathbf{r}}\,\frac{\rho_\ell}{2\pi\epsilon_0 r}\,. \tag{4.33}$$

(infinite line charge)

Note that Eq. (4.33) is applicable for any infinite line of charge, regardless of its location and direction, as long as $\hat{\mathbf{r}}$ is properly defined as the radial distance vector from the line charge to the observation point (i.e., $\hat{\mathbf{r}}$ is perpendicular to the line of charge).

Concept Question 4-7: Explain Gauss's law. Under what circumstances is it useful?

Concept Question 4-8: How should one choose a Gaussian surface?

Exercise 4-7: Two infinite lines, each carrying a uniform charge density ρ_ℓ, reside in free space parallel to the z axis at $x = 1$ and $x = -1$. Determine $\mathbf{E}$ at an arbitrary point along the y axis.

Answer: $\mathbf{E} = \hat{\mathbf{y}}\rho_\ell y / \left[\pi\epsilon_0(y^2+1)\right]$. (See EM.)

Exercise 4-8: A thin spherical shell of radius a carries a uniform surface charge density ρ_s. Use Gauss's law to determine $\mathbf{E}$ everywhere in free space.

Answer: $\mathbf{E} = 0$ for $R < a$;
$\mathbf{E} = \hat{\mathbf{R}}\rho_s a^2/(\epsilon R^2)$ for $R > a$. (*See* EM.)

Exercise 4-9: A spherical volume of radius a contains a uniform volume charge density ρ_v. Use Gauss's law to determine $\mathbf{D}$ for (a) $R \le a$ and (b) $R \ge a$.

Answer: (a) $\mathbf{D} = \hat{\mathbf{R}}\rho_v R/3$,
(b) $\mathbf{D} = \hat{\mathbf{R}}\rho_v a^3/(3R^2)$. (*See* EM.)

4-5 Electric Scalar Potential

The operation of an electric circuit usually is described in terms of the currents flowing through its branches and the voltages at its nodes. The voltage difference V between two points in a circuit represents the amount of work, or *potential* energy, required to move a unit charge from one to the other.

► The term "voltage" is short for "voltage potential" and synonymous with ***electric potential***. ◄

Even though when analyzing a circuit we may not consider the electric fields present in the circuit, it is in fact the existence of these fields that gives rise to voltage differences across circuit elements such as resistors or capacitors. The relationship between the electric field $\mathbf{E}$ and the electric potential V is the subject of this section.

4-5.1 Electric Potential as a Function of Electric Field

We begin by considering the simple case of a positive charge q in a uniform electric field $\mathbf{E} = -\hat{\mathbf{y}}E$, in the $-y$ direction (Fig. 4-11). The presence of the field $\mathbf{E}$ exerts a force $\mathbf{F}_e = q\mathbf{E}$

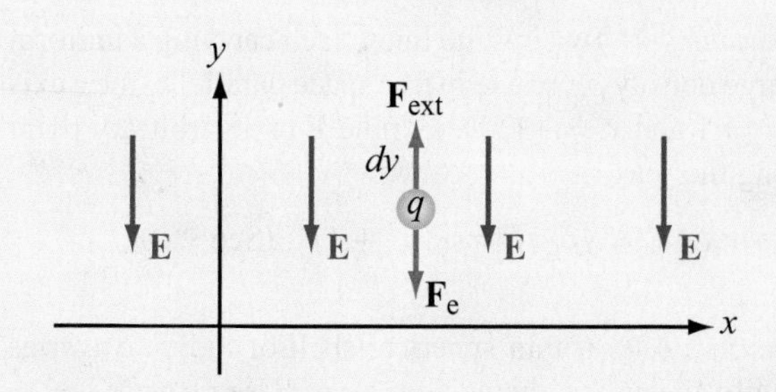

Figure 4-11 Work done in moving a charge q a distance dy against the electric field $\mathbf{E}$ is $dW = qE\ dy$.

on the charge in the $-y$ direction. To move the charge along the positive y direction (against the force $\mathbf{F}_e$), we need to provide an external force $\mathbf{F}_{ext}$ to counteract $\mathbf{F}_e$, which requires the expenditure of energy. To move q without acceleration (at constant speed), the net force acting on the charge must be zero, which means that $\mathbf{F}_{ext} + \mathbf{F}_e = 0$, or

$$\mathbf{F}_{ext} = -\mathbf{F}_e = -q\mathbf{E}. \tag{4.34}$$

The work done, or energy expended, in moving any object a vector differential distance $d\mathbf{l}$ while exerting a force $\mathbf{F}_{ext}$ is

$$dW = \mathbf{F}_{ext} \cdot d\mathbf{l} = -q\mathbf{E} \cdot d\mathbf{l} \qquad \text{(J)}. \tag{4.35}$$

Work, or energy, is measured in joules (J). If the charge is moved a distance dy along $\hat{\mathbf{y}}$, then

$$dW = -q(-\hat{\mathbf{y}}E) \cdot \hat{\mathbf{y}}\ dy = qE\ dy. \tag{4.36}$$

The differential electric potential energy dW per unit charge is called the ***differential electric potential*** (or differential voltage) dV. That is,

$$dV = \frac{dW}{q} = -\mathbf{E} \cdot d\mathbf{l} \qquad \text{(J/C or V)}. \tag{4.37}$$

The unit of V is the volt (V), with 1 V = 1 J/C, and since V is measured in volts, the electric field is expressed in volts per meter (V/m).

The ***potential difference*** corresponding to moving a point charge from point P_1 to point P_2 (**Fig. 4-12**) is obtained by integrating Eq. (4.37) *along any path* between them. That is,

$$\int_{P_1}^{P_2} dV = -\int_{P_1}^{P_2} \mathbf{E} \cdot d\mathbf{l}, \tag{4.38}$$

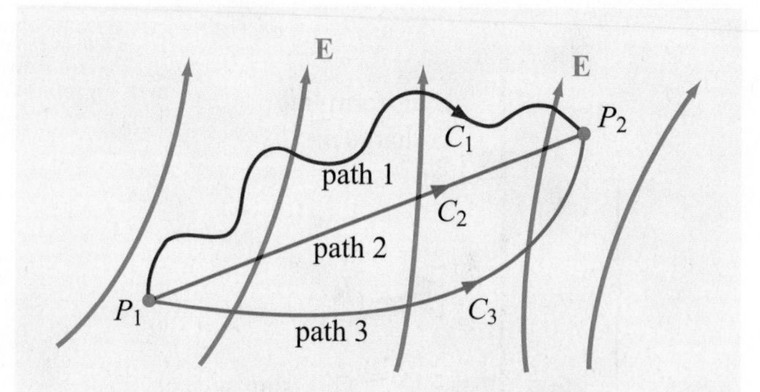

Figure 4-12 In electrostatics, the potential difference between P_2 and P_1 is the same irrespective of the path used for calculating the line integral of the electric field between them.

or

$$V_{21} = V_2 - V_1 = -\int_{P_1}^{P_2} \mathbf{E} \cdot d\mathbf{l}, \tag{4.39}$$

where V_1 and V_2 are the electric potentials at points P_1 and P_2, respectively. The result of the line integral on the right-hand side of Eq. (4.39) is independent of the specific integration path that connects points P_1 and P_2. This follows immediately from the law of conservation of energy. To illustrate with an example, consider a particle in Earth's gravitational field. If the particle is raised from a height h_1 above Earth's surface to height h_2, the particle gains potential energy in an amount proportional to $(h_2 - h_1)$. If, instead, we were to first raise the particle from height h_1 to a height h_3 greater than h_2, thereby giving it potential energy proportional to $(h_3 - h_1)$, and then let it drop back to height h_2 by expending an energy amount proportional to $(h_3 - h_2)$, its net gain in potential energy would again be proportional to $(h_2 - h_1)$. The same principle applies to the electric potential energy W and to the potential difference $(V_2 - V_1)$. The voltage difference between two nodes in an electric circuit has the same value regardless of which path in the circuit we follow between the nodes. Moreover, ***Kirchhoff's voltage law*** states that the net voltage drop around a closed loop is zero. If we go from P_1 to P_2 by path 1 in **Fig. 4-12** and then return from P_2 to P_1 by path 2, the right-hand side of Eq. (4.39) becomes a closed contour and the left-hand side vanishes. In fact, *the line integral of the electrostatic field* **E** *around any closed contour C is zero:*

$$\oint_C \mathbf{E} \cdot d\mathbf{l} = 0 \quad \textbf{(electrostatics).} \qquad (4.40)$$

▶ A vector field whose line integral along any closed path is zero is called a ***conservative*** or an ***irrotational*** field. Hence, the electrostatic field **E** is conservative. ◀

As we will see later in Chapter 6, if **E** is a time-varying function, it is no longer conservative, and its line integral along a closed path is not necessarily zero.

The conservative property of the electrostatic field can be deduced from Maxwell's second equation, Eq. (4.1b). If $\partial/\partial t = 0$, then

$$\nabla \times \mathbf{E} = 0. \qquad (4.41)$$

If we take the surface integral of $\nabla \times \mathbf{E}$ over an open surface S and then apply Stokes's theorem expressed by Eq. (3.107) to convert the surface integral into a line integral, we obtain

$$\int_S (\nabla \times \mathbf{E}) \cdot d\mathbf{s} = \oint_C \mathbf{E} \cdot d\mathbf{l} = 0, \qquad (4.42)$$

where C is a closed contour surrounding S. Thus, Eq. (4.41) is the differential-form equivalent of Eq. (4.40).

We now define what we mean by the electric potential V at a point in space. Before we do so, however, let us revisit our electric-circuit analogue. Just as a node in a circuit cannot be assigned an absolute voltage, a point in space cannot have an absolute electric potential. The voltage of a node in a circuit is measured relative to that of a conveniently chosen reference point to which we have assigned a voltage of zero, which we call *ground*. The same principle applies to the electric potential V. Usually (but not always), the reference point is chosen to be at infinity. That is, in Eq. (4.39) we assume that $V_1 = 0$ when P_1 is at infinity, and therefore the electric potential V at any point P is

$$V = -\int_{\infty}^{P} \mathbf{E} \cdot d\mathbf{l} \quad \text{(V)}. \qquad (4.43)$$

4-5.2 Electric Potential Due to Point Charges

The electric field due to a point charge q located at the origin is given by Eq. (4.32) as

$$\mathbf{E} = \hat{\mathbf{R}} \frac{q}{4\pi\epsilon R^2} \quad \text{(V/m)}. \qquad (4.44)$$

The field is radially directed and decays quadratically with the distance R from the observer to the charge.

As was stated earlier, the choice of integration path between the end points in Eq. (4.43) is arbitrary. Hence, we can conveniently choose the path to be along the radial direction $\hat{\mathbf{R}}$, in which case $d\mathbf{l} = \hat{\mathbf{R}}\, dR$ and

$$V = -\int_{\infty}^{R} \left(\hat{\mathbf{R}} \frac{q}{4\pi\epsilon R^2}\right) \cdot \hat{\mathbf{R}}\, dR = \frac{q}{4\pi\epsilon R} \quad \text{(V)}. \qquad (4.45)$$

If the charge q is at a location other than the origin, say at position vector $\mathbf{R}_1$, then V at observation position vector $\mathbf{R}$ becomes

$$V = \frac{q}{4\pi\epsilon|\mathbf{R} - \mathbf{R}_1|} \quad \text{(V)}, \qquad (4.46)$$

where $|\mathbf{R} - \mathbf{R}_1|$ is the distance between the observation point and the location of the charge q. The principle of superposition applied previously to the electric field **E** also applies to the electric potential V. Hence, for N discrete point charges $q_1, q_2, \ldots, q_N$ residing at position vectors $\mathbf{R}_1, \mathbf{R}_2, \ldots, \mathbf{R}_N$, the electric potential is

$$V = \frac{1}{4\pi\epsilon} \sum_{i=1}^{N} \frac{q_i}{|\mathbf{R} - \mathbf{R}_i|} \quad \text{(V)}. \qquad (4.47)$$

4-5.3 Electric Potential Due to Continuous Distributions

To obtain expressions for the electric potential V due to continuous charge distributions over a volume $\mathcal{V}'$, across a surface S', or along a line l', we (1) replace q_i in Eq. (4.47) with $\rho_{\mathrm{v}}\, d\mathcal{V}'$, $\rho_{\mathrm{s}}\, ds'$, and $\rho_\ell\, dl'$, respectively; (2) convert the summation into an integration; and (3) define $R' = |\mathbf{R} - \mathbf{R}_i|$ as the distance between the integration point and the observation point. These steps lead to the following expressions:

$$V = \frac{1}{4\pi\epsilon} \int_{\mathcal{V}'} \frac{\rho_v}{R'}\, d\mathcal{V}' \quad \textbf{(volume distribution)}, \tag{4.48a}$$

$$V = \frac{1}{4\pi\epsilon} \int_{S'} \frac{\rho_s}{R'}\, ds' \quad \textbf{(surface distribution)}, \tag{4.48b}$$

$$V = \frac{1}{4\pi\epsilon} \int_{l'} \frac{\rho_\ell}{R'}\, dl' \quad \textbf{(line distribution)}. \tag{4.48c}$$

4-5.4 Electric Field as a Function of Electric Potential

In Section 4-5.1, we expressed V in terms of a line integral over $\mathbf{E}$. Now we explore the inverse relationship by re-examining Eq. (4.37):

$$dV = -\mathbf{E} \cdot d\mathbf{l}. \tag{4.49}$$

For a scalar function V, Eq. (3.73) gives

$$dV = \nabla V \cdot d\mathbf{l}, \tag{4.50}$$

where ∇V is the gradient of V. Comparison of Eq. (4.49) with Eq. (4.50) leads to

$$\mathbf{E} = -\nabla V. \tag{4.51}$$

► This differential relationship between V and $\mathbf{E}$ allows us to determine $\mathbf{E}$ for any charge distribution by first calculating V and then taking the negative gradient of V to find $\mathbf{E}$. ◄

The expressions for V, given by Eqs. (4.47) to (4.48c), involve scalar sums and scalar integrals, and as such are usually much easier to evaluate than the vector sums and integrals in the expressions for $\mathbf{E}$ derived in Section 4-3 on the basis of Coulomb's law. Thus, even though the electric potential approach for finding $\mathbf{E}$ is a two-step process, it is conceptually and computationally simpler to apply than the direct method based on Coulomb's law.

Example 4-7: Electric Field of an Electric Dipole

An ***electric dipole*** consists of two point charges of equal magnitude but opposite polarity, separated by a distance d [**Fig. 4-13(a)**]. Determine V and $\mathbf{E}$ at any point P, given that P is at a distance $R \gg d$ from the dipole center, and the dipole resides in free space.

Solution: To simplify the derivation, we align the dipole along the z axis and center it at the origin [**Fig. 4-13(a)**]. For the two charges shown in **Fig. 4-13(a)**, application of Eq. (4.47) gives

$$V = \frac{1}{4\pi\epsilon_0}\left(\frac{q}{R_1} + \frac{-q}{R_2}\right) = \frac{q}{4\pi\epsilon_0}\left(\frac{R_2 - R_1}{R_1 R_2}\right).$$

Since $d \ll R$, the lines labeled R_1 and R_2 in **Fig. 4-13(a)**

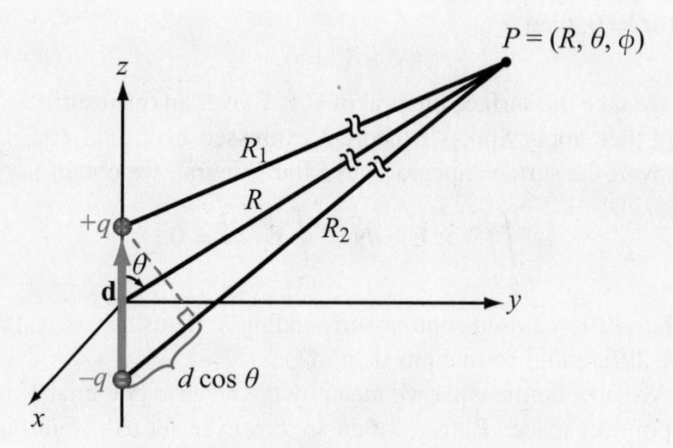

(a) Electric dipole

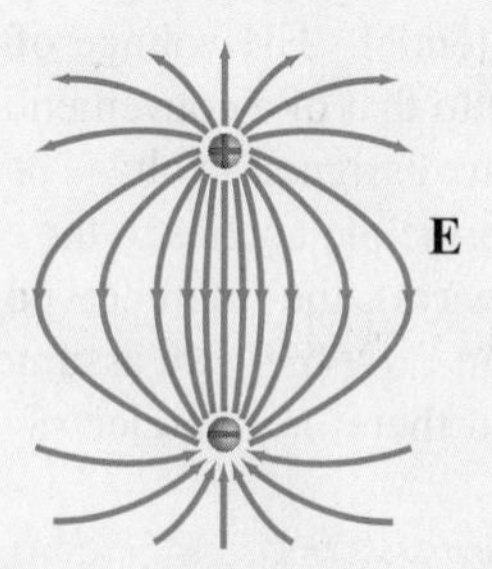

(b) Electric-field pattern

Figure 4-13 Electric dipole with dipole moment $\mathbf{p} = q\mathbf{d}$ (Example 4-7).

are approximately parallel to each other, in which case the following approximations apply:

$$R_2 - R_1 \approx d\cos\theta, \qquad R_1 R_2 \approx R^2.$$

Hence,

$$V = \frac{qd\cos\theta}{4\pi\epsilon_0 R^2}. \tag{4.52}$$

To generalize this result to an arbitrarily oriented dipole, note that the numerator of Eq. (4.52) can be expressed as the dot product of $q\mathbf{d}$ (where $\mathbf{d}$ is the distance vector from $-q$ to $+q$) and the unit vector $\hat{\mathbf{R}}$ pointing from the center of the dipole toward the observation point P. That is,

$$qd\cos\theta = q\mathbf{d}\cdot\hat{\mathbf{R}} = \mathbf{p}\cdot\hat{\mathbf{R}}, \tag{4.53}$$

where $\mathbf{p} = q\mathbf{d}$ is called the ***dipole moment***. Using Eq. (4.53) in Eq. (4.52) then gives

$$V = \frac{\mathbf{p}\cdot\hat{\mathbf{R}}}{4\pi\epsilon_0 R^2} \qquad \textbf{(electric dipole)}. \tag{4.54}$$

In spherical coordinates, Eq. (4.51) is given by

$$\begin{aligned}\mathbf{E} &= -\nabla V\\ &= -\left(\hat{\mathbf{R}}\frac{\partial V}{\partial R} + \hat{\boldsymbol{\theta}}\frac{1}{R}\frac{\partial V}{\partial \theta} + \hat{\boldsymbol{\phi}}\frac{1}{R\sin\theta}\frac{\partial V}{\partial \phi}\right),\end{aligned} \tag{4.55}$$

where we have used the expression for ∇V in spherical coordinates given on the inside back cover of the book. Upon taking the derivatives of the expression for V given by Eq. (4.52) with respect to R and θ and then substituting the results in Eq. (4.55), we obtain

$$\mathbf{E} = \frac{qd}{4\pi\epsilon_0 R^3}(\hat{\mathbf{R}}\,2\cos\theta + \hat{\boldsymbol{\theta}}\sin\theta) \quad \text{(V/m)}. \tag{4.56}$$

We stress that the expressions for V and $\mathbf{E}$ given by Eqs. (4.54) and (4.56) apply only when $R \gg d$. To compute V and $\mathbf{E}$ at points in the vicinity of the two dipole charges, it is necessary to perform all calculations without resorting to the far-distance approximations that led to Eq. (4.52). Such an exact calculation for $\mathbf{E}$ leads to the field pattern shown in **Fig. 4-13(b)**.

4-5.5 Poisson's Equation

With $\mathbf{D} = \epsilon\mathbf{E}$, the differential form of Gauss's law given by Eq. (4.26) may be cast as

$$\nabla\cdot\mathbf{E} = \frac{\rho_\text{v}}{\epsilon}. \tag{4.57}$$

Inserting Eq. (4.51) in Eq. (4.57) gives

$$\nabla\cdot(\nabla V) = -\frac{\rho_\text{v}}{\epsilon}. \tag{4.58}$$

Given Eq. (3.110) for the Laplacian of a scalar function V,

$$\nabla^2 V = \nabla\cdot(\nabla V) = \frac{\partial^2 V}{\partial x^2} + \frac{\partial^2 V}{\partial y^2} + \frac{\partial^2 V}{\partial z^2}, \tag{4.59}$$

Eq. (4.58) can be cast in the abbreviated form

$$\nabla^2 V = -\frac{\rho_\text{v}}{\epsilon} \qquad \textbf{(Poisson's equation)}. \tag{4.60}$$

This is known as ***Poisson's equation***. For a volume $\mathcal{V}'$ containing a volume charge density distribution ρ_v, the solution for V derived previously and expressed by Eq. (4.48a) as

$$V = \frac{1}{4\pi\epsilon}\int_{\mathcal{V}'} \frac{\rho_\text{v}}{R'}\,d\mathcal{V}' \tag{4.61}$$

satisfies Eq. (4.60). If the medium under consideration contains no charges, Eq. (4.60) reduces to

$$\nabla^2 V = 0 \qquad \textbf{(Laplace's equation)}, \tag{4.62}$$

and it is then referred to as ***Laplace's equation***. Poisson's and Laplace's equations are useful for determining the electrostatic potential V in regions with boundaries on which V is known, such as the region between the plates of a capacitor with a specified voltage difference across it.

Concept Question 4-9: What is a conservative field?

Concept Question 4-10: Why is the electric potential at a point in space always defined relative to the potential at some reference point?

Module 4.1 Fields due to Charges For any group of point charges, this module calculates and displays the electric field **E** and potential V across a 2-D grid. The user can specify the locations, magnitudes and polarities of the charges.

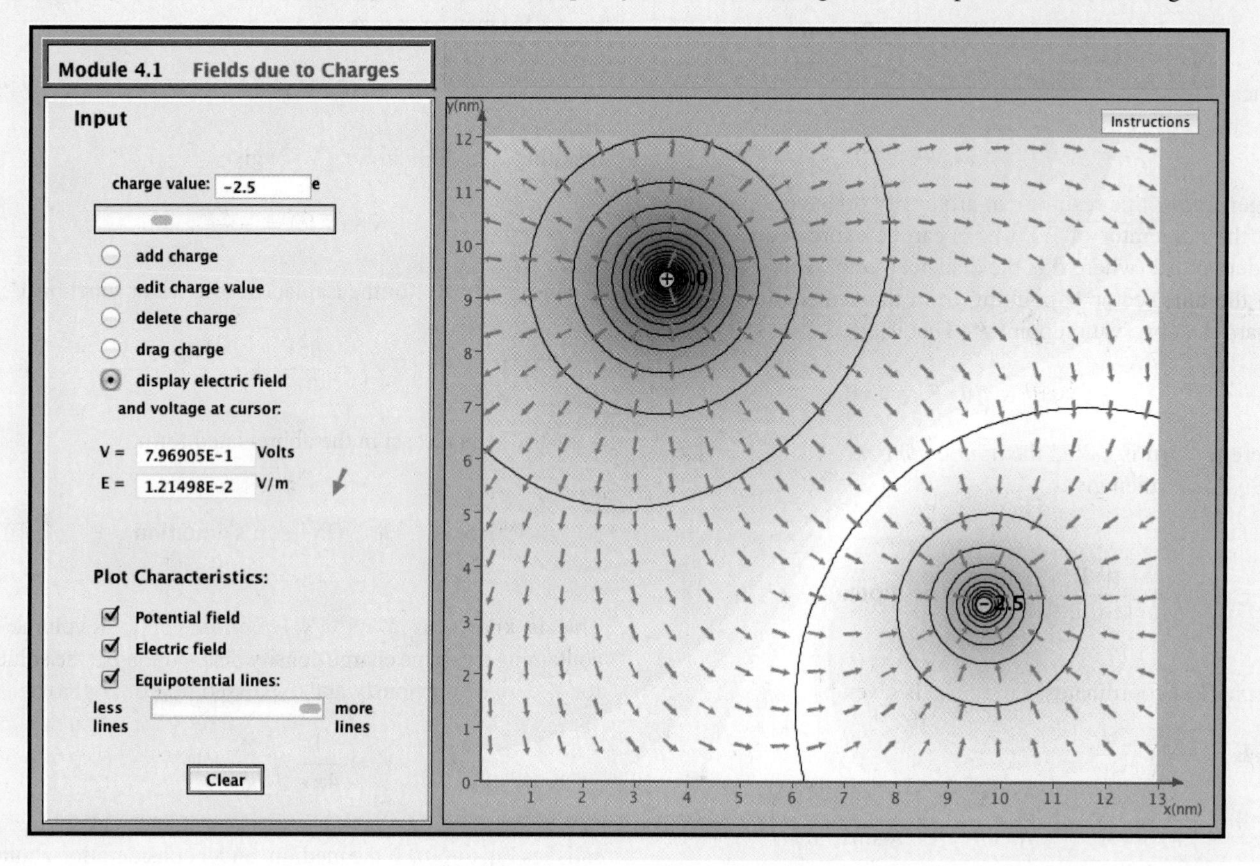

Concept Question 4-11: Explain why Eq. (4.40) is a mathematical statement of Kirchhoff's voltage law.

Concept Question 4-12: Why is it usually easier to compute V for a given charge distribution and then find **E** using $\mathbf{E} = -\nabla V$ than to compute **E** directly by applying Coulomb's law?

Concept Question 4-13: What is an electric dipole?

Exercise 4-10: Determine the electric potential at the origin due to four 20 μC charges residing in free space at the corners of a 2 m $\times$ 2 m square centered about the origin in the x–y plane.

Answer: $V = \sqrt{2} \times 10^{-5}/(\pi\epsilon_0)$ (V). (See EM.)

Exercise 4-11: A spherical shell of radius a has a uniform surface charge density ρ_s. Determine (a) the electric potential and (b) the electric field, both at the center of the shell.

Answer: (a) $V = \rho_s a/\epsilon$ (V), (b) $\mathbf{E} = 0$. (See EM.)

4-6 Conductors

The electromagnetic ***constitutive parameters*** of a material medium are its electrical permittivity ϵ, magnetic permeability μ, and conductivity σ. A material is said to be ***homogeneous*** if its constitutive parameters do not vary from point to point, and ***isotropic*** if they are independent of direction. Most materials are isotropic, but some crystals are not. Throughout this book, all materials are assumed to be homogeneous and isotropic. This section is concerned with σ, Section 4-7 examines ϵ, and discussion of μ is deferred to Chapter 5.

▶ The ***conductivity*** of a material is a measure of how easily electrons can travel through the material under the influence of an externally applied electric field. ◀

Materials are classified as ***conductors*** (metals) or ***dielectrics*** (insulators) according to the magnitudes of their conductivities. A conductor has a large number of loosely attached electrons in the outermost shells of its atoms. In the absence of an external electric field, these free electrons move in random directions and with varying speeds. Their random motion produces zero average current through the conductor. Upon applying an external electric field, however, the electrons migrate from one atom to the next in the direction opposite that of the external field. Their movement gives rise to a ***conduction current***

$$\mathbf{J} = \sigma \mathbf{E} \quad (\text{A/m}^2) \quad \textbf{(Ohm's law)}, \tag{4.63}$$

where σ is the material's conductivity with units of siemen per meter (S/m).

In yet other materials, called dielectrics, the electrons are tightly bound to the atoms, so much so that it is very difficult to detach them under the influence of an electric field. Consequently, no significant conduction current can flow through them.

▶ A ***perfect dielectric*** is a material with $\sigma = 0$. In contrast, a ***perfect conductor*** is a material with $\sigma = \infty$. Some materials, called superconductors, exhibit such a behavior. ◀

The conductivity σ of most metals is in the range from 10^6 to 10^7 S/m, compared with 10^{-10} to 10^{-17} S/m for good insulators

Table 4-1 Conductivity of some common materials at 20 °C.

Material	Conductivity, σ (S/m)
Conductors	
Silver	6.2×10^7
Copper	5.8×10^7
Gold	4.1×10^7
Aluminum	3.5×10^7
Iron	10^7
Mercury	10^6
Carbon	3×10^4
Semiconductors	
Pure germanium	2.2
Pure silicon	4.4×10^{-4}
Insulators	
Glass	10^{-12}
Paraffin	10^{-15}
Mica	10^{-15}
Fused quartz	10^{-17}

(Table 4-1). A class of materials called ***semiconductors*** allow for conduction currents even though their conductivities are much smaller than those of metals. The conductivity of pure germanium, for example, is 2.2 S/m. Tabulated values of σ at room temperature (20 °C) are given in Appendix B for some common materials, and a subset is reproduced in Table 4-1.

▶ The conductivity of a material depends on several factors, including temperature and the presence of impurities. In general, σ of metals increases with decreasing temperature. Most superconductors operate in the neighborhood of absolute zero. ◀

Concept Question 4-14: What are the electromagnetic constitutive parameters of a material?

Concept Question 4-15: What classifies a material as a conductor, a semiconductor, or a dielectric? What is a superconductor?

Concept Question 4-16: What is the conductivity of a perfect dielectric?

Technology Brief 7: Resistive Sensors

An ***electrical sensor*** is a device capable of responding to an applied ***stimulus*** by generating an electrical signal whose voltage, current, or some other attribute is related to the intensity of the stimulus.

> ▶ The family of possible stimuli encompasses a wide array of physical, chemical, and biological quantities, including temperature, pressure, position, distance, motion, velocity, acceleration, concentration (of a gas or liquid), blood flow, etc. ◀

The sensing process relies on measuring resistance, capacitance, inductance, induced electromotive force (emf), oscillation frequency or time delay, among others. Sensors are integral to the operation of just about every instrument that uses electronic systems, from automobiles and airplanes to computers and cell phones. This Technology Brief covers resistive sensors. ***Capacitive***, ***inductive***, and ***emf sensors*** are covered separately (here and in later chapters).

Piezoresistivity

According to Eq. (4.70), the resistance of a cylindrical resistor or wire conductor is given by $R = l/\sigma A$, where l is the cylinder's length, A is its cross-sectional area, and σ is the conductivity of its material. Stretching the wire by an applied external force causes l to increase and A to decrease. Consequently, R increases (**Fig. TF7-1**). Conversely, compressing the wire causes R to decrease. The Greek word ***piezein*** means to press, from which the term piezoresistivity is derived. *This should not be confused with piezoelectricity, which is an emf effect. (See EMF Sensors in Technology Brief 12.)*

The relationship between the resistance R of a piezoresistor and the ***applied force*** F can be modeled by the approximate linear equation

$$R = R_0 \left(1 + \frac{\alpha F}{A_0}\right),$$

where R_0 is the unstressed resistance (@ $F = 0$), A_0 is the unstressed cross-sectional area of the resistor, and α is the ***piezoresistive coefficient*** of the resistor material. The force F is positive if it is causing the resistor to stretch and negative if it is compressing it.

An elastic resistive sensor is well suited for measuring the deformation z of a surface (**Fig. TF7-2**), which can be related to the pressure applied to the surface; and if z is recorded as a function of time, it is possible to derive the velocity and acceleration of the surface's motion. To realize high longitudinal piezoresistive sensitivity (the ratio of the normalized change in resistance, $\Delta R/R_0$, to the corresponding change in length, $\Delta l/l_0$, caused by the applied force), the piezoresistor is often designed as a serpentine-shaped wire [**Fig. TF7-3(a)**] bonded on a flexible plastic substrate and glued onto the surface whose deformation is to be monitored. Copper and nickel alloys are commonly used for making the sensor wires, although in some applications silicon is used instead [**Fig. TF7-3(b)**] because it has a very high piezoresistive sensitivity.

> ▶ By connecting the piezoresistor to a Wheatstone bridge circuit (**Fig. TF7-4**) in which the other three resistors are all identical in value and equal to R_0 (the resistance of the piezoresistor when no external force is present), the voltage output becomes directly proportional to the normalized resistance change: $\Delta R/R_0$. ◀

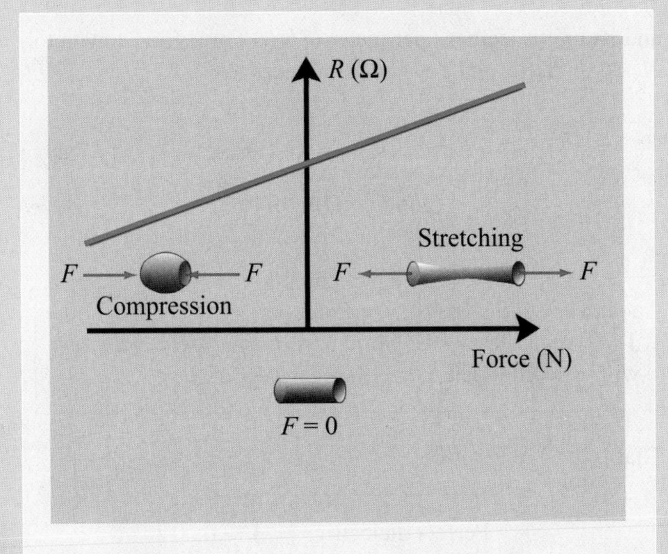

Figure TF7-1 Piezoresistance varies with applied force.

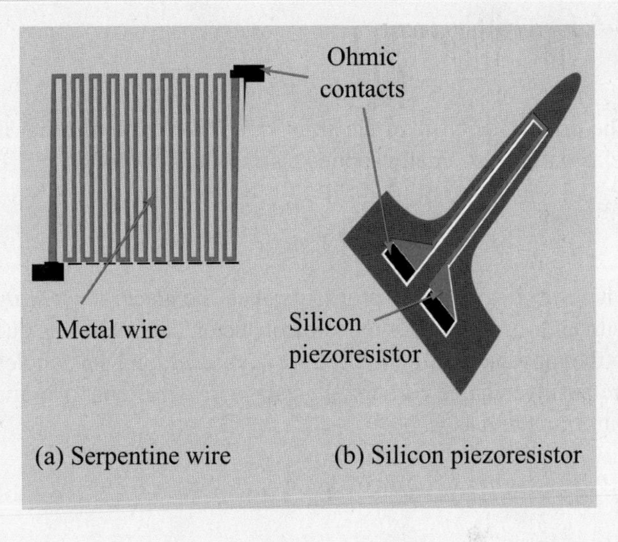

Figure TF7-3 Metal and silicon piezoresistors.

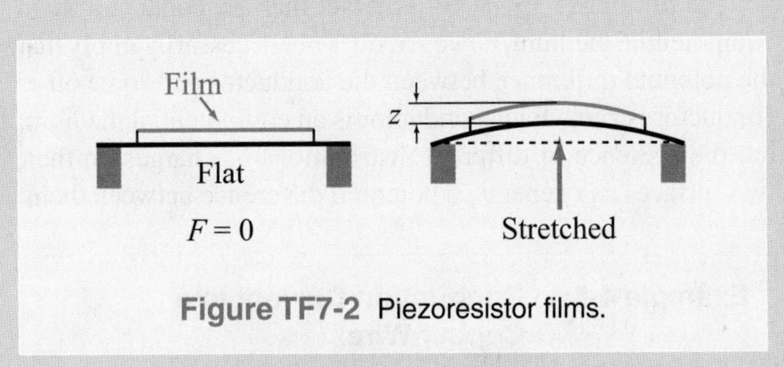

Figure TF7-2 Piezoresistor films.

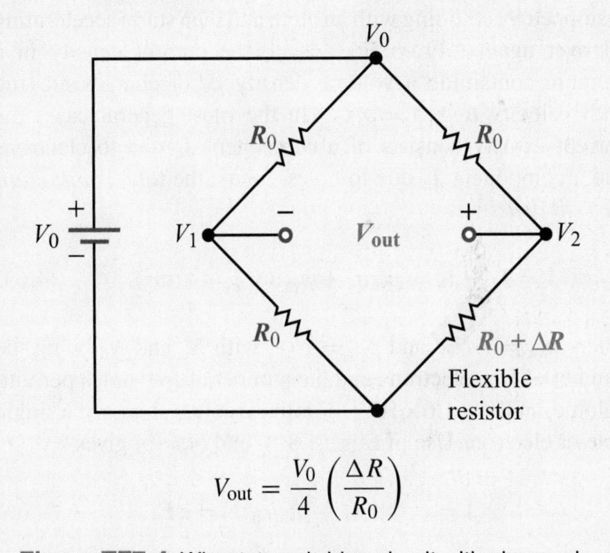

Figure TF7-4 Wheatstone bridge circuit with piezoresistor.

4-6.1 Drift Velocity

The drift velocity $\mathbf{u}_e$ of electrons in a conducting material is related to the externally applied electric field $\mathbf{E}$ through

$$\mathbf{u}_e = -\mu_e \mathbf{E} \qquad \text{(m/s)}, \tag{4.64a}$$

where μ_e is a material property called the ***electron mobility*** with units of ($\text{m}^2/\text{V}\cdot\text{s}$). In a semiconductor, current flow is due to the movement of both electrons and holes, and since holes are positive-charge carriers, the ***hole drift velocity*** $\mathbf{u}_h$ is in the same direction as $\mathbf{E}$,

$$\mathbf{u}_h = \mu_h \mathbf{E} \qquad \text{(m/s)}, \tag{4.64b}$$

where μ_h is the ***hole mobility***. The mobility accounts for the effective mass of a charged particle and the average distance over which the applied electric field can accelerate it before it is stopped by colliding with an atom and then starts accelerating all over again. From Eq. (4.11), the current density in a medium containing a volume density ρ_v of charges moving with velocity $\mathbf{u}$ is $\mathbf{J} = \rho_v \mathbf{u}$. In the most general case, the current density consists of a component $\mathbf{J}_e$ due to electrons and a component $\mathbf{J}_h$ due to holes. Thus, the total ***conduction current density*** is

$$\mathbf{J} = \mathbf{J}_e + \mathbf{J}_h = \rho_{ve}\mathbf{u}_e + \rho_{vh}\mathbf{u}_h \qquad (\text{A/m}^2), \tag{4.65}$$

where $\rho_{ve} = -N_e e$ and $\rho_{vh} = N_h e$, with N_e and N_h being the number of free electrons and the number of free holes per unit volume, and $e = 1.6\times10^{-19}$ C is the absolute charge of a single hole or electron. Use of Eqs. (4.64a) and (4.64b) gives

$$\mathbf{J} = (-\rho_{ve}\mu_e + \rho_{vh}\mu_h)\mathbf{E} = \sigma\mathbf{E}, \tag{4.66}$$

where the quantity inside the parentheses is defined as the ***conductivity*** of the material, σ. Thus,

$$\begin{aligned} \sigma &= -\rho_{ve}\mu_e + \rho_{vh}\mu_h \\ &= (N_e\mu_e + N_h\mu_h)e \qquad \text{(S/m)}, \end{aligned} \tag{4.67a}$$

(semiconductor)

and its unit is siemens per meter (S/m). For a good conductor, $N_h\mu_h \ll N_e\mu_e$, and Eq. (4.67a) reduces to

$$\sigma = -\rho_{ve}\mu_e = N_e\mu_e e \qquad \text{(S/m)}. \tag{4.67b}$$

(good conductor)

▶ In view of Eq. (4.66), in a ***perfect dielectric*** with $\sigma = 0$, $\mathbf{J} = 0$ regardless of $\mathbf{E}$. Similarly, in a ***perfect conductor*** with $\sigma = \infty$, $\mathbf{E} = \mathbf{J}/\sigma = 0$ regardless of $\mathbf{J}$. ◀

That is,

Perfect dielectric:	$\mathbf{J} = 0$,
Perfect conductor:	$\mathbf{E} = 0$.

Because σ is on the order of 10^6 S/m for most metals, such as silver, copper, gold, and aluminum (**Table 4-1**), it is common practice to treat them as perfect conductors and to set $\mathbf{E} = 0$ inside them.

A perfect conductor is an ***equipotential*** medium, meaning that the electric potential is the same at every point in the conductor. This property follows from the fact that V_{21}, the voltage difference between two points in the conductor equals the line integral of $\mathbf{E}$ between them, as indicated by Eq. (4.39), and since $\mathbf{E} = 0$ everywhere in the perfect conductor, the voltage difference $V_{21} = 0$. The fact that the conductor is an equipotential medium, however, does not necessarily imply that the potential difference between the conductor and some other conductor is zero. Each conductor is an equipotential medium, but the presence of different distributions of charges on their two surfaces can generate a potential difference between them.

Example 4-8: Conduction Current in a Copper Wire

A 2 mm diameter copper wire with conductivity of 5.8×10^7 S/m and electron mobility of 0.0032 ($\text{m}^2/\text{V}\cdot\text{s}$) is subjected to an electric field of 20 (mV/m). Find (a) the volume charge density of the free electrons, (b) the current density, (c) the current flowing in the wire, (d) the electron drift velocity, and (e) the volume density of the free electrons.

Solution:

(a)

$$\rho_{\text{ve}} = -\frac{\sigma}{\mu_{\text{e}}} = -\frac{5.8 \times 10^7}{0.0032} = -1.81 \times 10^{10} \text{ (C/m}^3\text{)}.$$

(b)

$$J = \sigma E = 5.8 \times 10^7 \times 20 \times 10^{-3} = 1.16 \times 10^6 \text{ (A/m}^2\text{)}.$$

(c)

$$I = JA$$
$$= J\left(\frac{\pi d^2}{4}\right) = 1.16 \times 10^6 \left(\frac{\pi \times 4 \times 10^{-6}}{4}\right) = 3.64 \text{ A}.$$

(d)

$$u_{\text{e}} = -\mu_{\text{e}} E = -0.0032 \times 20 \times 10^{-3} = -6.4 \times 10^{-5} \text{ m/s}.$$

The minus sign indicates that $\mathbf{u}_{\text{e}}$ is in the opposite direction of **E**.

(e)

$$N_{\text{e}} = -\frac{\rho_{\text{ve}}}{e} = \frac{1.81 \times 10^{10}}{1.6 \times 10^{-19}} = 1.13 \times 10^{29} \text{ electrons/m}^3.$$

Exercise 4-12: Determine the density of free electrons in aluminum, given that its conductivity is 3.5×10^7 (S/m) and its electron mobility is 0.0015 ($\text{m}^2/\text{V} \cdot \text{s}$).

Answer: $N_{\text{e}} = 1.46 \times 10^{29}$ electrons/m^3. (See EM.)

Exercise 4-13: The current flowing through a 100 m long conducting wire of uniform cross section has a density of 3×10^5 (A/m^2). Find the voltage drop along the length of the wire if the wire material has a conductivity of 2×10^7 (S/m).

Answer: $V = 1.5$ V. (See EM.)

4-6.2 Resistance

To demonstrate the utility of the point form of Ohm's law, we apply it to derive an expression for the resistance R of a conductor of length l and uniform cross section A, as shown in **Fig. 4-14**. The conductor axis is along the x direction and extends between points x_1 and x_2, with $l = x_2 - x_1$. A voltage V applied across the conductor terminals establishes an electric field $\mathbf{E} = \hat{\mathbf{x}} E_x$; the direction of **E** is from the point with higher potential (point 1 in **Fig. 4-14**) to the point with lower potential (point 2). The relation between V and E_x is obtained by applying Eq. (4.39):

$$\begin{aligned} V &= V_1 - V_2 \\ &= -\int_{x_2}^{x_1} \mathbf{E} \cdot d\mathbf{l} \\ &= -\int_{x_2}^{x_1} \hat{\mathbf{x}} E_x \cdot \hat{\mathbf{x}}\, dl = E_x l \qquad \text{(V)}. \end{aligned} \tag{4.68}$$

Using Eq. (4.63), the current flowing through the cross section A at x_2 is

$$I = \int_A \mathbf{J} \cdot d\mathbf{s} = \int_A \sigma \mathbf{E} \cdot d\mathbf{s} = \sigma E_x A \qquad \text{(A)}. \tag{4.69}$$

From $R = V/I$, the ratio of Eq. (4.68) to Eq. (4.69) gives

$$R = \frac{l}{\sigma A} \qquad (\Omega). \tag{4.70}$$

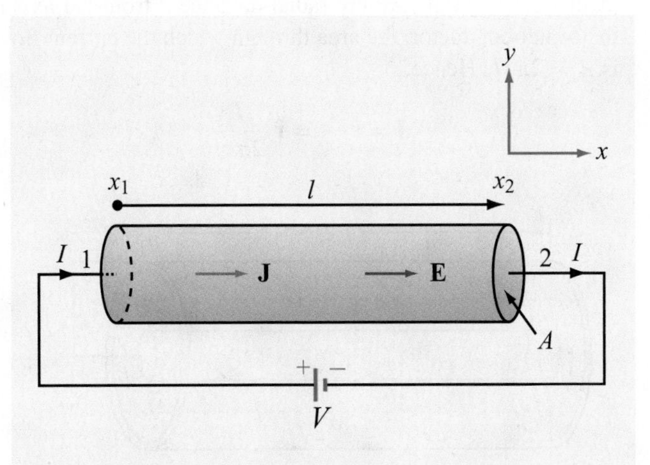

Figure 4-14 Linear resistor of cross section A and length l connected to a dc voltage source V.

We now generalize our result for R to any resistor of arbitrary shape by noting that the voltage V across the resistor is equal to the line integral of $\mathbf{E}$ over a path l between two specified points and the current I is equal to the flux of $\mathbf{J}$ through the surface S of the resistor. Thus,

$$R = \frac{V}{I} = \frac{-\int_l \mathbf{E} \cdot d\mathbf{l}}{\int_S \mathbf{J} \cdot d\mathbf{s}} = \frac{-\int_l \mathbf{E} \cdot d\mathbf{l}}{\int_S \sigma \mathbf{E} \cdot d\mathbf{s}} . \tag{4.71}$$

The reciprocal of R is called the ***conductance*** G, and the unit of G is (Ω^{-1}), or siemens (S). For the linear resistor,

$$G = \frac{1}{R} = \frac{\sigma A}{l} \quad \text{(S)}. \tag{4.72}$$

Example 4-9: Conductance of Coaxial Cable

The radii of the inner and outer conductors of a coaxial cable of length l are a and b, respectively (Fig. 4-15). The insulation material has conductivity σ. Obtain an expression for G', the conductance per unit length of the insulation layer.

Solution: Let I be the total current flowing radially (along $\hat{\mathbf{r}}$) from the inner conductor to the outer conductor through the insulation material. At any radial distance r from the axis of the center conductor, the area through which the current flows is $A = 2\pi rl$. Hence,

$$\mathbf{J} = \hat{\mathbf{r}} \frac{I}{A} = \hat{\mathbf{r}} \frac{I}{2\pi rl} , \tag{4.73}$$

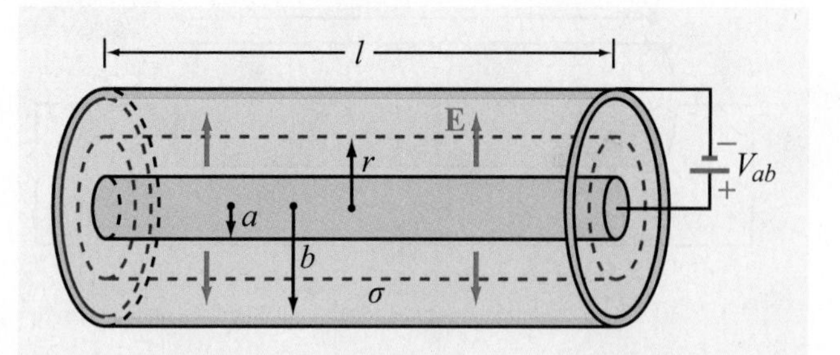

Figure 4-15 Coaxial cable of Example 4-9.

and from $\mathbf{J} = \sigma \mathbf{E}$,

$$\mathbf{E} = \hat{\mathbf{r}} \frac{I}{2\pi \sigma rl} . \tag{4.74}$$

In a resistor, the current flows from higher electric potential to lower potential. Hence, if $\mathbf{J}$ is in the $\hat{\mathbf{r}}$ direction, the inner conductor must be at a potential higher than that at the outer conductor. Accordingly, the voltage difference between the conductors is

$$\begin{aligned} V_{ab} &= -\int_b^a \mathbf{E} \cdot d\mathbf{l} = -\int_b^a \frac{I}{2\pi \sigma l} \frac{\hat{\mathbf{r}} \cdot \hat{\mathbf{r}}\, dr}{r} \\ &= \frac{I}{2\pi \sigma l} \ln\left(\frac{b}{a}\right). \end{aligned} \tag{4.75}$$

The conductance per unit length is then

$$G' = \frac{G}{l} = \frac{1}{Rl} = \frac{I}{V_{ab} l} = \frac{2\pi \sigma}{\ln(b/a)} \quad \text{(S/m)}. \tag{4.76}$$

4-6.3 Joule's Law

We now consider the power dissipated in a conducting medium in the presence of an electrostatic field $\mathbf{E}$. The medium contains free electrons and holes with volume charge densities ρ_{ve} and ρ_{vh}, respectively. The electron and hole charge contained in an elemental volume $\Delta \mathcal{V}$ is $q_{\text{e}} = \rho_{\text{ve}}\, \Delta \mathcal{V}$ and $q_{\text{h}} = \rho_{\text{vh}}\, \Delta \mathcal{V}$, respectively. The electric forces acting on q_{e} and q_{h} are $\mathbf{F}_{\text{e}} = q_{\text{e}} \mathbf{E} = \rho_{\text{ve}} \mathbf{E}\, \Delta \mathcal{V}$ and $\mathbf{F}_{\text{h}} = q_{\text{h}} \mathbf{E} = \rho_{\text{vh}} \mathbf{E}\, \Delta \mathcal{V}$. The work (energy) expended by the electric field in moving q_{e} a differential distance Δl_{e} and moving q_{h} a distance Δl_{h} is

$$\Delta W = \mathbf{F}_{\text{e}} \cdot \Delta \mathbf{l}_{\text{e}} + \mathbf{F}_{\text{h}} \cdot \Delta \mathbf{l}_{\text{h}}. \tag{4.77}$$

Power P, measured in watts (W), is defined as the time rate of change of energy. The power corresponding to ΔW is

$$\begin{aligned} \Delta P = \frac{\Delta W}{\Delta t} &= \mathbf{F}_{\text{e}} \cdot \frac{\Delta \mathbf{l}_{\text{e}}}{\Delta t} + \mathbf{F}_{\text{h}} \cdot \frac{\Delta \mathbf{l}_{\text{h}}}{\Delta t} \\ &= \mathbf{F}_{\text{e}} \cdot \mathbf{u}_{\text{e}} + \mathbf{F}_{\text{h}} \cdot \mathbf{u}_{\text{h}} \\ &= (\rho_{\text{ve}} \mathbf{E} \cdot \mathbf{u}_{\text{e}} + \rho_{\text{vh}} \mathbf{E} \cdot \mathbf{u}_{\text{h}})\, \Delta \mathcal{V} \\ &= \mathbf{E} \cdot \mathbf{J}\, \Delta \mathcal{V}, \end{aligned} \tag{4.78}$$

where $\mathbf{u}_{\text{e}} = \Delta \mathbf{l}_{\text{e}} / \Delta t$ and $\mathbf{u}_{\text{h}} = \Delta \mathbf{l}_{\text{h}} / \Delta t$ are the electron and hole drift velocities, respectively. Equation (4.65) was used in the last step of the derivation leading to Eq. (4.78). For a volume $\mathcal{V}$, the total dissipated power is

$$P = \int_{\mathcal{V}} \mathbf{E} \cdot \mathbf{J} \, d\mathcal{V} \quad \text{(W)} \quad \textbf{(Joule's law)}, \tag{4.79}$$

and in view of Eq. (4.63),

$$P = \int_{\mathcal{V}} \sigma |\mathbf{E}|^2 \, d\mathcal{V} \quad \text{(W)}. \tag{4.80}$$

Equation (4.79) is a mathematical statement of ***Joule's law***. For the resistor example considered earlier, $|\mathbf{E}| = E_x$ and its volume is $\mathcal{V} = lA$. Separating the volume integral in Eq. (4.80) into a product of a surface integral over A and a line integral over l, we have

$$\begin{aligned} P &= \int_{\mathcal{V}} \sigma |\mathbf{E}|^2 \, d\mathcal{V} \\ &= \int_A \sigma E_x \, ds \int_l E_x \, dl \\ &= (\sigma E_x A)(E_x l) = IV \quad \text{(W)}, \end{aligned} \tag{4.81}$$

where use was made of Eq. (4.68) for the voltage V and Eq. (4.69) for the current I. With $V = IR$, we obtain the familiar expression

$$P = I^2 R \quad \text{(W)}. \tag{4.82}$$

Concept Question 4-17: What is the fundamental difference between an insulator, a semiconductor, and a conductor?

Concept Question 4-18: Show that the power dissipated in the coaxial cable of Fig. 4-15 is $P = I^2 \ln(b/a)/(2\pi\sigma l)$.

Exercise 4-14: A 50 m long copper wire has a circular cross section with radius $r = 2$ cm. Given that the conductivity of copper is 5.8×10^7 S/m, determine (a) the resistance R of the wire and (b) the power dissipated in the wire if the voltage across its length is 1.5 mV.

Answer: (a) $R = 6.9 \times 10^{-4}$ Ω, (b) $P = 3.3$ mW. (See EM.)

Exercise 4-15: Repeat part (b) of Exercise 4.14 by applying Eq. (4.80). (See EM.)

4-7 Dielectrics

The fundamental difference between a conductor and a dielectric is that electrons in the outermost atomic shells of a conductor are only weakly tied to atoms and hence can freely migrate through the material, whereas in a dielectric they are strongly bound to the atom. In the absence of an electric field, the electrons in so-called ***nonpolar*** molecules form a symmetrical cloud around the nucleus, with the center of the cloud coinciding with the nucleus [Fig. 4-16(a)]. The electric field generated by the positively charged nucleus attracts and holds the electron cloud around it, and the mutual repulsion of the electron clouds of adjacent atoms shapes its form. When a conductor is subjected to an externally applied electric field, the most loosely bound electrons in each atom can jump from one atom to the next, thereby setting up an electric current. In a dielectric, however, an externally applied electric field $\mathbf{E}$ cannot effect mass migration of charges since none are able to move freely. Instead, $\mathbf{E}$ will ***polarize*** the atoms or molecules in the material by moving the center of the electron cloud away from the nucleus [Fig. 4-16(b)]. The polarized atom or molecule may be represented by an electric dipole consisting of charges $+q$ in the nucleus and $-q$ at the center of the electron cloud [Fig. 4-16(c)]. Each such dipole sets up a small electric field, pointing from the positively charged nucleus to the center of the equally but negatively charged electron cloud. This ***induced*** electric field, called a ***polarization*** field, generally is weaker than and opposite in direction to, $\mathbf{E}$. Consequently, the net electric field present in the

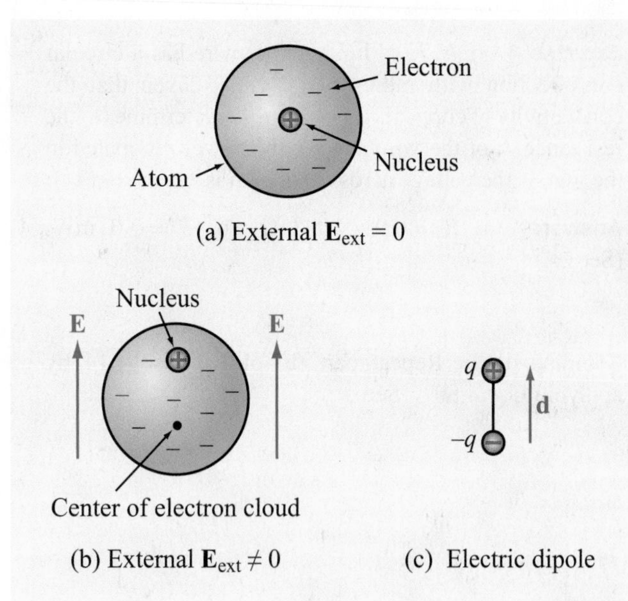

Figure 4-16 In the absence of an external electric field **E**, the center of the electron cloud is co-located with the center of the nucleus, but when a field is applied, the two centers are separated by a distance d.

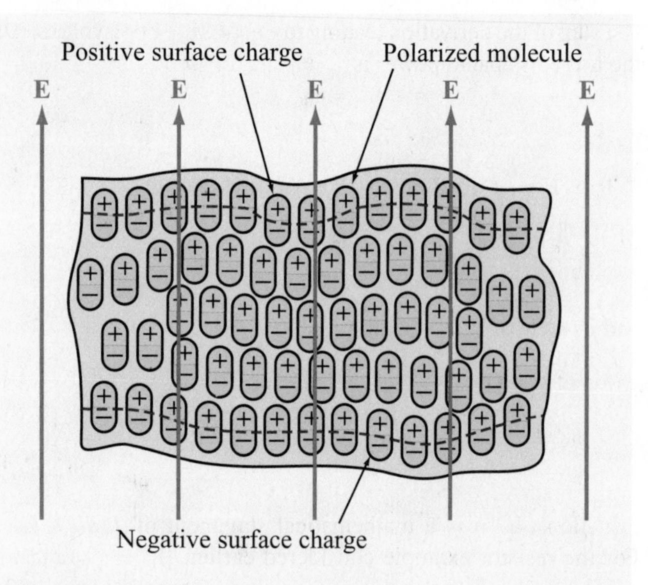

Figure 4-17 A dielectric medium polarized by an external electric field **E**.

dielectric material is smaller than **E**. At the microscopic level, each dipole exhibits a dipole moment similar to that described in Example 4-7. Within a block of dielectric material subject to a uniform external field, the dipoles align themselves linearly, as shown in Fig. 4-17. Along the upper and lower edges of the material, the dipole arrangement exhibits positive and negative surface charge densities, respectively.

It is important to stress that this description applies to only nonpolar molecules, which do not have permanent dipole moments. Nonpolar molecules become polarized only when an external electric field is applied, and when the field is removed, the molecules return to their original unpolarized state.

In polar materials, such as water, the molecules possess built-in ***permanent dipole moments*** that are randomly oriented in the absence of an applied electric field, and owing to their random orientations, the dipoles of polar materials produce no net macroscopic dipole moment (at the macroscopic scale, each point in the material represents a small volume containing thousands of molecules). Under the influence of an applied field, the permanent dipoles tend to align themselves along the direction of the electric field, in a manner similar to that shown in Fig. 4-17 for nonpolar materials.

4-7.1 Polarization Field

Whereas in free space $\mathbf{D} = \epsilon_0 \mathbf{E}$, the presence of microscopic dipoles in a dielectric material alters that relationship to

$$\mathbf{D} = \epsilon_0 \mathbf{E} + \mathbf{P}, \tag{4.83}$$

where **P**, called the ***electric polarization field***, accounts for the polarization properties of the material. The polarization field is produced by the electric field **E** and depends on the material properties. A dielectric medium is said to be ***linear*** if the magnitude of the induced polarization field **P** is directly proportional to the magnitude of **E**, and ***isotropic*** if **P** and **E** are in the same direction. Some crystals allow more polarization to take place along certain directions, such as the crystal axes, than along others. In such ***anisotropic*** dielectrics, **E** and **P** may have different directions. A medium is said to be ***homogeneous*** if its constitutive parameters (ϵ, μ, and σ) are constant throughout the medium. Our present treatment will be limited to media that are linear, isotropic, and homogeneous. For such media **P** is directly proportional to **E** and is expressed as

$$\mathbf{P} = \epsilon_0 \chi_e \mathbf{E}, \tag{4.84}$$

where χ_e is called the ***electric susceptibility*** of the material. Inserting Eq. (4.84) into Eq. (4.83), we have

$$\mathbf{D} = \epsilon_0 \mathbf{E} + \epsilon_0 \chi_e \mathbf{E} = \epsilon_0 (1 + \chi_e) \mathbf{E} = \epsilon \mathbf{E}, \tag{4.85}$$

which defines the permittivity ϵ of the material as

$$\epsilon = \epsilon_0 (1 + \chi_e). \tag{4.86}$$

It is often convenient to characterize the permittivity of a material relative to that of free space, ϵ_0; this is accommodated by the relative permittivity $\epsilon_r = \epsilon/\epsilon_0$. Values of ϵ_r are listed in **Table 4-2** for a few common materials, and a longer list is given in Appendix B. In free space $\epsilon_r = 1$, and *for most conductors* $\epsilon_r \approx 1$. The dielectric constant of air is approximately 1.0006 at sea level, and decreases toward unity with increasing altitude. Except in some special circumstances, such as when calculating electromagnetic wave refraction (bending) through the atmosphere over long distances, *air can be treated as if it were free space.*

4-7.2 Dielectric Breakdown

The preceding dielectric-polarization model presumes that the magnitude of **E** does not exceed a certain critical value, known as the ***dielectric strength*** E_{ds} of the material, beyond which electrons will detach from the molecules and accelerate through the material in the form of a conduction current. When this happens, sparking can occur, and the dielectric material can sustain permanent damage due to electron collisions with the molecular structure. This abrupt change in behavior is called ***dielectric breakdown***.

▶ The dielectric strength E_{ds} is the largest magnitude of **E** that the material can sustain without breakdown. ◀

Dielectric breakdown can occur in gases, liquids, and solids. The dielectric strength E_{ds} depends on the material composition, as well as other factors such as temperature and humidity. For air E_{ds} is roughly 3 (MV/m); for glass 25 to 40 (MV/m); and for mica 200 (MV/m) (see **Table 4-2**).

A charged thundercloud at electric potential V relative to the ground induces an electric field $E = V/d$ in the air beneath it, where d is the height of the cloud base above the ground. If V is sufficiently large so that E exceeds the dielectric strength of air, ionization occurs and a lightning discharge follows. The ***breakdown voltage*** V_{br} of a parallel-plate capacitor is discussed later in Example 4-11.

Concept Question 4-19: What is a polar material? A nonpolar material?

Concept Question 4-20: Do **D** and **E** always point in the same direction? If not, when do they not?

Concept Question 4-21: What happens when dielectric breakdown occurs?

4-8 Electric Boundary Conditions

A vector field is said to be spatially continuous if it does not exhibit abrupt changes in either magnitude or direction as a function of position. Even though the electric field may be continuous in adjoining dissimilar media, it may well be discontinuous at the boundary between them. Boundary conditions specify how the components of fields tangential and normal to an interface between two media relate across the interface. Here we derive a general set of boundary conditions for **E**, **D**, and **J**, applicable at the interface between *any* two dissimilar media, be they two dielectrics or a conductor and a dielectric. Of course, any of the dielectrics may be free space. Even though these boundary conditions are derived assuming electrostatic conditions, they remain valid for time-varying electric fields as well. **Figure 4-18** shows an interface between medium 1 with permittivity ϵ_1 and medium 2 with permittivity ϵ_2. In the general case, the interface may contain a surface charge density ρ_s (unrelated to the dielectric polarization charge density).

To derive the boundary conditions for the tangential components of **E** and **D**, we consider the closed rectangular loop $abcda$ shown in **Fig. 4-18** and apply the conservative property of the electric field expressed by Eq. (4.40), which states that the line integral of the electrostatic field around a closed path is always zero. By letting $\Delta h \to 0$, the contributions to the line

Table 4-2 Relative permittivity (dielectric constant) and dielectric strength of common materials.

Material	Relative Permittivity, ϵ_r	Dielectric Strength, E_{ds} (MV/m)
Air (at sea level)	1.0006	3
Petroleum oil	2.1	12
Polystyrene	2.6	20
Glass	4.5–10	25–40
Quartz	3.8–5	30
Bakelite	5	20
Mica	5.4–6	200

$\epsilon = \epsilon_r \epsilon_0$ and $\epsilon_0 = 8.854 \times 10^{-12}$ F/m.

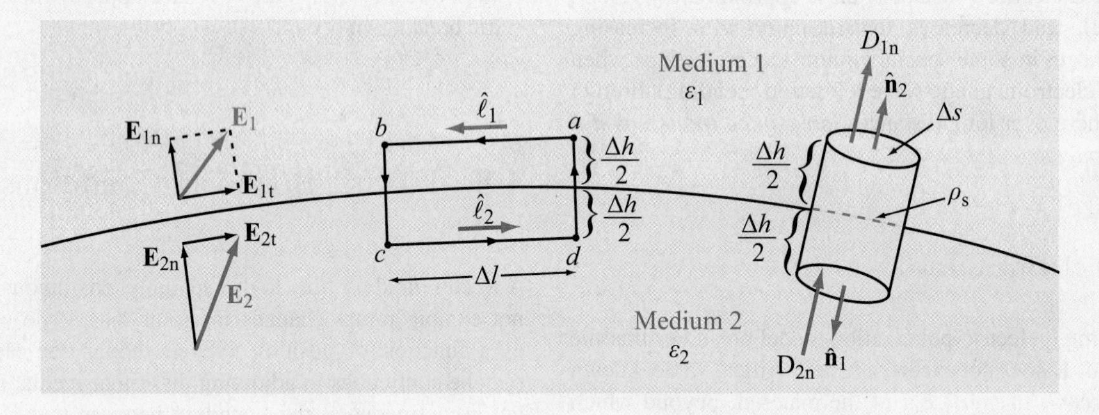

Figure 4-18 Interface between two dielectric media.

integral by segments bc and da vanish. Hence,

$$\oint_C \mathbf{E} \cdot d\mathbf{l} = \int_a^b \mathbf{E}_1 \cdot \hat{\boldsymbol{\ell}}_1 \, dl + \int_c^d \mathbf{E}_2 \cdot \hat{\boldsymbol{\ell}}_2 \, dl = 0, \quad (4.87)$$

where $\hat{\boldsymbol{\ell}}_1$ and $\hat{\boldsymbol{\ell}}_2$ are unit vectors along segments ab and cd, and $\mathbf{E}_1$ and $\mathbf{E}_2$ are the electric fields in media 1 and 2. Next, we decompose $\mathbf{E}_1$ and $\mathbf{E}_2$ into components tangential and normal to the boundary (Fig. 4-18),

$$\mathbf{E}_1 = \mathbf{E}_{1t} + \mathbf{E}_{1n}, \quad (4.88a)$$

$$\mathbf{E}_2 = \mathbf{E}_{2t} + \mathbf{E}_{2n}. \quad (4.88b)$$

Noting that $\hat{\boldsymbol{\ell}}_1 = -\hat{\boldsymbol{\ell}}_2$, it follows that

$$(\mathbf{E}_1 - \mathbf{E}_2) \cdot \hat{\boldsymbol{\ell}}_1 = 0. \quad (4.89)$$

In other words, the component of $\mathbf{E}_1$ along $\hat{\boldsymbol{\ell}}_1$ equals that of $\mathbf{E}_2$ along $\hat{\boldsymbol{\ell}}_1$, for all $\hat{\boldsymbol{\ell}}_1$ tangential to the boundary. Hence,

$$\mathbf{E}_{1t} = \mathbf{E}_{2t} \qquad \text{(V/m)}. \quad (4.90)$$

▶ Thus, the tangential component of the electric field is **continuous** across the boundary between any two media. ◀

Upon decomposing $\mathbf{D}_1$ and $\mathbf{D}_2$ into tangential and normal components (in the manner of Eq. (4.88)) and noting that

$\mathbf{D}_{1t} = \epsilon_1 \mathbf{E}_{1t}$ and $\mathbf{D}_{2t} = \epsilon_2 \mathbf{E}_{2t}$, the boundary condition on the tangential component of the electric flux density is

$$\frac{\mathbf{D}_{1t}}{\epsilon_1} = \frac{\mathbf{D}_{2t}}{\epsilon_2}. \qquad (4.91)$$

Next, we apply Gauss's law, as expressed by Eq. (4.29), to determine boundary conditions on the normal components of **E** and **D**. According to Gauss's law, the total outward flux of **D** through the three surfaces of the small cylinder shown in Fig. 4-18 must equal the total charge enclosed in the cylinder. By letting the cylinder's height $\Delta h \to 0$, the contribution to the total flux through the side surface goes to zero. Also, even if each of the two media happens to contain free charge densities, the only charge remaining in the collapsed cylinder is that distributed on the boundary. Thus, $Q = \rho_s \, \Delta s$, and

$$\oint_S \mathbf{D} \cdot d\mathbf{s} = \int_{\text{top}} \mathbf{D}_1 \cdot \hat{\mathbf{n}}_2 \, ds + \int_{\text{bottom}} \mathbf{D}_2 \cdot \hat{\mathbf{n}}_1 \, ds = \rho_s \, \Delta s, \quad (4.92)$$

where $\hat{\mathbf{n}}_1$ and $\hat{\mathbf{n}}_2$ are the outward normal unit vectors of the bottom and top surfaces, respectively. It is important to remember that ***the normal unit vector at the surface of any medium is always defined to be in the outward direction away from that medium.*** Since $\hat{\mathbf{n}}_1 = -\hat{\mathbf{n}}_2$, Eq. (4.92) simplifies to

$$\hat{\mathbf{n}}_2 \cdot (\mathbf{D}_1 - \mathbf{D}_2) = \rho_s \qquad (\text{C/m}^2). \qquad (4.93)$$

If D_{1n} and D_{2n} denote as the normal components of $\mathbf{D}_1$ and $\mathbf{D}_2$ along $\hat{\mathbf{n}}_2$, we have

$$D_{1n} - D_{2n} = \rho_s \qquad (\text{C/m}^2). \qquad (4.94)$$

▶ The normal component of **D** changes abruptly at a charged boundary between two different media in an amount equal to the surface charge density. ◀

The corresponding boundary condition for **E** is

$$\hat{\mathbf{n}}_2 \cdot (\epsilon_1 \mathbf{E}_1 - \epsilon_2 \mathbf{E}_2) = \rho_s, \qquad (4.95a)$$

or equivalently

$$\epsilon_1 E_{1n} - \epsilon_2 E_{2n} = \rho_s. \qquad (4.95b)$$

In summary, (1) the conservative property of **E**,

$$\nabla \times \mathbf{E} = 0 \quad \longleftrightarrow \quad \oint_C \mathbf{E} \cdot d\mathbf{l} = 0, \qquad (4.96)$$

led to the result that **E** has a continuous tangential component across a boundary, and (2) the divergence property of **D**,

$$\nabla \cdot \mathbf{D} = \rho_v \quad \longleftrightarrow \quad \oint_S \mathbf{D} \cdot d\mathbf{s} = Q, \qquad (4.97)$$

led to the result that the normal component of **D** changes by ρ_s across the boundary. A summary of the conditions that apply at the boundary between different types of media is given in **Table 4-3**.

Example 4-10: Application of Boundary Conditions

The x–y plane is a charge-free boundary separating two dielectric media with permittivities ϵ_1 and ϵ_2, as shown in **Fig. 4-19**. If the electric field in medium 1 is $\mathbf{E}_1 = \hat{\mathbf{x}}E_{1x} + \hat{\mathbf{y}}E_{1y} + \hat{\mathbf{z}}E_{1z}$, find (a) the electric field $\mathbf{E}_2$ in medium 2 and (b) the angles θ_1 and θ_2.

Solution: (a) Let $\mathbf{E}_2 = \hat{\mathbf{x}}E_{2x} + \hat{\mathbf{y}}E_{2y} + \hat{\mathbf{z}}E_{2z}$. Our task is to find the components of $\mathbf{E}_2$ in terms of the given components of $\mathbf{E}_1$. The normal to the boundary is $\hat{\mathbf{z}}$. Hence, the x and y components of the fields are tangential to the boundary and the z components are normal to the boundary. At a charge-

Table 4-3 **Boundary conditions for the electric fields.**

Field Component	Any Two Media	Medium 1 Dielectric ϵ_1	Medium 2 Conductor
Tangential E	$\mathbf{E}_{1t} = \mathbf{E}_{2t}$	$\mathbf{E}_{1t} = \mathbf{E}_{2t} = 0$	
Tangential D	$\mathbf{D}_{1t}/\epsilon_1 = \mathbf{D}_{2t}/\epsilon_2$	$\mathbf{D}_{1t} = \mathbf{D}_{2t} = 0$	
Normal E	$\epsilon_1 E_{1n} - \epsilon_2 E_{2n} = \rho_s$	$E_{1n} = \rho_s/\epsilon_1$	$E_{2n} = 0$
Normal D	$D_{1n} - D_{2n} = \rho_s$	$D_{1n} = \rho_s$	$D_{2n} = 0$
Notes: (1) ρ_s is the surface charge density at the boundary; (2) normal components of $\mathbf{E}_1$, $\mathbf{D}_1$, $\mathbf{E}_2$, and $\mathbf{D}_2$ are along $\hat{\mathbf{n}}_2$, the outward normal unit vector of medium 2.			

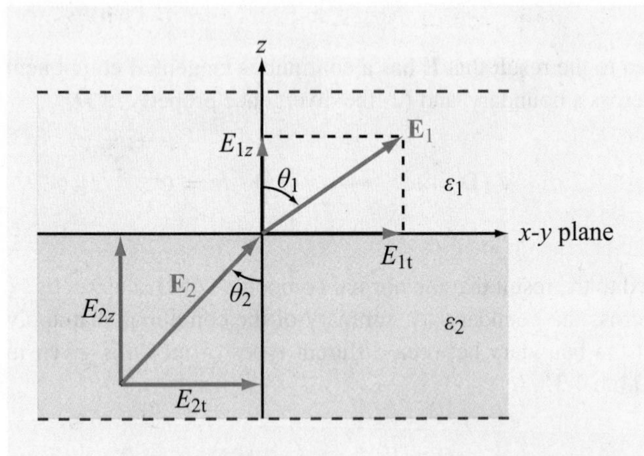

Figure 4-19 Application of boundary conditions at the interface between two dielectric media (Example 4-10).

free interface, the tangential components of **E** and the normal components of **D** are continuous. Consequently,

$$E_{2x} = E_{1x}, \qquad E_{2y} = E_{1y},$$

and

$$D_{2z} = D_{1z} \qquad \text{or} \qquad \epsilon_2 E_{2z} = \epsilon_1 E_{1z}.$$

Hence,

$$\mathbf{E}_2 = \hat{\mathbf{x}} E_{1x} + \hat{\mathbf{y}} E_{1y} + \hat{\mathbf{z}} \frac{\epsilon_1}{\epsilon_2} E_{1z}. \tag{4.98}$$

(b) The tangential components of $\mathbf{E}_1$ and $\mathbf{E}_2$ are $E_{1t} = \sqrt{E_{1x}^2 + E_{1y}^2}$ and $E_{2t} = \sqrt{E_{2x}^2 + E_{2y}^2}$. The angles θ_1 and θ_2 are then given by

$$\tan\theta_1 = \frac{E_{1t}}{E_{1z}} = \frac{\sqrt{E_{1x}^2 + E_{1y}^2}}{E_{1z}},$$

$$\tan\theta_2 = \frac{E_{2t}}{E_{2z}} = \frac{\sqrt{E_{2x}^2 + E_{2y}^2}}{E_{2z}} = \frac{\sqrt{E_{1x}^2 + E_{1y}^2}}{(\epsilon_1/\epsilon_2) E_{1z}},$$

and the two angles are related by

$$\frac{\tan\theta_2}{\tan\theta_1} = \frac{\epsilon_2}{\epsilon_1}. \tag{4.99}$$

Exercise 4-16: Find $\mathbf{E}_1$ in **Fig. 4-19** if

$$\mathbf{E}_2 = \hat{\mathbf{x}}2 - \hat{\mathbf{y}}3 + \hat{\mathbf{z}}3 \text{ (V/m)},$$

$$\epsilon_1 = 2\epsilon_0,$$

$$\epsilon_2 = 8\epsilon_0,$$

and the boundary is charge free.

Answer: $\mathbf{E}_1 = \hat{\mathbf{x}}2 - \hat{\mathbf{y}}3 + \hat{\mathbf{z}}12$ (V/m). (See EM.)

Exercise 4-17: Repeat Exercise 4.16 for a boundary with surface charge density $\rho_s = 3.54 \times 10^{-11}$ (C/m²).

Answer: $\mathbf{E}_1 = \hat{\mathbf{x}}2 - \hat{\mathbf{y}}3 + \hat{\mathbf{z}}14$ (V/m). (See EM.)

Module 4.2 Charges in Adjacent Dielectrics In two adjoining half-planes with selectable permittivities, the user can place point charges anywhere in space and select their magnitudes and polarities. The module then displays **E**, V, and the equipotential contours of V.

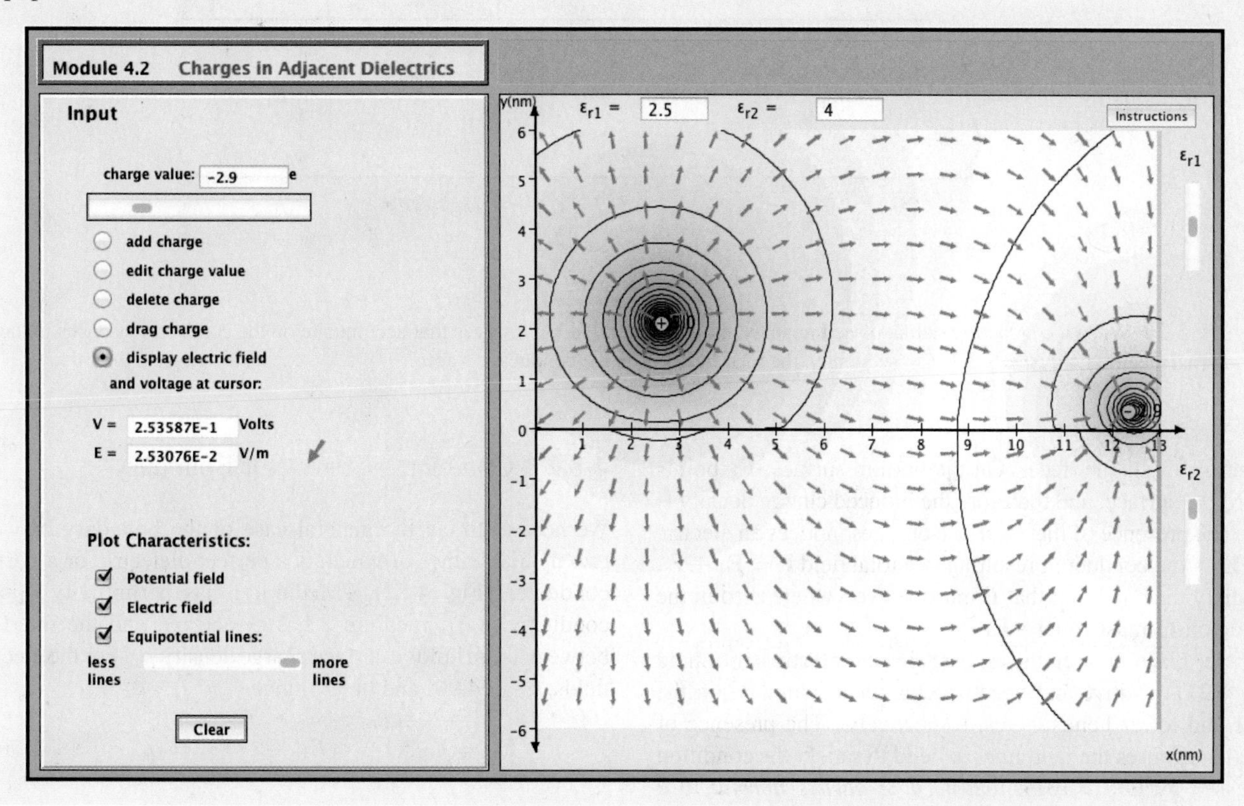

4-8.1 Dielectric-Conductor Boundary

Consider the case when medium 1 is a dielectric and medium 2 is a perfect conductor. Because in a perfect conductor, electric fields and fluxes vanish, it follows that $\mathbf{E}_2 = \mathbf{D}_2 = 0$, which implies that components of $\mathbf{E}_2$ and $\mathbf{D}_2$ tangential and normal to the interface are zero. Consequently, from Eq. (4.90) and Eq. (4.94), the fields in the dielectric medium, at the boundary with the conductor, satisfy

$$E_{1\text{t}} = D_{1\text{t}} = 0, \tag{4.100a}$$

$$D_{1\text{n}} = \epsilon_1 E_{1\text{n}} = \rho_\text{s}. \tag{4.100b}$$

These two boundary conditions can be combined into

$$\mathbf{D}_1 = \epsilon_1 \mathbf{E}_1 = \hat{\mathbf{n}} \rho_\text{s}, \tag{4.101}$$

(at conductor surface)

where $\hat{\mathbf{n}}$ is a unit vector directed normally outward from the conducting surface.

▶ The electric field lines point directly away from the conductor surface when ρ_s is positive and directly toward the conductor surface when ρ_s is negative. ◀

Figure 4-20 shows an infinitely long conducting slab placed in a uniform electric field $\mathbf{E}_1$. The media above and below the slab have permittivity ϵ_1. Because $\mathbf{E}_1$ points away from the upper surface, it induces a positive charge density $\rho_\text{s} = \epsilon_1 |\mathbf{E}_1|$

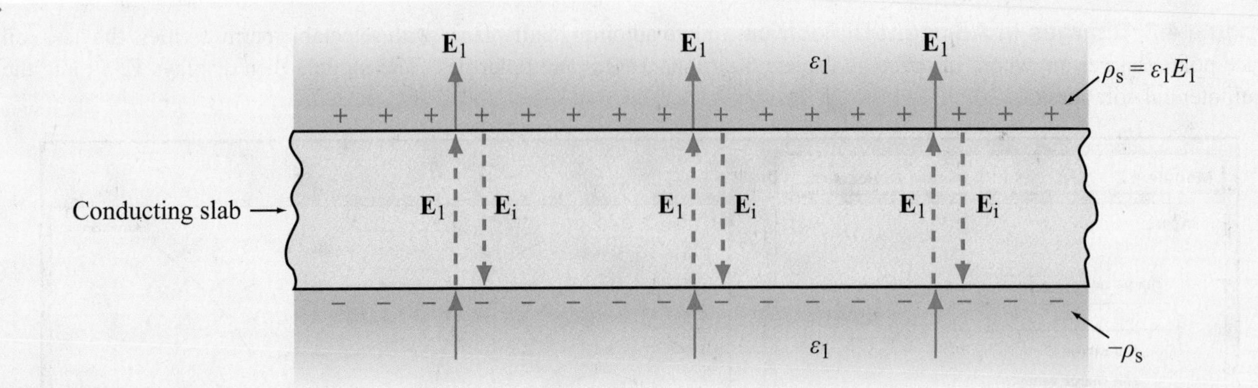

Figure 4-20 When a conducting slab is placed in an external electric field $\mathbf{E}_1$, charges that accumulate on the conductor surfaces induce an internal electric field $\mathbf{E}_i = -\mathbf{E}_1$. Consequently, the total field inside the conductor is zero.

on the upper slab surface. On the bottom surface, $\mathbf{E}_1$ points toward the surface, and therefore the induced charge density is $-\rho_s$. The presence of these surface charges induces an electric field $\mathbf{E}_i$ in the conductor, resulting in a total field $\mathbf{E} = \mathbf{E}_1 + \mathbf{E}_i$. To satisfy the condition that $\mathbf{E}$ must be everywhere zero in the conductor, $\mathbf{E}_i$ must equal $-\mathbf{E}_1$.

If we place a metallic sphere in an electrostatic field (**Fig. 4-21**), positive and negative charges accumulate on the upper and lower hemispheres, respectively. The presence of the sphere causes the field lines to bend to satisfy the condition expressed by Eq. (4.101); that is, **E** ***is always normal to a conductor boundary.***

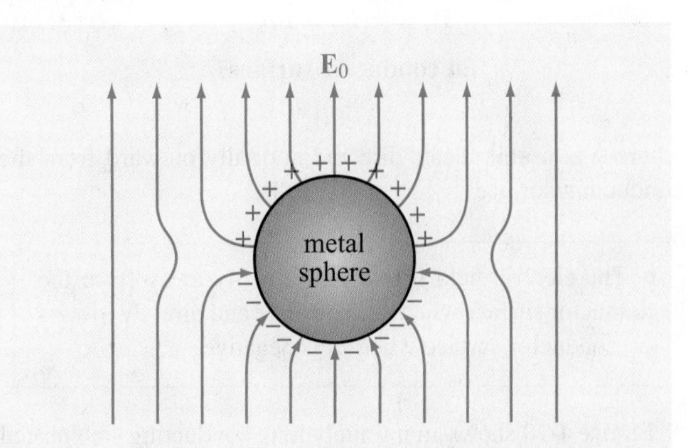

Figure 4-21 Metal sphere placed in an external electric field $\mathbf{E}_0$.

4-8.2 Conductor-Conductor Boundary

We now examine the general case of the boundary between two media neither of which is a perfect dielectric or a perfect conductor (**Fig. 4-22**). Medium 1 has permittivity ϵ_1 and conductivity σ_1, medium 2 has ϵ_2 and σ_2, and the interface between them holds a surface charge density ρ_s. For the electric fields, Eqs. (4.90) and (4.95b) give

$$\mathbf{E}_{1t} = \mathbf{E}_{2t}, \qquad \epsilon_1 E_{1n} - \epsilon_2 E_{2n} = \rho_s. \tag{4.102}$$

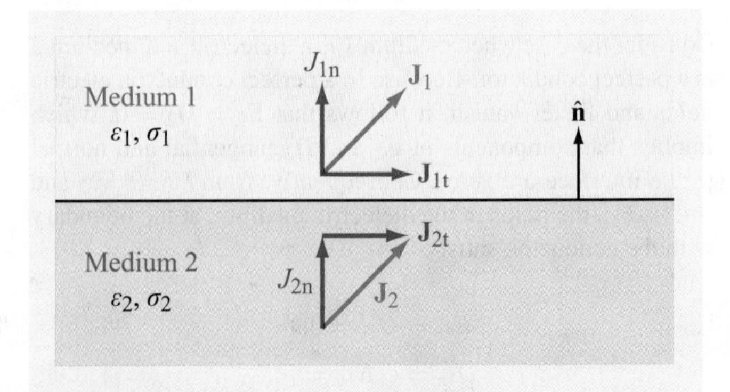

Figure 4-22 Boundary between two conducting media.

Module 4.3 Charges above Conducting Plane When electric charges are placed in a dielectric medium adjoining a conducting plane, some of the conductor's electric charges move to its surface boundary, thereby satisfying the boundary conditions outlined in Table 4-3. This module displays **E** and V everywhere and ρ_s along the dielectric-conductor boundary.

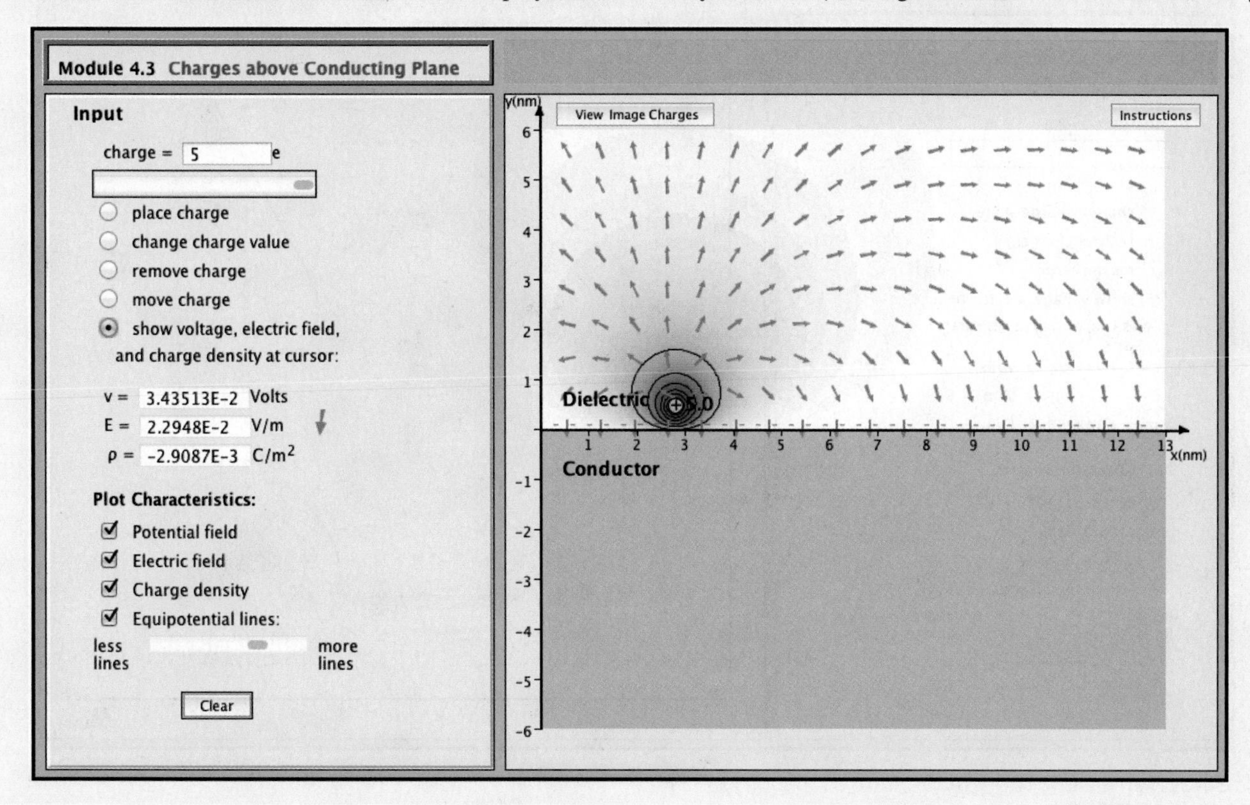

Since we are dealing with conducting media, the electric fields give rise to current densities $\mathbf{J}_1 = \sigma_1 \mathbf{E}_1$ and $\mathbf{J}_2 = \sigma_2 \mathbf{E}_2$. Hence

$$\frac{\mathbf{J}_{1t}}{\sigma_1} = \frac{\mathbf{J}_{2t}}{\sigma_2}, \qquad \epsilon_1 \frac{J_{1n}}{\sigma_1} - \epsilon_2 \frac{J_{2n}}{\sigma_2} = \rho_s. \tag{4.103}$$

The tangential current components $\mathbf{J}_{1t}$ and $\mathbf{J}_{2t}$ represent currents flowing in the two media in a direction parallel to the boundary, and hence there is no transfer of charge between them. This is not the case for the normal components. If $J_{1n} \neq J_{2n}$, then a different amount of charge arrives at the boundary than leaves it. Hence, ρ_s cannot remain constant in time, which violates the condition of electrostatics requiring all fields and charges to remain constant. Consequently, *the normal component of* **J** *has to be continuous across the boundary between two different media under electrostatic conditions.* Upon setting $J_{1n} = J_{2n}$ in Eq. (4.103), we have

$$J_{1n}\left(\frac{\epsilon_1}{\sigma_1} - \frac{\epsilon_2}{\sigma_2}\right) = \rho_s \quad \textbf{(electrostatics)}. \tag{4.104}$$

Concept Question 4-22: What are the boundary conditions for the electric field at a conductor–dielectric boundary?

Concept Question 4-23: Under electrostatic conditions, we require $J_{1n} = J_{2n}$ at the boundary between two conductors. Why?

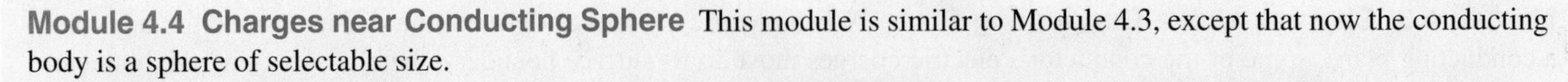

Module 4.4 Charges near Conducting Sphere This module is similar to Module 4.3, except that now the conducting body is a sphere of selectable size.

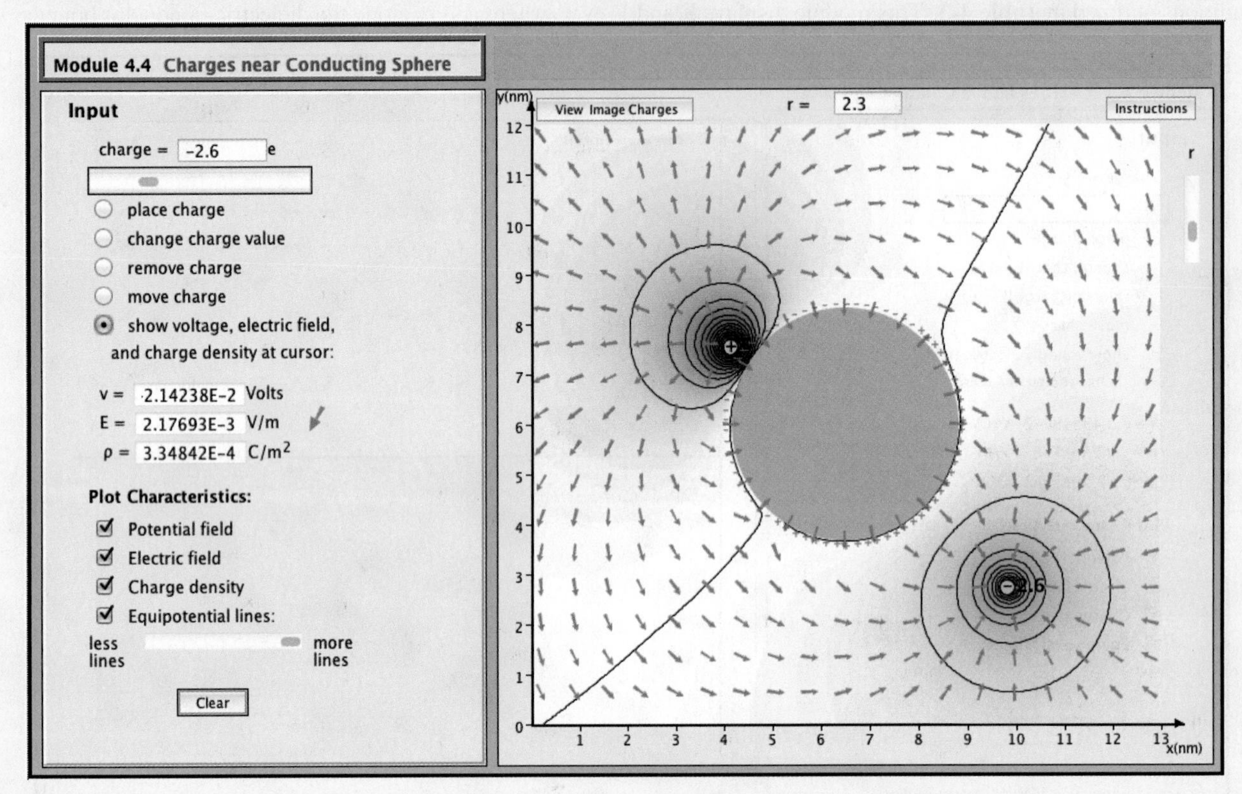

4-9 Capacitance

When separated by an insulating (dielectric) medium, any two conducting bodies, regardless of their shapes and sizes, form a ***capacitor***. If a dc voltage source is connected across them (**Fig. 4-23**) the surfaces of the conductors connected to the positive and negative source terminals accumulate charges $+Q$ and $-Q$, respectively.

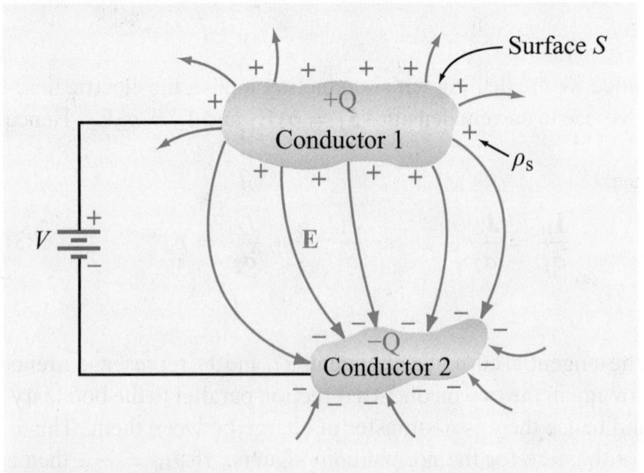

Figure 4-23 A dc voltage source connected to a capacitor composed of two conducting bodies.

▶ When a conductor has excess charge, it distributes the charge on its surface in such a manner as to maintain a zero electric field everywhere within the conductor, thereby ensuring that the electric potential is the same at every point in the conductor. ◀

The ***capacitance*** of a two-conductor configuration is defined as

$$C = \frac{Q}{V} \qquad \text{(C/V or F)}, \tag{4.105}$$

where V is the potential (voltage) difference between the conductors. Capacitance is measured in farads (F), which is equivalent to coulombs per volt (C/V).

The presence of free charges on the conductors' surfaces gives rise to an electric field $\mathbf{E}$ (**Fig. 4-23**) with field lines originating on the positive charges and terminating on the negative ones. Since the tangential component of $\mathbf{E}$ always vanishes at a conductor's surface, $\mathbf{E}$ is always perpendicular to the conducting surfaces. The normal component of $\mathbf{E}$ at any point on the surface of either conductor is given by

$$E_n = \hat{\mathbf{n}} \cdot \mathbf{E} = \frac{\rho_s}{\epsilon}, \qquad (4.106)$$

(at conductor surface)

where ρ_s is the surface charge density at that point, $\hat{\mathbf{n}}$ is the outward normal unit vector at the same location, and ϵ is the permittivity of the dielectric medium separating the conductors. The charge Q is equal to the integral of ρ_s over surface S (**Fig. 4-23**):

$$Q = \int_S \rho_s \, ds = \int_S \epsilon \hat{\mathbf{n}} \cdot \mathbf{E} \, ds = \int_S \epsilon \mathbf{E} \cdot d\mathbf{s}, \qquad (4.107)$$

where use was made of Eq. (4.106). The voltage V is related to $\mathbf{E}$ by Eq. (4.39):

$$V = V_{12} = -\int_{P_2}^{P_1} \mathbf{E} \cdot d\mathbf{l}, \qquad (4.108)$$

where points P_1 and P_2 are any two arbitrary points on conductors 1 and 2, respectively. Substituting Eqs. (4.107) and (4.108) into Eq. (4.105) gives

$$C = \frac{\displaystyle\int_S \epsilon \mathbf{E} \cdot d\mathbf{s}}{\displaystyle -\int_l \mathbf{E} \cdot d\mathbf{l}} \quad \text{(F)}, \qquad (4.109)$$

where l is the integration path from conductor 2 to conductor 1. To avoid making sign errors when applying Eq. (4.109), it is important to remember that surface S is the $+Q$ surface and P_1 is on S. (Alternatively, if you compute C and it comes out negative, just change its sign.) Because $\mathbf{E}$ appears in both the numerator and denominator of Eq. (4.109), *the value of C obtained for any specific capacitor configuration is always independent of $\mathbf{E}$'s magnitude.* In fact, C depends only on the capacitor geometry (sizes, shapes, and relative positions of the two conductors) and the permittivity of the insulating material.

If the material between the conductors is not a perfect dielectric (i.e., if it has a small conductivity σ), then current can flow through the material between the conductors, and the material exhibits a resistance R. The general expression for R for a resistor of arbitrary shape is given by Eq. (4.71):

$$R = \frac{\displaystyle -\int_l \mathbf{E} \cdot d\mathbf{l}}{\displaystyle \int_S \sigma \mathbf{E} \cdot d\mathbf{s}} \quad (\Omega). \qquad (4.110)$$

For a medium with uniform σ and ϵ, the product of Eqs. (4.109) and (4.110) gives

$$RC = \frac{\epsilon}{\sigma}. \qquad (4.111)$$

This simple relation allows us to find R if C is known, and vice versa.

Example 4-11: Capacitance and Breakdown Voltage of Parallel-Plate Capacitor

Obtain an expression for the capacitance C of a parallel-plate capacitor consisting of two parallel plates each of surface area A and separated by a distance d. The capacitor is filled with a dielectric material with permittivity ϵ. Also, determine the breakdown voltage if $d = 1$ cm and the dielectric material is quartz.

Solution: In **Fig. 4-24**, we place the lower plate of the capacitor in the x–y plane and the upper plate in the plane $z = d$. Because of the applied voltage difference V, charges $+Q$ and $-Q$ accumulate on the top and bottom capacitor plates. If the plate dimensions are much larger than the separation d, then these charges distribute themselves quasi-uniformly across the plates, giving rise to a quasi-uniform field between them pointing in the $-\hat{\mathbf{z}}$ direction. In addition, a ***fringing field*** will

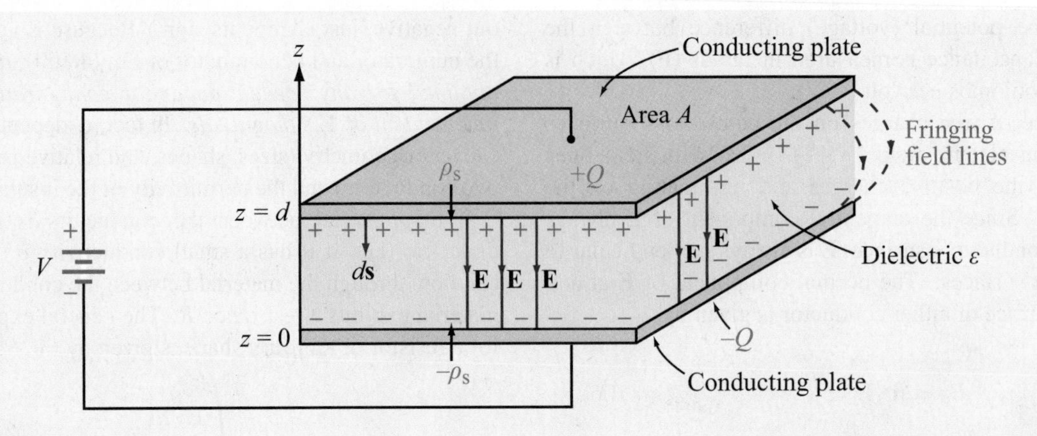

Figure 4-24 A dc voltage source connected to a parallel-plate capacitor (Example 4-11).

exist near the capacitor edges, but its effects may be ignored because the bulk of the electric field exists between the plates. The charge density on the upper plate is $\rho_s = Q/A$. Hence, in the dielectric medium

$$\mathbf{E} = -\hat{\mathbf{z}}E,$$

and from Eq. (4.106), the magnitude of $\mathbf{E}$ at the conductor–dielectric boundary is $E = \rho_s/\epsilon = Q/\epsilon A$. From Eq. (4.108), the voltage difference is

$$V = -\int_0^d \mathbf{E} \cdot d\mathbf{l} = -\int_0^d (-\hat{\mathbf{z}}E) \cdot \hat{\mathbf{z}}\, dz = Ed, \tag{4.112}$$

and the capacitance is

$$C = \frac{Q}{V} = \frac{Q}{Ed} = \frac{\epsilon A}{d}, \tag{4.113}$$

where use was made of the relation $E = Q/\epsilon A$.

From $V = Ed$, as given by Eq. (4.112), $V = V_{br}$ when $E = E_{ds}$, the dielectric strength of the material. According to **Table 4-2**, $E_{ds} = 30$ (MV/m) for quartz. Hence, the breakdown voltage is

$$V_{br} = E_{ds}d = 30 \times 10^6 \times 10^{-2} = 3 \times 10^5 \text{ V}.$$

Example 4-12: Capacitance per Unit Length of Coaxial Line

Obtain an expression for the capacitance of the coaxial line shown in **Fig. 4-25**.

Solution: For a given voltage V across the capacitor, charges $+Q$ and $-Q$ accumulate on the surfaces of the outer and inner conductors, respectively. We assume that these charges are uniformly distributed along the length and circumference of the conductors with surface charge density $\rho'_s = Q/2\pi bl$ on the outer conductor and $\rho''_s = -Q/2\pi al$ on the inner one. Ignoring fringing fields near the ends of the coaxial line, we can construct a cylindrical Gaussian surface in the dielectric in between the conductors, with radius r such that $a < r < b$. Symmetry implies that the $\mathbf{E}$-field is identical at all points on this surface, directed radially inward. From Gauss's law, it follows that the field magnitude equals the absolute value of the total charge enclosed, divided by the surface area. That is,

$$\mathbf{E} = -\hat{\mathbf{r}}\, \frac{Q}{2\pi\epsilon rl}. \tag{4.114}$$

The potential difference V between the outer and inner conductors is

$$V = -\int_a^b \mathbf{E} \cdot d\mathbf{l} = -\int_a^b \left(-\hat{\mathbf{r}}\, \frac{Q}{2\pi\epsilon rl}\right) \cdot (\hat{\mathbf{r}}\, dr)$$

$$= \frac{Q}{2\pi\epsilon l} \ln\left(\frac{b}{a}\right). \tag{4.115}$$

The capacitance C is then given by

$$C = \frac{Q}{V} = \frac{2\pi\epsilon l}{\ln(b/a)}, \tag{4.116}$$

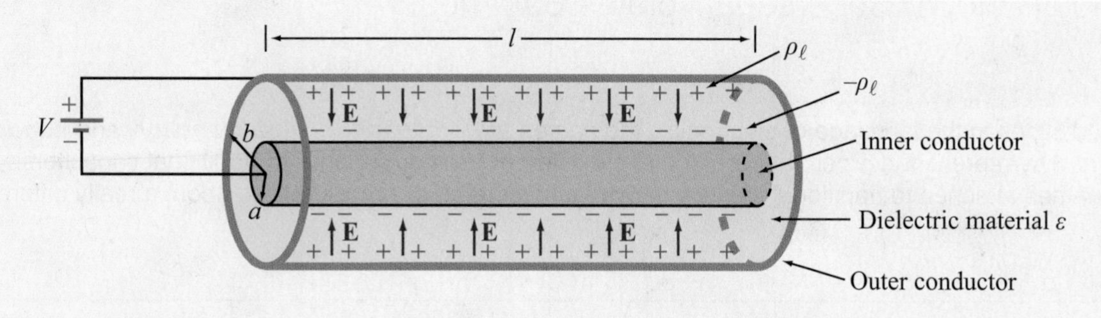

Figure 4-25 Coaxial capacitor filled with insulating material of permittivity ϵ (Example 4-12).

and the capacitance per unit length of the coaxial line is

$$C' = \frac{C}{l} = \frac{2\pi\epsilon}{\ln(b/a)} \quad \text{(F/m)}. \tag{4.117}$$

Concept Question 4-24: How is the capacitance of a two-conductor structure related to the resistance of the insulating material between the conductors?

Concept Question 4-25: What are fringing fields and when may they be ignored?

4-10 Electrostatic Potential Energy

A source connected to a capacitor expends energy in charging up the capacitor. If the capacitor plates are made of a good conductor with effectively zero resistance, and if the dielectric separating the two plates has negligible conductivity, then no real current can flow through the dielectric, and no ohmic losses occur anywhere in the capacitor. Where then does the energy expended in charging up the capacitor go? The energy ends up getting stored in the dielectric medium in the form of ***electrostatic potential energy***. The amount of stored energy W_e is related to Q, C, and V.

Suppose we were to charge up a capacitor by ramping up the voltage across it from $\upsilon = 0$ to $\upsilon = V$. During the process, charge $+q$ accumulates on one conductor, and $-q$ on the other. In effect, a charge q has been transferred from one of the conductors to the other. The voltage υ across the capacitor is related to q by

$$\upsilon = \frac{q}{C}. \tag{4.118}$$

From the definition of υ, the amount of work dW_e required to transfer an additional incremental charge dq from one conductor to the other is

$$dW_e = \upsilon\, dq = \frac{q}{C}\, dq. \tag{4.119}$$

If we transfer a total charge Q between the conductors of an initially uncharged capacitor, then the total amount of work performed is

$$W_e = \int_0^Q \frac{q}{C}\, dq = \frac{1}{2}\frac{Q^2}{C} \quad \text{(J)}. \tag{4.120}$$

Using $C = Q/V$, where V is the final voltage, W_e also can be expressed as

$$W_e = \tfrac{1}{2} C V^2 \quad \text{(J)}. \tag{4.121}$$

The capacitance of the parallel-plate capacitor discussed in Example 4-11 is given by Eq. (4.113) as $C = \epsilon A/d$, where A is the surface area of each of its plates and d is the separation between them. Also, the voltage V across the capacitor is

Technology Brief 8: Supercapacitors as Batteries

As recent additions to the language of electronics, the names ***supercapacitor***, ***ultracapacitor***, and ***nanocapacitor*** suggest that they represent devices that are somehow different from or superior to traditional capacitors. Are these just fancy names attached to traditional capacitors by manufacturers, or are we talking about a really different type of capacitor?

> ▶ The three aforementioned names refer to variations on an energy storage device known by the technical name ***electrochemical double-layer capacitor*** (EDLC), in which *energy storage is realized by a hybrid process that incorporates features from both the traditional electrostatic capacitor and the electrochemical voltaic battery.* ◀

For the purposes of this Technology Brief, we refer to this relatively new device as a supercapacitor: The battery is far superior to the traditional capacitor with regard to energy storage, but a capacitor can be charged and discharged much more rapidly than a battery. As a hybrid technology, the supercapacitor offers features that are intermediate between those of the battery and the traditional capacitor. The supercapacitor is now used to support a wide range of applications, from motor startups in large engines (trucks, locomotives, submarines, etc.) to flash lights in digital cameras, and its use is rapidly extending into consumer electronics (cell phones, MP3 players, laptop computers) and electric cars (**Fig. TF8-2**).

Figure TF8-1 Examples of electromechanical double-layer capacitors (EDLC), otherwise known as a supercapacitor.

Figure TF8-2 Examples of systems that use supercapacitors.

Capacitor Energy Storage Limitations

Energy density W' is often measured in watts-hours per kg (Wh/kg), with 1 Wh $= 3.6 \times 10^3$ joules. Thus, the energy capacity of a device is normalized to its mass. For batteries, W' extends between about 30 Wh/kg for a lead-acid battery to as high as 150 Wh/kg for a lithium-ion battery. In contrast, W' rarely exceeds 0.02 Wh/kg for a traditional capacitor. Let us examine what limits the value of W' for the capacitor by considering a small parallel-plate capacitor with plate area A and separation between plates d. For simplicity, we assign the capacitor a voltage rating of 1 V (maximum anticipated voltage across the capacitor). Our goal is to maximize the energy density W'. For a parallel-plate capacitor $C = \epsilon A/d$, where ϵ is the permittivity of the insulating material. Using Eq. (4.121) leads to

$$W' = \frac{W}{m} = \frac{1}{2m}\,CV^2 = \frac{\epsilon AV^2}{2md} \qquad \text{(J/kg)},$$

where m is the mass of the conducting plates and the insulating material contained in the capacitor. To keep the analysis simple, we assume that the plates can be made so thin as to ignore their masses relative to the mass of the insulating material. If the material's density is ρ (kg/m^3), then $m = \rho Ad$ and

$$W' = \frac{\epsilon V^2}{2\rho d^2} \qquad \text{(J/kg)}.$$

To maximize W', we need to select d to be the smallest possible, but we also have to be aware of the constraint associated with dielectric breakdown. To avoid sparking between the capacitor's two plates, the electric field strength should not exceed E_{ds}, the dielectric strength of the insulating material. Among the various types of materials commonly used in capacitors, mica has one of the highest values of E_{ds}, nearly 2×10^8 V/m. Breakdown voltage V_{br} is related to E_{ds} by $V_{\text{br}} = E_{\text{ds}}d$, so given that the capacitor is to have a voltage rating of 1 V, let us choose V_{br} to be 2 V, thereby allowing a 50% safety margin. With $V_{\text{br}} = 2$ V and $E_{\text{ds}} = 2 \times 10^8$ V/m, it follows that d should not be smaller than 10^{-8} m, or 10 nm. For mica, $\epsilon \simeq 6\epsilon_0$ and $\rho = 3 \times 10^3$ kg/m^3. Ignoring for the moment the practical issues associated with building a capacitor with a spacing of only 10 nm between conductors, the expression for energy density leads to $W' \simeq 90$ J/kg. Converting W' to Wh/kg (by dividing by 3.6×10^3 J/Wh) gives

$$W'(\text{max}) = 2.5 \times 10^{-2} \qquad \text{(Wh/kg)},$$

for a traditional capacitor at a voltage rating of 1 V.

► The energy storage capacity of a traditional capacitor is about four orders of magnitude smaller than the energy density capability of a lithium-ion battery. ◄

Energy Storage Comparison

The table in the upper part of **Fig. TF8-3** displays typical values or ranges of values for each of five attributes commonly used to characterize the performance of energy storage devices. In addition to the energy density W', they include the power density P', the charge and discharge rates, and the number of charge/discharge cycles that the device can withstand before deteriorating in performance. For most energy storage devices, the discharge rate usually is shorter than the charge rate, but for the purpose of the present discussion we treat them as equal. As a first-order approximation, the ***discharge rate*** is related to P' and W' by

$$T = \frac{W'}{P'}\,.$$

Energy Storage Devices

Feature	Traditional Capacitor	Supercapacitor	Battery
Energy density W' (Wh/kg)	$\sim 10^{-2}$	1 to 10	5 to 150
Power density P' (W/kg)	1,000 to 10,000	1,000 to 5,000	10 to 500
Charge and discharge rate T	10^{-3} sec	$\sim$ 1 sec to 1 min	$\sim$ 1 to 5 hrs
Cycle life N_c	∞	$\sim 10^6$	$\sim 10^3$

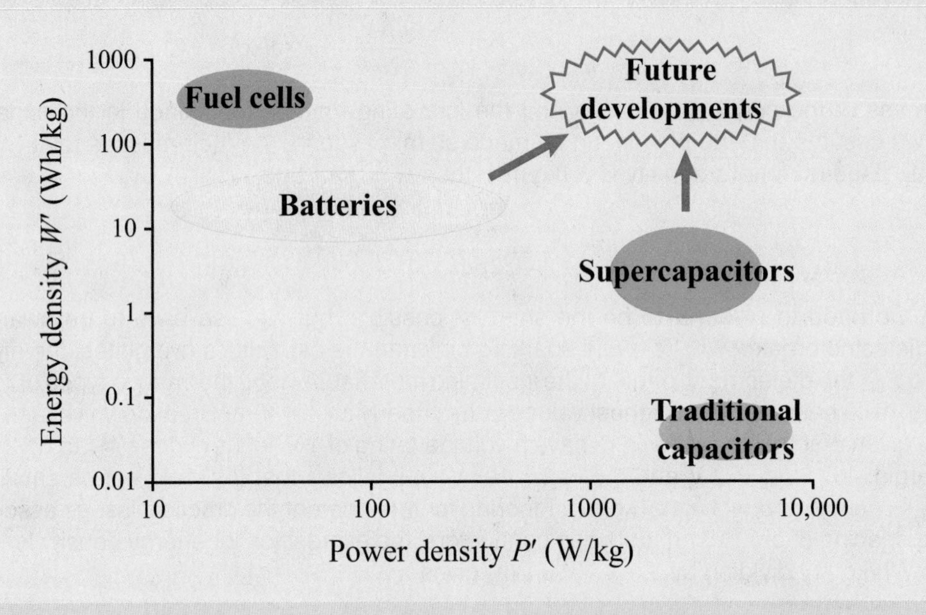

Figure TF8-3 Comparison of energy storage devices.

► Supercapacitors are capable of storing 100 to 1000 times more energy than a traditional capacitor, but 10 times less than a battery (**Fig. TF8-3**). On the other hand, *supercapacitors can discharge their stored energy in a matter of seconds, compared with hours for a battery.* ◄

Moreover, the supercapacitor's cycle life is on the order of 1 million, compared with only 1000 for a rechargeable battery. Because of these features, the supercapacitor has greatly expanded the scope and use of capacitors in electronic circuits and systems.

Future Developments

The upper right-hand corner of the plot in **Fig. TF8-3** represents the ideal energy storage device with $W' \simeq 100$–1000 Wh/kg and $P' \simeq 10^3$–10^4 W/kg. The corresponding discharge rate is $T \simeq 10$–100 ms. Current research aims to extend the capabilities of batteries and supercapacitors in the direction of this prized domain of the ***energy-power space***.

related to the magnitude of the electric field, E in the dielectric by $V = Ed$. Using these two expressions in Eq. (4.121) gives

$$W_e = \frac{1}{2}\,\frac{\epsilon A}{d}(Ed)^2 = \frac{1}{2}\,\epsilon E^2(Ad) = \frac{1}{2}\,\epsilon E^2 \mathcal{V}, \qquad (4.122)$$

where $\mathcal{V} = Ad$ is the volume of the capacitor. This expression affirms the assertion made at the beginning of this section, namely that the energy expended in charging up the capacitor is being stored in the electric field present in the dielectric material in between the two conductors.

The ***electrostatic energy density*** w_e is defined as the electrostatic potential energy W_e per unit volume:

$$w_e = \frac{W_e}{\mathcal{V}} = \frac{1}{2}\,\epsilon E^2 \quad \text{(J/m}^3\text{)}. \qquad (4.123)$$

Even though this expression was derived for a parallel-plate capacitor, it is equally valid for any dielectric medium containing an electric field $\mathbf{E}$, including vacuum. Furthermore, for any volume $\mathcal{V}$, the total electrostatic potential energy stored in it is

$$W_e = \frac{1}{2}\int_{\mathcal{V}} \epsilon E^2 \, d\mathcal{V} \quad \text{(J)}. \qquad (4.124)$$

Returning to the parallel-plate capacitor, the oppositely charged plates are attracted to each other by an electrical force $\mathbf{F}$. The force acting on any system of charges may be obtained from energy considerations. In the discussion that follows, we show how $\mathbf{F}$ can be determined from W_e, the electrostatic energy stored in the system by virtue of the presence of electric charges.

If two conductors comprising a capacitor are allowed to move closer to each other under the influence of the electrical force $\mathbf{F}$ by a differential distance $d\mathbf{l}$, while maintaining the charges on the plates constant, then the mechanical work done by the charged capacitor is

$$dW = \mathbf{F} \cdot d\mathbf{l}. \qquad (4.125)$$

The mechanical work is provided by *expending* electrostatic energy. Hence, dW equals the *loss* of energy stored in the dielectric insulating material of the capacitor, or

$$dW = -dW_e. \qquad (4.126)$$

From Eq. (3.73), dW_e may be written in terms of the gradient of W_e as

$$dW_e = \nabla W_e \cdot d\mathbf{l}. \qquad (4.127)$$

In view of Eq. (4.126), comparison of Eqs. (4.125) and (4.127) leads to

$$\mathbf{F} = -\nabla W_e \quad \text{(N)}. \qquad (4.128)$$

To apply Eq. (4.128) to the parallel-plate capacitor, we rewrite Eq. (4.120) in the form

$$W_e = \frac{1}{2}\,\frac{Q^2}{C} = \frac{Q^2 z}{2\epsilon A}, \qquad (4.129)$$

where we replaced d with the variable z, representing the vertical spacing between the conducting plates. Use of Eq. (4.129) in Eq. (4.128) gives

$$\mathbf{F} = -\nabla W_e = -\hat{\mathbf{z}}\,\frac{\partial}{\partial z}\left(\frac{Q^2 z}{2\epsilon A}\right) = -\hat{\mathbf{z}}\left(\frac{Q^2}{2\epsilon A}\right), \qquad (4.130)$$

and since $Q = \epsilon AE$, $\mathbf{F}$ can also be expressed as

$$\mathbf{F} = -\hat{\mathbf{z}}\,\frac{\epsilon A E^2}{2}. \qquad (4.131)$$

(parallel-plate capacitor)

Concept Question 4-26: To bring a charge q from infinity to a given point in space, a certain amount of work W is expended. Where does the energy corresponding to W go?

Concept Question 4-27: When a voltage source is connected across a capacitor, what is the direction of the electrical force acting on its two conducting surfaces?

Exercise 4-18: The radii of the inner and outer conductors of a coaxial cable are 2 cm and 5 cm, respectively, and the insulating material between them has a relative permittivity of 4. The charge density on the outer conductor is $\rho_\ell = 10^{-4}$ (C/m). Use the expression for $\mathbf{E}$ derived in Example 4-12 to calculate the total energy stored in a 20 cm length of the cable.

Answer: $W_e = 4.1$ J. (See EM.)

Technology Brief 9: Capacitive Sensors

To sense is to respond to a stimulus. (See Tech Brief 7 on resistive sensors.) A capacitor can function as a sensor if the stimulus changes the capacitor's ***geometry***—usually the spacing between its conductive elements—or the effective ***dielectric properties*** of the insulating material situated between them. Capacitive sensors are used in a multitude of applications. A few examples follow.

Fluid Gauge

The two metal electrodes in [**Fig. TF9-1(a)**], usually rods or plates, form a capacitor whose capacitance is directly proportional to the ***permittivity*** of the material between them. If the fluid section is of height $h_{\rm f}$ and the height of the empty space above it is $(h - h_{\rm f})$, then the overall capacitance is equivalent to two capacitors in parallel, or

$$C = C_{\rm f} + C_{\rm a} = \epsilon_{\rm f} w \, \frac{h_{\rm f}}{d} + \epsilon_{\rm a} w \, \frac{(h - h_{\rm f})}{d} \, ,$$

where w is the electrode plate width, d is the spacing between electrodes, and $\epsilon_{\rm f}$ and $\epsilon_{\rm a}$ are the permittivities of the fluid and air, respectively. Rearranging the expression as a linear equation yields

$$C = kh_{\rm f} + C_0,$$

where the constant coefficient is $k = (\epsilon_{\rm f} - \epsilon_{\rm a})w/d$ and $C_0 = \epsilon_{\rm a} wh/d$ is the capacitance of the tank when totally empty. Using the linear equation, the fluid height can be determined by measuring C with a bridge circuit [**Fig. TF9-1(b)**].

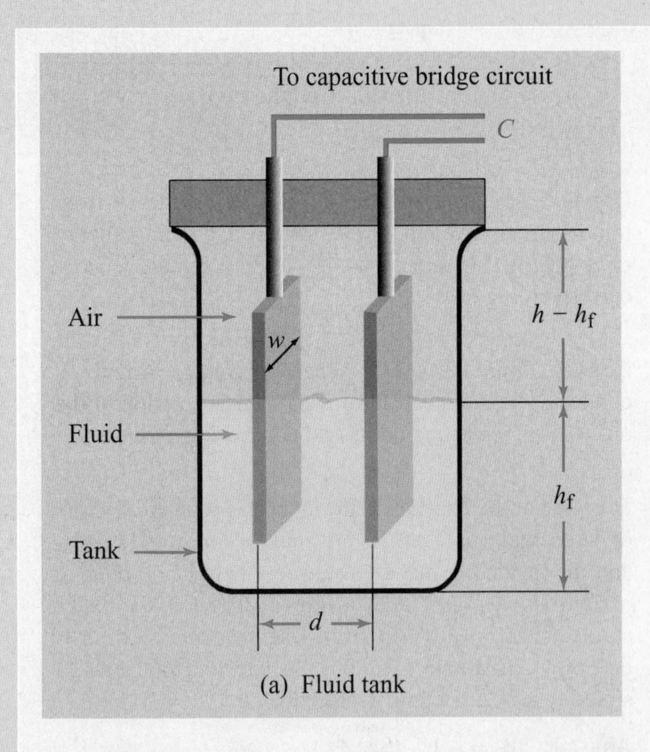

(a) Fluid tank

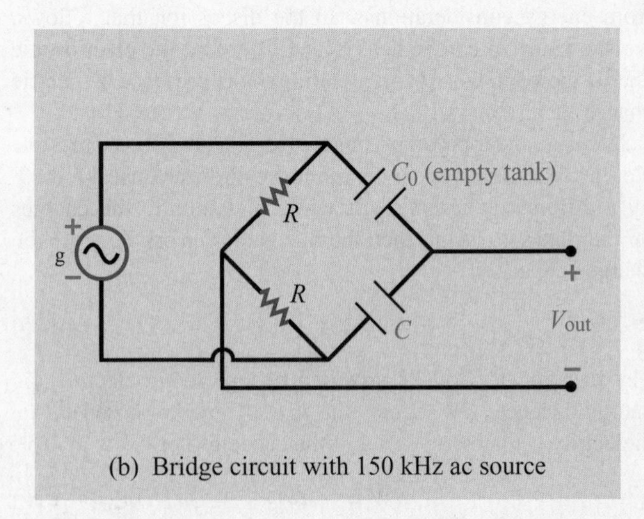

(b) Bridge circuit with 150 kHz ac source

Figure TF9-1 Fluid gauge and associated bridge circuit, with C_0 being the capacitance that an empty tank would have and C the capacitance of the tank under test.

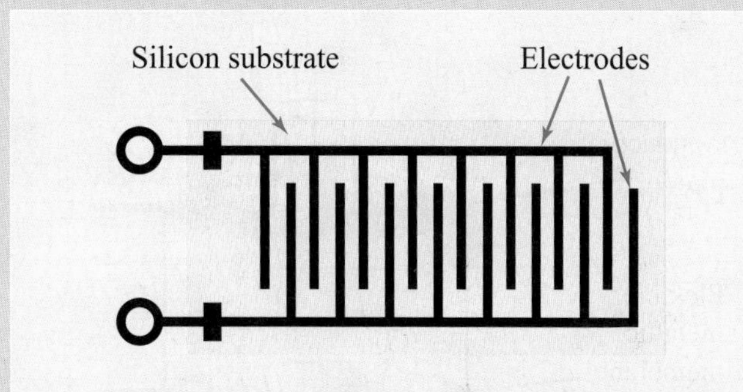

Figure TF9-2 Interdigital capacitor used as a humidity sensor.

► The output voltage V_{out} assumes a functional form that depends on the source voltage υ_{g}, the capacitance C_0 of the empty tank, and the unknown fluid height h_{f}. ◄

Humidity Sensor

Thin-film metal electrodes shaped in an ***interdigitized pattern*** (to enhance the ratio A/d) are fabricated on a silicon substrate (**Fig. TF9-2**). The spacing between digits is typically on the order of $0.2\ \mu\text{m}$. *The effective permittivity of the material separating the electrodes varies with the relative humidity of the surrounding environment.* Hence, the capacitor becomes a humidity sensor.

Pressure Sensor

A flexible metal ***diaphragm*** separates an oil-filled chamber with reference pressure P_0 from a second chamber exposed to the gas or fluid whose pressure P is to be measured by the sensor [**Fig. TF9-3(a)**]. The membrane is sandwiched, but electrically isolated, between two conductive parallel surfaces, forming two capacitors in series (**Fig. TF9-3(b)**). When $P > P_0$, the membrane bends in the direction of the lower plate. Consequently, d_1 increases and d_2 decreases, and in turn, C_1 decreases and C_2 increases [**Fig. TF9-3(c)**]. The converse happens when $P < P_0$. With the use of a capacitance bridge circuit, such as the one in **Fig. TF9-1(b)**, the sensor can be calibrated to measure the pressure P with good precision.

Noncontact Sensors

Precision positioning is a critical ingredient in semiconductor device fabrication, as well as in the operation and control of many mechanical systems. ***Noncontact capacitive sensors*** are used to sense the position of silicon wafers during the deposition, etching, and cutting processes, without coming in direct contact with the wafers.

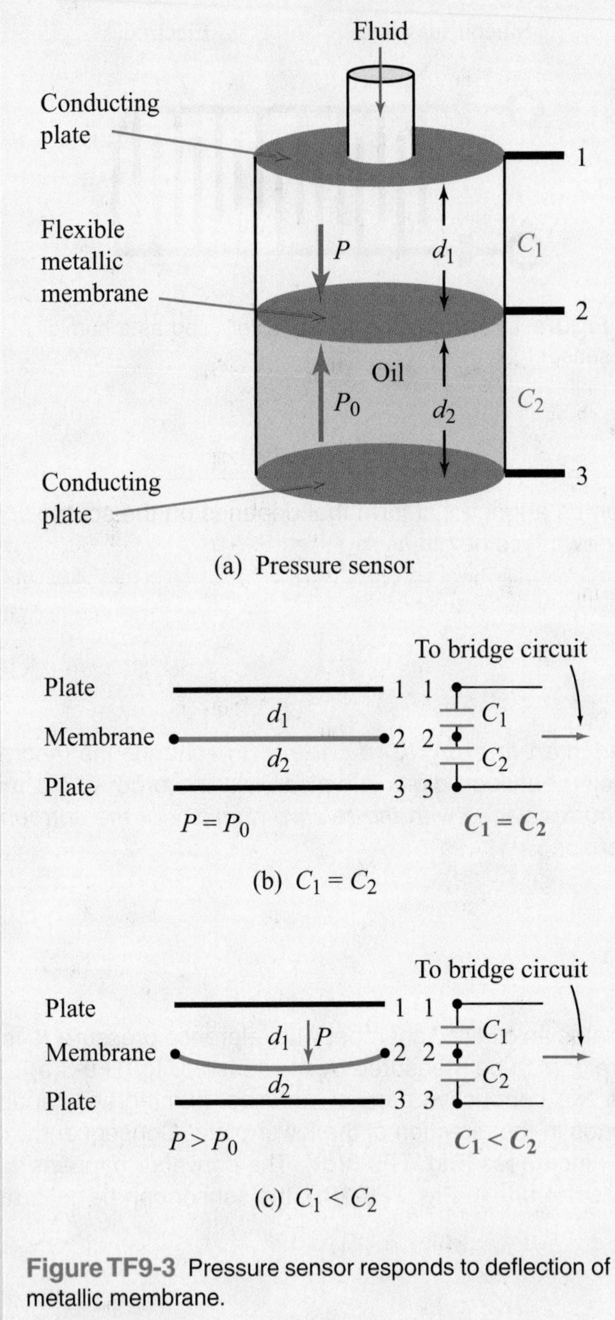

Figure TF9-3 Pressure sensor responds to deflection of metallic membrane.

► They are also used to sense and control robot arms in equipment manufacturing and to position hard disc drives, photocopier rollers, printing presses, and other similar systems. ◄

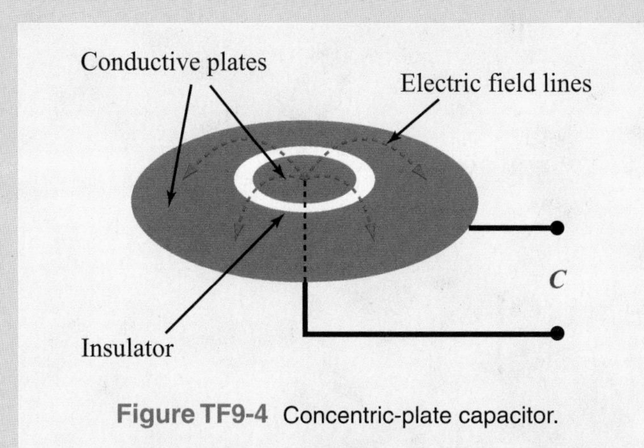

Figure TF9-4 Concentric-plate capacitor.

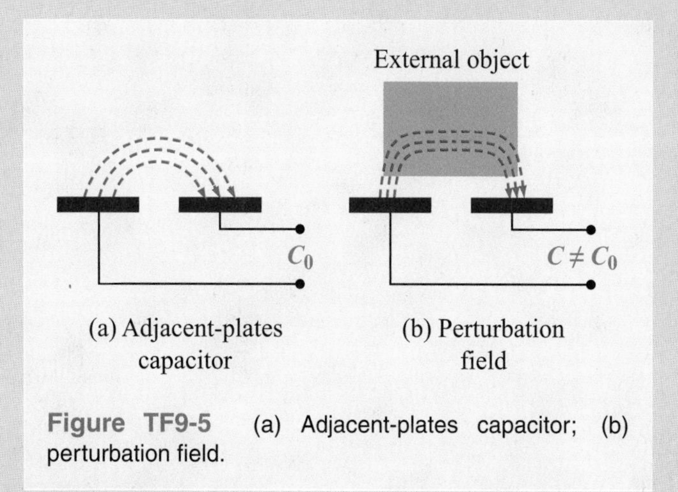

Figure TF9-5 (a) Adjacent-plates capacitor; (b) perturbation field.

The concentric plate capacitor in **Fig. TF9-4** consists of two metal plates, sharing the same plane, but electrically isolated from each other by an insulating material. When connected to a voltage source, charges of opposite polarity form on the two plates, resulting in the creation of electric-field lines between them. The same principle applies to the adjacent-plates capacitor in **Fig. TF9-5**. In both cases, the capacitance is determined by the shapes and sizes of the conductive elements and by the effective permittivity of the dielectric medium containing the electric field lines between them. Often, the capacitor surface is covered by a thin film of nonconductive material, the purpose of which is to keep the plate surfaces clean and dust free.

► The introduction of an external object into the proximity of the capacitor [**Fig. TF9-5(b)**] changes the effective permittivity of the medium, perturbs the electric field lines, and modifies the charge distribution on the plates. ◄

This, in turn, changes the value of the capacitance as would be measured by a capacitance meter or ***bridge circuit***. Hence, the capacitor becomes a ***proximity sensor***, and its sensitivity depends, in part, on how different the permittivity of the external object is from that of the unperturbed medium and on whether it is or is not made of a conductive material.

Fingerprint Imager

An interesting extension of noncontact capacitive sensors is the development of a fingerprint imager consisting of a two-dimensional array of capacitive sensor cells, constructed to record an electrical representation of a fingerprint (**Fig. TF9-6**). Each sensor cell is composed of an adjacent-plates capacitor connected to a capacitance measurement circuit (**Fig. TF9-7**). The entire surface of the imager is covered by a thin layer of nonconductive oxide. When the finger is placed on the oxide surface, it perturbs the field lines of the individual sensor cells to varying degrees, depending on the distance between the ridges and valleys of the finger's surface from the sensor cells.

► Given that the dimensions of an individual sensor are on the order of $65~\mu$m on the side, the imager is capable of recording a fingerprint image at a resolution corresponding to 400 dots per inch or better. ◄

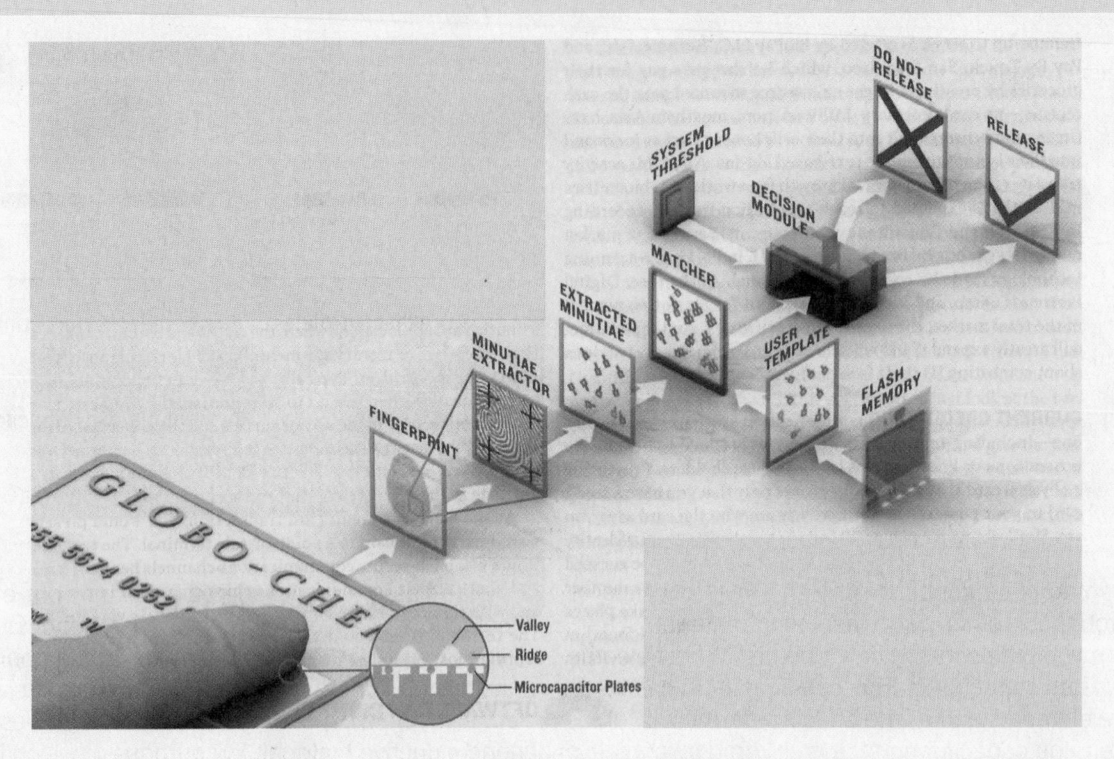

Figure TF9-6 Elements of a fingerprint matching system.

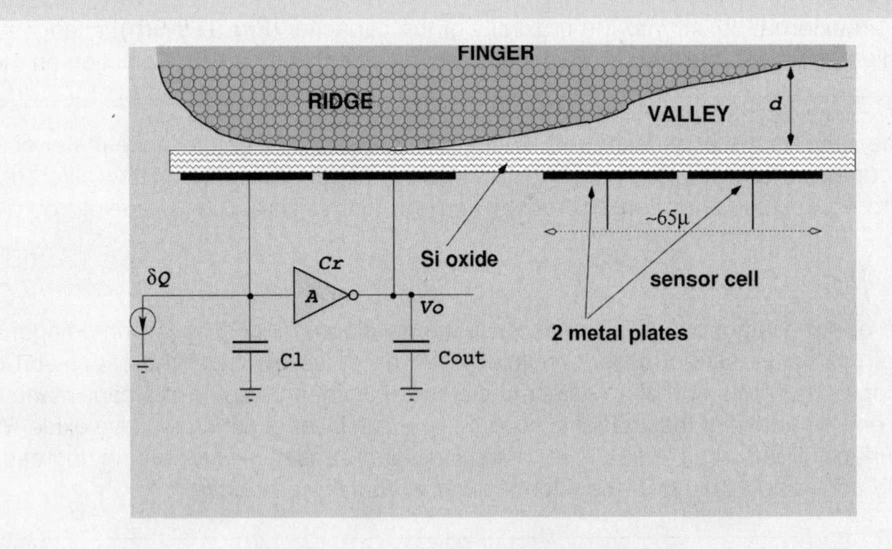

Figure TF9-7 Fingerprint representation.

4-11 Image Method

Consider a point charge Q at a distance d above a horizontally infinite, perfectly conducting plate [Fig. 4-26(a)]. We want to determine V and $\mathbf{E}$ at any point in the space above the plate, as well as the surface charge distribution on the plate. Three different methods for finding $\mathbf{E}$ have been introduced in this chapter The first method, based on Coulomb's law, requires knowledge of the magnitudes and locations of all the charges. In the present case, the charge Q induces an unknown and nonuniform distribution of charge on the plate. Hence, we cannot utilize Coulomb's method. The second method, based on Gauss's law, is equally difficult to use because it is not clear how to construct a Gaussian surface across which $\mathbf{E}$ is only tangential or only normal. The third method is based on evaluating the electric field using $\mathbf{E} = -\nabla V$ after solving Poisson's or Laplace's equation for V subject to the available boundary conditions, but it is mathematically involved.

Alternatively, the problem at hand can be solved using ***image theory***.

► Any given charge configuration above an infinite, perfectly conducting plane is electrically equivalent to the combination of the given charge configuration and its image configuration, with the conducting plane removed. ◄

The image-method equivalent of the charge Q above a conducting plane is shown in the right-hand section of Fig. 4-26. It consists of the charge Q itself and an image charge $-Q$ at a distance $2d$ from Q, with nothing else between them. The electric field due to the two isolated charges can now be easily found at any point (x, y, z) by applying Coulomb's method, as demonstrated by Example 4-13. By symmetry, the combination of the two charges always produces a potential $V = 0$ at every point in the plane previously occupied by the conducting surface. If the charge resides in the presence of more than one grounded plane, it is necessary to establish its images relative to each of the planes and then to establish images of each of those images against the remaining planes. The process is continued until the condition $V = 0$ is satisfied everywhere on all grounded planes. The image method applies not only to point charges, but also to distributions of charge, such as the line and volume distributions depicted in Fig. 4-27. Once $\mathbf{E}$ has been determined, the charge induced on the plate can be found from

$$\rho_s = (\hat{\mathbf{n}} \cdot \mathbf{E})\epsilon_0, \tag{4.132}$$

where $\hat{\mathbf{n}}$ is the normal unit vector to the plate [Fig. 4-26(a)].

Example 4-13: Image Method for Charge above Conducting Plane

Use image theory to determine $\mathbf{E}$ at an arbitrary point $P = (x, y, z)$ in the region $z > 0$ due to a charge Q in free space at a distance d above a grounded conducting plate residing in the $z = 0$ plane.

Solution: In Fig. 4-28, charge Q is at $(0, 0, d)$ and its image $-Q$ is at $(0, 0, -d)$. From Eq. (4.19), the electric field at point $P(x, y, z)$ due to the two charges is given by

$$\begin{aligned}\mathbf{E} &= \frac{1}{4\pi\epsilon_0}\left(\frac{Q\mathbf{R}_1}{R_1^3} + \frac{-Q\mathbf{R}_2}{R_2^3}\right)\\ &= \frac{Q}{4\pi\epsilon_0}\left[\frac{\hat{\mathbf{x}}x + \hat{\mathbf{y}}y + \hat{\mathbf{z}}(z-d)}{[x^2+y^2+(z-d)^2]^{3/2}}\right.\\ &\qquad\left. - \frac{\hat{\mathbf{x}}x + \hat{\mathbf{y}}y + \hat{\mathbf{z}}(z+d)}{[x^2+y^2+(z+d)^2]^{3/2}}\right]\end{aligned}$$

for $z \geq 0$.

Exercise 4-19: Use the result of Example 4-13 to find the surface charge density ρ_s on the surface of the conducting plane.

Answer: $\rho_s = -Qd/[2\pi(x^2 + y^2 + d^2)^{3/2}]$. (See EM.)

Concept Question 4-28: What is the fundamental premise of the image method?

Concept Question 4-29: Given a charge distribution, what are the various approaches described in this chapter for computing the electric field $\mathbf{E}$ at a given point in space?

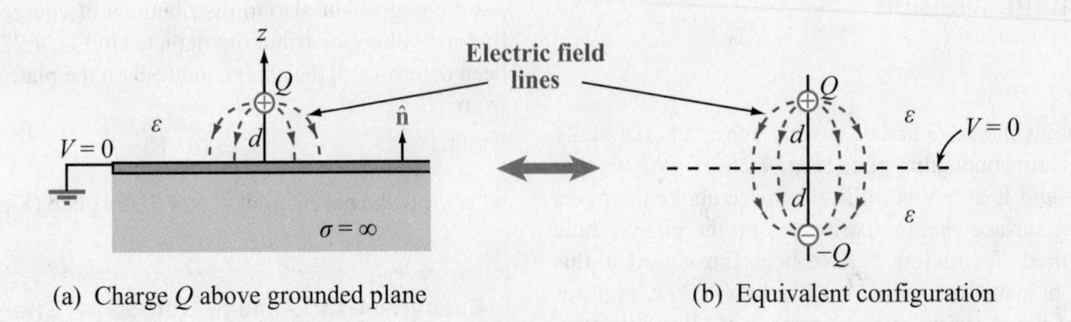

Figure 4-26 By image theory, a charge Q above a grounded perfectly conducting plane is equivalent to Q and its image $-Q$ with the ground plane removed.

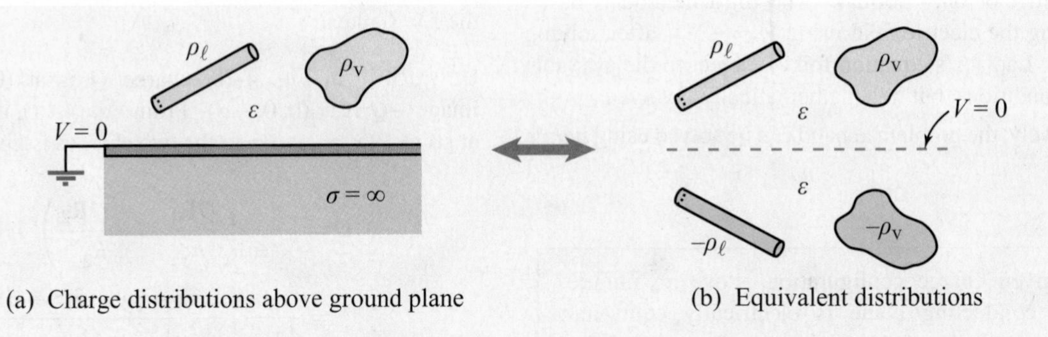

Figure 4-27 Charge distributions above a conducting plane and their image-method equivalents.

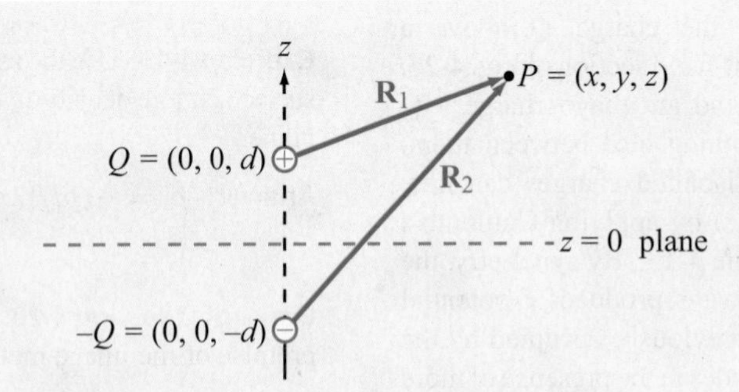

Figure 4-28 Application of the image method for finding $\mathbf{E}$ at point P (Example 4-13).

Chapter 4 Summary

Concepts

- Maxwell's equations are the fundamental tenets of electromagnetic theory.
- Under static conditions, Maxwell's equations separate into two uncoupled pairs, with one pair pertaining to electrostatics and the other to magnetostatics.
- Coulomb's law provides an explicit expression for the electric field due to a specified charge distribution.
- Gauss's law states that the total electric field flux through a closed surface is equal to the net charge enclosed by the surface.
- The electrostatic field **E** at a point is related to the electric potential V at that point by $\mathbf{E} = -\nabla V$, with V often being referenced to zero at infinity.
- Because most metals have conductivities on the order of 10^6 (S/m), they are treated in practice as perfect conductors. By the same token, insulators with conductivities smaller than 10^{-10} (S/m) often are treated as perfect dielectrics.
- Boundary conditions at the interface between two materials specify the relations between the normal and tangential components of **D**, **E**, and **J** in one of the materials to the corresponding components in the other.
- The capacitance of a two-conductor body and resistance of the medium between them can be computed from knowledge of the electric field in that medium.
- The electrostatic energy density stored in a dielectric medium is $w_{\mathrm{e}} = \frac{1}{2}\epsilon E^2$ (J/m^3).
- When a charge configuration exists above an infinite, perfectly conducting plane, the induced field **E** is the same as that due to the configuration itself and its image with the conducting plane removed.

Important Terms

Provide definitions or explain the meaning of the following terms:

boundary conditions
capacitance C
charge density
conductance G
conduction current
conductivity σ
conductor
conservative field
constitutive parameters
convection current
Coulomb's law
current density **J**
dielectric breakdown voltage V_{br}
dielectric material
dielectric strength E_{ds}
dipole moment **p**
electric dipole
electric field intensity **E**
electric flux density **D**
electric potential V
electric susceptibility χ_{e}
electron drift velocity $\mathbf{u}_{\mathrm{e}}$
electron mobility μ_{e}
electrostatic energy density w_{e}
electrostatic potential energy W_{e}
electrostatics
equipotential
Gaussian surface
Gauss's law
hole drift velocity $\mathbf{u}_{\mathrm{h}}$
hole mobility μ_{h}
homogeneous material
image method
isotropic material
Joule's law
Kirchhoff's voltage law
Laplace's equation
linear material
Ohm's law
perfect conductor
perfect dielectric
permittivity ϵ
Poisson's equation
polarization vector **P**
relative permittivity ϵ_{r}
semiconductor
static condition
superconductor
volume, surface, and line charge densities

Mathematical and Physical Models

Maxwell's Equations for Electrostatics

Name	Differential Form	Integral Form
Gauss's law	$\nabla \cdot \mathbf{D} = \rho_v$	$\oint_S \mathbf{D} \cdot d\mathbf{s} = Q$
Kirchhoff's law	$\nabla \times \mathbf{E} = 0$	$\oint_C \mathbf{E} \cdot d\mathbf{l} = 0$

Current density	$\mathbf{J} = \rho_v \mathbf{u}$
Poisson's equation	$\nabla^2 V = -\dfrac{\rho_v}{\epsilon}$
Laplace's equation	$\nabla^2 V = 0$
Resistance	$R = \dfrac{-\int_l \mathbf{E} \cdot d\mathbf{l}}{\int_S \sigma \mathbf{E} \cdot d\mathbf{s}}$
Boundary conditions	**Table 4-3**
Capacitance	$C = \dfrac{\int_S \epsilon \mathbf{E} \cdot d\mathbf{s}}{-\int_l \mathbf{E} \cdot d\mathbf{l}}$
RC relation	$RC = \dfrac{\epsilon}{\sigma}$
Energy density	$w_e = \frac{1}{2} \epsilon E^2$

Electric Field

Point charge	$\mathbf{E} = \hat{\mathbf{R}} \dfrac{q}{4\pi\epsilon R^2}$
Many point charges	$\mathbf{E} = \dfrac{1}{4\pi\epsilon} \sum_{i=1}^{N} \dfrac{q_i (\mathbf{R} - \mathbf{R}_i)}{\lvert\mathbf{R} - \mathbf{R}_i\rvert^3}$
Volume distribution	$\mathbf{E} = \dfrac{1}{4\pi\epsilon} \int_{\mathcal{V}'} \hat{\mathbf{R}}' \dfrac{\rho_v \, d\mathcal{V}'}{R'^2}$
Surface distribution	$\mathbf{E} = \dfrac{1}{4\pi\epsilon} \int_{S'} \hat{\mathbf{R}}' \dfrac{\rho_s \, ds'}{R'^2}$
Line distribution	$\mathbf{E} = \dfrac{1}{4\pi\epsilon} \int_{l'} \hat{\mathbf{R}}' \dfrac{\rho_\ell \, dl'}{R'^2}$
Infinite sheet of charge	$\mathbf{E} = \hat{\mathbf{z}} \dfrac{\rho_s}{2\epsilon_0}$
Infinite line of charge	$\mathbf{E} = \dfrac{\mathbf{D}}{\epsilon_0} = \hat{\mathbf{r}} \dfrac{D_r}{\epsilon_0} = \hat{\mathbf{r}} \dfrac{\rho_\ell}{2\pi\epsilon_0 r}$
Dipole	$\mathbf{E} = \dfrac{qd}{4\pi\epsilon_0 R^3} (\hat{\mathbf{R}}\, 2\cos\theta + \hat{\boldsymbol{\theta}} \sin\theta)$
Relation to V	$\mathbf{E} = -\nabla V$

PROBLEMS

Section 4-2: Charge and Current Distributions

4.1 Find the total charge contained in a cylindrical volume defined by $r \le 2$ m and $0 \le z \le 3$ m if $\rho_v = 30rz$ (mC/m^3).

***4.2** A cube 2 m on a side is located in the first octant in a Cartesian coordinate system, with one of its corners at the origin. Find the total charge contained in the cube if the charge density is given by $\rho_v = xy^2 e^{-2z}$ (mC/m^3).

***4.3** Find the total charge contained in a round-top cone defined by $R \le 2$ m and $0 \le \theta \le \pi/4$, given that $\rho_v = 30R^2 \cos^2\theta$ (mC/m^3).

4.4 If the line charge density is given by $\rho_l = 12y^2$ (mC/m), find the total charge distributed on the y axis from $y = -5$ to $y = 5$.

---
*Answer(s) available in Appendix D.

4.5 Find the total charge on a circular disk defined by $r \leq a$ and $z = 0$ if:

(a) $\rho_s = \rho_{s0} \cos\phi$ (C/m^2)

(b) $\rho_s = \rho_{s0} \sin^2\phi$ (C/m^2)

(c) $\rho_s = \rho_{s0} e^{-r}$ (C/m^2)

(d) $\rho_s = \rho_{s0} e^{-r} \sin^2\phi$ (C/m^2)

where ρ_{s0} is a constant.

4.6 If $\mathbf{J} = \hat{\mathbf{y}}6xz$ (A/m^2), find the current I flowing through a square with corners at (0, 0, 0), (2, 0, 0), (2, 0, 2), and (0, 0, 2).

***4.7** If $\mathbf{J} = \hat{\mathbf{R}}5/R$ (A/m^2), find I through the surface $R = 5$ m.

4.8 A circular beam of charge of radius a consists of electrons moving with a constant speed u along the $+z$ direction. The beam's axis is coincident with the z axis and the electron charge density is given by

$$\rho_v = -cr^2 \qquad \text{(c/m}^3\text{)}$$

where c is a constant and r is the radial distance from the axis of the beam.

***(a)** Determine the charge density per unit length.

(b) Determine the current crossing the z-plane.

4.9 An electron beam shaped like a circular cylinder of radius r_0 carries a charge density given by

$$\rho_v = \left(\frac{-\rho_0}{1+r^2}\right) \qquad \text{(C/m}^3\text{)}$$

where ρ_0 is a positive constant and the beam's axis is coincident with the z axis.

(a) Determine the total charge contained in length L of the beam.

(b) If the electrons are moving in the $+z$ direction with uniform speed u, determine the magnitude and direction of the current crossing the z-plane.

4.10 A line of charge of uniform density ρ_ℓ occupies a semicircle of radius b as shown in **Fig. P4.10**. Use the material presented in Example 4-4 to determine the electric field at the origin.

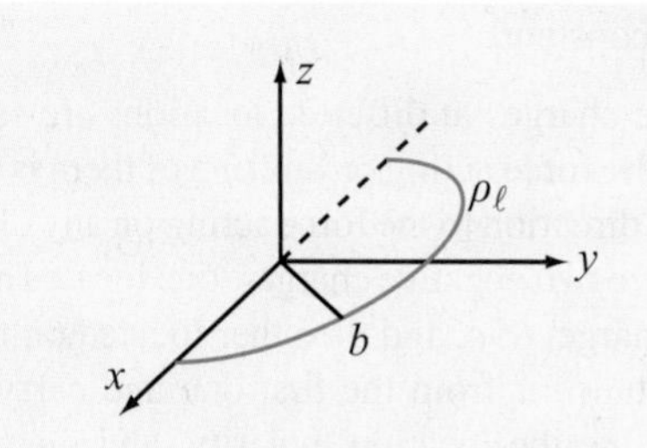

Figure P4.10 Problem 4.10.

Section 4-3: Coulomb's Law

***4.11** A square with sides of 2 m has a charge of 40 μC at each of its four corners. Determine the electric field at a point 5 m above the center of the square.

***4.12** Charge $q_1 = 6$ μC is located at (1 cm, 1 cm, 0) and charge q_2 is located at (0, 0, 4 cm). What should q_2 be so that $\mathbf{E}$ at (0, 2 cm, 0) has no y component?

4.13 Three point charges, each with $q = 3$ nC, are located at the corners of a triangle in the x–y plane, with one corner at the origin, another at (2 cm, 0, 0), and the third at (0, 2 cm, 0). Find the force acting on the charge located at the origin.

4.14 A line of charge with uniform density $\rho_\ell = 8$ (μC/m) exists in air along the z axis between $z = 0$ and $z = 5$ cm. Find $\mathbf{E}$ at (0,10 cm,0).

4.15 Electric charge is distributed along an arc located in the x–y plane and defined by $r = 2$ cm and $0 \leq \phi \leq \pi/4$. If $\rho_\ell = 5$ (μC/m), find $\mathbf{E}$ at $(0, 0, z)$ and then evaluate it at:

***(a)** The origin.

(b) $z = 5$ cm

(c) $z = -5$ cm

4.16 A line of charge with uniform density ρ_l extends between $z = -L/2$ and $z = L/2$ along the z axis. Apply Coulomb's law to obtain an expression for the electric field at any point $P(r, \phi, 0)$ on the x–y plane. Show that your result reduces to the expression given by (4.33) as the length L is extended to infinity.

***4.17** Repeat Example 4-5 for the circular disk of charge of radius a, but in the present case, assume the surface charge density to vary with r as

$$\rho_s = \rho_{s0} r^2 \qquad \text{(C/m}^2\text{)}$$

where ρ_{s0} is a constant.

4.18 Multiple charges at different locations are said to be in equilibrium if the force acting on any one of them is identical in magnitude and direction to the force acting on any of the others. Suppose we have two negative charges, one located at the origin and carrying charge $-9e$, and the other located on the positive x axis at a distance d from the first one and carrying charge $-36e$. Determine the location, polarity, and magnitude of a third charge whose placement would bring the entire system into equilibrium.

Section 4-4: Gauss's Law

4.19 Three infinite lines of charge, all parallel to the z axis, are located at the three corners of the kite-shaped arrangement shown in **Fig. P4.19**. If the two right triangles are symmetrical and of equal corresponding sides, show that the electric field is zero at the origin.

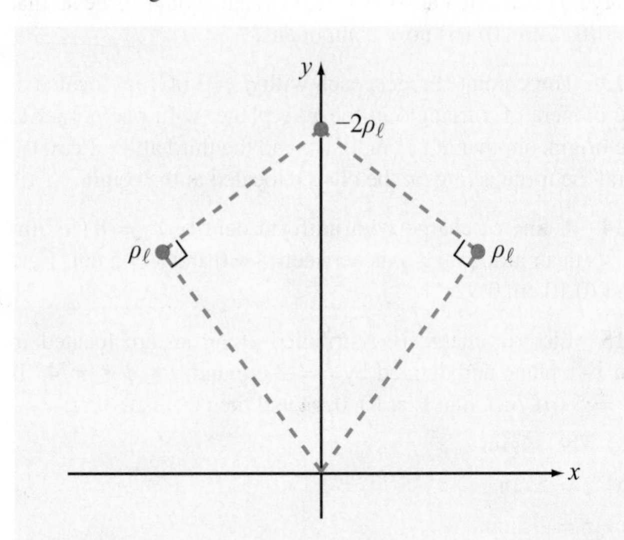

Figure P4.19 Kite-shaped arrangement of line charges for Problem 4.19.

*__4.20__ Three infinite lines of charge, $\rho_{l_1} = 3$ (nC/m), $\rho_{l_2} = -3$ (nC/m), and $\rho_{l_3} = 3$ (nC/m), are all parallel to the z axis. If they pass through the respective points $(0, -b)$, $(0, 0)$, and $(0, b)$ in the x–y plane, find the electric field at $(a, 0, 0)$. Evaluate your result for $a = 2$ cm and $b = 1$ cm.

4.21 Given the electric flux density

$$\mathbf{D} = \hat{\mathbf{x}}2(x + y) + \hat{\mathbf{y}}(3x - 2y) \qquad (\text{C/m}^2)$$

determine

(a) ρ_v by applying Eq. (4.26).

(b) The total charge Q enclosed in a cube 2 m on a side, located in the first octant with three of its sides coincident with the x-, y-, and z axes and one of its corners at the origin.

(c) The total charge Q in the cube, obtained by applying Eq. (4.29).

4.22 A horizontal strip lying in the x–y plane is of width d in the y direction and infinitely long in the x direction. If the strip is in air and has a uniform charge distribution ρ_s, use Coulomb's law to obtain an explicit expression for the electric field at a point P located at a distance h above the centerline of the strip. Extend your result to the special case where d is infinite and compare it with Eq. (4.25).

*__4.23__ Repeat Problem 4.21 for $\mathbf{D} = \hat{\mathbf{x}}xy^3z^3$ (C/m^2).

4.24 Charge Q_1 is uniformly distributed over a thin spherical shell of radius a, and charge Q_2 is uniformly distributed over a second spherical shell of radius b, with $b > a$. Apply Gauss's law to find $\mathbf{E}$ in the regions $R < a$, $a < R < b$, and $R > b$.

*__4.25__ The electric flux density inside a dielectric sphere of radius a centered at the origin is given by

$$\mathbf{D} = \hat{\mathbf{R}}\rho_0 R \qquad (\text{C/m}^2)$$

where ρ_0 is a constant. Find the total charge inside the sphere.

*__4.26__ An infinitely long cylindrical shell extending between $r = 1$ m and $r = 2$ m contains a uniform charge density ρ_{v0}. Apply Gauss's law to find $\mathbf{D}$ in all regions.

4.27 In a certain region of space, the charge density is given in cylindrical coordinates by the function:

$$\rho_v = 50re^{-r} \quad (\text{C/m}^3)$$

Apply Gauss's law to find $\mathbf{D}$.

4.28 If the charge density increases linearly with distance from the origin such that $\rho_v = 0$ at the origin and $\rho_v = 40$ C/m^3 at $R = 2$ m, find the corresponding variation of $\mathbf{D}$.

4.29 A spherical shell with outer radius b surrounds a charge-free cavity of radius $a < b$ (**Fig. P4.29**). If the shell contains a charge density given by

$$\rho_v = -\frac{\rho_{v0}}{R^2}, \qquad a \le R \le b,$$

where ρ_{v0} is a positive constant, determine $\mathbf{D}$ in all regions.

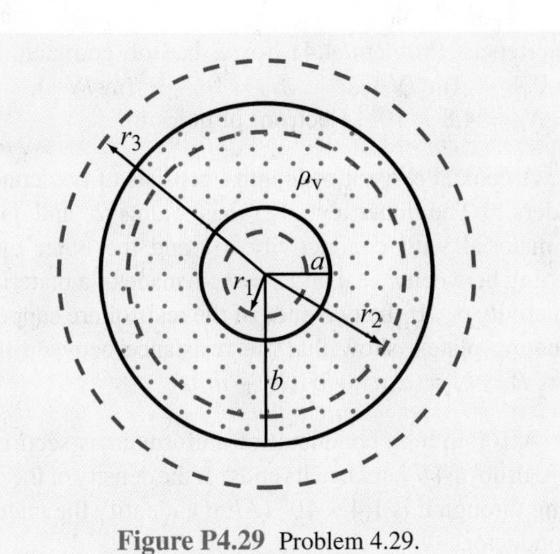

Figure P4.29 Problem 4.29.

Section 4-5: Electric Potential

*__4.30__ A square in the x–y plane in free space has a point charge of $+Q$ at corner $(a/2, a/2)$, the same at corner $(a/2, -a/2)$, and a point charge of $-Q$ at each of the other two corners.

(a) Find the electric potential at any point P along the x axis.

(b) Evaluate V at $x = a/2$.

4.31 The circular disk of radius a shown in **Fig. 4-7** has uniform charge density ρ_s across its surface.

(a) Obtain an expression for the electric potential V at a point $P(0, 0, z)$ on the z axis.

(b) Use your result to find **E** and then evaluate it for $z = h$. Compare your final expression with (4.24), which was obtained on the basis of Coulomb's law.

4.32 Show that the electric potential difference V_{12} between two points in air at radial distances r_1 and r_2 from an infinite line of charge with density ρ_ℓ along the z axis is $V_{12} = (\rho_\ell/2\pi\epsilon_0)\ln(r_2/r_1)$.

*__4.33__ A circular ring of charge of radius a lies in the x–y plane and is centered at the origin. Assume also that the ring is in air and carries a uniform density ρ_ℓ.

(a) Show that the electrical potential at $(0, 0, z)$ is given by $V = \rho_\ell a/[2\epsilon_0(a^2 + z^2)^{1/2}]$.

(b) Find the corresponding electric field **E**.

*__4.34__ Find the electric potential V at a location a distance b from the origin in the x–y plane due to a line charge with charge density ρ_ℓ and of length l. The line charge is coincident with the z axis and extends from $z = -l/2$ to $z = l/2$.

4.35 For the electric dipole shown in **Fig. 4-13**, $d = 1$ cm and $|\mathbf{E}| = 8$ (mV/m) at $R = 1$ m and $\theta = 0°$. Find **E** at $R = 2$ m and $\theta = 90°$.

4.36 For each of the distributions of the electric potential V shown in **Fig. P4.36**, sketch the corresponding distribution of **E** (in all cases, the vertical axis is in volts and the horizontal axis is in meters).

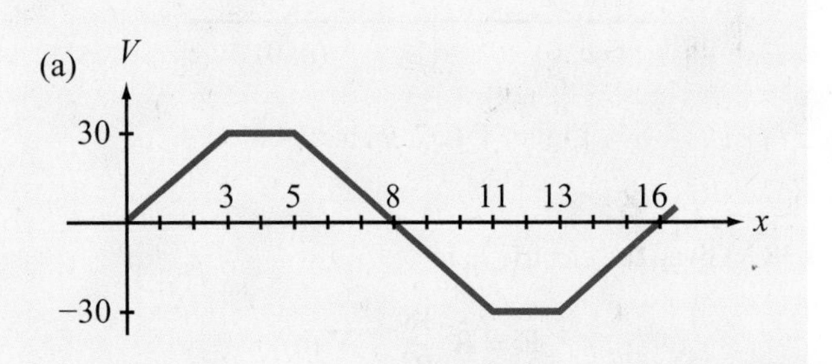

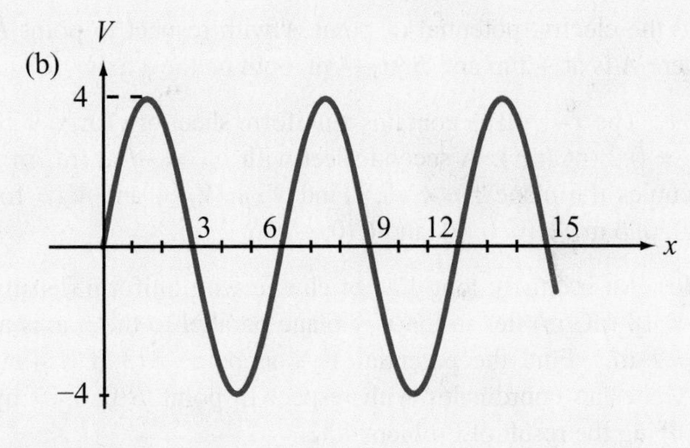

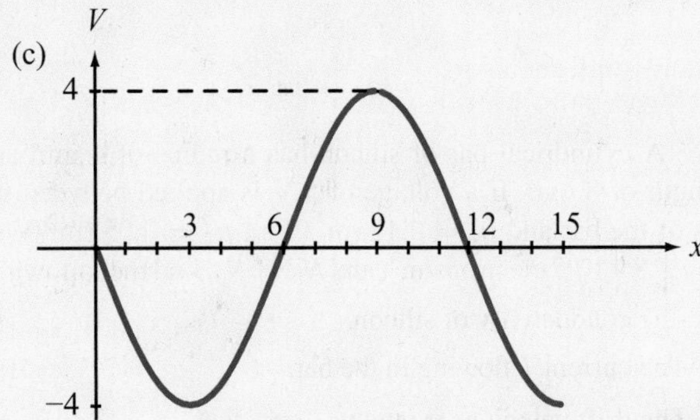

Figure P4.36 Electric potential distributions of Problem 4.36.

*4.37 Two infinite lines of charge, both parallel to the z axis, lie in the x–z plane, one with density ρ_ℓ and located at $x = a$ and the other with density $-\rho_\ell$ and located at $x = -a$. Obtain an expression for the electric potential $V(x, y)$ at a point $P = (x, y)$ relative to the potential at the origin.

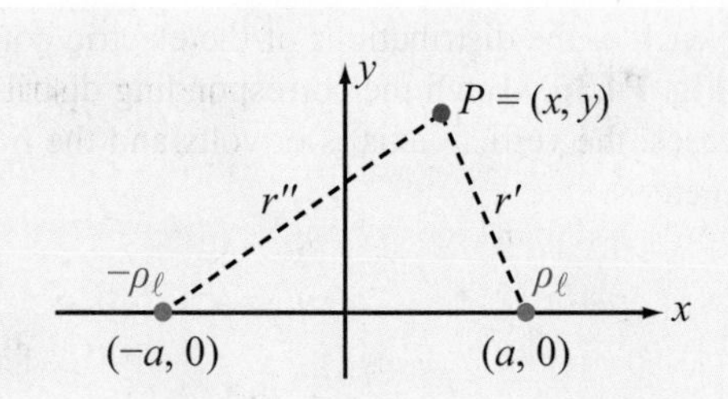

Figure P4.37 Problem 4.37.

4.38 Given the electric field

$$\mathbf{E} = \hat{\mathbf{R}}\,\frac{18}{R^2} \quad \text{(V/m)}$$

find the electric potential of point A with respect to point B where A is at $+2$ m and B at -4 m, both on the z axis.

4.39 The x–y plane contains a uniform sheet of charge with $\rho_{s_1} = 0.2$ (nC/m^2). A second sheet with $\rho_{s_2} = -0.2$ (nC/m^2) occupies the plane $z = 6$ m. Find V_{AB}, V_{BC}, and V_{AC} for $A(0, 0, 6\text{ m})$, $B(0, 0, 0)$, and $C(0, -2\text{ m}, 2\text{ m})$.

*4.40 An infinitely long line of charge with uniform density $\rho_l = 18$ (nC/m) lies in the x–y plane parallel to the y axis at $x = 2$ m. Find the potential V_{AB} at point $A(3\text{ m}, 0, 4\text{ m})$ in Cartesian coordinates with respect to point $B(0, 0, 0)$ by applying the result of Problem 4.32.

Section 4-6: Conductors

4.41 A cylindrical bar of silicon has a radius of 4 mm and a length of 8 cm. If a voltage of 5 V is applied between the ends of the bar and $\mu_e = 0.13$ (m^2/V·s), $\mu_h = 0.05$ (m^2/V·s), $N_e = 1.5 \times 10^{16}$ electrons/m^3, and $N_h = N_e$, find the following:

(a) The conductivity of silicon.

(b) The current I flowing in the bar.

***(c)** The drift velocities $\mathbf{u}_e$ and $\mathbf{u}_h$.

(d) The resistance of the bar.

(e) The power dissipated in the bar.

4.42 Repeat Problem 4.41 for a bar of germanium with $\mu_e = 0.4$ (m^2/V·s), $\mu_h = 0.2$ (m^2/V·s), and $N_e = N_h = 4.8 \times 10^{19}$ electrons or holes/m^3.

4.43 A coaxial resistor of length l consists of two concentric cylinders. The inner cylinder has radius a and is made of a material with conductivity σ_1, and the outer cylinder, extending between $r = a$ and $r = b$, is made of a material with conductivity σ_2. If the two ends of the resistor are capped with conducting plates, show that the resistance between the two ends is $R = l/[\pi(\sigma_1 a^2 + \sigma_2(b^2 - a^2))]$.

4.44 A 100 m long conductor of uniform cross-section has a voltage drop of 4 V between its ends. If the density of the current flowing through it is 1.4×10^6 (A/m^2), identify the material of the conductor.

*4.45 Apply the result of Problem 4.44 to find the resistance of a 20 cm long hollow cylinder (**Fig. P4.45**) made of carbon with $\sigma = 3 \times 10^4$ (S/m).

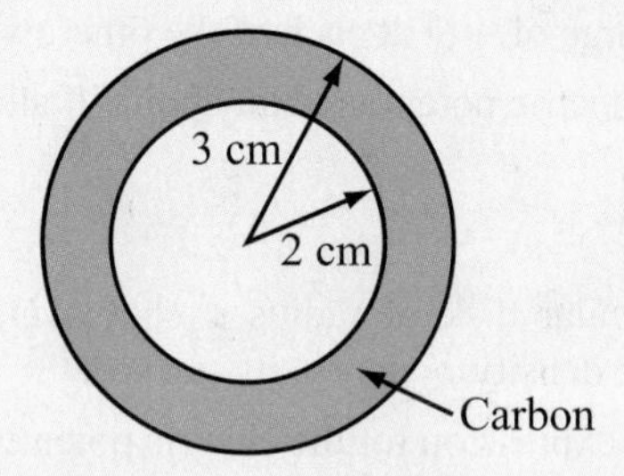

Figure P4.45 Cross section of hollow cylinder of Problem 4.45.

4.46 A cylinder-shaped carbon resistor is 8 cm in length and its circular cross section has a diameter $d = 1$ mm.

(a) Determine the resistance R.

(b) To reduce its resistance by 40%, the carbon resistor is coated with a layer of copper of thickness t. Use the result of Problem 4.43 to determine t.

4.47 A 4×10^{-3} mm thick square sheet of aluminum has 5 cm × 5 cm faces. Find the following:

(a) The resistance between opposite edges on a square face.

(b) The resistance between the two square faces. (See Appendix B for the electrical constants of materials.)

Section 4-8: Boundary Conditions

*4.48 With reference to Fig. 4-19, find $\mathbf{E}_1$ if $\mathbf{E}_2 = \hat{\mathbf{x}}3 - \hat{\mathbf{y}}2 + \hat{\mathbf{z}}2$ (V/m), $\epsilon_1 = 2\epsilon_0$, $\epsilon_2 = 18\epsilon_0$, and the boundary has a surface charge density $\rho_s = 3.54 \times 10^{-11}$ (C/m^2). What angle does $\mathbf{E}_2$ make with the z axis?

4.49 An infinitely long cylinder of radius a is surrounded by a dielectric medium that contains no free charges. If the tangential component of the electric field in the region $r \geq a$ is given by $\mathbf{E}_t = -\hat{\boldsymbol{\phi}} \cos^2\phi / r^2$, find $\mathbf{E}$ in that region.

*4.50 If $\mathbf{E} = \hat{\mathbf{R}}300$ (V/m) at the surface of a 5-cm conducting sphere centered at the origin, what is the total charge Q on the sphere's surface?

4.51 Figure P4.51 shows three planar dielectric slabs of equal thickness but with different dielectric constants. If $\mathbf{E}_0$ in air makes an angle of 45° with respect to the z axis, find the angle of $\mathbf{E}$ in each of the other layers.

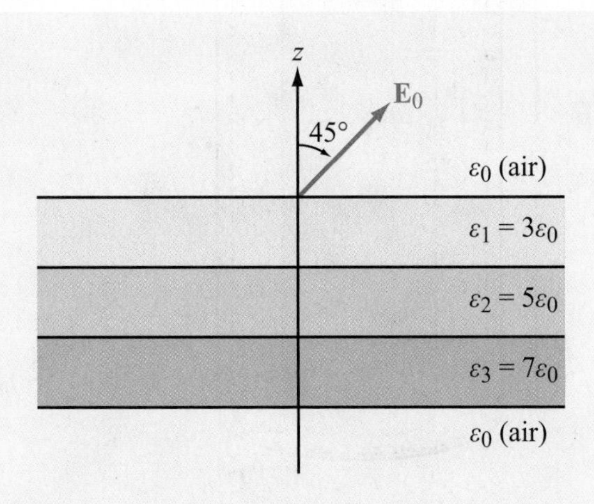

Figure P4.51 Dielectric slabs in Problem 4.51.

Sections 4-9 and 4-10: Capacitance and Electrical Energy

4.52 Dielectric breakdown occurs in a material whenever the magnitude of the field $\mathbf{E}$ exceeds the dielectric strength anywhere in that material. In the coaxial capacitor of Example 4-12,

*(a) At what value of r is $|E|$ maximum?

(b) What is the breakdown voltage if $a = 1$ cm, $b = 2$ cm, and the dielectric material is mica with $\epsilon_r = 6$?

4.53 Determine the force of attraction in a parallel-plate capacitor with $A = 5$ cm^2, $d = 2$ cm, and $\epsilon_r = 4$ if the voltage across it is 50 V.

4.54 An electron with charge $Q_e = -1.6 \times 10^{-19}$ C and mass $m_e = 9.1 \times 10^{-31}$ kg is injected at a point adjacent to the negatively charged plate in the region between the plates of an air-filled parallel-plate capacitor with separation of 1 cm and rectangular plates each 10 cm^2 in area (Fig. P4.54). If the voltage across the capacitor is 10 V, find the following:

(a) The force acting on the electron.

(b) The acceleration of the electron.

(c) The time it takes the electron to reach the positively charged plate, assuming that it starts from rest.

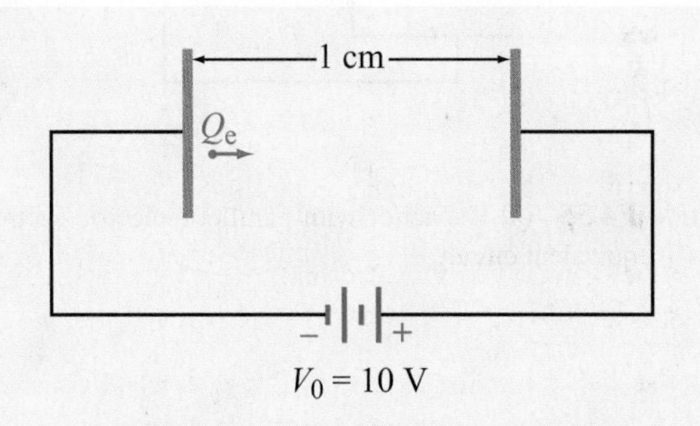

Figure P4.54 Electron between charged plates of Problem 4.54.

4.55 Figure P4.55(a) depicts a capacitor consisting of two parallel, conducting plates separated by a distance d. The space between the plates contains two adjacent dielectrics, one with permittivity ϵ_1 and surface area A_1 and another with ϵ_2 and A_2. The objective of this problem is to show that the capacitance C of the configuration shown in Fig. P4.55(a) is equivalent to two capacitances in parallel, as illustrated in Fig. P4.55(b), with

$$C = C_1 + C_2 \tag{4.133}$$

where

$$C_1 = \frac{\epsilon_1 A_1}{d} \tag{4.134}$$

$$C_2 = \frac{\epsilon_2 A_2}{d} \tag{4.135}$$

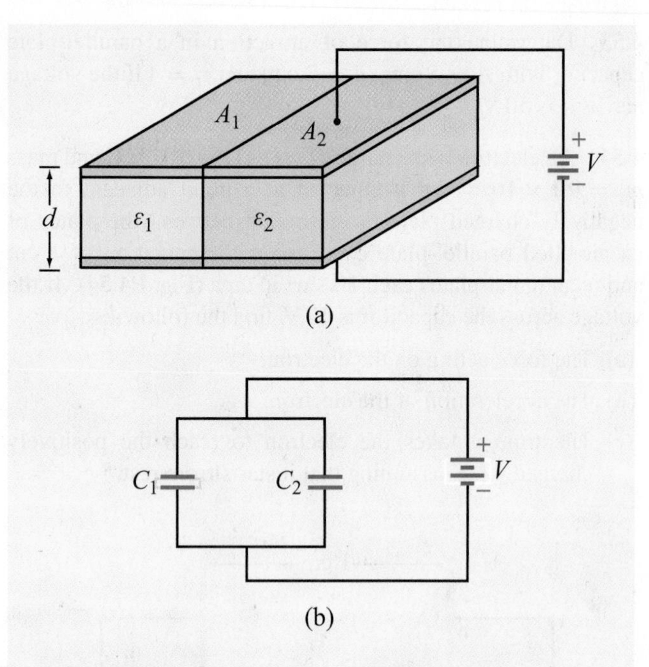

Figure P4.55 (a) Capacitor with parallel dielectric section, and (b) equivalent circuit.

To this end, proceed as follows:

(a) Find the electric fields $\mathbf{E}_1$ and $\mathbf{E}_2$ in the two dielectric layers.

(b) Calculate the energy stored in each section and use the result to calculate C_1 and C_2.

(c) Use the total energy stored in the capacitor to obtain an expression for C. Show that (4.133) is indeed a valid result.

*__4.56__ In a dielectric medium with $\epsilon_r = 4$, the electric field is given by

$$\mathbf{E} = \hat{\mathbf{x}}(x^2 + 2z) + \hat{\mathbf{y}}x^2 - \hat{\mathbf{z}}(y + z) \qquad \text{(V/m)}$$

Calculate the electrostatic energy stored in the region $-1 \text{ m} \le x \le 1 \text{ m}$, $0 \le y \le 2 \text{ m}$, and $0 \le z \le 3 \text{ m}$.

4.57 Use the result of Problem 4.55 to determine the capacitance for each of the following configurations:

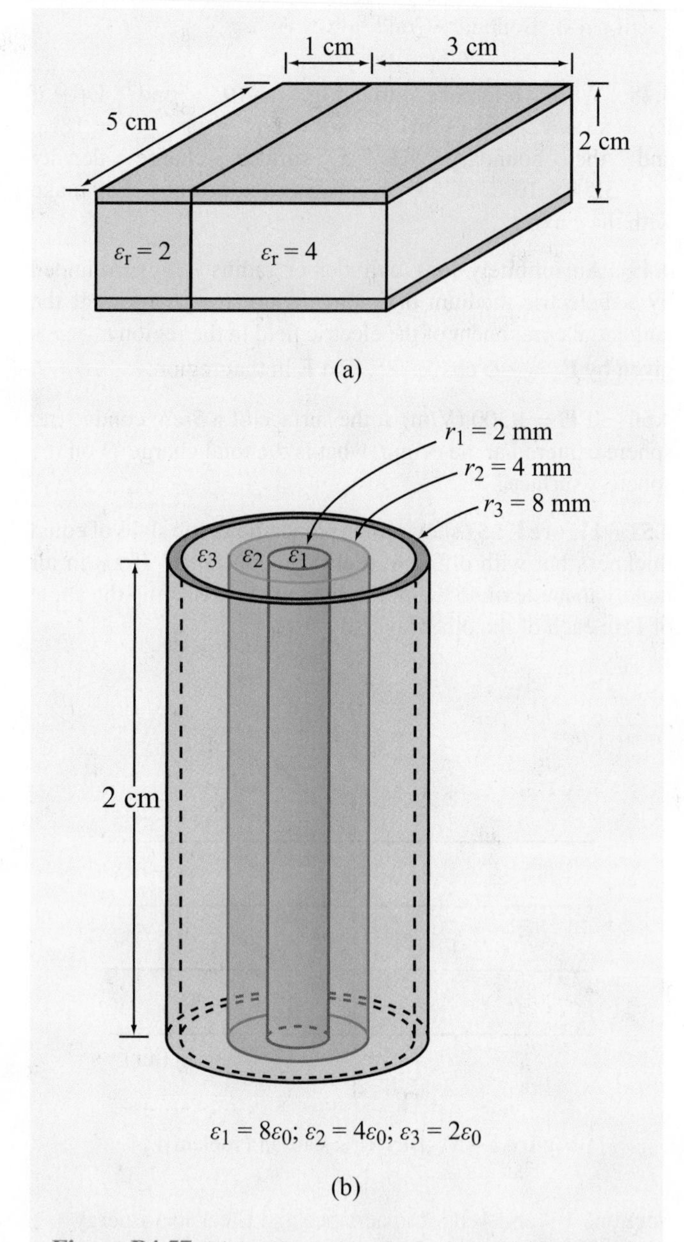

Figure P4.57 Dielectric sections for Problems 4.57 and 4.59.

*__(a)__ Conducting plates are on top and bottom faces of the rectangular structure in **Fig. P4.57(a)**.

(b) Conducting plates are on front and back faces of the structure in **Fig. P4.57(a)**.

(c) Conducting plates are on top and bottom faces of the cylindrical structure in **Fig. P4.57(b)**.

4.58 The capacitor shown in **Fig. P4.58** consists of two parallel dielectric layers. Use energy considerations to show that the equivalent capacitance of the overall capacitor, C, is equal to the series combination of the capacitances of the individual layers, C_1 and C_2, namely

$$C = \frac{C_1 C_2}{C_1 + C_2} \qquad (4.136)$$

where

$$C_1 = \epsilon_1 \frac{A}{d_1}, \qquad C_2 = \epsilon_2 \frac{A}{d_2}.$$

(a) Let V_1 and V_2 be the electric potentials across the upper and lower dielectrics, respectively. What are the corresponding electric fields E_1 and E_2? By applying the appropriate boundary condition at the interface between the two dielectrics, obtain explicit expressions for E_1 and E_2 in terms of ϵ_1, ϵ_2, V, and the indicated dimensions of the capacitor.

(b) Calculate the energy stored in each of the dielectric layers and then use the sum to obtain an expression for C.

(c) Show that C is given by Eq. (4.136).

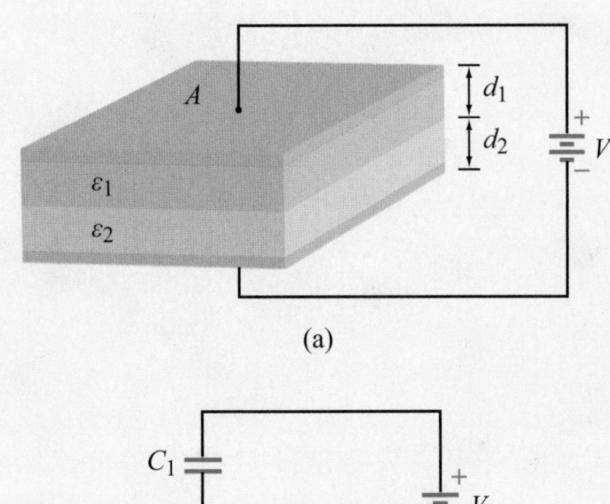

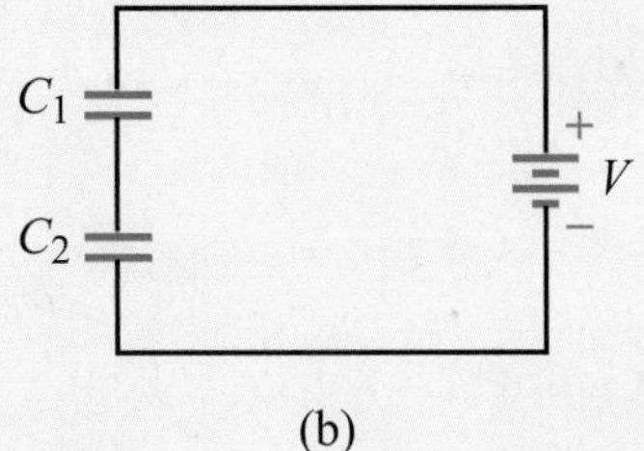

Figure P4.58 (a) Capacitor with parallel dielectric layers, and (b) equivalent circuit (Problem 4.58).

4.59 Use the expressions given in Problem 4.58 to determine the capacitance for the configurations in **Fig. P4.57(a)** when the conducting plates are placed on the right and left faces of the structure.

4.60 A coaxial capacitor consists of two concentric, conducting, cylindrical surfaces, one of radius a and another of radius b, as shown in **Fig. P4.60**. The insulating layer separating the two conducting surfaces is divided equally into two semi-cylindrical sections, one filled with dielectric ϵ_1 and the other filled with dielectric ϵ_2.

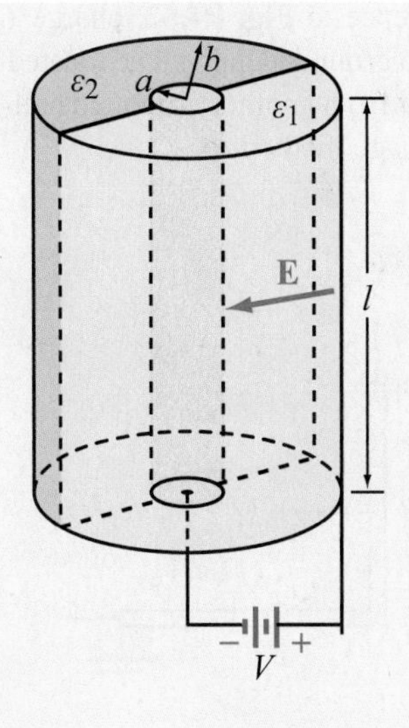

Figure P4.60 Problem 4.60.

(a) Develop an expression for C in terms of the length l and the given quantities.

*__(b)__ Evaluate the value of C for $a = 2$ mm, $b = 6$ mm, $\epsilon_{r_1} = 2$, $\epsilon_{r_2} = 4$, and $l = 4$ cm.

Section 4-12: Image Method

4.61 Conducting wires above a conducting plane carry currents I_1 and I_2 in the directions shown in **Fig. P4.61**. Keeping in mind that the direction of a current is defined in terms of the movement of positive charges, what are the directions of the image currents corresponding to I_1 and I_2?

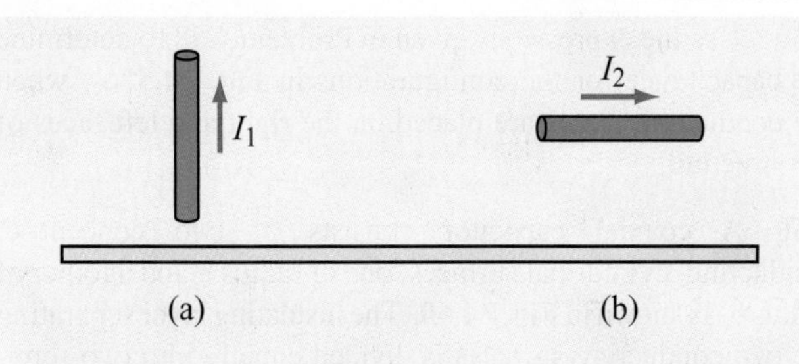

Figure P4.61 Currents above a conducting plane (Problem 4.61).

4.62 With reference to **Fig. P4.62**, charge Q is located at a distance d above a grounded half-plane located in the x–y plane and at a distance d from another grounded half-plane in the x–z plane. Use the image method to

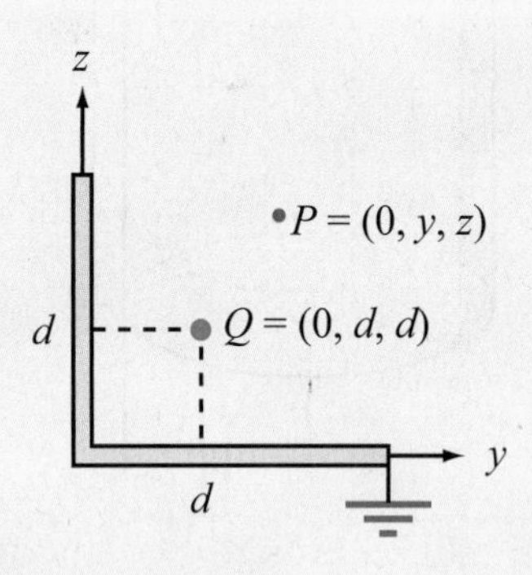

Figure P4.62 Charge Q next to two perpendicular, grounded, conducting half-planes.

(a) Establish the magnitudes, polarities, and locations of the images of charge Q with respect to each of the two ground planes (as if each is infinite in extent).

(b) Find the electric potential and electric field at an arbitrary point $P = (0, y, z)$.

*****4.63** Use the image method to find the capacitance per unit length of an infinitely long conducting cylinder of radius a situated at a distance d from a parallel conducting plane, as shown in **Fig. P4.63**.

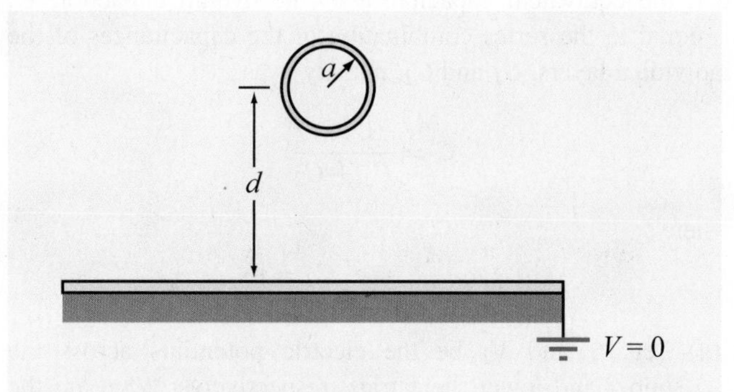

Figure P4.63 Conducting cylinder above a conducting plane (Problem 4.63).

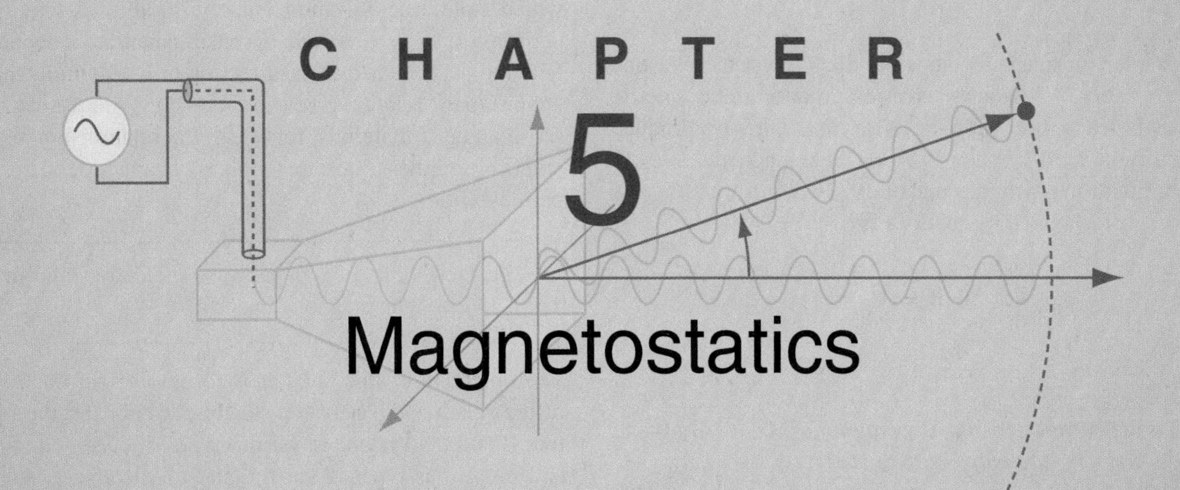

Chapter Contents

Overview, 258
5-1 Magnetic Forces and Torques, 259
5-2 The Biot–Savart Law, 266
5-3 Maxwell's Magnetostatic Equations, 273
TB10 Electromagnets, 278
5-4 Vector Magnetic Potential, 281
5-5 Magnetic Properties of Materials, 282
5-6 Magnetic Boundary Conditions, 286
5-7 Inductance, 287
TB11 Inductive Sensors, 290
5-8 Magnetic Energy, 293
Chapter 5 Summary, 294
Problems, 296

Objectives

Upon learning the material presented in this chapter, you should be able to:

1. Calculate the magnetic force on a current-carrying wire placed in a magnetic field and the torque exerted on a current loop.
2. Apply the Biot–Savart law to calculate the magnetic field due to current distributions.
3. Apply Ampère's law to configurations with appropriate symmetry.
4. Explain magnetic hysteresis in ferromagnetic materials.
5. Calculate the inductance of a solenoid, a coaxial transmission line, or other configurations.
6. Relate the magnetic energy stored in a region to the magnetic field distribution in that region.

Overview

This chapter on magnetostatics parallels the preceding one on electrostatics. Stationary charges produce static electric fields, and steady (i.e., non–time-varying) currents produce static magnetic fields. When $\partial/\partial t = 0$, the magnetic fields in a medium with magnetic permeability μ are governed by the second pair of Maxwell's equations [Eqs. (4.3a,b)]:

$$\nabla \cdot \mathbf{B} = 0, \tag{5.1a}$$

$$\nabla \times \mathbf{H} = \mathbf{J}, \tag{5.1b}$$

where $\mathbf{J}$ is the current density. The ***magnetic flux density*** $\mathbf{B}$ and the ***magnetic field intensity*** $\mathbf{H}$ are related by

$$\mathbf{B} = \mu \mathbf{H}. \tag{5.2}$$

When examining electric fields in a dielectric medium in Chapter 4, we noted that the relation $\mathbf{D} = \epsilon \mathbf{E}$ is valid only when the medium is linear and isotropic. These properties, which hold true for most materials, allow us to treat the permittivity ϵ as a constant, scalar quantity, independent of both the magnitude and the direction of $\mathbf{E}$. A similar statement applies to the relation given by Eq. (5.2). With the exception of ferromagnetic materials, for which the relationship between $\mathbf{B}$ and $\mathbf{H}$ is nonlinear, most materials are characterized by constant permeabilities.

► Furthermore, $\mu = \mu_0$ for most dielectrics and metals (excluding ferromagnetic materials). ◄

The objective of this chapter is to develop an understanding of the relationship between steady currents and the magnetic flux $\mathbf{B}$ and field $\mathbf{H}$ due to various types of current distributions and in various types of media, and to introduce a number of related quantities, such as the magnetic vector potential $\mathbf{A}$, the magnetic energy density w_m, and the inductance of a conducting structure, L. The parallelism that exists between these magnetostatic quantities and their electrostatic counterparts is elucidated in **Table 5-1**.

Table 5-1 Attributes of electrostatics and magnetostatics.

Attribute	Electrostatics	Magnetostatics
Sources	Stationary charges ρ_v	Steady currents $\mathbf{J}$
Fields and fluxes	$\mathbf{E}$ and $\mathbf{D}$	$\mathbf{H}$ and $\mathbf{B}$
Constitutive parameter(s)	ϵ and σ	μ
Governing equations		
• Differential form	$\nabla \cdot \mathbf{D} = \rho_v$ $\nabla \times \mathbf{E} = 0$	$\nabla \cdot \mathbf{B} = 0$ $\nabla \times \mathbf{H} = \mathbf{J}$
• Integral form	$\oint_S \mathbf{D} \cdot d\mathbf{s} = Q$ $\oint_C \mathbf{E} \cdot d\mathbf{l} = 0$	$\oint_S \mathbf{B} \cdot d\mathbf{s} = 0$ $\oint_C \mathbf{H} \cdot d\mathbf{l} = I$
Potential	Scalar V, with $\mathbf{E} = -\nabla V$	Vector $\mathbf{A}$, with $\mathbf{B} = \nabla \times \mathbf{A}$
Energy density	$w_e = \frac{1}{2}\epsilon E^2$	$w_m = \frac{1}{2}\mu H^2$
Force on charge q	$\mathbf{F}_e = q\mathbf{E}$	$\mathbf{F}_m = q\mathbf{u} \times \mathbf{B}$
Circuit element(s)	C and R	L

5-1 Magnetic Forces and Torques

The electric field $\mathbf{E}$ at a point in space was defined as the electric force $\mathbf{F}_e$ per unit charge acting on a charged test particle placed at that point. We now define the ***magnetic flux density*** $\mathbf{B}$ at a point in space in terms of the ***magnetic force*** $\mathbf{F}_m$ that acts on a charged test particle moving with velocity $\mathbf{u}$ through that point. Experiments revealed that a particle of charge q moving with velocity $\mathbf{u}$ in a magnetic field experiences a magnetic force $\mathbf{F}_m$ given by

$$\mathbf{F}_m = q\mathbf{u} \times \mathbf{B} \quad \text{(N)}. \tag{5.3}$$

Accordingly, the strength of $\mathbf{B}$ is measured in newtons/(C·m/s), also called the tesla (T). For a positively charged particle, the direction of $\mathbf{F}_m$ is that of the cross product $\mathbf{u} \times \mathbf{B}$, which is perpendicular to the plane containing $\mathbf{u}$ and $\mathbf{B}$ and governed by the right-hand rule. If q is negative, the direction of $\mathbf{F}_m$ is reversed (Fig. 5-1). The magnitude of $\mathbf{F}_m$ is given by

$$F_m = quB\sin\theta, \tag{5.4}$$

where θ is the angle between $\mathbf{u}$ and $\mathbf{B}$. We note that F_m is maximum when $\mathbf{u}$ is perpendicular to $\mathbf{B}$ ($\theta = 90°$), and zero when $\mathbf{u}$ is parallel to $\mathbf{B}$ ($\theta = 0$ or $180°$).

If a charged particle resides in the presence of both an electric field $\mathbf{E}$ and a magnetic field $\mathbf{B}$, then the total ***electromagnetic force*** acting on it is

$$\mathbf{F} = \mathbf{F}_e + \mathbf{F}_m = q\mathbf{E} + q\mathbf{u} \times \mathbf{B} = q(\mathbf{E} + \mathbf{u} \times \mathbf{B}). \tag{5.5}$$

The force expressed by Eq. (5.5) also is known as the ***Lorentz force***. Electric and magnetic forces exhibit a number of important differences:

1. Whereas the electric force is always in the direction of the electric field, the magnetic force is always perpendicular to the magnetic field.

2. Whereas the electric force acts on a charged particle whether or not it is moving, the magnetic force acts on it only when it is in motion.

3. Whereas the electric force expends energy in displacing a charged particle, the magnetic force does no work when a particle is displaced.

This last statement requires further elaboration. Because the magnetic force $\mathbf{F}_m$ is always perpendicular to $\mathbf{u}$, $\mathbf{F}_m \cdot \mathbf{u} = 0$.

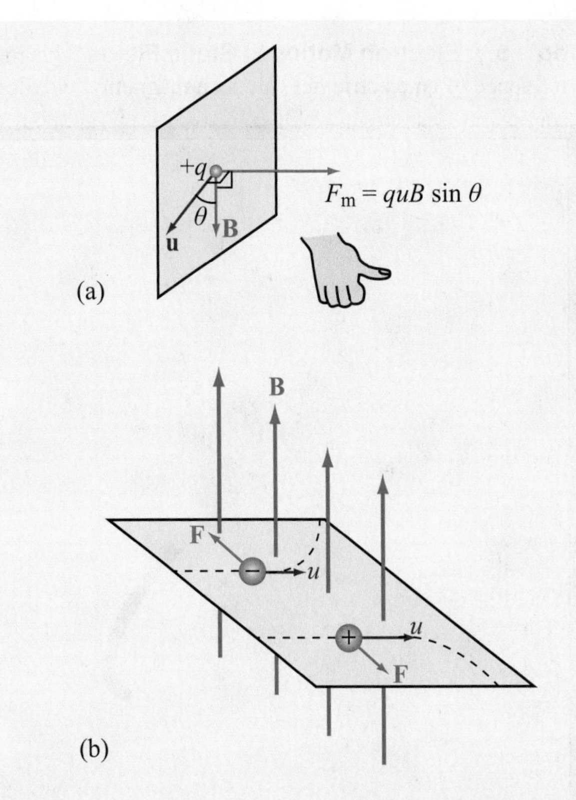

Figure 5-1 The direction of the magnetic force exerted on a charged particle moving in a magnetic field is (a) perpendicular to both $\mathbf{B}$ and $\mathbf{u}$ and (b) depends on the charge polarity (positive or negative).

Hence, the work performed when a particle is displaced by a differential distance $d\mathbf{l} = \mathbf{u}\,dt$ is

$$dW = \mathbf{F}_m \cdot d\mathbf{l} = (\mathbf{F}_m \cdot \mathbf{u})\,dt = 0. \tag{5.6}$$

► Since no work is done, a magnetic field cannot change the kinetic energy of a charged particle; the magnetic field can change the direction of motion of a charged particle, but not its speed. ◄

Exercise 5-1: An electron moving in the positive x direction perpendicular to a magnetic field is deflected in the negative z direction. What is the direction of the magnetic field?

Answer: Positive y direction. (See EM.)

Module 5.1 Electron Motion in Static Fields This module demonstrates the Lorentz force on an electron moving under the influence of an electric field alone, a magnetic field alone, or both acting simultaneously.

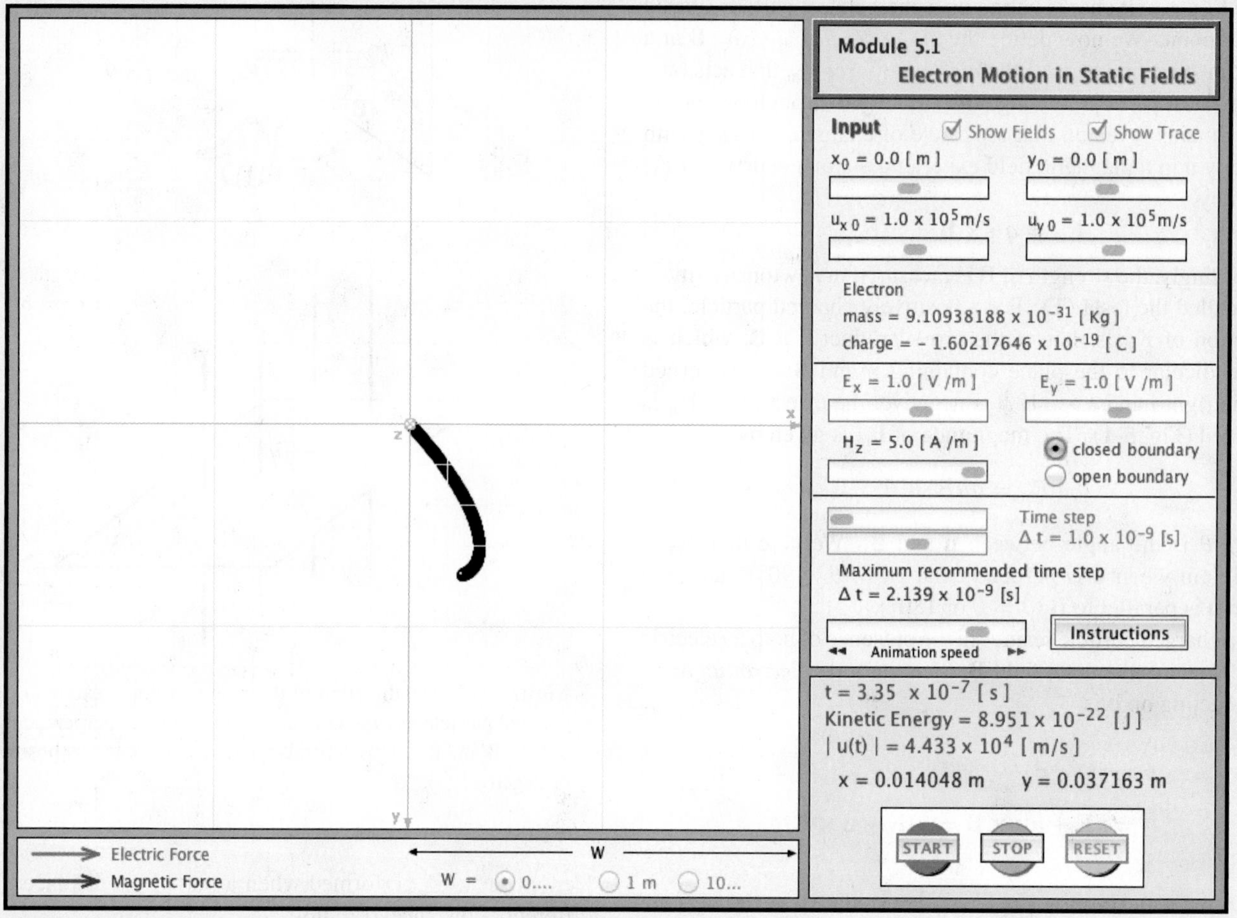

Exercise 5-2: A proton moving with a speed of 2×10^6 m/s through a magnetic field with magnetic flux density of 2.5 T experiences a magnetic force of magnitude 4×10^{-13} N. What is the angle between the magnetic field and the proton's velocity?

Answer: $\theta = 30°$ or $150°$. (See EM.)

Exercise 5-3: A charged particle with velocity **u** is moving in a medium with uniform fields $\mathbf{E} = \hat{\mathbf{x}}E$ and $\mathbf{B} = \hat{\mathbf{y}}B$. What should **u** be so that the particle experiences no net force?

Answer: $\mathbf{u} = \hat{\mathbf{z}}E/B$. [**u** may also have an arbitrary y component u_y]. (See EM.)

5-1.1 Magnetic Force on a Current-Carrying Conductor

A current flowing through a conducting wire consists of charged particles *drifting* through the material of the wire. Consequently, when a current-carrying wire is placed in a magnetic field, it experiences a force equal to the sum of the magnetic forces acting on the charged particles moving within it. Consider, for example, the arrangement shown in **Fig. 5-2**

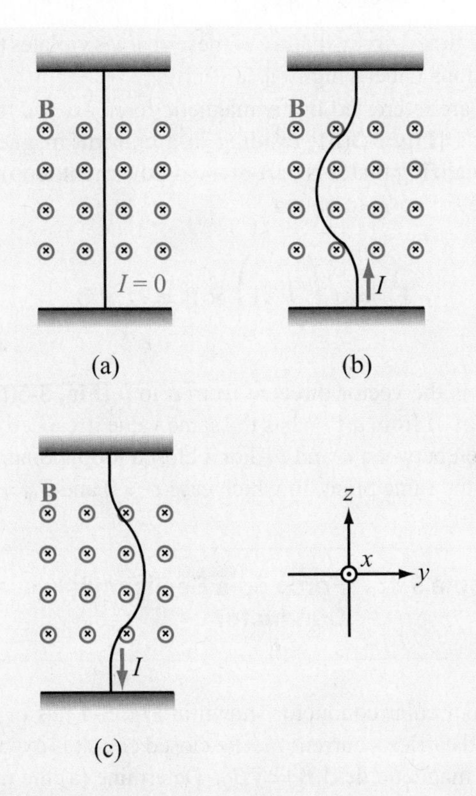

Figure 5-2 When a slightly flexible vertical wire is placed in a magnetic field directed into the page (as denoted by the crosses), it is (a) not deflected when the current through it is zero, (b) deflected to the left when I is upward, and (c) deflected to the right when I is downward.

in which a vertical wire oriented along the z direction is placed in a magnetic field $\mathbf{B}$ (produced by a magnet) oriented along the $-\hat{\mathbf{x}}$ direction (into the page). With no current flowing in the wire, $\mathbf{F}_\text{m} = 0$ and the wire maintains its vertical orientation [Fig. 5-2(a)], but when a current is introduced in the wire, the wire deflects to the left ($-\hat{\mathbf{y}}$ direction) if the current direction is upward ($+\hat{\mathbf{z}}$ direction), and to the right ($+\hat{\mathbf{y}}$ direction) if the current direction is downward ($-\hat{\mathbf{z}}$ direction). The directions of these deflections are in accordance with the cross product given by Eq. (5.3).

To quantify the relationship between $\mathbf{F}_\text{m}$ and the current I flowing in a wire, let us consider a small segment of the wire of cross-sectional area A and differential length $d\mathbf{l}$, with the direction of $d\mathbf{l}$ denoting the direction of the current. Without loss of generality, we assume that the charge carriers constituting the current I are exclusively electrons, which is always a valid assumption for a good conductor. If the wire contains a free-electron charge density $\rho_\text{ve} = -N_\text{e}e$, where N_e is the number of moving electrons per unit volume, then the total amount of moving charge contained in an elemental volume of the wire is

$$dQ = \rho_\text{ve} A\ dl = -N_\text{e} eA\ dl, \tag{5.7}$$

and the corresponding magnetic force acting on dQ in the presence of a magnetic field $\mathbf{B}$ is

$$d\mathbf{F}_\text{m} = dQ\ \mathbf{u}_\text{e} \times \mathbf{B} = -N_\text{e} eA\ dl\ \mathbf{u}_\text{e} \times \mathbf{B}, \tag{5.8a}$$

where $\mathbf{u}_\text{e}$ is the drift velocity of the electrons. Since the direction of a current is defined as the direction of flow of positive charges, the electron drift velocity $\mathbf{u}_\text{e}$ is parallel to $d\mathbf{l}$, but opposite in direction. Thus, $dl\ \mathbf{u}_\text{e} = -d\mathbf{l}\ u_\text{e}$ and Eq. (5.8a) becomes

$$d\mathbf{F}_\text{m} = N_\text{e} eAu_\text{e}\ d\mathbf{l} \times \mathbf{B}. \tag{5.8b}$$

From Eqs. (4.11) and (4.12), the current I flowing through a cross-sectional area A due to electrons with density $\rho_\text{ve} = -N_\text{e}e$, moving with velocity $-u_\text{e}$, is $I = \rho_\text{ve}(-u_\text{e})A = (-N_\text{e}e)(-u_\text{e})A = N_\text{e}eAu_\text{e}$. Hence, Eq. (5.8b) may be written in the compact form

$$d\mathbf{F}_\text{m} = I\ d\mathbf{l} \times \mathbf{B} \qquad \text{(N)}. \tag{5.9}$$

For a closed circuit of contour C carrying a current I, the total magnetic force is

$$\mathbf{F}_\text{m} = I \oint_C d\mathbf{l} \times \mathbf{B} \qquad \text{(N)}. \tag{5.10}$$

If the closed wire shown in Fig. 5-3(a) resides in a uniform external magnetic field $\mathbf{B}$, then $\mathbf{B}$ can be taken outside the integral in Eq. (5.10), in which case

$$\mathbf{F}_\text{m} = I \left(\oint_C d\mathbf{l} \right) \times \mathbf{B} = 0. \tag{5.11}$$

▶ This result, which is a consequence of the fact that the vector sum of the infinitesimal vectors $d\mathbf{l}$ over a closed path equals zero, states that the total magnetic force on any closed current loop in a uniform magnetic field is zero. ◀

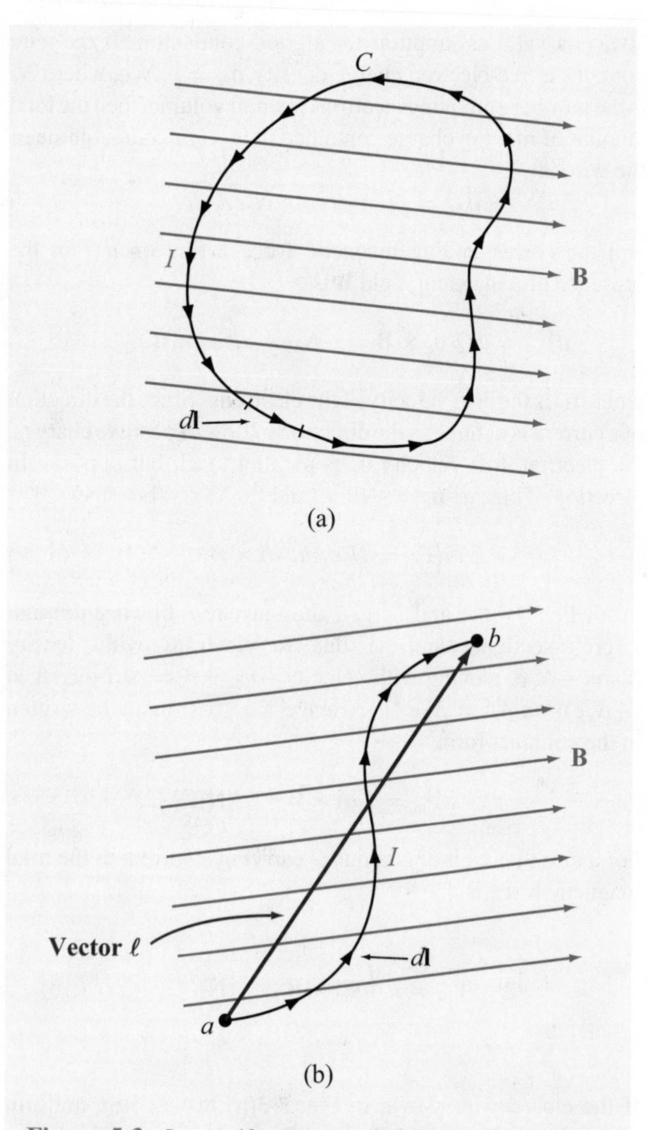

Figure 5-3 In a uniform magnetic field, (a) the net force on a closed current loop is zero because the integral of the displacement vector $d\mathbf{l}$ over a closed contour is zero, and (b) the force on a line segment is proportional to the vector between the end point ($\mathbf{F}_{\text{m}} = I\boldsymbol{\ell} \times \mathbf{B}$).

In the study of magnetostatics, all currents flow through closed paths. To understand why, consider the curved wire in **Fig. 5-3(b)** carrying a current I from point a to point b. In doing so, negative charges accumulate at a, and positive ones at b. The time-varying nature of these charges violates the static assumptions underlying Eqs. (5-1a,b).

If we are interested in the magnetic force exerted on a wire segment l [**Fig. 5-3(b)**] residing in a uniform magnetic field (while realizing that it is part of a closed current loop), we can integrate Eq. (5.9) to obtain

$$\mathbf{F}_{\text{m}} = I \left(\int_{\ell} d\mathbf{l} \right) \times \mathbf{B} = I\boldsymbol{\ell} \times \mathbf{B}, \tag{5.12}$$

where $\boldsymbol{\ell}$ is the vector directed from a to b [**Fig. 5-3(b)**]. The integral of $d\mathbf{l}$ from a to b has the same value irrespective of the path taken between a and b. For a closed loop, points a and b become the same point, in which case $\boldsymbol{\ell} = 0$ and $\mathbf{F}_{\text{m}} = 0$.

Example 5-1: Force on a Semicircular Conductor

The semicircular conductor shown in **Fig. 5-4** lies in the x–y plane and carries a current I. The closed circuit is exposed to a uniform magnetic field $\mathbf{B} = \hat{\mathbf{y}}B_0$. Determine (a) the magnetic

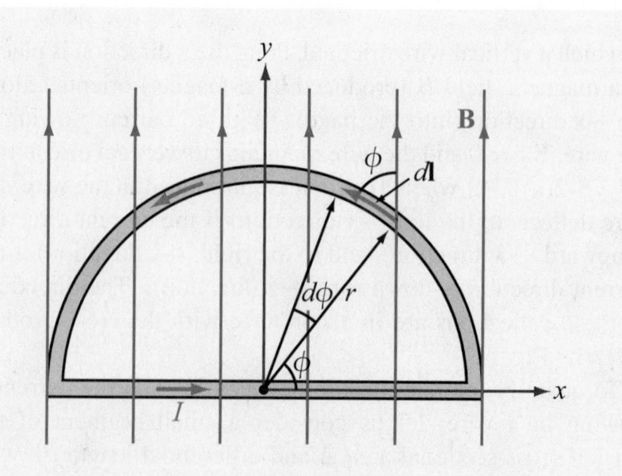

Figure 5-4 Semicircular conductor in a uniform field (Example 5-1).

force $\mathbf{F}_1$ on the straight section of the wire and (b) the force $\mathbf{F}_2$ on the curved section.

Solution: (a) To evaluate $\mathbf{F}_1$, consider that the straight section of the circuit is of length $2r$ and its current flows along the $+x$ direction. Application of Eq. (5.12) with $\boldsymbol{\ell} = \hat{\mathbf{x}}\,2r$ gives

$$\mathbf{F}_1 = \hat{\mathbf{x}}(2Ir) \times \hat{\mathbf{y}}B_0 = \hat{\mathbf{z}}\,2IrB_0 \qquad \text{(N)}.$$

The $\hat{\mathbf{z}}$ direction in Fig. 5-4 is *out of the page*.

(b) To evaluate $\mathbf{F}_2$, consider a segment of differential length $d\mathbf{l}$ on the curved part of the circle. The direction of $d\mathbf{l}$ is chosen to coincide with the direction of the current. Since $d\mathbf{l}$ and $\mathbf{B}$ are both in the x–y plane, their cross product $d\mathbf{l} \times \mathbf{B}$ points in the negative z direction, and the magnitude of $d\mathbf{l} \times \mathbf{B}$ is proportional to $\sin\phi$, where ϕ is the angle between $d\mathbf{l}$ and $\mathbf{B}$. Moreover, the magnitude of $d\mathbf{l}$ is $dl = r\,d\phi$. Hence,

$$\mathbf{F}_2 = I \int_{\phi=0}^{\pi} d\mathbf{l} \times \mathbf{B}$$

$$= -\hat{\mathbf{z}}I \int_{\phi=0}^{\pi} rB_0 \sin\phi\, d\phi = -\hat{\mathbf{z}}\,2IrB_0 \qquad \text{(N)}.$$

The $-\hat{\mathbf{z}}$ direction of the force acting on the curved part of the conductor is *into the page*. We note that $\mathbf{F}_2 = -\mathbf{F}_1$, implying that no net force acts on the closed loop, although opposing forces act on its two sections.

Concept Question 5-1: What are the major differences between the electric force $\mathbf{F}_e$ and the magnetic force $\mathbf{F}_m$?

Concept Question 5-2: The ends of a 10 cm long wire carrying a constant current I are anchored at two points on the x axis, namely $x = 0$ and $x = 6$ cm. If the wire lies in the x–y plane in a magnetic field $\mathbf{B} = \hat{\mathbf{y}}B_0$, which of the following arrangements produces a greater magnetic force on the wire: (a) wire is V-shaped with corners at (0, 0), (3, 4), and (6, 0), (b) wire is an open rectangle with corners at (0, 0), (0, 2), (6, 2), and (6, 0).

Exercise 5-4: A horizontal wire with a mass per unit length of 0.2 kg/m carries a current of 4 A in the $+x$ direction. If the wire is placed in a uniform magnetic flux density $\mathbf{B}$, what should the direction and minimum magnitude of $\mathbf{B}$ be in order to magnetically lift the wire vertically upward? [*Hint:* The acceleration due to gravity is $\mathbf{g} = -\hat{\mathbf{z}}\,9.8$ m/s^2.]

Answer: $\mathbf{B} = \hat{\mathbf{y}}0.49$ T. (See EM.)

5-1.2 Magnetic Torque on a Current-Carrying Loop

When a force is applied on a rigid body that can pivot about a fixed axis, the body will, in general, react by rotating about that axis. The angular acceleration depends on the cross product of the applied force vector $\mathbf{F}$ and the distance vector $\mathbf{d}$, measured *from* a point on the rotation axis (such that $\mathbf{d}$ is perpendicular to the axis) *to* the point of application of $\mathbf{F}$ (Fig. 5-5). The length of $\mathbf{d}$ is called the ***moment arm***, and the cross product

$$\mathbf{T} = \mathbf{d} \times \mathbf{F} \qquad \text{(N·m)} \tag{5.13}$$

is called the ***torque***. The unit for $\mathbf{T}$ is the same as that for work or energy, even though torque does not represent either. The force $\mathbf{F}$ applied on the disk shown in Fig. 5-5 lies in the x–y plane and makes an angle θ with $\mathbf{d}$. Hence,

$$\mathbf{T} = \hat{\mathbf{z}}rF\sin\theta, \tag{5.14}$$

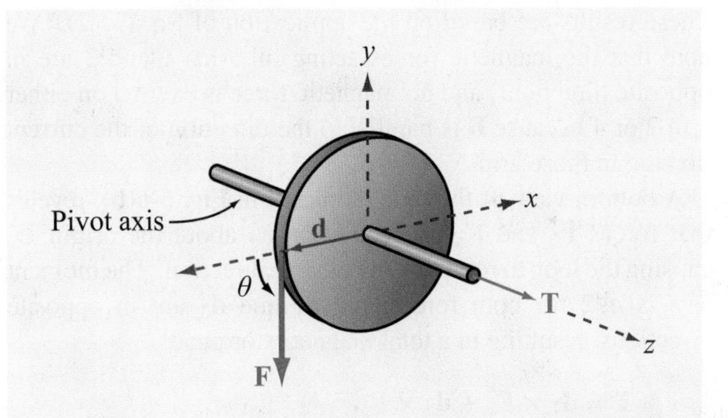

Figure 5-5 The force $\mathbf{F}$ acting on a circular disk that can pivot along the z axis generates a torque $\mathbf{T} = \mathbf{d} \times \mathbf{F}$ that causes the disk to rotate.

where $|\mathbf{d}| = r$, the radius of the disk, and $F = |\mathbf{F}|$. From Eq. (5.14) we observe that a torque along the positive z direction corresponds to a tendency for the cylinder to rotate counterclockwise and, conversely, a torque along the $-z$ direction corresponds to clockwise rotation.

▶ These directions are governed by the following ***right-hand rule:*** when the thumb of the right hand points along the direction of the torque, the four fingers indicate the direction that the torque tries to rotate the body. ◀

We now consider the ***magnetic torque*** exerted on a conducting loop under the influence of magnetic forces. We begin with the simple case where the magnetic field $\mathbf{B}$ is in the plane of the loop, and then we extend the analysis to the more general case where $\mathbf{B}$ makes an angle θ with the surface normal of the loop.

Magnetic Field in the Plane of the Loop

The rectangular conducting loop shown in **Fig. 5-6(a)** is constructed from rigid wire and carries a current I. The loop lies in the x–y plane and is allowed to pivot about the axis shown. Under the influence of an externally generated uniform magnetic field $\mathbf{B} = \hat{\mathbf{x}}B_0$, arms 1 and 3 of the loop are subjected to forces $\mathbf{F}_1$ and $\mathbf{F}_3$, given by

$$\mathbf{F}_1 = I(-\hat{\mathbf{y}}b) \times (\hat{\mathbf{x}}B_0) = \hat{\mathbf{z}}IbB_0, \qquad (5.15a)$$

and

$$\mathbf{F}_3 = I(\hat{\mathbf{y}}b) \times (\hat{\mathbf{x}}B_0) = -\hat{\mathbf{z}}IbB_0. \qquad (5.15b)$$

These results are based on the application of Eq. (5.12). We note that the magnetic forces acting on arms 1 and 3 are in opposite directions, and no magnetic force is exerted on either arm 2 or 4 because $\mathbf{B}$ is parallel to the direction of the current flowing in those arms.

A bottom view of the loop, depicted in **Fig. 5-6(b)**, reveals that forces $\mathbf{F}_1$ and $\mathbf{F}_3$ produce a torque about the origin O, causing the loop to rotate in a clockwise direction. The moment arm is $a/2$ for both forces, but $\mathbf{d}_1$ and $\mathbf{d}_3$ are in opposite directions, resulting in a total magnetic torque of

$$\begin{aligned}\mathbf{T} &= \mathbf{d}_1 \times \mathbf{F}_1 + \mathbf{d}_3 \times \mathbf{F}_3 \\ &= \left(-\hat{\mathbf{x}}\,\frac{a}{2}\right) \times (\hat{\mathbf{z}}IbB_0) + \left(\hat{\mathbf{x}}\,\frac{a}{2}\right) \times (-\hat{\mathbf{z}}IbB_0) \\ &= \hat{\mathbf{y}}IabB_0 = \hat{\mathbf{y}}IAB_0, \end{aligned} \qquad (5.16)$$

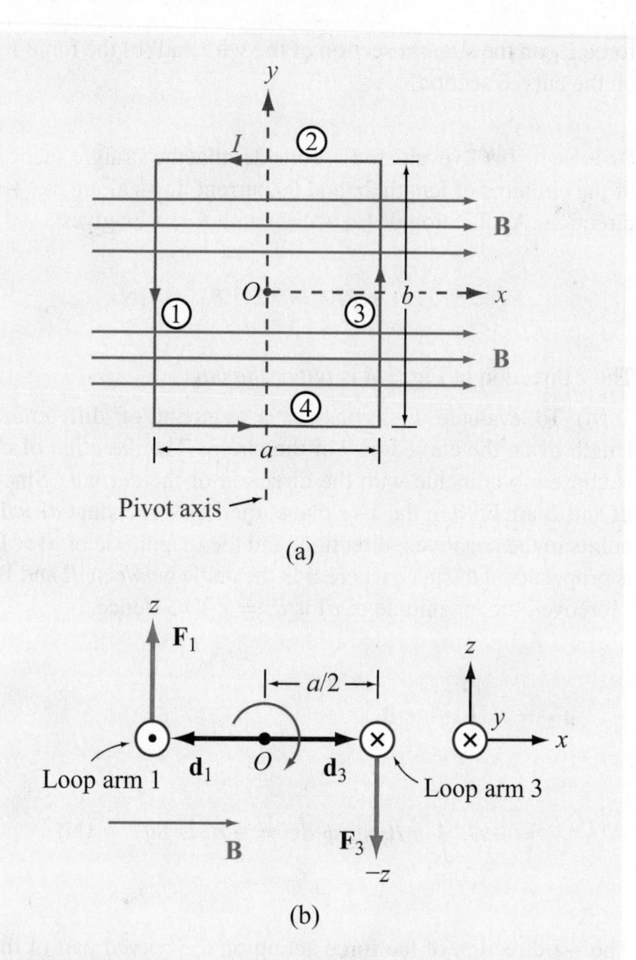

Figure 5-6 Rectangular loop pivoted along the y axis: (a) front view and (b) bottom view. The combination of forces $\mathbf{F}_1$ and $\mathbf{F}_3$ on the loop generates a torque that tends to rotate the loop in a clockwise direction as shown in (b).

where $A = ab$ is the area of the loop. The right-hand rule tells us that the sense of rotation is clockwise. The result given by Eq. (5.16) is valid only when the magnetic field $\mathbf{B}$ is parallel to the plane of the loop. As soon as the loop starts to rotate, the torque $\mathbf{T}$ decreases, and at the end of one quarter of a complete rotation, the torque becomes zero, as discussed next.

Magnetic Field Perpendicular to the Axis of a Rectangular Loop

For the situation represented by **Fig. 5-7**, where $\mathbf{B} = \hat{\mathbf{x}}B_0$, the field is still perpendicular to the loop's axis of rotation, but

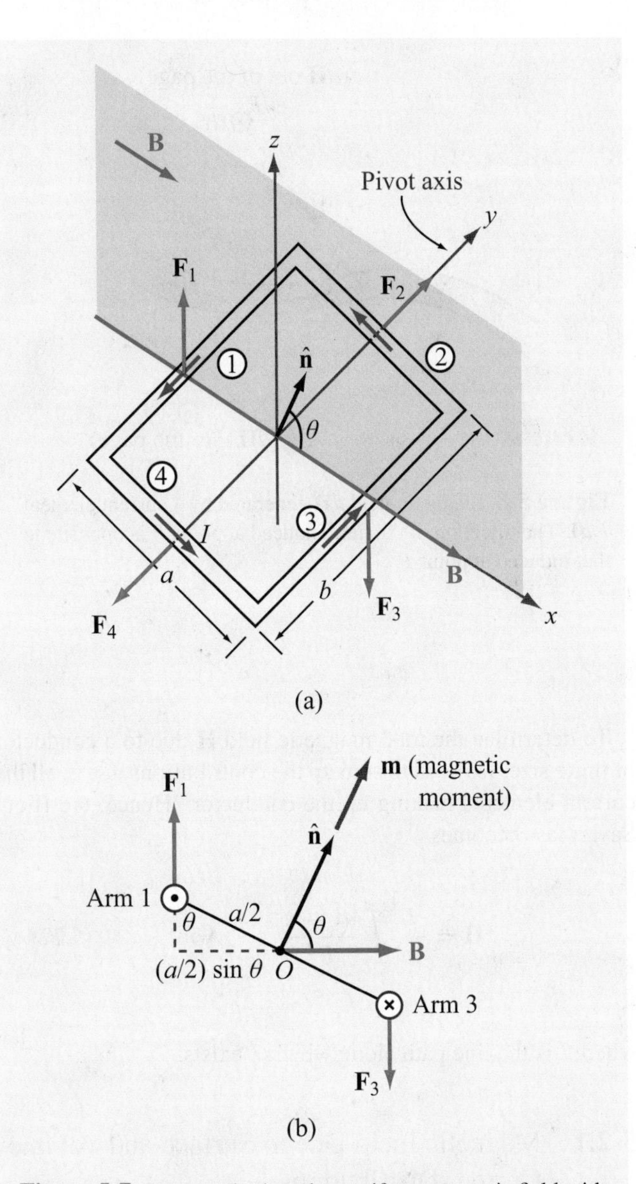

Figure 5-7 Rectangular loop in a uniform magnetic field with flux density **B** whose direction is perpendicular to the rotation axis of the loop but makes an angle θ with the loop's surface normal $\hat{\mathbf{n}}$.

because its direction may be at any angle θ with respect to the loop's surface normal $\hat{\mathbf{n}}$, we may now have nonzero forces on all four arms of the rectangular loop. However, forces $\mathbf{F}_2$ and $\mathbf{F}_4$ are equal in magnitude and opposite in direction and are along the rotation axis; hence, the net torque contributed by their combination is zero. The directions of the currents in arms 1 and 3 are always perpendicular to $\mathbf{B}$ regardless of the magnitude of θ. Hence, $\mathbf{F}_1$ and $\mathbf{F}_3$ have the same expressions given previously by Eqs. (5.15a,b), and for $0 \le \theta \le \pi/2$ their moment arms are of magnitude $(a/2)\sin\theta$, as illustrated in **Fig. 5-7(b)**. Consequently, the magnitude of the net torque exerted by the magnetic field about the axis of rotation is the same as that given by Eq. (5.16), but modified by $\sin\theta$:

$$T = IAB_0 \sin\theta. \qquad (5.17)$$

According to Eq. (5.17), the torque is maximum when the magnetic field is parallel to the plane of the loop ($\theta = 90°$) and zero when the field is perpendicular to the plane of the loop ($\theta = 0$). If the loop consists of N turns, each contributing a torque given by Eq. (5.17), then the total torque is

$$T = NIAB_0 \sin\theta. \qquad (5.18)$$

The quantity NIA is called the ***magnetic moment*** m of the loop. Now, consider the vector

$$\mathbf{m} = \hat{\mathbf{n}} NIA = \hat{\mathbf{n}} m \qquad (\text{A}\cdot\text{m}^2), \qquad (5.19)$$

where $\hat{\mathbf{n}}$ is the surface normal of the loop and governed by the following ***right-hand rule:*** *when the four fingers of the right hand advance in the direction of the current I, the direction of the thumb specifies the direction of* $\hat{\mathbf{n}}$. In terms of $\mathbf{m}$, the torque vector $\mathbf{T}$ can be written as

$$\mathbf{T} = \mathbf{m} \times \mathbf{B} \qquad (\text{N}\cdot\text{m}). \qquad (5.20)$$

Even though the derivation leading to Eq. (5.20) was obtained for $\mathbf{B}$ perpendicular to the axis of rotation of a rectangular loop, the expression is valid for any orientation of $\mathbf{B}$ and for a loop of any shape.

Concept Question 5-3: How is the direction of the magnetic moment of a loop defined?

Concept Question 5-4: If one of two wires of equal length is formed into a closed square loop and the other into a closed circular loop, and if both wires are carrying equal currents and both loops have their planes parallel to a uniform magnetic field, which loop would experience the greater torque?

Exercise 5-5: A square coil of 100 turns and 0.5 m long sides is in a region with a uniform magnetic flux density of 0.2 T. If the maximum magnetic torque exerted on the coil is 4×10^{-2} (N·m), what is the current flowing in the coil?

Answer: $I = 8$ mA. (See EM.)

5-2 The Biot–Savart Law

In the preceding section, we elected to use the magnetic flux density $\mathbf{B}$ to denote the presence of a magnetic field in a given region of space. We now work with the magnetic field intensity $\mathbf{H}$ instead. We do this in part to remind the reader that for most materials the flux and field are linearly related by $\mathbf{B} = \mu\mathbf{H}$, and therefore knowledge of one implies knowledge of the other (assuming that μ is known).

Through his experiments on the deflection of compass needles by current-carrying wires, Hans Oersted established that currents induce magnetic fields that form closed loops around the wires (see Section 1-3.3). Building upon Oersted's results, Jean Biot and Félix Savart arrived at an expression that relates the magnetic field $\mathbf{H}$ at any point in space to the current I that generates $\mathbf{H}$. The ***Biot–Savart law*** states that the differential magnetic field $d\mathbf{H}$ generated by a steady current I flowing through a differential length vector $d\mathbf{l}$ is

$$d\mathbf{H} = \frac{I}{4\pi} \frac{d\mathbf{l} \times \hat{\mathbf{R}}}{R^2} \quad \text{(A/m)}, \tag{5.21}$$

where $\mathbf{R} = \hat{\mathbf{R}}R$ is the distance vector between $d\mathbf{l}$ and the observation point P shown in Fig. 5-8. The SI unit for $\mathbf{H}$ is ampere·m/m^2 = (A/m). It is important to remember that Eq. (5.21) assumes that $d\mathbf{l}$ is along the direction of the current I and the unit vector $\hat{\mathbf{R}}$ points *from* the current element *to* the observation point.

According to Eq. (5.21), $d\mathbf{H}$ varies as R^{-2}, which is similar to the distance dependence of the electric field induced by an electric charge. However, unlike the electric field vector $\mathbf{E}$, whose direction is *along* the distance vector $\mathbf{R}$ joining the charge to the observation point, the magnetic field $\mathbf{H}$ is *orthogonal* to the plane containing the direction of the current element $d\mathbf{l}$ and the distance vector $\mathbf{R}$. At point P in Fig. 5-8, the direction of $d\mathbf{H}$ is out of the page, whereas at point P' the direction of $d\mathbf{H}$ is into the page.

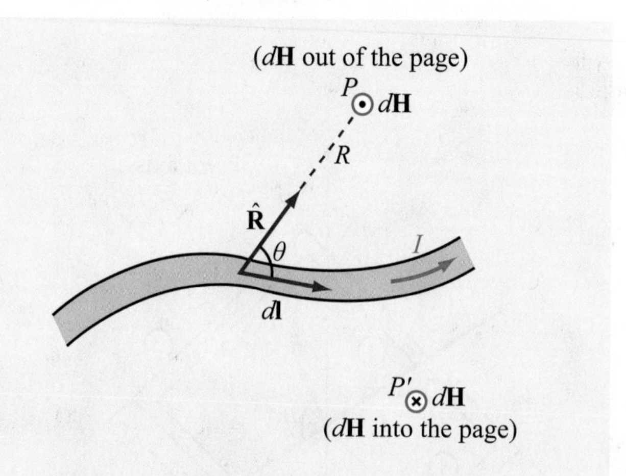

Figure 5-8 Magnetic field $d\mathbf{H}$ generated by a current element $I\,d\mathbf{l}$. The direction of the field induced at point P is opposite to that induced at point P'.

To determine the total magnetic field $\mathbf{H}$ due to a conductor of finite size, we need to sum up the contributions due to all the current elements making up the conductor. Hence, the Biot–Savart law becomes

$$\mathbf{H} = \frac{I}{4\pi} \int_l \frac{d\mathbf{l} \times \hat{\mathbf{R}}}{R^2} \quad \text{(A/m)}, \tag{5.22}$$

where l is the line path along which I exists.

5-2.1 Magnetic Field Due to Surface and Volume Current Distributions

The Biot–Savart law may also be expressed in terms of distributed current sources (Fig. 5-9) such as the ***volume current density*** $\mathbf{J}$, measured in (A/m^2), or the ***surface current density*** $\mathbf{J}_s$, measured in (A/m). The surface current density $\mathbf{J}_s$ applies to currents that flow on the surface of a conductor in the form of a sheet of effectively zero thickness. When current sources are specified in terms of $\mathbf{J}_s$ over a surface S or in terms of $\mathbf{J}$ over a volume $\mathcal{V}$, we can use the equivalence given by

$$I\,d\mathbf{l} \leftrightarrow \mathbf{J}_s\,ds \leftrightarrow \mathbf{J}\,d\mathcal{V} \tag{5.23}$$

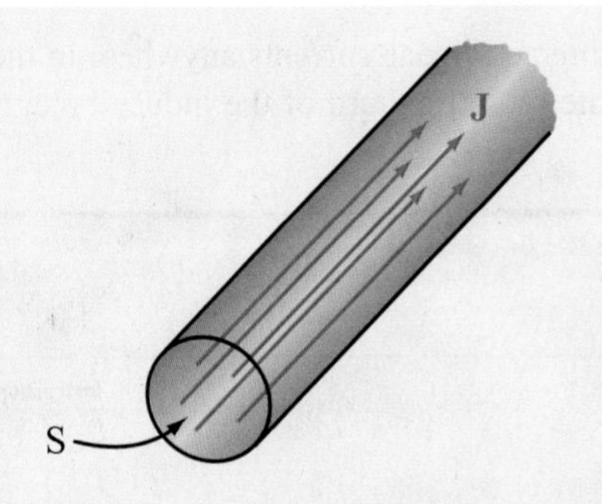

(a) Volume current density **J** in A/m²

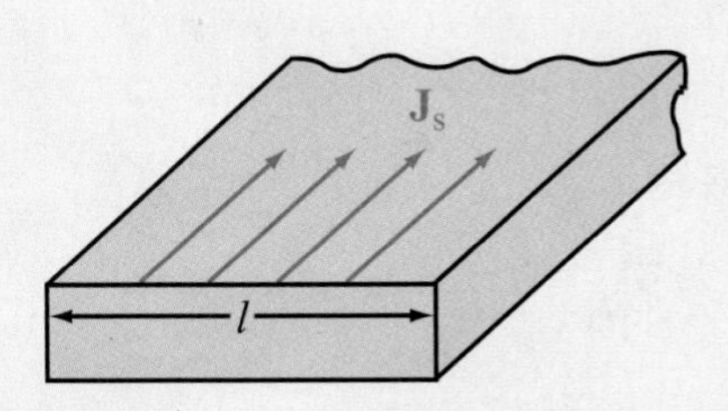

(b) Surface current density $\mathbf{J}_s$ in A/m

Figure 5-9 (a) The total current crossing the cross section S of the cylinder is $I = \int_S \mathbf{J} \cdot d\mathbf{s}$. (b) The total current flowing across the surface of the conductor is $I = \int_l J_s \, dl$.

to express the Biot–Savart law as

$$\mathbf{H} = \frac{1}{4\pi} \int_S \frac{\mathbf{J}_s \times \hat{\mathbf{R}}}{R^2} \, ds, \qquad (5.24a)$$

(surface current)

$$\mathbf{H} = \frac{1}{4\pi} \int_{\mathcal{V}} \frac{\mathbf{J} \times \hat{\mathbf{R}}}{R^2} \, d\mathcal{V}. \qquad (5.24b)$$

(volume current)

Example 5-2: Magnetic Field of a Linear Conductor

A free-standing linear conductor of length l carries a current I along the z axis as shown in Fig. 5-10. Determine the magnetic flux density **B** at a point P located at a distance r in the x–y plane.

Solution: From Fig. 5-10, the differential length vector $d\mathbf{l} = \hat{\mathbf{z}} \, dz$. Hence, $d\mathbf{l} \times \hat{\mathbf{R}} = dz \, (\hat{\mathbf{z}} \times \hat{\mathbf{R}}) = \hat{\boldsymbol{\phi}} \sin\hat{\boldsymbol{\theta}} \, dz$, where $\hat{\boldsymbol{\phi}}$ is the azimuth direction and θ is the angle between $d\mathbf{l}$ and $\hat{\mathbf{R}}$.

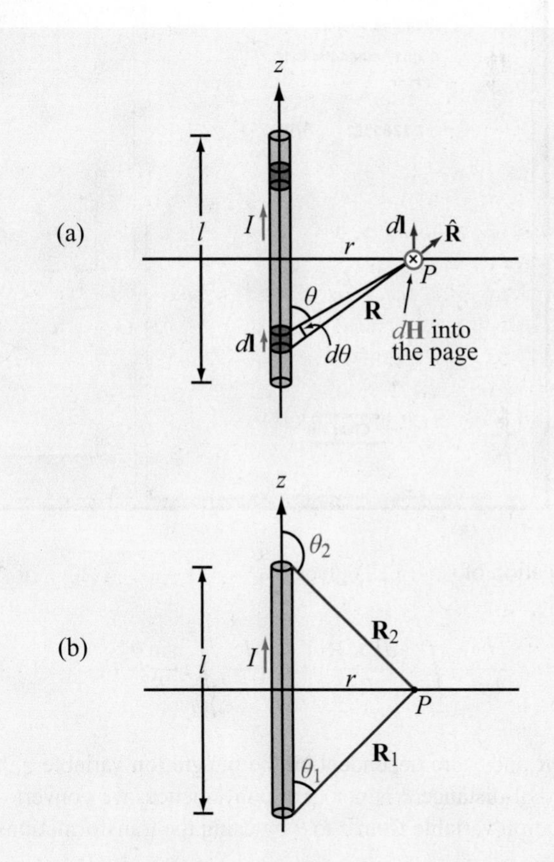

Figure 5-10 Linear conductor of length l carrying a current I. (a) The field $d\mathbf{H}$ at point P due to incremental current element $d\mathbf{l}$. (b) Limiting angles θ_1 and θ_2, each measured between vector $I \, d\mathbf{l}$ and the vector connecting the end of the conductor associated with that angle to point P (Example 5-2).

Module 5.2 Magnetic Fields due to Line Sources You can place z-directed linear currents anywhere in the display plane (x-y plane), select their magnitudes and directions, and then observe the spatial pattern of the induced magnetic flux $\mathbf{B}(x, y)$.

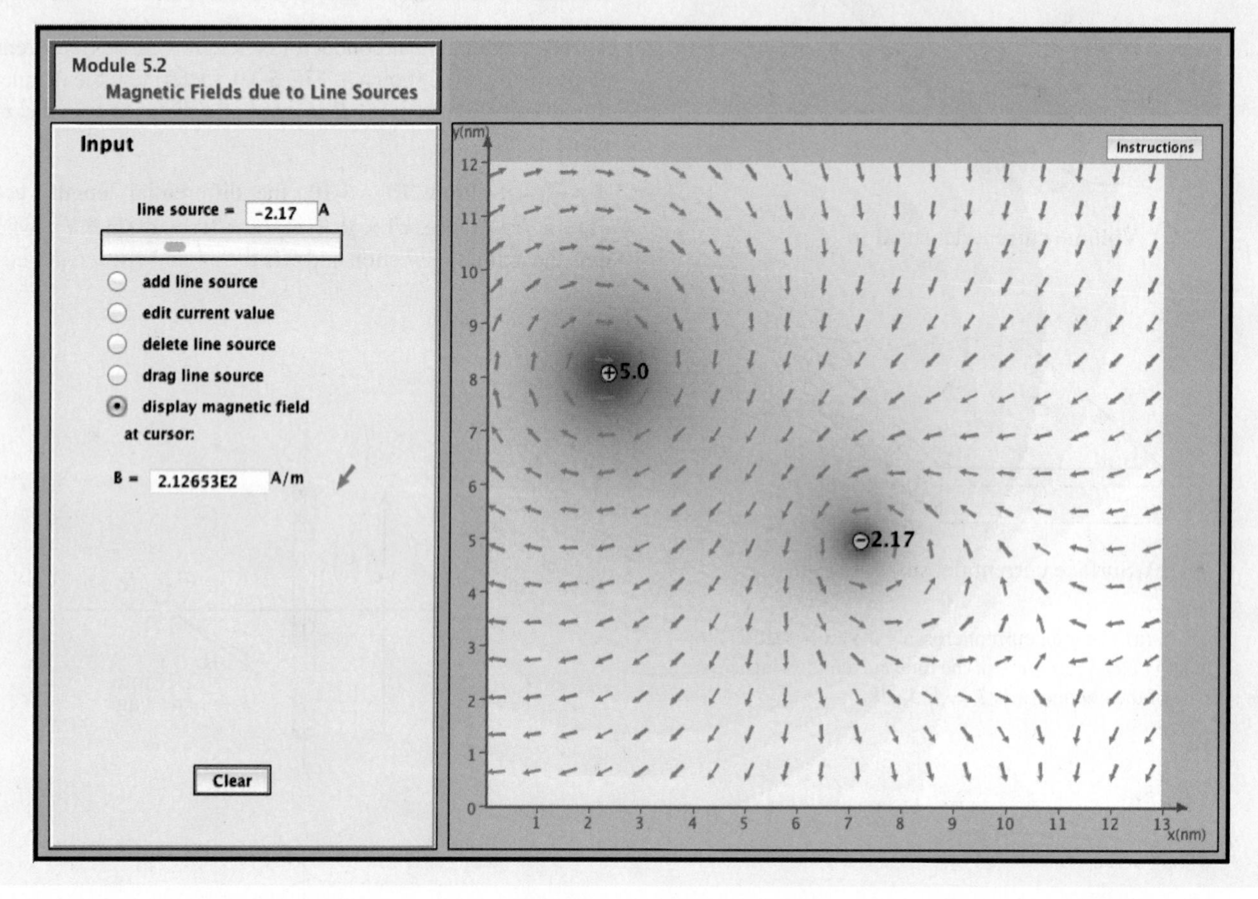

Application of Eq. (5.22) gives

$$\mathbf{H} = \frac{I}{4\pi} \int_{z=-l/2}^{z=l/2} \frac{d\mathbf{l} \times \hat{\mathbf{R}}}{R^2} = \hat{\boldsymbol{\phi}} \, \frac{I}{4\pi} \int_{-l/2}^{l/2} \frac{\sin\theta}{R^2} \, dz. \tag{5.25}$$

Both R and θ are dependent on the integration variable z, but the radial distance r is not. For convenience, we convert the integration variable from z to θ by using the transformations

$$R = r \csc\theta, \tag{5.26a}$$

$$z = -r \cot\theta, \tag{5.26b}$$

$$dz = r \csc^2\theta \, d\theta. \tag{5.26c}$$

Upon inserting Eqs. (5.26a) and (5.26c) into Eq. (5.25), we have

$$\begin{aligned} \mathbf{H} &= \hat{\boldsymbol{\phi}} \, \frac{I}{4\pi} \int_{\theta_1}^{\theta_2} \frac{\sin\theta \, r \csc^2\theta \, d\theta}{r^2 \csc^2\theta} \\ &= \hat{\boldsymbol{\phi}} \, \frac{I}{4\pi r} \int_{\theta_1}^{\theta_2} \sin\theta \, d\theta \\ &= \hat{\boldsymbol{\phi}} \, \frac{I}{4\pi r} (\cos\theta_1 - \cos\theta_2), \end{aligned} \tag{5.27}$$

where θ_1 and θ_2 are the limiting angles at $z = -l/2$ and $z = l/2$, respectively. From the right triangle in **Fig. 5-10(b)**, it follows that

$$\cos\theta_1 = \frac{l/2}{\sqrt{r^2 + (l/2)^2}}, \tag{5.28a}$$

$$\cos\theta_2 = -\cos\theta_1 = \frac{-l/2}{\sqrt{r^2 + (l/2)^2}}. \tag{5.28b}$$

Hence,

$$\mathbf{B} = \mu_0 \mathbf{H} = \hat{\boldsymbol{\phi}}\, \frac{\mu_0 I l}{2\pi r\sqrt{4r^2 + l^2}} \quad \text{(T)}. \tag{5.29}$$

For an infinitely long wire with $l \gg r$, Eq. (5.29) reduces to

$$\mathbf{B} = \hat{\boldsymbol{\phi}}\, \frac{\mu_0 I}{2\pi r} \quad \textbf{(infinitely long wire)}. \tag{5.30}$$

▶ This is a very important and useful expression to keep in mind. It states that in the neighborhood of a linear conductor carrying a current I, the induced magnetic field forms concentric circles around the wire (**Fig. 5-11**), and its intensity is directly proportional to I and inversely proportional to the distance r. ◀

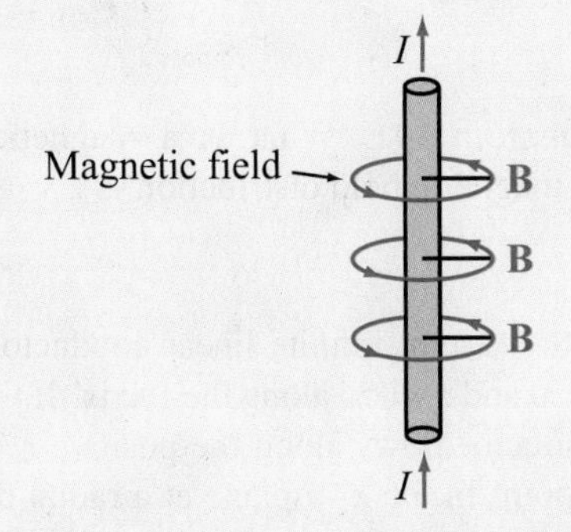

Figure 5-11 Magnetic field surrounding a long, linear current-carrying conductor.

Example 5-3: Magnetic Field of a Circular Loop

A circular loop of radius a carries a steady current I. Determine the magnetic field $\mathbf{H}$ at a point on the axis of the loop.

Solution: Let us place the loop in the x–y plane (**Fig. 5-12**). Our task is to obtain an expression for $\mathbf{H}$ at point $P(0, 0, z)$.

We start by noting that any element $d\mathbf{l}$ on the circular loop is perpendicular to the distance vector $\mathbf{R}$, and that all elements around the loop are at the same distance R from P, with $R = \sqrt{a^2 + z^2}$. From Eq. (5.21), the magnitude of $d\mathbf{H}$ due to current element $d\mathbf{l}$ is

$$dH = \frac{I}{4\pi R^2}\, |d\mathbf{l} \times \hat{\mathbf{R}}| = \frac{I\, dl}{4\pi(a^2 + z^2)}, \tag{5.31}$$

and the direction of $d\mathbf{H}$ is perpendicular to the plane containing $\mathbf{R}$ and $d\mathbf{l}$. $d\mathbf{H}$ is in the r–z plane (**Fig. 5-12**), and therefore it has components dH_r and dH_z. If we consider element $d\mathbf{l}'$, located diametrically opposite to $d\mathbf{l}$, we observe that the z components of the magnetic fields due to $d\mathbf{l}$ and $d\mathbf{l}'$ add because they are

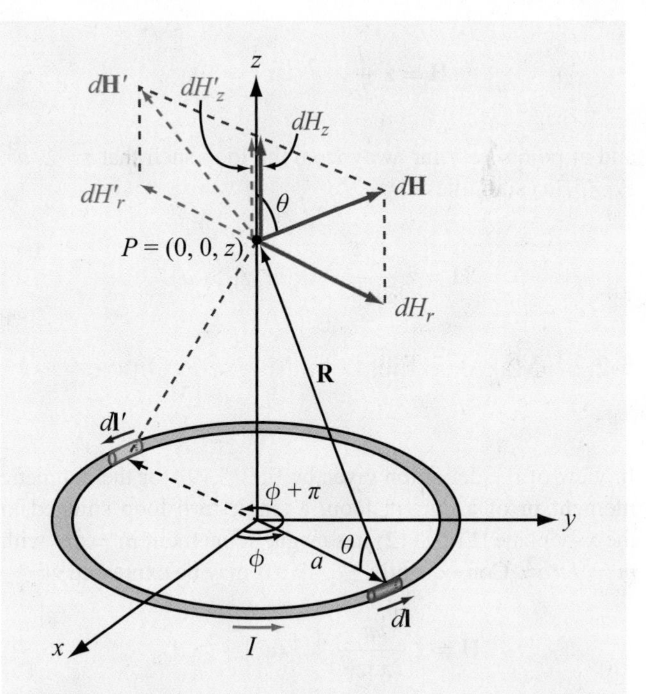

Figure 5-12 Circular loop carrying a current I (Example 5-3).

in the same direction, but their r components cancel because they are in opposite directions. Hence, the net magnetic field is along z only. That is,

$$d\mathbf{H} = \hat{\mathbf{z}}\, dH_z = \hat{\mathbf{z}}\, dH \cos\theta = \hat{\mathbf{z}} \frac{I\cos\theta}{4\pi(a^2+z^2)}\, dl. \quad (5.32)$$

For a fixed point $P(0, 0, z)$ on the axis of the loop, all quantities in Eq. (5.32) are constant, except for dl. Hence, integrating Eq. (5.32) over a circle of radius a gives

$$\mathbf{H} = \hat{\mathbf{z}} \frac{I\cos\theta}{4\pi(a^2+z^2)} \oint dl = \hat{\mathbf{z}} \frac{I\cos\theta}{4\pi(a^2+z^2)}(2\pi a). \quad (5.33)$$

Upon using the relation $\cos\theta = a/(a^2+z^2)^{1/2}$, we obtain

$$\mathbf{H} = \hat{\mathbf{z}} \frac{Ia^2}{2(a^2+z^2)^{3/2}} \quad \text{(A/m)}. \quad (5.34)$$

At the center of the loop ($z = 0$), Eq. (5.34) reduces to

$$\mathbf{H} = \hat{\mathbf{z}} \frac{I}{2a} \quad (\text{at } z = 0), \quad (5.35)$$

and at points very far away from the loop such that $z^2 \gg a^2$, Eq. (5.34) simplifies to

$$\mathbf{H} = \hat{\mathbf{z}} \frac{Ia^2}{2|z|^3} \quad (\text{at } |z| \gg a). \quad (5.36)$$

5-2.2 Magnetic Field of a Magnetic Dipole

In view of the definition given by Eq. (5.19) for the magnetic moment $\mathbf{m}$ of a current loop, a single-turn loop situated in the x–y plane (**Fig. 5-12**) has magnetic moment $\mathbf{m} = \hat{\mathbf{z}}m$ with $m = I\pi a^2$. Consequently, Eq. (5.36) may be expressed as

$$\mathbf{H} = \hat{\mathbf{z}} \frac{m}{2\pi|z|^3} \quad (\text{at } |z| \gg a). \quad (5.37)$$

This expression applies to a point P far away from the loop and on its axis. Had we solved for $\mathbf{H}$ at any distant point $P = (R, \theta, \phi)$ in a spherical coordinate system, with R the distance between the center of the loop and point P, we would have obtained the expression

$$\mathbf{H} = \frac{m}{4\pi R^3}(\hat{\mathbf{R}}\, 2\cos\theta + \hat{\boldsymbol{\theta}} \sin\theta) \quad (5.38)$$
$$(\text{for } R \gg a).$$

▶ A current loop with dimensions much smaller than the distance between the loop and the observation point is called a ***magnetic dipole***. This is because the pattern of its magnetic field lines is similar to that of a permanent magnet, as well as to the pattern of the electric field lines of the electric dipole (**Fig. 5-13**). ◀

Concept Question 5-5: Two infinitely long parallel wires carry currents of equal magnitude. What is the resultant magnetic field due to the two wires at a point midway between the wires, compared with the magnetic field due to one of them alone, if the currents are (a) in the same direction and (b) in opposite directions?

Concept Question 5-6: Devise a right-hand rule for the direction of the magnetic field due to a linear current-carrying conductor.

Concept Question 5-7: What is a magnetic dipole? Describe its magnetic field distribution.

Exercise 5-6: A semi-infinite linear conductor extends between $z = 0$ and $z = \infty$ along the z axis. If the current I in the conductor flows along the positive z direction, find $\mathbf{H}$ at a point in the x–y plane at a radial distance r from the conductor.

Answer: $\mathbf{H} = \hat{\boldsymbol{\phi}} \frac{I}{4\pi r}$ (A/m). (See EM.)

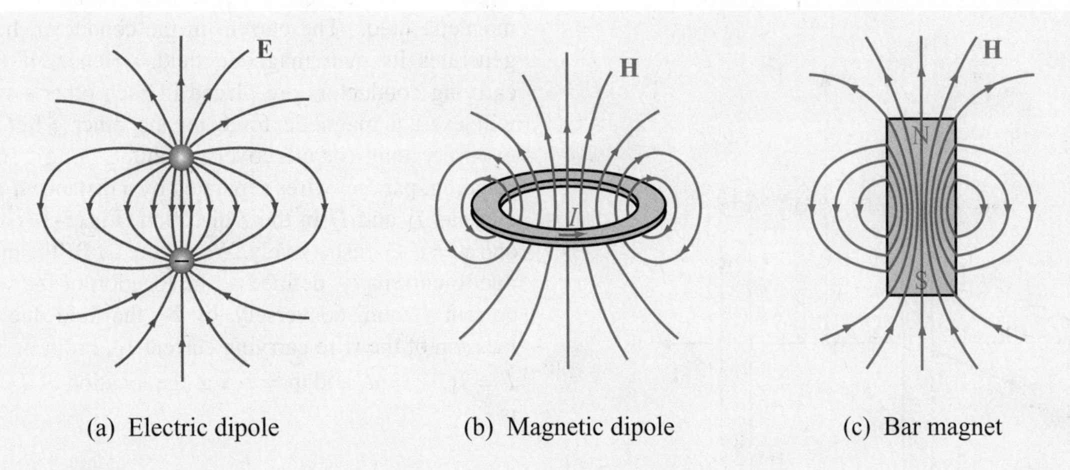

Figure 5-13 Patterns of (a) the electric field of an electric dipole, (b) the magnetic field of a magnetic dipole, and (c) the magnetic field of a bar magnet. Far away from the sources, the field patterns are similar in all three cases.

Module 5.3 Magnetic Field of a Current Loop Examine how the field along the loop axis changes with loop parameters.

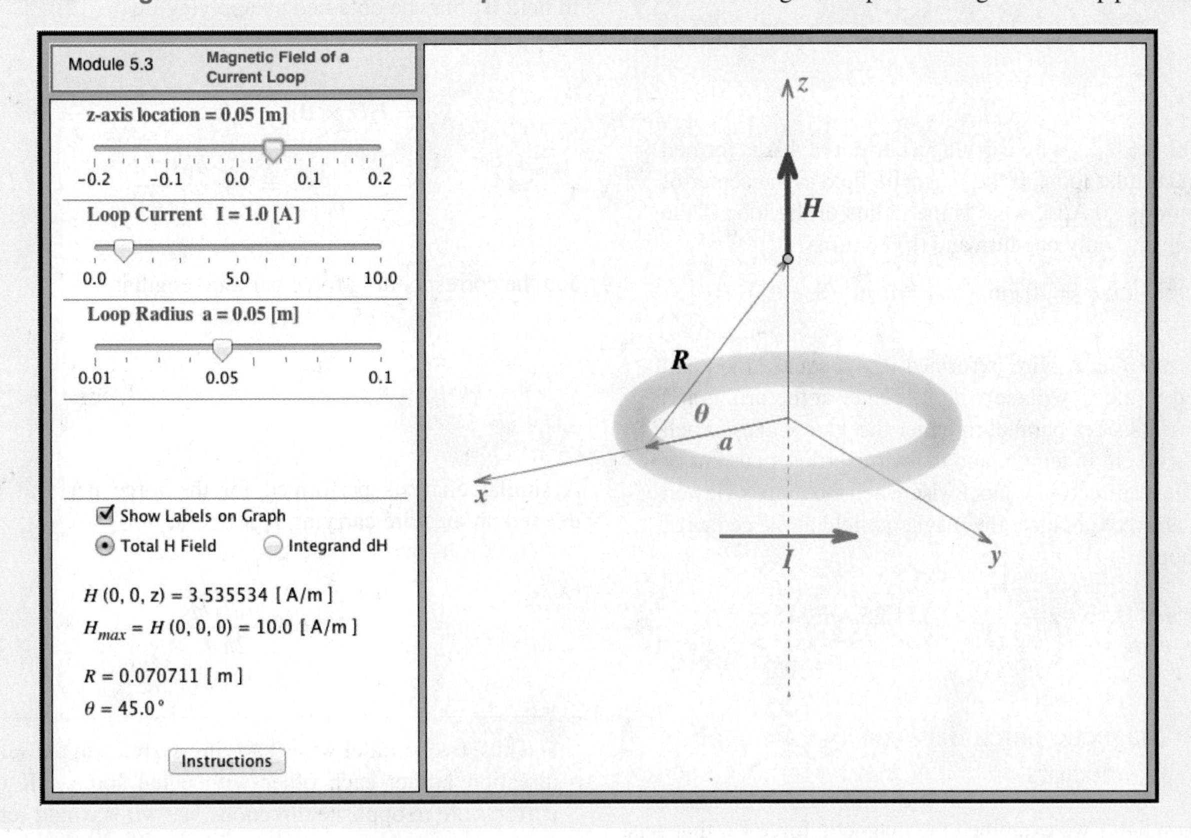

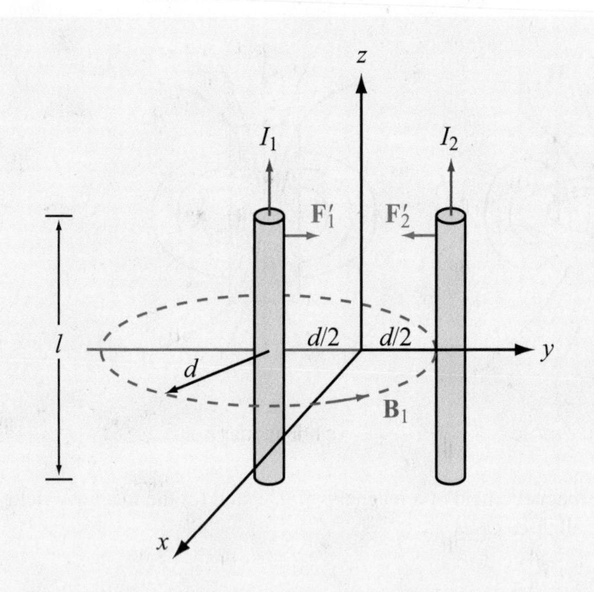

Figure 5-14 Magnetic forces on parallel current-carrying conductors.

Exercise 5-7: A wire carrying a current of 4 A is formed into a circular loop. If the magnetic field at the center of the loop is 20 A/m, what is the radius of the loop if the loop has (a) only one turn and (b) 10 turns?

Answer: (a) $a = 10$ cm, (b) $a = 1$ m. (See EM.)

Exercise 5-8: A wire is formed into a square loop and placed in the x–y plane with its center at the origin and each of its sides parallel to either the x or y axes. Each side is 40 cm in length, and the wire carries a current of 5 A whose direction is clockwise when the loop is viewed from above. Calculate the magnetic field at the center of the loop.

Answer: $\mathbf{H} = -\hat{\mathbf{z}}\dfrac{4I}{\sqrt{2}\pi l} = -\hat{\mathbf{z}}11.25$ A/m. (See EM.)

5-2.3 Magnetic Force Between Two Parallel Conductors

In Section 5-1.1 we examined the magnetic force $\mathbf{F}_m$ that acts on a current-carrying conductor when placed in an external magnetic field. The current in the conductor, however, also generates its own magnetic field. Hence, if two current-carrying conductors are placed in each other's vicinity, each will exert a magnetic force on the other. Let us consider two very long (or effectively infinitely long), straight, free-standing, parallel wires separated by a distance d and carrying currents I_1 and I_2 in the z direction (**Fig. 5-14**) at $y = -d/2$ and $y = d/2$, respectively. We denote by $\mathbf{B}_1$ the magnetic field due to current I_1, defined at the location of the wire carrying current I_2 and, conversely, by $\mathbf{B}_2$ the field due to I_2 at the location of the wire carrying current I_1. From Eq. (5.30), with $I = I_1$, $r = d$, and $\hat{\boldsymbol{\phi}} = -\hat{\mathbf{x}}$ at the location of I_2, the field $\mathbf{B}_1$ is

$$\mathbf{B}_1 = -\hat{\mathbf{x}}\,\frac{\mu_0 I_1}{2\pi d}\,. \tag{5.39}$$

The force $\mathbf{F}_2$ exerted on a length l of wire I_2 due to its presence in field $\mathbf{B}_1$ may be obtained by applying Eq. (5.12):

$$\begin{aligned}\mathbf{F}_2 = I_2 l\hat{\mathbf{z}} \times \mathbf{B}_1 &= I_2 l\hat{\mathbf{z}} \times (-\hat{\mathbf{x}})\,\frac{\mu_0 I_1}{2\pi d}\\ &= -\hat{\mathbf{y}}\,\frac{\mu_0 I_1 I_2 l}{2\pi d}\,,\end{aligned} \tag{5.40}$$

and the corresponding force per unit length is

$$\mathbf{F}'_2 = \frac{\mathbf{F}_2}{l} = -\hat{\mathbf{y}}\,\frac{\mu_0 I_1 I_2}{2\pi d}\,. \tag{5.41}$$

A similar analysis performed for the force per unit length exerted on the wire carrying I_1 leads to

$$\mathbf{F}'_1 = \hat{\mathbf{y}}\,\frac{\mu_0 I_1 I_2}{2\pi d}\,. \tag{5.42}$$

► Thus, two parallel wires carrying currents in the same direction attract each other with equal force. If the currents are in opposite directions, the wires would repel one another with equal force. ◄

Module 5.4 Magnetic Force Between Two Parallel Conductors Observe the direction and magnitude of the force exerted on parallel current-carrying wires.

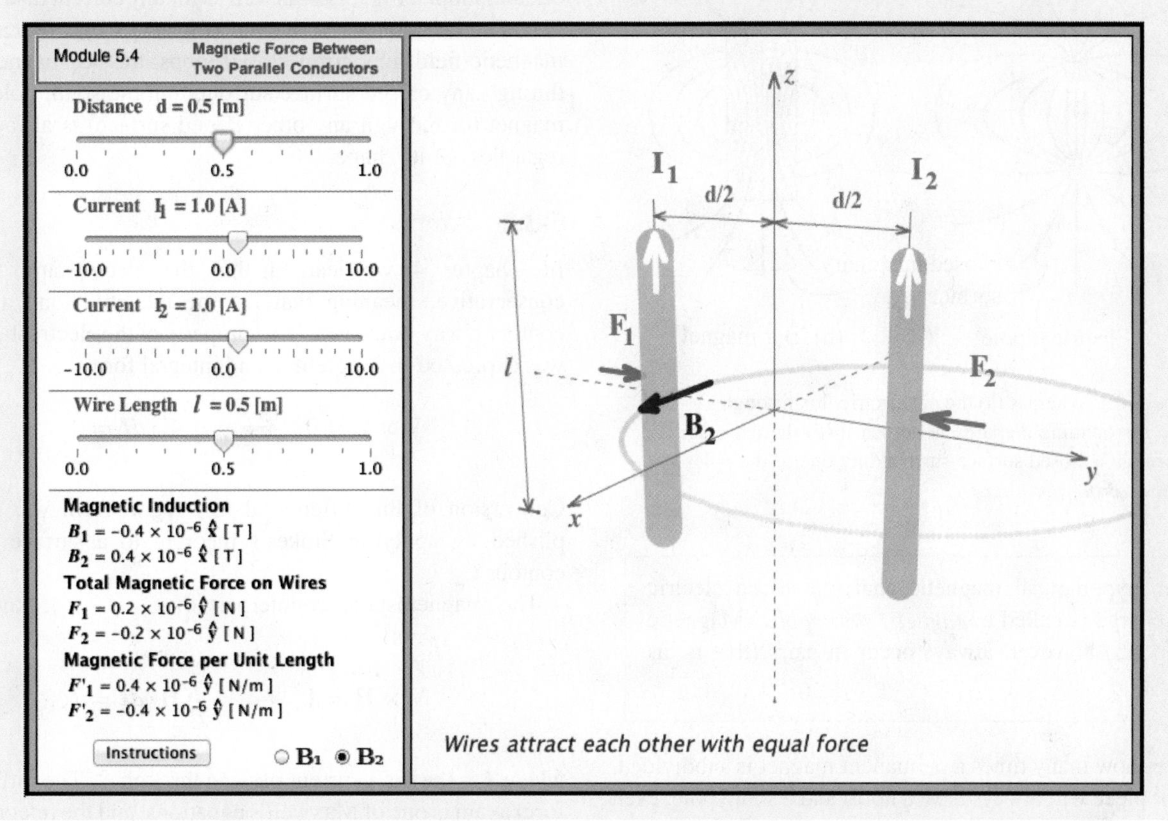

5-3 Maxwell's Magnetostatic Equations

Thus far, we have introduced the Biot–Savart law for finding the magnetic flux density **B** and field **H** due to any distribution of electric currents in free space, and we examined how magnetic fields can exert magnetic forces on moving charged particles and current-carrying conductors. We now examine two additional important properties of magnetostatic fields.

5-3.1 Gauss's Law for Magnetism

In Chapter 4 we learned that the net outward flux of the electric flux density **D** through a closed surface equals the enclosed net charge Q. We referred to this property as Gauss's law (for electricity), and expressed it mathematically in differential and integral forms as

$$\nabla \cdot \mathbf{D} = \rho_{\mathrm{v}} \quad \longleftrightarrow \quad \oint_S \mathbf{D} \cdot d\mathbf{s} = Q. \tag{5.43}$$

Conversion from differential to integral form was accomplished by applying the divergence theorem to a volume $\mathcal{V}$ that is enclosed by a surface S and contains a total charge $Q = \int_{\mathcal{V}} \rho_{\mathrm{v}}\, d\mathcal{V}$ (Section 4-4).

The magnetostatic counterpart of Eq. (5.43), often called ***Gauss's law for magnetism***, is

$$\nabla \cdot \mathbf{B} = 0 \quad \longleftrightarrow \quad \oint_S \mathbf{B} \cdot d\mathbf{s} = 0. \tag{5.44}$$

The differential form is one of Maxwell's four equations, and the integral form is obtained with the help of the divergence theorem. Note that the right-hand side of Gauss's law for magnetism is zero, reflecting the fact that the magnetic equivalence of an electric point charge does not exist in nature.

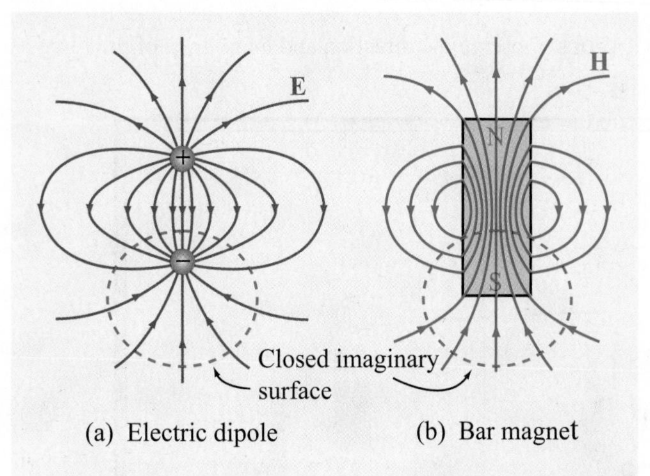

Figure 5-15 Whereas (a) the net electric flux through a closed surface surrounding a charge is not zero, (b) the net magnetic flux through a closed surface surrounding one of the poles of a magnet is zero.

▶ The hypothetical magnetic analogue to an electric point charge is called a ***magnetic monopole***. Magnetic monopoles, however, always occur in pairs (that is, as dipoles). ◀

No matter how many times a permanent magnet is subdivided, each new piece will always have a north and a south pole, even if the process were to be continued down to the atomic level. Consequently, there is no magnetic equivalence to an electric charge q or charge density ρ_v.

Formally, the name "Gauss's law" refers to the electric case, even when no specific reference to electricity is indicated. *The property described by Eq. (5.44) has been called "the law of nonexistence of isolated monopoles," "the law of conservation of magnetic flux," and "Gauss's law for magnetism," among others.* We prefer the last of the three cited names because it reminds us of the parallelism, as well as the differences, between the electric and magnetic laws of nature.

The difference between Gauss's law for electricity and its magnetic counterpart can be elucidated in terms of field lines. Electric field lines originate from positive electric charges and terminate on negative ones. Hence, for the electric field lines of the electric dipole shown in **Fig. 5-15(a)**, the electric flux through a closed surface surrounding one of the charges is nonzero. In contrast, *magnetic field lines always form continuous closed loops.* As we saw in Section 5-2, the magnetic field lines due to currents do not begin or end at any point; this is true for the linear conductor of **Fig. 5-11** and the circular loop of **Fig. 5-12**, as well as for any current distribution. It is also true for a bar magnet [**Fig. 5-15(b)**]. Because the magnetic field lines form closed loops, the net magnetic flux through any closed surface surrounding the south pole of the magnet (or through any other closed surface) is always zero, regardless of its shape.

5-3.2 Ampère's Law

In Chapter 4 we learned that the electrostatic field is conservative, meaning that its line integral along a closed contour always vanishes. This property of the electrostatic field was expressed in differential and integral forms as

$$\nabla \times \mathbf{E} = 0 \quad \longleftrightarrow \quad \oint_C \mathbf{E} \cdot d\boldsymbol{\ell} = 0. \tag{5.45}$$

Conversion of the differential to integral form was accomplished by applying Stokes's theorem to a surface S with contour C.

The magnetostatic counterpart of Eq. (5.45), known as ***Ampère's law***, is

$$\nabla \times \mathbf{H} = \mathbf{J} \quad \longleftrightarrow \quad \oint_C \mathbf{H} \cdot d\boldsymbol{\ell} = I, \tag{5.46}$$

where I is the total current passing through S. The differential form again is one of Maxwell's equations, and the integral form is obtained by integrating both sides of Eq. (5.46) over an open surface S,

$$\int_S (\nabla \times \mathbf{H}) \cdot d\mathbf{s} = \int_S \mathbf{J} \cdot d\mathbf{s}, \tag{5.47}$$

and then invoking Stokes's theorem with $I = \int \mathbf{J} \cdot d\mathbf{s}$.

▶ The sign convention for the direction of the contour path C in Ampère's law is taken so that I and $\mathbf{H}$ satisfy the right-hand rule defined earlier in connection with the Biot–Savart law. That is, if the direction of I is aligned with the direction of the thumb of the right hand, then the direction of the contour C should be chosen along that of the other four fingers. ◀

In words, *Ampère's circuital law states that the line integral of* **H** *around a closed path is equal to the current traversing the surface bounded by that path.* To apply Ampère's law, the

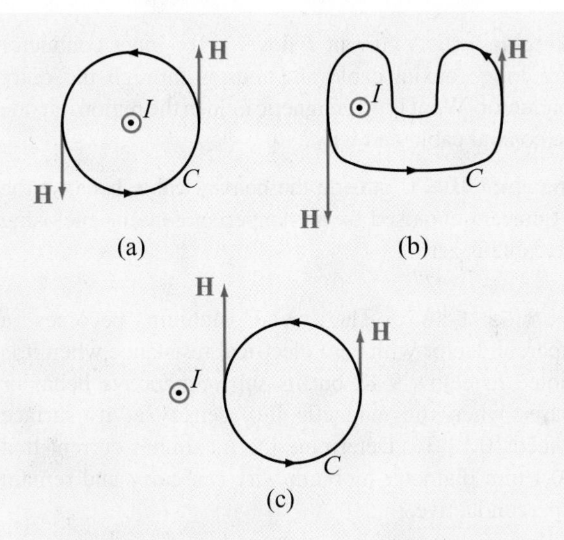

Figure 5-16 Ampère's law states that the line integral of **H** around a closed contour C is equal to the current traversing the surface bounded by the contour. This is true for contours (a) and (b), but the line integral of **H** is zero for the contour in (c) because the current I (denoted by the symbol ⊙) is not enclosed by the contour C.

current must flow through a closed path. By way of illustration, for both configurations shown in Figs. 5-16(a) and (b), the line integral of **H** is equal to the current I, even though the paths have very different shapes and the magnitude of **H** is not uniform along the path of configuration (b). By the same token, because path (c) in Fig. 5-16 does not enclose the current I, the line integral of **H** along it vanishes, even though **H** is not zero along the path.

When we examined Gauss's law in Section 4-4, we discovered that in practice its usefulness for calculating the electric flux density **D** is limited to charge distributions that possess a certain degree of symmetry and that the calculation procedure is subject to the proper choice of a Gaussian surface enclosing the charges. A similar restriction applies to Ampère's law: its usefulness is limited to symmetric current distributions that allow the choice of convenient ***Ampèrian contours*** around them, as illustrated by Examples 5-4 to 5-6.

Example 5-4: Magnetic Field of a Long Wire

A long (practically infinite) straight wire of radius a carries a steady current I that is uniformly distributed over its cross section. Determine the magnetic field **H** a distance r from the wire axis for (a) $r \le a$ (inside the wire) and (b) $r \ge a$ (outside the wire).

Solution: (a) We choose I to be along the $+z$ direction [Fig. 5-17(a)]. To determine $\mathbf{H}_1 = \mathbf{H}$ at a distance $r = r_1 \le a$, we choose the Ampèrian contour C_1 to be a circular path of radius $r = r_1$ [Fig. 5-17(b)]. In this case, Ampère's law takes the form

$$\oint_{C_1} \mathbf{H}_1 \cdot d\mathbf{l}_1 = I_1, \tag{5.48}$$

where I_1 is the fraction of the total current I flowing through C_1. From symmetry, $\mathbf{H}_1$ must be constant in magnitude and parallel to the contour at any point along the path. Furthermore, to satisfy the right-hand rule and given that I is along the z direction, $\mathbf{H}_1$ must be in the $+\phi$ direction. Hence, $\mathbf{H}_1 = \hat{\boldsymbol{\phi}} H_1$, $d\mathbf{l}_1 = \hat{\boldsymbol{\phi}} r_1 \, d\phi$, and the left-hand side of Eq. (5.48) becomes

$$\oint_{C_1} \mathbf{H}_1 \cdot d\mathbf{l}_1 = \int_0^{2\pi} H_1(\hat{\boldsymbol{\phi}} \cdot \hat{\boldsymbol{\phi}}) r_1 \, d\phi = 2\pi r_1 H_1.$$

The current I_1 flowing through the area enclosed by C_1 is equal to the total current I multiplied by the ratio of the area enclosed by C_1 to the total cross-sectional area of the wire:

$$I_1 = \left(\frac{\pi r_1^2}{\pi a^2}\right) I = \left(\frac{r_1}{a}\right)^2 I.$$

Equating both sides of Eq. (5.48) and then solving for $\mathbf{H}_1$ yields

$$\mathbf{H}_1 = \hat{\boldsymbol{\phi}} H_1 = \hat{\boldsymbol{\phi}} \frac{r_1}{2\pi a^2} I \qquad (\text{for } r_1 \le a). \tag{5.49a}$$

(b) For $r = r_2 \ge a$, we choose path C_2, which encloses all the current I. Hence, $\mathbf{H}_2 = \hat{\boldsymbol{\phi}} H_2$, $d\boldsymbol{\ell}_2 = \hat{\boldsymbol{\phi}} r_2 \, d\phi$, and

$$\oint_{C_2} \mathbf{H}_2 \cdot d\mathbf{l}_2 = 2\pi r_2 H_2 = I,$$

which yields

$$\mathbf{H}_2 = \hat{\boldsymbol{\phi}} H_2 = \hat{\boldsymbol{\phi}} \frac{I}{2\pi r_2} \qquad (\text{for } r_2 \ge a). \tag{5.49b}$$

Ignoring the subscript 2, we observe that Eq. (5.49b) provides the same expression for $\mathbf{B} = \mu_0 \mathbf{H}$ as Eq. (5.30), which was derived on the basis of the Biot–Savart law.

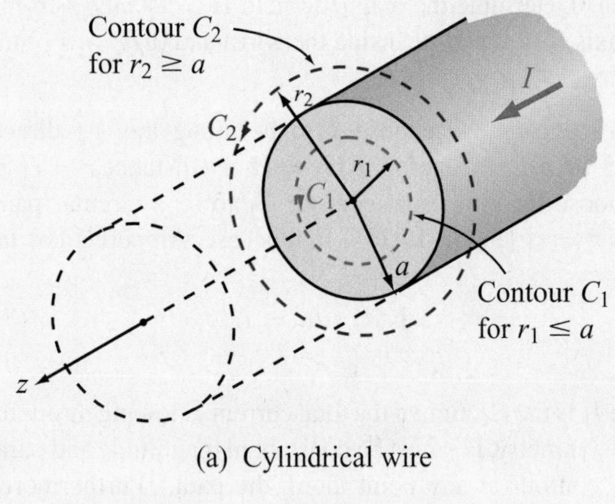

(a) Cylindrical wire

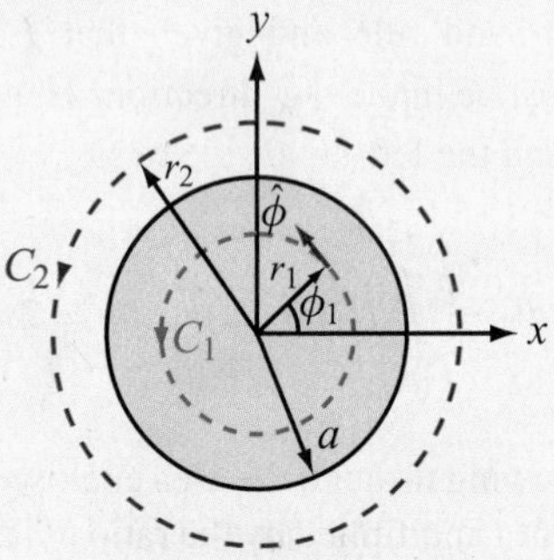

(b) Wire cross section

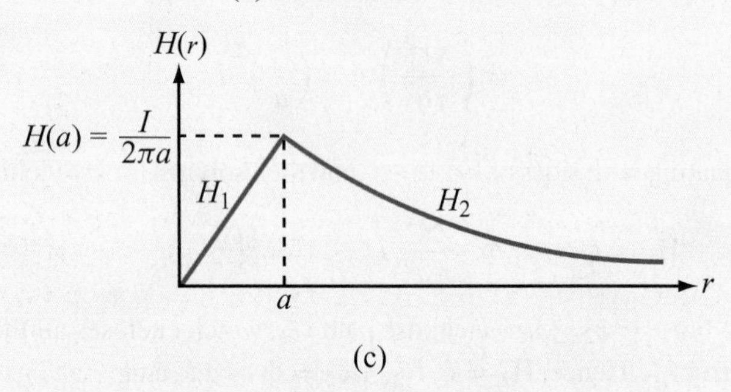

(c)

Figure 5-17 Infinitely long wire of radius a carrying a uniform current I along the $+z$ direction: (a) general configuration showing contours C_1 and C_2; (b) cross-sectional view; and (c) a plot of H versus r (Example 5-4).

The variation of the magnitude of H as a function of r is plotted in Fig. 5-17(c); H increases linearly between $r = 0$ and $r = a$ (inside the conductor), and then decreases as $1/r$ for $r > a$ (outside the conductor).

Exercise 5-9: A current I flows in the inner conductor of a long coaxial cable and returns through the outer conductor. What is the magnetic field in the region outside the coaxial cable and why?

Answer: $\mathbf{H} = 0$ outside the coaxial cable because the net current enclosed by an Ampèrian contour enclosing the cable is zero.

Exercise 5-10: The metal niobium becomes a superconductor with zero electrical resistance when it is cooled to below 9 K, but its superconductive behavior ceases when the magnetic flux density at its surface exceeds 0.12 T. Determine the maximum current that a 0.1 mm diameter niobium wire can carry and remain superconductive.

Answer: $I = 30$ A. (See EM.)

Example 5-5: Magnetic Field inside a Toroidal Coil

A ***toroidal coil*** (also called a torus or toroid) is a doughnut-shaped structure (called the core) wrapped in closely spaced turns of wire (Fig. 5-18). For clarity, we show the turns in the figure as spaced far apart, but in practice they are wound in

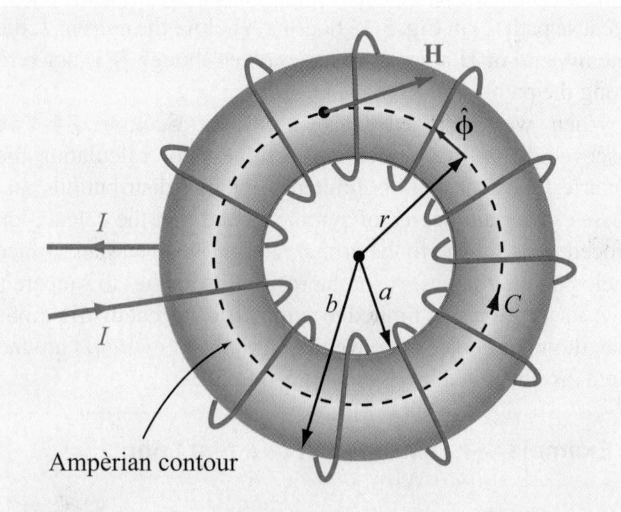

Figure 5-18 Toroidal coil with inner radius a and outer radius b. The wire loops usually are much more closely spaced than shown in the figure (Example 5-5).

a closely spaced arrangement to form approximately circular loops. The toroid is used to magnetically couple multiple circuits and to measure the magnetic properties of materials, as illustrated later in Fig. 5-30. For a toroid with N turns carrying a current I, determine the magnetic field $\mathbf{H}$ in each of the following three regions: $r < a$, $a < r < b$, and $r > b$, all in the azimuthal symmetry plane of the toroid.

Solution: From symmetry, it is clear that $\mathbf{H}$ is uniform in the azimuthal direction. If we construct a circular Ampèrian contour with center at the origin and radius $r < a$, there will be no current flowing through the surface of the contour, and therefore

$$\mathbf{H} = 0 \qquad \text{for } r < a.$$

Similarly, for an Ampèrian contour with radius $r > b$, the *net* current flowing through its surface is zero because an equal number of current coils cross the surface in both directions; hence,

$$\mathbf{H} = 0 \quad \text{for } r > b \text{ (region exterior to the toroidal coil).}$$

For the region inside the core, we construct a path of radius r (Fig. 5-18). For each loop, we know from Example 5-3 that the field $\mathbf{H}$ at the center of the loop points along the axis of the loop, which in this case is the ϕ direction, and in view of the direction of the current I shown in Fig. 5-18, the right-hand rule tells us that $\mathbf{H}$ must be in the $-\phi$ direction. Hence, $\mathbf{H} = -\hat{\boldsymbol{\phi}}H$. The total current crossing the surface of the contour with radius r is NI and its direction is into the page. According to the right-hand rule associated with Ampère's law, the current is positive if it crosses the surface of the contour in the direction of the four fingers of the right hand when the thumb is pointing along the direction of the contour C. Hence, the current through the surface spanned by the contour is $-NI$. Application of Ampère's law then gives

$$\oint_C \mathbf{H} \cdot d\mathbf{l} = \int_0^{2\pi} (-\hat{\boldsymbol{\phi}}H) \cdot \hat{\boldsymbol{\phi}} r \, d\phi = -2\pi r H = -NI.$$

Hence, $H = NI/(2\pi r)$ and

$$\mathbf{H} = -\hat{\boldsymbol{\phi}}H = -\hat{\boldsymbol{\phi}}\frac{NI}{2\pi r} \qquad \text{(for } a < r < b\text{).} \tag{5.50}$$

Example 5-6: Magnetic Field of an Infinite Current Sheet

The x–y plane contains an infinite current sheet with surface current density $\mathbf{J}_s = \hat{\mathbf{x}}J_s$ (Fig. 5-19). Find the magnetic field $\mathbf{H}$ everywhere in space.

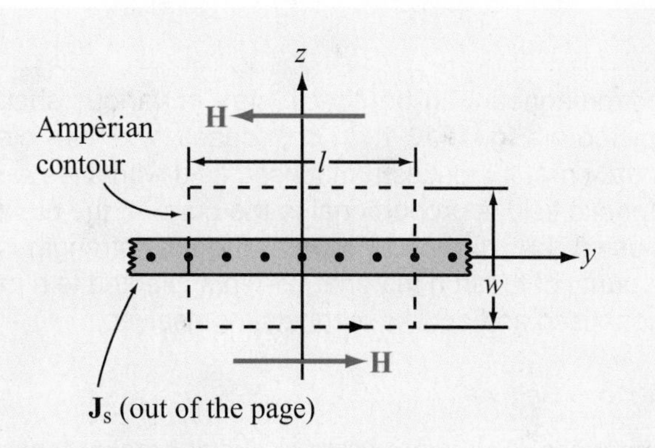

Figure 5-19 A thin current sheet in the x–y plane carrying a surface current density $\mathbf{J}_s = \hat{\mathbf{x}}J_s$ (Example 5-6).

Solution: From symmetry considerations and the right-hand rule, for $z > 0$ and $z < 0$ $\mathbf{H}$ must be in the directions shown in the figure. That is,

$$\mathbf{H} = \begin{cases} -\hat{\mathbf{y}}H & \text{for } z > 0, \\ \hat{\mathbf{y}}H & \text{for } z < 0. \end{cases}$$

To evaluate the line integral in Ampère's law, we choose a rectangular Ampèrian path around the sheet, with dimensions l and w (Fig. 5-19). Recalling that J_s represents current per unit length along the y direction, the total current crossing the surface of the rectangular loop is $I = J_s l$. Hence, applying Ampère's law over the loop, while noting that $\mathbf{H}$ is perpendicular to the paths of length w, we have

$$\oint_C \mathbf{H} \cdot d\mathbf{l} = 2Hl = J_s l,$$

from which we obtain the result

$$\mathbf{H} = \begin{cases} -\hat{\mathbf{y}}\dfrac{J_s}{2} & \text{for } z > 0, \\ \hat{\mathbf{y}}\dfrac{J_s}{2} & \text{for } z < 0. \end{cases} \tag{5.51}$$

Technology Brief 10: Electromagnets

William Sturgeon developed the first practical ***electromagnet*** in the 1820s. Today, the principle of the electromagnet is used in motors, relay switches in read/write heads for hard disks and tape drives, loud speakers, magnetic levitation, and many other applications.

Basic Principle

Electromagnets can be constructed in various shapes, including the linear ***solenoid*** and ***horseshoe*** geometries depicted in **Fig. TF10-1**. In both cases, when an electric current flows through the insulated wire coiled around the central core, it induces a magnetic field with lines resembling those generated by a bar magnet. The strength of the magnetic field is proportional to the current, the number of turns, and the magnetic permeability of the core material. By using a ***ferromagnetic core***, the field strength can be increased by several orders of magnitude, depending on the purity of the iron material. When subjected to a magnetic field, ferromagnetic materials, such as iron or nickel, get magnetized and act like magnets themselves.

Magnetic Relays

A magnetic relay is a ***switch*** or circuit breaker that can be activated into the "ON" and "OFF" positions magnetically. One example is the low-power ***reed relay*** used in telephone equipment, which consists of two flat nickel–iron blades separated by a small gap (**Fig. TF10-2**). The blades are shaped in such a way that in the absence of an external force, they remain apart and unconnected (OFF position). Electrical contact between the blades (ON position) is realized by applying a magnetic field along their length. The field, induced by a current flowing in the wire coiled around the glass envelope, causes the two blades to assume opposite magnetic polarities, thereby forcing them to attract together and close out the gap.

The Doorbell

In a doorbell circuit (**Fig. TF10-3**), the doorbell button is a switch; pushing and holding it down serves to connect the circuit to the household ac source through an appropriate ***step-down transformer***. The current from the source flows

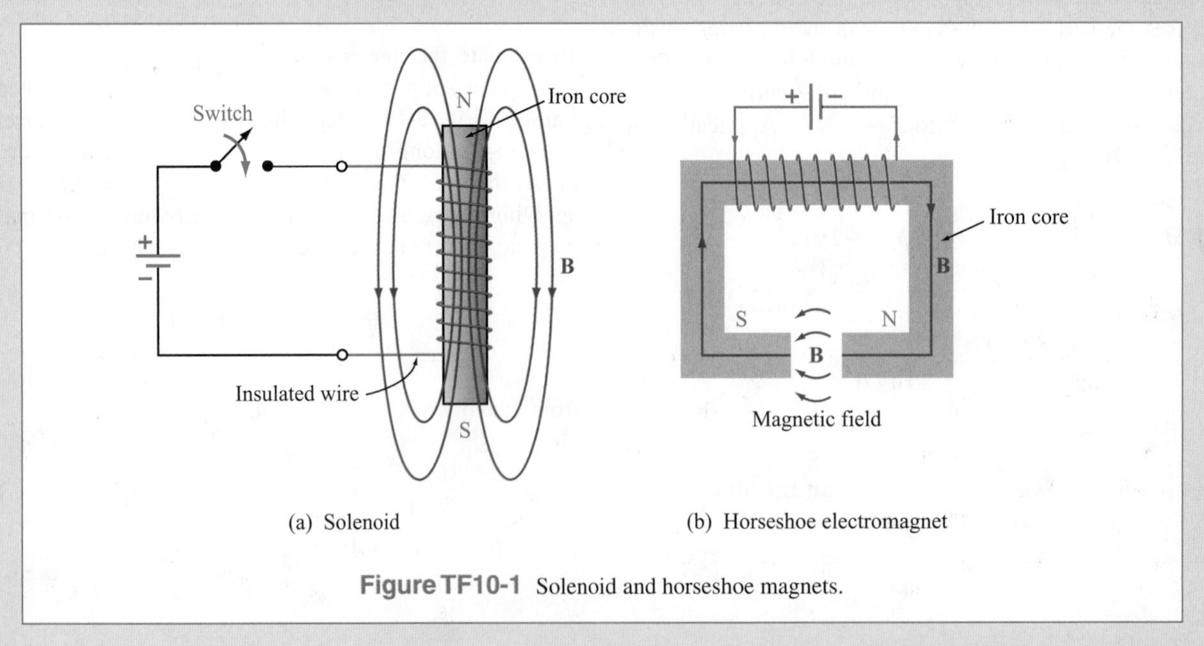

Figure TF10-1 Solenoid and horseshoe magnets.

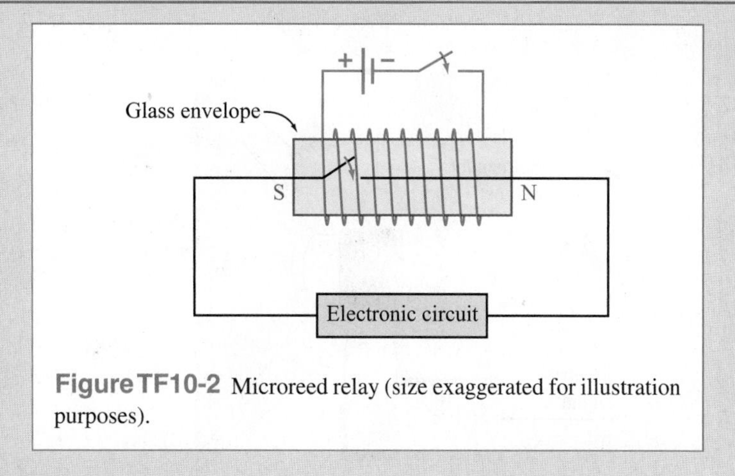

Figure TF10-2 Microreed relay (size exaggerated for illustration purposes).

through the electromagnet, via a contact arm with only one end anchored in place (and the other moveable), and onward to the switch. The magnetic field generated by the current flowing in the windings of the electromagnet pulls the unanchored end of the contact arm (which has an iron bar on it) closer in, in the direction of the electromagnet, thereby losing connection with the metal contact and severing current flow in the circuit. With no magnetic field to pull on the contact arm, it snaps back into its earlier position, re-establishing the current in the circuit. This back and forth cycle is repeated many times per second, so long as the doorbell button continues to be pushed down, and with every cycle, the clapper arm attached to the contact arm hits the metal bell and generates a ringing sound.

The Loudspeaker

By using a combination of a stationary, permanent magnet, and a moveable electromagnet, the electromagnet/speaker-cone of the loudspeaker (**Fig. TF10-4**) can be made to move back and forth in response to the electrical signal exciting the electromagnet. The vibrating movement of the cone generates sound waves with the same distribution of frequencies as contained in the spectrum of the electrical signal.

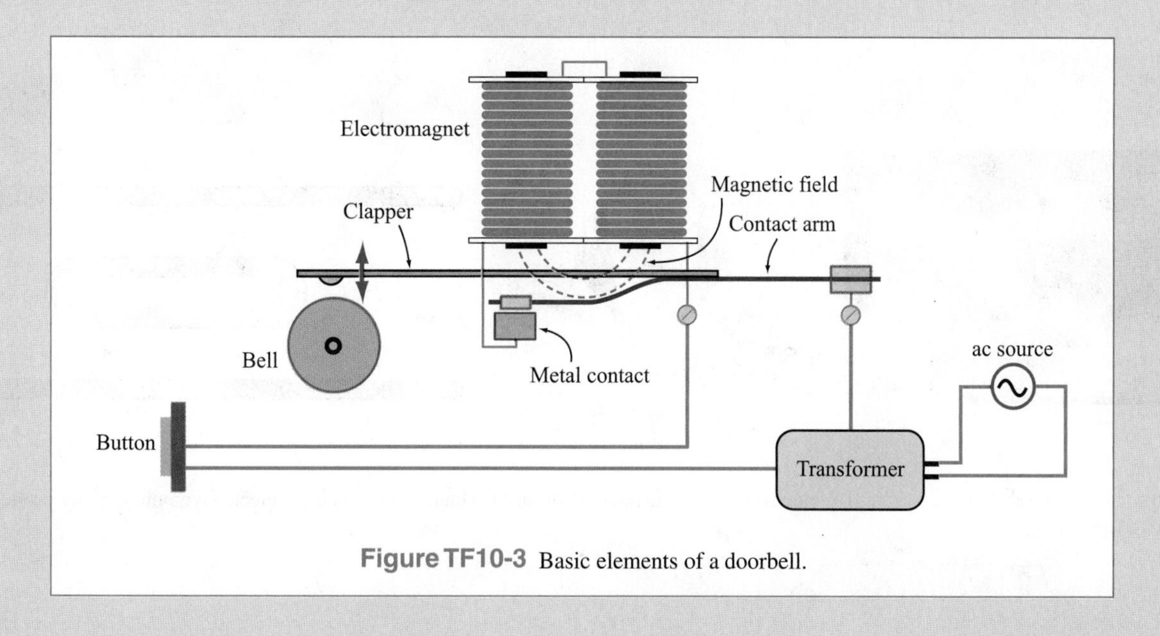

Figure TF10-3 Basic elements of a doorbell.

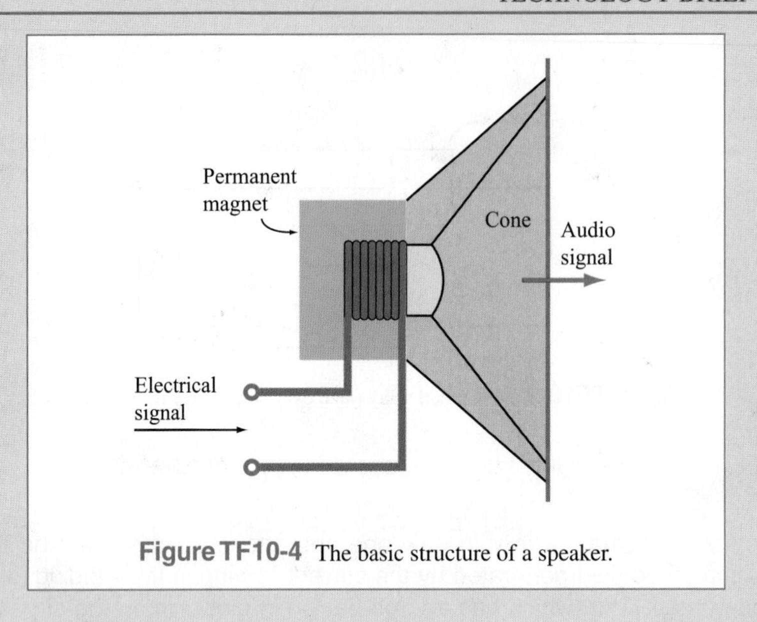

Figure TF10-4 The basic structure of a speaker.

Magnetic Levitation

► Magnetically levitated trains [**Fig. TF10-5(a)**], called ***maglevs*** for short, can achieve speeds as high as 500 km/hr, primarily because there is no friction between the train and the track. ◄

The train basically floats at a height of 1 or more centimeters above the track, made possible by magnetic levitation [**Fig. TF10-5(b)**]. The train carries superconducting electromagnets that induce currents in coils built into the guide rails alongside the train. The magnetic interaction between the train's superconducting electromagnets and the guide-rail coils serves not only to levitate the train, but also to propel it along the track.

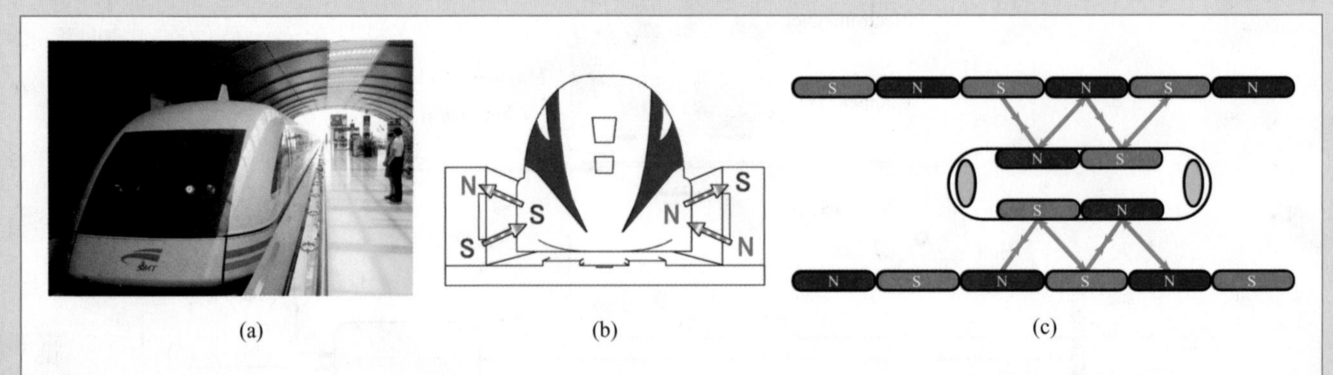

Figure TF10-5 (a) A maglev train, (b) electrodynamic suspension of an SCMaglev train, and (c) electrodynamic maglev propulsion via propulsion coils.

Concept Question 5-8: What are the fundamental differences between electric and magnetic fields?

Concept Question 5-9: If the line integral of $\mathbf{H}$ over a closed contour is zero, does it follow that $\mathbf{H} = 0$ at every point on the contour? If not, what then does it imply?

Concept Question 5-10: Compare the utility of applying the Biot–Savart law versus applying Ampère's law for computing the magnetic field due to current-carrying conductors.

Concept Question 5-11: What is a toroid? What is the magnetic field outside the toroid?

5-4 Vector Magnetic Potential

In our treatment of electrostatic fields in Chapter 4, we defined the electrostatic potential V as the line integral of the electric field $\mathbf{E}$, and found that V and $\mathbf{E}$ are related by $\mathbf{E} = -\nabla V$. This relationship proved useful not only in relating electric field distributions in circuit elements (such as resistors and capacitors) to the voltages across them, but also to determine $\mathbf{E}$ for a given charge distribution by first computing V using Eq. (4.48). We now explore a similar approach in connection with the magnetic flux density $\mathbf{B}$.

According to Eq. (5.44), $\nabla \cdot \mathbf{B} = 0$. We wish to define $\mathbf{B}$ in terms of a magnetic potential with the constraint that such a definition guarantees that the divergence of $\mathbf{B}$ is always zero. This can be realized by taking advantage of the vector identity given by Eq. (3.106b), which states that, for any vector $\mathbf{A}$,

$$\nabla \cdot (\nabla \times \mathbf{A}) = 0. \tag{5.52}$$

Hence, by introducing the ***vector magnetic potential*** $\mathbf{A}$ such that

$$\mathbf{B} = \nabla \times \mathbf{A} \qquad (\text{Wb/m}^2), \tag{5.53}$$

we are guaranteed that $\nabla \cdot \mathbf{B} = 0$. The SI unit for $\mathbf{B}$ is the tesla (T). An equivalent unit is webers per square meter (Wb/m^2). Consequently, the SI unit for $\mathbf{A}$ is (Wb/m).

With $\mathbf{B} = \mu \mathbf{H}$, the differential form of Ampère's law given by Eq. (5.46) can be written as

$$\nabla \times \mathbf{B} = \mu \mathbf{J}. \tag{5.54}$$

If we substitute Eq. (5.53) into Eq. (5.54), we obtain

$$\nabla \times (\nabla \times \mathbf{A}) = \mu \mathbf{J}. \tag{5.55}$$

For any vector $\mathbf{A}$, the Laplacian of $\mathbf{A}$ obeys the vector identity given by Eq. (3.113), that is,

$$\nabla^2 \mathbf{A} = \nabla(\nabla \cdot \mathbf{A}) - \nabla \times (\nabla \times \mathbf{A}), \tag{5.56}$$

where, by definition, $\nabla^2 \mathbf{A}$ in Cartesian coordinates is

$$\begin{aligned} \nabla^2 \mathbf{A} &= \left(\frac{\partial^2}{\partial x^2} + \frac{\partial^2}{\partial y^2} + \frac{\partial^2}{\partial z^2} \right) \mathbf{A} \\ &= \hat{\mathbf{x}} \nabla^2 A_x + \hat{\mathbf{y}} \nabla^2 A_y + \hat{\mathbf{z}} \nabla^2 A_z. \end{aligned} \tag{5.57}$$

Combining Eq. (5.55) with Eq. (5.56) gives

$$\nabla(\nabla \cdot \mathbf{A}) - \nabla^2 \mathbf{A} = \mu \mathbf{J}. \tag{5.58}$$

This equation contains a term involving $\nabla \cdot \mathbf{A}$. It turns out that we have a fair amount of latitude in specifying a value or a mathematical form for $\nabla \cdot \mathbf{A}$, without conflicting with the requirement represented by Eq. (5.53). The simplest among these allowed restrictions on $\mathbf{A}$ is

$$\nabla \cdot \mathbf{A} = 0. \tag{5.59}$$

Using this choice in Eq. (5.58) leads to the ***vector Poisson's equation***

$$\nabla^2 \mathbf{A} = -\mu \mathbf{J}. \tag{5.60}$$

Using the definition for $\nabla^2 \mathbf{A}$ given by Eq. (5.57), the vector Poisson's equation can be decomposed into three scalar Poisson's equations:

$$\nabla^2 A_x = -\mu J_x, \tag{5.61a}$$

$$\nabla^2 A_y = -\mu J_y, \tag{5.61b}$$

$$\nabla^2 A_z = -\mu J_z. \tag{5.61c}$$

In electrostatics, Poisson's equation for the scalar potential V is given by Eq. (4.60) as

$$\nabla^2 V = -\frac{\rho_v}{\epsilon}, \tag{5.62}$$

and its solution for a volume charge distribution ρ_v occupying a volume $\mathcal{V}'$ was given by Eq. (4.61) as

$$V = \frac{1}{4\pi\epsilon} \int_{\mathcal{V}'} \frac{\rho_v}{R'} \, d\mathcal{V}'. \quad (5.63)$$

Poisson's equations for A_x, A_y, and A_z are mathematically identical in form to Eq. (5.62). Hence, for a current density $\mathbf{J}$ with x component J_x distributed over a volume $\mathcal{V}'$, the solution for Eq. (5.61a) is

$$A_x = \frac{\mu}{4\pi} \int_{\mathcal{V}'} \frac{J_x}{R'} \, d\mathcal{V}' \quad \text{(Wb/m)}. \quad (5.64)$$

Similar solutions can be written for A_y in terms of J_y and for A_z in terms of J_z. The three solutions can be combined into a vector equation:

$$\mathbf{A} = \frac{\mu}{4\pi} \int_{\mathcal{V}'} \frac{\mathbf{J}}{R'} \, d\mathcal{V}' \quad \text{(Wb/m)}. \quad (5.65)$$

In view of Eq. (5.23), if the current distribution is specified over a surface S', then $\mathbf{J}\, d\mathcal{V}'$ should be replaced with $\mathbf{J}_s\, ds'$ and $\mathcal{V}'$ should be replaced with S'; similarly, for a line distribution, $\mathbf{J}\, d\mathcal{V}'$ should be replaced with $I\, d\mathbf{l}'$ and the integration should be performed over the associated path l'.

The vector magnetic potential provides a third approach for computing the magnetic field due to current-carrying conductors, in addition to the methods suggested by the Biot–Savart and Ampère laws. For a specified current distribution, Eq. (5.65) can be used to find $\mathbf{A}$, and then Eq. (5.53) can be used to find $\mathbf{B}$. Except for simple current distributions with symmetrical geometries that lend themselves to the application of Ampère's law, in practice we often use the approaches provided by the Biot–Savart law and the vector magnetic potential, and among these two the latter often is more convenient to apply because it is easier to perform the integration in Eq. (5.65) than that in Eq. (5.22).

The ***magnetic flux*** Φ linking a surface S is defined as the total magnetic flux density passing through it, or

$$\Phi = \int_S \mathbf{B} \cdot d\mathbf{s} \quad \text{(Wb)}. \quad (5.66)$$

If we insert Eq. (5.53) into Eq. (5.66) and then invoke Stokes's theorem, we obtain

$$\Phi = \int_S (\nabla \times \mathbf{A}) \cdot d\mathbf{s} = \oint_C \mathbf{A} \cdot d\mathbf{l} \quad \text{(Wb)}, \quad (5.67)$$

where C is the contour bounding the surface S. Thus, Φ can be determined by either Eq. (5.66) or Eq. (5.67), whichever is easier to integrate for the specific problem under consideration.

5-5 Magnetic Properties of Materials

Because of the similarity between the pattern of the magnetic field lines generated by a current loop and those exhibited by a permanent magnet, the loop can be regarded as a magnetic dipole with north and south poles (Section 5-2.2 and **Fig. 5-13**). The magnetic moment $\mathbf{m}$ of a loop of area A has magnitude $m = IA$ and a direction normal to the plane of the loop (in accordance with the right-hand rule). Magnetization in a material is due to atomic scale current loops associated with: (1) orbital motions of the electrons and protons around and inside the nucleus and (2) electron spin. The magnetic moment due to proton motion typically is three orders of magnitude smaller than that of the electrons, and therefore the total orbital and spin magnetic moment of an atom is dominated by the sum of the magnetic moments of its electrons.

► The magnetic behavior of a material is governed by the interaction of the magnetic dipole moments of its atoms with an external magnetic field. The nature of the behavior depends on the crystalline structure of the material and is used as a basis for classifying materials as ***diamagnetic***, ***paramagnetic***, or ***ferromagnetic***. ◄

The atoms of a diamagnetic material have no permanent magnetic moments. In contrast, both paramagnetic and ferromagnetic materials have atoms with permanent magnetic dipole moments, albeit with very different organizational structures.

5-5.1 Electron Orbital and Spin Magnetic Moments

This section presents a semiclassical, intuitive model of the atom, which provides quantitative insight into the origin of electron magnetic moments. An electron with charge of $-e$ moving at constant speed u in a circular orbit of radius r

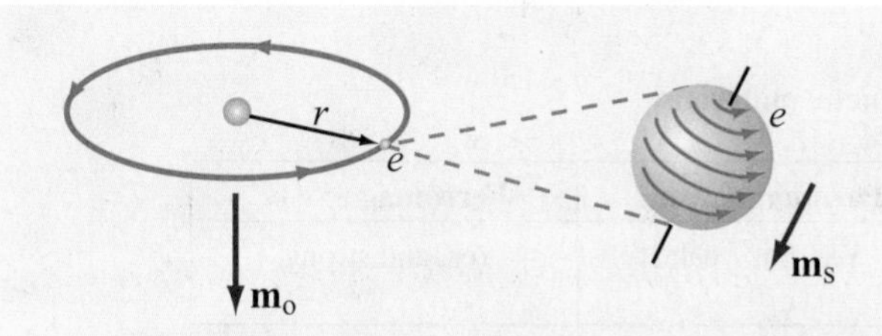

(a) Orbiting electron (b) Spinning electron

Figure 5-20 An electron generates (a) an orbital magnetic moment $\mathbf{m}_o$ as it rotates around the nucleus and (b) a spin magnetic moment $\mathbf{m}_s$, as it spins about its own axis.

[Fig. 5-20(a)] completes one revolution in time $T = 2\pi r/u$. This circular motion of the electron constitutes a tiny loop with current I given by

$$I = -\frac{e}{T} = -\frac{eu}{2\pi r} \,. \tag{5.68}$$

The magnitude of the associated ***orbital magnetic moment*** $\mathbf{m}_o$ is

$$\begin{aligned} m_o = IA &= \left(-\frac{eu}{2\pi r}\right)(\pi r^2) \\ &= -\frac{eur}{2} = -\left(\frac{e}{2m_e}\right) L_e, \end{aligned} \tag{5.69}$$

where $L_e = m_e ur$ is the angular momentum of the electron and m_e is its mass. According to quantum physics, the orbital angular momentum is quantized; specifically, L_e is always some integer multiple of $\hbar = h/2\pi$, where h is Planck's constant. That is, $L_e = 0, \hbar, 2\hbar, \ldots$. Consequently, the smallest nonzero magnitude of the orbital magnetic moment of an electron is

$$m_o = -\frac{e\hbar}{2m_e} \,. \tag{5.70}$$

Despite the fact that all materials contain electrons that exhibit magnetic dipole moments, most are effectively nonmagnetic. This is because, in the absence of an external magnetic field, the atoms of most materials are oriented *randomly*, as a result of which they exhibit a zero or very small net magnetic moment.

In addition to the magnetic moment due to its orbital motion, an electron has an intrinsic ***spin magnetic moment*** $\mathbf{m}_s$ due to its spinning motion about its own axis [Fig. 5-20(b)]. The magnitude of $\mathbf{m}_s$ predicted by quantum theory is

$$m_s = -\frac{e\hbar}{2m_e} \,, \tag{5.71}$$

which is equal to the minimum orbital magnetic moment m_o. The electrons of an atom with an even number of electrons usually exist in pairs, with the members of a pair having opposite spin directions, thereby canceling each others' spin magnetic moments. If the number of electrons is odd, the atom has a net nonzero spin magnetic moment due to its unpaired electron.

5-5.2 Magnetic Permeability

In Chapter 4, we learned that the relationship $\mathbf{D} = \epsilon_0 \mathbf{E}$, between the electric flux and field in free space, is modified to $\mathbf{D} = \epsilon_0 \mathbf{E} + \mathbf{P}$ in a dielectric material. Likewise, the relationship $\mathbf{B} = \mu_0 \mathbf{H}$ in free space is modified to

$$\mathbf{B} = \mu_0 \mathbf{H} + \mu_0 \mathbf{M} = \mu_0 (\mathbf{H} + \mathbf{M}), \tag{5.72}$$

where the ***magnetization vector*** $\mathbf{M}$ is defined as the vector sum of the magnetic dipole moments of the atoms contained in a unit volume of the material. Scale factors aside, the roles and interpretations of $\mathbf{B}$, $\mathbf{H}$, and $\mathbf{M}$ in Eq. (5.72) mirror those of $\mathbf{D}$, $\mathbf{E}$, and $\mathbf{P}$ in Eq. (4.83). Moreover, just as in most dielectrics $\mathbf{P}$ and $\mathbf{E}$ are linearly related, in most magnetic materials

$$\mathbf{M} = \chi_m \mathbf{H}, \tag{5.73}$$

where χ_m is a dimensionless quantity called the ***magnetic susceptibility*** of the material. For diamagnetic and paramagnetic materials, χ_m is a (temperature-dependent) constant, resulting in a linear relationship between $\mathbf{M}$ and $\mathbf{H}$ at a given temperature. This is not the case for ferromagnetic substances; the relationship between $\mathbf{M}$ and $\mathbf{H}$ not only is nonlinear, but also depends on the "history" of the material, as explained in the next section.

Keeping this fact in mind, we can combine Eqs. (5.72) and (5.73) to get

$$\mathbf{B} = \mu_0 (\mathbf{H} + \chi_m \mathbf{H}) = \mu_0 (1 + \chi_m) \mathbf{H}, \tag{5.74}$$

or

$$\mathbf{B} = \mu \mathbf{H}, \tag{5.75}$$

where μ, the ***magnetic permeability*** of the material, relates to χ_m as

$$\mu = \mu_0 (1 + \chi_m) \quad \text{(H/m)}. \tag{5.76}$$

Often it is convenient to define the magnetic properties of a material in terms of the ***relative permeability*** μ_r:

Table 5-2 **Properties of magnetic materials.**

	Diamagnetism	Paramagnetism	Ferromagnetism
Permanent magnetic dipole moment	No	Yes, but weak	Yes, and strong
Primary magnetization mechanism	Electron orbital magnetic moment	Electron spin magnetic moment	Magnetized domains
Direction of induced magnetic field (relative to external field)	Opposite	Same	Hysteresis [see **Fig. 5-22**]
Common substances	Bismuth, copper, diamond, gold, lead, mercury, silver, silicon	Aluminum, calcium, chromium, magnesium, niobium, platinum, tungsten	Iron, nickel, cobalt
Typical value of χ_m	$\approx -10^{-5}$	$\approx 10^{-5}$	$\lvert\chi_m\rvert \gg 1$ and hysteretic
Typical value of μ_r	≈ 1	≈ 1	$\lvert\mu_r\rvert \gg 1$ and hysteretic

$$\mu_r = \frac{\mu}{\mu_0} = 1 + \chi_m. \quad (5.77)$$

A material usually is classified as diamagnetic, paramagnetic, or ferromagnetic on the basis of the value of its χ_m (**Table 5-2**). Diamagnetic materials have negative susceptibilities whereas paramagnetic materials have positive ones. However, the absolute magnitude of χ_m is on the order of 10^{-5} for both classes of materials, which for most applications allows us to ignore χ_m relative to 1 in Eq. (5.77).

► Thus, $\mu_r \approx 1$ or $\mu \approx \mu_0$ for diamagnetic and paramagnetic substances, which include dielectric materials and most metals. In contrast, $|\mu_r| \gg 1$ for ferromagnetic materials; $|\mu_r|$ of purified iron, for example, is on the order of 2×10^5. ◄

Ferromagnetic materials are discussed next.

Exercise 5-11: The magnetic vector **M** is the vector sum of the magnetic moments of all the atoms contained in a unit volume (1m^3). If a certain type of iron with 8.5×10^{28} atoms/m^3 contributes one electron per atom to align its spin magnetic moment along the direction of the applied field, find (a) the spin magnetic moment of a single electron, given that $m_e = 9.1 \times 10^{-31}$ (kg) and $\hbar = 1.06 \times 10^{-34}$ (J·s), and (b) the magnitude of **M**.

Answer: (a) $m_s = 9.3 \times 10^{-24}$ (A·m^2), (b) $M = 7.9 \times 10^5$ (A/m). (See EM.)

5-5.3 Magnetic Hysteresis of Ferromagnetic Materials

Ferromagnetic materials, which include iron, nickel, and cobalt, exhibit unique magnetic properties due to the fact that their magnetic moments tend to readily align along the direction of an external magnetic field. Moreover, such materials remain partially magnetized even after the external field is removed. Because of these peculiar properties, ferromagnetic materials are used in the fabrication of permanent magnets.

A key to understanding the properties of ferromagnetic materials is the notion of ***magnetized domains***, microscopic regions (on the order of 10^{-10} m^3) within which the magnetic

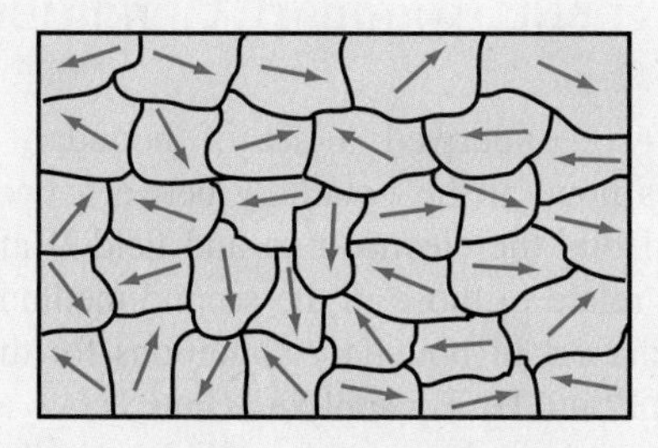

(a) Unmagnetized domains

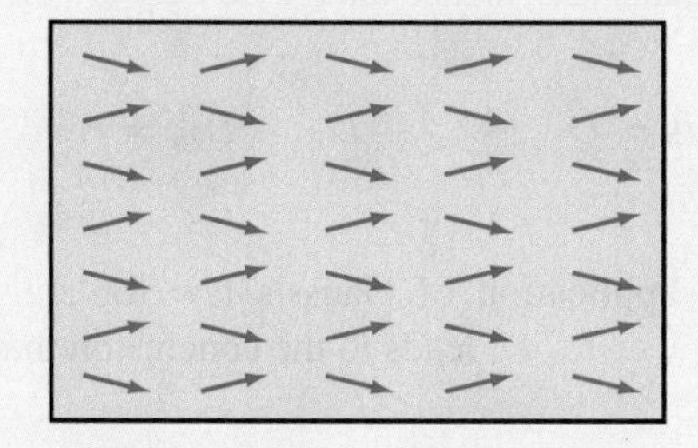

(b) Magnetized domains

Figure 5-21 Comparison of (a) unmagnetized and (b) magnetized domains in a ferromagnetic material.

moments of all atoms (typically on the order of 10^{19} atoms) are permanently aligned with each other. This alignment, which occurs in all ferromagnetic materials, is due to strong coupling forces between the magnetic dipole moments constituting an individual domain. In the absence of an external magnetic field, the domains take on random orientations relative to each other [**Fig. 5-21(a)**], resulting in zero net magnetization. The ***domain walls*** forming the boundaries between adjacent domains consist of thin transition regions. When an unmagnetized sample of a ferromagnetic material is placed in an external magnetic field, the domains partially align with the external field, as illustrated in **Fig. 5-21(b)**. A quantitative understanding of how the domains form and how they behave under the influence of an external magnetic field requires a heavy dose of quantum mechanics, and is outside the scope of the present treatment. Hence, we confine our discussion to a qualitative description of the magnetization process and its implications.

The magnetization behavior of a ferromagnetic material is described in terms of its ***B–H magnetization curve***, where B and H refer to the amplitudes of the **B** flux and **H** field in the material. Suppose that we start with an unmagnetized sample of iron, denoted by point O in **Fig. 5-22**. When we increase **H** continuously by, for example, increasing the current passing through a wire wound around the sample, **B** increases also along the B–H curve from point O to point A_1, at which nearly all the domains have become aligned with **H**. Point A_1 represents a saturation condition. If we then decrease **H** from its value at point A_1 back to zero (by reducing the current through the wire), the magnetization curve follows the path from A_1 to A_2. At point A_2, the external field **H** is zero (owing to the fact that the current through the wire is zero), but the flux density **B** in the material is not. The magnitude of **B** at A_2 is called the ***residual flux density*** B_r. The iron material is now magnetized and ready to be used as a permanent magnet owing to the fact that a large fraction of its magnetized domains have remained aligned. Reversing the direction of **H** and increasing its intensity causes **B** to decrease from B_r at point A_2 to zero at point A_3, and if the intensity of **H** is increased further while maintaining its direction, the magnetization moves to the saturation condition at point A_4. Finally, as **H** is made to return to zero and is then increased again in the positive direction, the curve follows the path from A_4 to A_1. This process is called ***magnetic hysteresis***. Hysteresis means "lag behind." The existence of a ***hysteresis loop*** implies that the magnetization process in ferromagnetic materials depends not only on the magnetic field **H**, but also on the magnetic history of the material. The shape and extent of the hysteresis loop depend on the properties of the ferromagnetic material and the peak-to-peak range over which **H** is made to vary. ***Hard ferromagnetic materials*** are characterized by wide hysteresis loops [**Fig. 5-23(a)**]. They cannot be easily demagnetized by

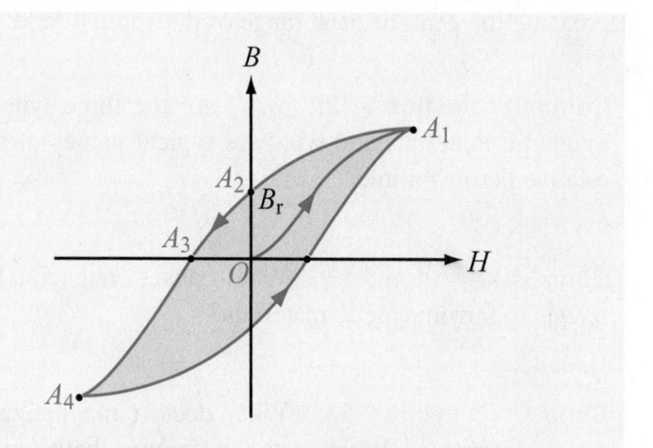

Figure 5-22 Typical hysteresis curve for a ferromagnetic material.

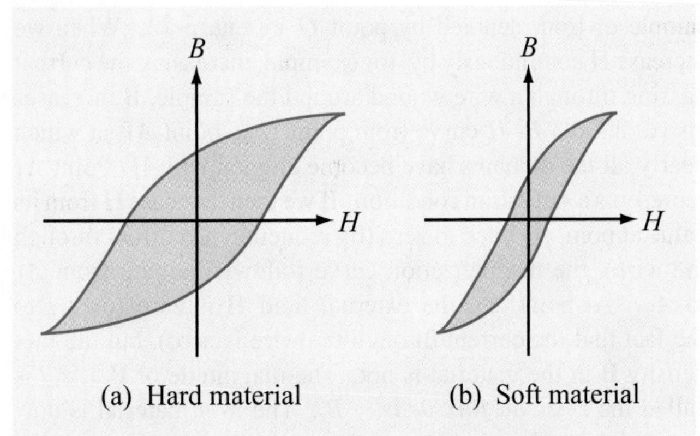

Figure 5-23 Comparison of hysteresis curves for (a) a hard ferromagnetic material and (b) a soft ferromagnetic material.

an external magnetic field because they have a large residual magnetization B_r. Hard ferromagnetic materials are used in the fabrication of permanent magnets for motors and generators. ***Soft ferromagnetic materials*** have narrow hysteresis loops [**Fig. 5-23(b)**], and hence can be more easily magnetized and demagnetized. To demagnetize any ferromagnetic material, the material is subjected to several hysteresis cycles while gradually decreasing the peak-to-peak range of the applied field.

Concept Question 5-12: What are the three types of magnetic materials and what are typical values of their relative permeabilities?

Concept Question 5-13: What causes magnetic hysteresis in ferromagnetic materials?

Concept Question 5-14: What does a magnetization curve describe? What is the difference between the magnetization curves of hard and soft ferromagnetic materials?

5-6 Magnetic Boundary Conditions

In Chapter 4, we derived a set of boundary conditions that describes how, at the boundary between two dissimilar contiguous media, the electric flux and field **D** and **E** in the first medium relate to those in the second medium. We now derive a similar set of boundary conditions for the magnetic flux and field **B** and **H**. By applying Gauss's law to a pill box that straddles the boundary, we determined that the difference between the normal components of the electric flux densities in two media equals the surface charge density ρ_s. That is,

$$\oint_S \mathbf{D} \cdot d\mathbf{s} = Q \quad \Rightarrow \quad D_{1n} - D_{2n} = \rho_s. \tag{5.78}$$

By analogy, application of Gauss's law for magnetism, as expressed by Eq. (5.44), leads to the conclusion that

$$\oint_S \mathbf{B} \cdot d\mathbf{s} = 0 \quad \Rightarrow \quad B_{1n} = B_{2n}. \tag{5.79}$$

▶ Thus the normal component of **B** is continuous across the boundary between two adjacent media. ◀

Because $\mathbf{B}_1 = \mu_1 \mathbf{H}_1$ and $\mathbf{B}_2 = \mu_2 \mathbf{H}_2$ for linear, isotropic media, the boundary condition for **H** corresponding to Eq. (5.79) is

$$\mu_1 H_{1n} = \mu_2 H_{2n}. \tag{5.80}$$

Comparison of Eqs. (5.78) and (5.79) reveals a striking difference between the behavior of the magnetic and electric fluxes across a boundary: *whereas the normal component of* **B** *is continuous across the boundary, the normal component of* **D** *is not (unless $\rho_s = 0$).* The reverse applies to the tangential components of the electric and magnetic fields **E** and **H**: *whereas the tangential component of* **E** *is continuous across the boundary, the tangential component of* **H** *is not (unless the surface current density $\mathbf{J}_s = 0$).* To obtain the boundary condition for the tangential component of **H**, we follow the same basic procedure used in Chapter 4 to establish the boundary condition for the tangential component of **E**. With reference to **Fig. 5-24**, we apply Ampère's law [Eq. (5.47)] to a closed

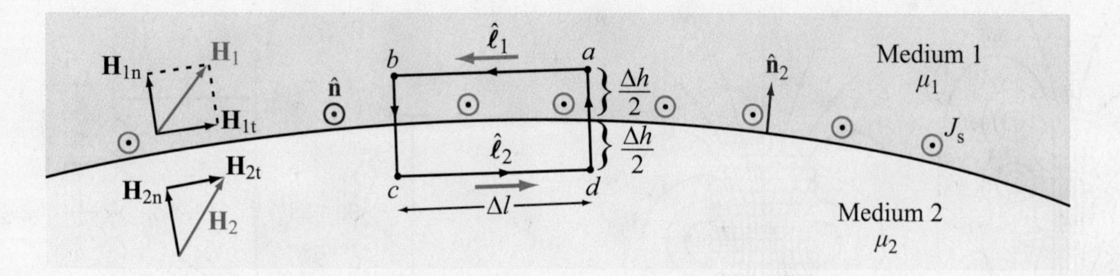

Figure 5-24 Boundary between medium 1 with μ_1 and medium 2 with μ_2.

rectangular path with sides of lengths Δl and Δh, and then let $\Delta h \to 0$, to obtain

$$\oint_C \mathbf{H} \cdot d\mathbf{l} = \int_a^b \mathbf{H}_1 \cdot \hat{\boldsymbol{\ell}}_1 \, d\ell + \int_c^d \mathbf{H}_2 \cdot \hat{\boldsymbol{\ell}}_2 \, d\ell = I, \tag{5.81}$$

where I is the net current crossing the surface of the loop in the direction specified by the right-hand rule (I is in the direction of the thumb when the fingers of the right hand extend in the direction of the loop C). As we let Δh of the loop approach zero, the surface of the loop approaches a thin line of length Δl. The total current flowing through this thin line is $I = J_s \, \Delta l$, where J_s is the magnitude of the component of the surface current density $\mathbf{J}_s$ normal to the loop. That is, $J_s = \mathbf{J}_s \cdot \hat{\mathbf{n}}$, where $\hat{\mathbf{n}}$ is the normal to the loop. In view of these considerations, Eq. (5.81) becomes

$$(\mathbf{H}_1 - \mathbf{H}_2) \cdot \hat{\boldsymbol{\ell}}_1 \, \Delta l = \mathbf{J}_s \cdot \hat{\mathbf{n}} \, \Delta l. \tag{5.82}$$

The vector $\hat{\boldsymbol{\ell}}_1$ can be expressed as $\hat{\boldsymbol{\ell}}_1 = \hat{\mathbf{n}} \times \hat{\mathbf{n}}_2$, where $\hat{\mathbf{n}}$ and $\hat{\mathbf{n}}_2$ are the normals to the loop and to the surface of medium 2 (Fig. 5-24), respectively. Using this relation in Eq. (5.82), and then applying the vector identity $\mathbf{A} \cdot (\mathbf{B} \times \mathbf{C}) = \mathbf{B} \cdot (\mathbf{C} \times \mathbf{A})$ leads to

$$\hat{\mathbf{n}} \cdot [\hat{\mathbf{n}}_2 \times (\mathbf{H}_1 - \mathbf{H}_2)] = \mathbf{J}_s \cdot \hat{\mathbf{n}}. \tag{5.83}$$

Since Eq. (5.83) is valid for any $\hat{\mathbf{n}}$, it follows that

$$\hat{\mathbf{n}}_2 \times (\mathbf{H}_1 - \mathbf{H}_2) = \mathbf{J}_s. \tag{5.84}$$

This equation implies that the tangential components of $\mathbf{H}$ parallel to $\mathbf{J}_s$ are continuous across the interface, whereas those orthogonal to $\mathbf{J}_s$ are discontinuous in the amount of $\mathbf{J}_s$.

Surface currents can exist only on the surfaces of perfect conductors and superconductors. Hence, *at the interface between media with finite conductivities,* $\mathbf{J}_s = 0$ and

$$H_{1t} = H_{2t}. \tag{5.85}$$

Exercise 5-12: With reference to Fig. 5-24, determine the angle between $\mathbf{H}_1$ and $\hat{\mathbf{n}}_2 = \hat{\mathbf{z}}$ if $\mathbf{H}_2 = (\hat{\mathbf{x}}3 + \hat{\mathbf{z}}2)$ (A/m), $\mu_{r_1} = 2$, and $\mu_{r_2} = 8$, and $\mathbf{J}_s = 0$.

Answer: $\theta = 20.6°$. (See EM.)

5-7 Inductance

An inductor is the magnetic analogue of an electric capacitor. Just as a capacitor can store energy in the electric field in the medium between its conducting surfaces, an inductor can store energy in the magnetic field near its current-carrying conductors. A typical inductor consists of multiple turns of wire helically coiled around a cylindrical core [Fig. 5-25(a)]. Such a structure is called a ***solenoid***. Its core may be air filled or may contain a magnetic material with magnetic permeability μ. If the wire carries a current I and the turns are closely spaced, the solenoid will produce a relatively uniform magnetic field within its interior with magnetic field lines resembling those of the permanent magnet [Fig. 5-25(b)].

5-7.1 Magnetic Field in a Solenoid

As a prelude to our discussion of inductance we derive an expression for the magnetic flux density $\mathbf{B}$ in the interior region of a tightly wound solenoid. The solenoid is of length l and radius a, and comprises N turns carrying current I. The number of turns per unit length is $n = N/l$, and the fact that the turns are tightly wound implies that the pitch of a single turn is small compared with the solenoid's radius. Even though the turns are slightly helical in shape, we can treat them as circular loops (Fig. 5-26). Let us start by considering the magnetic

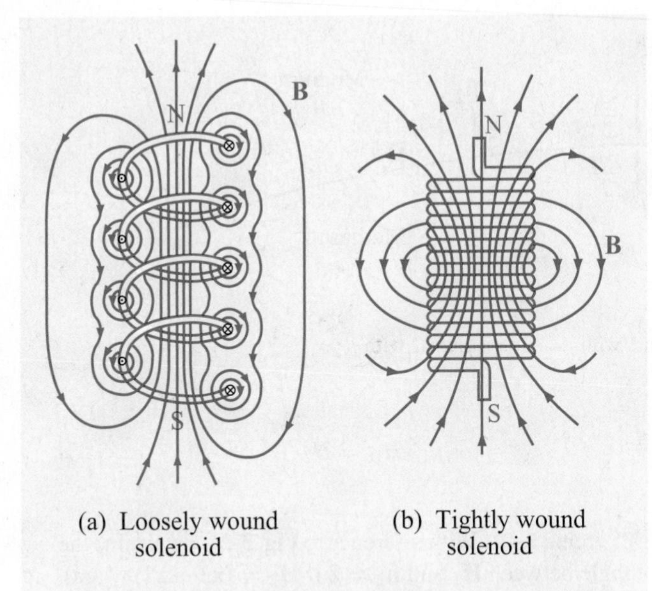

Figure 5-25 Magnetic field lines of (a) a loosely wound solenoid and (b) a tightly wound solenoid.

flux density **B** at point P on the axis of the solenoid. In Example 5-3, we derived the following expression for the magnetic field **H** along the axis of a circular loop of radius a, a distance z away from its center:

$$\mathbf{H} = \hat{\mathbf{z}} \, \frac{I' a^2}{2(a^2+z^2)^{3/2}}, \tag{5.86}$$

where I' is the current carried by the loop. If we treat an incremental length dz of the solenoid as an equivalent loop composed of $n\,dz$ turns carrying a current $I' = In\,dz$, then the induced field at point P is

$$d\mathbf{B} = \mu \, d\mathbf{H} = \hat{\mathbf{z}} \, \frac{\mu n I a^2}{2(a^2+z^2)^{3/2}} \, dz. \tag{5.87}$$

The total field **B** at P is obtained by integrating the contributions from the entire length of the solenoid. This is facilitated by expressing the variable z in terms of the angle θ, as seen from P to a point on the solenoid rim. That is,

$$z = a \tan\theta, \tag{5.88a}$$

$$a^2 + z^2 = a^2 + a^2\tan^2\theta = a^2\sec^2\theta, \tag{5.88b}$$

$$dz = a\sec^2\theta \, d\theta. \tag{5.88c}$$

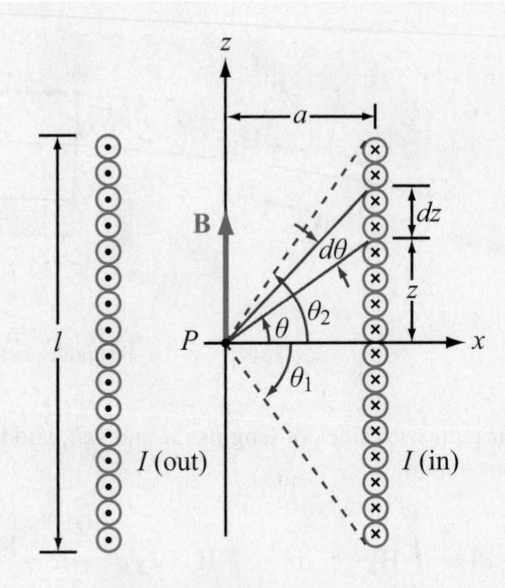

Figure 5-26 Solenoid cross section showing geometry for calculating **H** at a point P on the solenoid axis.

Upon substituting the last two expressions in Eq. (5.87) and integrating from θ_1 to θ_2, we have

$$\mathbf{B} = \hat{\mathbf{z}} \, \frac{\mu n I a^2}{2} \int_{\theta_1}^{\theta_2} \frac{a\sec^2\theta \, d\theta}{a^3\sec^3\theta}$$

$$= \hat{\mathbf{z}} \, \frac{\mu n I}{2} (\sin\theta_2 - \sin\theta_1). \tag{5.89}$$

If the solenoid length l is much larger than its radius a, then for points P away from the solenoid's ends, $\theta_1 \approx -90°$ and $\theta_2 \approx 90°$, in which case Eq. (5.89) reduces to

$$\mathbf{B} \approx \hat{\mathbf{z}}\mu n I = \frac{\hat{\mathbf{z}}\mu N I}{l} \quad \text{(long solenoid with } l/a \gg 1\text{)}. \tag{5.90}$$

Even though Eq. (5.90) was derived for the field **B** at the midpoint of the solenoid, it is approximately valid everywhere in the solenoid's interior, except near the ends.

We now return to a discussion of inductance, which includes the notion of ***self-inductance***, representing the magnetic flux linkage of a coil or circuit with itself, and ***mutual inductance***, which involves the magnetic flux linkage in a circuit due to the magnetic field generated by a current in another one. Usually, when the term ***inductance*** is used, the intended reference is to self-inductance.

Exercise 5-13: Use Eq. (5.89) to obtain an expression for **B** at a point on the axis of a very long solenoid but situated at its end points. How does **B** at the end points compare to **B** at the midpoint of the solenoid?

Answer: $\mathbf{B} = \hat{\mathbf{z}}(\mu NI/2l)$ at the end points, which is half as large as **B** at the midpoint. (See EM.)

5-7.2 Self-Inductance

From Eq. (5.66), the magnetic flux Φ linking a surface S is

$$\Phi = \int_S \mathbf{B} \cdot d\mathbf{s} \qquad \text{(Wb)}. \tag{5.91}$$

In a solenoid characterized by an approximately uniform magnetic field throughout its cross-section given by Eq. (5.90), the flux linking a single loop is

$$\Phi = \int_S \hat{\mathbf{z}} \left(\mu \frac{N}{l} I \right) \cdot \hat{\mathbf{z}}\, ds = \mu \frac{N}{l} IS, \tag{5.92}$$

where S is the cross-sectional area of the loop. ***Magnetic flux linkage*** Λ is defined as the total magnetic flux linking a given circuit or conducting structure. If the structure consists of a single conductor with multiple loops, as in the case of the solenoid, Λ equals the flux linking all loops of the structure. For a solenoid with N turns,

$$\Lambda = N\Phi = \mu \frac{N^2}{l} IS \qquad \text{(Wb)}. \tag{5.93}$$

If, on the other hand, the structure consists of two *separate* conductors, as in the case of the parallel-wire and coaxial transmission lines shown in **Fig. 5-27**, the flux linkage Λ associated with a length l of either line refers to the flux Φ through a closed surface between the two conductors, such as the shaded areas in **Fig. 5-27**. In reality, there is also some magnetic flux that passes through the conductors themselves, but it may be ignored by assuming that currents flow only on the surfaces of the conductors, in which case the magnetic field inside the conductors vanishes. This assumption is justified by the fact that our interest in calculating Λ is for the purpose of determining the inductance of a given structure, and inductance is of interest primarily in the ac case (i.e., time-varying currents, voltages, and fields). As we will see later in Section 7-5, the current flowing in a conductor under ac conditions is concentrated within a very thin layer on the skin of the conductor.

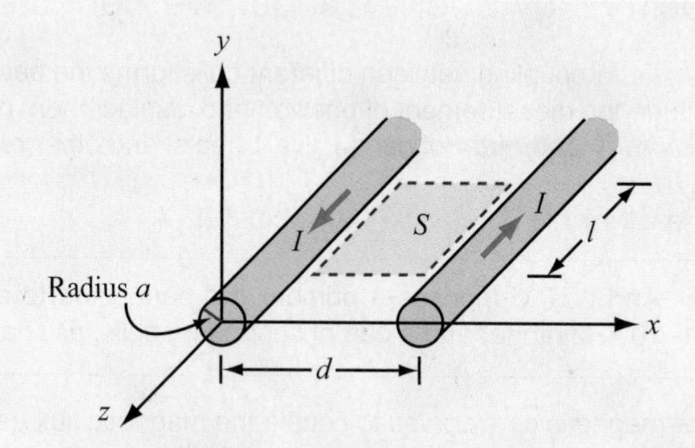

(a) Parallel-wire transmission line

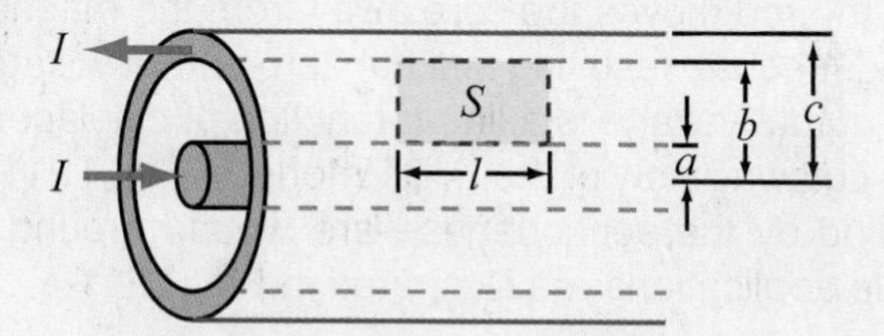

(b) Coaxial transmission line

Figure 5-27 To compute the inductance per unit length of a two-conductor transmission line, we need to determine the magnetic flux through the area S between the conductors.

▶ For the parallel-wire transmission line, ac currents flow on the outer surfaces of the wires, and for the coaxial line, the current flows on the outer surface of the inner conductor and on the inner surface of the outer one (the current-carrying surfaces are those adjacent to the electric and magnetic fields present in the region between the conductors). ◀

The ***self-inductance*** of any conducting structure is defined as the ratio of the magnetic flux linkage Λ to the current I flowing through the structure:

$$L = \frac{\Lambda}{I} \qquad \text{(H)}. \tag{5.94}$$

Technology Brief 11: Inductive Sensors

Magnetic coupling between different coils forms the basis of several different types of inductive sensors. Applications include the measurement of position and displacement (with submillimeter resolution) in device-fabrication processes, ***proximity detection*** of conductive objects, and other related applications.

Linear Variable Differential Transformer (LVDT)

► An LVDT comprises a primary coil connected to an ac source (typically a sine wave at a frequency in the 1–10 kHz range) and a pair of secondary coils, all sharing a common ***ferromagnetic core*** (**Fig. TF11-1**). ◄

The magnetic core serves to couple the magnetic flux generated by the primary coil into the two secondaries, thereby inducing an output voltage across each of them. The secondary coils are connected in opposition, so that when the core is positioned at the magnetic center of the LVDT, the individual output signals of the secondaries cancel each other out, producing a null output voltage. The core is connected to the outside world via a nonmagnetic push rod. When the rod moves the core away from the magnetic center, the magnetic fluxes induced in the secondary coils are no longer equal, resulting in a nonzero output voltage. The LVDT is called a "linear" transformer because the amplitude of the output voltage is a linear function of displacement over a wide operating range (**Fig. TF11-2**).

The cutaway view of the LVDT model in **Fig. TF11-3** depicts a configuration in which all three coils—with the primary straddled by the secondaries—are wound around a glass tube that contains the magnetic core and attached rod. Sample applications are illustrated in **Fig. TF11-4**.

Eddy-Current Proximity Sensor

The transformer principle can be applied to build a proximity sensor in which the output voltage of the secondary coil becomes a sensitive indicator of the presence of a conductive object in its immediate vicinity (**Fig. TF11-5**).

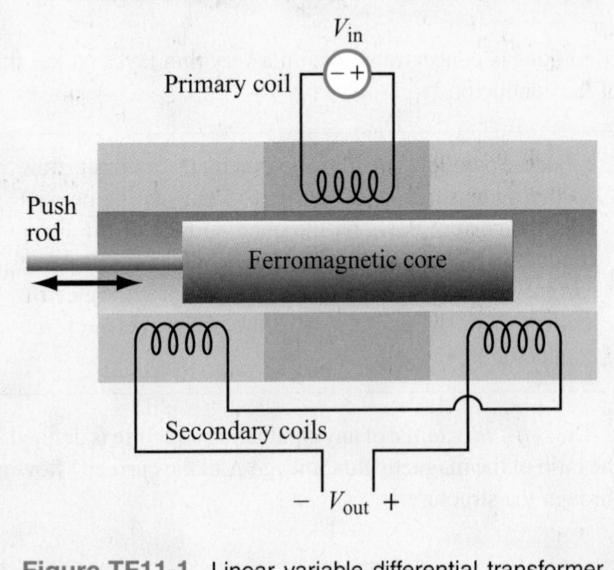

Figure TF11-1 Linear variable differential transformer (LVDT) circuit.

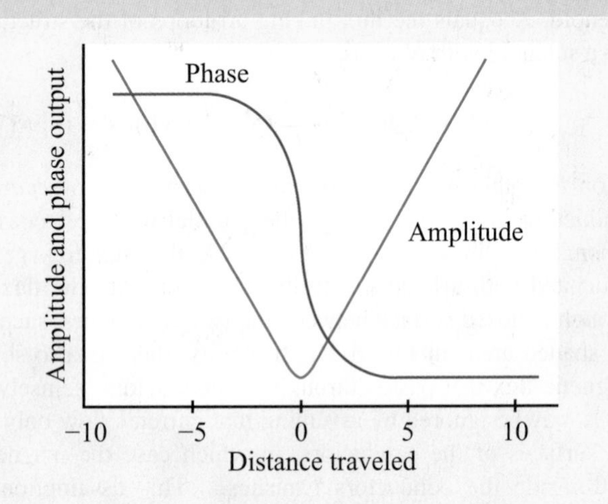

Figure TF11-2 Amplitude and phase responses as a function of the distance by which the magnetic core is moved away from the center position.

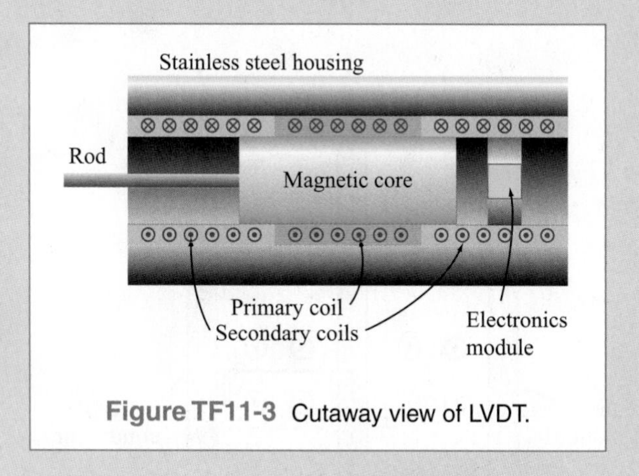

Figure TF11-3 Cutaway view of LVDT.

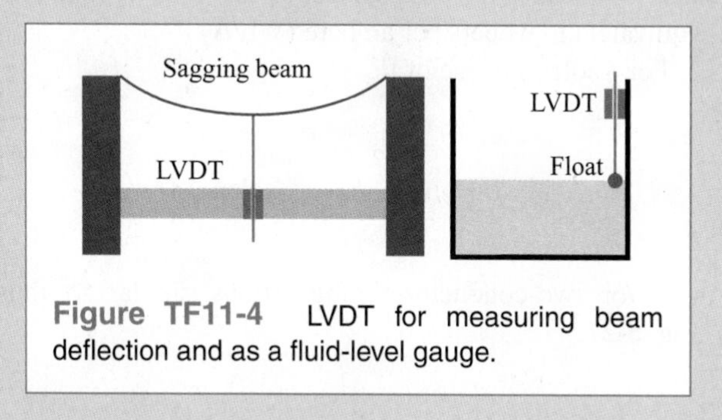

Figure TF11-4 LVDT for measuring beam deflection and as a fluid-level gauge.

► When an object is placed in front of the secondary coil, the magnetic field of the coil induces eddy (circular) currents in the object, which generate magnetic fields of their own having a direction that opposes the magnetic field of the secondary coil. ◄

The reduction in magnetic flux causes a drop in output voltage, with the magnitude of the change being dependent on the conductive properties of the object and its distance from the sensor.

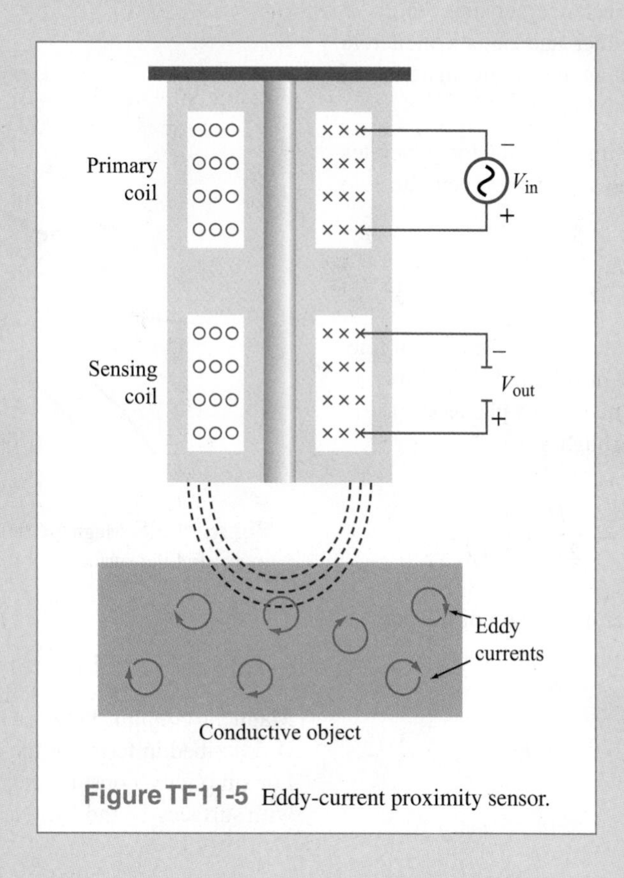

Figure TF11-5 Eddy-current proximity sensor.

The SI unit for inductance is the henry (H), which is equivalent to webers per ampere (Wb/A).

For a solenoid, use of Eq. (5.93) gives

$$L = \mu \frac{N^2}{l} S \qquad \textbf{(solenoid)}, \tag{5.95}$$

and for two-conductor configurations similar to those of **Fig. 5-27**,

$$L = \frac{\Lambda}{I} = \frac{\Phi}{I} = \frac{1}{I} \int_S \mathbf{B} \cdot d\mathbf{s}. \tag{5.96}$$

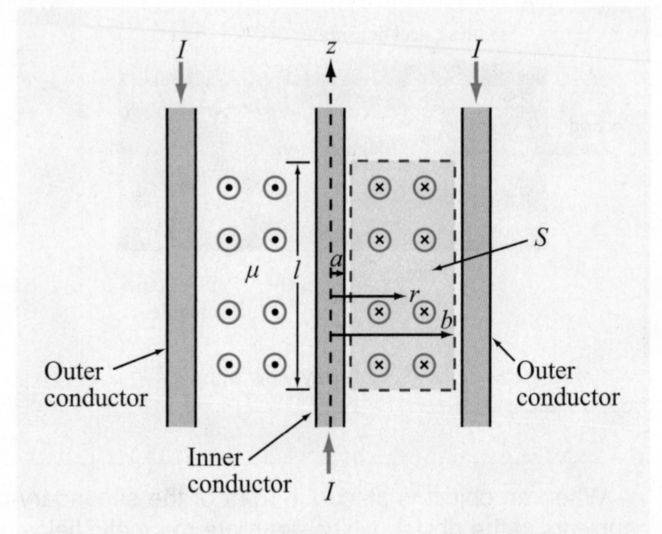

Figure 5-28 Cross-sectional view of coaxial transmission line (Example 5-7). ⊙ and ⊗ denote **H** field out of and into the page, respectively.

Example 5-7: Inductance of a Coaxial Transmission Line

Develop an expression for the inductance per unit length of a coaxial transmission line with inner and outer conductors of radii a and b (**Fig. 5-28**) and an insulating material of permeability μ.

Solution: The current I in the inner conductor generates a magnetic field **B** throughout the region between the two conductors. It is given by Eq. (5.30) as

$$\mathbf{B} = \hat{\boldsymbol{\phi}} \frac{\mu I}{2\pi r}, \tag{5.97}$$

where r is the radial distance from the axis of the coaxial line. Consider a transmission-line segment of length l as shown in **Fig. 5-28**. Because **B** is perpendicular to the planar surface S between the conductors, the flux through S is

$$\Phi = l \int_a^b B\, dr = l \int_a^b \frac{\mu I}{2\pi r}\, dr = \frac{\mu I l}{2\pi} \ln\left(\frac{b}{a}\right). \tag{5.98}$$

Using Eq. (5.96), the inductance per unit length of the coaxial transmission line is

$$L' = \frac{L}{l} = \frac{\Phi}{lI} = \frac{\mu}{2\pi} \ln\left(\frac{b}{a}\right). \tag{5.99}$$

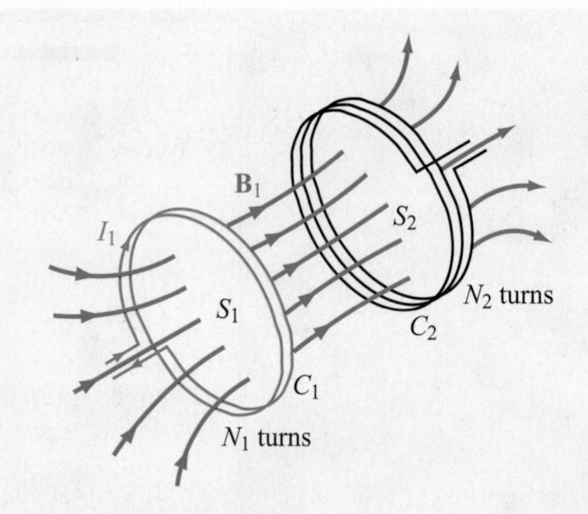

Figure 5-29 Magnetic field lines generated by current I_1 in loop 1 linking surface S_2 of loop 2.

5-7.3 Mutual Inductance

Magnetic coupling between two different conducting structures is described in terms of the mutual inductance between them. For simplicity, consider the case of two multiturn closed loops with surfaces S_1 and S_2. Current I_1 flows through the first loop

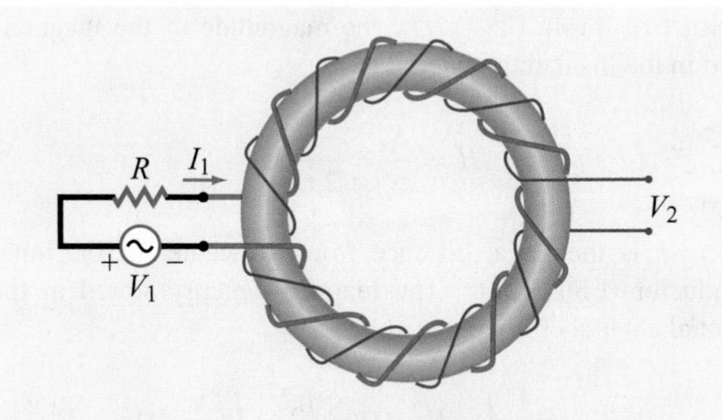

Figure 5-30 Toroidal coil with two windings used as a transformer.

(**Fig. 5-29**), and no current flows through the second one. The magnetic field $\mathbf{B}_1$ generated by I_1 results in a flux Φ_{12} through loop 2, given by

$$\Phi_{12} = \int_{S_2} \mathbf{B}_1 \cdot d\mathbf{s}, \qquad (5.100)$$

and if loop 2 consists of N_2 turns all coupled by $\mathbf{B}_1$ in exactly the same way, then the total magnetic flux linkage through loop 2 is

$$\Lambda_{12} = N_2\Phi_{12} = N_2 \int_{S_2} \mathbf{B}_1 \cdot d\mathbf{s}. \qquad (5.101)$$

The mutual inductance associated with this magnetic coupling is given by

$$L_{12} = \frac{\Lambda_{12}}{I_1} = \frac{N_2}{I_1} \int_{S_2} \mathbf{B}_1 \cdot d\mathbf{s} \quad \text{(H)}. \qquad (5.102)$$

Mutual inductance is important in transformers (as discussed in Chapter 6) wherein the windings of two or more circuits share a common magnetic core, as illustrated by the toroidal arrangement shown in **Fig. 5-30**.

Concept Question 5-15: What is the magnetic field like in the interior of a long solenoid?

Concept Question 5-16: What is the difference between self-inductance and mutual inductance?

Concept Question 5-17: How is the inductance of a solenoid related to its number of turns N?

5-8 Magnetic Energy

When we introduced electrostatic energy in Section 4-10, we did so by examining what happens to the energy expended in charging up a capacitor from zero voltage to some final voltage V. We introduce the concept of magnetic energy by considering an inductor with inductance L connected to a current source. Suppose that we were to increase the current i flowing through the inductor from zero to a final value I. From circuit theory, we know that the instantaneous voltage υ across the inductor is given by $\upsilon = L\,di/dt$. We will derive this relationship from Maxwell's equations in Chapter 6, thereby justifying the use of the i–υ relationship for the inductor. Power p equals the product of υ and i, and the time integral of power is work, or energy. Hence, the total energy in joules (J) expended in building up a current I in the inductor is

$$W_{\rm m} = \int p\,dt = \int i\upsilon\,dt = L\int_0^I i\,di = \tfrac{1}{2}LI^2 \quad \text{(J)}. \qquad (5.103)$$

We call this the ***magnetic energy*** stored in the inductor.

To justify this association, consider the solenoid inductor. Its inductance is given by Eq. (5.95) as $L = \mu N^2 S/l$, and the magnitude of the magnetic flux density in its interior is given by Eq. (5.90) as $B = \mu NI/l$, implying that $I = Bl/(\mu N)$. Using these expressions for L and I in Eq. (5.103), we obtain

$$\begin{aligned} W_{\rm m} &= \frac{1}{2}LI^2 \\ &= \frac{1}{2}\left(\mu\frac{N^2}{l}S\right)\left(\frac{Bl}{\mu N}\right)^2 \\ &= \frac{1}{2}\frac{B^2}{\mu}(lS) = \frac{1}{2}\mu H^2 \mathcal{V}, \end{aligned} \qquad (5.104)$$

where $\mathcal{V} = lS$ is the volume of the interior of the solenoid and $H = B/\mu$. The expression for $W_{\rm m}$ suggests that the energy expended in building up the current in the inductor is stored in the magnetic field with ***magnetic energy density*** $w_{\rm m}$, defined as the magnetic energy $W_{\rm m}$ per unit volume,

$$w_{\rm m} = \frac{W_{\rm m}}{\mathcal{V}} = \frac{1}{2}\mu H^2 \quad \text{(J/m}^3\text{)}. \qquad (5.105)$$

▶ Even though this expression was derived for a solenoid, it remains valid for any medium with a magnetic field **H**. ◀

Furthermore, for any volume $\mathcal{V}$ containing a material with permeability μ (including free space with permeability μ_0), the total magnetic energy stored in a magnetic field **H** is

$$W_m = \frac{1}{2} \int_{\mathcal{V}} \mu H^2 \, d\mathcal{V} \qquad \text{(J)}. \tag{5.106}$$

Example 5-8: Magnetic Energy in a Coaxial Cable

Derive an expression for the magnetic energy stored in a coaxial cable of length l and inner and outer radii a and b. The current flowing through the cable is I and its insulation material has permeability μ.

Solution: From Eq. (5.97), the magnitude of the magnetic field in the insulating material is

$$H = \frac{B}{\mu} = \frac{I}{2\pi r},$$

where r is the radial distance from the center of the inner conductor (**Fig. 5-28**). The magnetic energy stored in the coaxial cable therefore is

$$W_m = \frac{1}{2} \int_{\mathcal{V}} \mu H^2 \, d\mathcal{V} = \frac{\mu I^2}{8\pi^2} \int_{\mathcal{V}} \frac{1}{r^2} \, d\mathcal{V}.$$

Since H is a function of r only, we choose $d\mathcal{V}$ to be a cylindrical shell of length l, radius r, and thickness dr along the radial direction. Thus, $d\mathcal{V} = 2\pi rl \, dr$ and

$$W_m = \frac{\mu I^2}{8\pi^2} \int_a^b \frac{1}{r^2} \cdot 2\pi rl \, dr = \frac{\mu I^2 l}{4\pi} \ln\left(\frac{b}{a}\right) = \frac{1}{2} LI^2 \quad \text{(J)},$$

with L given by Eq. (5.99).

Chapter 5 Summary

Concepts

- The magnetic force acting on a charged particle q moving with a velocity **u** in a region containing a magnetic flux density **B** is $\mathbf{F}_m = q\mathbf{u} \times \mathbf{B}$.
- The total electromagnetic force, known as the Lorentz force, acting on a moving charge in the presence of both electric and magnetic fields is $\mathbf{F} = q(\mathbf{E} + \mathbf{u} \times \mathbf{B})$.
- Magnetic forces acting on current loops can generate magnetic torques.
- The magnetic field intensity induced by a current element is defined by the Biot–Savart law.
- Gauss's law for magnetism states that the net magnetic flux flowing out of any closed surface is zero.
- Ampère's law states that the line integral of **H** over a closed contour is equal to the net current crossing the surface bounded by the contour.
- The vector magnetic potential **A** is related to **B** by $\mathbf{B} = \nabla \times \mathbf{A}$.
- Materials are classified as diamagnetic, paramagnetic, or ferromagnetic, depending on their crystalline structure and the behavior under the influence of an external magnetic field.
- Diamagnetic and paramagnetic materials exhibit a linear behavior between **B** and **H**, with $\mu \approx \mu_0$ for both.
- Ferromagnetic materials exhibit a nonlinear hysteretic behavior between **B** and **H** and, for some, μ may be as large as $10^5 \mu_0$.
- At the boundary between two different media, the normal component of **B** is continuous, and the tangential components of **H** are related by $H_{2t} - H_{1t} = J_s$, where J_s is the surface current density flowing in a direction orthogonal to H_{1t} and H_{2t}.
- The inductance of a circuit is defined as the ratio of magnetic flux linking the circuit to the current flowing through it.
- Magnetic energy density is given by $w_m = \frac{1}{2}\mu H^2$.

Important Terms

Provide definitions or explain the meaning of the following terms:

Ampère's law
Ampèrian contour
Biot–Savart law
current density (volume) $\mathbf{J}$
diamagnetic
ferromagnetic
Gauss's law for magnetism
hard and soft ferromagnetic materials
inductance (self- and mutual)
Lorentz force $\mathbf{F}$
magnetic dipole
magnetic energy W_m
magnetic energy density w_m
magnetic flux Φ
magnetic flux density $\mathbf{B}$
magnetic flux linkage Λ
magnetic force $\mathbf{F}_m$
magnetic hysteresis
magnetic moment $\mathbf{m}$
magnetic potential $\mathbf{A}$
magnetic susceptibility χ_m
magnetization curve
magnetization vector $\mathbf{M}$
magnetized domains
moment arm $\mathbf{d}$
orbital and spin magnetic moments
paramagnetic
solenoid
surface current density $\mathbf{J}_s$
toroid
toroidal coil
torque $\mathbf{T}$
vector Poisson's equation

Mathematical and Physical Models

Maxwell's Magnetostatics Equations

Gauss's Law for Magnetism

$$\nabla \cdot \mathbf{B} = 0 \quad \longleftrightarrow \quad \oint_S \mathbf{B} \cdot d\mathbf{s} = 0$$

Ampère's Law

$$\nabla \times \mathbf{H} = \mathbf{J} \quad \longleftrightarrow \quad \oint_C \mathbf{H} \cdot d\boldsymbol{\ell} = I$$

Lorentz Force on Charge q

$$\mathbf{F} = q(\mathbf{E} + \mathbf{u} \times \mathbf{B})$$

Magnetic Force on Wire

$$\mathbf{F}_m = I \oint_C d\mathbf{l} \times \mathbf{B} \qquad \text{(N)}$$

Magnetic Torque on Loop

$$\mathbf{T} = \mathbf{m} \times \mathbf{B} \qquad \text{(N·m)}$$

$$\mathbf{m} = \hat{\mathbf{n}} NIA \qquad \text{(A·m}^2\text{)}$$

Biot–Savart Law

$$\mathbf{H} = \frac{I}{4\pi} \int_l \frac{d\mathbf{l} \times \hat{\mathbf{R}}}{R^2} \qquad \text{(A/m)}$$

Magnetic Field

Infinitely Long Wire $\quad \mathbf{B} = \hat{\boldsymbol{\phi}} \dfrac{\mu_0 I}{2\pi r} \qquad \text{(Wb/m}^2\text{)}$

Circular Loop $\quad \mathbf{H} = \hat{\mathbf{z}} \dfrac{Ia^2}{2(a^2 + z^2)^{3/2}} \qquad \text{(A/m)}$

Solenoid $\quad \mathbf{B} \approx \hat{\mathbf{z}} \mu n I = \dfrac{\hat{\mathbf{z}} \mu N I}{l} \qquad \text{(Wb/m}^2\text{)}$

Vector Magnetic Potential

$$\mathbf{B} = \nabla \times \mathbf{A} \qquad \text{(Wb/m}^2\text{)}$$

Vector Poisson's Equation

$$\nabla^2 \mathbf{A} = -\mu \mathbf{J}$$

Inductance

$$L = \frac{\Lambda}{I} = \frac{\Phi}{I} = \frac{1}{I} \int_S \mathbf{B} \cdot d\mathbf{s} \qquad \text{(H)}$$

Magnetic Energy Density

$$w_m = \frac{1}{2} \mu H^2 \qquad \text{(J/m}^3\text{)}$$

PROBLEMS

Section 5-1: Magnetic Forces and Torques

*5.1 An electron with a speed of 8×10^6 m/s is projected along the positive x direction into a medium containing a uniform magnetic flux density $\mathbf{B} = (\hat{\mathbf{x}}4 - \hat{\mathbf{z}}6)$ T. Given that $e = 1.6 \times 10^{-19}$ C and the mass of an electron is $m_e = 9.1 \times 10^{-31}$ kg, determine the initial acceleration vector of the electron (at the moment it is projected into the medium).

5.2 The circuit shown in Fig. P5.2 uses two identical springs to support a 10 cm long horizontal wire with a mass of 20 g. In the absence of a magnetic field, the weight of the wire causes the springs to stretch a distance of 0.2 cm each. When a uniform magnetic field is turned on in the region containing the horizontal wire, the springs are observed to stretch an additional 0.5 cm each. What is the intensity of the magnetic flux density $\mathbf{B}$? The force equation for a spring is $F = kd$, where k is the spring constant and d is the distance it has been stretched.

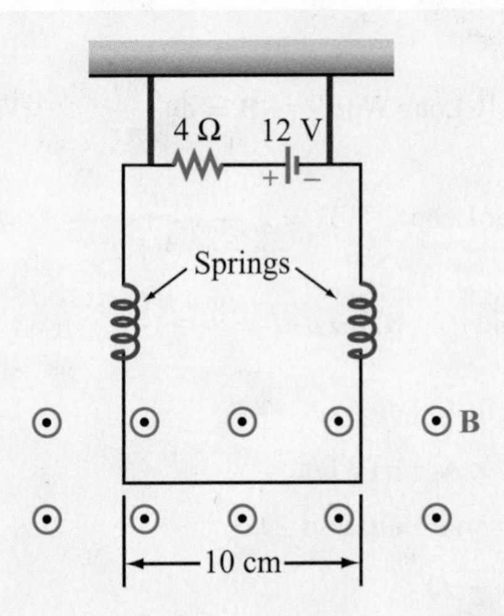

Figure P5.2 Configuration of Problem 5.2.

5.3 When a particle with charge q and mass m is introduced into a medium with a uniform field $\mathbf{B}$ such that the initial velocity of the particle $\mathbf{u}$ is perpendicular to $\mathbf{B}$ (Fig. P5.3), the magnetic force exerted on the particle causes it to move in a circle of radius a. By equating $\mathbf{F}_m$ to the centripetal force on the particle, determine a in terms of q, m, u, and $\mathbf{B}$.

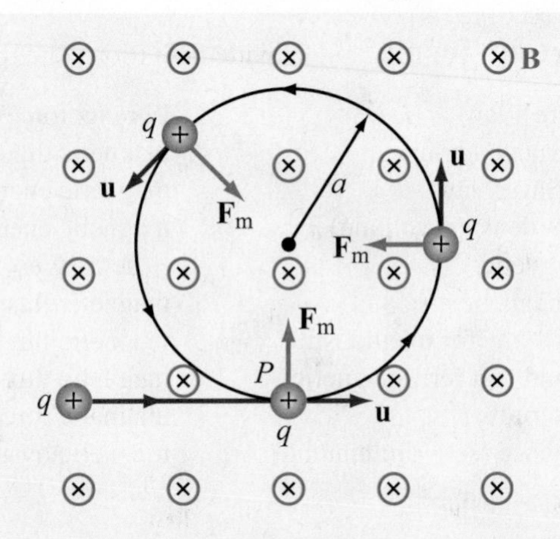

Figure P5.3 Particle of charge q projected with velocity $\mathbf{u}$ into a medium with a uniform field $\mathbf{B}$ perpendicular to $\mathbf{u}$ (Problem 5.3).

*5.4 The rectangular loop shown in Fig. P5.4 consists of 20 closely wrapped turns and is hinged along the z axis. The plane of the loop makes an angle of 30° with the y axis, and the current in the windings is 0.5 A. What is the magnitude of the torque exerted on the loop in the presence of a uniform field $\mathbf{B} = \hat{\mathbf{y}}\,2.4$ T? When viewed from above, is the expected direction of rotation clockwise or counterclockwise?

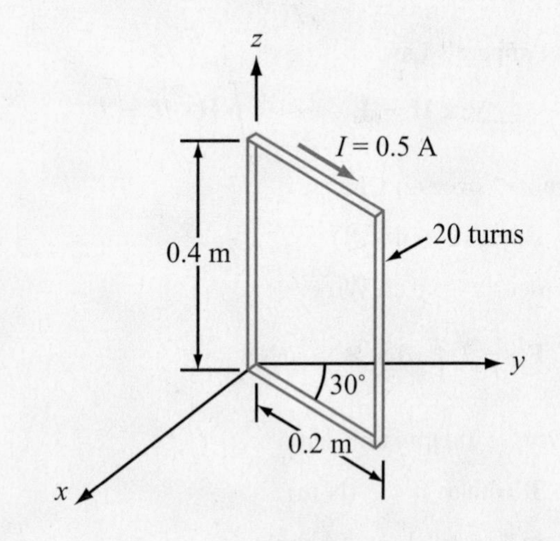

Figure P5.4 Hinged rectangular loop of Problem 5.4.

5.5 In a cylindrical coordinate system, a 2 m long straight wire carrying a current of 5 A in the positive z direction is located at $r = 4$ cm, $\phi = \pi/2$, and -1 m $\leq z \leq 1$ m.

*Answer(s) available in Appendix D.

*(a) If $\mathbf{B} = \hat{\mathbf{r}}\, 0.6 \cos\phi$ (T), what is the magnetic force acting on the wire?

(b) How much work is required to rotate the wire once about the z axis in the negative ϕ direction (while maintaining $r = 4$ cm)?

(c) At what angle ϕ is the force a maximum?

5.6 A 20-turn rectangular coil with sides $l = 30$ cm and $w = 10$ cm is placed in the y–z plane as shown in **Fig. P5.6**.

(a) If the coil, which carries a current $I = 10$ A, is in the presence of a magnetic flux density

$$\mathbf{B} = 2 \times 10^{-2}(\hat{\mathbf{x}} + \hat{\mathbf{y}}2) \quad \text{(T)},$$

determine the torque acting on the coil.

(b) At what angle ϕ is the torque zero?

(c) At what angle ϕ is the torque maximum? Determine its value.

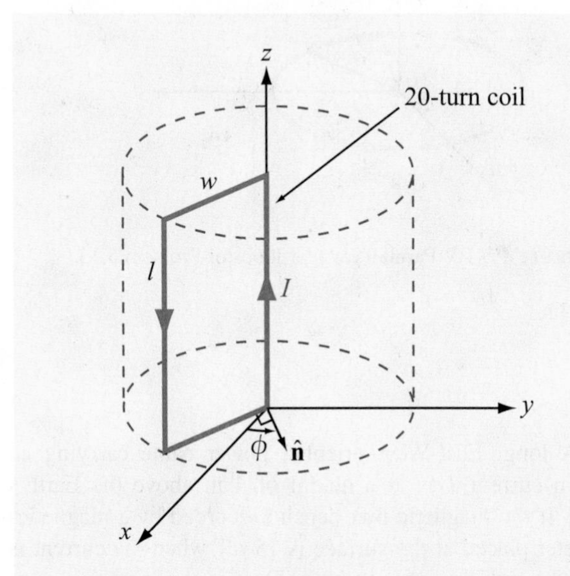

Figure P5.6 Rectangular loop of Problem 5.6.

Section 5-2: The Biot–Savart Law

5.7 An infinitely long, thin conducting sheet defined over the space $0 \le x \le w$ and $-\infty \le y \le \infty$ is carrying a current with a uniform surface current density $\mathbf{J}_s = \hat{\mathbf{y}}5$ (A/m). Obtain an expression for the magnetic field at point $P = (0, 0, z)$ in Cartesian coordinates.

5.8 Use the approach outlined in Example 5-2 to develop an expression for the magnetic field **H** at an arbitrary point P due to the linear conductor defined by the geometry shown in **Fig. P5.8**. If the conductor extends between $z_1 = 3$ m and $z_2 = 7$ m and carries a current $I = 15$ A, find **H** at $P = (2, \phi, 0)$.

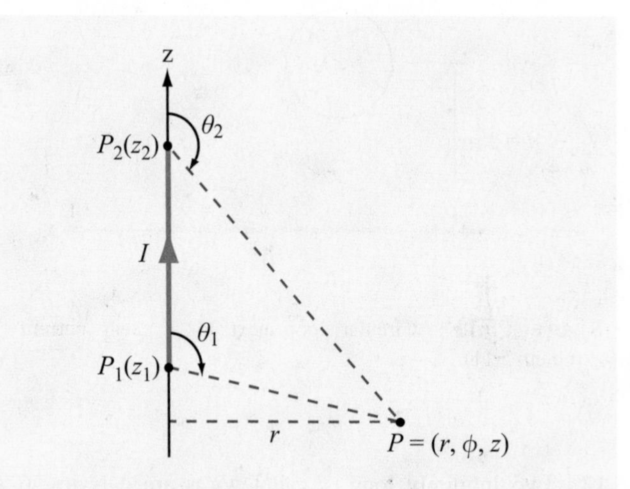

Figure P5.8 Current-carrying linear conductor of Problem 5.8.

***5.9** The loop shown in **Fig. P5.9** consists of radial lines and segments of circles whose centers are at point P. Determine the magnetic field **H** at P in terms of a, b, θ, and I.

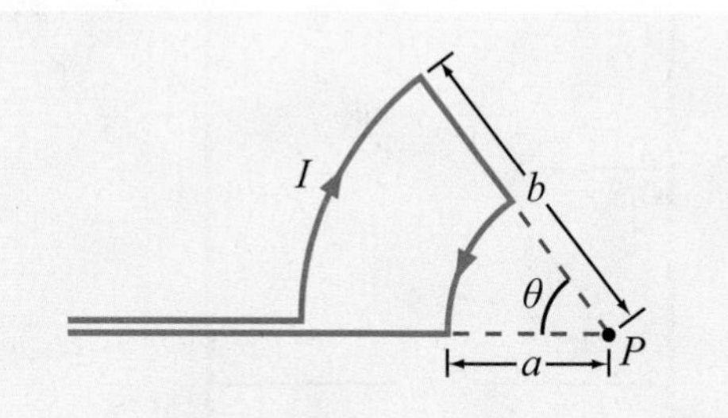

Figure P5.9 Configuration of Problem 5.9.

***5.10** An 8 cm × 12 cm rectangular loop of wire is situated in the x–y plane with the center of the loop at the origin and its long sides parallel to the x axis. The loop has a current of 50 A flowing clockwise (when viewed from above). Determine the magnetic flux density at the center of the loop.

*5.11 An infinitely long wire carrying a 50 A current in the positive x direction is placed along the x axis in the vicinity of a 20-turn circular loop located in the x–y plane (**Fig. P5.11**). If the magnetic field at the center of the loop is zero, what is the direction and magnitude of the current flowing in the loop?

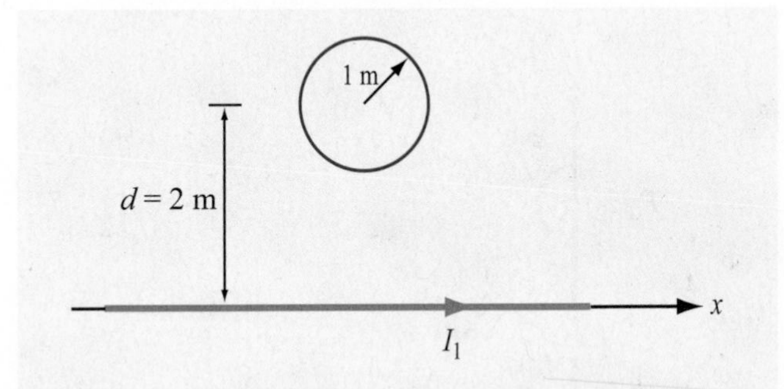

Figure P5.11 Circular loop next to a linear current (Problem 5.11).

5.12 Two infinitely long, parallel wires are carrying 6 A currents in opposite directions. Determine the magnetic flux density at point P in **Fig. P5.12**.

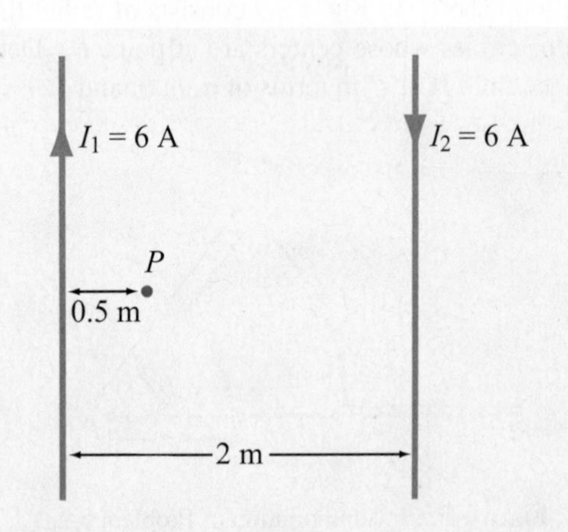

Figure P5.12 Arrangement for Problem 5.12.

5.13 Two parallel, circular loops carrying a current of 40 A each are arranged as shown in **Fig. P5.13**. The first loop is situated in the x–y plane with its center at the origin, and the second loop's center is at $z = 2$ m. If the two loops have the same radius $a = 3$ m, determine the magnetic field at:

(a) $z = 0$

(b) $z = 1$ m

(c) $z = 2$ m

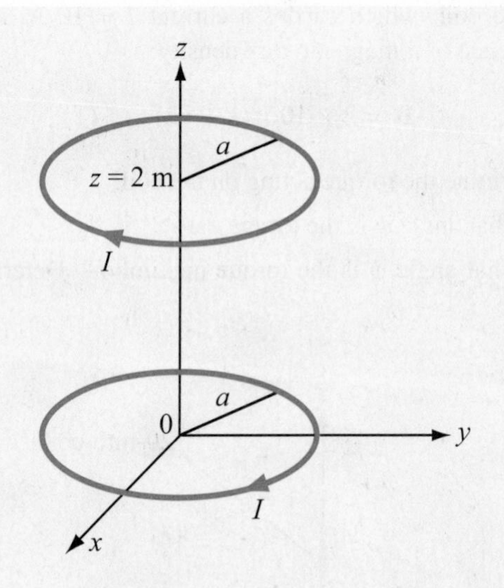

Figure P5.13 Parallel circular loops of Problem 5.13.

*5.14 A long, East-West–oriented power cable carrying an unknown current I is at a height of 4 m above the Earth's surface. If the magnetic flux density recorded by a magnetic-field meter placed at the surface is 15 μT when the current is flowing through the cable and 20 μT when the current is zero, what is the magnitude of I?

5.15 A circular loop of radius a carrying current I_1 is located in the x–y plane as shown in **Fig. P5.15**. In addition, an infinitely long wire carrying current I_2 in a direction parallel with the z axis is located at $y = y_0$.

(a) Determine **H** at $P = (0, 0, h)$.

(b) Evaluate **H** for $a = 3$ cm, $y_0 = 10$ cm, $h = 4$ cm, $I_1 = 10$ A, and $I_2 = 20$ A.

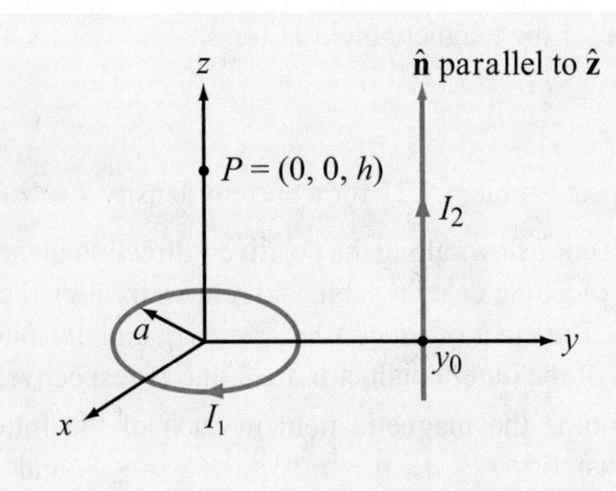

Figure P5.15 Problem 5.15.

*__5.16__ The long, straight conductor shown in **Fig. P5.16** lies in the plane of the rectangular loop at a distance $d = 0.1$ m. The loop has dimensions $a = 0.2$ m and $b = 0.5$ m, and the currents are $I_1 = 40$ A and $I_2 = 30$ A. Determine the net magnetic force acting on the loop.

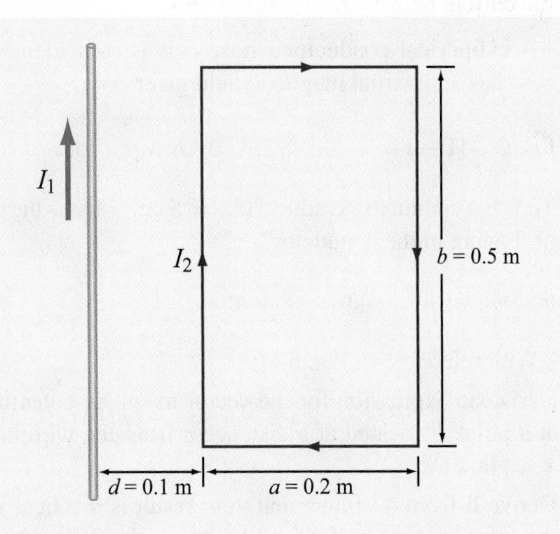

Figure P5.16 Current loop next to a conducting wire (Problem 5.16).

5.17 In the arrangement shown in **Fig. P5.17**, each of the two long, parallel conductors carries a current I, is supported by 8 cm long strings, and has a mass per unit length of 1.2 g/cm. Due to the repulsive force acting on the conductors, the angle θ between the supporting strings is 10°. Determine the magnitude of I and the relative directions of the currents in the two conductors.

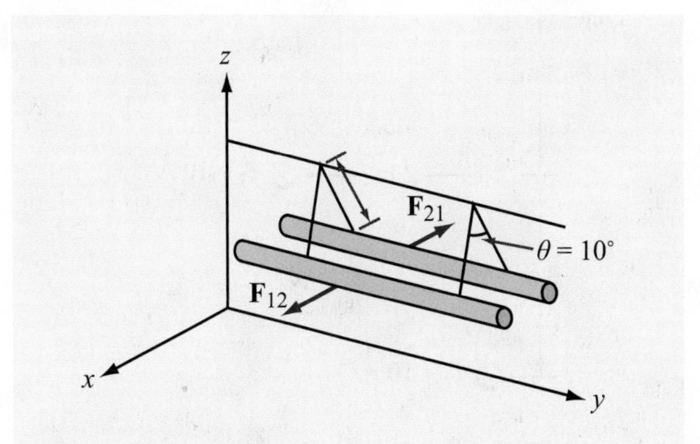

Figure P5.17 Parallel conductors supported by strings (Problem 5.17).

5.18 An infinitely long, thin conducting sheet of width w along the x direction lies in the x–y plane and carries a current I in the $-y$ direction. Determine the following:

*__(a)__ The magnetic field at a point P midway between the edges of the sheet and at a height h above it (**Fig. P5.18**).

(b) The force per unit length exerted on an infinitely long wire passing through point P and parallel to the sheet if the current through the wire is equal in magnitude but opposite in direction to that carried by the sheet.

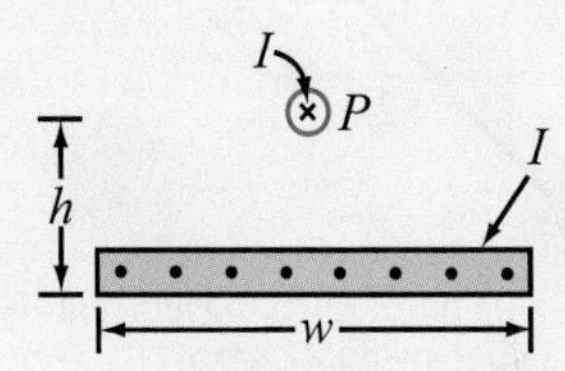

Figure P5.18 A linear current source above a current sheet (Problem 5.18).

5.19 Three long, parallel wires are arranged as shown in **Fig. P5.19**. Determine the force per unit length acting on the wire carrying I_3.

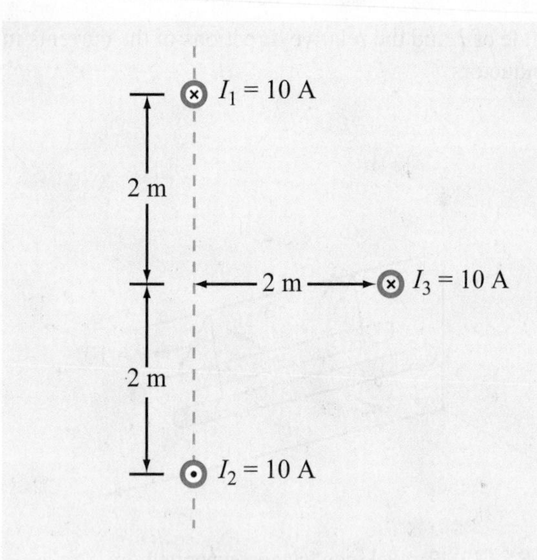

Figure P5.19 Three parallel wires of Problem 5.19.

*5.20 A square loop placed as shown in **Fig. P5.20** has 2 m sides and carries a current $I_1 = 5$ A. If a straight, long conductor carrying a current $I_2 = 20$ A is introduced and placed just above the midpoints of two of the loop's sides, determine the net force acting on the loop.

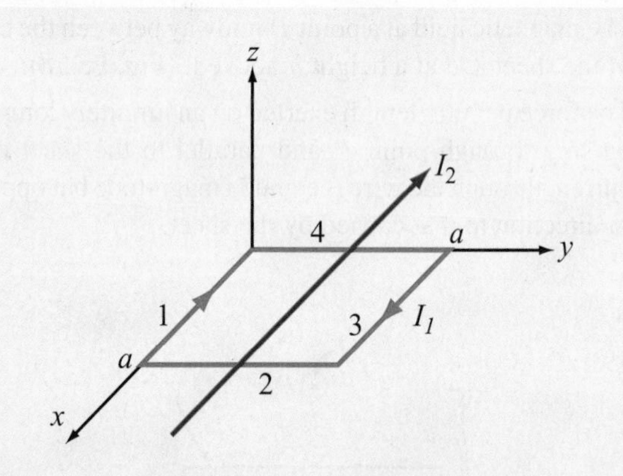

Figure P5.20 Long wire carrying current I_2, just above a square loop carrying I_1 (Problem 5.20).

Section 5-3: Maxwell's Magnetostatic Equations

5.21 A long cylindrical conductor whose axis is coincident with the z axis has a radius a and carries a current characterized by a current density $\mathbf{J} = \hat{\mathbf{z}}J_0/r$, where J_0 is a constant and r is the radial distance from the cylinder's axis. Obtain an expression for the magnetic field **H** for

(a) $0 \le r \le a$

(b) $r > a$

5.22 Repeat Problem 5.21 for a current density $\mathbf{J} = \hat{\mathbf{z}}J_0 e^{-r}$.

5.23 Current I flows along the positive z direction in the inner conductor of a long coaxial cable and returns through the outer conductor. The inner conductor has radius a, and the inner and outer radii of the outer conductor are b and c, respectively.

(a) Determine the magnetic field in each of the following regions: $0 \le r \le a$, $a \le r \le b$, $b \le r \le c$, and $r \ge c$.

(b) Plot the magnitude of **H** as a function of r over the range from $r = 0$ to $r = 10$ cm, given that $I = 10$ A, $a = 2$ cm, $b = 4$ cm, and $c = 5$ cm.

*5.24 In a certain conducting region, the magnetic field is given in cylindrical coordinates by

$$\mathbf{H} = \hat{\boldsymbol{\phi}}\frac{1}{r}[1 - (1 + 3r)e^{-3r}]$$

Find the current density **J**.

5.25 A cylindrical conductor whose axis is coincident with the z axis has an internal magnetic field given by

$$\mathbf{H} = \hat{\boldsymbol{\phi}}\,\frac{2}{r}[1 - (4r + 1)e^{-4r}] \qquad \text{(A/m)} \qquad \text{for } r \le a$$

where a is the conductor's radius. If $a = 5$ cm, what is the total current flowing in the conductor?

Section 5-4: Vector Magnetic Potential

5.26 With reference to **Fig. 5-10**:

*(a) Derive an expression for the vector magnetic potential **A** at a point P located at a distance r from the wire in the x–y plane.

(b) Derive **B** from **A**. Show that your result is identical with the expression given by Eq. (5.29), which was derived by applying the Biot–Savart law.

5.27 A uniform current density given by

$$\mathbf{J} = \hat{\mathbf{z}}J_0 \qquad \text{(A/m}^2\text{)}$$

gives rise to a vector magnetic potential

$$\mathbf{A} = -\hat{\mathbf{z}}\frac{\mu_0 J_0}{4}(x^2 + y^2) \qquad \text{(Wb/m).}$$

(a) Apply the vector Poisson's equation to confirm the above statement.

(b) Use the expression for **A** to find **H**.

(c) Use the expression for **J** in conjunction with Ampère's law to find **H**. Compare your result with that obtained in part (b).

5.28 In a given region of space, the vector magnetic potential is given by $\mathbf{A} = \hat{\mathbf{x}}5\cos\pi y + \hat{\mathbf{z}}(2 + \sin\pi x)$ (Wb/m).

*(a) Determine **B**.

(b) Use Eq. (5.66) to calculate the magnetic flux passing through a square loop with 0.25 m long edges if the loop is in the x–y plane, its center is at the origin, and its edges are parallel to the x and y axes.

(c) Calculate Φ again using Eq. (5.67).

***5.29** A thin current element extending between $z = -L/2$ and $z = L/2$ carries a current I along $+\hat{\mathbf{z}}$ through a circular cross-section of radius a.

(a) Find **A** at a point P located very far from the origin (assume R is so much larger than L that point P may be considered to be at approximately the same distance from every point along the current element).

(b) Determine the corresponding **H**.

Section 5-5: Magnetic Properties of Materials

***5.30** Iron contains 8.5×10^{28} atoms/m^3. At saturation, the alignment of the electrons' spin magnetic moments in iron can contribute 1.5 T to the total magnetic flux density **B**. If the spin magnetic moment of a single electron is 9.27×10^{-24} (A·m^2), how many electrons per atom contribute to the saturated field?

5.31 In the model of the hydrogen atom proposed by Bohr in 1913, the electron moves around the nucleus at a speed of 2×10^6 m/s in a circular orbit of radius 5×10^{-11} m. What is the magnitude of the magnetic moment generated by the electron's motion?

Section 5-6: Magnetic Boundary Conditions

5.32 The x–y plane separates two magnetic media with magnetic permeabilities μ_1 and μ_2 (**Fig. P5.32**). If there is no surface current at the interface and the magnetic field in medium 1 is

$$\mathbf{H}_1 = \hat{\mathbf{x}}H_{1x} + \hat{\mathbf{y}}H_{1y} + \hat{\mathbf{z}}H_{1z}$$

find:

(a) $\mathbf{H}_2$

(b) θ_1 and θ_2

(c) Evaluate $\mathbf{H}_2$, θ_1, and θ_2 for $H_{1x} = 2$ (A/m), $H_{1y} = 0$, $H_{1z} = 4$ (A/m), $\mu_1 = \mu_0$, and $\mu_2 = 4\mu_0$

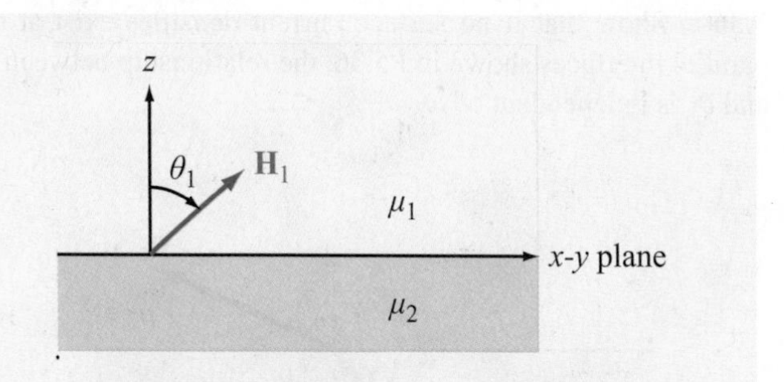

Figure P5.32 Adjacent magnetic media (Problem 5.32).

***5.33** Given that a current sheet with surface current density $\mathbf{J}_s = \hat{\mathbf{x}}4$ (A/m) exists at $y = 0$, the interface between two magnetic media, and $\mathbf{H}_1 = \hat{\mathbf{z}}11$ (A/m) in medium 1 ($y > 0$), determine $\mathbf{H}_2$ in medium 2 ($y < 0$).

5.34 In **Fig. P5.34**, the plane defined by $x - y = 1$ separates medium 1 of permeability μ_1 from medium 2 of permeability μ_2. If no surface current exists on the boundary and

$$\mathbf{B}_1 = \hat{\mathbf{x}}2 + \hat{\mathbf{y}}3 \qquad \text{(T)},$$

find $\mathbf{B}_2$ and then evaluate your result for $\mu_1 = 5\mu_2$. *Hint:* Start by deriving the equation for the unit vector normal to the given plane.

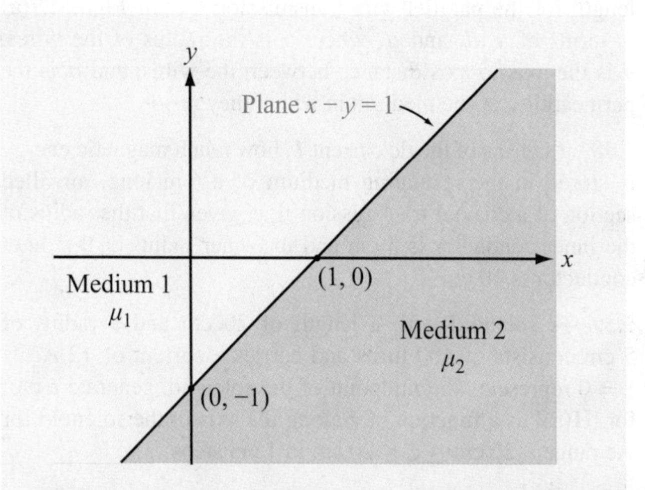

Figure P5.34 Magnetic media separated by the plane $x - y = 1$ (Problem 5.34).

*5.35 The plane boundary defined by $z = 0$ separates air from a block of iron. If $\mathbf{B}_1 = \hat{\mathbf{x}}4 - \hat{\mathbf{y}}6 + \hat{\mathbf{z}}12$ in air ($z \geq 0$), find $\mathbf{B}_2$ in iron ($z \leq 0$), given that $\mu = 5000\mu_0$ for iron.

5.36 Show that if no surface current densities exist at the parallel interfaces shown in P5.36, the relationship between θ_4 and θ_1 is independent of μ_2.

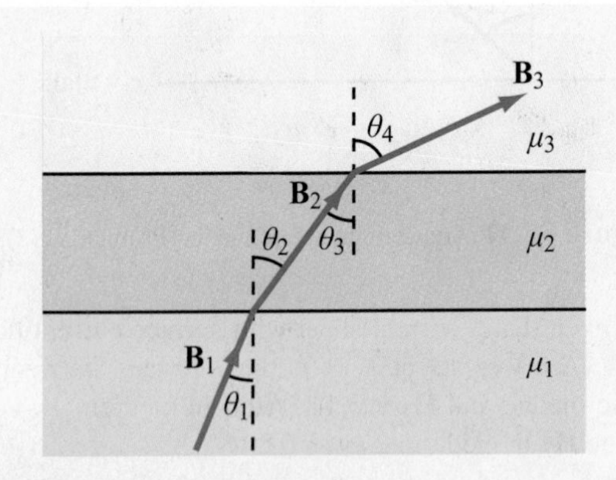

Figure P5.36 Three magnetic media with parallel interfaces (Problem 5.36).

Sections 5-7 and 5-8: Inductance and Magnetic Energy

*5.37 Obtain an expression for the self-inductance per unit length for the parallel wire transmission line of **Fig. 5-27(a)** in terms of a, d, and μ, where a is the radius of the wires, d is the axis-to-axis distance between the wires, and μ is the permeability of the medium in which they reside.

5.38 In terms of the dc current I, how much magnetic energy is stored in the insulating medium of a 6 m long, air-filled section of a coaxial transmission line, given that the radius of the inner conductor is 5 cm and the inner radius of the outer conductor is 10 cm?

5.39 A solenoid with a length of 20 cm and a radius of 5 cm consists of 400 turns and carries a current of 12 A. If $z = 0$ represents the midpoint of the solenoid, generate a plot for $|\mathbf{H}(z)|$ as a function of z along the axis of the solenoid for the range -20 cm $\leq z \leq 20$ cm in 1 cm steps.

*5.40 The rectangular loop shown in **Fig. P5.40** is coplanar with the long, straight wire carrying the current $I = 20$ A. Determine the magnetic flux through the loop.

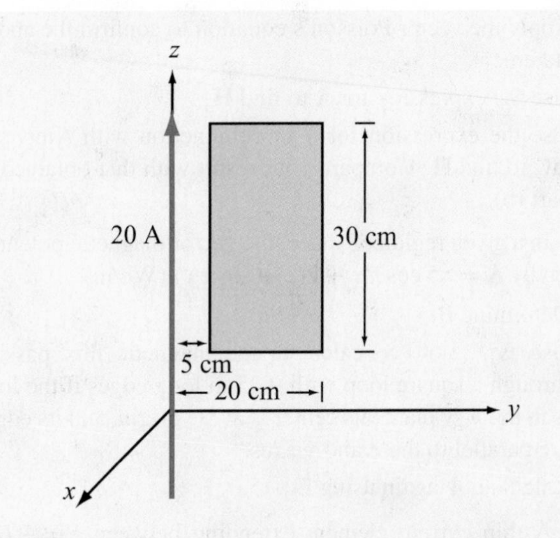

Figure P5.40 Loop and wire arrangement for Problem 5.40.

5.41 Determine the mutual inductance between the circular loop and the linear current shown in **Fig. P5.41**.

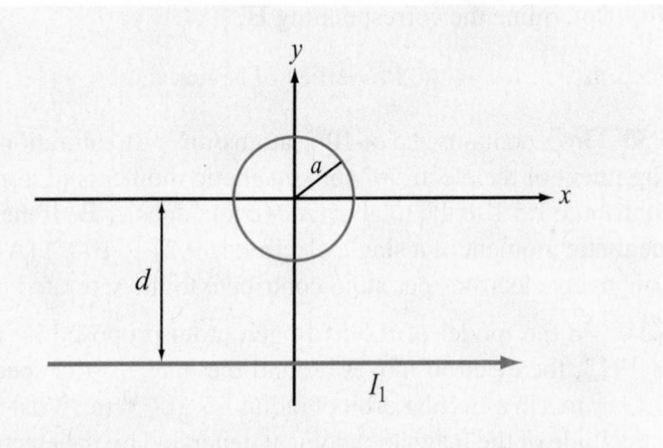

Figure P5.41 Linear conductor with current I_1 next to a circular loop of radius a at distance d (Problem 5.41).

CHAPTER 6

Maxwell's Equations for Time-Varying Fields

Chapter Contents

Dynamic Fields, 304
6-1 Faraday's Law, 304
6-2 Stationary Loop in a Time-Varying Magnetic Field, 306
6-3 The Ideal Transformer, 310
6-4 Moving Conductor in a Static Magnetic Field, 311
TB12 EMF Sensors, 314
6-5 The Electromagnetic Generator, 316
6-6 Moving Conductor in a Time-Varying Magnetic Field, 318
6-7 Displacement Current, 319
6-8 Boundary Conditions for Electromagnetics, 321
6-9 Charge-Current Continuity Relation, 321
6-10 Free-Charge Dissipation in a Conductor, 324
6-11 Electromagnetic Potentials, 324
Chapter 6 Summary, 329
Problems, 330

Objectives

Upon learning the material presented in this chapter, you should be able to:

1. Apply Faraday's law to compute the voltage induced by a stationary coil placed in a time-varying magnetic field or moving in a medium containing a magnetic field.
2. Describe the operation of the electromagnetic generator.
3. Calculate the displacement current associated with a time-varying electric field.
4. Calculate the rate at which charge dissipates in a material with known ϵ and σ.

Dynamic Fields

Electric charges induce electric fields, and electric currents induce magnetic fields. As long as the charge and current distributions remain constant in time, so will the fields they induce. If the charges and currents vary in time, the electric and magnetic fields vary accordingly. Moreover, the electric and magnetic fields become coupled and travel through space in the form of electromagnetic waves. Examples of such waves include light, x-rays, infrared rays, gamma rays, and radio waves (see **Fig. 1-16**).

To study time-varying electromagnetic phenomena, we need to consider the entire set of Maxwell's equations simultaneously. These equations, first introduced in the opening section of Chapter 4, are given in both differential and integral form in **Table 6-1**. In the static case ($\partial/\partial t = 0$), we use the first pair of Maxwell's equations to study electric phenomena (Chapter 4) and the second pair to study magnetic phenomena (Chapter 5). In the dynamic case ($\partial/\partial t \neq 0$), the coupling that exists between the electric and magnetic fields, as expressed by the second and fourth equations in **Table 6-1**, prevents such decomposition. The first equation represents Gauss's law for electricity, and it is equally valid for static and dynamic fields. The same is true for the third equation, Gauss's law for magnetism. By contrast, the second and fourth equations—Faraday's and Ampère's laws—are of a totally different nature. Faraday's law expresses the fact that a time-varying magnetic field gives rise to an electric field. Conversely, Ampère's law states that a time-varying electric field must be accompanied by a magnetic field.

Some statements in this and succeeding chapters contradict conclusions reached in Chapter 4 and 5 as those pertained to the special case of static charges and dc currents. The behavior of dynamic fields reduces to that of static ones when $\partial/\partial t$ is set to zero.

We begin this chapter by examining Faraday's and Ampère's laws and some of their practical applications. We then combine Maxwell's equations to obtain relations among the charge and current sources, ρ_v and $\mathbf{J}$, the scalar and vector potentials, V and $\mathbf{A}$, and the electromagnetic fields, $\mathbf{E}$, $\mathbf{D}$, $\mathbf{H}$, and $\mathbf{B}$, for the most general time-varying case and for the specific case of sinusoidal-time variations.

6-1 Faraday's Law

The close connection between electricity and magnetism was established by Oersted, who demonstrated that a wire carrying an electric current exerts a force on a compass needle and that the needle always turns so as to point in the $\hat{\boldsymbol{\phi}}$ direction when the current is along the $\hat{\mathbf{z}}$ direction. The force acting on the compass needle is due to the magnetic field produced by the current in the

Table 6-1 **Maxwell's equations.**

Reference	Differential Form	Integral Form	
Gauss's law	$\nabla \cdot \mathbf{D} = \rho_v$	$\oint_S \mathbf{D} \cdot d\mathbf{s} = Q$	(6.1)
Faraday's law	$\nabla \times \mathbf{E} = -\dfrac{\partial \mathbf{B}}{\partial t}$	$\oint_C \mathbf{E} \cdot d\mathbf{l} = -\int_S \dfrac{\partial \mathbf{B}}{\partial t} \cdot d\mathbf{s}$	(6.2)*
No magnetic charges (Gauss's law for magnetism)	$\nabla \cdot \mathbf{B} = 0$	$\oint_S \mathbf{B} \cdot d\mathbf{s} = 0$	(6.3)
Ampère's law	$\nabla \times \mathbf{H} = \mathbf{J} + \dfrac{\partial \mathbf{D}}{\partial t}$	$\oint_C \mathbf{H} \cdot d\mathbf{l} = \int_S \left(\mathbf{J} + \dfrac{\partial \mathbf{D}}{\partial t}\right) \cdot d\mathbf{s}$	(6.4)

*For a stationary surface S.

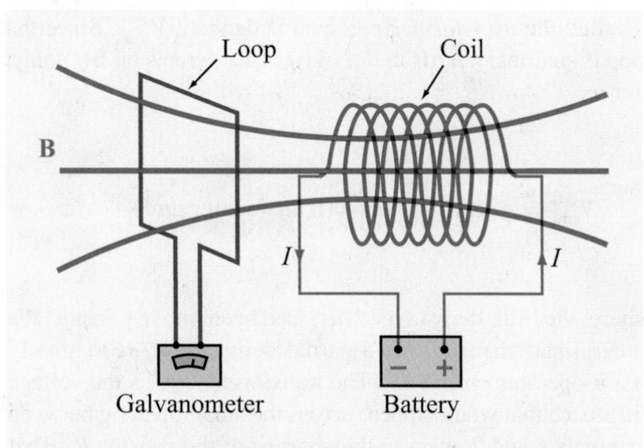

Figure 6-1 The galvanometer (predecessor of the ammeter) shows a deflection whenever the magnetic flux passing through the square loop changes with time.

wire. Following this discovery, *Faraday hypothesized that if a current produces a magnetic field, then the converse should also be true: a magnetic field should produce a current in a wire.* To test his hypothesis, he conducted numerous experiments in his laboratory in London over a period of about 10 years, all aimed at making magnetic fields induce currents in wires. Similar work was being carried out by Henry in Albany, New York. Wires were placed next to permanent magnets or current-carrying loops of all different sizes, but no currents were ever detected. Eventually, these experiments led to the discovery by both Faraday and Henry that:

► Magnetic fields can produce an electric current in a closed loop, but *only* if the magnetic flux linking the surface area of the loop changes with time. The key to the induction process is *change*. ◄

To elucidate the induction process, consider the arrangement shown in **Fig. 6-1**. A conducting loop connected to a galvanometer, a sensitive instrument used in the 1800s to detect current flow, is placed next to a conducting coil connected to a battery. The current in the coil produces a magnetic field **B** whose lines pass through the loop. In Section 5-4, we defined the magnetic flux Φ passing through a loop as the integral of the normal component of the magnetic flux density over the surface area of the loop, S, or

$$\Phi = \int_S \mathbf{B} \cdot d\mathbf{s} \quad \text{(Wb)}. \tag{6.5}$$

Under stationary conditions, the dc current in the coil produces a constant magnetic field **B**, which in turn produces a constant flux through the loop. When the flux is constant, no current is detected by the galvanometer. However, when the battery is disconnected, thereby interrupting the flow of current in the coil, the magnetic field drops to zero, and the consequent change in magnetic flux causes a momentary deflection of the galvanometer needle. When the battery is reconnected, the galvanometer again exhibits a momentary deflection, but in the opposite direction. Thus, current is induced in the loop when the magnetic flux changes, and the direction of the current depends on whether the flux increases (when the battery is being connected) or decreases (when the battery is being disconnected). It was further discovered that current can also flow in the loop while the battery is connected to the coil if the loop turns or moves closer to, or away from, the coil. The physical movement of the loop changes the amount of flux linking its surface S, even though the field **B** due to the coil has not changed.

A galvanometer is a predecessor of the voltmeter and ammeter. When a galvanometer detects the flow of current through the coil, it means that a voltage has been induced across the galvanometer terminals. This voltage is called the ***electromotive force*** (emf), V_{emf}, and the process is called ***electromagnetic induction***. The emf induced in a closed conducting loop of N turns is given by

$$V_{\text{emf}} = -N \frac{d\Phi}{dt} = -N \frac{d}{dt} \int_S \mathbf{B} \cdot d\mathbf{s} \quad \text{(V)}. \tag{6.6}$$

Even though the results leading to Eq. (6.6) were also discovered independently by Henry, Eq. (6.6) is attributed to Faraday and known as ***Faraday's law***. The significance of the negative sign in Eq. (6.6) is explained in the next section.

We note that the derivative in Eq. (6.6) is a total time derivative that operates on the magnetic field **B**, as well as the differential surface area $d\mathbf{s}$. Accordingly, an emf can be generated in a closed conducting loop under any of the following three conditions:

1. A time-varying magnetic field linking a stationary loop; the induced emf is then called the ***transformer emf***, $V_{\text{emf}}^{\text{tr}}$.

2. A moving loop with a time-varying area (relative to the normal component of **B**) in a static field **B**; the induced emf is then called the ***motional emf***, $V^{\text{m}}_{\text{emf}}$.

3. A moving loop in a time-varying field **B**.

The total emf is given by

$$V_{\text{emf}} = V^{\text{tr}}_{\text{emf}} + V^{\text{m}}_{\text{emf}}, \tag{6.7}$$

with $V^{\text{m}}_{\text{emf}} = 0$ if the loop is stationary [case (1)] and $V^{\text{tr}}_{\text{emf}} = 0$ if **B** is static [case (2)]. For case (3), both terms are important. Each of the three cases is examined separately in the following sections.

6-2 Stationary Loop in a Time-Varying Magnetic Field

The stationary, single-turn, conducting, circular loop with contour C and surface area S shown in **Fig. 6-2(a)** is exposed to a time-varying magnetic field $\mathbf{B}(t)$. As stated earlier, the emf induced when S is stationary and the field is time varying is called the ***transformer emf*** and is denoted $V^{\text{tr}}_{\text{emf}}$. Since the loop is stationary, d/dt in Eq. (6.6) now operates on $\mathbf{B}(t)$ only. Hence,

$$V^{\text{tr}}_{\text{emf}} = -N \int_S \frac{\partial \mathbf{B}}{\partial t} \cdot d\mathbf{s} \quad \textbf{(transformer emf)}, \tag{6.8}$$

where the full derivative d/dt has been moved inside the integral and changed into a partial derivative $\partial/\partial t$ to signify that it operates on **B** only. The transformer emf is the voltage difference that would appear across the small opening between terminals 1 and 2, even in the absence of the resistor R. That is, $V^{\text{tr}}_{\text{emf}} = V_{12}$, where V_{12} is the open-circuit voltage across the open ends of the loop. Under dc conditions, $V^{\text{tr}}_{\text{emf}} = 0$. For the loop shown in **Fig. 6-2(a)** and the associated definition for $V^{\text{tr}}_{\text{emf}}$ given by Eq. (6.8), the direction of $d\mathbf{s}$, the loop's differential surface normal, can be chosen to be either upward or downward. The two choices are associated with opposite designations of the polarities of terminals 1 and 2 in **Fig. 6-2(a)**.

► The connection between the direction of $d\mathbf{s}$ and the polarity of $V^{\text{tr}}_{\text{emf}}$ is governed by the following right-hand rule: if $d\mathbf{s}$ points along the thumb of the right hand, then the direction of the contour C indicated by the four fingers is such that it always passes across the opening from the positive terminal of $V^{\text{tr}}_{\text{emf}}$ to the negative terminal. ◄

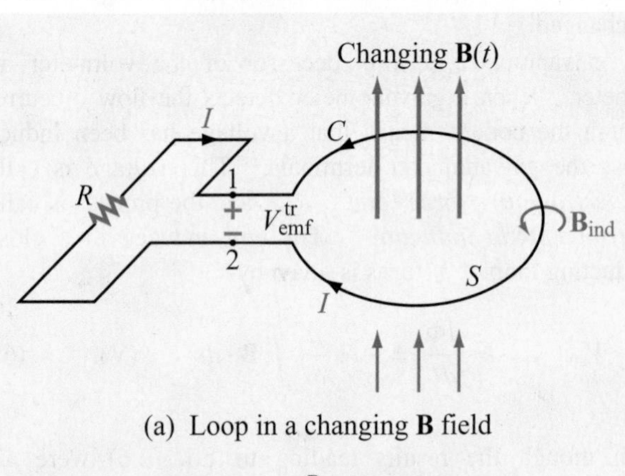

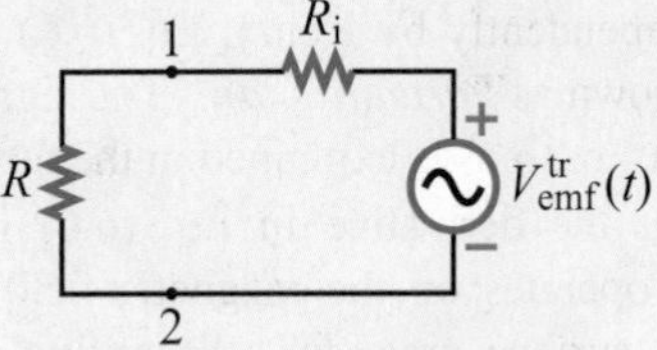

Figure 6-2 (a) Stationary circular loop in a changing magnetic field $\mathbf{B}(t)$, and (b) its equivalent circuit.

If the loop has an internal resistance R_{i}, the circuit in **Fig. 6-2(a)** can be represented by the equivalent circuit shown in **Fig. 6-2(b)**, in which case the current I flowing through the circuit is given by

$$I = \frac{V^{\text{tr}}_{\text{emf}}}{R + R_{\text{i}}} . \tag{6.9}$$

For good conductors, R_{i} usually is very small, and it may be ignored in comparison with practical values of R.

► The polarity of $V^{\text{tr}}_{\text{emf}}$ and hence the direction of I is governed by ***Lenz's law***, which states that the current in the loop is always in a direction that opposes the change of magnetic flux $\Phi(t)$ that produced I. ◄

The current I induces a magnetic field of its own, $\mathbf{B}_{\text{ind}}$, with a corresponding flux Φ_{ind}. The direction of $\mathbf{B}_{\text{ind}}$ is governed by the right-hand rule; if I is in a clockwise direction, then $\mathbf{B}_{\text{ind}}$ points downward through S and, conversely, if I is in a counterclockwise direction, then $\mathbf{B}_{\text{ind}}$ points upward through S. If the original field $\mathbf{B}(t)$ is increasing, which means that $d\Phi/dt > 0$, then according to Lenz's law, I has to be in the direction shown in Fig. 6-2(a) in order for $\mathbf{B}_{\text{ind}}$ to be in opposition to $\mathbf{B}(t)$. Consequently, terminal 2 would be at a higher potential than terminal 1, and $V_{\text{emf}}^{\text{tr}}$ would have a negative value. However, if $\mathbf{B}(t)$ were to remain in the same direction but to decrease in magnitude, then $d\Phi/dt$ would become negative, the current would have to reverse direction, and its induced field $\mathbf{B}_{\text{ind}}$ would be in the same direction as $\mathbf{B}(t)$ so as to oppose the change (decrease) of $\mathbf{B}(t)$. In that case, $V_{\text{emf}}^{\text{tr}}$ would be positive.

► It is important to remember that $\mathbf{B}_{\text{ind}}$ serves to oppose the *change* in $\mathbf{B}(t)$, and not necessarily $\mathbf{B}(t)$ itself. ◄

Despite the presence of the small opening between terminals 1 and 2 of the loop in Fig. 6-2(a), we shall treat the loop as a closed path with contour C. We do this in order to establish the link between $\mathbf{B}$ and the electric field $\mathbf{E}$ associated with the induced emf, $V_{\text{emf}}^{\text{tr}}$. Also, at any point along the loop, the field E is related to the current I flowing through the loop. For contour C, $V_{\text{emf}}^{\text{tr}}$ is related to $\mathbf{E}$ by

$$V_{\text{emf}}^{\text{tr}} = \oint_C \mathbf{E} \cdot d\mathbf{l}. \tag{6.10}$$

For $N = 1$ (a loop with one turn), equating Eqs. (6.8) and (6.10) gives

$$\oint_C \mathbf{E} \cdot d\mathbf{l} = -\int_S \frac{\partial \mathbf{B}}{\partial t} \cdot d\mathbf{s}, \tag{6.11}$$

which is the integral form of Faraday's law given in Table 6-1. We should keep in mind that the direction of the contour C and the direction of $d\mathbf{s}$ are related by the right-hand rule.

By applying Stokes's theorem to the left-hand side of Eq. (6.11), we have

$$\int_S (\nabla \times \mathbf{E}) \cdot d\mathbf{s} = -\int_S \frac{\partial \mathbf{B}}{\partial t} \cdot d\mathbf{s}, \tag{6.12}$$

and in order for the two integrals to be equal for all possible choices of S, their integrands must be equal, which gives

$$\nabla \times \mathbf{E} = -\frac{\partial \mathbf{B}}{\partial t} \qquad \textbf{(Faraday's law)}. \tag{6.13}$$

This differential form of Faraday's law states that a time-varying magnetic field induces an electric field $\mathbf{E}$ whose curl is equal to the negative of the time derivative of $\mathbf{B}$. Even though the derivation leading to Faraday's law started out by considering the field associated with a physical circuit, Eq. (6.13) applies at any point in space, whether or not a physical circuit exists at that point.

Example 6-1: Inductor in a Changing Magnetic Field

An inductor is formed by winding N turns of a thin conducting wire into a circular loop of radius a. The inductor loop is in the x–y plane with its center at the origin, and connected to a resistor R, as shown in Fig. 6-3. In the presence of a magnetic field $\mathbf{B} = B_0(\hat{\mathbf{y}}2 + \hat{\mathbf{z}}3) \sin \omega t$, where ω is the angular frequency, find

(a) the magnetic flux linking a single turn of the inductor,

(b) the transformer emf, given that $N = 10$, $B_0 = 0.2$ T, $a = 10$ cm, and $\omega = 10^3$ rad/s,

(c) the polarity of $V_{\text{emf}}^{\text{tr}}$ at $t = 0$, and

(d) the induced current in the circuit for $R = 1$ kΩ (assume the wire resistance to be much smaller than R).

Solution: (a) The magnetic flux linking each turn of the

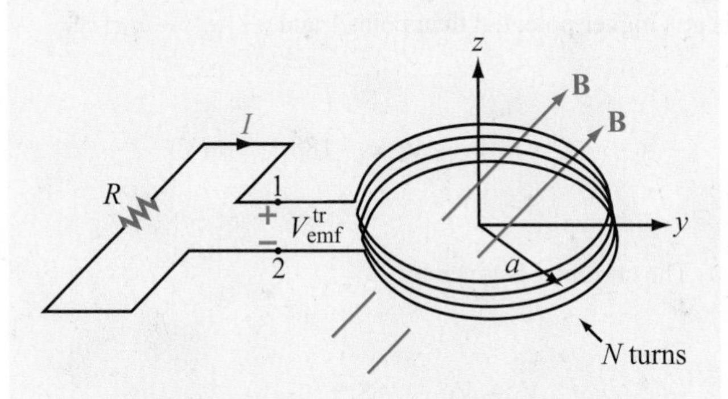

Figure 6-3 Circular loop with N turns in the x–y plane. The magnetic field is $\mathbf{B} = B_0(\hat{\mathbf{y}}2 + \hat{\mathbf{z}}3) \sin \omega t$ (Example 6-1).

inductor is

$$\Phi = \int_S \mathbf{B} \cdot d\mathbf{s}$$

$$= \int_S [B_0(\hat{\mathbf{y}}\,2 + \hat{\mathbf{z}}\,3) \sin \omega t] \cdot \hat{\mathbf{z}}\; ds = 3\pi a^2 B_0 \sin \omega t.$$

(b) To find $V_{\text{emf}}^{\text{tr}}$, we can apply Eq. (6.8) or we can apply the general expression given by Eq. (6.6) directly. The latter approach gives

$$V_{\text{emf}}^{\text{tr}} = -N \frac{d\Phi}{dt}$$

$$= -\frac{d}{dt}(3\pi N a^2 B_0 \sin \omega t) = -3\pi N \omega a^2 B_0 \cos \omega t.$$

For $N = 10$, $a = 0.1$ m, $\omega = 10^3$ rad/s, and $B_0 = 0.2$ T,

$$V_{\text{emf}}^{\text{tr}} = -188.5 \cos 10^3 t \qquad \text{(V)}.$$

(c) At $t = 0$, $d\Phi/dt > 0$ and $V_{\text{emf}}^{\text{tr}} = -188.5$ V. Since the flux is increasing, the current I must be in the direction shown in Fig. 6-3 in order to satisfy Lenz's law. Consequently, point 2 is at a higher potential than point 1 and

$$V_{\text{emf}}^{\text{tr}} = V_1 - V_2 = -188.5 \qquad \text{(V)}.$$

(d) The current I is given by

$$I = \frac{V_2 - V_1}{R} = \frac{188.5}{10^3} \cos 10^3 t$$

$$= 0.19 \cos 10^3 t \qquad \text{(A)}.$$

Exercise 6-1: For the loop shown in Fig. 6-3, what is $V_{\text{emf}}^{\text{tr}}$ if $\mathbf{B} = \hat{\mathbf{y}} B_0 \cos \omega t$?

Answer: $V_{\text{emf}}^{\text{tr}} = 0$ because $\mathbf{B}$ is orthogonal to the loop's surface normal $d\mathbf{s}$. (See EM.)

Exercise 6-2: Suppose that the loop of Example 6-1 is replaced with a 10-turn square loop centered at the origin and having 20 cm sides oriented parallel to the x and y axes. If $\mathbf{B} = \hat{\mathbf{z}} B_0 x^2 \cos 10^3 t$ and $B_0 = 100$ T, find the current in the circuit.

Answer: $I = -133 \sin 10^3 t$ (mA). (See EM.)

Example 6-2: Lenz's Law

Determine voltages V_1 and V_2 across the 2 Ω and 4 Ω resistors shown in Fig. 6-4. The loop is located in the x–y plane, its area is 4 m², the magnetic flux density is $\mathbf{B} = -\hat{\mathbf{z}}\,0.3t$ (T), and the internal resistance of the wire may be ignored.

Solution: The flux flowing through the loop is

$$\Phi = \int_S \mathbf{B} \cdot d\mathbf{s} = \int_S (-\hat{\mathbf{z}}\,0.3t) \cdot \hat{\mathbf{z}}\; ds$$

$$= -0.3t \times 4 = -1.2t \qquad \text{(Wb)},$$

and the corresponding transformer emf is

$$V_{\text{emf}}^{\text{tr}} = -\frac{d\Phi}{dt} = 1.2 \qquad \text{(V)}.$$

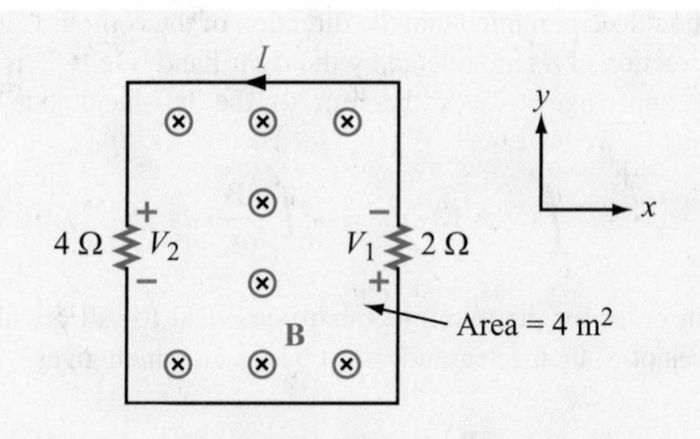

Figure 6-4 Circuit for Example 6-2.

Module 6.1 Circular Loop in Time-varying Magnetic Field Faraday's law of induction is demonstrated by simulating the current induced in a loop in response to the change in magnetic flux flowing through it.

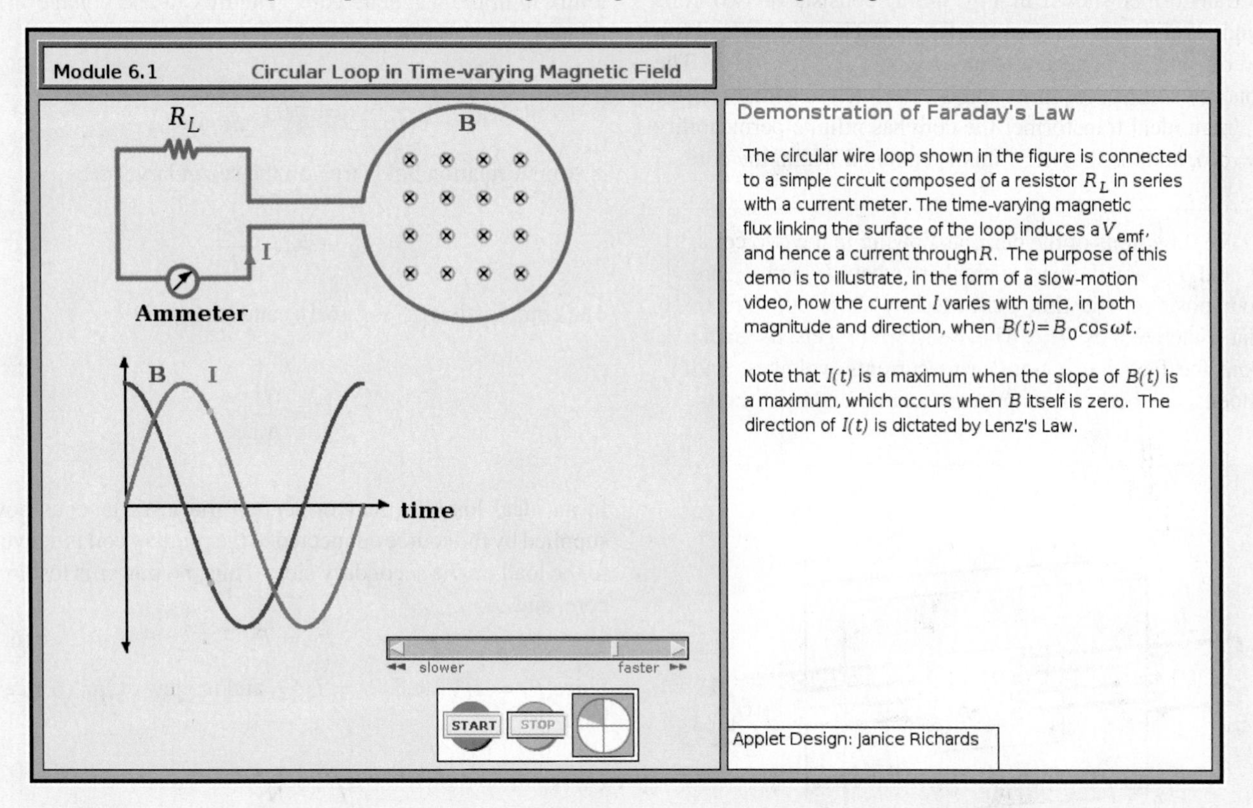

Since the magnetic flux through the loop is along the $-z$ direction (into the page) and increases in magnitude with time t, Lenz's law states that the induced current I should be in a direction such that the magnetic flux density $\mathbf{B}_{\text{ind}}$ it induces counteracts the direction of change of Φ. Hence, I has to be in the direction shown in the circuit because the corresponding $\mathbf{B}_{\text{ind}}$ is along the $+z$ direction in the region inside the loop area. This, in turn, means that V_1 and V_2 are positive voltages.

The total voltage of 1.2 V is distributed across two resistors in series. Consequently,

$$I = \frac{V_{\text{emf}}^{\text{tr}}}{R_1 + R_2} = \frac{1.2}{2+4} = 0.2 \text{ A},$$

and

$$V_1 = IR_1 = 0.2 \times 2 = 0.4 \text{ V},$$
$$V_2 = IR_2 = 0.2 \times 4 = 0.8 \text{ V}.$$

Concept Question 6-1: Explain Faraday's law and the function of Lenz's law.

Concept Question 6-2: Under what circumstances is the net voltage around a closed loop equal to zero?

Concept Question 6-3: Suppose the magnetic flux density linking the loop of Fig. 6-4 (Example 6-2) is given by $\mathbf{B} = -\hat{\mathbf{z}}\,0.3e^{-t}$ (T). What would the direction of the current be, relative to that shown in Fig. 6-4, for $t \geq 0$?

6-3 The Ideal Transformer

The transformer shown in **Fig. 6-5(a)** consists of two coils wound around a common magnetic core. The primary coil has N_1 turns and is connected to an ac voltage source $V_1(t)$. The secondary coil has N_2 turns and is connected to a load resistor R_L. In an ideal transformer the core has infinite permeability ($\mu = \infty$), and the magnetic flux is confined within the core.

▶ The directions of the currents flowing in the two coils, I_1 and I_2, are defined such that, when I_1 and I_2 are both positive, the flux generated by I_2 is opposite to that generated by I_1. The transformer gets its name from the fact that it transforms currents, voltages, and impedances between its primary and secondary circuits, and vice versa. ◀

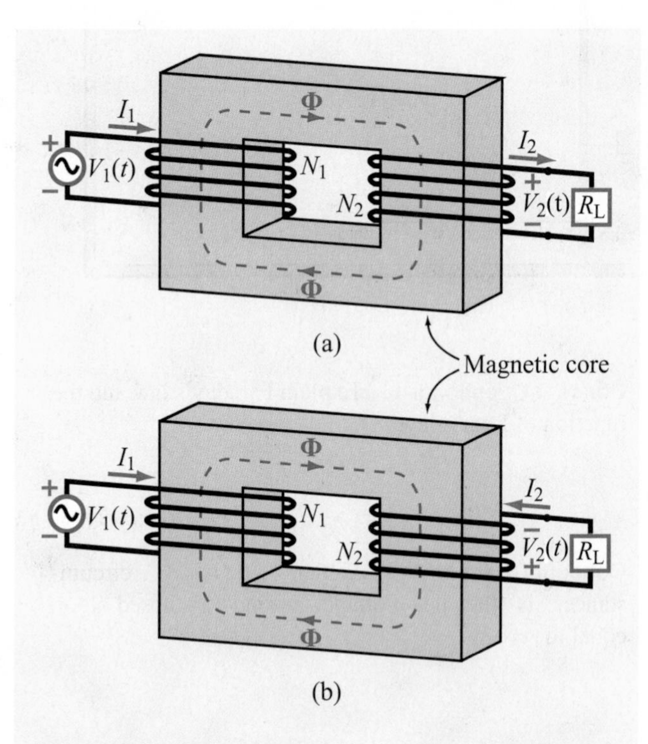

Figure 6-5 In a transformer, the directions of I_1 and I_2 are such that the flux Φ generated by one of them is opposite to that generated by the other. The direction of the secondary winding in (b) is opposite to that in (a), and so are the direction of I_2 and the polarity of V_2.

On the primary side of the transformer, the voltage source V_1 generates current I_1 in the primary coil, which establishes a flux Φ in the magnetic core. The flux Φ and voltage V_1 are related by Faraday's law:

$$V_1 = -N_1 \frac{d\Phi}{dt}. \tag{6.14}$$

A similar relation holds true on the secondary side:

$$V_2 = -N_2 \frac{d\Phi}{dt}. \tag{6.15}$$

The combination of Eqs. (6.14) and (6.15) gives

$$\frac{V_1}{V_2} = \frac{N_1}{N_2}. \tag{6.16}$$

In an ideal lossless transformer, all the instantaneous power supplied by the source connected to the primary coil is delivered to the load on the secondary side. Thus, no power is lost in the core, and

$$P_1 = P_2. \tag{6.17}$$

Since $P_1 = I_1 V_1$ and $P_2 = I_2 V_2$, and in view of Eq. (6.16), we have

$$\frac{I_1}{I_2} = \frac{N_2}{N_1}. \tag{6.18}$$

Thus, whereas the ratio of the voltages given by Eq. (6.16) is proportional to the corresponding turns ratio, the ratio of the currents is equal to the inverse of the turns ratio. If $N_1/N_2 = 0.1$, V_2 of the secondary circuit would be 10 times V_1 of the primary circuit, but I_2 would be only $I_1/10$.

The transformer shown in **Fig. 6-5(b)** is identical to that in **Fig. 6-5(a)** except for the direction of the windings of the secondary coil. Because of this change, the direction of I_2 and the polarity of V_2 in **Fig. 6-5(b)** are the reverse of those in **Fig. 6-5(a)**.

The voltage and current in the secondary circuit in **Fig. 6-5(a)** are related by $V_2 = I_2 R_L$. To the input circuit, the transformer may be represented by an equivalent input resistance R_{in}, as shown in **Fig. 6-6**, defined as

$$R_{in} = \frac{V_1}{I_1}. \tag{6.19}$$

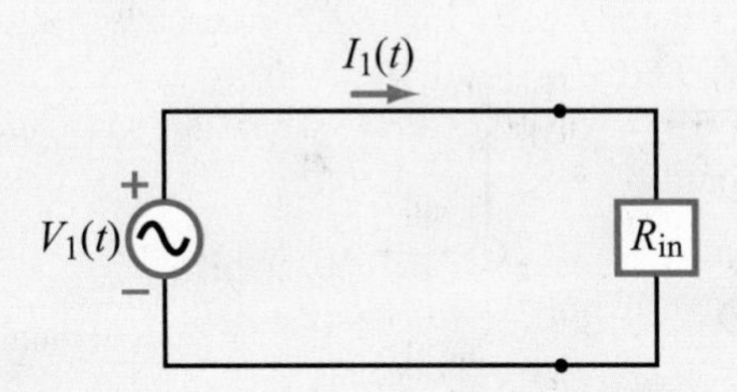

Figure 6-6 Equivalent circuit for the primary side of the transformer.

Use of Eqs. (6.16) and (6.18) gives

$$R_{in} = \frac{V_2}{I_2}\left(\frac{N_1}{N_2}\right)^2 = \left(\frac{N_1}{N_2}\right)^2 R_L. \quad (6.20)$$

When the load is an impedance Z_L and V_1 is a sinusoidal source, the phasor-domain equivalent of Eq. (6.20) is

$$Z_{in} = \left(\frac{N_1}{N_2}\right)^2 Z_L. \quad (6.21)$$

6-4 Moving Conductor in a Static Magnetic Field

Consider a wire of length l moving across a static magnetic field $\mathbf{B} = \hat{\mathbf{z}}B_0$ with constant velocity $\mathbf{u}$, as shown in Fig. 6-7. The conducting wire contains free electrons. From Eq. (5.3), the magnetic force $\mathbf{F}_m$ acting on a particle with charge q moving with velocity $\mathbf{u}$ in a magnetic field $\mathbf{B}$ is

$$\mathbf{F}_m = q(\mathbf{u} \times \mathbf{B}). \quad (6.22)$$

This magnetic force is equivalent to the electrical force that would be exerted on the particle by the electric field $\mathbf{E}_m$ given by

$$\mathbf{E}_m = \frac{\mathbf{F}_m}{q} = \mathbf{u} \times \mathbf{B}. \quad (6.23)$$

The field $\mathbf{E}_m$ generated by the motion of the charged particle is called a ***motional electric field*** and is in the direction perpendicular to the plane containing $\mathbf{u}$ and $\mathbf{B}$. For the wire shown in Fig. 6-7, $\mathbf{E}_m$ is along $-\hat{\mathbf{y}}$. The magnetic force acting on the (negatively charged) electrons in the wire causes them to drift in the direction of $-\mathbf{E}_m$; that is, toward the wire end

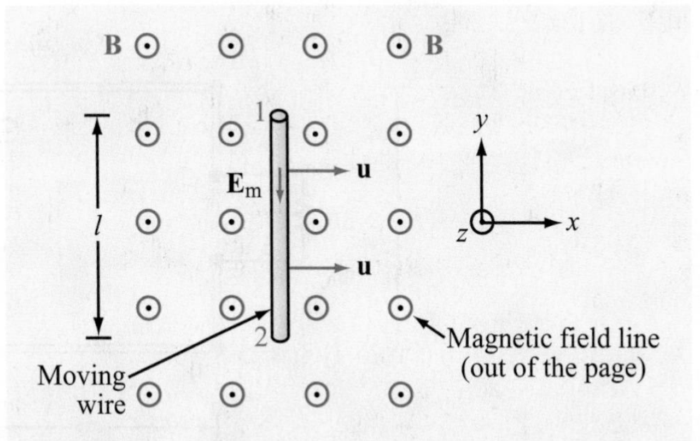

Figure 6-7 Conducting wire moving with velocity **u** in a static magnetic field.

labeled 1 in Fig. 6-7. This, in turn, induces a voltage difference between ends 1 and 2, with end 2 being at the higher potential. The induced voltage is called a ***motional emf***, V_{emf}^m, and is defined as the line integral of $\mathbf{E}_m$ between ends 2 and 1 of the wire,

$$V_{emf}^m = V_{12} = \int_2^1 \mathbf{E}_m \cdot d\mathbf{l} = \int_2^1 (\mathbf{u} \times \mathbf{B}) \cdot d\mathbf{l}. \quad (6.24)$$

For the conducting wire, $\mathbf{u} \times \mathbf{B} = \hat{\mathbf{x}}u \times \hat{\mathbf{z}}B_0 = -\hat{\mathbf{y}}uB_0$ and $d\mathbf{l} = \hat{\mathbf{y}}\, dl$. Hence,

$$V_{emf}^m = V_{12} = -uB_0 l. \quad (6.25)$$

In general, if any segment of a closed circuit with contour C moves with a velocity $\mathbf{u}$ across a static magnetic field $\mathbf{B}$, then the induced motional emf is given by

$$V_{emf}^m = \oint_C (\mathbf{u} \times \mathbf{B}) \cdot d\mathbf{l} \qquad \textbf{(motional emf)}. \quad (6.26)$$

► Only those segments of the circuit that cross magnetic field lines contribute to V_{emf}^m. ◄

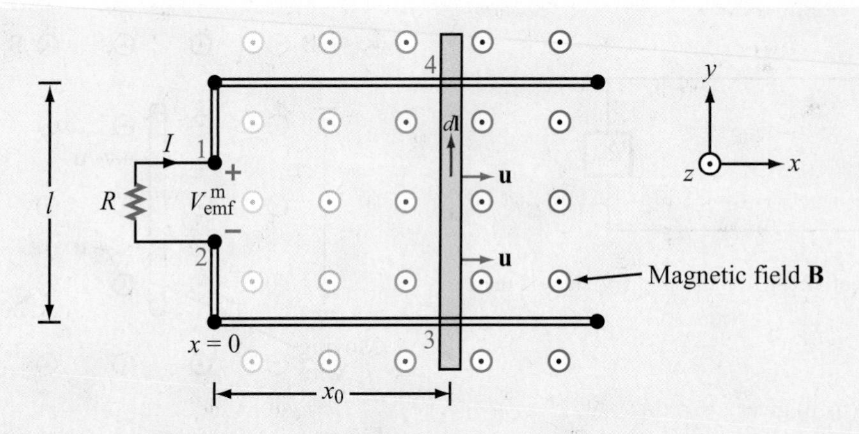

Figure 6-8 Sliding bar with velocity **u** in a magnetic field that increases linearly with x; that is, $\mathbf{B} = \hat{\mathbf{z}}B_0x$ (Example 6-3).

Example 6-3: Sliding Bar

The rectangular loop shown in Fig. 6-8 has a constant width l, but its length x_0 increases with time as a conducting bar slides with uniform velocity **u** in a static magnetic field $\mathbf{B} = \hat{\mathbf{z}}B_0x$. Note that **B** increases linearly with x. The bar starts from $x = 0$ at $t = 0$. Find the motional emf between terminals 1 and 2 and the current I flowing through the resistor R. Assume that the loop resistance $R_i \ll R$.

Solution: This problem can be solved by using the motional emf expression given by Eq. (6.26) or by applying the general formula of Faraday's law. We now show that the two approaches yield the same result.

The sliding bar, being the only part of the circuit that crosses the lines of the field **B**, is the only part of contour 2341 that contributes to V^{m}_{emf}. Hence, at $x = x_0$, for example,

$$V^{m}_{emf} = V_{12} = V_{43} = \int_3^4 (\mathbf{u} \times \mathbf{B}) \cdot d\mathbf{l} = \int_3^4 (\hat{\mathbf{x}}u \times \hat{\mathbf{z}}B_0x_0) \cdot \hat{\mathbf{y}}\, dl$$
$$= -uB_0x_0l.$$

The length of the loop is related to u by $x_0 = ut$. Hence,

$$V^{m}_{emf} = -B_0u^2lt \qquad \text{(V)}. \tag{6.27}$$

Since **B** is static, $V^{tr}_{emf} = 0$ and $V_{emf} = V^{m}_{emf}$ only. To verify that the same result can be obtained by the general form of Faraday's law, we evaluate the flux Φ through the surface of the loop. Thus,

$$\Phi = \int_S \mathbf{B} \cdot d\mathbf{s} = \int_S (\hat{\mathbf{z}}B_0x) \cdot \hat{\mathbf{z}}\, dx\, dy$$
$$= B_0l \int_0^{x_0} x\, dx = \frac{B_0lx_0^2}{2}. \tag{6.28}$$

Substituting $x_0 = ut$ in Eq. (6.28) and then evaluating the negative of the derivative of the flux with respect to time gives

$$V_{emf} = -\frac{d\Phi}{dt} = -\frac{d}{dt}\left(\frac{B_0lu^2t^2}{2}\right) = -B_0u^2lt \qquad \text{(V)}, \tag{6.29}$$

which is identical with Eq. (6.27). Since V_{12} is negative, the current $I = B_0u^2lt/R$ flows in the direction shown in Fig. 6-8.

Example 6-4: Moving Loop

The rectangular loop shown in Fig. 6-9 is situated in the x–y plane and moves away from the origin with velocity $\mathbf{u} = \hat{\mathbf{y}}5$ (m/s) in a magnetic field given by

$$\mathbf{B}(y) = \hat{\mathbf{z}}0.2e^{-0.1y} \qquad \text{(T)}.$$

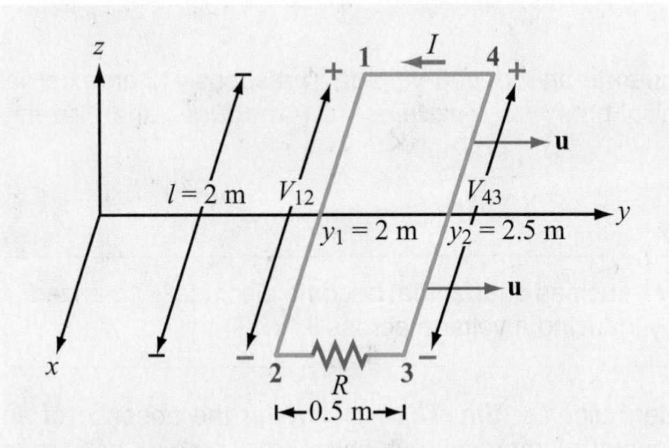

Figure 6-9 Moving loop of Example 6-4.

If $R = 5\ \Omega$, find the current I at the instant that the loop sides are at $y_1 = 2$ m and $y_2 = 2.5$ m. The loop resistance may be ignored.

Solution: Since $\mathbf{u} \times \mathbf{B}$ is along $\hat{\mathbf{x}}$, voltages are induced across only the sides oriented along $\hat{\mathbf{x}}$, namely the sides linking points 1 and 2 and points 3 and 4. Had $\mathbf{B}$ been uniform, the induced voltages would have been the same and the net voltage across the resistor would have been zero. In the present case, however, $\mathbf{B}$ decreases exponentially with y, thereby assuming a different value over side 1-2 than over side 3-4. Side 1-2 is at $y_1 = 2$ m, and the corresponding magnetic field is

$$\mathbf{B}(y_1) = \hat{\mathbf{z}}\,0.2e^{-0.1y_1} = \hat{\mathbf{z}}\,0.2e^{-0.2} \qquad \text{(T)}.$$

The induced voltage V_{12} is then given by

$$V_{12} = \int_2^1 [\mathbf{u} \times \mathbf{B}(y_1)] \cdot d\mathbf{l} = \int_{l/2}^{-l/2} (\hat{\mathbf{y}}5 \times \hat{\mathbf{z}}0.2e^{-0.2}) \cdot \hat{\mathbf{x}}\, dx$$

$$= -e^{-0.2}l = -2e^{-0.2} = -1.637 \qquad \text{(V)}.$$

Similarly,

$$V_{43} = -u\, B(y_2)\, l = -5 \times 0.2e^{-0.25} \times 2 = -1.558 \qquad \text{(V)}.$$

Consequently, the current is in the direction shown in the figure and its magnitude is

$$I = \frac{V_{43} - V_{12}}{R} = \frac{0.079}{5} = 15.8 \text{ (mA)}.$$

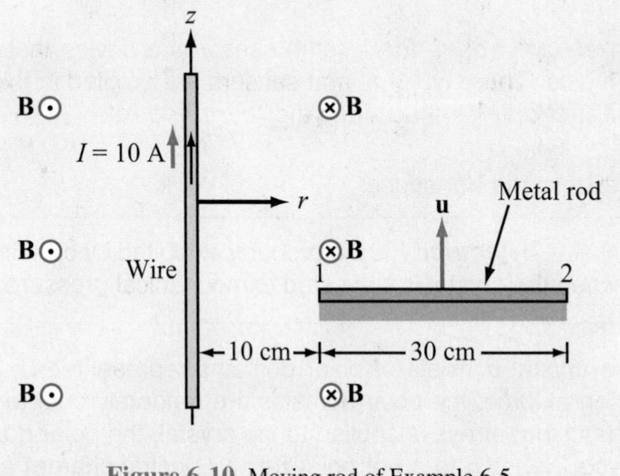

Figure 6-10 Moving rod of Example 6-5.

Example 6-5: Moving Rod Next to a Wire

The wire shown in Fig. 6-10 carries a current $I = 10$ A. A 30 cm long metal rod moves with a constant velocity $\mathbf{u} = \hat{\mathbf{z}}5$ m/s. Find V_{12}.

Solution: The current I induces a magnetic field

$$\mathbf{B} = \hat{\boldsymbol{\phi}}\,\frac{\mu_0 I}{2\pi r},$$

where r is the radial distance from the wire and the direction of $\hat{\boldsymbol{\phi}}$ is into the page on the rod side of the wire. The movement of the rod in the presence of the field $\mathbf{B}$ induces a motional emf given by

$$V_{12} = \int_{40\text{ cm}}^{10\text{ cm}} (\mathbf{u} \times \mathbf{B}) \cdot d\mathbf{l}$$

$$= \int_{40\text{ cm}}^{10\text{ cm}} \left(\hat{\mathbf{z}}\,5 \times \hat{\boldsymbol{\phi}}\,\frac{\mu_0 I}{2\pi r}\right) \cdot \hat{\mathbf{r}}\, dr$$

$$= -\frac{5\mu_0 I}{2\pi} \int_{40\text{ cm}}^{10\text{ cm}} \frac{dr}{r}$$

$$= -\frac{5 \times 4\pi \times 10^{-7} \times 10}{2\pi} \times \ln\left(\frac{10}{40}\right)$$

$$= 13.9 \qquad (\mu\text{V}).$$

Technology Brief 12: EMF Sensors

An ***electromotive force*** (emf) sensor is a device that can generate an induced voltage in response to an external stimulus. Three types of emf sensors are profiled in this technical brief: the ***piezoelectric transducer***, the ***Faraday magnetic flux sensor***, and the ***thermocouple***.

Piezoelectric Transducers

► ***Piezoelectricity*** is the property exhibited by certain crystals, such as quartz, that become electrically polarized when the crystal is subjected to mechanical pressure, thereby inducing a voltage across it. ◄

The crystal consists of polar domains represented by equivalent dipoles (**Fig. TF12-1**). Under the absence of an external force, the polar domains are randomly oriented throughout the material, but when ***compressive*** or ***tensile*** (stretching) stress is applied to the crystal, the polar domains align themselves along one of the principal axes of the crystal, leading to a net polarization (electric charge) at the crystal surfaces. Compression and stretching generate voltages of opposite polarity. The piezoelectric effect (***piezein*** means to press or squeeze in Greek) was discovered by the ***Curie brothers***, Pierre and Paul-Jacques, in 1880, and a year later, Lippmann predicted the converse property, namely that, if subjected to an electric field, the crystal would change in shape.

► The piezoelectric effect is a ***reversible (bidirectional)*** electromechanical process; application of force induces a voltage across the crystal, and conversely, application of a voltage changes the shape of the crystal. ◄

Piezoelectric crystals are used in ***microphones*** to convert mechanical vibrations (of the crystal surface) caused by acoustic waves into a corresponding electrical signal, and the converse process is used in ***loudspeakers*** to convert electrical signals into sound. In addition to having stiffness values comparable to that of steel, some piezoelectric materials exhibit very high sensitivity to the force applied upon them, with excellent linearity over a wide dynamic range. They can be used to measure surface deformations as small as ***nanometers*** (10^{-9} m), making them particularly attractive as positioning sensors in ***scanning tunneling microscopes***. As ***accelerometers***, they can measure acceleration levels as low as 10^{-4} g to as high as 100 g (where g is the acceleration due to gravity). Piezoelectric crystals and ceramics are used in cigarette lighters and gas grills as spark generators, in clocks and electronic circuitry as precision oscillators, in medical ***ultrasound*** diagnostic equipment as transducers (**Fig. TF12-2**), and in numerous other applications.

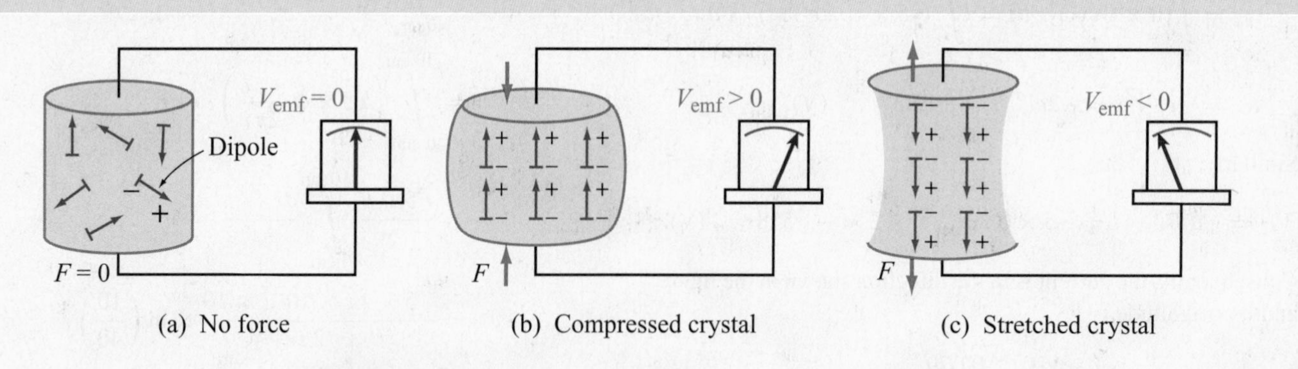

Figure TF12-1 Response of a piezoelectric crystal to an applied force.

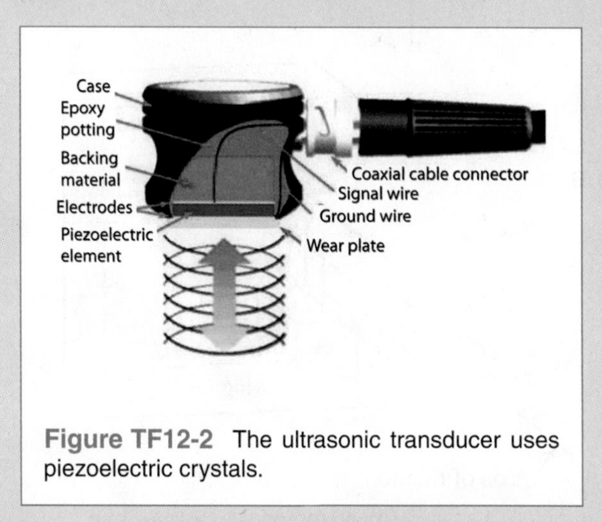

Figure TF12-2 The ultrasonic transducer uses piezoelectric crystals.

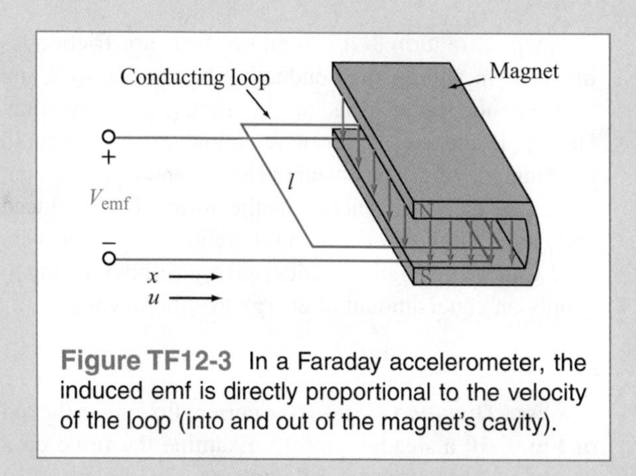

Figure TF12-3 In a Faraday accelerometer, the induced emf is directly proportional to the velocity of the loop (into and out of the magnet's cavity).

Faraday Magnetic Flux Sensor

According to Faraday's law [Eq. (6.6)], the emf voltage induced across the terminals of a conducting loop is directly proportional to the time rate of change of the magnetic flux passing through the loop. For the configuration in **Fig. TF12-3**,

$$V_{emf} = -uB_0 l,$$

where $u = dx/dt$ is the ***velocity*** of the loop (into or out of the magnet's cavity), with the direction of u defined as positive when the loop is moving inward into the cavity, B_0 is the magnetic field of the magnet, and l is the loop width. With B_0 and l being constant, the variation of $V_{emf}(t)$ with time t becomes a direct indicator of the time variation of $u(t)$. The time derivative of $u(t)$ provides the ***acceleration*** $a(t)$.

Thermocouple

In 1821, ***Thomas Seebeck*** discovered that when a junction made of two different conducting materials, such as bismuth and copper, is heated, it generates a thermally induced emf, which we now call the ***Seebeck potential*** V_S (**Fig. TF12-4**). When connected to a resistor, a current flows through the resistor, given by $I = V_S/R$.

This feature was advanced by ***A. C. Becquerel*** in 1826 as a means to measure the unknown temperature T_2 of a junction relative to a temperature T_1 of a (cold) reference junction. Today, such a generator of ***thermoelectricity*** is called a ***thermocouple***. Initially, an ice bath was used to maintain T_1 at 0°C, but in today's temperature sensor designs, an artificial cold junction is used instead. The artificial junction is an electric circuit that generates a potential equal to that expected from a reference junction at temperature T_1.

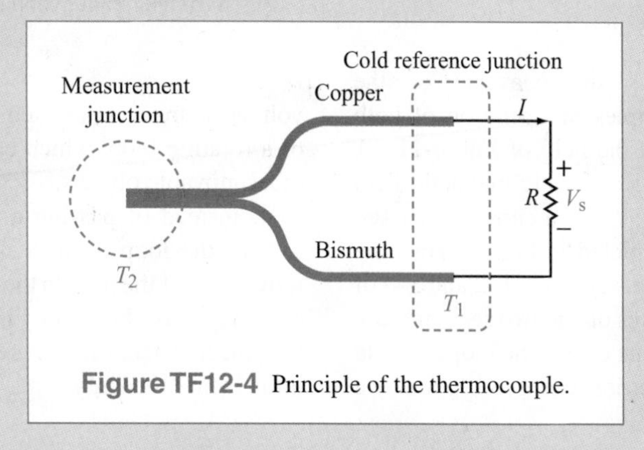

Figure TF12-4 Principle of the thermocouple.

Concept Question 6-4: Suppose that no friction is involved in sliding the conducting bar of Fig. 6-8 and that the horizontal arms of the circuit are very long. Hence, if the bar is given an initial push, it should continue moving at a constant velocity, and its movement generates electrical energy in the form of an induced emf, indefinitely. Is this a valid argument? If not, why not? Can we generate electrical energy without having to supply an equal amount of energy by other means?

Concept Question 6-5: Is the current flowing in the rod of Fig. 6-10 a steady current? Examine the force on a charge q at ends 1 and 2 and compare.

Exercise 6-3: For the moving loop of Fig. 6-9, find I when the loop sides are at $y_1 = 4$ m and $y_2 = 4.5$ m. Also, reverse the direction of motion such that $\mathbf{u} = -\hat{\mathbf{y}}5$ (m/s).

Answer: $I = -13$ (mA). (See EM.)

Exercise 6-4: Suppose that we turn the loop of Fig. 6-9 so that its surface is parallel to the x–z plane. What would I be in that case?

Answer: $I = 0$. (See EM.)

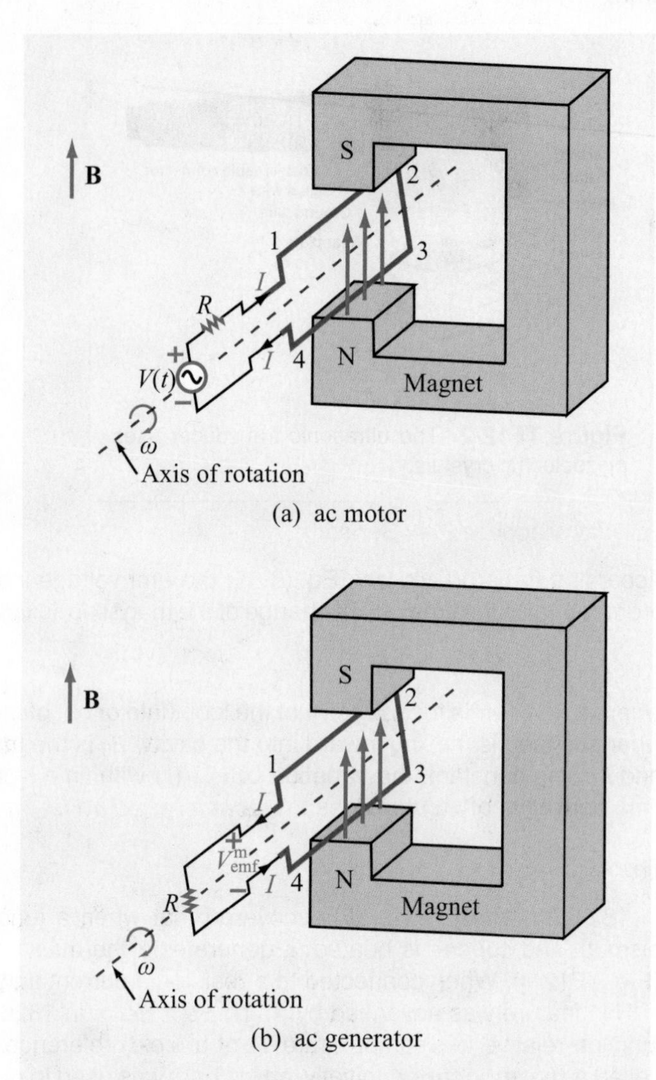

Figure 6-11 Principles of the ac motor and the ac generator. In (a) the magnetic torque on the wires causes the loop to rotate, and in (b) the rotating loop generates an emf.

6-5 The Electromagnetic Generator

The electromagnetic generator is the converse of the electromagnetic motor. The principles of operation of both instruments may be explained with the help of Fig. 6-11. A permanent magnet is used to produce a static magnetic field **B** in the slot between its two poles. When a current is passed through the conducting loop, as depicted in Fig. 6-11(a), the current flows in opposite directions in segments 1–2 and 3–4 of the loop. The induced magnetic forces on the two segments are also opposite, resulting in a torque that causes the loop to rotate about its axis. Thus, in a motor, electrical energy supplied by a voltage source is converted into mechanical energy in the form of a rotating loop, which can be coupled to pulleys, gears, or other movable objects.

If, instead of passing a current through the loop to make it turn, the loop is made to rotate by an external force, the movement of the loop in the magnetic field produces a motional emf, V^{m}_{emf}, as shown in Fig. 6-11(b). Hence, the motor has become a generator, and mechanical energy is being converted into electrical energy.

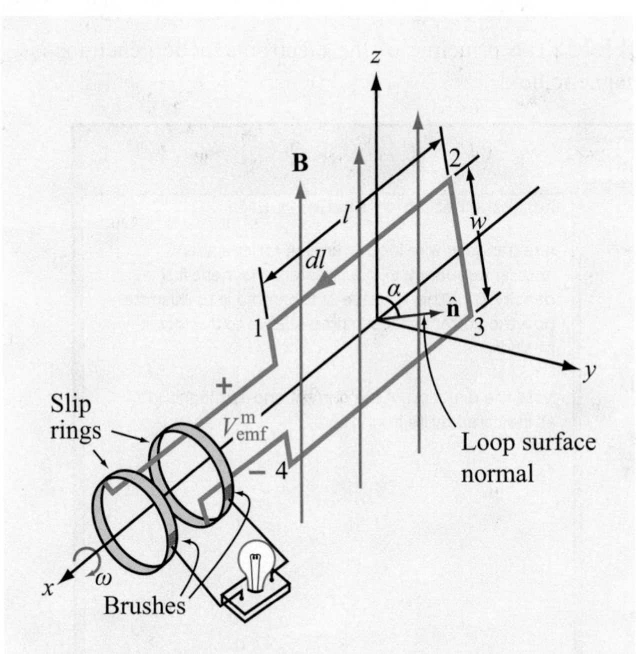

Figure 6-12 A loop rotating in a magnetic field induces an emf.

Let us examine the operation of the electromagnetic generator in more detail using the coordinate system shown in Fig. 6-12. The magnetic field is

$$\mathbf{B} = \hat{\mathbf{z}} B_0, \tag{6.30}$$

and the axis of rotation of the conducting loop is along the x axis. Segments 1–2 and 3–4 of the loop are of length l each, and both cross the magnetic flux lines as the loop rotates. The other two segments are each of width w, and neither crosses the **B** lines when the loop rotates. Hence, only segments 1–2 and 3–4 contribute to the generation of the motional emf, V^{m}_{emf}.

As the loop rotates with an angular velocity ω about its own axis, segment 1–2 moves with velocity **u** given by

$$\mathbf{u} = \hat{\mathbf{n}} \omega \, \frac{w}{2}\,, \tag{6.31}$$

where $\hat{\mathbf{n}}$, the surface normal to the loop, makes an angle α with the z axis. Hence,

$$\hat{\mathbf{n}} \times \hat{\mathbf{z}} = \hat{\mathbf{x}} \sin\alpha. \tag{6.32}$$

Segment 3–4 moves with velocity $-\mathbf{u}$. Application of Eq. (6.26), consistent with our choice of $\hat{\mathbf{n}}$, gives

$$\begin{aligned} V^{m}_{emf} = V_{14} &= \int_2^1 (\mathbf{u} \times \mathbf{B}) \cdot d\mathbf{l} + \int_4^3 (\mathbf{u} \times \mathbf{B}) \cdot d\mathbf{l} \\ &= \int_{-l/2}^{l/2} \left[\left(\hat{\mathbf{n}} \omega \frac{w}{2} \right) \times \hat{\mathbf{z}} B_0 \right] \cdot \hat{\mathbf{x}}\, dx \\ &\quad + \int_{l/2}^{-l/2} \left[\left(-\hat{\mathbf{n}} \omega \frac{w}{2} \right) \times \hat{\mathbf{z}} B_0 \right] \cdot \hat{\mathbf{x}}\, dx. \end{aligned} \tag{6.33}$$

Using Eq. (6.32) in Eq. (6.33), we obtain the result

$$V^{m}_{emf} = wl\omega B_0 \sin\alpha = A\omega B_0 \sin\alpha, \tag{6.34}$$

where $A = wl$ is the surface area of the loop. The angle α is related to ω by

$$\alpha = \omega t + C_0, \tag{6.35}$$

where C_0 is a constant determined by initial conditions. For example, if $\alpha = 0$ at $t = 0$, then $C_0 = 0$. In general,

$$V^{m}_{emf} = A\omega B_0 \sin(\omega t + C_0) \qquad \text{(V)}. \tag{6.36}$$

This same result can also be obtained by applying the general form of Faraday's law given by Eq. (6.6). The flux linking the surface of the loop is

$$\begin{aligned} \Phi = \int_S \mathbf{B} \cdot d\mathbf{s} &= \int_S \hat{\mathbf{z}} B_0 \cdot \hat{\mathbf{n}}\, ds \\ &= B_0 A \cos\alpha \\ &= B_0 A \cos(\omega t + C_0), \end{aligned} \tag{6.37}$$

and

$$\begin{aligned} V_{emf} = -\frac{d\Phi}{dt} &= -\frac{d}{dt}[B_0 A \cos(\omega t + C_0)] \\ &= A\omega B_0 \sin(\omega t + C_0), \end{aligned} \tag{6.38}$$

which is identical with the result given by Eq. (6.36).

► The voltage induced by the rotating loop is sinusoidal in time with an angular frequency ω equal to that of the rotating loop, and its amplitude is equal to the product of the surface area of the loop, the magnitude of the magnetic field generated by the magnet, and the angular frequency ω. ◄

Module 6.2 Rotating Wire Loop in Constant Magnetic Field The principle of the electromagnetic generator is demonstrated by a rectangular loop rotating in the presence of a magnetic field.

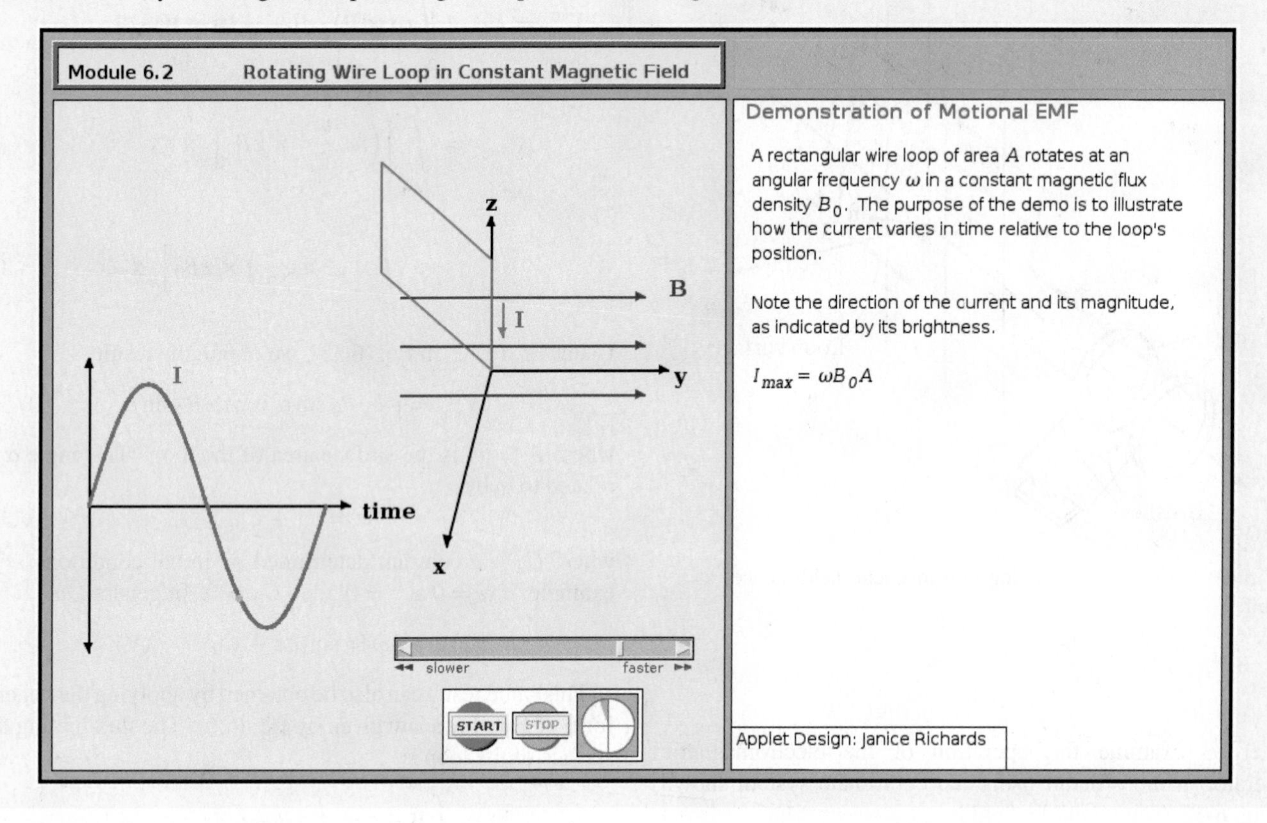

Concept Question 6-6: Contrast the operation of an ac motor with that of an ac generator.

Concept Question 6-7: The rotating loop of Fig. 6-12 had a single turn. What would be the emf generated by a loop with 10 turns?

Concept Question 6-8: The magnetic flux linking the loop shown in Fig. 6-12 is maximum when $\alpha = 0$ (loop in x–y plane), and yet according to Eq. (6.34), the induced emf is zero when $\alpha = 0$. Conversely, when $\alpha = 90°$, the flux linking the loop is zero, but $V_{\text{emf}}^{\text{m}}$ is at a maximum. Is this consistent with your expectations? Why?

6-6 Moving Conductor in a Time-Varying Magnetic Field

For the general case of a single-turn conducting loop moving in a time-varying magnetic field, the induced emf is the sum of a transformer component and a motional component. Thus, the sum of Eqs. (6.8) and (6.26) gives

$$\begin{aligned} V_{\text{emf}} &= V_{\text{emf}}^{\text{tr}} + V_{\text{emf}}^{\text{m}} \\ &= \oint_C \mathbf{E} \cdot d\mathbf{l} \\ &= -\int_S \frac{\partial \mathbf{B}}{\partial t} \cdot d\mathbf{s} + \oint_C (\mathbf{u} \times \mathbf{B}) \cdot d\mathbf{l}. \end{aligned} \tag{6.39}$$

V_{emf} is also given by the general expression of Faraday's law:

$$V_{\text{emf}} = -\frac{d\Phi}{dt} = -\frac{d}{dt} \int_S \mathbf{B} \cdot d\mathbf{s} \quad \text{(total emf)}. \qquad (6.40)$$

In fact, it can be shown mathematically that the right-hand side of Eq. (6.39) is equivalent to the right-hand side of Eq. (6.40). For a particular problem, the choice between using Eq. (6.39) or Eq. (6.40) is usually made on the basis of which is the easier to apply. In either case, for an N-turn loop, the right-hand sides of Eqs. (6.39) and (6.40) should be multiplied by N.

Example 6-6: Electromagnetic Generator

Find the induced voltage when the rotating loop of the electromagnetic generator of Section 6-5 is in a magnetic field $\mathbf{B} = \hat{\mathbf{z}} B_0 \cos \omega t$. Assume that $\alpha = 0$ at $t = 0$.

Solution: The flux Φ is given by Eq. (6.37) with B_0 replaced with $B_0 \cos \omega t$. Thus,

$$\Phi = B_0 A \cos^2 \omega t,$$

and

$$\begin{aligned} V_{\text{emf}} &= -\frac{\partial \Phi}{\partial t} \\ &= -\frac{\partial}{\partial t} (B_0 A \cos^2 \omega t) \\ &= 2B_0 A\omega \cos \omega t \sin \omega t = B_0 A \omega \sin 2\omega t. \end{aligned}$$

6-7 Displacement Current

Ampère's law in differential form is given by

$$\nabla \times \mathbf{H} = \mathbf{J} + \frac{\partial \mathbf{D}}{\partial t} \qquad \textbf{(Ampère's law)}. \qquad (6.41)$$

Integrating both sides of Eq. (6.41) over an arbitrary open surface S with contour C, we have

$$\int_S (\nabla \times \mathbf{H}) \cdot d\mathbf{s} = \int_S \mathbf{J} \cdot d\mathbf{s} + \int_S \frac{\partial \mathbf{D}}{\partial t} \cdot d\mathbf{s}. \qquad (6.42)$$

The surface integral of $\mathbf{J}$ equals the conduction current I_c flowing through S, and the surface integral of $\nabla \times \mathbf{H}$ can be converted into a line integral of $\mathbf{H}$ over the contour C bounding C by invoking Stokes's theorem. Hence,

$$\oint_C \mathbf{H} \cdot d\mathbf{l} = I_c + \int_S \frac{\partial \mathbf{D}}{\partial t} \cdot d\mathbf{s}. \qquad (6.43)$$

(Ampère's law)

The second term on the right-hand side of Eq. (6.43) of course has the same unit (amperes) as the current I_c, and because it is proportional to the time derivative of the electric flux density $\mathbf{D}$, which is also called the electric displacement, it is called the ***displacement current*** I_d. That is,

$$I_d = \int_S \mathbf{J}_d \cdot d\mathbf{s} = \int_S \frac{\partial \mathbf{D}}{\partial t} \cdot d\mathbf{s}, \qquad (6.44)$$

where $\mathbf{J}_d = \partial \mathbf{D}/\partial t$ represents a ***displacement current density***. In view of Eq. (6.44),

$$\oint_C \mathbf{H} \cdot d\mathbf{l} = I_c + I_d = I, \qquad (6.45)$$

where I is the total current. In electrostatics, $\partial \mathbf{D}/\partial t = 0$ and therefore $I_d = 0$ and $I = I_c$. The concept of displacement current was first introduced in 1873 by James Clerk Maxwell when he formulated the unified theory of electricity and magnetism under time-varying conditions.

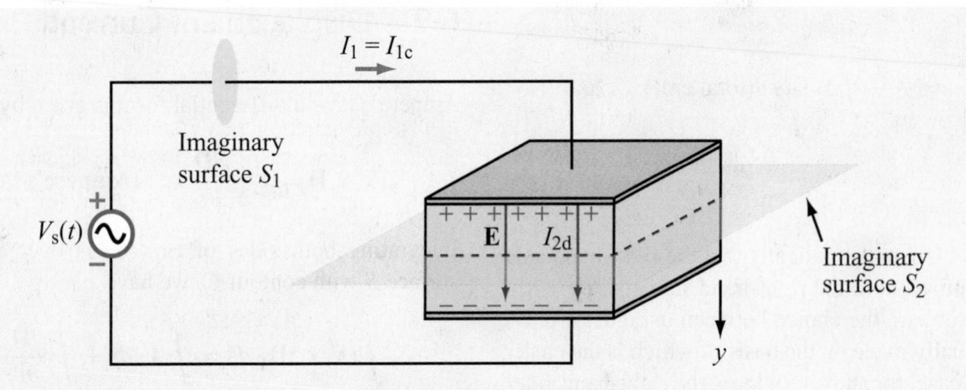

Figure 6-13 The displacement current I_{2d} in the insulating material of the capacitor is equal to the conducting current I_{1c} in the wire.

The parallel-plate capacitor is commonly used as an example to illustrate the physical meaning of the displacement current I_d. The simple circuit shown in **Fig. 6-13** consists of a capacitor and an ac source with voltage $V_s(t)$ given by

$$V_s(t) = V_0 \cos \omega t \qquad \text{(V)}. \tag{6.46}$$

According to Eq. (6.45), the total current flowing through any surface consists, in general, of a conduction current I_c and a displacement current I_d. Let us find I_c and I_d through each of the following two imaginary surfaces: (1) the cross section of the conducting wire, S_1, and (2) the cross section of the capacitor S_2 (**Fig. 6-13**). We denote the conduction and displacement currents in the wire as I_{1c} and I_{1d} and those through the capacitor as I_{2c} and I_{2d}.

In the perfectly conducting wire, $\mathbf{D} = \mathbf{E} = 0$; hence, Eq. (6.44) gives $I_{1d} = 0$. As for I_{1c}, we know from circuit theory that it is related to the voltage across the capacitor V_C by

$$I_{1c} = C \frac{dV_C}{dt} = C \frac{d}{dt} (V_0 \cos \omega t) = -CV_0\omega \sin \omega t, \tag{6.47}$$

where we used the fact that $V_C = V_s(t)$. With $I_{1d} = 0$, the total current in the wire is simply $I_1 = I_{1c} = -CV_0\omega \sin \omega t$.

In the perfect dielectric with permittivity ϵ between the capacitor plates, $\sigma = 0$. Hence, $I_{2c} = 0$ because no conduction current exists there. To determine I_{2d}, we need to apply Eq. (6.44). From Example 4-11, the electric field $\mathbf{E}$ in the dielectric spacing is related to the voltage V_c across its plates by

$$\mathbf{E} = \hat{\mathbf{y}} \frac{V_c}{d} = \hat{\mathbf{y}} \frac{V_0}{d} \cos \omega t, \tag{6.48}$$

where d is the spacing between the plates and $\hat{\mathbf{y}}$ is the direction from the higher-potential plate toward the lower-potential plate at $t = 0$. The displacement current I_{2d} is obtained by applying Eq. (6.44) with $d\mathbf{s} = \hat{\mathbf{y}}\, ds$:

$$\begin{aligned} I_{2d} &= \int_S \frac{\partial \mathbf{D}}{\partial t} \cdot d\mathbf{s} \\ &= \int_A \left[\frac{\partial}{\partial t} \left(\hat{\mathbf{y}} \frac{\epsilon V_0}{d} \cos \omega t \right) \right] \cdot (\hat{\mathbf{y}}\, ds) \\ &= -\frac{\epsilon A}{d} V_0 \omega \sin \omega t = -CV_0\omega \sin \omega t, \end{aligned} \tag{6.49}$$

where we used the relation $C = \epsilon A/d$ for the capacitance of the parallel-plate capacitor with plate area A. The expression for I_{2d} in the dielectric region between the conducting plates is identical with that given by Eq. (6.47) for the conduction current I_{1c} in the wire. The fact that these two currents are equal ensures the continuity of total current flow through the circuit.

► Even though the displacement current does not transport free charges, it nonetheless behaves like a real current. ◄

In the capacitor example, we treated the wire as a perfect conductor, and we assumed that the space between the capacitor plates was filled with a perfect dielectric. If the wire has a finite conductivity σ_w, then $\mathbf{D}$ in the wire would not be zero, and therefore the current I_1 would consist of a conduction current I_{1c}

as well as a displacement current I_{1d}; that is, $I_1 = I_{1c} + I_{1d}$. By the same token, if the dielectric spacing material has a nonzero conductivity σ_d, then free charges would flow between the two plates, and I_{2c} would not be zero. In that case, the total current flowing through the capacitor would be $I_2 = I_{2c} + I_{2d}$. No matter the circumstances, the total capacitor current remains equal to the total current in the wire. That is, $I_1 = I_2$.

Example 6-7: Displacement Current Density

The conduction current flowing through a wire with conductivity $\sigma = 2 \times 10^7$ S/m and relative permittivity $\epsilon_r = 1$ is given by $I_c = 2 \sin \omega t$ (mA). If $\omega = 10^9$ rad/s, find the displacement current.

Solution: The conduction current $I_c = JA = \sigma EA$, where A is the cross section of the wire. Hence,

$$E = \frac{I_c}{\sigma A} = \frac{2 \times 10^{-3} \sin \omega t}{2 \times 10^7 A} = \frac{1 \times 10^{-10}}{A} \sin \omega t \qquad \text{(V/m)}.$$

Application of Eq. (6.44), with $D = \epsilon E$, leads to

$$I_d = J_d A = \epsilon A \frac{\partial E}{\partial t} = \epsilon A \frac{\partial}{\partial t}\left(\frac{1 \times 10^{-10}}{A} \sin \omega t\right)$$
$$= \epsilon \omega \times 10^{-10} \cos \omega t = 0.885 \times 10^{-12} \cos \omega t \quad \text{(A)},$$

where we used $\omega = 10^9$ rad/s and $\epsilon = \epsilon_0 = 8.85 \times 10^{-12}$ F/m. Note that I_c and I_d are in phase quadrature (90° phase shift between them). Also, I_d is about nine orders of magnitude smaller than I_c, which is why the displacement current usually is ignored in good conductors.

Exercise 6-5: A poor conductor is characterized by a conductivity $\sigma = 100$ (S/m) and permittivity $\epsilon = 4\epsilon_0$. At what angular frequency ω is the amplitude of the conduction current density **J** equal to the amplitude of the displacement current density $\mathbf{J}_d$?

Answer: $\omega = 2.82 \times 10^{12}$ (rad/s). (See EM.)

6-8 Boundary Conditions for Electromagnetics

In Chapters 4 and 5 we applied the integral form of Maxwell's equations under static conditions to obtain boundary conditions applicable to the tangential and normal components of **E**, **D**, **B**, and **H** on interfaces between contiguous media (Section 4-8 for **E** and **D** and in Section 5-6 for **B** and **H**). In the dynamic case, Maxwell's equations (Table 6-1) include two new terms not accounted for in electrostatics and magnetostatics, namely, $\partial\mathbf{B}/\partial t$ in Faraday's law and $\partial\mathbf{D}/\partial t$ in Ampère's law.

▶ Nevertheless, the boundary conditions derived previously for electrostatic and magnetostatic fields remain valid for time-varying fields as well. ◀

This is because, if we were to apply the procedures outlined in the above-referenced sections for time-varying fields, we would find that the combination of the aforementioned terms vanish as the areas of the rectangular loops in Figs. 4-18 and 5-24 are made to approach zero.

The combined set of electromagnetic boundary conditions is summarized in Table 6-2.

Concept Question 6-9: When conduction current flows through a material, a certain number of charges enter the material on one end and an equal number leave on the other end. What's the situation like for the displacement current through a perfect dielectric?

Concept Question 6-10: Verify that the integral form of Ampère's law given by Eq. (6.43) leads to the boundary condition that the tangential component of **H** is continuous across the boundary between two dielectric media.

6-9 Charge-Current Continuity Relation

Under static conditions, the charge density ρ_v and the current density **J** at a given point in a material are totally independent of one another. This is no longer true in the time-varying case. To show the connection between ρ_v and **J**, we start by considering an arbitrary volume $\mathcal{V}$ bounded by a closed surface S (Fig. 6-14). The net positive charge contained in $\mathcal{V}$ is Q. Since, according to the law of conservation of

Table 6-2 **Boundary conditions for the electric and magnetic fields.**

Field Components	General Form	Medium 1 Dielectric	Medium 2 Dielectric	Medium 1 Dielectric	Medium 2 Conductor
Tangential E	$\hat{\mathbf{n}}_2 \times (\mathbf{E}_1 - \mathbf{E}_2) = 0$	$E_{1t} = E_{2t}$		$E_{1t} = E_{2t} = 0$	
Normal D	$\hat{\mathbf{n}}_2 \cdot (\mathbf{D}_1 - \mathbf{D}_2) = \rho_s$	$D_{1n} - D_{2n} = \rho_s$		$D_{1n} = \rho_s$	$D_{2n} = 0$
Tangential H	$\hat{\mathbf{n}}_2 \times (\mathbf{H}_1 - \mathbf{H}_2) = \mathbf{J}_s$	$H_{1t} = H_{2t}$		$H_{1t} = J_s$	$H_{2t} = 0$
Normal B	$\hat{\mathbf{n}}_2 \cdot (\mathbf{B}_1 - \mathbf{B}_2) = 0$	$B_{1n} = B_{2n}$		$B_{1n} = B_{2n} = 0$	

Notes: (1) ρ_s is the surface charge density at the boundary; (2) $\mathbf{J}_s$ is the surface current density at the boundary; (3) normal components of all fields are along $\hat{\mathbf{n}}_2$, the outward unit vector of medium 2; (4) $E_{1t} = E_{2t}$ implies that the tangential components are equal in magnitude and parallel in direction; (5) direction of $\mathbf{J}_s$ is orthogonal to $(\mathbf{H}_1 - \mathbf{H}_2)$.

Module 6.3 Displacement Current Observe the displacement current through a parallel plate capacitor.

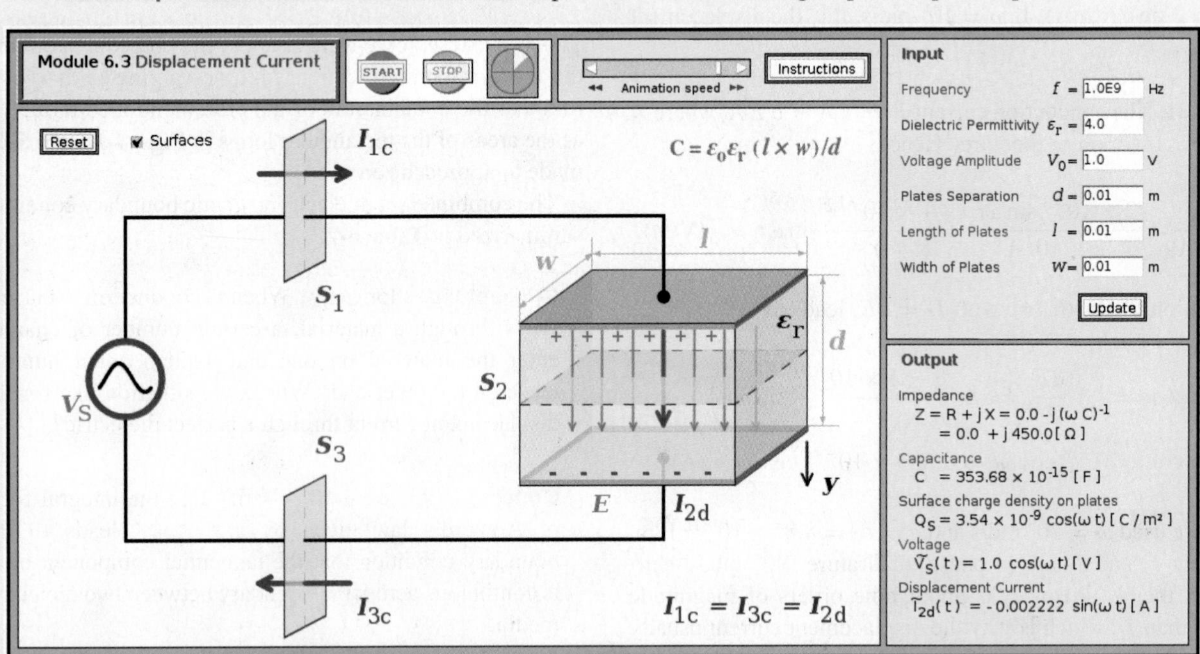

electric charge (Section 1-3.2), charge can neither be created nor destroyed, the only way Q can increase is as a result of a net inward flow of positive charge into the volume $\mathcal{V}$. By the same token, for Q to decrease there has to be a net outward flow of charge from $\mathcal{V}$. The inward and outward flow of charge constitute currents flowing across the surface S into and out of $\mathcal{V}$, respectively. We define I as the *net current flowing across S out of* $\mathcal{V}$. Accordingly, I is equal to the *negative* rate of change of Q:

$$I = -\frac{dQ}{dt} = -\frac{d}{dt} \int_{\mathcal{V}} \rho_v \, d\mathcal{V}, \tag{6.50}$$

where ρ_v is the volume charge density in $\mathcal{V}$. According to Eq. (4.12), the current I is also defined as the outward flux of

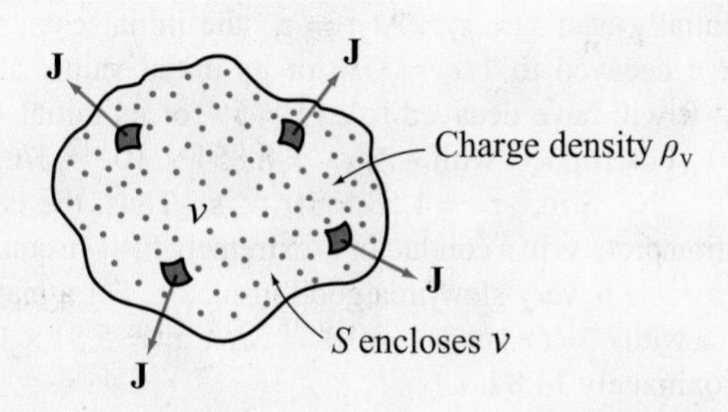

Figure 6-14 The total current flowing out of a volume $\mathcal{V}$ is equal to the flux of the current density **J** through the surface S, which in turn is equal to the rate of decrease of the charge enclosed in $\mathcal{V}$.

the current density **J** through the surface S. Hence,

$$\oint_S \mathbf{J} \cdot d\mathbf{s} = -\frac{d}{dt} \int_{\mathcal{V}} \rho_v \, d\mathcal{V}. \quad (6.51)$$

By applying the divergence theorem given by Eq. (3.98), we can convert the surface integral of **J** into a volume integral of its divergence $\nabla \cdot \mathbf{J}$, which then gives

$$\oint_S \mathbf{J} \cdot d\mathbf{s} = \int_{\mathcal{V}} \nabla \cdot \mathbf{J} \, d\mathcal{V} = -\frac{d}{dt} \int_{\mathcal{V}} \rho_v \, d\mathcal{V}. \quad (6.52)$$

For a stationary volume $\mathcal{V}$, the time derivative operates on ρ_v only. Hence, we can move it inside the integral and express it as a partial derivative of ρ_v:

$$\int_{\mathcal{V}} \nabla \cdot \mathbf{J} \, d\mathcal{V} = -\int_{\mathcal{V}} \frac{\partial \rho_v}{\partial t} \, d\mathcal{V}. \quad (6.53)$$

In order for the volume integrals on both sides of Eq. (6.53) to be equal for any volume $\mathcal{V}$, their integrands have to be equal at every point within $\mathcal{V}$. Hence,

$$\nabla \cdot \mathbf{J} = -\frac{\partial \rho_v}{\partial t}, \quad (6.54)$$

which is known as the ***charge-current continuity relation***, or simply the ***charge continuity equation***.

If the volume charge density within an elemental volume $\Delta\mathcal{V}$ (such as a small cylinder) is not a function of time (i.e., $\partial \rho_v / \partial t = 0$), it means that the net current flowing out of $\Delta\mathcal{V}$ is zero or, equivalently, that the current flowing into $\Delta\mathcal{V}$ is equal to the current flowing out of it. In this case, Eq. (6.54) implies

$$\nabla \cdot \mathbf{J} = 0, \quad (6.55)$$

and its integral-form equivalent [from Eq. (6.51)] is

$$\oint_S \mathbf{J} \cdot d\mathbf{s} = 0 \quad \textbf{(Kirchhoff's current law).} \quad (6.56)$$

Let us examine the meaning of Eq. (6.56) by considering a junction (or node) connecting two or more branches in an electric circuit. No matter how small, the junction has a volume $\mathcal{V}$ enclosed by a surface S. The junction shown in **Fig. 6-15** has been drawn as a cube, and its dimensions have been artificially enlarged to facilitate the present discussion. The junction has six faces (surfaces), which collectively constitute the surface S associated with the closed-surface integration given by Eq. (6.56). For each face, the integration represents the current flowing out through that face. Thus, Eq. (6.56) can be cast as

$$\sum_i I_i = 0 \quad \textbf{(Kirchhoff's current law),} \quad (6.57)$$

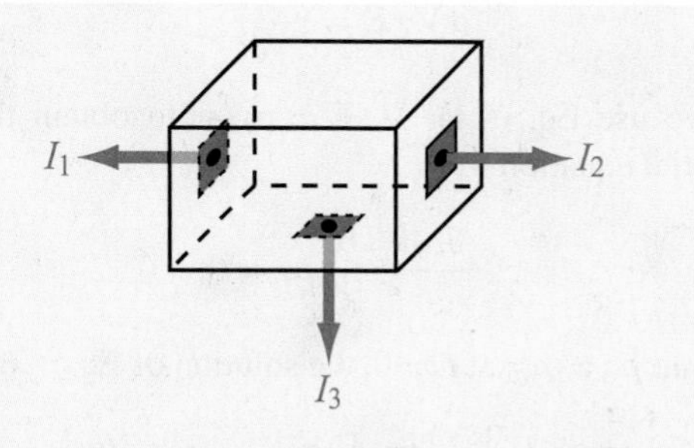

Figure 6-15 Kirchhoff's current law states that the algebraic sum of all the currents flowing out of a junction is zero.

where I_i is the current flowing outward through the ith face. For the junction of **Fig. 6-15**, Eq. (6.57) translates into $(I_1 + I_2 + I_3) = 0$. In its general form, Eq. (6.57) is an expression of ***Kirchhoff's current law***, which states that in an electric circuit *the sum of all the currents flowing out of a junction is zero.*

6-10 Free-Charge Dissipation in a Conductor

We stated earlier that current flow in a conductor is realized by the movement of loosely attached electrons under the influence of an externally applied electric field. These electrons, however, *are not excess charges*; their charge is balanced by an equal amount of positive charge in the atoms' nuclei. In other words, the conductor material is electrically neutral, and the net charge density in the conductor is zero ($\rho_v = 0$). What happens then if an excess free charge q is introduced at some interior point in a conductor? The excess charge gives rise to an electric field, which forces the charges of the host material nearest to the excess charge to rearrange their locations, which in turn cause other charges to move, and so on. The process continues until neutrality is reestablished in the conductor material and a charge equal to q resides on the conductor's surface.

How fast does the excess charge dissipate? To answer this question, let us introduce a volume charge density ρ_{vo} at the interior of a conductor and then find out the rate at which it decays down to zero. From Eq. (6.54), the continuity equation is given by

$$\nabla \cdot \mathbf{J} = -\frac{\partial \rho_v}{\partial t} . \tag{6.58}$$

In a conductor, the point form of Ohm's law, given by Eq. (4.63), states that $\mathbf{J} = \sigma \mathbf{E}$. Hence,

$$\sigma \nabla \cdot \mathbf{E} = -\frac{\partial \rho_v}{\partial t} . \tag{6.59}$$

Next, we use Eq. (6.1), $\nabla \cdot \mathbf{E} = \rho_v / \epsilon$, to obtain the partial differential equation

$$\frac{\partial \rho_v}{\partial t} + \frac{\sigma}{\epsilon} \rho_v = 0. \tag{6.60}$$

Given that $\rho_v = \rho_{vo}$ at $t = 0$, the solution of Eq. (6.60) is

$$\rho_v(t) = \rho_{vo} e^{-(\sigma/\epsilon)t} = \rho_{vo} e^{-t/\tau_r} \qquad (\text{C/m}^3), \tag{6.61}$$

where $\tau_r = \epsilon/\sigma$ is called the ***relaxation time constant***. We see from Eq. (6.61) that the initial excess charge ρ_{vo} decays exponentially at a rate τ_r. At $t = \tau_r$, the initial charge ρ_{vo} will have decayed to $1/e \approx 37\%$ of its initial value, and at $t = 3\tau_r$, it will have decayed to $e^{-3} \approx 5\%$ of its initial value at $t = 0$. For copper, with $\epsilon \approx \epsilon_0 = 8.854 \times 10^{-12}$ F/m and $\sigma = 5.8 \times 10^7$ S/m, $\tau_r = 1.53 \times 10^{-19}$ s. Thus, the charge dissipation process in a conductor is extremely fast. In contrast, the decay rate is very slow in a good insulator. For a material like mica with $\epsilon = 6\epsilon_0$ and $\sigma = 10^{-15}$ S/m, $\tau_r = 5.31 \times 10^4$ s, or approximately 14.8 hours.

Concept Question 6-11: Explain how the charge continuity equation leads to Kirchhoff's current law.

Concept Question 6-12: How long is the relaxation time constant for charge dissipation in a perfect conductor? In a perfect dielectric?

Exercise 6-6: Determine (a) the relaxation time constant and (b) the time it takes for a charge density to decay to 1% of its initial value in quartz, given that $\epsilon_r = 5$ and $\sigma = 10^{-17}$ S/m.

Answer: (a) $\tau_r = 51.2$ days, (b) 236 days. (See EM.)

6-11 Electromagnetic Potentials

Our discussion of Faraday's and Ampère's laws revealed two aspects of the link between time-varying electric and magnetic fields. We now examine the implications of this interconnection on the electric scalar potential V and the vector magnetic potential $\mathbf{A}$.

In the static case, Faraday's law reduces to

$$\nabla \times \mathbf{E} = 0 \qquad \textbf{(static case)}, \tag{6.62}$$

which states that the electrostatic field $\mathbf{E}$ is conservative. According to the rules of vector calculus, if a vector field $\mathbf{E}$ is conservative, it can be expressed as the gradient of a scalar. Hence, in Chapter 4 we defined $\mathbf{E}$ as

$$\mathbf{E} = -\nabla V \qquad \textbf{(electrostatics)}. \tag{6.63}$$

In the dynamic case, Faraday's law is

$$\nabla \times \mathbf{E} = -\frac{\partial \mathbf{B}}{\partial t}. \tag{6.64}$$

In view of the relation $\mathbf{B} = \nabla \times \mathbf{A}$, Eq. (6.64) can be expressed as

$$\nabla \times \mathbf{E} = -\frac{\partial}{\partial t}(\nabla \times \mathbf{A}), \tag{6.65}$$

which can be rewritten as

$$\nabla \times \left(\mathbf{E} + \frac{\partial \mathbf{A}}{\partial t}\right) = 0 \qquad \textbf{(dynamic case)}. \tag{6.66}$$

Let us for the moment define

$$\mathbf{E}' = \mathbf{E} + \frac{\partial \mathbf{A}}{\partial t}. \tag{6.67}$$

Using this definition, Eq. (6.66) becomes

$$\nabla \times \mathbf{E}' = 0. \tag{6.68}$$

Following the same logic that led to Eq. (6.63) from Eq. (6.62), we define

$$\mathbf{E}' = -\nabla V. \tag{6.69}$$

Upon substituting Eq. (6.67) for $\mathbf{E}'$ in Eq. (6.69) and then solving for $\mathbf{E}$, we have

$$\mathbf{E} = -\nabla V - \frac{\partial \mathbf{A}}{\partial t} \qquad \textbf{(dynamic case)}. \tag{6.70}$$

Equation (6.70) reduces to Eq. (6.63) in the static case.

When the scalar potential V and the vector potential $\mathbf{A}$ are known, $\mathbf{E}$ can be obtained from Eq. (6.70), and $\mathbf{B}$ can be obtained from

$$\mathbf{B} = \nabla \times \mathbf{A}. \tag{6.71}$$

Next we examine the relations between the potentials, V and $\mathbf{A}$, and their sources, the charge and current distributions ρ_v and $\mathbf{J}$, in the time-varying case.

6-11.1 Retarded Potentials

Consider the situation depicted in **Fig. 6-16**. A charge distribution ρ_v exists over a volume $\mathcal{V}'$ embedded in a perfect dielectric with permittivity ϵ. Were this a static charge distribution, then from Eq. (4.48a), the electric potential $V(\mathbf{R})$ at an observation point in space specified by the position vector $\mathbf{R}$ would be

$$V(\mathbf{R}) = \frac{1}{4\pi\epsilon} \int_{\mathcal{V}'} \frac{\rho_v(\mathbf{R}_i)}{R'}\, d\mathcal{V}', \tag{6.72}$$

where $\mathbf{R}_i$ denotes the position vector of an elemental volume $\Delta\mathcal{V}'$ containing charge density $\rho_v(\mathbf{R}_i)$, and $R' = |\mathbf{R} - \mathbf{R}_i|$ is the distance between $\Delta\mathcal{V}'$ and the observation point. If the charge distribution is time-varying, we may be tempted to rewrite Eq. (6.72) for the dynamic case as

$$V(\mathbf{R}, t) = \frac{1}{4\pi\epsilon} \int_{\mathcal{V}'} \frac{\rho_v(\mathbf{R}_i, t)}{R'}\, d\mathcal{V}', \tag{6.73}$$

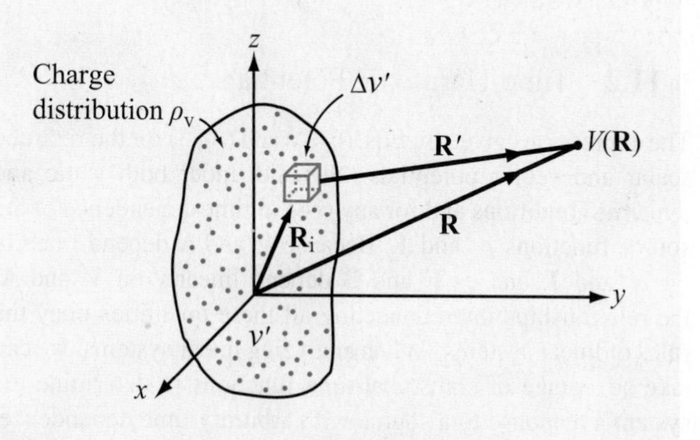

Figure 6-16 Electric potential $V(\mathbf{R})$ due to a charge distribution ρ_v over a volume $\mathcal{V}'$.

but such a form does not account for "reaction time." If V_1 is the potential due to a certain distribution ρ_{v1}, and if ρ_{v1} were to suddenly change to ρ_{v2}, it will take a finite amount of time before V_1 a distance R' away changes to V_2. *In other words, $V(\mathbf{R}, t)$ cannot change instantaneously.* The delay time is equal to $t' = R'/u_p$, where u_p is the velocity of propagation in the medium between the charge distribution and the observation point. Thus, $V(\mathbf{R}, t)$ at time t corresponds to ρ_v at an earlier time, that is, $(t - t')$. Hence, Eq. (6.73) should be rewritten as

$$V(\mathbf{R}, t) = \frac{1}{4\pi\epsilon} \int_{\mathcal{V}'} \frac{\rho_v(\mathbf{R}_i, t - R'/u_p)}{R'} \, d\mathcal{V}' \quad \text{(V)}, \tag{6.74}$$

and $V(\mathbf{R}, t)$ is appropriately called the ***retarded scalar potential***. If the propagation medium is vacuum, u_p is equal to the velocity of light c.

Similarly, the ***retarded vector potential*** $\mathbf{A}(\mathbf{R}, t)$ is related to the distribution of current density $\mathbf{J}$ by

$$\mathbf{A}(\mathbf{R}, t) = \frac{\mu}{4\pi} \int_{\mathcal{V}'} \frac{\mathbf{J}(\mathbf{R}_i, t - R'/u_p)}{R'} \, d\mathcal{V}' \quad \text{(Wb/m)}. \tag{6.75}$$

This expression is obtained by extending the expression for the magnetostatic vector potential $\mathbf{A}(\mathbf{R})$ given by Eq. (5.65) to the time-varying case.

6-11.2 Time-Harmonic Potentials

The expressions given by Eqs. (6.74) and (6.75) for the retarded scalar and vector potentials are valid under both static and dynamic conditions and for any type of time dependence of the source functions ρ_v and $\mathbf{J}$. Because V and $\mathbf{A}$ depend linearly on ρ_v and $\mathbf{J}$, and as $\mathbf{E}$ and $\mathbf{B}$ depend linearly on V and $\mathbf{A}$, the relationships interconnecting all these quantities obey the rules of linear systems. When analyzing linear systems, we can take advantage of sinusoidal-time functions to determine the system's response to a source with arbitrary time dependence. As was noted in Section 1-7, if the time dependence is described by a (nonsinusoidal) periodic time function, it can always be expanded into a Fourier series of sinusoidal components, and if the time function is nonperiodic, it can be represented by a Fourier integral. In either case, if the response of the linear system is known for all steady-state sinusoidal excitations, the principle of superposition can be used to determine its response to an excitation with arbitrary time dependence. Thus, the sinusoidal response of the system constitutes a fundamental building block that can be used to determine the response due to a source described by an arbitrary function of time. The term ***time-harmonic*** is often used in this context as a synonym for "steady-state sinusoidal time-dependent."

In this subsection, we derive expressions for the scalar and vector potentials due to time-harmonic sources. Suppose that $\rho_v(\mathbf{R}_i, t)$ is a sinusoidal-time function with angular frequency ω, given by

$$\rho_v(\mathbf{R}_i, t) = \rho_v(\mathbf{R}_i) \cos(\omega t + \phi). \tag{6.76}$$

Phasor analysis, which was first introduced in Section 1-7 and then used extensively in Chapter 2 to study wave propagation on transmission lines, is a useful tool for analyzing time-harmonic scenarios. A time harmonic charge distribution $\rho_v(\mathbf{R}_i, t)$ is related to its phasor $\tilde{\rho}_v(\mathbf{R}_i)$ as

$$\rho_v(\mathbf{R}_i, t) = \mathfrak{Re}\left[\tilde{\rho}_v(\mathbf{R}_i)\, e^{j\omega t}\right], \tag{6.77}$$

Comparison of Eqs. (6.76) and (6.77) shows that in the present case $\tilde{\rho}_v(\mathbf{R}_i) = \rho_v(\mathbf{R}_i)\, e^{j\phi}$.

Next, we express the retarded charge density $\rho_v(\mathbf{R}_i, t - R'/u_p)$ in phasor form by replacing t with $(t - R'/u_p)$ in Eq. (6.77):

$$\begin{aligned} \rho_v(\mathbf{R}_i, t - R'/u_p) &= \mathfrak{Re}\left[\tilde{\rho}_v(\mathbf{R}_i)\, e^{j\omega(t - R'/u_p)}\right] \\ &= \mathfrak{Re}\left[\tilde{\rho}_v(\mathbf{R}_i)\, e^{-j\omega R'/u_p} e^{j\omega t}\right] \\ &= \mathfrak{Re}\left[\tilde{\rho}_v(\mathbf{R}_i)\, e^{-jkR'} e^{j\omega t}\right], \end{aligned} \tag{6.78}$$

where

$$k = \frac{\omega}{u_p} \tag{6.79}$$

is called the ***wavenumber*** or phase constant of the propagation medium. (In general, the phase constant is denoted by the symbol "β", but for lossless dielectric media, it is commonly denoted by the symbol "k" and called the wavenumber.) Similarly, we define the phasor $\widetilde{V}(\mathbf{R})$ of the time function $V(\mathbf{R}, t)$ according to

$$V(\mathbf{R}, t) = \mathfrak{Re}\left[\widetilde{V}(\mathbf{R})\, e^{j\omega t}\right]. \tag{6.80}$$

Using Eqs. (6.78) and (6.80) in Eq. (6.74) gives

$$\Re\mathfrak{e}\left[\widetilde{V}(\mathbf{R})\,e^{j\omega t}\right] = \Re\mathfrak{e}\left[\frac{1}{4\pi\epsilon}\int\limits_{\mathcal{V}'}\frac{\widetilde{\rho}_{\mathrm{v}}(\mathbf{R}_{\mathrm{i}})\,e^{-jkR'}}{R'}\,e^{j\omega t}\,d\mathcal{V}'\right]. \quad (6.81)$$

By equating the quantities inside the square brackets on both sides of Eq. (6.81) and cancelling the common $e^{j\omega t}$ factor, we obtain the phasor-domain expression

$$\widetilde{V}(\mathbf{R}) = \frac{1}{4\pi\epsilon}\int\limits_{\mathcal{V}'}\frac{\widetilde{\rho}_{\mathrm{v}}(\mathbf{R}_{\mathrm{i}})\,e^{-jkR'}}{R'}\,d\mathcal{V}' \quad \text{(V)}. \quad (6.82)$$

For any given charge distribution, Eq. (6.82) can be used to compute $\widetilde{V}(\mathbf{R})$, and then the resultant expression can be used in Eq. (6.80) to find $V(\mathbf{R}, t)$. Similarly, the expression for $\mathbf{A}(\mathbf{R}, t)$ given by Eq. (6.75) can be transformed into

$$\mathbf{A}(\mathbf{R}, t) = \Re\mathfrak{e}\left[\widetilde{\mathbf{A}}(\mathbf{R})\,e^{j\omega t}\right] \quad (6.83)$$

with

$$\widetilde{\mathbf{A}}(\mathbf{R}) = \frac{\mu}{4\pi}\int\limits_{\mathcal{V}'}\frac{\widetilde{\mathbf{J}}(\mathbf{R}_{\mathrm{i}})\,e^{-jkR'}}{R'}\,d\mathcal{V}', \quad (6.84)$$

where $\widetilde{\mathbf{J}}(\mathbf{R}_{\mathrm{i}})$ is the phasor function corresponding to $\mathbf{J}(\mathbf{R}_{\mathrm{i}}, t)$.

The magnetic field phasor $\widetilde{\mathbf{H}}$ corresponding to $\widetilde{\mathbf{A}}$ is given by

$$\widetilde{\mathbf{H}} = \frac{1}{\mu}\,\nabla\times\widetilde{\mathbf{A}}. \quad (6.85)$$

Recalling that differentiation in the time domain is equivalent to multiplication by $j\omega$ in the phasor domain, in a nonconducting medium ($\mathbf{J} = 0$), Ampère's law given by Eq. (6.41) becomes

$$\nabla\times\widetilde{\mathbf{H}} = j\omega\epsilon\widetilde{\mathbf{E}} \quad \text{or} \quad \widetilde{\mathbf{E}} = \frac{1}{j\omega\epsilon}\,\nabla\times\widetilde{\mathbf{H}}. \quad (6.86)$$

Hence, given a time-harmonic current-density distribution with phasor $\widetilde{\mathbf{J}}$, Eqs. (6.84) to (6.86) can be used successively to determine both $\widetilde{\mathbf{E}}$ and $\widetilde{\mathbf{H}}$. The phasor vectors $\widetilde{\mathbf{E}}$ and $\widetilde{\mathbf{H}}$ also are related by the phasor form of Faraday's law:

$$\nabla\times\widetilde{\mathbf{E}} = -j\omega\mu\widetilde{\mathbf{H}}$$

$$\text{or} \quad \widetilde{\mathbf{H}} = -\frac{1}{j\omega\mu}\,\nabla\times\widetilde{\mathbf{E}}. \quad (6.87)$$

Example 6-8: Relating E to H

In a nonconducting medium with $\epsilon = 16\epsilon_0$ and $\mu = \mu_0$, the electric field intensity of an electromagnetic wave is

$$\mathbf{E}(z, t) = \hat{\mathbf{x}}\, 10 \sin(10^{10}t - kz) \qquad \text{(V/m)}. \tag{6.88}$$

Determine the associated magnetic field intensity $\mathbf{H}$ and find the value of k.

Solution: We begin by finding the phasor $\widetilde{\mathbf{E}}(z)$ of $\mathbf{E}(z, t)$. Since $\mathbf{E}(z, t)$ is given as a sine function and phasors are defined in this book with reference to the cosine function, we rewrite Eq. (6.88) as

$$\begin{aligned}\mathbf{E}(z, t) &= \hat{\mathbf{x}}\, 10 \cos(10^{10}t - kz - \pi/2) \qquad \text{(V/m)}\\ &= \mathfrak{Re}\left[\widetilde{\mathbf{E}}(z)\, e^{j\omega t}\right],\end{aligned} \tag{6.89}$$

with $\omega = 10^{10}$ (rad/s) and

$$\widetilde{\mathbf{E}}(z) = \hat{\mathbf{x}}\, 10 e^{-jkz} e^{-j\pi/2} = -\hat{\mathbf{x}} j 10 e^{-jkz}. \tag{6.90}$$

To find both $\widetilde{\mathbf{H}}(z)$ and k, we will perform a "circle": we will use the given expression for $\widetilde{\mathbf{E}}(z)$ in Faraday's law to find $\widetilde{\mathbf{H}}(z)$; then we will use $\widetilde{\mathbf{H}}(z)$ in Ampère's law to find $\widetilde{\mathbf{E}}(z)$, which we will then compare with the original expression for $\widetilde{\mathbf{E}}(z)$; and the comparison will yield the value of k. Application of Eq. (6.87) gives

$$\begin{aligned}\widetilde{\mathbf{H}}(z) &= -\frac{1}{j\omega\mu}\,\nabla \times \widetilde{\mathbf{E}}\\ &= -\frac{1}{j\omega\mu}\begin{vmatrix}\hat{\mathbf{x}} & \hat{\mathbf{y}} & \hat{\mathbf{z}}\\ \partial/\partial x & \partial/\partial y & \partial/\partial z\\ -j10e^{-jkz} & 0 & 0\end{vmatrix}\\ &= -\frac{1}{j\omega\mu}\left[\hat{\mathbf{y}}\,\frac{\partial}{\partial z}(-j10e^{-jkz})\right]\\ &= -\hat{\mathbf{y}} j\,\frac{10k}{\omega\mu}e^{-jkz}.\end{aligned} \tag{6.91}$$

So far, we have used Eq. (6.90) for $\widetilde{\mathbf{E}}(z)$ to find $\widetilde{\mathbf{H}}(z)$, but k remains unknown. To find k, we use $\widetilde{\mathbf{H}}(z)$ in Eq. (6.86) to find $\widetilde{\mathbf{E}}(z)$:

$$\begin{aligned}\widetilde{\mathbf{E}}(z) &= \frac{1}{j\omega\epsilon}\,\nabla \times \widetilde{\mathbf{H}}\\ &= \frac{1}{j\omega\epsilon}\left[-\hat{\mathbf{x}}\,\frac{\partial}{\partial z}\left(-j\,\frac{10k}{\omega\mu}e^{-jkz}\right)\right]\\ &= -\hat{\mathbf{x}} j\,\frac{10k^2}{\omega^2\mu\epsilon}e^{-jkz}.\end{aligned} \tag{6.92}$$

Equating Eqs. (6.90) and (6.92) leads to

$$k^2 = \omega^2\mu\epsilon,$$

or

$$\begin{aligned}k &= \omega\sqrt{\mu\epsilon}\\ &= 4\omega\sqrt{\mu_0\epsilon_0}\\ &= \frac{4\omega}{c} = \frac{4 \times 10^{10}}{3 \times 10^8} = 133 \qquad \text{(rad/m)}.\end{aligned} \tag{6.93}$$

With k known, the instantaneous magnetic field intensity is then given by

$$\begin{aligned}\mathbf{H}(z, t) &= \mathfrak{Re}\left[\widetilde{\mathbf{H}}(z)\, e^{j\omega t}\right]\\ &= \mathfrak{Re}\left[-\hat{\mathbf{y}} j\,\frac{10k}{\omega\mu}e^{-jkz}e^{j\omega t}\right]\\ &= \hat{\mathbf{y}}\, 0.11 \sin(10^{10}t - 133z) \qquad \text{(A/m)}.\end{aligned} \tag{6.94}$$

We note that k has the same expression as the phase constant of a lossless transmission line [Eq. (2.49)].

Exercise 6-7: The magnetic field intensity of an electromagnetic wave propagating in a lossless medium with $\epsilon = 9\epsilon_0$ and $\mu = \mu_0$ is

$$\mathbf{H}(z, t) = \hat{\mathbf{x}}\, 0.3 \cos(10^8 t - kz + \pi/4) \qquad \text{(A/m)}.$$

Find $\mathbf{E}(z, t)$ and k.

Answer: $\mathbf{E}(z, t) = -\hat{\mathbf{y}}\, 37.7 \cos(10^8 t - z + \pi/4)$ (V/m); $k = 1$ (rad/m). (See EM.)

Chapter 6 Summary

Concepts

- Faraday's law states that a voltage is induced across the terminals of a loop if the magnetic flux linking its surface changes with time.
- In an ideal transformer, the ratios of the primary to secondary voltages, currents, and impedances are governed by the turns ratio.
- Displacement current accounts for the "apparent" flow of charges through a dielectric. In reality, charges of opposite polarity accumulate along the two ends of a dielectric, giving the appearance of current flow through it.
- Boundary conditions for the electromagnetic fields at the interface between two different media are the same for both static and dynamic conditions.
- The charge continuity equation is a mathematical statement of the law of conservation of electric charge.
- Excess charges in the interior of a good conductor dissipate very quickly; through a rearrangement process, the excess charge is transferred to the surface of the conductor.
- In the dynamic case, the electric field **E** is related to both the scalar electric potential V and the magnetic vector potential **A**.
- The retarded scalar and vector potentials at a given observation point take into account the finite time required for propagation between their sources, the charge and current distributions, and the location of the observation point.

Mathematical and Physical Models

Faraday's Law

$$V_{\text{emf}} = -\frac{d\Phi}{dt} = -\frac{d}{dt}\int_S \mathbf{B} \cdot d\mathbf{s} = V_{\text{emf}}^{\text{tr}} + V_{\text{emf}}^{\text{m}}$$

Transformer

$$V_{\text{emf}}^{\text{tr}} = -N\int_S \frac{\partial \mathbf{B}}{\partial t} \cdot d\mathbf{s} \qquad (N \text{ loops})$$

Motional

$$V_{\text{emf}}^{\text{m}} = \oint_C (\mathbf{u} \times \mathbf{B}) \cdot d\mathbf{l}$$

Charge-Current Continuity

$$\nabla \cdot \mathbf{J} = -\frac{\partial \rho_{\text{v}}}{\partial t}$$

EM Potentials

$$\mathbf{E} = -\nabla V - \frac{\partial \mathbf{A}}{\partial t}$$

$$\mathbf{B} = \nabla \times \mathbf{A}$$

Current Density

Conduction $\mathbf{J}_{\text{c}} = \sigma \mathbf{E}$

Displacement $\mathbf{J}_{\text{d}} = \dfrac{\partial \mathbf{D}}{\partial t}$

Conductor Charge Dissipation

$$\rho_{\text{v}}(t) = \rho_{\text{vo}} e^{-(\sigma/\epsilon)t} = \rho_{\text{vo}} e^{-t/\tau_{\text{r}}}$$

Important Terms Provide definitions or explain the meaning of the following terms:

charge continuity equation
charge dissipation
displacement current I_d
electromagnetic induction
electromotive force V_{emf}
Faraday's law
Kirchhoff's current law
Lenz's law
motional emf V^{m}_{emf}
relaxation time constant
retarded potential
transformer emf V^{tr}_{emf}
wavenumber k

PROBLEMS

Sections 6-1 to 6-6: Faraday's Law and its Applications

***6.1** The switch in the bottom loop of **Fig. P6.1** is closed at $t = 0$ and then opened at a later time t_1. What is the direction of the current I in the top loop (clockwise or counterclockwise) at each of these two times?

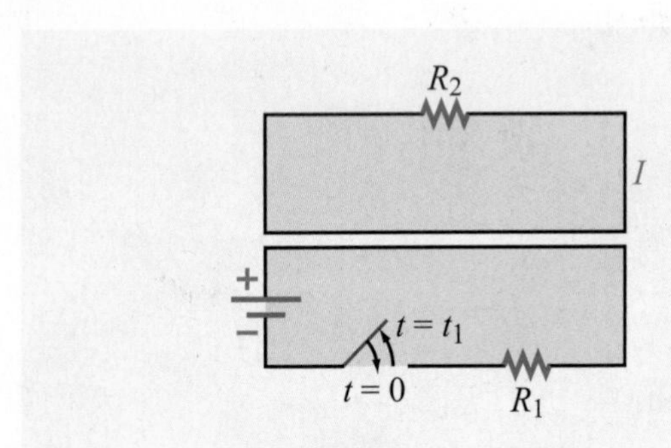

Figure P6.1 Loops of Problem 6.1.

6.2 The loop in **Fig. P6.2** is in the x–y plane and $\mathbf{B} = \hat{\mathbf{z}} B_0 \sin \omega t$ with B_0 positive. What is the direction of I ($\hat{\boldsymbol{\phi}}$ or $-\hat{\boldsymbol{\phi}}$) at:

(a) $t = 0$

(b) $\omega t = \pi/4$

(c) $\omega t = \pi/2$

*Answer(s) available in Appendix D.

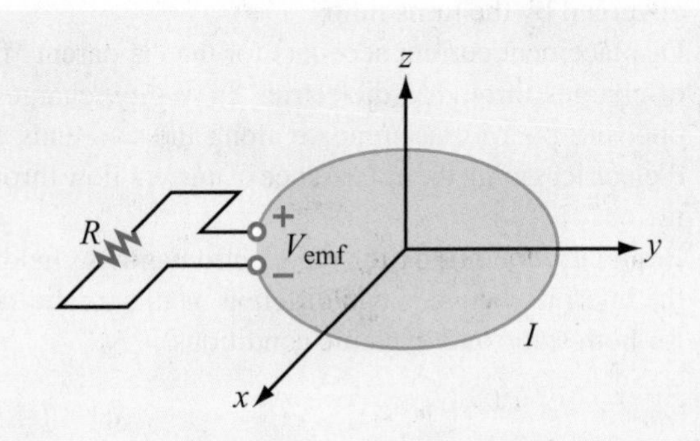

Figure P6.2 Loop of Problem 6.2.

6.3 A stationary conducting loop with an internal resistance of 0.5 Ω is placed in a time-varying magnetic field. When the loop is closed, a current of 5 A flows through it. What will the current be if the loop is opened to create a small gap and a 4.5 Ω resistor is connected across its open ends?

6.4 A coil consists of 200 turns of wire wrapped around a square frame of sides 0.25 m. The coil is centered at the origin with each of its sides parallel to the x- or y axis. Find the induced emf across the open-circuited ends of the coil if the magnetic field is given by

***(a)** $\mathbf{B} = \hat{\mathbf{z}}\, 20 e^{-3t}$ (T)

(b) $\mathbf{B} = \hat{\mathbf{z}}\, 20 \cos x \, \cos 10^3 t$ (T)

(c) $\mathbf{B} = \hat{\mathbf{z}}\, 20 \cos x \, \sin 2y \, \cos 10^3 t$ (T)

6.5 A rectangular conducting loop 5 cm × 10 cm with a small air gap in one of its sides is spinning at 7200 revolutions per minute. If the field **B** is normal to the loop axis and its magnitude is 3×10^{-6} T, what is the peak voltage induced across the air gap?

6.6 The square loop shown in **Fig. P6.6** is coplanar with a long, straight wire carrying a current

$$I(t) = 5\cos(2\pi \times 10^4 t) \qquad \text{(A)}.$$

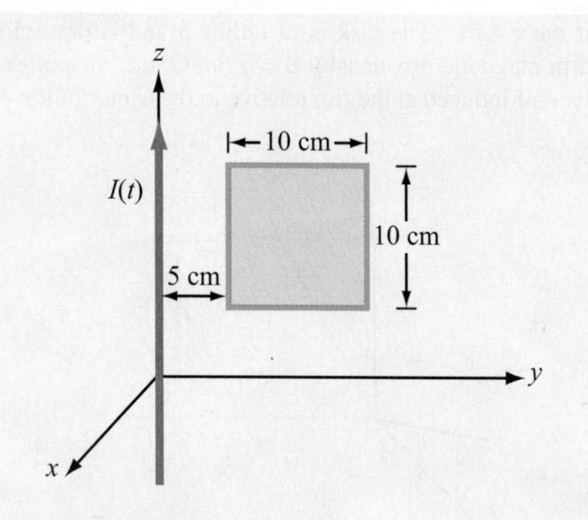

Figure P6.6 Loop coplanar with long wire (Problem 6.6).

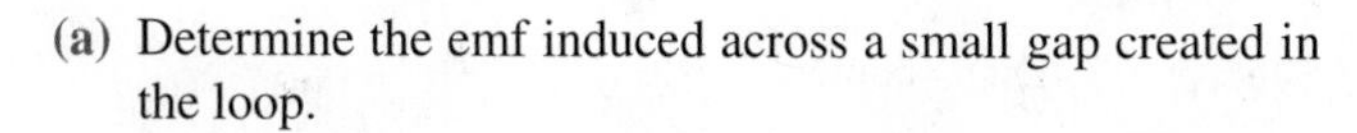

(a) Determine the emf induced across a small gap created in the loop.

(b) Determine the direction and magnitude of the current that would flow through a 4 Ω resistor connected across the gap. The loop has an internal resistance of 1 Ω.

6.7* The rectangular conducting loop shown in **Fig. P6.7 rotates at 3,000 revolutions per minute in a uniform magnetic flux density given by

$$\mathbf{B} = \hat{\mathbf{y}}\, 50 \qquad \text{(mT)}.$$

Determine the current induced in the loop if its internal resistance is 0.5 Ω.

6.8 The transformer shown in **Fig. P6.8** consists of a long wire coincident with the z axis carrying a current $I = I_0 \cos \omega t$, coupling magnetic energy to a toroidal coil situated in the x–y plane and centered at the origin. The toroidal core uses iron material with relative permeability μ_r, around which 100 turns of a tightly wound coil serves to induce a voltage V_{emf}, as shown in the figure.

(a) Develop an expression for V_{emf}.

(b) Calculate V_{emf} for $f = 60$ Hz, $\mu_r = 4000$, $a = 5$ cm, $b = 6$ cm, $c = 2$ cm, and $I_0 = 50$ A.

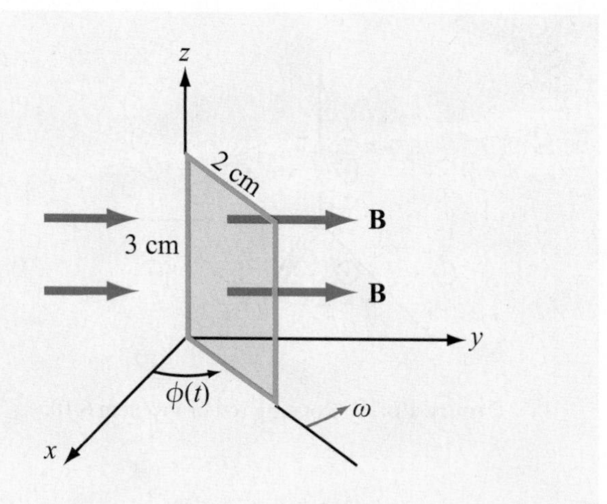

Figure P6.7 Rotating loop in a magnetic field (Problem 6.7).

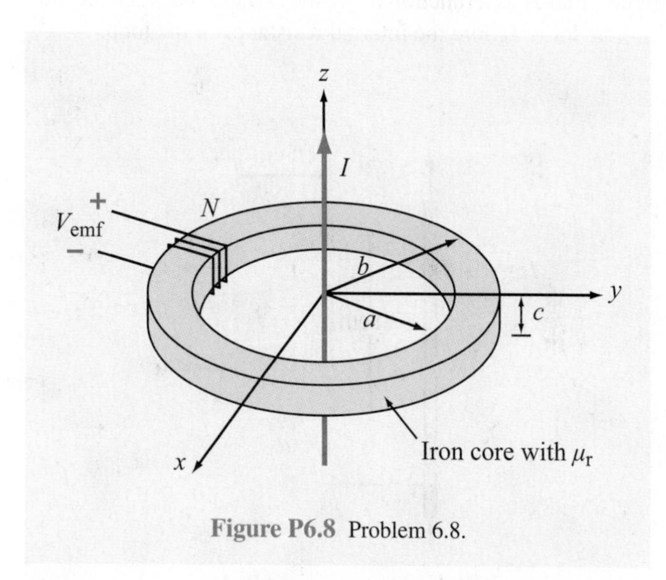

Figure P6.8 Problem 6.8.

6.9* A circular-loop TV antenna with 0.04 m^2 area is in the presence of a uniform-amplitude 300 Mhz signal. When oriented for maximum response, the loop develops an emf with a peak value of 30 (mV). What is the peak magnitude of **B of the incident wave?

6.10* A 50 cm long metal rod rotates about the z axis at 90 revolutions per minute, with end 1 fixed at the origin as shown in **Fig. P6.10. Determine the induced emf V_{12} if $\mathbf{B} = \hat{\mathbf{z}}\, 2 \times 10^{-4}$ T.

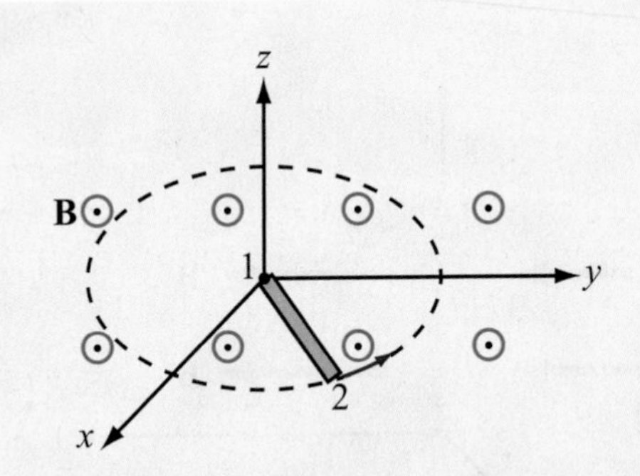

Figure P6.10 Rotating rod of Problem 6.10.

6.11 The loop shown in P6.11 moves away from a wire carrying a current $I_1 = 10$ A at a constant velocity $\mathbf{u} = \hat{\mathbf{y}}7.5$ (m/s). If $R = 10\ \Omega$ and the direction of I_2 is as defined in the figure, find I_2 as a function of y_0, the distance between the wire and the loop. Ignore the internal resistance of the loop.

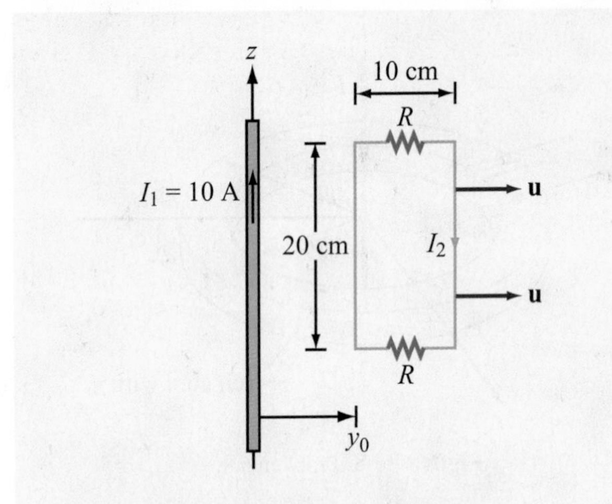

Figure P6.11 Moving loop of Problem 6.11.

***6.12** The electromagnetic generator shown in **Fig. 6-12** is connected to an electric bulb with a resistance of 50 Ω. If the loop area is 0.1 m^2 and it rotates at 3,600 revolutions per minute in a uniform magnetic flux density $B_0 = 0.4$ T, determine the amplitude of the current generated in the light bulb.

6.13 The circular, conducting, disk shown in **Fig. P6.13** lies in the x–y plane and rotates with uniform angular velocity ω about the z axis. The disk is of radius a and is present in a uniform magnetic flux density $\mathbf{B} = \hat{\mathbf{z}}B_0$. Obtain an expression for the emf induced at the rim relative to the center of the disk.

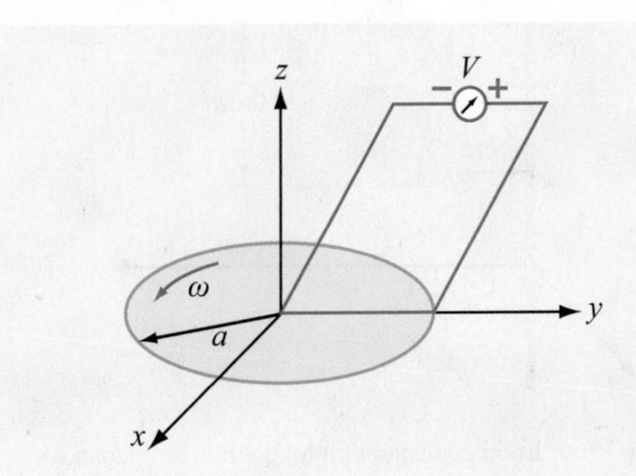

Figure P6.13 Rotating circular disk in a magnetic field (Problem 6.13).

Section 6-7: Displacement Current

***6.14** A coaxial capacitor of length $l = 6$ cm uses an insulating dielectric material with $\epsilon_r = 9$. The radii of the cylindrical conductors are 0.5 cm and 1 cm. If the voltage applied across the capacitor is

$$V(t) = 50\sin(120\pi t) \qquad \text{(V)},$$

what is the displacement current?

6.15 The plates of a parallel-plate capacitor have areas of 10 cm^2 each and are separated by 1 cm. The capacitor is filled with a dielectric material with $\epsilon = 4\epsilon_0$, and the voltage across it is given by $V(t) = 30\cos 2\pi \times 10^6 t$ (V). Find the displacement current.

6.16 The parallel-plate capacitor shown in **Fig. P6.16** is filled with a lossy dielectric material of relative permittivity ϵ_r and conductivity σ. The separation between the plates is d and each plate is of area A. The capacitor is connected to a time-varying voltage source $V(t)$.

(a) Obtain an expression for I_c, the conduction current flowing between the plates inside the capacitor, in terms of the given quantities.

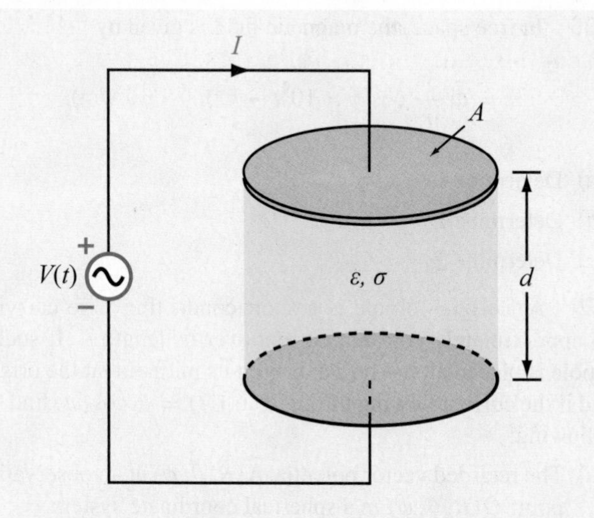

Figure P6.16 Parallel-plate capacitor containing a lossy dielectric material (Problem 6.16).

(b) Obtain an expression for I_d, the displacement current flowing inside the capacitor.

(c) Based on your expressions for parts (a) and (b), give an equivalent-circuit representation for the capacitor.

(d) Evaluate the values of the circuit elements for $A = 4$ cm^2, $d = 0.5$ cm, $\epsilon_\text{r} = 4$, $\sigma = 2.5$ (S/m), and $V(t) = 10\cos(3\pi \times 10^3 t)$ (V).

6.17 An electromagnetic wave propagating in seawater has an electric field with a time variation given by $\mathbf{E} = \hat{\mathbf{z}} E_0 \cos \omega t$. If the permittivity of water is $81\epsilon_0$ and its conductivity is 4 (S/m), find the ratio of the magnitudes of the conduction current density to displacement current density at each of the following frequencies:

(a) 1 kHz

*__(b)__ 1 MHz

(c) 1 GHz

(d) 100 GHz

*__6.18__ In wet soil, characterized by $\sigma = 10^{-2}$ (S/m), $\mu_\text{r} = 1$, and $\epsilon_\text{r} = 36$, at what frequency is the conduction current density equal in magnitude to the displacement current density?

Sections 6-9 and 6-10: Continuity Equation and Charge Dissipation

6.19 At $t = 0$, charge density $\rho_{\text{v}0}$ was introduced into the interior of a material with a relative permittivity $\epsilon_\text{r} = 6$. If at $t = 1$ μs the charge density has dissipated down to $10^{-3}\rho_{\text{v}0}$, what is the conductivity of the material?

*__6.20__ If the current density in a conducting medium is given by

$$\mathbf{J}(x, y, z; t) = (\hat{\mathbf{x}}z - \hat{\mathbf{y}}4y^2 + \hat{\mathbf{z}}2x) \cos \omega t$$

determine the corresponding charge distribution $\rho_\text{v}(x, y, z; t)$.

6.21 If we were to characterize how good a material is as an insulator by its resistance to dissipating charge, which of the following two materials is the better insulator?

Dry Soil:	$\epsilon_\text{r} = 2.5$,	$\sigma = 10^{-4}$ (S/m)
Fresh Water:	$\epsilon_\text{r} = 80$,	$\sigma = 10^{-3}$ (S/m)

6.22 In a certain medium, the direction of current density **J** points in the radial direction in cylindrical coordinates and its magnitude is independent of both ϕ and z. Determine **J**, given that the charge density in the medium is

$$\rho_\text{v} = \rho_0 r \cos \omega t \quad (\text{C/m}^3).$$

Sections 6-11: Electromagnetic Potentials

6.23 The electric field of an electromagnetic wave propagating in air is given by

$$\mathbf{E}(z, t) = \hat{\mathbf{x}}4\cos(6 \times 10^8 t - 2z) + \hat{\mathbf{y}}3\sin(6 \times 10^8 t - 2z) \qquad (\text{V/m}).$$

Find the associated magnetic field $\mathbf{H}(z, t)$.

*__6.24__ The magnetic field in a dielectric material with $\epsilon = 4\epsilon_0$, $\mu = \mu_0$, and $\sigma = 0$ is given by

$$\mathbf{H}(y, t) = \hat{\mathbf{x}}5\cos(2\pi \times 10^7 t + ky) \qquad (\text{A/m}).$$

Find k and the associated electric field **E**.

6.25 Given an electric field

$$\mathbf{E} = \hat{\mathbf{x}} E_0 \sin ay \cos(\omega t - kz),$$

where E_0, a, ω, and k are constants, find **H**.

*6.26 The electric field radiated by a short dipole antenna is given in spherical coordinates by

$$\mathbf{E}(R, \theta; t) = \hat{\boldsymbol{\theta}}\, \frac{2 \times 10^{-2}}{R} \sin\theta \, \cos(6\pi \times 10^8 t - 2\pi R) \quad \text{(V/m)}.$$

Find $\mathbf{H}(R, \theta; t)$.

6.27 The magnetic field in a given dielectric medium is given by

$$\mathbf{H} = \hat{\mathbf{y}}\, 6 \cos 2z \, \sin(2 \times 10^7 t - 0.1x) \qquad \text{(A/m)},$$

where x and z are in meters. Determine:

(a) $\mathbf{E}$,

(b) the displacement current density $\mathbf{J}_d$, and

(c) the charge density ρ_v.

6.28 In free space, the magnetic field is given by

$$\mathbf{H} = \hat{\boldsymbol{\phi}}\, \frac{36}{r} \cos(6 \times 10^9 t - kz) \qquad \text{(mA/m)}.$$

*(a) Determine k.

(b) Determine $\mathbf{E}$.

(c) Determine $\mathbf{J}_d$.

6.29 A Hertzian dipole is a short conducting wire carrying an approximately constant current over its length l. If such a dipole is placed along the z axis with its midpoint at the origin, and if the current flowing through it is $i(t) = I_0 \cos \omega t$, find the following:

(a) The retarded vector potential $\widetilde{\mathbf{A}}(R, \theta, \phi)$ at an observation point $Q(R, \theta, \phi)$ in a spherical coordinate system.

(b) The magnetic field phasor $\widetilde{\mathbf{H}}(R, \theta, \phi)$.

Assume l to be sufficiently small so that the observation point is approximately equidistant to all points on the dipole; that is, assume $R' \approx R$.

CHAPTER 7

Plane-Wave Propagation

Chapter Contents

Unbounded EM Waves, 336

7-1 Time-Harmonic Fields, 337

7-2 Plane-Wave Propagation in Lossless Media, 338

TB13 RFID Systems, 344

7-3 Wave Polarization, 346

7-4 Plane-Wave Propagation in Lossy Media, 353

TB14 Liquid Crystal Display (LCD), 358

7-5 Current Flow in a Good Conductor, 361

7-6 Electromagnetic Power Density, 365

Chapter 7 Summary, 368

Problems, 370

Objectives

Upon learning the material presented in this chapter, you should be able to:

1. Describe mathematically the electric and magnetic fields of TEM waves.
2. Describe the polarization properties of an EM wave.
3. Relate the propagation parameters of a wave to the constitutive parameters of the medium.
4. Characterize the flow of current in conductors and use it to calculate the resistance of a coaxial cable.
5. Calculate the rate of power carried by an EM wave, in both lossless and lossy media.

Unbounded EM Waves

It was established in Chapter 6 that a time-varying electric field produces a magnetic field and, conversely, a time-varying magnetic field produces an electric field. This cyclic pattern often results in electromagnetic (EM) waves propagating through free space and in material media. When a wave propagates through a homogeneous medium without interacting with obstacles or material interfaces, it is said to be ***unbounded***. Light waves emitted by the sun and radio transmissions by antennas are good examples. Unbounded waves may propagate in both lossless and lossy media. Waves propagating in a ***lossless medium*** (e.g., air and perfect dielectrics) are similar to those on a lossless transmission line in that they do not attenuate. When propagating in a ***lossy medium*** (material with nonzero conductivity, such as water), part of the power carried by an EM wave gets converted into heat. A wave produced by a localized source, such as an antenna, expands outwardly in the form of a ***spherical wave***, as depicted in **Fig. 7-1(a)**. Even though an antenna may radiate more energy along some directions than along others, the spherical wave travels at the same speed in all directions. To an observer very far away from the source, however, the ***wavefront*** of the spherical wave appears approximately *planar*, as if it were part of a ***uniform plane wave*** with identical properties at all points in the plane tangent to the wavefront [**Fig. 7-1(b)**]. Plane-waves are easily described using a Cartesian coordinate system, which is mathematically easier to work with than the spherical coordinate system needed to describe spherical waves.

When a wave propagates along a material structure, it is said to be ***guided***. Earth's surface and ionosphere constitute parallel boundaries of a natural structure capable of guiding short-wave radio transmissions in the HF band† (3 to 30 MHz); indeed, the ionosphere is a good reflector at these frequencies, thereby allowing the waves to zigzag between the two boundaries (**Fig. 7-2**). When we discussed wave propagation on a transmission line in Chapter 2, we dealt with voltages and currents. For a transmission-line circuit such as that shown in **Fig. 7-3**, the ac voltage source excites an incident wave that travels down the coaxial line toward the load, and unless the load is matched to the line, part (or all) of the incident wave is reflected back toward the generator. At any point on the line, the instantaneous total voltage $v(z,t)$ is the sum of the incident and reflected waves, both of which vary sinusoidally with time. Associated with the voltage difference between the inner and outer conductors of the coaxial line is a radial electric field $\mathbf{E}(z,t)$ that exists in the dielectric material between the conductors, and since $v(z,t)$ varies sinusoidally with time, so does $\mathbf{E}(z,t)$. Furthermore, the current flowing through the inner conductor induces an azimuthal magnetic field $\mathbf{H}(z,t)$ in the dielectric material surrounding it. These coupled fields, $\mathbf{E}(z,t)$ and $\mathbf{H}(z,t)$, constitute an electromagnetic wave. Thus, we can model wave propagation on a transmission line either in terms of the voltages across the line and the currents in its conductors,

†See Fig. 1-17.

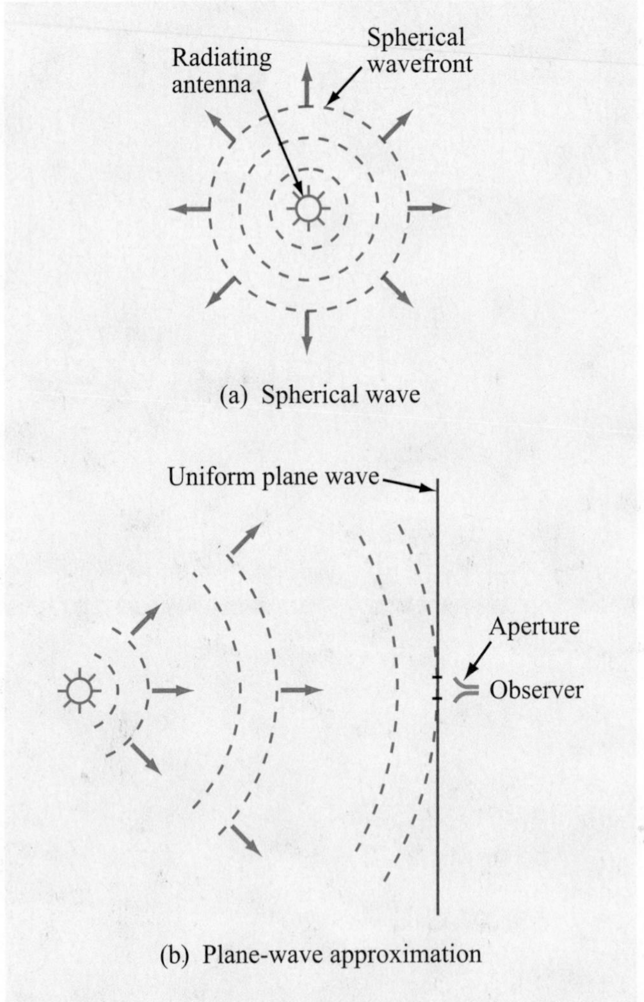

Figure 7-1 Waves radiated by an EM source, such as a light bulb or an antenna, have spherical wavefronts, as in (a); to a distant observer, however, the wavefront across the observer's aperture appears approximately planar, as in (b).

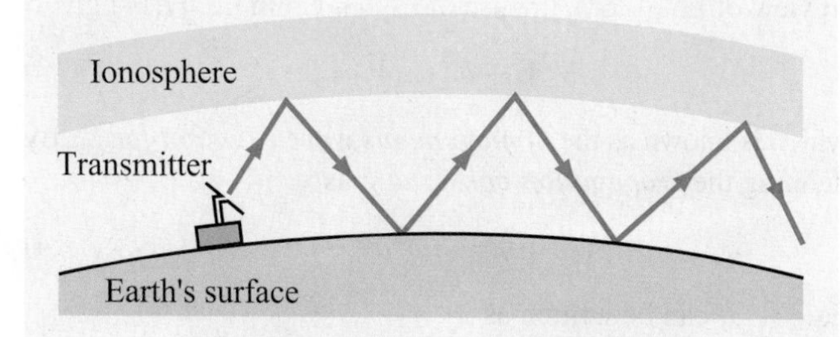

Figure 7-2 The atmospheric layer bounded by the ionosphere at the top and Earth's surface at the bottom forms a guiding structure for the propagation of radio waves in the HF band.

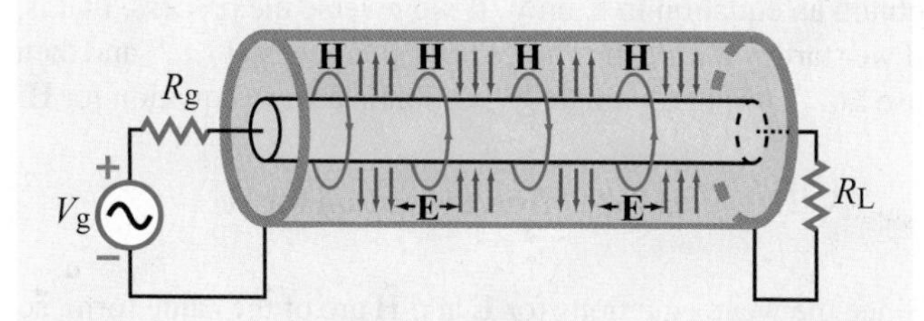

Figure 7-3 A guided electromagnetic wave traveling in a coaxial transmission line consists of time-varying electric and magnetic fields in the dielectric medium between the inner and outer conductors.

or in terms of the electric and magnetic fields in the dielectric medium between the conductors.

In this chapter we focus our attention on wave propagation in unbounded media. Unbounded waves have many practical applications in science and engineering. We consider both lossless and lossy media. Even though strictly speaking uniform plane waves cannot exist, we study them in this chapter to develop a physical understanding of wave propagation in lossless and lossy media. In Chapter 8 we examine how waves, both planar and spherical, are reflected by, and transmitted through, boundaries between dissimilar media. The processes of radiation and reception of waves by antennas are treated in Chapter 9.

7-1 Time-Harmonic Fields

Time-varying electric and magnetic fields (**E**, **D**, **B**, and **H**) and their sources (the charge density $\rho_{\rm v}$ and current density **J**) generally depend on the spatial coordinates (x, y, z) and the time variable t. However, if their time variation is sinusoidal with angular frequency ω, then these quantities can be represented by a phasor that depends on (x, y, z) only. The vector phasor $\widetilde{\mathbf{E}}(x, y, z)$ and the instantaneous field $\mathbf{E}(x, y, z; t)$ it describes are related as

$$\mathbf{E}(x, y, z; t) = \mathfrak{Re}\left[\widetilde{\mathbf{E}}(x, y, z)\, e^{j\omega t}\right]. \tag{7.1}$$

Similar definitions apply to **D**, **B**, and **H**, as well as to $\rho_{\rm v}$ and **J**. For a linear, isotropic, and homogeneous medium with electrical permittivity ϵ, magnetic permeability μ, and conductivity σ, Maxwell's equations (6.1) to (6.4) assume the following form in the phasor domain:

$$\nabla \cdot \widetilde{\mathbf{E}} = \tilde{\rho}_{\rm v}/\epsilon, \tag{7.2a}$$

$$\nabla \times \widetilde{\mathbf{E}} = -j\omega\mu\widetilde{\mathbf{H}}, \tag{7.2b}$$

$$\nabla \cdot \widetilde{\mathbf{H}} = 0, \tag{7.2c}$$

$$\nabla \times \widetilde{\mathbf{H}} = \widetilde{\mathbf{J}} + j\omega\epsilon\widetilde{\mathbf{E}}. \tag{7.2d}$$

To derive these equations we used $\mathbf{D} = \epsilon\mathbf{E}$ and $\mathbf{B} = \mu\mathbf{H}$, and the fact that for time-harmonic quantities, differentiation in the time domain corresponds to multiplication by $j\omega$ in the phasor domain. These equations are the starting point for the subject matter treated in this chapter.

7-1.1 Complex Permittivity

In a medium with conductivity σ, the conduction current density $\widetilde{\mathbf{J}}$ is related to $\widetilde{\mathbf{E}}$ by $\widetilde{\mathbf{J}} = \sigma\widetilde{\mathbf{E}}$. Assuming no other current flows in the medium, Eq. (7.2d) may be written as

$$\nabla \times \widetilde{\mathbf{H}} = \widetilde{\mathbf{J}} + j\omega\epsilon\widetilde{\mathbf{E}} = (\sigma + j\omega\epsilon)\widetilde{\mathbf{E}} = j\omega\left(\epsilon - j\frac{\sigma}{\omega}\right)\widetilde{\mathbf{E}}. \tag{7.3}$$

By defining the ***complex permittivity*** $\epsilon_{\rm c}$ as

$$\epsilon_{\rm c} = \epsilon - j\frac{\sigma}{\omega}, \tag{7.4}$$

Eq. (7.3) can be rewritten as

$$\nabla \times \widetilde{\mathbf{H}} = j\omega\epsilon_{\rm c}\widetilde{\mathbf{E}}. \tag{7.5}$$

Taking the divergence of both sides of Eq. (7.5), and recalling that the divergence of the curl of any vector field vanishes (i.e.,

$\nabla \cdot \nabla \times \widetilde{\mathbf{H}} = 0$), it follows that $\nabla \cdot (j\omega\epsilon_c \widetilde{\mathbf{E}}) = 0$, or $\nabla \cdot \widetilde{\mathbf{E}} = 0$. Comparing this with Eq. (7.2a) implies that $\tilde{\rho}_v = 0$. Upon replacing Eq. (7.2d) with Eq. (7.5) and setting $\tilde{\rho}_v = 0$ in Eq. (7.2a), Maxwell's equations become

$$\nabla \cdot \widetilde{\mathbf{E}} = 0, \tag{7.6a}$$

$$\nabla \times \widetilde{\mathbf{E}} = -j\omega\mu \widetilde{\mathbf{H}}, \tag{7.6b}$$

$$\nabla \cdot \widetilde{\mathbf{H}} = 0, \tag{7.6c}$$

$$\nabla \times \widetilde{\mathbf{H}} = j\omega\epsilon_c \widetilde{\mathbf{E}}. \tag{7.6d}$$

The complex permittivity ϵ_c given by Eq. (7.4) is often written in terms of a real part ϵ' and an imaginary part ϵ''. Thus,

$$\epsilon_c = \epsilon - j\frac{\sigma}{\omega} = \epsilon' - j\epsilon'', \tag{7.7}$$

with

$$\epsilon' = \epsilon, \tag{7.8a}$$

$$\epsilon'' = \frac{\sigma}{\omega}. \tag{7.8b}$$

For a lossless medium with $\sigma = 0$, it follows that $\epsilon'' = 0$ and $\epsilon_c = \epsilon' = \epsilon$.

7-1.2 Wave Equations

Next, we derive wave equations for $\widetilde{\mathbf{E}}$ and $\widetilde{\mathbf{H}}$ and then solve them to obtain explicit expressions for $\widetilde{\mathbf{E}}$ and $\widetilde{\mathbf{H}}$ as a function of the spatial variables (x, y, z). To that end, we start by taking the curl of both sides of Eq. (7.6b) to get

$$\nabla \times (\nabla \times \widetilde{\mathbf{E}}) = -j\omega\mu(\nabla \times \widetilde{\mathbf{H}}). \tag{7.9}$$

Upon substituting Eq. (7.6d) into Eq. (7.9) we obtain

$$\nabla \times (\nabla \times \widetilde{\mathbf{E}}) = -j\omega\mu(j\omega\epsilon_c \widetilde{\mathbf{E}}) = \omega^2\mu\epsilon_c \widetilde{\mathbf{E}}. \tag{7.10}$$

From Eq. (3.113), we know that the curl of the curl of $\widetilde{\mathbf{E}}$ is

$$\nabla \times (\nabla \times \widetilde{\mathbf{E}}) = \nabla(\nabla \cdot \widetilde{\mathbf{E}}) - \nabla^2 \widetilde{\mathbf{E}}, \tag{7.11}$$

where $\nabla^2 \widetilde{\mathbf{E}}$ is the Laplacian of $\widetilde{\mathbf{E}}$, which in Cartesian coordinates is given by

$$\nabla^2 \widetilde{\mathbf{E}} = \left(\frac{\partial^2}{\partial x^2} + \frac{\partial^2}{\partial y^2} + \frac{\partial^2}{\partial z^2}\right) \widetilde{\mathbf{E}}. \tag{7.12}$$

In view of Eq. (7.6a), the use of Eq. (7.11) in Eq. (7.10) gives

$$\nabla^2 \widetilde{\mathbf{E}} + \omega^2\mu\epsilon_c \widetilde{\mathbf{E}} = 0, \tag{7.13}$$

which is known as the ***homogeneous wave equation for*** $\widetilde{\mathbf{E}}$. By defining the ***propagation constant*** γ as

$$\gamma^2 = -\omega^2\mu\epsilon_c, \tag{7.14}$$

Eq. (7.13) can be written as

$$\nabla^2 \widetilde{\mathbf{E}} - \gamma^2 \widetilde{\mathbf{E}} = 0 \quad \textbf{(wave equation for } \widetilde{\mathbf{E}}\textbf{)}. \tag{7.15}$$

To derive Eq. (7.15), we took the curl of both sides of Eq. (7.6b) and then we used Eq. (7.6d) to eliminate $\widetilde{\mathbf{H}}$ and obtain an equation in $\widetilde{\mathbf{E}}$ only. If we reverse the process, that is, if we start by taking the curl of both sides of Eq. (7.6d) and then use Eq. (7.6b) to eliminate $\widetilde{\mathbf{E}}$, we obtain a wave equation for $\widetilde{\mathbf{H}}$:

$$\nabla^2 \widetilde{\mathbf{H}} - \gamma^2 \widetilde{\mathbf{H}} = 0 \quad \textbf{(wave equation for } \widetilde{\mathbf{H}}\textbf{)}. \tag{7.16}$$

Since the wave equations for $\widetilde{\mathbf{E}}$ and $\widetilde{\mathbf{H}}$ are of the same form, so are their solutions.

7-2 Plane-Wave Propagation in Lossless Media

The properties of an electromagnetic wave, such as its phase velocity u_p and wavelength λ, depend on the angular frequency ω and the medium's three constitutive parameters: ϵ, μ, and σ. If the medium is ***nonconducting*** ($\sigma = 0$), the wave does not suffer any attenuation as it travels and hence the medium is said to be ***lossless***. Because in a lossless medium $\epsilon_c = \epsilon$, Eq. (7.14) becomes

$$\gamma^2 = -\omega^2\mu\epsilon. \tag{7.17}$$

For lossless media, it is customary to define the ***wavenumber*** k as

$$k = \omega\sqrt{\mu\epsilon}\,. \tag{7.18}$$

In view of Eq. (7.17), $\gamma^2 = -k^2$ and Eq. (7.15) becomes

$$\nabla^2 \widetilde{\mathbf{E}} + k^2 \widetilde{\mathbf{E}} = 0. \tag{7.19}$$

7-2.1 Uniform Plane Waves

For an electric field phasor defined in Cartesian coordinates as

$$\widetilde{\mathbf{E}} = \hat{\mathbf{x}}\widetilde{E}_x + \hat{\mathbf{y}}\widetilde{E}_y + \hat{\mathbf{z}}\widetilde{E}_z, \tag{7.20}$$

substitution of Eq. (7.12) into Eq. (7.19) gives

$$\left(\frac{\partial^2}{\partial x^2} + \frac{\partial^2}{\partial y^2} + \frac{\partial^2}{\partial z^2}\right)(\hat{\mathbf{x}}\widetilde{E}_x + \hat{\mathbf{y}}\widetilde{E}_y + \hat{\mathbf{z}}\widetilde{E}_z) + k^2(\hat{\mathbf{x}}\widetilde{E}_x + \hat{\mathbf{y}}\widetilde{E}_y + \hat{\mathbf{z}}\widetilde{E}_z) = 0. \tag{7.21}$$

To satisfy Eq. (7.21), each vector component on the left-hand side of the equation must vanish. Hence,

$$\left(\frac{\partial^2}{\partial x^2} + \frac{\partial^2}{\partial y^2} + \frac{\partial^2}{\partial z^2} + k^2\right)\widetilde{E}_x = 0, \tag{7.22}$$

and similar expressions apply to $\widetilde{E}_y$ and $\widetilde{E}_z$.

► A ***uniform plane wave*** is characterized by electric and magnetic fields that have uniform properties at all points across an infinite plane. ◄

If this happens to be the x–y plane, then $\mathbf{E}$ and $\mathbf{H}$ do not vary with x or y. Hence, $\partial\widetilde{E}_x/\partial x = 0$ and $\partial\widetilde{E}_x/\partial y = 0$, and Eq. (7.22) reduces to

$$\frac{d^2\widetilde{E}_x}{dz^2} + k^2\widetilde{E}_x = 0. \tag{7.23}$$

Similar expressions apply to $\widetilde{E}_y$, $\widetilde{H}_x$, and $\widetilde{H}_y$. The remaining components of $\widetilde{\mathbf{E}}$ and $\widetilde{\mathbf{H}}$ are zero; that is, $\widetilde{E}_z = \widetilde{H}_z = 0$. To show that $\widetilde{E}_z = 0$, let us consider the z component of Eq. (7.6d),

$$\hat{\mathbf{z}}\left(\frac{\partial\widetilde{H}_y}{\partial x} - \frac{\partial\widetilde{H}_x}{\partial y}\right) = \hat{\mathbf{z}}j\omega\epsilon\widetilde{E}_z. \tag{7.24}$$

Since $\partial\widetilde{H}_y/\partial x = \partial\widetilde{H}_x/\partial y = 0$, it follows that $\widetilde{E}_z = 0$. A similar examination involving Eq. (7.6b) reveals that $\widetilde{H}_z = 0$.

► This means that a plane wave has no electric-field or magnetic-field components along its direction of propagation. ◄

For the phasor quantity $\widetilde{E}_x$, the general solution of the ordinary differential equation given by Eq. (7.23) is

$$\widetilde{E}_x(z) = \widetilde{E}_x^+(z) + \widetilde{E}_x^-(z) = E_{x0}^+e^{-jkz} + E_{x0}^-e^{jkz}, \tag{7.25}$$

where E_{x0}^+ and E_{x0}^- are constants to be determined from boundary conditions. The solution given by Eq. (7.25) is similar in form to the solution for the phasor voltage $\widetilde{V}(z)$ given by Eq. (2.54a) for the lossless transmission line. The first term in Eq. (7.25), containing the negative exponential e^{-jkz}, represents a wave with amplitude E_{x0}^+ traveling in the $+z$ direction. Likewise, the second term (with e^{jkz}) represents a wave with amplitude E_{x0}^- traveling in the $-z$ direction. Assume for the time being that $\widetilde{\mathbf{E}}$ only has a component along x (i.e., $\widetilde{E}_y = 0$) and that $\widetilde{E}_x$ is associated with a wave traveling in the $+z$ direction only (i.e., $E_{x0}^- = 0$). Under these conditions,

$$\widetilde{\mathbf{E}}(z) = \hat{\mathbf{x}}\widetilde{E}_x^+(z) = \hat{\mathbf{x}}E_{x0}^+e^{-jkz}. \tag{7.26}$$

To find the magnetic field $\widetilde{\mathbf{H}}$ associated with this wave, we apply Eq. (7.6b) with $\widetilde{E}_y = \widetilde{E}_z = 0$:

$$\nabla \times \widetilde{\mathbf{E}} = \begin{vmatrix} \hat{\mathbf{x}} & \hat{\mathbf{y}} & \hat{\mathbf{z}} \\ \dfrac{\partial}{\partial x} & \dfrac{\partial}{\partial y} & \dfrac{\partial}{\partial z} \\ \widetilde{E}_x^+(z) & 0 & 0 \end{vmatrix} = -j\omega\mu(\hat{\mathbf{x}}\widetilde{H}_x + \hat{\mathbf{y}}\widetilde{H}_y + \hat{\mathbf{z}}\widetilde{H}_z). \tag{7.27}$$

For a uniform plane wave traveling in the $+z$ direction,

$$\partial E_x^+(z)/\partial x = \partial E_x^+(z)/\partial y = 0.$$

Hence, Eq. (7.27) gives

$$\widetilde{H}_x = 0, \tag{7.28a}$$

$$\widetilde{H}_y = \frac{1}{-j\omega\mu}\frac{\partial\widetilde{E}_x^+(z)}{\partial z}, \tag{7.28b}$$

$$\widetilde{H}_z = \frac{1}{-j\omega\mu}\frac{\partial E_x^+(z)}{\partial y} = 0. \tag{7.28c}$$

Use of Eq. (7.26) in Eq. (7.28b) gives

$$\widetilde{H}_y(z) = \frac{k}{\omega\mu}E_{x0}^+e^{-jkz} = H_{y0}^+e^{-jkz}, \tag{7.29}$$

where H_{y0}^+ is the amplitude of $\widetilde{H}_y(z)$ and is given by

$$H_{y0}^+ = \frac{k}{\omega\mu}E_{x0}^+. \tag{7.30}$$

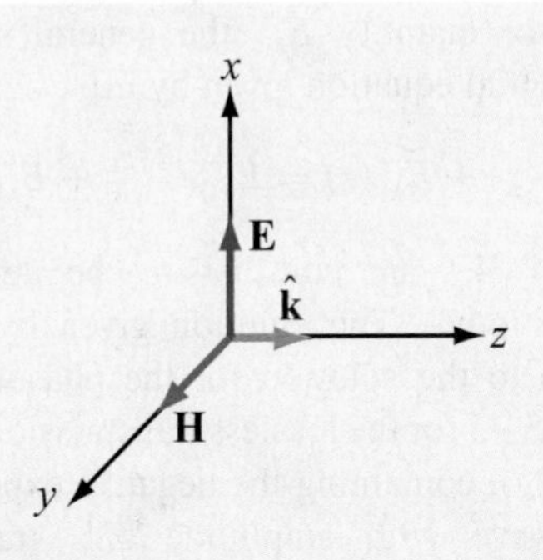

Figure 7-4 A transverse electromagnetic (TEM) wave propagating in the direction $\hat{\mathbf{k}} = \hat{\mathbf{z}}$. For all TEM waves, $\hat{\mathbf{k}}$ is parallel to $\mathbf{E} \times \mathbf{H}$.

For a wave traveling from the source toward the load on a transmission line, the amplitudes of its voltage and current phasors, V_0^+ and I_0^+, are related by the characteristic impedance of the line, Z_0. A similar connection exists between the electric and magnetic fields of an electromagnetic wave. The ***intrinsic impedance*** of a lossless medium is defined as

$$\eta = \frac{\omega\mu}{k} = \frac{\omega\mu}{\omega\sqrt{\mu\epsilon}} = \sqrt{\frac{\mu}{\epsilon}} \quad (\Omega), \tag{7.31}$$

where we used the expression for k given by Eq. (7.18).

In view of Eq. (7.31), the electric and magnetic fields of a $+z$-propagating plane wave with $\mathbf{E}$ field along $\hat{\mathbf{x}}$ are:

$$\widetilde{\mathbf{E}}(z) = \hat{\mathbf{x}}\widetilde{E}_x^+(z) = \hat{\mathbf{x}}E_{x0}^+ e^{-jkz}, \tag{7.32a}$$

$$\widetilde{\mathbf{H}}(z) = \hat{\mathbf{y}}\frac{\widetilde{E}_x^+(z)}{\eta} = \hat{\mathbf{y}}\frac{E_{x0}^+}{\eta}e^{-jkz}. \tag{7.32b}$$

► The electric and magnetic fields of a plane wave are perpendicular to each other, and both are perpendicular to the direction of wave travel (Fig. 7-4). These attributes qualify the wave as a ***transverse electromagnetic (TEM)***. ◄

Other examples of TEM waves include waves traveling on coaxial transmission lines ($\mathbf{E}$ is along $\hat{\mathbf{r}}$, $\mathbf{H}$ is along $\hat{\boldsymbol{\phi}}$, and the direction of travel is along $\hat{\mathbf{z}}$) and spherical waves radiated by antennas.

In the general case, E_{x0}^+ is a complex quantity with magnitude $|E_{x0}^+|$ and phase angle ϕ^+. That is,

$$E_{x0}^+ = |E_{x0}^+|e^{j\phi^+}. \tag{7.33}$$

The instantaneous electric and magnetic fields therefore are

$$\mathbf{E}(z,t) = \mathfrak{Re}\left[\widetilde{\mathbf{E}}(z)\, e^{j\omega t}\right] = \hat{\mathbf{x}}|E_{x0}^+|\cos(\omega t - kz + \phi^+) \quad \text{(V/m)}, \tag{7.34a}$$

and

$$\mathbf{H}(z,t) = \mathfrak{Re}\left[\widetilde{\mathbf{H}}(z)\, e^{j\omega t}\right] = \hat{\mathbf{y}}\frac{|E_{x0}^+|}{\eta}\cos(\omega t - kz + \phi^+) \quad \text{(A/m)}. \tag{7.34b}$$

Because $\mathbf{E}(z,t)$ and $\mathbf{H}(z,t)$ exhibit the same functional dependence on z and t, they are said to be ***in phase***; when the amplitude of one of them reaches a maximum, the amplitude of the other does so too. The fact that $\widetilde{\mathbf{E}}$ and $\widetilde{\mathbf{H}}$ are in phase is characteristic of waves propagating in lossless media.

From the material on wave motion presented in Section 1-4, we deduce that the ***phase velocity*** of the wave is

$$u_p = \frac{\omega}{k} = \frac{\omega}{\omega\sqrt{\mu\epsilon}} = \frac{1}{\sqrt{\mu\epsilon}} \quad \text{(m/s)}, \tag{7.35}$$

and its wavelength is

$$\lambda = \frac{2\pi}{k} = \frac{u_p}{f} \quad \text{(m)}. \tag{7.36}$$

In vacuum, $\epsilon = \epsilon_0$ and $\mu = \mu_0$, and the phase velocity u_p and the intrinsic impedance η given by Eq. (7.31) are

$$u_p = c = \frac{1}{\sqrt{\mu_0\epsilon_0}} = 3 \times 10^8 \quad \text{(m/s)}, \tag{7.37}$$

$$\eta = \eta_0 = \sqrt{\frac{\mu_0}{\epsilon_0}} = 377\ (\Omega) \approx 120\pi \quad (\Omega), \tag{7.38}$$

where c is the velocity of light and η_0 is called the ***intrinsic impedance of free space***.

Example 7-1: EM Plane Wave in Air

This example is analogous to the "Sound Wave in Water" problem given by Example 1-1.

The electric field of a 1 MHz plane wave traveling in the $+z$ direction in air points along the x direction. If this field reaches a peak value of 1.2π (mV/m) at $t = 0$ and $z = 50$ m, obtain expressions for $\mathbf{E}(z, t)$ and $\mathbf{H}(z, t)$ and then plot them as a function of z at $t = 0$.

Solution: At $f = 1$ MHz, the wavelength in air is

$$\lambda = \frac{c}{f} = \frac{3 \times 10^8}{1 \times 10^6} = 300 \text{ m},$$

and the corresponding wavenumber is $k = (2\pi/300)$ (rad/m). The general expression for an x-directed electric field traveling in the $+z$ direction is given by Eq. (7.34a) as

$$\begin{aligned}\mathbf{E}(z, t) &= \hat{\mathbf{x}}|E_{x0}^{+}| \cos(\omega t - kz + \phi^{+}) \\ &= \hat{\mathbf{x}}\, 1.2\pi \cos\left(2\pi \times 10^6 t - \frac{2\pi z}{300} + \phi^{+}\right) \quad \text{(mV/m)}.\end{aligned}$$

The field $\mathbf{E}(z, t)$ is maximum when the argument of the cosine function equals zero or a multiple of 2π. At $t = 0$ and $z = 50$ m, this condition yields

$$-\frac{2\pi \times 50}{300} + \phi^{+} = 0 \quad \text{or} \quad \phi^{+} = \frac{\pi}{3}.$$

Hence,

$$\mathbf{E}(z, t) = \hat{\mathbf{x}}\, 1.2\pi \cos\left(2\pi \times 10^6 t - \frac{2\pi z}{300} + \frac{\pi}{3}\right) \quad \text{(mV/m)},$$

and from Eq. (7.34b) we have

$$\begin{aligned}\mathbf{H}(z, t) &= \hat{\mathbf{y}}\, \frac{E(z, t)}{\eta_0} \\ &= \hat{\mathbf{y}}\, 10 \cos\left(2\pi \times 10^6 t - \frac{2\pi z}{300} + \frac{\pi}{3}\right) \quad (\mu\text{A/m}),\end{aligned}$$

where we have used the approximation $\eta_0 \approx 120\pi$ (Ω).

At $t = 0$,

$$\mathbf{E}(z, 0) = \hat{\mathbf{x}}\, 1.2\pi \cos\left(\frac{2\pi z}{300} - \frac{\pi}{3}\right) \quad \text{(mV/m)},$$

$$\mathbf{H}(z, 0) = \hat{\mathbf{y}}\, 10 \cos\left(\frac{2\pi z}{300} - \frac{\pi}{3}\right) \quad (\mu\text{A/m}).$$

Plots of $\mathbf{E}(z, 0)$ and $\mathbf{H}(z, 0)$ as a function of z are shown in Fig. 7-5.

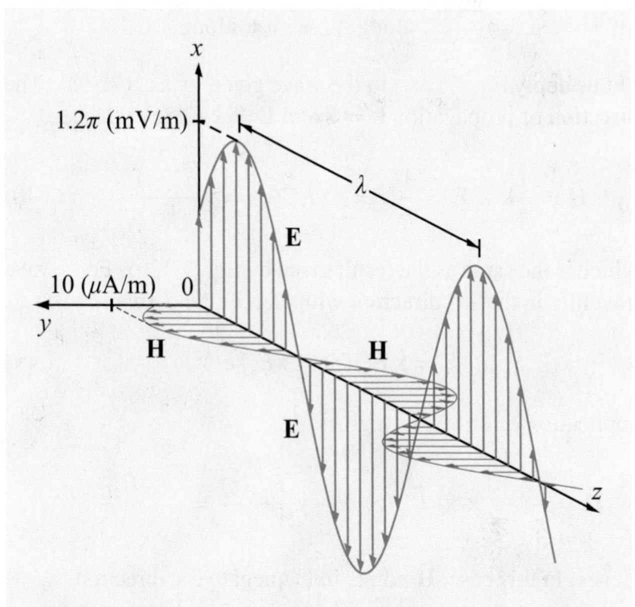

Figure 7-5 Spatial variations of **E** and **H** at $t = 0$ for the plane wave of Example 7-1.

7-2.2 General Relation between E and H

It can be shown that, for any uniform plane wave traveling in an arbitrary direction denoted by the unit vector $\hat{\mathbf{k}}$, the electric and magnetic field phasors $\widetilde{\mathbf{E}}$ and $\widetilde{\mathbf{H}}$ are related as

$$\widetilde{\mathbf{H}} = \frac{1}{\eta}\, \hat{\mathbf{k}} \times \widetilde{\mathbf{E}}, \quad (7.39a)$$

$$\widetilde{\mathbf{E}} = -\eta\, \hat{\mathbf{k}} \times \widetilde{\mathbf{H}}. \quad (7.39b)$$

► The following right-hand rule applies: when we rotate the four fingers of the right hand from the direction of **E** toward that of **H**, the thumb points in the direction of wave travel, $\hat{\mathbf{k}}$. ◄

The relations given by Eqs. (7.39a) and (b) are valid not only for lossless media but for lossy ones as well. As we see later in Section 7-4, the expression for η of a lossy medium is different from that given by Eq. (7.31). As long as the expression used for η is appropriate for the medium in which the wave is traveling, the relations given by Eqs. (7.39a) and (b) always hold.

(a) Wave propagating along $+z$ with $\mathbf{E}$ along $\hat{\mathbf{x}}$

Let us apply Eq. (7.39a) to the wave given by Eq. (7.32a). The direction of propagation $\hat{\mathbf{k}} = \hat{\mathbf{z}}$ and $\widetilde{\mathbf{E}} = \hat{\mathbf{x}}\,\widetilde{E}_x^+(z)$. Hence,

$$\widetilde{\mathbf{H}} = \frac{1}{\eta}\hat{\mathbf{k}} \times \widetilde{\mathbf{E}} = \frac{1}{\eta}(\hat{\mathbf{z}} \times \hat{\mathbf{x}})\,\widetilde{E}_x^+(z) = \hat{\mathbf{y}}\frac{\widetilde{E}_x^+(z)}{\eta}, \quad (7.40)$$

which is the same as the result given by Eq. (7.32b). For a wave traveling in the $-z$ direction with electric field given by

$$\widetilde{\mathbf{E}} = \hat{\mathbf{x}}\,\widetilde{E}_x^-(z) = \hat{\mathbf{x}}E_{x0}^- e^{jkz}, \quad (7.41)$$

application of Eq. (7.39a) gives

$$\widetilde{\mathbf{H}} = \frac{1}{\eta}(-\hat{\mathbf{z}} \times \hat{\mathbf{x}})\,\widetilde{E}_x^-(z) = -\hat{\mathbf{y}}\,\frac{\widetilde{E}_x^-(z)}{\eta} = -\hat{\mathbf{y}}\,\frac{E_{x0}^-}{\eta}\,e^{jkz}. \quad (7.42)$$

Hence, in this case, $\widetilde{\mathbf{H}}$ points in the negative y direction.

(b) Wave propagating along $+z$ with $\mathbf{E}$ along $\hat{\mathbf{x}}$ and $\hat{\mathbf{y}}$

In general, a uniform plane wave traveling in the $+z$ direction may have both x and y components, in which case $\widetilde{\mathbf{E}}$ is given by

$$\widetilde{\mathbf{E}} = \hat{\mathbf{x}}\,\widetilde{E}_x^+(z) + \hat{\mathbf{y}}\,\widetilde{E}_y^+(z), \quad (7.43a)$$

and the associated magnetic field is

$$\widetilde{\mathbf{H}} = \hat{\mathbf{x}}\,\widetilde{H}_x^+(z) + \hat{\mathbf{y}}\,\widetilde{H}_y^+(z). \quad (7.43b)$$

Application of Eq. (7.39a) gives

$$\widetilde{\mathbf{H}} = \frac{1}{\eta}\,\hat{\mathbf{z}} \times \widetilde{\mathbf{E}} = -\hat{\mathbf{x}}\,\frac{\widetilde{E}_y^+(z)}{\eta} + \hat{\mathbf{y}}\,\frac{\widetilde{E}_x^+(z)}{\eta}. \quad (7.44)$$

By equating Eq. (7.43b) to Eq. (7.44), we have

$$\widetilde{H}_x^+(z) = -\frac{\widetilde{E}_y^+(z)}{\eta}, \qquad \widetilde{H}_y^+(z) = \frac{\widetilde{E}_x^+(z)}{\eta}. \quad (7.45)$$

These results are illustrated in **Fig. 7-6**. The wave may be considered the sum of two waves, one with electric and magnetic components (E_x^+, H_y^+), and another with components (E_y^+, H_x^+). In general, a TEM wave may have an electric field in any direction in the plane orthogonal to the direction of wave travel, and the associated magnetic field is also in the same plane and its direction is dictated by Eq. (7.39a).

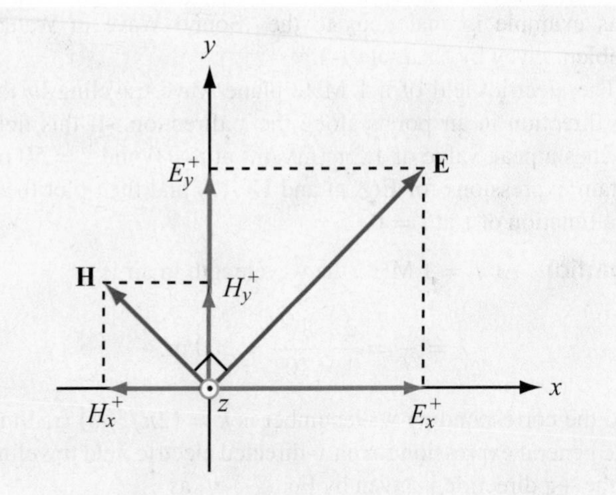

Figure 7-6 The wave (**E**, **H**) is equivalent to the sum of two waves, one with fields (E_x^+, H_y^+) and another with (E_y^+, H_x^+), with both traveling in the $+z$ direction.

Concept Question 7-1: What is a uniform plane wave? Describe its properties, both physically and mathematically. Under what conditions is it appropriate to treat a spherical wave as a plane wave?

Concept Question 7-2: Since $\widetilde{\mathbf{E}}$ and $\widetilde{\mathbf{H}}$ are governed by wave equations of the same form [Eqs. (7.15) and (7.16)], does it follow that $\widetilde{\mathbf{E}} = \widetilde{\mathbf{H}}$? Explain.

Module 7.1 Linking E to H Select the directions and magnitudes of E and H and observe the resultant wave vector.

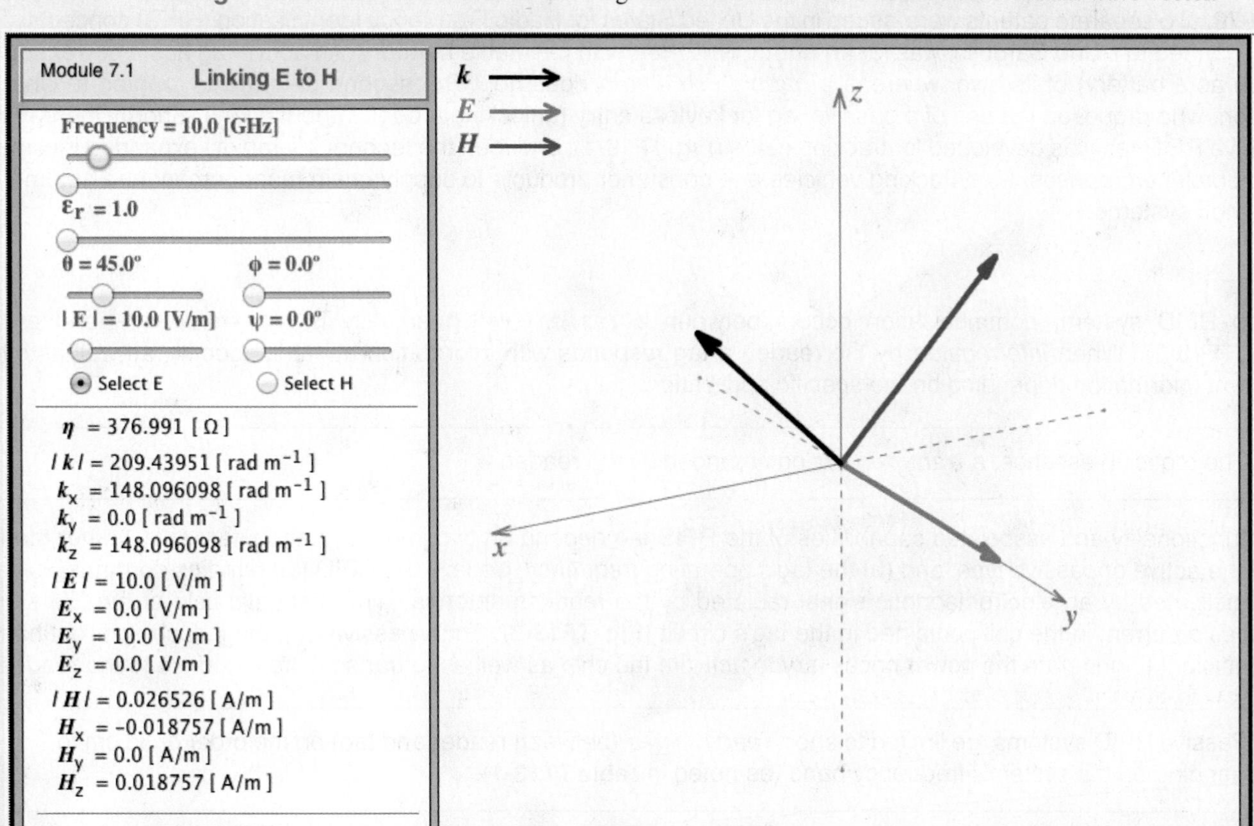

Concept Question 7-3: If a TEM wave is traveling in the $\hat{\mathbf{y}}$ direction, can its electric field have components along $\hat{\mathbf{x}}$, $\hat{\mathbf{y}}$, and $\hat{\mathbf{z}}$? Explain.

Exercise 7-1: A 10 MHz uniform plane wave is traveling in a nonmagnetic medium with $\mu = \mu_0$ and $\epsilon_r = 9$. Find (a) the phase velocity, (b) the wavenumber, (c) the wavelength in the medium, and (d) the intrinsic impedance of the medium.

Answer: (a) $u_p = 1 \times 10^8$ m/s, (b) $k = 0.2\pi$ rad/m, (c) $\lambda = 10$ m, (d) $\eta = 125.67\ \Omega$. (See EM.)

Exercise 7-2: The electric field phasor of a uniform plane wave traveling in a lossless medium with an intrinsic impedance of 188.5 Ω is given by $\widetilde{\mathbf{E}} = \hat{\mathbf{z}}\, 10e^{-j4\pi y}$ (mV/m). Determine (a) the associated magnetic field phasor and (b) the instantaneous expression for $\mathbf{E}(y, t)$ if the medium is nonmagnetic ($\mu = \mu_0$).

Answer: (a) $\widetilde{\mathbf{H}} = \hat{\mathbf{x}}\, 53e^{-j4\pi y}$ (μA/m),
(b) $\mathbf{E}(y, t) = \hat{\mathbf{z}}\, 10 \cos(6\pi \times 10^8 t - 4\pi y)$ (mV/m). (See EM.)

Technology Brief 13: RFID Systems

In 1973, two separate patents were issued in the United States for Radio Frequency Identification (RFID) concepts. The first, granted to Mario Cardullo, was for an ***active RFID tag*** with rewritable memory. An active tag has a power source (such as a battery) of its own, whereas a ***passive RFID tag*** does not. The second patent was granted to Charles Walton, who proposed the use of a passive tag for keyless entry (unlocking a door without a key). Shortly thereafter a passive RFID tag was developed for tracking cattle (**Fig. TF13-1**), and then the technology rapidly expanded into many commercial enterprises, from tracking vehicles and consumer products to supply chain management and automobile anti-theft systems.

RFID System Overview

In an RFID system, communication occurs between a ***reader***—which actually is a ***transceiver***—and a ***tag*** (**Fig. TF13-2**). When *interrogated* by the reader, a tag responds with information about its identity, as well as other relevant information depending on the specific application.

▶ The tag is, in essence, a ***transponder*** commanded by the reader. ◀

The functionality and associated capabilities of the RFID tag depend on two important attributes: (a) whether the tag is of the active or passive type, and (b) the tag's operating frequency. Usually the RFID tag remains dormant (*asleep*) until activated by an electromagnetic signal radiated by the reader's antenna. The magnetic field of the EM signal induces a current in the coil contained in the tag's circuit (**Fig. TF13-3**). For a passive tag, the induced current has to be sufficient to generate the power necessary to activate the chip as well as to transmit the response to the reader.

▶ Passive RFID systems are limited to short ***read ranges*** (between reader and tag) on the order of 30 cm to 3 m, depending on the system's frequency band (as noted in **Table TT13-1**). ◀

The obvious advantage of active RFID systems is that they can operate over greater distances and do not require reception of a signal from the reader's antenna to get activated. However, active tags are significantly more expensive to fabricate than their passive cousins.

RFID Frequency Bands

Table TT13-1 provides a comparison among the four frequency bands commonly used for RFID systems. Generally speaking, the higher-frequency tags can operate over longer read ranges and can carry higher data rates, but they are more expensive to fabricate.

Figure TF13-1 Passive RFID tags were developed in the 1970s for tracking cows.

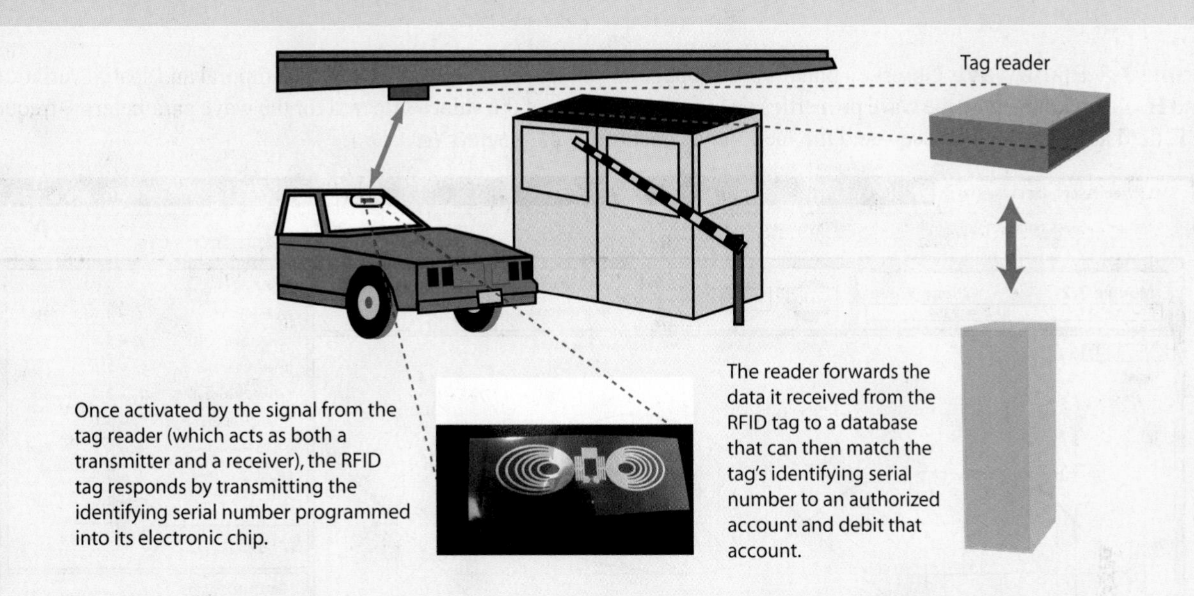

Figure TF13-2 How an RFID system works is illustrated through this EZ-Pass example. The UHF RFID shown is courtesy of Prof. C. F. Huang of Tatung University, Taiwan.

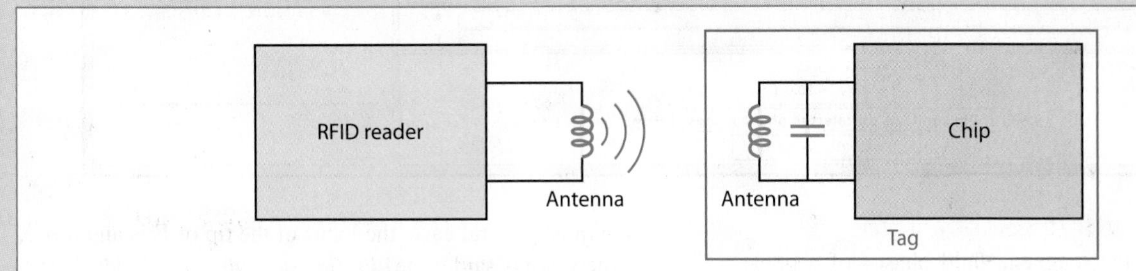

Figure TF13-3 Simplified diagram for how the RFID reader communicates with the tag. At the two lower carrier frequencies commonly used for RFID communication, namely 125 kHz and 13.56 MHz, coil inductors act as magnetic antennas. In systems designed to operate at higher frequencies (900 MHz and 2.54 GHz), dipole antennas are used instead.

Table TT13-1 Comparison of RFID frequency bands.

Band	LF	HF	UHF	Microwave
RFID frequency	125–134 kHz	13.56 MHz	865–956 MHz	2.45 GHz
Read range	≤ 0.5 m	≤ 1.5 m	≤ 5 m	≤ 10 m
Data rate	1 kbit/s	25 kbit/s	30 kbit/s	100 kbit/s
Typical applications	• Animal ID • Automobile key/antitheft • Access control	• Smart cards • Article surveillance • Airline baggage tracking • Library book tracking	• Supply chain management • Logistics	• Vehicle toll collection • Railroad car monitoring

Module 7.2 Plane Wave Observe a plane wave propagating along the z direction; note the temporal and spatial variations of **E** and **H**, and examine how the wave properties change as a function of the values selected for the wave parameters—frequency and **E** field amplitude and phase—and the medium's constitutive parameters (ϵ, μ, σ).

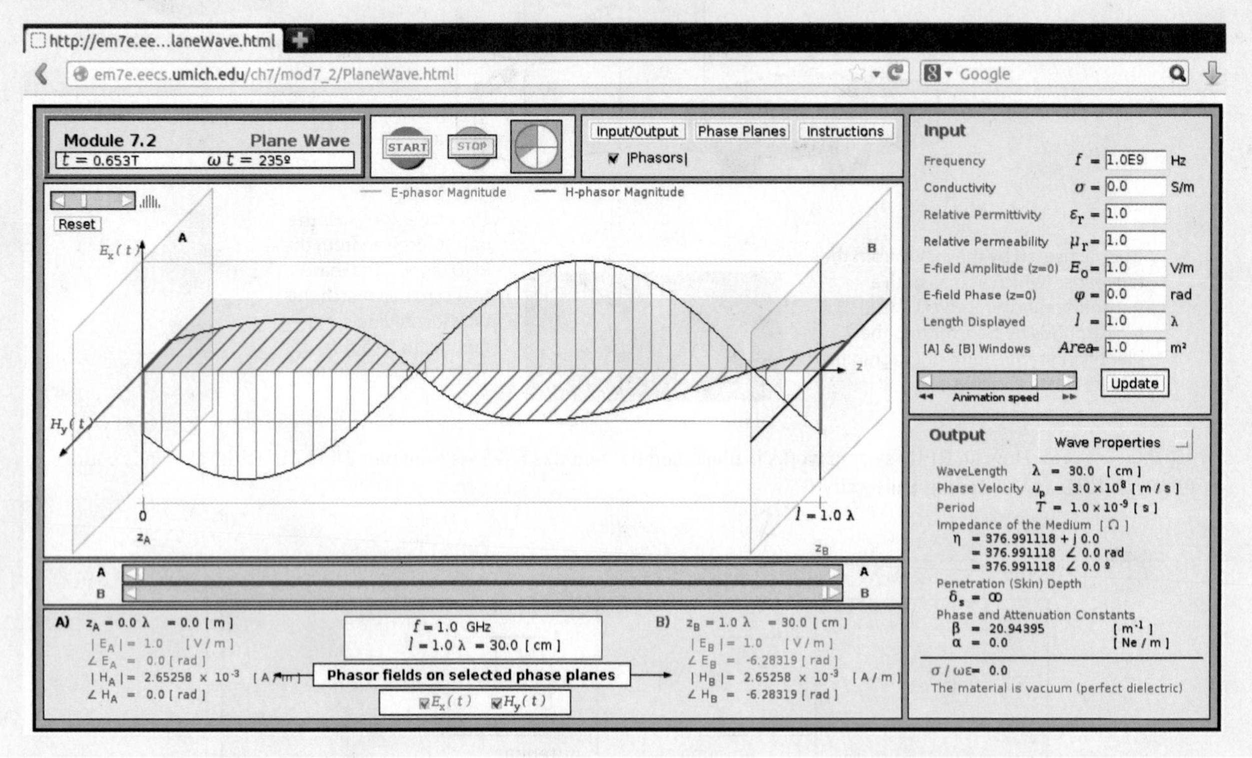

Exercise 7-3: If the magnetic field phasor of a plane wave traveling in a medium with intrinsic impedance $\eta = 100\ \Omega$ is given by $\widetilde{\mathbf{H}} = (\hat{\mathbf{y}}\, 10 + \hat{\mathbf{z}}\, 20)e^{-j4x}$ (mA/m), find the associated electric field phasor.

Answer: $\widetilde{\mathbf{E}} = (-\hat{\mathbf{z}} + \hat{\mathbf{y}}\, 2)e^{-j4x}$ (V/m). (See EM.)

Exercise 7-4: Repeat Exercise 7-3 for a magnetic field given by $\widetilde{\mathbf{H}} = \hat{\mathbf{y}}(10e^{-j3x} - 20e^{j3x})$ (mA/m).

Answer: $\widetilde{\mathbf{E}} = -\hat{\mathbf{z}}(e^{-j3x} + 2e^{j3x})$ (V/m). (See EM.)

7-3 Wave Polarization

► The ***polarization*** of a uniform plane wave describes the locus traced by the tip of the **E** vector (in the plane orthogonal to the direction of propagation) at a given point in space as a function of time. ◄

In the most general case, the locus of the tip of **E** is an ellipse, and the wave is said to be ***elliptically polarized***. Under certain conditions, the ellipse may degenerate into a circle or a straight line, in which case the ***polarization state*** is called ***circular*** or ***linear***, respectively.

It was shown in Section 7-2 that the z components of the electric and magnetic fields of a z-propagating plane wave are both zero. Hence, in the most general case, the electric field phasor $\widetilde{\mathbf{E}}(z)$ of a $+z$-propagating plane wave may consist of an x component, $\hat{\mathbf{x}}\, \widetilde{E}_x(z)$, and a y component, $\hat{\mathbf{y}}\, \widetilde{E}_y(z)$, or

$$\widetilde{\mathbf{E}}(z) = \hat{\mathbf{x}}\widetilde{E}_x(z) + \hat{\mathbf{y}}\widetilde{E}_y(z), \tag{7.46}$$

with

$$\widetilde{E}_x(z) = E_{x0}e^{-jkz}, \tag{7.47a}$$

$$\widetilde{E}_y(z) = E_{y0}e^{-jkz}, \tag{7.47b}$$

where E_{x0} and E_{y0} are the amplitudes of $\widetilde{E}_x(z)$ and $\widetilde{E}_y(z)$, respectively. For the sake of simplicity, the plus sign superscript has been suppressed; the negative sign in e^{-jkz} is sufficient to remind us that the wave is traveling in the positive z direction.

The two amplitudes E_{x0} and E_{y0} are, in general, complex quantities, each characterized by a magnitude and a phase angle. The phase of a wave is defined relative to a reference state, such as $z = 0$ and $t = 0$ or any other combination of z and t. As will become clear from the discussion that follows, the polarization of the wave described by Eqs. (7.46) and (7.47) depends on the phase of E_{y0} relative to that of E_{x0}, but not on the absolute phases of E_{x0} and E_{y0}. Hence, for convenience, we assign E_{x0} a phase of zero and denote the phase of E_{y0}, relative to that of E_{x0}, as δ. Thus, δ is the *phase difference* between the y and x components of $\widetilde{\mathbf{E}}$. Accordingly, we define E_{x0} and E_{y0} as

$$E_{x0} = a_x, \tag{7.48a}$$

$$E_{y0} = a_y e^{j\delta}, \tag{7.48b}$$

where $a_x = |E_{x0}| \geq 0$ and $a_y = |E_{y0}| \geq 0$ are the magnitudes of E_{x0} and E_{y0}, respectively. Thus, by definition, a_x and a_y may not assume negative values. Using Eqs. (7.48a) and (7.48b) in Eqs. (7.47a) and (7.47b), the total electric field phasor is

$$\widetilde{\mathbf{E}}(z) = (\hat{\mathbf{x}} a_x + \hat{\mathbf{y}} a_y e^{j\delta}) e^{-jkz}, \tag{7.49}$$

and the corresponding instantaneous field is

$$\begin{aligned}\mathbf{E}(z,t) &= \mathfrak{Re}\left[\widetilde{\mathbf{E}}(z)\, e^{j\omega t}\right] \\ &= \hat{\mathbf{x}} a_x \cos(\omega t - kz) \\ &\quad + \hat{\mathbf{y}} a_y \cos(\omega t - kz + \delta).\end{aligned} \tag{7.50}$$

When characterizing an electric field at a given point in space, two of its attributes that are of particular interest are its magnitude and direction. The magnitude of $\mathbf{E}(z,t)$ is

$$\begin{aligned}|\mathbf{E}(z,t)| &= [E_x^2(z,t) + E_y^2(z,t)]^{1/2} \\ &= [a_x^2 \cos^2(\omega t - kz) \\ &\quad + a_y^2 \cos^2(\omega t - kz + \delta)]^{1/2}.\end{aligned} \tag{7.51}$$

The electric field $\mathbf{E}(z,t)$ has components along the x and y directions. At a specific position z, the direction of $\mathbf{E}(z,t)$ is characterized by its ***inclination angle*** ψ, defined with respect to the x axis and given by

$$\psi(z,t) = \tan^{-1}\left[\frac{E_y(z,t)}{E_x(z,t)}\right]. \tag{7.52}$$

In the general case, both the intensity of $\mathbf{E}(z,t)$ and its direction are functions of z and t. Next, we examine some special cases.

7-3.1 Linear Polarization

► A wave is said to be linearly polarized if for a fixed z, the tip of $\mathbf{E}(z,t)$ traces a straight line segment as a function of time. This happens when $E_x(z,t)$ and $E_y(z,t)$ are ***in phase*** (i.e., $\delta = 0$) or ***out of phase*** ($\delta = \pi$). ◄

Under these conditions Eq. (7.50) simplifies to

$$\mathbf{E}(0,t) = (\hat{\mathbf{x}} a_x + \hat{\mathbf{y}} a_y)\cos(\omega t - kz) \quad \textbf{(in phase)}, \tag{7.53a}$$

$$\mathbf{E}(0,t) = (\hat{\mathbf{x}} a_x - \hat{\mathbf{y}} a_y)\cos(\omega t - kz) \quad \textbf{(out of phase)}. \tag{7.53b}$$

Let us examine the out-of-phase case. The field's magnitude is

$$|\mathbf{E}(z,t)| = [a_x^2 + a_y^2]^{1/2} |\cos(\omega t - kz)|, \tag{7.54a}$$

and the inclination angle is

$$\psi = \tan^{-1}\left(\frac{-a_y}{a_x}\right) \quad \textbf{(out of phase)}. \tag{7.54b}$$

We note that ψ is independent of both z and t. **Figure 7-7** displays the line segment traced by the tip of $\mathbf{E}$ at $z = 0$ over a half of a cycle. The trace would be the same at any other value of z as well. At $z = 0$ and $t = 0$, $|E(0,0)| = [a_x^2 + a_y^2]^{1/2}$. The length of the vector representing $\mathbf{E}(0,t)$ decreases to zero at $\omega t = \pi/2$. The vector then reverses direction and increases in magnitude to $[a_x^2 + a_y^2]^{1/2}$ in the second quadrant of the x–y plane at $\omega t = \pi$. Since ψ is independent of both z and t, $\mathbf{E}(z,t)$ maintains a direction along the line making an angle ψ with the x axis, while oscillating back and forth across the origin.

If $a_y = 0$, then $\psi = 0°$ or $180°$, and the wave is x-polarized; conversely, if $a_x = 0$, then $\psi = 90°$ or $-90°$, and the wave is y-polarized.

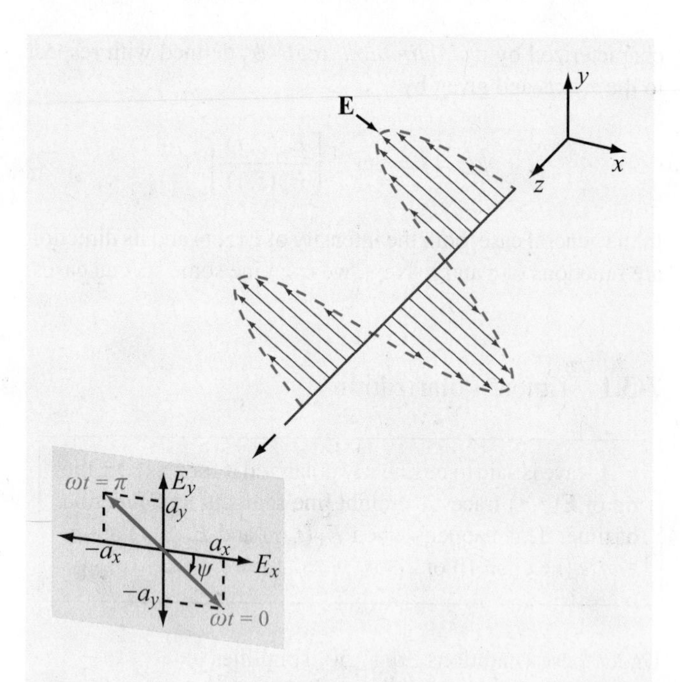

Figure 7-7 Linearly polarized wave traveling in the $+z$ direction (out of the page).

7-3.2 Circular Polarization

We now consider the special case when the magnitudes of the x and y components of $\widetilde{\mathbf{E}}(z)$ are equal, and the phase difference $\delta = \pm\pi/2$. For reasons that become evident shortly, the wave polarization is called ***left-hand circular*** when $\delta = \pi/2$, and ***right-hand circular*** when $\delta = -\pi/2$.

Left-hand circular (LHC) polarization

For $a_x = a_y = a$ and $\delta = \pi/2$, Eqs. (7.49) and (7.50) become

$$\widetilde{\mathbf{E}}(z) = (\hat{\mathbf{x}}a + \hat{\mathbf{y}}ae^{j\pi/2})e^{-jkz} = a(\hat{\mathbf{x}} + j\hat{\mathbf{y}})e^{-jkz}, \quad (7.55a)$$

$$\begin{aligned}\mathbf{E}(z,t) &= \mathfrak{Re}\left[\widetilde{\mathbf{E}}(z)\, e^{j\omega t}\right] \\ &= \hat{\mathbf{x}}a\cos(\omega t - kz) + \hat{\mathbf{y}}a\cos(\omega t - kz + \pi/2) \\ &= \hat{\mathbf{x}}a\cos(\omega t - kz) - \hat{\mathbf{y}}a\sin(\omega t - kz). \quad (7.55b)\end{aligned}$$

The corresponding field magnitude and inclination angle are

$$\begin{aligned}|\mathbf{E}(z,t)| &= \left[E_x^2(z,t) + E_y^2(z,t)\right]^{1/2} \\ &= [a^2\cos^2(\omega t - kz) + a^2\sin^2(\omega t - kz)]^{1/2} \\ &= a \quad (7.56a)\end{aligned}$$

and

$$\begin{aligned}\psi(z,t) &= \tan^{-1}\left[\frac{E_y(z,t)}{E_x(z,t)}\right] \\ &= \tan^{-1}\left[\frac{-a\sin(\omega t - kz)}{a\cos(\omega t - kz)}\right] = -(\omega t - kz). \quad (7.56b)\end{aligned}$$

We observe that the magnitude of $\mathbf{E}$ is independent of both z and t, whereas ψ depends on both variables. These functional dependencies are the converse of those for the linear polarization case.

At $z = 0$, Eq. (7.56b) gives $\psi = -\omega t$; the negative sign implies that the inclination angle decreases as time increases. As illustrated in **Fig. 7-8(a)**, the tip of $\mathbf{E}(t)$ traces a circle in the x–y plane and rotates in a clockwise direction as a function of time (when viewing the wave approaching). Such a wave is called ***left-hand circularly polarized*** because, when the thumb of the left hand points along the direction of propagation (the z direction in this case), the other four fingers point in the direction of rotation of $\mathbf{E}$.

Right-hand circular (RHC) polarization

For $a_x = a_y = a$ and $\delta = -\pi/2$, we have

$$|\mathbf{E}(z,t)| = a, \qquad \psi = (\omega t - kz). \quad (7.57)$$

The trace of $\mathbf{E}(0,t)$ as a function of t is shown in **Fig. 7-8(b)**. For RHC polarization, the fingers of the right hand point in the direction of rotation of $\mathbf{E}$ when the thumb is along the propagation direction. **Figure 7-9** depicts a right-hand circularly polarized wave radiated by a helical antenna.

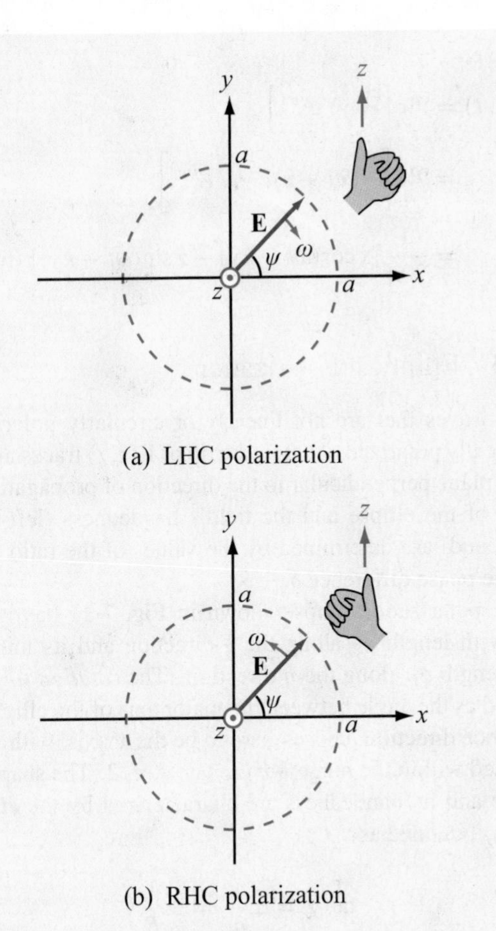

Figure 7-8 Circularly polarized plane waves propagating in the $+z$ direction (out of the page).

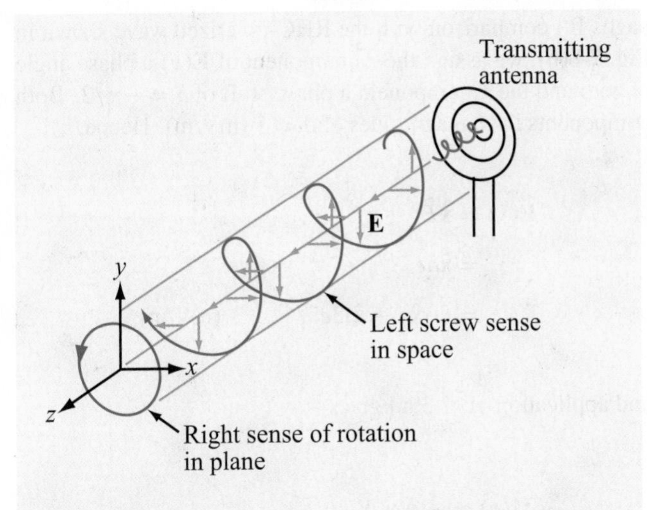

Figure 7-9 Right-hand circularly polarized wave radiated by a helical antenna.

▶ ***Polarization handedness*** is defined in terms of the rotation of **E** as a function of time in a fixed plane orthogonal to the direction of propagation, which is opposite of the direction of rotation of **E** as a function of distance at a fixed point in time. ◀

Example 7-2: RHC-Polarized Wave

An RHC-polarized plane wave with electric field magnitude of 3 (mV/m) is traveling in the $+y$ direction in a dielectric medium with $\epsilon = 4\epsilon_0$, $\mu = \mu_0$, and $\sigma = 0$. If the frequency is 100 MHz, obtain expressions for $\mathbf{E}(y, t)$ and $\mathbf{H}(y, t)$.

Solution: Since the wave is traveling in the $+y$ direction, its field must have components along the x and z directions. The rotation of $\mathbf{E}(y, t)$ is depicted in **Fig. 7-10**, where $\hat{\mathbf{y}}$ is out of the

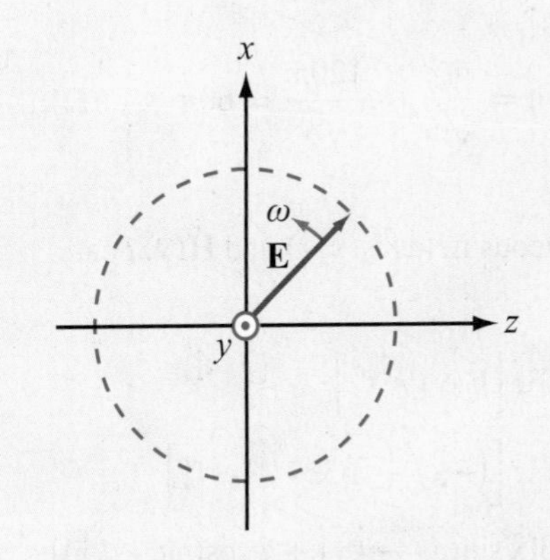

Figure 7-10 Right-hand circularly polarized wave of Example 7-2.

page. By comparison with the RHC-polarized wave shown in **Fig. 7-8(b)**, we assign the z component of $\widetilde{\mathbf{E}}(y)$ a phase angle of zero and the x component a phase shift of $\delta = -\pi/2$. Both components have magnitudes of $a = 3$ (mV/m). Hence,

$$\begin{aligned}\widetilde{\mathbf{E}}(y) &= \hat{\mathbf{x}}\widetilde{E}_x + \hat{\mathbf{z}}\widetilde{E}_z \\ &= \hat{\mathbf{x}}ae^{-j\pi/2}e^{-jky} + \hat{\mathbf{z}}ae^{-jky} \\ &= (-\hat{\mathbf{x}}j + \hat{\mathbf{z}})3e^{-jky} \qquad \text{(mV/m)},\end{aligned}$$

and application of (7.39a) gives

$$\begin{aligned}\widetilde{\mathbf{H}}(y) &= \frac{1}{\eta}\,\hat{\mathbf{y}} \times \widetilde{\mathbf{E}}(y) \\ &= \frac{1}{\eta}\,\hat{\mathbf{y}} \times (-\hat{\mathbf{x}}j + \hat{\mathbf{z}})3e^{-jky} \\ &= \frac{3}{\eta}(\hat{\mathbf{z}}j + \hat{\mathbf{x}})e^{-jky} \qquad \text{(mA/m)}.\end{aligned}$$

With $\omega = 2\pi f = 2\pi \times 10^8$ (rad/s), the wavenumber k is

$$k = \frac{\omega\sqrt{\epsilon_r}}{c} = \frac{2\pi \times 10^8\sqrt{4}}{3 \times 10^8} = \frac{4}{3}\pi \qquad \text{(rad/m)},$$

and the intrinsic impedance η is

$$\eta = \frac{\eta_0}{\sqrt{\epsilon_r}} \approx \frac{120\pi}{\sqrt{4}} = 60\pi \qquad (\Omega).$$

The instantaneous fields $\mathbf{E}(y,t)$ and $\mathbf{H}(y,t)$ are

$$\begin{aligned}\mathbf{E}(y,t) &= \mathfrak{Re}\left[\widetilde{\mathbf{E}}(y)\,e^{j\omega t}\right] \\ &= \mathfrak{Re}\left[(-\hat{\mathbf{x}}j + \hat{\mathbf{z}})3e^{-jky}e^{j\omega t}\right] \\ &= 3[\hat{\mathbf{x}}\sin(\omega t - ky) + \hat{\mathbf{z}}\cos(\omega t - ky)] \qquad \text{(mV/m)}\end{aligned}$$

and

$$\begin{aligned}\mathbf{H}(y,t) &= \mathfrak{Re}\left[\widetilde{\mathbf{H}}(y)\,e^{j\omega t}\right] \\ &= \mathfrak{Re}\left[\frac{3}{\eta}(\hat{\mathbf{z}}j + \hat{\mathbf{x}})e^{-jky}e^{j\omega t}\right] \\ &= \frac{1}{20\pi}[\hat{\mathbf{x}}\cos(\omega t - ky) - \hat{\mathbf{z}}\sin(\omega t - ky)] \text{ (mA/m)}.\end{aligned}$$

7-3.3 Elliptical Polarization

Plane waves that are not linearly or circularly polarized are elliptically polarized. That is, the tip of $\mathbf{E}(z,t)$ traces an ellipse in the plane perpendicular to the direction of propagation. The shape of the ellipse and the field's handedness (left-hand or right-hand) are determined by the values of the ratio (a_y/a_x) and the phase difference δ.

The polarization ellipse shown in **Fig. 7-11** has its major axis with length a_ξ along the ξ direction and its minor axis with length a_η along the η direction. The ***rotation angle*** γ is defined as the angle between the major axis of the ellipse and a reference direction, chosen here to be the x axis, with γ being bounded within the range $-\pi/2 \le \gamma \le \pi/2$. The shape of the ellipse and its handedness are characterized by the ***ellipticity angle*** χ, defined as

$$\tan\chi = \pm\frac{a_\eta}{a_\xi} = \pm\frac{1}{R}, \tag{7.58}$$

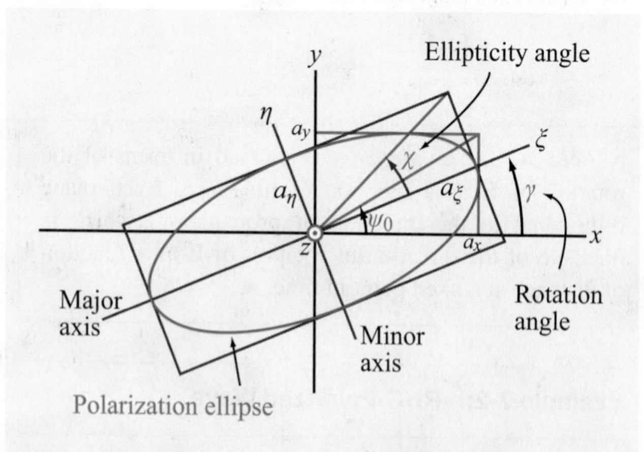

Figure 7-11 Polarization ellipse in the x–y plane, with the wave traveling in the z direction (out of the page).

with the plus sign corresponding to left-handed rotation and the minus sign corresponding to right-handed rotation. The limits for χ are $-\pi/4 \le \chi \le \pi/4$. The quantity $R = a_\xi/a_\eta$ is called the ***axial ratio*** of the polarization ellipse, and it varies between 1 for circular polarization and ∞ for linear polarization. The polarization angles γ and χ are related to the wave parameters a_x, a_y, and δ by†

$$\tan 2\gamma = (\tan 2\psi_0)\cos\delta \quad (-\pi/2 \le \gamma \le \pi/2), \tag{7.59a}$$

$$\sin 2\chi = (\sin 2\psi_0)\sin\delta \quad (-\pi/4 \le \chi \le \pi/4), \tag{7.59b}$$

where ψ_0 is an ***auxiliary angle*** defined by

$$\tan\psi_0 = \frac{a_y}{a_x} \qquad \left(0 \le \psi_0 \le \frac{\pi}{2}\right). \tag{7.60}$$

Sketches of the polarization ellipse are shown in Fig. 7-12 for various combinations of the angles (γ, χ). The ellipse reduces to a circle for $\chi = \pm 45°$ and to a line for $\chi = 0$.

► Positive values of χ, corresponding to $\sin\delta > 0$, are associated with left-handed rotation, and negative values of χ, corresponding to $\sin\delta < 0$, are associated with right-handed rotation. ◄

Since the magnitudes a_x and a_y are, by definition, nonnegative numbers, the ratio a_y/a_x may vary between zero for an x-polarized linear polarization and ∞ for a y-polarized linear polarization. Consequently, the angle ψ_0 is limited to the range $0 \le \psi_0 \le 90°$. Application of Eq. (7.59a) leads to two possible solutions for the value of γ, both of which fall within the defined range from $-\pi/2$ to $\pi/2$. The correct choice is governed by the following rule:

$$\gamma > 0 \text{ if } \cos\delta > 0,$$
$$\gamma < 0 \text{ if } \cos\delta < 0.$$

► In summary, the sign of the rotation angle γ is the same as the sign of $\cos\delta$ and the sign of the ellipticity angle χ is the same as the sign of $\sin\delta$. ◄

Example 7-3: Polarization State

Determine the polarization state of a plane wave with electric field

$$\mathbf{E}(z,t) = \hat{\mathbf{x}}\, 3\cos(\omega t - kz + 30°) - \hat{\mathbf{y}}\, 4\sin(\omega t - kz + 45°) \qquad \text{(mV/m)}.$$

Solution: We begin by converting the second term to a cosine reference,

$$\begin{aligned}\mathbf{E} &= \hat{\mathbf{x}}\, 3\cos(\omega t - kz + 30°) - \hat{\mathbf{y}}\, 4\cos(\omega t - kz + 45° - 90°)\\ &= \hat{\mathbf{x}}\, 3\cos(\omega t - kz + 30°) - \hat{\mathbf{y}}\, 4\cos(\omega t - kz - 45°).\end{aligned}$$

The corresponding field phasor $\widetilde{\mathbf{E}}(z)$ is

$$\begin{aligned}\widetilde{\mathbf{E}}(z) &= \hat{\mathbf{x}}\, 3e^{-jkz}e^{j30°} - \hat{\mathbf{y}}\, 4e^{-jkz}e^{-j45°}\\ &= \hat{\mathbf{x}}\, 3e^{-jkz}e^{j30°} + \hat{\mathbf{y}}\, 4e^{-jkz}e^{-j45°}e^{j180°}\\ &= \hat{\mathbf{x}}\, 3e^{-jkz}e^{j30°} + \hat{\mathbf{y}}\, 4e^{-jkz}e^{j135°},\end{aligned}$$

where we have replaced the negative sign of the second term with $e^{j180°}$ in order to have positive amplitudes for both terms, thereby allowing us to use the definitions given in Section 7-3.3. According to the expression for $\widetilde{\mathbf{E}}(z)$, the phase angles of the x and y components are $\delta_x = 30°$ and $\delta_y = 135°$, giving a phase difference $\delta = \delta_y - \delta_x = 135° - 30° = 105°$. The auxiliary angle ψ_0 is obtained from

$$\psi_0 = \tan^{-1}\left(\frac{a_y}{a_x}\right) = \tan^{-1}\left(\frac{4}{3}\right) = 53.1°.$$

From Eq. (7.59a),

$$\tan 2\gamma = (\tan 2\psi_0)\cos\delta = \tan 106.2° \cos 105° = 0.89,$$

†From M. Born and E. Wolf, *Principles of Optics*, New York: Macmillan, 1965, p. 27.

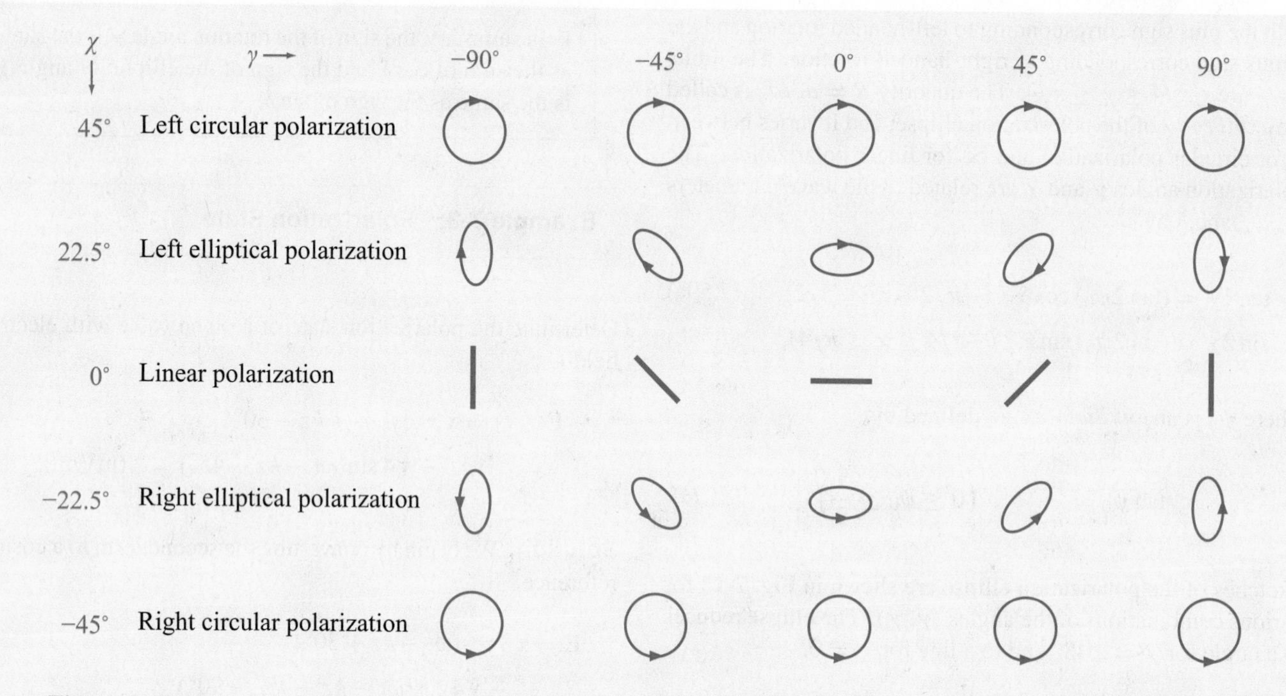

Figure 7-12 Polarization states for various combinations of the polarization angles (γ, χ) for a wave traveling out of the page.

which gives two solutions for γ, namely $\gamma = 20.8°$ and $\gamma = -69.2°$. Since $\cos\delta < 0$, the correct value of γ is $-69.2°$. From Eq. (7.59b),

$$\begin{aligned}\sin 2\chi &= (\sin 2\psi_0)\sin\delta \\ &= \sin 106.2° \sin 105° \\ &= 0.93 \quad \text{or} \quad \chi = 34.0°.\end{aligned}$$

The magnitude of χ indicates that the wave is elliptically polarized and its positive polarity specifies its rotation as left handed.

Concept Question 7-4: An elliptically polarized wave is characterized by amplitudes a_x and a_y and by the phase difference δ. If a_x and a_y are both nonzero, what should δ be in order for the polarization state to reduce to linear polarization?

Concept Question 7-5: Which of the following two descriptions defines an RHC-polarized wave: A wave incident upon an observer is RHC-polarized if its electric field appears to the observer to rotate in a counterclockwise direction (a) as a function of time in a fixed plane perpendicular to the direction of wave travel or (b) as a function of travel distance at a fixed time t?

Exercise 7-5: The electric field of a plane wave is given by

$$\mathbf{E}(z,t) = \hat{\mathbf{x}}\, 3\cos(\omega t - kz) + \hat{\mathbf{y}}\, 4\cos(\omega t - kz) \quad \text{(V/m)}.$$

Determine (a) the polarization state, (b) the modulus of $\mathbf{E}$, and (c) the auxiliary angle.

Answer: (a) Linear, (b) $|\mathbf{E}| = 5\cos(\omega t - kz)$ (V/m), (c) $\psi_0 = 53.1°$. (See EM.)

Module 7.3 Polarization I Upon specifying the amplitudes and phases of the x and y components of **E**, the user can observe the trace of **E** in the x–y plane.

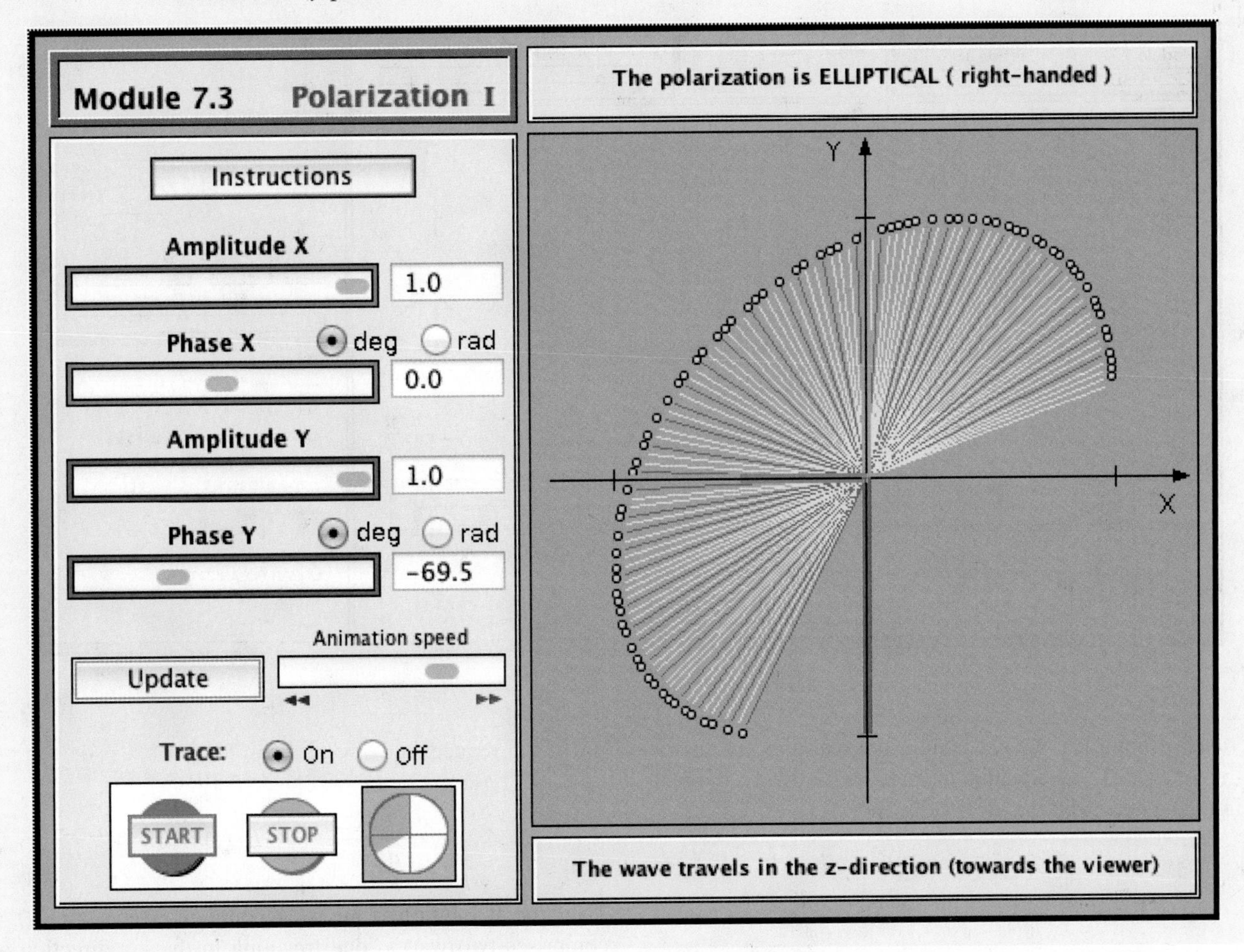

Exercise 7-6: If the electric field phasor of a TEM wave is given by $\widetilde{\mathbf{E}} = (\hat{\mathbf{y}} - \hat{\mathbf{z}}j)e^{-jkx}$, determine the polarization state.

Answer: RHC polarization. (See EM.)

7-4 Plane-Wave Propagation in Lossy Media

To examine wave propagation in a lossy (conducting) medium, we return to the wave equation given by Eq. (7.15),

$$\nabla^2\widetilde{\mathbf{E}} - \gamma^2\widetilde{\mathbf{E}} = 0, \tag{7.61}$$

with

$$\gamma^2 = -\omega^2\mu\epsilon_c = -\omega^2\mu(\epsilon' - j\epsilon''), \tag{7.62}$$

where $\epsilon' = \epsilon$ and $\epsilon'' = \sigma/\omega$. Since γ is complex, we express it as

$$\gamma = \alpha + j\beta, \tag{7.63}$$

where α is the medium's ***attenuation constant*** and β its ***phase constant***. By replacing γ with $(\alpha + j\beta)$ in Eq. (7.62), we obtain

$$\begin{aligned}(\alpha + j\beta)^2 &= (\alpha^2 - \beta^2) + j2\alpha\beta \\ &= -\omega^2\mu\epsilon' + j\omega^2\mu\epsilon''.\end{aligned} \tag{7.64}$$

Module 7.4 Polarization II Upon specifying the amplitudes and phases of the x and y components of **E**, the user can observe the 3-D profile of the **E** vector over a specified length span.

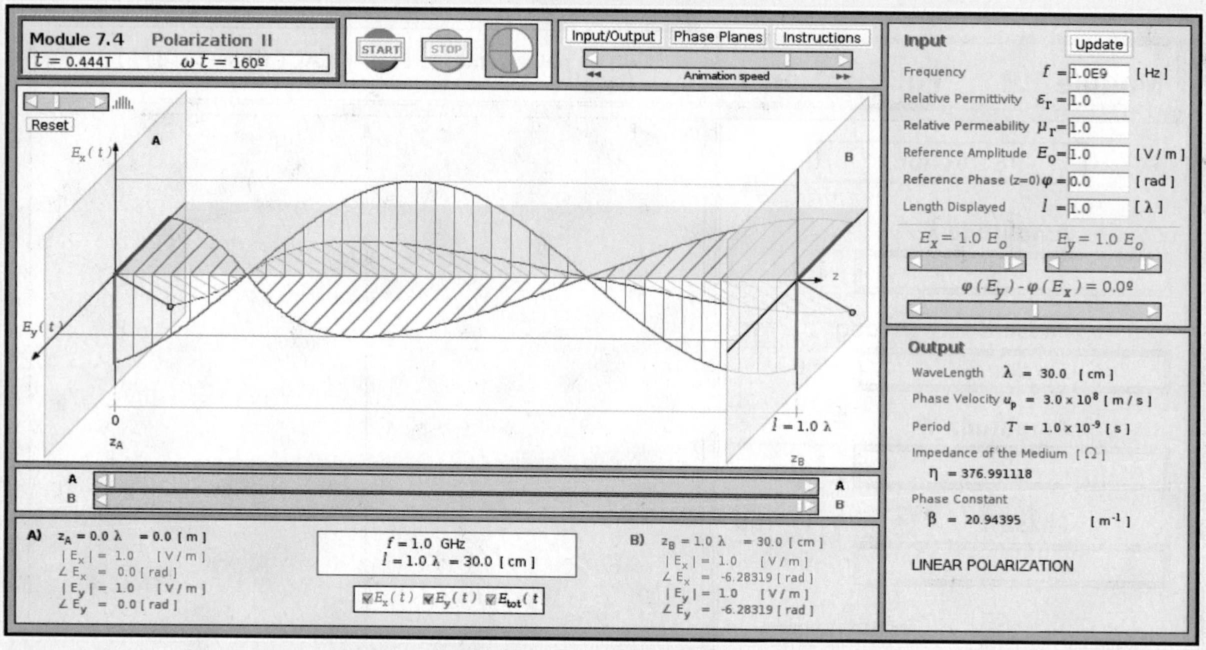

The rules of complex algebra require the real and imaginary parts on one side of an equation to equal, respectively, the real and imaginary parts on the other side. Hence,

$$\alpha^2 - \beta^2 = -\omega^2 \mu \epsilon', \tag{7.65a}$$

$$2\alpha\beta = \omega^2 \mu \epsilon''. \tag{7.65b}$$

Solving these two equations for α and β gives

$$\alpha = \omega \left\{ \frac{\mu\epsilon'}{2} \left[\sqrt{1 + \left(\frac{\epsilon''}{\epsilon'}\right)^2} - 1 \right] \right\}^{1/2} \quad \text{(Np/m)}, \tag{7.66a}$$

$$\beta = \omega \left\{ \frac{\mu\epsilon'}{2} \left[\sqrt{1 + \left(\frac{\epsilon''}{\epsilon'}\right)^2} + 1 \right] \right\}^{1/2} \quad \text{(rad/m)}. \tag{7.66b}$$

For a uniform plane wave with electric field $\widetilde{\mathbf{E}} = \hat{\mathbf{x}}\, \widetilde{E}_x(z)$ traveling along the z direction, the wave equation given by Eq. (7.61) reduces to

$$\frac{d^2\, \widetilde{E}_x(z)}{dz^2} - \gamma^2\, \widetilde{E}_x(z) = 0. \tag{7.67}$$

The general solution of the wave equation given by Eq. (7.67) comprises two waves, one traveling in the $+z$ direction and another traveling in the $-z$ direction. Assuming only the former is present, the solution of the wave equation leads to

$$\widetilde{\mathbf{E}}(z) = \hat{\mathbf{x}}\widetilde{E}_x(z) = \hat{\mathbf{x}}E_{x0}e^{-\gamma z} = \hat{\mathbf{x}}E_{x0}e^{-\alpha z}e^{-j\beta z}. \tag{7.68}$$

The associated magnetic field $\widetilde{\mathbf{H}}$ can be determined by applying Eq. (7.2b): $\nabla \times \widetilde{\mathbf{E}} = -j\omega\mu\widetilde{\mathbf{H}}$, or using Eq. (7.39a): $\widetilde{\mathbf{H}} = (\hat{\mathbf{k}} \times \widetilde{\mathbf{E}})/\eta_c$, where η_c is the ***intrinsic impedance of the lossy medium***. Both approaches give

$$\widetilde{\mathbf{H}}(z) = \hat{\mathbf{y}}\, \widetilde{H}_y(z) = \hat{\mathbf{y}}\, \frac{\widetilde{E}_x(z)}{\eta_c} = \hat{\mathbf{y}}\, \frac{E_{x0}}{\eta_c} e^{-\alpha z} e^{-j\beta z}, \tag{7.69}$$

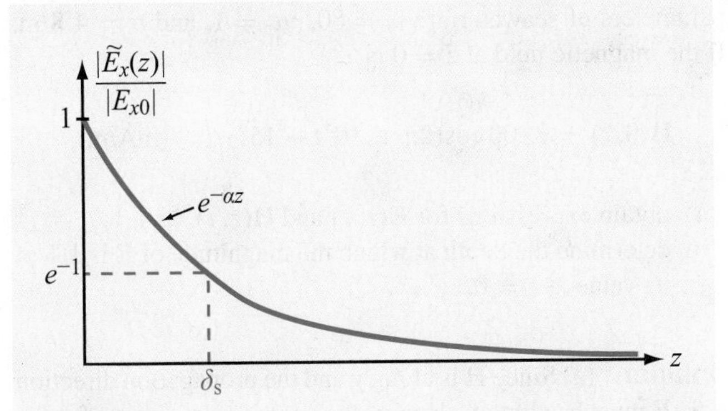

Figure 7-13 Attenuation of the magnitude of $\widetilde{E}_x(z)$ with distance z. The skin depth δ_s is the value of z at which $|\widetilde{E}_x(z)|/|E_{x0}| = e^{-1}$, or $z = \delta_s = 1/\alpha$.

where

$$\eta_c = \sqrt{\frac{\mu}{\epsilon_c}} = \sqrt{\frac{\mu}{\epsilon'}}\left(1 - j\frac{\epsilon''}{\epsilon'}\right)^{-1/2} \quad (\Omega). \quad (7.70)$$

We noted earlier that in a lossless medium, $\mathbf{E}(z,t)$ is in phase with $\mathbf{H}(z,t)$. This property no longer holds true in a lossy medium because η_c is complex. This fact is demonstrated in Example 7-4.

From Eq. (7.68), the magnitude of $\widetilde{E}_x(z)$ is given by

$$|\widetilde{E}_x(z)| = |E_{x0}e^{-\alpha z}e^{-j\beta z}| = |E_{x0}|e^{-\alpha z}, \quad (7.71)$$

which decreases exponentially with z at a rate dictated by the attenuation constant α. Since $\widetilde{H}_y = \widetilde{E}_x/\eta_c$, the magnitude of $\widetilde{H}_y$ also decreases as $e^{-\alpha z}$. As the field attenuates, part of the energy carried by the electromagnetic wave is converted into heat due to conduction in the medium. As the wave travels through a distance $z = \delta_s$ with

$$\delta_s = \frac{1}{\alpha} \quad \text{(m)}, \quad (7.72)$$

the wave magnitude decreases by a factor of $e^{-1} \approx 0.37$ (**Fig. 7-13**). At depth $z = 3\delta_s$, the field magnitude is less than 5% of its initial value, and at $z = 5\delta_s$, it is less than 1%.

► This distance δ_s, called the ***skin depth*** of the medium, characterizes how deep an electromagnetic wave can penetrate into a conducting medium. ◄

In a perfect dielectric, $\sigma = 0$ and $\epsilon'' = 0$; use of Eq. (7.66a) yields $\alpha = 0$ and therefore $\delta_s = \infty$. Thus, in free space, a plane wave can propagate indefinitely with no loss in magnitude. On the other extreme, in a perfect conductor, $\sigma = \infty$ and use of Eq. (7.66a) leads to $\alpha = \infty$ and hence $\delta_s = 0$. If the outer conductor of a coaxial cable is designed to be several skin depths thick, it prevents energy inside the cable from leaking outward and shields against penetration of electromagnetic energy from external sources into the cable.

The expressions given by Eqs. (7.66a), (7.66b), and (7.70) for α, β, and η_c are valid for any linear, isotropic, and homogeneous medium. For a perfect dielectric ($\sigma = 0$), these expressions reduce to those for the lossless case (Section 7-2), wherein $\alpha = 0$, $\beta = k = \omega\sqrt{\mu\epsilon}$, and $\eta_c = \eta$. For a lossy medium, the ratio $\epsilon''/\epsilon' = \sigma/\omega\epsilon$, which appears in all these expressions, plays an important role in classifying how lossy the medium is. When $\epsilon''/\epsilon' \ll 1$, the medium is considered a ***low-loss dielectric***, and when $\epsilon''/\epsilon' \gg 1$, it is considered a ***good conductor***. In practice, the medium may be regarded as a low-loss dielectric if $\epsilon''/\epsilon' < 10^{-2}$, as a good conductor if $\epsilon''/\epsilon' > 10^2$, and as a ***quasi-conductor*** if $10^{-2} \leq \epsilon''/\epsilon' \leq 10^2$. For low-loss dielectrics and good conductors, the expressions given by Eq. (7.66) can be significantly simplified, as shown next.

7-4.1 Low-Loss Dielectric

From Eq. (7.62), the general expression for γ is

$$\gamma = j\omega\sqrt{\mu\epsilon'}\left(1 - j\frac{\epsilon''}{\epsilon'}\right)^{1/2}. \quad (7.73)$$

For $|x| \ll 1$, the function $(1-x)^{1/2}$ can be approximated by the first two terms of its binomial series; that is, $(1-x)^{1/2} \approx 1 - x/2$. By applying this approximation to Eq. (7.73) for a low-loss dielectric with $x = j\epsilon''/\epsilon'$ and $\epsilon''/\epsilon' \ll 1$, we obtain

$$\gamma \approx j\omega\sqrt{\mu\epsilon'}\left(1 - j\frac{\epsilon''}{2\epsilon'}\right). \quad (7.74)$$

The real and imaginary parts of Eq. (7.74) are

$$\alpha \approx \frac{\omega\epsilon''}{2}\sqrt{\frac{\mu}{\epsilon'}} = \frac{\sigma}{2}\sqrt{\frac{\mu}{\epsilon}} \quad \text{(Np/m)}, \quad (7.75a)$$

$$\beta \approx \omega\sqrt{\mu\epsilon'} = \omega\sqrt{\mu\epsilon} \quad \text{(rad/m)}. \quad (7.75b)$$

(low-loss medium)

We note that the expression for β is the same as that for the wavenumber k of a lossless medium. Applying the binomial approximation $(1-x)^{-1/2} \approx (1+x/2)$ to Eq. (7.70) leads to

$$\eta_c \approx \sqrt{\frac{\mu}{\epsilon'}}\left(1+j\frac{\epsilon''}{2\epsilon'}\right) = \sqrt{\frac{\mu}{\epsilon}}\left(1+j\frac{\sigma}{2\omega\epsilon}\right). \quad (7.76a)$$

In practice, because $\epsilon''/\epsilon' = \sigma/\omega\epsilon < 10^{-2}$, the second term in Eq. (7.76a) often is ignored. Thus,

$$\eta_c \approx \sqrt{\frac{\mu}{\epsilon}}, \quad (7.76b)$$

which is the same as Eq. (7.31) for the lossless case.

7-4.2 Good Conductor

When $\epsilon''/\epsilon' > 100$, Eqs. (7.66a), (7.66b), and (7.70) can be approximated as

$$\alpha \approx \omega\sqrt{\frac{\mu\epsilon''}{2}} = \omega\sqrt{\frac{\mu\sigma}{2\omega}} = \sqrt{\pi f\mu\sigma} \quad \text{(Np/m)}, \quad (7.77a)$$

$$\beta = \alpha \approx \sqrt{\pi f\mu\sigma} \quad \text{(rad/m)}, \quad (7.77b)$$

$$\eta_c \approx \sqrt{j\frac{\mu}{\epsilon''}} = (1+j)\sqrt{\frac{\pi f\mu}{\sigma}} = (1+j)\frac{\alpha}{\sigma} \quad (\Omega).$$

(good conductor) (7.77c)

In Eq. (7.77c), we used the relation given by Eq. (1.53): $\sqrt{j} = (1+j)/\sqrt{2}$. For a perfect conductor with $\sigma = \infty$, these expressions yield $\alpha = \beta = \infty$, and $\eta_c = 0$. A perfect conductor is equivalent to a short circuit in a transmission line equivalent.

Expressions for the propagation parameters in various types of media are summarized in **Table 7-1**.

Example 7-4: Plane Wave in Seawater

A uniform plane wave is traveling in seawater. Assume that the x–y plane resides just below the sea surface and the wave travels in the $+z$ direction into the water. The constitutive parameters of seawater are $\epsilon_r = 80$, $\mu_r = 1$, and $\sigma = 4$ S/m. If the magnetic field at $z = 0$ is

$$\mathbf{H}(0,t) = \hat{\mathbf{y}}\, 100\cos(2\pi \times 10^3 t + 15^\circ) \qquad \text{(mA/m)},$$

(a) obtain expressions for $\mathbf{E}(z,t)$ and $\mathbf{H}(z,t)$, and

(b) determine the depth at which the magnitude of $\mathbf{E}$ is 1% of its value at $z = 0$.

Solution: (a) Since $\mathbf{H}$ is along $\hat{\mathbf{y}}$ and the propagation direction is $\hat{\mathbf{z}}$, $\mathbf{E}$ must be along $\hat{\mathbf{x}}$. Hence, the general expressions for the phasor fields are

$$\widetilde{\mathbf{E}}(z) = \hat{\mathbf{x}} E_{x0} e^{-\alpha z} e^{-j\beta z}, \quad (7.78a)$$

$$\widetilde{\mathbf{H}}(z) = \hat{\mathbf{y}}\, \frac{E_{x0}}{\eta_c}\, e^{-\alpha z} e^{-j\beta z}. \quad (7.78b)$$

To determine α, β, and η_c for seawater, we begin by evaluating the ratio ϵ''/ϵ'. From the argument of the cosine function of $\mathbf{H}(0,t)$, we deduce that $\omega = 2\pi \times 10^3$ (rad/s), and therefore $f = 1$ kHz. Hence,

$$\frac{\epsilon''}{\epsilon'} = \frac{\sigma}{\omega\epsilon} = \frac{\sigma}{\omega\epsilon_r\epsilon_0} = \frac{4}{2\pi \times 10^3 \times 80 \times (10^{-9}/36\pi)} = 9 \times 10^5.$$

This qualifies seawater as a good conductor at 1 kHz and allows us to use the good-conductor expressions given in **Table 7-1**:

$$\alpha = \sqrt{\pi f\mu\sigma} = \sqrt{\pi \times 10^3 \times 4\pi \times 10^{-7} \times 4} = 0.126 \quad \text{(Np/m)}, \quad (7.79a)$$

$$\beta = \alpha = 0.126 \quad \text{(rad/m)}, \quad (7.79b)$$

$$\eta_c = (1+j)\frac{\alpha}{\sigma} = (\sqrt{2}\, e^{j\pi/4})\frac{0.126}{4} = 0.044 e^{j\pi/4} \quad (\Omega). \quad (7.79c)$$

Table 7-1 Expressions for α, β, η_c, u_p, and λ for various types of media.

	Any Medium	Lossless Medium ($\sigma = 0$)	Low-loss Medium ($\epsilon''/\epsilon' \ll 1$)	Good Conductor ($\epsilon''/\epsilon' \gg 1$)	Units
$\alpha =$	$\omega \left[\frac{\mu\epsilon'}{2}\left[\sqrt{1+\left(\frac{\epsilon''}{\epsilon'}\right)^2} - 1\right]\right]^{1/2}$	0	$\frac{\sigma}{2}\sqrt{\frac{\mu}{\epsilon}}$	$\sqrt{\pi f \mu\sigma}$	(Np/m)
$\beta =$	$\omega \left[\frac{\mu\epsilon'}{2}\left[\sqrt{1+\left(\frac{\epsilon''}{\epsilon'}\right)^2} + 1\right]\right]^{1/2}$	$\omega\sqrt{\mu\epsilon}$	$\omega\sqrt{\mu\epsilon}$	$\sqrt{\pi f \mu\sigma}$	(rad/m)
$\eta_c =$	$\sqrt{\frac{\mu}{\epsilon'}}\left(1 - j\frac{\epsilon''}{\epsilon'}\right)^{-1/2}$	$\sqrt{\frac{\mu}{\epsilon}}$	$\sqrt{\frac{\mu}{\epsilon}}$	$(1+j)\frac{\alpha}{\sigma}$	(Ω)
$u_p =$	ω/β	$1/\sqrt{\mu\epsilon}$	$1/\sqrt{\mu\epsilon}$	$\sqrt{4\pi f/\mu\sigma}$	(m/s)
$\lambda =$	$2\pi/\beta = u_p/f$	u_p/f	u_p/f	u_p/f	(m)

Notes: $\epsilon' = \epsilon$; $\epsilon'' = \sigma/\omega$; in free space, $\epsilon = \epsilon_0$, $\mu = \mu_0$; in practice, a material is considered a low-loss medium if $\epsilon''/\epsilon' = \sigma/\omega\epsilon < 0.01$ and a good conducting medium if $\epsilon''/\epsilon' > 100$.

As no explicit information has been given about the electric field amplitude E_{x0}, we should assume it to be complex; that is, $E_{x0} = |E_{x0}|e^{j\phi_0}$. The wave's instantaneous electric and magnetic fields are given by

$$\mathbf{E}(z,t) = \mathfrak{Re}\left[\hat{\mathbf{x}}|E_{x0}|e^{j\phi_0}e^{-\alpha z}e^{-j\beta z}e^{j\omega t}\right]$$
$$= \hat{\mathbf{x}}|E_{x0}|e^{-0.126z}\cos(2\pi \times 10^3 t - 0.126z + \phi_0) \quad \text{(V/m)}, \tag{7.80a}$$

$$\mathbf{H}(z,t) = \mathfrak{Re}\left[\hat{\mathbf{y}}\,\frac{|E_{x0}|e^{j\phi_0}}{0.044e^{j\pi/4}}e^{-\alpha z}e^{-j\beta z}e^{j\omega t}\right]$$
$$= \hat{\mathbf{y}}22.5|E_{x0}|e^{-0.126z}\cos(2\pi \times 10^3 t - 0.126z + \phi_0 - 45^\circ) \quad \text{(A/m)}. \tag{7.80b}$$

At $z = 0$,

$$\mathbf{H}(0,t) = \hat{\mathbf{y}}\,22.5|E_{x0}|\cos(2\pi \times 10^3 t + \phi_0 - 45^\circ) \quad \text{(A/m)}. \tag{7.81}$$

By comparing Eq. (7.81) with the expression given in the problem statement,

$$\mathbf{H}(0,t) = \hat{\mathbf{y}}\,100\cos(2\pi \times 10^3 t + 15^\circ) \quad \text{(mA/m)},$$

we deduce that

$$22.5|E_{x0}| = 100 \times 10^{-3}$$

or

$$|E_{x0}| = 4.44 \quad \text{(mV/m)},$$

and

$$\phi_0 - 45^\circ = 15^\circ \quad \text{or} \quad \phi_0 = 60^\circ.$$

Hence, the final expressions for $\mathbf{E}(z,t)$ and $\mathbf{H}(z,t)$ are

$$\mathbf{E}(z,t) = \hat{\mathbf{x}}\,4.44e^{-0.126z}\cos(2\pi \times 10^3 t - 0.126z + 60^\circ) \quad \text{(mV/m)}, \tag{7.82a}$$

$$\mathbf{H}(z,t) = \hat{\mathbf{y}}\,100e^{-0.126z}\cos(2\pi \times 10^3 t - 0.126z + 15^\circ) \quad \text{(mA/m)}. \tag{7.82b}$$

Technology Brief 14: Liquid Crystal Display (LCD)

LCDs are used in digital clocks, cellular phones, desktop and laptop computers, and some televisions and other electronic systems. They offer a decided advantage over former display technologies, such as cathode ray tubes, in that they are much lighter and thinner and consume a lot less power to operate. LCD technology relies on special electrical and optical properties of a class of materials known as liquid crystals, first discovered in the 1880s by botanist ***Friedrich Reinitzer***.

Physical Principle

► ***Liquid crystals*** are neither a pure solid nor a pure liquid, but rather a hybrid of both. ◄

One particular variety of interest is the ***twisted nematic*** liquid crystal whose rod-shaped molecules have a natural tendency to assume a ***twisted spiral structure*** when the material is sandwiched between finely ***grooved glass substrates*** with orthogonal orientations (**Fig. TF14-1**). Note that the molecules in contact with the grooved surfaces align themselves in parallel along the grooves, from a y orientation at the entrance substrate into an x orientation at the

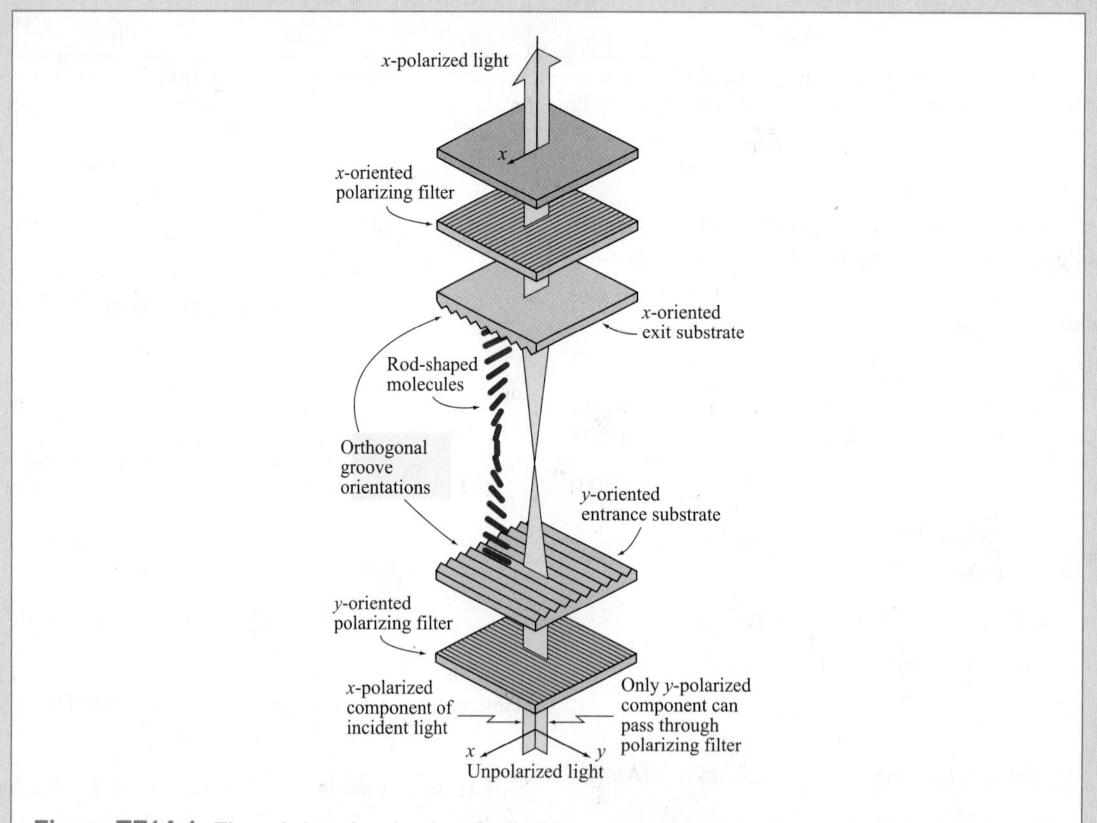

Figure TF14-1 The rod-shaped molecules of a liquid crystal sandwiched between grooved substrates with orthogonal orientations cause the electric field of the light passing through it to rotate by 90°.

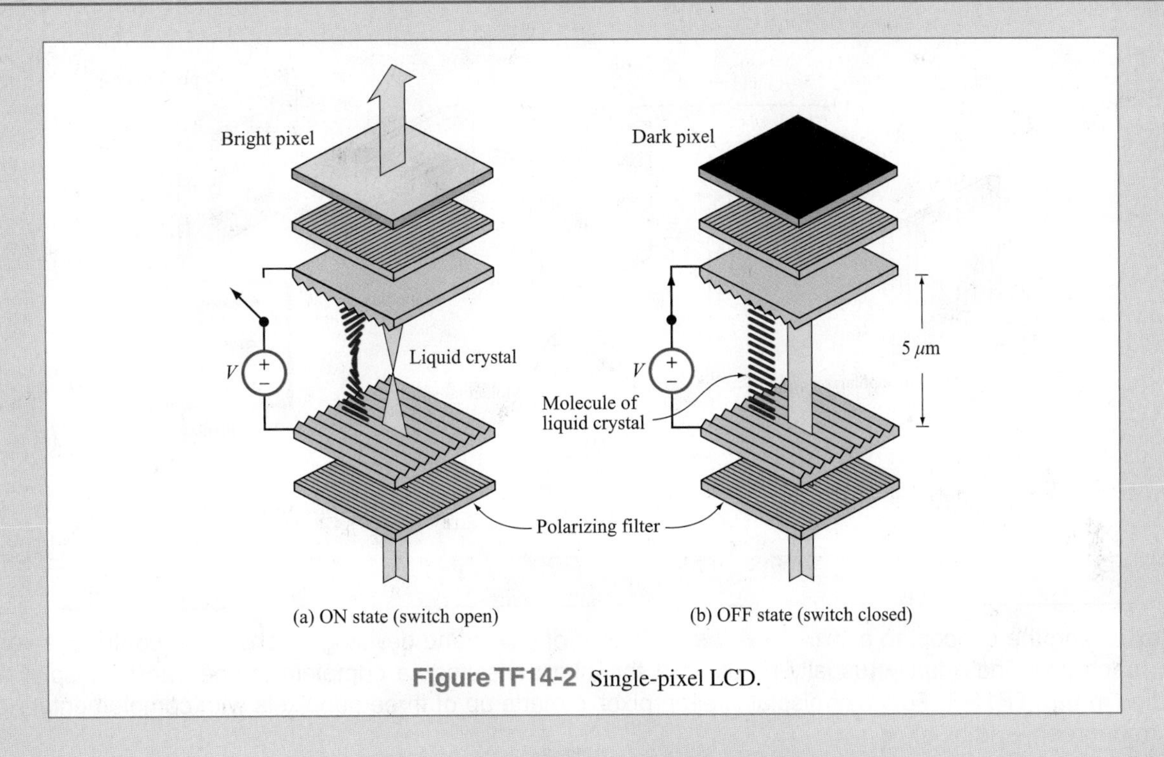

Figure TF14-2 Single-pixel LCD.

exit substrate. The molecular spiral causes the crystal to behave like a ***wave polarizer***: unpolarized light incident upon the entrance substrate follows the orientation of the spiral, emerging through the exit substrate with its polarization (direction of electric field) parallel to the groove's direction, which in **Fig. TF14-1** is along the x direction. Thus, of the x and y components of the incident light, only the y component is allowed to pass through the y-polarized filter, but as a consequence of the spiral action facilitated by the liquid crystal's molecules, the light that emerges from the LCD structure is x-polarized.

LCD Structure

A single-pixel LCD structure is shown in **Fig. TF14-2** for the OFF and ON states, with OFF corresponding to a bright-looking pixel and ON to a dark-looking pixel.

> ► The sandwiched liquid-crystal layer (typically on the order of ***5 microns in thickness***, or 1/20 of the width of a human hair) is straddled by a pair of optical filters with orthogonal polarizations. ◄

When no voltage is applied across the crystal layer [**Fig. TF14-2(a)**], incoming unpolarized light gets polarized as it passes through the entrance polarizer, then rotates by 90° as it follows the molecular spiral, and finally emerges from the exit polarizer, giving the exited surface a ***bright appearance***. A useful feature of nematic liquid crystals is that their spiral untwists [**Fig. TF14-2(b)**] under the influence of an electric field (induced by a voltage difference across the layer). The degree of untwisting depends on the strength of the electric field. With no spiral to rotate the wave polarization as the light travels through the crystal, the light polarization becomes orthogonal to that of the exit polarizer, allowing no light to pass through it. Hence, the pixel exhibits a ***dark appearance***.

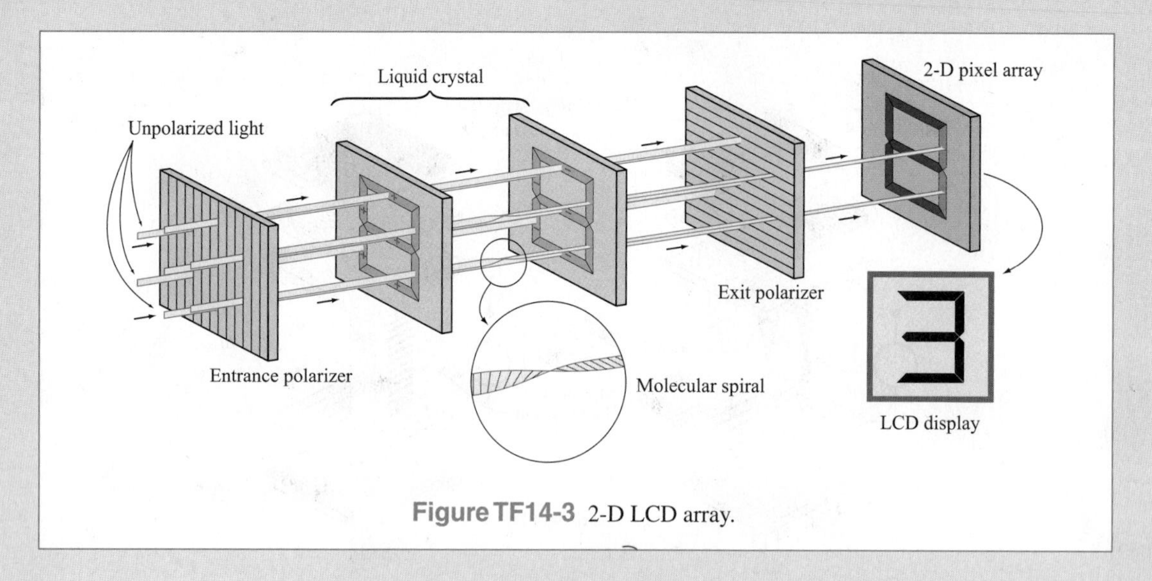

Figure TF14-3 2-D LCD array.

► By extending the concept to a ***two-dimensional array*** of pixels and devising a scheme to control the voltage across each pixel individually (usually by using a thin-film transistor), a complete image can be displayed as illustrated in **Fig. TF14-3**. For color displays, each pixel is made up of three subpixels with complementary color filters (red, green, and blue). ◄

Figure TF14-4 LCD display.

Module 7.5 Wave Attenuation Observe the profile of a plane wave propagating in a lossy medium. Determine the skin depth, the propagation parameters, and the intrinsic impedance of the medium.

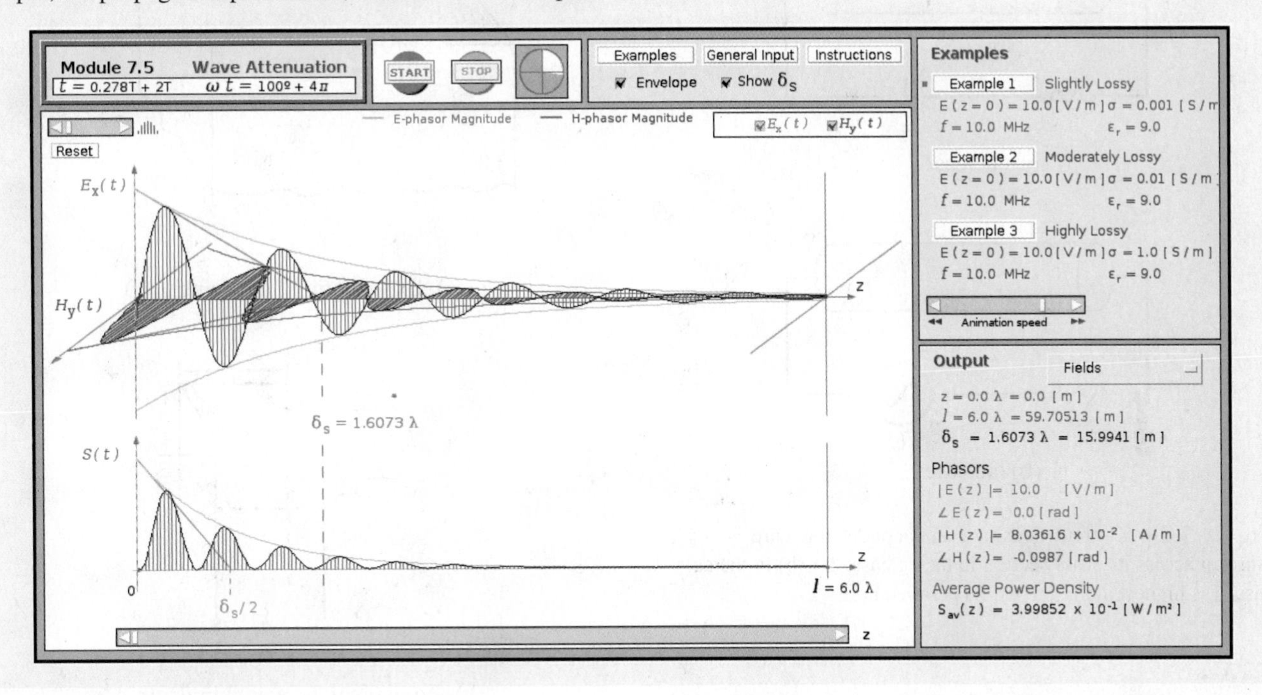

(b) The depth at which the amplitude of **E** has decreased to 1% of its initial value at $z = 0$ is obtained from

$$0.01 = e^{-0.126z}$$

or

$$z = \frac{\ln(0.01)}{-0.126} = 36.55 \text{ m} \approx 37 \text{ m}.$$

Exercise 7-7: The constitutive parameters of copper are $\mu = \mu_0 = 4\pi \times 10^{-7}$ (H/m), $\epsilon = \epsilon_0 \approx (1/36\pi) \times 10^{-9}$ (F/m), and $\sigma = 5.8 \times 10^7$ (S/m). Assuming that these parameters are frequency independent, over what frequency range of the electromagnetic spectrum (see **Fig. 1-16**) is copper a good conductor?

Answer: $f < 1.04 \times 10^{16}$ Hz, which includes the radio, infrared, visible, and part of the ultraviolet regions of the EM spectrum. (See EM.)

Exercise 7-8: Over what frequency range may dry soil, with $\epsilon_r = 3$, $\mu_r = 1$, and $\sigma = 10^{-4}$ (S/m), be regarded as a low-loss dielectric?

Answer: $f > 60$ MHz. (See EM.)

Exercise 7-9: For a wave traveling in a medium with a skin depth δ_s, what is the amplitude of **E** at a distance of $3\delta_s$ compared with its initial value?

Answer: $e^{-3} \approx 0.05$ or 5%. (See EM.)

7-5 Current Flow in a Good Conductor

When a dc voltage source is connected across the ends of a conducting wire, the current flowing through the wire is uniformly distributed over its cross section. That is, the current density **J** is the same along the axis of the wire and along its outer perimeter [**Fig. 7-14(a)**]. This is not true in the ac case. As we will see shortly, a time-varying current density is maximum along the perimeter of the wire and decreases exponentially as

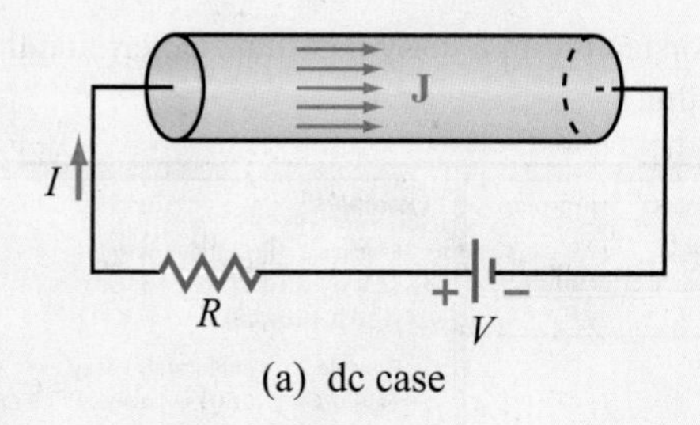

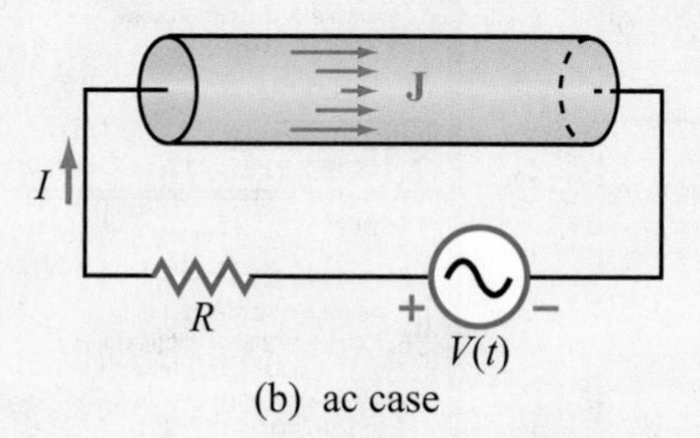

Figure 7-14 Current density **J** in a conducting wire is (a) uniform across its cross section in the dc case, but (b) in the ac case, **J** is highest along the wire's perimeter.

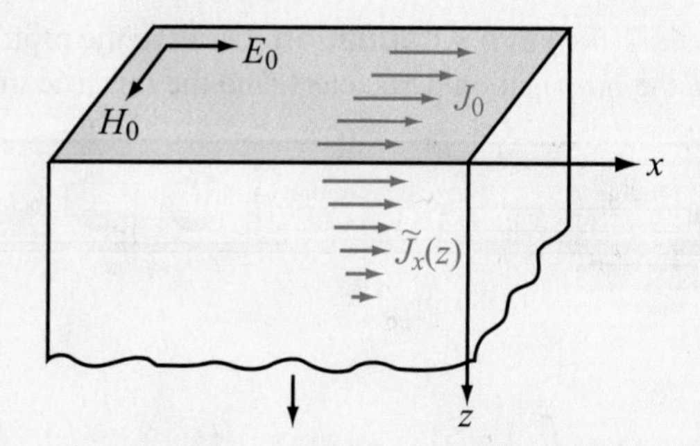

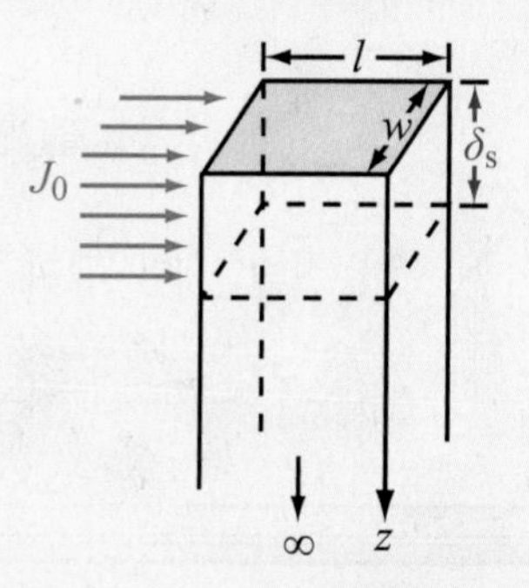

Figure 7-15 Exponential decay of current density $\widetilde{J}_x(z)$ with z in a solid conductor. The total current flowing through (a) a section of width w extending between $z = 0$ and $z = \infty$ is equivalent to (b) a constant current density J_0 flowing through a section of depth δ_s.

a function of distance toward the axis of the wire [Fig. 7-14(b)]. In fact, at very high frequencies most of the current flows in a thin layer near the wire surface, and if the wire material is a perfect conductor, the current flows entirely on the surface of the wire.

Before analyzing a wire with circular cross section, let us consider the simpler geometry of a semi-infinite conducting solid, as shown in Fig. 7-15(a). The solid's planar interface with a perfect dielectric is the x–y plane. If at $z = 0^-$ (just above the surface), an x-polarized electric field with $\widetilde{\mathbf{E}} = \hat{\mathbf{x}} E_0$ exists in the dielectric, a similarly polarized field is induced in the conducting medium and propagates as a plane wave along the $+z$ direction. As a consequence of the boundary condition mandating continuity of the tangential component of **E** across the boundary between any two contiguous media, the electric field at $z = 0^+$ (just below the boundary) is $\widetilde{\mathbf{E}}(0) = \hat{\mathbf{x}} E_0$ also. The EM fields at any depth z in the conductor are then given by

$$\widetilde{\mathbf{E}}(z) = \hat{\mathbf{x}} E_0 e^{-\alpha z} e^{-j\beta z}, \qquad (7.83a)$$

$$\widetilde{\mathbf{H}}(z) = \hat{\mathbf{y}} \frac{E_0}{\eta_c} e^{-\alpha z} e^{-j\beta z}. \qquad (7.83b)$$

From $\mathbf{J} = \sigma \mathbf{E}$, the current flows in the x direction, and its density is

$$\widetilde{\mathbf{J}}(z) = \hat{\mathbf{x}}\, \widetilde{J}_x(z), \qquad (7.84)$$

with

$$\widetilde{J}_x(z) = \sigma E_0 e^{-\alpha z} e^{-j\beta z} = J_0 e^{-\alpha z} e^{-j\beta z}, \qquad (7.85)$$

where $J_0 = \sigma E_0$ is the amplitude of the current density at the surface. In terms of the skin depth $\delta_s = 1/\alpha$ defined by Eq. (7.72) and using the fact that in a good conductor $\alpha = \beta$ as expressed by Eq. (7.77b), Eq. (7.85) can be written as

$$\widetilde{J}_x(z) = J_0 e^{-(1+j)z/\delta_s} \quad (\text{A/m}^2). \qquad (7.86)$$

The current flowing through a rectangular strip of width w along the y direction and extending between zero and ∞ in the z

direction is

$$\tilde{I} = w \int_0^\infty \tilde{J}_x(z)\, dz$$
$$= w \int_0^\infty J_0 e^{-(1+j)z/\delta_s}\, dz = \frac{J_0 w \delta_s}{(1+j)} \quad \text{(A)}. \tag{7.87}$$

The numerator of Eq. (7.87) is reminiscent of a uniform current density J_0 flowing through a thin surface of width w and depth δ_s. Because $\tilde{J}_x(z)$ decreases exponentially with depth z, a conductor of finite thickness d can be considered electrically equivalent to one of infinite depth as long as d exceeds a few skin depths. Indeed, if $d = 3\delta_s$ [instead of ∞ in the integral of Eq. (7.87)], the error incurred in using the result on the right-hand side of Eq. (7.87) is less than 5%; and if $d = 5\delta_s$, the error is less than 1%.

The voltage across a length l at the surface [Fig. 7-15(b)] is given by

$$\tilde{V} = E_0 l = \frac{J_0}{\sigma}\, l. \tag{7.88}$$

Hence, the impedance of a slab of width w, length l, and depth $d = \infty$ (or, in practice, $d > 5\delta_s$) is

$$Z = \frac{\tilde{V}}{\tilde{I}} = \frac{1+j}{\sigma\delta_s}\, \frac{l}{w} \quad (\Omega). \tag{7.89}$$

It is customary to represent Z as

$$Z = Z_s \frac{l}{w}, \tag{7.90}$$

where Z_s, the ***internal*** or ***surface impedance*** of the conductor, is defined as the impedance Z for a length $l = 1$ m and a width $w = 1$ m. Thus,

$$Z_s = \frac{1+j}{\sigma\delta_s} \quad (\Omega). \tag{7.91}$$

Since the reactive part of Z_s is positive, Z_s can be defined as

$$Z_s = R_s + j\omega L_s$$

with

$$R_s = \frac{1}{\sigma\delta_s} = \sqrt{\frac{\pi f \mu}{\sigma}} \quad (\Omega), \tag{7.92a}$$

$$L_s = \frac{1}{\omega\sigma\delta_s} = \frac{1}{2}\sqrt{\frac{\mu}{\pi f \sigma}} \quad \text{(H)}, \tag{7.92b}$$

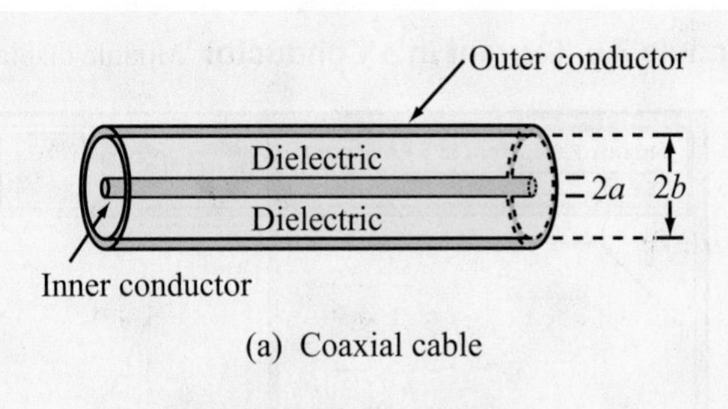

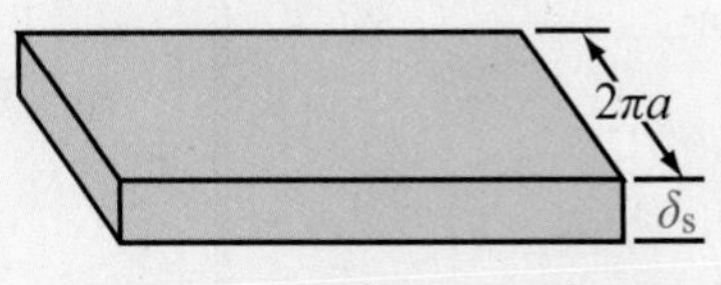

Figure 7-16 The inner conductor of the coaxial cable in (a) is represented in (b) by a planar conductor of width $2\pi a$ and depth δ_s, as if its skin has been cut along its length on the bottom side and then unfurled into a planar geometry.

where we used the relation $\delta_s = 1/\alpha \approx 1/\sqrt{\pi f \mu\sigma}$ given by Eq. (7.77a). In terms of the ***surface resistance*** R_s, the ***ac resistance*** of a slab of width w and length l is

$$R = R_s \frac{l}{w} = \frac{l}{\sigma\delta_s w} \quad (\Omega). \tag{7.93}$$

The expression for the ac resistance R is equivalent to the dc resistance of a plane conductor of length l and cross section $A = \delta_s w$.

The results obtained for the planar conductor can be extended to the coaxial cable shown in Fig. 7-16(a). If the conductors are made of copper with $\sigma = 5.8 \times 10^7$ S/m, the skin depth at 1 MHz is $\delta_s = 1/\sqrt{\pi f \mu\sigma} = 0.066$ mm, and since δ_s varies as $1/\sqrt{f}$, it becomes smaller at higher frequencies. As long as the inner conductor's radius a is greater than $5\delta_s$, or 0.33 mm at 1 MHz, its "depth" may be regarded as infinite. A similar criterion applies to the thickness of the outer conductor. To compute the resistance of the inner conductor, note that the current is concentrated near its outer surface and approximately equivalent to a uniform current flowing through a thin layer of depth δ_s and circumference $2\pi a$. In other words, the inner conductor's resistance is nearly the same as that of a planar conductor of depth δ_s and width $w = 2\pi a$, as shown in Fig. 7-16(b). The corresponding resistance per unit length is

Module 7.6 Current in a Conductor Module displays exponential decay of current density in a conductor.

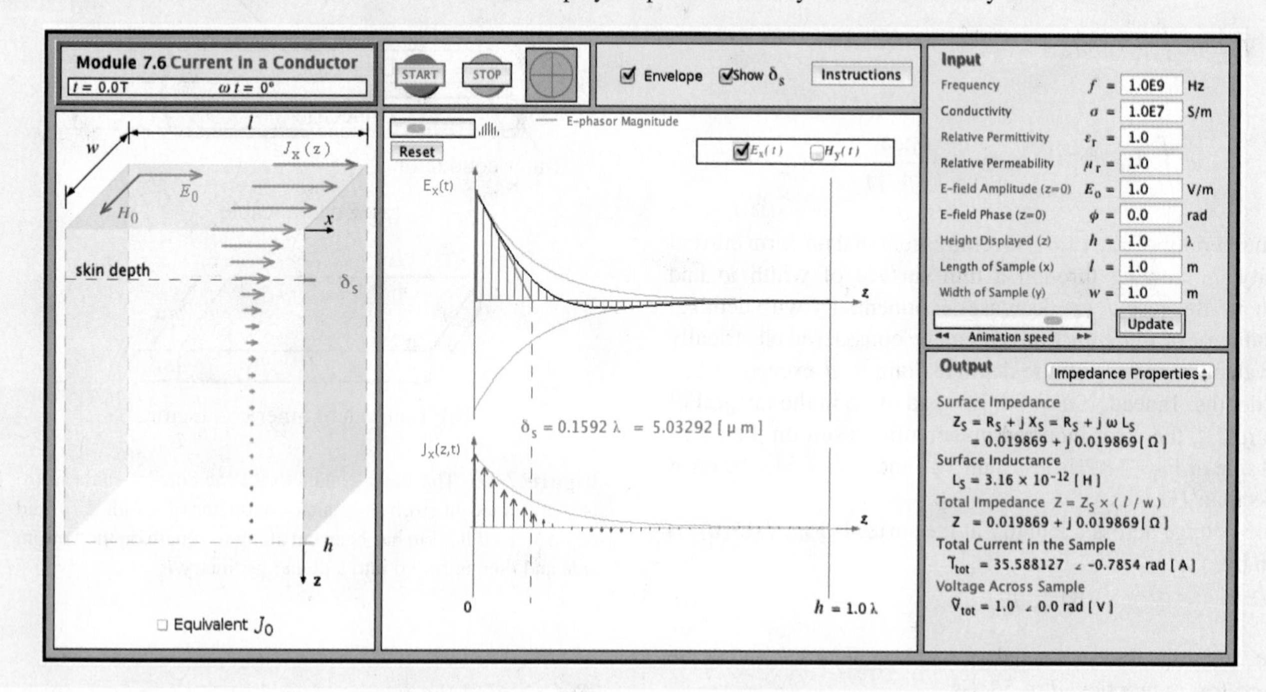

obtained by setting $w = 2\pi a$ and dividing by l in Eq. (7.93):

$$R'_1 = \frac{R}{l} = \frac{R_s}{2\pi a} \quad (\Omega/\text{m}). \tag{7.94}$$

Similarly, for the outer conductor, the current is concentrated within a thin layer of depth δ_s on the inside surface of the conductor adjacent to the insulating medium between the two conductors, which is where the EM fields exist. The resistance per unit length for the outer conductor with radius b is

$$R'_2 = \frac{R_s}{2\pi b} \quad (\Omega/\text{m}), \tag{7.95}$$

and the coaxial cable's total ac resistance per unit length is

$$R' = R'_1 + R'_2 = \frac{R_s}{2\pi}\left(\frac{1}{a} + \frac{1}{b}\right) \quad (\Omega/\text{m}). \tag{7.96}$$

This expression was used in Chapter 2 for characterizing the resistance per unit length of a coaxial transmission line.

Concept Question 7-6: How does β of a low-loss dielectric medium compare to that of a lossless medium?

Concept Question 7-7: In a good conductor, does the phase of **H** lead or lag that of **E** and by how much?

Concept Question 7-8: Attenuation means that a wave loses energy as it propagates in a lossy medium. What happens to the lost energy?

Concept Question 7-9: Is a conducting medium dispersive or dispersionless? Explain.

Concept Question 7-10: Compare the flow of current through a wire in the dc and ac cases. Compare the corresponding dc and ac resistances of the wire.

7-6 Electromagnetic Power Density

This section deals with the flow of power carried by an electromagnetic wave. For any wave with an electric field **E** and magnetic field **H**, the ***Poynting vector*** **S** is defined as

$$\mathbf{S} = \mathbf{E} \times \mathbf{H} \qquad (\text{W/m}^2). \qquad (7.97)$$

The unit of **S** is (V/m) × (A/m) = (W/m²), and the direction of **S** is along the wave's direction of propagation. Thus, **S** represents the power per unit area (or ***power density***) carried by the wave. If the wave is incident upon an aperture of area A with outward surface unit vector $\hat{\mathbf{n}}$ as shown in **Fig. 7-17**, then the total power that flows through or is intercepted by the aperture is

$$P = \int_A \mathbf{S} \cdot \hat{\mathbf{n}} \, dA \qquad (\text{W}). \qquad (7.98)$$

For a uniform plane wave propagating in a direction $\hat{\mathbf{k}}$ that makes an angle θ with $\hat{\mathbf{n}}$, $P = SA\cos\theta$, where $S = |\mathbf{S}|$.

Except for the fact that the units of **S** are per unit area, Eq. (7.97) is the vector analogue of the scalar expression for the instantaneous power $P(z,t)$ flowing through a transmission line,

$$P(z,t) = \upsilon(z,t)\; i(z,t), \qquad (7.99)$$

where $\upsilon(z,t)$ and $i(z,t)$ are the instantaneous voltage and current on the line.

Since both **E** and **H** are functions of time, so is the Poynting vector **S**. In practice, however, the quantity of greater interest is the ***average power density*** of the wave, $\mathbf{S}_{\text{av}}$, which is the time-average value of **S**:

$$\mathbf{S}_{\text{av}} = \tfrac{1}{2}\,\mathfrak{Re}\left[\widetilde{\mathbf{E}} \times \widetilde{\mathbf{H}}^*\right] \qquad (\text{W/m}^2). \qquad (7.100)$$

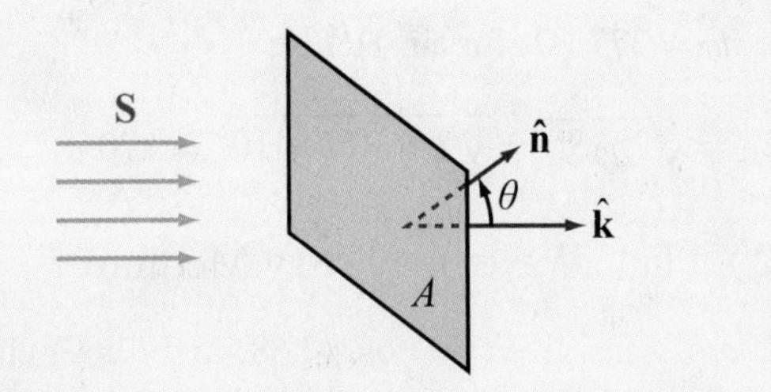

Figure 7-17 EM power flow through an aperture.

This expression may be regarded as the electromagnetic equivalent of Eq. (2.107) for the time-average power carried by a transmission line, namely

$$P_{\text{av}}(z) = \tfrac{1}{2}\,\mathfrak{Re}\left[\widetilde{V}(z)\;\widetilde{I}^*(z)\right], \qquad (7.101)$$

where $\widetilde{V}(z)$ and $\widetilde{I}(z)$ are the phasors corresponding to $\upsilon(z,t)$ and $i(z,t)$, respectively.

7-6.1 Plane Wave in a Lossless Medium

Recall that the general expression for the electric field of a uniform plane wave with arbitrary polarization traveling in the $+z$ direction is

$$\widetilde{\mathbf{E}}(z) = \hat{\mathbf{x}}\,\widetilde{E}_x(z) + \hat{\mathbf{y}}\,\widetilde{E}_y(z) = (\hat{\mathbf{x}}\,E_{x0} + \hat{\mathbf{y}}\,E_{y0})e^{-jkz}, \qquad (7.102)$$

where, in the general case, E_{x0} and E_{y0} may be complex quantities. The magnitude of $\widetilde{\mathbf{E}}$ is

$$|\widetilde{\mathbf{E}}| = (\widetilde{\mathbf{E}} \cdot \widetilde{\mathbf{E}}^*)^{1/2} = [|E_{x0}|^2 + |E_{y0}|^2]^{1/2}. \qquad (7.103)$$

The phasor magnetic field associated with $\widetilde{\mathbf{E}}$ is obtained by applying Eq. (7.39a):

$$\begin{aligned}\widetilde{\mathbf{H}}(z) &= (\hat{\mathbf{x}}\,\widetilde{H}_x + \hat{\mathbf{y}}\,\widetilde{H}_y)e^{-jkz} \\ &= \frac{1}{\eta}\,\hat{\mathbf{z}} \times \widetilde{\mathbf{E}} = \frac{1}{\eta}\,(-\hat{\mathbf{x}}\,E_{y0} + \hat{\mathbf{y}}\,E_{x0})e^{-jkz}. \qquad (7.104)\end{aligned}$$

The wave can be considered as the sum of two waves, one comprising fields $(\widetilde{E}_x, \widetilde{H}_y)$ and another comprising fields $(\widetilde{E}_y, \widetilde{H}_x)$. Use of Eqs. (7.102) and (7.104) in Eq. (7.100) leads to

$$\mathbf{S}_{\text{av}} = \hat{\mathbf{z}} \frac{1}{2\eta}(|E_{x0}|^2 + |E_{y0}|^2) = \hat{\mathbf{z}} \frac{|\widetilde{\mathbf{E}}|^2}{2\eta} \qquad (\text{W/m}^2),$$

(lossless medium) (7.105)

which states that power flows in the z direction with average power density equal to the sum of the average power densities of the $(\widetilde{E}_x, \widetilde{H}_y)$ and $(\widetilde{E}_y, \widetilde{H}_x)$ waves. Note that, because $\mathbf{S}_{\text{av}}$ depends only on η and $|\widetilde{\mathbf{E}}|$, waves characterized by different polarizations carry the same amount of average power as long as their electric fields have the same magnitudes.

Example 7-5: Solar Power

If solar illumination is characterized by a power density of 1 kW/m^2 on Earth's surface, find (a) the total power radiated by the sun, (b) the total power intercepted by Earth, and (c) the electric field of the power density incident upon Earth's surface, assuming that all the solar illumination is at a single frequency. The radius of Earth's orbit around the sun, R_s, is approximately 1.5×10^8 km, and Earth's mean radius R_e is 6,380 km.

Solution: (a) Assuming that the sun radiates isotropically (equally in all directions), the total power it radiates is $S_{\text{av}} A_{\text{sph}}$, where A_{sph} is the area of a spherical shell of radius R_s [Fig. 7-18(a)]. Thus,

$$P_{\text{sun}} = S_{\text{av}}(4\pi R_s^2) = 1 \times 10^3 \times 4\pi \times (1.5 \times 10^{11})^2$$
$$= 2.8 \times 10^{26} \text{ W}.$$

(b) With reference to Fig. 7-18(b), the power intercepted by Earth's cross section $A_e = \pi R_e^2$ is

$$P_{\text{int}} = S_{\text{av}}(\pi R_e^2) = 1 \times 10^3 \times \pi \times (6.38 \times 10^6)^2$$
$$= 1.28 \times 10^{17} \text{ W}.$$

(c) The power density S_{av} is related to the magnitude of the electric field $|\widetilde{E}| = E_0$ by

$$S_{\text{av}} = \frac{E_0^2}{2\eta_0},$$

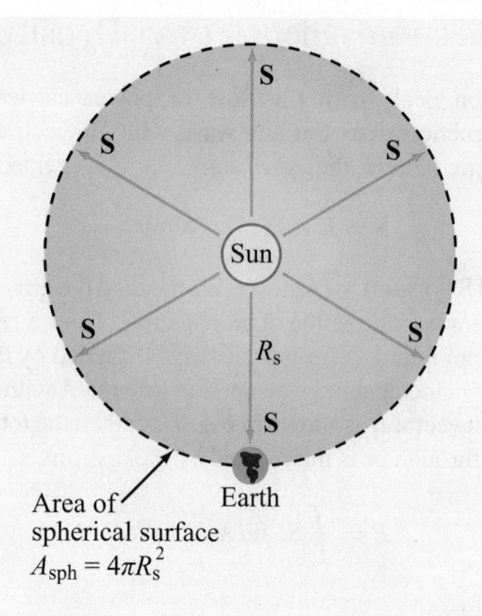

(a) Radiated solar power

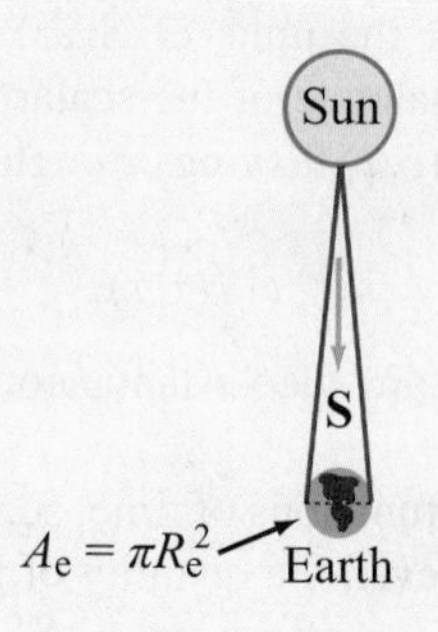

(b) Earth intercepted power

Figure 7-18 Solar radiation intercepted by (a) a spherical surface of radius R_s, and (b) Earth's surface (Example 7-5).

where $\eta_0 = 377$ (Ω) for air. Hence,

$$E_0 = \sqrt{2\eta_0 S_{\text{av}}} = \sqrt{2 \times 377 \times 10^3} = 870 \qquad (\text{V/m}).$$

7-6.2 Plane Wave in a Lossy Medium

The expressions given by Eqs. (7.68) and (7.69) characterize the electric and magnetic fields of an x-polarized plane wave propagating along the z direction in a lossy medium with propagation constant $\gamma = \alpha + j\beta$. By extending these expressions to the more general case of a wave with components along both x and y, we have

$$\widetilde{\mathbf{E}}(z) = \hat{\mathbf{x}}\,\widetilde{E}_x(z) + \hat{\mathbf{y}}\,\widetilde{E}_y(z) = (\hat{\mathbf{x}}\,E_{x0} + \hat{\mathbf{y}}\,E_{y0})e^{-\alpha z}e^{-j\beta z}, \quad (7.106\text{a})$$

$$\widetilde{\mathbf{H}}(z) = \frac{1}{\eta_c}(-\hat{\mathbf{x}}\,E_{y0} + \hat{\mathbf{y}}\,E_{x0})e^{-\alpha z}e^{-j\beta z}, \quad (7.106\text{b})$$

where η_c is the intrinsic impedance of the lossy medium. Application of Eq. (7.100) gives

$$\mathbf{S}_{\text{av}}(z) = \frac{1}{2}\,\Re\mathfrak{e}\left[\widetilde{\mathbf{E}} \times \widetilde{\mathbf{H}}^*\right] = \frac{\hat{\mathbf{z}}(|E_{x0}|^2 + |E_{y0}|^2)}{2}\,e^{-2\alpha z}\,\Re\mathfrak{e}\left(\frac{1}{\eta_c^*}\right). \quad (7.107)$$

By expressing η_c in polar form as

$$\eta_c = |\eta_c|e^{j\theta_\eta}, \quad (7.108)$$

Eq. (7.107) can be rewritten as

$$\mathbf{S}_{\text{av}}(z) = \hat{\mathbf{z}}\,\frac{|\widetilde{E}(0)|^2}{2|\eta_c|}\,e^{-2\alpha z}\cos\theta_\eta \quad (\text{W/m}^2), \quad (7.109)$$

(lossy medium)

where $|\widetilde{E}(0)|^2 = [|E_{x0}|^2 + |E_{y0}|^2]^{1/2}$ is the magnitude of $\widetilde{\mathbf{E}}(z)$ at $z = 0$.

▶ Whereas the fields $\widetilde{\mathbf{E}}(z)$ and $\widetilde{\mathbf{H}}(z)$ decay with z as $e^{-\alpha z}$, the power density $\mathbf{S}_{\text{av}}$ decreases as $e^{-2\alpha z}$. ◀

When a wave propagates through a distance $z = \delta_s = 1/\alpha$, the magnitudes of its electric and magnetic fields decrease to $e^{-1} \approx 37\%$ of their initial values, and its average power density decreases to $e^{-2} \approx 14\%$ of its initial value.

7-6.3 Decibel Scale for Power Ratios

The unit for power P is watts (W). In many engineering problems, the quantity of interest is the ratio of two power levels, P_1 and P_2, such as the incident and reflected powers on a transmission line, and often the ratio P_1/P_2 may vary over several orders of magnitude. The decibel (dB) scale is logarithmic, thereby providing a convenient representation of the power ratio, particularly when numerical values of P_1/P_2 are plotted against some variable of interest. If

$$G = \frac{P_1}{P_2}, \quad (7.110)$$

then

$$G\ [\text{dB}] = 10\log G = 10\log\left(\frac{P_1}{P_2}\right) \quad (\text{dB}). \quad (7.111)$$

Table 7-2 provides a comparison between values of G and the corresponding values of G [dB]. Even though decibels are defined for power ratios, they can sometimes be used to represent other quantities. For example, if $P_1 = V_1^2/R$ is the power dissipated in a resistor R with voltage V_1 across it at time t_1, and $P_2 = V_2^2/R$ is the power dissipated in the same resistor at time t_2, then

$$G\ [\text{dB}] = 10\log\left(\frac{P_1}{P_2}\right) = 10\log\left(\frac{V_1^2/R}{V_2^2/R}\right) = 20\log\left(\frac{V_1}{V_2}\right) = 20\log(g) = g\ [\text{dB}], \quad (7.112)$$

where $g = V_1/V_2$ is the voltage ratio. *Note that for voltage (or current) ratios the scale factor is 20 rather than 10,* which results in G [dB] $= g$ [dB].

The ***attenuation rate***, representing the rate of decrease of the magnitude of $\mathbf{S}_{\text{av}}(z)$ as a function of propagation distance, is

Table 7-2 Power ratios in natural numbers and in decibels.

G	G [dB]
10^x	$10x$ dB
4	6 dB
2	3 dB
1	0 dB
0.5	−3 dB
0.25	−6 dB
0.1	−10 dB
10^{-3}	−30 dB

defined as

$$A = 10\log\left[\frac{S_{\rm av}(z)}{S_{\rm av}(0)}\right]$$
$$= 10\log(e^{-2\alpha z})$$
$$= -20\alpha z \log e$$
$$= -8.68\alpha z = -\alpha\ [\text{dB/m}]\ z \qquad (\text{dB}), \qquad (7.113)$$

where

$$\alpha\ [\text{dB/m}] = 8.68\alpha \qquad (\text{Np/m}). \qquad (7.114)$$

We also note that, since $\mathbf{S}_{\rm av}(z)$ is directly proportional to $|\mathbf{E}(z)|^2$,

$$A = 10\log\left[\frac{|E(z)|^2}{|E(0)|^2}\right] = 20\log\left[\frac{|E(z)|}{|E(0)|}\right] \qquad (\text{dB}). \qquad (7.115)$$

Example 7-6: Power Received by a Submarine Antenna

A submarine at a depth of 200 m below the sea surface uses a wire antenna to receive signal transmissions at 1 kHz. Determine the power density incident upon the submarine antenna due to the EM wave of Example 7-4.

Solution: From Example 7-4, $|\widetilde{E}(0)| = |E_{x0}| = 4.44$ (mV/m), $\alpha = 0.126$ (Np/m), and $\eta_c = 0.044\angle 45^\circ$ (Ω). Application of Eq. (7.109) gives

$$\mathbf{S}_{\rm av}(z) = \hat{\mathbf{z}}\,\frac{|E_0|^2}{2|\eta_c|}\,e^{-2\alpha z}\cos\theta_\eta$$
$$= \hat{\mathbf{z}}\,\frac{(4.44\times10^{-3})^2}{2\times0.044}\,e^{-0.252z}\cos 45^\circ$$
$$= \hat{\mathbf{z}}\,0.16e^{-0.252z} \qquad (\text{mW/m}^2).$$

At $z = 200$ m, the incident power density is

$$\mathbf{S}_{\rm av} = \hat{\mathbf{z}}\,(0.16\times10^{-3}e^{-0.252\times200})$$
$$= 2.1\times10^{-26} \qquad (\text{W/m}^2).$$

Exercise 7-10: Convert the following values of the power ratio G to decibels: (a) 2.3, (b) 4×10^3, (c) 3×10^{-2}.

Answer: (a) 3.6 dB, (b) 36 dB, (c) −15.2 dB. (See EM.)

Exercise 7-11: Find the voltage ratio g corresponding to the following decibel values of the power ratio G: (a) 23 dB, (b) −14 dB, (c) −3.6 dB.

Answer: (a) 14.13, (b) 0.2, (c) 0.66. (See EM.)

Chapter 7 Summary

Concepts

- A spherical wave radiated by a source becomes approximately a uniform plane wave at large distances from the source.
- The electric and magnetic fields of a transverse electromagnetic (TEM) wave are orthogonal to each other, and both are perpendicular to the direction of wave travel.
- The magnitudes of the electric and magnetic fields of a TEM wave are related by the intrinsic impedance of the medium.
- Wave polarization describes the shape of the locus of the tip of the **E** vector at a given point in space as a function of time. The polarization state, which may be linear, circular, or elliptical, is governed by the ratio of the magnitudes of and the difference in phase between the two orthogonal components of the electric field vector.
- Media are classified as lossless, low-loss, quasi-conducting, or good-conducting on the basis of the ratio $\epsilon''/\epsilon' = \sigma/\omega\epsilon$.
- Unlike the dc case, wherein the current flowing through a wire is distributed uniformly across its cross section, in the ac case most of the current is concentrated along the outer perimeter of the wire.
- Power density carried by a plane EM wave traveling in an unbounded medium is akin to the power carried by the voltage/current wave on a transmission line.

Mathematical and Physical Models

Complex Permittivity

$$\epsilon_c = \epsilon' - j\epsilon''$$

$$\epsilon' = \epsilon$$

$$\epsilon'' = \frac{\sigma}{\omega}$$

Lossless Medium

$$k = \omega\sqrt{\mu\epsilon}$$

$$\eta = \sqrt{\frac{\mu}{\epsilon}} \qquad (\Omega)$$

$$u_p = \frac{\omega}{k} = \frac{1}{\sqrt{\mu\epsilon}} \qquad \text{(m/s)}$$

$$\lambda = \frac{2\pi}{k} = \frac{u_p}{f} \qquad \text{(m)}$$

Wave Polarization

$$\widetilde{\mathbf{H}} = \frac{1}{\eta}\,\hat{\mathbf{k}} \times \widetilde{\mathbf{E}}$$

$$\widetilde{\mathbf{E}} = -\eta\,\hat{\mathbf{k}} \times \widetilde{\mathbf{H}}$$

Maxwell's Equations for Time-Harmonic Fields

$$\nabla \cdot \widetilde{\mathbf{E}} = 0$$

$$\nabla \times \widetilde{\mathbf{E}} = -j\omega\mu\widetilde{\mathbf{H}}$$

$$\nabla \cdot \widetilde{\mathbf{H}} = 0$$

$$\nabla \times \widetilde{\mathbf{H}} = j\omega\epsilon_c\widetilde{\mathbf{E}}$$

Lossy Medium

$$\alpha = \omega\left\{\frac{\mu\epsilon'}{2}\left[\sqrt{1+\left(\frac{\epsilon''}{\epsilon'}\right)^2} - 1\right]\right\}^{1/2} \qquad \text{(Np/m)}$$

$$\beta = \omega\left\{\frac{\mu\epsilon'}{2}\left[\sqrt{1+\left(\frac{\epsilon''}{\epsilon'}\right)^2} + 1\right]\right\}^{1/2} \qquad \text{(rad/m)}$$

$$\eta_c = \sqrt{\frac{\mu}{\epsilon_c}} = \sqrt{\frac{\mu}{\epsilon'}}\left(1 - j\frac{\epsilon''}{\epsilon'}\right)^{-1/2} \qquad (\Omega)$$

$$\delta_s = \frac{1}{\alpha} \qquad \text{(m)}$$

Power Density

$$\mathbf{S}_{av} = \tfrac{1}{2}\,\mathfrak{Re}\left[\widetilde{\mathbf{E}} \times \widetilde{\mathbf{H}}^*\right] \qquad (\text{W/m}^2)$$

Important Terms

Provide definitions or explain the meaning of the following terms:

attenuation constant α
attenuation rate A
auxiliary angle ψ_0
average power density $\mathbf{S}_{av}$
axial ratio
circular polarization
complex permittivity ϵ_c
dc and ac resistances
elliptical polarization
ellipticity angle χ
good conductor
guided wave
homogeneous wave equation
in phase
inclination angle
internal or surface impedance
intrinsic impedance η
LHC and RHC polarizations
linear polarization
lossy medium
low-loss dielectric
out of phase
phase constant β
phase velocity
polarization state
Poynting vector $\mathbf{S}$
propagation constant γ
quasi-conductor
rotation angle γ
skin depth δ_s
spherical wave
surface resistance R_s
TEM wave
unbounded
unbounded wave
uniform plane wave
wave polarization
wavefront
wavenumber k

PROBLEMS

Section 7-2: Plane-Wave Propagation in Lossless Media

7.1 Write general expressions for the electric and magnetic fields of a 1 GHz sinusoidal plane wave traveling in the $+y$ direction in a lossless nonmagnetic medium with relative permittivity $\epsilon_r = 9$. The electric field is polarized along the x direction, its peak value is 6 V/m, and its intensity is 4 V/m at $t = 0$ and $y = 2$ cm.

7.2 The magnetic field of a wave propagating through a certain nonmagnetic material is given by

$$\mathbf{H} = \hat{\mathbf{z}}\,30\cos(10^8 t - 0.5y) \qquad \text{(mA/m)}.$$

Find the following:

*(a) The direction of wave propagation.

(b) The phase velocity.

*(c) The wavelength in the material.

(d) The relative permittivity of the material.

(e) The electric field phasor.

7.3 The electric field phasor of a uniform plane wave is given by $\widetilde{\mathbf{E}} = \hat{\mathbf{y}}\,10e^{j0.2z}$ (V/m). If the phase velocity of the wave is 1.5×10^8 m/s and the relative permeability of the medium is $\mu_r = 2.4$, find the following:

*(a) The wavelength.

(b) The frequency f of the wave.

(c) The relative permittivity of the medium.

(d) The magnetic field $\mathbf{H}(z, t)$.

7.4 The electric field of a plane wave propagating in a nonmagnetic material is given by

$$\mathbf{E} = [\hat{\mathbf{y}}\,6\sin(\pi \times 10^7 t - 0.2\pi x) + \hat{\mathbf{z}}\,4\cos(\pi \times 10^7 t - 0.2\pi x)] \qquad \text{(V/m)}.$$

Determine

(a) The wavelength.

(b) ϵ_r.

(c) $\mathbf{H}$.

*Answer(s) available in Appendix D.

*__7.5__ A wave radiated by a source in air is incident upon a soil surface, whereupon a part of the wave is transmitted into the soil medium. If the wavelength of the wave is 60 cm in air and 15 cm in the soil medium, what is the soil's relative permittivity? Assume the soil to be a very low-loss medium.

7.6 The magnetic field of a plane wave propagating in a nonmagnetic material is given by

$$\mathbf{H} = \hat{\mathbf{x}}\,60\cos(2\pi \times 10^7 t + 0.1\pi y)$$
$$\hat{\mathbf{z}}\,30\cos(2\pi \times 10^7 t + 0.1\pi y) \qquad \text{(mA/m)}.$$

Determine

*(a) The wavelength.

(b) ϵ_r.

(c) $\mathbf{E}$.

7.7 The electric field of a plane wave propagating in a lossless, nonmagnetic, dielectric material with $\epsilon_r = 2.56$ is given by

$$\mathbf{E} = \hat{\mathbf{y}}\,15\cos(6\pi \times 10^9 t - kz) \qquad \text{(V/m)}.$$

Determine:

(a) f, u_p, λ, k, and η.

(b) The magnetic field $\mathbf{H}$.

7.8 A 60 MHz plane wave traveling in the $-x$ direction in dry soil with relative permittivity $\epsilon_r = 4$ has an electric field polarized along the z direction. Assuming dry soil to be approximately lossless, and given that the magnetic field has a peak value of 10 (mA/m) and that its value was measured to be 7 (mA/m) at $t = 0$ and $x = -0.75$ m, develop complete expressions for the wave's electric and magnetic fields.

Section 7-3: Wave Polarization

7.9 An RHC-polarized wave with a modulus of 2 (V/m) is traveling in free space in the negative z direction. Write the expression for the wave's electric field vector, given that the wavelength is 3 cm.

7.10 For a wave characterized by the electric field

$$\mathbf{E}(z, t) = \hat{\mathbf{x}}\,a_x\cos(\omega t - kz) + \hat{\mathbf{y}}\,a_y\cos(\omega t - kz + \delta)$$

identify the polarization state, determine the polarization angles (γ, χ), and sketch the locus of $\mathbf{E}(0, t)$ for each of the following cases:

(a) $a_x = 3$ V/m, $a_y = 4$ V/m, and $\delta = 0$

(b) $a_x = 3$ V/m, $a_y = 4$ V/m, and $\delta = 180°$

(c) $a_x = 3$ V/m, $a_y = 3$ V/m, and $\delta = 45°$

(d) $a_x = 3$ V/m, $a_y = 4$ V/m, and $\delta = -135°$

7.11 The magnetic field of a uniform plane wave propagating in a dielectric medium with $\epsilon_r = 36$ is given by

$$\widetilde{\mathbf{H}} = 60(\hat{\mathbf{y}} + j\hat{\mathbf{z}})e^{-j\pi x/6} \qquad \text{(mA/m)}.$$

Specify the modulus and direction of the electric field intensity at the $x = 0$ plane at $t = 0$ and 5 ns.

7.12 The electric field of a uniform plane wave propagating in free space is given by

$$\widetilde{\mathbf{E}} = (\hat{\mathbf{x}} + j\hat{\mathbf{y}})20e^{-j\pi z/6} \qquad \text{(V/m)}.$$

Specify the modulus and direction of the electric field intensity at the $z = 0$ plane at $t = 0$, 5, and 10 ns.

7.13 Compare the polarization states of each of the following pairs of plane waves:

(a) Wave 1: $\mathbf{E}_1 = \hat{\mathbf{x}}\,2\cos(\omega t - kz) + \hat{\mathbf{y}}\,2\sin(\omega t - kz)$.
Wave 2: $\mathbf{E}_2 = \hat{\mathbf{x}}\,2\cos(\omega t + kz) + \hat{\mathbf{y}}\,2\sin(\omega t + kz)$.

(b) Wave 1: $\mathbf{E}_1 = \hat{\mathbf{x}}\,2\cos(\omega t - kz) - \hat{\mathbf{y}}\,2\sin(\omega t - kz)$.
Wave 2: $\mathbf{E}_2 = \hat{\mathbf{x}}\,2\cos(\omega t + kz) - \hat{\mathbf{y}}\,2\sin(\omega t + kz)$.

*__7.14__ The electric field of an elliptically polarized plane wave is given by

$$\mathbf{E}(z,t) = [-\hat{\mathbf{x}}\,10\sin(\omega t - kz - 60°) + \hat{\mathbf{y}}\,30\cos(\omega t - kz)] \qquad \text{(V/m)}.$$

Determine the following:

(a) The polarization angles (γ, χ).

(b) The direction of rotation.

7.15 A linearly polarized plane wave of the form $\widetilde{\mathbf{E}} = \hat{\mathbf{x}}\,a_x e^{-jkz}$ can be expressed as the sum of an RHC-polarized wave with magnitude a_R, and an LHC-polarized wave with magnitude a_L. Prove this statement by finding expressions for a_R and a_L in terms of a_x.

7.16 Plot the locus of $\mathbf{E}(0,t)$ for a plane wave with

$$\mathbf{E}(z,t) = \hat{\mathbf{x}}\sin(\omega t + kz) + \hat{\mathbf{y}}\,2\cos(\omega t + kz).$$

Determine the polarization state from your plot.

Section 7-4: Plane-Wave Propagation in Lossy Media

7.17 For each of the following combinations of parameters, determine if the material is a low-loss dielectric, a quasi-conductor, or a good conductor, and then calculate α, β, λ, u_p, and η_c:

*__(a)__ Glass with $\mu_r = 1$, $\epsilon_r = 5$, and $\sigma = 10^{-12}$ S/m at 10 GHz.

(b) Animal tissue with $\mu_r = 1$, $\epsilon_r = 12$, and $\sigma = 0.3$ S/m at 100 MHz.

(a) (c)] Wood with $\mu_r = 1$, $\epsilon_r = 3$, and $\sigma = 10^{-4}$ S/m at 1 kHz.

7.18 Dry soil is characterized by $\epsilon_r = 2.5$, $\mu_r = 1$, and $\sigma = 10^{-4}$ (S/m). At each of the following frequencies, determine if dry soil may be considered a good conductor, a quasi-conductor, or a low-loss dielectric, and then calculate α, β, λ, μ_p, and η_c:

(a) 60 Hz

(b) 1 kHz

(c) 1 MHz

(d) 1 GHz

*__7.19__ In a medium characterized by $\epsilon_r = 9$, $\mu_r = 1$, and $\sigma = 0.1$ S/m, determine the phase angle by which the magnetic field leads the electric field at 100 MHz.

7.20 Ignoring reflection at the air–water boundary, if the amplitude of a 1 GHz incident wave in air is 20 V/m at the water surface, at what depth will it be down to 1 μV/m? Water has $\mu_r = 1$, and at 1 GHz, $\epsilon_r = 80$ and $\sigma = 1$ S/m.

*__7.21__ Ignoring reflection at the air–soil boundary, if the amplitude of a 3 GHz incident wave is 10 V/m at the surface of a wet soil medium, at what depth will it be down to 1 mV/m? Wet soil is characterized by $\mu_r = 1$, $\epsilon_r = 9$, and $\sigma = 5 \times 10^{-4}$ S/m.

7.22 Generate a plot for the skin depth δ_s versus frequency for seawater for the range from 1 kHz to 10 GHz (use log-log scales). The constitutive parameters of seawater are $\mu_r = 1$, $\epsilon_r = 80$, and $\sigma = 4$ S/m.

*__7.23__ The skin depth of a certain nonmagnetic conducting material is 2 μm at 5 GHz. Determine the phase velocity in the material.

7.24 Based on wave attenuation and reflection measurements conducted at 1 MHz, it was determined that the intrinsic impedance of a certain medium is $28.1\angle 45°$ (Ω) and the skin depth is 2 m. Determine the following:

(a) The conductivity of the material.

(b) The wavelength in the medium.

(c) The phase velocity.

*__7.25__ The electric field of a plane wave propagating in a nonmagnetic medium is given by

$$\mathbf{E} = \hat{\mathbf{z}}\,25e^{-30x}\cos(2\pi \times 10^9 t - 40x) \qquad \text{(V/m)}.$$

Obtain the corresponding expression for **H**.

7.26 At 2 GHz, the conductivity of meat is on the order of 1 (S/m). When a material is placed inside a microwave oven and the field is activated, the presence of the electromagnetic fields in the conducting material causes energy dissipation in the material in the form of heat.

(a) Develop an expression for the time-average power per mm^3 dissipated in a material of conductivity σ if the peak electric field in the material is E_0.

(b) Evaluate the result for an electric field $E_0 = 4 \times 10^4$ (V/m).

7.27 The magnetic field of a plane wave propagating in a nonmagnetic medium is given by

$$\mathbf{H} = \hat{\mathbf{y}}\,60e^{-10z}\cos(2\pi \times 10^8 t - 12z) \qquad \text{(mA/m)}.$$

Obtain the corresponding expression for **E**.

Section 7-5: Current Flow in Conductors

7.28 In a nonmagnetic, lossy, dielectric medium, a 300 MHz plane wave is characterized by the magnetic field phasor

$$\tilde{\mathbf{H}} = (\hat{\mathbf{x}} - j4\hat{\mathbf{z}})e^{-2y}e^{-j9y} \qquad \text{(A/m)}.$$

Obtain time-domain expressions for the electric and magnetic field vectors.

*__7.29__ A rectangular copper block is 60 cm in height (along z). In response to a wave incident upon the block from above, a current is induced in the block in the positive x direction. Determine the ratio of the ac resistance of the block to its dc resistance at 1 kHz. The relevant properties of copper are given in Appendix B.

7.30 Repeat Problem 7.29 at 10 MHz.

7.31 The inner and outer conductors of a coaxial cable have radii of 0.5 cm and 1 cm, respectively. The conductors are made of copper with $\epsilon_r = 1$, $\mu_r = 1$, and $\sigma = 5.8 \times 10^7$ S/m, and the outer conductor is 0.5 mm thick. At 10 MHz:

(a) Are the conductors thick enough to be considered infinitely thick as far as the flow of current through them is concerned?

(b) Determine the surface resistance R_s.

(c) Determine the ac resistance per unit length of the cable.

7.32 Repeat Problem 7.31 at 1 GHz.

Section 7-6: Electromagnetic Power Density

7.33 A wave traveling in a nonmagnetic medium with $\epsilon_r = 9$ is characterized by an electric field given by

$$\mathbf{E} = [\hat{\mathbf{y}}\,3\cos(\pi \times 10^7 t + kx) - \hat{\mathbf{z}}\,2\cos(\pi \times 10^7 t + kx)] \qquad \text{(V/m)}.$$

Determine the direction of wave travel and average power density carried by the wave.

*__7.34__ The magnetic field of a plane wave traveling in air is given by $\mathbf{H} = \hat{\mathbf{x}}\,50\sin(2\pi \times 10^7 t - ky)$ (mA/m). Determine the average power density carried by the wave.

7.35 The electric-field phasor of a uniform plane wave traveling downward in water is given by

$$\tilde{\mathbf{E}} = \hat{\mathbf{x}}\,5e^{-0.2z}e^{-j0.2z} \qquad \text{(V/m)}.$$

where $\hat{\mathbf{z}}$ is the downward direction and $z = 0$ is the water surface. If $\sigma = 4$ S/m,

(a) Obtain an expression for the average power density.

(b) Determine the attenuation rate.

***(c)** Determine the depth at which the power density has been reduced by 40 dB.

*__7.36__ A wave traveling in a lossless, nonmagnetic medium has an electric field amplitude of 24.56 V/m and an average power density of 2.4 W/m^2. Determine the phase velocity of the wave.

7.37 The amplitudes of an elliptically polarized plane wave traveling in a lossless, nonmagnetic medium with $\epsilon_r = 4$ are $H_{y0} = 3$ (mA/m) and $H_{z0} = 4$ (mA/m). Determine the average power flowing through an aperture in the y–z plane if its area is 20 m^2.

7.38 At microwave frequencies, the power density considered safe for human exposure is 1 (mW/cm^2). A radar radiates a wave with an electric field amplitude E that decays with distance as

$E(R) = (3,000/R)$ (V/m), where R is the distance in meters. What is the radius of the unsafe region?

7.39 Consider the imaginary rectangular box shown in **Fig. P7.39**.

(a) Determine the net power flux $P(t)$ entering the box due to a plane wave in air given by

$$\mathbf{E} = \hat{\mathbf{x}} E_0 \cos(\omega t - ky) \qquad \text{(V/m)}.$$

***(b)** Determine the net time-average power entering the box.

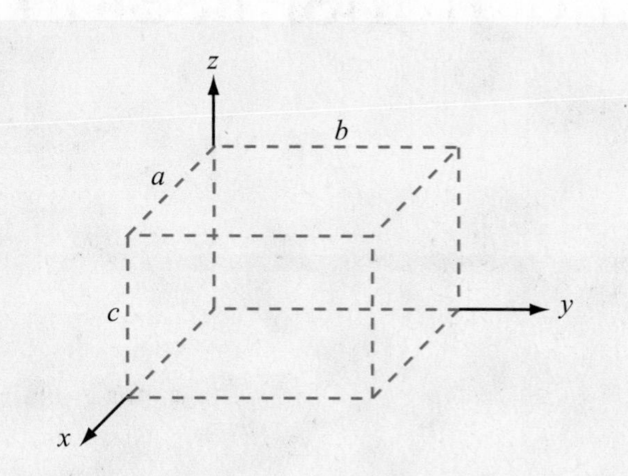

Figure P7.39 Imaginary rectangular box of Problems 7.39 and 7.40.

7.40 Repeat Problem 7.39 for a wave traveling in a lossy medium in which

$$\mathbf{E} = \hat{\mathbf{x}}\, 100e^{-20y} \cos(2\pi \times 10^9 t - 40y) \qquad \text{(V/m)},$$

$$\mathbf{H} = -\hat{\mathbf{z}}\, 0.64e^{-20y} \cdot \cos(2\pi \times 10^9 t - 40y - 36.85^\circ) \qquad \text{(A/m)}.$$

The box has dimensions $a = 1$ cm, $b = 2$ cm, and $c = 0.5$ cm.

7.41 A team of scientists is designing a radar as a probe for measuring the depth of the ice layer over the antarctic land mass. In order to measure a detectable echo due to the reflection by the ice-rock boundary, the thickness of the ice sheet should not exceed three skin depths. If $\epsilon_r' = 3$ and $\epsilon_r'' = 10^{-2}$ for ice and if the maximum anticipated ice thickness in the area under exploration is 1.2 km, what frequency range is useable with the radar?

7.42 Given a wave with $\mathbf{E} = \hat{\mathbf{x}} E_0 \cos(\omega t - kz)$:

***(a)** Calculate the time-average electric energy density

$$(w_e)_{av} = \frac{1}{T}\int_0^T w_e \, dt = \frac{1}{2T}\int_0^T \epsilon E^2 \, dt.$$

(b) Calculate the time-average magnetic energy density

$$(w_m)_{av} = \frac{1}{T}\int_0^T w_m \, dt = \frac{1}{2T}\int_0^T \mu H^2 \, dt.$$

(c) Show that $(w_e)_{av} = (w_m)_{av}$.

CHAPTER

8

Wave Reflection and Transmission

Chapter Contents

	EM Waves at Boundaries, 375
8-1	Wave Reflection and Transmission at Normal Incidence, 375
8-2	Snell's Laws, 384
8-3	Fiber Optics, 387
8-4	Wave Reflection and Transmission at Oblique Incidence, 389
TB15	Lasers, 390
8-5	Reflectivity and Transmissivity, 398
8-6	Waveguides, 402
TB16	Bar-Code Readers, 404
8-7	General Relations for **E** and **H**, 405
8-8	TM Modes in Rectangular Waveguide, 406
8-9	TE Modes in Rectangular Waveguide, 410
8-10	Propagation Velocities, 411
8-11	Cavity Resonators, 414
	Chapter 8 Summary, 417
	Problems, 419

Objectives

Upon learning the material presented in this chapter, you should be able to:

1. Characterize the reflection and transmission behavior of plane waves incident upon plane boundaries, for both normal and oblique incidence.
2. Calculate the transmission properties of optical fibers.
3. Characterize wave propagation in a rectangular waveguide.
4. Determine the behavior of resonant modes inside a rectangular cavity.

EM Waves at Boundaries

Figure 8-1 depicts the propagation path traveled by a signal transmitted by a shipboard antenna and received by an antenna on a submerged submarine. Starting from the transmitter (denoted Tx in **Fig. 8-1**), the signal travels along a transmission line to the transmitting antenna. The relationship between the transmitter (generator) output power, P_t, and the power supplied to the antenna is governed by the transmission-line equations of Chapter 2. If the transmission line is approximately lossless and properly matched to the transmitting antenna, then all of P_t is delivered to the antenna. If the antenna itself is lossless too, it will convert all of the power P_t in the guided wave provided by the transmission line into a spherical wave radiated outward into space. The radiation process is the subject of Chapter 9. From point 1, denoting the location of the shipboard antenna, to point 2, denoting the point of incidence of the wave onto the water's surface, the signal's behavior is governed by the equations characterizing wave propagation in lossless media, covered in Chapter 7. As the wave impinges upon the air–water boundary, part of it is reflected by the surface while another part gets transmitted across the boundary into the water. The transmitted wave is refracted, wherein its propagation direction moves closer toward the vertical, compared with that of the incident wave. Reflection and transmission processes are treated in this chapter. Wave travel from point 3, representing a point just below the water surface, to point 4, which denotes the location of the submarine antenna, is subject to the laws of wave propagation in lossy media, also treated in Chapter 7. Finally, some of the power carried by the wave traveling in water towards the submarine is intercepted by the receiving antenna. The received power, P_r, is then delivered to the receiver via a transmission line. The receiving properties of antennas are covered in Chapter 9. In summary, then, each wave-related aspect of the transmission process depicted in **Fig. 8-1**, starting with the transmitter and ending with the receiver, is treated in this book.

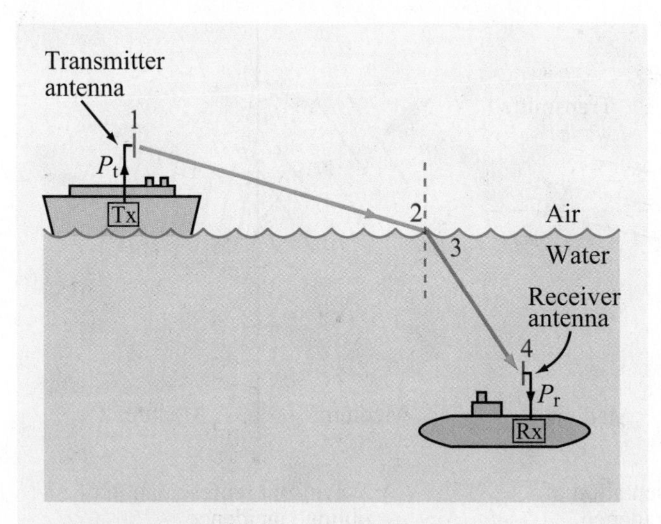

Figure 8-1 Signal path between a shipboard transmitter (Tx) and a submarine receiver (Rx).

This chapter begins by examining the reflection and transmission properties of plane waves incident upon planar boundaries and concludes with sections on waveguides and cavity resonators. Applications discussed along the way include fiber and laser optics.

8-1 Wave Reflection and Transmission at Normal Incidence

We know from Chapter 2 that, when a guided wave encounters a junction between two transmission lines with different characteristic impedances, the incident wave is partly reflected back toward the source and partly transmitted across the junction onto the other line. The same happens to a uniform plane wave when it encounters a boundary between two material half-spaces with different characteristic impedances. In fact, the situation depicted in **Fig. 8-2(b)** has an exact analogue in the transmission-line configuration of **Fig. 8-2(a)**. The boundary conditions governing the relationships between the electric and magnetic fields in **Fig. 8-2(b)** map one to one onto those we developed in Chapter 2 for the voltages and currents on the transmission line.

For convenience, we divide our treatment of wave reflection by, and transmission through, planar boundaries into two parts: in this section we confine our discussion to the normal-incidence case depicted in **Fig. 8-3(a)**, and in Sections 8-2 to 8-4 we examine the more general oblique-incidence case depicted in **Fig. 8-3(b)**. We will show the basis for the analogy between the transmission-line and plane-wave configurations so that we may use transmission-line equivalent models, tools (e.g., Smith chart), and techniques (e.g., quarter-wavelength matching) to expeditiously solve plane wave problems.

Before proceeding, however, we should explain the notion of rays and wavefronts and the relationship between them, as both are used throughout this chapter to represent electromagnetic waves. A *ray* is a line representing the direction of flow

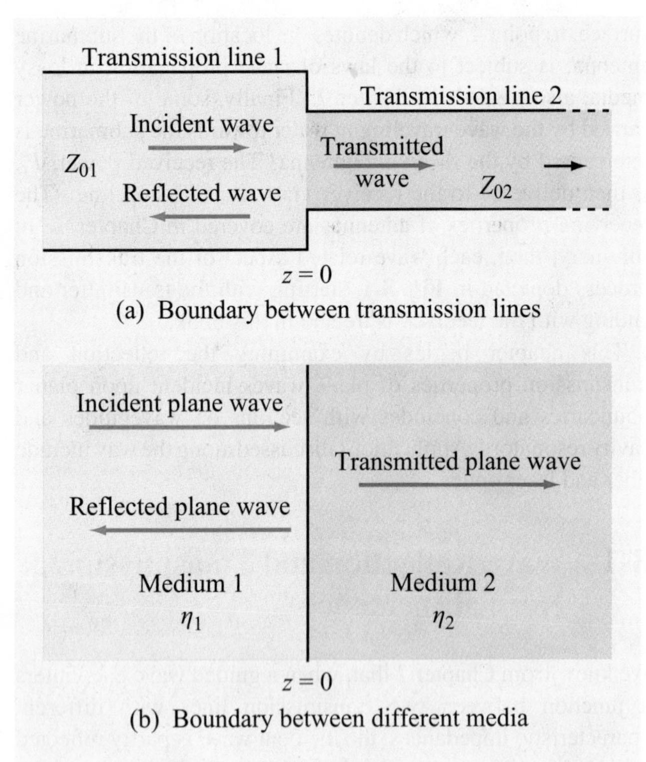

(a) Boundary between transmission lines

(b) Boundary between different media

Figure 8-2 Discontinuity between two different transmission lines is analogous to that between two dissimilar media.

of electromagnetic energy carried by a wave, and therefore it is parallel to the propagation unit vector $\hat{\mathbf{k}}$. A ***wavefront*** is a surface across which the phase of a wave is constant; it is perpendicular to the wavevector $\hat{\mathbf{k}}$. Hence, rays are perpendicular to wavefronts. The ray representation of wave incidence, reflection, and transmission shown in **Fig. 8-3(b)** is equivalent to the wavefront representation depicted in **Fig. 8-3(c)**. The two representations are complementary; the ray representation is easier to use in graphical illustrations, whereas the wavefront representation provides greater physical insight into what happens to a wave when it encounters a discontinuous boundary.

8-1.1 Boundary between Lossless Media

A planar boundary located at $z = 0$ [**Fig. 8-4(a)**] separates two lossless, homogeneous, dielectric media. Medium 1 has permittivity ϵ_1 and permeability μ_1 and fills the half-space $z \leq 0$. Medium 2 has permittivity ϵ_2 and permeability μ_2 and fills the half-space $z \geq 0$. An x-polarized plane wave with electric and magnetic fields $(\mathbf{E}^{\mathrm{i}}, \mathbf{H}^{\mathrm{i}})$ propagates in medium 1 along direction $\hat{\mathbf{k}}_{\mathrm{i}} = \hat{\mathbf{z}}$ toward medium 2. Reflection and transmission at the boundary at $z = 0$ result in a reflected wave, with electric and magnetic fields $(\mathbf{E}^{\mathrm{r}}, \mathbf{H}^{\mathrm{r}})$, traveling along direction

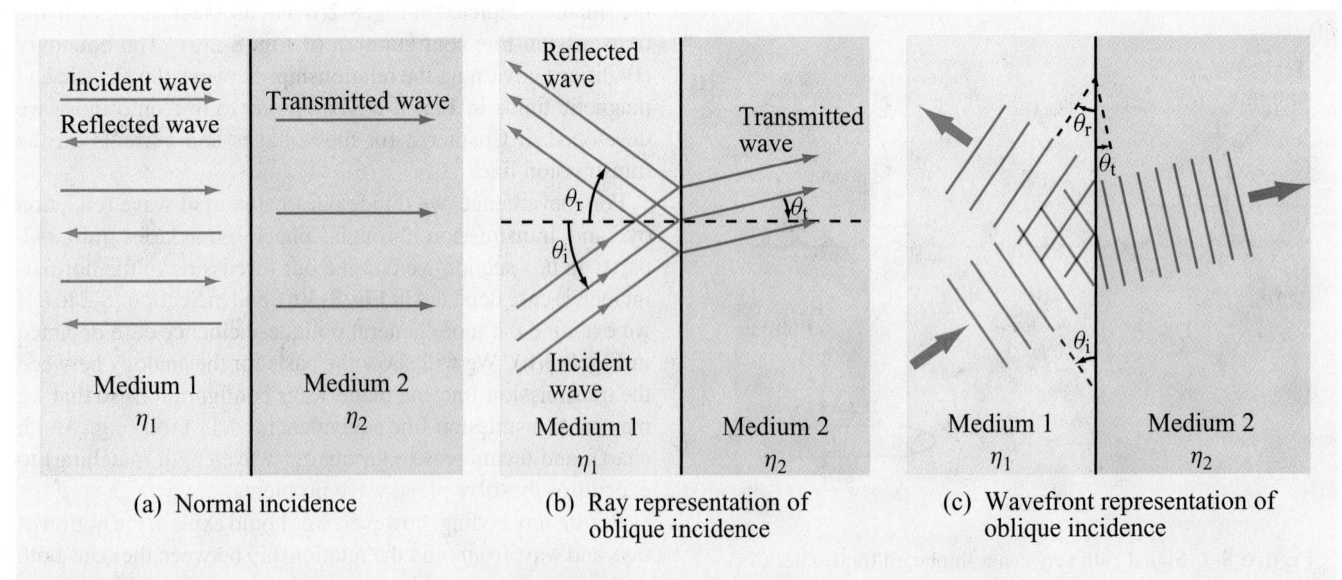

(a) Normal incidence

(b) Ray representation of oblique incidence

(c) Wavefront representation of oblique incidence

Figure 8-3 Ray representation of wave reflection and transmission at (a) normal incidence and (b) oblique incidence, and (c) wavefront representation of oblique incidence.

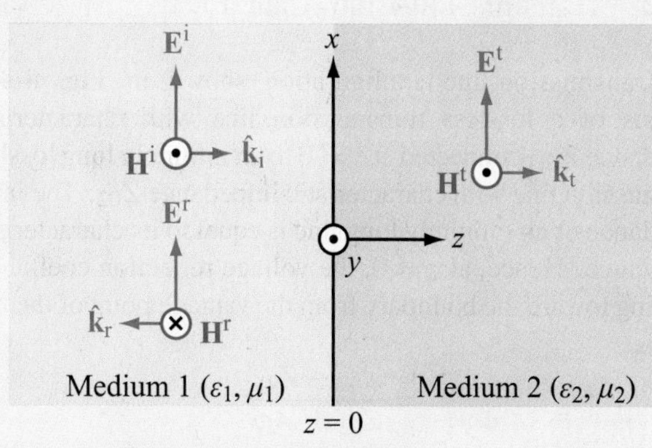

(a) Boundary between dielectric media

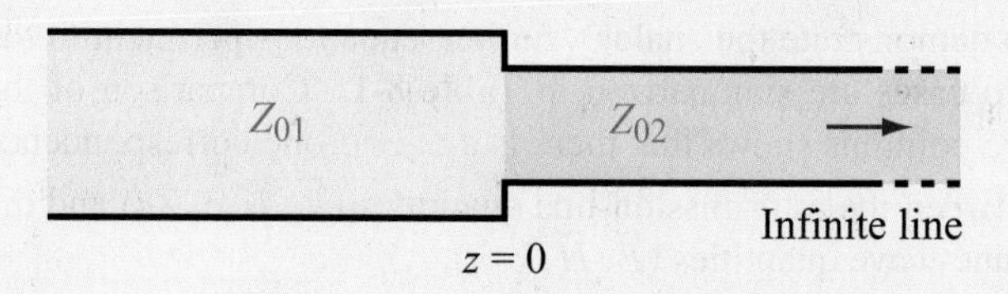

(b) Transmission-line analogue

Figure 8-4 The two dielectric media separated by the x–y plane in (a) can be represented by the transmission-line analogue in (b).

$\hat{\mathbf{k}}_r = -\hat{\mathbf{z}}$ in medium 1, and a transmitted wave, with electric and magnetic fields $(\mathbf{E}^t, \mathbf{H}^t)$, traveling along direction $\hat{\mathbf{k}}_t = \hat{\mathbf{z}}$ in medium 2. On the basis of the formulations developed in Sections 7-2 and 7-3 for plane waves, the three waves are described in phasor form by:

Incident Wave

$$\widetilde{\mathbf{E}}^i(z) = \hat{\mathbf{x}} E_0^i e^{-jk_1 z}, \tag{8.1a}$$

$$\widetilde{\mathbf{H}}^i(z) = \hat{\mathbf{z}} \times \frac{\widetilde{\mathbf{E}}^i(z)}{\eta_1} = \hat{\mathbf{y}} \, \frac{E_0^i}{\eta_1} \, e^{-jk_1 z}. \tag{8.1b}$$

Reflected Wave

$$\widetilde{\mathbf{E}}^r(z) = \hat{\mathbf{x}} E_0^r e^{jk_1 z}, \tag{8.2a}$$

$$\widetilde{\mathbf{H}}^r(z) = (-\hat{\mathbf{z}}) \times \frac{\widetilde{\mathbf{E}}^r(z)}{\eta_1} = -\hat{\mathbf{y}} \, \frac{E_0^r}{\eta_1} \, e^{jk_1 z}. \tag{8.2b}$$

Transmitted Wave

$$\widetilde{\mathbf{E}}^t(z) = \hat{\mathbf{x}} E_0^t e^{-jk_2 z}, \tag{8.3a}$$

$$\widetilde{\mathbf{H}}^t(z) = \hat{\mathbf{z}} \times \frac{\widetilde{\mathbf{E}}^t(z)}{\eta_2} = \hat{\mathbf{y}} \, \frac{E_0^t}{\eta_2} \, e^{-jk_2 z}. \tag{8.3b}$$

The quantities E_0^i, E_0^r, and E_0^t are, respectively, the amplitudes of the incident, reflected, and transmitted electric fields at $z = 0$ (the boundary between the two media). The wavenumber and intrinsic impedance of medium 1 are $k_1 = \omega\sqrt{\mu_1 \epsilon_1}$ and $\eta_1 = \sqrt{\mu_1/\epsilon_1}$, and those for medium 2 are $k_2 = \omega\sqrt{\mu_2 \epsilon_2}$ and $\eta_2 = \sqrt{\mu_2/\epsilon_2}$.

The amplitude E_0^i is imposed by the source responsible for generating the incident wave, and therefore is assumed known. Our goal is to relate E_0^r and E_0^t to E_0^i. We do so by applying boundary conditions for the total electric and magnetic fields at $z = 0$. According to **Table 6-2**, the tangential component of the total electric field is always continuous across a boundary between two contiguous media, and in the absence of current sources at the boundary, the same is true for the total magnetic field. In the present case, the electric and magnetic fields of the incident, reflected, and transmitted waves are all tangential to the boundary.

The total electric field $\widetilde{\mathbf{E}}_1(z)$ in medium 1 is the sum of the electric fields of the incident and reflected waves, and a similar statement applies to the magnetic field $\widetilde{\mathbf{H}}_1(z)$. Hence,

Medium 1

$$\widetilde{\mathbf{E}}_1(z) = \widetilde{\mathbf{E}}^i(z) + \widetilde{\mathbf{E}}^r(z)$$
$$= \hat{\mathbf{x}}(E_0^i e^{-jk_1 z} + E_0^r e^{jk_1 z}), \tag{8.4a}$$

$$\widetilde{\mathbf{H}}_1(z) = \widetilde{\mathbf{H}}^i(z) + \widetilde{\mathbf{H}}^r(z)$$
$$= \hat{\mathbf{y}} \, \frac{1}{\eta_1} \, (E_0^i e^{-jk_1 z} - E_0^r e^{jk_1 z}). \tag{8.4b}$$

With only the transmitted wave present in medium 2, the total fields are

Medium 2

$$\widetilde{\mathbf{E}}_2(z) = \widetilde{\mathbf{E}}^t(z) = \hat{\mathbf{x}} E_0^t e^{-jk_2 z}, \tag{8.5a}$$

$$\widetilde{\mathbf{H}}_2(z) = \widetilde{\mathbf{H}}^t(z) = \hat{\mathbf{y}} \, \frac{E_0^t}{\eta_2} e^{-jk_2 z}. \tag{8.5b}$$

At the boundary ($z = 0$), the tangential components of the

electric and magnetic fields are continuous. Hence,

$$\widetilde{\mathbf{E}}_1(0) = \widetilde{\mathbf{E}}_2(0) \quad \text{or} \quad E_0^{\mathrm{i}} + E_0^{\mathrm{r}} = E_0^{\mathrm{t}}, \tag{8.6a}$$

$$\widetilde{\mathbf{H}}_1(0) = \widetilde{\mathbf{H}}_2(0) \quad \text{or} \quad \frac{E_0^{\mathrm{i}}}{\eta_1} - \frac{E_0^{\mathrm{r}}}{\eta_1} = \frac{E_0^{\mathrm{t}}}{\eta_2}. \tag{8.6b}$$

Solving these equations for E_0^{r} and E_0^{t} in terms of E_0^{i} gives

$$E_0^{\mathrm{r}} = \left(\frac{\eta_2 - \eta_1}{\eta_2 + \eta_1}\right) E_0^{\mathrm{i}} = \Gamma E_0^{\mathrm{i}}, \tag{8.7a}$$

$$E_0^{\mathrm{t}} = \left(\frac{2\eta_2}{\eta_2 + \eta_1}\right) E_0^{\mathrm{i}} = \tau E_0^{\mathrm{i}}, \tag{8.7b}$$

where

$$\Gamma = \frac{E_0^{\mathrm{r}}}{E_0^{\mathrm{i}}} = \frac{\eta_2 - \eta_1}{\eta_2 + \eta_1} \quad \textbf{(normal incidence)}, \tag{8.8a}$$

$$\tau = \frac{E_0^{\mathrm{t}}}{E_0^{\mathrm{i}}} = \frac{2\eta_2}{\eta_2 + \eta_1} \quad \textbf{(normal incidence)}. \tag{8.8b}$$

The quantities Γ and τ are called the ***reflection*** and ***transmission coefficients***. For lossless dielectric media, η_1 and η_2 are real; consequently, both Γ and τ are real also. As we see in Section 8-1.4, the expressions given by Eqs. (8.8a) and (8.8b) are equally applicable when the media are conductive, even though in that case η_1 and η_2 may be complex, and hence Γ and τ may be complex as well. From Eqs. (8.8a) and (8.8b), it is easily shown that Γ and τ are interrelated as

$$\tau = 1 + \Gamma \quad \textbf{(normal incidence)}. \tag{8.9}$$

For nonmagnetic media,

$$\eta_1 = \frac{\eta_0}{\sqrt{\epsilon_{\mathrm{r}_1}}}, \qquad \eta_2 = \frac{\eta_0}{\sqrt{\epsilon_{\mathrm{r}_2}}},$$

where η_0 is the intrinsic impedance of free space, in which case Eq. (8.8a) may be expressed as

$$\Gamma = \frac{\sqrt{\epsilon_{\mathrm{r}_1}} - \sqrt{\epsilon_{\mathrm{r}_2}}}{\sqrt{\epsilon_{\mathrm{r}_1}} + \sqrt{\epsilon_{\mathrm{r}_2}}} \quad \textbf{(nonmagnetic media)}. \tag{8.10}$$

8-1.2 Transmission-Line Analogue

The transmission-line configuration shown in **Fig. 8-4(b)** consists of a lossless transmission line with characteristic impedance Z_{01}, connected at $z = 0$ to an infinitely long lossless transmission line with characteristic impedance Z_{02}. The input impedance of an infinitely long line is equal to its characteristic impedance. Hence, at $z = 0$, the voltage reflection coefficient (looking toward the boundary from the vantage point of the first line) is

$$\Gamma = \frac{Z_{02} - Z_{01}}{Z_{02} + Z_{01}},$$

which is identical in form to Eq. (8.8a). The analogy between plane waves and waves on transmission lines does not end here. To demonstrate the analogy further, equations pertinent to the two cases are summarized in **Table 8-1**. Comparison of the two columns shows that there is a one-to-one correspondence between the transmission-line quantities $(\widetilde{V}, \widetilde{I}, \beta, Z_0)$ and the plane-wave quantities $(\widetilde{E}, \widetilde{H}, k, \eta)$.

► This correspondence allows us to use the techniques developed in Chapter 2, including the Smith-chart method for calculating impedance transformations, to solve plane-wave propagation problems. ◄

The simultaneous presence of incident *and* reflected waves in medium 1 [**Fig. 8-4(a)**] gives rise to a standing-wave pattern. By analogy with the transmission-line case, the ***standing-wave ratio*** in medium 1 is defined as

$$S = \frac{|\widetilde{E}_1|_{\max}}{|\widetilde{E}_1|_{\min}} = \frac{1 + |\Gamma|}{1 - |\Gamma|}. \tag{8.15}$$

► If the two media have equal impedances ($\eta_1 = \eta_2$), then $\Gamma = 0$ and $S = 1$, and if medium 2 is a perfect conductor with $\eta_2 = 0$ (which is equivalent to a short-circuited transmission line), then $\Gamma = -1$ and $S = \infty$. ◄

The distance from the boundary to where the magnitude of the electric field intensity in medium 1 is a maximum, denoted $l_{\max}$, is described by the same expression as that given by Eq. (2.70) for the voltage maxima on a transmission line, namely

Table 8-1 **Analogy between plane-wave equations for normal incidence and transmission-line equations, both under lossless conditions.**

Plane Wave [Fig. 8-4(a)]		Transmission Line [Fig. 8-4(b)]	
$\widetilde{\mathbf{E}}_1(z) = \hat{\mathbf{x}} E_0^{\mathrm{i}}(e^{-jk_1z} + \Gamma e^{jk_1z})$	(8.11a)	$\widetilde{V}_1(z) = V_0^+(e^{-j\beta_1z} + \Gamma e^{j\beta_1z})$	(8.11b)
$\widetilde{\mathbf{H}}_1(z) = \hat{\mathbf{y}}\,\dfrac{E_0^{\mathrm{i}}}{\eta_1}(e^{-jk_1z} - \Gamma e^{jk_1z})$	(8.12a)	$\widetilde{I}_1(z) = \dfrac{V_0^+}{Z_{01}}(e^{-j\beta_1z} - \Gamma e^{j\beta_1z})$	(8.12b)
$\widetilde{\mathbf{E}}_2(z) = \hat{\mathbf{x}}\tau E_0^{\mathrm{i}} e^{-jk_2z}$	(8.13a)	$\widetilde{V}_2(z) = \tau V_0^+ e^{-j\beta_2z}$	(8.13b)
$\widetilde{\mathbf{H}}_2(z) = \hat{\mathbf{y}}\tau\,\dfrac{E_0^{\mathrm{i}}}{\eta_2}\,e^{-jk_2z}$	(8.14a)	$\widetilde{I}_2(z) = \tau\,\dfrac{V_0^+}{Z_{02}}\,e^{-j\beta_2z}$	(8.14b)
$\Gamma = (\eta_2 - \eta_1)/(\eta_2 + \eta_1)$		$\Gamma = (Z_{02} - Z_{01})/(Z_{02} + Z_{01})$	
$\tau = 1 + \Gamma$		$\tau = 1 + \Gamma$	
$k_1 = \omega\sqrt{\mu_1\epsilon_1}\,,\quad k_2 = \omega\sqrt{\mu_2\epsilon_2}$		$\beta_1 = \omega\sqrt{\mu_1\epsilon_1}\,,\quad \beta_2 = \omega\sqrt{\mu_2\epsilon_2}$	
$\eta_1 = \sqrt{\mu_1/\epsilon_1}\,,\quad \eta_2 = \sqrt{\mu_2/\epsilon_2}$		Z_{01} and Z_{02} depend on transmission-line parameters	

$$-z = l_{\max} = \frac{\theta_{\mathrm{r}} + 2n\pi}{2k_1} = \frac{\theta_{\mathrm{r}}\lambda_1}{4\pi} + \frac{n\lambda_1}{2}\,, \qquad (8.16)$$

$$\begin{cases} n = 1, 2, \ldots, & \text{if } \theta_{\mathrm{r}} < 0, \\ n = 0, 1, 2, \ldots, & \text{if } \theta_{\mathrm{r}} \geq 0, \end{cases}$$

where $\lambda_1 = 2\pi/k_1$ and θ_{r} is the phase angle of Γ (i.e., $\Gamma = |\Gamma|e^{j\theta_{\mathrm{r}}}$, and θ_{r} is bounded in the range $-\pi < \theta_{\mathrm{r}} \leq \pi$). The expression for $l_{\max}$ is valid not only when the two media are lossless dielectrics, but also when medium 1 is a low-loss dielectric. Moreover, medium 2 may be either a dielectric or a conductor. When both media are lossless dielectrics, $\theta_{\mathrm{r}} = 0$ if $\eta_2 > \eta_1$ and $\theta_{\mathrm{r}} = \pi$ if $\eta_2 < \eta_1$.

The spacing between adjacent maxima is $\lambda_1/2$, and the spacing between a maximum and the nearest minimum is $\lambda_1/4$. The electric-field minima occur at

$$l_{\min} = \begin{cases} l_{\max} + \lambda_1/4, & \text{if } l_{\max} < \lambda_1/4, \\ l_{\max} - \lambda_1/4, & \text{if } l_{\max} \geq \lambda_1/4. \end{cases} \qquad (8.17)$$

8-1.3 Power Flow in Lossless Media

Medium 1 in Fig. 8-4(a) is host to the incident and reflected waves, which together comprise the total electric and magnetic fields $\widetilde{\mathbf{E}}_1(z)$ and $\widetilde{\mathbf{H}}_1(z)$ given by Eqs. (8.11a) and (8.12a) of Table 8-1. Using Eq. (7.100), the net average power density flowing in medium 1 is

$$\begin{aligned} \mathbf{S}_{\mathrm{av}_1}(z) &= \tfrac{1}{2}\mathfrak{Re}[\widetilde{\mathbf{E}}_1(z) \times \widetilde{\mathbf{H}}_1^*(z)] \\ &= \tfrac{1}{2}\mathfrak{Re}\left[\hat{\mathbf{x}} E_0^{\mathrm{i}}(e^{-jk_1z} + \Gamma e^{jk_1z}) \right. \\ &\qquad \left. \times\, \hat{\mathbf{y}}\,\frac{E_0^{\mathrm{i}*}}{\eta_1}(e^{jk_1z} - \Gamma^* e^{-jk_1z})\right] \\ &= \hat{\mathbf{z}}\,\frac{|E_0^{\mathrm{i}}|^2}{2\eta_1}\,(1 - |\Gamma|^2), \end{aligned} \qquad (8.18)$$

which is analogous to Eq. (2.106) for the lossless transmission-line case. The first and second terms inside the bracket in Eq. (8.18) represent the average power density of the incident and reflected waves, respectively. Thus,

$$\mathbf{S}_{\mathrm{av}_1} = \mathbf{S}_{\mathrm{av}}^{\mathrm{i}} + \mathbf{S}_{\mathrm{av}}^{\mathrm{r}}, \qquad (8.19a)$$

with

$$\mathbf{S}^{\text{i}}_{\text{av}} = \hat{\mathbf{z}}\,\frac{|E^{\text{i}}_0|^2}{2\eta_1}\,, \tag{8.19b}$$

$$\mathbf{S}^{\text{r}}_{\text{av}} = -\hat{\mathbf{z}}|\Gamma|^2\,\frac{|E^{\text{i}}_0|^2}{2\eta_1} = -|\Gamma|^2\mathbf{S}^{\text{i}}_{\text{av}}. \tag{8.19c}$$

Even though Γ is purely real when both media are lossless dielectrics, we chose to treat it as complex, thereby providing in Eq. (8.19c) an expression that is also valid when medium 2 is conducting.

The average power density of the transmitted wave in medium 2 is

$$\begin{aligned}\mathbf{S}_{\text{av}_2}(z) &= \tfrac{1}{2}\Re\mathfrak{e}[\widetilde{\mathbf{E}}_2(z)\times\widetilde{\mathbf{H}}^*_2(z)]\\ &= \tfrac{1}{2}\Re\mathfrak{e}\left[\hat{\mathbf{x}}\tau E^{\text{i}}_0 e^{-jk_2z}\times\hat{\mathbf{y}}\tau^*\,\frac{E^{\text{i}*}_0}{\eta_2}\,e^{jk_2z}\right]\\ &= \hat{\mathbf{z}}|\tau|^2\,\frac{|E^{\text{i}}_0|^2}{2\eta_2}\,.\end{aligned} \tag{8.20}$$

Through the use of Eqs. (8.8a) and (8.8b), it can be easily shown that for lossless media

$$\frac{\tau^2}{\eta_2} = \frac{1-\Gamma^2}{\eta_1} \qquad \textbf{(lossless media)}, \tag{8.21}$$

which leads to

$$\mathbf{S}_{\text{av}_1} = \mathbf{S}_{\text{av}_2}.$$

This result is expected from considerations of power conservation.

Example 8-1: Radar Radome Design

A 10 GHz aircraft radar uses a narrow-beam scanning antenna mounted on a gimbal behind a dielectric radome, as shown in Fig. 8-5. Even though the radome shape is far from planar, it is approximately planar over the narrow extent of the radar beam. If the radome material is a lossless dielectric with $\epsilon_{\text{r}} = 9$ and $\mu_{\text{r}} = 1$, choose its thickness d such that the radome appears transparent to the radar beam. Structural integrity requires d to be greater than 2.3 cm.

Solution: **Figure 8-6(a)** shows a small section of the radome on an expanded scale. The incident wave can be approximated

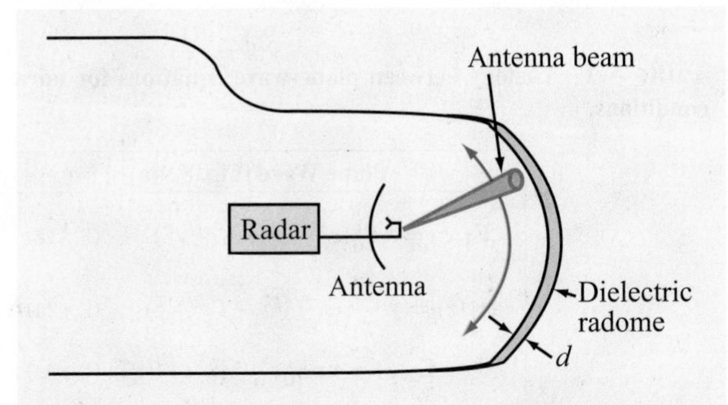

Figure 8-5 Antenna beam "looking" through an aircraft radome of thickness d (Example 8-1).

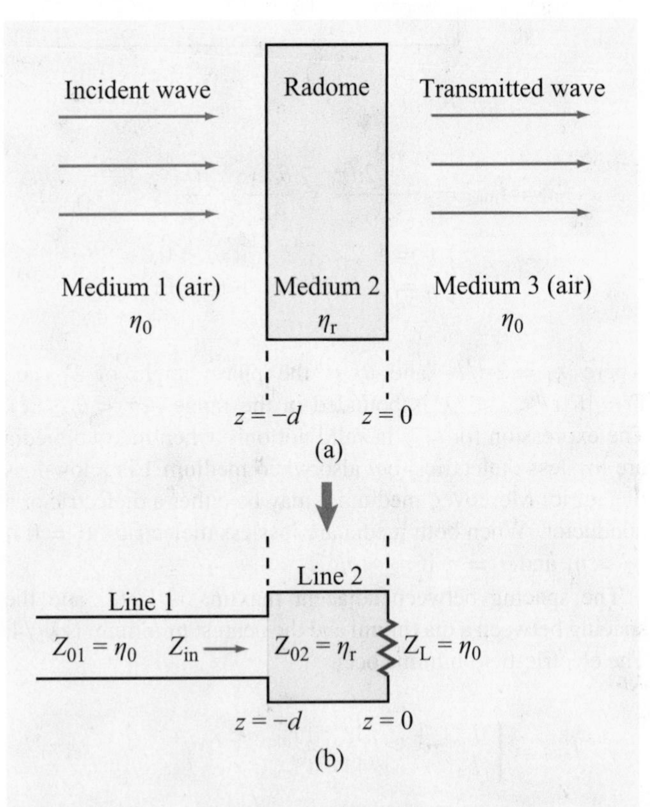

Figure 8-6 (a) Planar section of the radome of **Fig. 8-5** at an expanded scale and (b) its transmission-line equivalent model (Example 8-1).

as a plane wave propagating in medium 1 (air) with intrinsic impedance η_0. Medium 2 (the radome) is of thickness d and intrinsic impedance η_r, and medium 3 (air) is semi-infinite with intrinsic impedance η_0. **Figure 8-6(b)** shows an equivalent transmission-line model with $z = 0$ selected to coincide with the outside surface of the radome, and the load impedance $Z_L = \eta_0$ represents the input impedance of the semi-infinite air medium to the right of the radome.

For the radome to "appear" transparent to the incident wave, the reflection coefficient must be zero at $z = -d$, thereby guaranteeing total transmission of the incident power into medium 3. Since $Z_L = \eta_0$ in **Fig. 8-6(b)**, no reflection takes place at $z = -d$ if $Z_{in} = \eta_0$, which can be realized by choosing $d = n\lambda_2/2$ [see Section 2-8.4], where λ_2 is the wavelength in medium 2 and n is a positive integer. At 10 GHz, the wavelength in air is $\lambda_0 = c/f = 3$ cm, while in the radome material it is

$$\lambda_2 = \frac{\lambda_0}{\sqrt{\epsilon_r}} = \frac{3 \text{ cm}}{3} = 1 \text{ cm}.$$

Hence, by choosing $d = 5\lambda_2/2 = 2.5$ cm, the radome is both nonreflecting and structurally stable.

Example 8-2: Yellow Light Incident upon a Glass Surface

A beam of yellow light with wavelength 0.6 μm is normally incident in air upon a glass surface. Assume the glass is sufficiently thick as to ignore its back surface. If the surface is situated in the plane $z = 0$ and the relative permittivity of glass is 2.25, determine:

(a) the locations of the electric field maxima in medium 1 (air),
(b) the standing-wave ratio, and
(c) the fraction of the incident power transmitted into the glass medium.

Solution: (a) We begin by determining the values of η_1, η_2, and Γ:

$$\eta_1 = \sqrt{\frac{\mu_1}{\epsilon_1}} = \sqrt{\frac{\mu_0}{\epsilon_0}} \approx 120\pi \; (\Omega),$$

$$\eta_2 = \sqrt{\frac{\mu_2}{\epsilon_2}} = \sqrt{\frac{\mu_0}{\epsilon_0}} \cdot \frac{1}{\sqrt{\epsilon_r}} \approx \frac{120\pi}{\sqrt{2.25}} = 80\pi \; (\Omega),$$

$$\Gamma = \frac{\eta_2 - \eta_1}{\eta_2 + \eta_1} = \frac{80\pi - 120\pi}{80\pi + 120\pi} = -0.2.$$

Hence, $|\Gamma| = 0.2$ and $\theta_r = \pi$. From Eq. (8.16), the electric-field magnitude is maximum at

$$l_{max} = \frac{\theta_r \lambda_1}{4\pi} + n\frac{\lambda_1}{2} = \frac{\lambda_1}{4} + n\frac{\lambda_1}{2} \qquad (n = 0, 1, 2, \ldots)$$

with $\lambda_1 = 0.6$ μm.

(b)

$$S = \frac{1 + |\Gamma|}{1 - |\Gamma|} = \frac{1 + 0.2}{1 - 0.2} = 1.5.$$

(c) The fraction of the incident power transmitted into the glass medium is equal to the ratio of the transmitted power density, given by Eq. (8.20), to the incident power density, $S^i_{av} = |E^i_0|^2/2\eta_1$:

$$\frac{S_{av_2}}{S^i_{av}} = \tau^2 \frac{|E^i_0|^2}{2\eta_2} \bigg/ \left[\frac{|E^i_0|^2}{2\eta_1}\right] = \tau^2 \frac{\eta_1}{\eta_2}.$$

In view of Eq. (8.21),

$$\frac{S_{av_2}}{S^i_{av}} = 1 - |\Gamma|^2 = 1 - (0.2)^2 = 0.96, \text{ or } 96\%.$$

8-1.4 Boundary between Lossy Media

In Section 8-1.1 we considered a plane wave in a lossless medium incident normally on a planar boundary of another lossless medium. We now generalize our expressions to lossy media. In a medium with constitutive parameters (ϵ, μ, σ), the propagation constant $\gamma = \alpha + j\beta$ and the intrinsic impedance η_c are both complex. General expressions for α, β, and η_c are given by Eqs. (7.66a), (7.66b), and (7.70), respectively, and approximate expressions are given in **Table 7-2** for the special cases of low-loss media and good conductors. If media 1 and 2 have constitutive parameters $(\epsilon_1, \mu_1, \sigma_1)$ and $(\epsilon_2, \mu_2, \sigma_2)$ (**Fig. 8-7**), then expressions for the electric and magnetic fields in media 1 and 2 can be obtained from Eqs. (8.11) through (8.14) of **Table 8-1** by replacing jk with γ and η with η_c. Thus,

Medium 1

$$\widetilde{\mathbf{E}}_1(z) = \hat{\mathbf{x}} E^i_0 (e^{-\gamma_1 z} + \Gamma e^{\gamma_1 z}), \qquad (8.22a)$$

$$\widetilde{\mathbf{H}}_1(z) = \hat{\mathbf{y}} \frac{E^i_0}{\eta_{c_1}} (e^{-\gamma_1 z} - \Gamma e^{\gamma_1 z}), \qquad (8.22b)$$

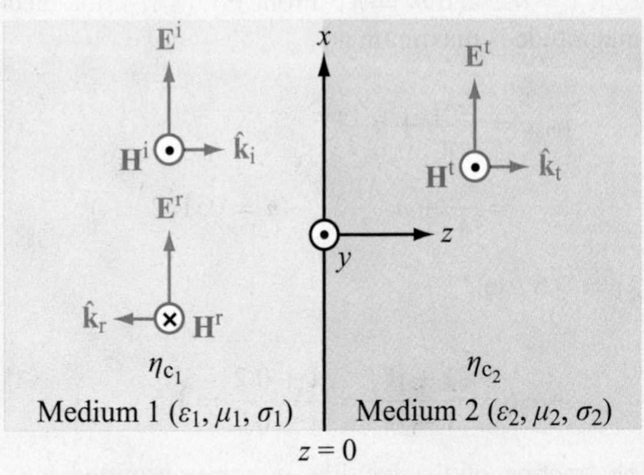

(a) Boundary between dielectric media

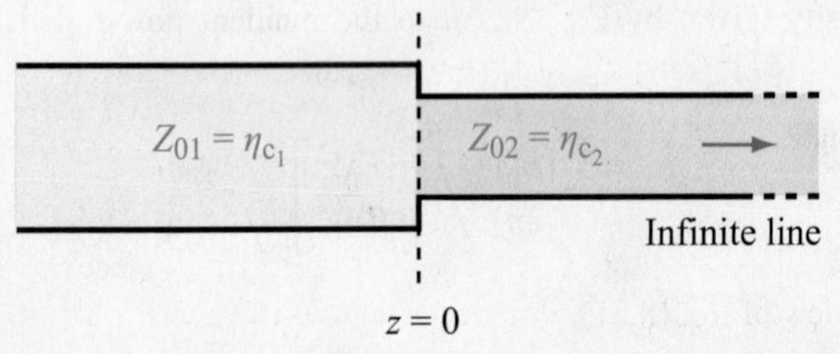

(b) Transmission-line analogue

Figure 8-7 Normal incidence at a planar boundary between two lossy media.

Medium 2

$$\widetilde{\mathbf{E}}_2(z) = \hat{\mathbf{x}}\tau E_0^{\mathrm{i}} e^{-\gamma_2 z}, \tag{8.23a}$$

$$\widetilde{\mathbf{H}}_2(z) = \hat{\mathbf{y}}\tau \frac{E_0^{\mathrm{i}}}{\eta_{\mathrm{c}_2}} e^{-\gamma_2 z}. \tag{8.23b}$$

Here, $\gamma_1 = \alpha_1 + j\beta_1$, $\gamma_2 = \alpha_2 + j\beta_2$, and

$$\Gamma = \frac{\eta_{\mathrm{c}_2} - \eta_{\mathrm{c}_1}}{\eta_{\mathrm{c}_2} + \eta_{\mathrm{c}_1}}, \tag{8.24a}$$

$$\tau = 1 + \Gamma = \frac{2\eta_{\mathrm{c}_2}}{\eta_{\mathrm{c}_2} + \eta_{\mathrm{c}_1}}. \tag{8.24b}$$

Because η_{c_1} and η_{c_2} are, in general, complex, Γ and τ may be complex as well.

Example 8-3: Normal Incidence on a Metal Surface

A 1 GHz x-polarized plane wave traveling in the $+z$ direction is incident from air upon a copper surface. The air-to-copper interface is at $z = 0$, and copper has $\epsilon_{\mathrm{r}} = 1$, $\mu_{\mathrm{r}} = 1$, and $\sigma = 5.8 \times 10^7$ S/m. If the amplitude of the electric field of the incident wave is 12 (mV/m), obtain expressions for the instantaneous electric and magnetic fields in the air medium. Assume the metal surface to be several skin depths deep.

Solution: In medium 1 (air), $\alpha = 0$,

$$\beta = k_1 = \frac{\omega}{c} = \frac{2\pi \times 10^9}{3 \times 10^8} = \frac{20\pi}{3} \quad \text{(rad/m)},$$

$$\eta_1 = \eta_0 = 377\ (\Omega), \qquad \lambda = \frac{2\pi}{k_1} = 0.3\ \text{m}.$$

At $f = 1$ GHz, copper is an excellent conductor because

$$\frac{\epsilon''}{\epsilon'} = \frac{\sigma}{\omega\epsilon_{\mathrm{r}}\epsilon_0} = \frac{5.8 \times 10^7}{2\pi \times 10^9 \times (10^{-9}/36\pi)} = 1 \times 10^9 \gg 1.$$

Use of Eq. (7.77c) gives

$$\begin{aligned}\eta_{\mathrm{c}_2} &= (1+j)\sqrt{\frac{\pi f \mu}{\sigma}} \\ &= (1+j)\left[\frac{\pi \times 10^9 \times 4\pi \times 10^{-7}}{5.8 \times 10^7}\right]^{1/2} \\ &= 8.25(1+j) \quad (\mathrm{m}\Omega).\end{aligned}$$

Since η_{c_2} is so small compared to $\eta_0 = 377$ (Ω) for air, the copper surface acts, in effect, like a short circuit. Hence,

$$\Gamma = \frac{\eta_{\mathrm{c}_2} - \eta_0}{\eta_{\mathrm{c}_2} + \eta_0} \approx -1.$$

Upon setting $\Gamma = -1$ in Eqs. (8.11) and (8.12) of **Table 8-1**, we obtain

$$\begin{aligned}\widetilde{\mathbf{E}}_1(z) &= \hat{\mathbf{x}} E_0^{\mathrm{i}} (e^{-jk_1 z} - e^{jk_1 z}) \\ &= -\hat{\mathbf{x}} j2E_0^{\mathrm{i}} \sin k_1 z,\end{aligned} \tag{8.25a}$$

$$\begin{aligned}\widetilde{\mathbf{H}}_1(z) &= \hat{\mathbf{y}} \frac{E_0^{\mathrm{i}}}{\eta_1} (e^{-jk_1 z} + e^{jk_1 z}) \\ &= \hat{\mathbf{y}} 2 \frac{E_0^{\mathrm{i}}}{\eta_1} \cos k_1 z.\end{aligned} \tag{8.25b}$$

With $E_0^{\mathrm{i}} = 12$ (mV/m), the instantaneous fields associated with these phasors are

$$\begin{aligned}
\mathbf{E}_1(z,t) &= \mathfrak{Re}[\widetilde{\mathbf{E}}_1(z)\, e^{j\omega t}] \\
&= \hat{\mathbf{x}}\, 2E_0^{\mathrm{i}} \sin k_1 z \sin \omega t \\
&= \hat{\mathbf{x}}\, 24 \sin(20\pi z/3) \sin(2\pi \times 10^9 t) \quad \text{(mV/m)}, \\
\mathbf{H}_1(z,t) &= \mathfrak{Re}[\widetilde{\mathbf{H}}_1(z)\, e^{j\omega t}] \\
&= \hat{\mathbf{y}}\, 2\, \frac{E_0^{\mathrm{i}}}{\eta_1} \cos k_1 z \cos \omega t \\
&= \hat{\mathbf{y}}\, 64 \cos(20\pi z/3) \cos(2\pi \times 10^9 t) \quad (\mu\text{A/m}).
\end{aligned}$$

Plots of the magnitude of $\mathbf{E}_1(z,t)$ and $\mathbf{H}_1(z,t)$ are shown in Fig. 8-8 as a function of negative z for various values of ωt. The wave patterns exhibit a repetition period of $\lambda/2$, and E and H are in phase quadrature (90° phase shift) in both space and time. This behavior is identical with that for voltage and current waves on a shorted transmission line.

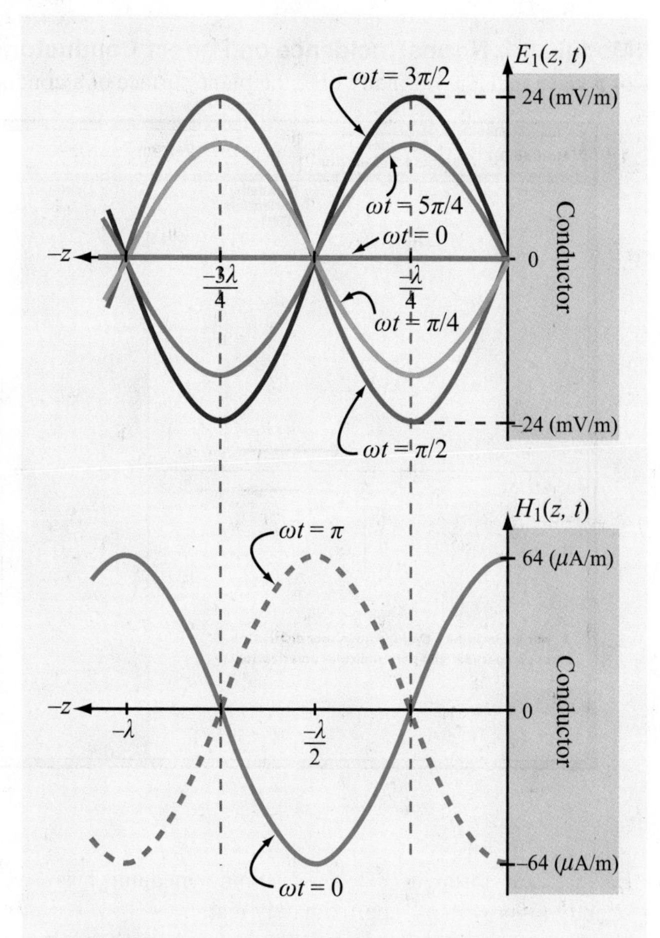

Figure 8-8 Wave patterns for fields $E_1(z,t)$ and $H_1(z,t)$ of Example 8-3.

Concept Question 8-1: What boundary conditions were used in the derivations of the expressions for Γ and τ?

Concept Question 8-2: In the radar radome design of Example 8-1, all the incident energy in medium 1 ends up getting transmitted into medium 3, and vice versa. Does this imply that no reflections take place within medium 2? Explain.

Concept Question 8-3: Explain on the basis of boundary conditions why it is necessary that $\Gamma = -1$ at the boundary between a dielectric and a perfect conductor.

Exercise 8-1: To eliminate reflections of normally incident plane waves, a dielectric slab of thickness d and relative permittivity ϵ_{r_2} is to be inserted between two semi-infinite media with relative permittivities $\epsilon_{r_1} = 1$ and $\epsilon_{r_3} = 16$. Use the quarter-wave transformer technique to select d and ϵ_{r_2}. Assume $f = 3$ GHz.

Answer: $\epsilon_{r_2} = 4$ and $d = (1.25 + 2.5n)$ (cm), with $n = 0, 1, 2, \ldots$. (See EM.)

Exercise 8-2: Express the normal-incidence reflection coefficient at the boundary between two nonmagnetic, conducting media in terms of their complex permittivities.

Answer: For incidence in medium 1 $(\epsilon_1, \mu_0, \sigma_1)$ onto medium 2 $(\epsilon_2, \mu_0, \sigma_2)$,

$$\Gamma = \frac{\sqrt{\epsilon_{c_1}} - \sqrt{\epsilon_{c_2}}}{\sqrt{\epsilon_{c_1}} + \sqrt{\epsilon_{c_2}}},$$

with $\epsilon_{c_1} = (\epsilon_1 - j\sigma_1/\omega)$ and $\epsilon_{c_2} = (\epsilon_2 - j\sigma_2/\omega)$. (See EM.)

Exercise 8-3: Obtain expressions for the average power densities in media 1 and 2 for the fields described by

Module 8.1 Normal Incidence on Perfect Conductor Observe the standing wave pattern created by the combination of a wave incident normally onto the plane surface of a conductor and its reflection.

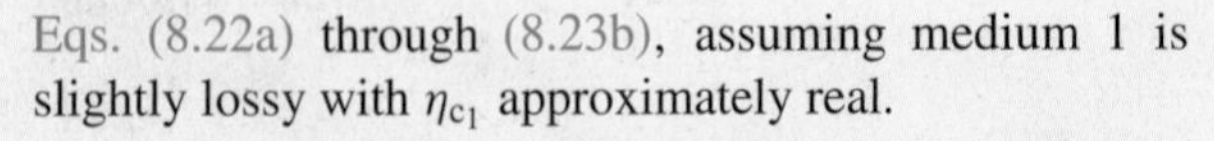

Eqs. (8.22a) through (8.23b), assuming medium 1 is slightly lossy with η_{c_1} approximately real.

Answer: (See EM.)

$$\mathbf{S}_{\text{av}_1} = \hat{\mathbf{z}}\,\frac{|E_0^{\text{i}}|^2}{2\eta_{c_1}}\left(e^{-2\alpha_1 z} - |\Gamma|^2 e^{2\alpha_1 z}\right),$$

$$\mathbf{S}_{\text{av}_2} = \hat{\mathbf{z}}|\tau|^2\,\frac{|E_0^{\text{i}}|^2}{2}\,e^{-2\alpha_2 z}\,\mathfrak{Re}\left(\frac{1}{\eta_{c_2}^*}\right).$$

8-2 Snell's Laws

In the preceding sections we examined reflection and transmission of plane waves that are normally incident upon a planar interface between two different media. We now consider the oblique-incidence case depicted in **Fig. 8-9**, and for simplicity we assume all media to be lossless. The $z = 0$ plane forms the boundary between media 1 and 2 with constitutive parameters (ϵ_1, μ_1) and (ϵ_2, μ_2), respectively. The two lines in **Fig. 8-9** with direction $\hat{\mathbf{k}}_{\text{i}}$ represent rays drawn normal to the wavefront of the incident wave, and those along directions $\hat{\mathbf{k}}_{\text{r}}$ and $\hat{\mathbf{k}}_{\text{t}}$ are similarly associated with the reflected and transmitted waves. The ***angles of incidence***, ***reflection***, and ***transmission*** (or ***refraction***), defined with respect to the normal to the boundary (the z axis), are θ_{i}, θ_{r}, and θ_{t}, respectively. These three angles are interrelated by ***Snell's laws***, which we derive shortly by considering the propagation of the wavefronts of the three waves. Rays of the incident wave intersect the boundary at O and O'. Here $A_{\text{i}}O$ represents a constant-phase wavefront of the incident wave. Likewise, $A_{\text{r}}O'$ and $A_{\text{t}}O'$ are constant-phase wavefronts of the reflected and transmitted waves, respectively (**Fig. 8-9**). The incident and reflected waves propagate in medium 1 with the same phase velocity $u_{\text{p}_1} = 1/\sqrt{\mu_1\epsilon_1}$, while the transmitted wave in medium 2 propagates with a velocity $u_{\text{p}_2} = 1/\sqrt{\mu_2\epsilon_2}$. The time it takes for the incident wave to travel from A_{i} to O' is the same as the time it takes for the reflected wave to travel from O to A_{r}, and also the time it takes the transmitted wave to travel from O to A_{t}. Since time equals

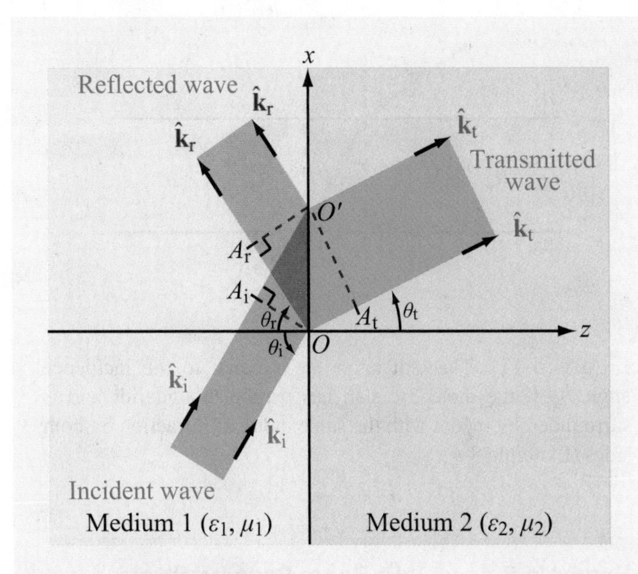

Figure 8-9 Wave reflection and refraction at a planar boundary between different media.

distance divided by velocity, it follows that

$$\frac{\overline{A_i O'}}{u_{p_1}} = \frac{\overline{OA_r}}{u_{p_1}} = \frac{\overline{OA_t}}{u_{p_2}}. \tag{8.26}$$

From the geometries of the three right triangles in **Fig. 8-9**, we deduce that

$$\overline{A_i O'} = \overline{OO'} \sin\theta_i, \tag{8.27a}$$

$$\overline{OA_r} = \overline{OO'} \sin\theta_r, \tag{8.27b}$$

$$\overline{OA_t} = \overline{OO'} \sin\theta_t. \tag{8.27c}$$

Use of these expressions in Eq. (8.26) leads to

$$\theta_i = \theta_r \quad \textbf{(Snell's law of reflection)}, \tag{8.28a}$$

$$\frac{\sin\theta_t}{\sin\theta_i} = \frac{u_{p_2}}{u_{p_1}} = \sqrt{\frac{\mu_1\epsilon_1}{\mu_2\epsilon_2}} \tag{8.28b}$$

(Snell's law of refraction).

► *Snell's law of reflection* states that the angle of reflection equals the angle of incidence, and *Snell's law of refraction* provides a relation between $\sin\theta_t$ and $\sin\theta_i$ in terms of the ratio of the phase velocities. ◄

The *index of refraction* of a medium, n, is defined as the ratio of the phase velocity in free space (i.e., the speed of light c) to the phase velocity in the medium. Thus,

$$n = \frac{c}{u_p} = \sqrt{\frac{\mu\epsilon}{\mu_0\epsilon_0}} = \sqrt{\mu_r\epsilon_r} \quad \textbf{(index of refraction)}. \tag{8.29}$$

In view of Eq. (8.29), Eq. (8.28b) may be rewritten as

$$\frac{\sin\theta_t}{\sin\theta_i} = \frac{n_1}{n_2} = \sqrt{\frac{\mu_{r_1}\epsilon_{r_1}}{\mu_{r_2}\epsilon_{r_2}}}. \tag{8.30}$$

For nonmagnetic materials, $\mu_{r_1} = \mu_{r_2} = 1$, in which case

$$\frac{\sin\theta_t}{\sin\theta_i} = \frac{n_1}{n_2} = \sqrt{\frac{\epsilon_{r_1}}{\epsilon_{r_2}}} = \frac{\eta_2}{\eta_1} \quad (\text{for } \mu_1 = \mu_2). \tag{8.31}$$

Usually, materials with higher densities have higher permittivities. Air, with $\mu_r = \epsilon_r = 1$, has an index of refraction $n_0 = 1$. Since for nonmagnetic materials $n = \sqrt{\epsilon_r}$, *a material is often referred to as more dense than another material if it has a greater index of refraction.*

At normal incidence ($\theta_i = 0$), Eq. (8.31) gives $\theta_t = 0$, as expected. At oblique incidence $\theta_t < \theta_i$ when $n_2 > n_1$ and $\theta_t > \theta_i$ when $n_2 < n_1$.

► If a wave is incident on a more dense medium [**Fig. 8-10(a)**], the transmitted wave refracts inwardly (toward the z axis), and the opposite is true if a wave is incident on a less dense medium [**Fig. 8-10(b)**]. ◄

A case of particular interest is when $\theta_t = \pi/2$, as shown in **Fig. 8-10(c)**; in this case, the refracted wave flows along the surface and no energy is transmitted into medium 2. The value

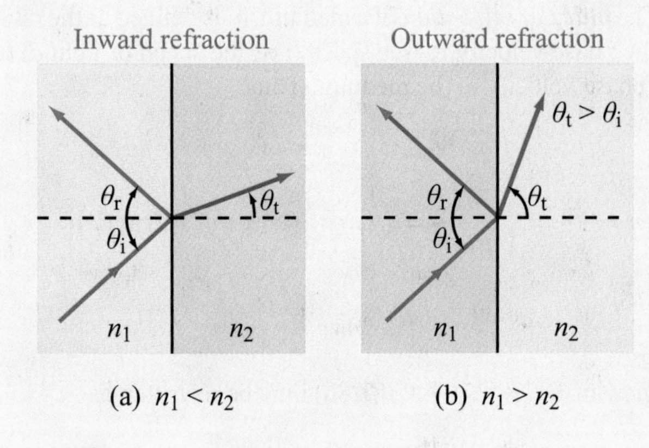

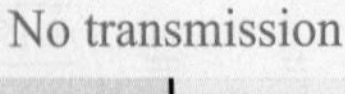

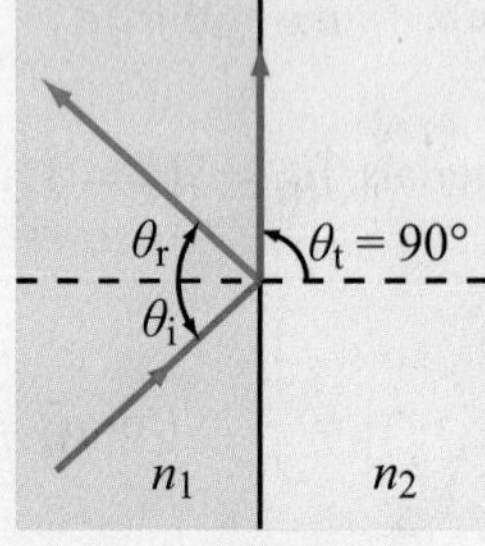

(c) $n_1 > n_2$ and $\theta_i = \theta_c$

Figure 8-10 Snell's laws state that $\theta_r = \theta_i$ and $\sin\theta_t = (n_1/n_2)\sin\theta_i$. Refraction is (a) inward if $n_1 < n_2$ and (b) outward if $n_1 > n_2$; and (c) the refraction angle is 90° if $n_1 > n_2$ and θ_i is equal to or greater than the critical angle $\theta_c = \sin^{-1}(n_2/n_1)$.

of the angle of incidence θ_i corresponding to $\theta_t = \pi/2$ is called the ***critical angle*** θ_c and is obtained from Eq. (8.30) as

$$\sin\theta_c = \frac{n_2}{n_1}\sin\theta_t\bigg|_{\theta_t=\pi/2} = \frac{n_2}{n_1} \tag{8.32a}$$

$$= \sqrt{\frac{\epsilon_{r_2}}{\epsilon_{r_1}}} \quad (\text{for } \mu_1 = \mu_2). \tag{8.32b}$$

(critical angle)

If θ_i exceeds θ_c, the incident wave is totally reflected, and the refracted wave becomes a nonuniform ***surface wave*** that travels along the boundary between the two media. This wave behavior is called ***total internal reflection***.

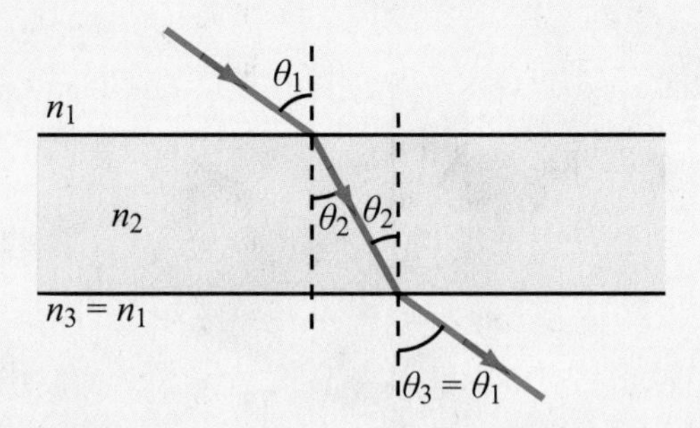

Figure 8-11 The exit angle θ_3 is equal to the incidence angle θ_1 if the dielectric slab has parallel boundaries and is surrounded by media with the same index of refraction on both sides (Example 8-4).

Example 8-4: Light Beam Passing through a Slab

A dielectric slab with index of refraction n_2 is surrounded by a medium with index of refraction n_1, as shown in **Fig. 8-11**. If $\theta_i < \theta_c$, show that the emerging beam is parallel to the incident beam.

Solution: At the slab's upper surface, Snell's law gives

$$\sin\theta_2 = \frac{n_1}{n_2}\sin\theta_1 \tag{8.33}$$

and, similarly, at the slab's lower surface,

$$\sin\theta_3 = \frac{n_2}{n_3}\sin\theta_2 = \frac{n_2}{n_1}\sin\theta_2. \tag{8.34}$$

Substituting Eq. (8.33) into Eq. (8.34) gives

$$\sin\theta_3 = \left(\frac{n_2}{n_1}\right)\left(\frac{n_1}{n_2}\right)\sin\theta_1 = \sin\theta_1.$$

Hence, $\theta_3 = \theta_1$. The slab displaces the beam's position, but the beam's direction remains unchanged.

Exercise 8-4: In the visible part of the electromagnetic spectrum, the index of refraction of water is 1.33. What is the critical angle for light waves generated by an upward-looking underwater light source?

Answer: $\theta_c = 48.8°$. (See EM.)

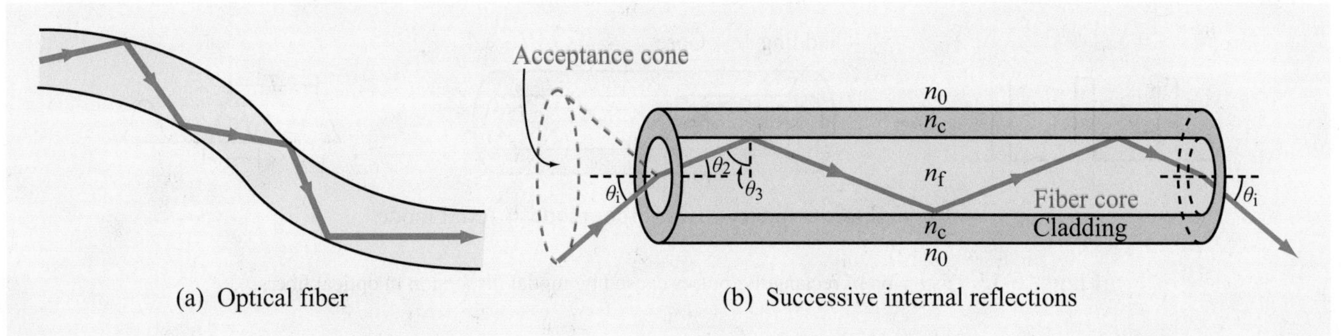

Figure 8-12 Waves can be guided along optical fibers as long as the reflection angles exceed the critical angle for total internal reflection.

Exercise 8-5: If the light source of Exercise 8-4 is situated at a depth of 1 m below the water surface and if its beam is isotropic (radiates in all directions), how large a circle would it illuminate when observed from above?

Answer: Circle's diameter = 2.28 m. (See EM.)

8-3 Fiber Optics

By successive total internal reflections, as illustrated in Fig. 8-12(a), light can be guided through thin dielectric rods made of glass or transparent plastic, known as ***optical fibers***. Because the light is confined to traveling within the rod, the only loss in power is due to reflections at the sending and receiving ends of the fiber and absorption by the fiber material (because it is not a perfect dielectric). Optical fibers are useful for the transmission of wide-band signals as well as many imaging applications.

An optical fiber usually consists of a cylindrical ***fiber core*** with an index of refraction n_f, surrounded by another cylinder of lower index of refraction, n_c, called the ***cladding*** [Fig. 8-12(b)]. The cladding layer serves to optically isolate the fiber when a large number of fibers are packed in close proximity, thereby avoiding leakage of light from one fiber into another. To ensure total internal reflection, the incident angle θ_3 in the fiber core must be equal to, or greater than, the critical angle θ_c for a wave in the fiber medium (with n_f) incident upon the cladding medium (with n_c). From Eq. (8.32a), we have

$$\sin\theta_c = \frac{n_c}{n_f}\,. \tag{8.35}$$

To meet the total reflection requirement $\theta_3 \geq \theta_c$, it is necessary that $\sin\theta_3 \geq n_c/n_f$. The angle θ_2 is the complement of angle θ_3; hence $\cos\theta_2 = \sin\theta_3$. The necessary condition therefore may be written as

$$\cos\theta_2 \geq \frac{n_c}{n_f}\,. \tag{8.36}$$

Moreover, θ_2 is related to the incidence angle on the face of the fiber, θ_i, by Snell's law:

$$\sin\theta_2 = \frac{n_0}{n_f}\sin\theta_i, \tag{8.37}$$

where n_0 is the index of refraction of the medium surrounding the fiber ($n_0 = 1$ for air and $n_0 = 1.33$ if the fiber is in water), or

$$\cos\theta_2 = \left[1 - \left(\frac{n_0}{n_f}\right)^2 \sin^2\theta_i\right]^{1/2}. \tag{8.38}$$

Using Eq. (8.38) on the left-hand side of Eq. (8.36) and then solving for $\sin\theta_i$ gives

$$\sin\theta_i \leq \frac{1}{n_0}\,(n_f^2 - n_c^2)^{1/2}. \tag{8.39}$$

The ***acceptance angle*** θ_a is defined as the maximum value of θ_i for which the condition of total internal reflection remains satisfied:

$$\sin\theta_a = \frac{1}{n_0}(n_f^2 - n_c^2)^{1/2}. \tag{8.40}$$

The angle θ_a is equal to half the angle of the acceptance cone of the fiber. Any ray of light incident upon the face of the core fiber at an incidence angle within the acceptance cone can propagate down the core. This means that there can be a large number of ray paths, called ***modes***, by which light energy can travel in the core. Rays characterized by large angles θ_i travel longer paths

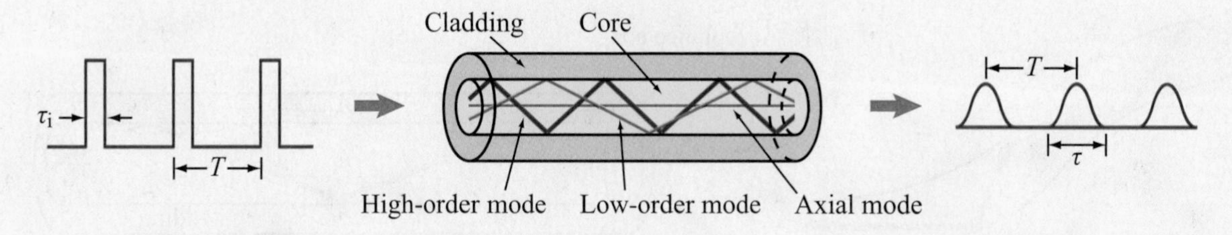

Figure 8-13 Distortion of rectangular pulses caused by modal dispersion in optical fibers.

than rays that propagate along the axis of the fiber, as illustrated by the three modes shown in **Fig. 8-13**. Consequently, different modes have different transit times between the two ends of the fiber. This property of optical fibers is called ***modal dispersion*** and has the undesirable effect of changing the shape of pulses used for the transmission of digital data. When a rectangular pulse of light incident upon the face of the fiber gets broken up into many modes and the different modes do not arrive at the other end of the fiber at the same time, the pulse gets distorted, both in shape and length. In the example shown in **Fig. 8-13**, the narrow rectangular pulses at the input side of the optical fiber are of width τ_i separated by a time duration T. After propagating through the fiber core, modal dispersion causes the pulses to look more like spread-out sine waves with spread-out temporal width τ. If the output pulses spread out so much that $\tau > T$, the output signals will smear out, making it impossible to decipher the transmitted message from the output signal. Hence, to ensure that the transmitted pulses remain distinguishable at the output side of the fiber, it is necessary that τ be shorter than T. As a safety margin, it is common practice to design the transmission system such that $T \geq 2\tau$.

The spread-out width τ is equal to the time delay Δt between the arrival of the slowest ray and the fastest ray. The slowest ray is the one traveling the longest distance and corresponds to the ray incident upon the input face of the fiber at the acceptance angle θ_a. From the geometry in **Fig. 8-12(b)** and Eq. (8.36), this ray corresponds to $\cos\theta_2 = n_c/n_f$. For an optical fiber of length l, the length of the path traveled by such a ray is

$$l_{\max} = \frac{l}{\cos\theta_2} = l\,\frac{n_f}{n_c}, \qquad (8.41)$$

and its travel time in the fiber at velocity $u_p = c/n_f$ is

$$t_{\max} = \frac{l_{\max}}{u_p} = \frac{l n_f^2}{c n_c}. \qquad (8.42)$$

The minimum time of travel is realized by the axial ray and is given by

$$t_{\min} = \frac{l}{u_p} = \frac{l}{c}\,n_f. \qquad (8.43)$$

The total time delay is therefore

$$\tau = \Delta t = t_{\max} - t_{\min} = \frac{l n_f}{c}\left(\frac{n_f - 1}{n_c}\right) \quad \text{(s)}. \qquad (8.44)$$

As we stated before, to retrieve the desired information from the transmitted signals, it is advisable that T, the interpulse period of the input train of pulses, be no shorter than 2τ. This, in turn, means that the data rate (in bits per second), or equivalently the number of pulses per second, that can be transmitted through the fiber is limited to

$$f_p = \frac{1}{T} = \frac{1}{2\tau} = \frac{c n_c}{2 l n_f (n_f - n_c)} \quad \text{(bits/s)}. \qquad (8.45)$$

Example 8-5: Transmission Data Rate on Optical Fibers

A 1 km long optical fiber (in air) is made of a fiber core with an index of refraction of 1.52 and a cladding with an index of refraction of 1.49. Determine

(a) the acceptance angle θ_a, and

(b) the maximum usable data rate of signals that can be transmitted through the fiber.

Solution: **(a)** From Eq. (8.40),

$$\sin\theta_a = \frac{1}{n_0}(n_f^2 - n_c^2)^{1/2} = [(1.52)^2 - (1.49)^2]^{1/2} = 0.3,$$

which corresponds to $\theta_a = 17.5°$.

Module 8.2 Multimode Step-Index Optical Fiber Choose the indices of refraction on the fibre core and cladding and then observe the zigzag pattern of the wave propagation inside the fiber.

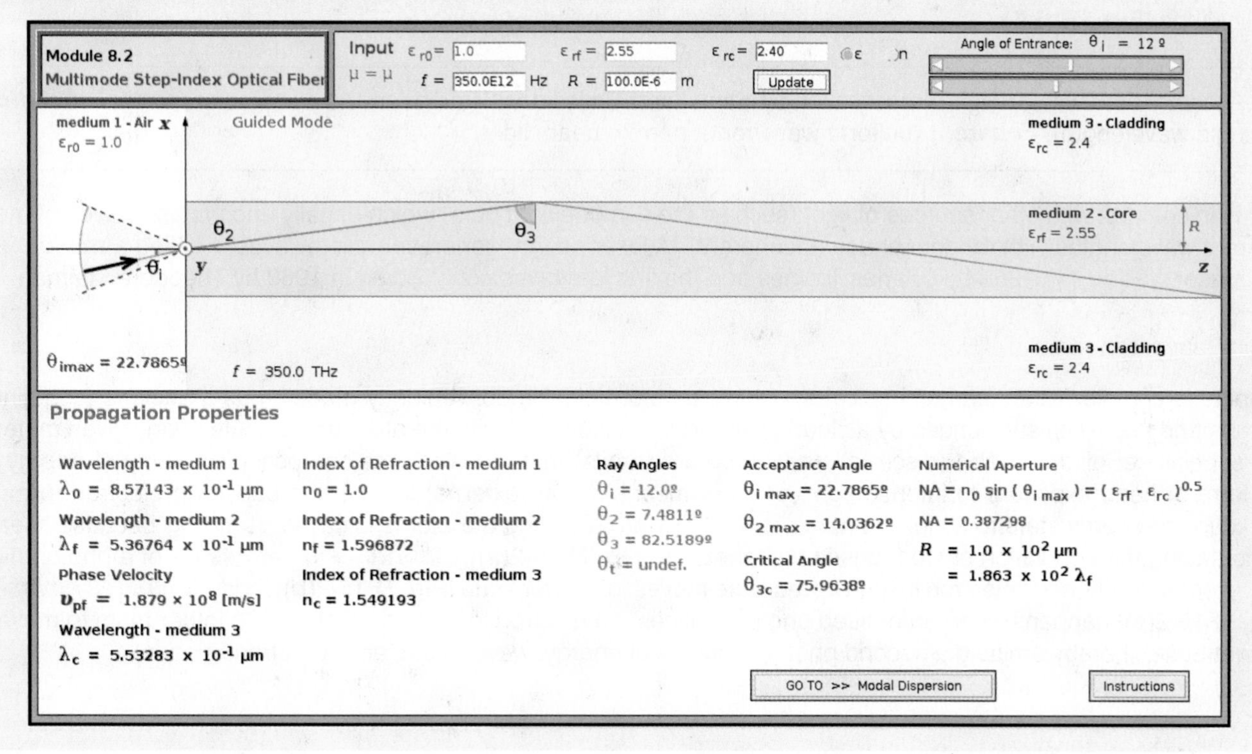

(b) From Eq. (8.45),

$$f_{\rm p} = \frac{cn_{\rm c}}{2ln_{\rm f}(n_{\rm f} - n_{\rm c})}$$

$$= \frac{3 \times 10^8 \times 1.49}{2 \times 10^3 \times 1.52(1.52 - 1.49)} = 4.9 \text{ (Mb/s)}.$$

Exercise 8-6: If the index of refraction of the cladding material in Example 8-5 is increased to 1.50, what would be the new maximum usable data rate?

Answer: 7.4 (Mb/s). (See EM.)

8-4 Wave Reflection and Transmission at Oblique Incidence

In this section we develop a rigorous theory of reflection and refraction of plane waves obliquely incident upon planar boundaries between different media. Our treatment parallels that in Section 8-1 for the normal-incidence case and goes beyond that in Section 8-2 on Snell's laws, which yielded information on only the angles of reflection and refraction.

For normal incidence, the reflection and transmission coefficients Γ and τ at a boundary between two media are independent of the polarization of the incident wave, as both the electric and magnetic fields of a normally incident plane wave are tangential to the boundary regardless of the wave polarization. This is not the case for obliquely incident waves travelling at an angle $\theta_{\rm i} \neq 0$ with respect to the normal to the interface.

► The ***plane of incidence*** is defined as the plane containing the normal to the boundary and the direction of propagation of the incident wave. ◄

Technology Brief 15: Lasers

Lasers are used in CD and DVD players, bar-code readers, eye surgery, and multitudes of other systems and applications (**Fig. TF15-1**).

> ► A laser—acronym for ***L****ight* ***A****mplification by* ***S****timulated* ***E****mission of* ***R****adiation*—is a source of ***monochromatic*** (single wavelength), ***coherent*** (uniform wavefront), narrow-beam light. ◄

This is in contrast with other sources of light (such as the sun or a light bulb) which usually encompass waves of many different wavelengths with random phase (incoherent). A laser source generating microwaves is called a ***maser***. The first maser was built in 1953 by Charles Townes and the first laser was constructed in 1960 by Theodore Maiman.

Basic Principles

Despite its complex quantum-mechanical structure, an atom can be conveniently modeled as a nucleus (containing protons and neutrons) surrounded by a cloud of electrons. Associated with the atom or molecule of any given material is a specific set of ***quantized*** (discrete) ***energy states*** (orbits) that the electrons can occupy. Supply of energy (in the form of heat, exposure to intense light, or other means) by an external source can cause an electron to move from a lower-energy state to a higher energy (***excited***) state. Exciting the atoms is called ***pumping*** because it leads to increasing the population of electrons in higher states [**Fig. TF15-2(a)**]. ***Spontaneous emission*** of a photon (light energy) occurs when the electron in the excited state moves to a lower state [**Fig. TF15-2(b)**], and ***stimulated emission*** [**Fig. TF15-2(c)**] happens when an emitted photon "entices" an electron in an excited state of another atom to move to a lower state, thereby emitting a second photon of identical energy, wavelength, and wavefront (phase).

Figure TF15-1 A few examples of laser applications.

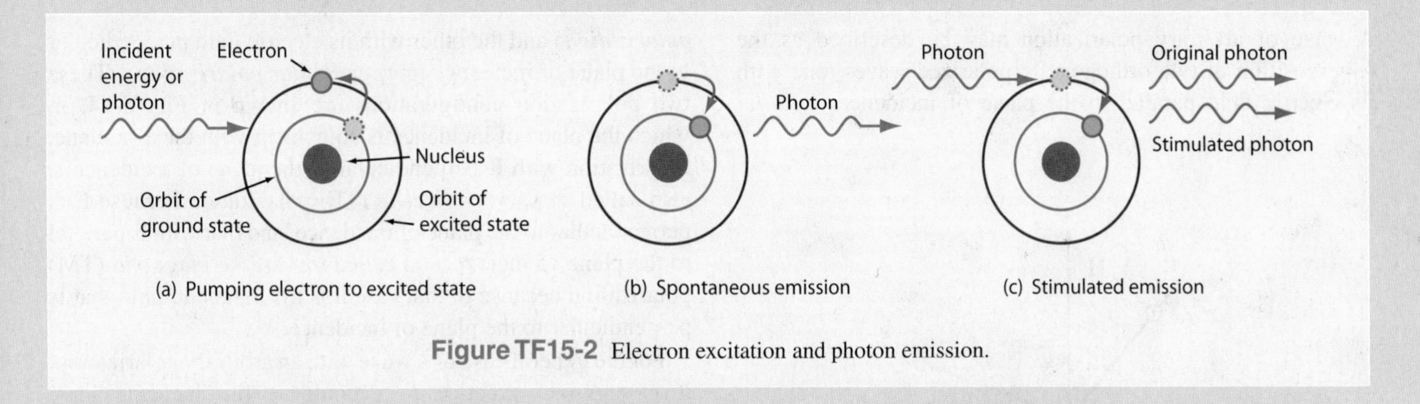

Figure TF15-2 Electron excitation and photon emission.

Principle of Operation

▶ Highly amplified stimulated emission is called ***lasing***. ◀

The lasing medium can be solid, liquid, or gas. Laser operation is illustrated in **Fig. TF15-3** for a ruby crystal surrounded by a flash tube (similar to a camera flash). A perfectly reflecting mirror is placed on one end of the crystal and a partially reflecting mirror on the other end. Light from the flash tube excites the atoms; some undergo spontaneous emission, generating photons that cause others to undergo stimulated emission; photons moving along the axis of the crystal bounce back and forth between the mirrors, causing additional stimulated emission (i.e., amplification), with only a fraction of the photons exiting through the partially reflecting mirror.

▶ Because all of the stimulated photons are identical, the light wave generated by the laser is of a single wavelength. ◀

Wavelength (Color) of Emitted Light

The atom of any given material has unique energy states. The difference in energy between the excited high-energy state and the stable lower-energy state determines the wavelength of the emitted photons (EM wave). Through proper choice of lasing material, monochromatic waves can be generated with wavelengths in the ultraviolet, visible, infrared or microwave bands.

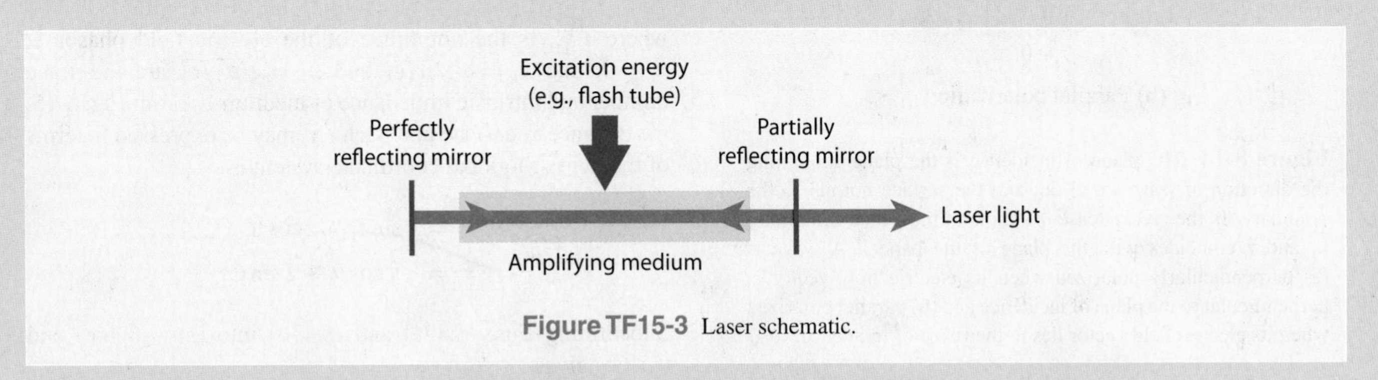

Figure TF15-3 Laser schematic.

A wave of arbitrary polarization may be described as the superposition of two orthogonally polarized waves, one with its electric field parallel to the plane of incidence (*parallel polarization*) and the other with its electric field perpendicular to the plane of incidence (*perpendicular polarization*). These two polarization configurations are shown in Fig. 8-14, in which the plane of incidence is coincident with the x–z plane. Polarization with **E** perpendicular to the plane of incidence is also called *transverse electric* (TE) polarization because **E** is perpendicular to the plane of incidence, and that with **E** parallel to the plane of incidence is called *transverse magnetic* (TM) polarization because in that case it is the magnetic field that is perpendicular to the plane of incidence.

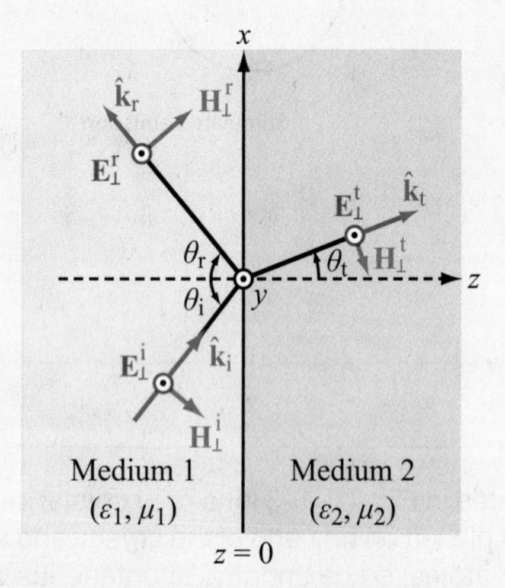

(a) Perpendicular polarization

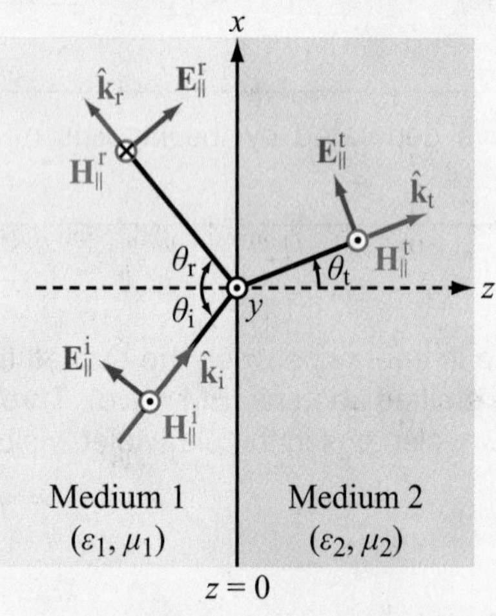

(b) Parallel polarization

Figure 8-14 The plane of incidence is the plane containing the direction of wave travel, $\hat{\mathbf{k}}_i$, and the surface normal to the boundary. In the present case the plane of incidence containing $\hat{\mathbf{k}}_i$ and $\hat{\mathbf{z}}$ coincides with the plane of the paper. A wave is (a) perpendicularly polarized when its electric field vector is perpendicular to the plane of incidence and (b) parallel polarized when its electric field vector lies in the plane of incidence.

For the general case of a wave with an arbitrary polarization, it is common practice to decompose the incident wave $(\mathbf{E}^i, \mathbf{H}^i)$ into a perpendicularly polarized component $(\mathbf{E}^i_\perp, \mathbf{H}^i_\perp)$ and a parallel polarized component $(\mathbf{E}^i_\parallel, \mathbf{H}^i_\parallel)$. Then, after determining the reflected waves $(\mathbf{E}^r_\perp, \mathbf{H}^r_\perp)$ and $(\mathbf{E}^r_\parallel, \mathbf{H}^r_\parallel)$ due to the two incident components, the reflected waves are added together to give the total reflected wave $(\mathbf{E}^r, \mathbf{H}^r)$ corresponding to the original incident wave. A similar process can be used to determine the total transmitted wave $(\mathbf{E}^t, \mathbf{H}^t)$.

8-4.1 Perpendicular Polarization

Figure 8-15 shows a perpendicularly polarized incident plane wave propagating along the x_i direction in dielectric medium 1. The electric field phasor $\widetilde{\mathbf{E}}^i_\perp$ points along the y direction, and the associated magnetic field phasor $\widetilde{\mathbf{H}}^i_\perp$ is along the y_i axis. The directions of $\widetilde{\mathbf{E}}^i_\perp$ and $\widetilde{\mathbf{H}}^i_\perp$ are such that $\widetilde{\mathbf{E}}^i_\perp \times \widetilde{\mathbf{H}}^i_\perp$ points along the propagation direction $\hat{\mathbf{x}}_i$. The electric and magnetic fields of such a plane wave are given by

$$\widetilde{\mathbf{E}}^i_\perp = \hat{\mathbf{y}} E^i_{\perp 0} e^{-jk_1 x_i}, \tag{8.46a}$$

$$\widetilde{\mathbf{H}}^i_\perp = \hat{\mathbf{y}}_i \frac{E^i_{\perp 0}}{\eta_1} e^{-jk_1 x_i}, \tag{8.46b}$$

where $E^i_{\perp 0}$ is the amplitude of the electric field phasor at $x_i = 0$, and $k_1 = \omega\sqrt{\mu_1 \epsilon_1}$ and $\eta_1 = \sqrt{\mu_1/\epsilon_1}$ are the wave number and intrinsic impedance of medium 1. From Fig. 8-15, the distance x_i and the unit vector $\hat{\mathbf{y}}_i$ may be expressed in terms of the (x, y, z) global coordinate system as

$$x_i = x \sin\theta_i + z\cos\theta_i, \tag{8.47a}$$

$$\hat{\mathbf{y}}_i = -\hat{\mathbf{x}}\cos\theta_i + \hat{\mathbf{z}}\sin\theta_i. \tag{8.47b}$$

Substituting Eqs. (8.47a) and (8.47b) into Eqs. (8.46a) and (8.46b) gives

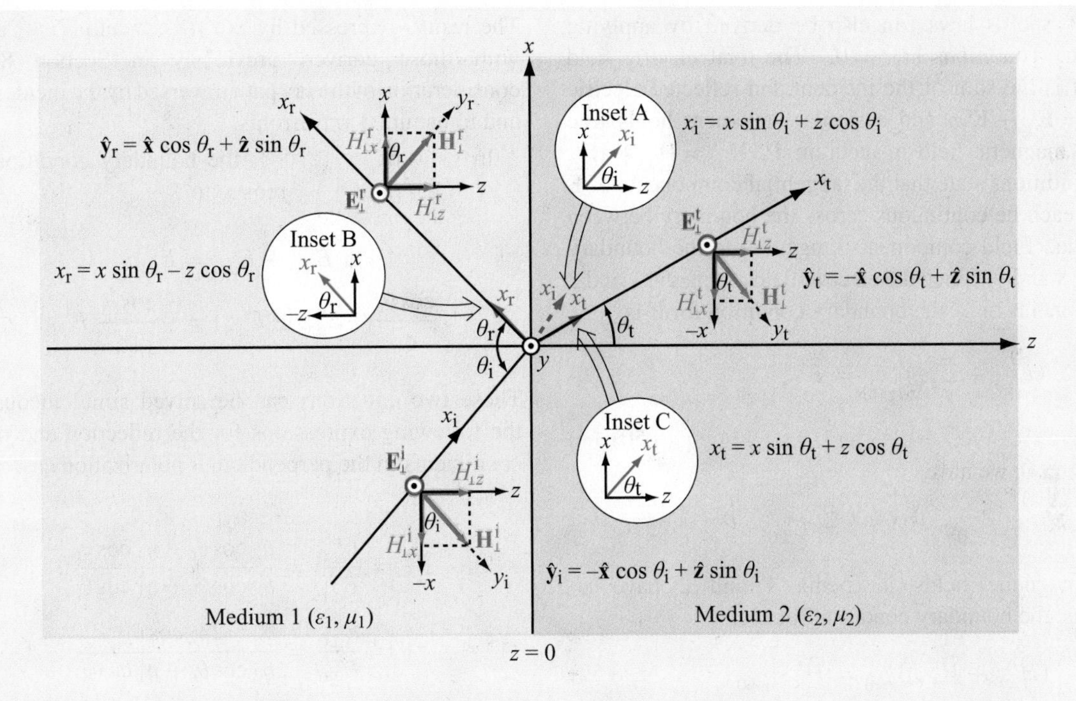

Figure 8-15 Perpendicularly polarized plane wave incident at an angle θ_i upon a planar boundary.

Incident Wave

$$\widetilde{\mathbf{E}}^i_\perp = \hat{\mathbf{y}} E^i_{\perp 0} e^{-jk_1(x\sin\theta_i + z\cos\theta_i)}, \tag{8.48a}$$

$$\widetilde{\mathbf{H}}^i_\perp = (-\hat{\mathbf{x}}\cos\theta_i + \hat{\mathbf{z}}\sin\theta_i) \times \frac{E^i_{\perp 0}}{\eta_1} e^{-jk_1(x\sin\theta_i + z\cos\theta_i)}. \tag{8.48b}$$

With the aid of the directional relationships given in **Fig. 8-15** for the reflected and transmitted waves, these fields are given by

Reflected Wave

$$\widetilde{\mathbf{E}}^r_\perp = \hat{\mathbf{y}} E^r_{\perp 0} e^{-jk_1 x_r} = \hat{\mathbf{y}} E^r_{\perp 0} e^{-jk_1(x\sin\theta_r - z\cos\theta_r)}, \tag{8.49a}$$

$$\widetilde{\mathbf{H}}^r_\perp = \hat{\mathbf{y}}_r \frac{E^r_{\perp 0}}{\eta_1} e^{-jk_1 x_r} = (\hat{\mathbf{x}}\cos\theta_r + \hat{\mathbf{z}}\sin\theta_r) \times \frac{E^r_{\perp 0}}{\eta_1} e^{-jk_1(x\sin\theta_r - z\cos\theta_r)}, \tag{8.49b}$$

Transmitted Wave

$$\widetilde{\mathbf{E}}^t_\perp = \hat{\mathbf{y}} E^t_{\perp 0} e^{-jk_2 x_t} = \hat{\mathbf{y}} E^t_{\perp 0} e^{-jk_2(x\sin\theta_t + z\cos\theta_t)}, \tag{8.49c}$$

$$\widetilde{\mathbf{H}}^t_\perp = \hat{\mathbf{y}}_t \frac{E^t_{\perp 0}}{\eta_2} e^{-jk_2 x_t} = (-\hat{\mathbf{x}}\cos\theta_t + \hat{\mathbf{z}}\sin\theta_t) \times \frac{E^t_{\perp 0}}{\eta_2} e^{-jk_2(x\sin\theta_t + z\cos\theta_t)}, \tag{8.49d}$$

where θ_r and θ_t are the reflection and transmission angles shown in **Fig. 8-15**, and k_2 and η_2 are the wavenumber and intrinsic impedance of medium 2. Our goal is to describe the reflected and transmitted fields in terms of the parameters that characterize the incident wave, namely the incidence angle θ_i and the amplitude $E^i_{\perp 0}$. The four expressions given by Eqs. (8.49a) through (8.49d) contain four unknowns: $E^r_{\perp 0}$, $E^t_{\perp 0}$, θ_r, and θ_t. Even though angles θ_r and θ_t are related to θ_i by Snell's laws (Eqs. (8.28a) and (8.28b)), here we choose to treat them as unknown for the time being, because we intend to show that Snell's laws can also be derived by applying field boundary conditions at $z = 0$. The total electric field in medium 1 is the sum of the incident and reflected electric fields: $\widetilde{\mathbf{E}}^1_\perp = \widetilde{\mathbf{E}}^i_\perp + \widetilde{\mathbf{E}}^r_\perp$; and a similar statement holds true for the total magnetic field in medium 1: $\widetilde{\mathbf{H}}^1_\perp = \widetilde{\mathbf{H}}^i_\perp + \widetilde{\mathbf{H}}^r_\perp$. Boundary conditions state that the tangential components of $\widetilde{\mathbf{E}}$ and $\widetilde{\mathbf{H}}$ must each be continuous across the boundary between the two media. Field components tangential to the boundary extend along $\hat{\mathbf{x}}$ and $\hat{\mathbf{y}}$. Since the electric fields in media 1 and 2 have $\hat{\mathbf{y}}$ components only, the boundary condition for $\widetilde{\mathbf{E}}$ is

$$(\widetilde{E}^i_{\perp y} + \widetilde{E}^r_{\perp y})\Big|_{z=0} = \widetilde{E}^t_{\perp y}\Big|_{z=0}. \tag{8.50}$$

Upon using Eqs. (8.48a), (8.49a), and (8.49c) in Eq. (8.50) and then setting $z = 0$, we have

$$E^i_{\perp 0} e^{-jk_1 x \sin\theta_i} + E^r_{\perp 0} e^{-jk_1 x \sin\theta_r} = E^t_{\perp 0} e^{-jk_2 x \sin\theta_t}. \tag{8.51}$$

Since the magnetic fields in media 1 and 2 have no $\hat{\mathbf{y}}$ components, the boundary condition for $\widetilde{\mathbf{H}}$ is

$$(\widetilde{H}^i_{\perp x} + \widetilde{H}^r_{\perp x})\big|_{z=0} = \widetilde{H}^t_{\perp x}\big|_{z=0}, \tag{8.52}$$

or

$$-\frac{E^i_{\perp 0}}{\eta_1} \cos\theta_i \, e^{-jk_1 x \sin\theta_i} + \frac{E^r_{\perp 0}}{\eta_1} \cos\theta_r \, e^{-jk_1 x \sin\theta_r} = -\frac{E^t_{\perp 0}}{\eta_2} \cos\theta_t \, e^{-jk_2 x \sin\theta_t}. \tag{8.53}$$

To satisfy Eqs. (8.51) and (8.53) for all possible values of x (i.e., all along the boundary), it follows that the arguments of all three exponentials must be equal. That is,

$$k_1 \sin\theta_i = k_1 \sin\theta_r = k_2 \sin\theta_t, \tag{8.54}$$

which is known as the ***phase-matching condition***. The first equality in Eq. (8.54) leads to

$$\theta_r = \theta_i \qquad \textbf{(Snell's law of reflection)}, \tag{8.55}$$

while the second equality leads to

$$\frac{\sin\theta_t}{\sin\theta_i} = \frac{k_1}{k_2} = \frac{\omega\sqrt{\mu_1\epsilon_1}}{\omega\sqrt{\mu_2\epsilon_2}} = \frac{n_1}{n_2}. \tag{8.56}$$

(Snell's law of refraction)

The results expressed by Eqs. (8.55) and (8.56) are identical with those derived previously in Section 8-2 through consideration of the ray path traversed by the incident, reflected, and transmitted wavefronts.

In view of Eq. (8.54), the boundary conditions given by Eqs. (8.51) and (8.53) reduce to

$$E^i_{\perp 0} + E^r_{\perp 0} = E^t_{\perp 0}, \tag{8.57a}$$

$$\frac{\cos\theta_i}{\eta_1}(-E^i_{\perp 0} + E^r_{\perp 0}) = -\frac{\cos\theta_t}{\eta_2} E^t_{\perp 0}. \tag{8.57b}$$

These two equations can be solved simultaneously to yield the following expressions for the reflection and transmission coefficients in the perpendicular polarization case:

$$\Gamma_\perp = \frac{E^r_{\perp 0}}{E^i_{\perp 0}} = \frac{\eta_2\cos\theta_i - \eta_1\cos\theta_t}{\eta_2\cos\theta_i + \eta_1\cos\theta_t}, \tag{8.58a}$$

$$\tau_\perp = \frac{E^t_{\perp 0}}{E^i_{\perp 0}} = \frac{2\eta_2\cos\theta_i}{\eta_2\cos\theta_i + \eta_1\cos\theta_t}. \tag{8.58b}$$

These two coefficients, which formally are known as the ***Fresnel reflection and transmission coefficients for perpendicular polarization***, are related by

$$\tau_\perp = 1 + \Gamma_\perp. \tag{8.59}$$

If medium 2 is a perfect conductor ($\eta_2 = 0$), Eqs. (8.58a) and (8.58b) reduce to $\Gamma_\perp = -1$ and $\tau_\perp = 0$, respectively, which means that the incident wave is totally reflected by the conducting medium.

For nonmagnetic dielectrics with $\mu_1 = \mu_2 = \mu_0$ and with the help of Eq. (8.56), the expression for $\Gamma_\perp$ can be written as

$$\Gamma_\perp = \frac{\cos\theta_i - \sqrt{(\epsilon_2/\epsilon_1) - \sin^2\theta_i}}{\cos\theta_i + \sqrt{(\epsilon_2/\epsilon_1) - \sin^2\theta_i}} \tag{8.60}$$

$$(\text{for } \mu_1 = \mu_2).$$

Since $(\epsilon_2/\epsilon_1) = (n_2/n_1)^2$, this expression can also be written in terms of the indices of refraction n_1 and n_2.

Example 8-6: Wave Incident Obliquely on a Soil Surface

Using the coordinate system of **Fig. 8-15**, a plane wave radiated by a distant antenna is incident in air upon a plane soil surface located at $z = 0$. The electric field of the incident wave is given by

$$\mathbf{E}^{\mathrm{i}} = \hat{\mathbf{y}}100\cos(\omega t - \pi x - 1.73\pi z) \qquad \text{(V/m)}, \qquad (8.61)$$

and the soil medium may be assumed to be a lossless dielectric with a relative permittivity of 4.

(a) Determine k_1, k_2, and the incidence angle θ_{i}.

(b) Obtain expressions for the total electric fields in air and in the soil.

(c) Determine the average power density carried by the wave traveling in soil.

Solution: **(a)** We begin by converting Eq. (8.61) into phasor form, akin to the expression given by Eq. (8.46a):

$$\widetilde{\mathbf{E}}^{\mathrm{i}} = \hat{\mathbf{y}}100e^{-j\pi x - j1.73\pi z} = \hat{\mathbf{y}}100e^{-jk_1 x_{\mathrm{i}}} \qquad \text{(V/m)}, \qquad (8.62)$$

where x_{i} is the axis along which the wave is traveling, and

$$k_1 x_{\mathrm{i}} = \pi x + 1.73\pi z. \qquad (8.63)$$

Using Eq. (8.47a), we have

$$k_1 x_{\mathrm{i}} = k_1 x \sin\theta_{\mathrm{i}} + k_1 z \cos\theta_{\mathrm{i}}. \qquad (8.64)$$

Hence,

$$k_1 \sin\theta_{\mathrm{i}} = \pi,$$

$$k_1 \cos\theta_{\mathrm{i}} = 1.73\pi,$$

which together give

$$k_1 = \sqrt{\pi^2 + (1.73\pi)^2} = 2\pi \qquad \text{(rad/m)},$$

$$\theta_{\mathrm{i}} = \tan^{-1}\left(\frac{\pi}{1.73\pi}\right) = 30^\circ.$$

The wavelength in medium 1 (air) is

$$\lambda_1 = \frac{2\pi}{k_1} = 1 \text{ m},$$

and the wavelength in medium 2 (soil) is

$$\lambda_2 = \frac{\lambda_1}{\sqrt{\epsilon_{\mathrm{r}_2}}} = \frac{1}{\sqrt{4}} = 0.5 \text{ m}.$$

The corresponding wave number in medium 2 is

$$k_2 = \frac{2\pi}{\lambda_2} = 4\pi \qquad \text{(rad/m)}.$$

Since $\widetilde{\mathbf{E}}^{\mathrm{i}}$ is along $\hat{\mathbf{y}}$, it is perpendicularly polarized ($\hat{\mathbf{y}}$ is perpendicular to the plane of incidence containing the surface normal $\hat{\mathbf{z}}$ and the propagation direction $\hat{\mathbf{x}}_{\mathrm{i}}$).

(b) Given that $\theta_{\mathrm{i}} = 30^\circ$, the transmission angle θ_{t} is obtained with the help of Eq. (8.56):

$$\sin\theta_{\mathrm{t}} = \frac{k_1}{k_2}\sin\theta_{\mathrm{i}} = \frac{2\pi}{4\pi}\sin 30^\circ = 0.25$$

or

$$\theta_{\mathrm{t}} = 14.5^\circ.$$

With $\epsilon_1 = \epsilon_0$ and $\epsilon_2 = \epsilon_{\mathrm{r}_2}\epsilon_0 = 4\epsilon_0$, the reflection and transmission coefficients for perpendicular polarization are determined with the help of Eqs. (8.59) and (8.60),

$$\Gamma_\perp = \frac{\cos\theta_{\mathrm{i}} - \sqrt{(\epsilon_2/\epsilon_1) - \sin^2\theta_{\mathrm{i}}}}{\cos\theta_{\mathrm{i}} + \sqrt{(\epsilon_2/\epsilon_1) - \sin^2\theta_{\mathrm{i}}}} = -0.38,$$

$$\tau_\perp = 1 + \Gamma_\perp = 0.62.$$

Using Eqs. (8.48a) and (8.49a) with $E^{\mathrm{i}}_{\perp 0} = 100$ V/m and $\theta_{\mathrm{i}} = \theta_{\mathrm{r}}$, the total electric field in medium 1 is

$$\begin{aligned}\widetilde{\mathbf{E}}^1_\perp &= \widetilde{\mathbf{E}}^{\mathrm{i}}_\perp + \widetilde{\mathbf{E}}^{\mathrm{r}}_\perp \\ &= \hat{\mathbf{y}}E^{\mathrm{i}}_{\perp 0}e^{-jk_1(x\sin\theta_{\mathrm{i}} + z\cos\theta_{\mathrm{i}})} \\ &\quad + \hat{\mathbf{y}}\Gamma E^{\mathrm{i}}_{\perp 0}e^{-jk_1(x\sin\theta_{\mathrm{i}} - z\cos\theta_{\mathrm{i}})} \\ &= \hat{\mathbf{y}}100e^{-j(\pi x + 1.73\pi z)} - \hat{\mathbf{y}}38e^{-j(\pi x - 1.73\pi z)},\end{aligned}$$

and the corresponding instantaneous electric field in medium 1 is

$$\mathbf{E}^1_\perp(x,z,t) = \mathfrak{Re}\left[\widetilde{\mathbf{E}}^1_\perp e^{j\omega t}\right]$$
$$= \hat{\mathbf{y}}[100\cos(\omega t - \pi x - 1.73\pi z) - 38\cos(\omega t - \pi x + 1.73\pi z)] \qquad \text{(V/m)}.$$

In medium 2, using Eq. (8.49c) with $E^t_{\perp 0} = \tau_\perp E^i_{\perp 0}$ gives

$$\widetilde{\mathbf{E}}^t_\perp = \hat{\mathbf{y}}\tau E^i_{\perp 0} e^{-jk_2(x\sin\theta_t + z\cos\theta_t)} = \hat{\mathbf{y}}62e^{-j(\pi x + 3.87\pi z)}$$

and, correspondingly,

$$\mathbf{E}^t_\perp(x,z,t) = \mathfrak{Re}\left[\widetilde{\mathbf{E}}^t_\perp e^{j\omega t}\right]$$
$$= \hat{\mathbf{y}}62\cos(\omega t - \pi x - 3.87\pi z) \qquad \text{(V/m)}.$$

(c) In medium 2, $\eta_2 = \eta_0/\sqrt{\epsilon_{r_2}} \approx 120\pi/\sqrt{4} = 60\pi$ (Ω), and the average power density carried by the wave is

$$S^t_{av} = \frac{|E^t_{\perp 0}|^2}{2\eta_2} = \frac{(62)^2}{2\times 60\pi} = 10.2 \qquad \text{(W/m}^2\text{)}.$$

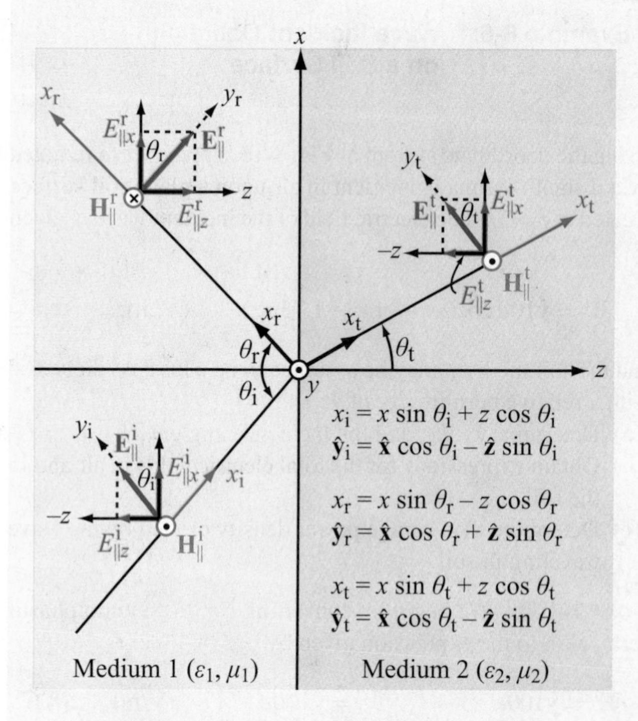

Figure 8-16 Parallel-polarized plane wave incident at an angle θ_i upon a planar boundary.

8-4.2 Parallel Polarization

If we interchange the roles played by **E** and **H** in the perpendicular polarization scenario covered in the preceding subsection, while keeping in mind the requirement that **E** × **H** must point in the direction of propagation for each of the incident, reflected, and transmitted waves, we end up with the parallel polarization scenario shown in Fig. 8-16. Now the electric fields lie in the plane of incidence, while the associated magnetic fields are perpendicular to the plane of incidence. With reference to the directions indicated in Fig. 8-16, the fields of the incident, reflected, and transmitted waves are given by

Incident Wave

$$\widetilde{\mathbf{E}}^i_\| = \hat{\mathbf{y}}_i E^i_{\|0} e^{-jk_1x_i}$$
$$= (\hat{\mathbf{x}}\cos\theta_i - \hat{\mathbf{z}}\sin\theta_i)E^i_{\|0}e^{-jk_1(x\sin\theta_i + z\cos\theta_i)}, \qquad (8.65a)$$

$$\widetilde{\mathbf{H}}^i_\| = \hat{\mathbf{y}}\,\frac{E^i_{\|0}}{\eta_1}\,e^{-jk_1x_i} = \hat{\mathbf{y}}\frac{E^i_{\|0}}{\eta_1}\,e^{-jk_1(x\sin\theta_i + z\cos\theta_i)}, \qquad (8.65b)$$

Reflected Wave

$$\widetilde{\mathbf{E}}^r_\| = \hat{\mathbf{y}}_r E^r_{\|0} e^{-jk_1x_r}$$
$$= (\hat{\mathbf{x}}\cos\theta_r + \hat{\mathbf{z}}\sin\theta_r)E^r_{\|0}e^{-jk_1(x\sin\theta_r - z\cos\theta_r)}, \qquad (8.65c)$$

$$\widetilde{\mathbf{H}}^r_\| = -\hat{\mathbf{y}}\,\frac{E^r_{\|0}}{\eta_1}\,e^{-jk_1x_r}$$
$$= -\hat{\mathbf{y}}\,\frac{E^r_{\|0}}{\eta_1}\,e^{-jk_1(x\sin\theta_r - z\cos\theta_r)}, \qquad (8.65d)$$

Transmitted Wave

$$\widetilde{\mathbf{E}}^t_\| = \hat{\mathbf{y}}_t E^t_{\|0} e^{-jk_2x_t}$$
$$= (\hat{\mathbf{x}}\cos\theta_t - \hat{\mathbf{z}}\sin\theta_t)E^t_{\|0}e^{-jk_2(x\sin\theta_t + z\cos\theta_t)}, \qquad (8.65e)$$

$$\widetilde{\mathbf{H}}^t_\| = \hat{\mathbf{y}}\,\frac{E^t_{\|0}}{\eta_2}\,e^{-jk_2x_t} = \hat{\mathbf{y}}\,\frac{E^t_{\|0}}{\eta_2}\,e^{-jk_2(x\sin\theta_t + z\cos\theta_t)}. \qquad (8.65f)$$

By matching the tangential components of $\widetilde{\mathbf{E}}$ and $\widetilde{\mathbf{H}}$ in both media at $z = 0$, we again obtain the relations defining

Snell's laws, as well as the following expressions for the ***Fresnel reflection and transmission coefficients for parallel polarization***:

$$\Gamma_\parallel = \frac{E^r_{\parallel 0}}{E^i_{\parallel 0}} = \frac{\eta_2 \cos\theta_t - \eta_1 \cos\theta_i}{\eta_2 \cos\theta_t + \eta_1 \cos\theta_i}, \tag{8.66a}$$

$$\tau_\parallel = \frac{E^t_{\parallel 0}}{E^i_{\parallel 0}} = \frac{2\eta_2 \cos\theta_i}{\eta_2 \cos\theta_t + \eta_1 \cos\theta_i}. \tag{8.66b}$$

The preceding expressions can be shown to yield the relation

$$\tau_\parallel = (1 + \Gamma_\parallel)\,\frac{\cos\theta_i}{\cos\theta_t}. \tag{8.67}$$

We noted earlier in connection with the perpendicular-polarization case that, when the second medium is a perfect conductor with $\eta_2 = 0$, the incident wave gets totally reflected at the boundary. The same is true for the parallel polarization case; setting $\eta_2 = 0$ in Eqs. (8.66a) and (8.66b) gives $\Gamma_\parallel = -1$ and $\tau_\parallel = 0$.

For nonmagnetic materials, Eq. (8.66a) becomes

$$\Gamma_\parallel = \frac{-(\epsilon_2/\epsilon_1)\cos\theta_i + \sqrt{(\epsilon_2/\epsilon_1) - \sin^2\theta_i}}{(\epsilon_2/\epsilon_1)\cos\theta_i + \sqrt{(\epsilon_2/\epsilon_1) - \sin^2\theta_i}} \quad (\text{for } \mu_1 = \mu_2). \tag{8.68}$$

To illustrate the angular variations of the magnitudes of $\Gamma_\perp$ and $\Gamma_\parallel$, **Fig. 8-17** shows plots for waves incident in air onto three different types of dielectric surfaces: dry soil ($\epsilon_r = 3$), wet soil ($\epsilon_r = 25$), and water ($\epsilon_r = 81$). For each of the surfaces, (1) $\Gamma_\perp = \Gamma_\parallel$ at normal incidence ($\theta_i = 0$), as expected, (2) $|\Gamma_\perp| = |\Gamma_\parallel| = 1$ at ***grazing incidence*** ($\theta_i = 90°$), and (3) $\Gamma_\parallel$ goes to zero at an angle called the ***Brewster angle*** in **Fig. 8-17**. Had the materials been magnetic too ($\mu_1 \neq \mu_2$), it would have been possible for $\Gamma_\perp$ to vanish at some angle as well. However, for nonmagnetic materials, the Brewster angle exists only for parallel polarization, and its value depends on the ratio (ϵ_2/ϵ_1), as we see shortly.

► At the Brewster angle, the parallel-polarized component of the incident wave is totally transmitted into medium 2. ◄

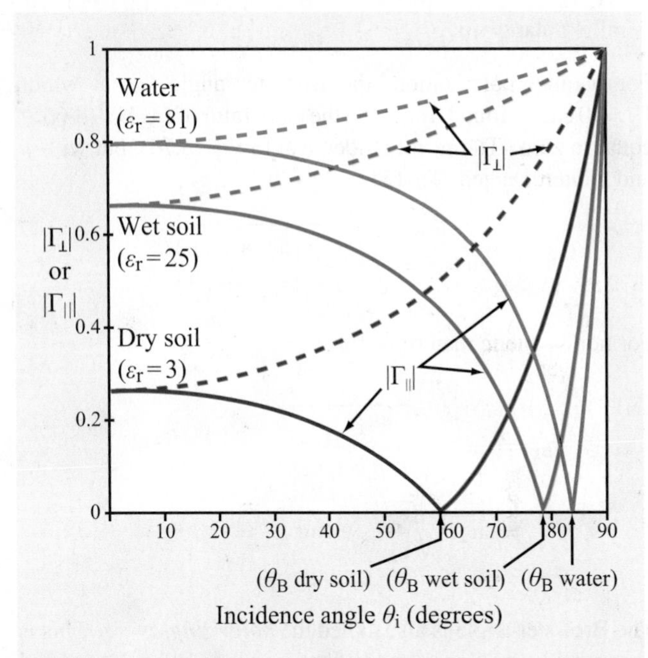

Figure 8-17 Plots for $|\Gamma_\perp|$ and $|\Gamma_\parallel|$ as a function of θ_i for a dry-soil surface, a wet-soil surface, and a water surface. For each surface, $|\Gamma_\parallel| = 0$ at the Brewster angle.

8-4.3 Brewster Angle

The Brewster angle θ_B is defined as the incidence angle θ_i at which the Fresnel reflection coefficient $\Gamma = 0$.

Perpendicular polarization

For perpendicular polarization, the Brewster angle $\theta_{B\perp}$ can be obtained by setting the numerator of the expression for $\Gamma_\perp$, given by Eq. (8.58a), equal to zero. This happens when

$$\eta_2 \cos\theta_i = \eta_1 \cos\theta_t. \tag{8.69}$$

By (1) squaring both sides of Eq. (8.69), (2) using Eq. (8.56), (3) solving for θ_i, and then denoting θ_i as $\theta_{B\perp}$, we obtain

$$\sin\theta_{B\perp} = \sqrt{\frac{1 - (\mu_1\epsilon_2/\mu_2\epsilon_1)}{1 - (\mu_1/\mu_2)^2}}. \tag{8.70}$$

Because the denominator of Eq. (8.70) goes to zero when $\mu_1 = \mu_2$, *$\theta_{B\perp}$ does not exist for nonmagnetic materials.*

Parallel polarization

For parallel polarization, the Brewster angle $\theta_{B\parallel}$ at which $\Gamma_\parallel = 0$ can be found by setting the numerator of $\Gamma_\parallel$, Eq. (8.66a), equal to zero. The result is identical to Eq. (8.70), but with μ and ϵ interchanged. That is,

$$\sin\theta_{B\parallel} = \sqrt{\frac{1-(\epsilon_1\mu_2/\epsilon_2\mu_1)}{1-(\epsilon_1/\epsilon_2)^2}}. \tag{8.71}$$

For nonmagnetic materials,

$$\theta_{B\parallel} = \sin^{-1}\sqrt{\frac{1}{1+(\epsilon_1/\epsilon_2)}} = \tan^{-1}\sqrt{\frac{\epsilon_2}{\epsilon_1}} \quad (\text{for } \mu_1 = \mu_2). \tag{8.72}$$

The Brewster angle is also called the ***polarizing angle***. This is because, if a wave composed of both perpendicular and parallel polarization components is incident upon a nonmagnetic surface at the Brewster angle $\theta_{B\parallel}$, the parallel polarized component is totally transmitted into the second medium, and only the perpendicularly polarized component is reflected by the surface. Natural light, including sunlight and light generated by most manufactured sources, is ***unpolarized*** because the direction of the electric field of the light waves varies randomly in angle over the plane perpendicular to the direction of propagation. Thus, on average half of the intensity of natural light is perpendicularly polarized and the other half is parallel polarized. When unpolarized light is incident upon a surface at the Brewster angle, the reflected wave is strictly perpendicularly polarized. Hence, the surface acts as a polarizer.

Concept Question 8-4: Can total internal reflection take place for a wave incident from medium 1 (with n_1) onto medium 2 (with n_2) when $n_2 > n_1$?

Concept Question 8-5: What is the difference between the boundary conditions applied in Section 8-1.1 for normal incidence and those applied in Section 8-4.1 for oblique incidence with perpendicular polarization?

Concept Question 8-6: Why is the Brewster angle also called the polarizing angle?

Concept Question 8-7: At the boundary, the vector sum of the tangential components of the incident and reflected electric fields has to equal the tangential component of the transmitted electric field. For $\epsilon_{r_1} = 1$ and $\epsilon_{r_2} = 16$, determine the Brewster angle and then verify the validity of the preceding statement by sketching to scale the tangential components of the three electric fields at the Brewster angle.

Exercise 8-7: A wave in air is incident upon a soil surface at $\theta_i = 50°$. If soil has $\epsilon_r = 4$ and $\mu_r = 1$, determine $\Gamma_\perp$, $\tau_\perp$, $\Gamma_\parallel$, and $\tau_\parallel$.

Answer: $\Gamma_\perp = -0.48$, $\tau_\perp = 0.52$, $\Gamma_\parallel = -0.16$, $\tau_\parallel = 0.58$. (See EM.)

Exercise 8-8: Determine the Brewster angle for the boundary of Exercise 8.7.

Answer: $\theta_B = 63.4°$. (See EM.)

Exercise 8-9: Show that the incident, reflected, and transmitted electric and magnetic fields given by Eqs. (8.65a) through (8.65f) all have the same exponential phase function along the x direction.

Answer: With the help of Eqs. (8.55) and (8.56), all six fields are shown to vary as $e^{-jk_1x\sin\theta_i}$. (See EM.)

8-5 Reflectivity and Transmissivity

The reflection and transmission coefficients derived earlier are ratios of the reflected and transmitted electric field amplitudes to the amplitude of the incident electric field. We now examine power ratios, starting with the perpendicular polarization case. **Figure 8-18** shows a circular beam of electromagnetic energy incident upon the boundary between two contiguous, lossless media. The area of the spot illuminated by the beam is A, and the incident, reflected, and transmitted beams have electric-field amplitudes $E^i_{\perp 0}$, $E^r_{\perp 0}$, and $E^t_{\perp 0}$, respectively. The average power densities carried by the incident, reflected, and

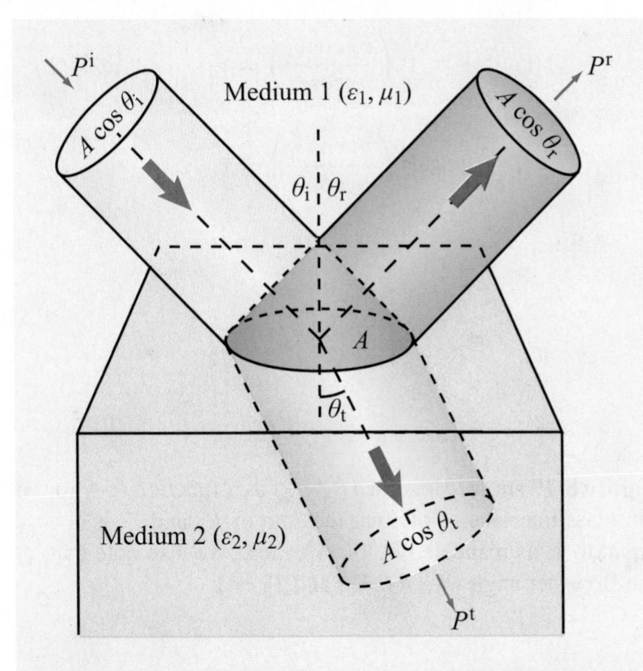

Figure 8-18 Reflection and transmission of an incident circular beam illuminating a spot of size A on the interface.

transmitted beams are

$$S^{\rm i}_\perp = \frac{|E^{\rm i}_{\perp 0}|^2}{2\eta_1}, \tag{8.73a}$$

$$S^{\rm r}_\perp = \frac{|E^{\rm r}_{\perp 0}|^2}{2\eta_1}, \tag{8.73b}$$

$$S^{\rm t}_\perp = \frac{|E^{\rm t}_{\perp 0}|^2}{2\eta_2}, \tag{8.73c}$$

where η_1 and η_2 are the intrinsic impedances of media 1 and 2, respectively. The cross-sectional areas of the incident, reflected, and transmitted beams are

$$A_{\rm i} = A\cos\theta_{\rm i}, \tag{8.74a}$$

$$A_{\rm r} = A\cos\theta_{\rm r}, \tag{8.74b}$$

$$A_{\rm t} = A\cos\theta_{\rm t}, \tag{8.74c}$$

and the corresponding average powers carried by the beams are

$$P^{\rm i}_\perp = S^{\rm i}_\perp A_{\rm i} = \frac{|E^{\rm i}_{\perp 0}|^2}{2\eta_1}\; A\cos\theta_{\rm i}, \tag{8.75a}$$

$$P^{\rm r}_\perp = S^{\rm r}_\perp A_{\rm r} = \frac{|E^{\rm r}_{\perp 0}|^2}{2\eta_1}\; A\cos\theta_{\rm r}, \tag{8.75b}$$

$$P^{\rm t}_\perp = S^{\rm t}_\perp A_{\rm t} = \frac{|E^{\rm t}_{\perp 0}|^2}{2\eta_2}\; A\cos\theta_{\rm t}. \tag{8.75c}$$

The ***reflectivity*** R (also called ***reflectance*** in optics) is defined as the ratio of the reflected to the incident power. The reflectivity for perpendicular polarization is then

$$R_\perp = \frac{P^{\rm r}_\perp}{P^{\rm i}_\perp} = \frac{|E^{\rm r}_{\perp 0}|^2\cos\theta_{\rm r}}{|E^{\rm i}_{\perp 0}|^2\cos\theta_{\rm i}} = \left|\frac{E^{\rm r}_{\perp 0}}{E^{\rm i}_{\perp 0}}\right|^2, \tag{8.76}$$

where we used the fact that $\theta_{\rm r} = \theta_{\rm i}$, in accordance with Snell's law of reflection. The ratio of the reflected to incident electric field amplitudes, $|E^{\rm r}_{\perp 0}/E^{\rm i}_{\perp 0}|$, is equal to the magnitude of the reflection coefficient $\Gamma_\perp$. Hence,

$$R_\perp = |\Gamma_\perp|^2, \tag{8.77}$$

and, similarly, for parallel polarization

$$R_\parallel = \frac{P^{\rm r}_\parallel}{P^{\rm i}_\parallel} = |\Gamma_\parallel|^2. \tag{8.78}$$

The ***transmissivity*** T (or ***transmittance*** in optics) is defined as the ratio of the transmitted power to incident power:

$$T_\perp = \frac{P^{\rm t}_\perp}{P^{\rm i}_\perp} = \frac{|E^{\rm t}_{\perp 0}|^2}{|E^{\rm i}_{\perp 0}|^2}\,\frac{\eta_1}{\eta_2}\,\frac{A\cos\theta_{\rm t}}{A\cos\theta_{\rm i}} = |\tau_\perp|^2\left(\frac{\eta_1\cos\theta_{\rm t}}{\eta_2\cos\theta_{\rm i}}\right), \tag{8.79a}$$

$$T_\parallel = \frac{P^{\rm t}_\parallel}{P^{\rm i}_\parallel} = |\tau_\parallel|^2\left(\frac{\eta_1\cos\theta_{\rm t}}{\eta_2\cos\theta_{\rm i}}\right). \tag{8.79b}$$

► The incident, reflected, and transmitted waves do not have to obey any such laws as conservation of electric field, conservation of magnetic field, or conservation of power density, but they do have to obey the law of conservation of power. ◄

In fact, in many cases the transmitted electric field is larger than the incident electric field. Conservation of power requires that the incident power equals the sum of the reflected and transmitted powers. That is, for perpendicular polarization,

$$P^{\rm i}_{\perp} = P^{\rm r}_{\perp} + P^{\rm t}_{\perp}, \tag{8.80}$$

or

$$\frac{|E^{\rm i}_{\perp 0}|^2}{2\eta_1} A\cos\theta_{\rm i} = \frac{|E^{\rm r}_{\perp 0}|^2}{2\eta_1} A\cos\theta_{\rm r} + \frac{|E^{\rm t}_{\perp 0}|^2}{2\eta_2} A\cos\theta_{\rm t}. \tag{8.81}$$

Use of Eqs. (8.76), (8.79a), and (8.79b) leads to

$$R_{\perp} + T_{\perp} = 1, \tag{8.82a}$$

$$R_{\parallel} + T_{\parallel} = 1, \tag{8.82b}$$

or

$$|\Gamma_{\perp}|^2 + |\tau_{\perp}|^2 \left(\frac{\eta_1 \cos\theta_{\rm t}}{\eta_2 \cos\theta_{\rm i}}\right) = 1, \tag{8.83a}$$

$$|\Gamma_{\parallel}|^2 + |\tau_{\parallel}|^2 \left(\frac{\eta_1 \cos\theta_{\rm t}}{\eta_2 \cos\theta_{\rm i}}\right) = 1. \tag{8.83b}$$

Figure 8-19 shows plots for $(R_{\parallel}, T_{\parallel})$ as a function of $\theta_{\rm i}$ for an air–glass interface. Note that the sum of $R_{\parallel}$ and $T_{\parallel}$ is always equal to 1, as mandated by Eq. (8.82b). We also note that, at the Brewster angle $\theta_{\rm B}$, $R_{\parallel} = 0$ and $T_{\parallel} = 1$.

Table 8-2 provides a summary of the general expressions for Γ, τ, R, and T for both normal and oblique incidence.

Table 8-2 Expressions for Γ, τ, R, and T for wave incidence from a medium with intrinsic impedance η_1 onto a medium with intrinsic impedance η_2. Angles $\theta_{\rm i}$ and $\theta_{\rm t}$ are the angles of incidence and transmission, respectively.

Property	Normal Incidence $\theta_{\rm i} = \theta_{\rm t} = 0$	Perpendicular Polarization	Parallel Polarization
Reflection coefficient	$\Gamma = \dfrac{\eta_2 - \eta_1}{\eta_2 + \eta_1}$	$\Gamma_{\perp} = \dfrac{\eta_2\cos\theta_{\rm i} - \eta_1\cos\theta_{\rm t}}{\eta_2\cos\theta_{\rm i} + \eta_1\cos\theta_{\rm t}}$	$\Gamma_{\parallel} = \dfrac{\eta_2\cos\theta_{\rm t} - \eta_1\cos\theta_{\rm i}}{\eta_2\cos\theta_{\rm t} + \eta_1\cos\theta_{\rm i}}$
Transmission coefficient	$\tau = \dfrac{2\eta_2}{\eta_2 + \eta_1}$	$\tau_{\perp} = \dfrac{2\eta_2\cos\theta_{\rm i}}{\eta_2\cos\theta_{\rm i} + \eta_1\cos\theta_{\rm t}}$	$\tau_{\parallel} = \dfrac{2\eta_2\cos\theta_{\rm i}}{\eta_2\cos\theta_{\rm t} + \eta_1\cos\theta_{\rm i}}$
Relation of Γ to τ	$\tau = 1 + \Gamma$	$\tau_{\perp} = 1 + \Gamma_{\perp}$	$\tau_{\parallel} = (1 + \Gamma_{\parallel})\dfrac{\cos\theta_{\rm i}}{\cos\theta_{\rm t}}$
Reflectivity	$R = \lvert\Gamma\rvert^2$	$R_{\perp} = \lvert\Gamma_{\perp}\rvert^2$	$R_{\parallel} = \lvert\Gamma_{\parallel}\rvert^2$
Transmissivity	$T = \lvert\tau\rvert^2\left(\dfrac{\eta_1}{\eta_2}\right)$	$T_{\perp} = \lvert\tau_{\perp}\rvert^2\dfrac{\eta_1\cos\theta_{\rm t}}{\eta_2\cos\theta_{\rm i}}$	$T_{\parallel} = \lvert\tau_{\parallel}\rvert^2\dfrac{\eta_1\cos\theta_{\rm t}}{\eta_2\cos\theta_{\rm i}}$
Relation of R to T	$T = 1 - R$	$T_{\perp} = 1 - R_{\perp}$	$T_{\parallel} = 1 - R_{\parallel}$

Notes: (1) $\sin\theta_{\rm t} = \sqrt{\mu_1\epsilon_1/\mu_2\epsilon_2}\,\sin\theta_{\rm i}$; (2) $\eta_1 = \sqrt{\mu_1/\epsilon_1}$; (3) $\eta_2 = \sqrt{\mu_2/\epsilon_2}$; (4) for nonmagnetic media, $\eta_2/\eta_1 = n_1/n_2$.

Module 8.3 Oblique Incidence Upon specifying the frequency, polarization, and incidence angle of a plane wave incident upon a planar boundary between two lossless media, this module displays vector information and plots of the reflection and transmission coefficients as a function of incidence angle.

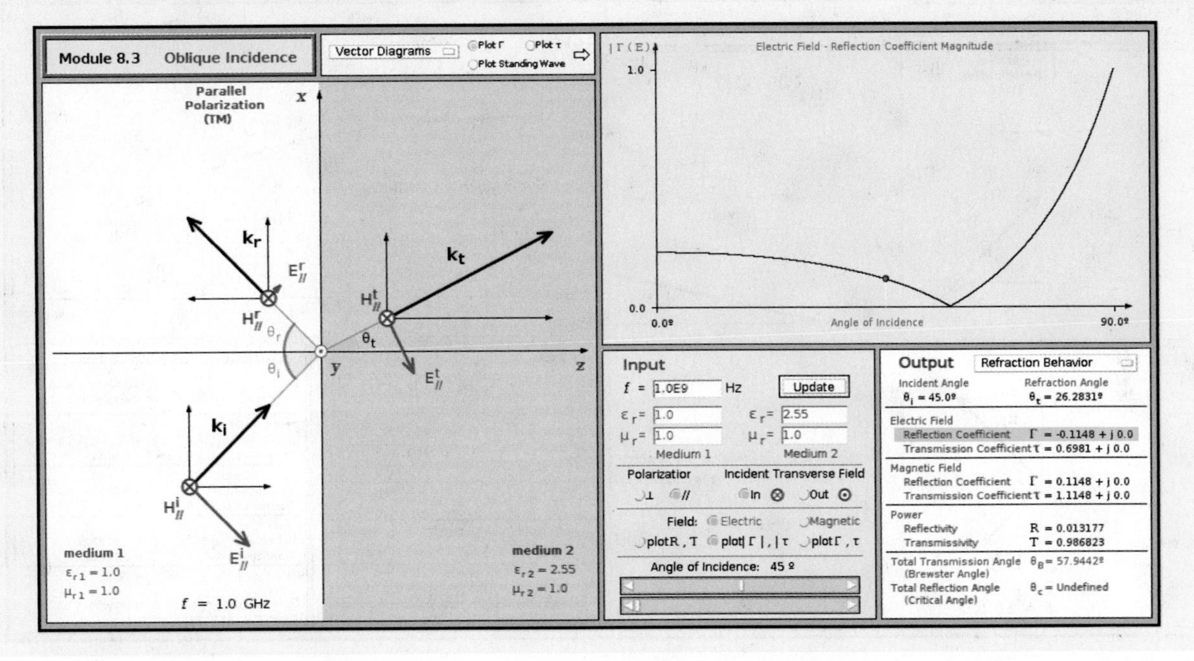

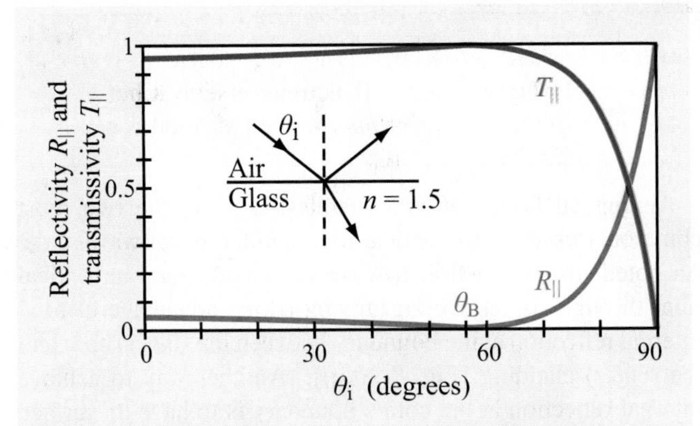

Figure 8-19 Angular plots for $(R_{\|}, T_{\|})$ for an air–glass interface.

Example 8-7: Beam of Light

A 5 W beam of light with circular cross section is incident in air upon the plane boundary of a dielectric medium with index of refraction of 5. If the angle of incidence is 60° and the incident wave is parallel polarized, determine the transmission angle and the powers contained in the reflected and transmitted beams.

Solution: From Eq. (8.56),

$$\sin\theta_t = \frac{n_1}{n_2}\sin\theta_i = \frac{1}{5}\sin 60^\circ = 0.17$$

or

$$\theta_t = 10^\circ.$$

With $\epsilon_2/\epsilon_1 = n_2^2/n_1^2 = (5)^2 = 25$, the reflection coefficient for parallel polarization follows from Eq. (8.68) as

$$\Gamma_{\|} = \frac{-(\epsilon_2/\epsilon_1)\cos\theta_i + \sqrt{(\epsilon_2/\epsilon_1) - \sin^2\theta_i}}{(\epsilon_2/\epsilon_1)\cos\theta_i + \sqrt{(\epsilon_2/\epsilon_1) - \sin^2\theta_i}}$$

$$= \frac{-25\cos 60^\circ + \sqrt{25 - \sin^2 60^\circ}}{25\cos 60^\circ + \sqrt{25 - \sin^2 60^\circ}} = -0.435.$$

Module 8.4 Oblique Incidence in Lossy Medium This module extends the capabilities of Module 8.1 to situations in which medium 2 is lossy.

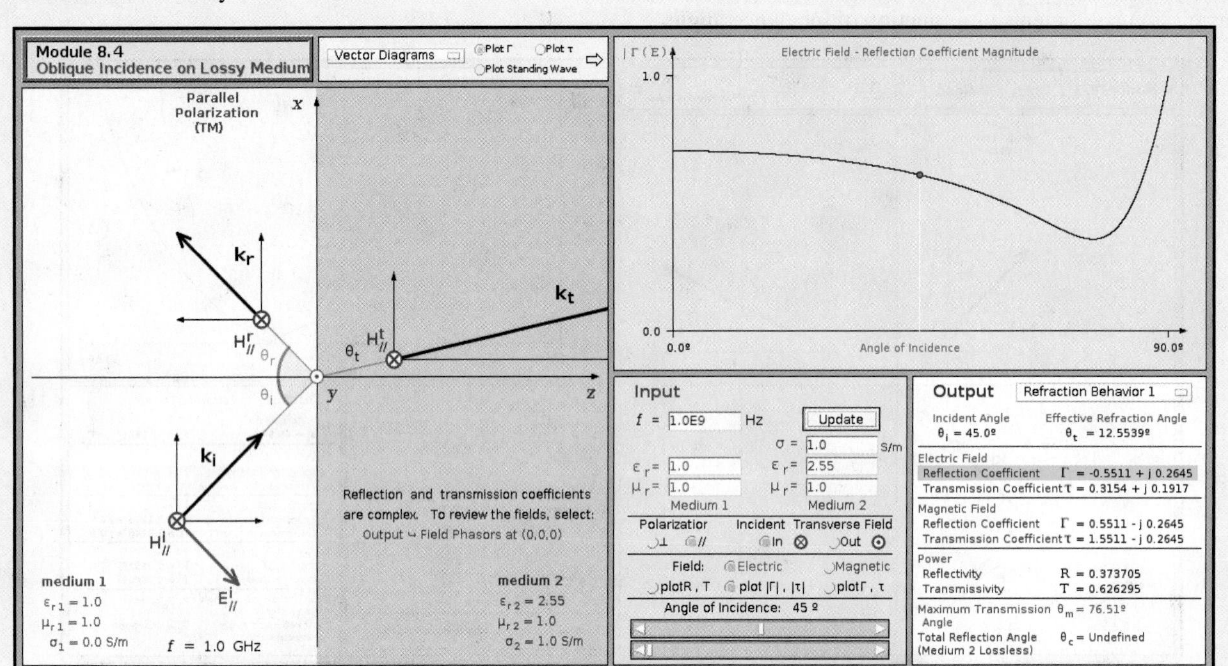

The reflected and transmitted powers therefore are

$$P^{\mathrm{r}}_{\parallel} = P^{\mathrm{i}}_{\parallel} |\Gamma_{\parallel}|^2 = 5(0.435)^2 = 0.95 \text{ W},$$

$$P^{\mathrm{t}}_{\parallel} = P^{\mathrm{i}}_{\parallel} - P^{\mathrm{r}}_{\parallel} = 5 - 0.95 = 4.05 \text{ W}.$$

8-6 Waveguides

Earlier in Chapter 2, we considered two families of transmission lines, namely those that support ***transverse-electromagnetic*** (TEM) modes and those that do not. Transmission lines belonging to the TEM family (Fig. 2-4), including coaxial, two-wire, and parallel-plate lines, support **E** and **H** fields that are orthogonal to the direction of propagation. Fields supported by lines in the other group, often called ***higher-order transmission lines***, may have **E** *or* **H** orthogonal to the direction of propagation $\hat{\mathbf{k}}$, but not both simultaneously. Thus, *at least one component of* **E** *or* **H** *is along* $\hat{\mathbf{k}}$.

► If **E** is transverse to $\hat{\mathbf{k}}$ but **H** is not, we call it a ***transverse electric*** (TE) mode, and if **H** is transverse to $\hat{\mathbf{k}}$ but **E** is not, we call it a ***transverse magnetic*** (TM) mode. ◄

Among all higher-order transmission lines, the two most commonly used are the optical fiber and the metal waveguide. As noted in Section 8-3, a wave is guided along an optical fiber through successive zigzags by taking advantage of total internal reflection at the boundary between the (inner) core and the (outer) cladding [Fig. 8-20(a)]. Another way to achieve internal reflection at the core's boundary is to have its surface coated by a conducting material. Under the proper conditions, on which we shall elaborate later, a wave excited in the interior of a hollow conducting pipe, such as the circular or rectangular waveguides shown in Figs. 8-20(b) and (c), undergoes a process similar to that of successive internal reflection in an optical fiber, resulting in propagation down the pipe. Most waveguide applications call for air-filled guides, but in some cases, the waveguide may be filled with a dielectric material so as to alter its propagation velocity or impedance, or it may be vacuum-

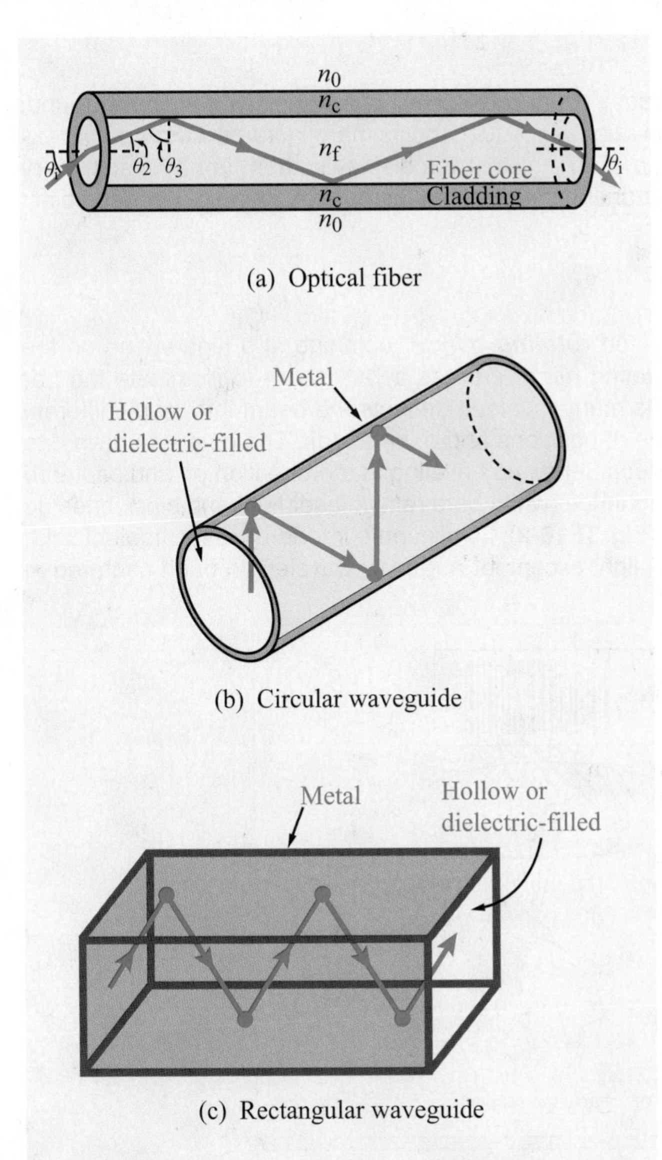

Figure 8-20 Wave travel by successive reflections in (a) an optical fiber, (b) a circular metal waveguide, and (c) a rectangular metal waveguide.

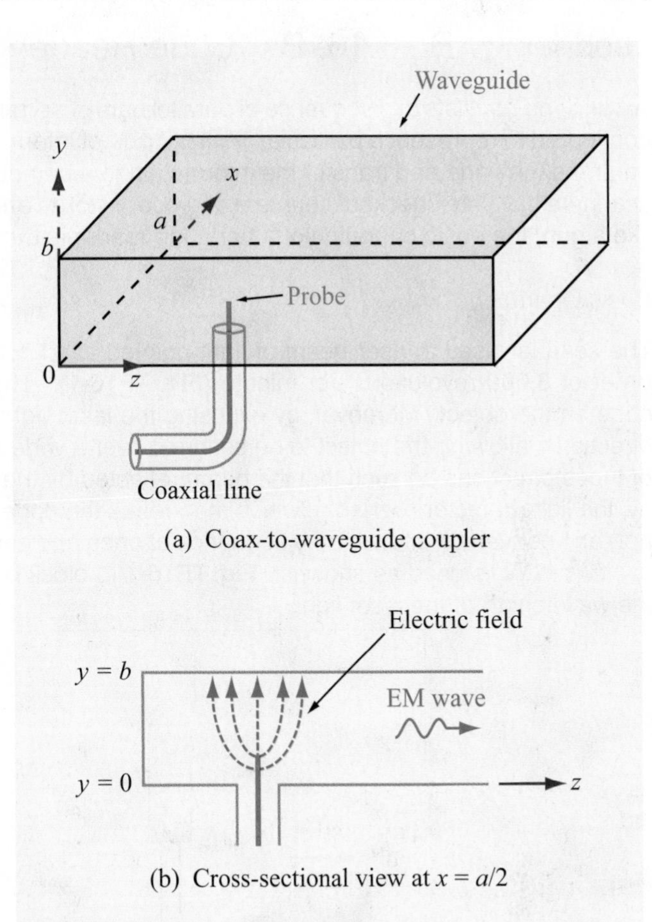

Figure 8-21 The inner conductor of a coaxial cable can excite an EM wave in the waveguide.

pumped to eliminate air molecules so as to prevent voltage breakdown, thereby increasing its power-handling capabilities.

Figure 8-21 illustrates how a coaxial cable can be connected to a rectangular waveguide. With its outer conductor connected to the metallic waveguide enclosure, the coaxial cable's inner conductor protrudes through a tiny hole into the waveguide's interior (without touching the conducting surface). Time-varying electric field lines extending between the protruding inner conductor and the inside surface of the guide provide the excitation necessary to transfer a signal from the coaxial line to the guide. Conversely, the center conductor can act like a probe, coupling a signal from the waveguide to the coaxial cable.

For guided transmission at frequencies below 30 GHz, the coaxial cable is by far the most widely used transmission line. At higher frequencies, however, the coaxial cable has a number of limitations: (a) in order for it to propagate only TEM modes, the cable's inner and outer conductors have to be reduced in size to satisfy a certain size-to-wavelength requirement, making it more difficult to fabricate; (b) the smaller cross section reduces the cable's power-handling capacity (limited by dielectric breakdown); and (c) the attenuation due to dielectric

Technology Brief 16: Bar-Code Readers

A ***bar code*** consists of a sequence of parallel bars of certain widths, usually printed in black against a white background, configured to represent a particular ***binary code*** of information about a product and its manufacturer. ***Laser scanners*** can read the code and transfer the information to a computer, a cash register, or a display screen. For both stationary scanners built into checkout counters at grocery stores and handheld units that can be pointed at the bar-coded object like a gun, the basic operation of a bar-code reader is the same.

Basic Operation

The scanner uses a laser beam of light pointed at a multifaceted ***rotating mirror***, spinning at a high speed on the order of 6,000 revolutions per minute (**Fig. TF16-1**). The rotating mirror creates a ***fan beam*** to illuminate the bar code on the object. Moreover, by exposing the laser light to its many facets, it deflects the beam into many different directions, allowing the object to be scanned over a wide range of positions and orientations. The goal is to have one of those directions be such that the beam reflected by the bar code ends up traveling in the direction of, and captured by, the light detector (***sensor***), which then reads the coded sequence (white bars reflect laser light and black ones do not) and converts it into a binary sequence of ones and zeros (**Fig. TF16-2**). To eliminate interference by ambient light, a ***glass filter*** is used as shown in **Fig. TF16-1** to block out all light except for a narrow wavelength band centered at the wavelength of the laser light.

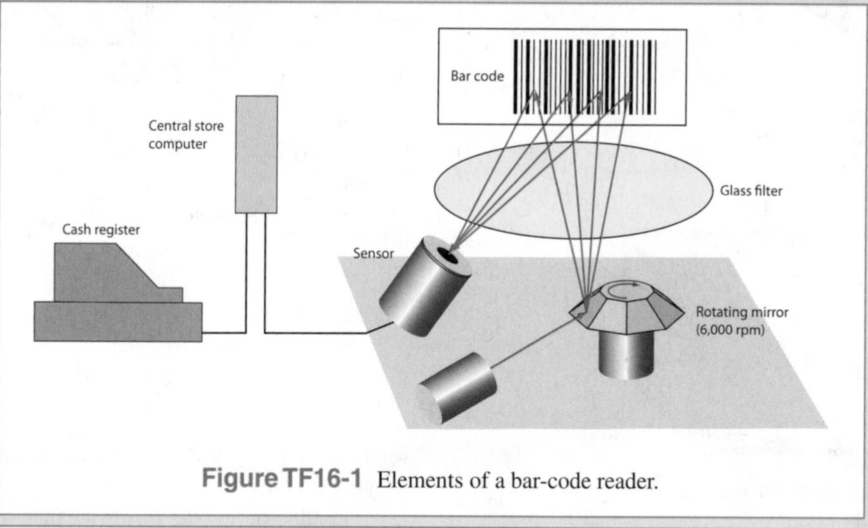

Figure TF16-1 Elements of a bar-code reader.

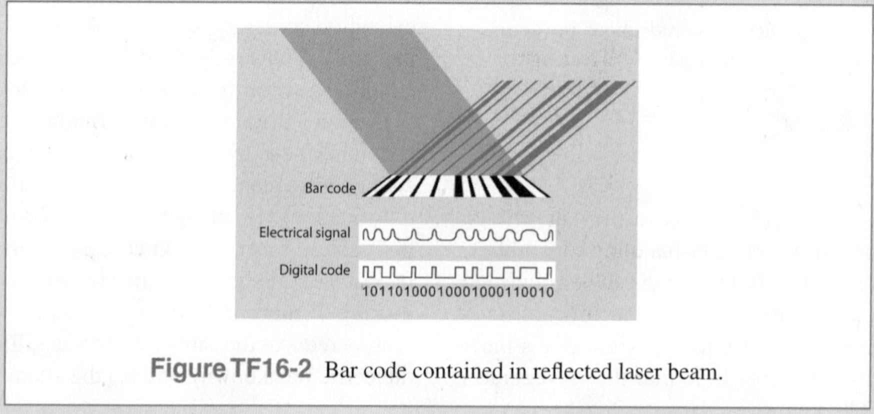

Figure TF16-2 Bar code contained in reflected laser beam.

losses increases with frequency. For all of these reasons, metal waveguides have been used as an alternative to coaxial lines for many radar and communication applications that operate at frequencies in the 5–100 GHz range, particularly those requiring the transmission of high levels of radio-frequency (RF) power. Even though waveguides with circular and elliptical cross sections have been used in some microwave systems, the rectangular shape has been the more prevalent geometry.

8-7 General Relations for E and H

The purpose of the next two sections is to derive expressions for **E** and **H** for the TE and TM modes in a rectangular waveguide, and to examine their wave properties. We choose the coordinate system shown in Fig. 8-22, in which propagation occurs along $\hat{\mathbf{z}}$. For TE modes, the electric field is transverse to the direction of propagation. Hence, **E** may have components along $\hat{\mathbf{x}}$ and $\hat{\mathbf{y}}$, but not along $\hat{\mathbf{z}}$. In contrast, **H** has a $\hat{\mathbf{z}}$-directed component and may have components along either $\hat{\mathbf{x}}$ or $\hat{\mathbf{y}}$, or both. The converse is true for TM modes.

Our solution procedure consists of four steps:

(1) Maxwell's equations are manipulated to develop general expressions for the phasor-domain transverse field components $\widetilde{E}_x$, $\widetilde{E}_y$, $\widetilde{H}_x$, and $\widetilde{H}_y$ in terms of $\widetilde{E}_z$ and $\widetilde{H}_z$. When specialized to the TE case, these expressions become functions of $\widetilde{H}_z$ only, and the converse is true for the TM case.

(2) The homogeneous wave equations given by Eqs. (7.15) and (7.16) are solved to obtain valid solutions for $\widetilde{E}_z$ (TM case) and $\widetilde{H}_z$ (TE case) in a waveguide.

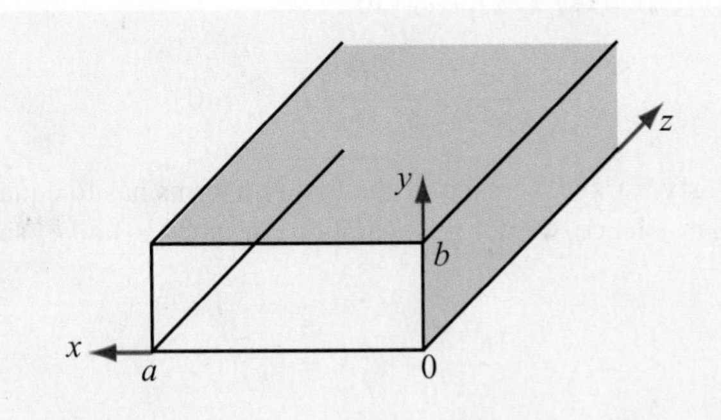

Figure 8-22 Waveguide coordinate system.

(3) The expressions derived in step 1 are then used to find $\widetilde{E}_x$, $\widetilde{E}_y$, $\widetilde{H}_x$, and $\widetilde{H}_y$.

(4) The solution obtained in step 3 are analyzed to determine the phase velocity and other properties of the TE and TM waves.

The intent of the present section is to realize the stated goals of step 1. We begin with a general form for the **E** and **H** fields in the phasor domain:

$$\widetilde{\mathbf{E}} = \hat{\mathbf{x}}\,\widetilde{E}_x + \hat{\mathbf{y}}\,\widetilde{E}_y + \hat{\mathbf{z}}\,\widetilde{E}_z, \tag{8.84a}$$

$$\widetilde{\mathbf{H}} = \hat{\mathbf{x}}\,\widetilde{H}_x + \hat{\mathbf{y}}\,\widetilde{H}_y + \hat{\mathbf{z}}\,\widetilde{H}_z. \tag{8.84b}$$

In general, all six components of $\widetilde{\mathbf{E}}$ and $\widetilde{\mathbf{H}}$ may depend on (x, y, z), and while we do not yet know how they functionally depend on (x, y), our prior experience suggests that $\widetilde{\mathbf{E}}$ and $\widetilde{\mathbf{H}}$ of a wave traveling along the $+z$ direction should exhibit a dependence on z of the form $e^{-j\beta z}$, where β is a yet-to-be-determined ***phase constant***. Hence, we adopt the form

$$\widetilde{E}_x(x, y, z) = \widetilde{e}_x(x, y)\, e^{-j\beta z}, \tag{8.85}$$

where $\widetilde{e}_x(x, y)$ describes the dependence of $\widetilde{E}_x(x, y, z)$ on (x, y) only. The form of Eq. (8.85) can be used for all other components of $\widetilde{\mathbf{E}}$ and $\widetilde{\mathbf{H}}$ as well. Thus,

$$\widetilde{\mathbf{E}} = (\hat{\mathbf{x}}\,\widetilde{e}_x + \hat{\mathbf{y}}\,\widetilde{e}_y + \hat{\mathbf{z}}\,\widetilde{e}_z)e^{-j\beta z}, \tag{8.86a}$$

$$\widetilde{\mathbf{H}} = (\hat{\mathbf{x}}\,\widetilde{h}_x + \hat{\mathbf{y}}\,\widetilde{h}_y + \hat{\mathbf{z}}\,\widetilde{h}_z)e^{-j\beta z}. \tag{8.86b}$$

The notation is intended to clarify that, in contrast to $\widetilde{E}$ and $\widetilde{H}$, which vary with (x, y, z), the lower-case $\widetilde{e}$ and $\widetilde{h}$ vary with (x, y) only.

In a lossless, source-free medium (such as the inside of a waveguide) characterized by permittivity ϵ and permeability μ (and conductivity $\sigma = 0$), Maxwell's curl equations are given by Eqs. (7.2b and d) with $\mathbf{J} = 0$,

$$\nabla \times \widetilde{\mathbf{E}} = -j\omega\mu\widetilde{\mathbf{H}}, \tag{8.87a}$$

$$\nabla \times \widetilde{\mathbf{H}} = j\omega\epsilon\widetilde{\mathbf{E}}. \tag{8.87b}$$

Upon inserting Eqs. (8.86a and b) into Eqs. (8.87a and b), and recalling that each of the curl equations actually consists of three separate equations—one for each of the unit vectors $\hat{\mathbf{x}}$, $\hat{\mathbf{y}}$, and $\hat{\mathbf{z}}$, we obtain the following relationships:

$$\frac{\partial \widetilde{e}_z}{\partial y} + j\beta\widetilde{e}_y = -j\omega\mu\widetilde{h}_x, \quad (8.88a)$$

$$-j\beta\widetilde{e}_x - \frac{\partial \widetilde{e}_z}{\partial x} = -j\omega\mu\widetilde{h}_y, \quad (8.88b)$$

$$\frac{\partial \widetilde{e}_y}{\partial x} - \frac{\partial \widetilde{e}_x}{\partial y} = -j\omega\mu\widetilde{h}_z, \quad (8.88c)$$

$$\frac{\partial \widetilde{h}_z}{\partial y} + j\beta\widetilde{h}_y = j\omega\epsilon\widetilde{e}_x, \quad (8.88d)$$

$$-j\beta\widetilde{h}_x - \frac{\partial \widetilde{h}_z}{\partial x} = j\omega\epsilon\widetilde{e}_y, \quad (8.88e)$$

$$\frac{\partial \widetilde{h}_y}{\partial x} - \frac{\partial \widetilde{h}_x}{\partial y} = j\omega\epsilon\widetilde{e}_z. \quad (8.88f)$$

Equations (8.88a–f) incorporate the fact that differentiation with respect to z is equivalent to multiplication by $-j\beta$. By manipulating these equations algebraically, we can obtain expressions for the x and y components of $\widetilde{\mathbf{E}}$ and $\widetilde{\mathbf{H}}$ in terms of their z components, namely

$$\widetilde{E}_x = \frac{-j}{k_c^2}\left(\beta\,\frac{\partial \widetilde{E}_z}{\partial x} + \omega\mu\,\frac{\partial \widetilde{H}_z}{\partial y}\right), \quad (8.89a)$$

$$\widetilde{E}_y = \frac{j}{k_c^2}\left(-\beta\,\frac{\partial \widetilde{E}_z}{\partial y} + \omega\mu\,\frac{\partial \widetilde{H}_z}{\partial x}\right), \quad (8.89b)$$

$$\widetilde{H}_x = \frac{j}{k_c^2}\left(\omega\epsilon\,\frac{\partial \widetilde{E}_z}{\partial y} - \beta\,\frac{\partial \widetilde{H}_z}{\partial x}\right), \quad (8.89c)$$

$$\widetilde{H}_y = \frac{-j}{k_c^2}\left(\omega\epsilon\,\frac{\partial \widetilde{E}_z}{\partial x} + \beta\,\frac{\partial \widetilde{H}_z}{\partial y}\right). \quad (8.89d)$$

Here

$$k_c^2 = k^2 - \beta^2 = \omega^2\mu\epsilon - \beta^2, \quad (8.90)$$

and k is the ***unbounded-medium wavenumber*** defined earlier as

$$k = \omega\sqrt{\mu\epsilon}\,. \quad (8.91)$$

For reasons that become clear later (in Section 8-8), the constant k_c is called the ***cutoff wavenumber***. In view of Eqs. (8.89a–d), the x and y components of $\widetilde{\mathbf{E}}$ and $\widetilde{\mathbf{H}}$ can now be found readily, so long as we have mathematical expressions for $\widetilde{E}_z$ and $\widetilde{H}_z$. For the TE mode, $\widetilde{E}_z = 0$, so all we need to know is $\widetilde{H}_z$, and the converse is true for the TM case.

8-8 TM Modes in Rectangular Waveguide

In the preceding section we developed expressions for $\widetilde{E}_x$, $\widetilde{E}_y$, $\widetilde{H}_x$, and $\widetilde{H}_y$, all in terms of $\widetilde{E}_z$ and $\widetilde{H}_z$. Since $\widetilde{H}_z = 0$ *for the TM mode*, our task reduces to obtaining a valid solution for $\widetilde{E}_z$. Our starting point is the ***homogeneous wave equation*** for $\widetilde{\mathbf{E}}$. For a lossless medium characterized by an unbounded-medium wavenumber k, the wave equation is given by Eq. (7.19) as

$$\nabla^2\widetilde{\mathbf{E}} + k^2\widetilde{\mathbf{E}} = 0. \quad (8.92)$$

To satisfy Eq. (8.92), each of its $\hat{\mathbf{x}}$, $\hat{\mathbf{y}}$, and $\hat{\mathbf{z}}$ components has to be satisfied independently. Its $\hat{\mathbf{z}}$ component is given by:

$$\frac{\partial^2 \widetilde{E}_z}{\partial x^2} + \frac{\partial^2 \widetilde{E}_z}{\partial y^2} + \frac{\partial^2 \widetilde{E}_z}{\partial z^2} + k^2\widetilde{E}_z = 0. \quad (8.93)$$

By adopting the mathematical form given by Eq. (8.85), namely

$$\widetilde{E}_z(x, y, z) = \widetilde{e}_z(x, y)\, e^{-j\beta z}, \quad (8.94)$$

Eq. (8.93) reduces to

$$\frac{\partial^2 \widetilde{e}_z}{\partial x^2} + \frac{\partial^2 \widetilde{e}_z}{\partial y^2} + k_c^2\widetilde{e}_z = 0, \quad (8.95)$$

where k_c^2 is as defined by Eq. (8.90).

The form of the partial differential equation (separate, uncoupled derivatives with respect to x and y) allows us to assume a product solution of the form

$$\widetilde{e}_z(x, y) = X(x)\, Y(y). \quad (8.96)$$

Substituting Eq. (8.96) into Eq. (8.95), followed with dividing all terms by $X(x)\,Y(y)$, leads to:

$$\frac{1}{X}\,\frac{d^2X}{dx^2} + \frac{1}{Y}\,\frac{d^2Y}{dy^2} + k_c^2 = 0. \quad (8.97)$$

To satisfy Eq. (8.97), each of the first two terms has to equal a constant. Hence, we define separation constants k_x and k_y such that

$$\frac{d^2X}{dx^2} + k_x^2 X = 0, \quad (8.98a)$$

$$\frac{d^2Y}{dy^2} + k_y^2 Y = 0, \quad (8.98b)$$

and

$$k_c^2 = k_x^2 + k_y^2. \quad (8.99)$$

Before proposing solutions for Eqs. (8.98a and b), we should consider the constraints that the solutions must meet. The electric field $\widetilde{E}_z$ is parallel to all four walls of the waveguide. Since $\mathbf{E} = 0$ in the conducting walls, *the boundary conditions require $\widetilde{E}_z$ in the waveguide cavity to go to zero as x approaches 0 and a, and as y approaches 0 and b* (**Fig. 8-22**). To satisfy these boundary conditions, sinusoidal solutions are chosen for $X(x)$ and $Y(y)$ as follows:

$$\begin{aligned}\widetilde{e}_z &= X(x)\,Y(y)\\ &= (A\cos k_x x + B\sin k_x x)(C\cos k_y y + D\sin k_y y).\end{aligned} \quad (8.100)$$

These forms for $X(x)$ and $Y(y)$ definitely satisfy the differential equations given by Eqs. (8.98a and b). The boundary conditions for $\widetilde{e}_z$ are:

$$\widetilde{e}_z = 0, \qquad \text{at } x = 0 \text{ and } a, \quad (8.101a)$$

$$\widetilde{e}_z = 0, \qquad \text{at } y = 0 \text{ and } b. \quad (8.101b)$$

Satisfying $\widetilde{e}_z = 0$ at $x = 0$ requires that we set $A = 0$, and similarly, satisfying $\widetilde{e}_z = 0$ at $y = 0$ requires $C = 0$. Satisfying $\widetilde{e}_z = 0$ at $x = a$ requires

$$k_x = \frac{m\pi}{a}, \qquad m = 1, 2, 3, \ldots \quad (8.102a)$$

and similarly, satisfying $\widetilde{e}_z = 0$ at $y = b$ requires

$$k_y = \frac{n\pi}{b}, \qquad n = 1, 2, 3, \ldots \quad (8.102b)$$

Consequently,

$$\widetilde{E}_z = \widetilde{e}_z e^{-j\beta z} = E_0 \sin\left(\frac{m\pi x}{a}\right) \sin\left(\frac{n\pi y}{b}\right) e^{-j\beta z}, \quad (8.103)$$

where $E_0 = BD$ is the amplitude of the wave in the guide. Keeping in mind that $\widetilde{H}_z = 0$ for the TM mode, the transverse components of $\widetilde{\mathbf{E}}$ and $\widetilde{\mathbf{H}}$ can now be obtained by applying Eq. (8.103) to (8.89a–d),

$$\widetilde{E}_x = \frac{-j\beta}{k_c^2}\left(\frac{m\pi}{a}\right) E_0 \cos\left(\frac{m\pi x}{a}\right) \sin\left(\frac{n\pi y}{b}\right) e^{-j\beta z}, \quad (8.104a)$$

$$\widetilde{E}_y = \frac{-j\beta}{k_c^2}\left(\frac{n\pi}{b}\right) E_0 \sin\left(\frac{m\pi x}{a}\right) \cos\left(\frac{n\pi y}{b}\right) e^{-j\beta z}, \quad (8.104b)$$

$$\widetilde{H}_x = \frac{j\omega\epsilon}{k_c^2}\left(\frac{n\pi}{b}\right) E_0 \sin\left(\frac{m\pi x}{a}\right) \cos\left(\frac{n\pi y}{b}\right) e^{-j\beta z}, \quad (8.104c)$$

$$\widetilde{H}_y = \frac{-j\omega\epsilon}{k_c^2}\left(\frac{m\pi}{a}\right) E_0 \cos\left(\frac{m\pi x}{a}\right) \sin\left(\frac{n\pi y}{b}\right) e^{-j\beta z}. \quad (8.104d)$$

Each combination of the integers m and n represents a viable solution, or a mode, denoted TM_{mn}. Associated with each mn mode are specific field distributions for the region inside the guide. **Figure 8-23** depicts the **E** and **H** field lines for the TM_{11} mode across two different cross sections of the guide.

According to Eqs. (8.103) and (8.104e), a rectangular waveguide with cross section $(a \times b)$ can support the propagation of waves with many different, but discrete, field configurations specified by the integers m and n. The only quantity in the fields' expressions that we have yet to determine is the propagation constant β, contained in the exponential $e^{-j\beta z}$. By combining Eqs. (8.90), (8.99), and (8.102), we obtain the following expression for β:

$$\begin{aligned}\beta &= \sqrt{k^2 - k_c^2}\\ &= \sqrt{\omega^2\mu\epsilon - \left(\frac{m\pi}{a}\right)^2 - \left(\frac{n\pi}{b}\right)^2}.\end{aligned} \quad (8.105)$$

(TE and TM)

Even though the expression for β was derived for TM modes, it is equally applicable to TE modes.

The exponential $e^{-j\beta z}$ describes a wave traveling in the $+z$ direction, provided that β is real, which corresponds to $k > k_c$. If $k < k_c$, β becomes imaginary: $\beta = -j\alpha$ with α real, in which case $e^{-j\beta z} = e^{-\alpha z}$, yielding ***evanescent waves*** characterized by amplitudes that decay rapidly with z due to the attenuation function $e^{-\alpha z}$. Corresponding to each mode

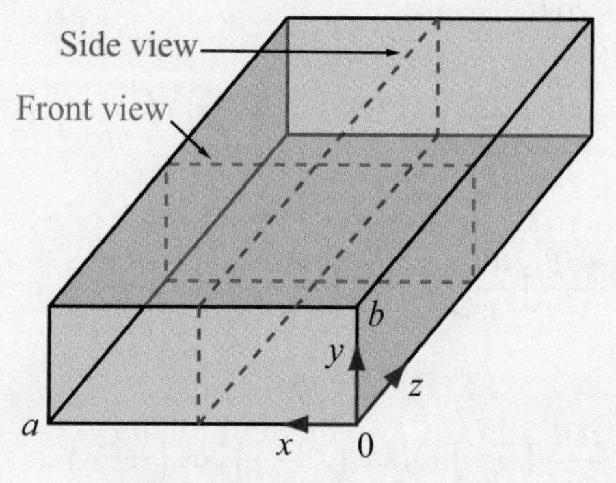

(a) Cross-sectional planes

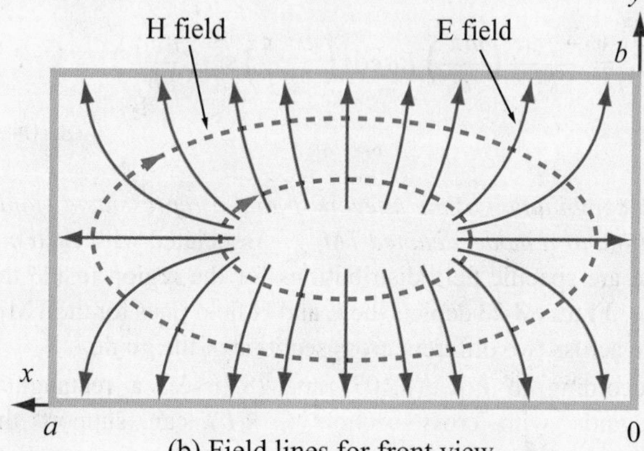

(b) Field lines for front view

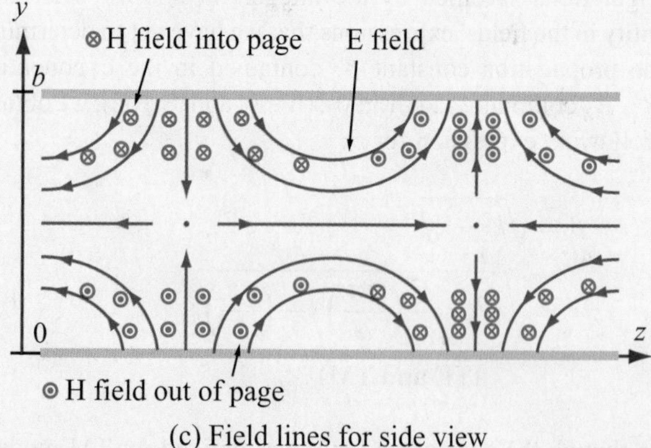

(c) Field lines for side view

Figure 8-23 TM_{11} electric and magnetic field lines across two cross-sectional planes.

(m,n), there is a ***cutoff frequency*** f_{mn} at which $\beta = 0$. By setting $\beta = 0$ in Eq. (8.105) and then solving for f, we have

$$f_{mn} = \frac{u_{p_0}}{2}\sqrt{\left(\frac{m}{a}\right)^2 + \left(\frac{n}{b}\right)^2}, \qquad (8.106)$$

(TE and TM)

where $u_{p_0} = 1/\sqrt{\mu\epsilon}$ is the phase velocity of a TEM wave in an unbounded medium with constitutive parameters ϵ and μ.

► A wave, in a given mode, can propagate through the guide only if its frequency $f > f_{mn}$, as only then $\beta =$ real. ◄

The mode with the lowest cutoff frequency is known as the ***dominant mode***. *The dominant mode is* TM_{11} *among TM modes and* TE_{10} *among TE modes* (whose solution is given in Section 8-8). *Whereas a value of zero for m or n is allowed for TE modes, it is not for TM modes* (because if either m or n is zero, $\widetilde{E}_z$ in Eq. (8.103) becomes zero and all other field components vanish as well).

By combining Eqs. (8.105) and (8.106), we can express β in terms of f_{mn},

$$\beta = \frac{\omega}{u_{p_0}}\sqrt{1 - \left(\frac{f_{mn}}{f}\right)^2} \quad \textbf{(TE and TM).} \qquad (8.107)$$

The phase velocity of a TE or TM wave in a waveguide is

$$u_p = \frac{\omega}{\beta} = \frac{u_{p_0}}{\sqrt{1-(f_{mn}/f)^2}}. \qquad (8.108)$$

(TE and TM)

The transverse electric field consists of components $\widetilde{E}_x$ and $\widetilde{E}_y$, given by Eqs. (8.104a and b). For a wave traveling in the $+z$ direction, the magnetic field associated with $\widetilde{E}_x$ is $\widetilde{H}_y$ [according to the right hand rule given by Eq. (7.39a)], and similarly, the magnetic field associated with $\widetilde{E}_y$ is $-\widetilde{H}_x$. The ratios, obtained by employing Eq. (8.104e), constitute the wave impedance in the guide,

$$Z_{\text{TM}} = \frac{\widetilde{E}_x}{\widetilde{H}_y} = -\frac{\widetilde{E}_y}{\widetilde{H}_x} = \frac{\beta\eta}{k} = \eta\sqrt{1 - \left(\frac{f_{mn}}{f}\right)^2}, \quad (8.109)$$

where $\eta = \sqrt{\mu/\epsilon}$ is the intrinsic impedance of the dielectric material filling the guide.

Example 8-8: Mode Properties

A TM wave propagating in a dielectric-filled waveguide of unknown permittivity has a magnetic field with y component given by

$$H_y = 6\cos(25\pi x)\sin(100\pi y) \times \sin(1.5\pi \times 10^{10} t - 109\pi z) \quad \text{(mA/m)}.$$

If the guide dimensions are $a = 2b = 4$ cm, determine: (a) the mode numbers, (b) the relative permittivity of the material in the guide, (c) the phase velocity, and (d) obtain an expression for E_x.

Solution: (a) By comparison with the expression for $\widetilde{H}_y$ given by Eq. (8.104d), we deduce that the argument of x is $(m\pi/a)$ and the argument of y is $(n\pi/b)$. Hence,

$$25\pi = \frac{m\pi}{4\times 10^{-2}}, \qquad 100\pi = \frac{n\pi}{2\times 10^{-2}},$$

which yield $m = 1$ and $n = 2$. Therefore, the mode is TM_{12}.

(b) The second sine function in the expression for H_y represents $\sin(\omega t - \beta z)$, which means that

$$\omega = 1.5\pi \times 10^{10} \text{ (rad/s), or } f = 7.5 \text{ GHz},$$

$$\beta = 109\pi \text{ (rad/m)}.$$

By rewriting Eq. (8.105) so as to obtain an expression for $\epsilon_r = \epsilon/\epsilon_0$ in terms of the other quantities, we have

$$\epsilon_r = \frac{c^2}{\omega^2}\left[\beta^2 + \left(\frac{m\pi}{a}\right)^2 + \left(\frac{n\pi}{b}\right)^2\right],$$

where c is the speed of light. Inserting the available values, we obtain

$$\epsilon_r = \frac{(3\times 10^8)^2}{(1.5\pi \times 10^{10})^2} \cdot \left[(109\pi)^2 + \left(\frac{\pi}{4\times 10^{-2}}\right)^2 + \left(\frac{2\pi}{2\times 10^{-2}}\right)^2\right] = 9.$$

(c)

$$u_p = \frac{\omega}{\beta} = \frac{1.5\pi \times 10^{10}}{109\pi} = 1.38 \times 10^8 \text{ m/s},$$

which is slower than the speed of light. However, as explained later in Section 8-10, the phase velocity in a waveguide may exceed c, but the velocity with which energy is carried down the guide is the group velocity u_g, which is never greater than c.

(d) From Eq. (8.109),

$$Z_{\text{TM}} = \eta\sqrt{1 - (f_{12}/f)^2}$$

Application of Eq. (8.106) yields $f_{12} = 5.15$ GHz for the TM_{12} mode. Using that in the expression for Z_{TM}, in addition to $f = 7.5$ GHz and $\eta = \sqrt{\mu/\epsilon} = (\sqrt{\mu_0/\epsilon_0})/\sqrt{\epsilon_r} = 377/\sqrt{9} = 125.67\ \Omega$, gives

$$Z_{\text{TM}} = 91.3\ \Omega.$$

Hence,

$$\begin{aligned} E_x &= Z_{\text{TM}} H_y \\ &= 91.3 \times 6\cos(25\pi x)\sin(100\pi y) \times \sin(1.5\pi \times 10^{10} t - 109\pi z) \quad \text{(mV/m)} \\ &= 0.55\cos(25\pi x)\sin(100\pi y) \times \sin(1.5\pi \times 10^{10} t - 109\pi z) \quad \text{(V/m)}. \end{aligned}$$

Concept Question 8-8: What are the primary limitations of coaxial cables at frequencies higher than 30 GHz?

Concept Question 8-9: Can a TE mode have a zero magnetic field along the direction of propagation?

Concept Question 8-10: What is the rationale for choosing a solution for $\widetilde{e}_z$ that involves sine and cosine functions?

Concept Question 8-11: What is an evanescent wave?

Exercise 8-10: For a square waveguide with $a = b$, what is the value of the ratio $\widetilde{E}_x/\widetilde{E}_y$ for the TM_{11} mode?

Answer: $\tan(\pi y/a)/\tan(\pi x/a)$.

Exercise 8-11: What is the cutoff frequency for the dominant TM mode in a waveguide filled with a material with $\epsilon_r = 4$? The waveguide dimensions are $a = 2b = 5$ cm.

Answer: For TM_{11}, $f_{11} = 3.35$ GHz.

Exercise 8-12: What is the magnitude of the phase velocity of a TE or TM mode at $f = f_{mn}$?

Answer: $u_p = \infty$! (See explanation in Section 8-10.)

8-9 TE Modes in Rectangular Waveguide

In the TM case, for which the wave has no magnetic field component along the z direction (i.e., $\widetilde{H}_z = 0$), we started our treatment in the preceding section by obtaining a solution for $\widetilde{E}_z$, and then we used it to derive expressions for the tangential components of $\widetilde{\mathbf{E}}$ and $\widetilde{\mathbf{H}}$. For the TE case, the same basic procedure can be applied, except for reversing the roles of $\widetilde{E}_z$ and $\widetilde{H}_z$. Such a process leads to:

$$\widetilde{E}_x = \frac{j\omega\mu}{k_c^2}\left(\frac{n\pi}{b}\right) H_0 \cos\left(\frac{m\pi x}{a}\right)\sin\left(\frac{n\pi y}{b}\right) e^{-j\beta z}, \tag{8.110a}$$

$$\widetilde{E}_y = \frac{-j\omega\mu}{k_c^2}\left(\frac{m\pi}{a}\right) H_0 \sin\left(\frac{m\pi x}{a}\right)\cos\left(\frac{n\pi y}{b}\right) e^{-j\beta z}, \tag{8.110b}$$

$$\widetilde{H}_x = \frac{j\beta}{k_c^2}\left(\frac{m\pi}{a}\right) H_0 \sin\left(\frac{m\pi x}{a}\right)\cos\left(\frac{n\pi y}{b}\right) e^{-j\beta z}, \tag{8.110c}$$

$$\widetilde{H}_y = \frac{j\beta}{k_c^2}\left(\frac{n\pi}{b}\right) H_0 \cos\left(\frac{m\pi x}{a}\right)\sin\left(\frac{n\pi y}{b}\right) e^{-j\beta z}, \tag{8.110d}$$

$$\widetilde{H}_z = H_0 \cos\left(\frac{m\pi x}{a}\right)\cos\left(\frac{n\pi y}{b}\right) e^{-j\beta z}, \tag{8.110e}$$

and, of course, $\widetilde{E}_z = 0$. The expressions for f_{mn}, β, and u_p given earlier by Eqs. (8.106), (8.107), and (8.108) remain unchanged.

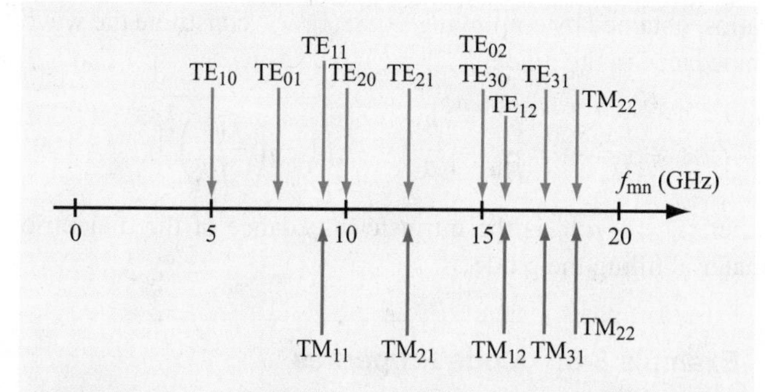

Figure 8-24 Cutoff frequencies for TE and TM modes in a hollow rectangular waveguide with $a = 3$ cm and $b = 2$ cm (Example 8-9).

► Because not all the fields vanish if m or n assume a value of zero, the lowest order TE mode is TE_{10} if $a > b$, or TE_{01} if $a < b$. It is customary to assign a to be the longer dimension, in which case the TE_{10} mode is the de facto dominant mode. ◄

Another difference between the TE and TM modes relates to the expression for the wave impedance. For TE,

$$Z_{\text{TE}} = \frac{\widetilde{E}_x}{\widetilde{H}_y} = -\frac{\widetilde{E}_y}{\widetilde{H}_x} = \frac{\eta}{\sqrt{1-(f_{mn}/f)^2}} . \tag{8.111}$$

A summary of the expressions for the various wave attributes of TE and TM modes is given in **Table 8-3**. By way of reference, corresponding expressions for the TEM mode on a coaxial transmission line are included as well.

Example 8-9: Cutoff Frequencies

For a hollow rectangular waveguide with dimensions $a = 3$ cm and $b = 2$ cm, determine the cutoff frequencies for all modes, up to 20 GHz. Over what frequency range will the guide support the propagation of a single dominant mode?

Solution: A hollow guide has $\mu = \mu_0$ and $\epsilon = \epsilon_0$. Hence, $u_{p_0} = 1/\sqrt{\mu_0\epsilon_0} = c$. Application of Eq. (8.106) gives the cutoff frequencies shown in **Fig. 8-24**, which start at 5 GHz for the TE_{10} mode. To avoid all other modes, the frequency of operation should be restricted to the 5–7.5 GHz range.

Table 8-3 **Wave properties for TE and TM modes in a rectangular waveguide with dimensions $a \times b$, filled with a dielectric material with constitutive parameters ϵ and μ. The TEM case, shown for reference, pertains to plane-wave propagation in an unbounded medium.**

Rectangular Waveguides		Plane Wave
TE Modes	TM Modes	TEM Mode
$\widetilde{E}_x = \frac{j\omega\mu}{k_c^2}\left(\frac{n\pi}{b}\right) H_0 \cos\left(\frac{m\pi x}{a}\right) \sin\left(\frac{n\pi y}{b}\right) e^{-j\beta z}$	$\widetilde{E}_x = \frac{-j\beta}{k_c^2}\left(\frac{m\pi}{a}\right) E_0 \cos\left(\frac{m\pi x}{a}\right) \sin\left(\frac{n\pi y}{b}\right) e^{-j\beta z}$	$\widetilde{E}_x = E_{x0} e^{-j\beta z}$
$\widetilde{E}_y = \frac{-j\omega\mu}{k_c^2}\left(\frac{m\pi}{a}\right) H_0 \sin\left(\frac{m\pi x}{a}\right) \cos\left(\frac{n\pi y}{b}\right) e^{-j\beta z}$	$\widetilde{E}_y = \frac{-j\beta}{k_c^2}\left(\frac{n\pi}{b}\right) E_0 \sin\left(\frac{m\pi x}{a}\right) \cos\left(\frac{n\pi y}{b}\right) e^{-j\beta z}$	$\widetilde{E}_y = E_{y0} e^{-j\beta z}$
$\widetilde{E}_z = 0$	$\widetilde{E}_z = E_0 \sin\left(\frac{m\pi x}{a}\right) \sin\left(\frac{n\pi y}{b}\right) e^{-j\beta z}$	$\widetilde{E}_z = 0$
$\widetilde{H}_x = -\widetilde{E}_y / Z_{TE}$	$\widetilde{H}_x = -\widetilde{E}_y / Z_{TM}$	$\widetilde{H}_x = -\widetilde{E}_y / \eta$
$\widetilde{H}_y = \widetilde{E}_x / Z_{TE}$	$\widetilde{H}_y = \widetilde{E}_x / Z_{TM}$	$\widetilde{H}_y = \widetilde{E}_x / \eta$
$\widetilde{H}_z = H_0 \cos\left(\frac{m\pi x}{a}\right) \cos\left(\frac{n\pi y}{b}\right) e^{-j\beta z}$	$\widetilde{H}_z = 0$	$\widetilde{H}_z = 0$
$Z_{TE} = \eta / \sqrt{1 - (f_c/f)^2}$	$Z_{TM} = \eta \sqrt{1 - (f_c/f)^2}$	$\eta = \sqrt{\mu/\epsilon}$
Properties Common to TE and TM Modes		
$f_c = \frac{u_{p0}}{2}\sqrt{\left(\frac{m}{a}\right)^2 + \left(\frac{n}{b}\right)^2}$		f_c = not applicable
$\beta = k\sqrt{1 - (f_c/f)^2}$		$k = \omega\sqrt{\mu\epsilon}$
$u_p = \frac{\omega}{\beta} = u_{p0} / \sqrt{1 - (f_c/f)^2}$		$u_{p0} = 1/\sqrt{\mu\epsilon}$

8-10 Propagation Velocities

When a wave is used to carry a message through a medium or along a transmission line, information is encoded into the wave's amplitude, frequency, or phase. A simple example is shown in Fig. 8-25, in which a high-frequency sinusoidal wave of frequency f is amplitude-modulated by a low-frequency Gaussian pulse. The waveform in (b) is the result of multiplying the Gaussian pulse shape in (a) by the carrier waveform.

By Fourier analysis, the waveform in (b) is equivalent to the superposition of a *group* of sinusoidal waves with specific amplitudes and frequencies. Exact equivalence may require a large, or infinite, number of frequency components, but in practice, it is often possible to represent the modulated waveform, to a fairly high degree of fidelity, with a wave group that extends over a relatively narrow bandwidth surrounding the high-frequency carrier f. The velocity with which the envelope—or equivalently the wave group—travels through the medium is called the ***group velocity*** u_g. As such, u_g is the velocity of the energy carried by the wave-group, and of the information encoded in it. Depending on whether or not the propagation medium is dispersive, u_g may or may not be equal to the phase velocity u_p. In Section 2-1.1, we described a "dispersive transmission line as one on which the phase velocity is not a constant as a function of frequency," a consequence of which is that the shape of a pulse transmitted through it gets progressively distorted as it moves down the line. A rectangular waveguide constitutes a dispersive transmission line because the phase velocity of a TE or TM mode propagating through it is a strong function of frequency [per Eq. (8.108)], particularly at frequencies close to the cutoff frequency f_{mn}. As we see shortly, if $f \gg f_{mn}$, the TE and TM modes become approximately TEM in character, not only in terms of the directional arrangement of the electric and magnetic fields, but also in terms of the frequency dependence of the phase velocity.

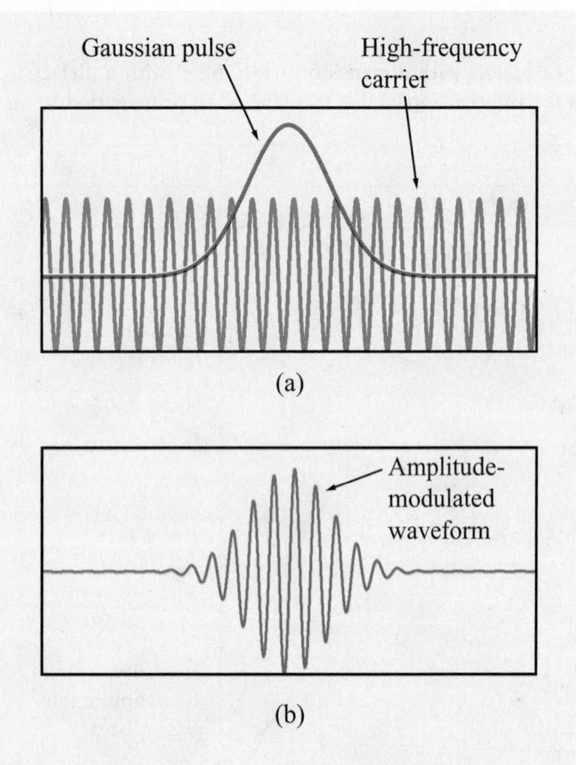

Figure 8-25 The amplitude-modulated high-frequency waveform in (b) is the product of the Gaussian-shaped pulse with the sinusoidal high-frequency carrier in (a).

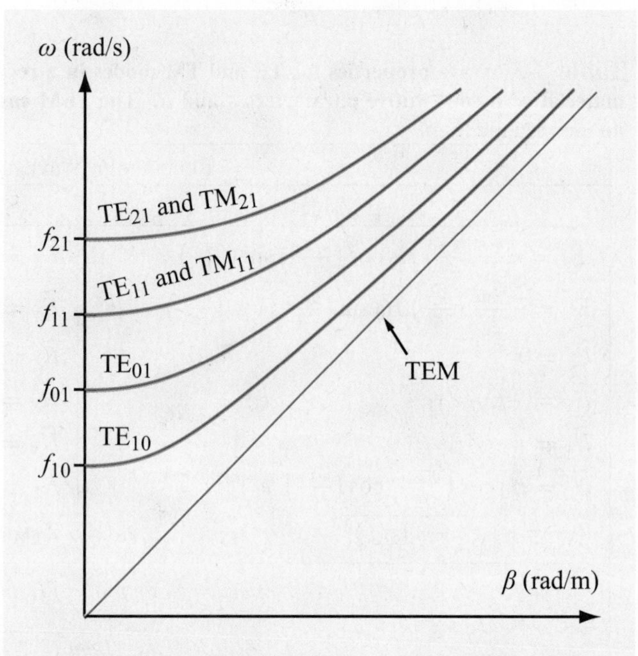

Figure 8-26 ω-β diagram for TE and TM modes in a hollow rectangular waveguide. The straight line pertains to propagation in an unbounded medium or on a TEM transmission line.

We now examine u_{p} and u_{g} in more detail. The phase velocity, defined as the velocity of the sinusoidal pattern of the wave, is given by

$$u_{\text{p}} = \frac{\omega}{\beta}, \tag{8.112}$$

while the group velocity u_{g} is given by

$$u_{\text{g}} = \frac{1}{d\beta/d\omega}. \tag{8.113}$$

Even though we will not derive Eq. (8.113) in this book, it is nevertheless important that we understand its properties for TE and TM modes in a metal waveguide. Using the expression for β given by Eq. (8.107),

$$u_{\text{g}} = \frac{1}{d\beta/d\omega} = u_{\text{p}0}\sqrt{1-(f_{mn}/f)^2}, \tag{8.114}$$

where, as before, $u_{\text{p}0}$ is the phase velocity in an unbounded dielectric medium. In view of Eq. (8.108) for the phase velocity u_{p},

$$u_{\text{p}}u_{\text{g}} = u_{\text{p}0}^2. \tag{8.115}$$

Above cutoff ($f > f_{mn}$), $u_{\text{p}} \geq u_{p0}$, and $u_{\text{g}} \leq u_{p0}$. As $f \to \infty$, or more precisely as $(f_{mn}/f) \to 0$, TE and TM modes approach the TEM case, for which $u_{\text{p}} = u_{\text{g}} = u_{\text{p}0}$.

A useful graphical tool for describing the propagation properties of a medium or transmission line is the ω-β ***diagram***. In Fig. 8-26, the straight line starting at the origin represents the ω-β relationship for a TEM wave propagating in an unbounded medium (or on a TEM transmission line). The TEM line provides a reference to which the ω-β curves of the TE/TM modes can be compared. At a given location on the ω-β line or curve, the ratio of the value of ω to that of β defines $u_{\text{p}} = \omega/\beta$, whereas it is the slope $d\omega/d\beta$ of the curve at that point that defines the group velocity u_g. For the TEM line, the ratio and the slope have identical values (hence, $u_p = u_g$), and the line starts at $\omega = 0$. In contrast, the curve for each of the indicated TE/TM modes starts at a cutoff frequency specific to that mode, below which the waveguide cannot support the propagation of a wave in that mode. At frequencies close to cutoff, u_p and u_g assume very different values; in fact, at cutoff $u_p = \infty$ and $u_g = 0$. On the other end of the frequency spectrum, at frequencies much higher than f_{mn}, the ω-β curves of the TE/TM modes approach the TEM line. We should note that for TE and TM modes, u_p may easily exceed the speed of light, but u_g will not, and since it is u_g that represents the actual transport of energy, Einstein's assertion that there is an upper bound on the speed of physical phenomena, is not violated.

So far, we have described the fields in the guide, but we have yet to interpret them in terms of plane waves that zigzag along the guide through successive reflections. To do just that, consider the simple case of a TE$_{10}$ mode. For $m = 1$ and $n = 0$, the only nonzero component of the electric field given by Eq. (8.110) is $\widetilde{E}_y$,

$$\widetilde{E}_y = -j\frac{\omega\mu}{k_c^2}\left(\frac{\pi}{a}\right) H_0 \sin\left(\frac{\pi x}{a}\right) e^{-j\beta z}. \tag{8.116}$$

Using the identity $\sin\theta = (e^{j\theta} - e^{-j\theta})/2j$ for any argument θ, we obtain

$$\begin{aligned}\widetilde{E}_y &= \left(\frac{\omega\mu\pi H_0}{2k_c^2 a}\right)(e^{-j\pi x/a} - e^{j\pi x/a})e^{-j\beta z} \\ &= E_0'(e^{-j\beta(z+\pi x/\beta a)} - e^{-j\beta(z-\pi x/\beta a)}) \\ &= E_0'(e^{-j\beta z'} - e^{-j\beta z''}),\end{aligned} \tag{8.117}$$

where we have consolidated the quantities multiplying the two exponential terms into the constant E_0'. The first exponential term represents a wave with propagation constant β traveling in the z' direction, where

$$z' = z + \frac{\pi x}{\beta a}, \tag{8.118a}$$

and the second term represents a wave travelling in the z'' direction, with

$$z'' = z - \frac{\pi x}{\beta a}. \tag{8.118b}$$

From the diagram shown in Fig. 8-27(a), it is evident that the z' direction is at an angle θ' relative to z and the z'' direction is at an angle $\theta'' = -\theta'$. This means that the electric field $\widetilde{E}_y$ (and its associated magnetic field $\widetilde{\mathbf{H}}$) of the TE$_{10}$ mode is composed of two TEM waves, as shown in Fig. 8-27(b), both traveling in the

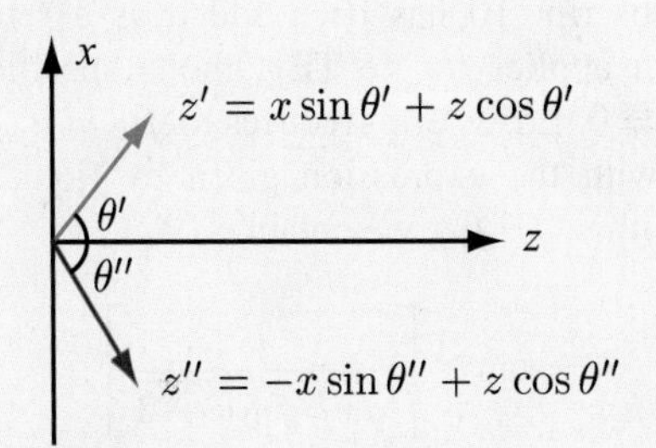

From Eq. (8.118a), $z' = \frac{\pi x}{\beta a} + z$.
Hence, $\theta' = \tan^{-1}(\pi/\beta a)$.

From Eq. (8.118b), $z'' = -\frac{\pi x}{\beta a} + z$.
Hence, $\theta'' = -\tan^{-1}(\pi/\beta a)$.

(a) z' and z'' propagation directions

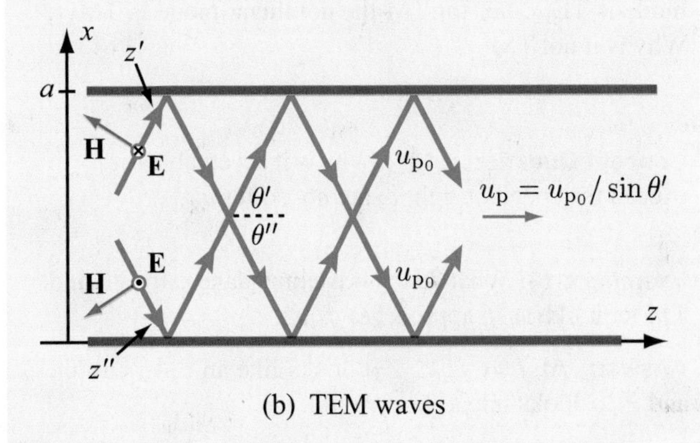

(b) TEM waves

Figure 8-27 The TE$_{10}$ mode can be constructed as the sum of two TEM waves.

$+z$ direction by zigzagging between the opposite walls of the waveguide. Along the zigzag directions (z' and z''), the phase velocity of the individual wave components is u_{p0}, but the phase velocity of the combination of the two waves along z is u_p.

Example 8-10: Zigzag Angle

For the TE$_{10}$ mode, express the zigzag angle θ' in terms of the ratio (f/f_{10}), and then evaluate it at $f = f_{10}$ and for $f \gg f_{10}$.

Solution: From Fig. 8-27,

$$\theta'_{10} = \tan^{-1}\left(\frac{\pi}{\beta_{10}a}\right),$$

where the subscript 10 has been added as a reminder that the expression applies to the TE_{10} mode specifically. For $m = 1$ and $n = 0$, Eq. (8.106) reduces to $f_{10} = u_{p_0}/2a$. After replacing β with the expression given by Eq. (8.107) and replacing a with $u_{p_0}/2f_{10}$, we obtain

$$\theta' = \tan^{-1}\left[\frac{1}{\sqrt{(f/f_{10})^2 - 1}}\right].$$

At $f = f_{10}$, $\theta' = 90°$, which means that the wave bounces back and forth at normal incidence between the two side walls of the waveguide, making no progress in the z direction. At the other end of the frequency spectrum, when $f \gg f_{10}$, θ' approaches 0 and the wave becomes TEM-like as it travels straight down the guide.

Concept Question 8-12: For TE waves, the dominant mode is TE_{10}, but for TM the dominant mode is TM_{11}. Why is it not TM_{10}?

Concept Question 8-13: Why is it acceptable for u_p to exceed the speed of light c, but not so for u_g?

Exercise 8-13: What do the wave impedances for TE and TM look like as f approaches f_{mn}?

Answer: At $f = f_{mn}$, Z_{TE} looks like an open circuit, and Z_{TM} looks like a short circuit.

Exercise 8-14: What are the values for (a) u_p, (b) u_g, and (c) the zigzag angle θ' at $f = 2f_{10}$ for a TE_{10} mode in a hollow waveguide?

Answer: (a) $u_p = 1.15c$, (b) $u_g = 0.87c$, (c) $\theta' = 30°$.

8-11 Cavity Resonators

A rectangular waveguide has metal walls on four sides. When the two remaining sides are terminated with conducting walls, the waveguide becomes a cavity. By designing cavities to *resonate* at specific frequencies, they can be used as circuit elements in ***microwave oscillators***, ***amplifiers***, and ***bandpass filters***.

The cavity shown in Fig. 8-28(a), with dimensions $(a \times b \times d)$, is connected to two coaxial cables that feed and extract signals into and from the cavity via input and output probes. As a bandpass filter, the function of a resonant cavity is to block all spectral components of the input signal except for those with frequencies that fall within a narrow band surrounding a specific center frequency f_0, the cavity's ***resonant frequency***. Comparison of the spectrum in Fig. 8-28(b), which describes the range of frequencies that might be contained in a typical input signal, with the narrow output spectrum in Fig. 8-28(c) demonstrates the filtering action imparted by the cavity.

In a rectangular waveguide, the fields constitute standing waves along the x and y directions, and a propagating wave along $\hat{\mathbf{z}}$. The terms TE and TM were defined relative to the propagation direction; TE meant that $\mathbf{E}$ was entirely transverse to $\hat{\mathbf{z}}$, and TM meant that $\mathbf{H}$ had no component along $\hat{\mathbf{z}}$. In a cavity, there is no unique propagation direction, as *no* fields propagate. Instead, standing waves exist along all three directions. Hence, the terms TE and TM need to be modified by defining the fields relative to one of the three rectangular axes. For the sake of consistency, *we will continue to define the transverse direction to be any direction contained in the plane whose normal is* $\hat{\mathbf{z}}$.

The TE mode in the rectangular waveguide consists of a single propagating wave whose $\widetilde{H}_z$ component is given by Eq. (8.110e) as

$$\widetilde{H}_z = H_0 \cos\left(\frac{m\pi x}{a}\right) \cos\left(\frac{n\pi y}{b}\right) e^{-j\beta z}, \qquad (8.119)$$

where the phase factor $e^{-j\beta z}$ signifies propagation along $+\hat{\mathbf{z}}$. Because the cavity has conducting walls at both $z = 0$ and $z = d$, it will contain two such waves, one with amplitude H_0 traveling along $+\hat{\mathbf{z}}$, and another with amplitude H_0^- traveling along $-\hat{\mathbf{z}}$. Hence,

$$\widetilde{H}_z = (H_0 e^{-j\beta z} + H_0^- e^{j\beta z}) \cos\left(\frac{m\pi x}{a}\right) \cos\left(\frac{n\pi y}{b}\right). \qquad (8.120)$$

Boundary conditions require the normal component of $\widetilde{\mathbf{H}}$ to be zero at a conducting boundary. Consequently, $\widetilde{H}_z$ must be zero at $z = 0$ and $z = d$. To satisfy these conditions, it is necessary that $H_0^- = -H_0$ and $\beta d = p\pi$, with $p = 1, 2, 3, \ldots$, in which case Eq. (8.120) becomes

$$\widetilde{H}_z = -2jH_0 \cos\left(\frac{m\pi x}{a}\right) \cos\left(\frac{n\pi y}{b}\right) \sin\left(\frac{p\pi z}{d}\right). \qquad (8.121)$$

Given that $\widetilde{E}_z = 0$ for the TE modes, all of the other components of $\widetilde{\mathbf{E}}$ and $\widetilde{\mathbf{H}}$ can be derived readily through the application of the relationships given by Eq. (8.89). A similar procedure can also be used to characterize cavity modes for the TM case.

Module 8.5 Rectangular Waveguide When givin the waveguide dimensions, the frequency f, and the mode type (TE or TM) and number, this module provides information about the wave impedance, cutoff frequency, and other wave attributes. It also displays the electric and magnetic field distributions inside the guide.

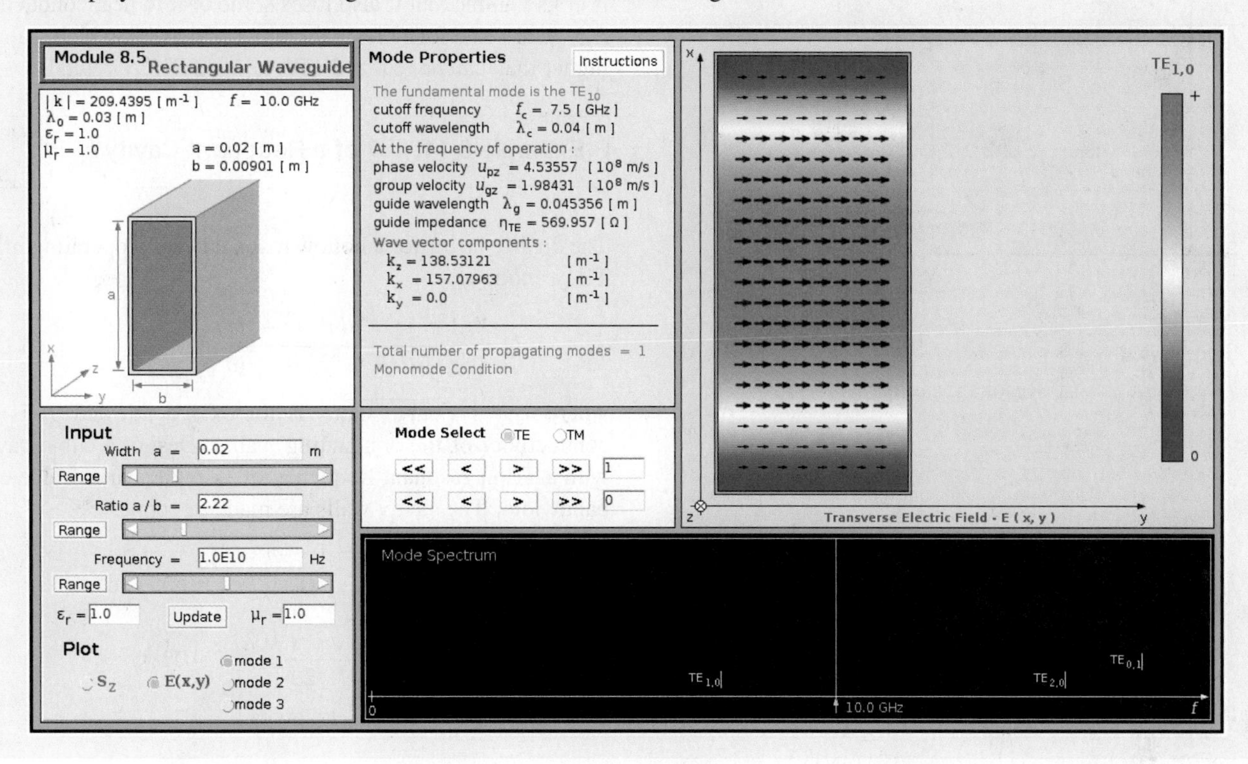

8-11.1 Resonant Frequency

The consequence of the quantization condition imposed on β, namely $\beta = p\pi/d$ with p assuming only integer values, is that for any specific set of integer values of (m, n, p), the wave inside the cavity can exist at only a single ***resonant frequency***, f_{mnp}, whose value has to satisfy Eq. (8.105). The resulting expression for f_{mnp} is

$$f_{mnp} = \frac{u_{\mathrm{p}0}}{2}\sqrt{\left(\frac{m}{a}\right)^2 + \left(\frac{n}{b}\right)^2 + \left(\frac{p}{d}\right)^2}\,. \qquad (8.122)$$

For TE, the indices m and n start at 0, but p starts at 1. The exact opposite applies to TM. By way of an example, the resonant frequency for a TE_{101} mode in a hollow cavity with dimensions $a = 2$ cm, $b = 3$ cm, and $d = 4$ cm is $f_{101} = 8.38$ GHz.

8-11.2 Quality Factor

In the ideal case, if a group of frequencies is introduced into the cavity to excite a certain TE or TM mode, only the frequency component at exactly f_{mnp} of that mode will survive, and all others will attenuate. If a probe is used to couple a sample of the resonant wave out of the cavity, the output signal will be a monochromatic sinusoidal wave at f_{mnp}. In practice, the cavity exhibits a frequency response similar to that shown in **Fig. 8-28(c)**, which is very narrow, but not a perfect spike. The bandwidth Δf of the cavity is defined as the frequency range between the two frequencies (on either side of f_{mnp}) at which the amplitude is $1/\sqrt{2}$ of the maximum amplitude (at f_{mnp}). The ***normalized bandwidth***, defined as $\Delta f / f_{mnp}$, is approximately equal to the reciprocal of the ***quality factor*** Q of the cavity,

$$Q \approx \frac{f_{mnp}}{\Delta f}\,. \qquad (8.123)$$

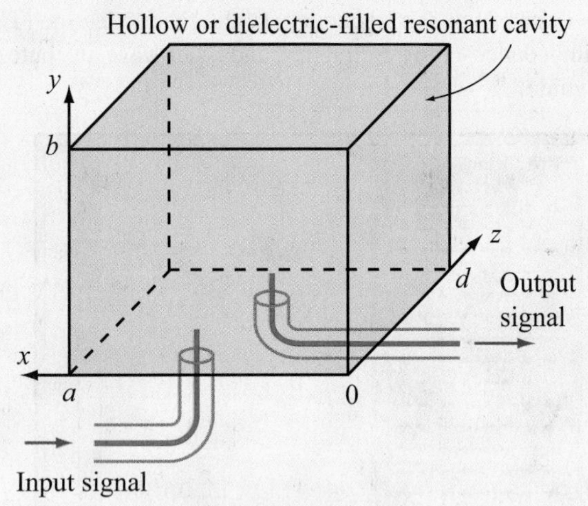

(a) Resonant cavity

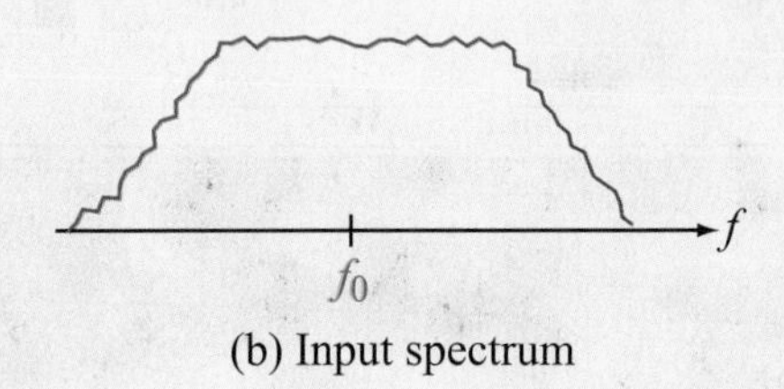

(b) Input spectrum

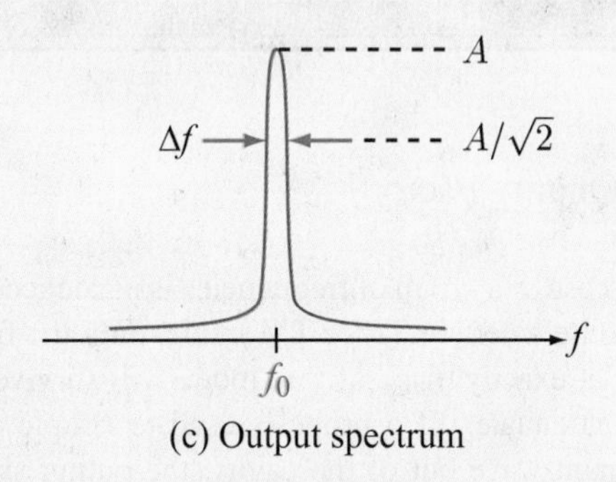

(c) Output spectrum

Figure 8-28 A resonant cavity supports a very narrow bandwidth around its resonant frequency f_0.

▶ The quality factor is defined in terms of the ratio of the energy stored in the cavity volume to the energy dissipated in the cavity walls through conduction. ◀

For an ideal cavity with perfectly conducting walls, no energy loss is incurred, as a result of which Q is infinite and $\Delta f \approx 0$. Metals have very high, but not infinite, conductivities, so a real cavity with metal walls stores most of the energy coupled into it in its volume, but it also loses some of it to heat conduction. A typical value for Q is on the order of 10,000, which is much higher than can be realized with lumped RLC circuits.

Example 8-11: Q of a Resonant Cavity

The quality factor for a hollow resonant cavity operating in the TE_{101} mode is

$$Q = \frac{1}{\delta_s} \frac{abd(a^2 + d^2)}{[a^3(d + 2b) + d^3(a + 2b)]}, \tag{8.124}$$

where $\delta_s = 1/\sqrt{\pi f_{mnp} \mu_0 \sigma_c}$ is the skin depth and σ_c is the conductivity of the conducting walls. Design a cubic cavity with a TE_{101} resonant frequency of 12.6 GHz and evaluate its bandwidth. The cavity walls are made of copper.

Solution: For $a = b = d$, $m = 1$, $n = 0$, $p = 1$, and $u_{p_0} = c = 3 \times 10^8$ m/s, Eq. (8.122) simplifies to

$$f_{101} = \frac{3\sqrt{2} \times 10^8}{2a} \qquad \text{(Hz)},$$

which, for $f_{101} = 12.6$ GHz, gives

$$a = 1.68 \text{ cm}.$$

At $f_{101} = 12.6$ GHz, the skin depth for copper (with $\sigma_c = 5.8 \times 10^7$ S/m) is

$$\delta_s = \frac{1}{[\pi f_{101} \mu_0 \sigma_c]^{1/2}} = \frac{1}{[\pi \times 12.6 \times 10^9 \times 4\pi \times 10^{-7} \times 5.8 \times 10^7]^{1/2}} = 5.89 \times 10^{-7} \text{ m}.$$

Upon setting $a = b = d$ in Eq. (8.124), the expression for Q of a cubic cavity becomes

$$Q = \frac{a}{3\delta_s} = \frac{1.68 \times 10^{-2}}{3 \times 5.89 \times 10^{-7}} \approx 9,500.$$

Hence, the cavity bandwidth is

$$\Delta f \approx \frac{f_{101}}{Q} \approx \frac{12.6 \times 10^9}{9,500} \approx 1.3 \text{ MHz}.$$

Chapter 8 Summary

Concepts

- The relations describing the reflection and transmission behavior of a plane EM wave at the boundary between two different media are the consequence of satisfying the conditions of continuity of the tangential components of **E** and **H** across the boundary.
- Snell's laws state that $\theta_i = \theta_r$ and

$$\sin\theta_t = (n_1/n_2)\sin\theta_i.$$

  For media such that $n_2 < n_1$, the incident wave is reflected totally by the boundary when $\theta_i \geq \theta_c$, where θ_c is the critical angle given by $\theta_c = \sin^{-1}(n_2/n_1)$.
- By successive multiple reflections, light can be guided through optical fibers. The maximum data rate of digital pulses that can be transmitted along optical fibers is dictated by modal dispersion.
- At the Brewster angle for a given polarization, the incident wave is transmitted totally across the boundary. For nonmagnetic materials, the Brewster angle exists for parallel polarization only.
- Any plane wave incident on a plane boundary can be synthesized as the sum of a perpendicularly polarized wave and a parallel polarized wave.
- Transmission-line equivalent models can be used to characterize wave propagation and reflection by and transmission through boundaries between different media.
- Waves can travel through a metal waveguide in the form of transverse electric (TE) and transverse magnetic (TM) modes. For each mode, the waveguide has a cutoff frequency below which a wave cannot propagate.
- A cavity resonator can support standing waves at specific resonant frequencies.

Important Terms

Provide definitions or explain the meaning of the following terms:

ω-β diagram
acceptance angle θ_a
angles of incidence, reflection, and transmission
Brewster angle θ_B
cladding
critical angle θ_c
cutoff frequency f_{mn}
cutoff wavenumber k_c
dominant mode
evanescent wave
fiber core
grazing incidence
group velocity u_g
index of refraction n
modal dispersion
modes
optical fibers
parallel polarization
perpendicular polarization
phase-matching condition
plane of incidence
polarizing angle
quality factor Q
reflection coefficient Γ
reflectivity (reflectance) R
refraction angle
resonant cavity
resonant frequency
Snell's laws
standing-wave ratio S
surface wave
total internal reflection
transmission coefficient τ
transmissivity (transmittance) T
transverse electric (TE) polarization
transverse magnetic (TM) polarization
unbounded-medium wavenumber
unpolarized
wavefront

Mathematical and Physical Models

Normal Incidence

$$\Gamma = \frac{E_0^{\rm r}}{E_0^{\rm i}} = \frac{\eta_2 - \eta_1}{\eta_2 + \eta_1}$$

$$\tau = \frac{E_0^{\rm t}}{E_0^{\rm i}} = \frac{2\eta_2}{\eta_2 + \eta_1}$$

$$\tau = 1 + \Gamma$$

$$\Gamma = \frac{\sqrt{\epsilon_{\rm r_1}} - \sqrt{\epsilon_{\rm r_2}}}{\sqrt{\epsilon_{\rm r_1}} + \sqrt{\epsilon_{\rm r_2}}} \quad (\text{if } \mu_1 = \mu_2)$$

Snell's Laws

$$\theta_{\rm i} = \theta_{\rm r}$$

$$\frac{\sin\theta_{\rm t}}{\sin\theta_{\rm i}} = \frac{u_{\rm p_2}}{u_{\rm p_1}} = \sqrt{\frac{\mu_1\epsilon_1}{\mu_2\epsilon_2}}$$

Oblique Incidence

Perpendicular Polarization

$$\Gamma_\perp = \frac{E_{\perp 0}^{\rm r}}{E_{\perp 0}^{\rm i}} = \frac{\eta_2\cos\theta_{\rm i} - \eta_1\cos\theta_{\rm t}}{\eta_2\cos\theta_{\rm i} + \eta_1\cos\theta_{\rm t}}$$

$$\tau_\perp = \frac{E_{\perp 0}^{\rm t}}{E_{\perp 0}^{\rm i}} = \frac{2\eta_2\cos\theta_{\rm i}}{\eta_2\cos\theta_{\rm i} + \eta_1\cos\theta_{\rm t}}$$

Parallel Polarization

$$\Gamma_\parallel = \frac{E_{\parallel 0}^{\rm r}}{E_{\parallel 0}^{\rm i}} = \frac{\eta_2\cos\theta_{\rm t} - \eta_1\cos\theta_{\rm i}}{\eta_2\cos\theta_{\rm t} + \eta_1\cos\theta_{\rm i}}$$

$$\tau_\parallel = \frac{E_{\parallel 0}^{\rm t}}{E_{\parallel 0}^{\rm i}} = \frac{2\eta_2\cos\theta_{\rm i}}{\eta_2\cos\theta_{\rm t} + \eta_1\cos\theta_{\rm i}}$$

Brewster Angle

$$\theta_{\rm B\parallel} = \sin^{-1}\sqrt{\frac{1}{1 + (\epsilon_1/\epsilon_2)}} = \tan^{-1}\sqrt{\frac{\epsilon_2}{\epsilon_1}}$$

Waveguides

$$\beta = \sqrt{\omega^2\mu\epsilon - \left(\frac{m\pi}{a}\right)^2 - \left(\frac{n\pi}{b}\right)^2}$$

$$f_{mn} = \frac{u_{\rm p_0}}{2}\sqrt{\left(\frac{m}{a}\right)^2 + \left(\frac{n}{b}\right)^2}$$

$$u_{\rm p} = \frac{\omega}{\beta} = \frac{u_{\rm p_0}}{\sqrt{1 - (f_{mn}/f)^2}}$$

$$u_{\rm p}u_{\rm g} = u_{\rm p_0}^2$$

$$Z_{\rm TE} = \frac{\eta}{\sqrt{1 - (f_{mn}/f)^2}}$$

$$Z_{\rm TM} = \eta\sqrt{1 - \left(\frac{f_{mn}}{f}\right)^2}$$

Resonant Cavity

$$f_{mnp} = \frac{u_{\rm p_0}}{2}\sqrt{\left(\frac{m}{a}\right)^2 + \left(\frac{n}{b}\right)^2 + \left(\frac{p}{d}\right)^2}$$

$$Q \approx \frac{f_{mnp}}{\Delta f}$$

PROBLEMS

Section 8-1: Wave Reflection and Transmission at Normal Incidence

8.1 A plane wave traveling in medium 1 with $\epsilon_{r1} = 2.25$ is normally incident upon medium 2 with $\epsilon_{r2} = 4$. Both media are made of nonmagnetic, nonconducting materials. If the electric field of the incident wave is given by

$$\mathbf{E}^i = \hat{\mathbf{y}} 8 \cos(6\pi \times 10^9 t - 30\pi x) \qquad \text{(V/m)}.$$

(a) Obtain time-domain expressions for the electric and magnetic fields in each of the two media.

(b) Determine the average power densities of the incident, reflected, and transmitted waves.

*__8.2__ A plane wave in air with an electric field amplitude of 20 V/m is incident normally upon the surface of a lossless, nonmagnetic medium with $\epsilon_r = 25$. Determine the following:

(a) The reflection and transmission coefficients.

(b) The standing-wave ratio in the air medium.

(c) The average power densities of the incident, reflected, and transmitted waves.

8.3 A plane wave traveling in a medium with $\epsilon_{r_1} = 9$ is normally incident upon a second medium with $\epsilon_{r_2} = 4$. Both media are made of nonmagnetic, nonconducting materials. If the magnetic field of the incident plane wave is given by

$$\mathbf{H}^i = \hat{\mathbf{z}} 2 \cos(2\pi \times 10^9 t - ky) \qquad \text{(A/m)}.$$

(a) Obtain time-domain expressions for the electric and magnetic fields in each of the two media.

*__(b)__ Determine the average power densities of the incident, reflected, and transmitted waves.

8.4 A 200 MHz, left-hand circularly polarized plane wave with an electric field modulus of 5 V/m is normally incident in air upon a dielectric medium with $\epsilon_r = 4$ and occupies the region defined by $z \geq 0$.

(a) Write an expression for the electric field phasor of the incident wave, given that the field is a positive maximum at $z = 0$ and $t = 0$.

(b) Calculate the reflection and transmission coefficients.

(c) Write expressions for the electric field phasors of the reflected wave, the transmitted wave, and the total field in the region $z \leq 0$.

(d) Determine the percentages of the incident average power reflected by the boundary and transmitted into the second medium.

8.5 Repeat Problem 8.4, but replace the dielectric medium with a poor conductor characterized by $\epsilon_r = 2.25$, $\mu_r = 1$, and $\sigma = 10^{-4}$ S/m.

8.6 A 50 MHz plane wave with electric field amplitude of 50 V/m is normally incident in air onto a semi-infinite, perfect dielectric medium with $\epsilon_r = 36$. Determine the following:

*__(a)__ Γ

(b) The average power densities of the incident and reflected waves.

(c) The distance in the air medium from the boundary to the nearest minimum of the electric field intensity, $|\mathbf{E}|$.

8.7 Repeat Problem 8.6, but replace the dielectric medium with a conductor with $\epsilon_r = 1$, $\mu_r = 1$, and $\sigma = 2.78 \times 10^{-3}$ S/m.

*__8.8__ What is the maximum amplitude of the total electric field in the air medium of Problem 8.6, and at what nearest distance from the boundary does it occur?

*__8.9__ The three regions shown in **Fig. P8.9** contain perfect dielectrics. For a wave in medium 1, incident normally upon the boundary at $z = -d$, what combination of ϵ_{r_2} and d produces

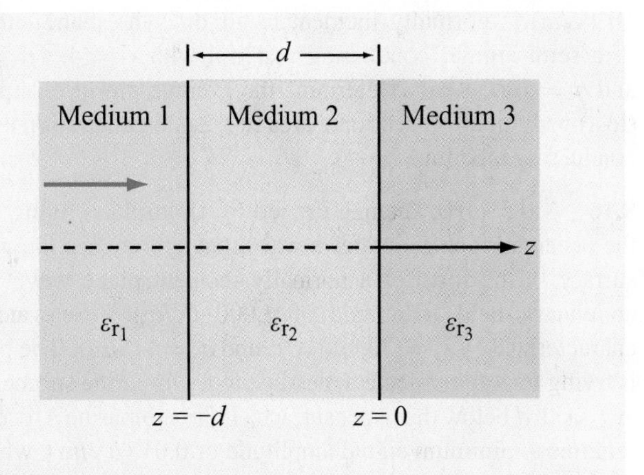

Figure P8.9 Dielectric layers for Problems 8.9 to 8.11.

*Answer(s) available in Appendix D.

no reflection? Express your answers in terms of ϵ_{r_1}, ϵ_{r_3}, and the oscillation frequency of the wave, f.

8.10 For the configuration shown in **Fig. P8.9**, use transmission-line equations (or the Smith chart) to calculate the input impedance at $z = -d$ for $\epsilon_{r_1} = 1$, $\epsilon_{r_2} = 9$, $\epsilon_{r_3} = 4$, $d = 1.2$ m, and $f = 50$ MHz. Also determine the fraction of the incident average power density reflected by the structure. Assume all media are lossless and nonmagnetic.

*__8.11__ Repeat Problem 8.10, but interchange ϵ_{r_1} and ϵ_{r_3}.

*__8.12__ A plane wave of unknown frequency is normally incident in air upon the surface of a perfect conductor. Using an electric-field meter, it was determined that the total electric field in the air medium is always zero when measured at a distance of 3 m from the conductor surface. Moreover, no such nulls were observed at distances closer to the conductor. What is the frequency of the incident wave?

8.13 Orange light of wavelength 0.61 μm in air enters a block of glass with $\epsilon_r = 1.44$. What color would it appear to a sensor embedded in the glass? The wavelength ranges of colors are violet (0.39 to 0.45 μm), blue (0.45 to 0.49 μm), green (0.49 to 0.58 μm), yellow (0.58 to 0.60 μm), orange (0.60 to 0.62 μm), and red (0.62 to 0.78 μm).

8.14 Consider a thin film of soap in air under illumination by yellow light with $\lambda = 0.6$ μm in vacuum. If the film is treated as a planar dielectric slab with $\epsilon_r = 1.72$, surrounded on both sides by air, what film thickness would produce strong reflection of the yellow light at normal incidence?

*__8.15__ A 5 MHz plane wave with electric field amplitude of 10 (V/m) is normally incident in air onto the plane surface of a semi-infinite conducting material with $\epsilon_r = 4$, $\mu_r = 1$, and $\sigma = 100$ (S/m). Determine the average power dissipated (lost) per unit cross-sectional area in a 2 mm penetration of the conducting medium.

8.16 A 0.5 MHz antenna carried by an airplane flying over the ocean surface generates a wave that approaches the water surface in the form of a normally incident plane wave with an electric-field amplitude of 3,000 (V/m). Seawater is characterized by $\epsilon_r = 72$, $\mu_r = 1$, and $\sigma = 4$ (S/m). The plane is trying to communicate a message to a submarine submerged at a depth d below the water surface. If the submarine's receiver requires a minimum signal amplitude of 0.01 (μV/m), what is the maximum depth d to which successful communication is still possible?

Sections 8-2 and 8-3: Snell's Laws and Fiber Optics

*__8.17__ A light ray is incident on a prism in air at an angle θ as shown in **Fig. P8.17**. The ray is refracted at the first surface and again at the second surface. In terms of the apex angle ϕ of the prism and its index of refraction n, determine the smallest value of θ for which the ray will emerge from the other side. Find this minimum θ for $n = 1.6$ and $\phi = 60°$.

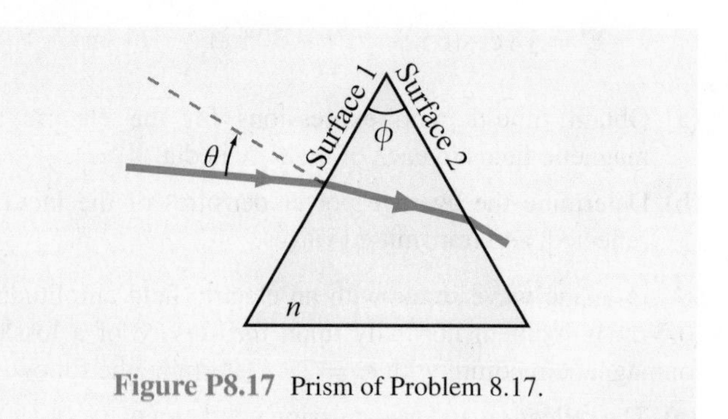

Figure P8.17 Prism of Problem 8.17.

8.18 For some types of glass, the index of refraction varies with wavelength. A prism made of a material with

$$n = 1.71 - \frac{4}{30}\lambda_0 \qquad (\lambda_0 \text{ in } \mu\text{m}),$$

where λ_0 is the wavelength in vacuum, was used to disperse white light as shown in **Fig. P8.18**. The white light is incident at an angle of 50°, the wavelength λ_0 of red light is 0.7 μm, and that of violet light is 0.4 μm. Determine the angular dispersion in degrees.

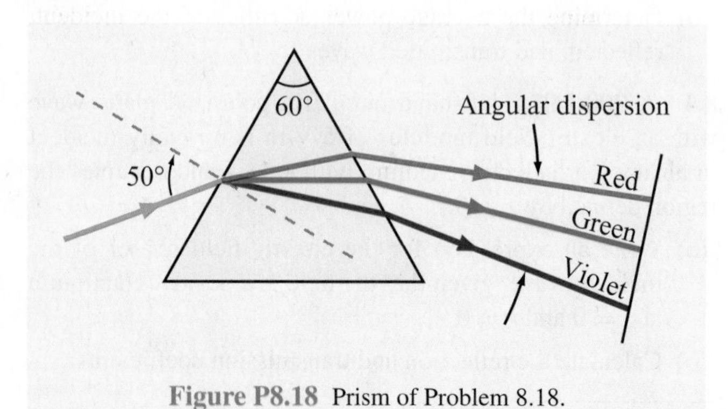

Figure P8.18 Prism of Problem 8.18.

8.19 A parallel-polarized plane wave is incident from air at an angle $\theta_i = 30°$ onto a pair of dielectric layers as shown in **Fig. P8.19**.

(a) Determine the angles of transmission θ_2, θ_3, and θ_4.

(b) Determine the lateral distance d.

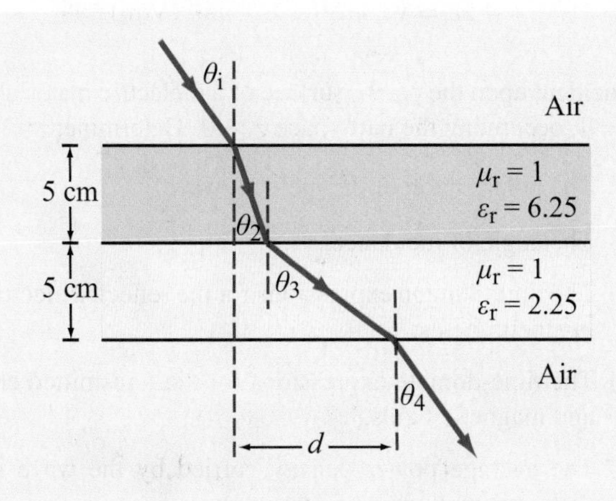

Figure P8.19 Problem P8.19.

*__8.20__ The two prisms in **Fig. P8.20** are made of glass with $n = 1.5$. What fraction of the power density carried by the ray incident upon the top prism emerges from the bottom prism? Neglect multiple internal reflections.

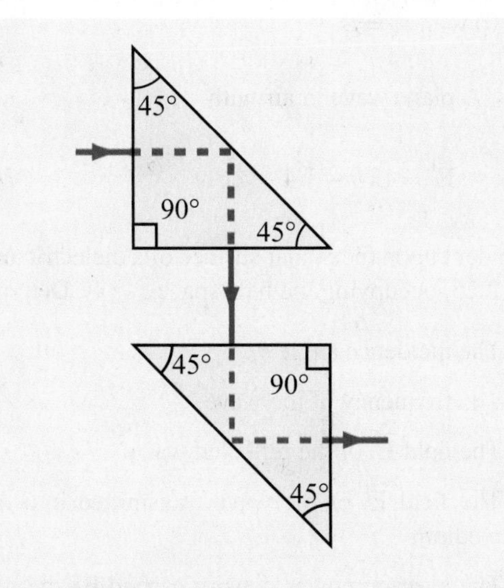

Figure P8.20 Periscope prisms of Problem 8.20.

8.21 A light ray incident at 45° passes through two dielectric materials with the indices of refraction and thicknesses given in **Fig. P8.21**. If the ray strikes the surface of the first dielectric at a height of 2 cm, at what height will it strike the screen?

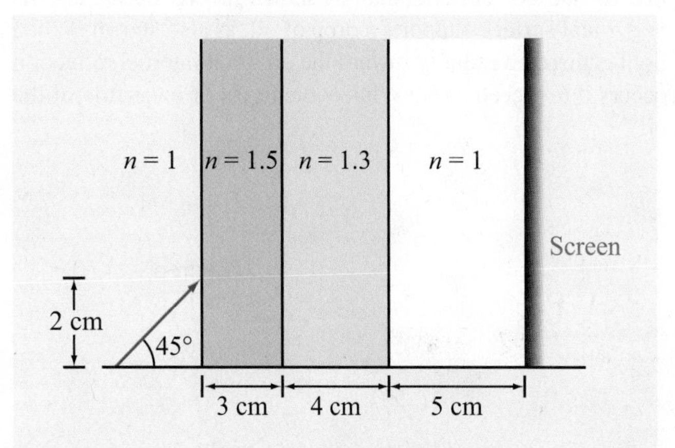

Figure P8.21 Light incident on a screen through a multilayered dielectric (Problem 8.21).

*__8.22__ **Figure P8.22** depicts a beaker containing a block of glass on the bottom and water over it. The glass block contains a small

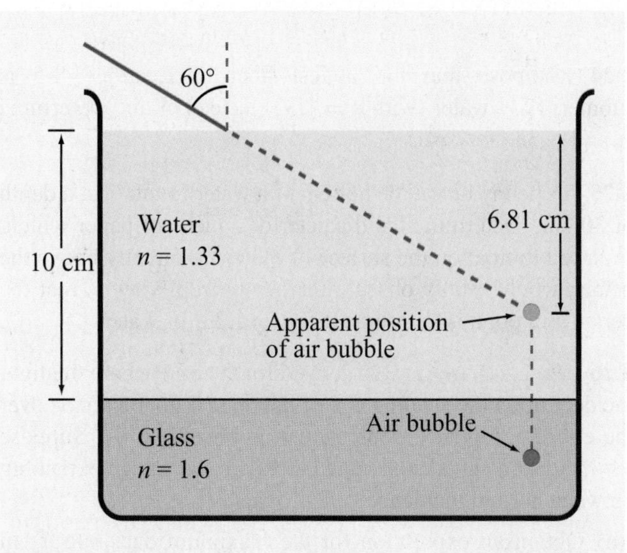

Figure P8.22 Apparent position of the air bubble in Problem 8.22.

air bubble at an unknown depth below the water surface. When viewed from above at an angle of 60°, the air bubble appears at a depth of 6.81 cm. What is the true depth of the air bubble?

8.23 A glass semicylinder with $n = 1.4$ is positioned such that its flat face is horizontal, as shown in **Fig. P8.23**, and its horizontal surface supports a drop of oil, as also shown. When light is directed radially toward the oil, total internal reflection occurs if θ exceeds 53°. What is the index of refraction of the oil?

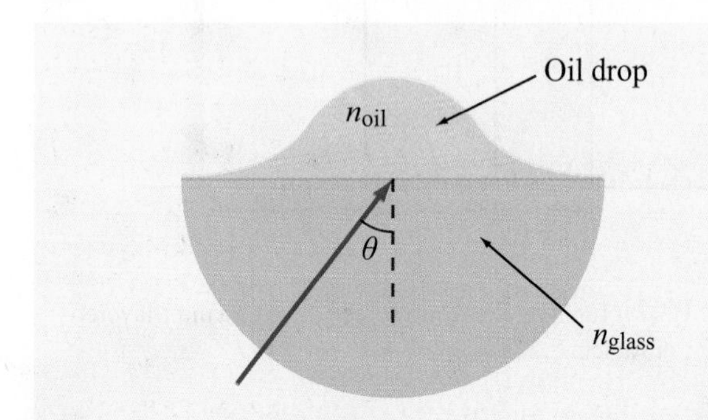

Figure P8.23 Oil drop on the flat surface of a glass semicylinder (Problem 8.23).

8.24 Suppose that the optical fiber of Example 8-5 is submerged in water (with $n = 1.33$) instead of air. Determine θ_a and f_p in that case.

***8.25** A penny lies at the bottom of a water fountain at a depth of 30 cm. Determine the diameter of a piece of paper which, if placed to float on the surface of the water directly above the penny, would totally obscure the penny from view. Treat the penny as a point and assume that $n = 1.33$ for water.

***8.26** Equation (8.45) was derived for the case where the light incident upon the sending end of the optical fiber extends over the entire acceptance cone shown in **Fig. 8-12(b)**. Suppose the incident light is constrained to a narrower range extending between normal incidence and θ', where $\theta' < \theta_a$.

(a) Obtain an expression for the maximum data rate f_p in terms of θ'.

(b) Evaluate f_p for the fiber of Example 8-5 when $\theta' = 3°$.

Sections 8-4 and 8-5: Reflection and Transmission at Oblique Incidence

8.27 A plane wave in air with

$$\widetilde{\mathbf{E}}^{\text{i}} = \hat{\mathbf{y}}\, 20e^{-j(3x+4z)} \qquad \text{(V/m)}$$

is incident upon the planar surface of a dielectric material, with $\epsilon_r = 4$, occupying the half-space $z \geq 0$. Determine:

(a) The polarization of the incident wave.

***(b)** The angle of incidence.

(c) The time-domain expressions for the reflected electric and magnetic fields.

(d) The time-domain expressions for the transmitted electric and magnetic fields.

(e) The average power density carried by the wave in the dielectric medium.

8.28 Repeat Problem 8.27 for a wave in air with

$$\widetilde{\mathbf{H}}^{\text{i}} = \hat{\mathbf{y}}\, 2 \times 10^{-2} e^{-j(8x+6z)} \qquad \text{(A/m)}$$

incident upon the planar boundary of a dielectric medium ($z \geq 0$) with $\epsilon_r = 9$.

8.29 A plane wave in air with

$$\widetilde{\mathbf{E}}^{\text{i}} = (\hat{\mathbf{x}}\, 9 - \hat{\mathbf{y}}\, 4 - \hat{\mathbf{z}}\, 6)e^{-j(2x+3z)} \qquad \text{(V/m)}$$

is incident upon the planar surface of a dielectric material, with $\epsilon_r = 2.25$, occupying the half-space $z \geq 0$. Determine

***(a)** The incidence angle θ_i.

(b) The frequency of the wave.

(c) The field $\widetilde{\mathbf{E}}^{\text{r}}$ of the reflected wave.

(d) The field $\widetilde{\mathbf{E}}^{\text{t}}$ of the wave transmitted into the dielectric medium.

(e) The average power density carried by the wave into the dielectric medium.

*8.30 A parallel-polarized plane wave is incident from air onto a dielectric medium with $\epsilon_r = 9$ at the Brewster angle. What is the refraction angle?

8.31 Natural light is randomly polarized, which means that, on average, half the light energy is polarized along any given direction (in the plane orthogonal to the direction of propagation) and the other half of the energy is polarized along the direction orthogonal to the first polarization direction. Hence, when treating natural light incident upon a planar boundary, we can consider half of its energy to be in the form of parallel-polarized waves and the other half as perpendicularly polarized waves. Determine the fraction of the incident power reflected by the planar surface of a piece of glass with $n = 1.5$ when illuminated by natural light at 70°.

8.32 Show that the reflection coefficient $\Gamma_\perp$ can be written in the following form:

$$\Gamma_\perp = \frac{\sin(\theta_t - \theta_i)}{\sin(\theta_t + \theta_i)} .$$

8.33 A perpendicularly polarized wave in air is obliquely incident upon a planar glass–air interface at an incidence angle of 30°. The wave frequency is 600 THz (1 THz = 10^{12} Hz), which corresponds to green light, and the index of refraction of the glass is 1.6. If the electric field amplitude of the incident wave is 50 V/m, determine the following:

(a) The reflection and transmission coefficients.

(b) The instantaneous expressions for **E** and **H** in the glass medium.

8.34 Show that for nonmagnetic media, the reflection coefficient $\Gamma_\parallel$ can be written in the following form:

$$\Gamma_\parallel = \frac{\tan(\theta_t - \theta_i)}{\tan(\theta_t + \theta_i)} .$$

8.35 A parallel-polarized beam of light with an electric field amplitude of 10 (V/m) is incident in air on polystyrene with $\mu_r = 1$ and $\epsilon_r = 2.6$. If the incidence angle at the air–polystyrene planar boundary is 60°, determine the following:

(a) The reflectivity and transmissivity.

(b) The power carried by the incident, reflected, and transmitted beams if the spot on the boundary illuminated by the incident beam is 1 m^2 in area.

8.36 A 50 MHz right-hand circularly polarized plane wave with an electric field modulus of 30 V/m is normally incident in air upon a dielectric medium with $\epsilon_r = 9$ and occupying the region defined by $z \geq 0$.

(a) Write an expression for the electric field phasor of the incident wave, given that the field is a positive maximum at $z = 0$ and $t = 0$.

(b) Calculate the reflection and transmission coefficients.

(c) Write expressions for the electric field phasors of the reflected wave, the transmitted wave, and the total field in the region $z \leq 0$.

(d) Determine the percentages of the incident average power reflected by the boundary and transmitted into the second medium.

8.37 Consider a flat 5 mm thick slab of glass with $\epsilon_r = 2.56$.

*(a) If a beam of green light ($\lambda_0 = 0.52$ μm) is normally incident upon one of the sides of the slab, what percentage of the incident power is reflected back by the glass?

(b) To eliminate reflections, it is desired to add a thin layer of antireflection coating material on each side of the glass. If you are at liberty to specify the thickness of the antireflection material as well as its relative permittivity, what would these specifications be?

Sections 8-6 to 8-11: Waveguides and Resonators

8.38 A TE wave propagating in a dielectric-filled waveguide of unknown permittivity has dimensions $a = 5$ cm and $b = 3$ cm. If the x component of its electric field is given by

$$E_x = -36\cos(40\pi x)\sin(100\pi y) \cdot \sin(2.4\pi \times 10^{10} t - 52.9\pi z), \qquad \text{(V/m)}$$

determine:

(a) the mode number,

(b) ϵ_r of the material in the guide,

(c) the cutoff frequency, and

(d) the expression for H_y.

*8.39 A hollow rectangular waveguide is to be used to transmit signals at a carrier frequency of 10 GHz. Choose its dimensions so that the cutoff frequency of the dominant TE mode is lower than the carrier by 25% and that of the next mode is at least 25% higher than the carrier.

8.40 Derive Eq. (8.89b).

*8.41 A waveguide filled with a material whose $\epsilon_r = 2.25$ has dimensions $a = 2$ cm and $b = 1.4$ cm. If the guide is to transmit 10.5 GHz signals, what possible modes can be used for the transmission?

8.42 A narrow rectangular pulse superimposed on a carrier with a frequency of 9.5 GHz was used to excite all possible modes in a hollow guide with $a = 3$ cm and $b = 2.0$ cm. If the guide is 100 m in length, how long will it take each of the excited modes to arrive at the receiving end?

*8.43 A waveguide, with dimensions $a = 2$ cm and $b = 1.4$ cm, is to be used at 10 GHz. Determine the wave impedance for the dominant mode when

(a) the guide is empty, and

(b) the guide is filled with polyethylene (whose $\epsilon_r = 2.25$).

8.44 For a rectangular waveguide operating in the TE_{10} mode, obtain expressions for the surface charge density $\widetilde{\rho}_s$ and surface current density $\widetilde{\mathbf{J}}_s$ on each of the four walls of the guide.

*8.45 If the zigzag angle θ' is 20° for the TE_{10} mode, what would it be for the TE_{20} mode?

8.46 A hollow cavity made of aluminum has dimensions $a = 4$ cm and $d = 3$ cm. Calculate Q of the TE_{101} mode for

*(a) $b = 2$ cm, and

(b) $b = 3$ cm.

8.47 Measurement of the TE_{101} frequency response of an air-filled cubic cavity revealed that its Q is 2401. If its volume is 8 mm^3, what material are its sides made of? (Hint: See Appendix B.)

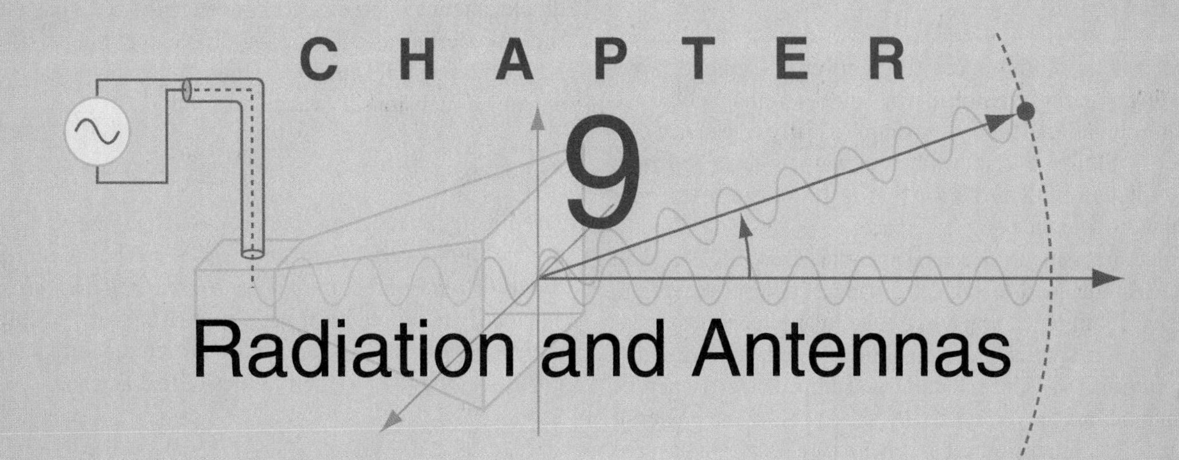

Chapter Contents

	Overview, 426
9-1	The Hertzian Dipole, 428
9-2	Antenna Radiation Characteristics, 432
9-3	Half-Wave Dipole Antenna, 439
9-4	Dipole of Arbitrary Length, 442
9-5	Effective Area of a Receiving Antenna, 444
TB17	Health Risks of EM Fields, 446
9-6	Friis Transmission Formula, 449
9-7	Radiation by Large-Aperture Antennas, 451
9-8	Rectangular Aperture with Uniform Aperture Distribution, 454
9-9	Antenna Arrays, 456
9-10	N-Element Array with Uniform Phase Distribution, 464
9-11	Electronic Scanning of Arrays, 466
	Chapter 9 Summary, 472
	Problems, 474

Objectives

Upon learning the material presented in this chapter, you should be able to:

1. Calculate the electric and magnetic fields of waves radiated by a dipole antenna.
2. Characterize the radiation of an antenna in terms of its radiation pattern, directivity, beamwidth, and radiation resistance.
3. Apply the Friis transmission formula to a free-space communication system.
4. Calculate the electric and magnetic fields of waves radiated by aperture antennas.
5. Calculate the radiation pattern of multi-element antenna arrays.

Overview

An *antenna* is a transducer that converts a guided wave propagating on a transmission line into an electromagnetic wave propagating in an unbounded medium (usually free space), or vice versa. **Figure 9-1** shows how a wave is launched by a hornlike antenna, with the horn acting as the transition segment between the waveguide and free space.

Antennas are made in various shapes and sizes (**Fig. 9-2**) and are used in radio and television broadcasting and reception, radio-wave communication systems, cellular telephones, radar systems, and anticollision automobile sensors, among many other applications. The radiation and impedance properties of an antenna are governed by its shape, size, and material properties. The dimensions of an antenna are usually measured in units of λ of the wave it is launching or receiving; a 1 m long dipole antenna operating at a wavelength $\lambda = 2$ m exhibits the same properties as a 1 cm long dipole operating at $\lambda = 2$ cm. Hence, in most of our discussions in this chapter, we refer to antenna dimensions in wavelength units.

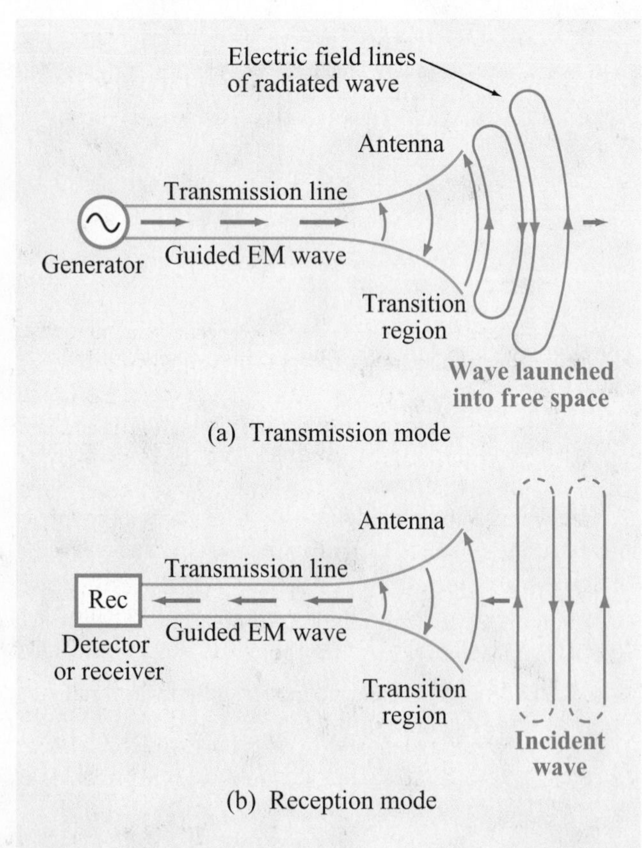

Figure 9-1 Antenna as a transducer between a guided electromagnetic wave and a free-space wave, for both transmission and reception.

Reciprocity

The directional function characterizing the *relative* distribution of power radiated by an antenna is known as the ***antenna radiation pattern***, or simply the ***antenna pattern***. An ***isotropic*** antenna is a hypothetical antenna that radiates equally in all directions, and it is often used as a reference radiator when describing the radiation properties of real antennas.

▶ Most antennas are ***reciprocal*** devices, exhibiting the same radiation pattern for transmission as for reception. ◀

Reciprocity means that, if in the transmission mode a given antenna transmits in direction A 100 times the power it transmits in direction B, then when used in the reception mode it is 100 times more sensitive to electromagnetic radiation incident from direction A than from B. All the antennas shown in **Fig. 9-2** obey the reciprocity law, but not all antennas are reciprocal devices. Reciprocity may not hold for some solid-state antennas composed of nonlinear semiconductors or ferrite materials. Such nonreciprocal antennas are beyond the scope of this chapter, and hence reciprocity is assumed throughout. The reciprocity property is very convenient because it allows us to compute the radiation pattern of an antenna in the transmission mode, even when the antenna is intended to operate as a receiver.

To fully characterize an antenna, one needs to study its radiation properties and impedance. The radiation properties include its directional radiation pattern and the associated polarization state of the radiated wave when the antenna is used in the transmission mode, also called the ***antenna polarization***.

▶ Being a reciprocal device, an antenna, when operating in the receiving mode, can extract from an incident wave only that component of the wave whose electric field matches the antenna polarization state. ◀

The second aspect, the ***antenna impedance***, pertains to the transfer of power from a generator to the antenna when the antenna is used as a transmitter and, conversely, the transfer of power from the antenna to a load when the antenna is used

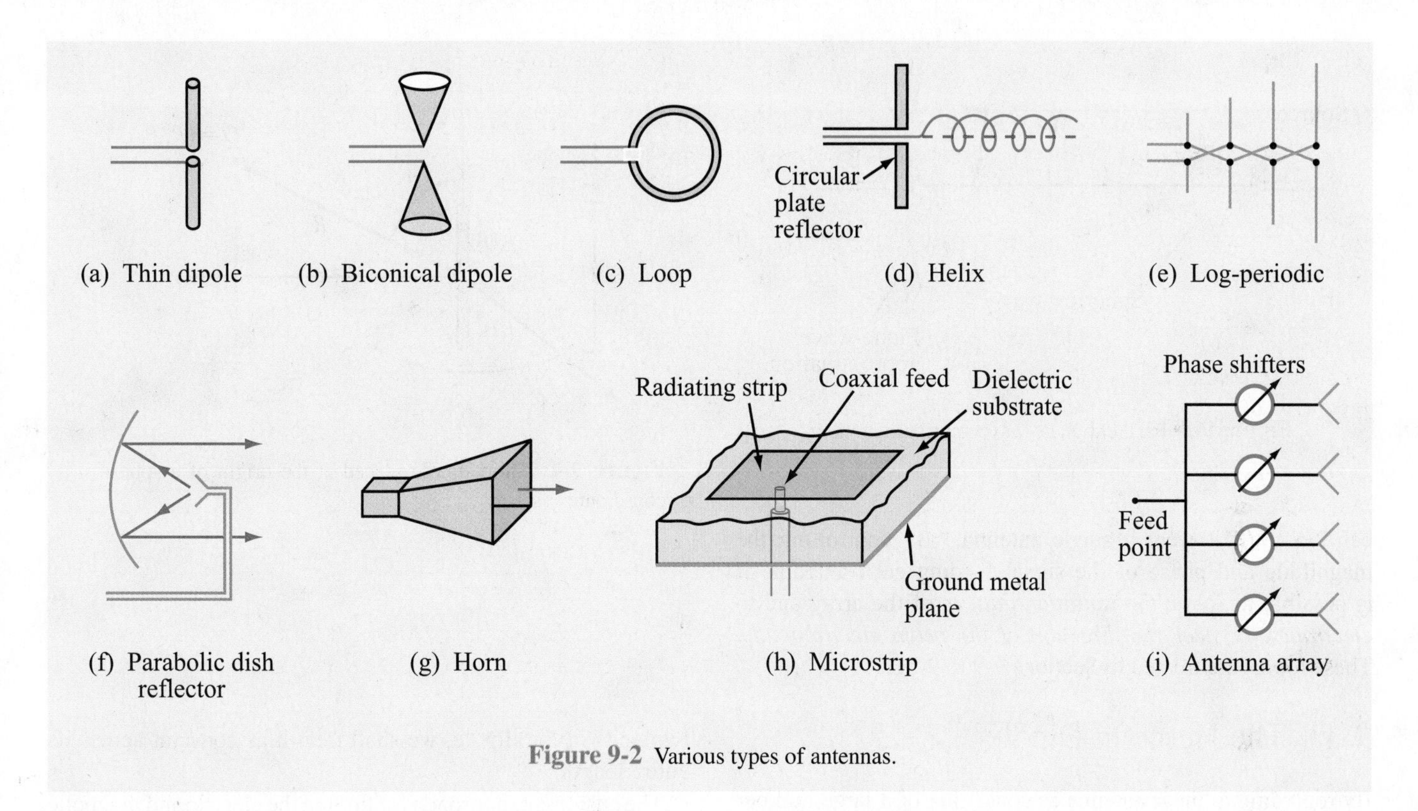

Figure 9-2 Various types of antennas.

as a receiver, as will be discussed later in Section 9-5. It should be noted that throughout our discussions in this chapter it will be assumed that the antenna is properly matched to the transmission line connected to its terminals, thereby avoiding reflections and their associated problems.

Radiation sources

Radiation sources fall into two categories: currents and aperture fields. The dipole and loop antennas [**Fig. 9-2(a)** and **(c)**] are examples of current sources; the time-varying currents flowing in the conducting wires give rise to the radiated electromagnetic fields. A horn antenna [**Fig. 9-2(g)**] is an example of the second group because the electric and magnetic fields across the horn's aperture serve as the sources of the radiated fields. The aperture fields are themselves induced by time-varying currents on the surfaces of the horn's walls, and therefore ultimately all radiation is due to time-varying currents. The choice of currents or apertures as the sources is merely a computational convenience arising from the structure of the antenna. We will examine the radiation processes associated with both types of sources.

Far-field region

The wave radiated by a point source is spherical in nature, with the wavefront expanding outward at a rate equal to the phase velocity u_{p} (or the velocity of light c if the medium is free space). If R, the distance between the transmitting antenna and the receiving antenna, is sufficiently large such that the wavefront across the receiving aperture may be considered planar (**Fig. 9-3**), then the receiving aperture is said to be in the ***far-field*** (or ***far-zone***) region of the transmitting point source. This region is of particular significance because for most applications, the location of the observation point is indeed in the far-field region of the antenna. The far-field plane-wave approximation allows the use of certain mathematical approximations that simplify the computation of the radiated field and, conversely, provide convenient techniques for synthesizing the appropriate antenna structure that would give rise to the desired far-field antenna pattern.

Antenna arrays

When multiple antennas operate together, the combination is called an ***antenna array*** [**Fig. 9-2(i)**], and the array as a whole

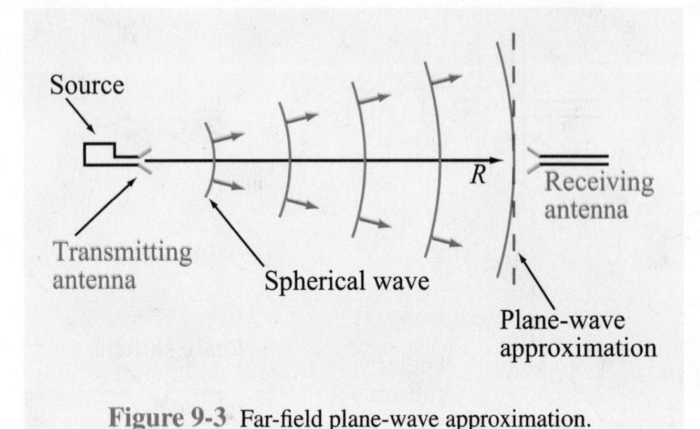

Figure 9-3 Far-field plane-wave approximation.

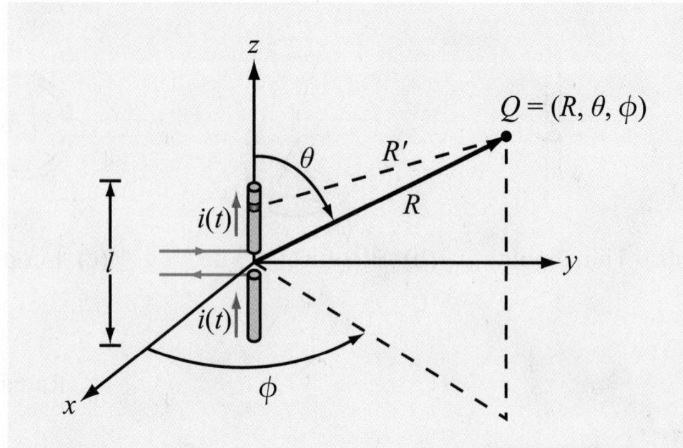

Figure 9-4 Short dipole placed at the origin of a spherical coordinate system.

behaves as if it were a single antenna. By controlling the magnitude and phase of the signal feeding each antenna, it is possible to *shape the radiation pattern* of the array and to *electronically steer the direction of the beam electronically.* These topics are treated in Sections 9-9 to 9-11.

9-1 The Hertzian Dipole

By regarding a linear antenna as consisting of a large number of infinitesimally short conducting elements, each of which is so short that current may be considered uniform over its length, the field of the entire antenna may be obtained by integrating the fields from all these differential antennas, with the proper magnitudes and phases taken into account. We shall first examine the radiation properties of such a differential antenna, known as a ***Hertzian dipole***, and then in Section 9-3 we will extend the results to compute the fields radiated by a half-wave dipole, which is commonly used as a standard antenna for many applications.

▶ A ***Hertzian dipole*** is a thin, linear conductor whose length l is very short compared with the wavelength λ; l should not exceed $\lambda/50$. ◀

The wire, oriented along the z direction in **Fig. 9-4**, carries a sinusoidally varying current given by

$$i(t) = I_0 \cos \omega t = \mathfrak{Re}[I_0 e^{j\omega t}] \qquad \text{(A)}, \tag{9.1}$$

where I_0 is the current amplitude. From Eq. (9.1), the phasor current $\widetilde{I} = I_0$. Even though the current has to go to zero at the two ends of the dipole, we shall treat it as constant across its entire length.

The customary approach for finding the electric and magnetic fields at a point Q in space (**Fig. 9-4**) due to radiation by a current source is through the retarded vector potential $\mathbf{A}$. From Eq. (6.84), the phasor retarded vector potential $\widetilde{\mathbf{A}}(R)$ at a distance vector $\mathbf{R}$ from a volume $\mathcal{V}'$ containing a phasor current distribution $\widetilde{\mathbf{J}}$ is given by

$$\widetilde{\mathbf{A}}(\mathbf{R}) = \frac{\mu_0}{4\pi} \int_{\mathcal{V}'} \frac{\widetilde{\mathbf{J}} e^{-jkR'}}{R'} \, d\mathcal{V}', \tag{9.2}$$

where μ_0 is the magnetic permeability of free space (because the observation point is in air) and $k = \omega/c = 2\pi/\lambda$ is the wavenumber. For the dipole, the current density is simply $\widetilde{\mathbf{J}} = \hat{\mathbf{z}}(I_0/s)$, where s is the cross-sectional area of the dipole wire. Also, $d\mathcal{V}' = s\,dz$ and the limits of integration are from $z = -l/2$ to $z = l/2$. In **Fig. 9-4**, the distance R' between the observation point and a given point along the dipole is not the same as the distance to its center, R, but because we are dealing with a very short dipole, we can set $R' \approx R$. Hence,

$$\widetilde{\mathbf{A}} = \frac{\mu_0}{4\pi} \frac{e^{-jkR}}{R} \int_{-l/2}^{l/2} \hat{\mathbf{z}} I_0 \, dz = \hat{\mathbf{z}} \frac{\mu_0}{4\pi} I_0 l \left(\frac{e^{-jkR}}{R} \right), \tag{9.3}$$

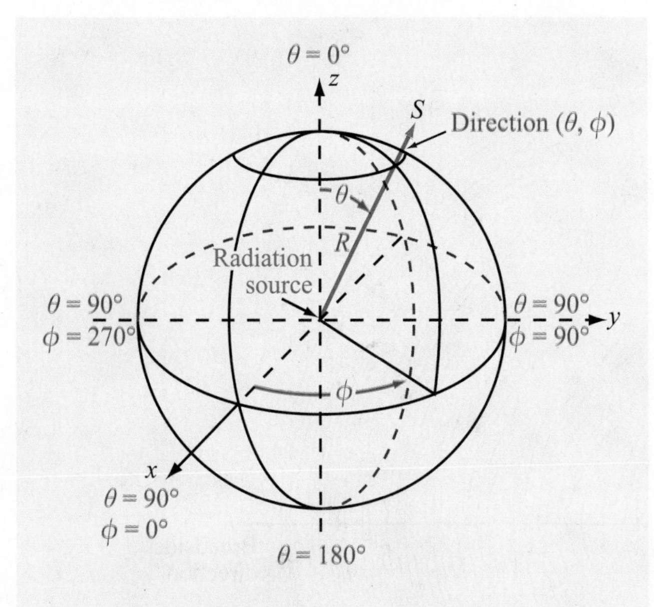

Figure 9-5 Spherical coordinate system.

► The function (e^{-jkR}/R) is called the ***spherical propagation factor***. It accounts for the $1/R$ decay of the magnitude with distance as well as the phase change represented by e^{-jkR}. ◄

The direction of $\widetilde{\mathbf{A}}$ is the same as that of the current (z direction).

Because our objective is to characterize the directional character of the radiated power at a fixed distance R from the antenna, antenna pattern plots are presented in a spherical coordinate system (Fig. 9-5). Its variables, R, θ, and ϕ, are called the ***range***, ***zenith angle***, and ***azimuth angle***, respectively. To that end, we need to write $\widetilde{\mathbf{A}}$ in terms of its spherical coordinate components, which is realized (with the help of Eq. (3.65c)) by expressing $\hat{\mathbf{z}}$ in terms of spherical coordinates:

$$\hat{\mathbf{z}} = \hat{\mathbf{R}}\cos\theta - \hat{\boldsymbol{\theta}}\sin\theta. \tag{9.4}$$

Upon substituting Eq. (9.4) into Eq. (9.3), we obtain

$$\begin{aligned}\widetilde{\mathbf{A}} &= (\hat{\mathbf{R}}\cos\theta - \hat{\boldsymbol{\theta}}\sin\theta)\,\frac{\mu_0 I_0 l}{4\pi}\left(\frac{e^{-jkR}}{R}\right) \\ &= \hat{\mathbf{R}}\widetilde{A}_R + \hat{\boldsymbol{\theta}}\widetilde{A}_\theta + \hat{\boldsymbol{\phi}}\widetilde{A}_\phi,\end{aligned} \tag{9.5}$$

with

$$\widetilde{A}_R = \frac{\mu_0 I_0 l}{4\pi}\cos\theta\left(\frac{e^{-jkR}}{R}\right), \tag{9.6a}$$

$$\widetilde{A}_\theta = -\frac{\mu_0 I_0 l}{4\pi}\sin\theta\left(\frac{e^{-jkR}}{R}\right), \tag{9.6b}$$

$$\widetilde{A}_\phi = 0.$$

With the spherical components of $\widetilde{\mathbf{A}}$ known, the next step is straightforward; we simply apply the free-space relationships given by Eqs. (6.85) and (6.86),

$$\widetilde{\mathbf{H}} = \frac{1}{\mu_0}\nabla\times\widetilde{\mathbf{A}}, \tag{9.7a}$$

$$\widetilde{\mathbf{E}} = \frac{1}{j\omega\epsilon_0}\nabla\times\widetilde{\mathbf{H}}, \tag{9.7b}$$

to obtain the expressions

$$\widetilde{H}_\phi = \frac{I_0 l k^2}{4\pi}\,e^{-jkR}\left[\frac{j}{kR} + \frac{1}{(kR)^2}\right]\sin\theta, \tag{9.8a}$$

$$\widetilde{E}_R = \frac{2 I_0 l k^2}{4\pi}\,\eta_0 e^{-jkR}\left[\frac{1}{(kR)^2} - \frac{j}{(kR)^3}\right]\cos\theta, \tag{9.8b}$$

$$\widetilde{E}_\theta = \frac{I_0 l k^2}{4\pi}\,\eta_0 e^{-jkR}\left[\frac{j}{kR} + \frac{1}{(kR)^2} - \frac{j}{(kR)^3}\right]\sin\theta, \tag{9.8c}$$

where $\eta_0 = \sqrt{\mu_0/\epsilon_0} \simeq 120\pi$ (Ω) is the intrinsic impedance of free space. The remaining components ($\widetilde{H}_R$, $\widetilde{H}_\theta$, and $\widetilde{E}_\phi$) are everywhere zero. **Figure 9-6** depicts the electric field lines of the wave radiated by the short dipole.

9-1.1 Far-Field Approximation

As was stated earlier, in most antenna applications we are primarily interested in the radiation pattern of the antenna at great distances from the source. For the electric dipole, this corresponds to distances R such that $R \gg \lambda$ or, equivalently, $kR = 2\pi R/\lambda \gg 1$. This condition allows us to neglect the terms varying as $1/(kR)^2$ and $1/(kR)^3$ in Eqs. (9.8a) to (9.8c) in favor of the terms varying as $1/kR$, which yields the far-field expressions

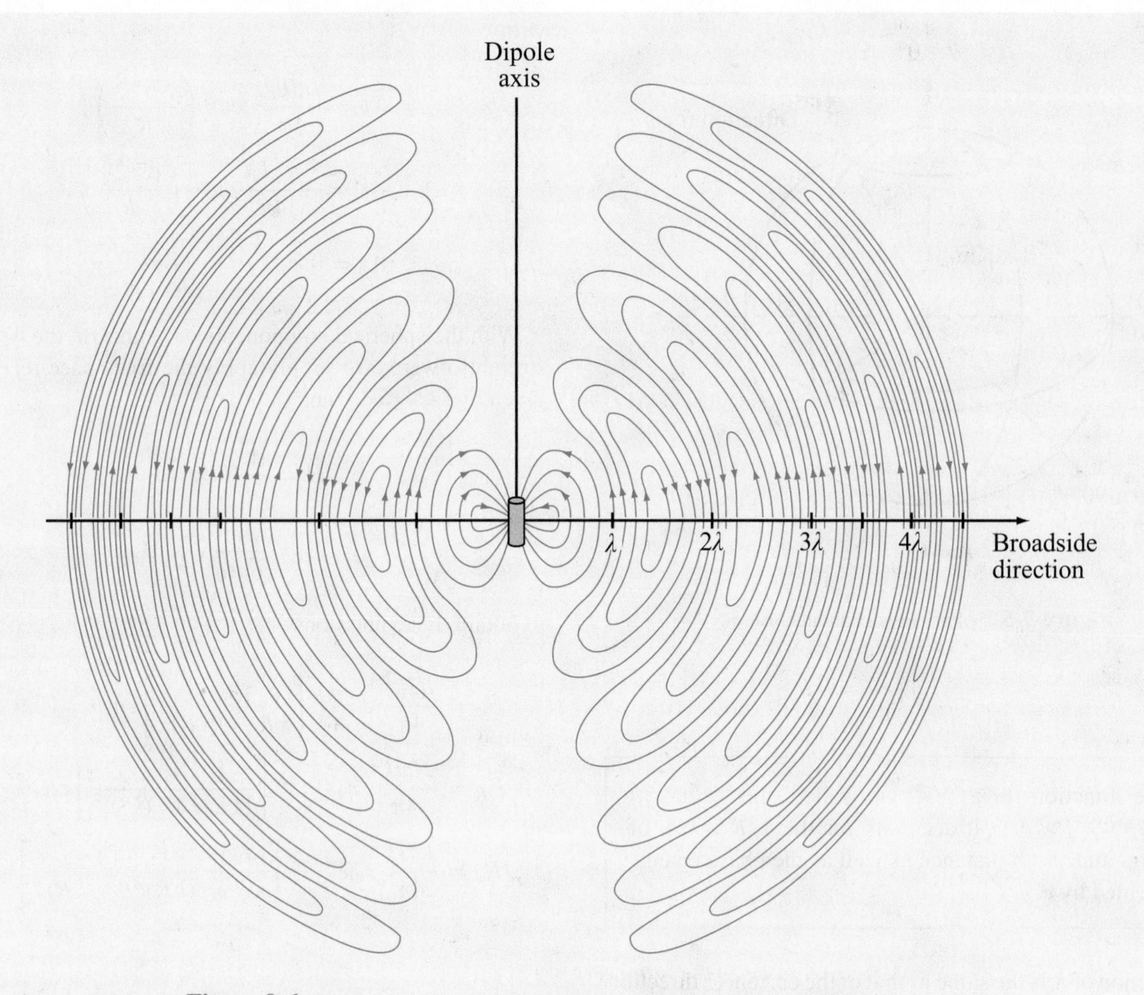

Figure 9-6 Electric field lines surrounding an oscillating dipole at a given instant.

$$\widetilde{E}_\theta = \frac{jI_0 l k \eta_0}{4\pi}\left(\frac{e^{-jkR}}{R}\right)\sin\theta \quad \text{(V/m)}, \tag{9.9a}$$

$$\widetilde{H}_\phi = \frac{\widetilde{E}_\theta}{\eta_0} \quad \text{(A/m)}, \tag{9.9b}$$

and $\widetilde{E}_R$ is negligible. At the observation point Q (Fig. 9-4), the wave now appears similar to a uniform plane wave with its electric and magnetic fields in phase, related by the intrinsic impedance of the medium η_0, and their directions orthogonal to each other and to the direction of propagation ($\hat{\mathbf{R}}$). Both fields are proportional to $\sin\theta$ and independent of ϕ (which is expected from symmetry considerations).

9-1.2 Power Density

Given $\widetilde{\mathbf{E}}$ and $\widetilde{\mathbf{H}}$, the ***time-average Poynting vector*** of the radiated wave, which is also called the ***power density***, can be obtained by applying Eq. (7.100); that is,

$$\mathbf{S}_{\text{av}} = \tfrac{1}{2}\mathfrak{Re}\left(\widetilde{\mathbf{E}} \times \widetilde{\mathbf{H}}^*\right) \quad \text{(W/m}^2\text{)}. \tag{9.10}$$

For the short dipole, use of Eqs. (9.9a) and (9.9b) yields

$$\mathbf{S}_{av} = \hat{\mathbf{R}}\, S(R, \theta), \tag{9.11}$$

with

$$S(R, \theta) = \left(\frac{\eta_0 k^2 I_0^2 l^2}{32\pi^2 R^2} \right) \sin^2\theta$$

$$= S_0 \sin^2\theta \qquad (\text{W/m}^2). \tag{9.12}$$

The directional pattern of any antenna is described in terms of the ***normalized radiation intensity*** $F(\theta, \phi)$, defined as the ratio of the power density $S(R, \theta, \phi)$ at a specified range R to S_{max}, the maximum value of $S(R, \theta, \phi)$ at the same range,

$$F(\theta, \phi) = \frac{S(R, \theta, \phi)}{S_{max}} \qquad \text{(dimensionless)}. \tag{9.13}$$

For the Hertzian dipole, the $\sin^2\theta$ dependence in Eq. (9.12) indicates that the radiation is maximum in the broadside direction ($\theta = \pi/2$), corresponding to the azimuth plane, and is given by

$$S_{max} = S_0 = \frac{\eta_0 k^2 I_0^2 l^2}{32\pi^2 R^2}$$

$$= \frac{15\pi I_0^2}{R^2} \left(\frac{l}{\lambda} \right)^2 \qquad (\text{W/m}^2), \tag{9.14}$$

where use was made of the relations $k = 2\pi/\lambda$ and $\eta_0 \approx 120\pi$. We observe that S_{max} is directly proportional to I_0^2 and l^2 (with l measured in wavelengths), and that it decreases with distance as $1/R^2$.

From the definition of the normalized radiation intensity given by Eq. (9.13), it follows that

$$F(\theta, \phi) = F(\theta) = \sin^2\theta. \tag{9.15}$$

Plots of $F(\theta)$ are shown in **Fig. 9-7** in both the elevation plane (the θ plane) and the azimuth plane (ϕ plane).

► No energy is radiated by the short dipole along the direction of the dipole axis, and maximum radiation ($F = 1$) occurs in the ***broadside direction*** ($\theta = 90°$). Since $F(\theta)$ is independent of ϕ, the pattern is doughnut-shaped in θ–ϕ space. ◄

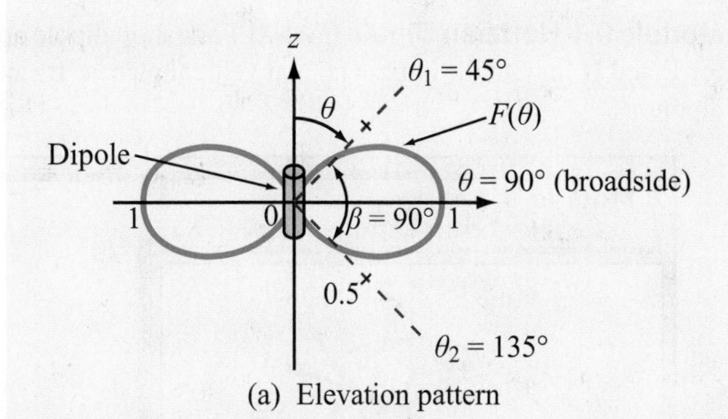

(a) Elevation pattern

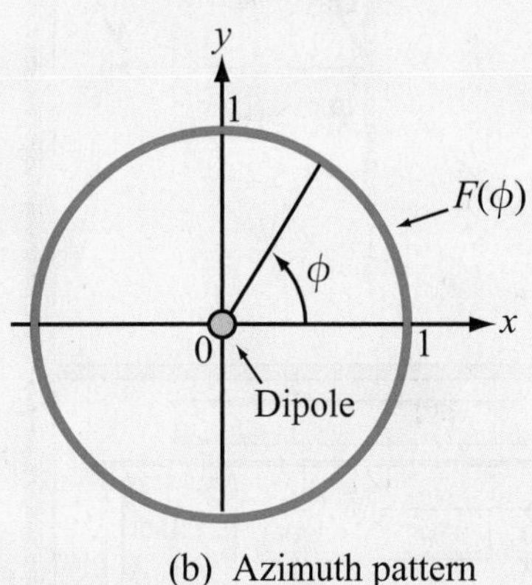

(b) Azimuth pattern

Figure 9-7 Radiation patterns of a short dipole.

Concept Question 9-1: What does it mean to say that most antennas are reciprocal devices?

Concept Question 9-2: What is the radiated wave like in the far-field region of the antenna?

Concept Question 9-3: In a Hertzian dipole, what is the underlying assumption about the current flowing through the wire?

Concept Question 9-4: Outline the basic steps used to relate the current in a wire to the radiated power density.

Module 9.1 Hertzian Dipole ($l \ll \lambda$) For a short dipole oriented along the z axis, this module displays the field distributions for **E** and **H** in both the horizontal and vertical planes. It can also animate the radiation process and current flow through the dipole.

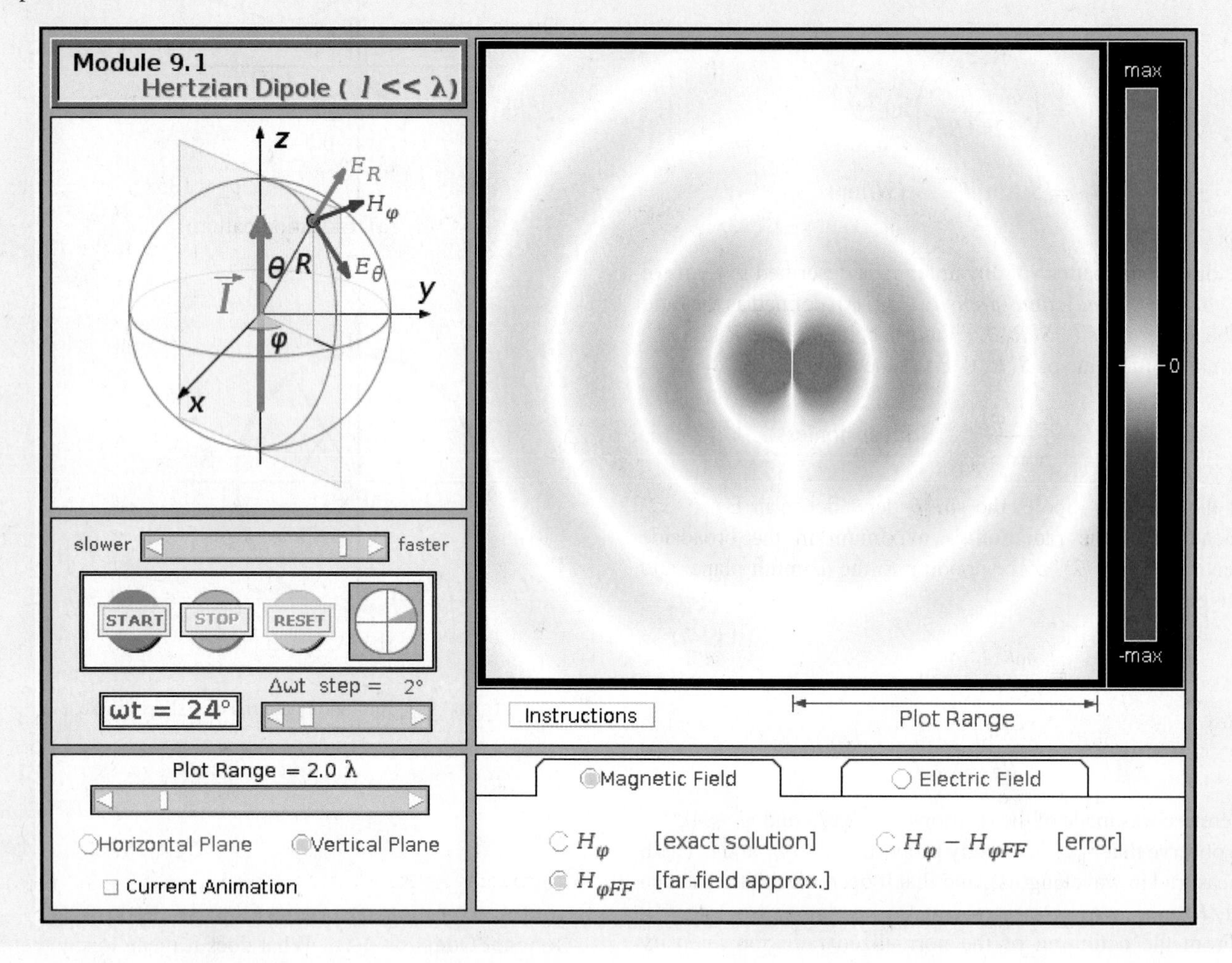

Exercise 9-1: A 1 m long dipole is excited by a 5 MHz current with an amplitude of 5 A. At a distance of 2 km, what is the power density radiated by the antenna along its broadside direction?

Answer: $S_0 = 8.2 \times 10^{-8}$ W/m^2. (See EM.)

9-2 Antenna Radiation Characteristics

An ***antenna pattern*** describes the far-field directional properties of an antenna when measured at a fixed distance from the antenna. In general, the antenna pattern is a three-dimensional plot that displays the strength of the radiated field or power density as a function of direction, with direction being specified by the zenith angle θ and the azimuth angle ϕ.

▶ By virtue of reciprocity, a receiving antenna has the same directional antenna pattern as the pattern that it exhibits when operated in the transmission mode. ◀

Consider a transmitting antenna placed at the origin of the

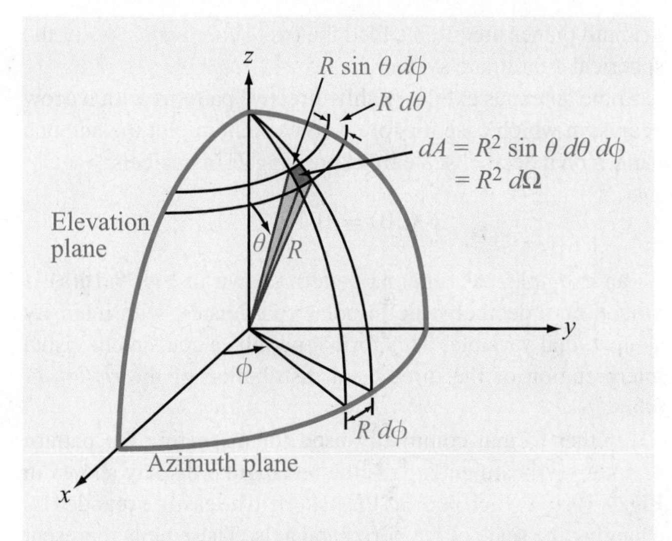

Figure 9-8 Definition of solid angle $d\Omega = \sin\theta \; d\theta \; d\phi$.

observation sphere shown in **Fig. 9-8**. The differential power radiated by the antenna through an elemental area dA is

$$dP_{\text{rad}} = \mathbf{S}_{\text{av}} \cdot d\mathbf{A} = \mathbf{S}_{\text{av}} \cdot \hat{\mathbf{R}} \, dA = S \, dA \qquad \text{(W)}, \qquad (9.16)$$

where S is the radial component of the time-average Poynting vector $\mathbf{S}_{\text{av}}$. In the far-field region of any antenna, $\mathbf{S}_{\text{av}}$ is always in the radial direction. In a spherical coordinate system,

$$dA = R^2 \sin\theta \; d\theta \; d\phi, \qquad (9.17)$$

and the ***solid angle*** $d\Omega$ associated with dA, defined as the subtended area divided by R^2, is given by

$$d\Omega = \frac{dA}{R^2} = \sin\theta \; d\theta \; d\phi \qquad \text{(sr)}. \qquad (9.18)$$

Note that, whereas a planar angle is measured in radians and the angular measure of a complete circle is 2π (rad), a solid angle is measured in ***steradians*** (sr), and the angular measure for a spherical surface is $\Omega = (4\pi R^2)/R^2 = 4\pi$ (sr). The solid angle of a hemisphere is 2π (sr).

Using the relation $dA = R^2 \, d\Omega$, dP_{rad} can be rewritten as

$$dP_{\text{rad}} = R^2 \, S(R, \theta, \phi) \, d\Omega. \qquad (9.19)$$

The total power radiated by an antenna through a spherical surface at a fixed distance R is obtained by integrating Eq. (9.19) over that surface:

$$\begin{aligned} P_{\text{rad}} &= R^2 \int_{\phi=0}^{2\pi} \int_{\theta=0}^{\pi} S(R, \theta, \phi) \sin\theta \; d\theta \; d\phi \\ &= R^2 S_{\max} \int_{\phi=0}^{2\pi} \int_{\theta=0}^{\pi} F(\theta, \phi) \sin\theta \; d\theta \; d\phi \\ &= R^2 S_{\max} \iint_{4\pi} F(\theta, \phi) \, d\Omega \qquad \text{(W)}, \qquad (9.20) \end{aligned}$$

where $F(\theta, \phi)$ is the normalized radiation intensity defined by Eq. (9.13). The 4π symbol under the integral sign is used as an abbreviation for the indicated limits on θ and ϕ. Formally, P_{rad} is called the ***total radiated power***.

9-2.1 Antenna Pattern

Each specific combination of the zenith angle θ and the azimuth angle ϕ denotes a specific direction in the spherical coordinate system of **Fig. 9-8**. The normalized radiation intensity $F(\theta, \phi)$

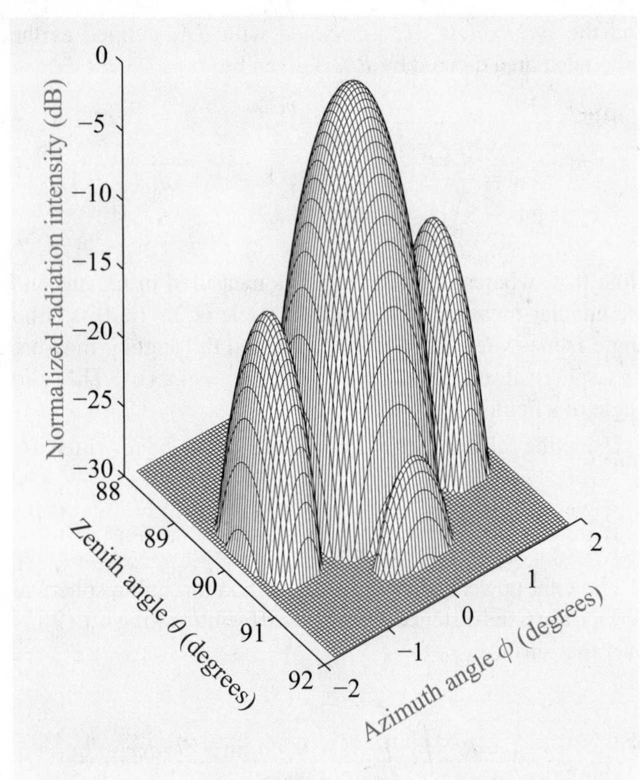

Figure 9-9 Three-dimensional pattern of a narrow-beam antenna.

characterizes the directional pattern of the energy radiated by an antenna, and a plot of $F(\theta, \phi)$ as a function of both θ and ϕ constitutes a three-dimensional pattern, an example of which is shown in **Fig. 9-9**.

Often, it is of interest to characterize the variation of $F(\theta, \phi)$ in the form of two-dimensional plots in specific planes in the spherical coordinate system. The two planes most commonly specified for this purpose are the elevation and azimuth planes. The ***elevation plane***, also called the θ plane, is a plane corresponding to a constant value of ϕ. For example, $\phi = 0$ defines the x–z plane and $\phi = 90°$ defines the y–z plane, both of which are elevation planes (**Fig. 9-8**). A plot of $F(\theta, \phi)$ versus θ in either of these planes constitutes a two-dimensional pattern in the elevation plane. This is not to imply, however, that the elevation-plane pattern is necessarily the same in all elevation planes.

The ***azimuth plane***, also called the ϕ plane, is specified by $\theta = 90°$ and corresponds to the x–y plane. The elevation and azimuth planes are often called the two ***principal planes*** of the spherical coordinate system.

Some antennas exhibit highly directive patterns with narrow beams, in which case it is often convenient to plot the antenna pattern on a decibel scale by expressing F in decibels:

$$F\ (\text{dB}) = 10 \log F.$$

As an example, the antenna pattern shown in **Fig. 9-10(a)** is plotted on a decibel scale in polar coordinates, with intensity as the radial variable. This format permits a convenient visual interpretation of the directional distribution of the ***radiation lobes***.

Another format commonly used for inspecting the pattern of a narrow-beam antenna is the rectangular display shown in **Fig. 9-10(b)**, which permits the pattern to be easily expanded by changing the scale of the horizontal axis. These plots represent the variation in only one plane in the observation sphere, the $\phi = 0$ plane. Unless the pattern is symmetrical in ϕ, additional patterns are required to define the overall variation of $F(\theta, \phi)$ with θ and ϕ.

Strictly speaking, the polar angle θ is always positive, being defined over the range from 0° (z direction) to 180° ($-z$ direction), and yet the θ axis in **Fig. 9-10(b)** is shown to have both positive and negative values. This is not a contradiction, but rather a different form of plotting antenna patterns. The right-hand half of the plot represents the variation of F (dB) with θ as θ is increased in a clockwise direction in the x–z plane [see inset in **Fig. 9-10(b)**], corresponding to $\phi = 0$, whereas the left-hand half of the plot represents the variation of F (dB) with θ as θ is increased in a counterclockwise direction at $\phi = 180°$. Thus, a negative θ value simply denotes that the direction (θ, ϕ) is in the left-hand half of the x–z plane.

The pattern shown in **Fig. 9-10(a)** indicates that the antenna is fairly directive, since most of the energy is radiated through a narrow sector called the ***main lobe***. In addition to the main lobe, the pattern exhibits several ***side lobes*** and ***back lobes*** as well. For most applications, these extra lobes are considered undesirable because they represent wasted energy for transmitting antennas and potential interference directions for receiving antennas.

9-2.2 Beam Dimensions

For an antenna with a single main lobe, the ***pattern solid angle*** Ω_p describes the equivalent width of the main lobe of the

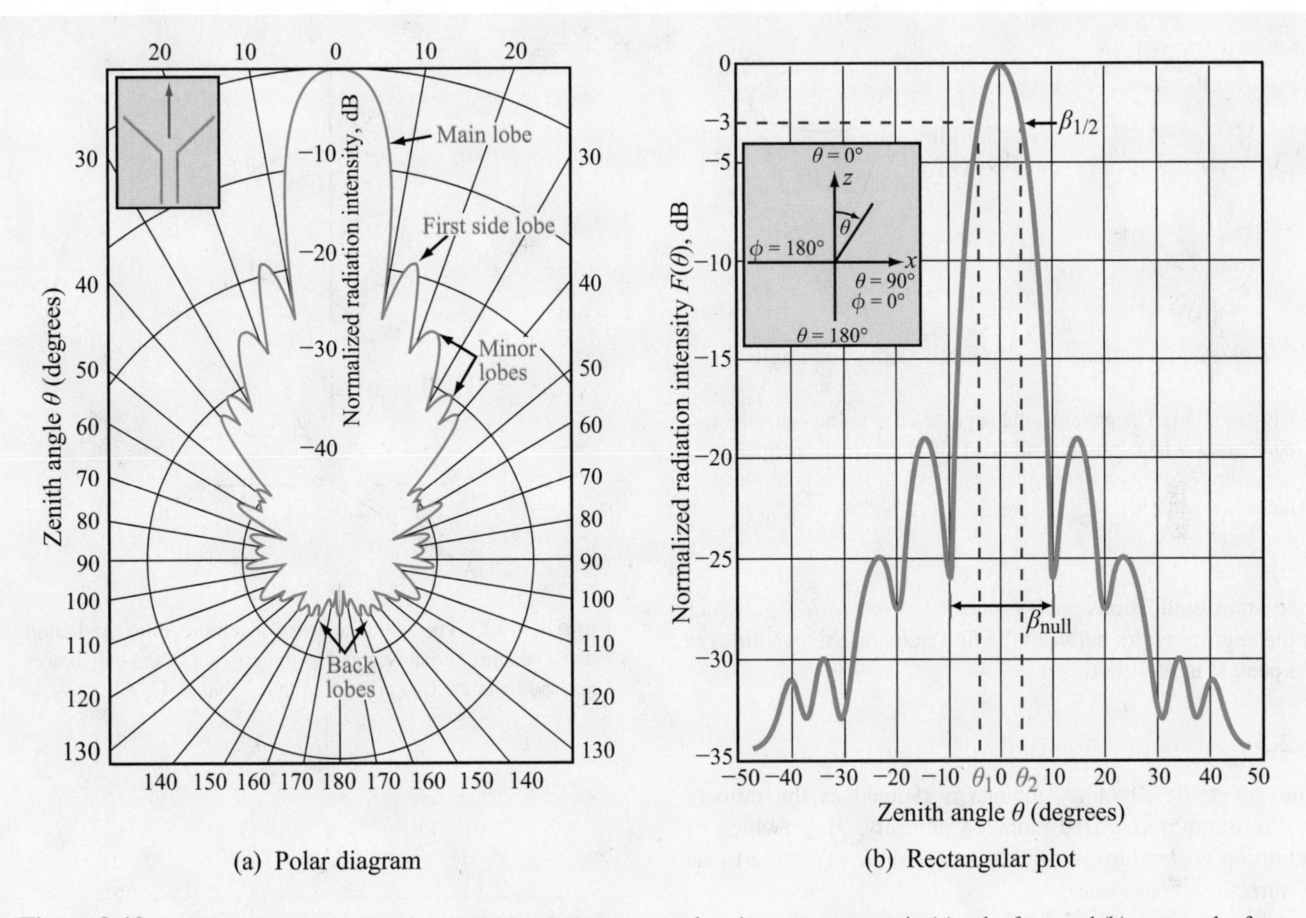

Figure 9-10 Representative plots of the normalized radiation pattern of a microwave antenna in (a) polar form and (b) rectangular form.

antenna pattern (**Fig. 9-11**). It is defined as the integral of the normalized radiation intensity $F(\theta, \phi)$ over a sphere:

$$\Omega_p = \iint_{4\pi} F(\theta, \phi)\, d\Omega \qquad \text{(sr)}. \qquad (9.21)$$

► For an isotropic antenna with $F(\theta, \phi) = 1$ in all directions, $\Omega_p = 4\pi$ (sr). ◄

The pattern solid angle characterizes the directional properties of the three-dimensional radiation pattern. To characterize the width of the main lobe in a given plane, the term used is ***beamwidth***. The ***half-power beamwidth***, or simply the beamwidth β, is defined as the angular width of the main lobe between the two angles at which the magnitude of $F(\theta, \phi)$ is equal to half of its peak value (or -3 dB on a decibel scale). For example, for the pattern displayed in **Fig. 9-10(b)**, β is given by

$$\beta = \theta_2 - \theta_1, \qquad (9.22)$$

where θ_1 and θ_2 are the ***half-power angles*** at which $F(\theta, 0) = 0.5$ (with θ_2 denoting the larger value and θ_1 denoting the smaller one, as shown in the figure). If the pattern is symmetrical and the peak value of $F(\theta, \phi)$ is at $\theta = 0$, then $\beta = 2\theta_2$. For the short-dipole pattern shown earlier in **Fig. 9-7(a)**, $F(\theta)$ is maximum at $\theta = 90°$, θ_2 is at 135°, and θ_1 is at 45°. Hence, $\beta = 135° - 45° = 90°$. The beamwidth β is also known as the ***3 dB beamwidth***. In addition to the half-power beamwidth, other beam dimensions may be of interest

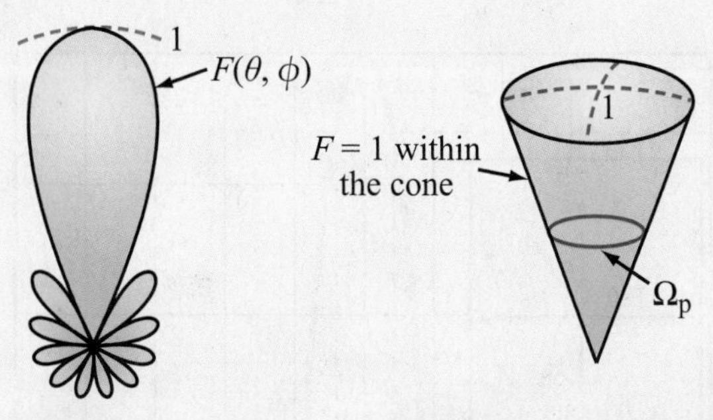

Figure 9-11 The pattern solid angle Ω_p defines an equivalent cone over which all the radiation of the actual antenna is concentrated with uniform intensity equal to the maximum of the actual pattern.

for certain applications, such as the ***null beamwidth*** β_{null}, which is the angular width between the first nulls on the two sides of the peak [Fig. 9-10(b)].

9-2.3 Antenna Directivity

The ***directivity*** D of an antenna is defined as the ratio of its maximum normalized radiation intensity, F_{max} (which by definition is equal to 1), to the average value of $F(\theta, \phi)$ over all directions (4π space):

$$D = \frac{F_{max}}{F_{av}} = \frac{1}{\dfrac{1}{4\pi} \displaystyle\iint_{4\pi} F(\theta, \phi)\, d\Omega} = \frac{4\pi}{\Omega_p} \quad \text{(dimensionless)}. \tag{9.23}$$

Here Ω_p is the pattern solid angle defined by Eq. (9.21). Thus, the narrower Ω_p of an antenna pattern is, the greater is the directivity. For an isotropic antenna, $\Omega_p = 4\pi$; hence, its directivity $D_{iso} = 1$.

By using Eq. (9.20) in Eq. (9.23), D can be expressed as

$$D = \frac{4\pi R^2 S_{max}}{P_{rad}} = \frac{S_{max}}{S_{av}}, \tag{9.24}$$

where $S_{av} = P_{rad}/(4\pi R^2)$ is the average value of the radiated power density and is equal to the total power radiated by the antenna, P_{rad}, divided by the surface area of a sphere of radius R.

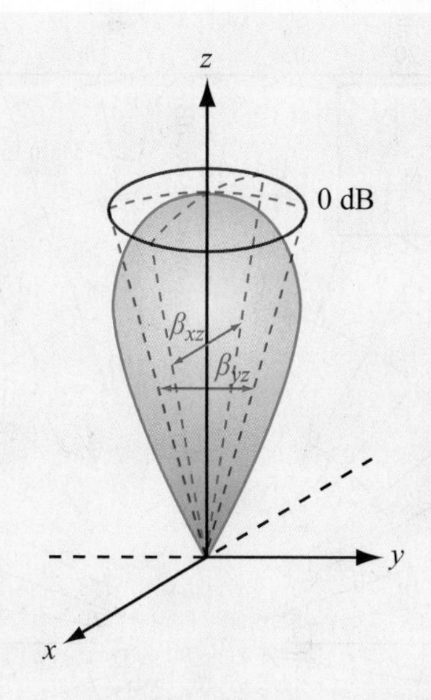

Figure 9-12 The solid angle of a unidirectional radiation pattern is approximately equal to the product of the half-power beamwidths in the two principal planes; that is, $\Omega_p \approx \beta_{xz}\beta_{yz}$.

► Since $S_{av} = S_{iso}$, where S_{iso} is the power density radiated by an isotropic antenna, D represents the ratio of the maximum power density radiated by the antenna to the power density radiated by an isotropic antenna, both measured at the same range R and excited by the same amount of input power. ◄

Usually, D is expressed in decibels:† D (dB) $= 10 \log D$.

For an antenna with a single main lobe pointing in the z direction as shown in Fig. 9-12, Ω_p may be approximated as the product of the half-power beamwidths β_{xz} and β_{yz} (in radians):

$$\Omega_p \approx \beta_{xz}\beta_{yz}, \tag{9.25}$$

†A note of caution: Even though we often express certain dimensionless quantities in decibels, we should always convert their decibel values to natural values before using them in the relations given in this chapter.

and therefore

$$D = \frac{4\pi}{\Omega_p} \approx \frac{4\pi}{\beta_{xz}\beta_{yz}} \quad \textbf{(single main lobe).} \qquad (9.26)$$

Although approximate, this relation provides a useful method for estimating the antenna directivity from measurements of the beamwidths in the two orthogonal planes whose intersection is the axis of the main lobe.

Example 9-1: Antenna Radiation Properties

Determine (a) the direction of maximum radiation, (b) pattern solid angle, (c) directivity, and (d) half-power beamwidth in the y–z plane for an antenna that radiates only into the upper hemisphere with normalized radiation intensity given by $F(\theta, \phi) = \cos^2\theta$.

Solution: The statement that the antenna radiates through only the upper hemisphere is equivalent to

$$F(\theta,\phi) = F(\theta) = \begin{cases} \cos^2\theta & \text{for } 0 \le \theta \le \pi/2 \\ & \text{and } 0 \le \phi \le 2\pi, \\ 0 & \text{elsewhere.} \end{cases}$$

(a) The function $F(\theta) = \cos^2\theta$ is independent of ϕ and is maximum when $\theta = 0°$. A polar plot of $F(\theta)$ is shown in Fig. 9-13.

(b) From Eq. (9.21), the pattern solid angle Ω_p is given by

$$\Omega_p = \iint_{4\pi} F(\theta,\phi)\, d\Omega$$

$$= \int_{\phi=0}^{2\pi} \left[\int_{\theta=0}^{\pi/2} \cos^2\theta \sin\theta \, d\theta \right] d\phi$$

$$= \int_{\phi=0}^{2\pi} \left[-\frac{\cos^3\theta}{3} \right]_0^{\pi/2} d\phi$$

$$= \int_0^{2\pi} \frac{1}{3}\, d\phi = \frac{2\pi}{3} \qquad \text{(sr)}.$$

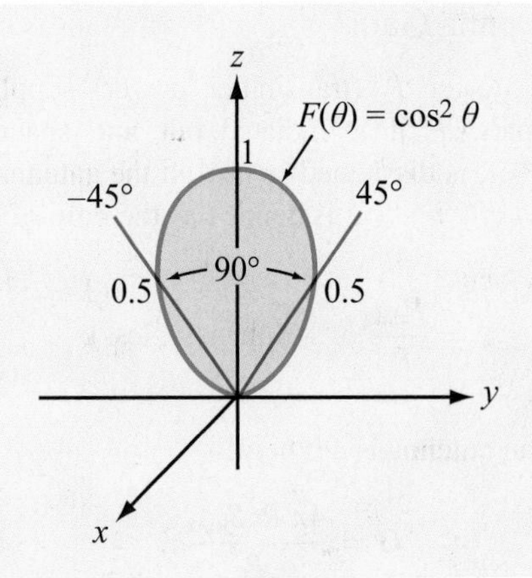

Figure 9-13 Polar plot of $F(\theta) = \cos^2\theta$.

(c) Application of Eq. (9.23) gives

$$D = \frac{4\pi}{\Omega_p} = 4\pi\left(\frac{3}{2\pi}\right) = 6,$$

which corresponds to D (dB) $= 10 \log 6 = 7.78$ dB.

(d) The half-power beamwidth β is obtained by setting $F(\theta) = 0.5$. That is,

$$F(\theta) = \cos^2\theta = 0.5,$$

which gives the half-power angles $\theta_1 = -45°$ and $\theta_2 = 45°$. Hence,

$$\beta = \theta_2 - \theta_1 = 90°.$$

Example 9-2: Directivity of a Hertzian Dipole

Calculate the directivity of a Hertzian dipole.

Solution: Application of Eq. (9.23) with $F(\theta) = \sin^2\theta$ [from Eq. (9.15)] gives

$$D = \frac{4\pi}{\iint_{4\pi} F(\theta,\phi) \sin\theta \, d\theta \, d\phi}$$

$$= \frac{4\pi}{\int_{\phi=0}^{2\pi}\int_{\theta=0}^{\pi} \sin^3\theta \, d\theta \, d\phi} = \frac{4\pi}{8\pi/3} = 1.5$$

or, equivalently, 1.76 dB.

9-2.4 Antenna Gain

Of the total power P_t (transmitter power) supplied to the antenna, a part, P_{rad}, is radiated out into space, and the remainder, P_{loss}, is dissipated as heat in the antenna structure. The ***radiation efficiency*** ξ is defined as the ratio of P_{rad} to P_t:

$$\xi = \frac{P_{rad}}{P_t} \quad \text{(dimensionless)}. \qquad (9.27)$$

The ***gain*** of an antenna is defined as

$$G = \frac{4\pi R^2 S_{max}}{P_t}, \qquad (9.28)$$

which is similar in form to the expression given by Eq. (9.24) for the directivity D except that it is referenced to the input power supplied to the antenna, P_t, rather than to the radiated power P_{rad}. In view of Eq. (9.27),

$$G = \xi D \quad \text{(dimensionless)}. \qquad (9.29)$$

► The gain accounts for ohmic losses in the antenna material, whereas the directivity does not. For a lossless antenna, $\xi = 1$, and $G = D$. ◄

9-2.5 Radiation Resistance

To a transmission line connected between a generator supplying power P_t on one end and an antenna on the other end, the antenna is merely a load with ***input impedance*** Z_{in}. If the line is lossless and properly matched to the antenna, all of P_t is transferred to the antenna. In general, Z_{in} consists of a resistive component R_{in} and a reactive component X_{in}:

$$Z_{in} = R_{in} + jX_{in}. \qquad (9.30)$$

The resistive component is defined as equivalent to a resistor R_{in} that would consume an average power P_t when the amplitude of the ac current flowing through it is I_0,

$$P_t = \tfrac{1}{2} I_0^2 R_{in}. \qquad (9.31)$$

Since $P_t = P_{rad} + P_{loss}$, it follows that R_{in} can be defined as the sum of a ***radiation resistance*** R_{rad} and a ***loss resistance*** R_{loss},

$$R_{in} = R_{rad} + R_{loss}, \qquad (9.32)$$

with

$$P_{rad} = \tfrac{1}{2} I_0^2 R_{rad}, \qquad (9.33a)$$

$$P_{loss} = \tfrac{1}{2} I_0^2 R_{loss}, \qquad (9.33b)$$

where I_0 is the amplitude of the sinusoidal current exciting the antenna. As defined earlier, the ***radiation efficiency*** is the ratio of P_{rad} to P_t, or

$$\xi = \frac{P_{rad}}{P_t} = \frac{P_{rad}}{P_{rad} + P_{loss}} = \frac{R_{rad}}{R_{rad} + R_{loss}}. \qquad (9.34)$$

The radiation resistance R_{rad} can be calculated by integrating the far-field power density over a sphere to obtain P_{rad} and then equating the result to Eq. (9.33a).

Example 9-3: Radiation Resistance and Efficiency of a Hertzian Dipole

A 4 cm long center-fed dipole is used as an antenna at 75 MHz. The antenna wire is made of copper and has a radius $a = 0.4$ mm. From Eqs. (7.92a) and (7.94), the loss resistance of a circular wire of length l is given by

$$R_{loss} = \frac{l}{2\pi a}\sqrt{\frac{\pi f \mu_c}{\sigma_c}}, \qquad (9.35)$$

where μ_c and σ_c are the magnetic permeability and conductivity of the wire, respectively. Calculate the radiation resistance and the radiation efficiency of the dipole antenna.

Solution: At 75 MHz,

$$\lambda = \frac{c}{f} = \frac{3 \times 10^8}{7.5 \times 10^7} = 4 \text{ m}.$$

The length to wavelength ratio is $l/\lambda = 4$ cm/4 m $= 10^{-2}$. Hence, this is a short dipole. From Eq. (9.24),

$$P_{rad} = \frac{4\pi R^2}{D} S_{max}. \qquad (9.36)$$

For the Hertzian dipole, S_{max} is given by Eq. (9.14), and from Example 9-2 we established that $D = 1.5$. Hence,

$$P_{rad} = \frac{4\pi R^2}{1.5} \times \frac{15\pi I_0^2}{R^2}\left(\frac{l}{\lambda}\right)^2 = 40\pi^2 I_0^2 \left(\frac{l}{\lambda}\right)^2. \qquad (9.37)$$

Equating this result to Eq. (9.33a) and then solving for the radiation resistance R_{rad} leads to

$$R_{rad} = 80\pi^2(l/\lambda)^2 \quad (\Omega) \quad \textbf{(short dipole)}. \tag{9.38}$$

For $l/\lambda = 10^{-2}$, $R_{rad} = 0.08\ \Omega$.

Next, we determine the loss resistance R_{loss}. For copper, Appendix B gives $\mu_c \approx \mu_0 = 4\pi \times 10^{-7}$ H/m and $\sigma_c = 5.8 \times 10^7$ S/m. Hence,

$$\begin{aligned} R_{loss} &= \frac{l}{2\pi a}\sqrt{\frac{\pi f \mu_c}{\sigma_c}} \\ &= \frac{4 \times 10^{-2}}{2\pi \times 4 \times 10^{-4}}\left(\frac{\pi \times 75 \times 10^6 \times 4\pi \times 10^{-7}}{5.8 \times 10^7}\right)^{1/2} \\ &= 0.036\ \Omega. \end{aligned}$$

Therefore, the radiation efficiency is

$$\xi = \frac{R_{rad}}{R_{rad} + R_{loss}} = \frac{0.08}{0.08 + 0.036} = 0.69.$$

Thus, the dipole is 69% efficient.

Concept Question 9-5: What does the pattern solid angle represent?

Concept Question 9-6: What is the magnitude of the directivity of an isotropic antenna?

Concept Question 9-7: What physical and material properties affect the radiation efficiency of a fixed-length Hertzian dipole antenna?

Exercise 9-2: An antenna has a conical radiation pattern with a normalized radiation intensity $F(\theta) = 1$ for θ between 0° and 45° and zero for θ between 45° and 180°. The pattern is independent of the azimuth angle ϕ. Find (a) the pattern solid angle and (b) the directivity.

Answer: (a) $\Omega_p = 1.84$ sr, (b) $D = 6.83$ or, equivalently, 8.3 dB. (See EM.)

Exercise 9-3: The maximum power density radiated by a short dipole at a distance of 1 km is 60 (nW/m^2). If $I_0 = 10$ A, find the radiation resistance.

Answer: $R_{rad} = 10$ mΩ. (See EM.)

9-3 Half-Wave Dipole Antenna

In Section 9-1 we developed expressions for the electric and magnetic fields radiated by a Hertzian dipole of length $l \ll \lambda$. We now use these expressions as building blocks to obtain expressions for the fields radiated by a half-wave dipole antenna, so named because its length $l = \lambda/2$. As shown in **Fig. 9-14**, the half-wave dipole consists of a thin wire fed at its center by a generator connected to the antenna terminals via a transmission line. The current flowing through the wire has a symmetrical distribution with respect to the center of the dipole, and the current is zero at its ends. Mathematically, $i(t)$ is given by

$$i(t) = I_0 \cos\omega t \cos kz = \mathfrak{Re}\left[I_0 \cos kz\, e^{j\omega t}\right], \tag{9.39a}$$

whose phasor is

$$\widetilde{I}(z) = I_0 \cos kz, \qquad -\lambda/4 \le z \le \lambda/4\,, \tag{9.39b}$$

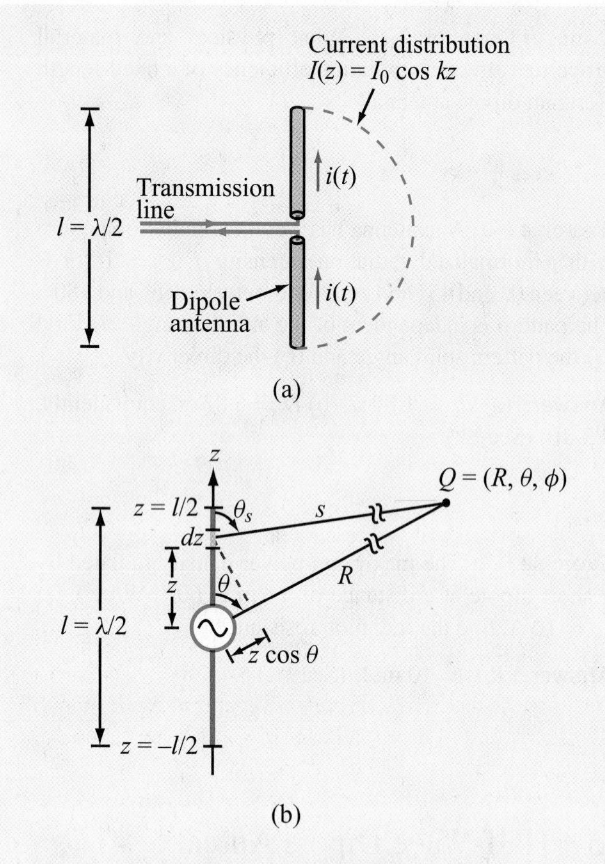

Figure 9-14 Center-fed half-wave dipole.

and $k = 2\pi/\lambda$. Equation (9.9a) gives an expression for $\widetilde{E}_\theta$, the far field radiated by a Hertzian dipole of length l when excited by a current I_0. Let us adapt that expression to an infinitesimal dipole segment of length dz, excited by a current $\widetilde{I}(z)$ and located at a distance s from the observation point Q [**Fig. 9-14(b)**]. Thus,

$$d\widetilde{E}_\theta(z) = \frac{jk\eta_0}{4\pi}\,\widetilde{I}(z)\,dz\left(\frac{e^{-jks}}{s}\right)\sin\theta_s, \tag{9.40a}$$

and the associated magnetic field is

$$d\widetilde{H}_\phi(z) = \frac{d\widetilde{E}_\theta(z)}{\eta_0}. \tag{9.40b}$$

The far field due to radiation by the entire antenna is obtained by integrating the fields from all of the Hertzian dipoles making up the antenna:

$$\widetilde{E}_\theta = \int_{z=-\lambda/4}^{\lambda/4} d\widetilde{E}_\theta. \tag{9.41}$$

Before we calculate this integral, we make the following two approximations. The first relates to the magnitude part of the spherical propagation factor, $1/s$. In **Fig. 9-14(b)**, the distance s between the current element and the observation point Q is considered so large in comparison with the length of the dipole that the difference between s and R may be neglected in terms of its effect on $1/s$. Hence, we may set $1/s \approx 1/R$, and by the same argument we set $\theta_s \approx \theta$. The error Δ between s and R is a maximum when the observation point is along the z axis and it is equal to $\lambda/4$ (corresponding to half of the antenna length). If $R \gg \lambda$, this error will have an insignificant effect on $1/s$. The second approximation is associated with the phase factor e^{-jks}. An error in distance Δ corresponds to an error in phase $k\Delta = (2\pi/\lambda)(\lambda/4) = \pi/2$. As a rule of thumb, a phase error greater than $\pi/8$ is considered unacceptable because it may lead to a significant error in the computed value of the field $\widetilde{E}_\theta$. Hence, the approximation $s \approx R$ is too crude for the phase factor and cannot be used. A more tolerable option is to use the ***parallel-ray approximation*** given by

$$s \approx R - z\cos\theta, \tag{9.42}$$

as illustrated in **Fig. 9-14(b)**.

Substituting Eq. (9.42) for s in the phase factor of Eq. (9.40a) and replacing s with R and θ_s with θ elsewhere in the expression, we obtain

$$d\widetilde{E}_\theta = \frac{jk\eta_0}{4\pi}\,\widetilde{I}(z)\,dz\left(\frac{e^{-jkR}}{R}\right)\sin\theta\, e^{jkz\cos\theta}. \tag{9.43}$$

After (1) inserting Eq. (9.43) into Eq. (9.41), (2) using the expression for $\widetilde{I}(z)$ given by Eq. (9.39b), and (3) carrying out the integration, the following expressions are obtained:

$$\widetilde{E}_\theta = j\,60I_0\left\{\frac{\cos[(\pi/2)\cos\theta]}{\sin\theta}\right\}\left(\frac{e^{-jkR}}{R}\right), \tag{9.44a}$$

$$\widetilde{H}_\phi = \frac{\widetilde{E}_\theta}{\eta_0}. \tag{9.44b}$$

The corresponding time-average power density is

$$S(R,\theta) = \frac{|\widetilde{E}_\theta|^2}{2\eta_0} = \frac{15I_0^2}{\pi R^2}\left\{\frac{\cos^2[(\pi/2)\cos\theta]}{\sin^2\theta}\right\} = S_0\left\{\frac{\cos^2[(\pi/2)\cos\theta]}{\sin^2\theta}\right\} \quad (\text{W/m}^2). \tag{9.45}$$

Examination of Eq. (9.45) reveals that $S(R,\theta)$ is maximum at $\theta = \pi/2$, and its value is

$$S_{\max} = S_0 = \frac{15I_0^2}{\pi R^2}.$$

Hence, the normalized radiation intensity is

$$F(\theta) = \frac{S(R,\theta)}{S_0} = \left\{\frac{\cos[(\pi/2)\cos\theta]}{\sin\theta}\right\}^2. \tag{9.46}$$

The radiation pattern of the half-wave dipole exhibits roughly the same doughnut-like shape shown earlier in **Fig. 9-7** for the short dipole. *Its directivity is slightly larger (1.64 compared with 1.5 for the short dipole), but its radiation resistance is 73 Ω (as shown later in Section 9-3.2), which is orders of magnitude larger than that of a short dipole.*

9-3.1 Directivity of $\lambda/2$ Dipole

To evaluate both the directivity D and the radiation resistance R_{rad} of the half-wave dipole, we first need to calculate the total radiated power P_{rad} by applying Eq. (9.20):

$$P_{\text{rad}} = R^2 \iint_{4\pi} S(R,\theta)\,d\Omega = \frac{15I_0^2}{\pi}\int_0^{2\pi}\int_0^{\pi}\left\{\frac{\cos[(\pi/2)\cos\theta]}{\sin\theta}\right\}^2 \sin\theta\,d\theta\,d\phi. \tag{9.47}$$

The integration over ϕ is equal to 2π, and numerical evaluation of the integration over θ gives the value 1.22. Consequently,

$$P_{\text{rad}} = 36.6\,I_0^2 \quad (\text{W}). \tag{9.48}$$

From Eq. (9.45), we found that $S_{\max} = 15I_0^2/(\pi R^2)$. Using this in Eq. (9.24) gives the following result for the directivity D of the half-wave dipole:

$$D = \frac{4\pi R^2 S_{\max}}{P_{\text{rad}}} = \frac{4\pi R^2}{36.6I_0^2}\left(\frac{15I_0^2}{\pi R^2}\right) = 1.64 \tag{9.49}$$

or, equivalently, 2.15 dB.

9-3.2 Radiation Resistance of $\lambda/2$ Dipole

From Eq. (9.33a),

$$R_{\text{rad}} = \frac{2P_{\text{rad}}}{I_0^2} = \frac{2\times 36.6I_0^2}{I_0^2} \approx 73\ \Omega. \tag{9.50}$$

As was noted earlier in Example 9-3, because the radiation resistance of a Hertzian dipole is comparable in magnitude to that of its loss resistance R_{loss}, its radiation efficiency ξ is rather small. For the 4 cm long dipole of Example 9-3, $R_{\text{rad}} = 0.08\ \Omega$ (at 75 MHz) and $R_{\text{loss}} = 0.036\ \Omega$. If we keep the frequency the same and increase the length of the dipole to 2 m ($\lambda = 4$ m at $f = 75$ MHz), R_{rad} becomes 73 Ω and R_{loss} increases to 1.8 Ω. The radiation efficiency increases from 69% for the short dipole to 98% for the half-wave dipole. More significant is the fact that it is practically impossible to match a transmission line to an antenna with a resistance on the order of 0.1 Ω, while it is quite easy to do so when $R_{\text{rad}} = 73\ \Omega$.

Moreover, since $R_{\text{loss}} \ll R_{\text{rad}}$ for the half-wave dipole, $R_{\text{in}} \approx R_{\text{rad}}$ and Eq. (9.30) becomes

$$Z_{\text{in}} \approx R_{\text{rad}} + jX_{\text{in}}. \tag{9.51}$$

Deriving an expression for X_{in} for the half-wave dipole is fairly complicated and beyond the scope of this book. However, it is significant to note that X_{in} is a strong function of l/λ, and that it decreases from 42 Ω at $l/\lambda = 0.5$ to zero at $l/\lambda = 0.48$, whereas R_{rad} remains approximately unchanged. Hence, by reducing the length of the half-wave dipole by 4%, Z_{in} becomes purely real and equal to 73 Ω, thereby making it possible to match the dipole to a 75 Ω transmission line without resorting to the use of a matching network.

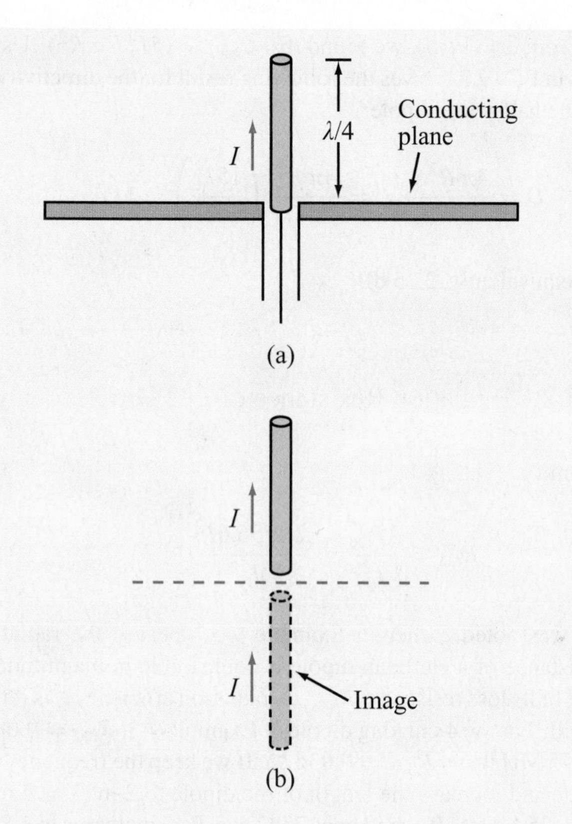

Figure 9-15 A quarter-wave monopole above a conducting plane is equivalent to a full half-wave dipole in free space.

9-3.3 Quarter-Wave Monopole Antenna

▶ When placed over a conducting ground plane, a quarter-wave monopole antenna excited by a source at its base [Fig. 9-15(a)] exhibits the same radiation pattern in the region above the ground plane as a half-wave dipole in free space. ◀

This is because, from image theory (Section 4-11), the conducting plane can be replaced with the image of the $\lambda/4$ monopole, as illustrated in **Fig. 9-15(b)**. Thus, the $\lambda/4$ monopole radiates an electric field identical to that given by Eq. (9.44a), and its normalized radiation intensity is given by Eq. (9.46); but the radiation is limited to the upper half-space defined by $0 \le \theta \le \pi/2$. Hence, a monopole radiates only half as much power as the dipole. Consequently, for a $\lambda/4$ monopole, $P_{\text{rad}} = 18.3 I_0^2$ and its radiation resistance is $R_{\text{rad}} = 36.5\ \Omega$.

The approach used with the quarter-wave monopole is also valid for any vertical wire antenna placed above a conducting plane, including a Hertzian monopole.

Concept Question 9-8: What is the physical length of a half-wave dipole operating at (a) 1 MHz (in the AM broadcast band), (b) 100 MHz (FM broadcast band), and (c) 10 GHz (microwave band)?

Concept Question 9-9: How does the radiation pattern of a half-wave dipole compare with that of a Hertzian dipole? How do their directivities, radiation resistances, and radiation efficiencies compare?

Concept Question 9-10: How does the radiation efficiency of a quarter-wave monopole compare with that of a half-wave dipole, assuming that both are made of the same material and have the same cross section?

Exercise 9-4: For the half-wave dipole antenna, evaluate $F(\theta)$ versus θ to determine the half-power beamwidth in the elevation plane (the plane containing the dipole axis).

Answer: $\beta = 78°$. (See EM.)

Exercise 9-5: If the maximum power density radiated by a half-wave dipole is 50 μW/m^2 at a range of 1 km, what is the current amplitude I_0?

Answer: $I_0 = 3.24$ A. (See EM.)

9-4 Dipole of Arbitrary Length

So far, we examined the radiation properties of the Hertzian and half-wave dipoles. We now consider the more general case of a linear dipole of arbitrary length l, relative to λ. For a center-fed dipole, as depicted in **Fig. 9-16**, the currents flowing through its two halves are symmetrical and must go to zero at

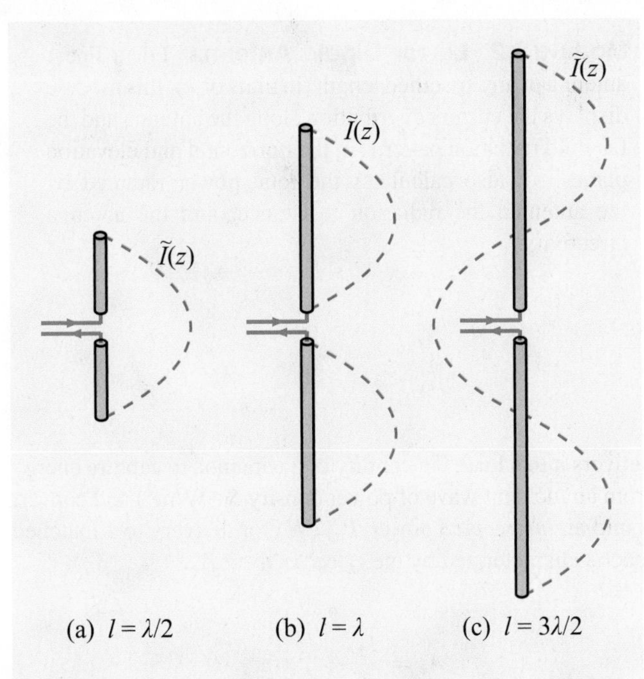

Figure 9-16 Current distribution for three center-fed dipoles.

its ends. Hence, the current phasor $\tilde{I}(z)$ can be expressed as a sine function with an argument that goes to zero at $z = \pm l/2$:

$$\tilde{I}(z) = \begin{cases} I_0 \sin\left[k\left(l/2 - z\right)\right], & \text{for } 0 \le z \le l/2, \\ I_0 \sin\left[k\left(l/2 + z\right)\right], & \text{for } -l/2 \le z < 0, \end{cases} \tag{9.52}$$

where I_0 is the current amplitude. The procedure for calculating the electric and magnetic fields and the associated power density of the wave radiated by such an antenna is basically the same as that used previously in connection with the half-wave dipole antenna. The only difference is the current distribution $\tilde{I}(z)$. If we insert the expression for $\tilde{I}(z)$ given by Eq. (9.52) into Eq. (9.43), we obtain the following expression for the differential electric field $d\widetilde{E}_\theta$ of the wave radiated by an elemental length dz at location z along the dipole:

$$d\widetilde{E}_\theta = \frac{jk\eta_0 I_0}{4\pi}\left(\frac{e^{-jkR}}{R}\right)\sin\theta\; e^{jkz\cos\theta}\; dz \times \begin{cases} \sin\left[k\left(l/2 - z\right)\right] & \text{for } 0 \le z \le l/2, \\ \sin\left[k\left(l/2 + z\right)\right] & \text{for } -l/2 \le z < 0. \end{cases} \tag{9.53}$$

The total field radiated by the dipole is

$$\begin{aligned} \widetilde{E}_\theta &= \int_{-l/2}^{l/2} d\widetilde{E}_\theta \\ &= \int_0^{l/2} d\widetilde{E}_\theta + \int_{-l/2}^{0} d\widetilde{E}_\theta \\ &= \frac{jk\eta_0 I_0}{4\pi}\left(\frac{e^{-jkR}}{R}\right)\sin\theta \\ &\quad \times \left\{ \int_0^{l/2} e^{jkz\cos\theta} \sin[k(l/2 - z)]\; dz \right. \\ &\quad \left. + \int_{-l/2}^{0} e^{jkz\cos\theta} \sin[k(l/2 + z)]\; dz \right\}. \end{aligned} \tag{9.54}$$

If we apply Euler's identity to express $e^{jkz\cos\theta}$ as $[\cos(kz\cos\theta) + j\sin(kz\cos\theta)]$, we can integrate the two integrals and obtain the result

$$\widetilde{E}_\theta = j60 I_0\left(\frac{e^{-jkR}}{R}\right) \cdot \left[\frac{\cos\left(\frac{kl}{2}\cos\theta\right) - \cos\left(\frac{kl}{2}\right)}{\sin\theta}\right]. \tag{9.55}$$

The corresponding time-average power density radiated by the dipole antenna is given by

$$\begin{aligned} S(\theta) &= \frac{|\widetilde{E}_\theta|^2}{2\eta_0} \\ &= \frac{15 I_0^2}{\pi R^2}\left[\frac{\cos\left(\frac{\pi l}{\lambda}\cos\theta\right) - \cos\left(\frac{\pi l}{\lambda}\right)}{\sin\theta}\right]^2, \end{aligned} \tag{9.56}$$

where we have used the relations $\eta_0 \approx 120\pi$ (Ω) and $k = 2\pi/\lambda$. For $l = \lambda/2$, Eq. (9.56) reduces to the expression given by Eq. (9.45) for the half-wave dipole. Plots of the normalized radiation intensity, $F(\theta) = S(R,\theta)/S_{\text{max}}$, are shown in **Fig. 9-17** for dipoles of lengths $\lambda/2$, λ, and $3\lambda/2$. The dipoles with $l = \lambda/2$ and $l = \lambda$ have similar radiation patterns with maxima along $\theta = 90°$, but the half-power beamwidth of the wavelength-long dipole is narrower than that of the half-wave dipole, and $S_{\text{max}} = 60 I_0^2/(\pi R^2)$ for the wavelength-long

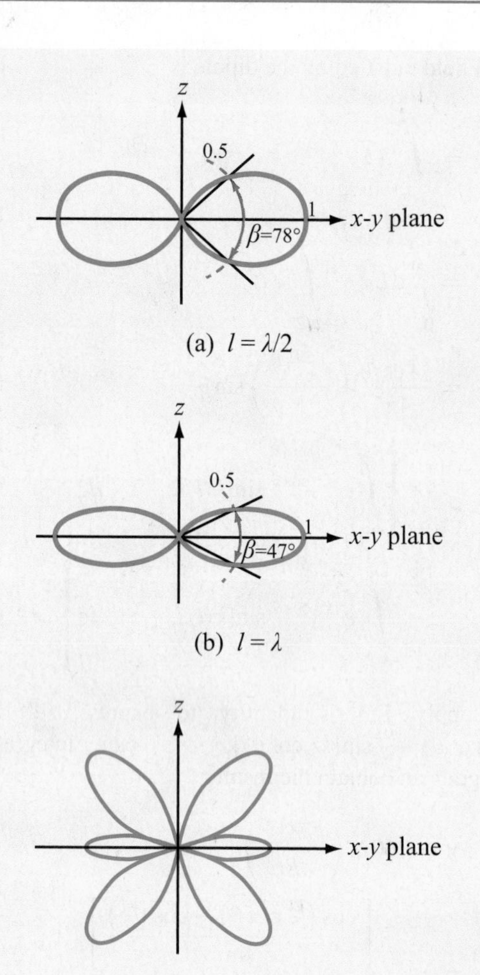

Figure 9-17 Radiation patterns of dipoles with lengths of $\lambda/2$, λ, and $3\lambda/2$.

dipole, which is four times that for the half-wave dipole. The pattern of the dipole with length $l = 3\lambda/2$ exhibits a structure with multiple lobes, and its direction of maximum radiation is not along $\theta = 90°$.

9-5 Effective Area of a Receiving Antenna

So far, antennas have been treated as directional radiators of energy. Now, we examine the reverse process, namely how a receiving antenna extracts energy from an incident wave and delivers it to a load. The ability of an antenna to capture energy from an incident wave of power density S_i (W/m^2) and convert it into an ***intercepted power*** P_int (W) for delivery to a matched load is characterized by the ***effective area*** A_e:

$$A_\mathrm{e} = \frac{P_\mathrm{int}}{S_\mathrm{i}} \qquad (\mathrm{m}^2). \tag{9.57}$$

Other commonly used names for A_e include ***effective aperture*** and ***receiving cross section***. The antenna receiving process may be modeled in terms of a Thévenin equivalent circuit (Fig. 9-18) consisting of a voltage $\widetilde{V}_\mathrm{oc}$ in series with the antenna input impedance Z_in. Here, $\widetilde{V}_\mathrm{oc}$ is the open-circuit voltage induced by the incident wave at the antenna terminals and Z_L is the impedance of the load connected to the antenna (representing a receiver or some other circuit). In general, both Z_in and Z_L are complex:

$$Z_\mathrm{in} = R_\mathrm{rad} + jX_\mathrm{in}, \tag{9.58a}$$

$$Z_\mathrm{L} = R_\mathrm{L} + jX_\mathrm{L}, \tag{9.58b}$$

where R_rad denotes the radiation resistance of the antenna (assuming $R_\mathrm{loss} \ll R_\mathrm{rad}$). To maximize power transfer to the load, the load impedance must be chosen such that $Z_\mathrm{L} = Z_\mathrm{in}^*$, or $R_\mathrm{L} = R_\mathrm{rad}$ and $X_\mathrm{L} = -X_\mathrm{in}$. In that case, the circuit reduces to a source $\widetilde{V}_\mathrm{oc}$ connected across a resistance equal to $2R_\mathrm{rad}$. Since $\widetilde{V}_\mathrm{oc}$ is a sinusoidal voltage phasor, the time-average power delivered to the load is

$$P_\mathrm{L} = \frac{1}{2}|\widetilde{I}_\mathrm{L}|^2 R_\mathrm{rad} = \frac{1}{2}\left[\frac{|\widetilde{V}_\mathrm{oc}|}{2R_\mathrm{rad}}\right]^2 R_\mathrm{rad} = \frac{|\widetilde{V}_\mathrm{oc}|^2}{8R_\mathrm{rad}}, \tag{9.59}$$

Module 9.2 Linear Dipole Antenna For a linear antenna of any specified length (in units of λ), this module displays the current distribution along the antenna and the far-field radiation patterns in the horizontal and elevation planes. It also calculates the total power radiated by the antenna, the radiation resistance, and the antenna directivity.

Module 9.3 Detailed Analysis of Linear Antenna This module complements Module 9.2 by offering extensive information about the specified linear antenna, including its directivity and plots of its current and field distributions.

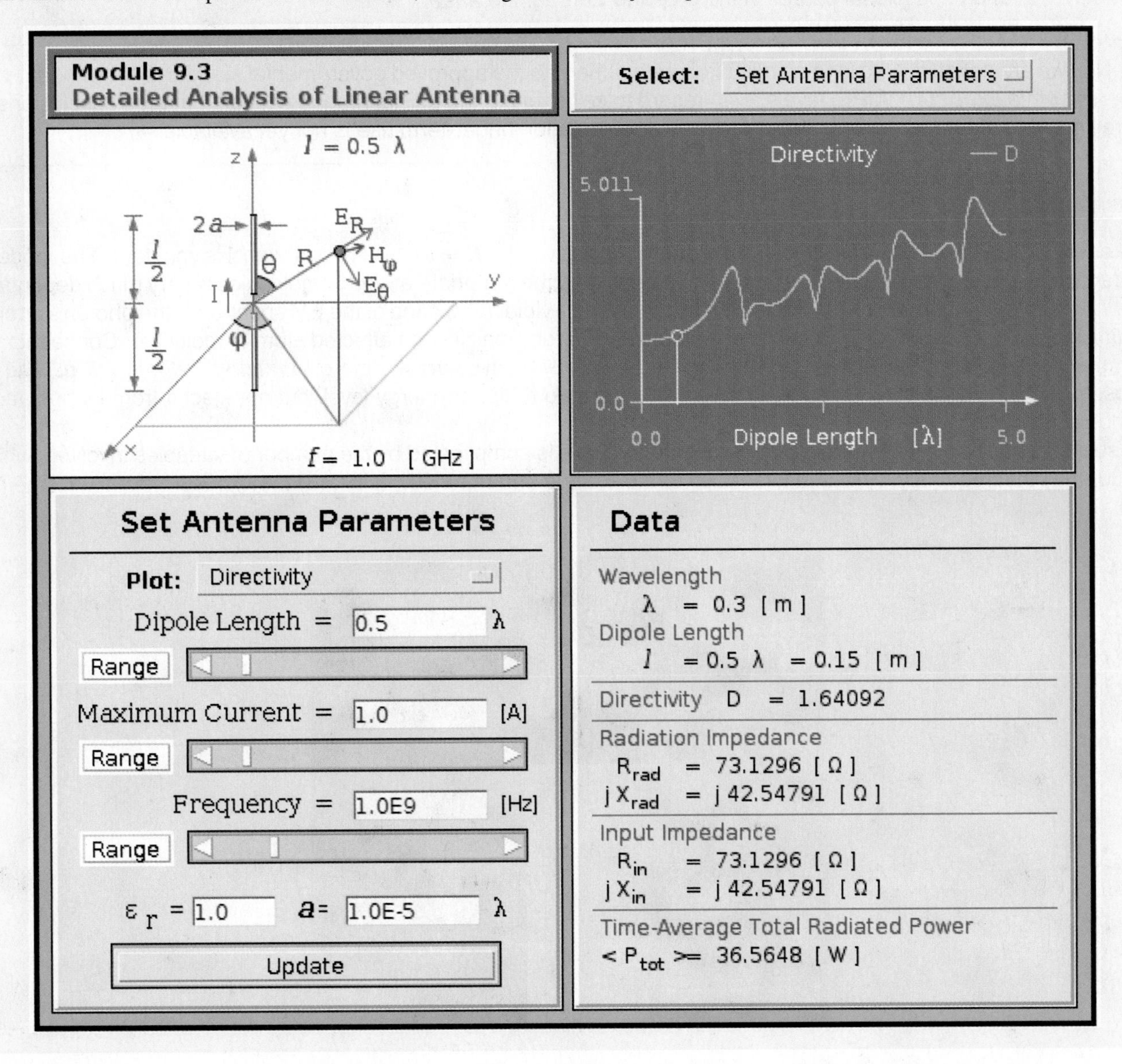

Technology Brief 17: Health Risks of EM Fields

Can the use of cell phones cause cancer? Does exposure to the electromagnetic fields (EMFs) associated with power lines pose health risks to humans? Are we endangered by EMFs generated by home appliances, telephones, electrical wiring, and the myriad of electronic gadgets we use every day (**Fig. TF17-1**)? Despite reports in some of the popular media alleging a causative relationship between low-level EMFs and many diseases, according to reports issued by governmental and professional boards in the U.S. and Europe, the answer is:

► NO, we are not at risk, so long as manufacturers adhere to the approved governmental standards for ***maximum permissible exposure*** (MPE) levels. With regard to cell phones, the official reports caution that their conclusions are limited to phone use of less than 15 years, since data for longer-term use is not yet available. ◄

Physiological Effects of EMFs

The energy carried by a photon with an EM frequency f is given by $E = hf$, where h is Planck's constant. The mode of interaction between a photon passing through a material and the material's atoms or molecules is very much dependent on f. If f is greater than about 10^{15} Hz (which falls in the ultraviolet (UV) band of the EM spectrum), the photon's energy is sufficient to free an electron and remove it completely, thereby ionizing the affected atom or molecule. Consequently, the energy carried by such EM waves is called ***ionizing radiation***, in contrast with ***non-ionizing radiation*** (**Fig. TF17-2**) whose photons may be able to cause an electron to move to a higher energy level, but not eject it from its host atom or molecule.

Assessing health risks associated with exposure to EMFs is complicated by the number of variables involved, which include: (1) the frequency f, (2) the intensities of the electric and magnetic fields, (3) the exposure duration, whether

Figure TF17-1 Electromagnetic fields are emitted by power lines, cell phones, TV towers, and many other electronic circuits and devices.

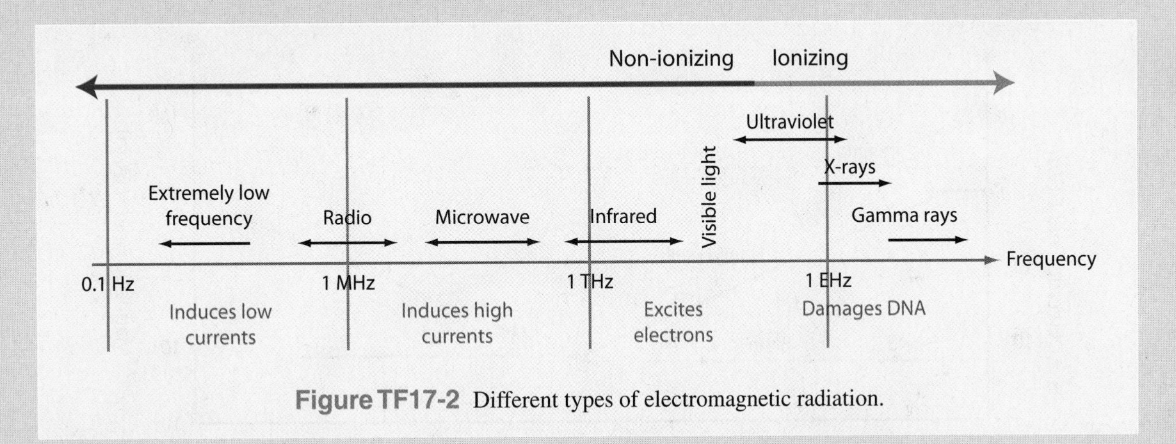

Figure TF17-2 Different types of electromagnetic radiation.

continuous or discontinuous, and whether pulsed or uniform, and (4) the specific part of the body that is getting exposed. We know that intense laser illumination can cause corneal burn, high-level X-rays can damage living tissue and cause cancer and, in fact, any form of EM energy can be dangerous if the exposure level and/or duration exceed certain safety limits. Governmental and professional safety boards are tasked with establishing maximum permissible exposure (MPE) levels that protect human beings against adverse health effects associated with EMFs. In the U.S., the relevant standards are IEEE Std C95.6 (dated 2002), which addresses EM fields in the 1 Hz to 3 kHz range, and IEEE Std 95.1 (dated 2005), which deals with the frequency range from 3 kHz to 300 GHz. On the European side of the Atlantic, responsibility for establishing MPE levels resides with the Scientific Committee on Emerging and Newly Identified Health Risks (SCENIHR) of the European Commission.

▶ At frequencies below 100 kHz, the goal is to minimize adverse effects of exposure to electric fields that can cause ***electrostimulation*** of nerve and muscle cells. Above 5 MHz, the main concern is excessive tissue heating, and in the transition region of 100 kHz to 5 MHz, safety standards are designed to protect against both electrostimulation and excessive heating. ◀

Frequency Range $0 \leq f \leq 3$ kHz: The plots in **Fig. TF17-3** display the values of MPE for electric and magnetic fields over the frequency range below 3 kHz. According to IEEE Std C95.6, it is sufficient to demonstrate compliance with the MPE levels for either the electric field E or the magnetic field H. According to the plot for H, exposure at 60 Hz should not exceed 720 A/m. The magnetic field due to power lines is typically in the range of 2–6 A/m underneath the lines, which is at least two orders of magnitude smaller than the established safe level for H.

Frequency Range 3 kHz $\leq f \leq 300$ GHz: At frequencies below 500 MHz, MPE is specified in terms of the electric and magnetic field strengths of the EM energy (**Fig. TF17-4**). From 100 MHz to 300 GHz (and beyond), MPE is specified in terms of the product of E and H, namely the power density S. Cell phones operate in the 1–2 GHz band; the specified MPE is 1 W/m^2, or equivalently 0.1 mW/cm^2.

Bottom Line

We are constantly bombarded by EM energy, from solar illumination to blackbody radiation emitted by all matter. Our bodies absorb, reflect, and emit EM energy all the time. Living organisms, including humans, require exposure to EM radiation to survive, but excessive exposure can cause adverse effects. The term *excessive exposure* connotes a complicated set of relationships among such variables as field strength, exposure duration and mode (continuous,

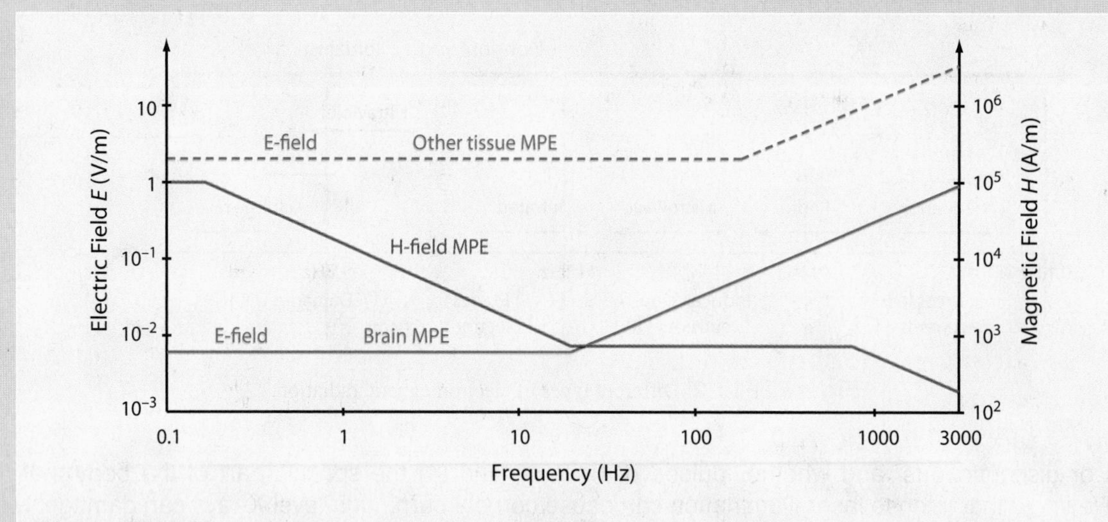

Figure TF17-3 Maximum permissible exposure (MPE) levels for E and H over the frequency range from 0.1 Hz to 3 kHz.

pulsed, etc.), body part, etc. The emission standards established by the Federal Communications Commission in the U.S. and similar governmental bodies in other countries are based on a combination of epidemiological studies, experimental observations, and theoretical understanding of how EM energy interacts with biological material. Generally speaking, the maximum permissible exposure levels specified by these standards are typically two orders of magnitude lower than the levels known to cause adverse effects, but in view of the multiplicity of variables involved, there is no guarantee that adhering to the standards will avoid health risks absolutely. The bottom line is: use common sense!

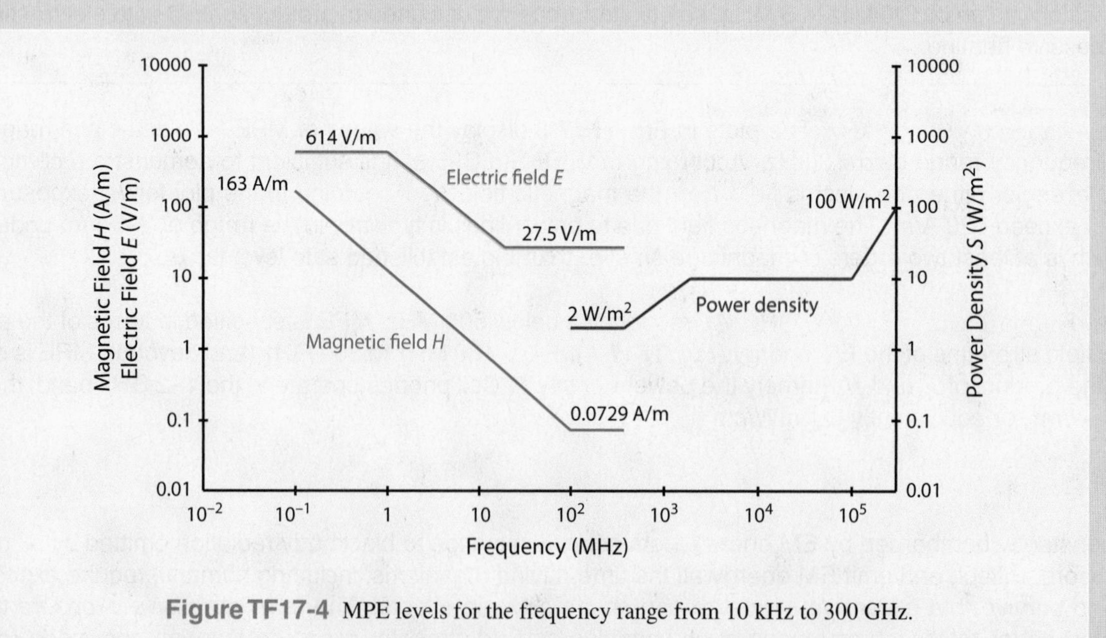

Figure TF17-4 MPE levels for the frequency range from 10 kHz to 300 GHz.

where $\widetilde{I}_L = \widetilde{V}_{oc}/(2R_{rad})$ is the phasor current flowing through the circuit. Since the antenna is lossless, all the intercepted power P_{int} ends up in the load resistance R_L. Hence,

$$P_{int} = P_L = \frac{|\widetilde{V}_{oc}|^2}{8R_{rad}}. \tag{9.60}$$

For an incident wave with electric field $\widetilde{E}_i$ parallel to the antenna polarization direction, the power density carried by the wave is

$$S_i = \frac{|\widetilde{E}_i|^2}{2\eta_0} = \frac{|\widetilde{E}_i|^2}{240\pi}. \tag{9.61}$$

The ratio of the results provided by Eqs. (9.60) and (9.61) gives

$$A_e = \frac{P_{int}}{S_i} = \frac{30\pi|\widetilde{V}_{oc}|^2}{R_{rad}|\widetilde{E}_i|^2}. \tag{9.62}$$

The open-circuit voltage $\widetilde{V}_{oc}$ induced in the receiving antenna is due to the incident field $\widetilde{E}_i$, but the relation between them depends on the specific antenna under consideration. By way of illustration, let us consider the case of the short-dipole antenna of Section 9-1. Because the length l of the short dipole is small compared with λ, the current induced by the incident field is uniform across its length, and the open-circuit voltage is simply $\widetilde{V}_{oc} = \widetilde{E}_i l$. Noting that $R_{rad} = 80\pi^2(l/\lambda)^2$ for the short dipole [see Eq. (9.38)] and using $\widetilde{V}_{oc} = \widetilde{E}_i l$, Eq. (9.62) simplifies to

$$A_e = \frac{3\lambda^2}{8\pi} \quad (\text{m}^2) \quad \textbf{(short dipole)}. \tag{9.63}$$

In Example 9-2 it was shown that for the Hertzian dipole the directivity $D = 1.5$. In terms of D, Eq. (9.63) can be rewritten in the form

$$A_e = \frac{\lambda^2 D}{4\pi} \quad (\text{m}^2) \quad \textbf{(any antenna)}. \tag{9.64}$$

▶ Despite the fact that the relation between A_e and D given by Eq. (9.64) was derived for a Hertzian dipole, it can be shown that it is also valid *for any antenna* under matched-impedance conditions. ◀

Exercise 9-6: The effective area of an antenna is 9 m^2. What is its directivity in decibels at 3 GHz?

Answer: $D = 40.53$ dB. (See EM.)

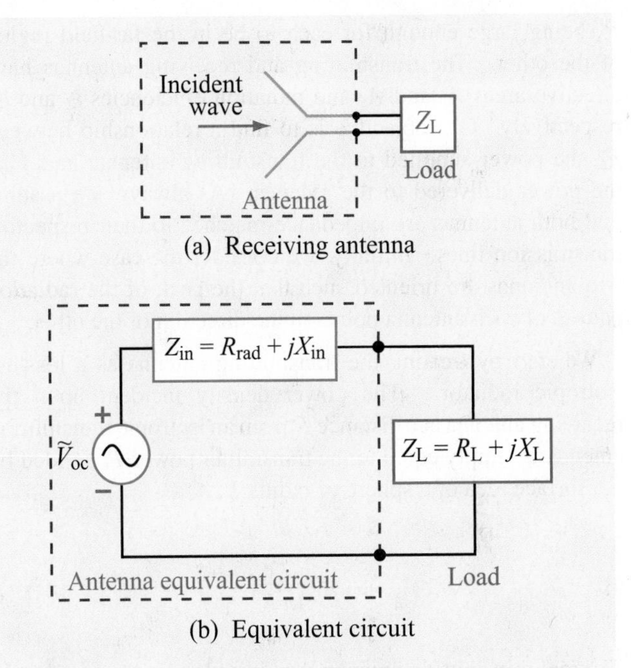

Figure 9-18 Receiving antenna represented by an equivalent circuit.

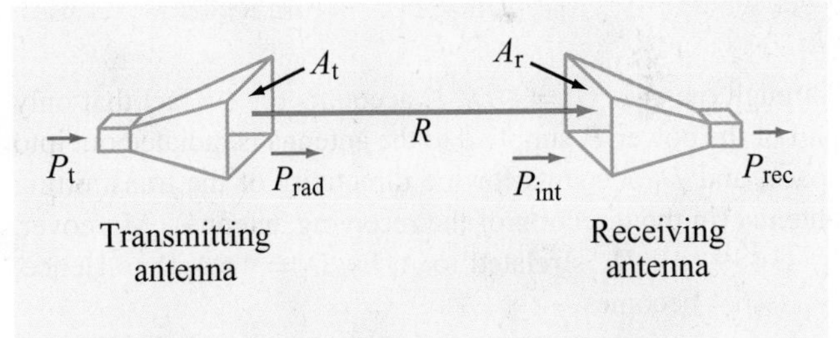

Figure 9-19 Transmitter–receiver configuration.

Exercise 9-7: At 100 MHz, the pattern solid angle of an antenna is 1.3 sr. Find (a) the antenna directivity D and (b) its effective area A_e.

Answer: (a) $D = 9.67$, (b) $A_e = 6.92$ m^2. (See EM.)

9-6 Friis Transmission Formula

The two antennas shown in **Fig. 9-19** are part of a free-space communication link, with the separation between the antennas,

R, being large enough for each to be in the far-field region of the other. The transmitting and receiving antennas have effective areas A_t and A_r and radiation efficiencies ξ_t and ξ_r, respectively. Our objective is to find a relationship between P_t, the power supplied to the transmitting antenna, and P_{rec}, the power delivered to the receiver. As always, we assume that both antennas are impedance-matched to their respective transmission lines. Initially, we consider the case where the two antennas are oriented such that the peak of the radiation pattern of each antenna points in the direction of the other.

We start by treating the transmitting antenna as a lossless isotropic radiator. The power density incident upon the receiving antenna at a distance R from an isotropic transmitting antenna is simply equal to the transmitter power P_t divided by the surface area of a sphere of radius R:

$$S_{iso} = \frac{P_t}{4\pi R^2} . \tag{9.65}$$

The real transmitting antenna is neither lossless nor isotropic. Hence, the power density S_r due to the real antenna is

$$S_r = G_t S_{iso} = \xi_t D_t S_{iso} = \frac{\xi_t D_t P_t}{4\pi R^2} . \tag{9.66}$$

Through the gain $G_t = \xi_t D_t$, ξ_t accounts for the fact that only part of the power P_t supplied to the antenna is radiated out into space, and D_t accounts for the directivity of the transmitting antenna (in the direction of the receiving antenna). Moreover, by Eq. (9.64), D_t is related to A_t by $D_t = 4\pi A_t/\lambda^2$. Hence, Eq. (9.66) becomes

$$S_r = \frac{\xi_t A_t P_t}{\lambda^2 R^2} . \tag{9.67}$$

On the receiving-antenna side, the power intercepted by the receiving antenna is equal to the product of the incident power density S_r and the effective area A_r:

$$P_{int} = S_r A_r = \frac{\xi_t A_t A_r P_t}{\lambda^2 R^2} . \tag{9.68}$$

The power delivered to the receiver, P_{rec}, is equal to the intercepted power P_{int} multiplied by the radiation efficiency of the receiving antenna, ξ_r. Hence, $P_{rec} = \xi_r P_{int}$, which leads to the result

$$\frac{P_{rec}}{P_t} = \frac{\xi_t \xi_r A_t A_r}{\lambda^2 R^2} = G_t G_r \left(\frac{\lambda}{4\pi R}\right)^2 . \tag{9.69}$$

► This relation is known as the ***Friis transmission formula***, and P_{rec}/P_t is called the ***power transfer ratio***. ◄

If the two antennas are not oriented in the direction of maximum power transfer, Eq. (9.69) assumes the general form

$$\frac{P_{rec}}{P_t} = G_t G_r \left(\frac{\lambda}{4\pi R}\right)^2 F_t(\theta_t, \phi_t)\, F_r(\theta_r, \phi_r), \tag{9.70}$$

where $F_t(\theta_t, \phi_t)$ is the normalized radiation intensity of the transmitting antenna at angles (θ_t, ϕ_t) corresponding to the direction of the receiving antenna (as seen by the antenna pattern of the transmitting antenna), and a similar definition applies to $F_r(\theta_r, \phi_r)$ for the receiving antenna.

Example 9-4: Satellite Communication System

A 6 GHz direct-broadcast TV satellite system transmits 100 W through a 2 m diameter parabolic dish antenna from a distance of approximately 40,000 km above Earth's surface. Each TV channel occupies a bandwidth of 5 MHz. Due to electromagnetic noise picked up by the antenna as well as noise generated by the receiver electronics, a home TV receiver has a noise level given by

$$P_n = K T_{sys} B \qquad \text{(W)}, \tag{9.71}$$

where T_{sys} [measured in kelvins (K)] is a figure of merit called the ***system noise temperature*** that characterizes the noise performance of the receiver–antenna combination, K is Boltzmann's constant [1.38×10^{-23} (J/K)], and B is the receiver bandwidth in Hz.

The ***signal-to-noise ratio*** S_n (which should not be confused with the power density S) is defined as the ratio of P_{rec} to P_n:

$$S_n = P_{rec}/P_n \qquad \text{(dimensionless)}. \tag{9.72}$$

For a receiver with $T_{sys} = 580$ K, what minimum diameter of a parabolic dish receiving antenna is required for high-quality TV reception with $S_n = 40$ dB? The satellite and ground receiving antennas may be assumed lossless, and their effective areas may be assumed equal to their physical apertures.

Solution: The following quantities are given:

$$P_t = 100 \text{ W}, \quad f = 6 \text{ GHz} = 6 \times 10^9 \text{ Hz}, \quad S_n = 10^4,$$

$$\text{Transmit antenna diameter } d_t = 2 \text{ m},$$

$$T_{sys} = 580 \text{ K}, \quad R = 40,000 \text{ km} = 4 \times 10^7 \text{ m},$$

$$B = 5 \text{ MHz} = 5 \times 10^6 \text{ Hz}.$$

The wavelength $\lambda = c/f = 5 \times 10^{-2}$ m, and the area of the transmitting satellite antenna is $A_t = (\pi d_t^2/4) = \pi$ (m^2). From Eq. (9.71), the receiver noise power is

$$P_n = KT_{sys}B = 1.38 \times 10^{-23} \times 580 \times 5 \times 10^6$$
$$= 4 \times 10^{-14} \text{ W}.$$

Using Eq. (9.69) with $\xi_t = \xi_r = 1$,

$$P_{rec} = \frac{P_t A_t A_r}{\lambda^2 R^2} = \frac{100\pi A_r}{(5 \times 10^{-2})^2 (4 \times 10^7)^2}$$
$$= 7.85 \times 10^{-11} A_r.$$

The area of the receiving antenna, A_r, can now be determined by equating the ratio P_{rec}/P_n to $S_n = 10^4$:

$$10^4 = \frac{7.85 \times 10^{-11} A_r}{4 \times 10^{-14}},$$

which yields the value $A_r = 5.1$ m^2. The required minimum diameter is $d_r = \sqrt{4A_r/\pi} = 2.55$ m.

Exercise 9-8: If the operating frequency of the communication system described in Example 9-4 is doubled to 12 GHz, what would then be the minimum required diameter of a home receiving TV antenna?

Answer: $d_r = 1.27$ m. (See EM.)

Exercise 9-9: A 3 GHz microwave link consists of two identical antennas each with a gain of 30 dB. Determine the received power, given that the transmitter output power is 1 kW and the two antennas are 10 km apart.

Answer: $P_{rec} = 6.33 \times 10^{-4}$ W. (See EM.)

Exercise 9-10: The effective area of a parabolic dish antenna is approximately equal to its physical aperture. If its directivity is 30 dB at 10 GHz, what is its effective area? If the frequency is increased to 30 GHz, what will be its new directivity?

Answer: $A_e = 0.07$ m^2, $D = 39.44$ dB. (See EM.)

9-7 Radiation by Large-Aperture Antennas

For wire antennas, the sources of radiation are the infinitesimal current elements comprising the current distribution along the wire, and the total radiated field at a given point in space is equal to the sum, or integral, of the fields radiated by all the elements. A parallel scenario applies to aperture antennas, except that now the source of radiation is the electric-field distribution across the aperture. Consider the horn antenna shown in Fig. 9-20. It is connected to a source through a coaxial transmission line, with the outer conductor of the line connected to the metal body of the horn and the inner conductor made to protrude, through a small hole, partially into the throat end of the horn. The protruding conductor acts as a monopole antenna, generating waves that radiate outwardly toward the horn's aperture. The electric field of the wave arriving at the aperture, which may

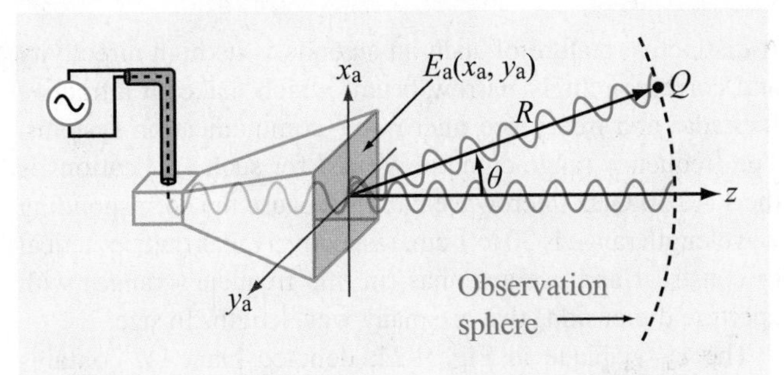

Figure 9-20 A horn antenna with aperture field distribution $E_a(x_a, y_a)$.

vary as a function of x_a and y_a over the horn's aperture, is called the ***electric-field aperture distribution*** or ***illumination***, $E_a(x_a, y_a)$. Inside the horn, wave propagation is guided by the horn's geometry; but as the wave transitions from a guided wave into an unbounded wave, every point of its wavefront serves as a source of spherical secondary wavelets. The aperture may then be represented as a distribution of isotropic radiators. At a distant point Q, the combination of all the waves arriving from all of these radiators constitutes the total wave that would be observed by a receiver placed at that point.

The radiation process described for the horn antenna is equally applicable to any aperture upon which an electromagnetic wave is incident. For example, if a light source is used to illuminate an opening in an opaque screen through a collimating lens, as shown in **Fig. 9-21(a)**, the opening becomes a source of secondary spherical wavelets, much like the aperture of the horn antenna. In the case of the parabolic reflector shown in **Fig. 9-21(b)**, it can be described in terms of an imaginary aperture representing the electric-field distribution across a plane in front of the reflector.

Two types of mathematical formulations are available for computing the electromagnetic fields of waves radiated by apertures. The first is a ***scalar formulation*** based on Kirchhoff's work and the second is a ***vector formulation*** based on Maxwell's equations. In this section, we limit our presentation to the scalar diffraction technique, in part because of its inherent simplicity and also because it is applicable across a wide range of practical applications.

► The key requirement for the validity of the scalar formulation is that the antenna aperture be at least several wavelengths long along each of its principal dimensions. ◄

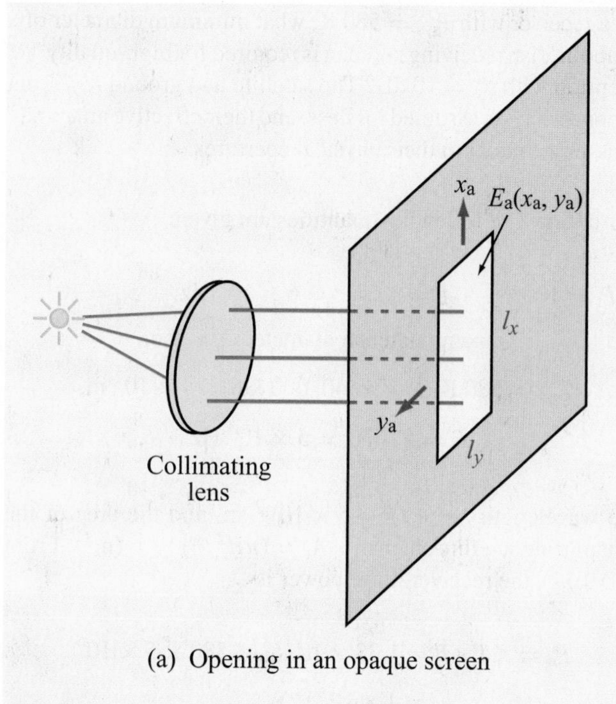

(a) Opening in an opaque screen

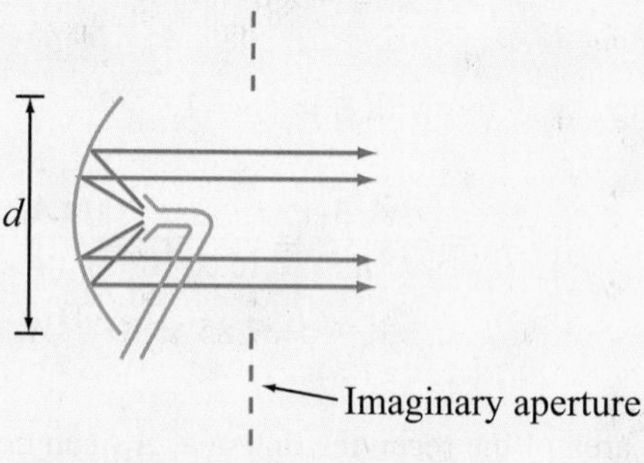

(b) Parabolic reflector antenna

Figure 9-21 Radiation by apertures: (a) an opening in an opaque screen illuminated by a light source through a collimating lens and (b) a parabolic dish reflector illuminated by a small horn antenna.

A distinctive feature of such an antenna is its high directivity and correspondingly narrow beam, which makes it attractive for radar and free-space microwave communication systems. The frequency range commonly used for such applications is the 1- to 30 GHz microwave band. Because the corresponding wavelength range is 30 to 1 cm, respectively, it is quite practical to construct and use antennas (in this frequency range) with aperture dimensions that are many wavelengths in size.

The x_a–y_a plane in **Fig. 9-22**, denoted plane A, contains an aperture with an electric field distribution $E_a(x_a, y_a)$. For the sake of convenience, the opening has been chosen to be rectangular in shape, with dimensions l_x along x_a and l_y along y_a, even though the formulation we are about to discuss is general enough to accommodate any two-dimensional aperture distribution, including those associated with circular and elliptical apertures. At a distance z from the aperture plane A in **Fig. 9-22**, we have an observation plane O with axes (x, y). The two planes have parallel axes and are separated by a distance z. Moreover, z is sufficiently large that any point Q

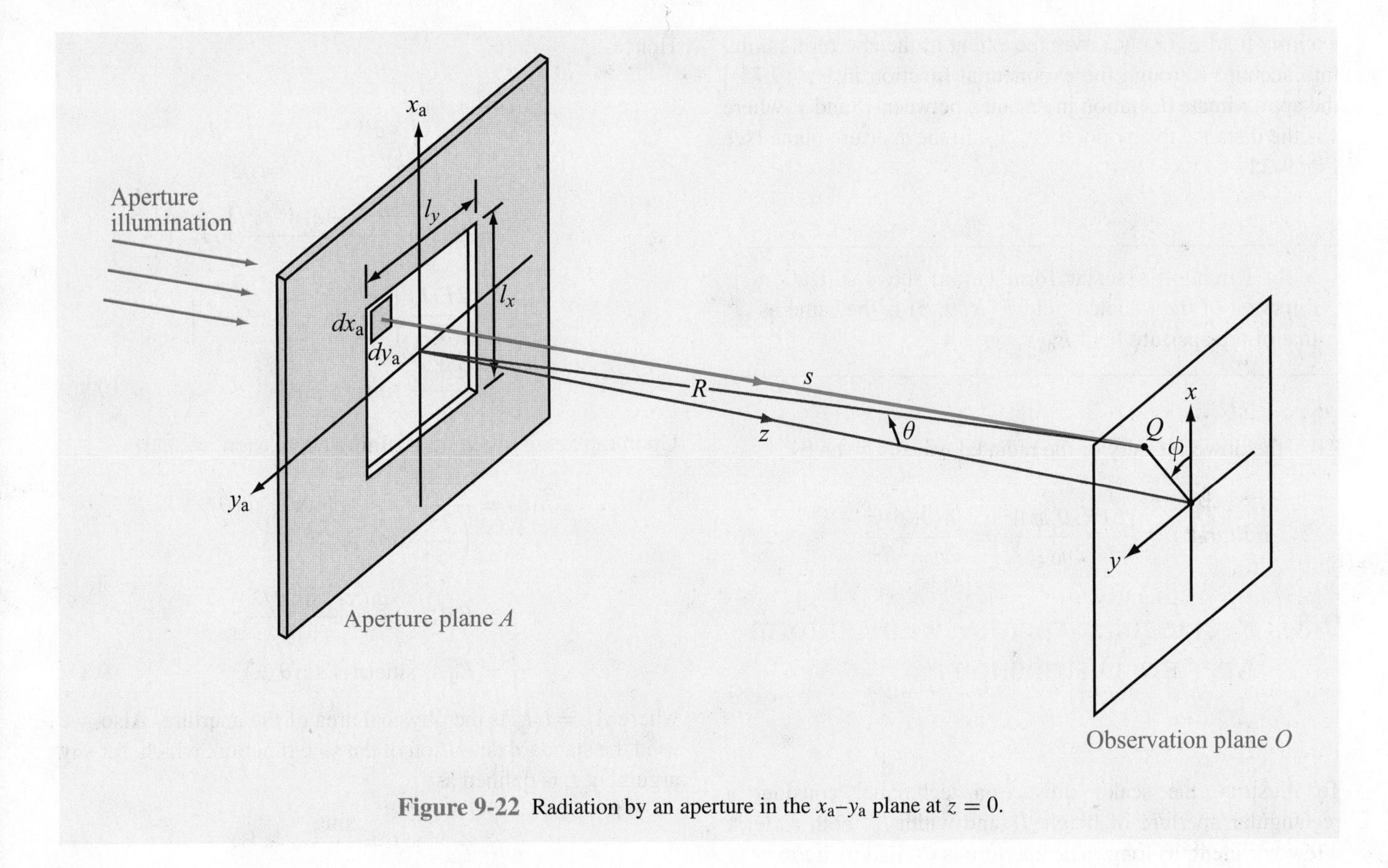

Figure 9-22 Radiation by an aperture in the x_a–y_a plane at $z = 0$.

in the observation plane is in the far-field region of the aperture. To satisfy the far-field condition, it is necessary that

$$R \geq 2d^2/\lambda \qquad \textbf{(far-field range)}, \tag{9.73}$$

where d is the longest linear dimension of the radiating aperture.

The position of observation point Q is specified by the range R between the center of the aperture and point Q and by the angles θ and ϕ (**Fig. 9-22**), which jointly define the direction of the observation point relative to the coordinate system of the aperture. In our treatment of the dipole antenna, we oriented the dipole along the z axis and we called θ the zenith angle. In the present context, the z axis is orthogonal to the plane containing the antenna aperture. Also, θ usually is called the ***elevation angle***. The electric field phasor of the wave incident upon point Q is denoted $\widetilde{E}(R, \theta, \phi)$. Kirchhoff's scalar diffraction theory provides the following relationship between the radiated field $\widetilde{E}(R, \theta, \phi)$ and the aperture illumination $\widetilde{E}_a(x_a, y_a)$:

$$\widetilde{E}(R, \theta, \phi) = \frac{j}{\lambda} \left(\frac{e^{-jkR}}{R} \right) \widetilde{h}(\theta, \phi), \tag{9.74}$$

where

$$\widetilde{h}(\theta, \phi) = \iint_{-\infty}^{\infty} \widetilde{E}_a(x_a, y_a) \cdot \exp\left[jk \sin\theta (x_a \cos\phi + y_a \sin\phi) \right] \, dx_a \, dy_a. \tag{9.75}$$

We shall refer to $\widetilde{h}(\theta, \phi)$ as the ***form factor*** of $\widetilde{E}(R, \theta, \phi)$. Its integral is written with infinite limits, with the understanding that $\widetilde{E}_a(x_a, y_a)$ is identically zero outside the aperture. The spherical propagation factor (e^{-jkR}/R) accounts for wave propagation between the center of the aperture and the observation point, and $\widetilde{h}(\theta, \phi)$ represents an integration of the exciting field $\widetilde{E}_a(x_a, y_a)$ over the extent of the aperture, taking into account [through the exponential function in Eq. (9.75)] the approximate deviation in distance between R and s, where s is the distance to any point (x_a, y_a) in the aperture plane (see Fig. 9-22).

► In Kirchhoff's scalar formulation, the polarization direction of the radiated field $\widetilde{E}(R, \theta, \phi)$ is the same as that of the aperture field $\widetilde{E}_a(x_a, y_a)$. ◄

Also, the power density of the radiated wave is given by

$$S(R, \theta, \phi) = \frac{|\widetilde{E}(R, \theta, \phi)|^2}{2\eta_0} = \frac{|\widetilde{h}(\theta, \phi)|^2}{2\eta_0 \lambda^2 R^2}. \quad (9.76)$$

9-8 Rectangular Aperture with Uniform Aperture Distribution

To illustrate the scalar diffraction technique, consider a rectangular aperture of height l_x and width l_y, both at least a few wavelengths long. The aperture is excited by a ***uniform field distribution*** (i.e., constant value) given by

$$\widetilde{E}_a(x_a, y_a) = \begin{cases} E_0 & \text{for } -l_x/2 \le x_a \le l_x/2 \\ & \text{and } -l_y/2 \le y_a \le l_y/2, \\ 0 & \text{otherwise.} \end{cases} \quad (9.77)$$

To keep the mathematics simple, let us confine our examination to the radiation pattern at a fixed range R in the x–z plane, which corresponds to $\phi = 0$. In this case, Eq. (9.75) simplifies to

$$\widetilde{h}(\theta) = \int_{y_a=-l_y/2}^{l_y/2} \int_{x_a=-l_x/2}^{l_x/2} E_0 \exp[jkx_a \sin\theta] \, dx_a \, dy_a. \quad (9.78)$$

In preparation for performing the integration in Eq. (9.78), we introduce the intermediate variable u defined as

$$u = k\sin\theta = \frac{2\pi \sin\theta}{\lambda}. \quad (9.79)$$

Hence,

$$\begin{aligned}\widetilde{h}(\theta) &= E_0 \int_{-l_x/2}^{l_x/2} e^{jux_a} \, dx_a \cdot \int_{-l_y/2}^{l_y/2} dy_a \\ &= E_0 \left[\frac{e^{jul_x/2} - e^{-jul_x/2}}{ju}\right] \cdot l_y \\ &= \frac{2E_0 l_y}{u} \left[\frac{e^{jul_x/2} - e^{-jul_x/2}}{2j}\right] \\ &= \frac{2E_0 l_y}{u} \sin(ul_x/2). \end{aligned} \quad (9.80)$$

Upon replacing u with its defining expression, we have

$$\begin{aligned}\widetilde{h}(\theta) &= \frac{2E_0 l_y}{\left(\dfrac{2\pi}{\lambda}\sin\theta\right)} \sin(\pi l_x \sin\theta/\lambda) \\ &= E_0 l_x l_y \frac{\sin(\pi l_x \sin\theta/\lambda)}{\pi l_x \sin\theta/\lambda} \\ &= E_0 A_p \operatorname{sinc}(\pi l_x \sin\theta/\lambda), \end{aligned} \quad (9.81)$$

where $A_p = l_x l_y$ is the physical area of the aperture. Also, we used the standard definition of the sinc function, which, for any argument t, is defined as

$$\operatorname{sinc} t = \frac{\sin t}{t}. \quad (9.82)$$

Using Eq. (9.76), we obtain the following expression for the power density at the observation point:

$$S(R, \theta) = S_0 \operatorname{sinc}^2(\pi l_x \sin\theta/\lambda) \quad (x\text{–}z \text{ plane}), \quad (9.83)$$

where $S_0 = E_0^2 A_p^2/(2\eta_0 \lambda^2 R^2)$.

► The sinc function is maximum when its argument is zero; $\operatorname{sinc}(0) = 1$. ◄

This occurs when $\theta = 0$. Hence, at a fixed range R, $S_{max} = S(\theta = 0) = S_0$. The normalized radiation intensity is then given by

$$\begin{aligned} F(\theta) &= \frac{S(R, \theta)}{S_{max}} \\ &= \operatorname{sinc}^2(\pi l_x \sin\theta/\lambda) \\ &= \operatorname{sinc}^2(\pi\gamma) \qquad (x\text{–}z \text{ plane}). \end{aligned} \quad (9.84)$$

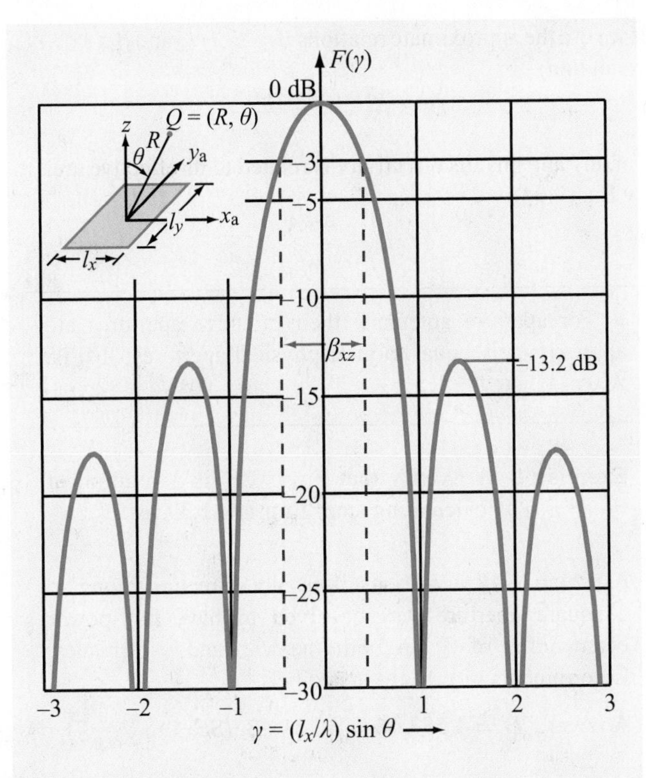

Figure 9-23 Normalized radiation pattern of a uniformly illuminated rectangular aperture in the x–z plane ($\phi = 0$).

Figure 9-23 shows $F(\theta)$ plotted (on a decibel scale) as a function of the intermediate variable $\gamma = (l_x/\lambda)\sin\theta$. The pattern exhibits nulls at nonzero integer values of γ.

9-8.1 Beamwidth

The normalized radiation intensity $F(\theta)$ is symmetrical in the x–z plane, and its maximum is along the boresight direction ($\theta = 0$, in this case). Its half-power beamwidth $\beta_{xz} = \theta_2 - \theta_1$, where θ_1 and θ_2 are the values of θ at which $F(\theta, 0) = 0.5$ (or -3 dB on a decibel scale), as shown in **Fig. 9-23**. Since the pattern is symmetrical with respect to $\theta = 0$, $\theta_1 = -\theta_2$ and $\beta_{xz} = 2\theta_2$. The angle θ_2 can be obtained from a solution of

$$F(\theta_2) = \text{sinc}^2(\pi l_x \sin\theta/\lambda) = 0.5. \tag{9.85}$$

From tabulated values of the sinc function, it is found that Eq. (9.85) yields the result

$$\frac{\pi l_x}{\lambda}\sin\theta_2 = 1.39, \tag{9.86}$$

or

$$\sin\theta_2 = 0.44\,\frac{\lambda}{l_x}. \tag{9.87}$$

Because $\lambda/l_x \ll 1$ (a fundamental condition of scalar diffraction theory is that the aperture dimensions be much larger than the wavelength λ), θ_2 is a small angle, in which case we can use the approximation $\sin\theta_2 \approx \theta_2$. Hence,

$$\beta_{xz} = 2\theta_2 \approx 2\sin\theta_2 = 0.88\frac{\lambda}{l_x} \quad \text{(rad)}. \tag{9.88a}$$

A similar solution for the y–z plane ($\phi = \pi/2$) gives

$$\beta_{yz} = 0.88\,\frac{\lambda}{l_y} \quad \text{(rad)}. \tag{9.88b}$$

► The uniform aperture distribution ($\widetilde{E}_\text{a} = E_0$ across the aperture) gives a far-field pattern with the narrowest possible beamwidth. ◄

The first sidelobe level is 13.2 dB below the peak value (see **Fig. 9-23**), which is equivalent to 4.8% of the peak value. If the intended application calls for a pattern with a lower sidelobe level (to avoid interference with signals from sources along directions outside the main beam of the antenna pattern), this can be accomplished by using a ***tapered aperture distribution***, one that is a maximum at the center of the aperture and decreases toward the edges.

► A tapered distribution provides a pattern with lower side lobes, but the main lobe becomes wider. ◄

The steeper the taper, the lower are the side lobes and the wider is the main lobe. In general, the beamwidth in a given plane, say the x–z plane, is given by

$$\beta_{xz} = k_x\frac{\lambda}{l_x}, \tag{9.89}$$

where k_x is a constant related to the steepness of the taper. For a uniform distribution with no taper, $k_x = 0.88$, and for a highly tapered distribution, $k_x \approx 2$. In the typical case, $k_x \approx 1$.

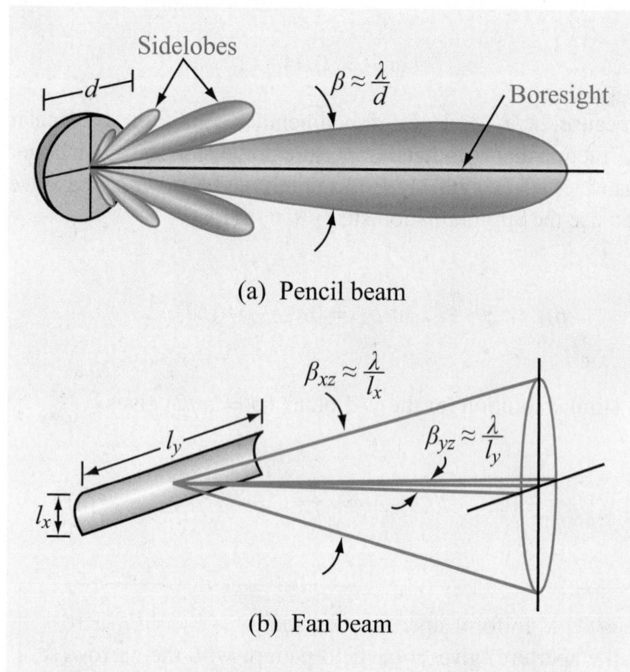

Figure 9-24 Radiation patterns of (a) a circular reflector and (b) a cylindrical reflector (side lobes not shown).

To illustrate the relationship between the antenna dimensions and the corresponding beam shape, we show in **Fig. 9-24** the radiation patterns of a circular reflector and a cylindrical reflector. The circular reflector has a circularly symmetric pattern, whereas the pattern of the cylindrical reflector has a narrow beam in the azimuth plane corresponding to its long dimension and a wide beam in the elevation plane corresponding to its narrow dimension. For a circularly symmetric antenna pattern, the beamwidth β is related to the diameter d by the approximate relation $\beta \approx \lambda/d$.

9-8.2 Directivity and Effective Area

In Section 9-2.3, we derived an approximate expression [Eq. (9.26)] for the antenna directivity D in terms of the half-power beamwidths β_{xz} and β_{yz} for antennas characterized by a single major lobe whose boresight is along the z direction:

$$D \approx \frac{4\pi}{\beta_{xz}\beta_{yz}} . \tag{9.90}$$

If we use the approximate relations $\beta_{xz} \approx \lambda/l_x$ and $\beta_{yz} \approx \lambda/l_y$, we obtain

$$D \approx \frac{4\pi l_x l_y}{\lambda^2} = \frac{4\pi A_p}{\lambda^2} . \tag{9.91}$$

For any antenna, its directivity is related to its effective area A_e by Eq. (9.64):

$$D = \frac{4\pi A_e}{\lambda^2} . \tag{9.92}$$

► For aperture antennas, their effective apertures are approximately equal to their physical apertures; that is, $A_e \approx A_p$. ◄

Exercise 9-11: Verify that Eq. (9.86) is a solution of Eq. (9.85) by calculating $\text{sinc}^2 t$ for $t = 1.39$.

Exercise 9-12: With its boresight direction along z, a square aperture was observed to have half-power beamwidths of 3° in both the x–z and y–z planes. Determine its directivity in decibels.

Answer: $D = 4{,}583.66 = 36.61$ dB. (See EM.)

Exercise 9-13: What condition must be satisfied in order to use scalar diffraction to compute the field radiated by an aperture antenna? Can we use it to compute the directional pattern of the eye's pupil ($d \simeq 0.2$ cm) in the visible part of the spectrum ($\lambda = 0.35$ to $0.7\ \mu$m)? What would the beamwidth of the eye's directional pattern be at $\lambda = 0.5\ \mu$m?

Answer: $\beta \approx \lambda/d = 2.5 \times 10^{-4}$ rad $= 0.86'$ (arc minute, with $60' = 1°$). (See EM.)

9-9 Antenna Arrays

AM broadcast services operate in the 535 to 1605 kHz band. The antennas they use are vertical dipoles mounted along tall towers. The antennas range in height from $\lambda/6$ to $5\lambda/8$, depending on the operating characteristics desired and other considerations. Their physical heights vary from 46 m (150 ft) to 274 m (900 ft); the wavelength at 1 MHz, approximately in the middle of the AM band, is 300 m. Because the field radiated by a single dipole is uniform in the horizontal plane (as discussed in Sections 9-1 and 9-3), it is not possible to direct the horizontal pattern along specific directions of interest, unless

Module 9.4 Large Parabolic Reflector For any specified reflector diameter d (such that $d \geq 2\lambda$) and illumination taper factor α, this module displays the pattern of the radiated field and computes the associated beamwidth and directivity.

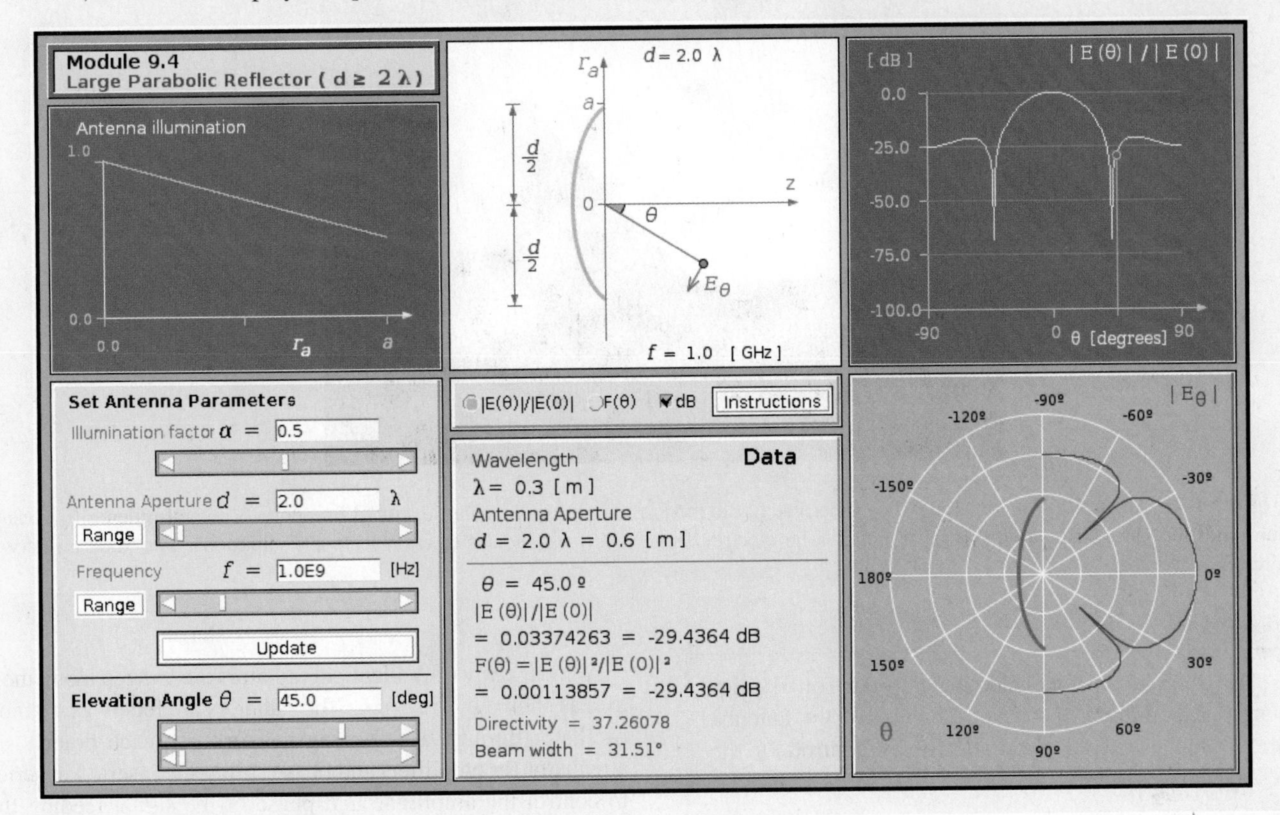

two or more antenna towers are used simultaneously. Directions of interest may include cities serviced by the AM station, and directions to avoid may include areas serviced by another station operating at the same frequency (thereby avoiding interference effects). When two or more antennas are used together, the combination is called an ***antenna array***.

The AM broadcast antenna array is only one example of the many antenna arrays used in communication systems and radar applications. Antenna arrays provide the antenna designer the flexibility to obtain high directivity, narrow beams, low side lobes, steerable beams, and shaped antenna patterns starting from very simple antenna elements. **Figure 9-25** shows a very large radar system consisting of a transmitter array composed of 5,184 individual dipole antenna elements and a receiver array composed of 4,660 elements. The radar system, part of the Space Surveillance Network operated by the U.S. Air Force, operates at 442 MHz and transmits a combined peak power of 30 MW!

Although an array need not consist of similar radiating elements, most arrays actually use identical elements, such as dipoles, slots, horn antennas, or parabolic dishes. The antenna elements composing an array may be arranged in various configurations, but the most common are the linear one-dimensional configuration—wherein the elements are arranged along a straight line—and the two-dimensional lattice configuration in which the elements sit on a planar grid. The desired shape of the far-field radiation pattern of the array can be synthesized by controlling the relative amplitudes of the array elements' excitations.

Figure 9-25 The AN/FPS-85 Phased Array Radar Facility in the Florida panhandle, near the city of Freeport. A several-mile no-fly zone surrounds the radar installation as a safety concern for electroexplosive devices, such as ejection seats and munitions, carried on military aircraft.

▶ Also, through the use of electronically controlled solid-state phase shifters, the beam direction of the antenna array can be steered electronically by controlling the relative phases of the array elements. ◀

This flexibility of the array antenna has led to numerous applications, including ***electronic steering*** and ***multiple-beam generation***.

The purpose of this and the next two sections is to introduce the reader to the basic principles of array theory and design techniques used in shaping the antenna pattern and steering the main lobe. The presentation is confined to the one-dimensional linear array with equal spacing between adjacent elements.

A linear array of N identical radiators is arranged along the z axis as shown in Fig. 9-26. The radiators are fed by a common oscillator through a branching network. In each branch, an attenuator (or amplifier) and phase shifter are inserted in series to control the amplitude and phase of the signal feeding the antenna element in that branch.

In the far-field region of any radiating element, the ***element electric-field intensity*** $\widetilde{E}_e(R,\theta,\phi)$ may be expressed as a product of two functions, the spherical propagation factor e^{-jkR}/R, which accounts for the dependence on the range R, and $\widetilde{f}_e(\theta,\phi)$, which accounts for the directional dependence of the element's electric field. Thus, for an isolated element, the radiated field is

$$\widetilde{E}_e(R,\theta,\phi) = \frac{e^{-jkR}}{R}\,\widetilde{f}_e(\theta,\phi), \tag{9.93}$$

and the corresponding power density S_e is

$$S_e(R,\theta,\phi) = \frac{1}{2\eta_0}|\widetilde{E}_e(R,\theta,\phi)|^2 = \frac{1}{2\eta_0 R^2}|\widetilde{f}_e(\theta,\phi)|^2. \tag{9.94}$$

Hence, for the array shown in Fig. 9-26(b), the far-zone field due to element i at range R_i from observation point Q is

$$\widetilde{E}_i(R_i,\theta,\phi) = A_i\,\frac{e^{-jkR_i}}{R_i}\,\widetilde{f}_e(\theta,\phi), \tag{9.95}$$

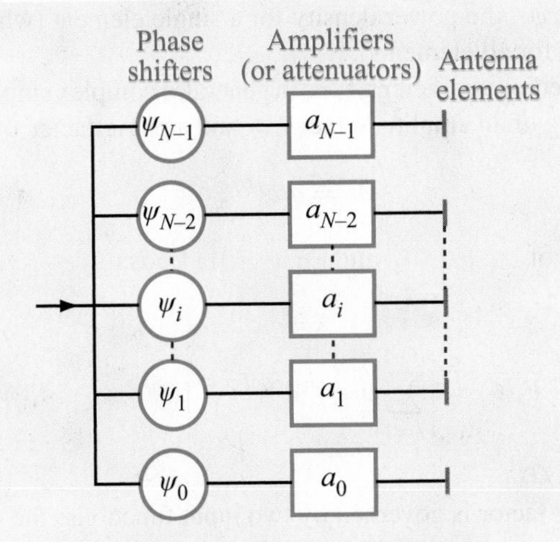

(a) Array elements with individual amplitude and phase control

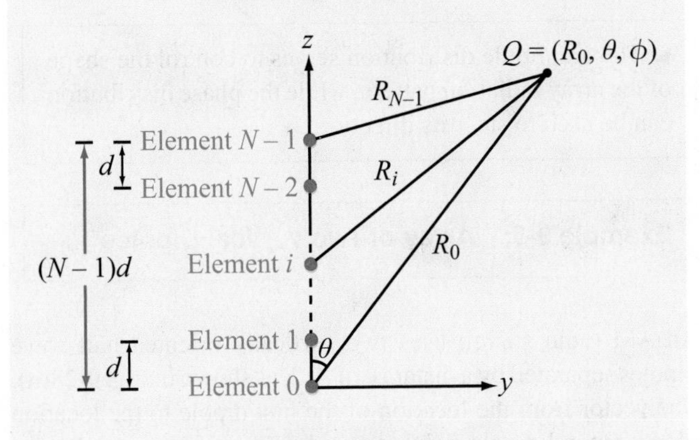

(b) Array geometry relative to observation point

Figure 9-26 Linear-array configuration and geometry.

where $A_i = a_i e^{j\psi_i}$ is a ***complex feeding coefficient*** representing the amplitude a_i and phase ψ_i of the excitation giving rise to $\widetilde{E}_i$, relative to a reference excitation. In practice, the excitation of one of the elements is used as reference. Note that R_i and A_i may be different for different elements in the array, but $\widetilde{f}_{\mathrm{e}}(\theta, \phi)$ is the same for all of them because they are all identical and hence exhibit identical directional patterns.

The total field at the observation point $Q(R_0, \theta, \phi)$ is the sum of the fields due to the N elements:

$$\widetilde{E}(R_0, \theta, \phi) = \sum_{i=0}^{N-1} \widetilde{E}_i(R_i, \theta, \phi)$$
$$= \left[\sum_{i=0}^{N-1} A_i \frac{e^{-jkR_i}}{R_i} \right] \widetilde{f}_{\mathrm{e}}(\theta, \phi), \tag{9.96}$$

where R_0 denotes the range of Q from the center of the coordinate system, chosen to be at the location of the zeroth element. To satisfy the far-field condition given by Eq. (9.73) for an array of length $l = (N-1)d$, where d is the interelement spacing, the range R_0 should be sufficiently large to satisfy

$$R_0 \geq \frac{2l^2}{\lambda} = \frac{2(N-1)^2 d^2}{\lambda}. \tag{9.97}$$

This condition allows us to ignore differences in the distances from Q to the individual elements as far as the magnitudes of the radiated fields are concerned. Thus, we can set $R_i = R_0$ in the denominator in Eq. (9.96) for all i. With regard to the phase part of the propagation factor, we can use the parallel-ray approximation given by

$$R_i \approx R_0 - z_i \cos\theta = R_0 - id \cos\theta, \tag{9.98}$$

where $z_i = id$ is the distance between the ith element and the zeroth element (**Fig. 9-27**). Employing these two approximations in Eq. (9.96) leads to

$$\widetilde{E}(R_0, \theta, \phi) = \widetilde{f}_{\mathrm{e}}(\theta, \phi) \left(\frac{e^{-jkR_0}}{R_0} \right) \cdot \left[\sum_{i=0}^{N-1} A_i e^{jikd\cos\theta} \right], \tag{9.99}$$

and the corresponding array-antenna power density is given by

$$S(R_0, \theta, \phi) = \frac{1}{2\eta_0} |\widetilde{E}(R_0, \theta, \phi)|^2$$
$$= \frac{1}{2\eta_0 R_0^2} |\widetilde{f}_{\mathrm{e}}(\theta, \phi)|^2 \left| \sum_{i=0}^{N-1} A_i e^{jikd\cos\theta} \right|^2$$
$$= S_{\mathrm{e}}(R_0, \theta, \phi) \left| \sum_{i=0}^{N-1} A_i e^{jikd\cos\theta} \right|^2, \tag{9.100}$$

where use was made of Eq. (9.94). This expression is a product of two factors. The first factor, $S_{\mathrm{e}}(R_0, \theta, \phi)$, is the power

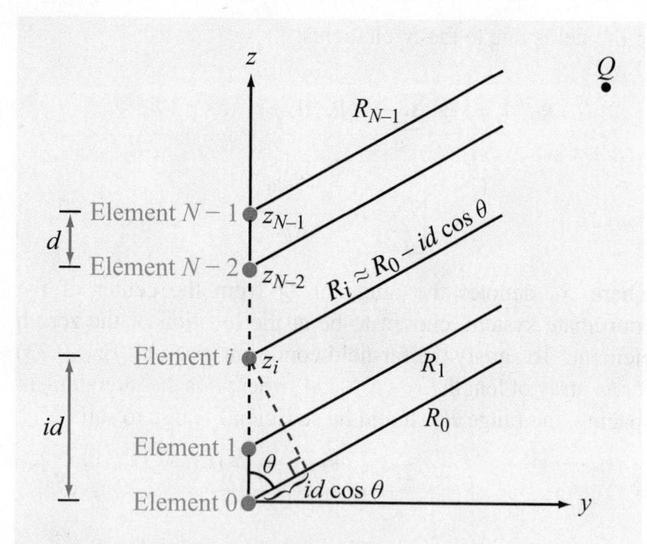

Figure 9-27 The rays between the elements and a faraway observation point are approximately parallel lines. Hence, the distance $R_i \approx R_0 - id\cos\theta$.

density of the energy radiated by an individual element, and the second, called the ***array factor***, is a function of the positions of the individual elements and their feeding coefficients, but not a function of the specific type of radiators used.

▶ The array factor represents the far-field radiation intensity of the N elements, had the elements been isotropic radiators. ◀

Denoting the array factor by

$$F_a(\theta) = \left| \sum_{i=0}^{N-1} A_i e^{jikd\cos\theta} \right|^2, \tag{9.101}$$

the power density of the antenna array is then written as

$$S(R_0, \theta, \phi) = S_e(R_0, \theta, \phi)\; F_a(\theta). \tag{9.102}$$

This equation demonstrates the ***pattern multiplication principle***. It allows us to find the far-field power density of the antenna array by first computing the far-field power pattern with the array elements replaced with isotropic radiators, which yields the array factor $F_a(\theta)$, and then multiplying the result by $S_e(R_0, \theta, \phi)$, the power density for a single element (which is the same for all elements).

The feeding coefficient A_i is, in general, a complex amplitude consisting of an amplitude factor a_i and a phase factor ψ_i:

$$A_i = a_i e^{j\psi_i}. \tag{9.103}$$

Insertion of Eq. (9.103) into Eq. (9.101) leads to

$$F_a(\theta) = \left| \sum_{i=0}^{N-1} a_i e^{j\psi_i} e^{jikd\cos\theta} \right|^2. \tag{9.104}$$

The array factor is governed by two input functions: the ***array amplitude distribution*** given by the a_i's and the ***array phase distribution*** given by the ψ_i's.

▶ The amplitude distribution serves to control the shape of the array radiation pattern, while the phase distribution can be used to steer its direction. ◀

Example 9-5: Array of Two Vertical Dipoles

An AM radio station uses two vertically oriented half-wave dipoles separated by a distance of $\lambda/2$, as shown in **Fig. 9-28(a)**. The vector from the location of the first dipole to the location of the second dipole points toward the east. The two dipoles are fed with equal-amplitude excitations, and the dipole farther east is excited with a phase shift of $-\pi/2$ relative to the other one. Find and plot the antenna pattern of the antenna array in the horizontal plane.

Solution: The array factor given by Eq. (9.104) was derived for radiators arranged along the z axis. To keep the coordinate system the same, we choose the easterly direction to be the z axis as shown in **Fig. 9-28(b)**, and we place the first dipole at $z = -\lambda/4$ and the second at $z = \lambda/4$. A dipole radiates uniformly in the plane perpendicular to its axis, which in this case is the horizontal plane. Hence, $S_e = S_0$ for all angles θ in **Fig. 9-28(b)**, where S_0 is the maximum value of the power density radiated by each dipole individually. Consequently, the power density radiated by the two-dipole array is

$$S(R, \theta) = S_0\, F_a(\theta).$$

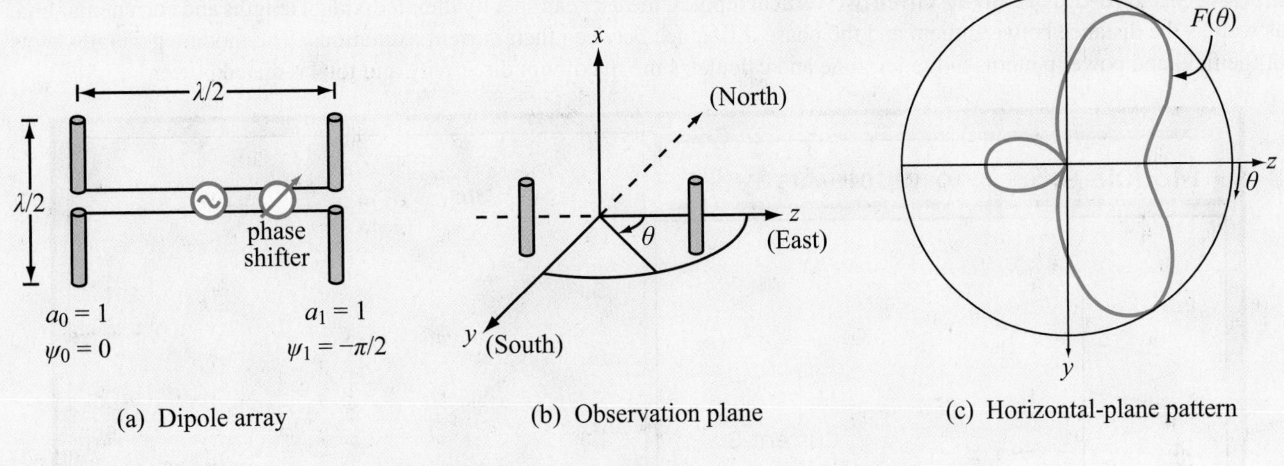

Figure 9-28 Two half-wave dipole array of Example 9-5.

For two elements separated by $d = \lambda/2$ and excited with equal amplitudes ($a_0 = a_1 = 1$) and with phase angles $\psi_0 = 0$ and $\psi_1 = -\pi/2$, Eq. (9.104) becomes

$$F_a(\theta) = \left| \sum_{i=0}^{1} a_i e^{j\psi_i} e^{jikd\cos\theta} \right|^2$$

$$= \left| 1 + e^{-j\pi/2} e^{j(2\pi/\lambda)(\lambda/2)\cos\theta} \right|^2$$

$$= \left| 1 + e^{j(\pi\cos\theta - \pi/2)} \right|^2 .$$

A function of the form $|1 + e^{jx}|^2$ can be evaluated by factoring out $e^{jx/2}$ from both terms:

$$|1 + e^{jx}|^2 = |e^{jx/2}(e^{-jx/2} + e^{jx/2})|^2$$

$$= |e^{jx/2}|^2 \, |e^{-jx/2} + e^{jx/2}|^2$$

$$= |e^{jx/2}|^2 \left| 2 \, \frac{[e^{-jx/2} + e^{jx/2}]}{2} \right|^2 .$$

The absolute value of $e^{jx/2}$ is 1, and we recognize the function inside the square bracket as $\cos(x/2)$. Hence,

$$|1 + e^{jx}|^2 = 4\cos^2\left(\frac{x}{2}\right).$$

Applying this result to the expression for $F_a(\theta)$, we have

$$F_a(\theta) = 4\cos^2\left(\frac{\pi}{2}\cos\theta - \frac{\pi}{4}\right).$$

The power density radiated by the array is then

$$S(R,\theta) = S_0 F_a(\theta) = 4S_0 \cos^2\left(\frac{\pi}{2}\cos\theta - \frac{\pi}{4}\right).$$

This function has a maximum value $S_{max} = 4S_0$, and it occurs when the argument of the cosine function is equal to zero. Thus,

$$\frac{\pi}{2}\cos\theta - \frac{\pi}{4} = 0,$$

which leads to the solution: $\theta = 60°$. Upon normalizing $S(R,\theta)$ by its maximum value, we obtain the normalized radiation intensity given by

$$F(\theta) = \frac{S(R,\theta)}{S_{max}} = \cos^2\left(\frac{\pi}{2}\cos\theta - \frac{\pi}{4}\right).$$

The pattern of $F(\theta)$ is shown in Fig. 9-28(c).

Example 9-6: Pattern Synthesis

In Example 9-5, we were given the array parameters a_0, a_1, ψ_0, ψ_1, and d, and we were then asked to determine the pattern of the two-element dipole array. We now consider the reverse process; given specifications on the desired pattern, we specify the array parameters to meet those specifications.

Given two vertical dipoles, as depicted in Fig. 9-28(b), specify the array parameters such that the array exhibits

Module 9.5 Two-dipole Array Given two vertical dipoles, the user can specify their individual lengths and current maxima, as well as the distance between them and the phase difference between their current excitations. The module generates plots of the field and power patterns in the far-zone and calculates the maximum directivity and total radiated power.

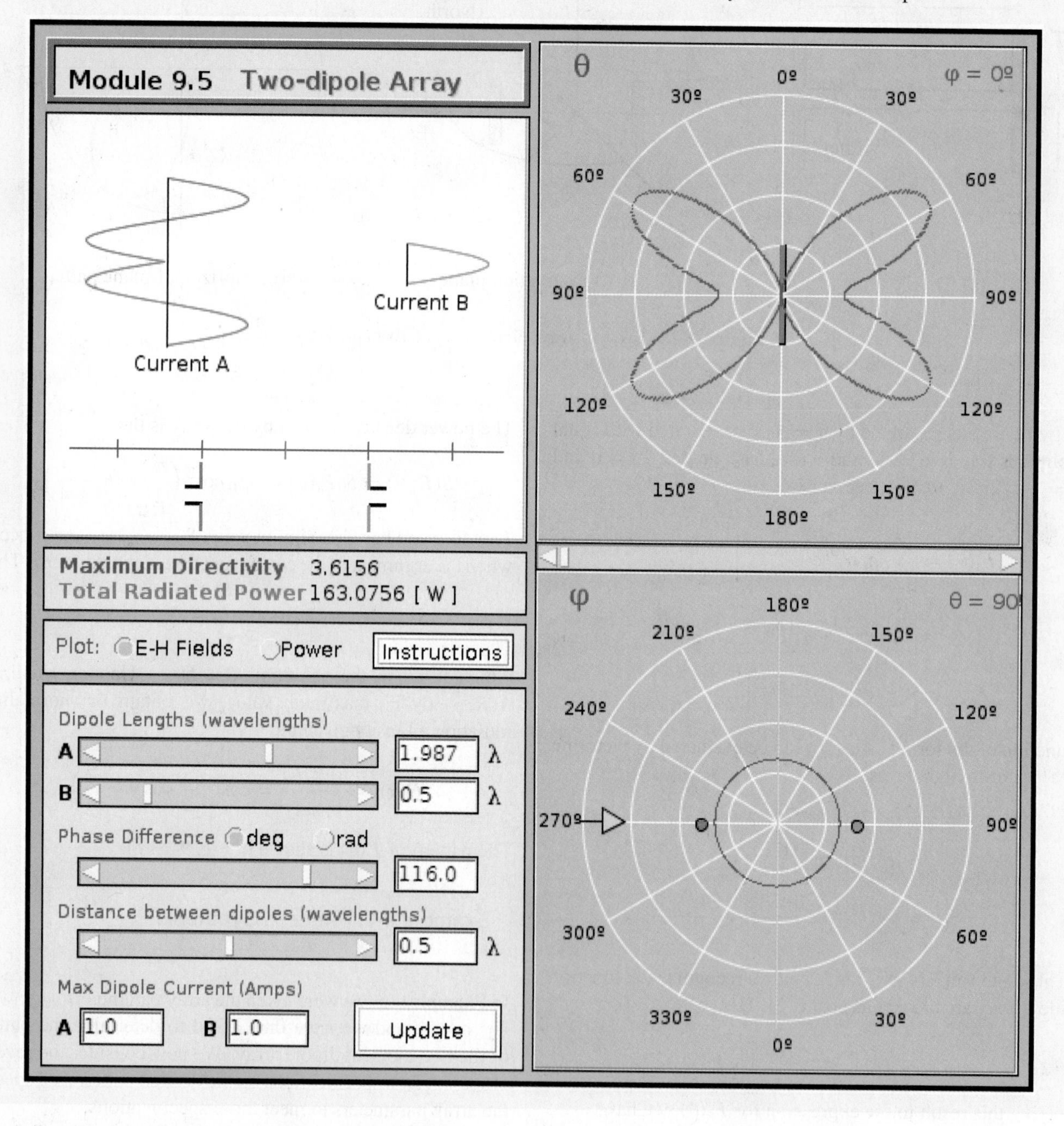

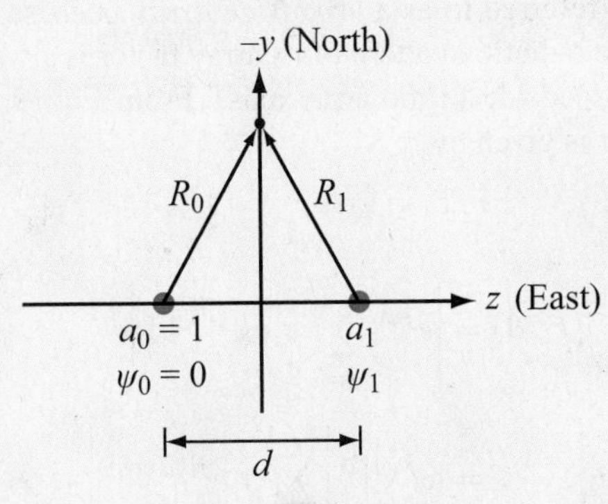

(a) Array arrangement

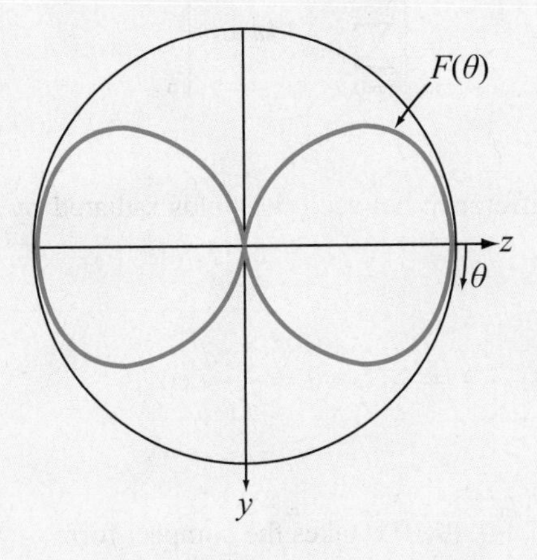

(b) Array pattern

Figure 9-29 (a) Two vertical dipoles separated by a distance d along the z axis; (b) normalized array pattern in the y–z plane for $a_0 = a_1 = 1$, $\psi_1 = \psi_0 = -\pi$, and $d = \lambda/2$.

maximum radiation toward the east and no radiation toward the north or south.

Solution: From Example 9-5, we established that because each dipole radiates equally along all directions in the y–z plane, the radiation pattern of the two-dipole array in that plane is governed solely by the array factor $F_a(\theta)$. The shape of the pattern of the array factor depends on three parameters: the amplitude ratio a_1/a_0, the phase difference $\psi_1 - \psi_0$, and the spacing d [**Fig. 9-29(a)**]. For convenience, we choose $a_0 = 1$ and $\psi_0 = 0$. Accordingly, Eq. (9.101) becomes

$$\begin{aligned} F_a(\theta) &= \left| \sum_{i=0}^{1} a_i e^{j\psi_i} e^{jikd\cos\theta} \right|^2 \\ &= |1 + a_1 e^{j\psi_1} e^{j(2\pi d/\lambda)\cos\theta}|^2. \end{aligned}$$

Next, we consider the specification that F_a be equal to zero when $\theta = 90°$ [north and south directions in **Fig. 9-29(a)**]. For any observation point on the y axis, the ranges R_0 and R_1 shown in **Fig. 9-29(a)** are equal, which means that the propagation phases associated with the time travel of the waves radiated by the two dipoles to that point are identical. Hence, to satisfy the stated condition, we need to choose $a_1 = a_0$ and $\psi_1 = \pm\pi$. With these choices, the signals radiated by the two dipoles have equal amplitudes and opposite phases, thereby interfering destructively. This conclusion can be ascertained by evaluating the array factor at $\theta = 90°$, with $a_0 = a_1 = 1$ and $\psi_1 = \pm\pi$:

$$\begin{aligned} F_a(\theta = 90°) &= |1 + 1e^{\pm j\pi}|^2 \\ &= |1 - 1| \\ &= 0. \end{aligned}$$

The two values of ψ_1, namely π and $-\pi$, lead to the same solution for the value of the spacing d to meet the specification that the array radiation pattern is maximum toward the east, corresponding to $\theta = 0°$. Let us choose $\psi_1 = -\pi$ and examine the array factor at $\theta = 0°$:

$$\begin{aligned} F_a(\theta = 0) &= |1 + e^{-j\pi} e^{j2\pi d/\lambda}|^2 \\ &= |1 + e^{j(-\pi + 2\pi d/\lambda)}|^2. \end{aligned}$$

Module 9.6 Detailed Analysis of Two-Dipole Array This module extends the display and computational capabilities of Module 9.5 by offering plots for individual components of **E** and **H** at any range from the antenna, including the near-field.

For $F_a(\theta = 0)$ to be a maximum, we require the phase angle of the second term to be zero or a multiple of 2π. That is,

$$-\pi + \frac{2\pi d}{\lambda} = 2n\pi,$$

or

$$d = (2n+1)\frac{\lambda}{2}, \qquad n = 0, 1, 2, \dots$$

In summary, the two-dipole array will meet the given specifications if $a_0 = a_1$, $\psi_1 - \psi_0 = -\pi$, and $d = (2n+1)\lambda/2$.

For $d = \lambda/2$, the array factor is

$$\begin{aligned} F_a(\theta) &= |1 + e^{-j\pi} e^{j\pi\cos\theta}|^2 \\ &= |1 - e^{j\pi\cos\theta}|^2 \\ &= \left| 2je^{-j(\pi/2)\cos\theta} \left[\frac{e^{j(\pi/2)\cos\theta} - e^{-j(\pi/2)\cos\theta}}{2j} \right] \right|^2 \\ &= 4\sin^2\left(\frac{\pi}{2}\cos\theta\right). \end{aligned}$$

The array factor has a maximum value of 4, which is the maximum level attainable from a two-element array with unit amplitudes. The directions along which $F_a(\theta)$ is a maximum are those corresponding to $\theta = 0$ (east) and $\theta = 180°$ (west), as shown in **Fig. 9-29(b)**.

Exercise 9-14: Derive an expression for the array factor of a two-element array excited in phase with $a_0 = 1$ and $a_1 = 3$. The elements are positioned along the z axis and are separated by $\lambda/2$.

Answer: $F_a(\theta) = [10 + 6\cos(\pi\cos\theta)]$. (See EM.)

Exercise 9-15: An equally spaced N-element array arranged along the z axis is fed with equal amplitudes and phases; that is, $A_i = 1$ for $i = 0, 1, \dots, (N-1)$. What is the magnitude of the array factor in the broadside direction?

Answer: $F_a(\theta = 90°) = N^2$. (See EM.)

9-10 *N*-Element Array with Uniform Phase Distribution

We now consider an array of N elements with equal spacing d and equal-phase excitations; that is, $\psi_i = \psi_0$ for $i = 1, 2, \dots, (N-1)$. Such an array of in-phase elements is sometimes referred to as a ***broadside array*** because the main beam of the radiation pattern of its array factor is always in the direction broadside to the array axis. From Eq. (9.104), the array factor is given by

$$\begin{aligned} F_a(\theta) &= \left| e^{j\psi_0} \sum_{i=0}^{N-1} a_i e^{jikd\cos\theta} \right|^2 \\ &= |e^{j\psi_0}|^2 \left| \sum_{i=0}^{N-1} a_i e^{jikd\cos\theta} \right|^2 \\ &= \left| \sum_{i=0}^{N-1} a_i e^{jikd\cos\theta} \right|^2. \end{aligned} \tag{9.105}$$

The phase difference between the fields radiated by adjacent elements is

$$\gamma = kd\cos\theta = \frac{2\pi d}{\lambda}\cos\theta. \tag{9.106}$$

In terms of γ, Eq. (9.105) takes the compact form

$$F_a(\gamma) = \left| \sum_{i=0}^{N-1} a_i e^{ji\gamma} \right|^2 \quad \text{(uniform phase)}. \tag{9.107}$$

For a uniform amplitude distribution with $a_i = 1$ for $i = 0, 1, \dots, (N-1)$, Eq. (9.107) becomes

$$F_a(\gamma) = |1 + e^{j\gamma} + e^{j2\gamma} + \dots + e^{j(N-1)\gamma}|^2. \tag{9.108}$$

This geometric series can be rewritten in a more compact form by applying the following recipe. First, we define

$$F_{\mathrm{a}}(\gamma) = |f_{\mathrm{a}}(\gamma)|^2, \qquad (9.109)$$

with

$$f_{\mathrm{a}}(\gamma) = [1 + e^{j\gamma} + e^{j2\gamma} + \cdots + e^{j(N-1)\gamma}]. \qquad (9.110)$$

Next, we multiply $f_{\mathrm{a}}(\gamma)$ by $e^{j\gamma}$ to obtain

$$f_{\mathrm{a}}(\gamma)\, e^{j\gamma} = (e^{j\gamma} + e^{j2\gamma} + \cdots + e^{jN\gamma}). \qquad (9.111)$$

Subtracting Eq. (9.111) from Eq. (9.110) gives

$$f_{\mathrm{a}}(\gamma)\,(1 - e^{j\gamma}) = 1 - e^{jN\gamma}, \qquad (9.112)$$

which, in turn, gives

$$\begin{aligned} f_{\mathrm{a}}(\gamma) &= \frac{1 - e^{jN\gamma}}{1 - e^{j\gamma}} \\ &= \frac{e^{jN\gamma/2}}{e^{j\gamma/2}} \, \frac{(e^{-jN\gamma/2} - e^{jN\gamma/2})}{(e^{-j\gamma/2} - e^{j\gamma/2})} \\ &= e^{j(N-1)\gamma/2} \, \frac{\sin(N\gamma/2)}{\sin(\gamma/2)} . \end{aligned} \qquad (9.113)$$

After multiplying $f_{\mathrm{a}}(\gamma)$ by its complex conjugate, we obtain the result:

$$F_{\mathrm{a}}(\gamma) = \frac{\sin^2(N\gamma/2)}{\sin^2(\gamma/2)} . \qquad (9.114)$$

(uniform amplitude and phase)

From Eq. (9.108), $F_{\mathrm{a}}(\gamma)$ is maximum when all terms are 1, which occurs when $\gamma = 0$ (or equivalently, $\theta = \pi/2$). Moreover, $F_{\mathrm{a}}(0) = N^2$. Hence, the normalized array factor is given by

$$F_{\mathrm{an}}(\gamma) = \frac{F_{\mathrm{a}}(\gamma)}{F_{\mathrm{a,max}}} nn \qquad (9.115)$$

$$\begin{aligned} &= \frac{\sin^2(N\gamma/2)}{N^2 \sin^2(\gamma/2)} \\ &= \frac{\sin^2\left(\dfrac{N\pi d}{\lambda}\cos\theta\right)}{N^2 \sin^2\left(\dfrac{\pi d}{\lambda}\cos\theta\right)} . \end{aligned} \qquad (9.116)$$

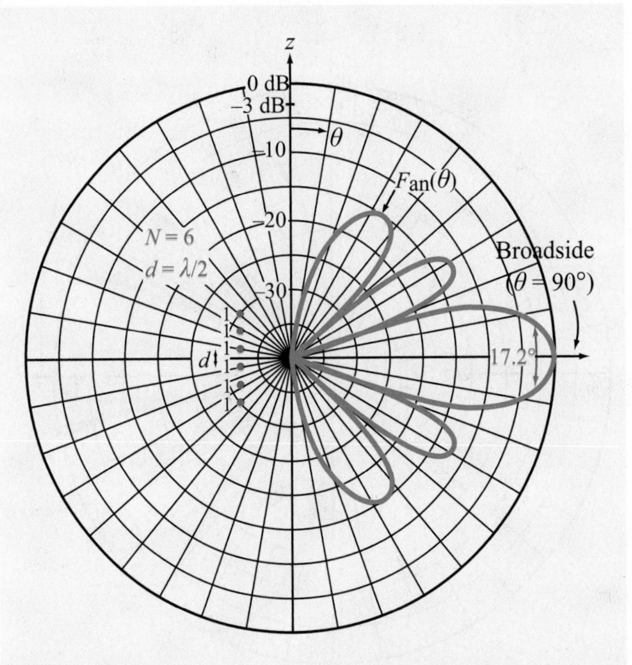

Figure 9-30 Normalized array pattern of a uniformly excited six-element array with interelement spacing $d = \lambda/2$.

A polar plot of $F_{\mathrm{an}}(\theta)$ is shown in Fig. 9-30 for $N = 6$ and $d = \lambda/2$. The reader is reminded that this is a plot of the radiation pattern of the array factor alone; the pattern for the antenna array is equal to the product of this pattern and that of a single element, as discussed earlier in connection with the pattern multiplication principle.

Example 9-7: Multiple-Beam Array

Obtain an expression for the array factor of a two-element array with equal excitation and a separation $d = 7\lambda/2$, and then plot the array pattern.

Solution: The array factor of a two-element array ($N = 2$)

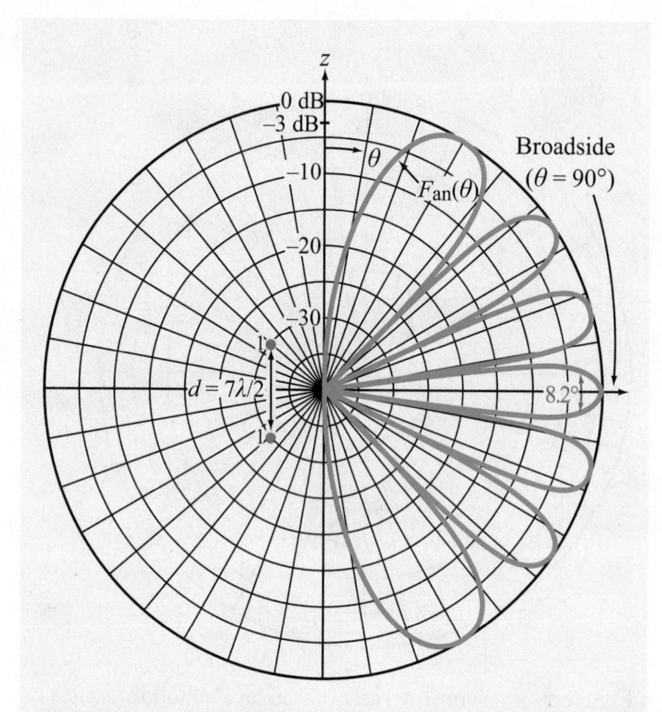

Figure 9-31 Normalized array pattern of a two-element array with spacing $d = 7\lambda/2$.

with equal excitation ($a_0 = a_1 = 1$) is given by

$$
\begin{aligned}
F_a(\gamma) &= \left| \sum_{i=0}^{1} a_i e^{ji\gamma} \right|^2 \\
&= |1 + e^{j\gamma}|^2, \\
&= |e^{j\gamma/2}(e^{-j\gamma/2} + e^{j\gamma/2})|^2 \\
&= |e^{j\gamma/2}|^2 \, |e^{-j\gamma/2} + e^{j\gamma/2}|^2 = 4\cos^2(\gamma/2),
\end{aligned}
$$

where $\gamma = (2\pi d/\lambda)\cos\theta$. The normalized array pattern, shown in Fig. 9-31, consists of seven beams, all with the same peak value, but not the same angular width. The number of beams in the angular range between $\theta = 0$ and $\theta = \pi$ is equal to the separation between the array elements, d, measured in units of $\lambda/2$.

9-11 Electronic Scanning of Arrays

The discussion in the preceding section was concerned with uniform-phase arrays, in which the phases of the feeding coefficients, ψ_0 to ψ_{N-1}, are all equal. In this section, we examine the use of phase delay between adjacent elements as a tool to ***electronically steer*** the direction of the array-antenna beam from broadside at $\theta = 90°$ to any desired angle θ_0. In addition to eliminating the need to mechanically steer an antenna to change its beam's direction, electronic steering allows beam scanning at very fast rates.

▶ Electronic steering is achieved by applying a ***linear phase distribution*** across the array: $\psi_0 = 0$, $\psi_1 = -\delta$, $\psi_2 = -2\delta$, etc. ◀

As shown in Fig. 9-32, the phase of the ith element, relative to that of the zeroth element, is

$$\psi_i = -i\delta, \tag{9.117}$$

where δ is the ***incremental phase delay*** between adjacent elements. Use of Eq. (9.117) in Eq. (9.104) leads to

$$
\begin{aligned}
F_a(\theta) &= \left| \sum_{i=0}^{N-1} a_i e^{-ji\delta} e^{jikd\cos\theta} \right|^2 \\
&= \left| \sum_{i=0}^{N-1} a_i e^{ji(kd\cos\theta - \delta)} \right|^2 \\
&= \left| \sum_{i=0}^{N-1} a_i e^{ji\gamma'} \right|^2 = F_a(\gamma'),
\end{aligned}
\tag{9.118}
$$

where we introduced a new variable given by

$$\gamma' = kd\cos\theta - \delta. \tag{9.119}$$

For reasons that become clear later, we define the phase shift δ in terms of an angle θ_0, which we call the ***scan angle***, as follows:

$$\delta = kd\cos\theta_0. \tag{9.120}$$

Hence, γ' becomes

$$\gamma' = kd(\cos\theta - \cos\theta_0). \tag{9.121}$$

The array factor given by Eq. (9.118) has the same functional form as the array factor developed earlier for the uniform-phase array [see Eq. (9.107)], except that γ is replaced with γ'. Hence:

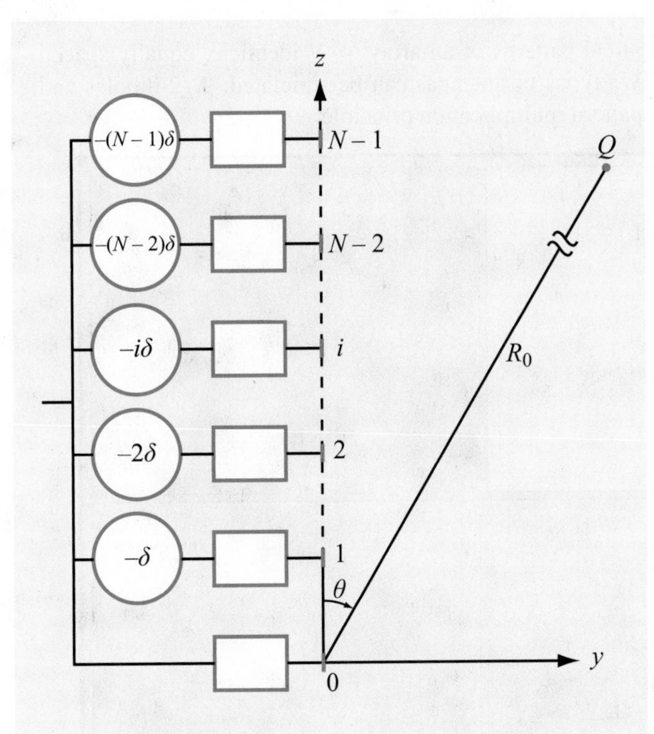

Figure 9-32 The application of linear phase.

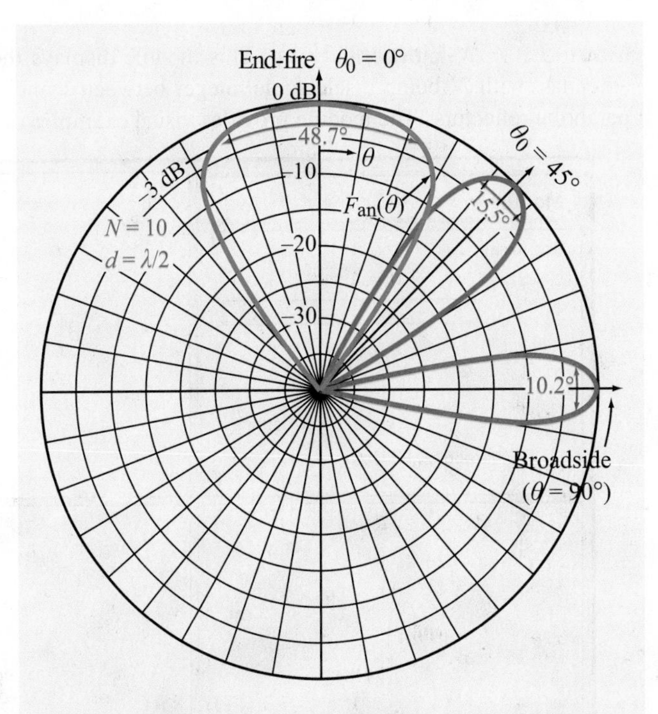

Figure 9-33 Normalized array pattern of a 10-element array with $\lambda/2$ spacing between adjacent elements. All elements are excited with equal amplitude. Through the application of linear phase across the array, the main beam can be steered from the broadside direction ($\theta_0 = 90°$) to any scan angle θ_0. Equiphase excitation corresponds to $\theta_0 = 90°$.

► Regardless of the amplitude distribution across an array, its array factor $F_a(\gamma')$ when excited by a linear-phase distribution can be obtained from $F_a(\gamma)$, the expression developed for the array assuming a uniform-phase distribution, by replacing γ with γ'. ◄

If the amplitude distribution is symmetrical with respect to the array center, the array factor $F_a(\gamma')$ is maximum when its argument $\gamma' = 0$. When the phase is uniform ($\delta = 0$), this condition corresponds to the direction $\theta = 90°$, which is why the uniform-phase arrangement is called a broadside array. According to Eq. (9.121), in a linearly phased array, $\gamma' = 0$ when $\theta = \theta_0$. Thus, by applying linear phase across the array, the array pattern is shifted along the $\cos\theta$ axis by an amount $\cos\theta_0$, and the direction of maximum radiation is ***steered*** from the broadside direction ($\theta = 90°$) to the direction $\theta = \theta_0$. To steer the beam all the way to the ***end-fire direction*** ($\theta = 0$), the incremental phase shift δ should be equal to kd radians.

9-11.1 Uniform-Amplitude Excitation

To illustrate the process with an example, consider the case of the N-element array excited by a uniform-amplitude distribution. Its normalized array factor is given by Eq. (9.116). Upon replacing γ with γ', we have

$$F_{\text{an}}(\gamma') = \frac{\sin^2(N\gamma'/2)}{N^2 \sin^2(\gamma'/2)}, \tag{9.122}$$

with γ' as defined by Eq. (9.121). For an array with $N = 10$ and $d = \lambda/2$, plots of the main lobe of $F_{\text{an}}(\theta)$ are shown in Fig. 9-33 for $\theta_0 = 0°$, 45°, and 90°. We note that the half-power beamwidth increases as the array beam is steered from broadside to end fire.

Module 9.7 ***N*-Element Array** This module displays the far-field patterns of an array of N identical, equally spaced antennas, with N being a selectable integer between 1 and 6. Two types of antennas can be simulated: $\lambda/2$-dipoles and parabolic reflectors. The module provides visual examples of the pattern multiplication principle.

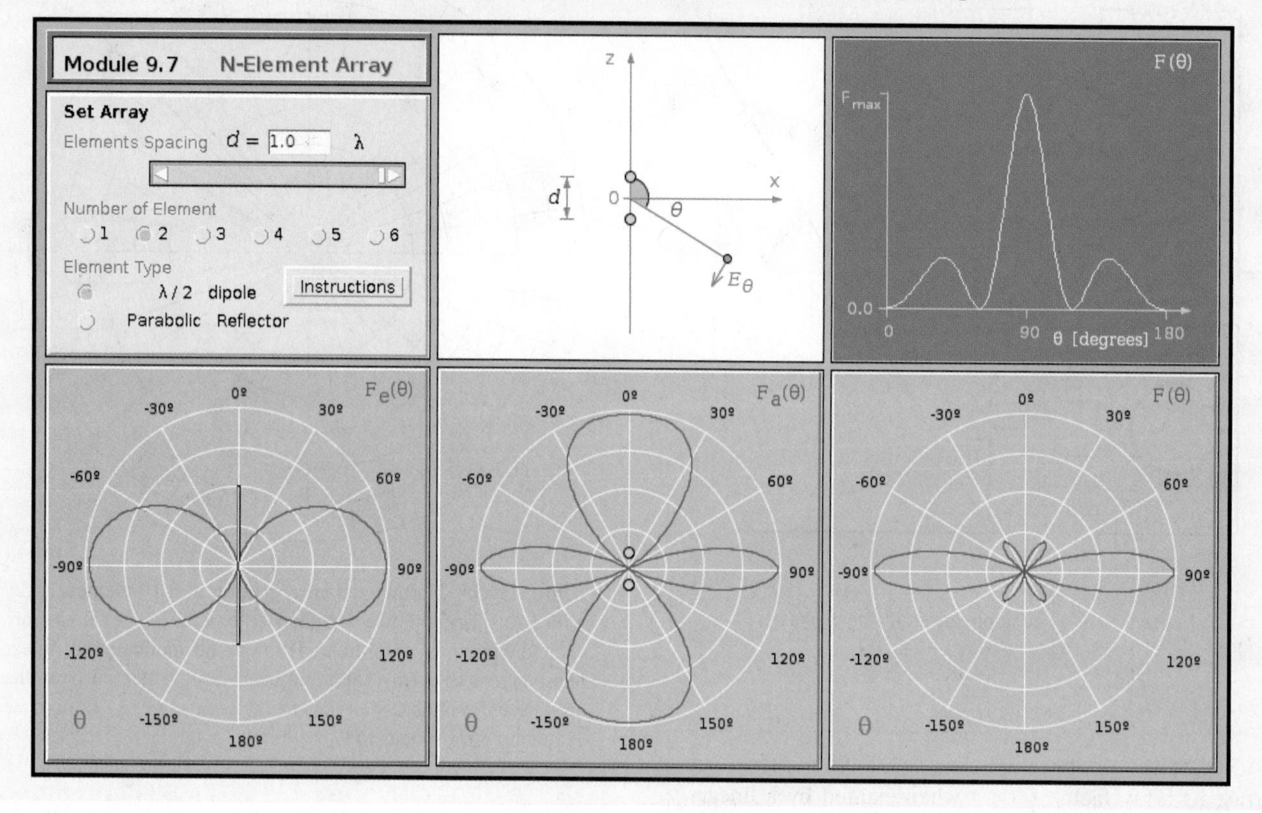

9-11.2 Array Feeding

According to the foregoing discussion, to steer the antenna beam to an angle θ_0, two conditions must be met: (1) the phase distribution must be linear across the array, and (2) the magnitude of the incremental phase delay δ must satisfy Eq. (9.120). The combination of these two conditions provides the necessary tilting of the beam from $\theta = 90°$ (broadside) to $\theta = \theta_0$. This can be accomplished by controlling the excitation of each radiating element individually through the use of electronically controlled phase shifters. Alternatively, a technique known as ***frequency scanning*** can be used to provide control of the phases of all the elements simultaneously. **Figure 9-34** shows an example of a simple feeding arrangement employed in frequency scanning arrays. A common feed point is connected to the radiating elements through transmission lines of varying lengths. Relative to the zeroth element, the path between the common feed point and a radiating element is longer by l for the first element, by $2l$ for the second, and by $3l$ for the third. Thus, the path length for the ith element is

$$l_i = il + l_0, \tag{9.123}$$

where l_0 is the path length of the zeroth element. Waves of frequency f propagating through a transmission line of length l_i are characterized by a phase factor $e^{-j\beta l_i}$, where $\beta = 2\pi f/u_\text{p}$ is the phase constant of the line and u_p is its phase velocity. Hence, the *incremental* phase delay of the ith element, relative to the phase of the zeroth element, is

$$\psi_i(f) = -\beta(l_i - l_0) = -\frac{2\pi}{u_\text{p}}\, f(l_i - l_0) = -\frac{2\pi i}{u_\text{p}}\, fl. \tag{9.124}$$

Suppose that at a given reference frequency f_0 we choose the incremental length l such that

$$l = \frac{n_0 u_{\rm p}}{f_0}, \qquad (9.125)$$

where n_0 is a *specific* positive integer. In this case, the phase delay $\psi_1(f_0)$ becomes

$$\psi_1(f_0) = -2\pi\left(\frac{f_0 l}{u_{\rm p}}\right) = -2n_0\pi \qquad (9.126)$$

and, similarly, $\psi_2(f_0) = -4n_0\pi$ and $\psi_3(f_0) = -6n_0\pi$. That is, at f_0 all the elements have equal phase (within multiples of 2π) and the array radiates in the broadside direction. If f is changed to $f_0 + \Delta f$, the new phase shift of the first element relative to the zeroth element is

$$\begin{aligned}\psi_1(f_0+\Delta f) &= -\frac{2\pi}{u_{\rm p}}(f_0+\Delta f)l \\ &= -\frac{2\pi f_0 l}{u_{\rm p}} - \left(\frac{2\pi l}{u_{\rm p}}\right)\Delta f \\ &= -2n_0\pi - 2n_0\pi\left(\frac{\Delta f}{f_0}\right) \\ &= -2n_0\pi - \delta, \end{aligned} \qquad (9.127)$$

where use was made of Eq. (9.125) and δ is defined as

$$\delta = 2n_0\pi\left(\frac{\Delta f}{f_0}\right). \qquad (9.128)$$

Similarly, $\psi_2(f_0+\Delta f) = 2\psi_1$ and $\psi_3(f_0+\Delta f) = 3\psi_1$. Ignoring the factor of 2π and its multiples (since they exercise no influence on the relative phases of the radiated fields), we see that the incremental phase shifts are directly proportional to the fractional frequency deviation $(\Delta f/f_0)$. Thus, in an array with N elements, controlling Δf provides a direct control of δ, which in turn controls the scan angle θ_0 according to Eq. (9.120). Equating Eq. (9.120) to Eq. (9.128) and then solving for $\cos\theta_0$ leads to

$$\cos\theta_0 = \frac{2n_0\pi}{kd}\left(\frac{\Delta f}{f_0}\right) \qquad (9.129)$$

As f is changed from f_0 to $f_0+\Delta f$, $k = 2\pi/\lambda = 2\pi f/c$ also changes with frequency. However, if $\Delta f/f_0$ is small, we may treat k as a constant equal to $2\pi f_0/c$; the error in $\cos\theta_0$ resulting from the use of this approximation in Eq. (9.129) is on the order of $\Delta f/f_0$.

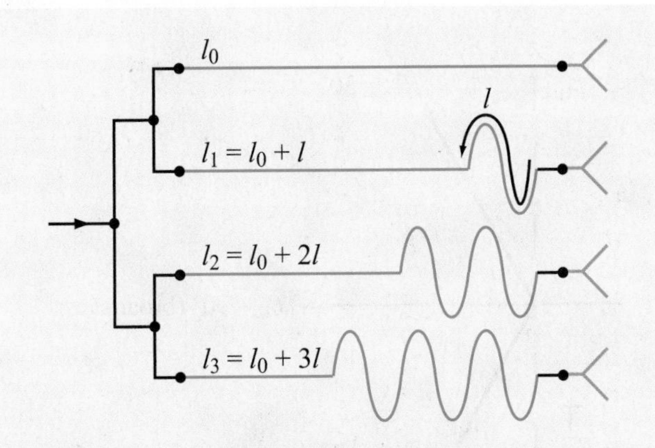

Figure 9-34 An example of a feeding arrangement for frequency-scanned arrays.

Example 9-8: Electronic Steering

Design a steerable six-element array with the following specifications:

1. All elements are excited with equal amplitudes.

2. At $f_0 = 10$ GHz, the array radiates in the broadside direction, and the interelement spacing $d = \lambda_0/2$, where $\lambda_0 = c/f_0 = 3$ cm.

3. The array pattern is to be electronically steerable in the elevation plane over the angular range extending between $\theta_0 = 30°$ and $\theta_0 = 150°$.

4. The antenna array is fed by a voltage-controlled oscillator whose frequency can be varied over the range from 9.5 to 10.5 GHz.

5. The array uses a feeding arrangement of the type shown in Fig. 9-34, and the transmission lines have a phase velocity $u_{\rm p} = 0.8$c.

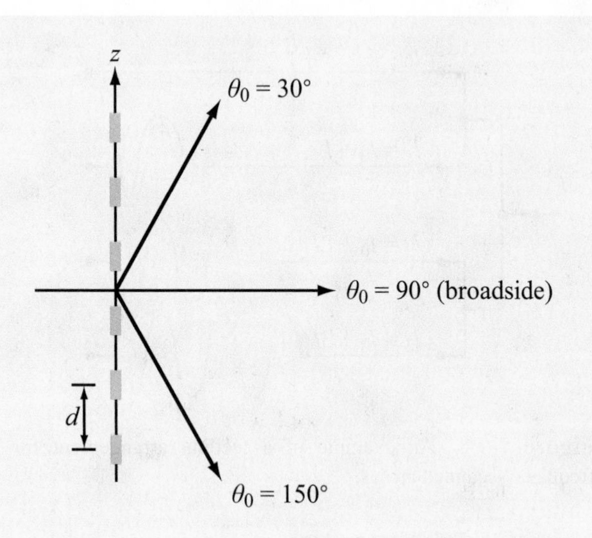

Figure 9-35 Steerable six-element array (Example 9-8).

Solution: The array is to be steerable from $\theta_0 = 30°$ to $\theta_0 = 150°$ (**Fig. 9-35**). For $\theta_0 = 30°$ and $kd = (2\pi/\lambda_0)(\lambda_0/2) = \pi$, Eq. (9.129) gives

$$0.87 = 2n_0\left(\frac{\Delta f}{f_0}\right). \tag{9.130}$$

We are given that $f_0 = 10$ GHz and the oscillator frequency can be varied between $(f_0 - 0.5\text{ GHz})$ and $(f_0 + 0.5\text{ GHz})$. Thus, $\Delta f_{\text{max}} = 0.5$ GHz. To satisfy Eq. (9.130), we need to choose n_0 such that Δf is as close as possible to, but not larger than, Δf_{max}. Solving Eq. (9.130) for n_0 with $\Delta f = \Delta f_{\text{max}}$ gives

$$n_0 = \frac{0.87}{2}\frac{f_0}{\Delta f_{\text{max}}} = 8.7.$$

Since n_0 is not an integer, we need to modify its value by rounding it upward to the next whole-integer value. Hence, we set $n_0 = 9$.

Application of Eq. (9.125) specifies the magnitude of the incremental length l:

$$l = \frac{n_0 u_{\text{p}}}{f_0} = \frac{9 \times 0.8 \times 3 \times 10^8}{10^{10}} = 21.6 \text{ cm}.$$

In summary, with $N = 6$ and $kd = \pi$, Eq. (9.122) becomes:

$$F_{\text{an}}(\gamma') = \frac{\sin^2(3\gamma')}{36\sin^2(\gamma'/2)},$$

with

$$\gamma' = kd(\cos\theta - \cos\theta_0) = \pi(\cos\theta - \cos\theta_0),$$

and

$$\cos\theta_0 = \frac{2n_0\pi}{kd}\left(\frac{\Delta f}{f_0}\right) = 18\left(\frac{f - 10\text{ GHz}}{10\text{ GHz}}\right). \tag{9.131}$$

The shape of the array pattern is similar to that shown in **Fig. 9-30**, and its main-beam direction is along $\theta = \theta_0$. For $f = f_0 = 10$ GHz, $\theta_0 = 90°$ (broadside direction); for $f = 10.48$ GHz, $\theta_0 = 30°$; and for $f = 9.52$ GHz, $\theta_0 = 150°$. For any other value of θ_0 between 30° and 150°, Eq. (9.131) provides the means for calculating the required value of the oscillator frequency f.

Concept Question 9-11: Why are antenna arrays useful? Give examples of typical applications.

Concept Question 9-12: Explain how the pattern multiplication principle is used to compute the radiation pattern of an antenna array.

Concept Question 9-13: For a linear array, what roles do the array amplitudes and phases play?

Concept Question 9-14: Explain how electronic beam steering is accomplished.

Concept Question 9-15: Why is frequency scanning an attractive technique for steering the beam of an antenna array?

Module 9.8 Uniform Dipole Array For an array of up to 50 identical vertical dipoles of selectable length and current maximum, excited with incremental phase delay δ between adjacent elements, the module displays the elevation and azimuthal patterns of the array. By varying δ, the array pattern can be steered in the horizontal plane.

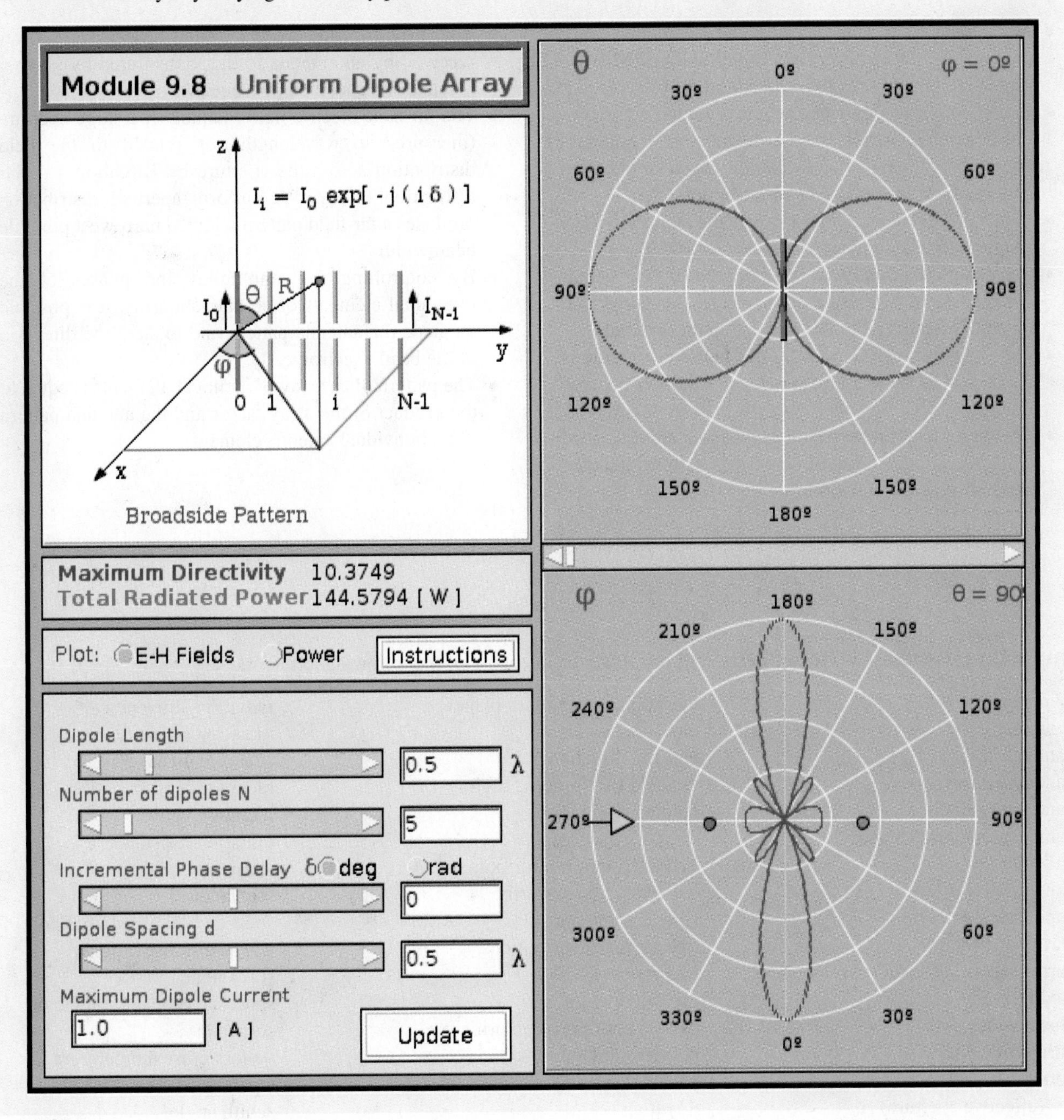

Chapter 9 Summary

Concepts

- An antenna is a transducer between a guided wave propagating on a transmission line and an EM wave propagating in an unbounded medium, or vice versa.
- Except for some solid-state antennas composed of non-linear semiconductors or ferrite materials, antennas are reciprocal devices; they exhibit the same radiation patterns for transmission as for reception.
- In the far-field region of an antenna, the radiated energy is approximately a plane wave.
- The electric field radiated by current antennas, such as wires, is equal to the sum of the electric fields radiated by all the Hertzian dipoles making up the antenna.
- The radiation resistance R_{rad} of a half-wave dipole is 73 Ω, which can be easily matched to a transmission line.
- The directional properties of an antenna are described by its radiation pattern, directivity, pattern solid angle, and half-power beamwidth.
- The Friis transmission formula relates the power received by an antenna to that transmitted by another antenna at a specified distance away.
- The far-zone electric field radiated by a large aperture (measured in wavelengths) is related to the field distribution across the aperture by Kirchhoff's scalar diffraction theory. A uniform aperture distribution produces a far-field pattern with the narrowest possible beamwidth.
- By controlling the amplitudes and phases of the individual elements of an antenna array, it is possible to shape the antenna pattern and to steer the direction of the beam electronically.
- The pattern of an array of identical elements is equal to the product of the array factor and the antenna pattern of an individual antenna element.

Important Terms

Provide definitions or explain the meaning of the following terms:

3 dB beamwidth
antenna
antenna array
antenna directivity D
antenna gain G
antenna input impedance
antenna pattern
antenna polarization
aperture distribution
array distribution
array factor $F_{\text{a}}(\theta, \phi)$
azimuth angle
beamwidth β
broadside direction
effective area (effective aperture) A_{e}
electronic steering
elevation and azimuth planes
elevation angle
end-fire direction
far-field (or far-zone) region
feeding coefficient
frequency scanning
Friis transmission formula
half-power beamwidth
isotropic antenna
linear phase distribution
loss resistance R_{loss}
null beamwidth
pattern multiplication principle
pattern solid angle Ω_{p}
power density $\mathbf{S}(R, \theta, \phi)$
Poynting vector
principal planes
radiation efficiency ξ
radiation intensity (normalized) $F(\theta, \phi)$
radiation lobes
radiation pattern
radiation resistance R_{rad}
reciprocal
scan angle
short dipole (Hertzian dipole)
signal-to-noise ratio S_{n}
solid angle
spherical propagation factor
steradian
system noise temperature T_{sys}
tapered aperture distribution
zenith angle

Mathematical and Physical Models

Antenna Properties

Pattern solid angle $\Omega_p = \iint_{4\pi} F(\theta, \phi)\, d\Omega$

Effective area $A_e = \dfrac{\lambda^2 D}{4\pi}$

Directivity $D = \dfrac{4\pi}{\Omega_p}$

Far-field distance $R > \dfrac{2d^2}{\lambda}$

Gain $G = \xi D, \quad \xi = \dfrac{P_{rad}}{P_{rad} + P_{loss}}$

Short Dipole ($l \ll \lambda$)

$$\widetilde{E}_\theta = \frac{j I_0 l k \eta_0}{4\pi}\left(\frac{e^{-jkR}}{R}\right)\sin\theta$$

$$\widetilde{H}_\phi = \frac{\widetilde{E}_\theta}{\eta_0}$$

$$S(R,\theta) = \left(\frac{\eta_0 k^2 I_0^2 l^2}{32\pi^2 R^2}\right)\sin^2\theta$$

$D = 1.5$

$\beta = 90°$

$R_{rad} = 80\pi^2 (l/\lambda)^2$

$\lambda/2$ Dipole

$$\widetilde{E}_\theta = j\,60 I_0 \left\{\frac{\cos[(\pi/2)\cos\theta]}{\sin\theta}\right\}\left(\frac{e^{-jkR}}{R}\right)$$

$$\widetilde{H}_\phi = \frac{\widetilde{E}_\theta}{\eta_0}$$

$$S(R,\theta) = \frac{15 I_0^2}{\pi R^2}\left\{\frac{\cos^2[(\pi/2)\cos\theta]}{\sin^2\theta}\right\}$$

$D = 1.64$

$\beta = 78°$

$R_{rad} \approx 73\ \Omega$

Friis Transmission Formula

$$\frac{P_{rec}}{P_t} = G_t G_r \left(\frac{\lambda}{4\pi R}\right)^2 F_t(\theta_t, \phi_t)\ F_r(\theta_r, \phi_r)$$

Rectangular Aperture (Uniform)

$S(R,\theta) = S_0\, \mathrm{sinc}^2(\pi l_x \sin\theta/\lambda)$, x-z plane

$S(R,\theta) = S_0\, \mathrm{sinc}^2(\pi l_y \sin\theta/\lambda)$, y-z plane

$$\beta_{xz} = 0.88\,\frac{\lambda}{l_x}, \quad \beta_{yz} = 0.88\,\frac{\lambda}{l_y}$$

$$D = \frac{4\pi A_e}{\lambda^2} \approx \frac{4\pi A_p}{\lambda^2}$$

Antenna Arrays

Multiplication Principle

$$S(R_0, \theta, \phi) = S_e(R_0, \theta, \phi)\ F_a(\theta)$$

Uniform Phase $\quad F_a(\gamma) = \left|\sum_{i=0}^{N-1} a_i e^{ji\gamma}\right|^2$, with $\gamma = kd\cos\theta = \dfrac{2\pi d}{\lambda}\cos\theta$

Linear Phase $\quad F_a(\theta) = \left|\sum_{i=0}^{N-1} a_i e^{ji\gamma'}\right|^2$, with $\gamma' = kd\cos\theta - \delta$

PROBLEMS

Sections 9-1 and 9-2: Hertizan Dipole and Antenna Radiation Characteristics

9.1 A 50 cm long center-fed dipole directed along the z direction and located at the origin is excited by a 1 MHz source. If the current amplitude is $I_0 = 20$ A, determine:

(a) The power density radiated at 2 km along the broadside of the antenna pattern.

(b) The fraction of the total power radiated within the sector between $\theta = 85°$ and $\theta = 95°$?

*__9.2__ A center-fed Hertzian dipole is excited by a current $I_0 = 50$ A. If the dipole is $\lambda/50$ in length, determine the maximum radiated power density at a distance of 1 km.

9.3 A 1 m long dipole is excited by a 1 MHz current with an amplitude of 24 A. What is the average power density radiated by the dipole at a distance of 5 km in a direction that is 45° from the dipole axis?

*__9.4__ Determine the following:

(a) The direction of maximum radiation.

(b) Directivity.

(c) Beam solid angle.

(d) Half-power beamwidth in the x–z plane.

for an antenna whose normalized radiation intensity is given by

$$F(\theta, \phi) = \begin{cases} 1 & \text{for } 0 \le \theta \le 60° \text{and } 0 \le \phi \le 2\pi \\ 0 & \text{elsewhere.} \end{cases}$$

Suggestion: Sketch the pattern prior to calculating the desired quantities.

9.5 Repeat Problem 9.4 for an antenna with

$$F(\theta, \phi) = \begin{cases} \sin^2\theta \cos^2\phi & \text{for } 0 \le \theta \le \pi \\ & \text{and } -\pi/2 \le \phi \le \pi/2 \\ 0 & \text{elsewhere} \end{cases}$$

9.6 A 2 m long center-fed dipole antenna operates in the AM broadcast band at 1 MHz. The dipole is made of copper wire with a radius of 1 mm.

(a) Determine the radiation efficiency of the antenna.

*__(b)__ What is the antenna gain in decibels?

(c) What antenna current is required so that the antenna will radiate 80 W, and how much power will the generator have to supply to the antenna?

9.7 Repeat Problem 9.6 for a 20 cm long antenna operating at 5 MHz.

9.8 Determine the frequency dependence of the radiation efficiency of the short dipole, and plot it over the range from 600 kHz to 60 MHz. The dipole is made of copper, its length is 10 cm, and its circular cross section has a radius of 1 mm.

*__9.9__ An antenna with a pattern solid angle of 1.5 (sr) radiates 90 W of power. At a range of 1 km, what is the maximum power density radiated by the antenna?

9.10 An antenna with a radiation efficiency of 90% has a directivity of 6.0 dB. What is its gain in decibels?

9.11 The normalized radiation intensity of a certain antenna is given by

$$F(\theta) = \exp(-20\theta^2) \qquad \text{for } 0 \le \theta \le \pi$$

where θ is in radians. Determine:

(a) The half-power beamwidth.

(b) The pattern solid angle.

(c) The antenna directivity.

*__9.12__ The radiation pattern of a circular parabolic-reflector antenna consists of a circular major lobe with a half-power beamwidth of 2° and a few minor lobes. Ignoring the minor lobes, obtain an estimate for the antenna directivity in dB.

Sections 9-3 and 9-4: Dipole Antennas

9.13 Repeat Problem 9.6 for a 1 m long half-wave dipole that operates in the FM/TV broadcast band at 150 MHz.

*__9.14__ Assuming the loss resistance of a half-wave dipole antenna to be negligibly small and ignoring the reactance component of its antenna impedance, calculate the standing-wave ratio on a 50 Ω transmission line connected to the dipole antenna.

9.15 For a short dipole with length l such that $l \ll \lambda$, instead of treating the current $\tilde{I}(z)$ as constant along the dipole, as was done in Section 9-1, a more realistic approximation that ensures the current goes to zero at the dipole ends is to describe $\tilde{I}(z)$ by the triangular function

$$\tilde{I}(z) = \begin{cases} I_0(1 - 2z/l) & \text{for } 0 \le z \le l/2 \\ I_0(1 + 2z/l) & \text{for } -l/2 \le z \le 0 \end{cases}$$

*Answer(s) available in Appendix D.

as shown in **Fig. P9.15**. Use this current distribution to determine the following:

*(a) The far-field $\widetilde{\mathbf{E}}(R, \theta, \phi)$.

(b) The power density $S(R, \theta, \phi)$.

(c) The directivity D.

(d) The radiation resistance R_{rad}.

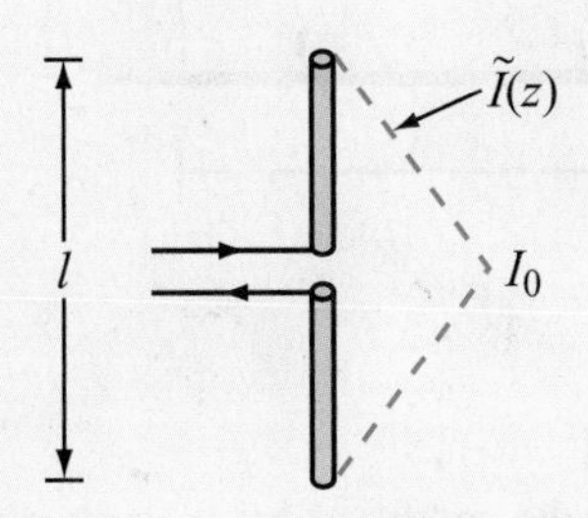

Figure P9.15 Triangular current distribution on a short dipole (Problem 9.15).

9.16 A 50 cm long dipole is excited by a sinusoidally varying current with an amplitude $I_0 = 5$ A. Determine the total radiated power if the oscillating frequency is:

(a) 1 MHz,

(b) 300 MHz.

9.17 For a dipole antenna of length $l = 3\lambda/2$,

*(a) Determine the directions of maximum radiation.

(b) Obtain an expression for $S_{\max}$.

(c) Generate a plot of the normalized radiation pattern $F(\theta)$.

(d) Compare your pattern with that shown in **Fig. 9-17(c)**.

9.18 For a dipole antenna of length $l = \lambda/4$,

(a) Determine the directions of maximum radiation.

(b) Obtain an expression for $S_{\max}$.

(c) Generate a plot of the normalized radiation pattern $F(\theta)$.

9.19 Repeat parts (a)–(c) of Problem 9.17 for a dipole of length $l = 3\lambda/4$.

***9.20** Repeat parts (a)–(c) of Problem 9.17 for a dipole of length $l = \lambda$.

9.21 A car antenna is a vertical monopole over a conducting surface. Repeat Problem 9.6 for a 1 m long car antenna operating at 1 MHz. The antenna wire is made of aluminum with $\mu_c = \mu_0$ and $\sigma_c = 3.5 \times 10^7$ S/m, and its diameter is 1 cm.

Sections 9-5 and 9-6: Effective Area and Friis Formula

9.22 Determine the effective area of a half-wave dipole antenna at 200 MHz, and compare it with its physical cross-section if the wire diameter is 2 cm.

9.23 A 3 GHz line-of-sight microwave communication link consists of two lossless parabolic dish antennas, each 1 m in diameter. If the receive antenna requires 1 nW of receive power for good reception and the distance between the antennas is 40 km, how much power should be transmitted?

9.24 A half-wave dipole TV broadcast antenna transmits 10 kW at 50 MHz. What is the power received by a home television antenna with 3 dB gain if located at a distance of 30 km?

9.25 A 150 MHz communication link consists of two vertical half-wave dipole antennas separated by 2 km. The antennas are lossless, the signal occupies a bandwidth of 3 MHz, the system noise temperature of the receiver is 900 K, and the desired signal-to-noise ratio is 17 dB. What transmitter power is required?

9.26 Consider the communication system shown in **Fig. P9.26**, with all components properly matched. If $P_t = 10$ W and $f = 6$ GHz:

(a) What is the power density at the receiving antenna (assuming proper alignment of antennas)?

(b) What is the received power?

(c) If $T_{\text{sys}} = 1{,}000$ K and the receiver bandwidth is 20 MHz, what is the signal-to-noise ratio in decibels?

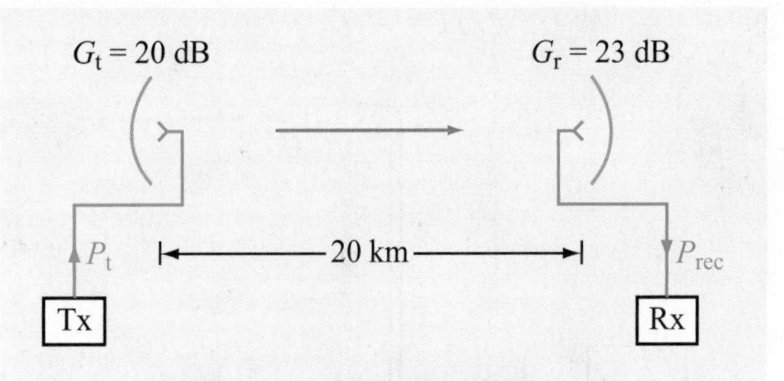

Figure P9.26 Communication system of Problem 9.26.

9.27 The configuration shown in **Fig. P9.27** depicts two vertically oriented half-wave dipole antennas pointed towards each other, with both positioned on 100 m tall towers separated

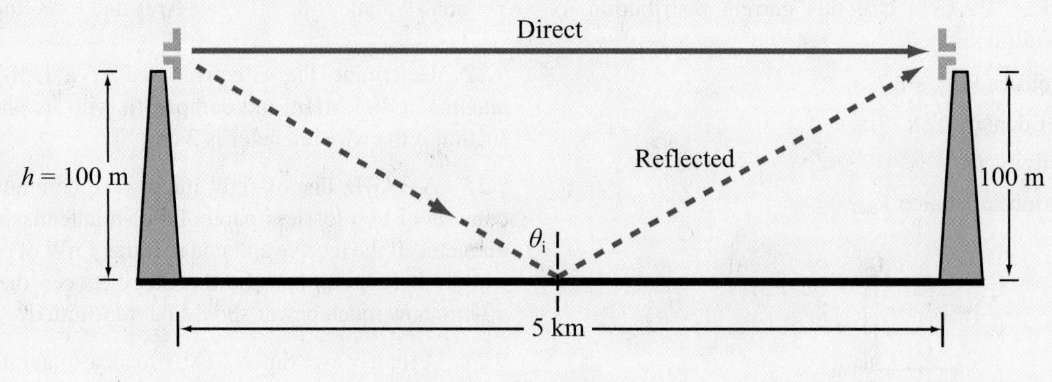

Figure P9.27 Problem 9.27.

by a distance of 5 km. If the transit antenna is driven by a 50 MHz current with amplitude $I_0 = 2$ A, determine:

*(a) The power received by the receive antenna in the absence of the surface. (Assume both antennas to be lossless.)

(b) The power received by the receive antenna after incorporating reflection by the ground surface, assuming the surface to be flat and to have $\epsilon_r = 9$ and conductivity $\sigma = 10^{-3}$ (S/m).

9.28 **Fig. P9.28** depicts a half-wave dipole connected to a generator through a matched transmission line. The directivity of the dipole can be modified by placing a reflecting rod a distance d behind the dipole. What would its reflectivity in the forward direction be if:

(a) $d = \lambda/4$,

(b) $d = \lambda/2$.

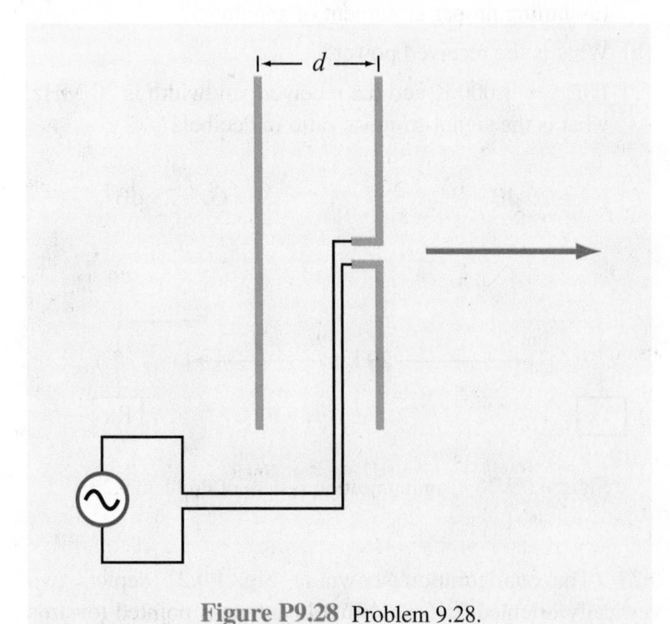

Figure P9.28 Problem 9.28.

9.29 The configuration shown in **Fig. P9.29** depicts a satellite repeater with two antennas, one pointed towards the antenna of ground station 1 and the other towards the antenna of ground station 2. All antennas are parabolic dishes, antennas A_1 and A_4 are each 4 m in diameter, antennas A_2 and A_3 are each 2 m in diameter, and the distance between the satellite and each of the ground stations is 40,000 km. Upon receiving the signal by its antenna A_2, the satellite transponder boosts the power gain by 80 dB and then retransmits the signal to A_4. The system operates at 10 GHz with $P_t = 1$ kW. Determine the received power P_r. Assume all antennas to be lossless.

Sections 9-7 and 9-8: Radiation by Apertures

9.30 The 10 dB beamwidth is the beam size between the angles at which $F(\theta)$ is 10 dB below its peak value. Determine the 10 dB beamwidth in the x–z plane for a uniformly illuminated aperture with length $l_x = 10\lambda$.

***9.31** A uniformly illuminated aperture is of length $l_x = 20\lambda$. Determine the beamwidth between first nulls in the x–z plane.

***9.32** A uniformly illuminated rectangular aperture situated in the x–y plane is 2 m high (along x) and 1 m wide (along y). If $f = 5$ GHz, determine the following:

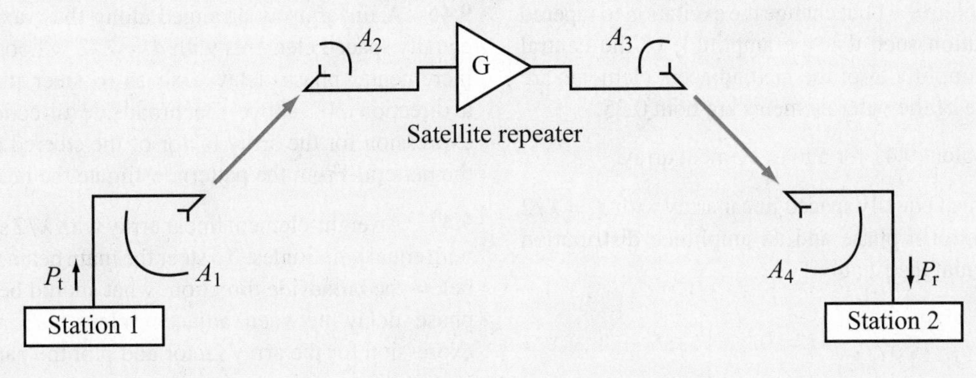

Figure P9.29 Satellite repeater system.

(a) The beamwidths of the radiation pattern in the elevation plane (x–z plane) and the azimuth plane (y–z plane).

(b) The antenna directivity D in decibels.

9.33 An antenna with a circular aperture has a circular beam with a beamwidth of 3° at 20 GHz.

(a) What is the antenna directivity in dB?

(b) If the antenna area is doubled, what will be the new directivity and new beamwidth?

(c) If the aperture is kept the same as in (a), but the frequency is doubled to 40 GHz, what will the directivity and beamwidth become then?

*__9.34__ A 94 GHz automobile collision-avoidance radar uses a rectangular-aperture antenna placed above the car's bumper. If the antenna is 1 m in length and 10 cm in height, determine the following:

(a) Its elevation and azimuth beamwidths.

(b) The horizontal extent of the beam at a distance of 300 m.

9.35 Compare directivity D_{ant} of a 1 m diameter antenna aperture operating at 10 GHz with directivity D_{eye} of the eye's pupil operating in the middle of the visible spectrum at $\lambda = 0.5$ μm. Treat the pupil as a circular aperture with a diameter of 4 mm.

9.36 A microwave telescope consisting of a very sensitive receiver connected to a 100 m parabolic-dish antenna is used to measure the energy radiated by astronomical objects at 20 GHz. If the antenna beam is directed toward the moon and the moon extends over a planar angle of 0.5° from Earth, what fraction of the moon's cross-section will be occupied by the beam?

Sections 9-9 through 9-11: Antenna Arrays

9.37 A two-element array consisting of two isotropic antennas separated by a distance d along the z axis is placed in a coordinate system whose z axis points eastward and whose x axis points toward the zenith. If a_0 and a_1 are the amplitudes of the excitations of the antennas at $z = 0$ and at $z = d$, respectively, and if δ is the phase of the excitation of the antenna at $z = d$ relative to that of the other antenna, find the array factor and plot the pattern in the x–z plane for the following:

*__(a)__ $a_0 = a_1 = 1$, $\delta = \pi/4$, and $d = \lambda/2$

(b) $a_0 = 1$, $a_1 = 2$, $\delta = 0$, and $d = \lambda$

(c) $a_0 = a_1 = 1$, $\delta = -\pi/2$, and $d = \lambda/2$

(d) $a_0 = 1$, $a_1 = 2$, $\delta = \pi/4$, and $d = \lambda/2$

(e) $a_0 = 1$, $a_1 = 2$, $\delta = \pi/2$, and $d = \lambda/4$

9.38 If the antennas in part (a) of Problem 9.37 are parallel, vertical, Hertzian dipoles with axes along the x direction, determine the normalized radiation intensity in the x–z plane and plot it.

*__9.39__ Consider the two-element dipole array of Fig. 9-29(a). If the two dipoles are excited with identical feeding coefficients ($a_0 = a_1 = 1$ and $\psi_0 = \psi_1 = 0$), choose (d/λ) such that the array factor has a maximum at $\theta = 45°$.

9.40 Choose (d/λ) so that the array pattern of the array of Problem 9.39 has a null, rather than a maximum, at $\theta = 45°$.

9.41 Find and plot the normalized array factor and determine the half-power beamwidth for a five-element linear array excited with equal phase and a uniform amplitude distribution. The interelement spacing is $3\lambda/4$.

9.42 Repeat Problem 9.41 but change the excitation to tapered amplitude distribution such that the amplitude of the central element is 1, the amplitudes of the next adjacent elements are both 0.5, and those of the outer elements are both 0.25.

9.43 Repeat Problem 9.41 for a nine-element array.

*__9.44__ A five-element equally spaced linear array with $d = \lambda/2$ is excited with uniform phase and an amplitude distribution given by the binomial distribution

$$a_i = \frac{(N-1)!}{i!(N-i-1)!}, \qquad i = 0, 1, \ldots, (N-1),$$

where N is the number of elements. Develop an expression for the array factor.

9.45 A three-element linear array of isotropic sources aligned along the z axis has an interelement spacing of $\lambda/4$ (**Fig. P9.45**). The amplitude excitation of the center element is twice that of the bottom and top elements, and the phases are $-\pi/2$ for the bottom element and $\pi/2$ for the top element, relative to that of the center element. Determine the array factor and plot it in the elevation plane.

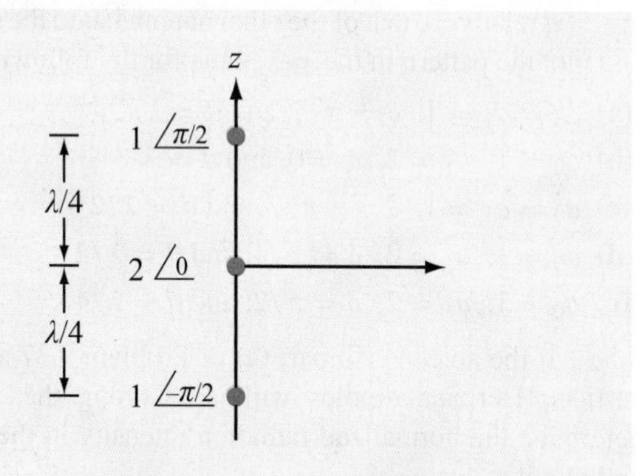

Figure P9.45 Three-element array of Problem 9.48.

9.46 A linear array arranged along the z axis consists of 12 equally spaced elements with $d = \lambda/2$. Choose an appropriate incremental phase delay δ so as to steer the main beam to a direction 30° above the broadside direction. Provide an expression for the array factor of the steered antenna and plot the pattern. From the pattern, estimate the beamwidth.

*__9.47__ An eight-element linear array with $\lambda/2$ spacing is excited with equal amplitudes. To steer the main beam to a direction 60° below the broadside direction, what should be the incremental phase delay between adjacent elements? Also, give the expression for the array factor and plot the pattern.

C H A P T E R

10

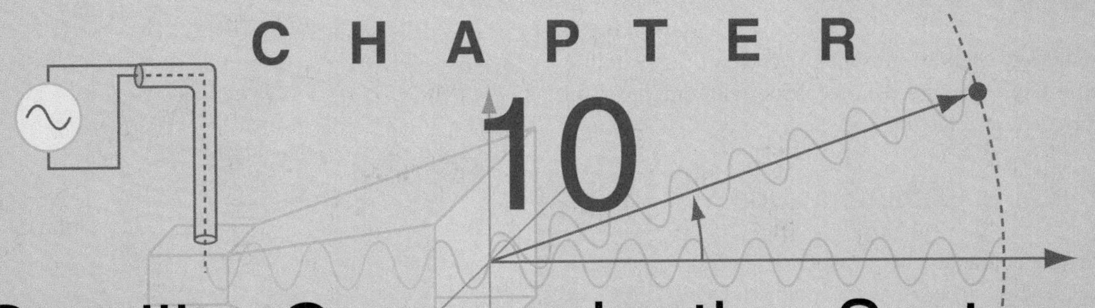

Satellite Communication Systems and Radar Sensors

Chapter Contents

Application Examples, 480
10-1 Satellite Communication Systems, 480
10-2 Satellite Transponders, 482
10-3 Communication-Link Power Budget, 484
10-4 Antenna Beams, 485
10-5 Radar Sensors, 486
10-6 Target Detection, 489
10-7 Doppler Radar, 491
10-8 Monopulse Radar, 492
Chapter 10 Summary, 495
Problems, 496

Objectives

Upon learning the material presented in this chapter, you should be able to:

1. Describe the basic operation of satellite transponders.
2. Calculate the power budget for a communication link.
3. Describe how radar attains spatial and angular resolutions, calculate the maximum detectable range, and explain the tradeoff between the probabilities of detection and false alarm.
4. Calculate the Doppler frequency shift observed by a radar.
5. Describe the monopulse-radar technique.

Application Examples

This concluding chapter presents overviews of satellite communication systems and radar sensors, with emphasis on their electromagnetic-related aspects.

10-1 Satellite Communication Systems

Today's world is connected by a vast communication network that provides a wide array of voice, data, and video services to both fixed and mobile terminals (**Fig. 10-1**). The viability and effectiveness of the network are attributed in large measure to the use of orbiting satellite systems that function as relay stations with wide area coverage of Earth's surface. From a geostationary orbit at 35,786 km above the equator, a satellite can view over one-third of Earth's surface and can connect any pair of points within its coverage (**Fig. 10-2**). The history of communication satellite engineering dates back to the late 1950s when the U.S. navy used the moon as a passive reflector to relay low-data-rate communications between Washington, D.C., and Hawaii. The first major development involving artificial Earth satellites took place in October of 1957 when the Soviet Union launched ***Sputnik I*** and used it for 21 days to transmit (one-way) telemetry information to a ground receiving station. This was followed by another telemetry satellite, ***Explorer I***, launched by the United States in January 1958. An important development took place in December of that year when the United States launched the ***Score*** satellite and used it to broadcast President Eisenhower's Christmas message, marking the first instance of two-way voice communication via an artificial satellite.

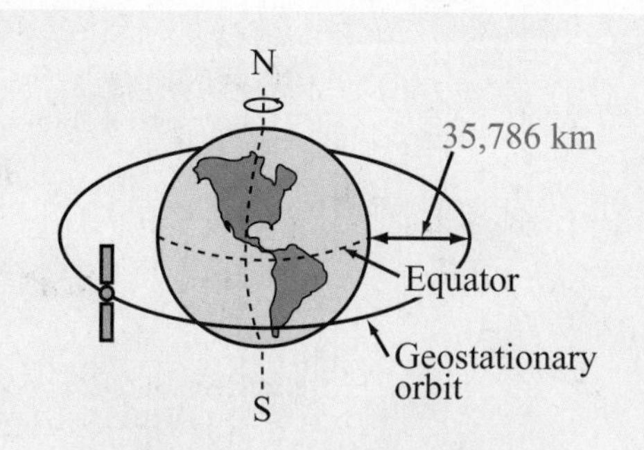

(a) Geostationary satellite orbit

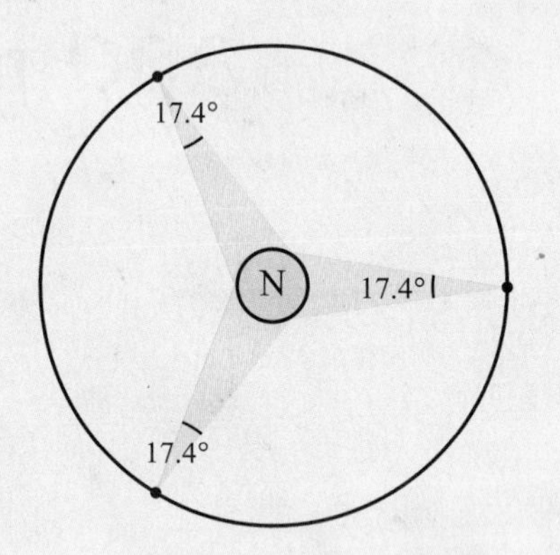

(b) Worldwide coverage by three satellites spaced 120° apart

Figure 10-2 Orbits of geostationary satellites.

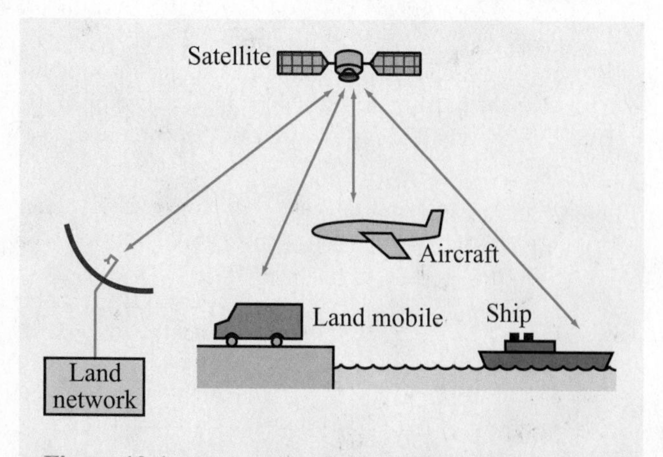

Figure 10-1 Elements of a satellite communication network.

These achievements were followed by a flurry of space activity, leading to the development of operational communication satellites by many countries for both commercial and governmental services. This section describes satellite communications links with emphasis on transmitter–receiver power calculations, propagation aspects, frequency allocations, and antenna design considerations.

A satellite is said to be in a ***geostationary orbit*** around Earth when it is in a circular orbit in a plane identical with Earth's equatorial plane at an altitude where the orbital period is identical with Earth's rotational period, thereby appearing stationary relative to Earth's surface. A satellite of mass M_s in circular orbit around Earth (**Fig. 10-3**) is subject to two forces, the attractive gravitational force F_g and the repelling centrifugal

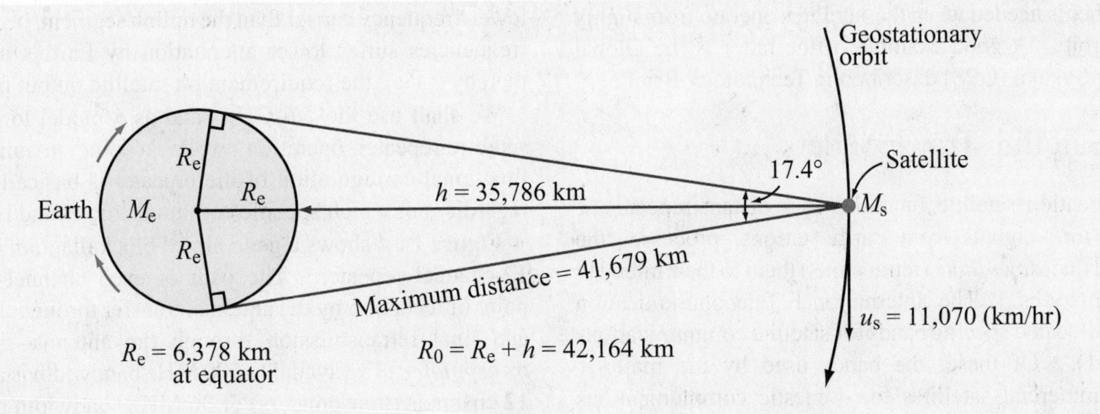

Figure 10-3 Satellite of mass m_s in orbit around Earth. For the orbit to be geostationary, the distance R_0 between the satellite and Earth's center should be 42,164 km. At the equator, this corresponds to an altitude of 35,786 km above Earth's surface.

force F_c. The magnitudes of these two forces are given by

$$F_g = \frac{GM_sM_e}{R_0^2}, \tag{10.1}$$

$$F_c = \frac{M_su_s^2}{R_0} = M_s\omega^2 R_0, \tag{10.2}$$

where $G = 6.67\times10^{-11}$ N·m²/kg² is the universal gravitational constant, $M_e = 5.98 \times 10^{24}$ kg is Earth's mass, R_0 is the distance between the satellite and the center of Earth, and u_s is the satellite velocity. For a rotating object, $u_s = \omega R_0$, where ω is its angular velocity. In order for the satellite to remain in orbit, the two opposing forces acting on it have to be equal in magnitude, or

$$G\frac{M_sM_e}{R_0^2} = M_s\omega^2 R_0, \tag{10.3}$$

which yields a solution for R_0 given by

$$R_0 = \left(\frac{GM_e}{\omega^2}\right)^{1/3}. \tag{10.4}$$

To remain stationary with respect to Earth's surface, the satellite's angular velocity has to be the same as that of Earth's own angular velocity around its own axis. Thus,

$$\omega = \frac{2\pi}{T}, \tag{10.5}$$

where T is the period of one sidereal day in seconds. A sidereal day, which takes into account Earth's rotation around the sun, is equal to 23 hours, 56 minutes, and 4.1 seconds. Using Eq. (10.5) in Eq. (10.4) gives

$$R_0 = \left(\frac{GM_eT^2}{4\pi^2}\right)^{1/3}, \tag{10.6}$$

and upon using the numerical values for T, M_e, and G, we obtain the result $R_0 = 42{,}164$ km. Subtracting 6,378 km for Earth's mean radius at the equator gives an altitude of $h = 35,786$ km above Earth's surface.

From a geostationary orbit, Earth subtends an angle of 17.4°, covering an arc of about 18,000 km along the equator, which corresponds to a longitude angle of about 160°. With three equally spaced satellites in geostationary orbit over Earth's equator, it is possible to achieve complete global coverage of the entire equatorial plane, with significant overlap between the beams of the three satellites. As far as coverage toward the poles, a global beam can reach Earth stations up to 81° of latitude on either side of the equator.

Not all satellite communication systems use spacecraft that are in geostationary orbits. Indeed, because of transmitter power limitations or other considerations, it is sometimes necessary to operate from much lower altitudes, in which case the satellite is placed in a ***highly elliptical orbit*** (to satisfy Kepler's law) such that for part of the orbit (near its perigee) it is at a range of only a few hundred kilometers from Earth's surface. Whereas only three geostationary satellites are needed to provide near-global coverage of Earth's surface, a much larger number is needed when the satellites operate from highly elliptical orbits. A good example of the latter is the Global Positioning System (GPS) described in Technology Brief 5.

10-2 Satellite Transponders

A communication satellite functions as a distant repeater; it receives ***uplink*** signals from Earth stations, processes the signals, and then ***downlinks*** (retransmits) them to their intended Earth destinations. The International Telecommunication Union has allocated specific bands for satellite communications (**Table 10-1**). Of these, the bands used by the majority of U.S. commercial satellites for domestic communications are the ***4/6 GHz band*** (3.7 to 4.2 GHz downlink and 5.925 to 6.425 GHz uplink) and the ***12/14 GHz band*** (11.7 to 12.2 GHz downlink and 14.0 to 14.5 GHz uplink). Each uplink and downlink segment has been allocated 500 MHz of bandwidth. By using different frequency bands for Earth-to-satellite uplink segments and for satellite-to-Earth downlink segments, the same antennas can be used for both functions while simultaneously guarding against interference between the two signals. The downlink segment commonly uses a lower-frequency carrier than the uplink segment, because lower frequencies suffer lower attenuation by Earth's atmosphere, thereby easing the requirement on satellite output power.

Table 10-1 Communications satellite frequency allocations.

Use	Downlink Frequency (MHz)	Uplink Frequency (MHz)
	Fixed Service	
Commercial (C-band)	3,700–4,200	5,925–6,425
Military (X-band)	7,250–7,750	7,900–8,400
Commercial (K-band)		
Domestic (USA)	11,700–12,200	14,000–14,500
International	10,950–11,200	27,500–31,000
	Mobile Service	
Maritime	1,535–1,542.5	1,635–1,644
Aeronautical	1,543.5–1,558.8	1,645–1,660
	Broadcast Service	
	2,500–2,535	2,655–2,690
	11,700–12,750	
	Telemetry, Tracking, and Command	
	137–138, 401–402, 1,525–1,540	

We shall use the 4/6 GHz band as a model to discuss the satellite-repeater operation, while keeping in mind that the functional configuration of the repeater is basically the same regardless of which specific communication band is used.

Figure 10-4 shows a generalized block diagram of a typical 12-channel repeater. The path of each channel—from the point of reception by the antenna, transfer through the repeater, and final retransmission through the antenna—is called a ***transponder***. The available 500 MHz bandwidth is allocated to 12 channels (transponders) of 36 MHz bandwidth per channel and 4 MHz separation between channels. The basic functions of a transponder are: (a) isolation of neighboring radio frequency (RF) channels, (b) frequency translation, and (c) amplification. With ***frequency-division multiple access*** (FDMA)—one of the schemes commonly used for information transmission—each transponder can accommodate thousands of individual telephone channels within its 36 MHz of bandwidth (telephone speech signals require a minimum bandwidth of 3 kHz, so frequency spacing is nominally 4 kHz per telephone channel), several TV channels (each requiring a bandwidth of 6 MHz), millions of bits of digital data, or combinations of all three.

When the same antenna is used for both transmission and reception, a ***duplexer*** is used to perform the signal separation. Many types of duplexers are available, but among the simplest to understand is the circulator shown in **Fig. 10-5**. A ***circulator*** is a three-port device that uses a ferrite material placed in a magnetic field induced by a permanent magnet to achieve power flow from ports 1 to 2, 2 to 3, and 3 to 1, but not in the reverse directions. With the antenna connected to port 1, the received signal is channeled only to port 2; if port 2 is properly matched to the band-pass filter, no part of the received signal is reflected from port 2 to 3. Similarly, the transmitted signal connected to port 3 is channeled by the circulator to port 1 for transmission by the antenna.

Following the duplexer shown in **Fig. 10-4**, the received signal passes through a receiver band-pass filter that ensures isolation of the received signal from the transmitted signal. The receiver filter covers the bandwidth from 5.925 to 6.425 GHz, which encompasses the cumulative bandwidths of all 12 channels; the first received channel extends from 5,927 to 5,963 MHz, the second one from 5,967 to 6,003 MHz, and so on until the twelfth channel, which covers the range from 6,367 to 6,403 MHz. Tracing the signal path, the next subsystem is the wideband receiver, which consists of three elements:

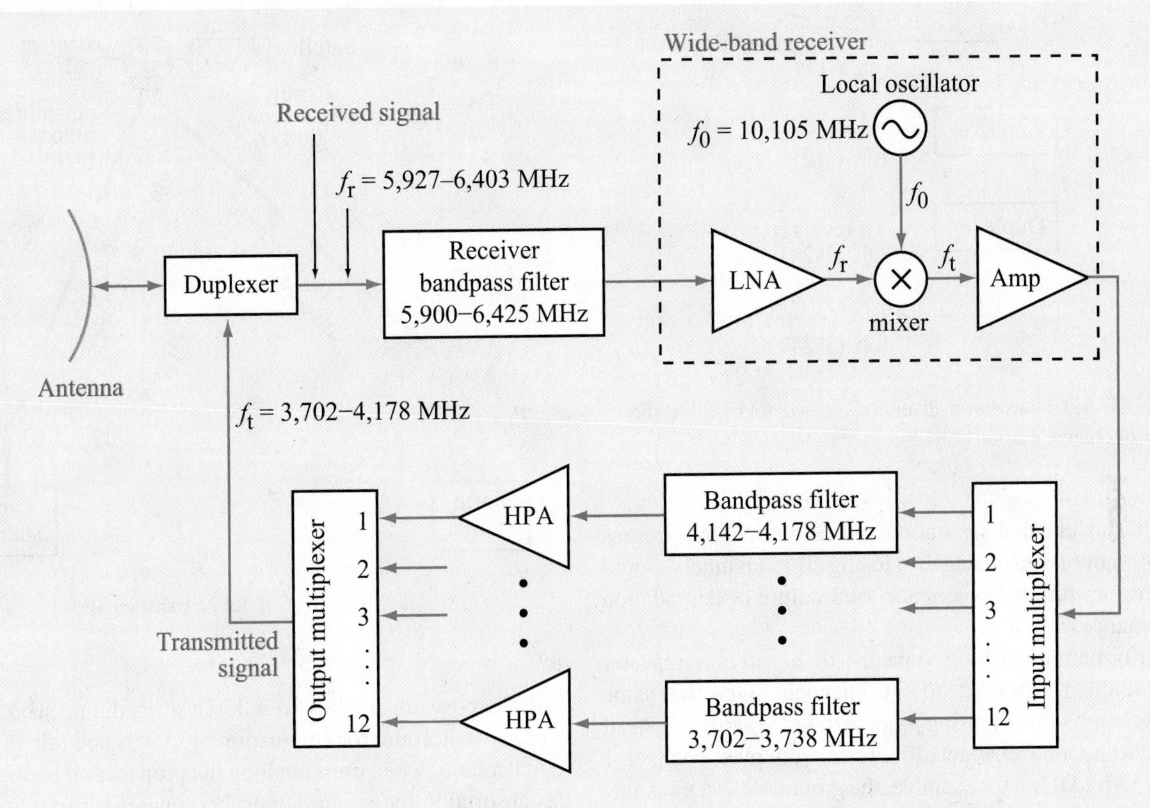

Figure 10-4 Elements of a 12-channel (transponder) communications system.

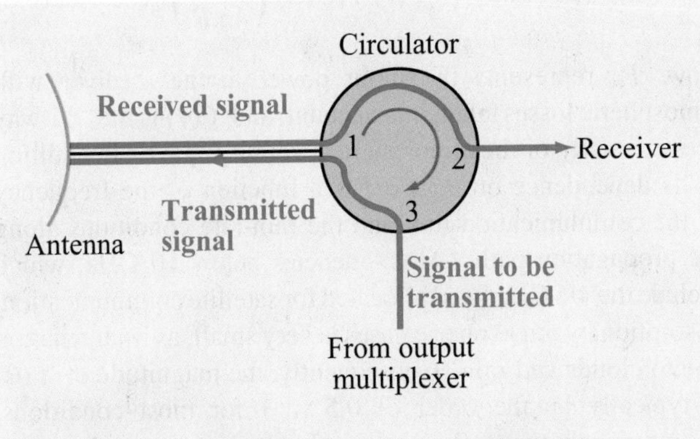

Figure 10-5 Basic operation of a ferrite circulator.

a low-noise wideband amplifier, a frequency translator, and an output amplifier. The frequency translator consists of a stable local oscillator, which generates a signal at frequency $f_0 = 10{,}105$ MHz, connected to a nonlinear microwave mixer. The mixer serves to convert the frequency f_r of the received signal (which covers the range from 5,927 to 6,403 MHz) to a lower-frequency signal $f_t = f_0 - f_r$. Thus, the lower end of the received signal frequency band gets converted from 5,927 to 4,178 MHz and the upper end gets converted from 6,403 to 3,702 MHz. This translation results in 12 channels with new frequency ranges, but whose signals carry the same information (modulation) that was present in the received signals. In principle, the receiver output signal can now be further amplified and then channeled to the antenna through the duplexer for transmission back to Earth. Instead, the receiver output signal is separated into the 12 transponder channels through a ***multiplexer*** followed by a bank of narrow band-pass filters, each covering the bandwidth of one transponder channel. Each of the 12 channels is amplified by its own ***high-power amplifier*** (HPA), and then the 12 channels are combined by another multiplexer that feeds the combined spectrum into the

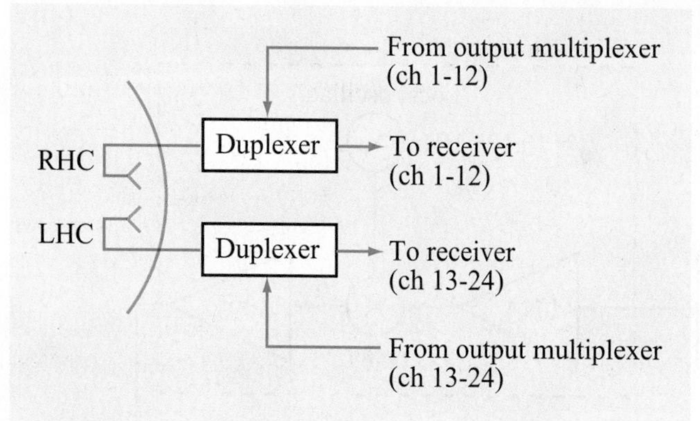

Figure 10-6 Polarization diversity is used to increase the number of channels from 12 to 24.

duplexer. This channel separation and recombination process is used as a safety measure against losing all 12 channels should a high-power amplifier experience total failure or degradation in performance.

The information-carrying capacity of a satellite repeater can be doubled from 12 to 24 channels over the same 500 MHz bandwidth by using ***polarization diversity***. Instead of transmitting one channel of information over channel 1 (5,927 to 5,963 MHz), for example, the ground station transmits to the satellite two signals carrying different information and covering the same frequency band, but with different antenna polarization configurations, such as right-hand circular (RHC) and left-hand circular (LHC) polarizations. The satellite antenna is equipped with a feed arrangement that can receive each of the two circular polarization signals individually with negligible interference between them. Two duplexers are used in this case, one connected to the RHC polarization feed and another connected to the LHC polarization feed, as illustrated in **Fig. 10-6**.

10-3 Communication-Link Power Budget

The uplink and downlink segments of a satellite communication link (**Fig. 10-7**) are each governed by the ***Friis transmission formula*** (Section 9-6), which states that the power P_r received by an antenna with gain G_r due to the transmission of power P_t by an antenna with gain G_t at a range R is given by

$$P_r = P_t G_t G_r \left(\frac{\lambda}{4\pi R}\right)^2. \tag{10.7}$$

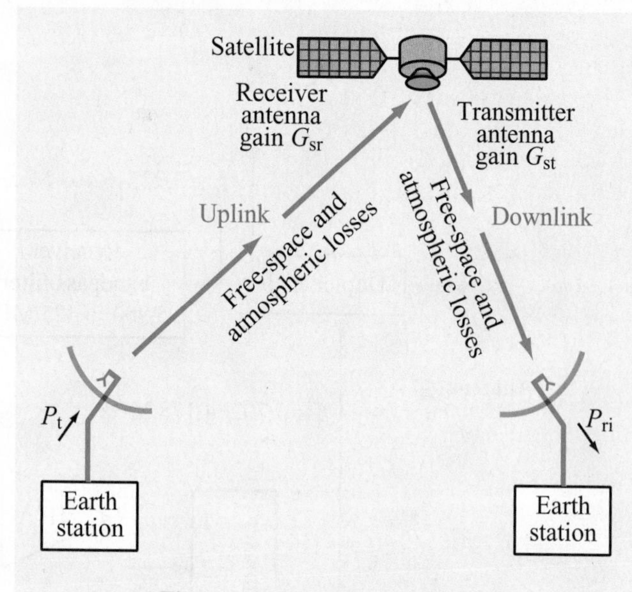

Figure 10-7 Satellite transponder.

This expression applies to a lossless medium, such as free space. To account for attenuation by clouds and rain in Earth's atmosphere (when present along the propagation path), as well as absorption by certain atmospheric gases (primarily oxygen and water vapor), we rewrite Eq. (10.7) as

$$P_{ri} = \Upsilon(\theta)\ P_r = \Upsilon(\theta)\ P_t G_t G_r \left(\frac{\lambda}{4\pi R}\right)^2. \tag{10.8}$$

Now, P_{ri} represents the input power at the receiver with atmospheric losses taken into account, and $\Upsilon(\theta)$ is the one-way ***transmissivity*** of the atmosphere at zenith angle θ. In addition to its dependence on θ, $\Upsilon(\theta)$ is a function of the frequency of the communication link and the rain-rate conditions along the propagation path. At frequencies below 10 GHz, which include the 4/6 GHz band allocated for satellite communication, absorption by atmospheric gases is very small, as is attenuation due to clouds and rain. Consequently, the magnitude of $\Upsilon(\theta)$ is typically on the order of 0.5 to 1 for most conditions. A transmissivity of 0.5 means that twice as much power needs to be transmitted (compared to the free-space case) in order to receive a specified power level. Among the various sources of atmospheric attenuation, the most serious is rainfall, and its attenuation coefficient increases rapidly with increasing frequency. Consequently, atmospheric attenuation assumes greater importance with regard to transmitter power requirements as the communication-system frequency is increased toward higher bands in the microwave region.

The noise appearing at the receiver output, P_{no}, consists of three contributions: (1) noise internally generated by the receiver electronics, (2) noise picked up by the antenna due to external sources, including emission by the atmosphere, and (3) noise due to thermal emission by the antenna material. The combination of all noise sources can be represented by an equivalent ***system noise temperature***, T_{sys}, defined such that

$$P_{no} = G_{rec} K T_{sys} B, \tag{10.9}$$

where K is Boltzmann's constant and G_{rec} and B are the receiver power gain and bandwidth. This output noise level is the same as would appear at the output of a noise-free receiver with input noise level

$$P_{ni} = \frac{P_{no}}{G_{rec}} = K T_{sys} B. \tag{10.10}$$

The ***signal-to-noise ratio*** is defined as the ratio of the signal power to the noise power *at the input of an equivalent noise-free receiver*. Hence,

$$S_n = \frac{P_{ri}}{P_{ni}} = \frac{\Upsilon(\theta)\, P_t G_t G_r}{K T_{sys} B} \left(\frac{\lambda}{4\pi R}\right)^2. \tag{10.11}$$

The performance of a communication system is governed by two sets of issues. The first encompasses the signal-processing techniques used to encode, modulate, combine, and transmit the signal at the transmitter end and to receive, separate, demodulate, and decode the signal at the receiver end. The second set encompasses the gains and losses in the communication link, and they are represented by the signal-to-noise ratio S_n. For a given set of signal-processing techniques, S_n determines the quality of the received signal, such as the bit error rate in digital data transmission and sound and picture quality in audio and video transmissions. Very high quality signal transmission requires very high values of S_n; in broadcast-quality television by satellite, some systems are designed to provide values of S_n exceeding 50 dB (or a factor of 10^5).

The performance of a satellite link depends on the composite performance of the uplink and downlink segments. If either segment performs poorly, the composite performance will be poor, regardless of how good the performance of the other segment is.

10-4 Antenna Beams

Whereas most Earth-station antennas are designed to provide highly directive beams (to avoid interference effects), the satellite antenna system is designed to produce beams tailored to match the areas served by the satellite. For global coverage, beamwidths of 17.4° are required. In contrast, for transmission to and reception from a small area, beamwidths on the order of 1° or less may be needed (**Fig. 10-8**).

An antenna with a beamwidth β of 1° would produce a spot beam on Earth covering an area approximately 630 km in diameter.

Beam size has a direct connection to antenna gain and, in turn, to transmitter power requirements. Antenna gain G

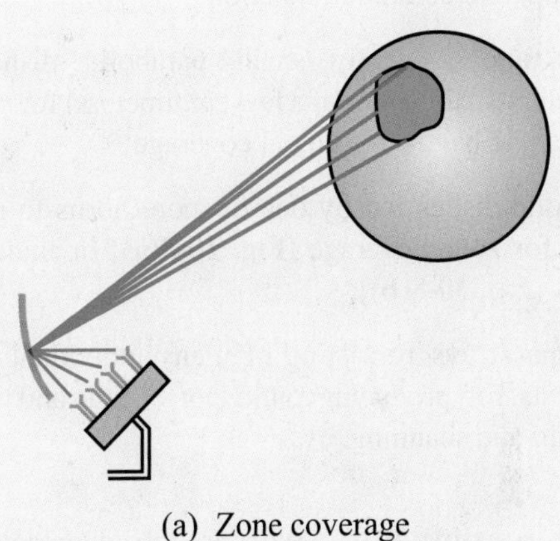

(a) Zone coverage

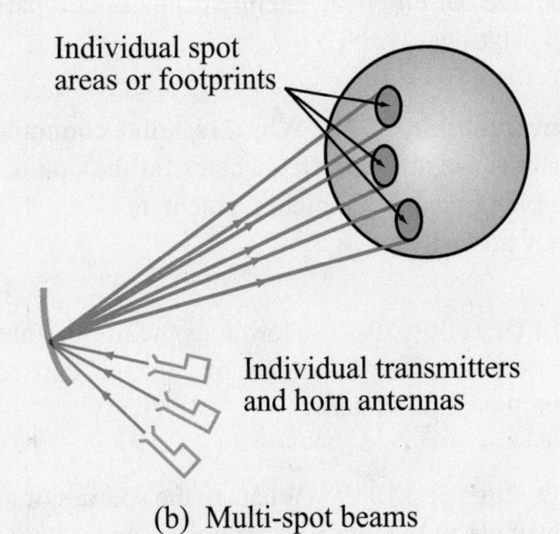

(b) Multi-spot beams

Figure 10-8 Spot and multibeam satellite antenna systems for coverage of defined areas on Earth's surface.

is related to the directivity D by $G = \xi D$, where ξ is the radiation efficiency, and D is related to the beamwidth β by the approximate expression given by Eq. (9.26). For a circular beam,

$$G = \xi \frac{4\pi}{\beta^2}, \quad (10.12)$$

where β is in radians. For a lossless antenna ($\xi = 1$), a global beam with $\beta = 17.4°$ ($= 0.3$ rad) corresponds to a gain $G = 136$, or 21.3 dB. A narrow 1° beam, on the other hand, corresponds to an antenna gain of 41,253, or 46.2 dB.

To accommodate the various communication functions associated with satellite systems, four main types of antennas are used[†]:

1. Dipoles and helices at VHF and UHF for telemetry, tracking, and command functions;

2. Horns and relatively small parabolic dishes (with diameters on the order of a few centimeters) for producing wide-angle beams for global coverage;

3. Parabolic dishes fed by one or more horns to provide a beam for zone coverage [Fig. 10-8(a)] or multiple spot beams [Fig. 10-8(b)];

4. Antenna arrays consisting of many individual radiating elements for producing multispot beams and for beam steering and scanning.

Concept Question 10-1: What are the advantages and disadvantages of elliptical satellite orbits in comparison to the geostationary orbit?

Concept Question 10-2: Why do satellite communication systems use different frequencies for the uplink and downlink segments? Which segment uses the higher frequency and why?

Concept Question 10-3: How does the use of antenna polarization increase the number of channels carried by the communication system?

Concept Question 10-4: What are the sources of noise that contribute to the total system noise temperature of a receiver?

[†]R. G. Meadows and A. J. Parsons, *Satellite Communications*, Hutchinson Publishers, London, 1989.

10-5 Radar Sensors

The term ***radar*** is a contracted form of the phrase ***radio detection and ranging***, which conveys some, but not all, of the features of a modern radar system. Historically, radar systems were first developed and used at *radio* frequencies, including the microwave band, but we now also have light radars, or ***lidars***, that operate at optical wavelengths. Over the years, the name radar has lost its original meaning and has come to signify any *active* electromagnetic sensor that uses its own source to illuminate a region of space and then measure the echoes generated by reflecting objects contained in that region. In addition to detecting the presence of a reflecting object and determining its *range* by measuring the time delay of short-duration pulses transmitted by the radar, a radar is also capable of specifying the *position* of the target and its radial velocity. Measurement of the ***radial velocity*** of a moving object is realized by measuring the ***Doppler frequency shift*** produced by the object. Also, the strength and shape of the reflected pulse carry information about the shape and material properties of the reflecting object.

Radar is used for a wide range of civilian and military applications, including air traffic control, aircraft navigation, law enforcement, control and guidance of weapon systems, remote sensing of Earth's environment, weather observation, astronomy, and collision avoidance for automobiles. The frequency bands used for the various types of radar applications extend from the megahertz region to frequencies as high as 225 GHz.

10-5.1 Basic Operation of a Radar System

The block diagram shown in Fig. 10-9 contains the basic functional elements of a pulse radar system. The ***synchronizer–modulator*** unit serves to synchronize the operation of the transmitter and the ***videoprocessor–display unit*** by generating a train of direct-current (dc) narrow-duration, evenly spaced pulses. These pulses, which are supplied to both the transmitter and the videoprocessor–display unit, specify the times at which radar pulses are transmitted. The transmitter contains a high-power radio-frequency (RF) oscillator with an on/off control voltage actuated by the pulses supplied by the synchronizer–modulator unit. Hence, the transmitter generates pulses of RF energy equal in duration and spacing to the dc pulses generated by the synchronizer–modulator unit. Each pulse is supplied to the antenna through a ***duplexer***, which allows the antenna to be shared between the transmitter and the receiver.

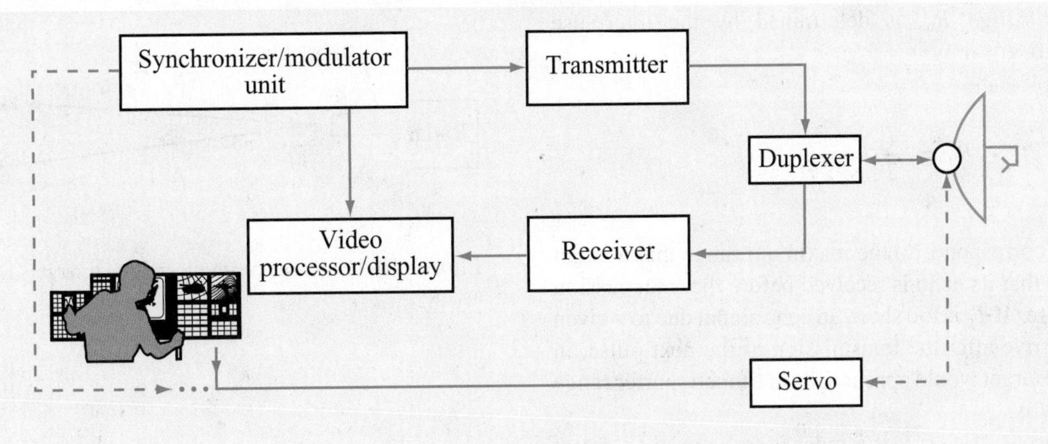

Figure 10-9 Basic block diagram of a radar system.

The duplexer, which often is called the ***transmitter/receiver (T/R) switch***, connects the transmitter to the antenna for the duration of the pulse, and then connects the antenna to the receiver for the remaining period until the start of a new pulse. Some duplexers, however, are passive devices that perform the sharing and isolation functions continuously. The circulator shown in **Fig. 10-5** is an example of a passive duplexer. After transmission by the antenna, a portion of the transmitted signal is intercepted by a reflecting object (often called a ***target***) and scattered in many directions. The energy reradiated by the target back toward the radar is collected by the antenna and delivered to the receiver, which processes the signal to detect the presence of the target and to extract information on its location and velocity. The receiver converts the reflected RF signals into lower-frequency video signals and supplies them to the videoprocessor–display unit, which displays the extracted information in a format suitable for the intended application. The servo unit positions the orientation of the antenna beam in response to control signals provided by either an operator, a control unit with preset functions, or a control unit commanded by another system. The control unit of an air-traffic-control radar, for example, commands the servo to rotate the antenna in azimuth continuously. In contrast, the radar antenna placed in the nose of an aircraft is made to scan back and forth over only a specified angular sector.

10-5.2 Unambiguous Range

The collective features of the energy transmitted by a radar are called the ***signal waveform***. For a pulse radar, these features

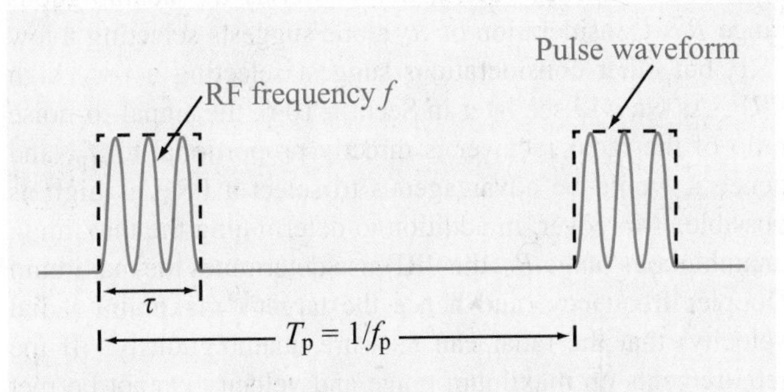

Figure 10-10 A pulse radar transmits a continuous train of RF pulses at a repetition frequency f_{p}.

include (1) the carrier frequency f, (2) the ***pulse length*** τ, (3) the ***pulse repetition frequency*** f_{p} (number of pulses per second), or equivalently the ***interpulse period*** $T_{\text{p}} = 1/f_{\text{p}}$, and (4) the modulation (if any) within the pulses. Three of these features are illustrated in **Fig. 10-10**. Modulation, which refers to control of the amplitude, frequency, or phase of the signal, is beyond the level of the present treatment.

The range to a target is determined by measuring the time delay T taken by the pulse to travel to the target and back. For a target at range R,

$$T = \frac{2R}{c}, \tag{10.13}$$

where $c = 3 \times 10^8$ m/s is the speed of light, and the factor 2 accounts for the two-way propagation. The maximum target range that a radar can measure unambiguously, called the

unambiguous range R_u, is determined by the interpulse period T_p and is given by

$$R_u = \frac{cT_p}{2} = \frac{c}{2f_p}. \quad (10.14)$$

The range R_u corresponds to the maximum range that a target can have such that its echo is received before the transmission of the next pulse. If T_p is too short, an echo signal due to a given pulse might arrive after the transmission of the next pulse, in which case the target would appear to be at a much shorter range than it actually is.

According to Eq. (10.14), if a radar is to be used to detect targets that are as far away as 100 km, for example, then f_p should be less than 1.5 kHz, and the higher the pulse repetition frequency (PRF), the shorter is the unambiguous range R_u. Consideration of R_u alone suggests selecting a low PRF, but other considerations suggest selecting a very high PRF. As we will see later in Section 10-6, the signal-to-noise ratio of the radar receiver is directly proportional to f_p, and hence it would be advantageous to select a PRF as high as possible. Moreover, in addition to determining the maximum unambiguous range R_u, the PRF also determines the maximum Doppler frequency (and hence the target's maximum radial velocity) that the radar can measure unambiguously. If the requirements on maximum range and velocity cannot be met by the same PRF, then some compromise may be necessary. Alternatively, it is possible to use a ***multiple-PRF*** radar system that transmits a few pulses at one PRF followed by another series of pulses at another PRF, and then the two sets of received pulses are processed together to remove the ambiguities that would have been present with either PRF alone.

10-5.3 Range and Angular Resolutions

Consider a radar observing two targets located at ranges R_1 and R_2, as shown in **Fig. 10-11**. Let $t = 0$ denote the time corresponding to the start of the transmitted pulse. The pulse length is τ. The return due to target 1 will arrive at $T_1 = 2R_1/c$ and will have a length τ (assuming that the pulse length in space is much greater than the radial extent of the target). Similarly, the return due to target 2 will arrive at $T_2 = 2R_2/c$. The two targets are resolvable as distinct targets so long as $T_2 \geq T_1 + \tau$ or, equivalently,

$$\frac{2R_2}{c} \geq \frac{2R_1}{c} + \tau. \quad (10.15)$$

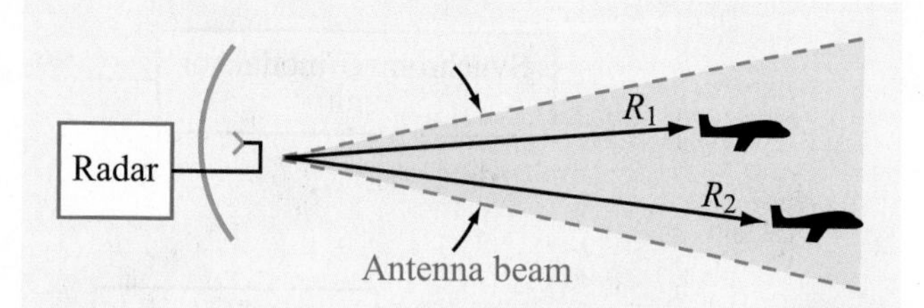

Figure 10-11 Radar beam viewing two targets at ranges R_1 and R_2.

The ***range resolution*** of the radar, ΔR, is defined as the minimum spacing between two targets necessary to avoid overlap between the echoes from the two targets. From Eq. (10.15), this occurs when

$$\Delta R = R_2 - R_1 = c\tau/2. \quad (10.16)$$

Some radars are capable of transmitting pulses as short as 1 ns in duration or even shorter. For $\tau = 1$ ns, $\Delta R = 15$ cm.

The basic angular resolution of a radar system is determined by its antenna beamwidth β, as shown in **Fig. 10-12**. The corresponding ***azimuth resolution*** Δx at a range R is given by

$$\Delta x = \beta R, \quad (10.17)$$

where β is in radians. In some cases, special techniques are used to improve the angular resolution down to a fraction of the beamwidth. One example is the monopulse tracking radar described in Section 10-8.

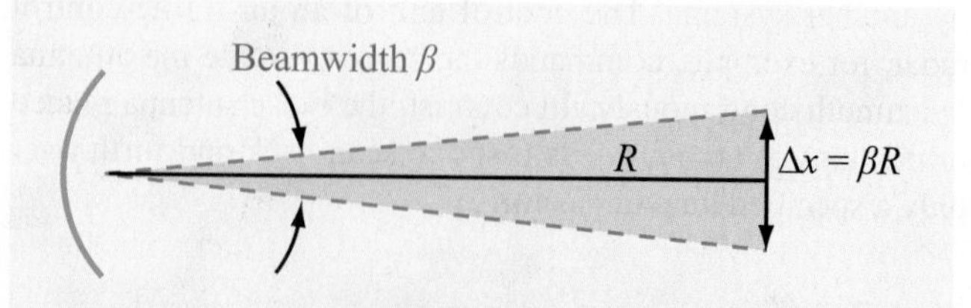

Figure 10-12 The azimuth resolution Δx at a range R is equal to βR.

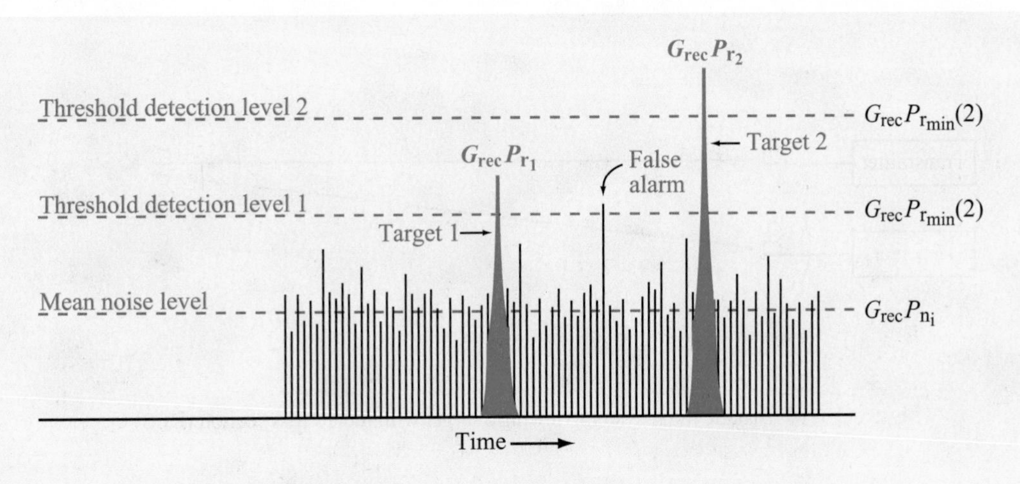

Figure 10-13 The output of a radar receiver as a function of time.

10-6 Target Detection

Target detection by radar is governed by two factors: (1) the ***signal energy*** received by the radar receiver due to reflection of part of the transmitted energy by the target, and (2) the ***noise energy*** generated by the receiver. **Figure 10-13**, which depicts the output response of a radar receiver as a function of time, shows the signals due to two targets displayed against the noise contributed by external sources as well as by the devices making up the receiver. The random variations exhibited by the noise may at times make it difficult to distinguish the signal reflected by the target from a noise spike. In **Fig. 10-13**, the mean noise-power level at the receiver output is denoted by $P_{no} = G_{rec}P_{ni}$, where G_{rec} is the receiver gain and P_{ni} is the noise level referred to the receiver's input terminals. The power levels P_{r_1} and P_{r_2} represent the echoes of the two targets observed by the radar. Because of the random nature of noise, it is necessary to set a threshold level, $P_{r_{min}}$, for detection. For threshold level 1 indicated in **Fig. 10-13**, the radar will produce the presence of both targets, but it will also detect a ***false alarm***. The chance of this occurring is called the ***false-alarm probability***. On the other hand, if the threshold level is raised to level 2 to avoid the false alarm, the radar will not detect the presence of the first target. A radar's ability to detect the presence of a target is characterized by a ***detection probability***. The setting of the threshold signal level relative to the mean noise level is thus made on the basis of a compromise that weighs both probabilities.

To keep the noise level at a minimum, the receiver is designed such that its bandwidth B is barely wide enough to pass most of the energy contained in the received pulse. Such a design, called a ***matched filter***, requires that B be equal to the reciprocal of the pulse length τ (i.e., $B = 1/\tau$). Hence, for a matched-filter receiver, Eq. (10.10) becomes

$$P_{ni} = KT_{sys}B = \frac{KT_{sys}}{\tau}. \tag{10.18}$$

The signal power received by the radar, P_r, is related to the transmitted power level, P_t, through the radar equation. We will first derive the radar equation for the general case of a ***bistatic radar*** configuration in which the transmitter and receiver are not necessarily at the same location, and then we will specialize the results to the ***monostatic radar*** case wherein the transmitter and receiver are colocated. In **Fig. 10-14**, the target is at range R_t from the transmitter and at range R_r from the receiver. The power density illuminating the target is given by

$$S_t = \frac{P_t}{4\pi R_t^2}\, G_t \qquad (\text{W/m}^2), \tag{10.19}$$

where $(P_t/4\pi R_t^2)$ represents the power density that would have been radiated by an isotropic radiator, and G_t is the gain of the transmitting antenna in the direction of the target. The target is characterized by a ***radar cross section*** (RCS) σ_t (m^2), defined such that the power intercepted and then reradiated by the target is

$$P_{rer} = S_t\sigma_t = \frac{P_tG_t\sigma_t}{4\pi R_t^2} \qquad (\text{W}). \tag{10.20}$$

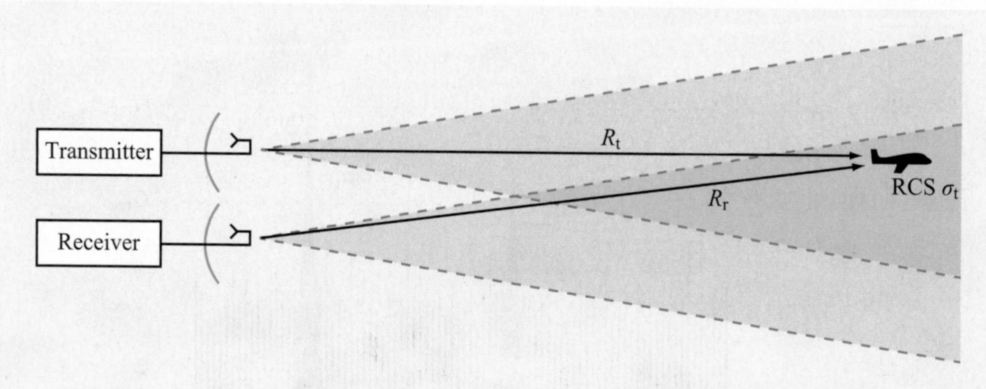

Figure 10-14 Bistatic radar system viewing a target with radar cross section (RCS) σ_t.

This reradiated power spreads out over a spherical surface, resulting in a power density S_r incident upon the receiving radar antenna. Hence,

$$S_r = \frac{P_{rer}}{4\pi R_r^2} = \frac{P_t G_t \sigma_t}{(4\pi R_t R_r)^2} \quad (\text{W/m}^2). \tag{10.21}$$

With an effective area A_r and radiation efficiency ξ_r, the receiving radar antenna intercepts and delivers (to the receiver) power P_r given by

$$P_r = \xi_r A_r S_r = \frac{P_t G_t \xi_r A_r \sigma_t}{(4\pi R_t R_r)^2} = \frac{P_t G_t G_r \lambda^2 \sigma_t}{(4\pi)^3 R_t^2 R_r^2}, \tag{10.22}$$

where we have used Eqs. (9.29) and (9.64) to relate the effective area of the receiving antenna, A_r, to its gain G_r. For a monostatic antenna that uses the same antenna for the transmit and receive functions, $G_t = G_r = G$ and $R_t = R_r = R$. Hence,

$$P_r = \frac{P_t G^2 \lambda^2 \sigma_t}{(4\pi)^3 R^4} \quad \textbf{(radar equation)}. \tag{10.23}$$

Unlike the one-way communication system for which the dependence on R is as $1/R^2$, the range dependence given by the ***radar equation*** goes as $1/R^4$, the product of two one-way propagation processes.

The detection process may be based on the echo from a single pulse or on the addition (integration) of echoes from several pulses. We will consider only the single-pulse case here. A target is said to be detectable if its echo signal power P_r exceeds $P_{r_{min}}$, the ***threshold detection level*** indicated in **Fig. 10-13**. The ***maximum detectable range*** R_{max} is the range beyond which the target cannot be detected, corresponding to the range at which $P_r = P_{r_{min}}$ in Eq. (10.23). Thus,

$$R_{max} = \left[\frac{P_t G^2 \lambda^2 \sigma_t}{(4\pi)^3 P_{r_{min}}} \right]^{1/4}. \tag{10.24}$$

The signal-to-noise ratio is equal to the ratio of the received signal power P_r to the mean input noise power P_{ni} given by Eq. (10.18):

$$S_n = \frac{P_r}{P_{ni}} = \frac{P_r \tau}{K T_{sys}}, \tag{10.25}$$

and the ***minimum signal-to-noise ratio*** S_{min} corresponds to when $P_r = P_{r_{min}}$:

$$S_{min} = \frac{P_{r_{min}} \tau}{K T_{sys}}. \tag{10.26}$$

Use of Eq. (10.26) in Eq. (10.24) gives

$$R_{max} = \left[\frac{P_t \tau G^2 \lambda^2 \sigma_t}{(4\pi)^3 K T_{sys} S_{min}} \right]^{1/4}. \tag{10.27}$$

The product $P_t\tau$ is equal to the energy of the transmitted pulse. Hence, according to Eq. (10.27), *it is the energy of the transmitted pulse rather than the transmitter power level alone that determines the maximum detectable range.* A high-power narrow pulse and an equal-energy, low-power long pulse will yield the same radar performance as far as maximum detectable range is concerned. However, the range-resolution capability

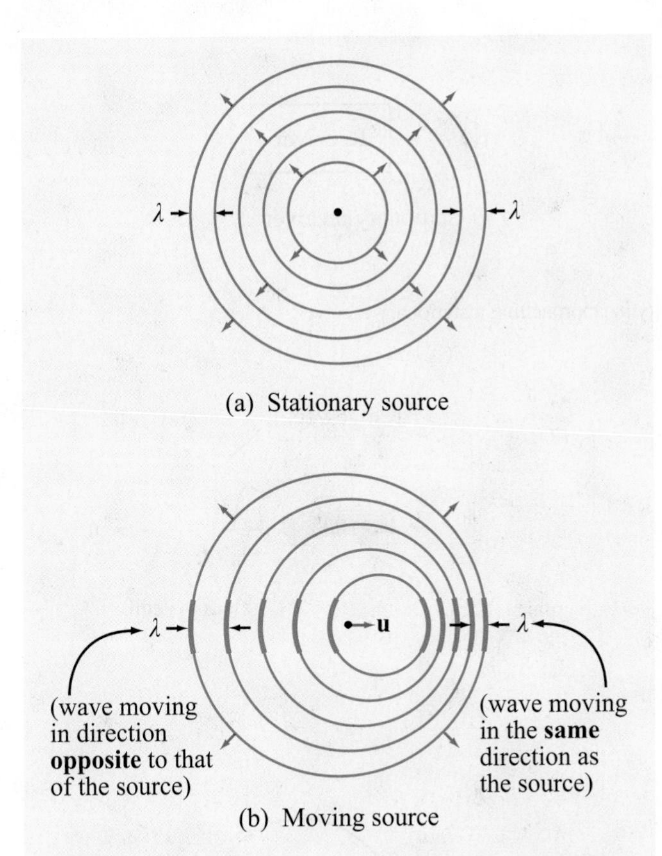

Figure 10-15 A wave radiated from a point source when (a) stationary and (b) moving. The wave is compressed in the direction of motion, spread out in the opposite direction, and unaffected in the direction normal to motion.

of the long pulse is much poorer than that of the short pulse [see Eq. (10.16)].

The maximum detectable range $R_{\max}$ can also be increased by improving the signal-to-noise ratio. This can be accomplished by integrating the echoes from multiple pulses in order to increase the total amount of energy received from the target. The number of pulses available for integration over a specified integration time is proportional to the PRF. Hence, from the standpoint of maximizing target detection, it is advantageous to use as high a PRF as allowed by other considerations.

10-7 Doppler Radar

The Doppler effect is a shift in the frequency of a wave caused by the motion of the transmitting source, the reflecting object, or the receiving system. As illustrated in **Fig. 10-15**, a wave radiated by a stationary isotropic point source forms equally spaced concentric circles as a function of time travel from the source. In contrast, a wave radiated by a moving source is compressed in the direction of motion and is spread out in the opposite direction. Compressing a wave shortens its wavelength, which is equivalent to increasing its frequency. Conversely, spreading it out decreases its frequency. The change in frequency is called the ***Doppler frequency shift*** f_d. That is, if f_t is the frequency of the wave radiated by the moving source, then the frequency f_r of the wave that would be observed by a stationary receiver is

$$f_r = f_t + f_d. \tag{10.28}$$

The magnitude and sign of f_d depend on the direction of the velocity vector relative to the direction of the range vector connecting the source to the receiver.

Consider a source transmitting an electromagnetic wave with frequency f_t (**Fig. 10-16**). At a distance R from the source, the electric field of the radiated wave is given by

$$E(R) = E_0 e^{j(\omega_t t - kR)} = E_0 e^{j\phi}, \tag{10.29}$$

where E_0 is the wave's magnitude, $\omega_t = 2\pi f_t$, and $k = 2\pi/\lambda_t$, where λ_t is the wavelength of the transmitted wave. The magnitude depends on the distance R and the gain of the source antenna, but it is not of concern as far as the Doppler effect is concerned. The quantity

$$\phi = \omega_t t - kR = 2\pi f_t t - \frac{2\pi}{\lambda_t} R \tag{10.30}$$

is the phase of the radiated wave relative to its phase at $R = 0$ and reference time $t = 0$. If the source is moving toward the receiver, as in **Fig. 10-16**, or vice versa, at a radial velocity u_r, then

$$R = R_0 - u_r t, \tag{10.31}$$

where R_0 is the distance between the source and the receiver at $t = 0$. Hence,

$$\phi = 2\pi f_t t - \frac{2\pi}{\lambda_t}(R_0 - u_r t). \tag{10.32}$$

This is the phase of the signal detected by the receiver. The frequency of a wave is defined as the time derivative of the phase ϕ divided by 2π. Thus,

$$f_r = \frac{1}{2\pi}\frac{d\phi}{dt} = f_t + \frac{u_r}{\lambda_t}. \tag{10.33}$$

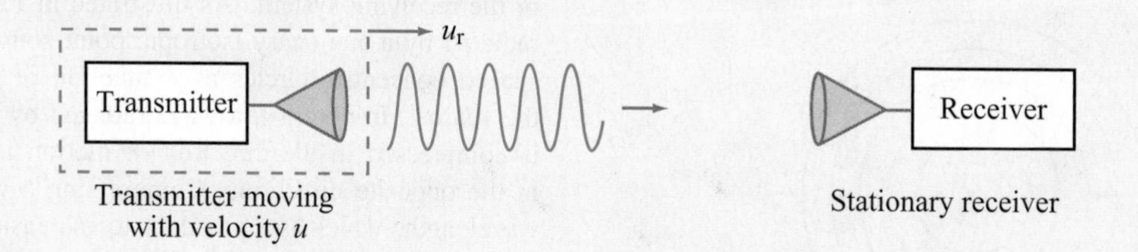

Figure 10-16 Transmitter with radial velocity u_r approaching a stationary receiver.

Comparison of Eq. (10.33) with Eq. (10.28) leads to $f_d = u_r/\lambda_t$. For radar, the Doppler shift happens twice, once for the wave from the radar to the target and again for the wave reflected by the target back to the radar. Hence, $f_d = 2u_r/\lambda_t$. The dependence of f_d on direction is given by the dot product of the velocity and range unit vectors, which leads to

$$f_d = -2\frac{u_r}{\lambda_t} = -\frac{2u}{\lambda_t}\cos\theta, \qquad (10.34)$$

where u_r is the radial velocity component of u and θ is the angle between the range vector and the velocity vector (**Fig. 10-17**), with the direction of the range vector defined to be *from* the radar *to* the target. For a receding target (relative to the radar), $0 \le \theta \le 90°$, and for an approaching target, $90° \le \theta \le 180°$.

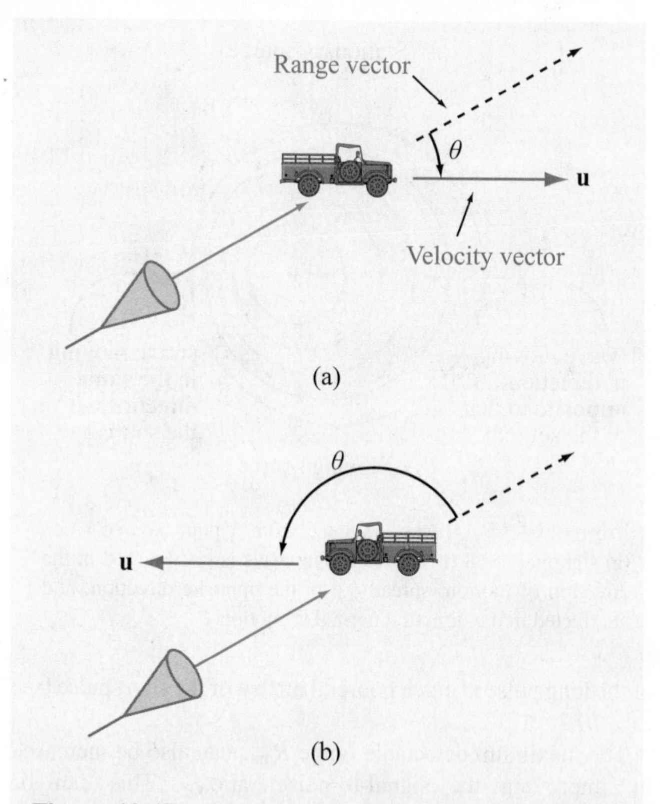

Figure 10-17 The Doppler frequency shift is negative for a receding target ($0 \le \theta \le 90°$), as in (a), and positive for an approaching target ($90° \le \theta \le 180°$), as in (b).

10-8 Monopulse Radar

On the basis of information extracted from the echo due to a single pulse, a ***monopulse radar*** can track the direction of a target with an angular accuracy equal to a fraction of its antenna beamwidth. To track a target in both elevation and azimuth, a monopulse radar uses an antenna (such as a parabolic dish), with four separate small horns at its focal point (**Fig. 10-18**). Monopulse systems are of two types. The first is called ***amplitude-comparison monopulse*** because the tracking information is extracted from the amplitudes of the echoes received by the four horns, and the second is called ***phase-comparison monopulse*** because it relies on the phases of the received signals. We shall limit our present discussion to the amplitude-comparison scheme.

Individually, each horn would produce its own beam, with the four beams pointing in slightly different directions. **Figure 10-19** shows the beams of two adjacent horns. The basic principle of the amplitude-comparison monopulse is to measure the amplitudes of the echo signals received through the two beams and then apply the difference between them to repoint the antenna boresight direction toward the target. Using computer-controlled phase shifters, the ***phasing network*** shown in **Fig. 10-18** can combine the signal delivered to the four-element horn array by the transmitter or by the echo signals received by them in different ways. Upon transmission, the network excites all four feeds in phase, thereby producing a single main beam called the ***sum beam***. The phasing network uses special microwave devices that allow it to provide the desired

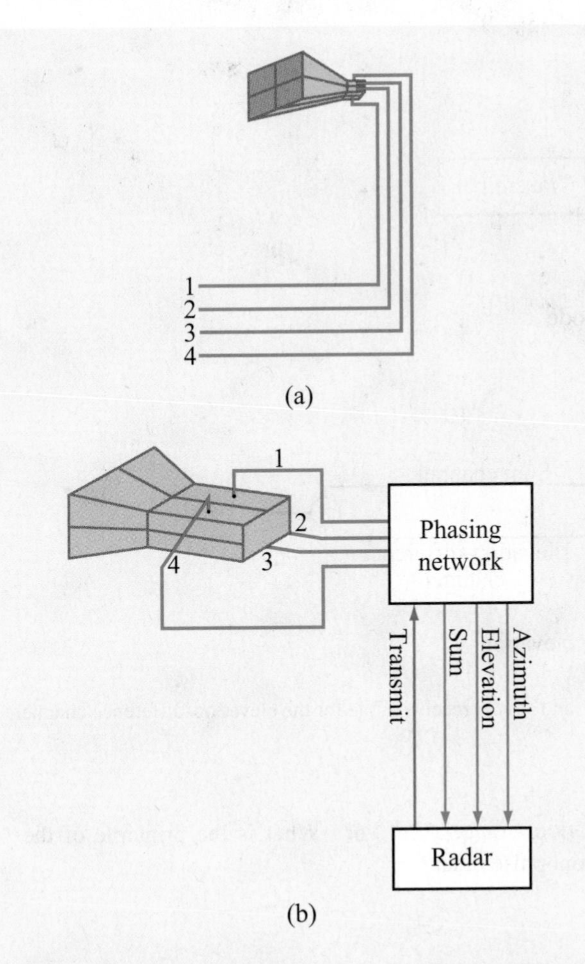

Figure 10-18 Antenna feeding arrangement for an amplitude-comparison monopulse radar: (a) feed horns and (b) connection to phasing network.

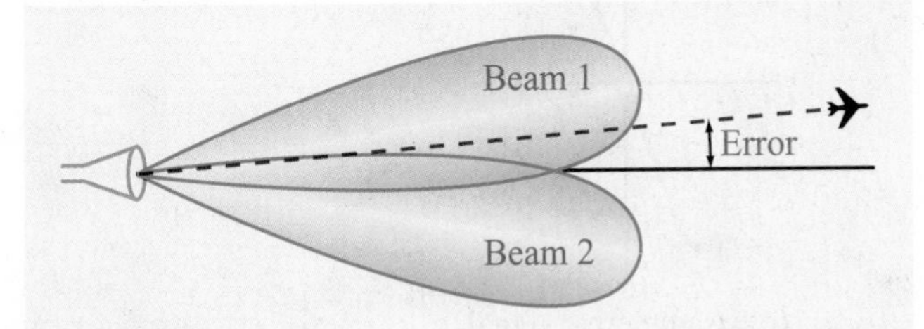

Figure 10-19 A target observed by two overlapping beams of a monopulse radar.

functionality during both the transmit and receive modes. Its equivalent functionality is described by the circuits shown in **Fig. 10-20**. During the receive period, the phasing network uses power dividers, power combiners, and phase shifters so as to generate three different output channels. One of these is the sum channel, corresponding to adding all four horns in phase, and its radiation pattern is depicted in **Fig. 10-21(a)**. The second channel, called the ***elevation-difference channel***, is obtained by first adding the outputs of the top-right and top-left horns [**Fig. 10-20(b)**], then adding the outputs of the bottom-right and bottom-left horns, and then subtracting the second sum from the first. The subtraction process is accomplished by adding a 180° phase shifter in the path of the second sum before adding it to the first sum. The beam pattern of the elevation-difference channel is shown in **Fig. 10-21(b)**. If the observed target is centered between the two elevation beams, the receiver echoes will have the same strength for both beams, thereby producing a zero output from the elevation-difference channel. If it is not, the amplitude of the elevation-difference channel will be proportional to the angular deviation of the target from the boresight direction, and its sign will denote the direction of the deviation. The third channel (not shown in **Fig. 10-20**) is the ***azimuth-difference channel***, and it is accomplished through a similar process that generates a beam corresponding to the difference between the sum of the two right horns and the sum of the two left horns.

In practice, the output of the difference channel is multiplied by the output of the sum channel to increase the strength of the difference signal and to provide a phase reference for extracting the sign of the angle. This product, called the ***angle error signal***, is displayed in **Fig. 10-21(c)** as a function of the angle error. The error signal activates a servo-control system to reposition the antenna direction. By applying a similar procedure along the azimuth direction using the product of the azimuth-difference channel and the sum channel, a monopulse radar provides automatic tracking in both directions. The range to the target is obtained by measuring the round-trip delay of the signal.

Concept Question 10-5: How is the PRF related to unambiguous range?

Concept Question 10-6: Explain how the false-alarm probability and the detection probability are related to the noise level of the receiver.

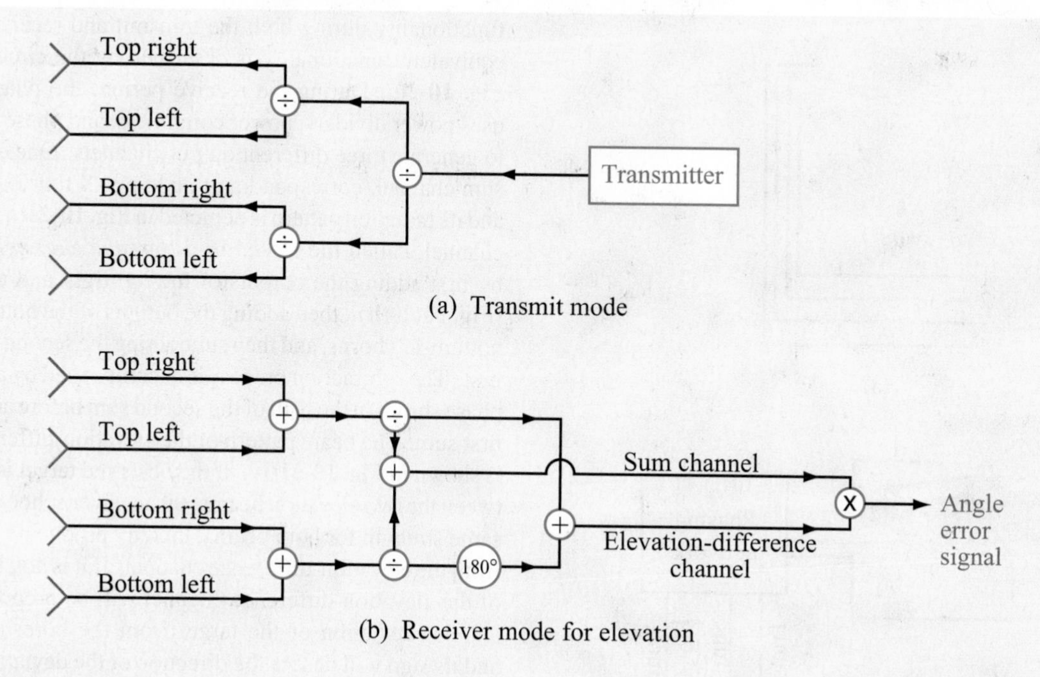

Figure 10-20 Functionality of the phasing network in (a) the transmit mode and (b) the receive mode for the elevation-difference channel.

Concept Question 10-7: In terms of the geometry shown in **Fig. 10-17**, when is the Doppler shift a maximum?

Concept Question 10-8: What is the principle of the monopulse radar?

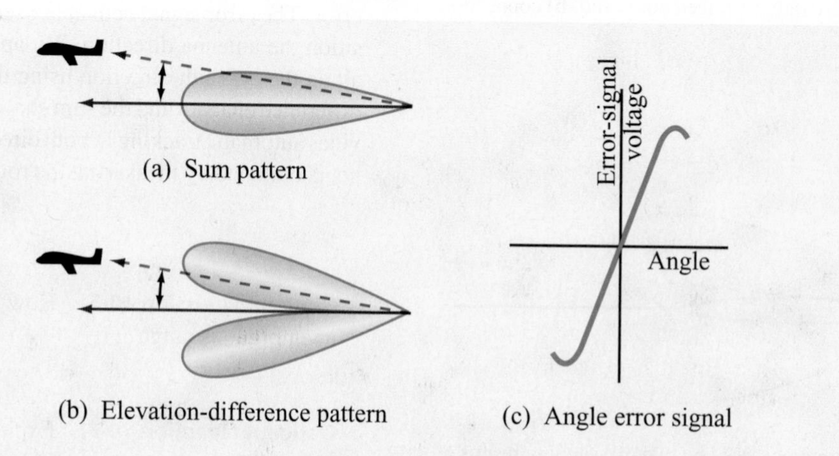

Figure 10-21 Monopulse antenna (a) sum pattern, (b) elevation-difference pattern, and (c) angle error signal.

Chapter 10 Summary

Concepts

- Three equally spaced satellites in geostationary orbit can provide coverage of most of Earth's surface.
- The use of polarization diversity makes it possible to double the number of channels per unit bandwidth carried by a satellite repeater.
- A satellite antenna system is designed to produce beams tailored to match the areas served by the satellite. Antenna arrays are particularly suitable for this purpose.
- A radar is an electromagnetic sensor that illuminates a region of space and then measures the echoes due to reflecting objects. From the echoes, information can be extracted about the range of a target, its radial velocity, direction of motion, and other characteristics.
- Due to the random nature of receiver noise, target detection is a statistical process characterized by detection and false-alarm probabilities.
- A moving object produces a Doppler frequency shift proportional to the radial velocity of the object (relative to the radar) and inversely proportional to λ.
- A monopulse radar uses multiple beams to track the direction of a target, with an angular accuracy equal to a fraction of its antenna beamwidth.

Mathematical and Physical Models

Satellite Communication Systems

Radius of geostationary orbit

$$R_0 = \left(\frac{GM_e T^2}{4\pi^2}\right)^{1/3}$$

Received power

$$P_{ri} = \Upsilon(\theta)\; P_r = \Upsilon(\theta)\; P_t G_t G_r \left(\frac{\lambda}{4\pi R}\right)^2$$

Noise power

$$P_{ni} = K T_{sys} B$$

Signal-to-noise ratio

$$S_n = \frac{P_{ri}}{P_{ni}} = \frac{\Upsilon(\theta)\; P_t G_t G_r}{K T_{sys} B}\left(\frac{\lambda}{4\pi R}\right)^2$$

Radar Sensors

Unambiguous range

$$R_u = \frac{cT_p}{2} = \frac{c}{2f_p}$$

Range resolution

$$\Delta R = R_2 - R_1 = c\tau/2$$

Azimuth resolution

$$\Delta x = \beta R$$

Radar equation

$$P_r = \frac{P_t G^2 \lambda^2 \sigma_t}{(4\pi)^3 R^4}$$

Doppler frequency shift

$$f_d = -2\frac{u_r}{\lambda_t} = -\frac{2u}{\lambda_t}\cos\theta$$

Important Terms Provide definitions or explain the meaning of the following terms:

atmospheric transmissivity Υ
azimuth resolution
bistatic radar
circulator
detection probability
Doppler frequency shift f_d
duplexer
Explorer I
false-alarm probability
FDMA
geostationary orbit
interpulse period T_p
lidar
matched filter
maximum detectable range R_{max}
monopulse radar
monostatic radar
multiplexer
polarization diversity
pulse length τ
pulse repetition frequency (PRF) f_p
radar
radar cross section σ_t
radar equation
radial velocity u_r
range resolution
Score
signal-to-noise ratio
Sputnik I
sum and difference channels
synchronizer
system noise temperature
threshold detection level
transponder
unambiguous range R_u
uplink and downlink

PROBLEMS

Sections 10-1 to 10-4: Satellite Communication Systems

*__10.1__ A remote sensing satellite is in circular orbit around Earth at an altitude of 1,500 km above Earth's surface. What is its orbital period?

10.2 A transponder with a bandwidth of 400 MHz uses polarization diversity. If the bandwidth allocated to transmit a single telephone channel is 4 kHz, how many telephone channels can be carried by the transponder?

*__10.3__ Repeat Problem 10.2 for TV channels, each requiring a bandwidth of 6 MHz.

10.4 A geostationary satellite is at a distance of 40,000 km from a ground receiving station. The satellite transmitting antenna is a circular aperture with a 1 m diameter, and the ground station uses a parabolic dish antenna with an effective diameter of 20 cm. If the satellite transmits 1 kW of power at 12 GHz and the ground receiver is characterized by a system noise temperature of 1,000 K, what would be the signal-to-noise ratio of a received TV signal with a bandwidth of 6 MHz? The antennas and the atmosphere may be assumed lossless.

Sections 10-5 to 10-8: Radar Sensors

10.5 A 10 GHz weather radar uses a 15 cm diameter lossless antenna. At a distance of 1 km, what are the dimensions of the volume resolvable by the radar if the pulse length is 1 μs?

*__10.6__ A collision-avoidance automotive radar is designed to detect the presence of vehicles up to a range of 1 km. What is the maximum usable PRF?

*__10.7__ A radar system is characterized by the following parameters: $P_t = 1$ kW, $\tau = 0.1$ μs, $G = 30$ dB, $\lambda = 3$ cm, and $T_{sys} = 1{,}500$ K. The radar cross section of a car is typically 5 m^2. How far away can the car be and remain detectable by the radar with a minimum signal-to-noise ratio of 13 dB?

10.8 A 3 cm wavelength radar is located at the origin of an x–y coordinate system. A car located at $x = 100$ m and $y = 200$ m is heading east (x direction) at a speed of 120 km/hr. What is the Doppler frequency measured by the radar?

*Answer(s) available in Appendix D.

APPENDIX A

Symbols, Quantities, and Units

Symbol	Quantity	SI Unit	Abbreviation
$\mathbf{A}$	Magnetic potential (vector)	webers/meter	Wb/m
B	Susceptance	siemens	S
$\mathbf{B}$	Magnetic flux density	teslas or webers/meter^2	T or W/m^2
C	Capacitance	farads	F
D	Directivity (antenna)	(dimensionless)	—
$\mathbf{D}$	Electric flux density	coulombs/meter^2	C/m^2
$\mathbf{d}$	Moment arm	meters	m
$\mathbf{E}$	Electric field intensity	volts/meter	V/m
E_{ds}	Dielectric strength	volts/meter	V/m
F	Radiation intensity (normalized)	(dimensionless)	—
$\mathbf{F}$	Force	newtons	N
f	Frequency	hertz	Hz
f_{d}	Doppler frequency	hertz	Hz
f_{mn}	Cutoff frequency	hertz	Hz
G	Conductance	siemens	S
G	Gain (power)	(dimensionless)	—
$\mathbf{H}$	Magnetic field intensity	amperes/meter	A/m
I	Current	amperes	A
$\mathbf{J}$	Current density (volume)	amperes/meter^2	A/m^2
$\mathbf{J}_{\mathrm{s}}$	Current density (surface)	amperes/meter	A/m
k	Wavenumber	radians/meter	rad/m
k_{c}	Cutoff wavenumber	radians/second	rad/s
L	Inductance	henrys	H
l	Length	meters	m

Symbol	Quantity	SI Unit	Abbreviation
M, m	Mass	kilograms	kg
$\mathbf{M}$	Magnetization vector	amperes/meter	A/m
$\mathbf{m}$	Magnetic dipole moment	ampere-meters^2	A·m^2
n	Index of refraction	(dimensionless)	—
P	Power	watts	W
$\mathbf{P}$	Electric polarization vector	coulombs/meter^2	C/m^2
p	Pressure	newtons/meter^2	N/m^2
$\mathbf{p}$	Electric dipole moment	coulomb-meters	C·m
Q	Quality factor	(dimensionless)	—
Q, q	Charge	coulombs	C
R	Reflectivity (reflectance)	(dimensionless)	—
R	Resistance	ohms	Ω
R	Range	meters	m
r	Radial distance	meters	m
S	Standing-wave ratio	(dimensionless)	—
$\mathbf{S}$	Poynting vector	watts/meter^2	W/m^2
$\mathbf{S}_{\text{av}}$	Power density	watts/meter^2	W/m^2
T	Temperature	kelvins	K
T	Transmissivity (transmittance)	(dimensionless)	—
$\mathbf{T}$	Torque	newton-meters	N·m
t	Time	seconds	s
T	period	seconds	s
$\mathbf{u}$	Velocity	meters/second	m/s
u_{g}	Group velocity	meters/second	m/s
u_{p}	Phase velocity	meters/second	m/s
V	Electric potential	volts	V
V	Voltage	volts	V
V_{bv}	Voltage breakdown	volts	V
V_{emf}	Electromotive force (emf)	volts	V
W	Energy (work)	joules	J
w	Energy density	joules/meter^3	J/m^3
X	Reactance	ohms	Ω
Y	Admittance	siemens	S
Z	Impedance	ohms	Ω
α	Attenuation constant	nepers/meter	Np/m
β	Beamwidth	degrees	°
β	Phase constant (wavenumber)	radians/meter	rad/m
Γ	Reflection coefficient	(dimensionless)	—
γ	Propagation constant	meters^{-1}	m^{-1}
δ_{s}	Skin depth	meters	m
ϵ, ϵ_0	Permittivity	farads/meter	F/m
ϵ_{r}	Relative permittivity	(dimensionless)	—
η	Impedance	ohms	Ω
λ	Wavelength	meters	m

Symbol	Quantity	SI Unit	Abbreviation
μ, μ_0	Permeability	henrys/meter	H/m
μ_r	Relative permeability	(dimensionless)	—
μ_e, μ_h	Mobility (electron, hole)	meters^2/volt·second	m^2/V·s
ρ_l	Charge density (linear)	coulombs/meter	C/m
ρ_s	Charge density (surface)	coulombs/meter^2	C/m^2
ρ_v	Charge density (volume)	coulombs/meter^3	C/m^3
σ	Conductivity	siemens/meter	S/m
σ_t	Radar cross section	meters^2	m^2
τ	Transmission coefficient	(dimensionless)	—
τ	Pulse length	seconds	s
Υ	Atmospheric transmissivity	(dimensionless)	—
Φ	Magnetic flux	webers	Wb
$\boldsymbol{\psi}$	Gravitational field	newtons/kilogram	N/kg
χ_e	Electric susceptibility	(dimensionless)	—
χ_m	Magnetic susceptibility	(dimensionless)	—
Ω	Solid angle	steradians	sr
ω	Angular frequency	radians/second	rad/s
ω	Angular velocity	radians/second	rad/s

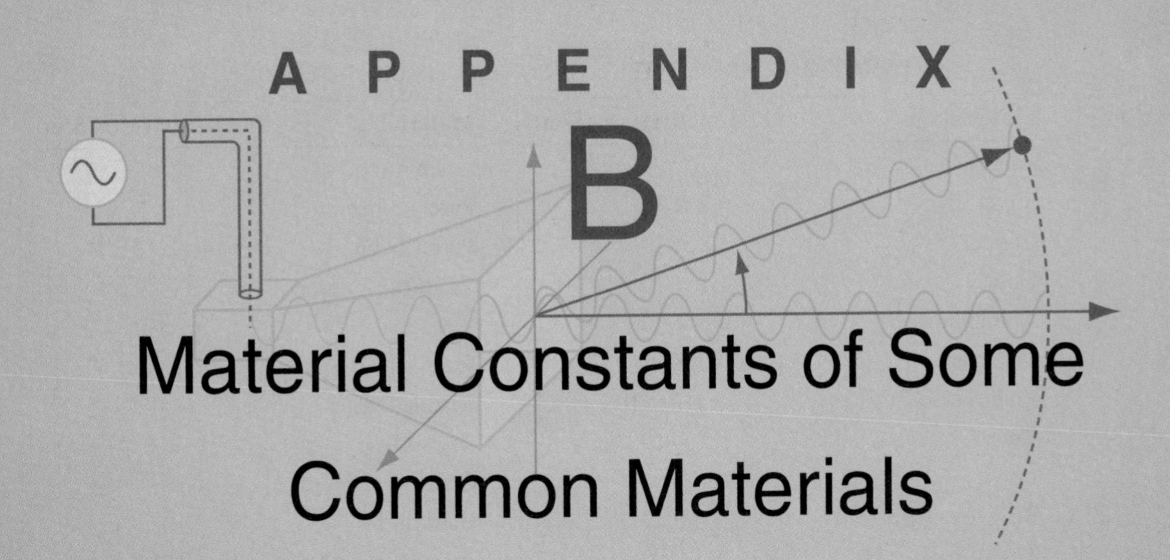

Material Constants of Some Common Materials

Table B-1 RELATIVE PERMITTIVITY ϵ_r OF COMMON MATERIALS[a]

$\epsilon = \epsilon_r \epsilon_0$ and $\epsilon_0 = 8.854 \times 10^{-12}$ F/m.

Material	Relative Permittivity, ϵ_r	Material	Relative Permittivity, ϵ_r
Vacuum	1	Dry soil	2.5–3.5
Air (at sea level)	1.0006	Plexiglass	3.4
Styrofoam	1.03	Glass	4.5–10
Teflon	2.1	Quartz	3.8–5
Petroleum oil	2.1	Bakelite	5
Wood (dry)	1.5–4	Porcelain	5.7
Paraffin	2.2	Formica	6
Polyethylene	2.25	Mica	5.4–6
Polystyrene	2.6	Ammonia	22
Paper	2–4	Seawater	72–80
Rubber	2.2–4.1	Distilled water	81

[a]These are low-frequency values at room temperature (20° C).

Note: For most metals, $\epsilon_r \simeq 1$.

Table B-2 CONDUCTIVITY σ OF SOME COMMON MATERIALS[a]

Material	Conductivity, σ (S/m)	Material	Conductivity, σ (S/m)
Conductors		**Semiconductors**	
Silver	6.2×10^7	Pure germanium	2.2
Copper	5.8×10^7	Pure silicon	4.4×10^{-4}
Gold	4.1×10^7	**Insulators**	
Aluminum	3.5×10^7	Wet soil	$\sim 10^{-2}$
Tungsten	1.8×10^7	Fresh water	$\sim 10^{-3}$
Zinc	1.7×10^7	Distilled water	$\sim 10^{-4}$
Brass	1.5×10^7	Dry soil	$\sim 10^{-4}$
Iron	10^7	Glass	10^{-12}
Bronze	10^7	Hard rubber	10^{-15}
Tin	9×10^6	Paraffin	10^{-15}
Lead	5×10^6	Mica	10^{-15}
Mercury	10^6	Fused quartz	10^{-17}
Carbon	3×10^4	Wax	10^{-17}
Seawater	4		
Animal body (average)	0.3 (poor cond.)		

[a]These are low-frequency values at room temperature (20° C).

Table B-3 RELATIVE PERMEABILITY μ_r OF SOME COMMON MATERIALS[a]

$\mu = \mu_r \mu_0$ and $\mu_0 = 4\pi \times 10^{-7}$ H/m.

Material	Relative Permeability, μ_r
Diamagnetic	
Bismuth	$0.99983 \simeq 1$
Gold	$0.99996 \simeq 1$
Mercury	$0.99997 \simeq 1$
Silver	$0.99998 \simeq 1$
Copper	$0.99999 \simeq 1$
Water	$0.99999 \simeq 1$
Paramagnetic	
Air	$1.000004 \simeq 1$
Aluminum	$1.00002 \simeq 1$
Tungsten	$1.00008 \simeq 1$
Titanium	$1.0002 \simeq 1$
Platinum	$1.0003 \simeq 1$
Ferromagnetic (nonlinear)	
Cobalt	250
Nickel	600
Mild steel	2,000
Iron (pure)	4,000–5,000
Silicon iron	7,000
Mumetal	$\sim$ 100, 000
Purified iron	$\sim$ 200, 000

[a]These are typical values; actual values depend on material variety.

Note: Except for ferromagnetic materials, $\mu_r \simeq 1$ for all dielectrics and conductors.

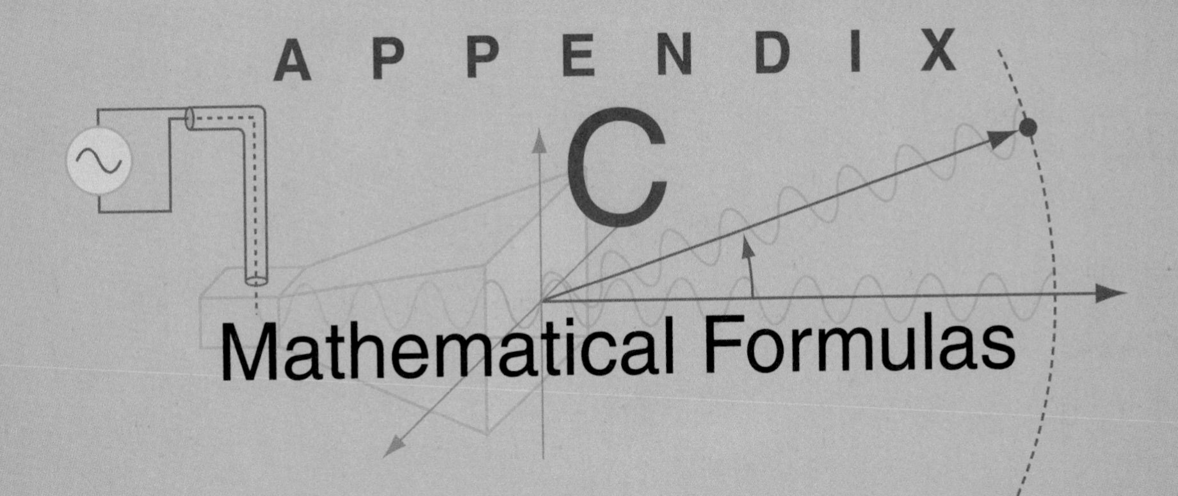

APPENDIX C
Mathematical Formulas

Trigonometric Relations

$$\sin(x \pm y) = \sin x \cos y \pm \cos x \sin y$$

$$\cos(x \pm y) = \cos x \cos y \mp \sin x \sin y$$

$$2 \sin x \sin y = \cos(x - y) - \cos(x + y)$$

$$2 \sin x \cos y = \sin(x + y) + \sin(x - y)$$

$$2 \cos x \cos y = \cos(x + y) + \cos(x - y)$$

$$\sin 2x = 2 \sin x \cos x$$

$$\cos 2x = 1 - 2 \sin^2 x$$

$$\sin x + \sin y = 2 \sin\left(\frac{x + y}{2}\right) \cos\left(\frac{x - y}{2}\right)$$

$$\sin x - \sin y = 2 \cos\left(\frac{x + y}{2}\right) \sin\left(\frac{x - y}{2}\right)$$

$$\cos x + \cos y = 2 \cos\left(\frac{x + y}{2}\right) \cos\left(\frac{x - y}{2}\right)$$

$$\cos x - \cos y = -2 \sin\left(\frac{x + y}{2}\right) \sin\left(\frac{x - y}{2}\right)$$

$$\cos(x \pm 90^\circ) = \mp \sin x$$

$$\cos(-x) = \cos x$$

$$\sin(x \pm 90^\circ) = \pm \cos x$$

$$\sin(-x) = -\sin x$$

$$e^{jx} = \cos x + j \sin x \qquad \text{(Euler's identity)}$$

$$\sin x = \frac{e^{jx} - e^{-jx}}{2j}$$

$$\cos x = \frac{e^{jx} + e^{-jx}}{2}$$

Approximations for Small Quantities

For $|x| \ll 1$,

$$(1 \pm x)^n \simeq 1 \pm nx$$

$$(1 \pm x)^2 \simeq 1 \pm 2x$$

$$\sqrt{1 \pm x} \simeq 1 \pm \frac{x}{2}$$

$$\frac{1}{\sqrt{1 \pm x}} \simeq 1 \mp \frac{x}{2}$$

$$e^x = 1 + x + \frac{x^2}{2!} + \cdots \simeq 1 + x$$

$$\ln(1 + x) \simeq x$$

$$\sin x = x - \frac{x^3}{3!} + \frac{x^5}{5!} + \cdots \simeq x$$

$$\cos x = 1 - \frac{x^2}{2!} + \frac{x^4}{4!} + \cdots \simeq 1 - \frac{x^2}{2}$$

$$\lim_{x \to 0} \frac{\sin x}{x} = 1$$

APPENDIX D

Answers to Selected Problems

Chapter 1

1.1 5 cm

1.3 $p(x, t) = 51.04 \cos(4\pi \times 10^3 t - 12.12\pi x + 36°)$ (N/m^2)

1.6 $u_p = 0.83$ (m/s); $\lambda = 10.47$ m

1.8 **(a)** $y_1(x, t)$ is traveling in positive x direction. $y_2(x, t)$ is traveling in negative x direction.

1.10 $y_2(t)$ lags $y_1(t)$ by 54°.

1.12 $T = 1.25$ s; $u_p = 0.28$ m/s; $\lambda = 0.35$ m

1.14 $\alpha = 2 \times 10^{-3}$ (Np/m)

1.16 **(c)** $z_1 z_2 = 18e^{j109.4°}$

1.17 **(b)** $z_2 = \sqrt{3}\, e^{j3\pi/4}$

1.19 **(c)** $|z|^2$,

1.20 **(d)** $t = 0$; $s = 6\, e^{j30°}$

1.22 $\ln(z) = 1.76 + j1.03$

1.25 $v_c(t) = 15.57 \cos(2\pi \times 10^3 t - 81.5°)$ V

1.26 **(d)** $i(t) = 3.61 \cos(\omega t + 146.31°)$ A

1.27 **(d)** $\tilde{I} = 2e^{j\pi/4}$ A

Chapter 2

2.2 **(a)** $l/\lambda = 2 \times 10^{-5}$; transmission line may be ignored.

(c) $l/\lambda = 0.6$; transmission line effects should be included.

2.4 $R' = 0.69$ (Ω/m), $L' = 1.57 \times 10^{-7}$ (H/m), $G' = 0$, $C' = 1.84 \times 10^{-10}$ (F/m)

2.7 $\alpha = 0.109$ Np/m; $\beta = 44.5$ rad/m; $Z_0 = (19.6 + j0.030)$ Ω; $u_p = 1.41 \times 10^8$ m/s

2.10 $w = 0.613$ mm, $\lambda = 0.044$ m

2.14 $R' = 0.5$ (Ω/m); $L' = 200$ (nH/m); $G' = 200$ (μS/m); $C' = 80$ (pF/m); $\lambda = 2.5$ m

2.16 $R' = 0.4$ Ω/m, $L' = 38.2$ nH/m, $G' = 0.25$ mS/m, $C' = 23.9$ pF/m

2.17 **(a)** $b = 4.2$ mm

(b) $u_p = 2 \times 10^8$ m/s

2.22 $Z_L = (120.5 - j89.3)$ Ω

2.23 $Z_0 = 70.7$ Ω

2.29 $Z_{in} = (40 + j20)$ Ω

2.31 **(b)** $\Gamma = 0.16\,e^{-j80.54^\circ}$.

2.32 **(a)** $\Gamma = 0.62e^{-j29.7^\circ}$

2.33 **(a)** $Z_{in_1} = (35.20 - j8.62)$ Ω

2.35 $L = 8.3 \times 10^{-9}$ H

2.36 $l = \lambda/4 + n\lambda/2$

2.39 $Z_{in} = \dfrac{100^2}{33.33} = 300$ Ω

2.41 **(b)** $i_L(t) = 3\cos(6\pi \times 10^8 t - 135^\circ)$ (A)

2.42 **(a)** $Z_{in} = (41.25 - j16.35)$ Ω

2.44 $P^i_{av} = 10.0$ mW; $P^r_{av} = -1.1$ mW; $P^t_{av} = 8.9$ mW

2.45 **(a)** $P_{av} = 0.29$ W

2.48 **(b)** $\Gamma = 0.62 \exp -29.7^\circ$

2.50 $Z_{in} = (66 - j125)$ Ω

2.52 **(b)** $S = 1.64$

2.53 $Z_{01} = 40$ Ω; $Z_{02} = 250$ Ω

2.55 **(a)** $Z_{in} = -j154$ Ω

(b) $0.074\lambda + (n\lambda/2),\ n = 0, 1, 2, \ldots$

2.57 The reciprocal of point Z is at point Y, which is at $0.55 + j0.26$.

2.61 $Z_L = (41 - j19.5)$ Ω

2.63 $Z_{in} = (95 - j70)$ Ω

2.69 First solution: Stub at $d = 0.199\lambda$ from antenna and stub length $l = 0.125\lambda$. Second solution: $d = 0.375\lambda$ from antenna and stub length $l = 0.375\lambda$.

2.73 $Z_{in} = 100$ Ω

2.78 $V_g = 19.2$ V; $R_g = 30$ Ω; $l = 700$ m

2.80 **(a)** $l = 600$ m

(b) $Z_L = 0$

(c) $R_g = \left(\dfrac{1+\Gamma_g}{1-\Gamma_g}\right) Z_0 = \left(\dfrac{1+0.25}{1-0.25}\right) 50 = 83.3$ Ω

(d) $V_g = 32$ V

Chapter 3

3.2 $\hat{\mathbf{a}} = \hat{\mathbf{x}}0.32 - \hat{\mathbf{z}}0.95$

3.3 Area = 36

3.5 **(a)** $A = \sqrt{14}$; $\hat{\mathbf{a}}_A = (\hat{\mathbf{x}} + \hat{\mathbf{y}}2 - \hat{\mathbf{z}}3)/\sqrt{14}$

(e) $\mathbf{A} \cdot (\mathbf{B} \times \mathbf{C}) = 20$

(h) $(\mathbf{A} \times \hat{\mathbf{y}}) \cdot \hat{\mathbf{z}} = 1$

3.9 $\hat{\mathbf{a}} = \dfrac{\mathbf{A}}{|\mathbf{A}|} = \dfrac{-\hat{\mathbf{x}} - \hat{\mathbf{y}}y - \hat{\mathbf{z}}2}{\sqrt{5 + y^2}}$

3.10 $\hat{\mathbf{a}} = (\hat{\mathbf{x}}3 - \hat{\mathbf{z}}6)/\sqrt{45}$

3.12 $\mathbf{A} = \hat{\mathbf{x}}0.8 + \hat{\mathbf{y}}1.6$

3.15 $\hat{\mathbf{c}} = \hat{\mathbf{x}}0.37 + \hat{\mathbf{y}}0.56 + \hat{\mathbf{z}}0.74$

3.17 $\mathbf{G} = \pm\left(-\hat{\mathbf{x}}\tfrac{8}{3} + \hat{\mathbf{y}}\tfrac{8}{3} + \hat{\mathbf{z}}\tfrac{4}{3}\right)$

3.23 **(a)** $P_1 = (2.24, 63.4^\circ, 0)$ in cylindrical; $P_1 = (2.24, 90^\circ, 63.4^\circ)$ in spherical

(d) $P_4 = (2.83, 135^\circ, -2)$ in cylindrical; $P_4 = (3.46, 125.3^\circ, 135^\circ)$ in spherical

3.24 **(a)** $P_1 = (0, 0, 5)$

3.25 **(c)** $A = 12$

3.27 **(a)** $V = 21\pi/2$

3.30 **(a)** $\theta_{AB} = 90^\circ$

(b) $\pm(\hat{\mathbf{r}}0.487 + \hat{\boldsymbol{\phi}}0.228 + \hat{\mathbf{z}}0.843)$

3.31 (a) $d = \sqrt{3}$

3.34 (c) $\mathbf{C}(P_3) = \hat{\mathbf{r}}0.707 + \hat{\mathbf{z}}4$

(e) $\mathbf{E}(P_5) = -\hat{\mathbf{r}} + \hat{\boldsymbol{\phi}}$

3.35 (c) $\mathbf{C}(P_3) = \hat{\mathbf{R}}0.854 + \hat{\boldsymbol{\theta}}0.146 - \hat{\boldsymbol{\phi}}0.707$

3.36 (e) $\nabla S = \hat{\mathbf{x}}8xe^{-z} + \hat{\mathbf{y}}3y^2 - \hat{\mathbf{z}}4x^2e^{-z}$

3.37 (b) $\nabla T = \hat{\mathbf{x}}\,2x$

(g) $\nabla T = -\hat{\mathbf{x}}\,\frac{2\pi}{6}\sin\left(\frac{\pi x}{3}\right)$

3.38 $T(z) = 10 + (1 - e^{-4z})/4$

3.39 $\left(\frac{dV}{dl}\right)\Big|_{(1,-1,2)} = 1.34$

3.42 $dU/dl = -0.02$

3.46 $\mathbf{E} = \hat{\mathbf{R}}2R$

3.48 (a) $\oint \mathbf{D}\cdot d\mathbf{s} = 150\pi$

(b) $\iiint \nabla\cdot\mathbf{D}\,d\mathcal{V} = 150\pi$

3.56 (a) **A** is solenoidal, but not conservative.

(d) **D** is conservative, but not solenoidal.

(h) **H** is conservative, but not solenoidal.

3.58 (c) $\nabla^2\left(\frac{3}{x^2+y^2}\right) = \frac{12}{(x^2+y^2)^2}$

Chapter 4

4.2 $Q = 2.62$ (mC)

4.3 $Q = 260$ (mC)

4.7 $I = 314.2$ A

4.8 (a) $\rho_l = -\frac{\pi c a^4}{2}$ (C/m)

4.11 $\mathbf{E} = \hat{\mathbf{z}}\,51.2$ kV/m

4.12 $q_2 \approx -94.69$ (μC)

4.15 (a) $\mathbf{E} = -\hat{\mathbf{x}}\,1.6 - \hat{\mathbf{y}}\,0.66$ (MV/m)

4.17 $\mathbf{E} = \hat{\mathbf{z}}\,(\rho_{s0}h/2\epsilon_0)\left[\sqrt{a^2+h^2} + h^2/\sqrt{a^2+h^2} - 2h\right]$

4.20 $\mathbf{E} = -\hat{\mathbf{y}}\,\frac{\rho_l}{\pi\epsilon_0 R_1}\,\frac{R_1}{R_2} + \hat{\mathbf{y}}\,\frac{\rho_l}{\pi\epsilon_0 R_2} = 0$

4.23 (a) $\rho_v = y^3 z^3$

(b) $Q = 32$ (C)

(c) $Q = 32$ (C)

4.25 $Q = 4\pi\rho_0 a^3$ (C)

4.26 $\mathbf{D} = \hat{\mathbf{r}}\,\frac{\rho_{v0}(r^2-1)}{2r}, \qquad 1 \le r \le 2$ m

$\mathbf{D} = \hat{\mathbf{r}}\,D_r = \hat{\mathbf{r}}\,\frac{3\rho_{v0}}{2r}, \qquad r \ge 2$ m

4.30 $R_1 = \frac{a}{2},\ R_3 = \frac{a\sqrt{5}}{2},\ V = \frac{0.55Q}{\pi\epsilon_0 a}$

4.33 (b) $\mathbf{E} = \hat{\mathbf{z}}(\rho_l a/2\epsilon_0)[z/(a^2+z^2)^{3/2}]$ (V/m)

4.34 $V(b) = (\rho_l/4\pi\epsilon)$
$\times \ln\left[\frac{l+\sqrt{l^2+4b^2}}{-l+\sqrt{l^2+4b^2}}\right]$ (V)

4.37 $V = \frac{\rho_l}{2\pi\epsilon_0}\left[\ln\left(\frac{a}{\sqrt{(x-a)^2+y^2}}\right) - \ln\left(\frac{a}{\sqrt{(x+a)^2+y^2}}\right)\right]$

4.40 $V_{AB} = -234.18$ V

4.41 (c) $\mathbf{u}_e = -8.125\mathbf{E}/|\mathbf{E}|$ (m/s); $\mathbf{u}_h = 3.125\mathbf{E}/|\mathbf{E}|$ (m/s)

4.45 $R = 4.2$ (mΩ)

4.48 $\theta = 61°$

4.50 $Q = 3\pi\epsilon_0$ (C)

4.52 (a) $|E|$ is maximum at $r = a$.

4.56 $W_e = 4.62 \times 10^{-9}$ (J)

4.57 (a) $C = 3.1$ pF

4.60 (b) $C = 6.07$ pF

4.63 $C' = \frac{\pi\epsilon_0}{\ln[(2d/a)-1]}$ (C/m)

Chapter 5

5.1 $\mathbf{a} = -\hat{\mathbf{y}}8.44 \times 10^{18}$ (m/s^2)

5.4 $\mathbf{T} = -\hat{\mathbf{z}}1.66$ (N·m); clockwise

5.5 (a) $\mathbf{F} = 0$

5.9 $\mathbf{H} = \hat{\mathbf{z}}\dfrac{I\theta\,(b-a)}{4\pi ab}$

5.10 $\mathbf{B} = -\hat{\mathbf{z}}0.6$ (mT)

5.11 $I_2 = \dfrac{2aI_1}{2\pi Nd} = \dfrac{1 \times 50}{\pi \times 20 \times 2} = 0.4$ A

5.14 $I = 100$ A

5.16 $\mathbf{F} = -\hat{\mathbf{x}}0.8$ (mN)

5.18 (a) $\mathbf{H}(0,0,h) = -\hat{\mathbf{x}}\,\dfrac{I}{\pi w}\tan^{-1}\left(\dfrac{w}{2h}\right)$ (A/m)

5.20 $\mathbf{F} = \hat{\mathbf{y}}\,8 \times 10^{-5}$ N

5.24 $\mathbf{J} = \hat{\mathbf{z}}\,9e^{-3r}$ A/m^2

5.26 (a) $\mathbf{A} = \hat{\mathbf{z}}\,\dfrac{\mu_0 I}{4\pi}\ln\left(\dfrac{\ell + \sqrt{\ell^2 + 4r^2}}{-\ell + \sqrt{\ell^2 + 4r^2}}\right)$

5.28 (a) $\mathbf{B} = \hat{\mathbf{z}}5\pi\sin\pi y - \hat{\mathbf{y}}\pi\cos\pi x$ (T)

5.29 (a) $\mathbf{A} = \hat{\mathbf{z}}\mu_0 IL/(4/piR)$

(b) $\mathbf{H} = (IL/4\pi)[(-\hat{\mathbf{x}}y + \hat{\mathbf{y}}x)/(x^2+y^2+z^2)^{3/2}]$

5.30 $n_e = 1.5$ electrons/atom

5.33 $\mathbf{H}_2 = \hat{\mathbf{z}}\,7$

5.35 $\vec{B}_2 = \hat{\mathbf{x}}20000 - \hat{\mathbf{y}}30000 + \hat{\mathbf{z}}12$

5.37 $L' = (\mu/\pi)\ln[(d-a)/a]$ (H)

5.40 $\Phi = 1.66 \times 10^{-6}$ (Wb)

Chapter 6

6.1 At $t = 0$, current in top loop is momentarily clockwise. At $t = t_1$, current in top loop is momentarily counterclockwise.

6.4 (a) $V_{\text{emf}} = 750e^{-3t}$ (V)

6.7 $I_{\text{ind}} = 18.85\sin(200\pi t)$ mA

6.9 $B_0 = 0.4$ (nA/m)

6.10 $V_{12} = -236$ (μV)

6.12 $I = 0.3$ (A)

6.14 $I = 0.82\cos(120\pi t)$ (μA)

6.17 (b) 888

6.18 $f = 5$ MHz

6.20 $\rho_v = (8y/\omega)\sin\omega t + C_0$, where C_0 is a constant of integration.

6.24 $k = (4\pi/30)$ rad/m;
$\mathbf{E} = -\hat{\mathbf{z}}941\cos(2\pi \times 10^7 t + 4\pi y/30)$ (V/m)

6.26 $\mathbf{H}(R,\theta;t) = \hat{\boldsymbol{\phi}}\,(53/R)\sin\theta\cos(6\pi \times 10^8 t - 2\pi R)$ (μA/m)

6.28 (a) $k = 20$ (rad/m)

Chapter 7

7.2 (a) Positive y-direction

(c) $\lambda = 12.6$ m

7.3 (a) $\lambda = 31.42$ m

7.5 $\epsilon_r = 16$

7.6 (a) $\lambda = 10$ m

7.14 (a) $\gamma = 73.5°$ and $\chi = -8.73°$

(b) Right-hand elliptically polarized

7.17 (a) Low-loss dielectric. $\alpha = 8.42 \times 10^{-11}$ Np/m, $\beta = 468.3$ rad/m, $\lambda = 1.34$ cm, $u_p = 1.34 \times 10^8$ m/s, $\eta_c \approx 168.5\ \Omega$

7.19 **H** lags **E** by 31.72°

7.21 $z = 287.82$ m

7.23 $u_p = 6.28 \times 10^4$ (m/s)

7.25 $\mathbf{H} = -\hat{\mathbf{y}}0.16\,e^{-30x}\cos(2\pi \times 10^9 t - 40x - 36.85°)$ (A/m)

7.29 $(R_{ac}/R_{dc}) = 287.1$

7.34 $\mathbf{S}_{av} = \hat{\mathbf{y}}0.48$ (W/m^2)

7.35 (c) $z = 23.03$ m

7.36 $u_p = 1 \times 10^8$ (m/s)

7.39 (b) $P_{av} = 0$

7.42 (a) $(w_e)_{av} = \dfrac{\epsilon E_0^2}{4}$

Chapter 8

8.2 (a) $\Gamma = -0.67;\ \tau = 0.33$

(b) $S = 5$

(c) $S^i_{av} = 0.52$ (W/m^2); $S^r_{av} = 0.24$ (W/m^2); $S^t_{av} = 0.28$ (W/m^2)

8.3 (b) $\mathbf{S}^i_{av} = \hat{\mathbf{y}}\,251.34$, $\mathbf{S}^r_{av} = \hat{\mathbf{y}}\,10.05$, $\mathbf{S}^t_{av} = \hat{\mathbf{y}}\,241.29$ (W/m^2)

8.6 (a) $\Gamma = -0.71$

8.8 $|\widetilde{\mathbf{E}}_1|_{max} = 85.5$ (V/m); $l_{max} = 1.5$ m

8.9 $\epsilon_{r_2} = \sqrt{\epsilon_{r_1}\epsilon_{r_3}}$; $d = c/[4f(\epsilon_{r_1}\epsilon_{r_3})^{1/4}]$

8.11 $Z_{in}(-d) = 0.43\eta_0\angle{-51.7°}$
$|\Gamma|^2 = 0.24$

8.12 $f = 50$ MHz

8.15 $P' = (3.3 \times 10^{-3})^2 \frac{10^2}{2} \times 1.14\,[1 - e^{-2\times 44.43\times 2\times 10^{-3}}] = 1.01 \times 10^{-4}$ (W/m^2)

8.17 $\theta_{min} = 35.57°$

8.20 $\dfrac{S^t}{S^i} = 0.85$

8.22 $d = 15$ cm

8.25 $d = 68.42$ cm

8.26 $f_p = 166.33$ (Mb/s)

8.27 (b) $\theta_i = 36.87°$

8.29 (a) $\theta_i = 33.7$o

8.30 $\theta_t = 18.44°$

8.37 (a) 9.4%

8.39 $a = 2$ cm; $b = 1.6$ cm

8.41 Any one of the first four modes.

8.43 570 Ω (empty); 290 Ω (filled)

8.45 $\theta'_{20} = 43.16°$

8.46 (a) $Q = 8367$

Chapter 9

9.2 $S_{max} = 47.5$ (μW/m^2)

9.4 (a) Direction of maximum radiation is a circular cone 120° wide, centered around the $+z$ axis.

(b) $D = 4 = 6$ dB

(c) $\Omega_p = \pi$ (sr) $= 3.14$ (sr)

(d) $\beta = 120°$

9.6 (b) $G = -3.5$ dB

9.9 $S_{max} = 6 \times 10^{-5}$ (W/m^2)

9.12 $D = 40.11$ dB

9.14 $S = 1.46$

9.15 (a) $\widetilde{\mathbf{E}}(R, \theta, \phi) = \hat{\boldsymbol{\theta}} \widetilde{E}_\theta = \hat{\boldsymbol{\theta}} j \dfrac{I_0 l k \eta_0}{8\pi} \left(\dfrac{e^{-jkR}}{R} \right) \sin\theta$ (V/m)

9.17 (a) $\theta_{\max_1} = 42.6°$, $\theta_{\max_2} = 137.4°$

9.20 (a) $\theta_{\max_1} = 90°$, $\theta_{\max_2} = 270°$

(b) $S_{\max} = \dfrac{60 I_0^2}{\pi R^2}$

(c) $F(\theta) = \dfrac{1}{4} \left[\dfrac{\cos(\pi \cos\theta) + 1}{\sin\theta} \right]^2$

9.23 $P_t = 25.9$ (mW)

9.27 (a) $P_{rec} = 3.6 \times 10^{-6}$ W

9.31 $\beta_{null} = 5.73°$

9.32 $D = 39.96$ dB

9.34 (a) $\beta_e = 1.8°$; $\beta_a = 0.18°$

(b) $\Delta y = \beta_a R = 0.96$ m

9.37 (a) $F_a(\theta) = 4\cos^2\left[\frac{\pi}{8}(4\cos\theta + 1)\right]$

9.39 $d/\lambda = 1.414$

9.44 $F_a(\theta) = [6 + 8\cos(\pi\cos\theta) + 2\cos(2\pi\cos\theta)]^2$

9.47 $\delta = -2.72$ (rad) $= -155.9°$

Chapter 10

10.1 $T = 89.72$ minutes

10.3 133.3 ≈ 133 channels

10.6 $(f_p)_{\max} = 150$ kHz

10.7 $R_{\max} = 4.84$ km

Bibliography

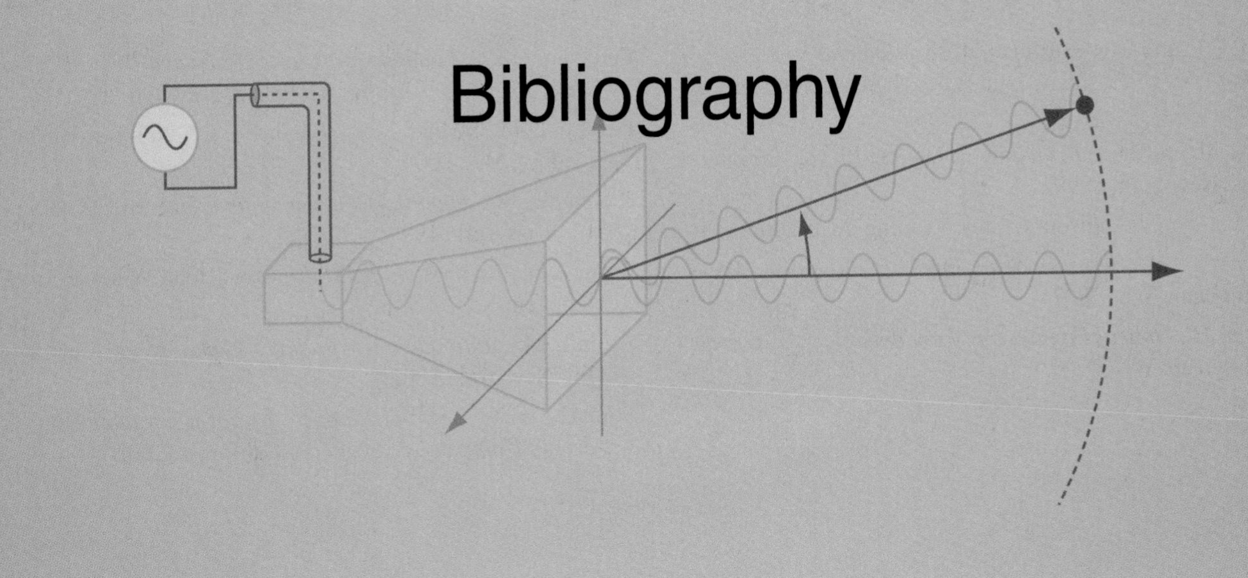

The following list of books, arranged alphabetically by the last name of the first author, provides references for further reading.

Electromagnetics

Balanis, C.A., *Advanced Engineering Electromagnetics,* John Wiley & Sons, Hoboken, NJ, 1989.

Cheng, D.K., *Fundamentals of Engineering Electromagnetics,* Addison Wesley, Reading, MA, 1993.

Hayt, W.H., Jr. and J.A. Buck, *Engineering Electromagnetics,* 7th ed., McGraw-Hill, New York, 2005.

Iskander, M.F., *Electromagnetic Fields & Waves*, Prentice Hall, Upper Saddle River, NJ, 2000.

King, R.W.P. and S. Prasad, *Fundamental Electromagnetic Theory and Applications,* Prentice Hall, Englewood Cliffs, NJ, 1986.

Ramo, S., J.R. Whinnery, and T. Van Duzer, *Fields and Waves in Communication Electronics,* 3rd ed., John Wiley & Sons, Hoboken, NJ, 1994.

Rao, N.N., *Elements of Engineering Electromagnetics*, Prentice Hall, Upper Saddle River, NJ, 2004.

Shen, L.C. and J.A. Kong, *Applied Electromagnetism,* 3rd ed., PWS Engineering, Boston, MA, 1995.

Antennas and Radiowave Propagation

Balanis, C.A., *Antenna Theory: Analysis and Design,* John Wiley & Sons, Hoboken, NJ, 2005.

Ishimaru, A., *Electromagnetic Wave Propagation, Radiation, and Scattering,* Prentice Hall, Upper Saddle River, NJ, 1991.

Stutzman, W.L. and G.A. Thiele, *Antenna Theory and Design*, John Wiley & Sons, Hoboken, NJ, 1997.

Optical Engineering

Bohren, C.F. and D.R. Huffman, *Absorption and Scattering of Light by Small Particles*, John Wiley & Sons, Hoboken, NJ, 1998.

Born, M. and E. Wolf, *Principles of Optics*, 7th ed., Pergamon Press, New York, 1999.

Hecht, E., *Optics,* Addison-Wesley, Reading, MA, 2001.

Smith, W.J., *Modern Optical Engineering,* SPIE Press, Bellingham, WA, 2007.

Walker, B.H., *Optical Engineering Fundamentals*, SPIE Press, Bellingham, WA, 2009.

Microwave Engineering

Freeman, J.C., *Fundamentals of Microwave Transmission Lines,* John Wiley & Sons, Hoboken, NJ, 1996.

Pozar, D.M., *Microwave Engineering,* Addison-Wesley, Reading, MA, 2004.

Richharia, M., *Satellite Communication Systems,* McGraw-Hill, New York, 1999.

Scott, A.W., *Understanding Microwaves,* John Wiley & Sons, Hoboken, NJ, 2005.

Skolnik, M.I., *Introduction to Radar Systems,* 3rd ed., McGraw-Hill, New York, 2002.

Stimson, G.W., *Introduction to Airborne Radar,* Hughes Aircraft Company, El Segundo, California, 2001.

Index

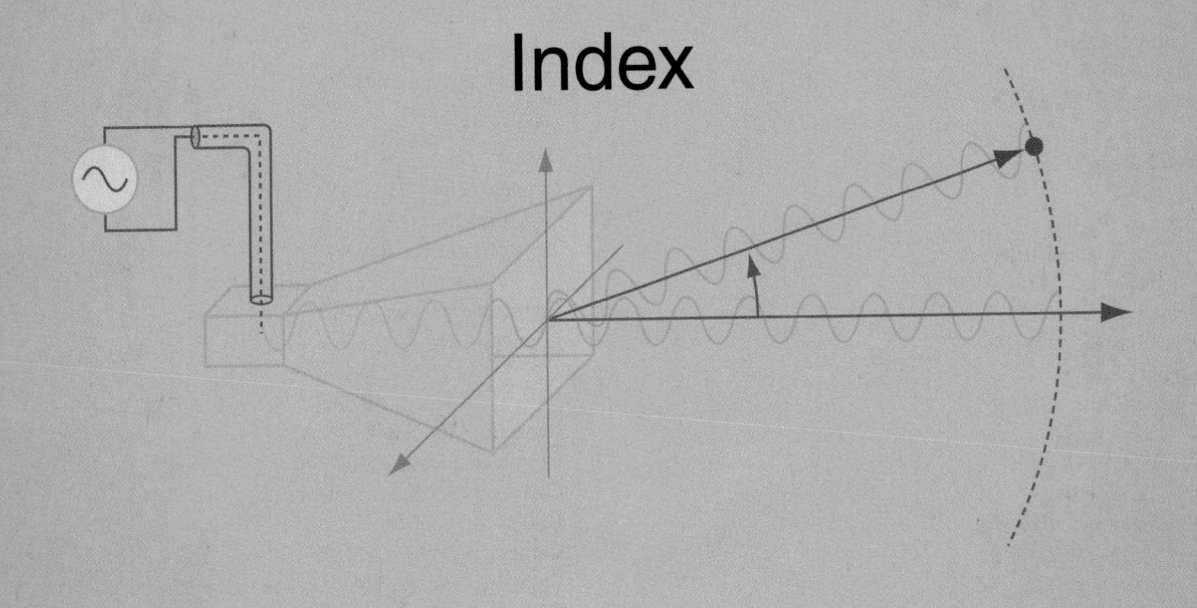

3-dB beamwidth, 436

A

Abacus, 30
Ablation, 32, 134
ac motor, 25, 27
ac resistance R, 363
Acceptance angle θ_a, 387
Adding machine, 30
Admittance Y, 118
Alternating current (ac), 27
AM radio, 28
Ampère, André-Marie, 26
Ampère's law, 274–277, 295
Amplitude-comparison monopulse radar, 492
Amplitude modulation (AM), 28
Analog computer, 30
Angle error signal, 493
Angle of incidence θ_i, 385
Angle of reflection θ_r, 385
Angle of transmission θ_t, 385
Angular frequency ω, 47, 80
Angular velocity ω, 47
Antennas, 426–471, 485–486
    aperture, 451
        rectangular, 454–456
        scalar formulation, 452
        vector formulation, 452
    arrays, 457–464
        linear phase, 468
        pattern multiplication principle, 460
        scanning, 466–471
        uniform phase, 464–465
    broadside direction, 431
    directivity D, 436, 456
    effective area, 456
    far-field (far-zone) region, 427, 430–431
    gain, 438–439
    half-wave dipole, 439–444
    input impedance, 426
    isotropic, 426, 435
    large aperture, 451–457
    multiplication principle, 460
    normalized radiation intensity, 431
    pattern solid angle Ω_p, 434
    patterns, 426, 433
        beam dimensions, 434
        beamwidth β, 435–436
        directivity D, 436–437
    polarization, 426
    receiving, 444–449
    reciprocal, 426
    types, 486
        arrays, 486
        dipoles, 486
        helices, 486

horns, 486
parabolic dishes, 486
Antenna radiation pattern, 426
Arithmometer, 30
Armstrong, Edwin, 28, 29
ARPANET, 29
Array factor $F_a(\theta)$, 460
array amplitude distribution, 460
array phase distribution, 460
Atmospheric transmissivity Υ, 484
Attenuation constant α, 79, 353
Average power $\mathbf{S}_{av}$, 365
Average power density $\mathbf{S}_{av}$, 365
Auxiliary angle ψ_0, 351
Axial ratio R, 351
Azimuth angle ϕ, 429
Azimuth-difference channel, 493
Azimuth plane (ϕ-plane), 434
Azimuth resolution Δx, 488

B

bac-cab rule, 161
Backus, John, 30
Band gap energy, 61
Bar-code readers, 404–405
Bardeen, John, 29
Base vector, 156
BASIC, 30
Beam dimensions, 434
Beamwidth β, 435, 436, 455–456
Becquerel, Alexandre-Edmond, 60, 315
Bell, Alexander, 28
Berliner, Emil, 28
Berners–Lee, Tim, 31
Bhatia, Sabeer, 31
Bioelectrics, 135
Biot, Jean-Baptiste, 26, 38
Biot–Savart law, 26, 38, 266–273, 295
current distributions, 266–270
surface current density $\mathbf{J}_s$, 266
volume current density $\mathbf{J}$, 266
volume distributions, 266–270
Bistatic radar, 489
Bounce diagram, 140
Boundary conditions, 225–232
Brattain, Walter, 29
Braun, Karl, 28
Brewster (polarizing) angle, 397–398, 418
Broadside array, 464
Broadside direction, 431
Bush, Vannevar, 30

C

Capacitance C, 232–235
capacitor, 232
of a coaxial line, 234
of a parallel-plate capacitor, 233–235
Capacitive sensors, 218, 240–244
Capacitor, 26,
as batteries, 236–238
electrochemical double-layer (EDLC), 236
Cardullo, Mario, 344
Carrier frequency f, 487
Cartesian coordinate system x, y, z, 163, 164
CAT (CT) scan, 186
Cathode ray tube (CRT), 28
Cavity resonators, 414–416, 418
Cell phone, 29
Charge continuity equation, 323, 329
Charge dissipation, 324
Charge distribution, 202–203, 206
surface distribution, 207
Circular polarization, 346, 348–350
Circulation, 184
Circulator, 482
Cladding, 387
Coaxial line, 73
Complex conjugate, 56
Complex feeding coefficient A_i, 459
Complex numbers, 54–58
complex conjugate, 56
Euler's identity, 54, 65
polar form, 54
properties, 56
rectangular form, 54
rectangular-polar relations, 54, 65
Complex permittivity ϵ_c, 337
Compressive stress, 314
Conductance G, 118
Conductivity σ, 30, 40, 220, 499
Conductors, 217–223
conduction current, 217
conduction current density $\mathbf{J}$, 217
conductivity, 220, 499
equipotential medium, 220
resistance, 221–222
semiconductors, 217, 220
Conservative (irrotational) field, 188, 213

Constitutive parameters, 217
Convection current, 204
Conversion efficiency, 60
Coordinate systems, 162–176
  Cartesian x, y, z, 163, 164
  cylindrical r, ϕ, z, 162, 164–167
  spherical R, θ, ϕ, 162, 167–169
Coplanar waveguide, 73
Cormack, Allan, 186
Coulomb (C), 35
Coulomb, Charles-Augustin de, 25, 26, 35
Coulomb's law, 35, 204–209
  charge distribution, 206
  circular disk of charge, 208
  infinite sheet of charge, 209
  line distribution, 207
  relative permittivity (dielectric constant) ϵ_r, 205
  ring of charge, 207
  surface distribution, 207
  two-point charges, 206
  volume distribution, 207
Critical angle θ_c, 386
Cross (vector) product, 160–161
CT (CAT) scan, 186
Curie, Paul-Jacques, 314
Curie, Pierre, 314
Curl operator, 184, 185
Current density, 217, 266, 319
Cutoff frequency f_{mn}, 408
Cutoff wavenumber k_c, 406
Cylindrical coordinate system r, ϕ, z, 162, 164–167

D

dc motor, 25
De Forest, Lee, 28
Deep Blue, 31
Del (gradient operator) ∇, 177
Detection, 489–491
  maximum detectable range R_{max}, 490
  threshold detection level $P_{r_{min}}$, 490
Diamagnetic, 282
Dielectric constant (relative permittivity) ϵ_r, 37, 205, 224, 501
Dielectrics, 217, 223–225
  anisotropic, 224
  breakdown, 225–225
  breakdown voltage V_{br}, 225
  electric polarization field **P**, 224
  electric susceptibility χ_e, 225
  homogeneous, 224
  isotropic, 224
  linear, 224
  nonpolar, 223
  perfect, 217, 220
  permanent dipole moments, 224
  polar materials, 223
  polarization, 223
  strength E_{ds}, 225
  tables, 226, 501
Difference channel, 493
Digital computer, 30
Dimensions, 33
Dipole, 36, 104, 214, 270, 274
  electric, 36, 104, 214
  half-wave, 439–444, 473
  Hertzian, 428–431
  linear, 442–444
  moment, 215
  short, 449, 473
  vertical, 457
Direct current (dc), 25
Directional derivative dT/dl, 177
Directivity D, 436, 456
Dispersive, 72
Displacement current I_d, 319–321
Displacement current density $\mathbf{J}_d$, 319
Distance vector, 158
Divergence operator, 180–184
Divergence theorem, 181
Dominant mode, 408
Doppler frequency shift f_d, 486, 491
Doppler radar, 491–492
Dot (scalar) product, 158–159
Downlink, 482
Drift velocity $\mathbf{u}_e$, 220
du Fay, Charles François, 25, 26
Duplexer (T/R switch), 482, 487

E

e electron charge, 35
Echo satellite, 29
Eckert, J. Presper, 30
Edison, Thomas, 28, 42
Effective aperture, 444, *See also* Effective area
Effective area A_e, 444
Einstein, Albert, 25, 27, 60
Electric, 25, 26

Electric charge, 25, 26, 35–36
law of conservation of electric charge, 36
principle of linear superposition, 36
Electric dipole, 36, 104, 214
moment, 215
Electric-field aperture distribution $E_a(x_a, y_a)$, 452
Electric field intensity **E**, 36, 201
Electric field phasor $\widetilde{\mathbf{E}}$, 341
Electric fields, 35–37, 201, 205–209
dipole, 36, 214
e charge, 35
polarization, 36, 223
Electric flux density **D**, 37, 201
Electric generator, 25
Electric potential V, 211
Electric scalar potential, 211–216
as a function of electric field, 211–213
due to continuous distributions, 213
due to point charges, 213, 245
electric dipole, 214
Kirchhoff's voltage law, 212
Laplace's equation, 215
line distribution, 213
Poisson's equation, 215
potential energy, 211
Electric susceptibility χ_e, 225
Electric typewriter, 28
Electrical force $\mathbf{F}_e$, 35
Electrical permittivity ϵ, 35, 88, 205–206, 225
of free space ϵ_0, 35
Electrical sensors, 218
capacitive, 218
emf, 218
inductive, 218
resistive, 218–219
Electromagnetic (EM) force, 34, 259
Electromagnetic (EM) spectrum, 52–54
gamma rays, 52, 54
infrared, 52, 54
microwave band, 54, 54
EHF, 54
millimeter-wave band, 54
SHF, 54
UHF, 54
monochromatic, 52
properties, 52
radio spectrum, 52, 54, 54
ultraviolet, 54, 54
visible, 54, 54
X-rays, 52, 54
Electromagnetic generator, 316–318
Electromagnetic induction, 305
Electromagnetic telegraph, 28
Electromagnetic waves, 27, 104, 375–416
Electromagnets, 278–280
ferromagnetic core, 278
horseshoe, 278
loudspeaker, 279–280
magnetic levitation, 280
magnetically levitated trains (maglevs), 280–280
reed relay, 278
step-down transformer, 278
switch, 278
Electromotive force (emf) V_{emf}, 27, 305
Electron, 25, 27, 35
Electronic beeper, 29
Electronic steering, 458
EM, 25
Electrostatics, 39, 201
Elevation angle (θ-plane), 434
Elevation-difference channel, 493
Elevation plane (θ-plane), 434
Elliptical polarization, 346, 350–352
Ellipticity angle χ, 350
Emf sensor, 218
End-fire direction, 467
Engelbart, Douglas, 31
ENIAC, 30
Equipotential, 220
Euler's identity, 54, 65
Evanescent wave, 407
Explorer I satellite, 480

F

Faraday, Michael, 25, 27, 305
Faraday's law, 304–306, 329
motional emf, 311, 329
transformer emf, 306, 329
Far-field (far-zone) region, 427
approximation, 430–431
power density, 431
False alarm probability, 489
Feeding coefficient A_i, 459
Felt, Dorr, 30
Ferromagnetic, 282, 284–286
Fessenden, Reginald, 28
Fiber, 29, 73, 387
Fiber optics, 387–389

Field lines, 180
Floppy disk, 30
Fluorescence, 42
Fluorescent bulb, 42–45
Flux density, 180
Flux sensor, 315
FORTRAN, 30
Franklin, Benjamin, 25, 26
Free space, 35
  velocity of light c, 38
  magnetic permeability μ_0, 38
  electric permittivity ϵ, 35
Frequency, 47
Frequency-division multiple access (FDMA), 482
Frequency modulation (FM), 29
Frequency scanning, 467–471
Friis transmission formula, 449–451, 484
Fundamental forces
  electromagnetic, 34, 201
  nuclear, 34
  weak-interaction, 34
  gravitational, 34

G

Gamma rays, 52, 54
Gauss, Carl Friedrich, 27
Gauss's law, 27, 209–211
  differential form, 209
  of infinite line charge, 211
  integral form, 209
  Gaussian surface, 209
Gauss's law for magnetism, 273, 274, 295
Geostationary orbit, 480
Gilbert, William, 25, 26
Global Positioning System (GPS), 172–173
Grad (gradient) ∇T, 177
Gradient operator, 177–180
Gravitational force, 34
  gravitational field $\boldsymbol{\psi}$, 34
Grazing incidence, 397
Group velocity u_g, 409

H

Half-power angle, 435
Half-power beamwidth, 435
Half-wave dipole, 439–444
Henry, Joseph, 25, 27, 305
Hertz, Heinrich, 25, 27, 28, 47
Hertzian dipole, 428–431
High-power amplifier, 483
Hoff, Ted, 31
Hole drift velocity $\mathbf{u}_h$, 220
Hole mobility μ_h, 220
Homogeneous material, 217
Homogeneous wave equation, 338
Horn antenna, 427
Hotmail, 31
Hounsfield, Godfrey, 186
Humidity sensor, 241

I

Illumination $E_a(x_a, y_a)$, 452
Image method, 245–246
Imaginary part $\mathfrak{Im}$, 54
Impedance, 71, 80, 88, 90, 97, 98, 115
Impedance matching, 123–132
  lumped element matching, 124–130
  matching points, 129
  network, 124
  shunt stub, 130
  single-stub matching, 130–133
  stub, 130
Impulse period T_p, 487
In-phase, 91
Incandescence, 42
Incandescent bulb, 42–45
Inclination angle ψ, 347
Incremental phase delay δ, 468
Index of refraction, 385
Inductance, 27, 287–293, 295
  of a coaxial line, 289
  mutual, 288, 292–293
  self, 288, 289
  solenoid, 287
Inductive sensors, 218, 290–291
  eddy-current proximity sensor, 290
  ferromagnetic core, 290
  linear variable differential transformer (LVDT), 290
  proximity detection, 290
Infrared rays, 52, 54
In-phase, 91
Input impedance Z_{in}, 438
Integrated circuit (IC), 29
Intercepted power P_{int}, 444
Internal (surface) impedance Z_s, 363
International System of Units (SI), 33
Internet, 29, 31

Intrinsic impedance η, 340
Isotropic, 217
Isotropic antenna, 426, 435
Isotropic material, 217

J

Java, 31
Joule's law, 223

K

Kapany, Narinder, 29
Kemeny, John, 30
Kilby, Jack, 29
Kirchhoff's laws, 71
  current, 71, 323, 324
  voltage, 71, 212
Kurtz, Thomas, 30

L

Laplace's equation, 215
Laplacian operator, 189–191
Lasers, 390–391
Law of conservation of electric charge, 36
LED bulb, 42–45
LED lighting, 42–45
Left-hand circular (LHC) polarization, 348
Leibniz, Gottfried von, 30
Lenz's law, 307, 308–309
Leyden Jar, 25
Lidars, 486
Light emitting diode (LED), 44
Lightning rod, 26
Line charge, 202
Line charge density ρ_ℓ, 202
Linear phase distribution, 466
Liquid crystal display (LCD), 24, 358–360
Liquid crystals, 24
Logarithm, 30
Lorentz force, 259, 295
Loss resistance R_{loss}, 438
Lossless media, 380, 398–402
Lossy media, 50
Loudspeaker, 279–280
Low-loss dielectric, 355
Luminous efficacy (LE), 45

M

Macintosh, 31
Maiman, Theodore, 390
Maglevs, 280–280
Magnetic dipole, 270
Magnetic energy W_{m}, 293–294
Magnetic field intensity **H**, 38, 258
Magnetic field phasor $\widetilde{\mathbf{H}}$, 341
Magnetic field, 266–272
  between two parallel conductors, 272–273
  in a solenoid, 287
  inside a toroidal coil, 276–277
  of a circular loop, 269–270, 295
  of a linear conductor, 266–269
  of a long wire, 275–276, 295
  of a magnetic dipole, 270
  of an infinite current sheet, 277
Magnetic flux Φ, 282
Magnetic flux density **B**, 37, 258
Magnetic flux linkage Λ, 289
Magnetic force $\mathbf{F}_{\text{m}}$, 38, 258–263
Magnetic hysteresis, 284
Magnetic levitation, 280
Magnetic moment **m**, 283–284
Magnetic monopole, 274
Magnetic permeability μ, 38, 284
Magnetic potential **A**, 281–282
Magnetic properties of materials, 282–286
Magnetic sound recorder, 28
Magnetic susceptibility χ_{m}, 283
Magnetic torque, 263–266
Magnetite, 25, 37
Magnetized domains, 284
Magnetron tube, 105
Magnus, 26
Marconi, Guglielmo, 28
Mars Pathfinder, 29
Maser, 390
Matched filter, 489
Matched line, 93, 107
Maximum detectable range R_{max}, 490
Maxwell, James Clerk, 25, 27, 201
Maxwell's equations, 273–277, 295, 304
Mauchley, John, 30
Microprocessor, 31
Microstrip line, 73
Microwave band, 54, 54
Mobility μ_{e}, 220
Modal dispersion, 388

Mode, 387, 408
Modem, 30
Moment, 215, 224, 283–284
Monochromatic, 52, 390
Monopulse radar, 492–494, 495
  amplitude-comparison monopulse, 492
  phase-comparison monopulse, 492
Monostatic radar, 489
Morse, Samuel, 27, 28
Motional emf $V_{\text{emf}}^{\text{m}}$, 306, 311, 329
MS-DOS, 31
Multiple-beam generation, 458
Multiple-PRF, 488
Multiplexer, 483

N

n-type layer, 60
Nakama, Yoshiro, 30
Nanocapacitor, 236
Napier, John, 30
Negative electric charge, 25
Neutrons, 35
Newton, Isaac, 26
Noise power, 490, 495
Normal incidence, 378, 418
Normalized load impedance z_{L}, 90
Normalized load reactance x_{L}, 112
Normalized load resistance r_{L}, 112
Notation, 33
Noyce, Robert, 29
Nuclear force, 34
Null beamwidth, 436

O

Oblique incidence, 384–386, 418
Oersted, Hans Christian, 26, 37, 304
Ohm, Georg Simon, 27
Ohm's law, 27, 217
Optical fiber, 29, 73, 387–389
Orbital magnetic moment, 283–284

P

p–n junction, 60
p-type layer, 60
Pager, 29
Parallel-plate transmission line, 73
Parallel polarization, 396–398
Paramagnetic, 282
Pascal, Blaise, 30
Pattern multiplication principle, 460
Pattern solid angle Ω_{p}, 435
Perfect conductor, 217, 220
Perfect dielectric, 217, 220
Permittivity ϵ, 205, 225, 499
Perpendicular polarization, 392–396
Phase, 46
Phase constant β, 79, 353
Phase constant (wavenumber) k, 327
Phase lag, 48
Phase lead, 48
Phase-matching condition, 394
Phase velocity (propagation velocity) u_{p}, 340
Phasor representation, 33
Phasors 58–65
Photoelectric effect, 25, 27, 60
Photovoltaic (PV), 60
Photovoltaic effect, 60
Piezein, 218, 314
Piezoelectric transducer, 314
Piezoresistivity, 218–219
Planck, Max, 25
Plane-wave propagation, 335–368
  attenuation rate A, 368
  circular polarization, 346, 348–350
    left-hand circular (LHC), 348
    right-hand circular (RHC), 348–350
  complex permittivity ϵ_{c}, 337
    imaginary part ϵ'', 338
    real part ϵ', 338
  elliptical polarization, 346, 350–352
    auxiliary angle ψ_0, 351
    axial ratio R, 351
    ellipticity angle χ, 350
    rotation angle γ, 350
  electromagnetic power density, 365
  linear polarization, 346, 347–348
  lossy medium, 336, 353–361
    attenuation constant α, 353
    skin depth δ_{s}, 355
  low-loss dielectric, 355
Pocket calculator, 31
Poisson's equation, 215
Polarization, 36, 346, 392
  parallel polarization, 392, 396–398
  perpendicular polarization, 392–396
  transverse electric (TE) polarization, 392

transverse magnetic (TM) polarization, 392
unpolarized, 398
Polarization diversity, 484
Polarization field **P**, 224
Polarization state, 346
Position vector, 158
Potential energy W_e, 235, 239
Poulsen, Valdemar, 28
Power density $S(R, \theta, \phi)$, 431
Power transfer ratio P_{rec}/P_t, 450
Poynting vector (power density) **S**, 365, 431
Pressure sensor, 241
Principle of linear superposition, 36
Principal planes, 434
Propagation constant γ, 338
Propagation velocity (phase velocity) u_p, 47
Pulse code modulation (PCM), 29
Pulse length τ, 487
Pulse repetition frequency (PRF) f_p, 487

Q

Quality factor Q, 415
Quarter-wavelength transformer, 106
Quasi-conductor, 355

R

Radar (radio detection and ranging), 29, 489–491
azimuth resolution Δx, 488
cross-section, 489
bistatic, 489
detection, 489–491
Doppler, 491–492
monopulse, 492–494, 495
monostatic, 489
multiple-PRF, 488
operation, 486
pulse, 487
range, 487
range resolution ΔR, 488
unambiguous range R_u, 488
Radar cross-section, 489
Radar equation, 490
Radial distance, 38, 164, 486
Radial velocity u_r, 486
Radiation efficiency ξ, 438
Radiation intensity, 431
Radiation pattern, 426
Radiation resistance R_{rad}, 438
Radio frequency identification (RFID) systems, 344–345
Radio telegraphy, 28
Radio waves, 28, 54,
Radius of geostationary orbit, 481, 495
Range R, 167
Range resolution ΔR, 488
RC relation, 233, 248
Real part $\mathfrak{Re}$, 54
Received power, 485, 495
Receiving cross section, 444, *See also* Effective area
Rectangular aperture, 454–456
Rectangular waveguide, 73
Reeves, H. A., 29
Reflection coefficient, 88–90
Reflectivity R, 399–402
Refraction angle, 385
Reinitzer, Friedrich, 358
Relaxation time constant τ_r, 324
Resistive sensor, 218–219
Resonant frequency f_0, 414, 415–416
Retarded potentials, 325–326
Right-hand circular (RHC) polarization, 348–350
Röntgen, Wilhelm, 25, 27
Rotation angle γ, 350

S

Satellite, 480–491
antennas, 485–486
elliptical orbit, 481
geostationary, 480
transponders, 482–484
Savart, Félix, 26, 38
Scalar (dot) product, 158–159
Scalar quantity, 33
Scan angle δ, 468
Score satellite, 480
Seebeck, Thomas, 315
Seebeck potential V_s, 315
Semiconductor, 217, 220
Sensors, 218
capacitive, 218, 240–244
emf, 218, 314–315
inductive, 218, 290–291
resistive, 218–219
Shockley, William, 29
Signal-to-noise ratio S_n, 450, 490, 495
Signal waveform, 487
Skin depth δ_s, 355
Smith chart, 74, 110–123

admittance Y, 118
admittance transformation, 118–122
angle of reflected coefficient, 113
characteristic admittance Y_0, 118
conductance G, 118
constant-SWR (-$|\Gamma|$) circle, 115
matching points, 129
normalized admittance y, 118
normalized conductance g, 118
normalized susceptance b, 118
normalized load admittance y_L, 118
normalized load impedance z_L, 112
normalized load reactance x_L, 112
normalized load resistance r_L, 112
normalized wave impedance $z(d)$, 114
parametric equations, 111–113
phase-shifted coefficient Γ_d, 114
standing-wave ratio (SWR), 115–117
susceptance B, 118
unit circle, 112
voltage maxima $|\widetilde{V}|_{max}$, 115–118
voltage minima $|\widetilde{V}|_{min}$, 115–118
wavelengths toward generator (WTG), 115
wavelengths toward load (WTL), 115
Smith, Jack, 31
Smith, P.H., 110
Snell's laws, 384–386
of reflection, 385, 394, 418
of refraction, 385, 394, 418
Solar cell, 60
Solenoid, 278
Solid angle $d\Omega$, 433
Spherical propagation factor (e^{-jkR}/R), 429
Spherical wave, 336
Spin magnetic moment, 283
Spontaneous emission, 390
Sputnik I satellite, 480
Standing wave, 81, 92–97
first voltage maximum, 94
first voltage minimum, 94
in-phase, 91
interference, 93
minimum value, 93
maximum value, 93
pattern, 93, 105
phase-opposition, 93
properties, 107
voltage standing wave ratio [(VSWR) or (SWR)] S, 94
Static conditions, 201
Steradians (sr), 433
Stimulated emission, 390
Stokes's theorem, 188–189
Strip line, 73
Sturgeon, William, 28, 29, 278
Sum channel, 493
Sun beam, 492
Supercapacitor, 236
Superconductor, 220
Superheterodyne radio receiver, 28
Surface charge density ρ_s, 202
Surface current density $\mathbf{J}_s$, 266
Surface (internal) impedance Z_s, 363
Surface resistance R_s, 363
SWR (standing-wave ratio), 115–117
Synchronizer–modulator, 486
System noise temperature T_{sys}, 450, 485

T

Tapered aperture distribution, 455
Telegraph, 27
Telephone, 28
Television (TV), 29
TEM (transverse electromagnetic), 73–74
Tensile stress, 314
Tesla, Nikola, 25, 27, 38
Thales of Miletus, 25, 26
Thermocouple, 314, 315
Thomas de Colmar, Charles Xavier, 30
Thompson, Joseph, 25, 27
Threshold detection level $P_{r_{min}}$, 490
Tomography, 186
Toroidal coil, 276–277
Torque, 263–266
Total internal reflection, 386
Townes, Charles, 390
Transformer emf V^{tr}_{emf}, 305
Transient response, 133–137
Transistor, 29
Transmission coefficients τ, 378
Transmission lines, 70–143
admittance Y, 118
air line, 77, 81
bounce diagram, 140
characteristic impedance Z_0, 80
characteristic parameters, 89
coaxial line, 73, 75, 83
complex propagation constant γ, 79

attenuation constant α, 79
phase constant β, 79
conductance G, 118
current maxima and minima, 94
definition, 71
dispersive transmission line, 74
distortionless line, 74
effective relative permittivity ϵ_{eff}, 84
load impedance Z_{L}, 88
guide wavelength λ, 81
input impedance Z_{in}, 98, 115
input reactance X_{in}, 101
input resistance R_{in}, 101
lossless line, 87–97
lossless microstrip line, 82–87
lumped-element model, 74–75
matched load, 90
matching network, 124
microstrip line, 73, 82–87
nondispersive, 88
open-circuited line, 103
parallel-plate line, 73
parameters, 74–75
phase-shifted coefficient Γ_d, 114
power loss, 72
power flow, 108–110
quarter-wavelength transformer, 106
slotted line, 96
Smith chart, 74, 110–123
standing wave, 81, 92–97
Transmission lines (*continued*)
standing wave pattern, 93, 105
SWR circle, 115
TEM (transverse electromagnetic) transmission lines, 73–74
transient response, 133–137
transmission line parameters, 74
capacitance C', 75
conductance G', 75
inductance L', 75
resistance R', 74
voltage maxima $|\widetilde{V}|_{\text{max}}$, 115–118
voltage minima $|\widetilde{V}|_{\text{min}}$, 115–118
voltage reflection coefficient Γ, 88–90
voltage standing wave ratio [(VSWR) or (SWR)] S, 94
wave impedance $Z(d)$, 97–100
Transmissivity $\Upsilon(\theta)$, 399–402, 484
Transmitter/receiver (T/R) switch, 487
Transponder, 482–484
Transverse electric (TE), 392
Transverse electric (TE) polarization, 392
Transverse electromagnetic (TEM) wave, 340
Transverse magnetic (TM), 392
Transverse magnetic (TM) polarization, 392
Travelling waves, 40–54, *See also* Waves
Triode tube, 28
Two-wire line, 73

U

Ultracapacitor, 236
Ultraviolet rays, 53, 54
Unambiguous range R_{u}, 488
Uniform field, 184–185
Uniform field distribution, 454
Units, 33
Unit vectors, 33, 156
Uplink, 482

V

van Musschenbroek, Pieter, 26
Vector analysis, 155–191
transformations between coordinate systems, 169–176
Vector magnetic potential, 281–282, 295
Vector Poisson's equation, 281, 295
Vector (cross) product, 160–161
Vector quantities, 33
Velocity of light in free space c, 38
Video processor/display, 486
Visible light, 54, 54
Volta, Alessandro, 25, 26
VSWR (voltage standing wave ratio) S, 94. See also SWR
Volume charge density ρ_{v}, 202
Volume current density $\mathbf{J}$, 266

W

Walton, Charles, 344
Watson-Watt, Robert, 29
Wave polarization, 346
circular, 346, 348–350
elliptically, 346, 350–352
electric field phasor $\widetilde{\mathbf{E}}$, 347
inclination angle ψ, 347
linear, 346, 347–348

Wave polarizer, 359
Wavefront, 336
Waveguides 402–405, 418
Wavelength, 47, 53
Wavenumber (phase constant) k, 327, 338
Waves, 40–54
Weak-interaction force, 34
White light, 26
Wireless transmission, 28
World Wide Web (WWW), 31

X

X-rays, 25, 27, 52, 54

Z

Zenith angle θ, 167, 429
Zuse, Konrad, 30
Zworykin, Vladimir, 29

ω-β diagram, 412

FUNDAMENTAL PHYSICAL CONSTANTS

CONSTANT	SYMBOL	VALUE
speed of light in vacuum	c	$2.998 \times 10^8 \approx 3 \times 10^8$ m/s
gravitational constant	G	6.67×10^{-11} N·m^2/kg^2
Boltzmann's constant	K	1.38×10^{-23} J/K
elementary charge	e	1.60×10^{-19} C
permittivity of free space	ε_0	$8.85 \times 10^{-12} \approx \frac{1}{36\pi} \times 10^{-9}$ F/m
permeability of free space	μ_0	$4\pi \times 10^{-7}$ H/m
electron mass	m_e	9.11×10^{-31} kg
proton mass	m_p	1.67×10^{-27} kg
Planck's constant	h	6.63×10^{-34} J·s
intrinsic impedance of free space	η_0	$376.7 \approx 120\pi$ Ω

FUNDAMENTAL SI UNITS

DIMENSION	UNIT	SYMBOL
Length	meter	m
Mass	kilogram	kg
Time	second	s
Electric current	ampere	A
Temperature	kelvin	K
Amount of substance	mole	mol
Luminous Intensity	candela	cd

MULTIPLE & SUBMULTIPLE PREFIXES

PREFIX	SYMBOL	MAGNITUDE	PREFIX	SYMBOL	MAGNITUDE
exa	E	10^{18}	milli	m	10^{-3}
peta	P	10^{15}	micro	μ	10^{-6}
tera	T	10^{12}	nano	n	10^{-9}
giga	G	10^{9}	pico	p	10^{-12}
mega	M	10^{6}	femto	f	10^{-15}
kilo	k	10^{3}	atto	a	10^{-18}

Book Website: www.pearsonglobaleditions.com/Ulaby

GRADIENT, DIVERGENCE, CURL, & LAPLACIAN OPERATORS

CARTESIAN (RECTANGULAR) COORDINATES (x, y, z)

$$\nabla V = \hat{\mathbf{x}}\frac{\partial V}{\partial x} + \hat{\mathbf{y}}\frac{\partial V}{\partial y} + \hat{\mathbf{z}}\frac{\partial V}{\partial z}$$

$$\nabla \cdot \mathbf{A} = \frac{\partial A_x}{\partial x} + \frac{\partial A_y}{\partial y} + \frac{\partial A_z}{\partial z}$$

$$\nabla \times \mathbf{A} = \begin{vmatrix} \hat{\mathbf{x}} & \hat{\mathbf{y}} & \hat{\mathbf{z}} \\ \dfrac{\partial}{\partial x} & \dfrac{\partial}{\partial y} & \dfrac{\partial}{\partial z} \\ A_x & A_y & A_z \end{vmatrix} = \hat{\mathbf{x}}\left(\frac{\partial A_z}{\partial y} - \frac{\partial A_y}{\partial z}\right) + \hat{\mathbf{y}}\left(\frac{\partial A_x}{\partial z} - \frac{\partial A_z}{\partial x}\right) + \hat{\mathbf{z}}\left(\frac{\partial A_y}{\partial x} - \frac{\partial A_x}{\partial y}\right)$$

$$\nabla^2 V = \frac{\partial^2 V}{\partial x^2} + \frac{\partial^2 V}{\partial y^2} + \frac{\partial^2 V}{\partial z^2}$$

CYLINDRICAL COORDINATES (r, ϕ, z)

$$\nabla V = \hat{\mathbf{r}}\frac{\partial V}{\partial r} + \hat{\boldsymbol{\phi}}\frac{1}{r}\frac{\partial V}{\partial \phi} + \hat{\mathbf{z}}\frac{\partial V}{\partial z}$$

$$\nabla \cdot \mathbf{A} = \frac{1}{r}\frac{\partial}{\partial r}(rA_r) + \frac{1}{r}\frac{\partial A_\phi}{\partial \phi} + \frac{\partial A_z}{\partial z}$$

$$\nabla \times \mathbf{A} = \frac{1}{r}\begin{vmatrix} \hat{\mathbf{r}} & \hat{\boldsymbol{\phi}}r & \hat{\mathbf{z}} \\ \dfrac{\partial}{\partial r} & \dfrac{\partial}{\partial \phi} & \dfrac{\partial}{\partial z} \\ A_r & rA_\phi & A_z \end{vmatrix} = \hat{\mathbf{r}}\left(\frac{1}{r}\frac{\partial A_z}{\partial \phi} - \frac{\partial A_\phi}{\partial z}\right) + \hat{\boldsymbol{\phi}}\left(\frac{\partial A_r}{\partial z} - \frac{\partial A_z}{\partial r}\right) + \hat{\mathbf{z}}\frac{1}{r}\left[\frac{\partial}{\partial r}(rA_\phi) - \frac{\partial A_r}{\partial \phi}\right]$$

$$\nabla^2 V = \frac{1}{r}\frac{\partial}{\partial r}\left(r\frac{\partial V}{\partial r}\right) + \frac{1}{r^2}\frac{\partial^2 V}{\partial \phi^2} + \frac{\partial^2 V}{\partial z^2}$$

SPHERICAL COORDINATES (R, θ, ϕ)

$$\nabla V = \hat{\mathbf{R}}\frac{\partial V}{\partial R} + \hat{\boldsymbol{\theta}}\frac{1}{R}\frac{\partial V}{\partial \theta} + \hat{\boldsymbol{\phi}}\frac{1}{R\sin\theta}\frac{\partial V}{\partial \phi}$$

$$\nabla \cdot \mathbf{A} = \frac{1}{R^2}\frac{\partial}{\partial R}(R^2 A_R) + \frac{1}{R\sin\theta}\frac{\partial}{\partial \theta}(A_\theta \sin\theta) + \frac{1}{R\sin\theta}\frac{\partial A_\phi}{\partial \phi}$$

$$\nabla \times \mathbf{A} = \frac{1}{R^2\sin\theta}\begin{vmatrix} \hat{\mathbf{R}} & \hat{\boldsymbol{\theta}}R & \hat{\boldsymbol{\phi}}R\sin\theta \\ \dfrac{\partial}{\partial R} & \dfrac{\partial}{\partial \theta} & \dfrac{\partial}{\partial \phi} \\ A_R & RA_\theta & (R\sin\theta)A_\phi \end{vmatrix}$$

$$= \hat{\mathbf{R}}\frac{1}{R\sin\theta}\left[\frac{\partial}{\partial \theta}(A_\phi \sin\theta) - \frac{\partial A_\theta}{\partial \phi}\right] + \hat{\boldsymbol{\theta}}\frac{1}{R}\left[\frac{1}{\sin\theta}\frac{\partial A_R}{\partial \phi} - \frac{\partial}{\partial R}(RA_\phi)\right] + \hat{\boldsymbol{\phi}}\frac{1}{R}\left[\frac{\partial}{\partial R}(RA_\theta) - \frac{\partial A_R}{\partial \theta}\right]$$

$$\nabla^2 V = \frac{1}{R^2}\frac{\partial}{\partial R}\left(R^2\frac{\partial V}{\partial R}\right) + \frac{1}{R^2\sin\theta}\frac{\partial}{\partial \theta}\left(\sin\theta\frac{\partial V}{\partial \theta}\right) + \frac{1}{R^2\sin^2\theta}\frac{\partial^2 V}{\partial \phi^2}$$

SOME USEFUL VECTOR IDENTITIES

$\mathbf{A} \cdot \mathbf{B} = AB \cos \theta_{AB}$ Scalar (or dot) product

$\mathbf{A} \times \mathbf{B} = \hat{\mathbf{n}} AB \sin \theta_{AB}$ Vector (or cross) product, $\hat{\mathbf{n}}$ normal to plane containing $\mathbf{A}$ and $\mathbf{B}$

$$\mathbf{A} \cdot (\mathbf{B} \times \mathbf{C}) = \mathbf{B} \cdot (\mathbf{C} \times \mathbf{A}) = \mathbf{C} \cdot (\mathbf{A} \times \mathbf{B})$$

$$\mathbf{A} \times (\mathbf{B} \times \mathbf{C}) = \mathbf{B}(\mathbf{A} \cdot \mathbf{C}) - \mathbf{C}(\mathbf{A} \times \mathbf{B})$$

$$\nabla(U + V) = \nabla U + \nabla V$$

$$\nabla(UV) = U \nabla V + V \nabla U$$

$$\nabla \cdot (\mathbf{A} + \mathbf{B}) = \nabla \cdot \mathbf{A} + \nabla \cdot \mathbf{B}$$

$$\nabla \cdot (U\mathbf{A}) = U \nabla \cdot \mathbf{A} + \mathbf{A} \cdot \nabla U$$

$$\nabla \times (U\mathbf{A}) = U \nabla \times \mathbf{A} + \nabla U \times \mathbf{A}$$

$$\nabla \times (\mathbf{A} + \mathbf{B}) = \nabla \times \mathbf{A} + \nabla \times \mathbf{B}$$

$$\nabla \cdot (\mathbf{A} \times \mathbf{B}) = \mathbf{B} \cdot (\nabla \times \mathbf{A}) - \mathbf{A} \cdot (\nabla \times \mathbf{B})$$

$$\nabla \cdot (\nabla \times \mathbf{A}) = 0$$

$$\nabla \times \nabla V = 0$$

$$\nabla \cdot \nabla V = \nabla^2 V$$

$$\nabla \times \nabla \times \mathbf{A} = \nabla(\nabla \cdot \mathbf{A}) - \nabla^2 \mathbf{A}$$

$\int_{\mathcal{V}} (\nabla \cdot \mathbf{A})\, d\mathcal{V} = \oint_S \mathbf{A} \cdot d\mathbf{s}$ Divergence theorem (S encloses $\mathcal{V}$)

$\int_S (\nabla \times \mathbf{A}) \cdot d\mathbf{s} = \oint_C \mathbf{A} \cdot d\mathbf{l}$ Stokes's theorem (S bounded by C)